Principles
of Anatomy
and Physiology

Principles
of Anatomy
and Physiology

Fourth Edition

Gerard J. Tortora

Bergen Community College

Nicholas P. Anagnostakos

late, Bergen Community College

1817

HARPER & ROW, PUBLISHERS, New York

Cambridge, Philadelphia, San Francisco,
London, Mexico City, São Paulo, Sydney

Sponsoring Editor: Claudia M. Wilson
Development Editor: Robert Ginsberg
Project Editor: Robert Greiner
Design Supervision: Helen Iranyi
Production Assistant: Debi Forrest-Bochner
Manager, New Book Production: Kewal Sharma
Photo Researcher: Mira Schachne
Compositor: Kingsport Press
Printer and Binder: Kingsport Press
Art Studio: Vantage Art, Inc.
Medical Illustrators: Leonard Dank, Charles Bridgman, Paul Melloni,
         Marsha J. Dohrmann, Helen Gee Jeung
Cover Design and Page Layout: Caliber Design Planning, Inc.

PRINCIPLES OF ANATOMY AND PHYSIOLOGY, Fourth Edition

Library of Congress Cataloging in Publication Data

Tortora, Gerard J.
  Principles of anatomy and physiology.

  Bibliography: p.
  Includes index.
  1. Human physiology. 2. Anatomy, Human.
I. Anagnostakos, Nicholas Peter, 1924–1981. II. Title.
[DNLM: 1. Anatomy. 2. Homeostasis. 3. Physiology.
QS 4 T712p]
QP34.5.T67 1984   612   83–18426
ISBN 0–06–046656–1

Harper International Edition
ISBN 0–06–350734X
Australian Edition
ISBN 0–06–3507315

COVER ILLUSTRATIONS

Front: The heart by Leonard Dank
Back: Diagram of a lobule of the lung by Leonard Dank

The fourth edition of *Principles of Anatomy and Physiology* is dedicated to
NICHOLAS P. ANAGNOSTAKOS
Friend, Coauthor, and Colleague

Contents in Brief

Contents in Detail

Preface

AUDIENCE

Principles of Anatomy and Physiology, Fourth Edition, is designed for the introductory course in anatomy and physiology. The text is geared to students in health-oriented, medical, and biological programs. Among the students specifically served by this text are those aiming for careers as nurses, medical assistants, medical laboratory technologists, radiologic technologists, inhalation therapists, dental hygienists, physical therapists, morticians, and medical record keepers. Because of its scope, the text is also useful for students in the biological sciences, science technology, liberal arts, physical education, and in premedical, predental, and prechiropractic programs.

OBJECTIVES

The objectives of the fourth edition are the same as those of previous editions: (1) to provide a basic understanding and working knowledge of the human body and (2) to present this essential material at a level that average students can handle.

Throughout, our goal has been to eliminate barriers to a ready comprehension of the structure and function of the human body. We recognize, however, that some technical vocabulary and difficult concepts are vital to the course. Such material is developed in step-by-step, easy-to-understand explanations that avoid needlessly difficult nontechnical vocabulary and syntax.

THEMES

Two major themes still dominate the book—*homeostasis* and *pathology.* Throughout, the book shows students how dynamic counterbalancing forces maintain normal anatomy and physiology. Pathology is viewed as a disruption in homeostasis. Accordingly, we present a number of clinical topics and contrast them with specific normal processes.

ORGANIZATION

The book follows the same unit and topic sequence as its three earlier editions. It is divided into five principal areas of concentration: organization, support and move-ment, control systems, maintenance, and continuity. Unit I, Organization of the Human Body, provides an understanding of the structural and functional levels of the body, from molecules to organ systems. Unit II, Principles of Support and Movement, analyzes the anatomy and physiology of the skeletal system, articulations, and the muscular system. Unit III, Control Systems of the Human Body, emphasizes the importance of the nerve impulse in the immediate maintenance of homeostasis, the role of receptors in providing information about the internal and external environment, and the significance of hormones in maintaining long-range homeostasis. Unit IV, Maintenance of the Human Body, illustrates how the body maintains itself on a day-to-day basis through the mechanisms of circulation, respiration, digestion, cellular metabolism, urine production, and buffer systems. Unit V, Continuity, covers the anatomy and physiology of the reproductive systems, sexual development, and the basic concepts of genetics.

In the fourth edition, the strengths of the previous editions have been maintained. Revisions made for the new edition have focused on updating certain topics and strengthening the coverage of physiology. Among the specific changes made in topic coverage in the fourth edition are the following.

UNIT I. ORGANIZATION OF THE HUMAN BODY

Chapter 1 has been updated to include basic principles and clinical applications of the dynamic spatial reconstructor (DSR). In Chapter 2, new sections have been added on nuclear magnetic resonance (NMR); positron emission tomography (PET); hydrogen bonding; saturated, unsaturated, and polyunsaturated fats; structural levels of protein organization; and left-handed DNA. Chapter 3 includes revised discussions of the structure, chemistry, and functions of plasma membranes; facilitated diffusion; active transport; ribosomes; lysosomes; cell division; and cells and cancer. New to Chapter 3 are discussions on nucleosomes, peroxisomes, the cellular cytoskeleton, protein synthesis, DNA repair, recombinant DNA, and laetrile and cancer. In Chapter 4, the discussion of the basement membrane has been revised, and the section on the inflammatory response has been updated. Also in Chapter 4, the exhibits on epithelial and connective tissue have again been improved by the addition of new color photomicrographs. Changes in Chapter 5 consist of new material on lines of cleavage; epidermal ridges and grooves; hair color, growth, and replacement; apocrine and eccrine sweat glands; ceruminous glands; collagen implants; pruritis; and over-the-counter corticosteroids. The discussions on skin color, systemic lupus erythematosus (SLE), and sun-screening agents have been revised.

UNIT II. PRINCIPLES OF SUPPORT AND MOVEMENT

New to Chapter 6 is a section on pulsating electromagnetic fields (PEMFs) and fracture repair. In Chapter 7 there is a new section on sternal puncture. Chapter 9 has been expanded to include discussions of arthroscopy and DMSO and arthritis. In Chapter 10 there are new sections on rigor mortis and recruitment. The section on the histology of cardiac muscle tissue has been revised. Chapter 11 includes new exhibits that describe and illustrate the muscles of the pharynx and larynx and sports injuries related to muscles.

UNIT III. CONTROL SYSTEMS OF THE HUMAN BODY

Chapter 12 has been updated to include a new section on neuronal intracellular transport and revised sections on factors that alter synaptic transmission and membrane potentials. In Chapter 14, discussions of drugs and the blood–brain barrier, brain injuries, brain electrical activity mapping (BEAM), rabies, Reye's syndrome (RS), and senility have been added. A new exhibit in this chapter summarizes cranial nerves, and material on the thalamus, brain lateralization, the cerebellum, and multiple sclerosis (MS) has been revised. Chapter 15 now contains a revised discussion of cutaneous reception and new sections on the somatosensory cortex and motor cortex. In Chapter 17 coverage has been expanded to include corneal transplants, color blindness, light and dark adaptation, and deafness. The treatment of static and dynamic equilibrium, olfactory and gustatory sensations, and sound waves has been revised. Chapter 18 contains revised material on hormone chemistry, prostaglandins, the mechanism of hormone action, the control of hormone secretion through feedback systems, and the histology of the adenohypophysis. Endocrine disorders are now integrated throughout the chapter as clinical applications and several new summary exhibits are included

UNIT IV. MAINTENANCE OF THE HUMAN BODY

Chapter 19 contains a revised discussion of hemostasis and new sections on blood substitutes and hemophilia. The section on the cardiac cycle and stroke volume in Chapter 20 has been extensively expanded; new material on the artificial heart, the Holter monitor, and defibrillation have been added. Updated and expanded topics in Chapter 21 are the histology of capillaries, atherosclerosis, hypertension, and hemodynamics. New topics in the chapter are coronary artery disease (CAD), coronary artery spasm, and the cardiovascular effects of exercise.

The section on capillary exchange has been moved from Chapter 27 to Chapter 21. Chapter 22 has been revised by the addition of discussions of immunology and cancer, monoclonal antibodies, and acquired immune deficiency syndrome (AIDS). The discussion of nonspecific resistance, hypersensitivity, and autoimmune disease have been revised. Chapter 23 contains additional coverage on pulmonary ventilation, the nervous control of respiration, and factors that modify respiration. There are now new sections on pulmonary embolism, compliance, and airway resistance, nitrogen narcosis, hyperbaric oxygenation, bronchitis, the common cold, and influenza. In Chapter 24, the sections on deglutition, the control of digestive secretions and motility, and the absorption of nutrients have been expanded. New material has been included on the absorption of water, electrolytes, and vitamins; hepatitis; bulimia; mumps; heartburn; vomiting; diarrhea and constipation; and liver biopsy. Several new summary exhibits have been added and the discussion of the regulation of food intake has been moved from Chapter 25 to Chapter 24. Chapter 25 contains expanded and reorganized discussions of metabolism and updated mineral and vitamin exhibits. The chapter also includes new material on absorptive and postabsorptive states. Chapter 26 provides new material on tests for renal function, bilirubin, urobilinogen, microbes in urine, and continuous ambulatory peritoneal dialysis (CAPD).

UNIT V. CONTINUITY

New or expanded topics in Chapter 28 include actions of inhibin and relaxin, diagnosis of pregnancy, premenstrual syndrome (PMS), and toxic shock syndrome (TSS). In Chapter 29, the sections on embryonic development, hormones of pregnancy, and lactation have been expanded. There are also new sections on morning sickness and twinning.

SPECIAL FEATURES

As in previous editions, the book contains numerous learning aids. Users of the book have cited the pedagogical aids as one of the book's many strengths. All of the tested and successful learning aids of previous editions have been kept in the fourth edition, and several new ones have been added. These special features are:

1. Student Objectives. Each chapter opens with a comprehensive list of Student Objectives. Each objective describes a knowledge or skill students should acquire while studying the chapter. (See **Note to the Student** for an explanation of how the objectives can be used.)

2. Study Outline. A Study Outline at the end of each chapter provides a brief summary of major topics. This section consolidates the essential points covered in the chapter so that students can recall and relate the points to one another. Page numbers have been added to the outlines so that topics within chapters can be located more easily.

3. Review Questions. Review Questions at the end of each chapter provide a check to see if the objectives stated at the beginning of the chapter have been met. After answering the questions, students should reread the objectives to determine whether they have met the goals.

4. Exhibits. Health-science students are generally expected to learn a great deal about the anatomy of certain organ systems, specifically, skeletal muscles, articulations, blood vessels, and nerves. In order to avoid interrupting the discussion of concepts and to organize the data, anatomical details have been presented in tabular form in exhibits, most of which are accompanied by illustrations. New summary exhibits have been added to the fourth edition, especially in relation to physiological principles.

5. Disorders: Homeostatic Imbalances. Abnormalities of structure or function are grouped at the end of appropriate chapters in sections entitled "Disorders: Homeostatic Imbalances." These sections provide a review of normal body processes as well as demonstrate the importance of the study of anatomy and physiology to a career in any of the health fields. All disorders have been updated and many new ones have been added.

6. Phonetic Pronunciations. Throughout the text, phonetic pronunciations are provided in parentheses for selected anatomical and physiological terms. These pronunciations are given at the point where the terms are introduced and are repeated in the Glossary of Terms. The **Note to the Student** explains the pronunciation key.

7. Medical Terminology. Glossaries of selected medical terms appear at the end of appropriate chapters; these listings are entitled "Medical Terminology." All of the glossaries have been revised for the fourth edition.

8. Drugs. Drugs associated with various body systems are discussed at the end of selected chapters. They have been revised for the fourth edition.

9. Clinical Applications. Throughout the text, Clinical Applications are boxed off for greater emphasis. Many new ones have been added.

10. Line Art. The line drawings in the book are large so that details are easily seen. In the fourth edition, many new illustrations have been added and full color is used throughout to differentiate structures and regions.

11. Photographs. The photographs amplify the narrative and the line drawings. Numerous photomicrographs (in full color), newly added scanning electron micrographs, and transmission electron micrographs enhance the histological discussions. Color photographs of specimens clarify gross anatomy discussions and have been added throughout, especially in Chapter 13 (spinal cord), Chapter 14 (brain), and Chapter 17 (special senses). Photographs of regional dissections, some in full color, have been added to the fourth edition in chapters dealing with the spinal cord, brain, heart, respiratory, digestive, urinary, and reproductive systems. A unique feature of the photographic component of the fourth edition is the addition of colored cross sections of various regions of the body in Chapter 11 (muscular system).

12. Appendixes. Appendix A, Measurements, summarizes U.S., metric, and apothecary units of length, mass, volume, and time. Appendix B, Abbreviations, is new to the fourth edition. It is an alphabetical list of commonly encountered medical abbreviations.

13. Glossary. Two glossaries appear at the end of the book. The first deals with prefixes, suffixes, and combining forms. The second is a comprehensive glossary of terms and has been greatly expanded for the fourth edition.

14. Bibliography. The extensive bibliography lists current suggested readings that correspond to the unit organization of the text.

SUPPLEMENTS

The following supplementary items are available to accompany the fourth edition of *Principles of Anatomy and Physiology:*

1. Learning Guide. By Kathleen S. Prezbindowski and Gerard J. Tortora, the *Learning Guide* is designed to help students *learn* anatomy and physiology. Rather than asking for mere repetition of the text material, the *Learning Guide* requires students to perform activities such as labeling a drawing or coloring parts of a diagram, as well as to answer multiple-choice, matching, and fill-in questions. The book's emphasis is on *using* new information in order to learn it. To enhance the effectiveness of the exercises, answers are provided for key-concept exercises so that the student has immediate feedback. A Mastery Test at the end of each chapter provides the student with a means of evaluating his or her learning of the chapter material and also gives practice for classroom testing situations.

2. Test Bank. The complimentary *Test Bank* contains 55 test items—multiple-choice, true-false, and matching questions—for each of the 29 chapters in the book. The test bank is available in standard printed format as well as on ACCESS (a computer-based test generator) and MICROTEST (for use with a microcomputer).

3. Instructor's Manual. Each chapter in the complimentary *Instructor's Manual*, prepared by Professor Tortora, consists of a chapter overview, a list of instructional concepts relating to the chapter, suggested problem-solving essay questions, and lists of audiovisual materials relating to the chapter topic.

4. Transparencies. Seventy-five full-color transparencies will be available. Illustrations from the text that are often shown and discussed in class are the subjects for the transparencies, and special care has been taken to make them clear and usable as overhead projections.

ACKNOWLEDGMENTS

I wish to thank the following people for their contribution to the fourth edition of *Principles of Anatomy and Physiology:*

Michael J. Autuori, University of Bridgeport
Frank Baker, Golden West College
Anita Been, University of Wisconsin
Mary Jane Burge, Cuyahoga Community College
Kenneth Carpenter, Shelby State Community College
Daniel Fertig, East Los Angeles College
Elvis Holt, Indiana University–Purdue University
Emron A. Jensen, Weber State College
R. Bruce Judd, Edison Community College
Michael C. Kennedy, Hahnemann Medical College
Donald Kisiel, Suffolk County Community College
Katherine Klein, Community College of Allegheny County
Gloria Jeanne Rogillio, Hinds Junior College
Paul Spannbauer, Hudson Valley Community College
Kent Van De Graaff, Brigham Young University
Edward P. Wallen, University of Wisconsin
Donald Wheeler, Cuyahoga Community College

Continuing thanks go to those people who worked on the three previous editions. Gratitude is also extended for the contributions of those people whose names appear with the photographs in the text. Special thanks to Dr. Michael H. Ross of the University of Florida for providing the excellent color photomicrographs. The exceedingly complex task of coordinating the art program was carried out smoothly and efficiently by Phil Alkana and Ellen Meek Tweedy, both of whom have my deepest thanks. Finally, for typing drafts of the manuscript and numerous other clerical duties associated with the task of putting together a textbook, thanks to Geraldine C. Tortora.

As the acknowledgments indicate, the participation of many individuals of diverse talent and expertise is required in the production of a textbook of this scope and complexity. For this reason, readers and users of the fourth edition are invited to send their reactions and suggestions so that plans can be formulated for subsequent editions.

Gerard J. Tortora

Biology Department
Bergen Community College
400 Paramus Road
Paramus, NJ 07652

Note to the Student

At the beginning of each chapter is a listing of **Student Objectives.** Before you read the chapter, please read the objectives carefully. Each objective is a statement of a skill or knowledge that you should acquire. To meet these objectives, you will have to perform several activities. Obviously, you must read the chapter carefully. If there are sections of the chapter that you do not understand after one reading, you should reread those sections before continuing. In conjunction with your reading, pay particular attention to the figures and exhibits; they have been carefully coordinated with the textual narrative.

At the end of each chapter are two and sometimes three other learning guides that you may find useful. The first, **Study Outline,** is a concise summary of important topics discussed in the chapter. This section is designed to consolidate the essential points covered in the chapter, so that you may recall and relate them to one another. The second guide, **Review Questions,** is a series of questions designed specifically to help you master the objectives. A third aid, **Medical Terminology,** appears in some chapters. This is a listing of terms designed to build your medical vocabulary. After you have answered the review questions, you should return to the beginning of the chapter and reread the objectives to determine whether you have achieved the goals.

As a further aid, we have included pronunciations for many terms that may be new to you. These appear in parentheses immediately following the new words, and they are repeated in the glossary of terms at the back of the book. (Of course, since there will always be some conflict among medical personnel and dictionaries about pronunciation, you will come across variations in different sources.) Look at the words carefully and say them out loud several times. Learning to pronounce a new word will help you remember it and make it a useful part of

your medical vocabulary. Take a few minutes now to read the following pronunciation key, so it will be familiar as you encounter new words. The key is repeated at the beginning of the glossary of terms.

PRONUNCIATION KEY

1. The strongest accented syllable appears in capital letters, for example, bilateral (bī-LAT-er-al) and diagnosis (dī-ag-NŌ-sis).
2. If there is a secondary accent, it is noted by a single quote mark('), for example, constitution (kon'-sti-TOO-shun) and physiology (fiz'-ē-OL-ō-jē). Any additional secondary accents are also noted by a single quote mark, for example, decarboxylation (dē'-kar-bok'-si-LĀ-shun).
3. Vowels marked with a line above the letter are pronounced with the long sound as in the following common words.

   ā as in *māke*
   ē as in *bē*
   ī as in *īvy*
   ō as in *pōle*

4. Vowels not so marked are pronounced with the short sound as in the following words.

   e as in *bet*
   i as in *sip*
   o as in *not*
   u as in *bud*

5. Other phonetic symbols are used to indicate the following sounds:

   a as in *above*
   oo as in *sue*
   yoo as in *cute*
   oy as in *oil*

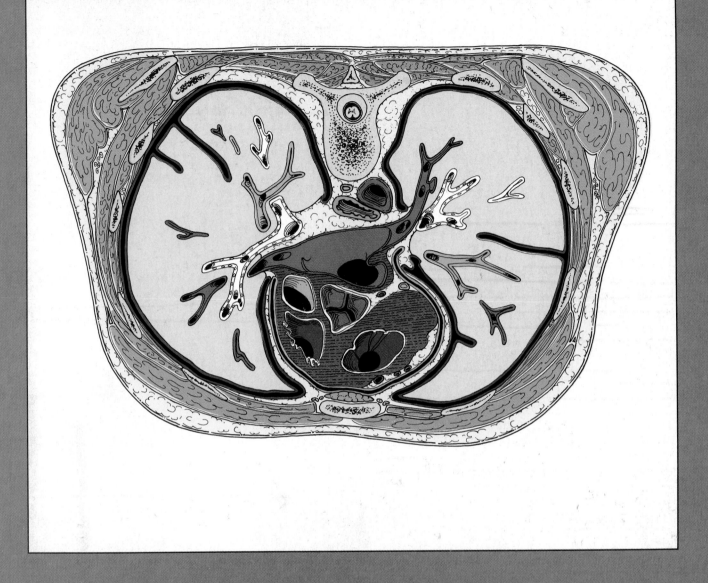

UNIT I

Organization of the Human Body

This unit is designed to show you how your body is organized at different levels. After you study the various regions and parts of your body, you will discover the importance of the chemicals that make it up. You will then find out how your cells, tissues, and organs form the systems that keep you alive and healthy.

- Define anatomy, with its subdivisions, and physiology.

- Explain the relationship between structure and function.

- Define each of the following levels of structural organization that make up the human body: chemical, cellular, tissue, organ, system, and organismic.

- Identify the principal systems of the human body, list the representative organs of each system, and describe the function of each system.

- Describe the general anatomical characteristics of the structural plan of the human body.

- Define the anatomical position.

- Compare common and anatomical terms used to describe various regions of the human body.

- Define several directional terms used in association with the human body.

- Define the common anatomical planes that may be passed through the human body.

- Distinguish a cross section, frontal section, and midsagittal section.

- Define the relationship of radiographic anatomy to the diagnosis of disease.

- List by name and location the principal body cavities and their major organs.

- Explain how the abdominopelvic cavity is divided into nine regions and four quadrants.

- Contrast the principles employed in conventional radiography and in computed tomography (CT) scanning.

- Point out the differences between a roentgenogram and a computed tomography (CT) scan.

- Explain how the dynamic spatial reconstructor (DSR) operates and its diagnostic importance in radiographic anatomy.

- Define homeostasis and explain why homeostasis is a state that results in normal body activities and why the inability to achieve homeostasis leads to disorders.

- Define a stress and identify the effects of stress on homeostasis.

- Describe the interrelationships of body systems in maintaining homeostasis.

- Compare the role of the endocrine and nervous systems in maintaining homeostasis.

- Contrast the homeostasis of blood pressure (BP) through nervous control and of blood sugar (BS) level through hormonal control.

- Define a feedback system and explain its role in homeostasis.

- Compare negative and positive feedback systems.

1 An Introduction to the Human Body

You are about to begin a study of the human body in order to learn how your body is organized and how it functions. The study of the human body involves many branches of science. Each contributes to a comprehensive understanding of how your body normally works and what happens when it is injured, diseased, or placed under stress.

ANATOMY AND PHYSIOLOGY DEFINED

Two branches of science that will help you understand your body parts and functions are anatomy and physiology. **Anatomy** (*anatome* = to dissect) refers to the study of *structure* and the relationships among structures. Anatomy is a broad science, and the study of structure becomes more meaningful when specific aspects of the science are considered. **Surface anatomy** is the study of the form (morphology) and markings of the surface of the body. **Gross** or **macroscopic anatomy** deals with structures that can be studied without a microscope. Another kind of anatomy, **systemic** or **systematic anatomy,** covers specific systems of the body, such as the system of nerves, spinal cord, and brain or the system of heart, blood vessels, and blood. **Regional anatomy** deals with a specific region of the body, such as the head, neck, chest, or abdomen. **Developmental anatomy** is the study of development from fertilized egg to adult form. **Embryology** is generally restricted to the study of development from the fertilized egg through the eighth week in utero. Other branches of anatomy are **pathological** (*patho* = disease) **anatomy,** the study of structural changes caused by disease; **histology** (*histio* = tissue), the microscopic study of the structure of tissues; and **cytology** (*cyt, cyto, cyte* = cell), the study of cells. **Radiographic anatomy,** the study of the structure of the body using radiographic (x-ray) techniques, is discussed in detail later in the chapter.

Whereas anatomy and its branches deal with structures of the body, **physiology** (fiz'-ē-OL-ō-jē) deals with *functions* of the body parts, that is, how the body parts work. Since physiology cannot be completely separated from anatomy, you will learn about the human body by studying its structures and functions together, and you will see how each structure of the body is custom-modeled to carry out a particular function. The structure of a part often determines the functions it will perform. In turn, body functions often influence the size, shape, and health of the structures.

LEVELS OF STRUCTURAL ORGANIZATION

The human body consists of several levels of structural organization that are associated with one another in several ways. Here we will consider the principal levels that will help you to understand how your body is organized (Figure 1-1). The lowest level of organization, the **chemical level,** includes all chemical substances essential for maintaining life. All these chemicals are made up of atoms joined together in various ways.

The chemicals, in turn, are put together to form the next higher level of organization: the **cellular level. Cells** are the basic structural and functional units of an organism. Among the many kinds of cells in your body are muscle cells, nerve cells, and blood cells. Figure 1-1 shows several isolated cells from the lining of the stomach. Each has a different structure, and each performs a different function.

The next higher level of structural organization is the **tissue level. Tissues** are made up of groups of similar cells and their intercellular material (substance between cells) that perform certain special functions. When the isolated cells shown in Figure 1-1 are joined together, they form a tissue called epithelium, which lines the stomach. Each cell in the tissue has a specific function. Mucous cells produce mucus, a secretion that lubricates food as it passes through the stomach. Parietal cells produce acid in the stomach. Chief cells produce enzymes needed to digest proteins. Other examples of tissues in your body are muscle tissue, connective tissue, and nervous tissue.

In many places in the body, different kinds of tissues are joined together to form an even higher level of organization: the **organ level. Organs** are structures of definite form and function composed of two or more different tissues. Organs usually have a recognizable shape. Examples of organs are the heart, liver, lungs, brain, and stomach. Figure 1-1 shows three of the tissues that make up the stomach. The serosa is a layer of connective tissue and epithelium around the outside that protects the stomach and reduces friction when the stomach moves and rubs against other organs. The muscle tissue layers of the stomach contract to mix food and pass it on to the next digestive organ. The epithelial tissue layer lining the stomach produces mucus, acid, and enzymes.

The next higher level of structural organization in the body is the **system level. A system** consists of an association of organs that have a common function. The digestive system, which functions in the breakdown of food, is composed of the mouth, saliva-producing glands called salivary glands, pharynx (throat), esophagus (gullet), stomach, small intestine, large intestine, rectum, liver, gallbladder, and pancreas.

The highest level is the **organismic level.** All the parts of the body functioning with one another constitute the total **organism**—one living individual.

In the chapters that follow, you will examine the anatomy and physiology of the major body systems. Exhibit 1-1 illustrates these systems, their representative organs, and their general functions. Unless otherwise stated, male and female bodies are similar. The systems are presented in the exhibit in the order in which they are discussed in later chapters.

STRUCTURAL PLAN

The human body has certain general **anatomical characteristics** that will help you to understand its overall structural plan. For example, humans have a **backbone,** or **vertebral column,** a characteristic that places them in a large group of organisms called *vertebrates.* Another characteristic is the body's **tube within a tube** construction. The outer tube is formed by the body wall; the inner tube is the digestive tract. Moreover, humans are for the most part **bilaterally symmetrical,** that is, externally the left and right sides of the body are mirror images.

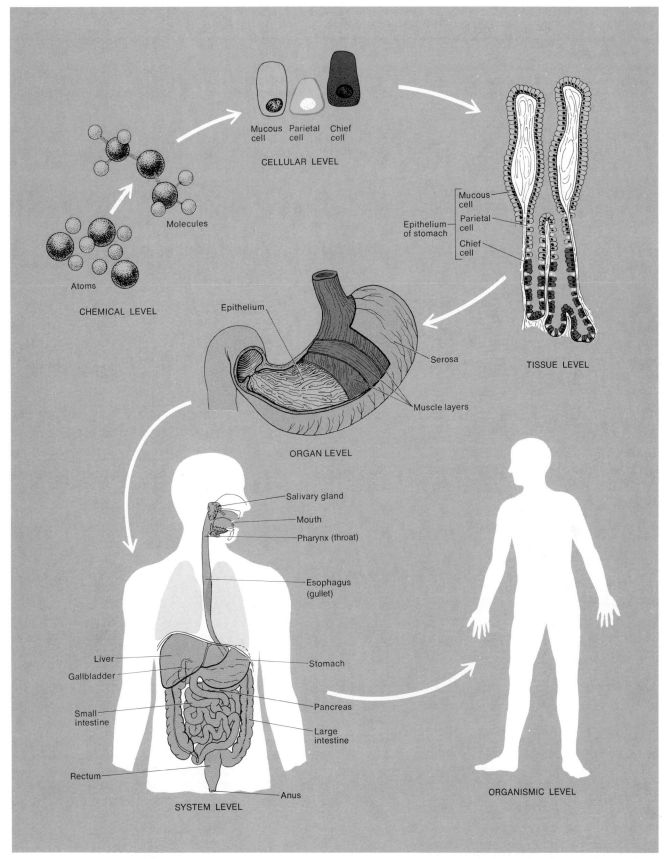

FIGURE 1-1 Levels of structural organization that compose the human body.

EXHIBIT 1-1
PRINCIPAL SYSTEMS OF HUMAN BODY, REPRESENTATIVE ORGANS, AND FUNCTIONS

1. Integumentary

Definition: The skin and structures derived from it, such as hair, nails, and sweat and oil glands.

Function: Helps regulate body temperature, protects the body, eliminates wastes, synthesizes vitamin D, and receives certain stimuli such as temperature, pressure, and pain.

2. Skeletal

Definition: All the bones of the body, their associated cartilages, and the joints of the body.

Function: Supports and protects the body, gives leverage, produces blood cells, and stores minerals.

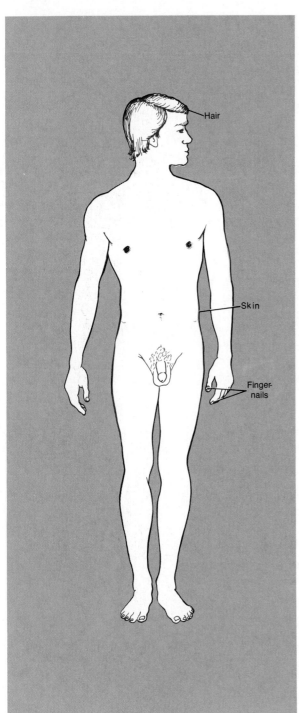

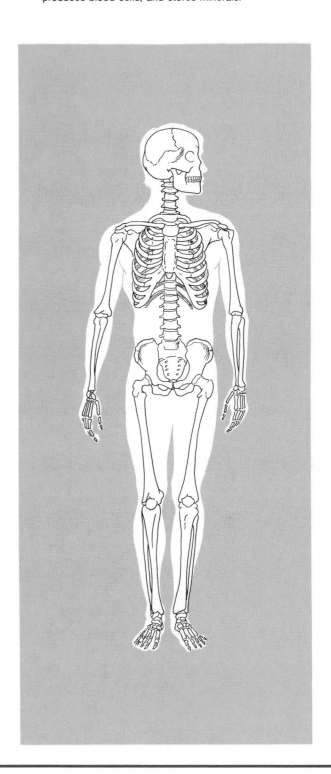

3. Muscular
Definition: All the muscle tissue of the body, including skeletal (shown in the illustration), visceral, and cardiac.
Function: Participates in bringing about movement, maintains posture, and produces heat.

4. Nervous
Definition: Brain, spinal cord, nerves, and sense organs, such as the eye and ear.
Function: Regulates body activities through nerve impulses.

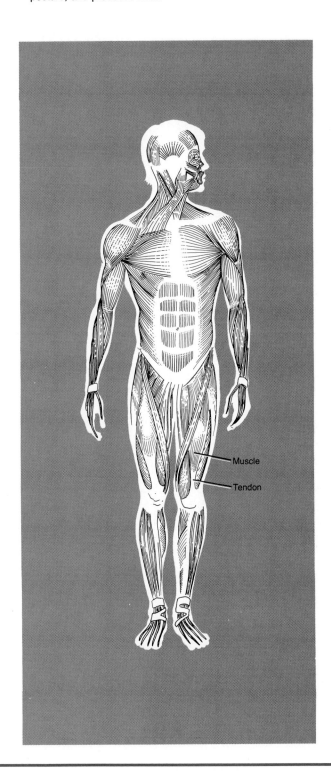

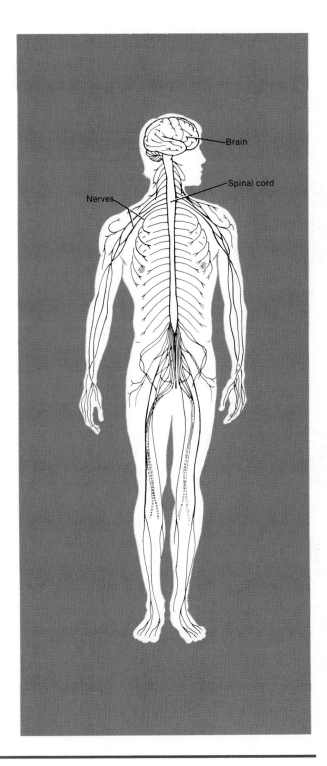

EXHIBIT 1-1
(Continued)

5. Endocrine
Definition: All glands that produce hormones.
Function: Regulates body activities through hormones transported by the cardiovascular system.

6. Cardiovascular
Definition: Blood, heart, and blood vessels.
Function: Distributes oxygen and nutrients to cells, carries carbon dioxide and wastes from cells, maintains the acid–base balance of the body, protects against disease, prevents hemorrhage by forming blood clots, and helps regulate body temperature.

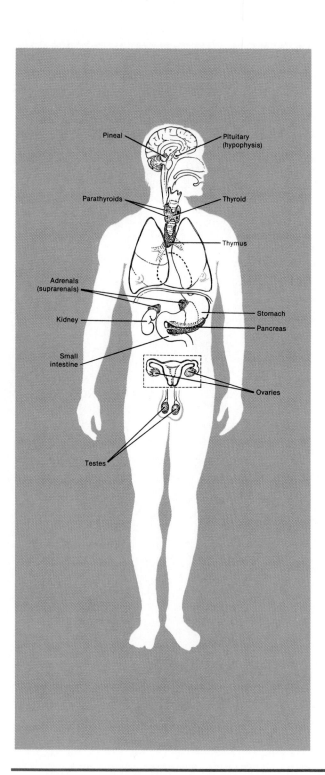

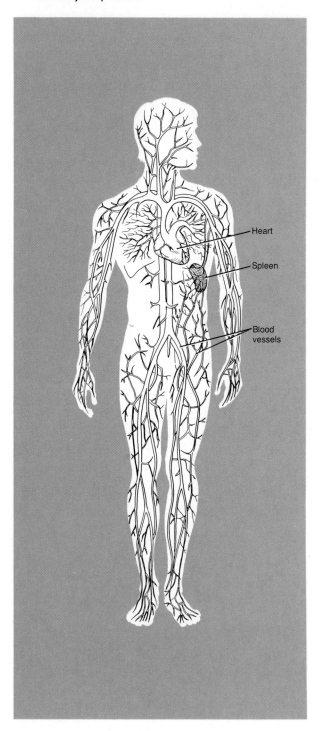

7. Lymphatic

Definition: Lymph, lymph nodes, lymph vessels, and lymph glands, such as the spleen, thymus gland, and tonsils.

Function: Returns proteins and plasma to the cardiovascular system, transports fats from the digestive system to the cardiovascular system, filters the blood, produces white blood cells, and protects against disease.

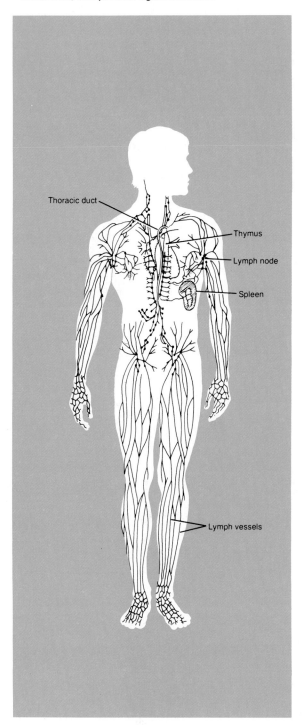

8. Respiratory

Definition: The lungs and a series of passageways leading into and out of them.

Function: Supplies oxygen, eliminates carbon dioxide, and helps regulate the acid–base balance of the body.

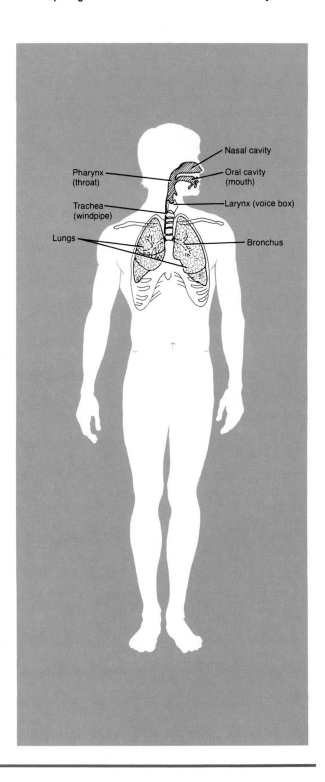

EXHIBIT 1-1
(*Continued*)

9. Digestive

Definition: A long tube and associated organs such as the salivary glands, liver, gallbladder, and pancreas.

Function: Performs the physical and chemical breakdown of food for use by cells and eliminates solid and other wastes.

10. Urinary

Definition: Organs that produce, collect, and eliminate urine.

Function: Regulates the chemical composition of blood, eliminates wastes, regulates fluid and electrolyte balance and volume, and helps maintain the acid–base balance of the body.

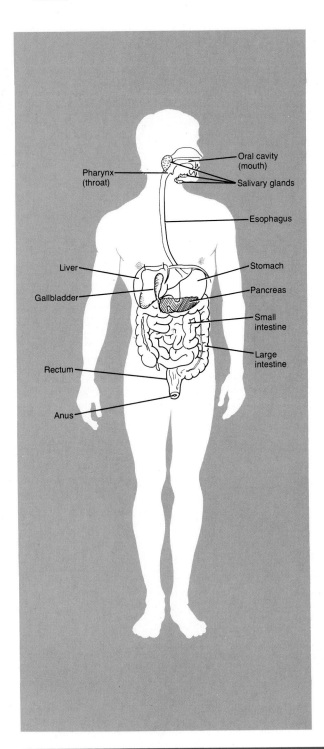

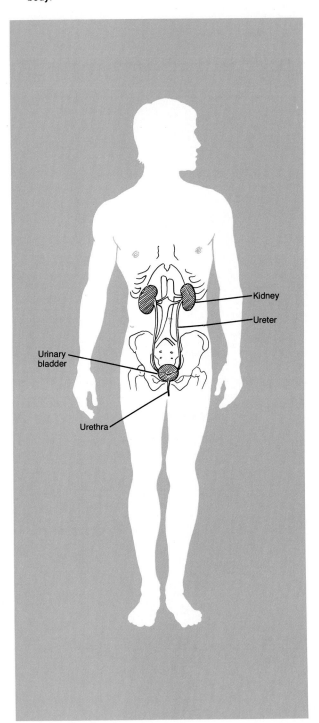

11. Reproductive

Definition: Organs (testes and ovaries) that produce reproductive cells (sperm and ova) and organs that transport and store reproductive cells.

Function: Reproduces the organism.

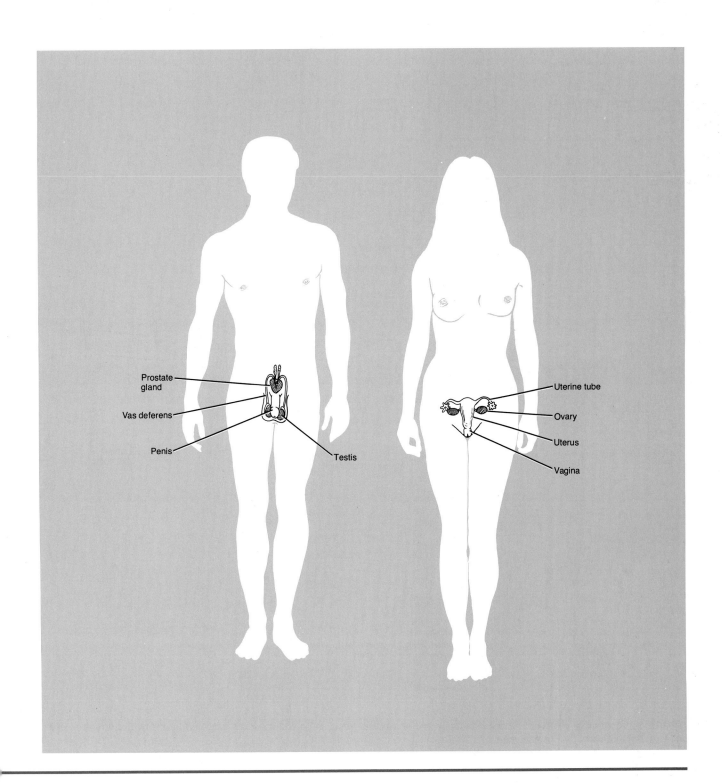

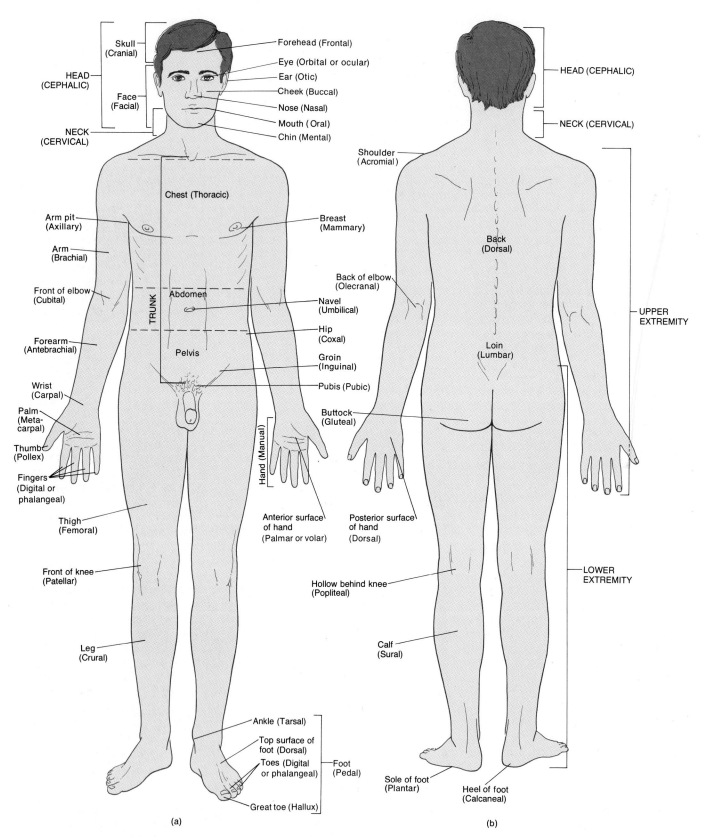

FIGURE 1-2 Anatomical position. The common name and anatomical terms, in parentheses, are indicated for many of the regions of the body. (a) Anterior view. (b) Posterior view.

ANATOMICAL POSITION AND ANATOMICAL NAMES

In all anatomical texts and charts, descriptions of any region or part of the human body assume that the body is in a specific position, called the **anatomical position,** so that directional terms are clear and any part can be related to any other part. In the anatomical position, the subject is standing erect facing the observer, the upper extremities are placed at the sides, and the palms of the hands are turned forward (Figure 1-2). Once the body is in the anatomical position, it is easier to visualize and understand how it is organized into various regions. The common and anatomical terms, in parentheses, of the principal body regions are also presented in Figure 1-2.

DIRECTIONAL TERMS

In order to explain exactly where various body structures are located in relationship to each other, anatomists use certain **directional terms.** If you want to point out the sternum (breastbone) to someone who knows where the clavicle (collarbone) is, you can say that the sternum is inferior (farther away from the head) and medial (toward the middle of the body) to the clavicle. As you can see, using the terms *inferior* and *medial* avoids a great deal of complicated description. Many directional terms are defined in Exhibit 1-2 and the parts of the body referred to in the examples are labeled in Figure 1-3. Studying the exhibit and the figure together should make clear to you the directional relationships among various body parts.

PLANES

The structural plan of the human body may also be discussed with respect to **planes** (imaginary flat surfaces) that pass through it. Several of the commonly used planes are illustrated in Figure 1-4. A **midsagittal (median) plane** is a vertical plane that passes through the midline of the body and divides the body or organs into equal right and left sides. A **sagittal** (SAJ-i-tal) **plane,** also called a **parasagittal** (*para* = near) **plane,** is a plane parallel to a midsagittal plane that divides the body or organs into unequal left and right portions. A **frontal (coronal;** kō-RŌ-nal) **plane** is a plane at a right angle to a midsagittal (or sagittal) plane that divides the body or organs into anterior and posterior portions. Finally, a **horizontal (transverse) plane** is a plane that is parallel to the ground, that is, at a right angle to the midsagittal, sagittal, and frontal planes. It divides the body or organs into superior and inferior portions.

When you study a body structure, you will often view it in section, that is, look at the flat surface resulting from a cut made through the three-dimensional structure. It is important to know the plane of the section so that you can understand the anatomical relationship of one part to another. Figure 1-5 indicates how three different sections—a *cross section,* a *frontal section,* and a *midsagittal section*—are made through different parts of the brain.

BODY CAVITIES

Spaces within the body that contain internal organs are called **body cavities.** Specific cavities may be distinguished if the body is divided into right and left halves. Figure 1-6 shows the two principal body cavities. The **dorsal body cavity** is located near the dorsal (posterior) surface of the body. It is further subdivided into a **cranial cavity,** which is a bony cavity formed by the cranial (skull) bones and contains the brain, and a **vertebral (spinal) canal,** which is a bony cavity formed by the vertebrae of the backbone and contains the spinal cord and the beginnings of spinal nerves.

The other principal body cavity is the **ventral body cavity** or **coelom** (SĒ-lōm). This cavity is located on the ventral (anterior) aspect of the body. The organs inside the ventral body cavity are called the **viscera** (VIS-er-a). Its walls are composed of skin, connective tissue, bone, muscles, and a membrane called the peritoneum. Like the dorsal body cavity, the ventral body cavity has two principal subdivisions—an upper portion, called the **thoracic** (thō-RAS-ik) **cavity** (or chest cavity), and a lower portion, called the **abdominopelvic** (ab-dom'-i-nō-PEL-vik) **cavity.** The anatomical landmark that divides the ventral body cavity into the thoracic and abdominopelvic cavities is the muscular diaphragm.

The thoracic cavity contains several divisions. There are two **pleural cavities** (Figure 1-7). Each is a small potential space between the visceral pleura and parietal pleura, the membranes covering the lungs. The **mediastinum** (mē'-dē-as-TĪ-num), is a mass of tissue between the pleurae of the lungs that extends from the sternum to the vertebral column (Figure 1-7a, b). The mediastinum includes all of the contents of the thoracic cavity, except the lungs themselves. The **pericardial** (per'-ē-KAR-dē-al; *peri* = around, *cardi* = heart) **cavity** is a small potential space between the visceral pericardium and parietal pericardium, the membranes covering the heart (Figure 1-7b).

The abdominopelvic cavity, as the name suggests, is divided into two portions, although no wall separates them (see Figure 1-6). The upper portion, the **abdominal cavity,** contains the stomach, spleen, liver, gallbladder, pancreas, small intestine, most of the large intestine, the kidneys, and the ureters. The lower portion, the **pelvic cavity,** contains the urinary bladder, sigmoid colon, rectum, and the internal male or female reproductive organs. One way to mark the division between the abdominal and pelvic cavities is to draw an imaginary line from the symphysis pubis (anterior joint between hipbones) to the superior border of the sacrum (sacral promontory).

ABDOMINOPELVIC REGIONS

To describe the location of organs easily, the abdominopelvic cavity may be divided into the **nine regions** shown in Figure 1-8. Although some unfamiliar terms are used in describing the nine regions and their contents, follow the descriptions as well as you can. When the organs are studied in detail in later chapters, they will have more meaning. The **epigastric** (*epi* = above; *gaster* = stomach) **region** contains the left lobe and medial part of the right

EXHIBIT 1-2
DIRECTIONAL TERMS

TERM	DEFINITION	EXAMPLE
Superior (cephalad or **craniad)**	Toward the head or the upper part of a structure; generally refers to structures in the trunk.	The heart is superior to the liver.
Inferior (caudad)	Away from the head or toward the lower part of a structure; generally refers to structures in the trunk.	The stomach is inferior to the lungs.
Anterior (ventral)	Nearer to or at the front of the body.	The sternum is anterior to the heart.
Posterior (dorsal)	Nearer to or at the back of the body.	The esophagus is posterior to the trachea.
Medial	Nearer to the midline of the body or a structure.	The ulna is on the medial side of the forearm.
Lateral	Farther from the midline of the body or a structure.	The ascending colon of the large intestine is lateral to the urinary bladder.
Intermediate	Between two structures, one of which is medial and one of which is lateral.	The ring finger is intermediate between the little (medial) and middle (lateral) fingers.
Ipsilateral	On the same side of the body.	The gallbladder and ascending colon of the large intestine are ipsilateral.
Contralateral	On the opposite side of the body.	The ascending and descending colons of the large intestine are contralateral.

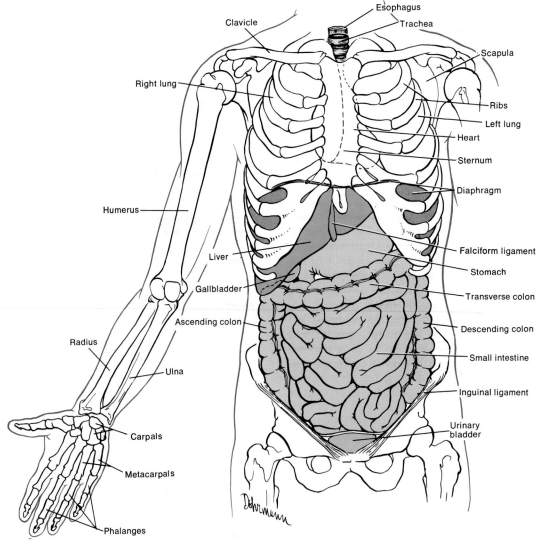

FIGURE 1-3 Anatomical and directional terms. By studying Exhibit 1-2 with this figure, you should gain an understanding of the meanings of the terms *superior, inferior, anterior, posterior, medial, lateral, intermediate, ipsilateral, contralateral, proximal,* and *distal.*

Proximal	Nearer to the attachment of an extremity to the trunk or a structure; nearer to the point of origin.	The humerus is proximal to the radius.
Distal	Farther from the attachment of an extremity to the trunk or a structure; farther from the point of origin.	The phalanges are distal to the carpals (wrist bones).
Superficial	Toward or on the surface of the body.	The muscles of the thoracic wall are superficial to the viscera in the thoracic cavity. (See Figure 1-7b.)
Deep	Away from the surface of the body.	The muscles of the arm are deep to the skin of the arm.
Parietal	Pertaining to or forming the outer wall of a body cavity.	The parietal pleura forms the outer layer of the pleural sacs that surround the lungs. (See Figure 1-7b.)
Visceral	Pertaining to the covering of an organ (viscus).	The visceral pleura forms the inner layer of the pleural sacs and covers the external surface of the lungs. (See Figure 1-7b.)

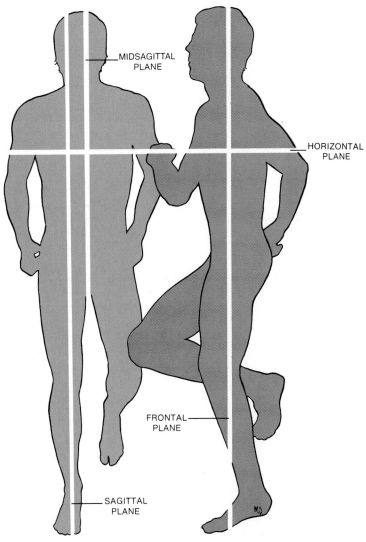

FIGURE 1-4 Planes of the human body.

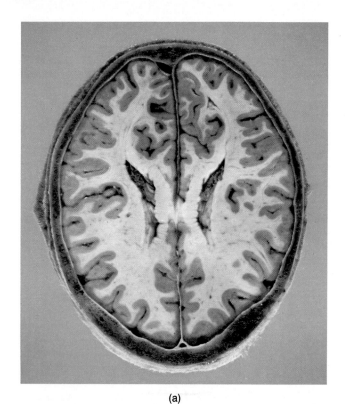

(a)

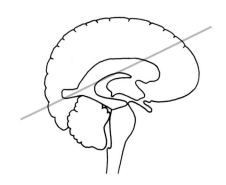

FIGURE 1-5 Sections through different parts of the brain. (a) Cross section. (Courtesy of Stephen A. Kieffer and E. Robert Heitzman, *An Atlas of Cross-Sectional Anatomy,* Harper & Row, Publishers, Inc., New York, 1979.) (b) Frontal section. (Courtesy of C. Yokochi and J. W. Rohen, *Photographic Anatomy of the Human Body,* 2nd ed., 1979, IGAKU-SHOIN, Ltd., Tokyo, New York.) (c) Midsagittal section. (Courtesy of C. Yokochi and J. W. Rohen, *Photographic Anatomy of the Human Body,* 2nd ed., 1979, IGAKU-SHOIN, Ltd., Tokyo, New York.)

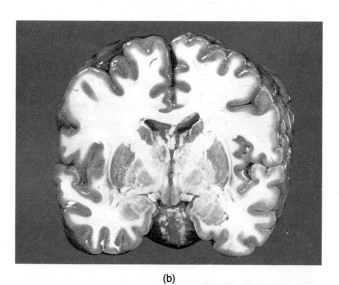

(b)

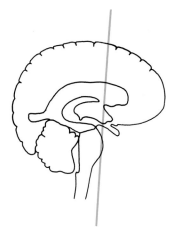

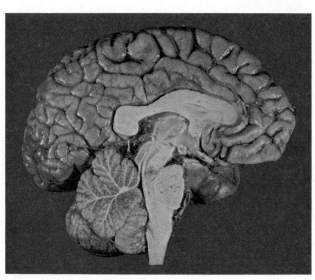

(c)

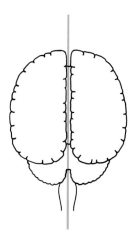

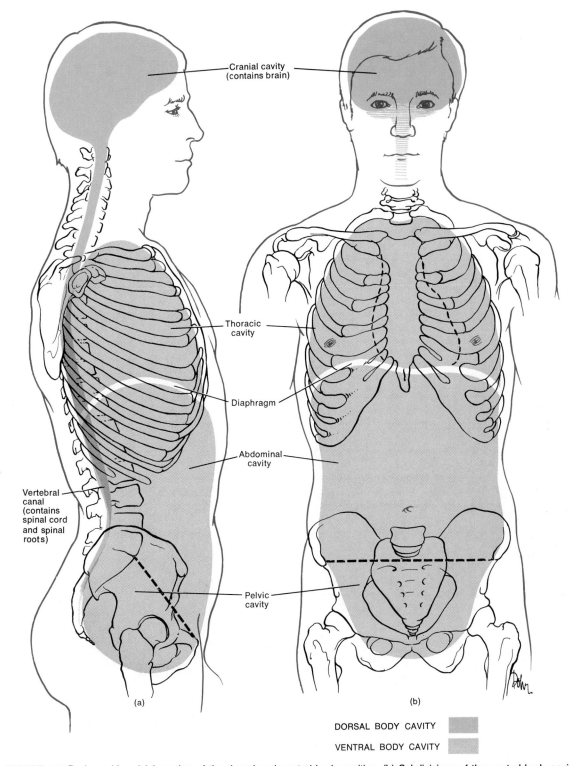

Cranial cavity
(contains brain)

Thoracic
cavity

Diaphragm

Abdominal
cavity

Vertebral
canal
(contains
spinal cord
and spinal
roots)

Pelvic
cavity

(a)

(b)

DORSAL BODY CAVITY

VENTRAL BODY CAVITY

FIGURE 1-6 Body cavities. (a) Location of the dorsal and ventral body cavities. (b) Subdivisions of the ventral body cavity.

lobe of the liver, the pyloric part and lesser curvature of the stomach, the superior and descending portions of the duodenum, the body and upper part of the head of the pancreas, and the two adrenal (suprarenal) glands. The **right hypochondriac** (*hypo* = under; *chondro* = cartilage, of ribs) **region** contains the right lobe of the liver, the gallbladder, and the upper third of the right kidney. The **left hypochondriac region** contains the body and fun-

dus of the stomach, the spleen, the left colic (splenic) flexure, the upper two-thirds of the left kidney, and the tail of the pancreas. The **umbilical region** contains the middle of the transverse colon, the inferior part of the duodenum, the jejunum, the ileum, the hilar regions of the kidneys, and the bifurcations (branching) of the abdominal aorta and inferior vena cava. The **right lumbar** (*lumbus* = loin) **region** contains the superior part of the

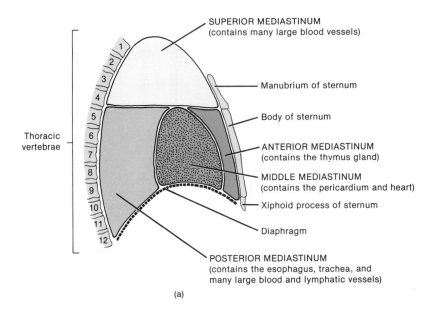

(a)

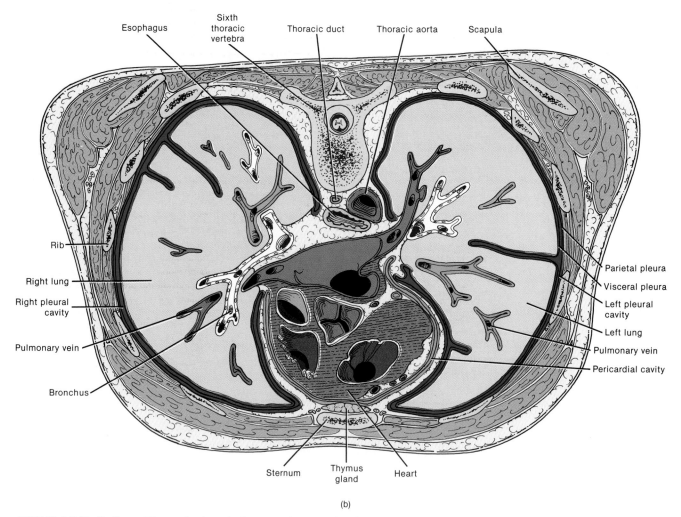

(b)

FIGURE 1-7 Mediastinum. The mediastinum is the space between the pleurae of the lungs that extends from the sternum to the vertebral column. (a) Subdivisions of the mediastinum seen in right lateral view. (b) Mediastinum seen in a cross section of the thorax.

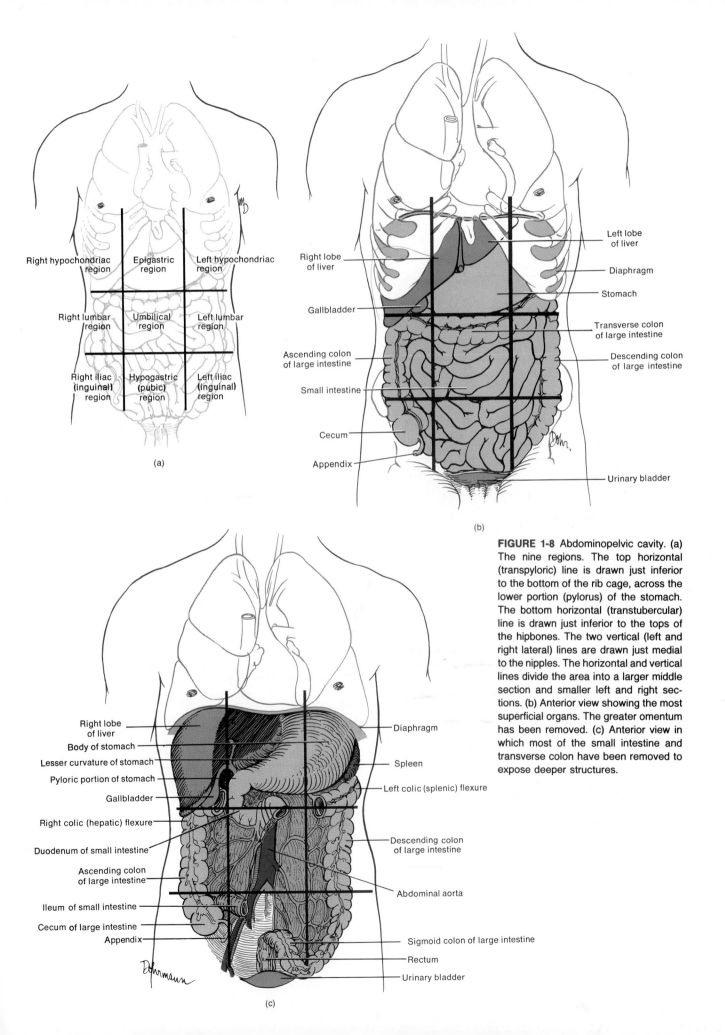

(a)

Right hypochondriac region

Epigastric region

Left hypochondriac region

Right lumbar region

Umbilical region

Left lumbar region

Right iliac (inguinal) region

Hypogastric (pubic) region

Left iliac (inguinal) region

(b)

Right lobe of liver

Gallbladder

Ascending colon of large intestine

Small intestine

Cecum

Appendix

Left lobe of liver

Diaphragm

Stomach

Transverse colon of large intestine

Descending colon of large intestine

Urinary bladder

(c)

Right lobe of liver

Body of stomach

Lesser curvature of stomach

Pyloric portion of stomach

Gallbladder

Right colic (hepatic) flexure

Duodenum of small intestine

Ascending colon of large intestine

Ileum of small intestine

Cecum of large intestine

Appendix

Diaphragm

Spleen

Left colic (splenic) flexure

Descending colon of large intestine

Abdominal aorta

Sigmoid colon of large intestine

Rectum

Urinary bladder

FIGURE 1-8 Abdominopelvic cavity. (a) The nine regions. The top horizontal (transpyloric) line is drawn just inferior to the bottom of the rib cage, across the lower portion (pylorus) of the stomach. The bottom horizontal (transtubercular) line is drawn just inferior to the tops of the hipbones. The two vertical (left and right lateral) lines are drawn just medial to the nipples. The horizontal and vertical lines divide the area into a larger middle section and smaller left and right sections. (b) Anterior view showing the most superficial organs. The greater omentum has been removed. (c) Anterior view in which most of the small intestine and transverse colon have been removed to expose deeper structures.

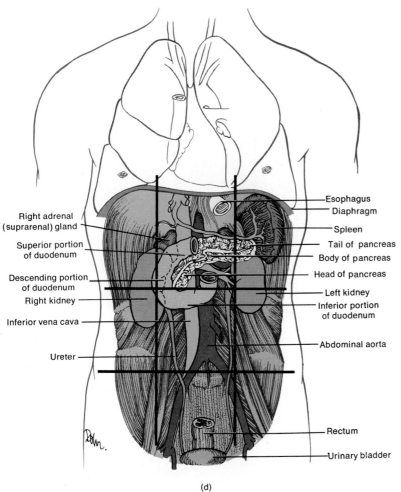

Right adrenal (suprarenal) gland

Superior portion of duodenum

Descending portion of duodenum

Right kidney

Inferior vena cava

Ureter

Esophagus
Diaphragm
Spleen
Tail of pancreas
Body of pancreas
Head of pancreas
Left kidney
Inferior portion of duodenum
Abdominal aorta

Rectum
Urinary bladder

(d)

FIGURE 1-8 (*Continued*) Abdominopelvic cavity. (d) Anterior view in which many organs have been removed, exposing the posterior structures. (e) Left lateral view.

cecum, the ascending colon, the right colic (hepatic) flexure, the lower lateral portion of the right kidney, and the small intestine. The **left lumbar region** contains the descending colon, the lower third of the left kidney, and the small intestine. The **hypogastric (pubic) region** contains the urinary bladder when full, the small intestine, and part of the sigmoid colon. The **right iliac (right inguinal) region** contains the lower end of the cecum, the appendix, and the small intestine. The term *iliacus* refers to the superior portion of the pelvic bone (hipbone). The **left iliac (left inguinal) region** contains the junction of the descending and sigmoid parts of the colon and the small intestine.

ABDOMINOPELVIC QUADRANTS

The abdominopelvic cavity may be divided more simply into four **quadrants** (*quad* = four). These are shown in Figure 1-9. In this method, frequently used by clinicians, a horizontal line and a vertical line are passed through the umbilicus. These two lines divide the abdomen into a **right upper quadrant (RUQ), left upper quadrant (LUQ), right lower quadrant (RLQ),** and **left lower quadrant (LLQ).** Whereas the nine-region designation is more widely used for anatomical studies, the four-quad-

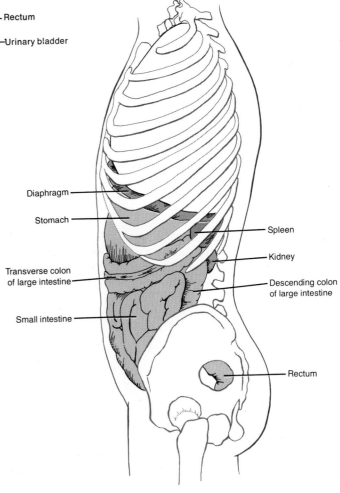

Diaphragm

Stomach

Transverse colon of large intestine

Small intestine

Spleen

Kidney

Descending colon of large intestine

Rectum

(e)

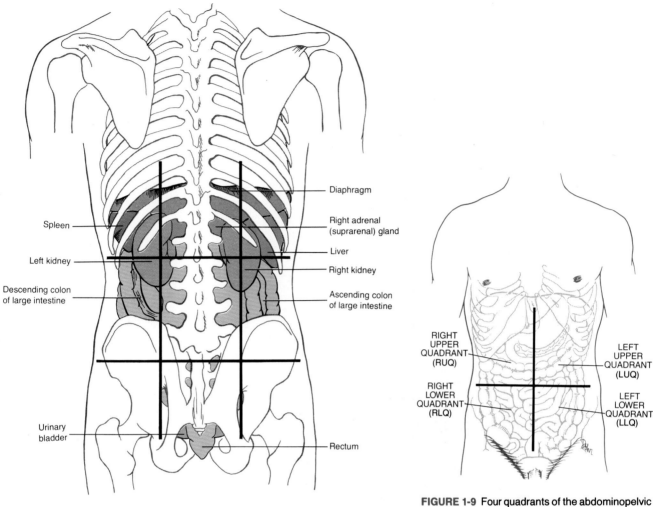

(f)

FIGURE 1-8 (Continued) Abdominopelvic cavity. (f) Posterior view.

FIGURE 1-9 Four quadrants of the abdominopelvic cavity. The two lines intersect at right angles at the umbilicus.

rant designation is better suited for locating the site of an abdominopelvic pain, tumor, or other abnormality.

RADIOGRAPHIC ANATOMY

A specialized branch of anatomy that is essential for the diagnosis of many disorders is **radiographic** (*radius* = ray) **anatomy,** which includes the use of several radiographic techniques (x-rays).

CONVENTIONAL RADIOGRAPHY

The most common and familiar type of radiographic anatomy employs the use of a single barrage of x-rays. The x-rays pass through the body and expose an x-ray film, producing a photographic image called a **roentgenogram** (RENT-gen-ō-gram). A roentgenogram provides a two-dimensional shadow image of the interior of the body (Figure 1-10a). As valuable as they are in diagnosis, conventional x-rays compress the body image onto a flat sheet of film, often resulting in an overlap of organs and tissues that could make diagnosis difficult. Moreover, x-rays do not always differentiate between subtle differences in tissue density.

COMPUTED TOMOGRAPHY (CT) SCANNING

These diagnostic difficulties have been virtually eliminated by the use of an x-ray technique called **computed tomography (CT) scanning** or **computerized axial tomography (CAT) scanning.** First introduced in 1971, CT scanning combines the principles of x-ray and advanced computer technologies. An x-ray source moves in an arc around the part of the body being scanned and repeatedly sends out x-ray beams. As the beams pass through the body, the tissues absorb small amounts of radiation, depending on their densities. Once the beams pass through, they are converted by light-sensitive crystal detectors to electronic signals that are transmitted to the scanner's computer.

Through mathematical analysis of the difference between the total radiation emitted and the amount striking the detectors, the computer reconstructs what happens to the x-ray beams. It determines where and how much of the beams are absorbed as they pass through a tissue. And, like slicing an orange and turning the slice to see the pits, the scanner's computer rotates the mathematical information 90°. It then projects an image, called a **CT scan,** onto a television screen (physician's console). The

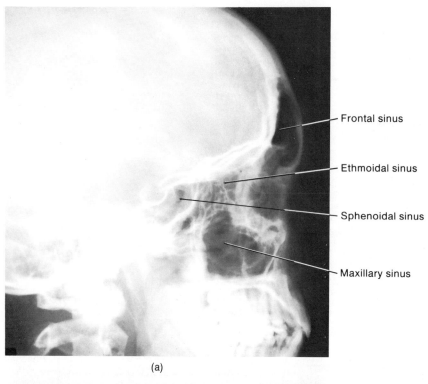

(a)

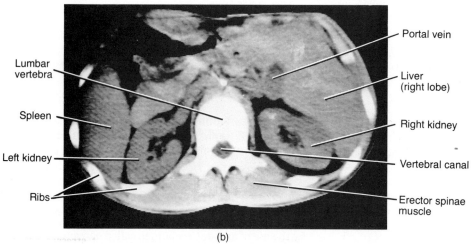

Lumbar vertebra

Spleen

Left kidney

Ribs

Portal vein

Liver (right lobe)

Right kidney

Vertebral canal

Erector spinae muscle

(b)

FIGURE 1-10 Radiologic anatomy. (a) Roentgenogram of the skull showing the paranasal sinuses. (Courtesy of Eastman Kodak Company.) (b) CT scan of the abdomen. (Courtesy of General Electric Company.)

CT scan provides a very accurate cross-sectional picture of any area of the body (Figure 1-10b). A series of scans permits a physician to examine layer after layer of a patient's tissues. Additional scans taken from other angles can be used to pinpoint extremely small abnormalities in a tissue.

The CT scan permits a significant differentiation of body parts that was never possible with conventional x-rays. The entire CT scanning process takes only seconds, it is completely painless, and the x-ray dose is equal to or less than that of many other diagnostic procedures.

DYNAMIC SPATIAL RECONSTRUCTOR (DSR)

A recent development in radiographic anatomy is a highly sophisticated x-ray machine called the **dynamic spatial reconstructor (DSR).** It is the most complex and versatile medical instrument built to date. It resembles a giant pencil sharpener, weighs over 15 tons, and has the ability to construct *moving, three-dimensional,* life-size images of all or part of an internal organ from any view desired. The image produced by the DSR can be rotated and tipped and, like an electronic knife, the DSR can pictorially slice open an organ and expose its interior (Figure 1-11). The instrument also provides enlargements, stop-action, replay, high-speed, and slow-motion viewing.

During a DSR scan, 28 x-ray guns revolve around the patient, each firing beams of x-rays 60 times per second. In a period of 5 seconds, the DSR can produce an amazing 75,000 cross sections. In the same period of time, a CT scanner can make only one section. The level of radiation exposure of the DSR is about double that for a chest roentgenogram.

The DSR has been designed to provide three-dimensional imaging of the heart, lungs, and circulation. It can be used to measure the volumes and movements of

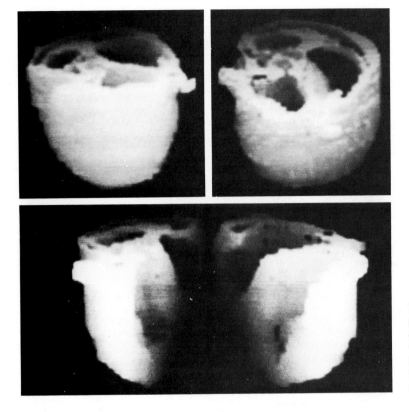

FIGURE 1-11 Dynamic spatial reconstructor (DSR) display of a living dog's heart. One of the unique properties of the DSR is its ability to act like an electronic knife by "dissecting" various organs, as shown in the photos. (Courtesy of Erik L. Ritman, M.D., Ph.D., Head, Biodynamics Research Unit, Mayo Foundation, Rochester, Minnesota.)

the heart and lungs and other internal organs, to detect cancer and heart defects, and to measure tissue damage following a heart attack, stroke, or other disease.

HOMEOSTASIS

Having considered some of the principal anatomical features of the body, we will now turn our attention to an important physiological feature of the body. This is homeostasis, one of the major themes of this textbook.

Homeostasis (hō'-mē-ō-STĀ-sis) is the condition in which the body's internal environment remains relatively constant, within limits (*homeo* = same; *stasis* = standing still). For the body's cells to survive, the composition of the surrounding fluids must be precisely maintained at all times. Fluid outside body cells is called **extracellular** (*extra* = outside) **fluid (ECF)** and is found in two principal places. The fluid filling the microscopic spaces between the cells of tissues is called *intercellular fluid, interstitial* (in'-ter-STISH-al) *fluid* (*inter* = between), or *tissue fluid*. The extracellular fluid in blood vessels is termed *plasma* (Figure 1-12). Fluid within cells is called **intracellular** (*intra* = within, inside) **fluid (ICF)**. Among the substances in extracellular fluid are gases, nutrients, and electrically charged chemical particles called ions—all needed for the maintenance of life. Extracellular fluid circulates through the blood and lymph vessels and from there moves into the spaces between the tissue cells. Thus it is in constant motion throughout the body. Essentially, all body cells are surrounded by the same fluid environment. For this reason, extracellular fluid is often called the body's internal environment.

An organism is said to be in homeostasis when its internal environment (1) contains exactly the optimum concentrations of gases, nutrients, ions, and water, (2) has an optimal temperature, and (3) has an optimal pressure for the health of the cells. When homeostasis is disturbed, ill health may result. If the body fluids are not eventually brought back into balance, death may occur.

STRESS AND HOMEOSTASIS

Homeostasis in all organisms is continually disturbed by **stress,** which is any stimulus that creates an imbalance in the internal environment. The stress may come from the external environment in the form of heat, cold, loud noises, or lack of oxygen. Or the stress may originate within the body in the form of high blood pressure, pain, tumors, or unpleasant thoughts. Most stresses are mild and routine. Poisoning, overexposure, severe infection, and surgical operations are examples of extreme stress.

Fortunately, the body has many regulating (homeostatic) devices that oppose the forces of stress and bring the internal environment back into balance. High resistance to stress is a striking feature of all organisms. Some people live in deserts where the daytime temperatures easily reach 49°C (120°F). Others work outside all day in subzero weather. Yet everyone's internal body temperature remains near 37°C (98.6°F). Mountain climbers exercise strenuously at high altitudes, where the oxygen content of the air is low. But once they adjust to the new altitude, they do not suffer from oxygen shortage. The extremes in temperature and in oxygen content of the air are external stresss, and the exercise performed is an internal stress, yet the body compensates and remains in homeostasis. Walter B. Cannon (1871–1945), an American physiologist who coined the term *homeostasis,* noted that the heat produced by the muscles during

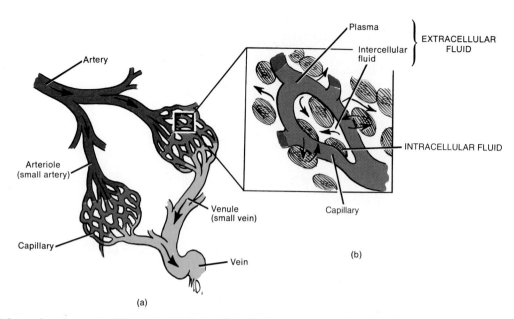

FIGURE 1-12 Internal environment of the body. (a) Extracellular fluid is found in two principal places: in blood vessels as plasma and between cells as interstitial fluid. Plasma circulates through arteries and arterioles and then into microscopic blood vessels called capillaries. From there it moves into the spaces between body cells where it is called interstitial fluid. This fluid then returns to capillaries as plasma and passes through the venules, and then into the veins. (b) Enlarged detail.

strenuous exercise would curdle and inactivate the body's proteins if the body did not dissipate heat quickly. Muscles that are being exercised also produce, in addition to heat, a great deal of lactic acid. If the body did not have a homeostatic mechanism for reducing the amount of the acid, the extracellular fluid would become acidic and destroy the cells.

Every body structure, from the cellular to the system level, contributes in some way to keeping the internal environment within normal limits. One homeostatic function of the cardiovascular system, for example, is to keep the fluids of all parts of the body constantly moving. When we are at rest, fresh blood is circulated throughout the entire body about once every minute. But when we are active and our muscles need nutrients rapidly, the heart quickens its pace and pumps fresh blood to the organs five times a minute. In this way, the cardiovascular system helps compensate for the stress of increased activity.

The respiratory system offers another example of a homeostatic mechanism in the body. The cells use more oxygen and produce more carbon dioxide when they are very active. Therefore, during periods of activity the respiratory system must work faster to keep the oxygen in the extracellular fluid from falling below normal limits and prevent excessive amounts of carbon dioxide from accumulating.

The digestive system and related organs help maintain the homeostasis involved in providing nutrients and removing wastes. As circulating blood passes through the organs of digestion, the products of digestion are transported to the body fluids so they can be used as nutrients by the cells. The liver, kidneys, endocrine glands, and other organs help to alter or store the products of digestion in various ways. The kidneys also help to remove

the wastes produced by cells after the cells have utilized the nutrients.

The homeostatic mechanisms of the body, such as those performed by the cardiovascular, respiratory, and digestive systems, are themselves subject to control by the nervous system and the endocrine system. The nervous system regulates homeostasis by detecting when the body is deviating from its balanced state and by sending messages to the proper organs to counteract the stress. For instance, when muscle cells are active, they take a great deal of oxygen from the blood. They also give off carbon dioxide, which is picked up by the blood. Certain nerve cells detect the chemical changes occurring in the blood and send a message to the brain. The brain then sends a message to the heart to pump blood more quickly to the lungs so the blood can give up its excess carbon dioxide and take on more oxygen. Simultaneously, the brain sends a message to the muscles that control breathing to contract faster. As a result, carbon dioxide can be exhaled and more oxygen can be inhaled.

Homeostasis is also controlled by the endocrine system—a series of glands that secrete chemical regulators, called hormones, into the blood. Whereas nerve impulses coordinate homeostasis rapidly, hormones work slowly. Both means of control are directed toward the same end.

HOMEOSTASIS OF BLOOD PRESSURE

Blood pressure (BP) is the force exerted by blood as it presses against and attempts to stretch the walls of the blood vessels, especially the arteries. It is determined primarily by three factors: the rate and strength of the heartbeat, the amount of blood, and the resistance offered by the arteries as blood passes through them. The resistance

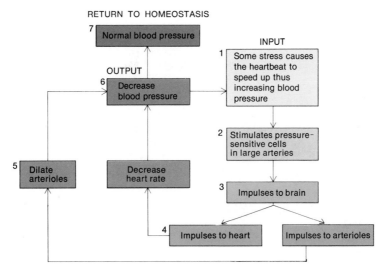

FIGURE 1-13 Homeostasis of blood pressure. Note that the output is fed back into the system, and the system continues to lower blood pressure until there is a return to homeostasis.

of the arteries results from the chemical properties of the blood and the size of the arteries.

If some stress, either internal or external, causes the heartbeat to speed up, the following sequence occurs (Figure 1-13). As the heart pumps faster, it pushes more blood into the arteries per minute, increasing pressure in the arteries. The higher pressure is detected by pressure-sensitive nerve cells in the walls of certain arteries, which send nerve impulses to the brain. The brain interprets the message and responds by sending impulses to the heart to slow the heart rate, thus decreasing blood pressure. The continual monitoring of blood pressure by the nervous system is an attempt to maintain a normal blood pressure and employs what is called a feedback system.

A **feedback system** is any circular situation in which information about the status of something is continually reported (fed back) to a central control region. The nervous control that results in a normal constant blood pressure is an example. In the case of regulating blood pressure, the **input (stimulus)** is the information picked up by the pressure-sensitive nerve cells (high blood pressure), and the **output (response)** is the return toward normal blood pressure due to decreased heartbeat. Figure 1-13 shows that the system runs in a circle. The pressure-sensitive nerve cells continue to monitor pressure and to feed this information back to the brain, even after the return to homeostasis has begun and blood pressure has begun to normalize. In other words, the pressure-sensitive nerve cells send impulses to the brain about the changed blood pressure, and, if the pressure is still too high, the brain continues to send out impulses to slow the heartbeat.

This type of feedback system reverses the direction of the initial condition from a rising to a falling blood pressure and is called a **negative feedback system.** The reaction of the body (output) counteracts the stress (input) in order to restore homeostasis. Such a system is therefore a stimulatory-inhibitory one. If instead the brain had signaled the heart to beat even faster and the blood pressure

had kept on rising, the system would have been a **positive feedback system.** In a positive feedback system the output **intensifies** the input. This system is therefore a stimulatory-stimulatory one. Most positive feedback systems are destructive and result in various disorders. Most of the feedback systems of the body are negative.

Figure 1-13 also shows that a second negative feedback control is involved in maintaining normal blood pressure. Small arteries, called arterioles, have muscular walls that can constrict or relax upon receiving an appropriate signal from the brain. When the blood pressure increases, pressure-sensitive nerve cells in certain arteries send messages to the brain. The brain interprets the messages and responds by sending impulses to the arterioles, causing them to relax (dilate). Thus the blood flowing through the arterioles is offered less resistance and blood pressure drops back to normal.

HOMEOSTASIS OF BLOOD SUGAR LEVEL

An example of homeostatic regulation by hormones is the maintenance of blood sugar (BS) level. Glucose, the sugar found in blood, is one of the body's principal sources of energy. Under normal circumstances, the concentration of sugar in the blood averages about 90 milligrams/100 milliliters of blood.* This level is maintained primarily by two hormones secreted by the pancreas: insulin and glucagon. Suppose you have just eaten some candy. The sugar in the candy is broken down by the organs of digestion and moves from the digestive tract into the blood. The sugar then becomes a stress because it raises the blood sugar level above normal. In response to this stress, the cells of the pancreas are stimulated to secrete insulin (Figure 1-14a). Once insulin enters the blood, it has two principal effects. First, it increases sugar uptake by cells.

* One milligram (mg) = 0.001 gram (g), 1 g = 0.035 ounce (oz), and 100 milliliters (ml) = 3.38 oz.

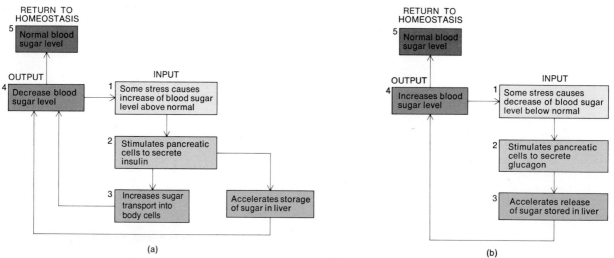

FIGURE 1-14 Homeostasis of blood sugar level. (a) Homeostatic mechanism that lowers high blood sugar level down to normal. (b) Homeostatic mechanism that raises low blood sugar level up to normal.

The blood sugar level is lowered by this action since the sugar moves from the blood into body cells. Second, insulin accelerates the process by which sugar is stored in the liver and muscles. Thus even more sugar is removed from the blood. In essence, insulin decreases blood sugar concentration until it returns to normal.

The other hormone produced by the pancreas, glucagon, has the opposite effect of insulin. Suppose you have not eaten for several hours and your blood sugar level is steadily decreasing. Lack of sugar is now the stress, and under this condition other cells of the pancreas are instead stimulated to secrete glucagon (Figure 1-14b). This hormone accelerates the process by which sugar stored in the liver is sent back into the bloodstream. The blood sugar level is thus increased until it returns to normal.

MEASURING THE HUMAN BODY

An important aspect of describing the body and understanding how it works is **measurement.** Examples of measurements involve the size of an organ, the weight of an organ, the time it takes for a physiological reaction to occur, and the amount of a medication to be administered. Measurements involving time, weight, temperature, size, length, and volume are routine to a medical science program. The metric system of measurement is standardly used in sciences. In this book measurements are given in metric units followed by the approximate U.S. equivalent in parentheses, where convenient. If you are not familiar with U.S.–metric conversions, consult Appendix A.

STUDY OUTLINE

Anatomy and Physiology Defined (p. 4)

1. Anatomy is the study of structure and the relationship among structures.
2. Subdivisions of anatomy include surface anatomy (form and markings of surface features), gross anatomy (macroscopic), systemic or systematic anatomy (systems), regional anatomy (regions), developmental anatomy (development from fertilization to adulthood), embryology (development from fertilized egg through eighth week in utero), pathological anatomy (disease), histology (tissues), cytology (cells), and radiographic anatomy (x-rays).
3. Physiology is the study of how body structures function.

Levels of Structural Organization (p. 4)

1. The human body consists of several levels of structural organization; among these are the chemical, cellular, tissue, organ, system, and organismic levels.
2. Cells are the basic structural and functional units of an organism.
3. Tissues consist of groups of similarly specialized cells and their intercellular material that perform certain special functions.

4. Organs are structures of definite form and function composed of two or more different tissues.
5. Systems consist of associations of organs that have a common function.
6. The human organism is a collection of structurally and functionally integrated systems.
7. The systems of the human body are the integumentary, skeletal, muscular, nervous, endocrine, cardiovascular, lymphatic, respiratory, digestive, urinary, and reproductive (see Exhibit 1-1).

Structural Plan (p. 4)

1. The human body has certain general characteristics.
2. Among the characteristics are a backbone, a tube within a tube organization, and bilateral symmetry.

Anatomical Position and Anatomical Names (p. 13)

1. When in the anatomical position, the subject stands erect facing the observer, the upper extremities are placed at the sides, and the palms of the hands are turned forward.
2. Regional names are terms given to specific regions of the body for reference. Examples of regional names include cra-

nial (skull), thoracic (chest), brachial (arm), patellar (knee), cephalic (head), and gluteal (buttock).

Directional Terms (p. 13)

1. Directional terms indicate the relationship of one part of the body to another.
2. Commonly used directional terms are superior (toward the head or upper part of a structure), inferior (away from the head or toward the lower part of a structure), anterior (near or at the front of the body), posterior (near or at the back of the body), medial (nearer the midline of the body or a structure), intermediate (between a medial and lateral structure), ipsilateral (on the same side of the body), contralateral (on the opposite side of the body), proximal (nearer the attachment of an extremity to the trunk or a structure), distal (farther from the attachment of an extremity to the trunk or a structure), superficial (toward or on the surface of the body), deep (away from the surface of the body), parietal (pertaining to the outer wall of a body cavity), and visceral (pertaining to the covering of an organ).

Planes (p. 13)

1. Planes are imaginary flat surfaces that are used to divide the body or organs into definite areas. A midsagittal (median) plane is a vertical plane through the midline of the body that divides the body or organs into equal right and left sides; a sagittal (parasagittal) plane is a plane parallel to the midsagittal plane that divides the body or organs into unequal right and left sides; a frontal (coronal) plane is a plane at a right angle to a midsagittal (or sagittal) plane that divides the body or organs into anterior and posterior portions; and a horizontal (transverse) plane is a plane parallel to the ground and at a right angle to the midsagittal, sagittal, and frontal planes that divides the body or organs into superior and inferior portions.
2. Sections are flat surfaces resulting from cuts through body structures. They are named according to the plane on which the cut is made and include cross sections, frontal sections, and midsagittal sections.

Body Cavities (p. 13)

1. Spaces in the body that contain internal organs are called cavities.
2. The dorsal and ventral cavities are the two principal body cavities. The dorsal cavity contains the brain and spinal cord. The organs of the ventral cavity are collectively called the viscera.
3. The dorsal cavity is subdivided into the cranial cavity, which contains the brain, and the vertebral or spinal canal, which contains the spinal cord and beginnings of spinal nerves.
4. The ventral body cavity is subdivided by the diaphragm into an upper thoracic cavity and a lower abdominopelvic cavity.
5. The thoracic cavity contains two pleural cavities and a mediastinum, which includes the pericardial cavity.
6. The mediastinum is a mass of tissue between the pleurae of the lungs that extends from the sternum to the vertebral column; it contains all contents of the thoracic cavity, except the lungs.
7. The abdominopelvic cavity is divided into a superior abdominal and an inferior pelvic cavity by an imaginary line extending from the symphysis pubis to the sacral promontory.
8. Viscera of the abdominal cavity include the stomach, spleen, pancreas, liver, gallbladder, kidneys, small intestine, and most of the large intestine.
9. Viscera of the pelvic cavity include the urinary bladder, sigmoid colon, rectum, and internal female and male reproductive structures.

Abdominopelvic Regions (p. 13)

1. To describe the location of organs easily, the abdominopelvic cavity may be divided into nine regions by drawing four imaginary lines (left lateral, right lateral, transpyloric, and transtubercular).
2. The names of the nine abdominopelvic regions are epigastric, right hypochondriac, left hypochondriac, umbilical, right lumbar, left lumbar, hypogastric (pubic), right iliac (inguinal), and left iliac (inguinal).

Abdominopelvic Quadrants (p. 20)

1. To locate the site of an abdominopelvic abnormality in clinical studies, the abdominopelvic cavity may be divided into four quadrants by passing imaginary horizontal and vertical lines through the umbilicus.
2. The names of the four abdominopelvic quadrants are right upper quadrant (RUQ), left upper quadrant (LUQ), right lower quadrant (RLQ), and left lower quadrant (LLQ).

Radiographic Anatomy (p. 21)

1. Radiographic anatomy is a specialized branch of anatomy that makes use of x-rays.
2. The value of radiographic anatomy is its application in the diagnosis of disease.

Conventional Radiography

1. Conventional radiography uses a single barrage of x-rays.
2. The photographic two-dimensional image produced is called a roentgenogram.
3. Conventional radiography has several diagnostic drawbacks including overlapping of organs and tissues and inability to differentiate subtle differences in tissue density.

Computed Tomography (CT) Scanning

1. CT scanning combines the principles of x-ray and advanced computer technology.
2. The image produced, called a CT scan, provides a very accurate cross-sectional picture of any area of the body.

Dynamic Spatial Reconstructor (DSR)

1. The DSR is a highly sophisticated x-ray machine that can produce moving, three-dimensional images of different organs of the body.
2. The DSR has been designed to provide imaging of the heart, lungs, and circulation.

Homeostasis (p. 23)

1. Homeostasis is a condition in which the internal environment (extracellular fluid) of the body remains relatively constant in terms of chemical composition, temperature, and pressure.
2. All body systems attempt to maintain homeostasis.
3. Homeostasis is controlled mainly by the nervous and endocrine systems.

Stress and Homeostasis

1. Stress is any external or internal stimulus that creates a change in the internal environment.
2. If a stress acts on the body, homeostatic mechanisms attempt to counteract the effects of the stress and bring the condition back to normal.

Homeostasis of Blood Pressure

1. Blood pressure (BP) is the force exerted by blood as it presses against and attempts to stretch the walls of arteries. It is determined by the rate and force of the heartbeat, the amount of blood, and arterial resistance.
2. If a stress causes the heartbeat to increase, blood pressure

also increases; pressure-sensitive nerve cells in certain arteries inform the brain, and the brain responds by sending impulses that decrease heartbeat, thus decreasing blood pressure back to normal; a rise in blood pressure also causes the brain to send impulses that dilate arterioles, thereby also helping to decrease blood pressure back to normal.

3. Any circular situation in which information about the status of something is continually fed back to a control region is called a feedback system.

4. A negative feedback system is one in which the reaction of the body (output) counteracts the stress (input) in order to maintain homeostasis; most feedback systems of the body are negative. A positive feedback system is one in which the output intensifies the input; the system is usually destructive.

Homeostasis of Blood Sugar Level

1. A normal blood sugar (BS) level is maintained by the actions of two different pancreatic hormones: insulin and glucagon.

2. Insulin lowers blood sugar level by increasing sugar uptake by cells and accelerating sugar storage as glycogen in the liver and skeletal muscles.

3. Glucagon raises blood sugar level by accelerating the rate of sugar released from glycogen by the liver.

Measuring the Human Body (p. 26)

1. Various kinds of measurements are important in understanding the human body.

2. Examples of such measurements include organ dimensions and weight, physiological response time, and amount of medication to be administered.

3. Measurements in this book are given in metric units followed by the approximate U.S. equivalents in parentheses, where convenient.

4. The principles and applications of the various metric and U.S. units of measurement are discussed in Appendix A.

REVIEW QUESTIONS

1. Define anatomy. List and define the various subdivisions of anatomy.
2. Define physiology.
3. Give several examples of how structure and function are related.
4. Construct a diagram to illustrate the levels of structural organization that characterize the body.
5. Define each of the following terms: cell, tissue, organ, system, and organism.
6. Using Exhibit 1-1 as a guide, outline the functions of each system of the body, and list several organs that compose each system.
7. What does bilateral symmetry mean? Why is the body considered to be a tube within a tube? What is a vertebrate?
8. Define the anatomical position. Why is the anatomical position used?
9. Review Figure 1-2. See if you can locate each region on your own body, and name each by its common and anatomical term.
10. What is a directional term? Why are these terms important?
11. Use each of the directional terms listed in Exhibit 1-2 in a complete sentence.
12. Define the various planes that may be passed through the body. Explain how each plane divides the body.
13. What is meant by the phrase "a part of the body has been sectioned"?
14. Define a body cavity. List the body cavities discussed, and tell which major organs are located in each. What landmarks separate the various body cavities from one another?
15. What is the mediastinum?
16. Describe how the abdominopelvic area is subdivided into nine regions. Name and locate each region and list the organs, or parts of organs, in each.
17. Describe how the abdominopelvic cavity is divided into four quadrants, and name each quadrant.

18. Explain the principle of computed tomography (CT) scanning. Contrast it with conventional radiography in terms of principle and diagnostic value.
19. Distinguish between a CT scan and a roentgenogram.
20. Explain the principle and clinical application of the dynamic spatial reconstructor (DSR).
21. Define homeostasis. What is extracellular fluid? Why is it called the internal environment of the body?
22. Under what conditions are the internal environment said to be in homeostasis?
23. What is a stress? Give several examples. How is stress related to homeostasis?
24. How is homeostasis related to normal and abnormal conditions in the body?
25. Substantiate this statement: "Homeostasis is a cooperative effort of all body parts."
26. What systems of the body control homeostasis? Explain.
27. Discuss briefly how the regulation of blood pressure and blood glucose level are examples of homeostasis.
28. Define a feedback system. Distinguish between a negative and a positive feedback system.
29. Describe several situations that involve measurement of the human body.
30. Using Appendix A as a guide, solve the following problems.
    a. A bacterial cell measures 100 μm in length. Give its length in nanometers.
    b. A person's arm measures 2 ft in length. How many centimeters is this?
    c. The indicated dosage of a drug is 50 μg. How many milligrams is this?
    d. A person weighs 110 lb. Give the weight in kilograms.
    e. If you excrete 1,300 ml of urine in a day, how many liters is this?
    f. If you remove 15 cm^3 of blood from a patient, how many milliliters have you removed?

- Identify by name and symbol the principal chemical elements of the human body.

- Explain, by diagraming, the structure of an atom.

- Describe the principle of nuclear magnetic resonance (NMR) and its diagnostic importance.

- Define a chemical reaction as a function of electrons in incomplete outer energy levels.

- Describe ionic bond formation in a molecule of sodium chloride (NaCl).

- Discuss covalent bond formation as the sharing of outer energy level electrons.

- Explain the nature and importance of hydrogen bonding.

- Define a radioisotope and explain the principle of positron emission tomography (PET) scanning and its diagnostic importance.

- Explain the basic differences among synthesis, decomposition, exchange, and reversible chemical reactions.

- Discuss how chemical reactions are related to metabolism.

- Define the type of energy involved when chemical bonds are formed and broken.

- Define and distinguish between inorganic and organic compounds.

- Discuss the functions of water as a solvent, suspending medium, chemical reactant, heat absorber, and lubricant.

- List and compare the properties of acids, bases, and salts.

- Define pH as the degree of acidity or alkalinity of a solution.

- Explain the role of a buffer system as a homeostatic mechanism that maintains the pH of a body fluid.

- Compare the structure and functions of carbohydrates, lipids, and proteins.

- Differentiate between dehydration synthesis and hydrolysis of organic molecules.

- Define the homeostatic role of enzymes as catalysts.

- Contrast the structure of deoxyribonucleic acid (DNA) and ribonucleic acid (RNA).

- Define the roles of DNA and RNA in heredity and protein synthesis.

- Identify the function and importance of adenosine triphosphate (ATP) and cyclic adenosine-3',5'-monophosphate (cyclic AMP).

2 The Chemical Level of Organization

Many of the common substances you eat and drink—water, sugar, table salt, cooking oil—play vital roles in keeping you alive. In this chapter you will learn something about how the molecules of these substances function in your body. Fundamental to this study is a knowledge of basic chemistry and chemical processes. To understand the nature of the matter you are made from and the changes this matter goes through in your body, you will need to know which chemical elements are present in the human organism and how they interact.

INTRODUCTION TO BASIC CHEMISTRY

CHEMICAL ELEMENTS

All living and nonliving things consist of **matter,** which is anything that occupies space and has mass. Matter may exist in a solid, liquid, or gaseous state. All forms of matter are made up of a limited number of building units called **chemical elements,** substances that cannot be decomposed into simpler substances by ordinary chemical reactions. At present scientists recognize 106 different elements, of which 92 are naturally occurring. Elements are designated by letter abbreviations, usually derived from the first or first and second letters of the Latin or English name for the element. Such letter abbreviations are called **chemical symbols.** Examples of chemical symbols are H (hydrogen), C (carbon), O (oxygen), N (nitrogen), Na (sodium), K (potassium), Fe (iron), and Ca (calcium).

Approximately 24 elements are found in the human organism. Carbon, hydrogen, oxygen, and nitrogen make up about 96 percent of the body's weight. These 4 elements together with phosphorus and calcium constitute approximately 99 percent of the total body weight. Eighteen other chemical elements, called trace elements, are found in low concentrations and compose the remaining 1 percent.

STRUCTURE OF ATOMS

Each element is made up of units of matter called **atoms.** An element is simply a quantity of matter composed of atoms all of the same type. A handful of the element carbon, such as pure coal, contains only carbon atoms. A tank of oxygen contains only oxygen atoms. Measurements indicate that the smallest atoms are less than 0.00000001 cm (1/250,000,000 inch) in diameter, and the largest atoms are 0.00000005 cm (1/50,000,000 inch) in diameter. In other words, if 50 million of the largest atoms were placed end to end, they would measure approximately 1 inch in length.

An atom consists of two basic parts: the nucleus and the electrons (Figure 2-1). The centrally located **nucleus** comprises most of the atomic mass and contains positively charged particles called **protons** (p^+) and uncharged (neutral) particles called **neutrons** (n^0). Because each proton has one positive charge, the nucleus itself is positively charged. The second basic part of an atom contains the **electrons** (e^-). These negatively charged particles spin around the nucleus. The number of electrons in an atom of an element always equals the number of protons. Since each electron carries one negative charge, the negatively

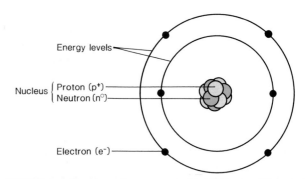

FIGURE 2-1 Structure of an atom. In this highly simplified version of a carbon atom, note the centrally located nucleus. The nucleus contains six neutrons and six protons, although all are not visible in this view since some are behind others. The six electrons orbit the nucleus at varying distances from its center.

charged electrons and the positively charged protons balance each other, and the atom is electrically neutral.

What makes the atoms of one element different from those of another? The answer lies in the number of protons. Figure 2-2 shows that the hydrogen atom contains one proton. The helium atom contains two. The carbon atom has six, and so on. Each different kind of atom has a different number of protons in its nucleus. The number of protons in an atom is called the atom's **atomic number.** Therefore we can say that each kind of atom, or element, has a different atomic number.

CLINICAL APPLICATION

New in the arsenal for diagnosing disease is **nuclear magnetic resonance (NMR).** It focuses on the nuclei of atoms of a single element in a tissue at a time and determines if the nuclei behave normally in response to an external force such as magnetism. In most studies to date, NMR imaging of hydrogen nuclei has been popular because of the body's large water content. The part of the body to be studied, ranging from a finger to the entire body, is placed in the scanner, exposing the nuclei to a uniform magnetic field. This causes the nuclei to line up in the direction of the magnetic field. Then the aligned nuclei are exposed to a brief burst of an alternating magnetic field at a 90° angle to the first magnetic field. When the magnetic field is cut off, the energy absorbed by the nuclei becomes a small electrical voltage. The voltage is picked up by detectors and relayed to a computer for analysis. The computerized reconstruction, called the **NMR image,** shows the density and energy loss (voltage) of the nuclei of a particular element and somewhat resembles a CT scan (Figure 2-3). The reconstructions can be displayed in color and as two- or three-dimensional images and indicate a biochemical blueprint of cellular activity.

The diagnostic advantage of NMR is that in addition to providing images of diseased organs and tissues, it also provides information about what chemicals are present in the organ and tissue. Such an analysis can indicate a particular disease is in progress, even before symptoms occur. NMR also offers the advantages of being noninvasive (not involving puncture or incision of the skin or insertion of an instrument or foreign material into the body), not utilizing radiation, and gathering biochemical information without time-consuming chemical analyses.

In clinical and preclinical trials, NMR studies have confirmed the findings of other diagnostic techniques such as CT scans and, in some cases, have added more detail. Thus, NMR has proved useful in identifying existing pathologies. Scientists hope that NMR can also be used to perform a "biopsy" on tumors without an operation, assess mental disorders, measure blood flow, study the evolution of hematomas and infarctions in the brain following a stroke, identify the potential for developing a stroke, assess treatment for conditions such as heart disease and stroke, monitor the progress and treatment of a disease, study the effects of toxic drugs on tissues, measure intracellular pH, study metabolism, and determine how donor organs from cadavers may function after transplantation.

ATOMS AND MOLECULES

When atoms combine with or break apart from other atoms, a **chemical reaction** occurs. In the process, new products with different properties are formed. Chemical reactions are the foundation of all life processes.

The electrons of the atom actively participate in chemical reactions. The electrons spin around the nucleus in orbits, shown in Figure 2-2 as concentric circles lying at varying distances from the nucleus. We call these orbits **energy levels.** Each orbit has a maximum number of electrons it can hold. For instance, the orbit nearest the nucleus never holds more than two electrons, no matter what the element. This orbit can be referred to as the first energy level. The second energy level holds a maximum of eight electrons. The third level of atoms whose atomic number is less than 20 also can hold a maximum of eight electrons. The third level of more complex atoms can hold a maximum of 18 electrons.

An atom always attempts to have its outermost orbit with the maximum number of electrons it can hold. To do this, the atom may give up an electron, take on an electron, or share an electron with another atom—whichever is easiest. Take a look at the chlorine atom. Its outermost energy level, which happens to be the third level, has seven electrons. Since the third level of an atom can hold a maximum of eight electrons, chlorine can be described as having a shortage of one electron. In fact, chlorine usually does try to pick up an extra electron. Sodium, by contrast, has only one electron in its outer level. This again happens to be the third energy level. It is much easier for sodium to get rid of the one electron than to fill the third level by taking on seven more electrons. Atoms of a few elements, like helium, have completely filled outer energy levels and do not need to gain or lose electrons. These are called **inert** elements.

Atoms with incompletely filled outer energy levels, like sodium and chlorine, tend to combine with other atoms

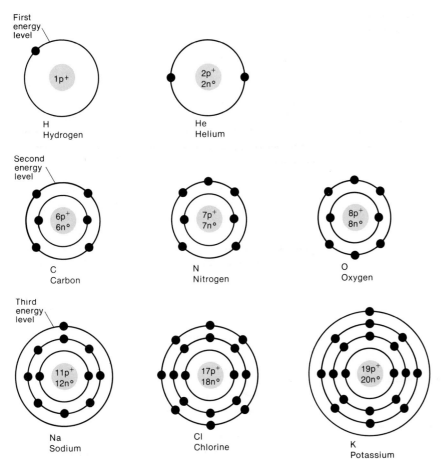

FIGURE 2-2 Atomic structures of some representative atoms.

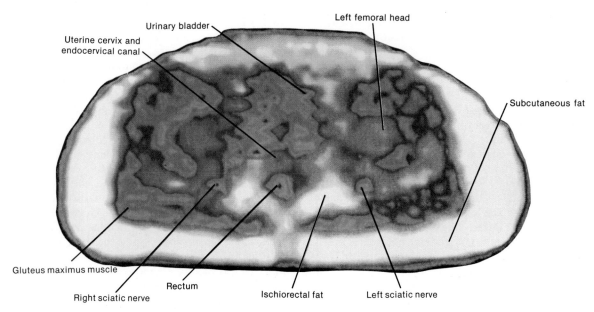

FIGURE 2-3 Nuclear magnetic resonance (NMR) image of the female pelvis. (Courtesy of John Cassese, Director of Marketing, Fonar Corporation, Melville, New York.)

in a chemical reaction. During the reaction, the atoms can trade off or share electrons and, thereby, fill their outer energy levels. Atoms that already have filled outer levels generally do not participate in chemical reactions for the simple reason that they do not need to gain or lose electrons. When two or more atoms combine in a chemical reaction, the resulting combination is called a **molecule** (MOL-e-kyool). A molecule may contain two atoms of the same kind, as in the hydrogen molecule: H_2. The subscript 2 indicates that there are two hydrogen atoms in the molecule. Molecules may also be formed by the reaction of two or more different kinds of atoms, as in the hydrochloric acid molecule: HCl. Here an atom of hydrogen is attached to an atom of chlorine. A molecule that contains at least two different kinds of atoms is called a **compound.** Hydrochloric acid, which is present in the digestive juices of the stomach, is a compound. A molecule of hydrogen is not.

The atoms in a molecule are held together by forces of attraction called **chemical bonds.** Here we will consider ionic bonds, covalent bonds, and hydrogen bonds.

Ionic Bonds

Atoms are electrically neutral because the number of positively charged protons equals the number of negatively charged electrons. But when an atom gains or loses electrons, this balance is destroyed. If the atom gains electrons, it acquires an overall negative charge. If the atom loses electrons, it acquires an overall positive charge. Such a negatively or positively charged atom or group of atoms is called an **ion** (Ī-on).

Consider the sodium ion (Figure 2-4a). The sodium atom (Na) has 11 protons and 11 electrons, with 1 electron in its outer energy level. When sodium gives up the single electron in its outer level, it is left with 11 protons and

only 10 electrons. The atom now has an overall positive charge of one (+1). This positively charged sodium atom is called a sodium ion (written Na^+).

Another example is the formation of the chloride ion (Figure 2-4b). Chlorine has a total of 17 electrons, 7 of them in the outer energy level. Since this energy level can hold 8 electrons, chlorine tends to pick up an electron that has been lost by another atom. By accepting an electron, chlorine acquires a total of 18 electrons. However, it still has only 17 protons in its nucleus. The chloride ion therefore has a negative charge of one (−1) and is written as Cl^-.

The positively charged sodium ion (Na^+) and the negatively charged chloride ion (Cl^-) attract each other—unlike charges attract each other. The attraction, called an **ionic bond,** holds the two atoms together, and a molecule is formed (Figure 2-4c). The formation of this molecule, sodium chloride (NaCl) or table salt, is one of the most common examples of ionic bonding. Thus an ionic bond is an attraction between atoms in which one atom loses electrons and another atom gains electrons. Generally, atoms whose outer energy level is less than half-filled lose electrons and form positively charged ions called **cations** (KAT-ī-ons). Examples of cations are potassium ion (K^+), calcium ion (Ca^{2+}), iron ion (Fe^{2+}), and sodium ion (Na^+). By contrast, atoms whose outer energy level is more than half-filled tend to gain electrons and form negatively charged ions called **anions** (AN-ī-ons). Examples of anions include iodine ion (I^-), chloride ion (Cl^-), and sulfur ion (S^{2-}).

Notice that an ion is always symbolized by writing the chemical abbreviation followed by the number of positive (+) or negative (−) charges the ion acquires.

Hydrogen is an example of an atom whose outer level is exactly half-filled. The first energy level can hold two electrons, but in hydrogen atoms it contains only one.

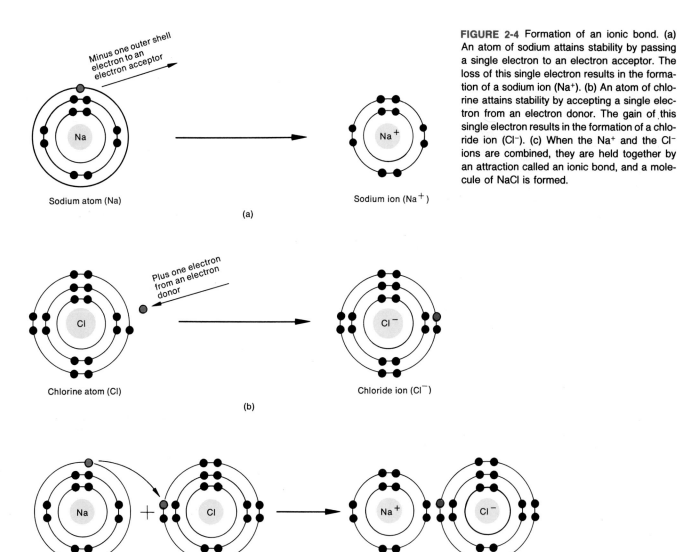

Sodium atom (Na)

Minus one outer shell electron to an electron acceptor

Sodium ion (Na⁺)

(a)

FIGURE 2-4 Formation of an ionic bond. (a) An atom of sodium attains stability by passing a single electron to an electron acceptor. The loss of this single electron results in the formation of a sodium ion (Na⁺). (b) An atom of chlorine attains stability by accepting a single electron from an electron donor. The gain of this single electron results in the formation of a chloride ion (Cl⁻). (c) When the Na⁺ and the Cl⁻ ions are combined, they are held together by an attraction called an ionic bond, and a molecule of NaCl is formed.

Chlorine atom (Cl)

Plus one electron from an electron donor

Chloride ion (Cl⁻)

(b)

Sodium atom (Na)   Chlorine atom (Cl)

Sodium ion   Chloride ion

(c)

Sodium chloride (NaCl)

Hydrogen may lose its electron and become a positive ion (H^+). This is precisely what happens when hydrogen combines with chlorine to form hydrochloric acid (H^+Cl^-). However, hydrogen is equally capable of forming another kind of bond altogether: a covalent bond.

Covalent Bonds

The second chemical bond to be considered is the **covalent bond.** This bond is far more common in organisms than is the ionic bond. When a covalent bond is formed, neither of the combining atoms loses or gains an electron. Instead, the two atoms share one, two, or three electron pairs. Look at the hydrogen atom again. One way a hydrogen atom can fill its outer energy level is to combine with another hydrogen atom to form the molecule H_2 (Figure 2-5a). In the H_2 molecule, the two atoms share a pair of electrons. Each hydrogen atom has its own electron plus one electron from the other atom. The shared pair

actually circles the nuclei of both atoms. Therefore the outer energy levels of both atoms are filled. When one pair of electrons is shared between atoms, as in the H_2 molecule, a *single covalent bond* is formed. A single covalent bond is expressed as a single line between the atoms (H—H). When two pairs of electrons are shared between two atoms, a *double covalent bond* is formed, which is expressed as two parallel lines (=) (Figure 2-5b). A *triple covalent bond,* expressed by three parallel lines (≡), occurs when three pairs of electrons are shared (Figure 2-5c).

The same principles that apply to covalent bonding between atoms of the same element also apply to atoms of different elements. Methane (CH_4) is an example of covalent bonding between atoms of different elements (Figure 2-5d). The outer energy level of the carbon atom can hold eight electrons, but has only four of its own. Each hydrogen atom can hold two electrons, but has only one of its own. In the methane molecule the carbon

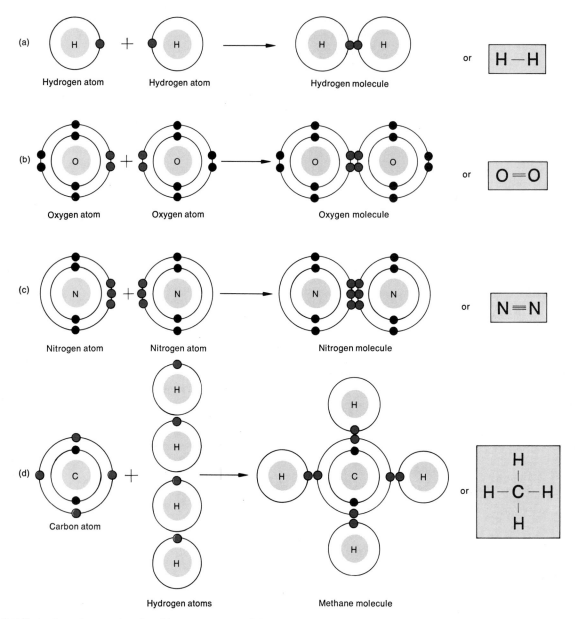

FIGURE 2-5 Formation of a covalent bond between atoms of the same element and between atoms of different elements. (a) A single covalent bond between two hydrogen atoms. (b) A double covalent bond between two oxygen atoms. (c) A triple covalent bond between two nitrogen atoms. (d) Single covalent bonds between a carbon atom and four hydrogen atoms. In the symbols on the right, each covalent bond is represented by a straight line between atoms.

atom shares four pairs of electrons. One pair is shared with each hydrogen atom. Each of the four carbon electrons orbits around both the carbon nucleus and a hydrogen nucleus. Each hydrogen electron circles around its own nucleus and the carbon nucleus.

Elements whose outer energy levels are half-filled, such as hydrogen and carbon, form covalent bonds quite easily. In fact, carbon always forms covalent bonds. It never becomes an ion. However, many atoms whose outer energy levels are more than half-filled also form covalent bonds. An example is oxygen. We won't go into the reasons why some atoms tend to form covalent bonds rather than ionic bonds.

Hydrogen Bonds

A **hydrogen bond** consists of a hydrogen atom covalently bonded to one oxygen atom or one nitrogen atom, but shared by another oxygen or nitrogen atom. Because hydrogen bonds are weak, only about 5 percent as strong as covalent bonds, they do not bind atoms into molecules. However, they do serve as bridges between different molecules or between various parts of the same molecule. The weak bonds may be formed and broken fairly easily. It is this property that accounts for the temporary bonding between certain atoms within large complex molecules such as proteins and nucleic acids. It should be noted that even though hydrogen bonds are relatively weak, such large molecules may contain several hundred of these bonds, resulting in considerable strength and stability.

Chemical reactions are nothing more than the making or breaking of bonds between atoms. And these reactions occur continually in all the cells of your body. As you will see again and again, reactions are the processes by which body structures are built and body functions carried out.

RADIOISOTOPES

Atoms of an element, although chemically alike, may have different nuclear masses because of one or more extra neutrons in some of the atoms, so the atomic weight assigned to an element is only an average. Each of the chemically identical atoms of an element with a particular nuclear mass is an **isotope** of that element. Isotopes are named by a number that indicates their atomic mass, the sum of their neutrons and protons. Certain isotopes called **radioisotopes** are unstable—they "decay" or change their nuclear structure to a more stable configuration. And in decaying they emit radiation that can be detected by instruments. These instruments estimate the amount of radioisotope present in a part of the body or in a sample of material and form an image of its distribution.

Radioisotopes of iodine were among the first discovered, and their specificity for the thyroid gland made them the cornerstone for the study of thyroid physiology. Nuclear medicine now uses ^{32}P to treat leukemia and ^{59}Fe for the study of red blood cell production. Moreover, short-lived agents such as technetium-99m pertechnetate (^{99m}Tc) have improved the quality of the images and reduced the patient's radiation dose.

CLINICAL APPLICATION

In Chapter 1 you learned that conventional radiographic techniques, CT scanners, and the DSR use radiation in the form of x-rays to produce images that are invaluable in the diagnosis of disease. In **radioisotopic scanning** the chemical activity of various tissues of the body can be recorded. Physicians can diagnose certain diseases by charting the course of an injected radioisotope through the body and by noting how much of the substance concentrates and where. Radioisotope scans, like the DSR, can provide early warning of disease.

Although radioisotope scans have been used in diagnosis since the 1950s, their value was limited by distorted images of the distribution of a radioisotope. Within the past few years more precise localization of radioisotopes in the body has been made possible by new computer imaging techniques similar to those used in CT scanning. The result is a sophisticated version of radioisotope scanning called **positron emission tomography (PET).**

The principle behind PET scanning is as follows. Short-lived radioisotopes such as ^{11}C, ^{13}N, or ^{15}O are produced and incorporated into a solution that can be injected into the body. As the radioisotope circulates

through the body, it emits positively charged electrons called *positrons*. Positrons collide with negatively charged electrons in body tissues, causing their annihilation and the release of gamma rays. The gamma rays travel in opposite directions and are detected and recorded by PET receptors. A computer then takes the information and constructs a colored **PET scan** that shows where the radioisotopes are being used in the body (Figure 2-6).

Although PET may not become a clinical tool for several years, it has already provided information that cannot be obtained by any other technique. Using PET, physicians can study the effects of drugs in body organs, measure blood flow through organs such as the brain and heart, identify the extent of damage as a result of strokes or heart attacks, and detect cancers and measure the effects of treatment. PET studies with schizophrenic and manic-depressive patients have revealed that schizophrenics tend to use less glucose in certain brain areas, whereas manic-depressive patients use more glucose during manic phases. PET technology is also now being used to study the chemical changes that occur during epileptic seizures and in persons suffering from senile dementia. It is hoped that PET could eventually become a routine part of psychiatric examinations, helping to diagnose patients who show symptoms of more than one kind of mental illness. As impressive as PET is in the diagnosis of disease, it is also being used by scientists to probe the healthy brain. For example, by detecting and recording changes in glucose consumption in the brain, scientists can identify which specific areas of the brain are involved in specific sensory and motor activities (see Figure 14-11).

CHEMICAL REACTIONS

In this section, we look at four basic types of chemical reactions. These reactions are simple, yet central to life processes. Once you have learned them, you will be able to understand the chemical reactions discussed later.

Synthesis Reactions—Anabolism

When two or more atoms, ions, or molecules combine to form new and larger molecules, the process is called **a synthesis reaction.** The word *synthesis* means "combination," and synthesis reactions involve the *forming of new bonds.* Synthesis reactions can be expressed in the following way:

$$A \quad + \quad B \quad \rightarrow \quad AB$$

Atom, ion,   Atom, ion,   Combine to form new
or molecule A   or molecule B   molecule AB

The combining substances, A and B, are called the **reactants;** the substance formed by the combination is the **end product.** The arrow indicates the direction in which the reaction is proceeding. An example of a synthesis reaction is:

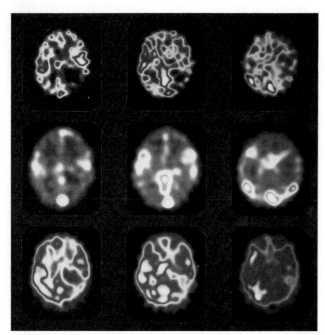

FIGURE 2-6 Positron emission tomography (PET) scans. (Top row) Oxygen metabolism in the brain. (Middle row) Blood volume in the brain. (Bottom row) Blood flow through the brain. (Courtesy of Dr. Michel M. Ter-Rogossian, Washington University, School of Medicine.)

$$N \ + \ 3H \ \rightarrow \ NH_3$$

Nitrogen   Hydrogen   Ammonia
  atom   atoms   molecule

All the synthesis reactions that occur in your body are collectively called anabolic reactions, or simply **anabolism** (a-NAB-ō-lizm). Combining glucose molecules to form glycogen and combining amino acids to form proteins are two examples of anabolism. The importance of anabolism is considered in detail in Chapter 25.

Decomposition Reactions—Catabolism

The reverse of a synthesis reaction is a **decomposition reaction.** The word *decompose* means to break down into smaller parts. In a decomposition reaction, the *bonds are broken.* Large molecules are broken down into smaller molecules, ions, or atoms. A decomposition reaction occurs in this way:

$$AB \quad \rightarrow \quad A \quad + \quad B$$

Molecule AB   Atom, ion, or   Atom, ion,
breaks down   molecule A   or molecule B
  into

Under the proper conditions, methane can decompose into carbon and hydrogen:

$$CH_4 \ \rightarrow \ C \ + \ 4H$$

Methane   Carbon   Hydrogen
molecule   atom   atoms

The subscript 4 on the left-hand side of the reaction equation indicates that four atoms of hydrogen are bonded to one carbon atom in the methane molecule. The number 4 on the right-hand side of the equation shows that four single hydrogen atoms have been set free.

All the decomposition reactions that occur in your body are collectively called catabolic reactions, or simply **catabolism** (ka-TAB-ō-lizm). The digestion and oxidation of food molecules are examples of catabolism. The importance of catabolism is also considered in detail in Chapter 25.

Exchange Reactions

All chemical reactions are based on synthesis or decomposition processes. In other words, chemical reactions are simply the making and/or breaking of ionic or covalent bonds. Many reactions, such as **exchange reactions,** are partly synthesis and partly decomposition. An exchange reaction works like this:

$$AB + CD \rightarrow AD + BC \quad or \quad AC + BD$$

The bonds between A and B and between C and D are broken in a decomposition process. New bonds are then formed between A and D and between B and C or between A and C and between B and D in a synthesis process.

Reversible Reactions

When chemical reactions are reversible, the end product can revert to the original combining molecules. A **reversible reaction** is indicated by two arrows:

$$A + B \underset{\text{breaks down to}}{\overset{\text{combines with}}{\rightleftharpoons}} AB$$

Some reversible reactions occur because neither the reactants nor the end products are stable. Other reactions reverse themselves only under special conditions:

$$A + B \underset{\text{water}}{\overset{\text{heat}}{\rightleftharpoons}} AB$$

Whatever is written above or below the arrows indicates the special condition under which the reaction occurs. In this case, A and B react to produce AB only when heat is applied, and AB breaks down into A and B only when water is added. Figure 2-7 summarizes the basic chemical reactions that can occur.

Metabolism

The word **metabolism** (me-TAB-ō-lizm) represents the sum of all the synthesis and decomposition reactions occurring in the body—the sum of all anabolic and catabolic reactions that go on inside you. When we say that a person has a high metabolism that is due to a disruption of homeostasis, we mean that the chemical reactions in the body are proceeding at a faster rate than normal. The decomposition reactions are occurring so quickly that foods are broken down completely before the body has a chance to store them. Consequently, people with a high metabolism can usually eat a great deal without gaining weight. Because rapid decomposition reactions generate large amounts of energy, including heat, such people appear to have a lot of "nervous" energy and often complain that they are too hot.

GENERAL NATURE   SPECIFIC EXAMPLE

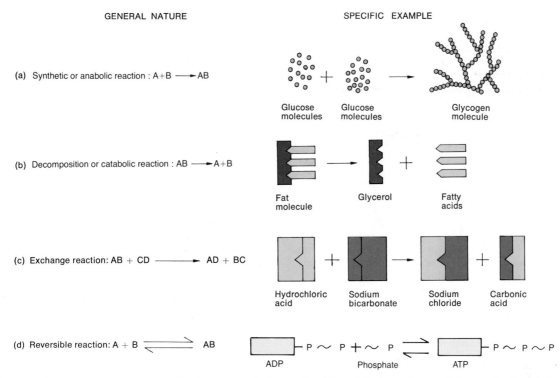

(a) Synthetic or anabolic reaction : A+B ⟶ AB

Glucose   Glucose   Glycogen
molecules   molecules   molecule

(b) Decomposition or catabolic reaction : AB ⟶ A+B

Fat   Glycerol   Fatty
molecule   acids

(c) Exchange reaction: AB + CD ⟶ AD + BC

Hydrochloric   Sodium   Sodium   Carbonic
acid   bicarbonate   chloride   acid

(d) Reversible reaction: A + B ⇌ AB

ADP   Phosphate   ATP

FIGURE 2-7 Kinds of chemical reactions. (a) Synthetic or anabolic reaction. When linked together as shown, molecules of glucose form a molecule of glycogen. Glucose is a sugar that is the primary source of energy. Glycogen is a storage form of that sugar found in the liver and skeletal muscles. (b) Decomposition or catabolic reaction. The example shown is a molecule of fat breaking down into glycerol and fatty acids. This reaction occurs whenever a food that contains fat is digested. (c) Exchange reaction. In this reaction, atoms of different molecules are exchanged with each other. Shown is a buffer reaction in which the body eliminates strong acids to help maintain homeostasis. (d) Reversible reaction. ATP (adenosine triphosphate) is an important source of stored energy. When the energy is needed, the ATP breaks down into ADP (adenosine diphosphate) and PO_4 (phosphate group), and energy is released in the reaction. The phosphate group is symbolized as P. The cells of the body reconstruct ATP by using the energy of foods to attach ADP to PO_4.

Chemical reactions in the bodies of people with low metabolism that is due to a disruption of homeostasis proceed more slowly than normal. Food is broken down slowly. Much of it is only partially broken down and then stored. Such people tend to gain weight easily, have little energy, and are often cold. Because their synthesis reactions are also slowed down, their bodies build up new structures very slowly. Wounds, for instance, often take a long time to heal.

Energy and Chemical Reactions

Energy is involved whenever bonds between atoms in molecules are formed or broken during the chemical reactions taking place in the body. When a chemical bond is formed, energy is required. When a bond is broken, energy is released. This means that synthesis reactions need energy in order to occur, whereas decomposition reactions give off energy. The building processes of the body—the construction of bones, the growth of hair and nails, the replacement of injured cells—occur basically through synthesis reactions. The breakdown of foods, on the other hand, occurs through decomposition reactions. When foods are decomposed, they release energy that can be used by the body for its building processes. Re-

leased energy can also be used to warm the body by taking the form of **heat energy.** Foods can be partially broken down into compounds that can be stored in the body. Later, when additional energy is required, the body finishes breaking down these reserve compounds. This stored energy is called **potential energy.**

Another form of energy that is important to life processes, **kinetic energy,** is discussed in Chapter 3.

CHEMICAL COMPOUNDS AND LIFE PROCESSES

Most of the chemicals in the body exist in the form of compounds. Biologists and chemists divide these compounds into two principal classes: inorganic compounds, which usually lack carbon, and organic compounds, which always contain carbon. **Inorganic compounds** are usually small, ionically bonded molecules that are vital to body functions. They include water, many salts, acids, and bases. **Organic compounds** are held together mostly or entirely by covalent bonds. They tend to be very large molecules and are therefore good building blocks for body structures. Organic compounds present in the body include carbohydrates, lipids, proteins, nucleic acids, and ATP.

INORGANIC COMPOUNDS

Water

One of the most important, as well as the most abundant, inorganic substances in the human organism is **water.** In fact, with a few exceptions, such as tooth enamel and bone tissue, water is by far the most abundant material in all tissues. About 60 percent of red blood cells, 75 percent of muscle tissue, and 92 percent of blood plasma is water. The following functions of water explain why it is such a vital compound in living systems:

1. Water is an excellent solvent and suspending medium. A *solvent* is a liquid or gas in which some other material (solid, liquid, or gas), called a *solute,* has been dissolved. The combination of solvent plus solute is called a **solution,** and one common example of a solution is salt water. A solute, such as salt in water, typically does not settle out of its solution. The solute can be retrieved through a chemical reaction or, in some cases, by boiling off the solvent. In a **suspension,** by contrast, the suspended material mixes with the liquid or suspending medium, but it will eventually settle out of the mixture. An example of a suspension is cornstarch and water. If the two materials are shaken together, a milky mixture forms. After the mixture sits for a while, however, the water clears at the top and the cornstarch settles to the bottom.

The solvating property of water is essential to health and survival. For example, water in the blood forms a solution with a small portion of the oxygen you inhale, allowing the oxygen to be carried to your body cells. Water in the blood also dissolves a small portion of the carbon dioxide that is carried from the cells to the lungs to be exhaled. Furthermore, if the surfaces of the air sacs in your lungs are not moist, oxygen cannot dissolve and, therefore, cannot move into your blood to be distributed throughout your body. Water, moreover, is the solvent that carries nutrients into and wastes out of your body cells.

As a suspending medium, water is also vital to your survival. Many large organic molecules are suspended in the water of your body cells. These molecules are consequently able to come in contact with other chemicals, allowing various essential chemical reactions to occur.

2. Water can participate in chemical reactions. During digestion, for example, water can be added to large nutrient molecules in order to break them down into smaller molecules. This kind of breakdown is necessary if the body is to utilize the energy in nutrients. Water molecules are also used in synthesis reactions. Such reactions occur in the production of hormones and enzymes.

EXHIBIT 2-1

DISSOCIATION OF REPRESENTATIVE SALTS INTO IONS THAT PROVIDE ESSENTIAL CHEMICAL ELEMENTS FOR THE BODY

SALT	DISSOCIATES INTO	CATION		ANION
NaCl Sodium chloride	$\longrightarrow$	Na^+ Sodium ion	+	Cl^- Chloride ion
KCl Potassium chloride	$\longrightarrow$	K^+ Potassium ion	+	Cl^- Chloride ion
$CaCl_2$ Calcium chloride	$\longrightarrow$	Ca^{2+} Calcium ion	+	$2Cl^-$ Chloride ions
$MgCl_2$ Magnesium chloride	$\longrightarrow$	Mg^{2+} Magnesium ion	+	$2Cl^-$ Chloride ions
$CaCO_3$ Calcium carbonate	$\longrightarrow$	Ca^{2+} Calcium ion	+	$CO_3{}^{2-}$ Carbonate ion
$Ca_3(PO_4)_2$ Calcium phosphate	$\longrightarrow$	$3Ca^{2+}$ Calcium ions	+	$2PO_4{}^{3-}$ Phosphate ions
Na_2SO_4 Sodium sulfate	$\longrightarrow$	$2Na^+$ Sodium ions	+	$SO_4{}^{2-}$ Sulfate ion

3. Water absorbs and releases heat very slowly. In comparison to other substances, water requires a large amount of heat to increase its temperature and a great loss of heat to decrease its temperature. Thus the presence of a large amount of water moderates the effects of fluctuations in environmental temperature and thereby helps to maintain a homeostatic body temperature.

4. Water requires a large amount of heat to change from a liquid to a gas. When water (perspiration) evaporates from the skin, it takes with it large quantities of heat and provides an excellent cooling mechanism.

5. Water serves as a lubricant in various regions of the body. It is a major part of mucus and other lubricating fluids. Lubrication is especially necessary in the chest and abdomen, where internal organs touch and slide over each other. It is also needed at joints, where bones, ligaments, and tendons rub against each other. In the digestive tract, water moistens foods to ensure their smooth passage.

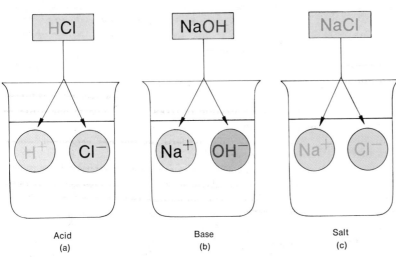

FIGURE 2-8 Ionization of acids, bases, and salts. (a) When placed in water, hydrochloric acid (HCl) dissociates into H^+ ions and Cl^- ions. Acids are proton donors. (b) When the base sodium hydroxide (NaOH) is placed in water, it dissociates into OH^- ions and Na^+ ions. Bases are proton acceptors. (c) When table salt (NaCl) is placed in water, it dissociates into positive and negative ions (Na^+ and Cl^-), neither of which is H^+ or OH^-.

Acids, Bases, and Salts

When molecules of inorganic acids, bases, or salts are dissolved in water in the body cells, they undergo **ionization** (ī'-on-i-ZĀ-shun) or **dissociation** (dis'-sō-sē-Ā-shun), that is, they break apart into ions. Such molecules are also called **electrolytes** (ē-LEK-trō-līts) because the solution will conduct an electric current (the chemistry and importance of electrolytes are discussed in detail in Chapter 27). An **acid** may be defined as a substance that dissociates into one or more *hydrogen ions* (H⁺) and one or more negative ions (anions). An acid may also be defined as a proton (H⁺) donor. A **base,** by contrast, dissociates into one or more *hydroxyl ions* (OH⁻) and one or more positive ions (cations). A base may also be viewed as a proton acceptor. Hydroxyl ions, as well as some other negative ions, have a strong attraction for protons. A **salt,** when dissolved in water, dissociates into cations and anions, neither of which is H⁺ or OH⁻ (Figure 2-8). Acids and bases react with one another to form salts. For example, the combination of hydrochloric acid (HCl), an acid, and sodium hydroxide (NaOH), a base, produces sodium chloride (NaCl), a salt, and water (H_2O).

Many salts are found in the body. Some are in cells, whereas others are in the body fluids, such as lymph, blood, and the extracellular fluid of tissues. The ions of salts are the source of many essential chemical elements. Exhibit 2-1 shows how salts dissociate into ions that provide these elements. Chemical analyses reveal that sodium and chloride ions are present in higher concentrations than other ions in extracellular body fluids. Inside the cells, phosphate and potassium ions are more abundant than other ions. Chemical elements such as sodium, phosphorus, potassium, or iodine are present in the body only in chemical combination with other elements or as ions. Their presence as free, un-ionized atoms could be instantly fatal. Exhibit 2-2 lists representative elements found in the body.

Acid–Base Balance: The Concept of pH

The fluids of your body must maintain a fairly constant balance of acids and bases. In solutions such as those found in body cells or in extracellular fluids, acids dissociate into hydrogen ions (H⁺) and anions. Bases, on the other hand, dissociate into hydroxyl ions (OH⁻) and cations. The more hydrogen ions that exist in a solution, the more acid the solution; conversely, the more hydroxyl ions, the more basic (alkaline) the solution. The term **pH** is used to describe the degree of *acidity* or *alkalinity* (*basicity*) of a solution. Biochemical reactions—reactions that occur in living systems—are extremely sensitive to even small changes in the acidity or alkalinity of the environment in which they occur. In fact, H⁺ and OH⁻ ions are involved in practically all biochemical processes, and the functions of cells are modified greatly by any departure from narrow limits of normal H⁺ and OH⁻ concentrations. For this reason, the acids and bases that are constantly formed in the body must be kept in balance.

A solution's acidity or alkalinity is expressed on a **pH scale** that runs from 0 to 14 (Figure 2-9). The pH scale is based on the number of H⁺ ions in a solution expressed

EXHIBIT 2-2

REPRESENTATIVE CHEMICAL ELEMENTS FOUND IN THE BODY

CHEMICAL ELEMENT	COMMENT
Oxygen (O)	Constituent of water and organic molecules; functions in cellular respiration.
Carbon (C)	Found in every organic molecule.
Hydrogen (H)	Constituent of water, all foods, and most organic molecules.
Nitrogen (N)	Component of all protein molecules and nucleic acid molecules.
Calcium (Ca)	Constituent of bone and teeth; required for blood clotting, intake (endocytosis) and output (exocytosis) of substances through plasma membranes, motility of cells, movement of chromosomes prior to cell division, glycogen metabolism, synthesis and release of transmitter substances, and contraction of muscle.
Phosphorus (P)	Component of many proteins, nucleic acids, ATP, and cyclic AMP; required for normal bone and tooth structure; found in nerve tissue.
Chlorine (Cl)	Cl⁻ is an anion of NaCl, a salt important in water movement between cells.
Sulfur (S)	Component of many proteins, especially the contractile protein of muscle.
Potassium (K)	Required for growth and important in conduction of nerve impulses and muscle contraction.
Sodium (Na)	Na⁺ is a cation of NaCl; structural component of bone; essential in blood to maintain water balance; needed for conduction of nerve impulses.
Magnesium (Mg)	Component of many enzymes.
Iodine (I)	Vital to functioning of thyroid gland.
Iron (Fe)	Essential component of hemoglobin and respiratory enzymes.

in chemical units called moles per liter. A pH of 7 means that a solution contains one ten-millionth (0.0000001) of a mole of H⁺ ions per liter. The number 0.0000001 is written 10^{-7} in exponential form. To convert this value to pH, the negative exponent (−7) is converted into the positive number 7. A solution with a concentration of 0.0001 (10^{-4}) of H⁺ ions per liter has a pH of 4, a solution with a concentration of 0.000000001 (10^{-9}) has a pH of 9, and so on.

A solution that is zero on the pH scale has many H⁺ ions and few OH⁻ ions. A solution that rates 14, by contrast, has many OH⁻ ions and few H⁺ ions. The midpoint in the scale is 7, where the concentration of H⁺ and OH⁻ ions is equal (see Exhibit 2-3). A substance with a pH of 7, such as pure water, is neutral. A solution that has more H⁺ ions than OH⁻ ions is an *acid solution* and has a pH below 7. A solution that has more OH⁻

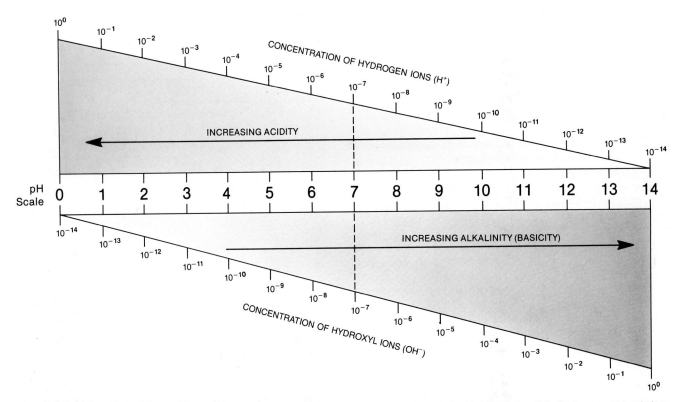

FIGURE 2-9 pH scale. At pH 7 (neutrality), the concentration of H^+ and OH^- ions is equal. A pH value below 7 indicates an acid solution; that is, there are more H^+ ions than OH^- ions. The lower the numerical value of the pH, the more acid the solution is because the H^+ ion concentration becomes progressively greater. A pH value above 7 indicates an alkaline (basic) solution, that is, there are more OH^- ions than H^+ ions. The higher the numerical value of the pH, the more alkaline the solution is because the OH^- ion concentration becomes progressively greater. A change in one whole number on the pH scale represents a tenfold change from the previous concentration.

ions than H^+ ions is a *basic* or *alkaline solution* and has a pH above 7. A change of one whole number on the pH scale represents a tenfold change from the previous concentration. That is, a pH of 2 indicates 10 times fewer H^+ ions than a pH of 1. A pH of 3 indicates 10 times fewer H^+ ions than a pH of 2 and 100 times fewer H^+ ions than a pH of 1.

Maintaining pH: Buffer Systems

Although the pH of body fluids may differ, the normal limits for the various fluids are generally quite specific and narrow. Exhibit 2-4 shows the pH values for certain body fluids compared with common substances. Even though strong acids and bases are continually taken into

EXHIBIT 2-3
RELATIONSHIP OF pH SCALE TO RELATIVE CONCENTRATIONS OF HYDROGEN (H^+) AND HYDROXYL (OH^-) IONS

CONCENTRATION OF H^+ IONS		pH	CONCENTRATION OF OH^- IONS	
1.0	10^0	0	10^{-14}	0.00000000000001
0.1	10^{-1}	1	10^{-13}	0.0000000000001
0.01	10^{-2}	2	10^{-12}	0.000000000001
0.001	10^{-3}	3	10^{-11}	0.00000000001
0.0001	10^{-4}	4	10^{-10}	0.0000000001
0.00001	10^{-5}	5	10^{-9}	0.000000001
0.000001	10^{-6}	6	10^{-8}	0.00000001
0.0000001	10^{-7}	Neutrality 7 Neutrality	10^{-7}	0.0000001
0.00000001	10^{-8}	8	10^{-6}	0.000001
0.000000001	10^{-9}	9	10^{-5}	0.00001
0.0000000001	10^{-10}	10	10^{-4}	0.0001
0.00000000001	10^{-11}	11	10^{-3}	0.001
0.000000000001	10^{-12}	12	10^{-2}	0.01
0.0000000000001	10^{-13}	13	10^{-1}	0.1
0.00000000000001	10^{-14}	14	10^0	1.0

EXHIBIT 2-4

NORMAL pH VALUES OF REPRESENTATIVE SUBSTANCES

SUBSTANCE	pH VALUE
Gastric juice (digestive juice of the stomach)	1.2–3.0
Lemon juice	2.2–2.4
Grapefruit juice	3.0
Cider	2.8–3.3
Pineapple juice	3.5
Tomato juice	4.2
Clam chowder	5.7
Urine	5.0–7.8
Saliva	6.35–6.85
Milk	6.6–6.9
Pure (distilled) water	7.0
Blood	7.35–7.45
Semen (fluid containing sperm)	7.35–7.50
Cerebrospinal fluid (fluid associated with nervous system)	7.4
Pancreatic juice (digestive juice of the pancreas)	7.1–8.2
Eggs	7.6–8.0
Bile (liver secretion that aids in fat digestion)	7.6–8.6
Milk of magnesia	10.0–11.0
Limewater	12.3

the body, the pH levels of these body fluids remain relatively constant. The mechanisms that maintain these homeostatic pH values in the body are called **buffer systems.**

The essential function of a buffer system is to react with strong acids or bases in the body and replace them with weak acids or bases that can change the normal pH values only slightly. Strong acids (or bases) ionize easily and contribute many H^+ (or OH^-) ions to a solution. They therefore change the pH drastically. Weak acids (or bases) do not ionize so easily. They contribute fewer H^+ (or OH^-) ions and have little effect on the pH. The chemicals that change strong acids or bases into weak ones are called **buffers** and are found in the body's fluids. Of the several buffer systems, we will examine only the *carbonic acid–bicarbonate buffer system*—the most important one found in extracellular fluid.

The carbonic acid–bicarbonate buffer system consists of a *pair* of compounds. One of them is a *weak acid*, and the other is a *weak base*. The weak acid of the buffer pair is *carbonic acid* (H_2CO_3); the weak base is *sodium bicarbonate* ($NaHCO_3$). The carbonic acid is a proton (H^+) donor and the bicarbonate ion of the sodium bicarbonate is a proton acceptor. In solution, the members of this buffer pair dissociate as follows:

Acidic component:
$$H_2CO_3 \rightleftharpoons H^+ + HCO_3^-$$

Carbonic acid   Hydrogen ion   Bicarbonate ion

Basic component:
$$NaHCO_3 \rightleftharpoons Na^+ + HCO_3^-$$

Sodium bicarbonate   Sodium ion   Bicarbonate ion

Each member of the buffer pair has a specific role in helping the body maintain a constant pH. If the body's pH is threatened by the presence of a strong acid, the weak base of the buffer pair goes into operation. If the body's pH is threatened by a strong base, the weak acid goes into play.

Consider the following situation. If a strong acid, such as HCl, is added to extracellular fluid, the weak base of the buffer system goes to work and the following acid-buffering reaction occurs:

$$HCl + NaHCO_3 \rightleftharpoons NaCl + H_2CO_3$$

Hydrochloric acid (strong acid)   Sodium bicarbonate (weak base of buffer system)   Sodium chloride (salt)   Carbonic acid (weak acid)

The chloride ion of HCl and the sodium ion of sodium bicarbonate combine to form NaCl, a substance that has no effect on pH. The hydrogen ion of the HCl could greatly lower pH by making the solution more acid, but this H^+ ion combines with the bicarbonate ion (HCO_3^-) of sodium bicarbonate to form carbonic acid, a weak acid that lowers pH only slightly. In other words, because of the action of the weak base of the buffer system, the strong acid (HCl) has been replaced by a weak acid and a salt, and the pH remains relatively constant.

Now suppose a strong base, such as sodium hydroxide (NaOH), is added to the extracellular fluid. In this instance, the weak acid of the buffer system goes to work and the following base-buffering reaction takes place:

$$NaOH + H_2CO_3 \rightleftharpoons H_2O + NaHCO_3$$

Sodium hydroxide (strong base)   Carbonic acid (weak acid of buffer system)   Water   Sodium bicarbonate (weak base)

In this reaction, the OH^- ion of sodium hydroxide could greatly raise the pH of the solution by making it more alkaline. However, the OH^- ion combines with an H^+ ion of carbonic acid and forms water, a substance that has no effect on pH. In addition, the Na^+ ion of sodium hydroxide combines with the bicarbonate ion (HCO_3^-) to form sodium bicarbonate, a weak base that has little effect on pH. Thus, because of the action of the buffer system, the strong base is replaced by water and a weak base, and the pH remains relatively constant.

Whenever a buffering reaction occurs, the concentration of one member of the buffer pair is increased while the concentration of the other decreases. When a strong acid is buffered, for example, the concentration of carbonic acid is increased, but the concentration of sodium bicarbonate is decreased. This happens because carbonic acid is produced and sodium bicarbonate is used up in the acid-buffering reaction. When a strong base is buffered, the concentration of sodium bicarbonate is increased, but the concentration of carbonic acid is decreased because sodium bicarbonate is produced and carbonic acid is used up in the base-buffering reaction. When

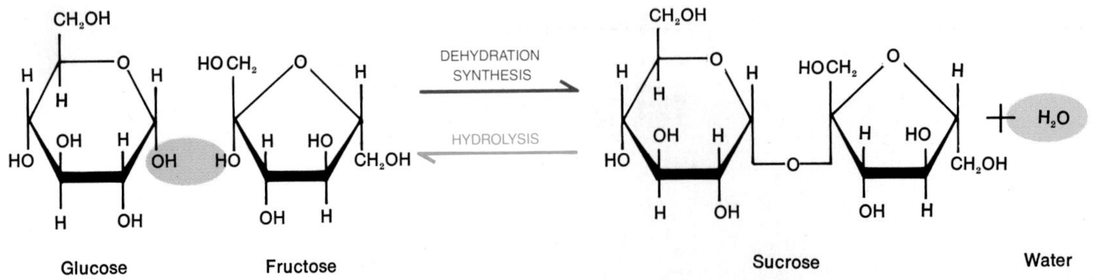

FIGURE 2-10 Dehydration synthesis and hydrolysis of a molecule of sucrose. In the dehydration synthesis reaction (read from left to right), the two smaller molecules, glucose and fructose, are joined to form a larger molecule of sucrose. Note the loss of a water molecule. In hydrolysis (read from right to left), the larger sucrose molecule is broken down into the two smaller molecules, glucose and fructose. Here, a molecule of water is added to sucrose for the reaction to occur.

the buffered substances, HCl and NaOH in this case, are removed from the body via the lungs or kidneys, the carbonic acid and sodium bicarbonate formed as products of the reactions function again as components of the buffer pair. Do you now understand why buffers are sometimes called "chemical sponges"?

ORGANIC COMPOUNDS

Organic compounds contain carbon and usually hydrogen and oxygen as well. Carbon has several properties that make it particularly useful to living organisms. For one thing, it can react with one to several hundred other carbon atoms to form large molecules of many different shapes. This means that the body can build many compounds out of carbon, hydrogen, and oxygen. Each compound can be especially suited for a particular structure or function. The relatively large size of most carbon-containing molecules and the fact that they do not dissolve easily in water make them useful materials for building body structures. Carbon compounds are mostly or entirely held together by covalent bonds and tend to decompose easily. This means that organic compounds are also a good source of energy. Ionic compounds are not good energy sources because they form new ionic bonds as soon as the old ones are broken.

Carbohydrates

A large and diverse group of organic compounds found in the body is the **carbohydrates,** also known as sugars and starches. The carbohydrates perform a number of major functions in living systems. A few even form structural units. For instance, one type of sugar (deoxyribose) is a building block of genes, the molecules that carry hereditary information. Some carbohydrates are converted to proteins and to fats or fatlike substances, which are used to build structures and provide an emergency source of energy. Other carbohydrates function as food reserves. One example is glycogen, which is stored in the liver and skeletal muscles. The principal function of carbohydrates, however, is to provide the most readily available source of energy to sustain life.

Carbon, hydrogen, and oxygen are the elements found in carbohydrates. The ratio of hydrogen to oxygen atoms is always 2 to 1, the same as in water. This ratio can be seen in the formulas for carbohydrates such as ribose

$(C_5H_{10}O_5)$, glucose $(C_6H_{12}O_6)$, and sucrose $(C_{12}H_{22}O_{11})$. Although there are exceptions, the general formula for carbohydrates is $(CH_2O)_n$, where n symbolizes three or more CH_2O units. Carbohydrates can be divided into three major groups: monosaccharides, disaccharides, and polysaccharides.

1. Monosaccharides. Monosaccharides (mon-ō-SAK-a-rīds), or simple sugars, are compounds containing from three to seven carbon atoms. Simple sugars with three carbons in the molecule are called trioses. The number of carbon atoms in the molecule is indicated by the prefix *tri*. There are also tetroses (four-carbon sugars), pentoses (five-carbon sugars), hexoses (six-carbon sugars), and heptoses (seven-carbon sugars). Pentoses and hexoses are exceedingly important to the human organism. The pentose called deoxyribose is a component of genes. The hexose called glucose is the main energy-supplying molecule of the body.

2. Disaccharides. A second group of carbohydrates, the **disaccharides** (dī-SAK-a-rīds), consists of two monosaccharides joined chemically. In the process of disaccharide formation, two monosaccharides combine to form a disaccharide molecule and a molecule of water is lost. This reaction is known as **dehydration synthesis** (*dehydration* = loss of water). The following reaction shows disaccharide formation. Molecules of the monosaccharides glucose and fructose combine to form a molecule of the disaccharide sucrose (table sugar):

$$C_6H_{12}O_6 \quad + \quad C_6H_{12}O_6 \quad \rightarrow \quad C_{12}H_{22}O_{11} + H_2O$$

Glucose   Fructose   Sucrose   Water
(monosaccharide)   (monosaccharide)   (disaccharide)

You may be puzzled to see that glucose and fructose have the same chemical formulas. Actually, they are different monosaccharides, since the relative positions of the oxygens and carbons vary in the two different molecules (see Figure 2-10). The formula for sucrose is $C_{12}H_{22}O_{11}$ and not $C_{12}H_{24}O_{12}$, since a molecule of H_2O is lost in the process of disaccharide formation. In every dehydration synthesis a molecule of water is lost. Along with this water loss, there is the synthesis of two small molecules, such as glucose and fructose, into one large, more complex molecule, such as sucrose (Figure 2-10). Similarly, the dehydration synthesis of the two monosaccharides glucose and galactose forms the disaccharide lactose (milk sugar).

Disaccharides can also be broken down into smaller, simpler molecules by adding water. This reverse chemical reaction is called **digestion** or **hydrolysis,** which means to split by using water. A molecule of sucrose, for example, may be digested into its components of glucose and fructose by the addition of water. The mechanism of this reaction also is represented in Figure 2-10.

EXHIBIT 2-5
RELATIONSHIPS OF REPRESENTATIVE LIPIDS TO HUMAN ORGANISM

LIPIDS	RELATIONSHIP
FATS	Protection, insulation, source of energy.
PHOSPHOLIPIDS	
Lecithin	Major lipid component of cell membranes; constituent of plasma.
Cephalin and sphingomyelin	Found in high concentrations in nerves and brain tissue.
STEROIDS	
Cholesterol	Constituent of all animal cells, blood, and nervous tissue; suspected relationship to heart disease and atherosclerosis; precursor of bile salts, vitamin D, and steroid hormones.
Bile salts	Substances that emulsify or suspend fats before their digestion and absorption; needed for absorption of fat-soluble vitamins (A, D, E, K).
Vitamin D	Produced in skin on exposure to ultraviolet radiation; necessary for bone growth, development, and repair.
Estrogens	Sex hormones produced in large quantities by females.
Androgens	Sex hormones produced in large quantities by males.
PORPHYRINS (lipid portions of organic molecules)	
Hemoglobin	Oxygen-transporting pigment in red blood cells.
Bile pigments	Bilirubin, a reddish pigment, and biliverdin, a greenish pigment, both formed from the heme group of hemoglobin; bilirubin responsible for the brown color of feces and both pigments contribute to the color changes associated with a bruise.
Cytochromes	Coenzymes involved in the respiration of all cells.
OTHER LIPOID SUBSTANCES	
Carotenes	Pigment in egg yolk, carrots, and tomatoes; vitamin A is formed from carotenes; retinene, formed from vitamin A, is a photoreceptor in the retina of the eye.
Vitamin E	May promote wound healing, prevent scarring, and contribute to the normal structure and functioning of the nervous system; deficiency causes sterility in rats and muscular dystrophy in monkeys; deficiency in humans believed to cause the oxidation of certain fats (unsaturated) resulting in abnormal structure and function of certain parts of cells (mitochondria, lysosomes, and plasma membranes); may reduce the severity of visual loss associated with retrolental fibroplasia (eye disease in premature infants caused by too much oxygen in incubators) by functioning as an antioxidant.
Vitamin K	Vitamin that promotes blood clotting and prevents excessive bleeding.
Prostaglandins	Membrane-associated lipids that stimulate uterine contractions, induce labor and abortions, regulate blood pressure, transmit nerve impulses, regulate metabolism, regulate stomach secretions, inhibit lipid breakdown, and regulate muscular contractions of the gastrointestinal tract.

3. Polysaccharides. The third major group of carbohydrates, the **polysaccharides** (pol'-ē-SAK-a-rīds), consists of three or more monosaccharides joined together through dehydration synthesis. Polysaccharides have the formula $(C_6H_{10}O_5)_n$. Like disaccharides, polysaccharides can be broken into their constituent sugars through hydrolysis reactions. Unlike monosaccharides or disaccharides, however, they usually lack the characteristic sweetness of sugars like fructose or sucrose and are usually not soluble in water. One of the chief polysaccharides is glycogen.

Lipids

A second group of organic compounds that is vital to the human organism is the **lipids.** Like carbohydrates, lipids are composed of carbon, hydrogen, and oxygen, but they do not have a 2:1 ratio of hydrogen to oxygen.

Most lipids are insoluble in water, but they readily dissolve in solvents such as alcohol, chloroform, and ether. Since lipids are a large and diverse group of compounds, we will only discuss two types in detail at this point: fats and prostaglandins. Pertinent information regarding other lipids is provided in Exhibit 2-5.

A molecule of **fat** (triglyceride) consists of two basic components: **glycerol** and **fatty acids** (Figure 2-11). A single molecule of fat is formed when a molecule of glycerol combines with three molecules of fatty acids. This reaction, like the one described for disaccharide formation, is a dehydration synthesis reaction. During hydrolysis, a single molecule of fat is broken down into fatty acids and glycerol.

In subsequent chapters, we will be talking about saturated, unsaturated, and polyunsaturated fats. These terms

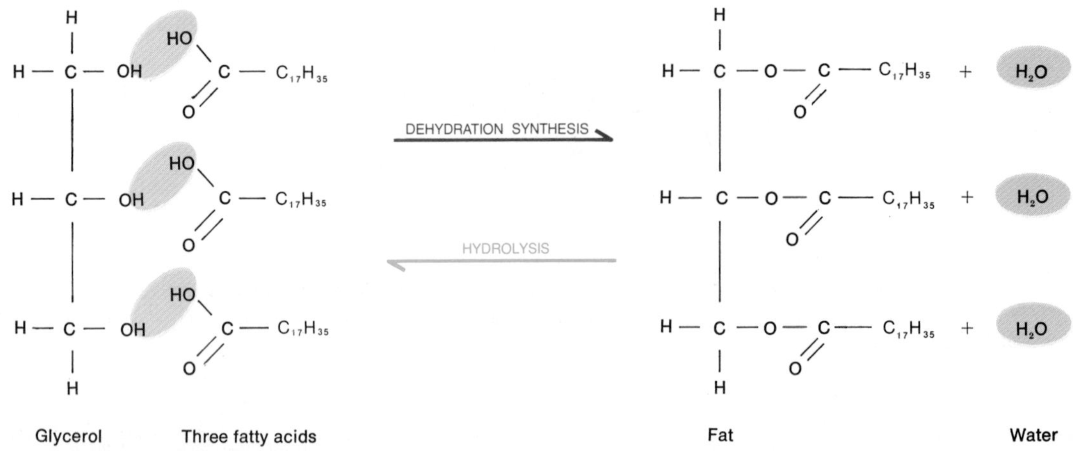

FIGURE 2-11 Dehydration synthesis and hydrolysis of a fat. In the dehydration synthesis reaction (read from left to right), one molecule of glycerol combines with three fatty acid molecules, and there is a loss of three molecules of water. In hydrolysis (read from right to left), a molecule of fat is broken down into a single molecule of glycerol and three fatty acid molecules upon the addition of three molecules of water. The fatty acid shown is stearic acid, a component of corn oil, coconut oil, beef fat, and pork fat.

have the following meanings. A *saturated fat* contains only single covalent bonds between its carbon atoms and all the carbon atoms are bonded to the maximum number of hydrogen atoms; it is saturated with hydrogen atoms. Saturated fats occur mostly in animal foods already high in cholesterol such as beef, pork, butter, whole milk, eggs, and cheese. They also occur in some plant products such as cocoa butter, palm oil, and coconut oil. An *unsaturated fat* contains at least one double covalent bond between its carbon atoms; it is not completely saturated with hydrogen atoms. Examples are olive oil and peanut oil and

they have no significant effect on cholesterol level. A *polyunsaturated fat* is one in which each carbon atom can still bond to two or more hydrogen atoms. Corn oil, safflower oil, sesame oil, and soybean oil are examples of polyunsaturated fats. Many researchers believe that they help to reduce cholesterol in the blood.

Fats represent the body's most highly concentrated source of energy. They provide more than twice as many calories (a measure of energy) per weight as either carbohydrates or proteins. In general, however, fats are about 10 to 12 percent less efficient as body fuels than are carbohydrates. A great amount of the fat calorie is wasted and thus not available for the body to use.

Prostaglandins (pros'-ta-GLAN-dins), also called **PG,** are a large group of membrane-associated lipids composed of 20-carbon fatty acids containing 5 carbon atoms joined to form a ring (cyclopentane ring). Prostaglandins were first discovered in prostate gland secretions, but they are now known to be produced in all nucleated cells in the body and to be able to influence the functioning of any type of cell.

Prostaglandins are produced in cell membranes and are rapidly decomposed by catabolic enzymes. Although synthesized in minute quantities, they are potent substances and exhibit a wide variety of effects on the body. Basically, prostaglandins mimic hormones. They are involved in the modulation of many hormonal responses (Chapter 18), inducing menstruation, and inducing second-trimester abortions. They are also involved in the inflammatory response (Chapter 4), preventing peptic ulcers, opening bronchial and nasal passages, and platelet aggregation and inhibition of aggregation.

Proteins

A third principal group of organic compounds is **proteins.** These compounds are much more complex in structure than the carbohydrates or lipids. They are also responsible for much of the structure of body cells and are related to many physiological activities. For example, proteins in the form of enzymes speed up many essential biochemi-

EXHIBIT 2-6

CLASSIFICATION OF PROTEINS BY FUNCTION

TYPE OF PROTEIN	DESCRIPTION
Structural	Proteins that form the structural framework of various parts of the body. Examples: keratin in the skin, hair, and fingernails and collagen in connective tissue.
Regulatory	Proteins that function as hormones and regulate various physiological processes. Examples: insulin, which regulates blood sugar, and adrenaline, which regulates the diameter of blood vessels.
Contractile	Proteins that serve as contractile elements in muscle tissue. Examples: myosin and actin.
Immunological	Proteins that serve as antibodies to protect the body against invading microbes. Example: gamma globulin.
Transport	Proteins that transport vital substances throughout the body. Example: hemoglobin, which transports oxygen and carbon dioxide in the blood.
Catalytic	Proteins that act as enzymes and function in controlling biochemical reactions. Examples: salivary amylase, pepsin, and lactase.

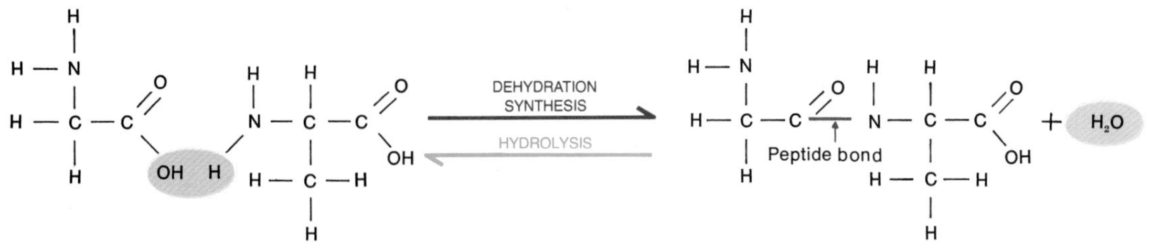

FIGURE 2-12 Protein formation. When two or more amino acids are chemically united, the resulting bond between them is called a peptide bond. In the example shown, glycine and alanine, the two amino acids, are joined to form the dipeptide glycylalanine. At the point where water is lost, the peptide bond is formed.

cal reactions. Other proteins assume a necessary role in muscular contraction. Antibodies are proteins that provide the human organism with defenses against invading microbes. And some hormones that regulate body functions are also proteins. A classification of proteins on the basis of function is shown in Exhibit 2-6.

Chemically, proteins always contain carbon, hydrogen, oxygen, and nitrogen. Many proteins also contain sulfur and phosphorus. Just as monosaccharides are the building units of sugars, and fatty acids and glycerol are the building units of fats, **amino acids** are the building blocks of proteins. In protein formation, amino acids combine to form more complex molecules, while water molecules are lost. The process is a dehydration synthesis reaction, and the bonds formed between amino acids are called *peptide bonds* (Figure 2-12).

When two amino acids combine, a **dipeptide** results. Adding another amino acid to a dipeptide produces a **tripeptide.** Further additions of amino acids result in the formation of **polypeptides,** which are large protein molecules. At least 20 different amino acids are found in proteins. A great variety of proteins is possible because each variation in the number or sequence of amino acids can produce a different protein. The situation is similar to using an alphabet of 20 letters to form words. Each letter could be compared to a different amino acid, and each word would be a different protein.

Proteins have four levels of structural organization. The *primary structure* is the sequence of amino acids making up the protein. An alteration in primary structure can have serious consequences. For example, a single substitution of an amino acid in a blood protein can result in a deformed hemoglobin molecule that produces sickle-cell anemia. The *secondary structure* of a protein is its coiling or zigzag arrangement along one dimension, such as clockwise spirals and pleated sheets. The *tertiary structure* refers to the bending or folding of a protein into a three-dimensional shape. The *quaternary structure* of a protein refers to two or more tertiary patterns bonded to each other (see Figure 19-3).

Enzymes and Homeostasis

Enzymes (EN-zīms) are specialized proteins that are produced by living cells to catalyze reactions in the body. A *catalyst* is a chemical substance that alters the speed of a chemical reaction without becoming part of the prod-

ucts of the reaction or being used up. Some enzymes consist entirely of proteins. Many also contain a nonprotein portion called a **cofactor,** without which the enzyme will not function. Some cofactors are ions; others are complex organic molecules called **coenzymes.** Vitamins function as coenzymes.

For a chemical or biochemical reaction to occur, a certain amount of energy is required: the **activation energy.** As stated earlier, energy can be transformed from one state to another. When heat energy is added to molecules, some of the heat is transformed to kinetic energy and the energy of the molecules is increased. One way to speed up molecules to activation-level energy is to heat them. Unfortunately, moderate heat denatures (coagulates and destroys the activity of) many body proteins. The role of an enzyme, then, is to decrease the amount of energy needed to start the reaction.

Exactly how enzymes lower activation energies is not fully understood. However, it is known that an enzyme attaches itself to one of the reacting molecules, called a **substrate,** and the two form a temporary **enzyme-substrate complex.** The surface of the substrate makes contact with a specific region on the surface of the enzyme called the **active site.** Thousands of enzymes exist, but each kind can attach to only one kind of substrate. Apparently, the enzyme molecule must fit perfectly with the substrate molecule like pieces of a jigsaw puzzle (Figure 2-13). If the enzyme and substrate do not fit properly, no reaction occurs.

After the reaction is completed, the products of the reaction move away from the enzyme and the enzyme is free to attach to another substrate molecule. The whole process of attachment, reaction, and detachment takes place very quickly. Most enzymes are capable of interacting with up to 5 million substrate molecules per minute at 0°C and, within limits, about double that for every 10°C increase in temperature above 0°C.

Many enzymes are named by adding the suffix *ase* to the name of their substrates. For example, the enzyme involved in breaking down sucrose is called *sucrase.* Enzymes that hydrolyze proteins are classified as *proteases.* Other enzymes are named by their action. *Dehydrogenases,* for example, are enzymes that remove hydrogen atoms from a substrate.

Enzymes are clearly essential to the body's overall homeostasis. They quickly catalyze chemical reactions, and they also govern the reactions that occur.

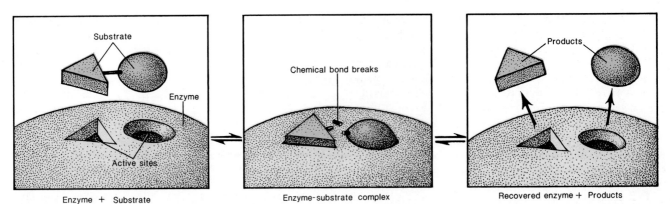

FIGURE 2-13 Enzyme action in a decomposition reaction. The enzyme and substrate molecules combine to form an enzyme-substrate complex. During combination, the substrate is changed into products. Once the products are formed, the enzyme is recovered and may be used again to catalyze a similar reaction.

Nucleic Acids: Deoxyribonucleic Acid (DNA) and Ribonucleic Acid (RNA)

Nucleic (nu-KLĒ-ic) **acids,** compounds first discovered in the nuclei of cells, are exceedingly large organic molecules containing carbon, hydrogen, oxygen, nitrogen,

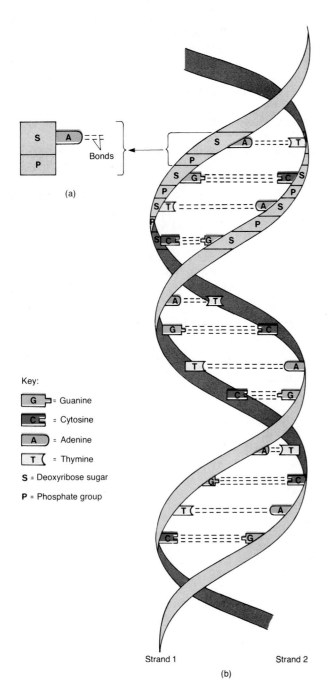

Key:

G = Guanine

C = Cytosine

A = Adenine

T = Thymine

S = Deoxyribose sugar

P = Phosphate group

Strand 1   Strand 2

(b)

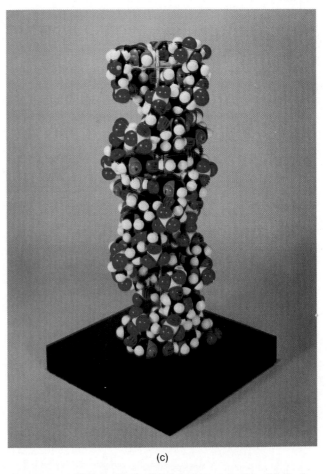

(c)

FIGURE 2-14 DNA molecule. (a) Adenine nucleotide. (b) Portion of an assembled DNA molecule. (c) Space-filling (three-dimensional) model to show relative size and location of atoms. (Courtesy of Ealing Corp.)

and phosphorus. They are divided into two principal kinds: **deoxyribonucleic** (dē-ok'-sē-rī'-bō-nyoo-KLĒ-ik) **acid (DNA)** and **ribonucleic acid (RNA).**

Whereas the basic structural units of proteins are amino acids, the basic units of nucleic acids are **nucleotides.** A molecule of DNA is a chain composed of repeating nucleotide units. Each nucleotide of DNA consists of three basic parts (Figure 2-14a):

1. It contains one of four possible *nitrogen bases,* which are ring-shaped structures containing atoms of C, H, O, and N. The nitrogen bases found in DNA are adenine, thymine, cytosine, and guanine.

2. It contains a pentose sugar called *deoxyribose.*

3. It also contains *phosphate groups.*

The nucleotides are named according to the nitrogen base that is present. Thus a nucleotide containing thymine is called a *thymine nucleotide.* One containing adenine is called an *adenine nucelotide,* and so on.

The chemical components of the DNA molecule were known before 1900, but it was not until 1953 that a model of the organization of the chemicals was constructed. This model was proposed by J. D. Watson and F. H. C. Crick on the basis of data from many investigations. Figure 2-14b shows the following structural characteristics of the DNA molecule.

1. The molecule consists of two strands with crossbars. The strands twist about each other in the form of a *double helix* so that the shape resembles a twisted ladder. For many years it was assumed that all DNA was in the form of a double helix that twisted smoothly to the right (B-DNA or right-handed DNA). It was discovered, however, that DNA can also twist jaggedly to the left (Z-DNA or left-handed DNA). This alternate form of DNA may help to explain how genes turn on and off and how some cells may become malignant.

2. The uprights of the DNA ladder consist of alternating phosphate groups and the deoxyribose portions of the nucleotides.

3. The rungs of the ladder contain paired nitrogen bases. As shown, adenine always pairs with thymine and cytosine always pairs with guanine.

Cells contain hereditary material called *genes,* each of which is a segment of a DNA molecule. Our genes determine which traits we inherit, and they control all

the activities that take place in our cells throughout a lifetime. When a cell divides, its hereditary information is passed on to the next generation of cells. The passing of information is possible because of DNA's unique structure.

RNA, the second principal kind of nucleic acid, differs from DNA in several respects. RNA is single-stranded; DNA is double-stranded. The sugar in the RNA nucleotide is the pentose ribose. And RNA does not contain the nitrogen base thymine. Instead of thymine, RNA has the nitrogen base uracil. At least three different kinds of RNA have been identified in cells. Each type has a specific role to perform with DNA in protein synthesis reactions (Chapter 3).

Adenosine Triphosphate (ATP)

A molecule that is indispensable to the life of the cell is **adenosine** (a-DEN-ō-sēn) **triphosphate (ATP).** This substance is found universally in living systems and performs the essential function of storing energy for various cellular activities. Structurally, ATP consists of three phosphate groups and an adenosine unit composed of adenine and the five-carbon sugar ribose (Figure 2-15). ATP is regarded as a high-energy molecule because of the total amount of usable energy it releases when it is broken down by the addition of a water molecule (hydrolysis).

When the terminal phosphate group is hydrolyzed, the reaction liberates a great deal of energy. This energy is used by the cell to perform its basic activities. Removal of the terminal phosphate group leaves a molecule called **adenosine diphosphate (ADP).** This reaction may be represented as follows.

$$\text{ATP} \quad \rightleftharpoons \quad \text{ADP} \quad + \quad \text{P} \quad + \quad \text{E}$$

| Adenosine triphosphate | Adenosine diphosphate | Phosphate | Energy |

The energy supplied by the catabolism of ATP into ADP is constantly being used by the cell. Since the supply of ATP at any given time is limited, a mechanism exists to replenish it—a phosphate group is added to ADP to manufacture more ATP. The reaction may be represented as follows:

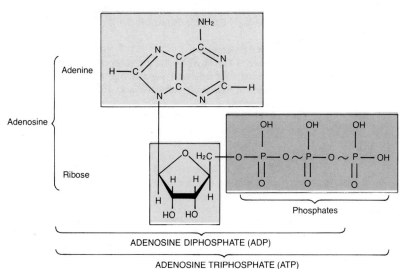

FIGURE 2-15 Structure of ATP and ADP.

$$ADP \; + \; P \; + \; E \; \rightleftharpoons \; ATP$$
Adenosine   Phosphate   Energy   Adenosine
diphosphate   triphosphate

Logically, energy is required to manufacture ATP. The energy required to attach a phosphate group to ADP is supplied by various decomposition reactions taking place in the cell, particularly by the decomposition of glucose. ATP can be stored in every cell, where it provides potential energy that is not released until needed.

Cyclic Adenosine-3′,5′-Monophosphate (Cyclic AMP)

A chemical substance closely related to ATP is **cyclic adenosine-3′,5′-monophosphate (cyclic AMP)**. Essentially it is a molecule of adenosine monophosphate with the phosphate attached to the ribose sugar at two places (Figure 2-16). The attachment forms a ring-shaped structure and thus the name cyclic AMP.

Cyclic AMP is formed from ATP by the action of a special enzyme, called *adenyl cyclase,* located in the cell

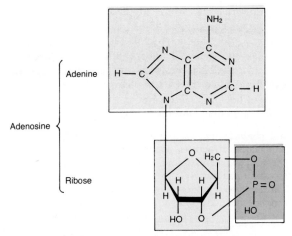

FIGURE 2-16 Structure of cyclic AMP.

membrane. Although cyclic AMP was discovered in 1958, only recently has its function in cells become clear. One function is related to the action of hormones, a topic we explore in detail in Chapter 18.

STUDY OUTLINE

Introduction to Basic Chemistry (p. 30)
Chemical Elements
1. Matter is anything that occupies space and has mass. It is made up of building units called chemical elements.
2. Carbon, hydrogen, oxygen, and nitrogen make up 96 percent of body weight. These elements together with phosphorus and calcium make up 99 percent of total body weight.

Structure of Atoms
1. Units of matter of all chemical elements are called atoms.
2. Atoms consist of a nucleus, which contains protons and neutrons, and orbiting electrons moving in energy levels.
3. The total number of protons of an atom is its atomic number. This number is equal to the number of electrons in the atom.
4. Nuclear magnetic resonance (NMR) is based on the reaction of atomic nuclei to magnetism.
5. NMR can identify existing pathologies, evaluate drug therapy, measure metabolism, and assess the potential for certain diseases.

Atoms and Molecules
1. The electrons are the part of an atom that actively participate in chemical reactions.
2. A molecule is the smallest unit of two or more combined atoms. A molecule containing two or more different kinds of atoms is a compound.
3. In an ionic bond, outer energy level electrons are transferred from one atom to another. The transfer forms ions, whose unlike charges attract each other and form ionic bonds.
4. In a covalent bond, there is a sharing of pairs of outer-energy-level electrons.
5. Hydrogen bonding provides temporary bonding between certain atoms within large complex molecules such as proteins and nucleic acids.

Radioisotopes
1. Radioisotopes emit radiation that can be detected by instruments.
2. A sophisticated version of radioisotopic scanning is called positron emission tomography (PET).
3. PET scans provide images of chemical activity in tissues.

Chemical Reactions
1. Synthesis reactions involve the combination of reactants to produce a new molecule. The reactions are anabolic: bonds are formed.
2. In decomposition reactions, a substance breaks down into other substances. The reactions are catabolic: bonds are broken.
3. Exchange reactions involve the replacement of one atom or atoms by another atom or atoms.
4. In reversible reactions, end products can revert to the original combining molecules.
5. The sum of all synthetic and decomposition reactions that occur within an organism is referred to as metabolism.
6. When chemical bonds are formed, energy is needed. When bonds are broken, energy is released. This is known as chemical bond energy.

Chemical Compounds and Life Processes (p. 37)
1. Inorganic substances usually lack carbon, contain ionic bonds, resist decomposition, and dissolve readily in water.
2. Organic substances always contain carbon and usually hydrogen. Most organic substances contain covalent bonds and many are insoluble in water.

Inorganic Compounds
1. Water is the most abundant substance in the body. It is an excellent solvent and suspending medium, participates in chemical reactions, absorbs and releases heat slowly, and lubricates.
2. Acids, bases, and salts dissociate into ions in water. An acid ionizes into H^+ ions; a base ionizes into OH^- ions. A salt ionizes into neither H^+ nor OH^- ions. Cations are positively charged ions; anions are negatively charged ions.
3. The pH of different parts of the body must remain fairly constant for the body to remain healthy. On the pH scale, 7 represents neutrality. Values below 7 indicate acid solutions, and values above 7 indicate alkaline solutions.
4. The pH values of different parts of the body are maintained by buffer systems, which usually consist of a weak acid and a weak base. Buffer systems eliminate excess H^+ ions and excess OH^- ions in order to maintain pH homeostasis.

Organic Compounds

1. Carbohydrates are sugars or starches that provide most of the energy needed for life. They may be monosaccharides, disaccharides, or polysaccharides. Carbohydrates, and other organic molecules, are joined together to form larger molecules with the loss of water by a process called dehydration synthesis. In the reverse process, called hydrolysis or digestion, large molecules are broken down into smaller ones upon the addition of water.

2. Lipids are a diverse group of compounds that includes fats, phospholipids, steroids, porphyrins, and prostaglandins (PG). Fats protect, insulate, provide energy, and are stored. Prostaglandins mimic the effects of hormones and are involved in the inflammatory response and the modulation of hormonal responses.

3. Proteins are constructed from amino acids. They give structure to the body, regulate processes, provide protection, help muscles to contract, and transport substances.

4. Enzymes are proteins produced by the body. They catalyze chemical reactions. Enzymes act on substrates by lowering the activation energy needed for reaction.

5. Deoxyribonucleic acid (DNA) and ribonucleic acid (RNA) are nucleic acids consisting of nitrogen bases, sugar, and phosphate groups. DNA is a double helix and is the primary chemical in genes. RNA differs in structure and chemical composition from DNA and is mainly concerned with protein synthesis reactions.

6. The principal energy-storing molecule in the body is adenosine triphosphate (ATP). When its energy is liberated, it is decomposed to adenosine diphosphate (ADP). ATP is manufactured from ADP using the energy supplied by various decomposition reactions, particularly of glucose.

7. Cyclic AMP is closely related to ATP and assumes a function in certain hormonal reactions.

REVIEW QUESTIONS

1. What is the relationship of matter to the body?
2. Define a chemical element. List the chemical symbols for 10 different chemical elements. Which chemical elements make up the bulk of the human organism?
3. What is an atom? Diagram the positions of the nucleus, protons, neutrons, and electrons in an atom of oxygen and an atom of nitrogen. What is an atomic number?
4. Explain the principle of nuclear magnetic resonance (NMR). How is NMR useful in the diagnosis of disease?
5. What is meant by an energy level?
6. How are chemical bonds formed? Distinguish between an ionic bond and a covalent bond. Give at least one example of each.
7. Can you determine how a molecule of $MgCl_2$ is ionically bonded? Magnesium has two electrons in its outer energy level. Construct a diagram to verify your answer.
8. Refer to Figure 2-5b and c. See if you can determine why there is a double covalent bond between atoms in an oxygen molecule (O_2) and a triple covalent bond between atoms in a nitrogen molecule (N_2).
9. Define a hydrogen bond. Why are hydrogen bonds important?
10. Define a radioisotope. Explain the principle of positron emission tomography (PET) scanning. What is the diagnostic importance of PET scanning?
11. What are the four principal kinds of chemical reactions? How are anabolism and catabolism related to synthesis and decomposition reactions, respectively? How is energy related to chemical reactions?
12. Identify what kind of reaction each of the following represents:
    a. $H_2 + Cl_2 \rightarrow 2HCl$
    b. $3\,NaOH + H_3PO_4 \rightarrow Na_3PO_4 + 3\,H_2O$
    c. $CaCO_3 + CO_2 + H_2O \rightarrow Ca(HCO_3)_2$
    d. $HNO_3 \rightarrow H^+ + NO_3^-$
    e. $NH_3 + H_2O \rightleftharpoons NH_4^+ + OH^-$
13. How do inorganic compounds differ from organic compounds? List and define the principal inorganic and organic compounds that are important to the human body.
14. What are the essential functions of water in the body? Distinguish between a solution and a suspension.
15. Define an acid, a base, and a salt. How does the body acquire some of these substances? List some functions of the chemical elements furnished as ions of salts.
16. What is pH? Why is it important to maintain a relatively constant pH? What is the pH scale? If there are 100 OH^- ions at a pH of 8.5, how many OH^- ions are there at a pH of 9.5?
17. List the normal pH values of some common fluids, biological solutions, and foods. Refer to Exhibit 2-4, and select the two substances whose pH values are closest to neutrality. Is the pH of milk or of cerebrospinal fluid closer to 7? Is the pH of bile or of urine farther from neutrality?
18. What are the components of a buffer system? What is the function of a buffer pair? Diagram and explain how the carbonic acid–bicarbonate buffer system of extracellular fluid maintains a constant pH even in the presence of a strong acid or strong base. How is this an example of homeostasis?
19. Why are the reactions of buffer pairs more important with strong acids and bases than with weak acids and bases?
20. Define a carbohydrate. Why are carbohydrates essential to the body? How are carbohydrates classified?
21. Compare dehydration synthesis and hydrolysis. Why are they significant?
22. How do lipids differ from carbohydrates? What are some relationships of lipids to the body?
23. Define a prostaglandin (PG). List some physiological effects of prostaglandins.
24. Define a protein. What is a peptide bond? Discuss the classification of proteins on the basis of function.
25. Distinguish between an enzyme and a substrate. List some principal characteristics of enzymes. Relate the concept of activation energy to enzyme action. How do enzymes maintain homeostasis?
26. What is a nucleic acid? How do deoxyribonucleic acid (DNA) and ribonucleic acid (RNA) differ with regard to chemical composition, structure, and function?
27. What is adenosine triphosphate (ATP)? What is the essential function of ATP in the human body? How is this function accomplished?
28. How is cyclic AMP related to ATP? What is the function of cyclic AMP?

STUDENT OBJECTIVES

- Define and list a cell's generalized parts.

- Explain the chemistry, structure, and functions of the plasma membrane.

- Describe how materials move across plasma membranes by diffusion, facilitated diffusion, osmosis, filtration, dialysis, active transport, phagocytosis, and pinocytosis.

- Define the structure and function of several modified plasma membranes.

- Identify the chemical composition and functions of cytoplasm.

- Describe the structure and two general functions of a cell nucleus.

- Define the structure and function of ribosomes.

- Distinguish between agranular and granular endoplasmic reticulum with regard to structure and function.

- Describe the structure and functions of the Golgi complex.

- Discuss the structure and function of mitochondria as "powerhouses of the cell."

- Explain why a lysosome in a cell is called a "suicide packet."

- Discuss the role of peroxisomes.

- Distinguish between the structure and function of microfilaments and microtubules as components of the cytoskeleton.

- Discuss the structure and function of centrioles in cellular reproduction.

- Differentiate between cilia and flagella in terms of function.

- Define a cell inclusion and give several examples.

- Define extracellular material and give several examples.

- Discuss the stages and events involved in cell division.

- Explain the significance of cell division.

- Define a gene and explain the sequence of events involved in protein synthesis.

- Describe the principle and importance of genetic engineering.

- Explain the relationship of aging to cells and the effects of aging on various body systems.

- Describe cancer as a homeostatic imbalance of cells.

- Define medical terminology associated with cells.

3 The Cellular Level of Organization

The study of the body at the cellular level of organization is important because many activities essential to life occur in cells and many disease processes originate there. A **cell** may be defined as the basic, living, structural and functional unit of the body and, in fact, of all organisms. **Cytology** is the branch of science concerned with the study of cells. This chapter concentrates on the structure, functions, and reproduction of cells.

A series of illustrations accompanies each cell structure that you study. A diagram of a generalized animal cell shows the location of the structure within the cell. An electron micrograph shows the actual appearance of the structure.* A diagram of the electron micrograph clarifies some of the small details by exaggerating their outlines. Finally, an enlarged diagram of the structure shows its details.

GENERALIZED ANIMAL CELL

A **generalized animal cell** is a composite of many different cells in the body. Examine the generalized cell illustrated in Figure 3-1, but keep in mind that no such single cell actually exists.

For convenience, we can divide the generalized cell into four principal parts:

1. Plasma (cell) membrane. The outer, limiting membrane separating the cell's internal parts from the extracellular fluid and external environment.
2. Cytoplasm. The substance between the nucleus and the plasma membrane.
3. Organelles. The cellular components that are highly specialized for specific cellular activities.
4. Inclusions. The secretions and storage areas of cells.

Extracellular materials, which are substances external to the cell surface, will also be examined in connection with cells.

PLASMA (CELL) MEMBRANE

The exceedingly thin structure that separates one cell from other cells and from the external environment is called the **plasma membrane** or **cell membrane.** The membrane ranges from 65 to 100 angstroms (Å) in thickness.

CHEMISTRY AND STRUCTURE

Proteins are the major component of nearly all plasma membranes, comprising from 50 to 70 percent by weight. Next in abundance are phospholipids. Other constituents in lesser amounts include cholesterol, water, carbohydrates, and ions. Recent studies of membrane structure suggest a new concept regarding the arrangement of the various molecules. This new concept is referred to as the **fluid mosaic model** (Figure 3-2b).

The phospholipid molecules are arranged in two parallel rows, forming a **phospholipid bilayer.** A phospholipid molecule consists of a polar, phosphate-containing "head" which mixes with water and nonpolar fatty acid "tails" that do not mix with water. The molecules are oriented in the bilayer so that "heads" face outward on either side and the "tails" face each other in the membrane's interior.

The membrane proteins are classified into two categories: integral and peripheral. **Integral proteins** are embedded in the phospholipid bilayer among the fatty acid "tails." Some of the integral proteins lie at or near the inner and outer membrane surfaces; others penetrate the membrane completely. Since the phospholipid bilayer is somewhat fluid and flexible and the integral proteins have been observed moving from one location to another in the membrane, the relationship has been compared to icebergs (proteins) floating in the sea (phospholipid bilayer). The subunits of some integral proteins form minute channels through which substances could be transported into and out of the cell (described shortly). Other integral proteins carry branching chains of carbohydrates. Such combinations of carbohydrate and protein, called *glycoproteins,* provide receptor sites that enable a cell to recognize other cells of its own kind so that they can associate to form a tissue, to recognize and respond to foreign cells that might be potentially dangerous, and to recognize and attach to hormones, nutrients, and other chemicals. Red blood cells also have glycoprotein receptors that prevent them from clumping and producing unwanted blood clots. Diabetes mellitus is a disease believed to be caused by faulty cell receptor sites.

Peripheral proteins are loosely bound to the membrane surface and easily separated from it. Far less is known about them than integral proteins and their functions are not yet completely understood. For example, some peripheral proteins are believed to serve as enzymes that catalyze cellular reactions. An example is cytochrome c, involved in cellular respiration. Other peripheral proteins, such as spectrin of red blood cells, are believed to have a mechanical function by serving as a scaffolding to support the plasma membrane. It is also believed that peripheral proteins may assume a role in changes in membrane shape during such processes as cell division, locomotion, and ingestion.

FUNCTIONS

Based upon the discussion of the chemistry and structure of the plasma membrane, we can now describe its several important functions. First, the plasma membrane provides a flexible boundary that encloses the cellular contents and separates them from the external environment. Second, the membrane facilitates contact with other body cells or with foreign cells or substances. Third, the membrane provides receptors for chemicals such as hormones, enzymes, nutrients, and antibodies. Fourth, the plasma membrane mediates the entrance and exit of materials. The ability of a plasma membrane to permit certain substances to enter and exit, but to restrict the passage of others is called **selective permeability.** Let us look at this mechanism in a bit more detail.

A membrane is said to be *permeable* to a substance if it allows the free passage of that substance. Although

---

* An **electron micrograph** (EM) is a photograph taken with an electron microscope. The electron microscope can magnify an object more than 200,000 times. In comparison, the light microscope that you probably use in your laboratory magnifies objects up to 1,000 times their size.

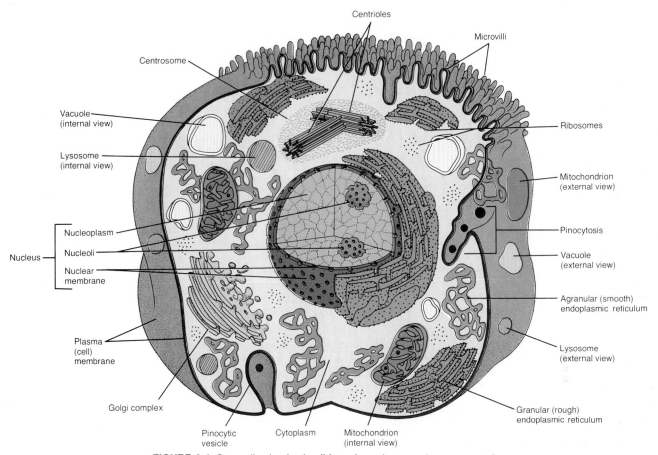

FIGURE 3-1 Generalized animal cell based on electron microscope studies.

plasma membranes are not actually freely permeable to any substance, they do permit some substances to pass more readily than others. For example, water passes more readily than most other substances. The permeability of a plasma membrane appears to be a function of several factors.

1. Size of molecules. Large molecules cannot pass through the plasma membrane. Water and amino acids are small molecules and can enter and exit the cell easily. However, most proteins, which consist of many amino acids linked together, seem to be too large to pass through the membrane. Many scientists believe that the giant-sized molecules do not enter the cell because they are larger than the minute channels in the integral proteins.

2. Solubility in lipids. Substances that dissolve easily in lipids pass through the membrane more readily than other substances, since a major part of the plasma membrane consists of lipid molecules. Examples of lipid-soluble substances are oxygen, carbon dioxide, and steroid hormones.

3. Charge on ions. The charge of an ion attempting to cross the plasma membrane can determine how easily the ion enters or leaves the cell. The protein portion of the membrane is capable of ionization. If an ion has a charge opposite that of the membrane, it is attracted to the membrane and passes through more readily. If the ion attempting to cross the membrane has the same charge as the membrane, it is repelled by the membrane and its passage is restricted. This phenomenon conforms to the rule of physics that opposite charges attract, whereas like charges repel each other.

4. Presence of carrier molecules. Some integral proteins

called carriers are capable of attracting and transporting substances across the membrane regardless of size, ability to dissolve in lipids, or membrane charge. Their purpose is to modify membrane permeability. The mechanism by which carriers do this will be described shortly.

MOVEMENT OF MATERIALS ACROSS PLASMA MEMBRANES

The mechanisms whereby substances move across the plasma membrane are essential to the life of the cell. Certain substances, for example, must move into the cell to support life, whereas waste materials or harmful substances must be moved out. Plasma membranes mediate the movements of such materials. The processes involved in these movements may be classed as either passive or active. In **passive processes,** substances move across plasma membranes without assistance from the cell. Their movement involves the kinetic energy of individual molecules. The substances move, on their own, down a concentration gradient from an area where their concentration is greater to an area where their concentration is less. The substances may also be forced across the plasma membrane by pressure from an area where the pressure is greater to an area where it is less. In **active processes,** the cell contributes energy in moving the substance across the membrane since the substance moves against a concentration gradient.

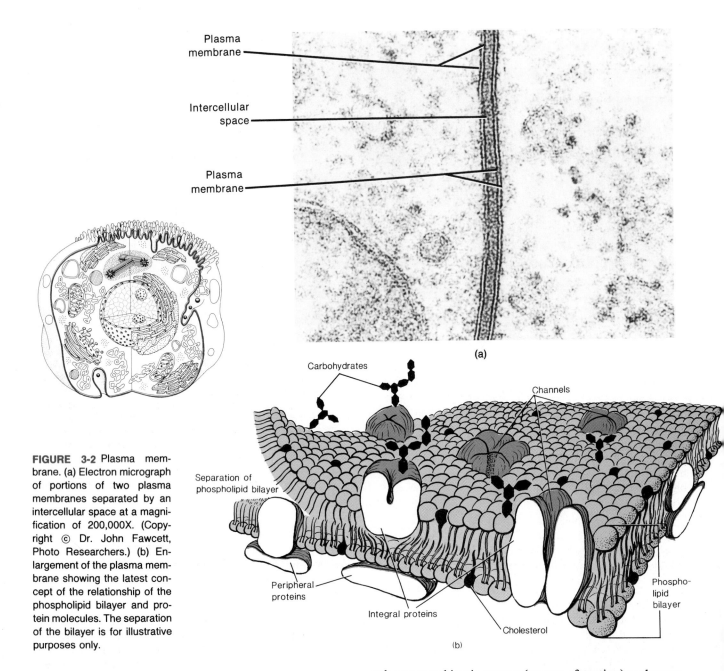

FIGURE 3-2 Plasma membrane. (a) Electron micrograph of portions of two plasma membranes separated by an intercellular space at a magnification of 200,000X. (Copyright © Dr. John Fawcett, Photo Researchers.) (b) Enlargement of the plasma membrane showing the latest concept of the relationship of the phospholipid bilayer and protein molecules. The separation of the bilayer is for illustrative purposes only.

Passive Processes

● *Diffusion* A passive process called **diffusion** occurs when there is a *net* or greater movement of molecules or ions from a region of high concentration to a region of low concentration. The movement from high to low concentration continues until the molecules are evenly distributed. At this point, they move in both directions at an equal rate. This point of even distribution is called *equilibrium.* The difference between high and low concentrations is called the *concentration gradient.* Molecules moving from the high-concentration area to the low-concentration area are said to move *down* or *with* the concentration gradient. If a dye pellet is placed in a beaker filled with water, the color of the dye is seen immediately around the pellet. At increasing distances from the pellet, the color becomes lighter (Figure 3-3). Later, however, the water solution will be a uniform color. The dye mole-

cules possess kinetic energy (energy of motion) and move about at random. The dye molecules move down the concentration gradient from an area of high dye concentration to an area of low dye concentration. The water molecules also move from their high-concentration to their low-concentration area. When dye molecules and water molecules are evenly distributed among themselves, equilibrium is reached and diffusion ceases, even though molecular movements continue. As another example of diffusion, consider what would happen if you opened a bottle of perfume in a room. The perfume molecules would diffuse until an equilibrium is reached between the perfume molecules and the air molecules in the room.

In the examples cited, no membranes were involved. Diffusion may occur, however, through selectively permeable membranes in the body. Large and small lipid-soluble molecules pass through the phospholipid bilayer of the membrane. A good example of this is the movement of

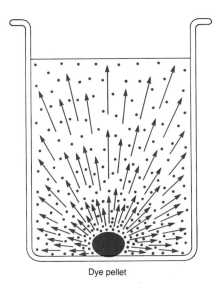

FIGURE 3-3 Principle of diffusion. Molecules of dye (solute) in a beaker of water (solvent) move down the concentration gradient from a region of high concentration to a region of low concentration.

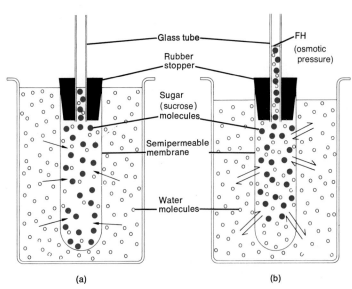

FIGURE 3-4 Principle of osmosis. (a) Apparatus at the start of the experiment. (b) Apparatus at equilibrium. In (a), the cellophane tube contains a 20-percent sugar solution and is immersed in a beaker of distilled water. The 20 percent sugar solution contains 20 parts sugar and 80 parts water, while the distilled water contains 0 parts sugar and 100 parts water. The arrows indicate that water molecules can pass freely into the tube, but that sugar (sucrose) molecules are held back by the selectively permeable membrane. As water moves into the tube by osmosis, the sugar solution is diluted and the volume of the solution in the cellophane tube increases. This increased volume is shown in (b), with the sugar solution moving up the glass tubing. The final height reached (FH) occurs at equilibrium and represents the osmotic pressure. At this point, the number of water molecules leaving the cellophane tube is equal to the number of water molecules entering the tube.

oxygen from the blood into the cells and the movement of carbon dioxide from the cells back into the blood. This is essential in order for body cells to maintain homeostasis. It ensures that cells receive adequate amounts of oxygen and eliminate carbon dioxide as part of their normal metabolism. Small molecules that are not lipid-soluble, such as certain ions (sodium, potassium, chloride), are able to diffuse through channels formed by integral proteins in the membrane.

● *Facilitated Diffusion* Another type of diffusion through a selectively permeable membrane occurs by a process called **facilitated diffusion.** This process is accomplished with the assistance of integral proteins in the membrane that serve as carriers. Although some chemical substances are large molecules and insoluble in lipids, they can still pass through the plasma membrane. Among these are different sugars, especially glucose. In the process of facilitated diffusion, it is believed that glucose is picked up by a carrier. The combined glucose-carrier is soluble in the phospholipid bilayer of the membrane, and the carrier moves the glucose to the inside of the membrane and then inside the cell. The carrier makes the glucose soluble in the phospholipid bilayer of the membrane so it can pass through the membrane. By itself, glucose is insoluble and cannot penetrate the membrane. In the process of facilitated diffusion, the cell does not expend energy, and the movement of the substance is from a region of its higher concentration to a region of its lower concentration.

The rate of facilitated diffusion is considerably faster than that of simple diffusion and depends on (1) the difference in concentration of the substance on either side of the membrane, (2) the amount of carrier available to transport the substance, and (3) how quickly the carrier and substance combine. The process is greatly accelerated by insulin, a hormone produced by the pancreas. One of insulin's functions is to lower the blood glucose level by accelerating the transportation of glucose from the blood into body cells. This transportation, as we have just seen, is by facilitated diffusion.

● *Osmosis* Another passive process by which materials move across membranes is **osmosis.** It is the net movement of water molecules through a selectively permeable membrane from an area of higher water concentration to an area of lower water concentration. The water molecules pass through channels in integrated proteins in the membrane. Once again, a simple apparatus may be used to demonstrate the process. The apparatus shown in Figure 3-4 consists of a tube constructed from cellophane, a selectively permeable membrane. The cellophane tube is filled with a colored, 20 percent sugar (sucrose) solution. Such a solution consists of 20 parts sugar and 80 parts water. The upper portion of the cellophane tube is plugged with a rubber stopper through which a glass tube is fitted. The cellophane tube is placed into a beaker containing distilled (pure) water. Pure water consists of 0 parts sugar and 100 parts water. Initially, the concentrations of water on either side of the selectively permeable membrane are different. There is a lower concentration of water inside the cellophane tube than outside. Because of this difference, water moves from the beaker into the cellophane tube. The force with which the water moves is called osmotic pressure. Very simply, **osmotic pressure** is the force under which a solvent, usually water, moves from a solution of higher water (lower solute) concentration to a solution of lower water (higher solute) concentration when the solutions are separated by a selectively permeable membrane. There is no movement of sugar from the cellophane tube inside the beaker, since the cellophane is impermeable to molecules of sugar—sugar molecules are too large to go through the pores of the membrane. As water moves into the cellophane tube, the sugar solution becomes increasingly diluted and the increased volume forces the mixture up the glass tubing. In time, the water that has accumulated in the cellophane tube and the glass tube exerts a downward pressure that forces water molecules back out of the cellophane tube and into the beaker. When water molecules leave and enter the cellophane tube at the same rate, equilibrium is reached. Osmotic pressure is an important force in the movement of water between various compartments of the body.

● *Isotonic, Hypotonic, and Hypertonic Solutions*   Osmosis may also be understood by considering the effects of different water concentrations on red blood cells. If the normal shape of a red blood cell is to be maintained, the cell must be placed in an **isotonic solution.** This is a solution in which the total concentrations of water molecules and solute molecules are the same on both sides of the semipermeable cell membrane. The concentrations of water and solute in the extracellular fluid outside the red blood cell must be the same as the concentration of the intracellular fluid. Under ordinary circumstances, a 0.85 percent NaCl solution is isotonic for red blood cells. In this condition, water molecules enter and exit the cell at the same rate, allowing the cell to maintain its normal shape.

A different situation results if red blood cells are placed in a solution that has a lower concentration of solutes and, therefore, a higher concentration of water. This is called a **hypotonic solution.** In this condition, water molecules enter the cells faster than they can leave, causing the red blood cells to swell and eventually burst. The rupture of red blood cells in this manner is called **hemolysis** (hē-MOL-i-sis) or **laking.** Distilled water is a strongly hypotonic solution.

A **hypertonic solution** has a higher concentration of solutes and a lower concentration of water than the red blood cells. One example of a hypertonic solution is a 10 percent NaCl solution. In such a solution, water molecules move out of the cells faster than they can enter. This situation causes the cells to shrink. The shrinkage of red blood cells in this manner is called **crenation** (krē-NĀ-shun). Red blood cells may be greatly impaired or destroyed if placed in solutions that deviate significantly from the isotonic state.

● *Filtration*   A third passive process involved in moving materials in and out of cells is **filtration.** This process involves the movement of solvents such as water and dissolved substances such as sugar across a selectively permeable membrane by gravity or mechanical pressure, usually hydrostatic (water) pressure. Such a movement is always from an area of higher pressure to an area of lower pressure and continues as long as a pressure difference exists. Most small to medium-sized molecules can be forced through a cell membrane.

An example of filtration occurs in the kidneys, where the blood pressure supplied by the heart forces water and small molecules like urea through thin cell membranes of tiny blood vessels and into the kidney tubules. In this basic process, protein molecules are retained by the body since they are too large to be forced through the cell membranes of the blood vessels. The molecules of harmful substances such as urea are small enough to be forced through and eliminated, however.

● *Dialysis*   The final passive process to be considered is **dialysis.** Dialysis is the diffusion of solute particles across a selectively permeable membrane and involves the separation of small molecules from large molecules. Suppose a solution containing molecules of various sizes is placed in a tube that is permeable only to the smaller molecules. The tube is then placed in a beaker of distilled water. Eventually, the smaller molecules will move from the tube into the water in the beaker and the larger molecules will be left behind.

The principle of dialysis is employed in artificial kidney machines. It does not occur within the human body. The patient's blood is passed across a dialysis membrane outside the body. The dialysis membrane takes the place of the kidneys. As the blood moves across the membrane, small-particle waste products pass from the blood into a solution surrounding the dialysis membrane. At the same time, certain nutrients can be passed from the solution into the blood. The blood is then returned to the body.

Active Processes

When cells actively participate in moving substances across membranes, they must expend energy. Cells can even move substances against a concentration gradient. The active processes considered here are active transport, phagocytosis, and pinocytosis.

● *Active Transport*   The process by which substances, usually ions, are transported across plasma membranes typically from an area of lower concentration to an area of higher concentration is called **active transport.** In order to transport a substance against a concentration gradient, the membrane uses energy supplied by ATP. In fact, a typical body cell probably expends up to 40 percent of its ATP for active transport. Although the molecular events involved in active transport are not completely understood, integral proteins in the plasma membrane do assume a role.

As just noted, glucose can be transported across cell membranes via facilitated diffusion from areas of higher to lower concentration. Glucose can also be moved by the cells lining the digestive tract from the cavity of the tract into the blood even though blood concentration of glucose is higher. This involves active transport. One proposed mechanism is that glucose (or other substance) enters a channel in an integral membrane protein (Figure

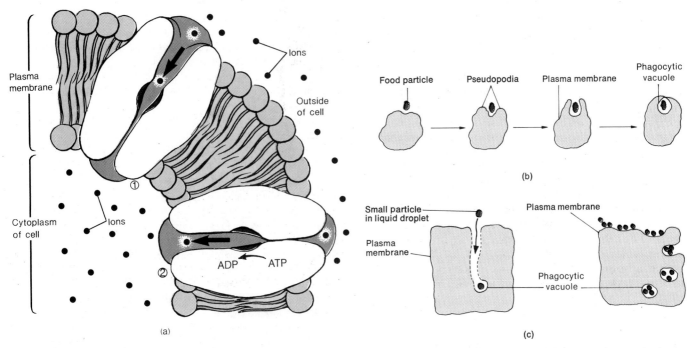

FIGURE 3-5 Active processes. (a) Proposed mechanism of active transport. (b) Phagocytosis. (c) Two variations of pinocytosis. In the variation on the left, the ingested substance enters a channel formed by the plasma membrane and becomes enclosed in a vacuole at the base of the channel. In the variation on the right, the ingested substance becomes enclosed in a vacuole that forms and detaches at the surface of the cell.

3-5a). When the glucose molecule makes contact with an active site in the channel, the energy from ATP induces a change in the membrane protein that expels the glucose on the opposite side of the membrane. As you will see later, kidney cells also have the ability to actively transport glucose back into the blood so that it is not lost in the urine.

Active transport is also an important process in maintaining the concentrations of some ions inside body cells and other ions outside body cells. For example, before a nerve cell can conduct an impulse, the concentration of potassium (K^+) ions must be considerably higher inside the cell than outside, even when the intracellular K^+ concentration is higher than the extracellular concentration. This is accomplished by a "potassium pump" which consists of an integral protein serving as the "pump," powered by ATP (see Figure 12-5a). Nerve impulse conduction also requires that nerve cells have a higher concentration of sodium (Na^+) ions outside than inside, even when the extracellular Na^+ concentration is higher than the intracellular concentration. This is accomplished by a "sodium pump," whose operation is coupled with that of the "potassium pump." The "sodium pump" is also an integral protein powered by ATP (see Figure 12-5a).

● *Phagocytosis*   Another active process by which cells take in substances across the plasma membrane is called **phagocytosis** (fag'-ō-sī-TŌ-sis), or "cell eating" (Figure 3-5b). In this process, projections of cytoplasm, called *pseudopodia* (soo'-dō-PŌ-dē-a), engulf solid particles external to the cell. Once the particle is surrounded, the membrane folds inwardly, forming a membrane sac around the particle. This newly formed sac, called a *phagocytic vacuole,* breaks off from the outer cell mem-

brane, and the solid material inside the vacuole is digested. Indigestible particles and cell products are removed from the cell by a reverse phagocytosis. This process is important because molecules and particles of material that would normally be restricted from crossing the plasma membrane can be brought into or removed from the cell. The phagocytic white blood cells of the body constitute a vital defense mechanism. Through phagocytosis, the white blood cells engulf and destroy bacteria and other foreign substances (see Figure 4-4).

● *Pinocytosis*   In **pinocytosis** (pi'-nō-sī-TŌ-sis), or "cell drinking," the engulfed material consists of a liquid rather than a solid (Figure 3-5c). Moreover, no cytoplasmic projections are formed. Instead, the liquid is attracted to the surface of the membrane. The membrane folds inwardly, surrounds the liquid, and detaches from the rest of the intact membrane. Few cells are capable of phagocytosis, but many cells carry on pinocytosis. Examples include cells in the kidneys and urinary bladder.

The movement of materials into the cell by either phagocytosis or pinocytosis is referred to as **endocytosis.** The export of material from the cell by the reverse processes is called **exocytosis.**

MODIFIED PLASMA MEMBRANES

Electron-microscope studies have revealed that plasma membranes of certain cells are modified in various ways for specific purposes. For example, the membranes of some cells lining the small intestine have small, cylindrical projections called **microvilli** (see Figure 3-1). These fingerlike projections enormously increase the absorbing area of the cell surface. A single cell may have as many as 3,000 microvilli, and a 1 sq mm (0.00155 sq inch) area of small intestine may contain as many as 200 million microvilli, which would increase the surface area for absorption by 20 times.

Another membrane modification is found in the rods and cones of the eye. They serve as photoreceptors, or light-receiving cells. The upper portion of each rod contains two-layered, disc-shaped membranes called **sacs** that contain the pigments involved in vision.

Another example of a membrane modification is the **stereocilia.** They are found only in cells lining a duct (ductus epididymis) of the male reproductive system. They appear by light microscopy as long, slender, branching processes at the free surfaces of the lining cells (see Figure 28-6). Electron micrographs show stereocilia to be microvilli.

A final example of a membrane modification is the **myelin sheath** that surrounds portions of certain nerve cells (see Figure 12-3). It is thought that the myelin sheath increases the velocity of impulse conduction, protects the portion of the nerve cell it surrounds, is related to the nutrition of the nerve cell, and is necessary for its repair and regeneration.

CYTOPLASM

The substance inside the cell's plasma membrane and external to the nucleus is called **cytoplasm** (Figure 3-6a, b). It is the matrix, or ground substance, in which various cellular components are found. Physically, cytoplasm may be described as a thick, semitransparent, elastic fluid containing suspended particles and a series of minute tubules and filaments that form a cytoskeleton. Chemically, cytoplasm is 75 to 90 percent water plus solid components. Proteins, carbohydrates, lipids, and inorganic substances compose the bulk of the solid components. The inorganic substances and most carbohydrates are soluble in water and are present as a true solution. The majority of organic compounds, however, are found as colloids—particles that remain suspended in the surrounding ground substance. Since the particles of a colloid bear electrical charges that repel each other, they remain suspended and separated from each other.

Functionally, cytoplasm is the substance in which chemical reactions occur. The cytoplasm receives raw materials from the external environment and converts them into usable energy by decomposition reactions. Cytoplasm is also the site where new substances are synthesized for

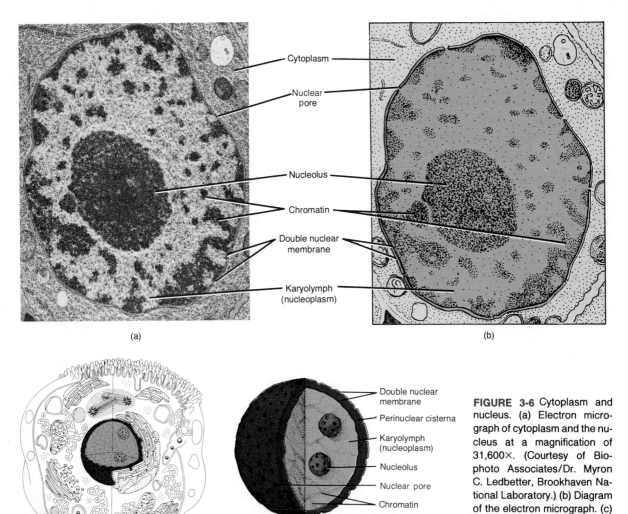

(a) (b) (c)

FIGURE 3-6 Cytoplasm and nucleus. (a) Electron micrograph of cytoplasm and the nucleus at a magnification of 31,600×. (Courtesy of Biophoto Associates/Dr. Myron C. Ledbetter, Brookhaven National Laboratory.) (b) Diagram of the electron micrograph. (c) Diagram of a nucleus with two nucleoli.

cellular use. It packages chemicals for transport to other parts of the cell or other cells of the body and facilitates the excretion of waste materials.

ORGANELLES

Despite the myriad chemical activities occurring simultaneously in the cell, there is little interference of one reaction with another. The cell has a system of compartmentalization provided by structures called **organelles.** These structures are specialized portions of the cell that assume specific roles in growth, maintenance, repair, and control.

NUCLEUS

The **nucleus** is generally a spherical or oval organelle and is the largest structure in the cell (Figure 3-6a–c). It contains hereditary factors of the cell, called genes, which control cellular structure and direct many cellular activities. Mature red blood cells do not have nuclei. These cells carry on only limited types of chemical activity and are not capable of growth or reproduction.

The nucleus is separated from the cytoplasm by a double membrane called the *nuclear membrane* or *envelope* (Figure 3-6c). Between the two layers of the nuclear membrane is a space called the *perinuclear cisterna*. Each of the nuclear membranes resembles the structure of the plasma membrane. Minute pores in the nuclear membrane allow the nucleus to communicate with a membranous network in the cytoplasm called the endoplasmic reticulum. Substances entering and exiting the nucleus are believed to pass through the tiny pores.

Three prominent structures are visible internal to the nuclear membrane. The first of these is a gel-like fluid that fills the nucleus called *karyolymph* (*nucleoplasm*). One or more spherical bodies called the *nucleoli* are also present. These structures are composed of protein, DNA, and RNA. DNA synthesizes the RNA which is stored in the nucleoli. The RNA, as you will see shortly, assumes a function in protein synthesis. Finally, there is the *genetic material* consisting principally of DNA. When the cell is not reproducing, the genetic material appears as a threadlike mass called *chromatin*. Prior to cellular reproduction the chromatin shortens and coils into rod-shaped bodies called *chromosomes*.

Studies of chromosomes have provided a great deal of information about their chemistry and detailed structure. Although it has been known for some time that chromosomes consist of DNA and proteins called *histones,* their organization and arrangement in chromosomes have only recently been determined. In specially prepared electron micrographs, it can be seen that chromosomes consist of elementary subunits of structure called **nucleosomes.** Each nucleosome has a diameter of about 100 Å. A nucleosome consists of two each of four different histone proteins (collectively called an *octamer*) associated with a relatively fixed length of DNA (about 200 nitrogenous base pairs). Present evidence indicates that the DNA is wrapped around the histones in some way. A fifth type of histone maintains adjacent nucleosomes in a helical coli (*solenoid*). Nucleosomes are separated by stretches of DNA only so that the arrangement resembles beads on a string. The functional significance of nucleosome structure is still speculative. It is believed that histones may facilitate changes in chromosomal structure that expose activated genes (DNA) to perform a specific task in the cell.

RIBOSOMES

Ribosomes are tiny granules, 250 Å at their largest dimension, that are composed of a type of RNA called ribosomal RNA (rRNA) and a number of specific ribosomal proteins. The rRNA is manufactured by DNA in the nucleolus. Ribosomes were so named because of their high content of rRNA. Structurally, a ribosome consists of two subunits, one about half the size of the other. Recently scientists have provided three-dimensional models of the structure of a ribosome (Figure 3-7d). Functionally, ribosomes are the sites of protein synthesis: they receive genetic instructions and translate them into protein. The mechanism of ribosomal function is discussed later in the chapter.

Some ribosomes, called *free ribosomes,* are scattered in the cytoplasm; they have no attachments to other parts of the cell. The free ribosomes occur singly or in clusters, and they are primarily concerned with synthesizing proteins for use inside the cell. Other ribosomes are attached to a cellular structure called the endoplasmic reticulum. These ribosomes are concerned with the synthesis of proteins for export from the cell.

ENDOPLASMIC RETICULUM (ER)

Within the cytoplasm, there is a system of pairs of parallel membranes enclosing narrow cavities of varying shapes. This system is known as the **endoplasmic reticulum,** or **ER** (Figure 3-7). The ER, in other words, is a network of channels (*cisternae*) running through the cytoplasm. These channels are continuous with the nuclear membrane.

On the basis of its association with ribosomes, ER is distinguished into two types. **Granular (rough) ER** is studded with ribosomes; **agranular (smooth) ER** is free of ribosomes. Recent studies using radioactive tracers suggest that agranular ER is synthesized from granular ER.

Numerous functions related to homeostasis are attributed to the ER. It contributes to the mechanical support and distribution of the cytoplasm. It also conducts intracellular nerve impulses, such as in muscle cells, where it is called sarcoplasmic reticulum. The ER is involved in the intracellular exchange of materials with the cytoplasm and provides a surface area for chemical reactions. Various products are transported from one portion of the cell to another via the ER, so the ER is considered an intracellular circulatory system. The ER also serves as a storage area for synthesized molecules. And, together with a cellular structure called the Golgi complex, the ER assumes a role in the synthesis and packaging of molecules.

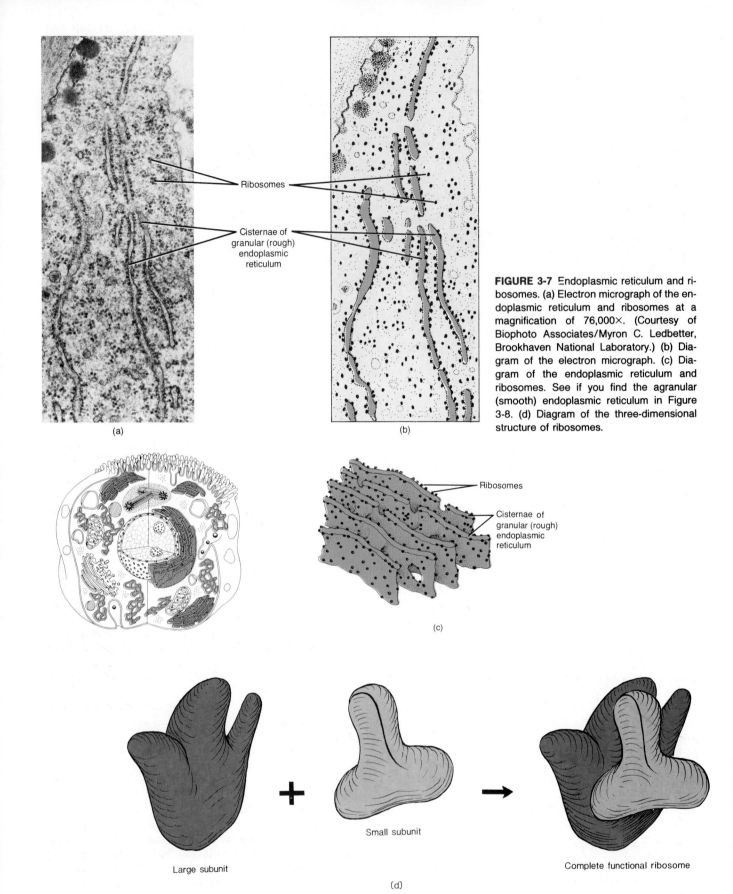

Ribosomes

Cisternae of
granular (rough)
endoplasmic
reticulum

(a)

(b)

FIGURE 3-7 Endoplasmic reticulum and ribosomes. (a) Electron micrograph of the endoplasmic reticulum and ribosomes at a magnification of 76,000×. (Courtesy of Biophoto Associates/Myron C. Ledbetter, Brookhaven National Laboratory.) (b) Diagram of the electron micrograph. (c) Diagram of the endoplasmic reticulum and ribosomes. See if you find the agranular (smooth) endoplasmic reticulum in Figure 3-8. (d) Diagram of the three-dimensional structure of ribosomes.

Ribosomes

Cisternae of
granular (rough)
endoplasmic
reticulum

(c)

Large subunit

+

Small subunit

Complete functional ribosome

(d)

GOLGI COMPLEX

Another structure found in the cytoplasm is the **Golgi** (GOL-jē) **complex.** This structure consists of four to eight flattened baglike channels, stacked upon each other with expanded areas at their ends. Like those of the ER, the stacked elements are called *cisternae,* and the expanded, terminal areas are *vesicles* (Figure 3-8). Generally, the Golgi complex is located near the nucleus.

One function of the Golgi complex is the packaging of secreted proteins. *Secretion* is the production and re-

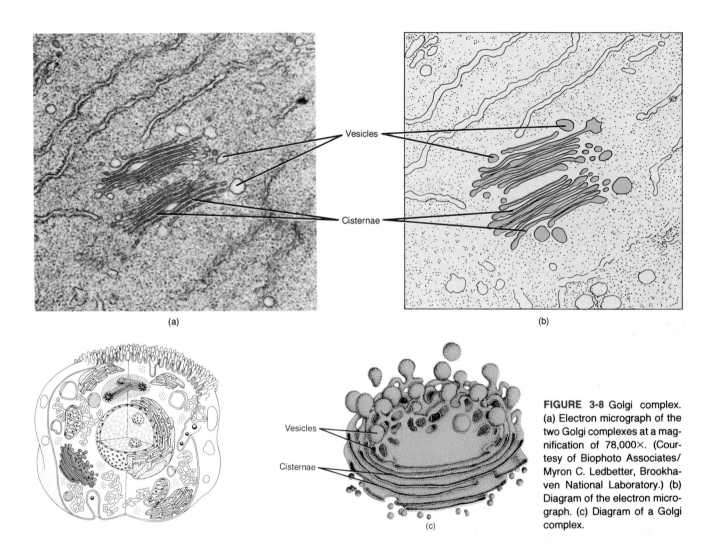

(a)

(b)

Vesicles

Cisternae

Vesicles

Cisternae

(c)

FIGURE 3-8 Golgi complex. (a) Electron micrograph of the two Golgi complexes at a magnification of 78,000×. (Courtesy of Biophoto Associates/ Myron C. Ledbetter, Brookhaven National Laboratory.) (b) Diagram of the electron micrograph. (c) Diagram of a Golgi complex.

lease from a gland cell of a fluid that usually contains a variety of substances. Proteins synthesized at ribosomes associated with granular ER are transported into the ER cisternae (Figure 3-9). The proteins then pass in membrane sacs pinched off from the ER into the Golgi complex. As proteins accumulate and concentrate in the cisternae of the Golgi complex, the cisternae expand to form vesicles. After a certain size is reached, the vesicles pinch off from the cisternae. The protein and its associated vesicle is referred to as a *secretory granule.* The secretory granule then moves toward the surface of the cell where the protein is secreted. The contents of the granule are discharged into the extracellular space, and the granule membrane is incorporated into the plasma membrane. Cells of the digestive tract that secrete protein enzymes utilize this mechanism. The secretory granule prevents "digestion" of the cytoplasm of the cells by the enzymes as it moves toward the cell surface. Other vesicles that pinch off from the Golgi complex are loaded with special digestive enzymes and remain within the cell. They become cellular structures called lysosomes.

Another function of the Golgi complex is associated with lipid secretion. It occurs in essentially the same way as protein secretion, except the lipids are synthesized by the agranular ER. The lipids pass through the ER into the Golgi complex. As in the mechanism just described,

the lipids migrate into the cisternae and vesicles and are discharged at the surface of the cell. In the course of moving through the cytoplasm, the vesicle may release lipids into the cytoplasm before being discharged from the cell. These appear in the cytoplasm as lipid droplets. Among the lipids secreted in this manner are the steroids (see Exhibit 2-5).

The Golgi complex also functions in the synthesis of carbohydrates. Recent evidence indicates that carbohydrates synthesized by the Golgi complex are combined with proteins synthesized by ribosomes associated with granular ER to form glycoproteins. As these carbohydrate-protein complexes are assembled, they accumulate in the flattened channels of the Golgi complex. The channels expand and form vesicles. After a critical size is reached, the vesicles pinch off from the channel, migrate through the cytoplasm, and discharge their contents through the plasma membrane. The Golgi complex is well developed and highly active in secretory cells such as those found in the pancreas and the salivary glands.

MITOCHONDRIA

Small, spherical, rod-shaped, or filamentous structures called **mitochondria** (mī'-tō-KON-drē-a) appear throughout the cytoplasm. When sectioned and viewed under

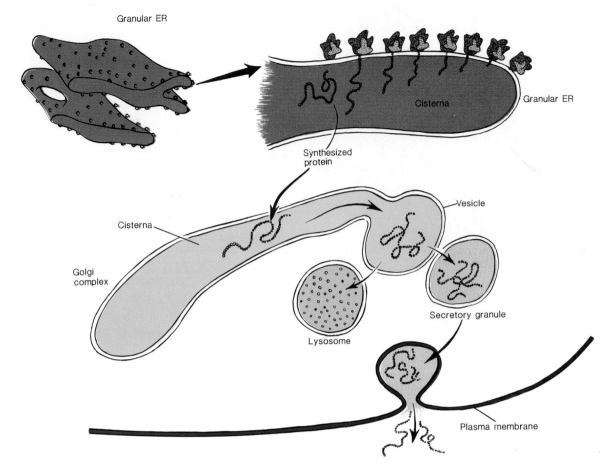

FIGURE 3-9 Packaging of synthesized protein for export from the cell.

an electron microscope, each reveals an elaborate internal organization (Figure 3-10). A mitochondrion consists of two membranes, each of which is similar in structure to the plasma membrane. The outer mitochondrial membrane is smooth, but the inner membrane is arranged in a series of folds called *cristae*. The center of the mitochondrion is called the *matrix*.

Because of the nature and arrangement of the cristae, the inner membrane provides an enormous surface area for chemical reactions. Enzymes involved in energy-releasing reactions that form ATP are located on the cristae. Mitochondria are frequently called the "powerhouses of the cell" because they are the sites for the production of ATP. Active cells, such as muscle and liver cells, have a large number of mitochondria because of their high energy expenditure.

Mitochondria are self-replicative, that is, they can divide to form new ones. The replication process is controlled by DNA which is incorporated into the mitochondrial structure. Self-replication usually occurs in response to increased cellular need for ATP.

LYSOSOMES

When viewed under the electron microscope, **lysosomes** (*lysis* = dissolution; *soma* = body) appear as membrane-enclosed spheres (Figure 3-11). They are formed from Golgi complexes, have a double membrane, and lack detailed structure. They contain powerful digestive enzymes capable of breaking down many kinds of molecules. You will see in Chapter 14 that Tay-Sachs disease results from a deficiency of a lysosomal enzyme. These enzymes are also capable of digesting bacteria that enter the cell (see Figure 22-8). White blood cells, which ingest bacteria by phagocytosis, contain large numbers of lysosomes.

Scientists have wondered why these powerful enzymes do not also destroy their own cells. Perhaps the lysosome membrane in a healthy cell is impermeable to enzymes so they cannot move out into the cytoplasm. However, when a cell is injured, the lysosomes release their enzymes. The enzymes then promote reactions that break the cell down into its chemical constituents. This process of self-destruction by cells is known as **autolysis.** The chemical remains are either reused by the body or excreted. Because of this function, lysosomes have been called "suicide packets."

Lysosomes are crucial in the removal of cell parts, whole cells, and even extracellular material, which may be the process underlying bone removal. In bone reshaping, especially during the growth process, special bone-destroying cells called osteoclasts, secrete extracellular enzymes that dissolve bone. Bone tissue cultures given excess amounts of vitamin A remove bone apparently through an activation process involving lysosomes.

CLINICAL APPLICATION

Animals overfed on vitamin A interestingly develop spontaneous fractures, suggesting greatly increased lysosomal activity. On the other hand, cortisone and hydrocortisone, steroid hormones produced by the adrenal gland, have a stabilizing effect on lysosomal mem-

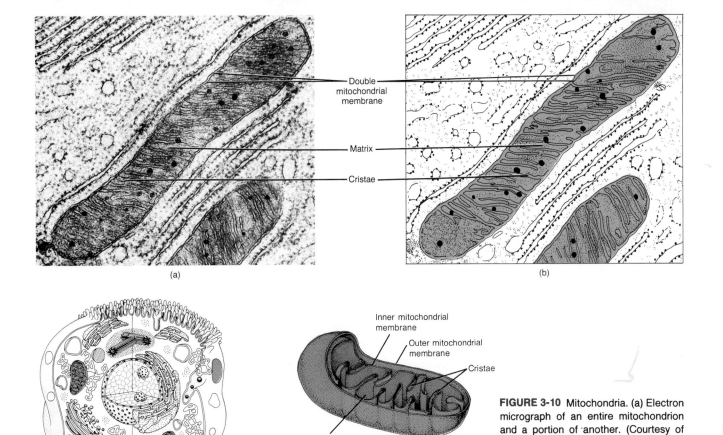

Double mitochondrial membrane

Matrix

Cristae

(a)

(b)

Inner mitochondrial membrane

Outer mitochondrial membrane

Cristae

Matrix

(c)

FIGURE 3-10 Mitochondria. (a) Electron micrograph of an entire mitochondrion and a portion of 'another. (Courtesy of Lester V. Bergman & Associates.) (b) Diagram of the electron micrograph. (c) Diagram of a mitochondrion.

branes. The steroid hormones are well known for their antiinflammatory properties, which suggests that they reduce destructive cellular activity by lysosomes.

PEROXISOMES

Organelles similar in structure to lysosomes, but smaller, are called **peroxisomes** (pe-ROKS-i-sōms). They are abundant in liver cells and contain several enzymes related to the metabolism of hydrogen peroxide (H_2O_2), a substance that is toxic to body cells. One of the enzymes

in peroxisomes, called *catalase,* immediately breaks down H_2O_2 into water and oxygen:

$$\underset{\substack{\text{Hydrogen}\\\text{peroxide}}}{H_2O_2} \xrightarrow{\text{Catalyase}} \underset{\text{Water}}{H_2O} + \underset{\text{Oxygen}}{O_2}$$

MICROFILAMENTS AND MICROTUBULES: THE CYTOSKELETON

In recent years, electron microscopy has revealed that the cytoplasm of cells is more than just a structureless

FIGURE 3-11 Lysosome. (a) Electron micrograph of a lysosome at a magnification of 55,000×. (Courtesy of F. Van Hoof, Université Catholique de Louvain.) (b) Diagram of the electron micrograph.

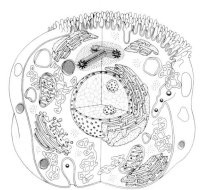

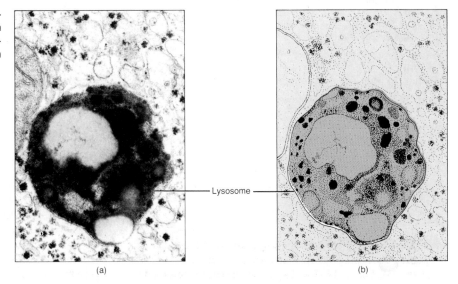

Lysosome

(a)

(b)

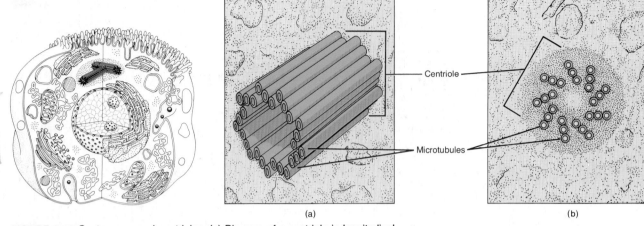

(a)   (b)

FIGURE 3-12 Centrosome and centrioles. (a) Diagram of a centriole in longitudinal section. (b) Diagram of a centriole in cross section. (c) Electron micrograph of a centriole in cross section at a magnification of 67,000×. (Courtesy of Biophoto Associates.)

medium in which the various cellular components are suspended. It has been shown that cytoplasm actually has a complex internal structure, consisting of a series of exceedingly small microfilaments and microtubules, together referred to as the **cytoskeleton.**

Microfilaments are rodlike structures ranging from 30 to 120 Å in diameter. They are of variable length and may occur in bundles, randomly scattered throughout the cytoplasm, or arranged in a meshwork, depending on the type of cell in which they are found. Some microfilaments consist of a protein called *actin;* others consist of a protein called *myosin.* In muscle tissue, actin microfilaments (thin myofilaments) and myosin microfilaments (thick myofilaments) are involved in the contraction of muscle cells. This mechanism is described in Chapter 10. In nonmuscle cells, microfilaments help to provide support and shape and assist in the movement of entire cells (phagocytes and cells of developing embryos) and movements within cells (secretion, phagocytosis, pinocytosis).

Microtubles are relatively straight, slender, cylindrical structures that range in diameter from 180 to 300 Å, usually averaging about 240 Å. They consist of a protein called *tubulin.* Microtubules dispersed in the cytoplasm, together with microfilaments, help to provide support and shape for cells. Some evidence suggests that microtubules may form conducting channels through which various substances can move throughout the cytoplasm. This mechanism has been studied extensively in nerve cells. Microtubules also assist in the movement of pseudopodia that are characteristic of phagocytes. As you will see shortly, microtubules form the structure of flagella and cilia (cellular appendages involved in motility), centrioles (organelles that may direct the assembly of microtubules), and the mitotic spindle (structures that are involved in cell division).

CENTROSOME AND CENTRIOLES

A dense area of cytoplasm, generally spherical and located near the nucleus, is called the **centrosome (centrosphere).** Within the centrosome is a pair of cylindrical structures:

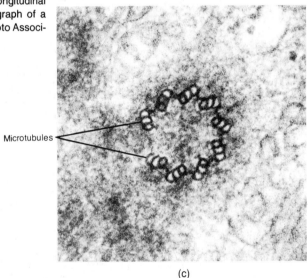

(c)

the **centrioles** (Figure 3-12). Each centriole is composed of a ring of nine evenly spaced bundles. Each bundle, in turn, consists of three microtubules. The two centrioles are situated so that the long axis of one is at right angles to the long axis of the other. Centrioles assume a role in cell reproduction by serving as centers about which microtubules involved in chromosome movement are organized. This will be described shortly as part of cell division. Certain cells, such as most mature nerve cells, do not have a centrosome and so do not reproduce. This is why they cannot be replaced if destroyed. Like mitochondria, centrioles contain DNA that controls their self-replication.

FLAGELLA AND CILIA

Some body cells possess projections for moving the entire cell or for moving substances along the surface of the cell. These projections contain cytoplasm and are bounded by the plasma membrane. If the projections are few and long in proportion to the size of the cell, they are called **flagella.** The only example of a flagellum in the human body is the tail of a sperm cell, used for locomotion (see Figure 28-4). If the projections are numerous and short, resembling many hairs, they are called **cilia.**

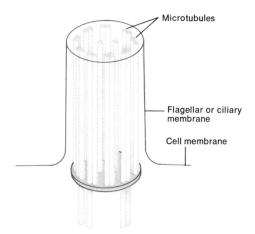

FIGURE 3-13 Structure of a flagellum or cilium.

Microtubules

Flagellar or ciliary membrane

Cell membrane

In humans, ciliated cells of the respiratory tract move mucus that has trapped foreign particles over the surface of the tissue (see Figure 23-5b). Electron microscopy has revealed no fundamental structural difference between cilia and flagella (Figure 3-13). Both consist of nine pairs of microtubules that form a ring around two microtubules in the center.

CELL INCLUSIONS

Cell inclusions are a large and diverse group of chemical substances, some of which have recognizable shapes. These products are principally organic and may appear or disappear at various times in the life of the cell. *Melanin* is a pigment stored in certain cells of the skin, hair, and eyes. It protects the body by screening out harmful ultraviolet rays from the sun. *Glycogen* is a polysaccharide that is stored in the liver, skeletal muscle cells, and the vaginal mucosa. When the body requires quick energy, liver cells can break down the glycogen into glucose and release it. *Lipids,* which are stored in fat cells, may be decomposed for producing energy. A final example of an inclusion is *mucus,* which is produced by cells that line organs. Its function is to provide lubrication and protection.

The major parts of the cell and their functions are summarized in Exhibit 3-1.

EXTRACELLULAR MATERIALS

The substances that lie outside cells are called **extracellular materials.** They include the body fluids, which provide a medium for dissolving, mixing, and transporting substances. Among the body fluids are interstitial fluid, the fluid that fills the microscopic spaces (interstitial spaces), and plasma, the liquid portion of blood. Extracellular materials also include secreted inclusions like mucus and special substances that form the matrix in which some cells are embedded.

The matrix materials are produced by certain cells and deposited outside their plasma membranes. The matrix supports the cells, binds them together, and gives strength and elasticity to the tissue. Some matrix materials

EXHIBIT 3-1
CELL PARTS AND THEIR FUNCTIONS

PART	FUNCTIONS
Plasma membrane	Protects cellular contents; makes contact with other cells; provides receptors for hormones, enzymes, and antibodies; mediates the entrance and exit of materials.
Cytoplasm	Serves as the ground substance in which chemical reactions occur.
Organelles Nucleus	Contains genes and controls cellular activities.
Ribosomes	Sites of protein synthesis.
Endoplasmic reticulum (ER)	Contributes to mechanical support; conducts intracellular nerve impulses in muscle cells; facilitates intracellular exchange of materials with cytoplasm; provides a surface area for chemical reactions; provides a pathway for transporting chemicals; serves as a storage area; together with Golgi complex synthesizes and packages molecules for export.
Golgi complex	Packages synthesized proteins for secretion in conjunction with endoplasmic reticulum; forms lysosomes; secretes lipids; synthesizes carbohydrates; combines carbohydrates with proteins to form glycoproteins for secretion.
Mitochondria	Sites for production of ATP.
Lysosomes	Digest substances and foreign microbes; may be involved in bone removal.
Peroxisomes	Contains several enzymes, such as catalase, related to hydrogen peroxide metabolism.
Microfilaments	Form cytoskeleton with microtubules; involved in muscle cell contraction; provide support and shape; assist in cellular and intracellular movement.
Microtubules	Form cytoskeleton with microfilaments; provide support and shape; form intracellular conducting channels; assist in cellular movement; form the structure of flagella, cilia, centrioles, and spindle fibers.
Centrioles	Help organize mitotic spindle during cell division.
Flagella and cilia	Allow movement of entire cell (flagella) or movement of particles along surface of cell (cilia).
Inclusions	Melanin (pigment in skin, hair, eyes) screens out ultraviolet rays; glycogen (stored glucose) can be decomposed to provide energy; lipids (stored in fat cells) can be decomposed to produce energy; mucus provides lubrication and protection.

are *amorphous* (*a* = without, *morpho* = shape); they have no specific shape. These include hyaluronic acid and chondroitin sulfate. **Hyaluronic** (hī'-a-loo-RON-ik) **acid** is a viscous, fluidlike substance that binds cells together, lubricates joints, and maintains the shape of the eyeballs. **Chondroitin** (kon-DROY-tin) **sulfate** is a jelly-like substance that provides support and adhesiveness in cartilage, bone, heart valves, the cornea of the eye, and the umbilical cord.

Other matrix materials are *fibrous,* or threadlike. Fibrous materials provide strength and support for tissues. Among these are **collagenous** (*kolla* = glue) **fibers** consisting of the protein *collagen.* These fibers are found in all types of connective tissue, especially in bones, cartilage, tendons, and ligaments. **Reticular** (*rete* = net) **fibers** consisting of the protein collagen and a coating of glycoprotein, form a network around fat cells, nerve fibers, muscle cells, and blood vessels. They also form the framework or stroma for many soft organs of the body such as the spleen. **Elastic fibers,** consisting of the protein *elastin,* give elasticity to skin and to tissues forming the walls of blood vessels.

CELL DIVISION

Most of the cell activities mentioned thus far maintain the life of the cell on a day-to-day basis. However, cells become damaged, diseased, or worn out and then die. New cells must be produced as replacements and for growth.

Cell division is the process by which cells reproduce themselves. It consists of a nuclear division and a cytoplasmic division. Because nuclear division can be of two types, two kinds of cell division are recognized. The first is the mechanism by which sperm and egg cells are produced, preliminary to the formation of a new organism. The process consists of a nuclear division called **meiosis** plus a cytoplasmic division called **cytokinesis.** It is often called reproductive cell division and is discussed in detail in Chapter 29.

In the second kind of division, often called somatic cell division, a single parent cell duplicates itself. This process consists of a nuclear division called **mitosis** and cytokinesis. The process ensures that each new daughter cell has the same *number* and *kind* of chromosomes as the original parent cell. After the process is complete, the two daughter cells have the same hereditary material and genetic potential as the parent cell. This kind of cell division results in an increase in the number of body cells. In a 24-hour period, the average adult loses trillions of cells from different parts of the body. Obviously, these cells must be replaced. Cells that have a short life span— the cells of the outer layer of skin, the cornea of the eye, the digestive tract—are continually being replaced. Mitosis and cytokinesis are the means by which dead or injured cells are replaced and new cells are added for body growth.

When a cell reproduces, it must replicate its chromosomes so its hereditary traits may be passed on to succeeding generations of cells. A **chromosome** is a highly coiled DNA molecule that is partly covered by protein. The protein causes changes in the length and thickness of the chromosome. Hereditary information is contained in the DNA portion of the chromosome in units called **genes.** Each human chromosome consists of about 20,000 genes.

When a cell is carrying on every life process except division, it is said to be in **interphase** or the **metabolic phase.** It is during this stage that the replication of chromosomes occurs.

When DNA replicates, its helical structure partially uncoils (Figure 3-14). Those portions of DNA that remain coiled stain darker than the uncoiled portions. This unequal distribution of stain causes the DNA to appear as a granular mass called **chromatin** (see Figure 3-15a). During uncoiling, DNA separates at the points where the nitrogen bases are connected. Each exposed nitrogen base then picks up a complementary nitrogen base (with associated sugar and phosphate group) from the cytoplasm of the cell. This uncoiling and complementary base pairing continues until each of the two original DNA strands is matched and joined with two newly formed DNA strands. The original DNA molecule has become two DNA molecules.

During interphase the cell is also synthesizing most of its RNA and proteins. It is producing chemicals so that all cellular components can be doubled during division. A microscopic view of a cell during interphase shows a clearly defined nuclear membrane, nucleoli, karyolymph, chromatin, and a pair of centrioles. Once a cell completes its replication of DNA and its synthesis of RNA and proteins during interphase, mitosis begins.

MITOSIS

The succession of events that takes place during mitosis and cytokinesis is plainly visible under a microscope after the cells have been stained in the laboratory.

The process called **mitosis** is the distribution of the two sets of chromosomes into two separate and equal nuclei following the replication of the chromosomes of the parent nucleus. For convenience, biologists divide the process into four stages: prophase, metaphase, anaphase, and telophase. These are arbitrary classifications. Mitosis is actually a continuous process, one stage merging imperceptibly into the next.

Prophase

During **prophase** (Figure 3-15b), the first stage of mitosis, chromatin shortens and coils into chromosomes. The nucleoli become less distinct and the nuclear membrane disappears. Each prophase "chromosome" is actually composed of a pair of structures called **chromatids.** Each chromatid is a complete chromosome consisting of a double-stranded DNA molecule and each is attached to its chromatid pair by a small spherical body called a **centromere.** During prophase, the chromatid pairs assemble near the equatorial plane region or equator of the cell.

Also during prophase, the paired centrioles separate and each pair moves to an opposite pole (end) of the cell. Between the centrioles, a series of microtubules is organized into two groups of fibers. The *continuous* (*inter-*

polar) *microtubules* originate from the vicinity of each pair of centrioles and grow toward each other. Thus, they extend from one pole of the cell to another. As they grow toward each other, the second group of microtubules develops. These are called *chromosomal microtubules* and grow out of the centromeres; they extend from a centromere to a pole of the cell. Together, the continuous and chromosomal microtubules constitute the **mitotic spindle** and, with the centrioles, are referred to as the **mitotic apparatus.**

Metaphase

During **metaphase** (Figure 3-15c), the second stage of mitosis, the centromeres of the chromatid pairs line up on the equatorial plane of the cell. The centromeres of each chromatid pair form a chromosomal microtubule that attaches the centromere to a pole of the cell.

Anaphase

The third stage of mitosis, **anaphase** (Figure 3-15d), is characterized by the division of the centromeres and the movement of complete identical sets of chromatids, now called chromosomes, to opposite poles of the cell. During this movement, the centromeres attached to the chromosomal microtubules seem to drag the trailing parts of the chromosomes toward opposite poles.

The mechanism by which the chromosomes move to opposite poles is not completely understood. Several theories have been presented and the key points of two of them may be summarized as follows. One suggests that the movement of chromosomes is caused by the assembly

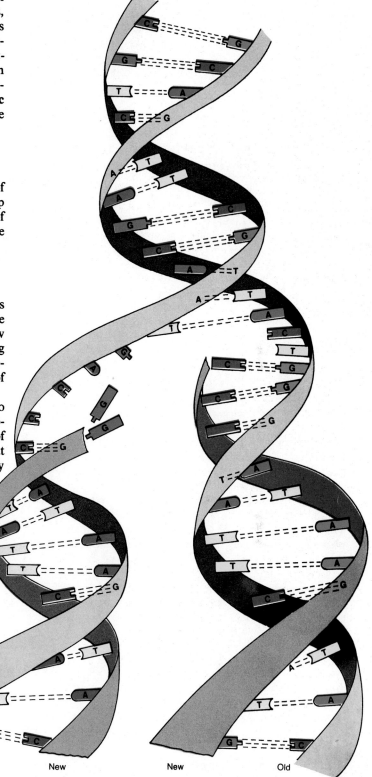

FIGURE 3-14 Replication of DNA. The two strands of the double helix separate by breaking the bonds between nucleotides. New nucleotides attach at the proper sites, and a new strand of DNA is paired off with each of the original strands. After replication, the two DNA molecules, each consisting of a new and an old strand, return to their helical structure.

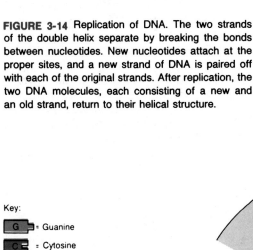

Key:

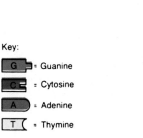

G = Guanine

C = Cytosine

A = Adenine

T = Thymine

Old   New   New   Old

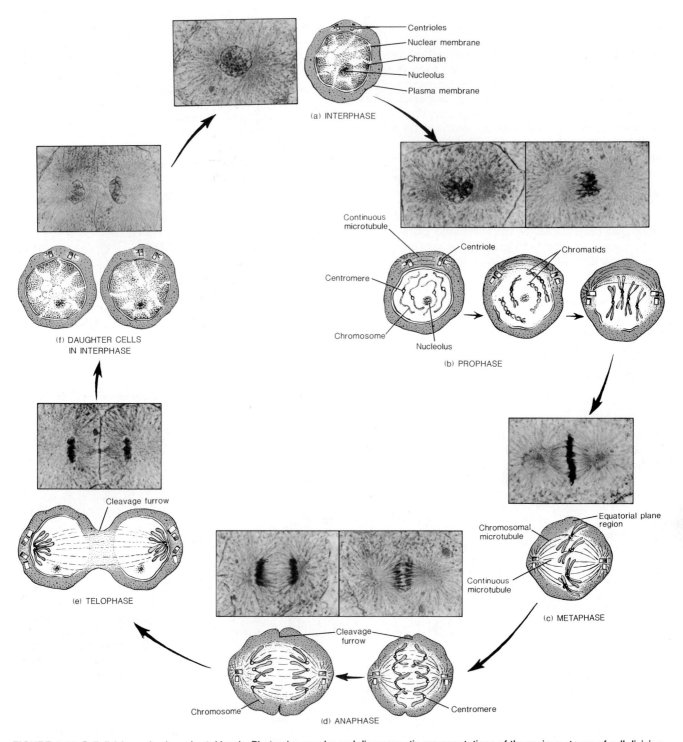

FIGURE 3-15 Cell division: mitosis and cytokinesis. Photomicrographs and diagrammatic representations of the various stages of cell division in whitefish eggs. Read the sequence starting at (a), and move clockwise until you complete the cycle. (Courtesy of Carolina Biological Supply Company.)

and disassembly of microtubules. As the chromosomal microtubules are shortened by the removal of tubulin subunits, the chromosomes are pulled toward the poles. Simultaneously, the continuous microtubules are lengthened by the addition of tubulin subunits, further contributing to chromosomal movement.

A second theory is that chromosomal movement and

pole separation are caused by mechanochemical interactions between continuous and chromosomal microtubules. This is accomplished by temporary cross bridges between microtubules. As a result of reactions that use energy, the cross bridges cause the continuous and chromosomal microtubules to slide past each other in a manner similar to the closing of an extension ladder.

Telophase

Telophase (Figure 3-15e), the final stage of mitosis, consists of a series of events nearly the reverse of prophase. By this time, two identical sets of chromosomes have reached opposite poles. As telophase progresses, new nuclear membranes begin to enclose the chromosomes, the chromosomes start to assume their chromatin form, nucleoli reappear, and the mitotic spindles disappear. The centrioles also replicate so that each cell has two centriole pairs. The formation of two nuclei identical to those of cells in interphase terminates telophase. A mitotic cycle has thus been completed (Figure 3-15f).

Time Required

The time required for mitosis varies with the kind of cell, its location, and the influence of factors such as temperature. Furthermore, the different stages of mitosis are not equal in duration. Prophase is usually the longest stage, lasting from one to several hours. Metaphase is considerably shorter, ranging from 5 to 15 minutes. Anaphase is the shortest stage, lasting from 2 to 10 minutes. Telophase lasts from 10 to 30 minutes. These lengths of time are only approximate, however.

CYTOKINESIS

Division of the cytoplasm, called **cytokinesis** (sī'-tō-ki-NĒ-sis), often begins during late anaphase and terminates at the same time as telophase. Cytokinesis begins with the formation of a *cleavage furrow* that extends around the cell's equator. The furrow progresses inward, resembling a constricting ring, and cuts completely through the cell to form two separate portions of cytoplasm (Figure 3-15d–f).

GENE ACTION

PROTEIN SYNTHESIS

We have already noted that genes consist of DNA and that genes determine the traits that a cell inherits and control the activities that take place in a cell. Mitosis and cytokinesis are vital processes because they ensure that daughter cells have identical sets of chromosomes and, therefore, identical genes. In this way, hereditary information passes from one generation of cells to another.

Although cells synthesize numerous chemicals in order to maintain homeostasis, much of the cellular machinery is concerned with producing proteins. Some of the proteins are structural, helping to form plasma membranes, microfilaments, microtubules, centrioles, flagella, cilia, the mitotic spindle, and other parts of cells. Other proteins serve as hormones, antibodies, and contractile elements in muscle tissue. Still other proteins serve as enzymes that regulate the myriad chemical reactions that occur in cells. Basically, cells are protein factories that constantly synthesize large numbers of diverse proteins that determine the physical and chemical characteristics of cells and, therefore, of organisms.

The genetic instructions for making proteins are found in DNA. Cells make proteins by translating the genetic information encoded in DNA into specific proteins. In the process, the genetic information in a region of DNA is copied to produce a specific molecule of RNA. Through a complex series of reactions, the information contained in RNA is translated into a corresponding specific sequence of amino acids in a newly produced protein molecule. Let us take a look at how DNA directs protein synthesis by considering the two principal steps in protein synthesis: transcription and translation.

Transcription

Transcription is the process by which genetic information encoded in DNA is copied by a strand of RNA called **messenger RNA (mRNA).** It is called transcription because it resembles the transcription of a sequence of words from one tape to another. By using a specific portion of the cell's DNA as a template, the genetic information stored in the sequence of nitrogen bases of DNA is rewritten so that the same information appears in the nitrogen bases of mRNA. As in DNA replication (see Figure 3-14), a cytosine (C) in the DNA template dictates a guanine (G) in the mRNA strand being made; a G in the DNA template dictates a C in the mRNA strand; and a thymine (T) in the DNA template dictates an adenine (A) in the mRNA. Since RNA contains uracil (U) instead of T, an A in the DNA template dictates a U in the mRNA. As an example, if the template portion of DNA has the base sequence ATGCAT, the transcribed mRNA strand would have the complementary base sequence UACGUA (Figure 3-16). Note that only one of the two DNA strands serves as a template for RNA synthesis. This strand is referred to as the *sense strand.* The other strand, the one not transcribed, is the complement of the sense strand and is called the *antisense strand.*

In addition to serving as the template for the synthesis of mRNA, DNA also synthesizes two other kinds of RNA. One is called **ribosomal RNA** or **rRNA** which, together with ribosomal proteins, makes up ribosomes. The other is called **transfer RNA (tRNA).** Once synthesized, mRNA, rRNA, and tRNA leave the nucleus of the cell. In the cytoplasm, they participate in the next principal step in protein synthesis—translation.

Translation

The process by which information in the nitrogen base sequence of mRNA is used to specify the amino acid sequence of a protein is called **translation.** The key events involved in translation are as follows (Figure 3-17).

1. In the cytoplasm, the small ribosomal subunit binds one end of the mRNA molecule (Figure 3-17a). In this respect, ribosomes are the sites of protein synthesis.

2. There are 20 different amino acids in the cytoplasm that may participate in protein synthesis. Whichever amino acids participate to form a particular protein are picked up by tRNAs (Figure 3-17b). For each different amino acid there is a different type of tRNA. One end of a tRNA molecule couples with a specific amino acid. This amino acid activation requires energy

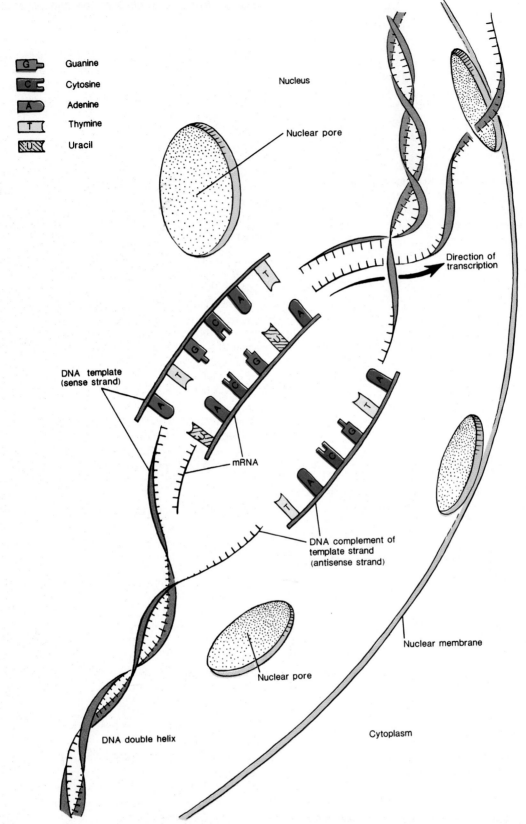

Legend:

G Guanine
C Cytosine
A Adenine
T Thymine
U Uracil

Nucleus

Nuclear pore

Direction of transcription

DNA template (sense strand)

mRNA

DNA complement of template strand (antisense strand)

Nuclear membrane

Nuclear pore

Cytoplasm

DNA double helix

FIGURE 3-16 Transcription. Transcription is the process whereby an mRNA molecule is produced that is complementary in base sequence to a section of one of the DNA strands (the sense strand) in a DNA double helix. The process occurs in the nucleus. The diagram shows a DNA double helix, partly unwound, with transcription occurring from the DNA template in the unwound region. The upper enlargement shows in more detail a six-base segment of the template and transcript; note the nitrogen-base pairing (A with T and G with C), although in RNA U replaces T. The lower enlargement shows the base sequence of the nontemplate DNA strand (antisense strand) for the same six-base segment as detailed in the upper enlargement. In addition to mRNA, DNA templates can also synthesize rRNA and tRNA. Note here, first, that the base sequence is complementary to the base sequence of the DNA template strand and, second, that the base sequence is the same as that found in the RNA transcript from the template strand (except, of course, that in RNA, U replaces T). The process of transcription is carried out by a special enzyme—RNA polymerase.

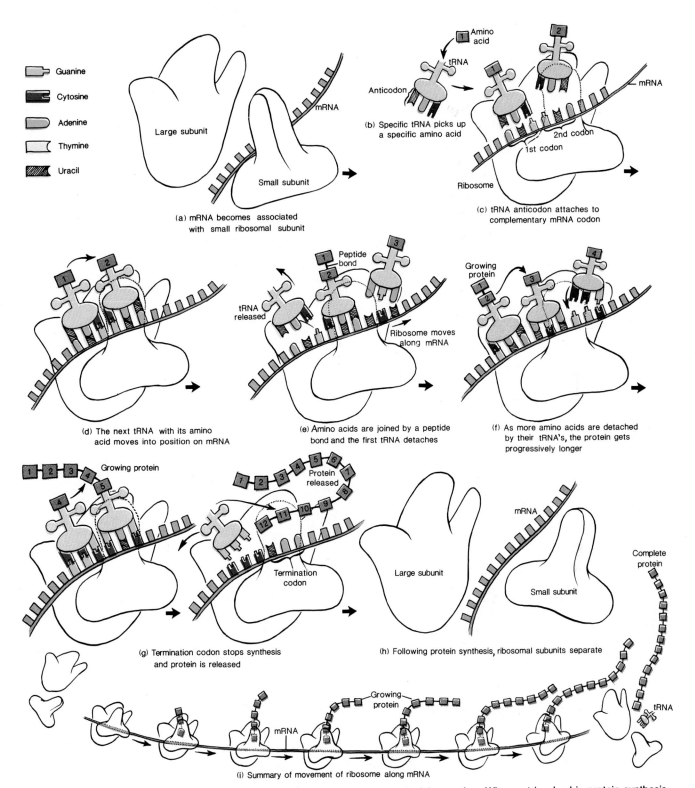

Guanine

Cytosine

Adenine

Thymine

Uracil

(a) mRNA becomes associated with small ribosomal subunit

(b) Specific tRNA picks up a specific amino acid

(c) tRNA anticodon attaches to complementary mRNA codon

(d) The next tRNA with its amino acid moves into position on mRNA

(e) Amino acids are joined by a peptide bond and the first tRNA detaches

(f) As more amino acids are detached by their tRNA's, the protein gets progressively longer

(g) Termination codon stops synthesis and protein is released

(h) Following protein synthesis, ribosomal subunits separate

(i) Summary of movement of ribosome along mRNA

FIGURE 3-17 Translation. Note that during protein synthesis the ribosomal subunits join together. When not involved in protein synthesis, the two subunits are separate entities.

EXHIBIT 3-2
EFFECTS OF AGING ON VARIOUS BODY SYSTEMS

SYSTEM	AGING EFFECTS
Integumentary	General changes include wrinkling; loss of subcutaneous (under the skin) fat; atrophy (wasting away) of sweat glands; reduction of blood flow to skin, decrease in number of functioning pigment-producing cells, resulting in gray hair and atypical skin color; increase in size of some pigment cells, producing pigmented blotching; atrophy of sebaceous (oil) glands, producing dry and broken skin that is susceptible to infection. Aged skin becomes susceptible to pathological conditions such as senile pruritus (itching), decubitus ulcers (bedsores), and herpes zoster (shingles).
Skeletomuscular	There is loss of muscle mass and diminishing of muscle reflexes. Common disorders include osteoarthritis, rheumatoid arthritis, gout, osteoporosis (decrease in bone mass), and Paget's disease.
Nervous	Nerve cells (neurons) are lost. Associated with this decline is a decreased capacity for sending nerve impulses to and from the brain. Conduction velocity decreases, voluntary motor movements slow down, and the reflex time for skeletal muscles is decreased. Degenerative changes and disease states involving the sense organs can alter vision, hearing, taste, smell, and touch. The disorders that represent the most common visual problems and may be responsible for serious loss of vision are presbyopia (inability to focus on nearby objects), cataracts (cloudiness of the lens), and glaucoma (excessive fluid pressure in the eyeball). Impaired hearing associated with aging, known as presbycusis, is usually the result of changes in important structures of the inner ear. Parkinson's disease is the most common movement disorder involving the central nervous system.
Endocrine	The endocrine system exhibits a variety of changes, and many researchers look to this system with the hope of finding the key to the aging process. Disorders of the endocrine system are not frequent, and when they do occur, most often they are related to pathological changes rather than age. Diabetes and thyroid disorders are the most important endocrine problems that have a significant effect on health and function.
Cardiovascular	General changes include loss of extensibility of the aorta, reduction in cardiac muscle cell size, progressive loss of cardiac muscular strength, and a reduced output of blood by the heart. There is an increase in blood pressure. Coronary artery disease increases and represents the major cause of heart disease and death in older Americans. Congestive heart failure occurs and is viewed as a set of symptoms associated with the impaired pumping performance of the heart. Changes in blood vessels such as hardening of the arteries and cholesterol deposits in arteries that serve brain tissue can reduce its nourishment and result in the malfunction or death of brain cells, which is called cerebrovascular disease.
Respiratory	The cardiovascular and respiratory systems operate as a unit, and damage or disease in one of these organ systems is often secondarily reflected in the other. In pulmonary heart disease the right side of the

from ATP breakdown. The other end has a specific sequence of three nitrogen bases (triplet) known as an *anticodon*. The complementary triplet on an mRNA strand is called a *codon.*

3. By base pairing, the anticodon of a specific tRNA recognizes the corresponding codon of mRNA and attaches to it. If the tRNA anticodon is UAC, the mRNA codon would be AUG (Figure 3-17c). In the process, the tRNA also brings along the specific amino acid. The pairing of anticodon and codon occurs only where mRNA is attached to a ribosome.

4. Once the first tRNA attaches to mRNA, the ribosome moves along the mRNA, and the next tRNA with its amino acid moves into position (Figure 3-17d).

5. The two amino acids are joined by a peptide bond and the first tRNA detaches itself from the mRNA strand. The larger ribosomal subunit contains the enzymes that join the amino acids together (Figure 3-17e). The released tRNA can now pick up another similar amino acid if necessary.

6. As the proper amino acids are brought into line, one by one, peptide bonds are formed between them and the protein gets progressively longer (Figure 3-17f).

7. When the specified protein is completed, further synthesis is stopped by a special *termination codon.* The assembled protein is released from the ribosome and the ribosome comes apart into its component subunits (Figure 3-17g,h).

As each ribosome moves along the mRNA strand, it "reads" the information coded in mRNA and synthesizes

a protein according to that information; the ribosome synthesizes the protein by translating the codon sequences into an amino acid sequence.

Protein synthesis progresses at the rate of about 15 amino acids per second. As the ribosome moves along the mRNA and before it completes translation of that gene, another ribosome may attach and begin translation of the same mRNA strand, so that several ribosomes may be attached to the same mRNA strand. Such an mRNA strand with its several ribosomes attached is called a **polyribosome.** Several ribosomes moving simultaneously in tandem along the same mRNA molecule permit the translation of a single mRNA strand into several identical proteins simultaneously.

Based upon the description of protein synthesis just presented, we can appropriately define a **gene** as a group of nucleotides on a DNA molecule that serves as the master mold for manufacturing a specific protein. Genes average about 1,000 pairs of nucleotides, which appear in a specific sequence on the DNA molecule. No two genes have exactly the same sequence of nucleotides and this is the key to heredity.

Remember that the base sequence of the gene determines the sequence of the bases in the mRNA. The sequence of the bases in the mRNA then determines the

heart enlarges in response to certain lung diseases. This condition is known as right ventricular hypertrophy or cor pulmonale. The airways and tissues of the respiratory tract, including the air sacs, become less elastic and more rigid. The threat of serious respiratory infections such as pneumonia and tuberculosis increases, as does the threat of the obstructive conditions such as chronic bronchitis, emphysema, and lung cancer.

Digestive   Overall general changes include atrophy of the secretion mechanisms, decreasing motility (muscular movement) of the digestive organs, loss of strength and tone of the muscular tissue and its supporting structures, changes in neurosensory feedback on enzyme and hormone release, and diminished response to pain and internal sensations. Specific changes include reduced sensitivity to mouth irritations and sores, loss of taste, pyorrhea, difficulty in swallowing, hiatus hernia, cancer of the esophagus, gastritis, peptic ulcer, and gastric cancer. Changes in the small intestine include duodenal ulcers, appendicitis, malabsorption, and maldigestion. Other pathologies that increase in incidence are gallbladder problems, jaundice, cirrhosis, and acute pancreatitis. Large-intestinal changes include constipation, cancer of the colon or rectum, hemorrhoids, and diverticular disease of the colon.

Urinary   Urinary incontinence (lack of voluntary control over urination) and urinary tract infections are major problems. Other pathologies include polyuria (excessive urine production), nocturia (excessive urination at night), increased frequency of urination, dysuria (painful urination), retention of urine (failure to produce urine), and hematuria (blood in the urine). The prostate gland is often implicated in various disorders of the urinary tract, and cancer of the prostate is the most frequent malignancy of old men. Changes and diseases in the kidney include acute and chronic kidney inflammations and renal calculi (kidney stones).

Reproductive   On the whole, very few age-specific disorders are associated with this system. In the male the decreasing production of the hormone testosterone produces less muscle strength, fewer viable sperm, and decreased sexual desire. However, abundant spermatozoa may be found even in old age. Most of the pathologies produce uneventful recoveries, except for prostate problems that could become serious and fatal. The female reproductive system becomes less efficient, possibly as a result of less frequent ovulation and the declining ability of the uterine tubes and uterus to support the young embryo. There is a decrease in the hormones progesterone and estrogen. Menopause is only one of a series of phases that leads to reduced fertility, irregular or absent menstruation, and a variety of physical changes. Uterine cancer peaks at about 65 years of age, but cervical cancer is more common in younger women, and breast cancer is the leading cause of death among women between the ages of 40 and 60. Prolapse (falling down or sinking) of the uterus is possibly the commonest complaint among female geriatric patients.

order and kind of amino acids that will form the protein. Thus each gene is responsible for making a particular protein as follows:

$$DNA \xrightarrow{\text{Transcription}} RNA \xrightarrow{\text{Translation}} protein$$

"SOS" GENES: DNA REPAIR

The normal daily activities of a cell and the continuity of life from one generation to another depend on the stable and precise replication of the DNA molecule in which hereditary information is encoded. Unfortunately, the architecture of DNA is such that it is vulnerable to damage by harmful radiations and various chemical agents. If the damage results in the incorporation of an incorrect or altered nitrogen base or other structural change that distorts the DNA molecule, its ability to replicate and synthesize proteins is impaired. If the damage is not repaired, the cell functions abnormally and could die. The damage seems sometimes to initiate a sequence of events leading to cancer.

In recent years, it has been learned that extensive damage to DNA can be repaired by certain enzymes. From what can be learned, it appears that damage to DNA switches on an **"SOS response."** As part of this response,

certain genes produce increased quantities of proteins (enzymes) that operate together to repair the genetic damage. Scientists are now trying to learn the mechanisms of different repair systems and under what conditions they operate. It is hoped that such information will provide clues to the nature of cancer and other diseases and the aging process.

CLINICAL APPLICATION

Different kinds of cells make different proteins following instructions encoded in the DNA of their genes. Since 1973 scientists have been able to alter those instructions in bacterial cells by adding genes from other organisms to the bacterial genes. This causes the bacterial cells to produce proteins that they normally do not synthesize. Bacteria so altered are called **recombinants** and their DNA, a combination of DNA from different sources, is called **recombinant DNA.** When recombinant DNA is introduced into a bacterium, the bacterium will synthesize the proteins of whatever new gene it has acquired. The new technology that has arisen from manipulating the genetic material is called **genetic engineering.** The bacterium that currently plays the most important role is *Escherichia coli,* a common inhabitant of the human intestine.

An important goal of recombinant DNA research is to increase our understanding of how genes are organized, how they function, and how they are regulated. This information will enable scientists to identify each of the 100,000 genes in the human cell and might be used to find a way to replace defective genes that are responsible for hemophilia, sickle-cell anemia, and other diseases. Recombinant DNA technology might also help scientists understand why normal cells turn malignant.

The practical applications of recombinant DNA technology are staggering. Strains of recombinant bacteria are presently producing several important therapeutic substances. These include *growth hormone (GH)*, required for growth during childhood; *somatostatin*, a brain hormone that helps regulate growth; *insulin*, a hormone that helps regulate blood sugar level and is used by diabetics; *interferon* (INF), an antiviral (and possibly anticancer) substance; and *beta-endorphin*, a brain peptide that has morphinelike properties that suppress pain. Some day bacteria will be programmed to produce other therapeutic substances like antibiotics, vaccines, and serums to fight an assortment of diseases. In addition to the medical benefits, recombinant DNA technology could be applied to producing proteins to help alleviate food shortages, to increase the alcohol yield from corn, to consume hazardous oil spills in the ocean, and to extract scarce minerals from soil.

CELLS AND AGING

Aging is a progressive failure of the body's homeostatic adaptive responses. It is a general response that produces observable changes in structure and function and increased vulnerability to environmental stress and disease. Disease and aging probably accelerate each other.

The obvious characteristics of aging are well known: graying and loss of hair, loss of teeth, wrinkling of skin, decreased muscle mass, and increased fat deposits. The physiological signs of aging are gradual deterioration in function and capacity to respond to environmental stress. Thus basic kidney and digestive metabolic rates decrease, as does the ability to respond effectively to changes in temperature, diet, and oxygen supply in order to maintain a constant internal environment. These manifestations of aging are related to a net decrease in the number of cells in the body (100,000 brain cells are lost each day) and to the disordered functioning of the cells that remain.

The extracellular components of tissues also change with age. Collagen fibers, responsible for the strength in tendons, increase in number and change in quality with aging. These changes in the collagen of arterial walls are as much responsible for their loss of extensibility as are the deposits associated with atherosclerosis. Elastin, another extracellular component, is responsible for the elasticity of blood vessels and skin. It thickens, fragments, and acquires a greater affinity for calcium with age—changes that may be associated with the development of atherosclerosis.

Several kinds of cells in the body—heart cells, skeletal muscle cells, neurons—are incapable of replacement. Recent experiments have proved that certain other cell types are limited when it comes to cell division. Cells grown outside the body divided only a certain number of times and then stopped. The number of divisions correlated with the donor's age. The number of divisions also correlated with the normal life span of the different species from which the cells were obtained—strong evidence for the hypothesis that cessation of mitosis is a normal, genetically programmed event. According to this view, an "aging" gene is part of the genetic blueprint at birth and it turns on at a preprogrammed time, slowing down or halting processes vital to life.

Some scientists believe that DNA and its protein-making apparatus wear out, gradually losing their capacity for self-repair, resulting in deterioration. Others feel that a mechanism exists for casting off a faulty gene and moving in a replacement. Longer-lived species presumably have more reserve DNA but, in time, all organisms deplete their reserves.

Whereas some theories of aging explain the process at the cellular level, others concentrate on regulatory mechanisms operating within the entire organism. For example, one such theory holds that the immune system, which manufactures antibodies against foreign invaders, turns on its own cells. This autoimmune response might be caused by changes in the surfaces of cells, causing antibodies to attack the body's own cells. As surface changes in cells increase, the autoimmune response intensifies, producing the well-known signs of aging. Another organismic theory suggests that aging is programmed in the pituitary gland, a hormone-producing gland attached to the undersurface of the brain. Supposedly, at a set time in life, the gland releases a hormone that triggers age-associated disruptions.

A detailed description of the aging process in humans, as it occurs in the various systems of the body, appears in Exhibit 3-2.

DISORDERS: HOMEOSTATIC IMBALANCES

Cancer

Definition

Cancer is not a single disease but many. The human body contains more than a hundred different types of cells, each of which can malfunction in its own distinctive way to cause cancer. When cells in some area of the body duplicate unusually quickly, the excess of tissue that develops is called a growth, or **tumor.** The study of tumors is called **oncology** (*onco* = swelling or

mass; *logos* = study of) and a physician who specializes in this field is called an **oncologist.** Tumors may be cancerous and sometimes fatal or they may be quite harmless. A cancerous growth is called a **malignant tumor,** or **malignancy.** A noncancerous growth is called a **benign growth.** Benign tumors are composed of cells that do not spread to other parts of the body and they may be removed if they interfere with a normal body function or are disfiguring.

Spread

Cells of malignant growths duplicate continuously and very often quickly without control. The majority of cancer patients are not killed by the **primary tumor** that develops. Rather, they are killed by **metastasis** (me-TAS-ta-sis), the spread of the disease to other parts of the body, since metastatic groups of cells are harder to detect and eliminate than primary tumors. One of the unique properties of a malignant tumor is its ability to metastasize.

In the process of metastasis, there is an initial invasion of the malignant cells into surrounding tissues. As the cancer grows, it expands and begins to compete with normal tissues for space and nutrients. Eventually, the normal tissue atrophies and dies. The invasiveness of the malignant cells may be related to mechanical pressure of the growing tumor, motility of the malignant cells, and enzymes produced by the malignant cells. Also, malignant cells lack what is called *contact inhibition*. When nonmalignant cells of the body divide and migrate (for example, skin cells that multiply to heal a superficial cut), their further migration is inhibited by contact on all sides with other skin cells. Unfortunately, malignant cells do not conform to the rules of contact inhibition; they have the ability to invade healthy body tissues with very few restrictions.

Following invasion, some of the malignant cells may detach from the primary tumor and invade a body cavity (abdominal or thoracic) or enter the blood or lymph. This latter condition can lead to widespread metastasis. In the next step in metastasis, those malignant cells that survive in the blood or lymph invade adjacent body tissues and establish **secondary tumors**. It is believed that some of the invading cells involved in metastasis have properties different from those of the primary tumor that enhance metastasis. These include appropriate mechanical, enzymatic, and surface properties. In the final stage of metastasis, the secondary tumors become vascularized, that is, they take on new networks of blood vessels that provide nutrients for their further growth. In all stages of metastasis, the malignant cells resist the antitumor defenses of the body. Usually death results from the atrophy of a vital organ. The pain associated with cancer develops when the growth puts pressure on nerves or blocks a passageway so that secretions build up pressure.

Types

At present, cancers are classified by their microscopic appearance and the body site from which they arise. At least 100 different cancers have been identified in this way. If finer details of appearance are taken into consideration, the number can be increased to 200 or more. The name of the cancer is derived from the type of tissue in which it develops. **Carcinoma** (*carc* = cancer; *oma* = tumor) refers to a malignant tumor consisting of epithelial cells. **Sarcoma** is a general term for any cancer arising from connective tissue. **Osteogenic sarcomas** (*osteo* = bone; *genic* = origin), the most frequent type of childhood cancer, destroy normal bone tissue and eventually spread to other areas of the body. **Myelomas** (*myelos* = marrow) are malignant tumors, occurring in middle-aged and older people, that interfere with the blood-cell-producing function of bone marrow and cause anemia. **Chondrosarcomas** are cancerous growths of cartilage (*chondro* = cartilage).

Possible Causes

What triggers a perfectly normal cell to lose control and become abnormal? Scientists are uncertain. First, there are environmental agents: substances in the air we breathe, the water we drink, the food we eat. A chemical or other environmental agent that produces cancer is called a **carcinogen.** The World Health Organization estimates that carcinogens may be associated with 60 to 90 percent of all human cancer. Examples of carcinogens are the hydrocarbons found in cigarette tar. Ninety percent of all lung cancer patients are smokers. Another environmental factor is radiation. Ultraviolet (UV) light from the sun, for example, may cause genetic mutations in exposed skin cells and lead to cancer, especially among light-skinned people.

Viruses are a second cause of cancer, at least in animals. These agents are tiny packages of nucleic acids, either DNA or RNA, that are capable of infecting cells and converting them to virus-producers. Virologists have linked tumor viruses with cancer in many species of birds and mammals, including primates. Since these experiments have not been performed on humans, there is no absolute proof that viruses cause human cancer. Nevertheless, with over 100 separate viruses identified as carcinogens in many species and tissues of animals, it is also probable that at least some cancers in humans are due to virus. In fact, the *Epstein-Barr* (*EB*) *virus,* the causative agent of infectious mononucleosis, has recently been linked as the causative agent of three human cancers—*Burkett's lymphoma* (a tumor of the jaw in African children and young adults), *nasopharyngeal carcinoma* (common in Chinese males), and *Hodgkin's disease* (a cancer of the lymphatic system). Also, the *hepatitis B virus* (*HBV*) has been associated with cancer of the liver.

One theory of the relationship between viruses and cancer holds that a set of viral genes, acquired early in evolution, is part of the genetic makeup of all vertebrate cells. Some of the genes, called **oncogenes,** have the ability to transform a normal cell into a cancerous cell. Such a transformation requires activation of the oncogenes by carcinogens. Some scientists feel that the oncogenes are not viral genes, but rather normal genes that become cancer-producing genes when activated by carcinogens.

Currently, scientists are trying to establish a relationship between stress and cancer. Some believe that stress may play a role not only in the development, but also in the metastasis of cancer.

Treatment

Treating cancer is difficult because it is not a single disease and because all the cells in a single population (tumor) do not behave in the same way. Even though many cancer researchers believe that a given cancer arises from a single malfunctioning cell, the same cancer may contain a diverse population of cells by the time it reaches a clinically detectable size. Although tumor cells look alike when stained and viewed under the microscope, they do not necessarily behave in the same manner in the body. For example, some metastasize and others do not. Some divide and others do not. Some are sensitive to drugs and some are resistant. As a consequence of differences in drug resistance, a single chemotherapeutic drug may destroy susceptible cells, but permit resistant cells to proliferate. This is probably one of the reasons that combination chemotherapy is usually more successful.

In recent years, there has been considerable debate over the use of **Laetrile** (LĀ-e-tril) in the treatment of human cancer. Laetrile, also known as amygdalin, is a naturally derived substance prepared from apricot pits. It has had many centuries of use for medical purposes for a variety of illnesses but was abandoned as a therapeutic agent until 1952 when it was revived as a drug for the treatment of cancer. Since that time, Laetrile has become the most controversial drug ever used for the treatment of a disease.

In response to public pressure, the National Cancer Institute (NCI) and Food and Drug Administration (FDA) conducted a clinical trial to determine the effectiveness of Laetrile in the treatment of advanced cancer. The results of the trial were published in 1982 and are summarized as follows:

1. Laetrile does not work. Even when combined with metabolic therapy (diet, vitamins, and enzymes), Laetrile produced no discernible benefit in patients with a variety of types of advanced cancer. More than 75 percent of the patients died by the end of the study and their survival times were the same as patients who received no treatment at all. The data also suggest that Laetrile would not have a beneficial effect in patients in earlier stages of cancer.

2. Laetrile is a toxic drug. Several patients in the study exhibited symptoms of cyanide toxicity or blood cyanide levels approaching the lethal range.

MEDICAL TERMINOLOGY

Note to the Student

Each chapter in this text that discusses a major system of the body is followed by a glossary of **medical terminology**. Both normal and pathological conditions of the system are included in these glossaries. You should familiarize yourself with the terms, since they will play an essential role in your medical vocabulary.

Some of these disorders, as well as disorders discussed in the text, are referred to as local or systemic. A **local disease** is one that affects one part or a limited area of the body. A **systemic disease** affects either the entire body or several parts.

Atrophy (*a* = without; *tropho* = nourish) A decrease in the size of cells with subsequent decrease in the size of the affected tissue or organ; wasting away.

Biopsy (*bio* = life; *opsis* = vision) The removal and microscopic examination of tissue from the living body for diagnosis.

Deterioration (*deterior* = worse or poorer) The process or state of growing worse; disintegration or wearing away.

Geriatrics (*geri* = aged) The branch of medicine devoted to the medical problems and care of elderly persons.

Gerontology (*gero* = aged; *logos* = study of) The study of old age.

Hyperplasia (*hyper* = over; *plas* = grow) Increase in the number of cells due to an increase in the frequency of cell division.

Hypertrophy Increase in the size of cells without cell division.

Insidious Hidden, not apparent, as a disease that does not exhibit distinct symptoms of its arrival.

Metaplasia (*meta* = change) The transformation of one cell into another.

Metastasis (*stasis* = standing still) The transfer of disease from one part of the body to another that is not directly connected with it.

Necrosis (*necros* = death; *osis* = condition) Death of a group of cells.

Neoplasm (*neo* = new) Any abnormal formation or growth, usually a malignant tumor.

Progeny (*progignere* = to bring forth) Offspring or descendants.

Senescence The process of growing old.

STUDY OUTLINE

Generalized Animal Cell (p. 52)

1. A cell is the basic, living, structural and functional unit of the body.
2. A generalized cell is a composite that represents various cells of the body.
3. Cytology is the science concerned with the study of cells.
4. The principal parts of a cell are the plasma (cell) membrane, cytoplasm, organelles, and inclusions. Extracellular materials are manufactured by the cell and deposited outside the plasma membrane.

Plasma (Cell) Membrane (p. 52)

Chemistry and Structure

1. The plasma (cell) membrane, surrounds the cell and separates it from other cells and the external environment.
2. It is composed primarily of proteins and phospholipids. According to the fluid mosaic model, the membrane consists of a phospholipid bilayer with integral and peripheral proteins.

Functions

1. Functionally, the plasma membrane facilitates contact with other cells, provides receptors, and mediates the passage of materials.
2. The membrane's selectively permeable nature restricts the passage of certain substances. Substances can pass through the membrane depending on their molecular size, lipid solubility, electrical charges, and the presence of carriers.

Movement of Materials Across Plasma Membranes

1. Passive processes involve the kinetic energy of individual molecules.
2. Diffusion is the net movement of molecules or ions from an area of higher concentration to an area of lower concentration until an equilibrium is reached.
3. In facilitated diffusion, certain molecules, such as glucose, combine with a carrier to become soluble in the phospholipid portion of the membrane.
4. Osmosis is the movement of water through a selectively permeable membrane from an area of higher water concentration to an area of lower water concentration.
5. Osmotic pressure is the force under which a solvent moves from a solution of lower solute concentration to a solution of higher solute concentration when the solutions are separated by a selectively permeable membrane.
6. In an isotonic solution, red blood cells maintain their normal shape; in a hypotonic solution, they undergo hemolysis; in a hypertonic solution, they undergo crenation.
7. Filtration is the movement of water and dissolved substances across a selectively permeable membrane by pressure.
8. Dialysis is the diffusion of solute particles across a selectively permeable membrane and involves the separation of small molecules from large molecules.
9. Active processes involve the use of ATP by the cell.
10. Active transport is the movement of ions across a cell membrane from lower to higher concentration.
11. Phagocytosis is the ingestion of solid particles by pseudopo-

dia. It is an important process used by white blood cells to destroy bacteria that enter the body.
12. Pinocytosis is the ingestion of a liquid by the plasma membrane. In this process, the liquid becomes surrounded by a vacuole.
13. When phagocytosis and pinocytosis involve the movement of substances into a cell, they are referred to as endocytosis; when they involve the movement of substances out of a cell, they are referred to as exocytosis.

Modified Plasma Membranes
1. Membranes of certain cells are modified for specific functions.
2. Microvilli are microscopic, fingerlike projections of the plasma membrane that increase the surface area for absorption.
3. Rods of the eye contain sacs of light-sensitive pigments.
4. Stereocilia are long, slender, branching cells which line the ductus epididymis.
5. The myelin sheath of nerve cells protects, aids impulse conduction, and provides nutrition.

Cytoplasm (p. 58)
1. Cytoplasm is the substance inside the cell that contains organelles and inclusions.
2. It is composed mostly of water plus proteins, carbohydrates, lipids, and inorganic substances. The chemicals in cytoplasm are either in solution or in a colloid (suspended) form.
3. Functionally, cytoplasm is the medium in which chemical reactions occur.

Organelles (p. 59)
1. Organelles are specialized portions of the cell that carry on specific activities.
2. They assume specific roles in cellular growth, maintenance, repair, and control.

Nucleus
1. Usually the largest organelle, the nucleus controls cellular activities and contains the genetic information.
2. Cells without nuclei, such as mature red blood cells, do not grow or reproduce.
3. The parts of the nucleus include the nuclear membrane, karyolymph, nucleoli, and genetic material (DNA), comprising the chromosomes.
4. Chromosomes consist of DNA and histones and consist of subunits called nucleosomes.

Ribosomes
1. Ribosomes are granular structures consisting of ribosomal RNA and ribosomal proteins.
2. They occur free (singly or in clusters) or in conjunction with endoplasmic reticulum.
3. Functionally, ribosomes are the sites of protein synthesis.

Endoplasmic Reticulum (ER)
1. The ER is a network of parallel membranes continuous with the plasma membrane and nuclear membrane.
2. Granular or rough ER has ribosomes attached to it. Agranular or smooth ER does not contain ribosomes.
3. The ER provides mechanical support, conducts intracellular nerve impulses in muscle cells, exchanges materials with cytoplasm, transports substances intracellularly, stores synthesized molecules, and helps export chemicals from the cell.

Golgi Complex
1. The Golgi complex consists of four to eight stacked, flattened channels (cisternae) with expanded terminal areas (vesicles).
2. In conjunction with the ER, the Golgi complex secretes proteins and lipids and synthesizes and secretes glycoproteins.
3. It is particularly prominent in secretory cells such as those in the pancreas or salivary glands.

Mitochondria
1. Mitochondria consist of a smooth outer membrane and a folded inner membrane surrounding the interior matrix. The inner folds are called cristae.
2. The mitochondria are called "powerhouses of the cell" because ATP is produced in them.

Lysosomes
1. Lysosomes are spherical structures that contain digestive enzymes. They are formed from Golgi complexes.
2. They are found in large numbers in white blood cells, which carry on phagocytosis.
3. If the cell is injured, lysosomes release enzymes and digest the cell. Thus they are called "suicide packets."
4. Lysosomes may be involved in bone removal.

Peroxisomes
1. Peroxisomes are similar to lysosomes, but smaller.
2. They contain enzymes (e.g., catalase) involved in the metabolism of hydrogen peroxide.

Microfilaments and Microtubules: The Cytoskeleton
1. Together microfilaments and microtubules form the cytoskeleton.
2. Microfilaments are rodlike structures consisting of the protein actin or myosin. They are involved in muscular contraction, support, and movement.
3. Microtubules are cylindrical structures consisting of the protein tubulin. They support, provide movement, and form the structure of flagella, cilia, centrioles, and the mitotic spindle.

Centrosome and Centrioles
1. The dense area of cytoplasm containing the centrioles is called a centrosome.
2. Centrioles are paired cylinders arranged at right angles to one another. They assume an important role in cell reproduction.

Flagella and Cilia
1. These cellular projections have the same basic structure and are used in movement.
2. If projections are few and long, they are called flagella. If they are numerous and hairlike, they are called cilia.
3. The flagellum on a sperm cell moves the entire cell. The cilia on cells of the respiratory tract move foreign matter trapped in mucus along the cell surfaces toward the throat for elimination.

Cell Inclusions (p. 65)
1. Cell inclusions are chemical substances produced by cells. They are usually organic and may have recognizable shapes.
2. Examples are melanin, glycogen, lipids, and mucus.

Extracellular Materials (p. 65)
1. These are all the substances that lie outside the cell membrane.
2. They provide support and a medium for the diffusion of nutrients and wastes.
3. Some, like hyaluronic acid and chondroitin sulfate, are amorphous. Others, like collagenous, reticular, and elastic fibers, are fibrous.

Cell Division (p. 66)

1. Cell division is the process by which cells reproduce themselves. It consists of nuclear division and cytoplasmic division (cytokinesis).
2. Cell division that results in the production of sperm and eggs consists of a nuclear division called meiosis and cytokinesis.
3. Cell division that results in an increase in body cells involves a nuclear division called mitosis and cytokinesis.
4. Prior to mitosis and cytokinesis, the DNA molecules, or chromosomes, replicate themselves so the same chromosomal complement can be passed on to future generations of cells.
5. A cell carrying on every life process except division is said to be in interphase or metabolic phase.

Mitosis

1. Mitosis is the distribution of two sets of chromosomes into separate and equal nuclei following their replication.
2. It consists of prophase, metaphase, anaphase, and telophase.

Cytokinesis

1. Cytokinesis begins in late anaphase and terminates in telophase.
2. A cleavage furrow forms at the cell's equator and progresses inward, cutting through the cell to form two separate portions of cytoplasm.

Gene Action (p. 69)

Protein Synthesis

1. Most of the cellular machinery is concerned with synthesizing proteins.
2. Cells make proteins by translating the genetic information encoded in DNA into specific proteins. This involves transcription and translation.
3. In transcription, genetic information encoded in DNA is copied by a strand of messenger RNA (mRNA); the DNA strand that serves as the template is called the sense strand.
4. DNA also synthesizes ribosomal RNA (rRNA) and template RNA (tRNA).

5. The process of using the information in the nitrogen base sequence of mRNA to dictate the amino acid sequence of a protein is known as translation.
6. mRNA associates with ribosomes, which consist of rRNA and protein.
7. Specific amino acids are attached to molecules of tRNA. Another portion of the tRNA has a triplet of bases called an anticodon; a codon is a segment of three bases of mRNA.
8. tRNA delivers a specific amino acid to the codon; the ribosome moves along an mRNA strand as amino acids are joined to form a growing polypeptide.

"SOS" Genes: DNA Repair

1. The structure of DNA is vulnerable to damage by harmful radiations and various chemicals.
2. Damage could lead to cellular malfunction that might lead to cancer.
3. In response to DNA damage, an "SOS response" occurs—certain genes produce enzymes that repair genetic damage.

Cells and Aging (p. 74)

1. Aging is a progressive failure of the body's homeostatic adaptive responses.
2. Many theories of aging have been proposed, including genetically programmed cessation of cell division and excessive immune responses, but none successfully answers all the experimental objections.
3. All of the various body systems exhibit definitive and sometimes extensive changes with aging (see Exhibit 3-2).

Disorders: Homeostatic Imbalances (p. 74)

1. Cancerous tumors are referred to as malignant; noncancerous tumors are called benign; the study of tumors is called oncology.
2. The spread of cancer from its primary site is called metastasis.
3. Carcinogens include environmental agents and viruses.
4. Treating cancer is difficult because all the cells in a single population do not behave the same way.

REVIEW QUESTIONS

1. Define a cell. What are the four principal portions of a cell? What is meant by a generalized cell?
2. Discuss the chemistry and structure of the plasma membrane with respect to the fluid mosaic model.
3. How do integral and peripheral membrane proteins differ in function?
4. Describe the various functions of the plasma membrane. What determines membrane permeability?
5. Describe the structure and function of microvilli, rod sacs, myelin sheaths, and stereocilia as membrane modifications.
6. What are the major differences between active processes and passive processes in moving substances across plasma membranes?
7. Define and give an example of each of the following: diffusion, facilitated diffusion, osmosis, filtration, active transport, phagocytosis, pinocytosis.
8. Distinguish between endocytosis and exocytosis.
9. Compare the effect on red blood cells of an isotonic, hypertonic, and hypotonic solution. What is osmotic pressure?
10. Discuss the chemical composition and physical nature of cytoplasm. What is its function?
11. What is an organelle? By means of a labeled diagram, indicate the parts of a generalized animal cell.
12. Describe the structure and functions of the nucleus of a cell. What are nucleosomes?
13. Discuss the distribution of ribosomes. What is their function?
14. Distinguish between granular (rough) and agranular (smooth) endoplasmic reticulum (ER). What are the functions of ER?
15. Describe the structure and functions of the Golgi complex.
16. Why are mitochondria referred to as "powerhouses of the cell"?
17. List and describe the various functions of lysosomes.
18. What is the importance of peroxisomes?
19. Contrast the structure and functions of microfilaments and microtubules.
20. Describe the structure and function of centrioles.
21. How are cilia and flagella distinguished on the basis of structure and function?
22. Define a cell inclusion. Provide examples and indicate their functions.
23. What is an extracellular material? Give examples and the functions of each.

24. How does DNA replicate itself?
25. Distinguish between the two types of cell division. Why is each important?
26. Describe the principal events of each stage of mitosis.
27. Summarize the steps involving gene action in protein synthesis.
28. How is the repair of DNA accomplished?
29. What is recombinant DNA? What is its clinical usefulness?
30. What is aging? List some of the characteristics of aging.
31. Briefly describe the various theories regarding aging.
32. List the major effects of aging on all of the body systems.
33. What is a tumor? Distinguish between malignant and benign tumors.
34. Define metastasis. What factors contribute to metastasis?
35. Discuss how certain carcinogens may be related to cancer.
36. What are some of the problems with respect to treating cancer?
37. Refer to the glossary of medical terminology associated with cells. Be sure that you can define each term.

- Define a tissue.

- Classify the tissues of the body into four major types and define each type.

- Discuss the distinguishing characteristics of epithelial tissues.

- Contrast the structural and functional differences between covering and lining epithelium and glandular epithelium.

- Compare the layering arrangements and cell shapes of covering and lining epithelium.

- List the structure, location, and function for the following types of epithelium: simple squamous, simple cuboidal, simple columnar (nonciliated and ciliated), stratified squamous, stratified cuboidal, stratified columnar, transitional, and pseudostratified.

- Define a gland and distinguish between exocrine and endocrine glands.

- Classify exocrine glands according to structural complexity and function and give an example of each.

- Identify the distinguishing characteristics of connective tissue.

- Contrast the structural and functional differences between embryonic and adult connective tissues.

- Discuss the ground substance, fibers, and cells that constitute connective tissue.

- List the structure, function, and location of loose (areolar) connective tissue, adipose tissue, and dense, elastic, and reticular connective tissue.

- List the structure, function, and location of the three types of cartilage.

- Distinguish between interstitial and appositional growth of cartilage.

- Define an epithelial membrane.

- List the location and function of mucous, serous, cutaneous, and synovial membranes.

- List and describe the symptoms of tissue inflammation.

- Outline the stages involved in the inflammatory response.

- Describe the conditions necessary for tissue repair.

- Explain the importance of nutrition, adequate circulation, and age to tissue repair.

4 The Tissue Level of Organization

Cells are highly organized units, but they do not function in isolation. They work together in a group of similar cells called a tissue.

TYPES OF TISSUES

A **tissue** is a group of similar cells and their intercellular substance functioning together to perform a specialized activity. Certain tissues function to move body parts. Others move food through body organs. Some tissues protect and support the body. Others produce chemicals such as enzymes and hormones. The various tissues of the body are classified into four principal types according to their function and structure:

1. **Epithelial** (ep'-i-THĒ-lē-al) **tissue,** which covers body surfaces or tissues, lines body cavities, and forms glands.
2. **Connective tissue,** which protects and supports the body and its organs and binds organs together.
3. **Muscular tissue,** which is responsible for movement.
4. **Nervous tissue,** which initiates and transmits nerve impulses that coordinate body activities.

EPITHELIAL TISSUE

The tissues in this principal type perform many activities in the body, ranging from protection to secretion. **Epithelial tissue,** or more simply **epithelium,** may be divided into two subtypes: (1) *covering and lining epithelium* and (2) *glandular epithelium.* Covering and lining epithelium forms the outer covering of external body surfaces and the outer covering of some internal organs. It lines the body cavities and the interiors of the respiratory and digestive tracts, blood vessels, and ducts. It makes up, along with nervous tissue, the parts of the sense organs that are sensitive to the stimuli that produce smell and hearing sensations. And, it is the tissue from which gametes (sperm and eggs) develop. Glandular epithelium constitutes the secreting portion of glands.

Both types of epithelium consist largely or entirely of closely packed cells with little or no intercellular material between adjacent cells. (Such intercellular material is also called the matrix.) Epithelial cells are arranged in continuous sheets that may be either single or multilayered. Nerves may extend through these sheets, but blood vessels do not. They are *avascular* (*a* = without, *vascular* = blood vessels). The vessels that supply nutrients and remove wastes are located in underlying connective tissue.

Both types of epithelium overlie and adhere firmly to the connective tissue, which holds the epithelium in position and prevents it from being torn. The surface of attachment between the epithelium and the connective tissue is a thin extracellular layer called the **basement membrane.** With few exceptions, epithelial cells secrete along their basal surfaces a material consisting of a special type of collagen and glycoproteins. This structure averages 500 to 800 Å in thickness and is referred to as the **basal lamina.** Frequently, the basal lamina is reinforced by an underlying *reticular lamina* consisting of reticular fibers and glycoproteins. This lamina is produced by cells in the underlying connective tissue. The combination of the basal lamina and reticular lamina constitutes the basement membrane.

Since all epithelium is subjected to a certain degree of wear, tear, and injury, its cells must divide and produce new cells to replace those that are destroyed.

COVERING AND LINING EPITHELIUM

Arrangement of Layers

Covering and lining epithelium is arranged in several different ways related to location and function. If the epithelium is specialized for absorption or filtration and is in an area that has minimal wear and tear, the cells of the tissue are arranged in a single layer. Such an arrangement is called **simple epithelium.** If the epithelium is not specialized for absorption or filtration and is found in an area with a high degree of wear and tear, then the cells are stacked in several layers. This tissue is referred to as **stratified epithelium.** A third, less common, arrangement of epithelium is called **pseudostratified.** Like simple epithelium, pseudostratified epithelium has only one layer of cells. However, some of the cells do not reach the surface—an arrangement that gives the tissue a multilayered, or stratified, appearance. The pseudostratified cells that do reach the surface either secrete mucus or contain cilia that move mucus and foreign particles for eventual elimination from the body.

Cell Shapes

In addition to classifying covering and lining epithelium according to the number of its layers, we may also categorize it by cell shape. The cells may be flat, cubelike, or columnar or may resemble a cross between shapes. **Squamous** (SKWĀ-mus) cells are flattened and scalelike. They are attached to each other like a mosaic. **Cuboidal** cells are usually cube-shaped in cross section. They sometimes appear as hexagons. **Columnar** cells are tall and cylindrical, appearing as rectangles set on end. **Transitional** cells often have a combination of shapes and are found where there is a great degree of distention or expansion in the body. Transitional cells in the bottom layer of an epithelial tissue may range in shape from cuboidal to columnar. In the intermediate layer, they may be cuboidal or polyhedral. In the superficial layer they may range from cuboidal to squamous, depending on how much they are pulled out of shape during certain body functions.

Classification

Considering layers and cell type in combination, we may classify covering and lining epithelium as follows:

Simple
1. Squamous
2. Cuboidal
3. Columnar

Stratified
1. Squamous
2. Cuboidal
3. Columnar
4. Transitional

Pseudostratified

Each of the epithelial tissues described in the following sections is illustrated in Exhibit 4-1.

EXHIBIT 4-1
EPITHELIAL TISSUES

COVERING AND LINING EPITHELIUM
Simple Squamous (250×)
Description: Single layer of flat, scalelike cells; large, centrally located nuclei.
Location: Lines air sacs of lungs, glomerular capsule of kidneys, and inner surfaces of the membranous labyrinth and tympanic membrane of the ear. Called endothelium when it lines heart, blood and lymphatic vessels, and forms capillaries. Called mesothelium when it lines the ventral body cavity and covers viscera as part of a serous membrane.
Function: Filtration, absorption, and secretion in serous membranes.

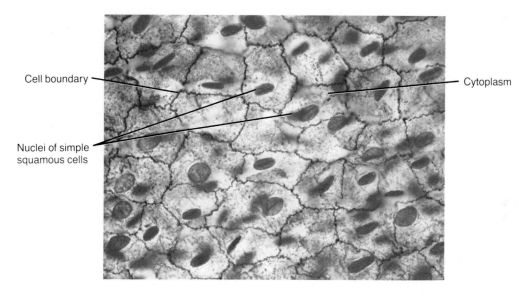

Simple Cuboidal (450×)
Description: Single layer of cube-shaped cells; centrally located nuclei.
Location: Covers surface of ovary, lines anterior surface of capsule of the lens of eyes, forms pigmented epithelium of retina of eye, and lines kidney tubules and smaller ducts of many glands.
Function: Secretion and absorption.

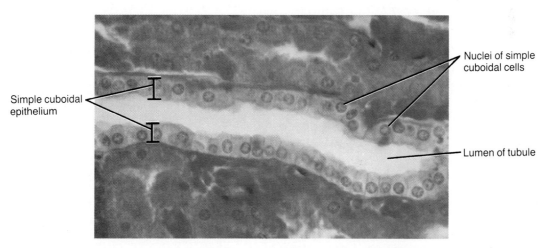

Simple Epithelium

● *Simple Squamous Epithelium* This type of simple epithelium consists of a single layer of flat, scalelike cells. Its surface resembles a tiled floor. The nucleus of each cell is centrally located and oval or spherical. Since simple squamous epithelium has only one layer of cells, it is highly adapted to diffusion, osmosis, and filtration. Thus it lines the air sacs of the lungs, where oxygen is exchanged with carbon dioxide. It is present in the part of the kidney that filters the blood. It also lines the inner surfaces of the membranous labyrinth and tympanic membrane of the ear. Simple squamous epithelium is found in body parts that have little wear and tear.

EXHIBIT 4-1
(*Continued*)

Simple Columnar (Nonciliated) (600×)
Description: Single layer of nonciliated rectangular cells; contains goblet cells; nuclei at bases of cells.
Location: Lines the digestive tract from the cardia of the stomach to the anus, excretory ducts of many glands, and gallbladder.
Function: Secretion and absorption.

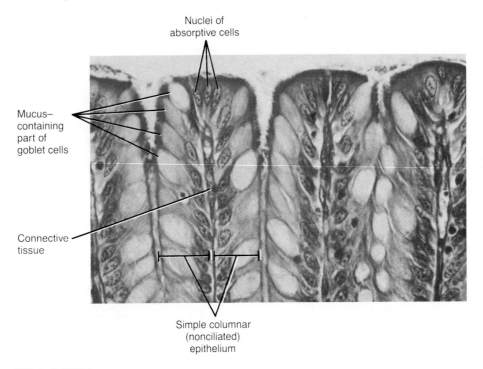

Simple Columnar (Ciliated) (400×)
Description: Single layer of ciliated columnar cells; contains goblet cells; nuclei at bases of cells.
Location: Lines a few portions of upper respiratory tract, uterine (fallopian) tubes, uterus, some paranasal sinuses, and central canal of spinal cord.
Function: Moves mucus by ciliary action.

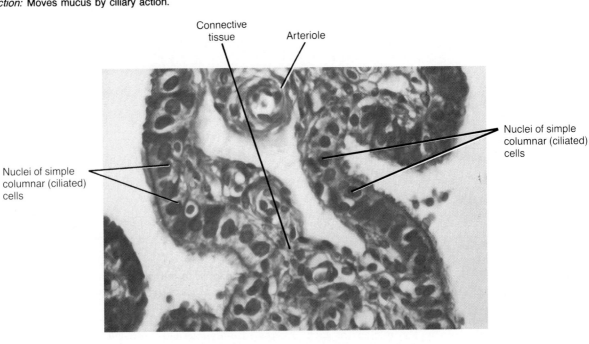

Stratified Squamous (185×)

Description: Several layers of cells; cuboidal to columnar shape in deep layers; scalelike shape in superficial layers; basal cells replace surface cells as they are lost.

Location: Nonkeratinized variety lines wet surfaces such as lining of the mouth, tongue, esophagus, part of epiglottis, and vagina.

Function: Protection.

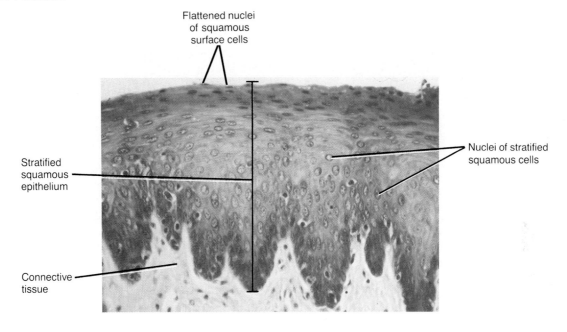

Stratified Cuboidal (185×)

Description: Two or more layers of cells in which the surface cells are cube-shaped.

Location: Ducts of adult sweat glands, fornix of conjunctiva of eye, cavernous urethra of male urogenital system, pharynx, and epiglottis.

Function: Protection.

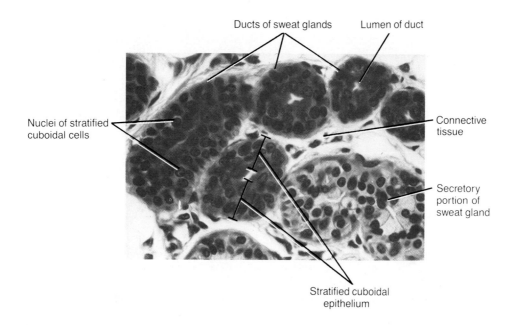

EXHIBIT 4-1
(*Continued*)

Stratified Columnar (600×)
Description: Several layers of polyhedral cells; columnar cells only in superficial layer.
Location: Lines part of male urethra, large excretory ducts of some glands, and small areas in anal mucous membrane.
Function: Protection and secretion.

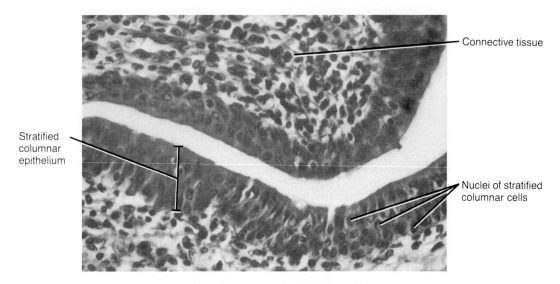

Stratified Transitional (185×)
Description: Resembles nonkeratinized stratified squamous tissue, except that superficial cells are larger and more rounded.
Location: Lines urinary bladder.
Function: Permits distention.

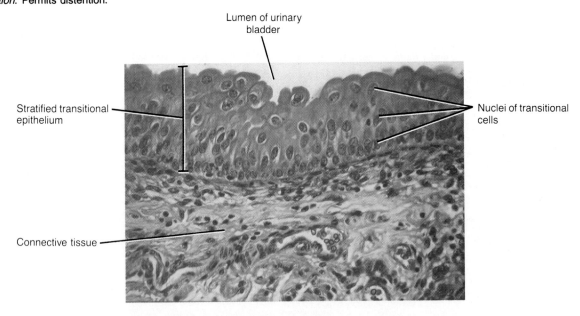

A tissue similar to simple squamous epithelium is endothelium. **Endothelium** lines the heart, blood vessels, and lymph vessels and forms the walls of capillaries. Another tissue similar to simple squamous epithelium that forms the epithelial layer of serous membranes is **mesothelium.** This tissue lines the thoracic and abdominopelvic cavities and covers the viscera within them.

● *Simple Cuboidal Epithelium*   Viewed from above, the cells of simple cuboidal epithelium appear as closely fitted polygons. The cuboidal nature of the cells is obvious only when the tissue is sectioned at right angles. Like simple squamous epithelium, these cells possess a central nucleus. Simple cuboidal epithelium covers the surface of the ovaries, lines the anterior surface of the capsule of the lens

Pseudostratified (500×)

Description: Not a true stratified tissue; nuclei of cells at different levels; all cells attached to basement membrane, but not all reach surface.

Location: Lines larger excretory ducts of many large glands, male urethra, and auditory tubes; ciliated variety with goblet cells lines most of the upper respiratory tract and some ducts of male reproductive system.

Function: Secretion and movement of mucus by ciliary action.

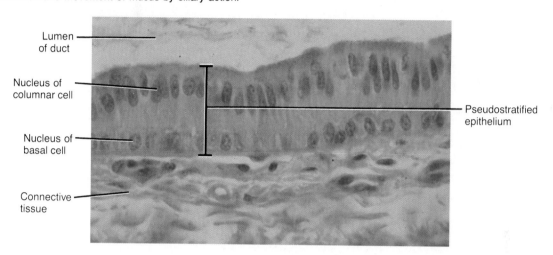

Lumen of duct

Nucleus of columnar cell

Nucleus of basal cell

Connective tissue

Pseudostratified epithelium

GLANDULAR EPITHELIUM

Exocrine Gland (300×)

Description: Secretes products into ducts.

Location: Sweat, oil, wax, and mammary glands of the skin; digestive glands such as salivary glands which secrete into mouth cavity.

Function: Produces mucus, perspiration, oil, wax, milk, or digestive enzymes.

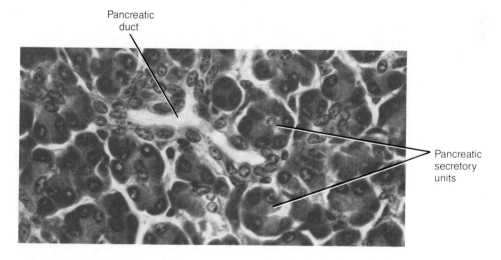

Pancreatic duct

Pancreatic secretory units

of the eye, and forms the pigmented epithelium of the retina of the eye. In the kidneys, where it forms the kidney tubules and contains microvilli, it functions in water reabsorption. It also lines the smaller ducts of some glands and the secreting units of glands, such as the thyroid.

This tissue performs the functions of secretion and absorption. **Secretion,** usually a function of epithelium, is the production and release by cells of a fluid that may contain a variety of substances such as mucus, perspiration, or enzymes. **Absorption** is the intake of fluids or other substances by cells of the skin or mucous membranes.

● *Simple Columnar Epithelium* The surface view of simple columnar epithelium is similar to that of simple cuboi-

EXHIBIT 4-1
(*Continued*)

Endocrine Gland (180×)

Description: Secretes hormones into blood.

Location: Pituitary at base of brain, thyroid and parathyroids near larynx, adrenals above kidneys, pancreas below stomach, ovaries in pelvic cavity and testes in scrotum, pineal at base of brain, and thymus between lungs.

Function: Produces hormones that regulate various body activities.

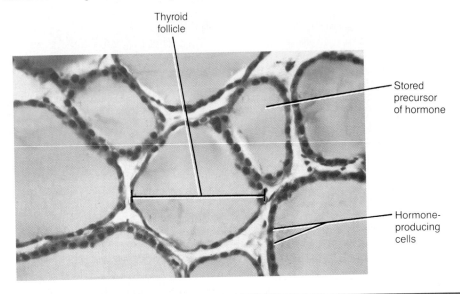

Thyroid follicle

Stored precursor of hormone

Hormone-producing cells

Photomicrographs © 1983 by Michael H. Ross. Used by permission.

dal tissue. When sectioned at right angles to the surface, however, the cells appear somewhat rectangular. The nuclei are located near the bases of the cells.

The luminal surfaces (surfaces adjacent to the lumen, or cavity of a hollow organ, vessel, or duct) of simple columnar epithelial cells are modified in several ways, depending on location and function. Simple columnar epithelium lines the digestive tract from the cardia of the stomach to the anus, the gallbladder, and excretory ducts of many glands. In such sites, the cells protect the underlying tissues. Many of them are also modified to aid in food-related activities. In the small intestine especially, the plasma membranes of the cells are folded into **microvilli** (see Figure 3-1). The microvilli arrangement increases the surface area of the plasma membrane and thereby allows larger amounts of digested nutrients and fluids to be absorbed into the body.

Interspersed among the typical columnar cells of the intestine are other modified columnar cells called **goblet cells.** These cells, which secrete mucus, are so named because the mucus accumulates in the upper half of the cell, causing the area to bulge out. The whole cell resembles a goblet or wine glass. The secreted mucus serves as a lubricant between the food and the walls of the digestive tract.

A third modification of columnar epithelium is found in cells with hairlike processes called **cilia.** In a few portions of the upper respiratory tract, ciliated columnar cells are interspersed with goblet cells. Mucus secreted by the goblet cells forms a film over the respiratory sur-

face. This film traps foreign particles that are inhaled. The cilia wave in unison and move the mucus, with any foreign particles, toward the throat, where it can be swallowed or eliminated. Air is filtered by this process before entering the lungs. Ciliated columnar epithelium is also found in the uterus and uterine tubes of the female reproductive system, some paranasal sinuses, and the central canal of the spinal cord.

Stratified Epithelium

In contrast to simple epithelium, stratified epithelium consists of at least two layers of cells. Thus it is durable and can protect underlying tissues from the external environment and from wear and tear. Some stratified epithelium cells also produce secretions. The name of the specific kind of stratified epithelium depends on the shape of the surface cells.

● *Stratified Squamous Epithelium*   In the more superficial layers of this type of epithelium, the cells are flat, whereas in the deep layers, cells vary in shape from cuboidal to columnar. The basal, or bottom, cells are continually multiplying by cell division. As new cells grow, they compress the cells on the surface and push them outward. The basal cells continually shift upward and outward. As they move farther from the deep layer and their blood supply, they become dehydrated, shrink, and grow harder. At the surface, the cells are rubbed off. New cells continually emerge, are sloughed off, and replaced.

One form of stratified squamous epithelium is called **nonkeratinized stratified squamous epithelium.** This tissue is found on wet surfaces that are subjected to considerable wear and tear and does not perform the function of absorption—the lining of the mouth, the tongue, the esophagus, and the vagina. Another form of stratified squamous epithelium is called **keratinized stratified squamous epithelium.** The surface cells of this type are modified into a tough layer of material containing keratin. **Keratin** is a protein that is waterproof and resistant to friction and helps to resist bacterial invasion. The outer layer of skin consists of keratinized tissue.

● *Stratified Cuboidal Epithelium*   This relatively rare type of epithelium is found in the ducts of the sweat glands of adults, fornix of the conjunctiva of the eye, cavernous urethra of the male urogenital system, pharynx, and epiglottis. It sometimes consists of more than two layers of cells. Its function is mainly protective.

● *Stratified Columnar Epithelium*   Like stratified cuboidal epithelium, this type of tissue is also uncommon in the body. Usually the basal layer or layers consist of shortened, irregularly polyhedral cells. Only the superficial cells are columnar in form. This kind of epithelium lines part of the male urethra, some larger excretory ducts such as lactiferous (milk) ducts in the mammary glands, and small areas in the anal mucous membrane. It functions in protection and secretion.

● *Transitional Epithelium*   This kind of epithelium is very much like nonkeratinized stratified squamous epithelium. The distinction is that cells of the outer layer in transitional epithelium tend to be large and rounded rather than flat. This feature allows the tissue to be stretched (distended) without the outer cells breaking apart from one another. When stretched, they are drawn out into squamouslike cells. Because of this arrangement, transitional epithelium lines hollow structures that are subjected to expansion from within, such as the urinary bladder. Its function is to help prevent a rupture of the organ.

Pseudostratified Epithelium

The third category of covering and lining epithelium consists of columnar cells, and is called pseudostratified epithelium. The nuclei of the cells are at varying depths. Even though all the cells are attached to the basement membrane in a single layer, some do not reach the surface. This feature gives the impression of a multilayered tissue, the reason for the designation *pseudo*stratified epithelium. It lines the larger excretory ducts of many glands, parts of the male urethra, and the auditory tubes. Pseudostratified epithelium may be ciliated and may contain goblet cells. In this form, it lines most of the upper respiratory tract and certain ducts of the male reproductive system.

GLANDULAR EPITHELIUM

The function of glandular epithelium is secretion, accomplished by glandular cells that lie in clusters deep to the covering and lining epithelium. A **gland** may consist of one cell or a group of highly specialized epithelial cells that secrete substances into ducts, onto a surface, or into the blood. The production of such substances always requires active work by the glandular cells and results in an expenditure of energy.

All glands of the body are classified as exocrine or endocrine according to whether they secrete substances into ducts (or directly onto a free surface) or into the blood. **Exocrine glands** secrete their products into ducts (tubes) that empty at the surface of covering and lining epithelium or directly onto a free surface. The product of an exocrine gland may be released at the skin surface or into the lumen of a hollow organ. The secretions of exocrine glands include mucus, perspiration, oil, wax, and digestive enzymes. Examples of exocrine glands are sweat glands, which eliminate perspiration to cool the skin, salivary glands, which secrete a digestive enzyme, and goblet cells, which produce mucus. **Endocrine glands** are ductless and secrete their products into the blood. The secretions of endocrine glands are always hormones, chemicals that regulate various physiological activities. The pituitary, thyroid, and adrenal glands are endocrine glands.

Structural Classification of Exocrine Glands

Exocrine glands are classified into two structural types: unicellular and multicellular. **Unicellular glands** are single-celled. A good example of a unicellular gland is a goblet cell (Exhibit 4-1). Goblet cells are found in the epithelial lining of the digestive, respiratory, urinary, and reproductive systems. They produce mucus to lubricate the free surfaces of these membranes.

Multicellular glands occur in several different forms (Figure 4-1). If the secretory portions of a gland are tubular, it is referred to as a **tubular gland.** If they are flasklike, it is called an **acinar** (AS-i-nar) **gland.** If the gland contains both tubular and flasklike secretory portions, it is called a **tubuloacinar gland.** Further, if the duct of the gland does not branch, it is referred to as a **simple gland;** if the duct does branch, it is called a **compound gland.** By combining the shape of the secretory portion with the degree of branching of the duct, we arrive at the following structural classification for exocrine glands:

I. **Unicellular.** Single-celled gland that secretes mucus. Example: goblet cell of the digestive and respiratory systems.

II. **Multicellular.** Many-celled glands.
   A. *Simple.* Single, nonbranched duct.
      1. **Tubular.** The secretory portion is straight and tubular. Example: intestinal glands.
      2. **Branched tubular.** The secretory portion is branched and tubular. Examples: gastric and uterine glands.
      3. **Coiled tubular.** The secretory portion is coiled. Example: sudoriferous (sweat) glands.
      4. **Acinar.** The secretory portion is flasklike. Example: seminal vesicle glands.
      5. **Branched acinar.** The secretory portion is branched and flasklike. Example: sebaceous (oil) glands.
   B. *Compound.* Branched duct.
      1. **Tubular.** The secretory portion is tubular. Examples: bulbourethral glands, testes, and liver.

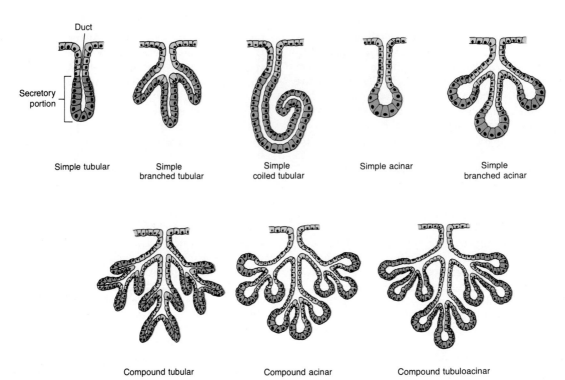

FIGURE 4-1 Structural types of multicellular exocrine glands. The secretory portions of the glands are indicated in purple. The blue areas represent the ducts of the glands.

2. **Acinar.** The secretory portion is flasklike. Examples: salivary glands (sublingual and submandibular).
3. **Tubuloacinar.** The secretory portion is both tubular and flasklike. Examples: salivary glands (parotid) and pancreas.

Functional Classification of Exocrine Glands

The functional classification of exocrine glands is based on how the gland releases its secretion. The three recognized categories are holocrine, merocrine, and apocrine glands. **Holocrine glands** accumulate a secretory product in their cytoplasm. The cell then dies and is discharged with its contents as the glandular secretion (Figure 4-2a). The discharged cell is replaced by a new cell. One example of a holocrine gland is a sebaceous gland of the skin. **Merocrine glands** simply form the secretory product and discharge it from the cell (Figure 4-2b). Examples of merocrine glands are the salivary glands and pancreas. **Apocrine glands** accumulate their secretory product at the apical (outer) margin of the secreting cell. That portion of the cell pinches off from the rest of the cell to form the secretion (Figure 4-2c). The remaining part of the cell repairs itself and repeats the process. An example of an apocrine gland is the mammary gland.

CONNECTIVE TISSUE

The most abundant tissue in the body is **connective tissue.** This binding and supporting tissue usually is highly vascular and thus has a rich blood supply. An exception is cartilage which is avascular. The cells are widely scattered, rather than closely packed, and there is considerable intercellular material, or matrix. In contrast to epithelium, connective tissues do not occur on free surfaces, such as the surfaces of a body cavity or the external surface of the body. The general functions of connective tissues are protection, support, and the binding together of various organs.

The intercellular substance in a connective tissue largely determines the tissue's qualities. These substances are nonliving and may consist of fluid, semifluid, or mucoid (mucuslike) material. In cartilage, the intercellular material is firm but pliable. In bone, it is considerably harder and not pliable. The cells of connective tissue produce the intercellular substances. The cells may also store fat, ingest bacteria and cell debris, form anticoagulants, or give rise to antibodies that protect against disease.

CLASSIFICATION

Connective tissue may be classified in several ways. We will classify them as follows:

I. **Embryonic connective tissue**
  A. Mesenchyme
  B. Mucous connective tissue
II. **Adult connective tissue**
  A. Connective tissue proper
    1. Loose (areolar) connective tissue
    2. Adipose tissue
    3. Dense (collagenous) connective tissue
    4. Elastic connective tissue
    5. Reticular connective tissue
  B. Cartilage
    1. Hyaline cartilage
    2. Fibrocartilage
    3. Elastic cartilage

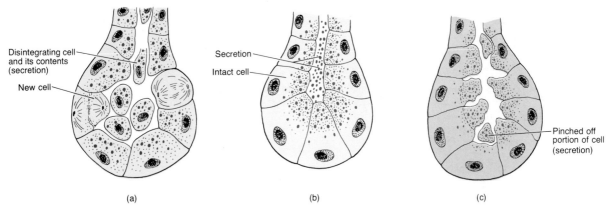

FIGURE 4-2 Functional classification of multicellular exocrine glands. (a) Holocrine gland. (b) Merocrine gland. (c) Apocrine gland.

C. Osseous (bone) tissue
D. Vascular (blood) tissue

Each of the connective tissues described in the following sections is illustrated in Exhibit 4-2.

EMBRYONIC CONNECTIVE TISSUE

Connective tissue that is present primarily in the embryo or fetus is called **embryonic connective tissue.** The term *embryo* refers to a developing human from fertilization through the first two months of pregnancy; a *fetus* refers to a developing human from the third month of pregnancy to birth.

One example of embryonic connective tissue found almost exclusively in the embryo is **mesenchyme** (MEZ-en-kīm)—the tissue from which all other connective tissues eventually arise. Mesenchyme may be observed beneath the skin and along the developing bones of the embryo. Some mesenchymal cells are scattered irregularly throughout adult connective tissue, most frequently around blood vessels. Here mesenchymal cells differentiate into fibroblasts that assist in wound healing.

Another kind of embryonic connective tissue is **mucous connective tissue,** found primarily in the fetus. This tissue, also called **Wharton's jelly,** is located in the umbilical cord of the fetus, where it supports the wall of the cord.

ADULT CONNECTIVE TISSUE

Adult connective tissue is connective tissue that exists in the newborn and that does not change after birth. It is subdivided into several kinds.

Connective Tissue Proper

Connective tissue that has a more or less fluid intercellular material and a fibroblast as the typical cell is termed **connective tissue proper.** Five examples of such tissues may be distinguished.

● *Loose (Areolar) Connective Tissue* Loose or areolar (a-RĒ-ō-lar) connective tissue is one of the most widely distributed connective tissues in the body. Structurally, it consists of fibers and several kinds of cells embedded

in a semifluid intercellular substance. The term "loose" refers to the loosely woven arrangement of fibers in the intercellular substance. The fibers are neither abundant nor arranged to prevent stretching.

The intercellular substance consists of a viscous material called **hyaluronic acid.** Hyaluronic acid normally facilitates the passage of nutrients from the blood vessels of the connective tissue into adjacent cells and tissues, although the thick consistency of this acid may impede the movement of some drugs. But if an enzyme called **hyaluronidase** is injected into the tissue, hyaluronic acid changes to a watery consistency. This feature is of clinical importance because the reduced viscosity hastens the absorption and diffusion of injected drugs and fluids through the tissue and thus can lessen tension and pain. Some bacteria, white blood cells, and sperm cells produce hyaluronidase.

The three types of fibers embedded between the cells of loose connective tissue are collagenous, elastic, and reticular fibers. **Collagenous (white) fibers** are very tough and resistant to a pulling force, yet are somewhat flexible because they are usually wavy. These fibers often occur in bundles. They are composed of many minute fibers called fibrils lying parallel to one another. The bundle arrangement affords a great deal of strength. Chemically, collagenous fibers consist of the protein collagen. **Elastic (yellow) fibers,** by contrast, are smaller than collagenous fibers and freely branch and rejoin one another. Elastic fibers consist of a protein called elastin. These fibers also provide strength and have great elasticity, up to 50 percent of their length. **Reticular fibers** also consist of collagen, plus some glycoprotein. They are very thin fibers that branch extensively and are not as strong as collagenous fibers. Some authorities believe that reticular fibers are immature collagenous fibers. Like collagenous fibers, reticular fibers provide support and strength and form the *stroma* (framework) of many soft organs.

The cells in loose connective tissue are numerous and varied. Most are **fibroblasts**—large, flat cells with branching processes. If the tissue is injured, the fibroblasts are believed to form collagenous fibers, elastic fibers, and the viscous ground substance. Mature fibroblasts are referred to as **fibrocytes.** The basic distinction between the two is that fibroblasts (or *"blasts"* of any form of cell) are involved in the formation of immature tissue or repair

EXHIBIT 4-2
CONNECTIVE TISSUES

EMBRYONIC
Messenchymal (180×)

Description: Consists of highly branched mesenchymal cells embedded in a fluid substance.

Location: Under skin and along developing bones of embryo; some mesenchymal cells found in adult connective tissue, especially along blood vessels.

Function: Forms all other kinds of connective tissue.

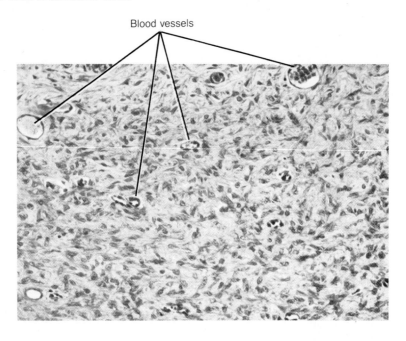

Blood vessels

Mucous (320×)

Description: Consists of flattened or spindle-shaped cells embedded in a mucuslike substance containing fine collagenous fibers.

Location: Umbilical cord of fetus.

Function: Support.

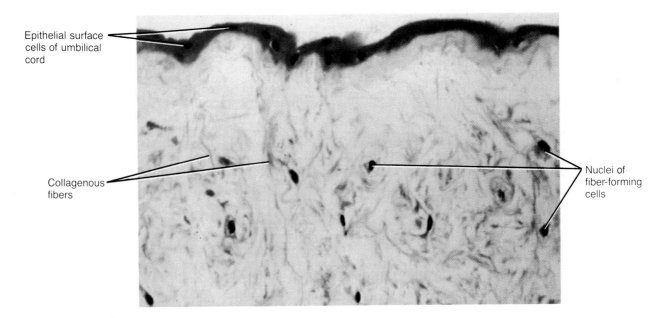

Epithelial surface cells of umbilical cord

Collagenous fibers

Nuclei of fiber-forming cells

ADULT

Loose or Areolar (180×)

Description: Consists of fibers (collagenous, elastic, and reticular) and several kinds of cells (fibroblasts, macrophages, plasma cells, and mast cells) embedded in a semifluid ground substance.

Location: Subcutaneous layer of skin, mucous membranes, blood vessels, nerves, and body organs.

Function: Strength, elasticity, and support.

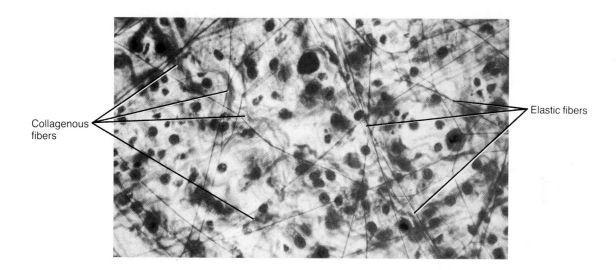

Collagenous fibers

Elastic fibers

Adipose (250×)

Description: Consists of adipocytes, "signet ring"-shaped cells with peripheral nuclei, that are specialized for fat storage.

Location: Subcutaneous layer of skin, around heart and kidneys, marrow of long bones, and padding around joints.

Function: Reduces heat loss through skin, serves as an energy reserve, supports, and protects.

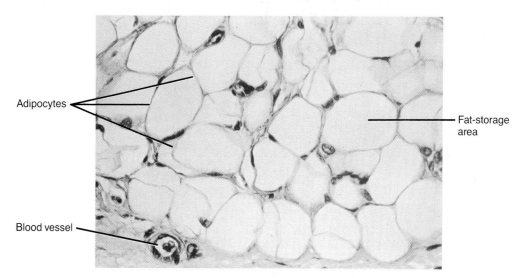

Adipocytes

Fat-storage area

Blood vessel

EXHIBIT 4-2
(Continued)

Dense or Collagenous (250×)
Description: Consists of predominately collagenous, or white, fibers arranged in bundles; fibroblasts present in rows between bundles.
Location: Forms tendons, ligaments, aponeuroses, membranes around various organs, and fasciae.
Function: Provides strong attachment between various structures.

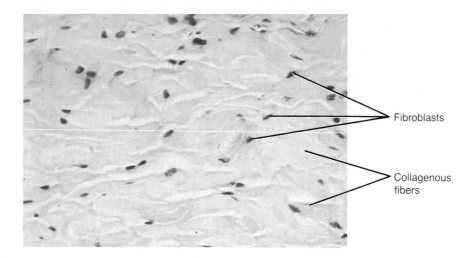

Fibroblasts

Collagenous fibers

Elastic (180×)
Description: Consists of predominately freely branching elastic, or yellow, fibers; fibroblasts present in spaces between fibers.
Location: Lung tissue, cartilage of larynx, wall of arteries, trachea, bronchial tubes, true vocal cords, and ligamenta flava of vertebrae.
Function: Allows stretching of various organs.

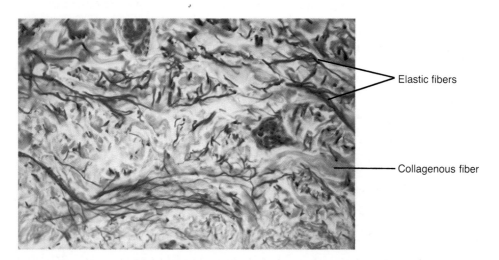

Elastic fibers

Collagenous fiber

of mature tissue, and fibrocytes (or *"cytes"* of any form of cell) are inactive cells that no longer produce fibers or matrix.

Other cells found in loose connective tissue are called **macrophages** (MAK-rō-fā-jēz; *macro* = large; *phagein* = to eat). They are irregular in form with short branching projections and are capable of engulfing bacteria and cellular debris by the process of phagocytosis. Thus they provide a vital defense for the body. The types of macrophages are discussed shortly in relation to tissue inflammation.

A third kind of cell in loose connective tissue is the **plasma cell.** These cells are small and either round or irregular. Plasma cells develop from a type of white blood cell called a lymphocyte (B cell). They give rise to antibodies and, accordingly, provide a defensive mechanism

Reticular (250×)
Description: Consists of a network of interlacing reticular fibers with thin, flat cells wrapped around fibers.
Location: Liver, spleen, and lymph nodes.
Function: Forms stroma of organs; binds together smooth muscle tissue cells.

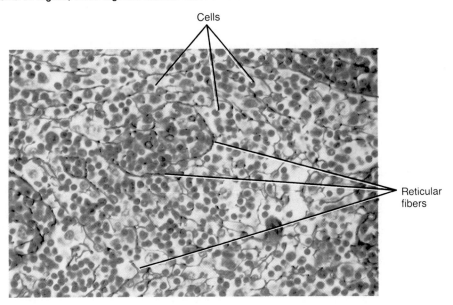

Cells

Reticular fibers

Hyaline Cartilage (350×)
Description: Also called gristle; appears as a bluish white, glossy mass; contains numerous chondrocytes; is the most abundant type of cartilage.
Location: Ends of long bones, ends of ribs, nose, parts of larynx, trachea, bronchi, bronchial tubes, and embryonic skeleton.
Function: Provides movement at joints, flexibility, and support.

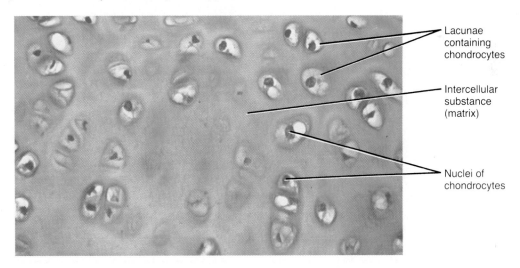

Lacunae containing chondrocytes

Intercellular substance (matrix)

Nuclei of chondrocytes

through immunity. Plasma cells are found in many places of the body, but most are found in connective tissue, especially that of the digestive tract and the mammary glands.

Another cell in loose connective tissue is the **mast cell.** It may develop from another type of white blood cell called a basophil. The mast cell is somewhat larger than a basophil and is found in abundance along blood vessels.

It forms heparin, an anticoagulant that prevents blood from clotting in the vessels. Mast cells are also believed to produce histamine and serotonin, chemicals that dilate small blood vessels.

Other cells in loose connective tissue include **melanocytes** (pigment cells), fat cells, and white blood cells.

Loose connective tissue is continuous throughout the body. It is present in all mucous membranes and around

EXHIBIT 4-2
(*Continued*)

Fibrocartilage (180×)
Description: Consists of chondrocytes scattered among bundles of collagenous fibers.
Location: Symphysis pubis and intervertebral discs.
Function: Support and fusion.

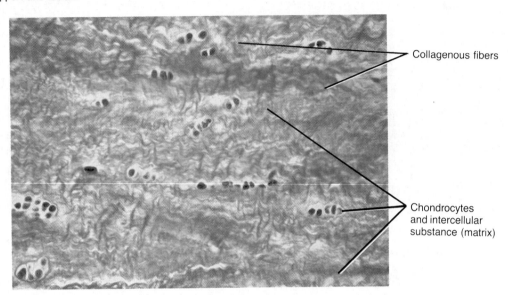

Collagenous fibers

Chondrocytes and intercellular substance (matrix)

Elastic Cartilage (350×)
Description: Consists of chondrocytes located in a threadlike network of elastic fibers.
Location: Epiglottis, parts of larynx, external ear, and auditory tubes.
Function: Gives support and maintains shape.

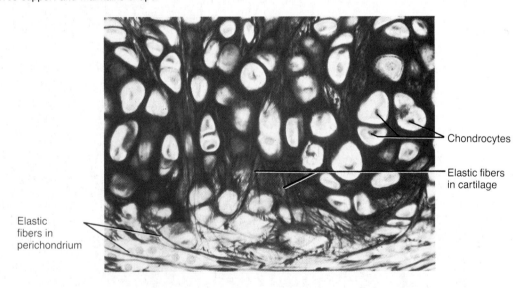

Chondrocytes

Elastic fibers in cartilage

Elastic fibers in perichondrium

Photomicrographs © 1983 by Michael H. Ross. Used by permission.

all blood vessels and nerves. And it occurs around body organs and in the papillary region of the dermis of the skin. Combined with adipose tissue, it forms the **subcutaneous** (sub'-kyoo-TĀ-nē-us; *sub* = under, *cut* = skin) **layer**—the layer of tissue that attaches the skin to underlying tissues and organs. The subcutaneous layer is also referred to as the **superficial fascia** (FASH-ē-a).

● *Adipose Tissue*   Adipose tissue is basically a form of loose connective tissue in which the cells, called **adipocytes,** are specialized for fat storage. Adipocytes are derived from fibroblasts and the cells have the shape of a "signet ring" because the cytoplasm and nucleus are pushed to the edge of the cell by a large droplet of fat. Adipose tissue is found wherever loose connective tissue is located. Specifically, it is in the subcutaneous layer below the skin, around the kidneys, at the base and on the surface of the heart, in the marrow of long bones, as a padding around joints, and behind the eyeball in the orbit. Adipose tissue is a poor conductor of heat and therefore reduces heat loss through the skin. It is also a major energy reserve and generally supports and protects various organs.

● *Dense (Collagenous) Connective Tissue*   Dense (collagenous) connective tissue is characterized by a close packing of fibers and less intercellular substance than in loose connective tissue. The fibers can be either irregularly arranged or regularly arranged. In areas of the body where tensions are exerted in various directions, the fiber bundles are interwoven and without regular orientation. Such a dense connective tissue is referred to as **irregularly arranged** and occurs in sheets. It forms most fasciae, the reticular region of the dermis of the skin, the periosteum of bone, the perichondrium of cartilage, and the *membrane (fibrous) capsules* around organs, such as the kidneys, liver, testes, and lymph nodes.

In other areas of the body, dense connective tissue is adapted for tension in one direction, and the fibers have an orderly, parallel arrangement. Such a dense connective tissue is known as **regularly arranged.** The most common variety of dense regularly arranged connective tissue has a predominance of collagenous (white) fibers arranged in bundles. Fibroblasts are placed in rows between the bundles. The tissue is silvery white, tough, yet somewhat pliable. Because of its great strength, it is the principal component of *tendons,* which attach muscles to bones; many *ligaments* (collagenous ligaments), which hold bones together at joints; and *aponeuroses* (ap'-ō-noo-RŌ-sēz), which are flat bands connecting one muscle with another or with bone.

CLINICAL APPLICATION

Scientists have developed a carbon fiber implant for **reconstructing severely torn ligaments and tendons.** The implant consists of carbon fibers coated with a plastic called polylactic acid. The coated fibers are sewn in and around torn ligaments and tendons to reinforce them and to provide a scaffolding around which the body's own collagenous fibers grow. Within two weeks the polylactic acid is absorbed by the body and the carbon fibers eventually fracture. By this time, the fibers are completely clad in collagen produced by fibroblasts.

● *Elastic Connective Tissue*   Unlike collagenous connective tissue, elastic connective tissue has a predominance of freely branching elastic fibers. These fibers give the tissue a yellowish color. Fibroblasts are present only in the spaces between the fibers. Elastic connective tissue can be stretched and will snap back into shape. It is a component of the cartilages of the larynx, the walls of elastic arteries, the trachea, the bronchial tubes to the lungs, and the lungs themselves. Elastic connective tissue provides stretch and strength, allowing structures to perform their functions efficiently. Yellow elastic ligaments, as contrasted with collagenous ligaments, are composed mostly of elastic fibers; they form the ligamenta flava of the vertebrae (ligaments between successive vertebrae), the suspensory ligament of the penis, and the true vocal cords.

● *Reticular Connective Tissue*   Reticular connective tissue consists of interlacing reticular fibers. It helps to form the stroma of many organs, including the liver, spleen, and lymph nodes. Reticular connective tissue also helps to bind together the cells of smooth muscle tissue. It is especially adapted to providing strength and support.

Cartilage

One type of connective tissue, called cartilage, is capable of enduring considerably more stress than the tissues just discussed. Unlike other connective tissues, cartilage has no blood vessels (except for the perichondrium) or nerves of its own. **Cartilage** consists of a dense network of collagenous fibers and elastic fibers firmly embedded in chondroitin sulfate, a jellylike substance. The cells of mature cartilage, called **chondrocytes** (KON-drō-sīts), occur singly or in groups within spaces called **lacunae** (la-KOO-nē) in the intercellular substance. The surface of cartilage is surrounded by irregularly arranged dense connective tissue called the **perichondrium** (per'-i-KON-drē-um; *chondro* = cartilage, *peri* = around). Three kinds of cartilage are recognized: hyaline cartilage, fibrocartilage, and elastic cartilage (Exhibit 4-2).

● *Growth of Cartilage*   The growth of cartilage follows two basic patterns. In **interstitial (endogenous) growth,** the cartilage increases rapidly in size through the division of existing chondrocytes and continuous deposition of increasing amounts of intercellular matrix by the chondrocytes. The formation of new chondrocytes and their production of new intercellular matrix causes the cartilage to expand from within, thus the term *interstitial growth.* This growth pattern occurs while the cartilage is young and pliable—during childhood and adolescence.

In **appositional (exogenous) growth,** the growth of cartilage occurs because of the activity of the inner chondrogenic layer of the perichondrium. The deeper cells of the perichondrium, the fibroblasts, divide. Some differentiate into chondroblasts (immature cells that develop into specialized cells) and then into chondrocytes. As differentiation occurs, the chondroblasts become surrounded with intercellular matrix and become chondrocytes. As a result, the matrix is deposited on the surface of the cartilage, increasing its size. The new layer of cartilage is added beneath the perichondrium on the surface of the cartilage, causing it to grow in width. Appositional growth starts later than interstitial growth and continues throughout life.

● *Hyaline Cartilage*   This cartilage, also called gristle, appears as a bluish-white, glossy, homogeneous mass. The collagenous fibers, although present, are not visible with ordinary staining techniques, and the prominent chondrocytes are found in lacunae. Hyaline cartilage is the most abundant kind of cartilage in the body. It is found at joints over the ends of long bones (where it is called *articular cartilage*) and forms the *costal cartilages* at the ventral ends of the ribs. Hyaline cartilage also helps to form the nose, larynx, trachea, bronchi, and bronchial tubes leading to the lungs. Most of the embryonic skeleton consists of hyaline cartilage. It affords flexibility and support.

● *Fibrocartilage*   Chondrocytes scattered through many bundles of visible collagenous fibers are found in this type of cartilage. Fibrocartilage is found at the symphysis pubis, the point where the coxal (hip) bones fuse anteriorly at the midline. It is also found in the discs between the vertebrae. This tissue combines strength and rigidity.

● *Elastic Cartilage*   In this tissue, chondrocytes are located in a threadlike network of elastic fibers. Elastic cartilage provides strength and maintains the shape of certain organs—the larynx, the external part of the ear (the pinna), and the auditory tubes (the internal connection between the middle ear cavity and the upper throat).

Osseous Tissue and Vascular Tissue

The details of **osseous tissue** (bone), another kind of connective tissue, are discussed in Chapter 6 as part of the skeletal system. **Vascular tissue,** also known as blood, is treated in Chapter 19 as a component of the cardiovascular system.

MUSCLE TISSUE AND NERVOUS TISSUE

Epithelial and connective tissue can take a variety of forms to provide a variety of body functions. They are all-purpose tissues. By contrast, **muscle tissue** consists of highly modified cells that perform one basic function: contraction (Chapter 10), and **nervous tissue** is specialized to generate and conduct electrical impulses (Chapter 12).

MEMBRANES

The combination of an epithelial layer and an underlying connective tissue layer constitutes an **epithelial membrane.** The principal epithelial membranes of the body are mucous membranes, serous membranes, and the cutaneous membrane, or skin. Another kind of membrane, a **synovial membrane,** does not contain epithelium.

MUCOUS MEMBRANES

A **mucous membrane,** or **mucosa,** lines a body cavity that opens directly to the exterior. Mucous membranes line the entire digestive, respiratory, excretory, and reproductive tracts (see Figure 24-10). The surface tissue of a mucous membrane may vary in type—it is stratified squamous epithelium in the esophagus and simple columnar epithelium in the intestine.

The epithelial layer of a mucous membrane secretes mucus, which prevents the cavities from drying out. It also traps dust in the respiratory passageways and lubricates food as it moves through the digestive tract. In addition, the epithelial layer is responsible for the secretion of digestive enzymes and the absorption of food.

The connective tissue layer of a mucous membrane is called the *lamina propria*. It binds the epithelium to the underlying structures and allows some flexibility of the membrane. It also holds the blood vessels in place, protects underlying muscles from abrasion or puncture, provides the epithelium covering it with oxygen and nutrients, and removes wastes.

SEROUS MEMBRANES

A **serous membrane,** or **serosa,** lines a body cavity that does not open directly to the exterior, and it covers the organs that lie within the cavity. Serous membranes consist of thin layers of loose connective tissue covered by a layer of mesothelium and they are in the form of invaginated double-walled sacs. The part attached to the cavity wall is called the *parietal* (pa-RĪ-i-tal) *portion*. The part that covers the organs inside these cavities is the *visceral portion*. The serous membrane lining the thoracic cavity and covering the lungs is called the *pleura* (see Figure 1-7b). The membrane lining the heart cavity and covering the heart is the *pericardium* (*cardio* = heart). The serous membrane lining the abdominal cavity and covering the abdominal organs and some pelvic organs is called the *peritoneum*.

The epithelial layer of a serous membrane secretes a lubricating fluid that allows the organs to glide easily against one another or against the walls of the cavities. The connective tissue layer of the serous membrane consists of a relatively thin layer of loose connective tissue.

CUTANEOUS MEMBRANE

The **cutaneous membrane,** or skin, constitutes an organ of the integumentary system and is discussed in the next chapter.

SYNOVIAL MEMBRANES

Synovial membranes line the cavities of the joints (see Figure 9-1a). Like serous membranes, they line structures that do not open to the exterior. Unlike mucous, serous, and cutaneous membranes, they do not contain epithelium and are therefore not epithelial membranes. They are composed of loose connective tissue with elastic fibers and varying amounts of fat. Synovial membranes secrete *synovial fluid,* which lubricates the ends of bones as they move at joints and nourishes the articular cartilage covering the bones that form the joints.

TISSUE INFLAMMATION: AN ATTEMPT TO RESTORE HOMEOSTASIS

The tissues of the body perform various roles in attempting to maintain homeostasis. Among these functions are protection, support, filtration, absorption, secretion, movement, transportation, and defense against disease. Damage to a tissue is a stress because it interferes with these vital functions. When cells are injured, many parts of the body work to overcome the stress and return the body to homeostasis. They do this by contributing to the inflammatory response and to the repair process.

When cells are damaged, the injury sets off an **inflammatory response.** The injury, which may be viewed as a form of stress, can have various causes. It could result from mechanical trauma, such as a clean knife incision during surgery. Bacteria that give off toxic chemicals could enter through the nose, pores in the skin, or burns, or by way of a splinter or nail. Cells can be injured if the blood supply is cut off, causing the cells to "starve."

SYMPTOMS

Inflammation is usually characterized by four fundamental symptoms: *redness, pain, heat,* and *swelling.* A fifth symptom can be the *loss of function* of the injured area. Whether loss of function occurs depends on the site and extent of the injury. The inflammatory response serves a protective and defensive role. It is an attempt to neutralize and destroy toxic agents at the site of injury and to prevent their spread to other organs. Thus the inflammatory response is an attempt to restore tissue homeostasis.

The immediate inflammatory response to tissue injury consists of a complicated sequence of physiological and anatomical adjustments. Various body components are involved in the initial response: blood vessels; intercellular fluid mixed with parts of injured cells, called the *exudate* (EKS-yoo-dāt); the cellular components of blood; and the surrounding epithelial and connective tissues. Other factors that affect the inflammatory response are the individual's age and general state of health. Healing processes of all types exert a great demand on the body's store of nutrients. Thus nutrition plays an essential role in healing.

STAGES

The inflammatory response is one of the body's internal systems of defense. The response of a tissue to a rusty nail wound is similar to other inflammatory responses, such as the sore throat that results from bacterial or viral invasion. These are the basic stages of response.

1. Vasodilation and increased permeability of blood vessels. Immediately following tissue damage, there is vasodilation and increased permeability of blood vessels in the area of the injury. **Vasodilation** is an increase in diameter in the blood vessels. **Increased permeability** means that substances normally retained in blood are permitted to pass from the blood vessels. Vasodilation allows more blood to go to the damaged area, and increased permeability permits defensive substances in the blood to enter the injured area. Such defensive substances include white blood cells and clot-forming chemicals. The increased blood supply also removes toxic products and dead cells, preventing them from complicating the injury. These toxic substances include waste products released by invading microorganisms. As the blood clots around the site of activity, the microbe and its toxins are prevented from spreading to other parts of the body.

Vasodilation and increased permeability are caused by the release of certain chemicals by damaged cells in response to injury. One such substance is *histamine* (Figure 4-3a). It is present in many tissues of the body, especially in mast cells in connective tissue, circulating basophils (one type of white blood cell), and blood platelets. Histamine is released in direct response to any injured cells containing it. Neutrophils, another type of white blood cell attracted to the site of injury, can also produce chemicals that cause the release of histamine. *Serotonin* (*5-HT*) also produced by mast cells, assumes a role in

vasodilation. Other substances that play a role in vasodilation and increased permeability of blood vessels are *kinins.* These are chemicals that attract neutrophils to the injured area. Inflammation also results in an increased synthesis of *prostaglandins* (*PG*), especially of the E series. Prostaglandins are potent vasodilators by themselves and they also intensify the effects of histamine, serotonin, and kinins. Prostaglandins also bring about increased permeability of blood vessels.

In addition to the responses in the area of injury, the body may also respond by increasing the metabolic rate and quickening the heartbeat so that more blood circulates to the injured area per minute. Within minutes after the injury, the quickened metabolism and circulation and, especially, the dilation and increased permeability of capillaries produce heat, redness, and swelling. The heat results from the large amount of warm blood that accumulates in the area and, to a certain extent, from the heat energy produced by the metabolic reactions. The large amounts of blood in the area are also responsible for the redness.

CLINICAL APPLICATION

The increased permeability of the capillary walls allows quantities of fluid to move out of the blood and into the intercellular spaces of the tissue. Since the fluid moves into the intercellular spaces faster than it can be drained off, it accumulates in the tissue, causing it to swell. The swelling is called **edema.**

Pain, whether immediate or delayed, is a cardinal symptom of inflammation. It can result from an injury to nerve fibers or from an irritation caused by the release of toxic chemicals from microorganisms. Pain may also be due to increased pressure from edema. Prostaglandins intensify and prolong the pain associated with inflammation. However, kinins affect some nerve endings, causing much of the pain associated with inflammation.

2. Phagocyte migration. Generally, within an hour after the inflammatory process is initiated, phagocytes appear on the scene. Phagocytes include neutrophils (microphages) and monocytes (develop into macrophages); both are types of white blood cells. As the flow of blood decreases, neutrophils begin to stick to the inner surface of the endothelium (lining) of blood vessels. This is called **margination.** Then, the neutrophils begin to squeeze through the wall of the blood vessel to reach the damaged area. This migration, which resembles amoeboid movement, is called **diapedesis** (di'-a-pe-DĒ-sis). It can take as little as two minutes (Figure 4-3b). The movement of neutrophils depends on **chemotaxis** (kē'-mō-TAK-sis; *chemo* = chemicals, *taxis* = arrangement), the attraction of neutrophils by certain chemicals. Neutrophils are attracted by microbes, kinins, and other neutrophils. A steady stream of neutrophils is ensured by the production and release of additional cells from bone marrow. This is brought about by a substance called **leucocytosis-promoting factor,** which is released from inflamed tissues. Neutrophils attempt to destroy the invading microbes by phagocytosis (Figure 4-4).

As the inflammatory response continues, monocytes follow the neutrophils into the infected area. Once in the tissue, monocytes become transformed into **wandering macrophages.** They are so named because they leave the blood and migrate through tissues to infected areas. Other macrophages, called **fixed macrophages** or **histiocytes** (HIS-tē-ō-sīts), are found in the liver (stellate reticuloendothelial cells), lungs (alveolar macrophages), brain (microglia), spleen, lymph nodes, subcutaneous tissue, and bone marrow. Since fixed macrophages can be either reticular (forming a supporting network) or endothelial (lining a sinus), they are referred to as cells of the **reticuloendothelial system (RES).** Wandering macrophages augment the phagocytic activ-

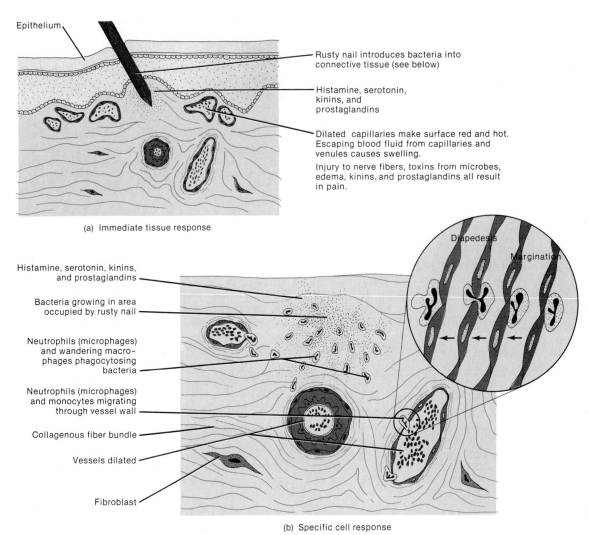

FIGURE 4-3 Tissue and cell response to an injury. (a) Immediate tissue response. (b) Specific cell response.

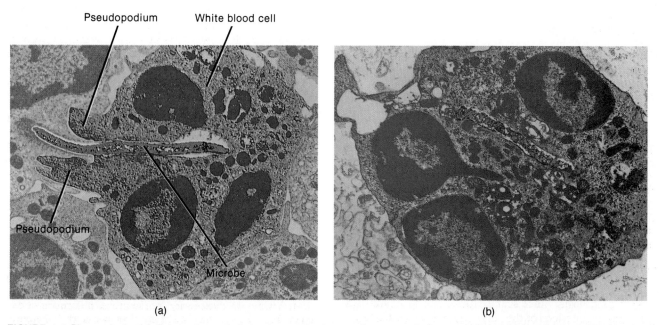

FIGURE 4-4 Phagocytosis. (a) Color-enhanced photomicrograph of an earlier stage of phagocytosis in which a human white blood cell (neutrophil) is engulfing a microbe. (b) Color-enhanced photomicrograph of a later stage of phagocytosis in which the engulfed microbe is being destroyed. (Courtesy, Abbott Laboratories.)

ity of fixed macrophages. In addition, fixed macrophages mobilize under the stimulus of inflammation and, as wandering macrophages, also migrate to the infected area. The neutrophils predominate in the early stages of infection, but tend to die off rapidly. Macrophages enter the picture during a later stage of the infection. They are several times more phagocytic than neutrophils and large enough to engulf tissue that has been destroyed, neutrophils that have been destroyed, and invading microbes.

3. Release of nutrients. Nutrients stored in the body are used to support the defensive cells. They are also used in the increased metabolic reactions of the cells under attack.

4. Fibrin formation. The blood contains a soluble protein called **fibrinogen** (fī-BRIN-ō-jen). The increased permeability of capillaries causes leakage of fibrinogen to tissues. Fibrinogen is then converted to an insoluble, thick network called **fibrin** (FĪ-brin), which localizes and traps the invading organisms, preventing their spread. This network eventually forms a fibrin clot that prevents hemorrhage and isolates the infected area.

5. Pus formation. In all but very mild inflammations, **pyogenesis** (*pyo* = pus; *genesis* = to produce) occurs. **Pus** is a thick fluid that contains living, as well as nonliving, white blood cells plus debris from other dead tissue.

CLINICAL APPLICATION

If the pus cannot drain out of the body, an abscess develops. An **abscess** is simply an excessive accumulation of pus in a confined space. Common examples are pimples and boils. When inflamed tissue is shed many times, it produces an open sore, called an **ulcer,** on the surface of an organ or tissue. Ulcers may result from a prolonged inflammatory response to a continuously injured tissue. For instance, overproduction of digestive acids in the stomach may cause a steady erosion of the epithelial tissue lining the stomach. People with poor circulation are susceptible to ulcers in the tissues of their legs. The ulcers develop when the tissues are continuously damaged by a shortage of oxygen and nutrients.

TISSUE REPAIR

Tissue repair, the process by which tissues replace dead or damaged cells, begins during the active phase of inflammation, but it can be completed only after all harmful substances have been neutralized or removed from the site of the injury. New cells originate by cell duplication from the **stroma,** the supporting connective tissue, or from the **parenchyma,** cells which form the organ's functioning part. The epithelial cells that secrete and absorb are the parenchymal cells of the intestine, for example. The restoration of an injured organ or tissue to normal structure and function depends entirely on which type of cell— parenchymal or stromal—is active in the repair. If only parenchymal elements accomplish the repair, a perfect or near-perfect reconstruction of the injured tissue may occur. However, if the fibroblast cells of the stroma are active in the repair, the tissue will be replaced with new connective tissue called *scar tissue.* This condition is known as **fibrosis.** Since scar tissue is not specialized to perform the functions of the parenchymal tissue, the function of the tissue is impaired.

CLINICAL APPLICATION

The scar tissue formed by fibrosis can cause the abnormal joining of tissues called **adhesions.** They are common in the abdomen and may occur around a site of previous inflammation such as an inflamed appendix, or they can follow surgery. Such adhesions make subsequent surgery more difficult.

The cardinal factor in tissue repair lies in the capacity of parenchymal tissue to regenerate. This capacity, in turn, depends on the ability of the parenchymal cells to replicate quickly.

REPAIR PROCESS

If injury to a tissue is slight, repair may sometimes be accomplished with the drainage and reabsorption of pus, followed by parenchymal regeneration. When the area of skin loss is great, fluid moves out of the capillaries and the area becomes dry. Fibrin seals the open tissue by hardening into a **scab.**

When tissue and cell damage are extensive and severe, as in large, open wounds, both the connective tissue stroma and the parenchymal cells are active in repair. This repair involves the rapid cell division of many fibroblasts, the manufacture of new collagenous fibers to provide strength, and an increase by cell division of the number of small blood vessels in the area. All these processes create an actively growing, connective tissue called **granulation tissue.** This new granulation tissue forms across a wound or surgical incision to provide a framework (stroma). The framework supports the epithelial cells that migrate into the open area and fill it. The newly formed granulation tissue also secretes a fluid that kills bacteria.

CONDITIONS AFFECTING REPAIR

Three factors affect tissue repair: nutrition, blood circulation, and age. Nutrition is vital in the healing process since a great demand is placed on the body's store of nutrients. Protein-rich diets are important since most of the cell structure is made from proteins. Vitamins also play a direct role in wound healing. Among the vitamins involved and their roles in wound healing are the following:

1. Vitamin A is essential in the replacement of epithelial tissues, especially in the respiratory tract.

2. The B vitamins—thiamine, nicotinic acid, riboflavin—are needed by many enzyme systems in cells. They are needed especially for the enzymes involved in decomposing glucose to CO_2 and H_2O which is crucial to both heart and nervous tissue. These vitamins may relieve pain in some cases.

3. Vitamin C directly affects the normal production and maintenance of intercellular substances. It is required for the manufacture of cementing elements of connective tissues, especially collagen. Vitamin C also strengthens and promotes the formation of new blood vessels. With vitamin C deficiency, even superficial wounds fail to heal, and the walls of the blood vessels become fragile and are easily ruptured.

4. Vitamin D is necessary for the proper absorption of calcium from the intestine. Calcium gives bones their hardness and is necessary for the healing of fractures.

5. Vitamin E is believed to promote healing of injured tissues and may prevent scarring.

6. Vitamin K assists in the clotting of blood and thus prevents the injured person from bleeding to death.

In tissue repair, proper blood circulation is indispensable. It is the blood that transports oxygen, nutrients, antibodies, and many defensive cells to the site of injury. The blood also plays an important role in the removal of tissue fluid, blood cells that have been depleted of oxygen, bacteria, foreign bodies, and debris. These elements would otherwise interfere with healing.

Generally, tissues heal faster and leave less obvious scars in the young than in the aged. The young body is generally in a much better nutritional state, its tissues have a better blood supply, and the cells of younger people have a faster metabolic rate. Thus cells can duplicate their materials and divide more quickly.

STUDY OUTLINE

Types of Tissues (p. 82)

1. A tissue is a group of similar cells and their intercellular substance specialized for a particular function.
2. Depending on their function and structure, the various tissues of the body are classified into four principal types: epithelial, connective, muscular, and nervous.

Epithelial Tissue (p. 82)

1. Epithelium has many cells, little intracellular material, and no blood vessels (avascular). It is attached to connective tissue by a basement membrane. It can replace itself.
2. The subtypes of epithelium include covering and lining epithelium and glandular epithelium.

Covering and Lining Epithelium

1. Layers are arranged as simple (one layer), stratified (several layers), and pseudostratified (one layer that appears as several); cell shapes include squamous (flat), cuboidal (cubelike), columnar (rectangular), and transitional (variable).
2. Simple squamous epithelium is adapted for diffusion and filtration and is found in lungs and kidneys. Endothelium lines the heart and blood vessels. Mesothelium lines the thoracic and abdominopelvic cavities and covers the organs within them.
3. Simple cuboidal epithelium is adapted for secretion and absorption. It is found covering ovaries, in kidneys and eyes, and lining some glandular ducts.
4. Nonciliated simple columnar epithelium lines most of the digestive tract. Specialized cells containing microvilli perform absorption. Goblet cells perform secretion of mucus. In a few portions of the respiratory tract, the cells are ciliated to move foreign particles trapped in mucus out of the body.
5. Stratified squamous epithelium is protective. It lines the upper digestive tract and vagina and forms the outer layer of skin.
6. Stratified cuboidal epithelium is found in adult sweat glands, portion of urethra, pharynx, and epiglottis.
7. Stratified columnar epithelium protects and secretes. It is found in the male urethra and large excretory ducts.
8. Transitional epithelium lines the urinary bladder and is capable of stretching.
9. Pseudostratified epithelium has only one layer but gives the appearance of many. It lines larger excretory ducts, parts of urethra, auditory tubes, and most upper respiratory structures, where it protects and secretes.

Glandular Epithelium

1. A gland is a single cell or a mass of epithelial cells adapted for secretion.
2. Exocrine glands (sweat, oil, and digestive glands) secrete into ducts or directly onto a free surface.
3. Structural classification includes unicellular and multicellular glands; multicellular glands are further classified as tubular, acinar, tubuloacinar, simple, and compound.
4. Functional classification includes holocrine, merocrine, and apocrine glands.
5. Endocrine glands secrete hormones directly into the blood.

Connective Tissue (p. 90)

1. Connective tissue is the most abundant body tissue. It has few cells, an extensive intercellular substance, and a rich blood supply (vascular), except for cartilage. It does not occur on free surfaces.
2. The intercellular substance determines the tissue's qualities.
3. Connective tissue protects, supports, and binds organs together.
4. Connective tissue is classified into two principal types: embryonic and adult.

Embryonic Connective Tissue

1. Mesenchyme forms all other connective tissues.
2. Mucous connective tissue is found in the umbilical cord of the fetus, where it gives support.

Adult Connective Tissue

1. Adult connective tissue is connective tissue that exists in the newborn and that does not change after birth. It is subdivided into several kinds: connective tissue proper, cartilage, bone tissue, and vascular tissue.
2. Connective tissue proper has a more or less fluid intercellular material, and a typical cell is the fibroblast. Five examples of such tissues may be distinguished.
3. Loose (areolar) connective tissue is one of the most widely distributed connective tissues in the body. Its intercellular substance (hyaluronic acid) contains fibers (collagenous, elastic, and reticular) and various cells (fibroblasts, macrophages, plasma, mast, and melanocytes). Loose connective tissue is found in all mucous membranes, around body organs, and in the subcutaneous layer.
4. Adipose tissue is a form of loose connective tissue in which the cells, called adipocytes, are specialized for fat storage. It is found in the subcutaneous layer and around various organs.
5. Dense (collagenous) connective tissue has a close packing of fibers (regularly or irregularly arranged). It is found as a component of fascia, membranes of organs, tendons, ligaments, and aponeuroses.
6. Elastic connective tissue has a predominance of freely branching elastic fibers that give it a yellow color. It is found in the cartilages of the larynx, elastic arteries, trachea, bronchial tubes, and true vocal cords.

7. Reticular connective tissue consists of interlacing reticular fibers and forms the stroma of the liver, spleen, and lymph nodes.
8. Cartilage has a jellylike matrix containing collagenous and elastic fibers and chondrocytes.
9. The growth of cartilage is accomplished by interstitial growth (from within) and appositional growth (from without).
10. Hyaline cartilage is found in the embryonic skeleton, at the ends of bones, in the nose, and in respiratory structures. It is flexible, allows movement, and provides support.
11. Fibrocartilage connects the pelvic bones and the vertebrae. It provides strength.
12. Elastic cartilage maintains the shape of organs such as the larynx, auditory tubes, and external ear.

Muscle Tissue and Nervous Tissue (p. 98)
1. Muscle tissue performs one major function: contraction.
2. Nervous tissue is specialized to conduct electrical impulses.

Membranes (p. 98)
1. An epithelial membrane is an epithelial layer overlying a connective tissue layer. Examples are mucous, serous, and cutaneous.
2. Mucous membranes line cavities that open to the exterior, such as the digestive tract.
3. Serous membranes (pleura, pericardium, peritoneum) line closed cavities and cover the organs in the cavities. These membranes consist of parietal and visceral portions.
4. The cutaneous membrane is the skin.
5. Synovial membranes line joint cavities and do not contain epithelium.

Tissue Inflammation: An Attempt to Restore Homeostasis (p. 98)
1. Damage to a tissue causes an inflammatory response characterized by redness, pain, heat, and swelling; sometimes loss of function occurs.
2. The inflammatory response is initiated by histamine, serotonin, kinins, and prostaglandins released by damaged tissue. They cause vasodilation and increased permeability of blood vessels.
3. Further cell injury is prevented by phagocytes. These include neutrophils (microphages) and macrophages.
4. The role of fibrin is to isolate the infected area.
5. In most inflammations, pus is produced; if it cannot drain out of the body, an abscess develops.

Tissue Repair (p. 101)
1. Tissue repair is the replacement of damaged or destroyed cells by healthy ones.
2. It begins during the active phase of inflammation and is not completed until after harmful substances in the inflamed area have been neutralized or removed.

Repair Process
1. If the injury is superficial, tissue repair involves pus removal (if pus is present), scab formation, and parenchymal regeneration.
2. If damage is extensive, granulation tissue is involved.

Conditions Affecting Repair
1. Nutrition is important to tissue repair. Various vitamins (A, some B, D, C, E, and K) and a protein-rich diet are needed.
2. Adequate circulation of blood is needed.
3. The tissues of young people repair rapidly and efficiently; the process slows down with aging.

REVIEW QUESTIONS

1. Define a tissue. What are the four basic kinds of human tissue?
2. Distinguish covering and lining epithelium from glandular epithelium. What characteristics are common to all epithelium?
3. Describe the origin and composition of the basement membrane.
4. Describe the various layering arrangements and cell shapes of epithelium.
5. How is epithelium classified? List the various types.
6. For each of the following kinds of epithelium, briefly describe the microscopic appearance, location in the body, and functions: simple squamous, simple cuboidal, simple columnar, stratified squamous, stratified cuboidal, stratified columnar, transitional, and pseudostratified.
7. Define the following terms: endothelium, mesothelium, secretion, absorption, goblet cell, and keratin.
8. What is a gland? Distinguish between endocrine and exocrine glands.
9. Describe the classification of exocrine glands according to structure and according to function and give at least one example of each.
10. Enumerate the ways in which connective tissue differs from epithelium.
11. How are connective tissues classified? List the various types.
12. How are embryonic connective tissue and adult connective tissue distinguished?
13. Describe the following connective tissues with regard to microscopic appearance, location in the body, and function: loose (areolar), adipose, dense (collagenous), elastic, reticular, hyaline cartilage, fibrocartilage, and elastic cartilage.
14. Distinguish between the interstitial and appositional growth of cartilage.
15. Define the following terms: hyaluronic acid, collagenous fiber, elastic fiber, reticular fiber, fibroblast, macrophage, plasma cell, mast cell, melanocyte, adipocyte, chondrocyte, and lacuna.
16. Define the following kinds of membranes: mucous, serous, cutaneous, and synovial. Where is each located in the body? What are their functions?
17. Following are some descriptions of various tissues of the body. For each description, name the tissue described.
    a. An epithelium that permits distention (stretching).
    b. A single layer of flat cells concerned with filtration and absorption.
    c. Forms all other kinds of connective tissue.
    d. Specialized for fat storage.
    e. An epithelium with waterproofing qualities.
    f. Forms the framework of many organs.
    g. Produces perspiration, wax, oil, or digestive enzymes.
    h. Cartilage that shapes the external ear.
    i. Contains goblet cells and lines the intestine.
    j. Most widely distributed connective tissue.
    k. Forms tendons, ligaments, and aponeuroses.
    l. Specialized for the secretion of hormones.
    m. Provides support in the umbilical cord.

n. Lines kidney tubules and is specialized for absorption and secretion.

o. Permits extensibility of lung tissue.

18. Define an inflammation. In what sense is the inflammatory response protective?

19. Describe the principal symptoms associated with inflammation.

20. How are vasodilation and increased permeability of blood vessels initiated during inflammation? Why are they important? What causes edema?

21. Describe the mechanism and importance of phagocytic migration during inflammation.

22. What is the role of fibrin in inflammation?

23. What is an abscess?

24. What is meant by tissue repair?

25. Distinguish between stromal and parenchymal repair.

26. What is the importance of granulation tissue?

27. What conditions affect tissue repair?

- Define the integumentary system.

- Describe the various functions of the skin.

- List the various layers of the epidermis and describe their structure and functions.

- Describe the composition and function of the dermis.

- Explain the basis for skin color.

- Explain the basic pattern of epidermal ridges and grooves.

- Describe the development, distribution, structure, color, growth, and replacement of hair.

- Compare the structure, distribution, and functions of sebaceous (oil), sudoriferous (sweat), and ceruminous glands.

- List the parts of a nail and describe their composition.

- Explain the role of the skin in helping to maintain normal body temperature.

- Describe the causes and effects for the following skin disorders: pruritus, acne, impetigo, systemic lupus erythematosus (SLE), psoriasis, decubitus ulcers, warts, cold sores, sunburn, and skin cancer.

- Define a burn and list the systemic effects of a burn.

- Classify burns into first, second, and third degrees.

- Describe how to estimate the extent of a burn.

- Define medical terminology associated with the integumentary system.

- Explain the actions of selected drugs associated with the integumentary system.

5 The Integumentary System

A aggregation of tissues that performs a specific function is an **organ.** The next higher level of organization is a **system**—a group of organs operating together to perform specialized functions. The skin and its derivatives, such as hair, nails, glands, and several specialized receptors, constitute the **integumentary** (in'-teg-yoo-MEN-tar-ē) **system** of the body.

SKIN

The **skin** or **cutis** is an organ because it consists of tissues structurally joined together to perform specific activities. It is not just a simple thin covering that keeps the body together and gives it protection. The skin is quite complex in structure and performs several functions essential for survival.

The skin is one of the larger organs of the body in terms of surface area. For the average adult, the skin occupies a surface area of approximately 19,355 sq cm (3,000 sq inches). It varies in thickness from 0.5 to 3 mm (0.02 to 0.12 inch). In general, skin is thicker on the dorsal surface of a part of the body than it is on the ventral surface, but this condition is reversed in the hand and the foot. The skin of the palmar surface of the hand and the plantar surface of the foot is thicker than on their dorsal aspects.

FUNCTIONS

The skin covers the body and protects the underlying tissues from bacterial invasion, drying out, and harmful light rays. It helps to maintain body temperature; prevents excessive loss of inorganic and organic materials; receives stimuli from the environment; stores chemical compounds; excretes water, salts, and several organic compounds; and synthesizes the hormone, vitamin D. The role of the skin in receiving various stimuli is considered in detail in Chapter 15.

STRUCTURE

Structurally, the skin consists of two principal parts (Figure 5-1). The outer, thinner portion, which is composed of epithelium, is called the **epidermis.** The epidermis is cemented to the inner, thicker, connective tissue part called the **dermis.** Thick skin has a relatively thick epidermis, whereas thin skin has a relatively thin epidermis. Beneath the dermis is a **subcutaneous layer.** This layer, also called the **superficial fascia** or **hypodermis,** consists of areolar and adipose tissues. Fibers from the dermis extend down into the superficial fascia and anchor the skin to the subcutaneous layer. The superficial fascia, in turn, is firmly attached to underlying tissues and organs.

EPIDERMIS

The **epidermis** is composed of *Langerhans' cells* (probably macrophages that have invaded the epidermis) and stratified squamous epithelium organized in four or five cell layers, depending on its location in the body (see Figure 5-1 and Figure 5-2). Where exposure to friction is greatest, such as the palms and soles, the epidermis has five layers.

In all other parts it has four layers. The names of the five layers from the deepest to the most superficial are as follows.

1. Stratum basale. This single layer of cuboidal to columnar cells is capable of continued cell division. As these cells multiply, they push up toward the surface and become part of the layers to be described next. Their nuclei degenerate, and the cells die. Eventually, the cells are shed from the top layer of the epidermis.

2. Stratum spinosum. This layer of the epidermis contains eight to ten rows of polyhedral (many-sided) cells that fit closely together. The surfaces of these cells may assume a prickly appearance when prepared for microscope examination (*spinosum* = prickly). The stratum basale and stratum spinosum are sometimes collectively referred to as the **stratum germinativum** (jer'-mi-na-TĒ-vum) to indicate the layers where new cells are germinated. The deeper layers of the epidermis of hairless skin contain nerve endings sensitive to touch called *Merkel's* (*tactile*) *discs* (see Figure 15-1).

3. Stratum granulosum. The third layer of the epidermis consists of three to five rows of flattened cells that contain darkly staining granules of a substance called *keratohyalin* (ker'-a-tō-HĪ-a-lin). This compound is involved in the first step of keratin formation. *Keratin* is a waterproofing protein found in the top layer of the epidermis. The nuclei of the cells in the stratum granulosum are in various stages of degeneration. As these nuclei break down, the cells are no longer capable of carrying out vital metabolic reactions and die.

4. Stratum lucidum. This layer is quite pronounced in the thick skin of the palms and soles. It consists of several rows of clear, flat, dead cells that contain droplets of a substance called *eleidin* (el-Ē-i-din). The layer is so named because eleidin is translucent (*lucidum* = clear). Eleidin is formed from keratohyalin and is eventually transformed to keratin.

5. Stratum corneum. This layer consists of 25 to 30 rows of flat, dead cells completely filled with keratin. These cells are continuously shed and replaced. The stratum corneum serves as an effective barrier against light and heat waves, bacteria, and many chemicals.

DERMIS

The second principal part of the skin, the **dermis,** is composed of connective tissue containing collagenous and elastic fibers (see Figure 5-1). The dermis is very thick in the palms and soles and very thin in the eyelids, penis, and scrotum. It also tends to be thicker on the dorsal aspects of the body than the ventral and thicker on the lateral aspects of extremities than medial aspects. Numerous blood vessels, nerves, glands, and hair follicles are embedded in the dermis.

The upper region of the dermis, about one-fifth of the thickness of the total layer, is named the **papillary region** or **layer.** Its surface area is greatly increased by small, fingerlike projections called **dermal papillae** (pa-PIL-ē). These structures project into the concavities between ridges in the deep surface of the epidermis and many contain loops of capillaries. Some dermal papillae contain *Meissner's* (MĪS-nerz) *corpuscles,* nerve endings sensitive to touch. The papillary region consists of loose connective tissue containing fine elastic fibers.

The remaining portion of the dermis is called the **reticular region** or **layer.** It consists of dense, irregularly arranged connective tissue containing interlacing bundles of collagenous and coarse elastic fibers. It is named the

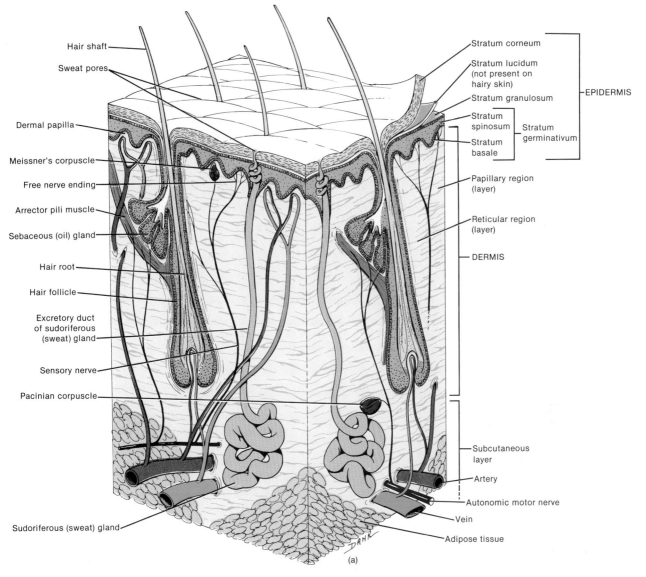

Hair shaft —

Sweat pores —

Dermal papilla —

Meissner's corpuscle —

Free nerve ending —

Arrector pili muscle —

Sebaceous (oil) gland —

Hair root —

Hair follicle —

Excretory duct of sudoriferous (sweat) gland —

Sensory nerve —

Pacinian corpuscle —

Sudoriferous (sweat) gland —

Stratum corneum

Stratum lucidum (not present on hairy skin)

Stratum granulosum

Stratum spinosum — Stratum germinativum

Stratum basale

EPIDERMIS

Papillary region (layer)

Reticular region (layer)

DERMIS

Subcutaneous layer

Artery

Autonomic motor nerve

Vein

Adipose tissue

(a)

FIGURE 5-1 Skin. (a) Structure of the skin and underlying subcutaneous layer. (b) Scanning electron micrograph of the skin and several hairs at a magnification of 260×. (From *Tissues and Organs: A Text-Atlas of Scanning Electron Microscopy,* by Richard K. Kessel and Randy H. Kardon. Copyright © 1979 by Scientific American, Inc.)

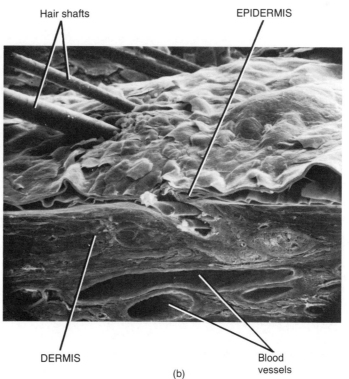

Hair shafts

EPIDERMIS

DERMIS

Blood vessels

(b)

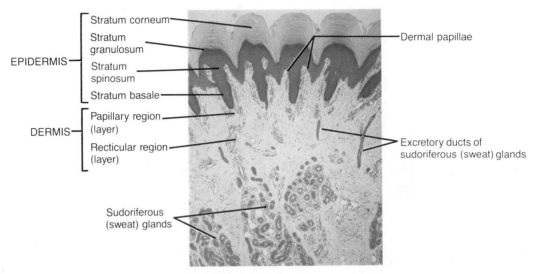

EPIDERMIS
- Stratum corneum
- Stratum granulosum
- Stratum spinosum
- Stratum basale

DERMIS
- Papillary region (layer)
- Recticular region (layer)

Dermal papillae

Excretory ducts of sudoriferous (sweat) glands

Sudoriferous (sweat) glands

Photomicrograph of thick skin at a magnification of 50×. The stratum lucidum is not shown here, but is illustrated in Figure 5-1a. (© 1983 by Michael H. Ross. Used by permission.)

reticular layer because the bundles of collagenous fibers interlace in a netlike manner. Spaces between the fibers are occupied by a small quantity of adipose tissue, hair follicles, nerves, oil glands, and the ducts of sweat glands. Varying thicknesses of the reticular region are responsible for differences in the thickness of the skin.

The combination of collagenous and elastic fibers in the reticular region provides the skin with strength, extensibility, and elasticity. (Extensibility is the ability to stretch; elasticity is the ability to return to original shape after extension or contraction.) The ability of the skin to stretch can readily be seen during conditions of pregnancy, obesity, and edema. The small tears that occur during extreme stretching are initially red and remain visible afterward as silvery white streaks called *striae* (STRĪ-ē).

The reticular region is attached to underlying organs, such as bone and muscle, by the subcutaneous layer. The subcutaneous layer also contains nerve endings called *Pacinian* (pa-SIN-ē-an) *corpuscles* that are sensitive to pressure (see Figure 15-1).

The collagenous fibers in the dermis run in all directions, but in particular regions of the body they tend to run more in one direction than another. The predominant direction of the underlying collagenous fibers is indicated in the skin by **lines of cleavage (tension lines).** The lines are especially evident on the palmar surfaces of the fingers where they run parallel to the long axis of the digit. The characteristic lines for each part of the body are shown in Figure 5-3.

Lines of cleavage are of particular interest to a surgeon because an incision running parallel to the collagen fibers will heal with only a fine scar. An incision made across the rows of fibers disrupts the collagen, and the wound tends to gape open and heal in a broad, thick scar.

The color of the skin is due to melanin, a pigment in the epidermis; carotene, a pigment mostly in the dermis; and blood in the capillaries in the dermis. The amount of **melanin** varies the skin color from pale yellow to black. This pigment is found primarily in the basale and spinosum layers. Melanin is synthesized in cells called **melanocytes** (me-LAN-ō-sīts), located either just beneath or between cells of the stratum basale. Since the number of melanocytes is about the same in all races, differences in skin color are due to the amount of pigment the melanocytes produce and disperse. An inherited inability of an individual in any race to produce melanin results in **albinism** (AL-bi-nizm). The pigment is absent in the hair and eyes as well as the skin. An individual affected with albinism is called an **albino** (al-BĪ-no). In some people, melanin tends to form in patches called **freckles.**

Melanocytes synthesize melanin from the amino acid *tyrosine* in the presence of an enzyme called *tyrosinase.* Exposure to ultraviolet radiation increases the enzymatic activity of melanocytes and leads to increased melanin production. The cell bodies of melanocytes send out long processes between epidermal cells. Upon contact with the processes, epidermal cells take up the melanin by phagocytosis. When the skin is again exposed to ultraviolet radiation, both the amount and the darkness of melanin increase, tanning and further protecting the body against radiation. Thus melanin serves a vital protective function. In mammals, the melanocyte-stimulating hormone (MSH) produced by the anterior pituitary gland causes increased melanin synthesis and dispersion through the epidermis. The exact role of MSH in humans is not clear.

Overexposure of the skin to the ultraviolet light of the sun may lead to skin cancer. Among the most malignant and lethal skin cancers is **melanoma** (*melano* = dark-colored; *oma* = tumor), cancer of the melanocytes. Fortunately, most skin cancers are operable.

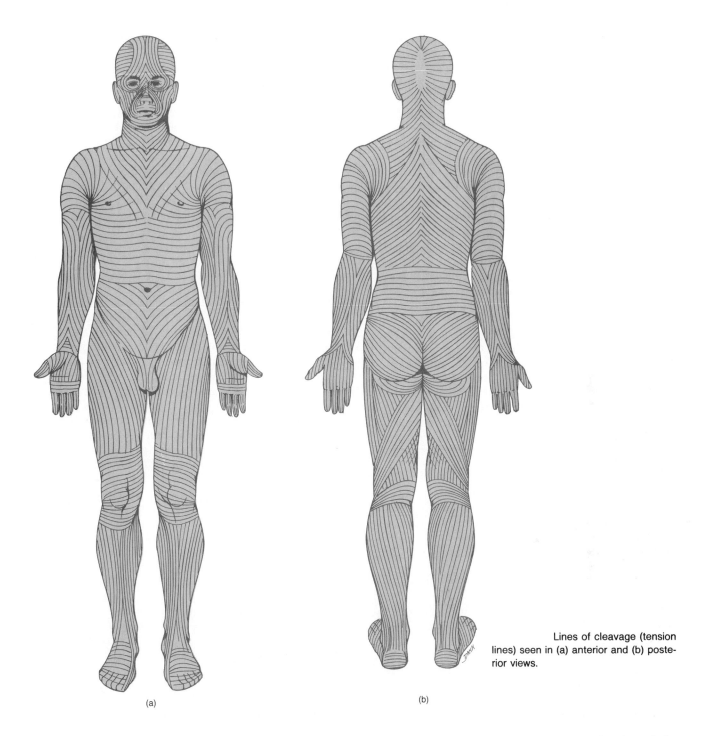

Lines of cleavage (tension lines) seen in (a) anterior and (b) posterior views.

(a)

(b)

Another pigment, called **carotene** (KAR-o-tēn), is found in the stratum corneum and fatty areas of the dermis in Oriental people. Together carotene and melanin account for the yellowish hue of their skin.

The pink color of Caucasian skin is due to blood in capillaries in the dermis. The redness of the vessels is not heavily masked by pigment. The epidermis has no blood vessels, a characteristic of all epithelia.

The outer surface of the skin of the palms and fingers and soles and toes is marked by a series of ridges and grooves which appear either as fairly straight lines or as a pattern of loops and whorls, as on the tips of the digits.

The **epidermal ridges** develop during the third and fourth fetal months as the epidermis conforms to the contours of the underlying dermal papillae (see Figure 5-1a). Since the ducts of sweat glands open on the summits of the epidermal ridges, fingerprints (or footprints) are left when a smooth object is touched. The ridge pattern, which is genetically determined, is unique for each individual. It does not change throughout life, except to enlarge, and thus can serve as the basis for identification through fingerprints or footprints. The function of the ridges is to increase the grip of the hand or foot by increasing friction.

Epidermal grooves on other parts of the skin divide the surface into a number of diamond-shaped areas. Examine the dorsum of the hand as an example. Note that hairs typically emerge at the points of intersection of the grooves. Note also that the grooves increase in frequency and depth as regions of free joint movement are approached.

CLINICAL APPLICATION

Scientists have now made it possible to improve in many cases the appearance of scars from deep acne, chickenpox, burns, cleft lip, and other disorders and to make age-related wrinkles disappear. The substance that makes this possible is called **Zyderm Collagen Implant.** Collagen is the body's principal structural material. It is found in skin, bone, cartilage, tendons, ligaments, and various viscera and accounts for almost one-third of the total protein of the body.

Zyderm Collagen Implant is prepared from cattle collagen that is suspended in a saline solution containing lidocaine, a local anesthetic. Once injected into the skin, the solution disappears and the collagen, under the influence of body temperature, becomes a stationary fleshlike substance that is incorporated into the skin. The collagen implant becomes colonized by blood vessels and cells and acts as a natural structural framework in the skin.

EPIDERMAL DERIVATIVES

Structures developed from the embryonic epidermis—hair, glands, nails—perform functions that are necessary and sometimes vital. Hair and nails protect the body. The sweat glands help regulate body temperature.

HAIR

Growths of the epidermis variously distributed over the body are **hairs** or **pili.** The primary function of hair is protection. Though the protection is limited, hair guards the scalp from injury and the sun's rays. Eyebrows and eyelashes protect the eyes from foreign particles. Hair in the nostrils and external ear canal protects these structures from insects and dust.

Development and Distribution

During the third and fourth months of fetal life, the epidermis develops downgrowths into the dermis called **hair follicles.** Originating from these follicles are small hairs. By the fifth or six month, the follicles produce delicate hairs called **lanugo** (lan-YOO-gō; *lana* = wool) that covers the fetus. The lanugo is usually shed prior to birth, except in the regions of the eyebrows, eyelids, and scalp. Here they persist and become stronger. Several months after birth, these hairs are shed and are replaced by still coarser ones, while the remainder of the body develops a new hair growth called the **vellus** (fleece). At puberty, coarse hairs develop in the axillary and pubic regions of both sexes and on the face and to a lesser extent on

other parts of the body in the male. The coarse hairs that develop at puberty, plus the hairs of the scalp and eyebrows, are referred to as **terminal hairs.**

Hairs are distributed on nearly all parts of the body. It has been estimated that an average adult has about five million hairs, of which about 100,000 are in the scalp. They are absent from the palms of the hands, the soles of the feet, the dorsal surfaces of the distal phalanges, lips, nipples, clitoris, glans penis, inner surface of the prepuce, inner surfaces of the labia majora and minora, and outer surfaces of the labia minora. Straight hairs are oval or cylindrical in cross section, whereas curly hairs are flattened. Straight hairs are stronger than curly ones.

Structure

Each hair consists of a shaft and a root (Figure 5-4a). The **shaft** is the superficial portion, most of which projects above the surface of the skin. The shaft of coarse hairs consists of three principal parts. The inner **medulla** is composed of rows of polyhedral cells containing granules of eleidin and air spaces. The medulla is poorly developed or not present at all in fine hairs. The second principal part of the shaft is the middle **cortex.** It forms the major part of the shaft and consists of elongated cells that contain pigment granules in dark hair, but mostly air in white hair. The **cuticle of the hair,** the outermost layer, consists of a single layer of thin, flat, scalelike cells that are the most heavily keratinized. They are arranged like shingles on the side of a house, but the free edges of the cuticle cells point upward rather than downward like shingles (Figure 5-4e). The **root** is the portion below the surface that penetrates into the dermis and even into the subcutaneous layer and, like the shaft, contains a medulla, cortex, and cuticle (Figure 5-4a,b).

Surrounding the root is the **hair follicle,** which is made up of an external zone of epithelium (the external root sheath) and an internal zone of epithelium (the internal root sheath). The **external root sheath** is a downward continuation of the basale and spinosum layers of the epidermis. Near the surface it contains all the epidermal layers. As it descends, it does not exhibit the superficial epidermal layers. At the bottom of the hair follicle, the external root sheath contains only the stratum basale. The **internal root sheath** is formed from proliferating cells of the matrix (described shortly) and takes the form of a cellular tubular sheath deep to the external root sheath. The internal root sheath extends only partway up the follicle and consists of (1) an inner layer, the **cuticle of the internal root sheath,** which is a single layer of flattened cells with atrophied nuclei, (2) a middle **granular layer,** which is one to three layers of flattened nucleated cells, and (3) an outer **pallid layer,** which is a single layer of cuboidal cells with flattened nuclei.

The base of each follicle is enlarged into an onion-shaped structure, the **bulb.** This structure contains an indentation, the **papilla of the hair,** filled with loose connective tissue. The papilla of the hair contains many blood vessels and provides nourishment for the growing hair. The bulb also contains a region of cells called the **matrix,** a germinal layer. The cells of the matrix produce new hairs by cell division when older hairs are shed. This replacement occurs within the same follicle.

Sebaceous glands and a bundle of smooth muscle are also associated with hair. Details of the sebaceous glands are discussed shortly. The smooth muscle is called **arrector pili**; it extends from the dermis of the skin to the side of the hair follicle (Figure 5-4a). In its normal position hair is arranged at an angle to the surface of the skin. The arrectores pilorum muscles contract under stresses of fright and cold and pull the hairs into a vertical position. This contraction results in "goosebumps" or "gooseflesh" because the skin around the shaft forms slight elevations.

Around each hair follicle are nerve endings, called *root hair plexuses,* that are sensitive to touch (see Figure 15-1). They respond if a hair shaft is moved.

Color

The color of hair is due primarily to melanin. It is formed by melanocytes distributed in the matrix of the bulb of the follicle. There are only three colors of hair pigment—black, brown, and yellow, due to the pigments black melanin, brown melanin, and pheomelanin, respectively. The many variations in hair color are combinations of different amounts of the three pigments. Graying of hair is the loss of pigment believed to be the result of a progressive inability of the melanocytes to make tyrosinase, the enzyme necessary for the synthesis of melanin. White hair results from air in the medullary shaft.

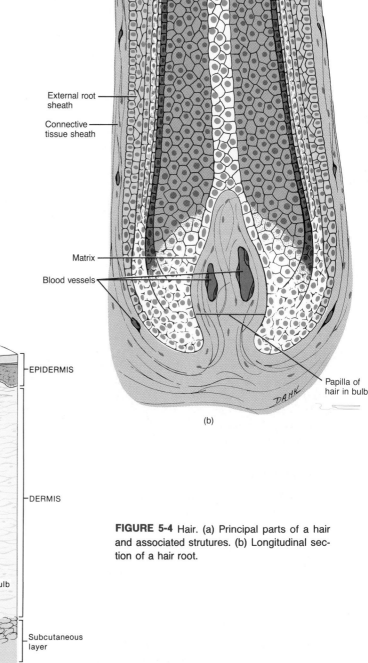

(b)

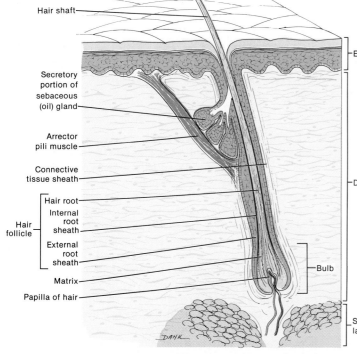

(a)

FIGURE 5-4 Hair. (a) Principal parts of a hair and associated strutures. (b) Longitudinal section of a hair root.

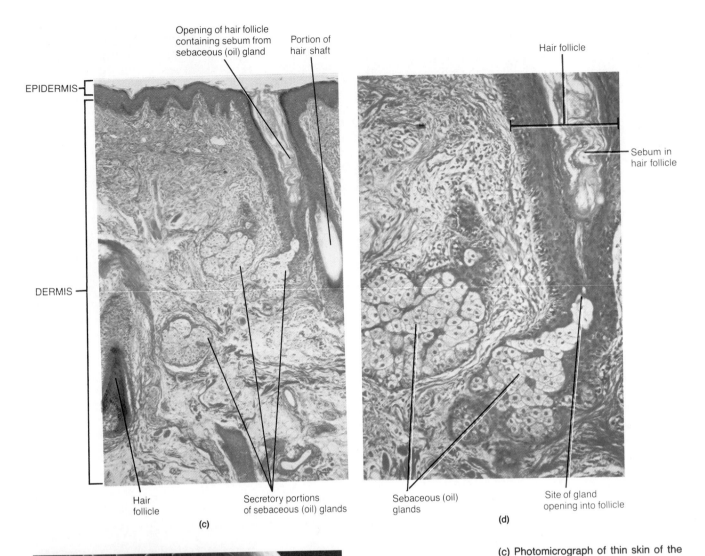

EPIDERMIS

DERMIS

Opening of hair follicle
containing sebum from
sebaceous (oil) gland

Portion of
hair shaft

Hair follicle

Sebum in
hair follicle

Hair
follicle

Secretory portions
of sebaceous (oil) glands

Sebaceous (oil)
glands

Site of gland
opening into follicle

(c)

(d)

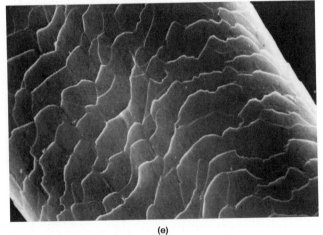

(e)

(c) Photomicrograph of thin skin of the face at a magnification of 60×. (© 1983 by Michael H. Ross. Used by permission.) (d) Photomicrograph of a sebaceous (oil) gland opening into a hair follicle at a magnification of 150×. (© 1983 by Michael H. Ross. Used by permission.) (e) Scanning electron micrograph of the surface of a hair shaft showing the shinglelike cuticular scales at a magnification of 1,000×. (Courtesy of Fisher Scientific Company and S.T.E.M. Laboratories, Inc., Copyright, 1975.)

During the resting phase, the matrix becomes inactive and undergoes atrophy. At this point the root of the hair detaches from the matrix and the hair slowly moves up the follicle. It may remain there for some time until pulled out, shed, or pushed up by a replacing hair.

Either before or after the hair comes out of the follicle, proliferation of cells by the external root sheath forms a new matrix. The new matrix undergoes cell division, forming a new hair that grows up the follicle and replaces the old hair. You might be interested to know that shaving or cutting the hair has no effect on its growth.

The cycle of hair growth varies in different parts of the body. In the scalp, each hair grows steadily and continuously for 2 to 6 years, growth then stops, and after 3 months, the hair is shed. After another 3 months of a resting phase, a new hair starts to grow from the same

Hair replacement occurs according to a cyclic pattern, alternating between growing and resting periods. During the growing phase of the cycle, the cells of the matrix are active. They increase in number by cell division and are pushed upward and eventually die, a situation similar to epidermal growth. The product is a hair, which is essentially dead protein tissue. Hair grows about 1 mm (0.04 inch) every 3 days.

follicle. Eyebrows, by contrast, have a growing phase of only about 10 weeks. It is for this reason that these hairs do not grow very long.

Normal hair loss in an adult scalp is about 70 to 100 hairs per day. Both the rate of growth and the replacement cycle may be altered by illness, diet, and other factors. For example, high fever, major illness, major surgery, blood loss, or severe emotional stress may increase the rate of shedding. Rapid weight-loss diets involving severe restriction of calories or protein also increase hair loss. An increase in the rate of shedding can also occur for 3 to 4 months after childbirth. Certain drugs and radiation therapy are also factors in increasing hair loss.

The age of onset, degree of thinning, and ultimate hair pattern associated with **common baldness** ("male-pattern" baldness) are determined by male hormones, called androgens, and heredity. Androgens are involved in promoting normal sexual development. Genetic baldness cannot be prevented or helped by local applications of various chemicals, injections, radiation, or any other treatment.

GLANDS

Three kinds of glands associated with the skin are sebaceous glands, sudoriferous glands, and ceruminous glands.

Sebaceous (Oil) Glands

Sebaceous (se-BĀ-shus) or **oil glands,** with few exceptions, are connected to hair follicles (see Figure 5-4a,c, d). The secreting portions of the glands lie in the dermis and those glands associated with hairs open into the necks of hair follicles. Sebaceous glands not associated with hair follicles open directly onto the surface of the skin (lips, glans penis, labia minora, and meibomian glands of the eyelids). Sebaceous glands are simple branched acinar glands. Absent in the palms and soles, they vary in size and shape in other regions of the body. For example, they are small in most areas of the trunk and extremities, but large in the skin of the breasts, face, neck, and upper chest.

The sebaceous glands secrete an oily substance called **sebum** (SĒ-bum), a mixture of fats, cholesterol, proteins, and inorganic salts. Sebum helps keep hair from drying and becoming brittle, forms a protective film that prevents excessive evaporation of water from the skin, and keeps the skin soft and pliable.

CLINICAL APPLICATION
When sebaceous glands of the face become enlarged because of accumulated sebum, **blackheads** develop. Since sebum is nutritive to certain bacteria, **pimples** or **boils** often result. The color of blackheads is due to melanin and oxidized oil, not dirt.

Sudoriferous (Sweat) Glands

Sudoriferous (soo'-dor-IF-er-us; *sudor* = sweat; *ferre* = to bear) or **sweat glands** are distinguished into two principal types on the basis of structure and location. **Apocrine** sweat glands are simple, branched tubular glands. Their distribution is limited primarily to the skin of the axilla, pubic region, and pigmented areas (areolae) of the breasts. The secretory portion of apocrine sweat glands is located in the dermis and the excretory duct opens into hair follicles. Apocrine sweat glands begin to function at puberty and produce a more viscous secretion than the other type of sweat gland.

Eccrine sweat glands are much more common than apocrine sweat glands and are simple, coiled tubular glands. They are distributed throughout the skin except for the margins of the lips, nail beds of the fingers and toes, glans penis, glans clitoris, and eardrums. Eccrine sweat glands are most numerous in the skin of the palms and the soles; their density can be as high as 3,000 per square inch in the palms. The secretory portion of eccrine sweat glands is located in the subcutaneous layer and the excretory duct projects upward through the dermis and epidermis to terminate at a pore at the surface of the epidermis (see Figure 5-1). Eccrine sweat glands function throughout life and produce a secretion that is more watery than that of apocrine sweat glands.

Perspiration, or **sweat,** is the substance produced by sudoriferous glands. It is a mixture of water, salts (mostly NaCl), urea, uric acid, amino acids, ammonia, sugar, lactic acid, and ascorbic acid. Its principal function is to help maintain body temperature. It also helps to eliminate wastes.

Ceruminous Glands

In certain parts of the skin, sudoriferous glands are modified as **ceruminous** (se-ROO-mi-nus) **glands.** Such modified glands are simple, coiled tubular glands present in the external auditory meatus (canal). The secretory portions of ceruminous glands lie in the submucosa, deep to sebaceous glands, and the excretory ducts open either directly onto the surface of the external auditory meatus or into sebaceous ducts. The combined secretion of the ceruminous and sebaceous glands is called **cerumen** (*cera* = wax). Cerumen may accumulate resulting in impacted cerumen (Chapter 17).

NAILS

Hard, keratinized cells of the epidermis are referred to as **nails.** The cells form a clear, solid covering over the dorsal surfaces of the terminal portions of the fingers and toes. Each nail (Figure 5-5) consists of a **nail body,** a **free edge,** and a **nail root.** The nail body is the portion of the nail that is visible, the free edge is the part that projects beyond the distal end of the digit, and the nail root is the portion that is hidden in the nail groove. Most of the nail body is pink because of the underlying vascular tissue. The whitish semilunar area at the proximal end of the body is called the **lunula** (LOO-nyoo-la). It appears whitish because the vascular tissue underneath does not show through.

The fold of skin that extends around the proximal and lateral borders of the nail is known as the **nail fold,** and the epidermis beneath the nail constitutes the **nail bed.** The furrow between the two is the **nail groove.**

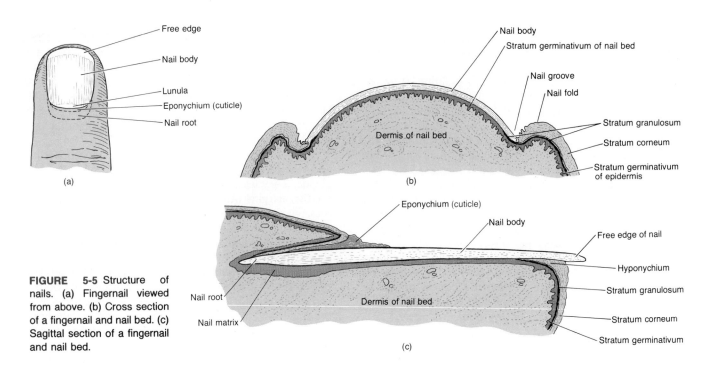

FIGURE 5-5 Structure of nails. (a) Fingernail viewed from above. (b) Cross section of a fingernail and nail bed. (c) Sagittal section of a fingernail and nail bed.

The **eponychium** (ep'-ō-NIK-ē-um) or **cuticle** is a narrow band of epidermis that extends from the margin of the nail wall (lateral border), adhering to it. It occupies the proximal border of the nail and consists of stratum corneum. The thickened area of stratum corneum below the free edge of the nail is referred to as the **hyponychium** (hī'-pō-NIK-ē-um).

The epithelium of the proximal part of the nail bed is known as the **nail matrix.** Its function is to bring about the growth of nails. Essentially, growth occurs by the transformation of superficial cells of the matrix into nail cells. In the process, the outer, harder layer is pushed forward over the stratum germinativum. The average growth in the length of fingernails is about 1 mm (0.04 inch) per week. The growth rate is somewhat slower in toenails.

HOMEOSTASIS

One of the best examples of homeostasis in humans is the maintenance of body temperature by the skin. Humans, like other mammals, are *homeotherms*—warm-blooded organisms. This means that we are able to maintain a remarkably constant body temperature of 37°C (98.6°F) even though the environmental temperature may vary over a broad range.

Suppose you are in an environment where the temperature is 37.8°C (100°F). A sequence of events is set into operation to counteract this above-normal temperature, which may be considered a stress. Sensing devices in the skin called receptors pick up the stimulus—in this case, heat—and activate nerves that send the message to your brain. A temperature-regulating area of the brain then sends nerve impulses to the sudoriferous glands, which produce more perspiration. As the perspiration evaporates from the surface of your skin, it is cooled and your body temperature is maintained. This sequence of events is shown in Figure 5-6.

Note that temperature regulation by the skin involves a feedback system because the output (cooling of the skin) is fed back to the skin receptors and becomes part of a new stimulus-response cycle. In other words, after the sudoriferous glands are activated, the skin receptors keep the brain informed about the external temperature. The brain, in turn, continues to send messages to the sudoriferous glands until the temperature returns to 37°C (98.6°F). Like most of the body's feedback systems, temperature regulation is a negative feedback system—the output, cooling, is the opposite of the original condition, overheating.

Temperature maintenance by perspiration represents only one mechanism by which we maintain normal body temperature. Other mechanisms include adjusting blood flow to the skin, regulating metabolic rate, and regulating skeletal muscle contractions. These other mechanisms are discussed in detail in Chapter 25.

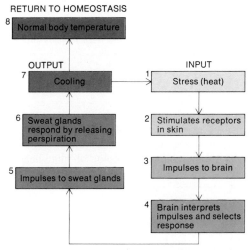

FIGURE 5-6 Role of the skin in maintaining the homeostasis of body temperature.

DISORDERS: HOMEOSTATIC IMBALANCES

Pruritus

Pruritus (*pruire* = to itch), or itching, is probably one of the most common of all dermatological conditions. The causes of pruritus may be placed in three general categories: (1) cutaneous disorders, such as *xerosis* (dry skin), *miliaria* (sweat retention in the skin), and infectious agents (bacteria, fungi, viruses, and body lice); (2) systemic disorders, such as cancer, chronic renal failure, liver diseases, and thyroid diseases; (3) psychogenic factors, as might occur during or just after an episode of emotional stress.

Acne

Acne is an inflammation of sebaceous glands and usually begins at puberty. In fact, a few comedones (blackheads or whiteheads) on the face may be the first signs of approaching puberty. At puberty the sebaceous glands, under the influence of androgens (male hormones) grow in size and increase production of their complex lipid product, sebum. Although testosterone, a male hormone, appears to be the most potent circulating androgen for sebaceous cell stimulation, adrenal and ovarian androgens can stimulate sebaceous secretions as well.

Acne occurs predominantly in sebaceous follicles. The four basic types of acne lesions in order of increasing severity are comedones, papules, pustules, and cysts. The sebaceous follicles are rapidly colonized by microorganisms such as staphylococcal species and *Proprionibacterium acne* bacteria that thrive in the lipid-rich environment of the follicles. When this occurs, the cyst or sac of connective tissue cells can destroy and displace epidermal cells, resulting in permanent scarring. This type of acne, called *cystic acne*, may be treated successfully with a new and powerful drug called Accutane, a synthetic form of vitamin A. Care must be taken to avoid squeezing, pinching, or scratching the lesions.

Impetigo

Impetigo (im'-pe-TĪ-gō) is a superficial skin infection caused by staphylococci or streptococci. It is characterized by isolated pustules that become crusted and rupture. Occurring principally around mouth, nose, and hands, the inflammation is located in the papillary layer of the skin and involves the capillary network and stratum corneum. The disease is most common in children, and epidemics in nurseries (due to staphylococci) may be serious.

Systemic Lupus Erythematosus (SLE)

Systemic lupus erythematosus (er-i'-them-a-TŌ-sus), **SLE,** or **lupus** is an autoimmune, inflammatory disease, occurring mostly in young women in their reproductive years. An *autoimmune disease* is one in which the body attacks its own tissues, failing to differentiate between what is foreign and what is not. In SLE, damage to blood vessel walls results in the release of chemicals that mediate the inflammatory response. The blood vessel damage can be associated with virtually every body system.

The cause of SLE is not known. It is not contagious, and although it is not thought to be hereditary, there seems to be a strong incidence of other connective tissue disorders—especially rheumatoid arthritis (RA) and rheumatic fever—in relatives of SLE victims. The disease may be triggered by medication, such as penicillin, sulfa, or tetracycline, exposure to excessive sunlight, injury, emotional upset, infection, or other stress. These triggering factors, once recognized, are to be avoided by the patient in the future.

Symptoms include low-grade fever, aches, fatigue, photosensitivity, rapid loss of large amounts of scalp hair, and sometimes an eruption across the bridge of the nose and cheeks called a "butterfly rash." Other skin lesions may occur with blistering and ulceration. The erosive nature of some of the SLE skin lesions was thought to resemble the damage inflicted by the bite of a wolf, thus the term *lupus*. The most serious complications of the disease involve inflammation of the kidneys, liver, spleen, lungs, heart, and the central nervous system.

Psoriasis

Psoriasis (sō-RĪ-a-sis) is a chronic, occasionally acute, relapsing skin disease borne by 6 to 8 million people in the United States. In its severe forms, psoriasis is a disabling and disfiguring affliction. It may begin at any age, although it is most severe between ages 10 and 50. It is thought to be hereditary in at least one-third of the patients. Recent studies have traced the complex cause of psoriasis to an abnormally high rate of mitosis in epidermal cells. Triggering factors such as trauma, infections, seasonal and hormonal changes, and emotional stress can initiate and intensify the skin eruptions.

Psoriasis is characterized by distinct, reddish, slightly raised plaques or papules (small, round skin elevations) covered with scales. Itching is seldom severe, and the lesions heal without scarring. Psoriasis ordinarily involves the scalp, the elbows and knees, the back, and the buttocks. Occasionally the disease is generalized.

Decubitus Ulcers

Decubitus (dē-KYOO-be-tus) **ulcers,** also known as **bedsores, pressure sores,** or **trophic ulcers** are caused by a constant deficiency of blood to tissues overlying a bony projection that has been subjected to prolonged pressure against an object such as a bed, cast, or splint. The deficiency results in tissue ulceration. Small breaks in the epidermis become infected, and the sensitive subcutaneous and deeper tissues are damaged. Eventually the tissue is destroyed.

Decubitus ulcers are seen most frequently in patients who are bedridden for long periods of time. The most common areas involved are the skin over the sacrum, heels, ankles, buttocks, and other large bony projections. The chief causes are pressure from infrequent turning of the patient, trauma and maceration of the skin, and malnutrition. Maceration of the skin often follows soaking of bed and clothing by perspiration, urine, or feces.

Warts

Certain viruses (papovaviruses) can cause epithelial skin cells to proliferate into uncontrolled growths called **warts.** Warts are generally benign (noncancerous), and most regress spontaneously. It is possible to spread warts by contact-transfer of the viruses, and sexual contact has been implicated in the transmission of warts on the genitals. After infection, there is an incubation period of several weeks before the appearance of the wart.

Cold Sores (Fever Blisters)

The herpes simplex virus (HSV) has the ability to lay dormant for extended periods of time without expressing any signs or symptoms of disease. Cold sores are caused by type 1 herpes simplex virus, which is transmitted by oral or respiratory routes. During its periods of dormancy, the virus probably infects only a few cells at a time. Although the initial infection usually occurs during infancy, it is for the most part subclinical. The few individuals who express the disease during infancy have characteristic lesions, usually in the oral mucous membrane. These infections subside, but recur as what are called **cold sores (fever blisters).** The appearance of cold sore lesions is associated with stimuli such as exposure to the ultraviolet (UV) radiation from the sun, hormonal changes related to the menstrual cycle, or even emotional upset.

Type 2 herpes simplex virus causes genital herpes infections and is transmitted primarily by sexual contact (see Chapter 28).

Sunburn

Sunburn is injury to the skin as a result of acute, prolonged exposure to the ultraviolet (UV) rays of sunlight. It usually occurs 2 to 8 hours after exposure. Acute redness and pain are maximal in about 12 hours and dissipate approximately 72 to 96 hours later. Sunburn is without doubt one of the most common and painful skin afflictions. The damage to skin cells caused by sunburn is due to inhibition of DNA and RNA synthesis which leads to cell death. There can also be damage to blood vessels as well as other structures in the dermis. Overexposure over a period of years results in a leathery skin texture, wrinkles, skin folds, sagging skin, warty growths called keratoses, freckling, a yellow discoloration due to abnormal elastic tissue, and premature aging of the skin.

The tendency to sunburn is determined by type of skin. There are basically four skin groups. Those who burn and never tan, those who burn but keep some tan, those who burn slightly and develop a good tan, and those who never burn and always tan.

Covering the body remains the best protection against the sun's rays, while a topical sunscreen will help protect uncovered parts. The relative degree of protection provided by a sunscreen-ing agent is discussed in the section on drugs at the end of the chapter.

Extreme and severe sunburn may produce *sunstroke*—a disruption of the body's heat-regulating mechanism. It is characterized by high fever and collapse and sometimes by convulsions, coma, and death (Chapter 25).

Skin Cancer

Excessive sun exposure can result in **skin cancer.** Everyone, regardless of skin pigmentation, is a potential victim of skin cancer if exposure to sunlight is sufficiently intense and continuous. Natural skin pigment can never give complete protection. If you must be in direct sunlight for long periods of time, use a suitable sunscreen. One of the best agents for protection against overexposure to ultraviolet rays of the sun is para-aminobenzoic acid (PABA). The alcohol preparations of PABA are best because the active ingredient binds to the stratum corneum of the skin.

A number of widely prescribed drugs (tetracyclines, sulfa drugs, and others), constituents of foods (such as riboflavins), and saccharin and cyclamates in dietetic drinks are potential photosensitizers. When activated in the body by light, they may produce substances that can damage the tissues in sensitive individuals. A typical response is the appearance of a rash on parts of the body exposed to the sun. Repeated and persistent exposure could produce permanent changes leading to skin cancer.

FIGURE 5-7 The Lund-Browder method for estimating the extent of burns. Relative proportions of various body regions in (a) a young child compared to (b) an adult.

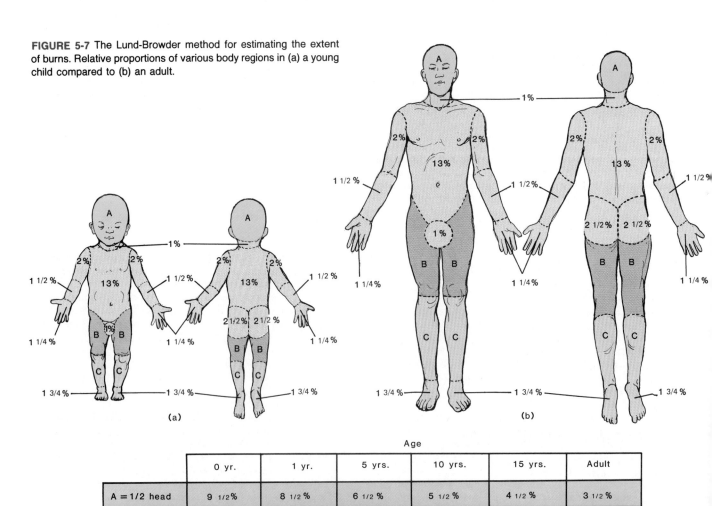

	Age					
	0 yr.	1 yr.	5 yrs.	10 yrs.	15 yrs.	Adult
A = 1/2 head	9 1/2 %	8 1/2 %	6 1/2 %	5 1/2 %	4 1/2 %	3 1/2 %
B = 1/2 thigh	2 3/4 %	3 1/4 %	4%	4 1/4 %	4 1/2 %	4 3/4 %
C = 1/2 calf	2 1/2 %	2 1/2 %	2 3/4 %	3%	3 1/4 %	3 1/2 %

Burns

Tissues may be damaged by thermal (heat), electrical, radioactive, or chemical agents. These agents can destroy the proteins in the exposed cells and cause cell injury or death. Such damage is a **burn.** The injury to tissues directly or indirectly in contact with the damaging agent, such as the skin or the linings of the respiratory and digestive tracts, is the local effect of a burn. Generally, however, the systemic effects of a burn are a greater threat to life than the local effects. The systemic effects of a burn may include (1) a large loss of water, plasma, and plasma proteins, which causes shock; (2) bacterial infection; (3) reduced circulation of blood; and (4) decreased production of urine.

Classification

A burn may extend through the entire thickness of the skin or it may damage or destroy only part of the skin. Clinically, the depth of a burn is determined by its color, the presence or absence of sensation, blister formation, or loss of elasticity. A **first-degree burn** is characterized by mild pain and erythema (redness) and involves only the surface epithelium. Generally, a first-degree burn will heal in about two to three days and may be accompanied by flaking or peeling. A typical sunburn is an example of a first-degree burn.

A **second-degree burn** involves the deeper layers of the epidermis or the upper levels of the dermis. In a superficial second-degree burn, the deeper layers of the epidermis are injured and there is characteristic erythema, blister formation, edema, and pain. Blisters beneath or within the epidermis are called *bullae* (BYOOL-ē), meaning bubbles. Such an injury usually heals within seven to ten days with only mild scarring. In a deep second-degree burn, there is destruction of the epidermis as well as the upper levels of the dermis. Epidermal derivatives, such as hair follicles, sebaceous glands, and sweat glands are usually not injured. If there is no infection, deep second-degree burns heal without grafting in about three to four weeks. Scarring may result.

First- and second-degree burns are collectively referred to as **partial-thickness burns.** A **third-degree burn** or **full-thickness burn** involves destruction of the epidermis, dermis, and the epidermal derivatives. Such burns vary in appearance from marble-white to mahogany colored to charred, dry wounds. There is little if any edema and such a burn is usually not painful to the touch due to destruction of nerve endings. Regeneration is slow and much granulation tissue forms before being covered by epithelium. Even if skin grafting is quickly begun, third-degree burns quickly contract and produce scarring.

The severity of a burn is measured in terms of the amount of surface area affected as well as in terms of the depth of injury. A fairly accurate method for estimating the extent of a burn is to apply the *Lund-Browder method.* This method estimates the extent by measuring the areas affected against the percentage of total surface area for body parts shown in Figure 5-7. For example, if the anterior of the head and neck of an adult is affected, the burn covers 4½ percent of the body surface. Because the proportions of the body change with growth, the percentages vary for different ages. Thus, the extent of burn damage can be made fairly accurately for any age group.

Treatment

A severely burned individual should be moved as quickly as possible to a hospital. Treatment may then include:

1. Cleansing the burn wounds thoroughly.
2. Removing all dead tissue (debridement) so antibacterial agents can directly contact the wound surface and thereby prevent infection.
3. Replacing lost body fluids and electrolytes (ions).
4. Covering wounds with temporary protection as soon as possible. One such covering, *human amniotic membrane,* is now easier to use because of recently developed methods of collection and storage. The membrane is stripped from the placenta ("afterbirth") up to 24 hours after delivery and stored in a saline-penicillin solution for up to 8 weeks. The membrane's desirable permeable properties keep wounds from drying while preventing infection from without. Investigators have found that the membrane is also useful for treating chronic decubitus ulcers. Another covering used as a temporary protection for burns is a newly developed semisynthetic membrane called *Biobrane.* It consists of nylon, Silastic, and a collagen derivative.
5. Removing a thin layer of skin from another part of the burn victim's body and transplanting it to the injured area (skin graft).

MEDICAL TERMINOLOGY

Albinism (*alb* = white; *ism* = condition)   Congenital (existing at birth) absence of pigment from the skin, hair, and parts of the eye.

Anhidrosis (*an* = without; *hidr* = sweating; *osis* = condition)   A rare genetic condition characterized by inability to sweat.

Antiperspirant (*anti* = against; *perspirare* = to breathe through)   An agent that inhibits or prevents perspiration and usually contains aluminum compounds as the active ingredient.

Athlete's foot   A superficial fungus infection of the skin of the foot.

Callus (callosity)   An area of hardened and thickened skin that is usually seen in palms and soles and is due to pressure and friction.

Carbuncle (*carbunculus* = little coal)   A hard, round, deep, painful inflammation of the subcutaneous tissue that causes necrosis (death) and pus formation (abscess).

Comedo (*comedo* = to eat up)   A collection of sebaceous material and dead cells in the hair follicle and excretory duct of the sebaceous gland. Usually found over the face, chest, and back, and more commonly during adolescence. Also called blackhead or whitehead.

Corn   A painful conical thickening of the skin found principally over toe joints and between the toes. It may be hard or soft, depending on the location. Hard corns are usually found over toe joints, and soft corns are usually found between the fourth and fifth toes.

Cyst (*cyst* = sac containing fluid)   A sac with a distinct connective tissue wall, containing a fluid or other material.

Deodorant (*de* = from; *odorare* = perfume)   An agent that masks offensive odors; usually contains aluminum compounds as the active ingredient. Body odor is not caused by perspiration itself as much as by the activities of bacteria that grow in perspiration.

Dermabrasion (*derm* = skin)   Removal of acne, scars, tattoos, or nevi (see below) by sandpaper or a high-speed brush.

Dermatome (*tomy* = cut)   An instrument for excising areas of skin to be used for grafting.

Detritus (*deterere* = to rub away)   Particulate matter produced by or remaining after the wearing away or disintegration of a substance or tissue; scales, crusts, and loosened skin.

Eczema (*ekzein* = to boil out)   An acute or chronic superficial inflammation of the skin, characterized by redness, oozing, crusting, and scaling. Also called chronic dermatitis.

Electrolysis   Method of hair removal in which the hair bulb is destroyed by an electric current so that the hair cannot regrow.

Erythema (*erythema* = redness)   Redness of the skin caused by engorgement of capillaries in lower layers of the skin. Erythema occurs with any skin injury, infection, or inflammation.

Furuncle   A boil; an abscess resulting from infection of a hair follicle.

Hypodermic (*hypo* = under)   Relating to the area beneath the skin. Also called subcutaneous.

Intradermal (*intra* = within)   Within the skin. Also called intracutaneous.

Keratosis (*kera* = horn)   Formation of a hardened growth of tissue.

Nevus   A round, pigmented, flat, or raised skin area varying in color from yellow-brown to black. It may be present at birth or develop later. Also called mole or birthmark.

Nodule (*nodulus* = little knot)   A large cluster of cells raised above the skin but extending deep into the tissues.

Papule   A small, round skin elevation varying in size from a pinpoint to that of a split pea. One example is a pimple.

Polyp   A tumor on a stem found especially on mucous membranes.

Pustule   A small, round elevation of the skin containing pus.

Subcutaneous (*sub* = under)   Beneath the skin. Also called hypodermic.

Topical   Pertaining to a definite area; local. Also, of a medication, applied to the surface rather than ingested or injected.

DRUGS ASSOCIATED WITH THE INTEGUMENTARY SYSTEM

In the sections on drugs associated with the various systems of the body, we have included old, firmly established drugs and new, recently approved medications. Drugs are referred to by generic and nonproprietary names and in some cases by brand names. The generic name indicates the general category of the drug. It usually describes the use or action of the drug, such as keratolytic agent. The nonproprietary name is the common name for the drug as approved by the U.S. Food and Drug Administration (FDA). Nonproprietary names appear in italic type. A brand name is a name assigned to a drug by a manufacturer and is a registered trademark. A specific drug may be manufactured by only one company, or it may be produced by several companies under different brand names. Brand names are given in parentheses following the nonproprietary names, for example, *triamcinolone* (Aristocort, Kenalog).

Disorders of the skin are treated primarily with liquid lotions, liniments, semisolid ointments, pastes, water-miscible bases, or wetting agents. Generally, the skin acts as a barrier to most drugs that are applied to it, and this has the advantage of minimizing the absorption of potentially toxic drugs.

Keratolytic Agents
Keratolytic agents are keratin-dissolving drugs that help speed the rate of peeling of keratin-containing cells. They are used to treat acne, athelete's foot, corns, and calluses. Examples are *salicylic acid, tretinoin,* and *benzoyl peroxide.*

Depilatories
Depilatories are substances that remove superfluous hair. They dissolve the protein in the hair shaft, turning it into a gelatinous mass which can be wiped away. Since the hair root is not affected, regrowth of the hair occurs. Examples of depilatories include calcium hydroxide, strontium hydroxide, and *calcium thioglycollate* (Neet, Nair).

Caustics
Caustics are substances that destroy excessive growths of keratin-containing cells, such as warts. Examples are *glacial acetic acid* and *trichloroacetic acid.*

Astringents
Astringents cause a slight protein-coagulating (tightening) effect on the skin and are used to treat poison ivy, athlete's foot, and acute eczema. An example is *aluminum salts.*

Dusting Powders
Dusting powders absorb moisture from the skin and are used to reduce friction between adjacent skin areas that rub together. Examples include *corn starch, talc,* and *magnesium oxide.*

Protective Dressings
Protective dressings are drugs used to seal small wounds, protect the skin from external irritants, and prevent diaper rash, bed sores, and chapping. Examples include *colloidion* and *dimethicone.*

Corticosteroids
Corticosteroids are antiinflammatory drugs that are useful in treatment of severe sunburn and certain dermatoses. Examples include *hydrocortisone* (Cortaid), *triamcinolone* (Aristocort, Kenalog), *prednisone, dexamethasone* (Decadron), and *betamethasone* (Celestone, Valisone).

Since 0.5 percent hydrocortisone (e.g., Cortaid) is now available over-the-counter (OTC), the harmful effects of topical corticosteroid overuse are likely to become more frequent. One of the primary dangers of continued topical corticosteroid use is masking the symptoms of certain bacterial, viral, and fungal infections. Another is actually worsening the infection, such as occurs in acne, scabies, impetigo, and lice. Topical corticosteroid overuse can also result in thinning and weakening of the epidermis and superficial dermis, especially in the axillae, groin, and face; stimulation of dermal structures causing acne or excessive growth of unwanted hair; and tissue atrophy. A very serious systemic complication is suppression of interactions between the pituitary and suprarenal glands (Chapter 18).

Sunscreening Agents
Presently there are over 50 products available that provide varying degrees of protection from direct ultraviolet light exposure as well as indirect or reflected light exposure from the sun. Such *sunscreening agents* fall into two principal classes: physical reflectors and chemical absorbers. Both classes retard premature aging of the skin and development of skin cancers caused by ultraviolet light. *Physical reflectors* include opaque white and tinted creams containing substances such as titanium dioxide or zinc oxide. Most people, however, prefer chemical absorbers in creams, oils, and lotions, which are more cosmetically attractive than physical reflectors. The most effective ingredient in

chemical absorbers is para-aminobenzoic acid (PABA) or a related compound.

The FDA has established guidelines to indicate the relative degree of protection provided by a sunscreening agent. The degree of protection is indicated by a numerical rating and is called a *sun protection factor* (*SPF*). The SPF is the time required to produce erythema (redness) when a particular sunscreen product is used divided by the time required to produce er-

ythema when the product is not used. For example, if you can remain exposed to the sun with protection for 30 minutes before your skin gets red, but you can remain exposed to the sun without protection for only 5 minutes, then the SPF of the product you are using is 6. In other words, the product allows you to stay in the sun 6 times longer than without protection. SPF ranges from 2 (practically no protection) to 15 or more (super protection).

STUDY OUTLINE

Skin (p. 106)

1. The skin and its derivatives (hair, glands, and nails) constitute the integumentary system.
2. The skin is one of the larger organs of the body. It performs the functions of protection, maintaining body temperature, preventing excessive loss of inorganic and organic materials, receiving stimuli, storage of chemical compounds, synthesis of vitamin D, and excretion of water, salts, and several organic compounds.
3. The principal parts of the skin are the outer epidermis and inner dermis. The dermis overlies the subcutaneous layer.
4. The epidermal layers, from deepest to most superficial, are the stratum basale, spinosum, granulosum, lucidum, and corneum. The basale and spinosum undergo continuous cell division and produce all other layers.
5. The dermis consists of a papillary region and a reticular region. The papillary region is loose connective tissue containing blood vessels, nerves, hair follicles, dermal papillae, and Meissner's corpuscles. The reticular region is dense, irregularly arranged connective tissue containing adipose tissue, hair follicles, nerves, oil glands, and ducts of sweat glands.
6. Lines of cleavage indicate the direction of collagenous fiber bundles in the dermis and are considered during surgery.
7. The color of skin is due to melanin, carotene, and blood in capillaries in the dermis.
8. Epidermal ridges increase friction for better grasping ability and provide the basis for fingerprints and footprints.

Epidermal Derivatives (p. 110)

1. Epidermal derivatives are structures developed from the embryonic epidermis.
2. Among the epidermal derivatives are hair, skin glands (sebaceous, sudoriferous, and ceruminous), and nails.

Hair

1. Hairs are epidermal growths that function in protection.
2. Hair consists of a shaft above the surface, a root that penetrates the dermis and subcutaneous layer, and a hair follicle.
3. Associated with hairs are sebaceous glands, arrectores pilorum muscles, and root hair plexuses.
4. Hair color is due to combinations of various amounts of the three hair pigments, black melanin, brown melanin, and pheomelanin (yellow). Graying is due to the loss of melanin.
5. New hairs develop from cell division of the matrix in the bulb; hair replacement and growth occurs in a cyclic pattern. "Male-pattern" baldness is caused by androgens and heredity.

Glands

1. Sebaceous (oil) glands are usually connected to hair follicles; they are absent in the palms and soles. Sebaceous glands produce sebum which moistens hairs and waterproofs the skin. Enlarged sebaceous glands may produce blackheads, pimples, and boils.
2. Sudoriferous (sweat) glands are distinguished into apocrine

and eccrine. Apocrine sweat glands are limited in distribution to the skin of the axilla, pubis, and areolae; their ducts open into hair follicles. Eccrine sweat glands have an extensive distribution; their ducts terminate at pores at the surface of the epidermis. Sudoriferous glands produce perspiration, which carries small amounts of wastes to the surface and assists in maintaining body temperature.
3. Ceruminous glands are modified sudoriferous glands that secrete cerumen. They are found in the external auditory meatus.

Nails

1. Nails are hard, keratinized epidermal cells over the dorsal surfaces of the terminal portions of the fingers and toes.
2. The principal parts of a nail are the body, free edge, root, lunula, eponychium, hyponychium, and matrix. Cell division of the matrix cells produces new nails.

Homeostasis (p. 114)

1. One of the functions of the skin is the maintenance of the normal body temperature of 37°C (98.6°F).
2. If environmental temperature is high, skin receptors sense the stimulus (heat) and generate impulses that are transmitted to the brain. The brain then causes the sweat glands to produce perspiration. As the perspiration evaporates, the skin is cooled.
3. The skin-cooling response is a negative feedback mechanism.
4. Temperature maintenance is also accomplished by adjusting blood flow to the skin, regulating metabolic rate, and regulating skeletal muscle contractions.

Disorders: Homeostatic Imbalances (p. 115)

1. Pruritus or itching is a common skin problem that may be related to skin disorders, systemic diseases, or psychogenic factors.
2. Acne is an inflammation of sebaceous glands.
3. Impetigo is a superficial skin infection caused by bacteria and characterized by pustules that become crusted and rupture.
4. Systemic lupus erythematosus (SLE) is an autoimmune disease of connective tissue.
5. Psoriasis is a chronic skin disease characterized by reddish, raised plaques or papules.
6. Decubitus ulcers are caused by a chronic deficiency of blood to tissues subjected to prolonged pressure.
7. Warts are uncontrolled growths of epithelial skin cells caused by a virus. Most warts are benign.
8. Cold sores (fever blisters) are lesions caused by type 1 herpes simplex virus. The dormant infection is triggered by certain stimuli.
9. Sunburn is a skin injury resulting from prolonged exposure to the ultraviolet (UV) rays of sunlight.
10. Skin cancer can be caused by excessive exposure to sunlight.
11. Tissue damage that destroys protein is called a burn. Depending on the depth of damage, skin burns are classified

as first-degree and second-degree (partial-thickness) and third-degree (full-thickness). One method employed for determining the extent of a burn is the Lund-Browder method.

Burn treatment may include cleansing the wound, removing dead tissue, replacing lost body fluids, and covering wounds with temporary protection, and skin grafting.

REVIEW QUESTIONS

1. Define an organ. In what respect is the skin an organ? What is the integumentary system?
2. List the principal functions of the skin.
3. Compare the structure of epidermis and dermis. What is the subcutaneous layer?
4. List and describe the epidermal layers from the deepest outward. What is the importance of each layer?
5. Contrast the structural differences between the papillary and reticular regions of the dermis.
6. What are lines of cleavage? What is their importance during surgery?
7. Explain the factors that produce skin color. What is an albino?
8. Describe how melanin is synthesized and distributed to epidermal cells.
9. How are epidermal ridges formed? Why are they important?
10. List the receptors in the epidermis, dermis, and subcutaneous layer, and indicate the location and role of each.
11. Describe the development and distribution of hair.
12. Describe the structure of a hair. How are hairs moistened? What produces "goosebumps" or "gooseflesh"?
13. Explain the basis for hair color. Why does hair turn gray?
14. Describe how hair grows and is replaced.
15. Contrast the locations and functions of sebaceous glands, sudoriferous glands, and ceruminous glands. What are the names and chemical components of the secretions of each?
16. Distinguish between apocrine and eccrine sweat glands.
17. From what layer of the skin do nails form? Describe the principal parts of a nail.
18. Explain with a labeled diagram how the skin helps maintain normal body temperature.
19. What is a feedback system? A negative feedback system? Relate these definitions to the maintenance of normal body temperature.
20. Define each of the following disorders of the integumentary system: pruritus, acne, impetigo, systemic lupus erythematosus (SLE), psoriasis, decubitus ulcers, warts, cold sores, sunburn, and skin cancer.
21. Define a burn. Classify burns according to degree.
22. Explain how the Lund-Browder method is used to estimate the extent of a burn.
23. How are burns treated?
24. Refer to the glossary of key medical terms associated with the integumentary system. Be sure that you can define each term.
25. Describe the actions of drugs associated with the integumentary system.

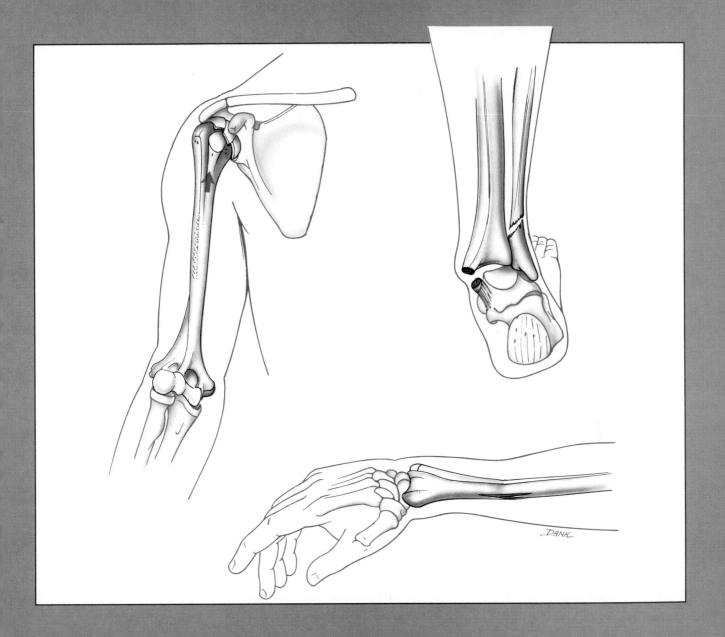

UNIT II

Principles of Support and Movement

This unit considers two primary themes—support and movement. You will study the various ways in which the body is supported and the different movements it can perform. Both support and movement are made possible by the cooperative effort of bones, joints, and muscles.

6 Skeletal Tissue

ithout the skeletal system we would be unable to perform movements, such as walking or grasping. The slightest jar to the head or chest could damage the brain or heart. It would even be impossible to chew food. The framework of bones and cartilage that protects our organs and allows us to move is called the **skeletal system.**

FUNCTIONS

The skeletal system performs several basic functions. First, it **supports** the soft tissues of the body so that the form of the body and an erect posture can be maintained. Second, the system **protects** delicate structures—the brain, the spinal cord, the lungs, the heart, the major blood vessels in the thoracic cavity. Third, the bones serve as levers to which the muscles of the body are attached. When the muscles contract, the bones acting as levers produce **movement.** Fourth, the bones serve as **storage areas** for mineral salts, especially calcium and phosphorus, and fat. Fifth, **blood cell production,** occurs in the

red marrow of the bones. This process is referred to as **hematopoiesis** (hē'-ma-tō-poy-Ē-sis) or **hemopoiesis** (hē'-mō-poy-Ē-sis). Red marrow consists of blood cells in immature stages, fat cells, and macrophages. It is responsible for producing red blood cells, some white blood cells, and platelets.

HISTOLOGY

Structurally, the skeletal system consists of two types of connective tissue: cartilage and bone. We described the microscopic structure of cartilage in Chapter 4. Here, our attention will be directed to the microscopic structure of bone tissue.

Like other connective tissues, **bone,** or **osseous** (OS-ē-us) **tissue,** contains a great deal of intercellular substance surrounding widely separated cells. Mature bone cells are called **osteocytes.** Unlike other connective tissues, the intercellular substance of bone contains abundant mineral salts, primarily calcium phosphate $(Ca_3(PO_4)_2 \cdot (OH)_2)$

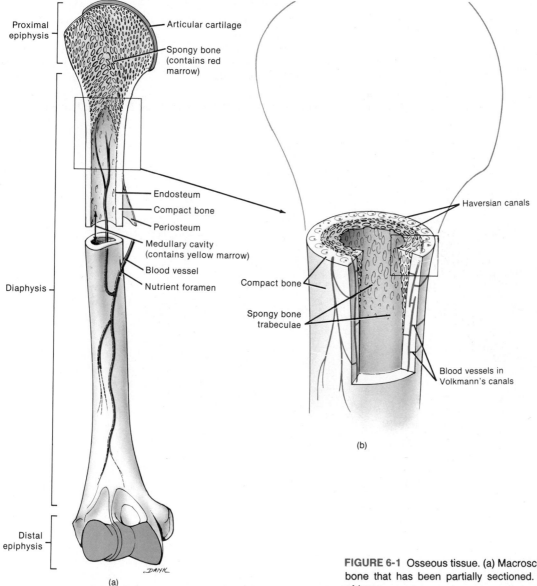

Proximal epiphysis

Articular cartilage

Spongy bone (contains red marrow)

Endosteum

Compact bone

Periosteum

Medullary cavity (contains yellow marrow)

Blood vessel

Nutrient foramen

Diaphysis

Distal epiphysis

(a)

Haversian canals

Compact bone

Spongy bone trabeculae

Blood vessels in Volkmann's canals

(b)

FIGURE 6-1 Osseous tissue. (a) Macroscopic appearance of a long bone that has been partially sectioned. (b) Histological structure of bone.

and some calcium carbonate ($CaCO_3$). Together these salts are referred to as *hydroxyapatites*. As these salts are deposited in the framework of collagenous fibers of the intercellular substance, the tissue hardens, that is, becomes ossified. The hydroxyapatites compose 67 percent of the weight of bone, and the collagenous fibers make up the remaining 33 percent.

The microscopic structure of bone may be analyzed by considering the anatomy of a long bone such as the humerus (arm bone) shown in Figure 6-1a. A typical long bone consists of the following parts.

    1. Diaphysis (dī-AF-i-sis). The shaft or long, main portion of the bone.

    2. Epiphyses (ē-PIF-i-sēz). The extremities or ends of the bone.

    3. Metaphysis (me-TAF-i-sis). The region in a mature bone where the diaphysis joins the epiphysis. In a growing bone, it is the region where calcified cartilage is reinforced and then replaced by bone (described later in the chapter).

    4. Articular cartilage. A thin layer of hyaline cartilage covering the epiphysis where the bone forms a joint with another bone.

    5. Periosteum (per'-ē-OS-tē-um). A dense, white, fibrous covering around the remaining surface of the bone. The periosteum (*peri* = around; *osteo* = bone) consists of two layers (Figure 6-1c). The outer **fibrous layer** is composed of connective tissue containing blood vessels, lymphatic vessels, and nerves that pass into the bone. The inner **osteogenic** (os'-tē-ō-JEN-ik) **layer** contains elastic fibers, blood vessels, and **osteoblasts**—cells responsible for forming new bone during growth and repair.

The word *blast* means a germ or bud. It denotes an immature cell or tissue that later develops into a specialized form. The periosteum is essential for bone growth, repair, and nutrition. It also serves as a point of attachment for ligaments and tendons. A photomicrograph of the periosteum is shown in Figure 6-2.

    6. Medullary (MED-yoo-lar'-ē) or **marrow cavity.** The space within the diaphysis that contains the fatty *yellow marrow* in adults. Yellow marrow consists primarily of fat cells and a few scattered blood cells.

    7. Endosteum. A layer of osteoblasts that lines the medullary cavity and contains scattered osteoclasts (cells that may assume a role in the removal of bone).

Bone is not a completely solid, homogenous substance. In fact, all bone has some spaces between its hard components. The spaces provide channels for blood vessels that supply bone cells with nutrients. The spaces also make bones lighter. Depending on the size and distribution of the spaces, the regions of a bone may be categorized as spongy or compact (Figures 6-1b and 6-3).

Spongy, or **cancellous** bone tissue contains many large spaces filled with red marrow. It makes up most of the bone tissue of short, flat, and irregularly shaped bones and most of the epiphyses of long bones. Spongy bone tissue provides some support and a storage area for marrow. **Compact,** or **dense** bone tissue, by contrast, contains few spaces. It is deposited in a layer over the spongy bone tissue. The layer of compact bone is thicker in the diaphyses than the epiphyses. Compact bone tissue provides protection and considerable support and helps the long bones resist the stress of weight placed on them.

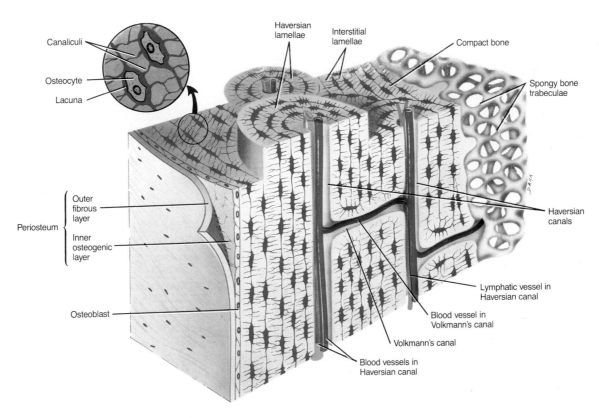

(c)

FIGURE 6-1 (*Continued*) Osseous tissue. (c) Enlarged aspect of Haversian systems in compact bone.

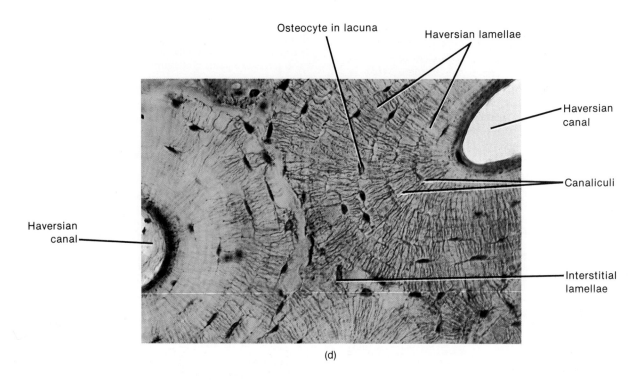

Osteocyte in lacuna   Haversian lamellae

Haversian canal

Canaliculi

Haversian canal

Interstitial lamellae

(d)

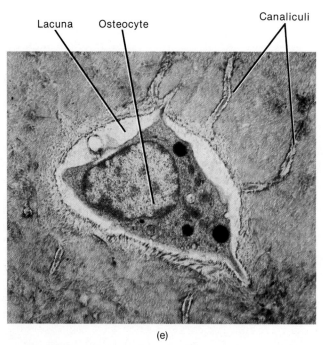

Lacuna   Osteocyte   Canaliculi

(e)

FIGURE 6-1 (*Continued*) Osseous tissue. (d) Photomicrograph of portions of several Haversian systems at a magnification of 150×. (Courtesy of Biophoto Associates.) (e) Electron micrograph of an osteocyte. (Courtesy of Biophoto Associates.)

COMPACT BONE

We can compare the differences between spongy and compact bone tissue by looking at the highly magnified section in Figure 6-1c. One main difference is that adult compact bone has a concentric-ring structure, whereas spongy bone does not. Blood vessels and nerves from the periosteum and endosteum penetrate the compact bone through **Volkmann's canals.** The blood vessels of these canals connect with blood vessels and nerves of the medullary cavity and those of the **Haversian** (ha-VER-shun) **canals.** The Haversian canals run longitudinally through the bone.

Around them are **lamellae** (la-MEL-ē)—concentric rings of hard, calcified, intercellular substance. Between the lamellae are small spaces called **lacunae** (la-KOO-nē), where osteocytes are found. **Osteocytes** are mature osteoblasts that have lost their ability to produce new bone tissue. Radiating in all directions from the lacunae are minute canals called **canaliculi** (kan'-a-LIK-yoo-lē), which contain slender processes of osteocytes (Figure 6-1e). The canaliculi connect with other lacunae and, eventually, with the Haversian canals. Thus an intricate network is formed throughout the bone. This branching network of canaliculi provides numerous routes so that

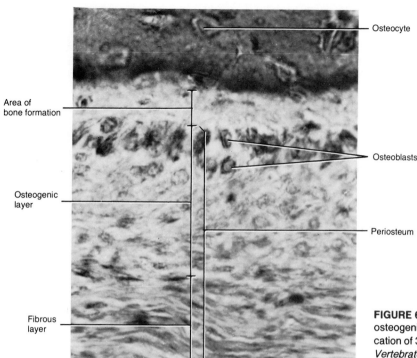

Osteocyte

Area of bone formation

Osteoblasts

Osteogenic layer

Periosteum

Fibrous layer

FIGURE 6-2 Photomicrograph of compact bone showing the osteogenic and fibrous layers of the periosteum at a magnification of 300×. (Courtesy of Donald I. Patt, from *Comparative Vertebrate Histology,* by Donald I. Patt and Gail R. Patt, Harper & Row, Publishers, Inc., New York, 1969.)

nutrients can reach the osteocytes and wastes can be removed. Each Haversian canal, with its surrounding lamellae, lacunae, osteocytes, and canaliculi, is called an **Haversian system** or **osteon.** Haversian systems are characteristic of adult bone. The areas between Haversian systems contain **interstitial lamellae.** These also possess lacunae with osteocytes and canaliculi, but their lamellae are usually not connected to the Haversian systems.

SPONGY BONE

In contrast to compact bone, spongy bone does not contain true Haversian systems. It consists of an irregular latticework of thin plates of bone called **trabeculae** (tra-BEK-yoo-lē). The spaces between the trabeculae of some bones are filled with red marrow. The cells of red marrow are responsible for producing blood cells. Within the trabeculae lie lacunae, which contain the osteocytes. Blood vessels from the periosteum penetrate through to the spongy bone, and osteocytes in the trabeculae are nourished directly from the blood circulating through the marrow cavities.

Most people think of all bone as a very hard, white material. Yet the bones of an infant are not hard at all, and a child's bones are generally more pliable than those of an adult. The final shape and hardness of adult bones require many years to develop and depend on a complex series of chemical changes. Let us now see how bones are formed and how they grow.

OSSIFICATION

The process by which bone forms in the body is called **ossification** or **osteogenesis.** The "skeleton" of a human embryo is composed of either fibrous membranes or hyaline cartilage. Both are shaped like bones and provide the medium for ossification. Ossification begins around the sixth or seventh week of embryonic life and continues throughout adulthood. Two kinds of bone formation occur. The first is called intramembranous (in'-tra-MEM-bra-nus) ossification. This term refers to the formation of bone directly on or within the fibrous membranes (*intra* = within; *membranous* = membrane). The second kind, endochondral (en'-dō-KON-dral) ossification, refers to the formation of bone in cartilage (*endo* = within; *chondro* = cartilage). These two kinds of ossification do *not* lead to differences in the structure of mature bones. They simply indicate difference methods of bone formation. Both mechanisms involve the replacement of a preexisting connective tissue with bone.

The first stage in the development of bone is the migration of mesenchymal cells (embryonic connective tissue cells) into the area where bone formation is about to begin. These cells increase in number and size. In some skeletal structures where capillaries are lacking they become chondroblasts; in others where capillaries are present some become osteoblasts. The **chondroblasts** will be responsible for cartilage formation. The osteoblasts will form bone tissue by intramembranous or endochondral ossification.

INTRAMEMBRANOUS OSSIFICATION

Of the two types of bone formation, the simpler and more direct is **intramembranous ossification.** The flat bones of the roof of the skull, parts of the mandible (lower jawbone), and probably part of the clavicles are formed in this way. The essentials of this process are as follows.

Osteoblasts formed from mesenchymal cells cluster in the fibrous membrane. The site of such a cluster is called

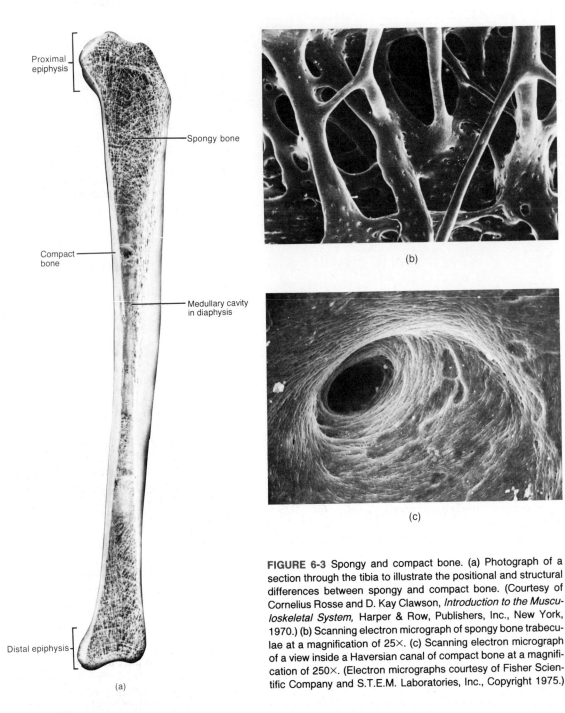

(a)

(b)

(c)

FIGURE 6-3 Spongy and compact bone. (a) Photograph of a section through the tibia to illustrate the positional and structural differences between spongy and compact bone. (Courtesy of Cornelius Rosse and D. Kay Clawson, *Introduction to the Musculoskeletal System,* Harper & Row, Publishers, Inc., New York, 1970.) (b) Scanning electron micrograph of spongy bone trabeculae at a magnification of 25×. (c) Scanning electron micrograph of a view inside a Haversian canal of compact bone at a magnification of 250×. (Electron micrographs courtesy of Fisher Scientific Company and S.T.E.M. Laboratories, Inc., Copyright 1975.)

a *center of ossification.* The osteoblasts then secrete intercellular substances partly composed of collagenous fibers that form a framework, or matrix, in which calcium salts are quickly deposited. The deposition of calcium salts is called *calcification.* When a cluster of osteoblasts is completely surrounded by the calcified matrix, it is called a *trabecula.* As trabeculae form in nearby ossification centers, they fuse into the open latticework characteristic of spongy bone. With the formation of successive layers of bone, some osteoblasts become trapped in the lacunae. The entrapped osteoblasts lose their ability to form bone and are called osteocytes. The spaces between the trabeculae fill with red marrow. The original connective tissue that surrounds the growing mass of bone then becomes the periosteum. The ossified area has now become true

spongy bone. Eventually, the surface layers of the spongy bone will be reconstructed into compact bone. Much of this newly formed bone will be destroyed and reformed so the bone may reach its final adult size and shape.

ENDOCHONDRAL OSSIFICATION

The replacement of cartilage by bone is called **endochondral** or **intracartilaginous ossification.** Most bones of the body, including parts of the skull, are formed in this way. Since this type of ossification is best observed in a long bone, we will investigate the tibia, or shinbone (Figure 6-4).

Early in embryonic life, a cartilage model or template of the future bone is laid down. This model is covered

Proximal epiphysis

Spongy bone

Compact bone

Medullary cavity in diaphysis

Distal epiphysis

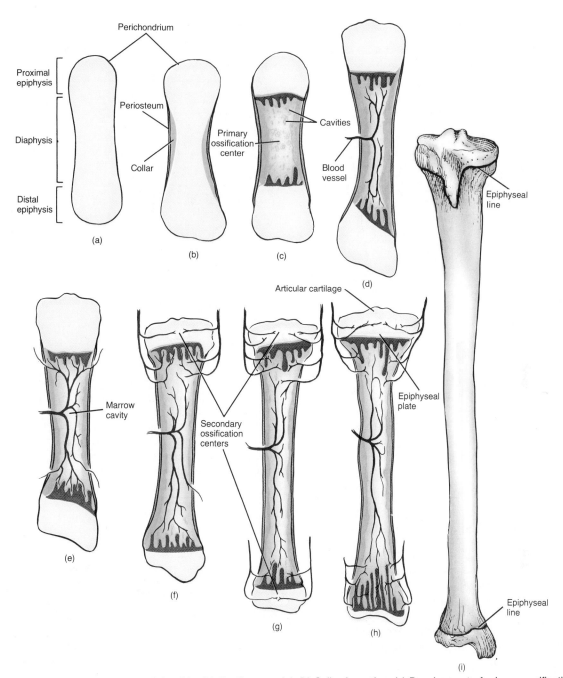

Endochondral ossification of the tibia. (a) Cartilage model. (b) Collar formation. (c) Development of primary ossification center. (d) Entrance of blood vessels. (e) Marrow-cavity formation. (f) Thickening and lengthening of collar. (g) Formation of secondary ossification centers. (h) Remains of cartilage as articular cartilage and epiphyseal plate. (i) Formation of epiphyseal lines.

by a membrane called the *perichondrium* (per-i-KON-drē-um). Midway along the shaft of this model a blood vessel penetrates the perichondrium, stimulating the cells in the internal layer of the perichondrium to enlarge and become osteoblasts. The cells begin to form a collar of compact bone around the middle of the diaphysis of the cartilage model. Once the perichondrium starts to form bone, it is called the *periosteum.* Simultaneously with the appearance of the bone collar and the penetration of blood vessels, changes occur in the cartilage in the center of the diaphysis. In this area, called the **primary ossification center,** cartilage cells hypertrophy (increase in size)—probably because they accumulate glycogen for energy and produce enzymes to catalyze future chemical reactions. When the hypertrophied cells burst, there is a change in extracellular pH to a more alkaline pH causing the intercellular substance to become *calcified,* that is, minerals are deposited within it. Once the cartilage becomes calcified, nutritive materials required by the cartilage cells can no longer diffuse through the intercellular substance and this may cause the cartilage cells to die. Then the intercellular substance begins to degenerate,

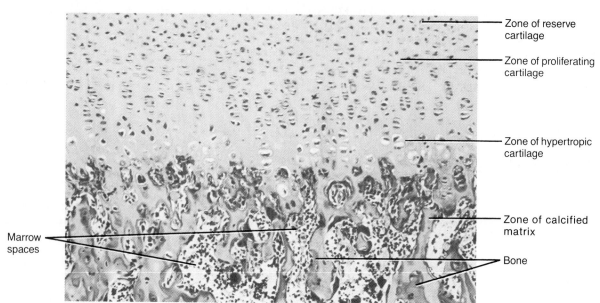

Zone of reserve cartilage

Zone of proliferating cartilage

Zone of hypertropic cartilage

Zone of calcified matrix

Bone

Marrow spaces

FIGURE 6-5 Epiphyseal plate. Photomicrograph of the epiphysis of a long bone showing the various zones of the epiphyseal plate at a magnification of 160×. (© 1983 by Michael H. Ross. Used by permission.)

leaving large cavities in the cartilage model. The blood vessels grow along the spaces where cartilage cells were previously located and enlarge the cavities further. Gradually, these spaces in the middle of the shaft join with each other, and the marrow cavity is formed.

As these developmental changes are occurring, the osteoblasts of the periosteum deposit successive layers of bone on the outer surface so that the collar thickens, becoming thickest in the diaphysis. The cartilage model continues to grow at its ends, steadily increasing in length. Eventually, blood vessels enter the epiphyses and **secondary ossification centers** appear in the epiphyses and also lay down spongy bone. In the tibia, one secondary ossification center develops in the proximal epiphysis soon after birth. The other center develops in the distal epiphysis during the child's second year.

After the two secondary ossification centers have formed, bone tissue has completely replaced cartilage, except in two regions. Cartilage continues to cover the articular surfaces of the epiphyses, where it is called **articular cartilage.** It also remains as a plate between the epiphysis and diaphysis, where it is called the **epiphyseal plate.**

The epiphyseal (ep'-i-FIZ-ē-al) plate consists of four zones (Figure 6-5). The *zone of reserve cartilage* is adjacent to the epiphysis of the bone. It consists of small chondrocytes that are scattered irregularly throughout the intercellular matrix. The cells of this zone do not function in bone growth. The zone of reserve cartilage functions in anchoring the epiphyseal plate to the bone of the epiphysis. Its blood vessels also provide nutrients for the other zones of the epiphyseal plate.

The second zone, the *zone of proliferating cartilage,* consists of slightly larger chondrocytes that are arranged like stacks of coins. The function of this zone is to make new chondrocytes by cell division to replace those that die at the diaphyseal surface of the epiphyseal plate.

The third zone, the *zone of hypertrophic* (hī-per-TRŌF-ik) *cartilage,* consists of even larger chondrocytes also arranged in columns. The cells are in various stages of maturation, with the more mature cells closer to the diaphysis. The lengthwise expansion of the epiphyseal plate is the result of cellular proliferation of the zone of proliferating cartilage and maturation of the cells in the zone of hypertrophic cartilage. Near the diaphyseal end of the bone, the intercellular matrix of the cells of the zone of hypertrophic cartilage becomes calcified and dies.

The fourth zone, the *zone of calcified matrix,* is only a few cells thick and consists mostly of dead cells because the intercellular matrix around them has calcified. The calcified matrix is taken up by osteoclasts, and the area is invaded by osteoblasts and capillaries from the bone in the diaphysis. These cells lay down bone on the calcified cartilage that persists. As a result, the diaphyseal border of the epiphyseal plate is firmly cemented to the bone of the diaphysis.

The region between the diaphysis and epiphysis of a bone where the calcified matrix is replaced by bone is called the *metaphysis* (me-TAF-i-sis). The activity of the epiphyseal plate is the only mechanism by which the diaphysis can increase in length. Unlike cartilage, which can grow by both interstitial and appositional growth, bone can grow only by appositional growth.

The epiphyseal plate allows the diaphysis of the bone to increase in length until early adulthood. As the child grows, cartilage cells are produced by mitosis on the epiphyseal side of the plate. Cartilage cells are then destroyed and the cartilage is replaced by bone on the diaphyseal side of the plate. In this way, the thickness of the epiphyseal plate remains fairly constant, but the bone on the diaphyseal side increases in length. Growth in diameter occurs along with growth in length. In this process, the bone lining the marrow cavity is destroyed so that the cavity increases in diameter. At the same time, osteoblasts from the periosteum add new osseous tissue around the outer surface of the bone. Initially, diaphyseal and epiphyseal ossification produce only spongy bone. Later, by reconstruction, the outer region of spongy bone is reorganized into compact bone.

CLINICAL APPLICATION

The epiphyseal cartilage cells stop dividing and the cartilage is replaced by bone at about age 18 in females and age 20 in males. The newly formed bony structure is called the **epiphyseal line,** a remnant of the once active epiphyseal plate. With the appearance of the epiphyseal line, appositional bone growth stops. The clavicle is the last bone to stop growing.

Ossification of all bones is usually completed by age 25. Bones undergoing either intramembranous or endochondral ossification are continually remodeled from the time that initial calcification occurs until the final structure appears. **Remodeling** is the replacement of old bone tissue by new bone tissue. Compact bone is formed by the transformation of spongy bone. The diameter of a long bone is increased by the destruction of the bone closest to the marrow cavity and the construction of new bone around the outside of the diaphysis. However, even after bones have reached their adult shapes and sizes, old bone is perpetually destroyed and new osseous tissue is formed in its place.

HOMEOSTASIS

Bone shares with skin the feature of replacing itself throughout adult life. Remodeling takes place at different rates in various body regions. The distal portion of the femur (thighbone) is replaced about every 4 months. By contrast, bone in certain areas of the shaft will not be completely replaced during the individual's life. Remodeling allows worn or injured bone to be removed and replaced with new tissue. It also allows bone to serve as the body's storage area for calcium. Many other tissues in the body need calcium in order to perform their functions. For example, nerve cells need calcium for their activities, muscle needs calcium in order to contract, and blood needs calcium in order to clot. The blood continually trades off calcium with the bones, removing calcium when it and other tissues are not receiving enough of this element and resupplying the bones with dietary calcium to keep them from losing too much bone mass.

The cells believed to be responsible for the resorption (loss of a substance through a physiological or pathological process) of bone tissue are called **osteoclasts** (*clast*

= break). In the healthy adult, a delicate homeostasis is maintained between the action of the osteoclasts in removing calcium and the action of the bone-making osteoblasts in depositing calcium. Should too much new tissue be formed, the bones become abnormally thick and heavy. If too much calcium is deposited in the bone, the surplus may form thick bumps, or spurs, on the bone that interfere with movement at joints. A loss of too much tissue or calcium weakens the bones and allows them to break easily or to become very flexible.

In the process of resorption, it is believed that osteoclasts send out projections that secrete proteolytic enzymes released from lysosomes and several acids (lactic and citric). The enzymes may function by digesting the collagen and other organic substances, while the acids may cause the bone salts to dissolve in solution. It is also presumed that the osteoclastic projections may phagocytose whole fragments of collagen and bone salts.

Normal bone growth in the young and bone replacement in the adult depend on several factors. First, sufficient quantities of calcium and phosphorus, components of the primary salt that makes bone hard, must be included in the diet.

Second, the individual must obtain sufficient amounts of vitamins A, C, and D, substances that are responsible for the proper utilization of calcium and phosphorus by the body. Vitamin D, for example, is required for the absorption of calcium by the kidneys and from the digestive tract into the blood. It also has important effects on bone deposition and resorption.

Third, the body must manufacture the proper amounts of the hormones responsible for bone tissue activity (Chapter 18). Growth hormone (GH), secreted by the pituitary gland, is responsible for the general growth of bones. Too much or too little of these hormones during childhood makes the adult abnormally tall or short. Other hormones specialize in regulating the osteoclasts. Calcitonin (CT), produced by the thyroid gland, inhibits osteoclastic activity, while parathormone (PTH), synthesized by the parathyroid glands, increases osteoclastic activity. And still others, especially the sex hormones, aid osteoblastic activity and thus promote the growth of new bone. The sex hormones act as a double-edged sword. They aid in the growth of new bone, but they also bring about the degeneration of all the cartilage cells in the epiphyseal plates. Because of the sex hormones, the typical adolescent experiences a spurt of growth during puberty, when sex hormone levels start to increase. The individual then quickly completes the growth process as the epiphyseal cartilage disappears. Premature puberty can actually prevent one from reaching an average adult height because of the simultaneous premature degeneration of the plates.

DISORDERS: HOMEOSTATIC IMBALANCES ▇▇▇▇▇▇▇▇▇▇▇▇

You have already learned how important certain vitamins, minerals, and hormones are to bone growth and development. Many bone disorders result from deficiencies in vitamins or minerals or from too much or too little of the hormones that regulate bone homeostasis. Infection and tumors which disrupt homeostasis are also responsible for certain bone disorders.

Vitamin Deficiencies

Vitamin D is important to normal bone growth and maintenance. It is essential for the synthesis of a protein that transports the calcium obtained from foods across the lining of the intestine and into the extracellular fluid. When the body lacks this vitamin, it is unable to absorb calcium and phosphorus. A deficiency

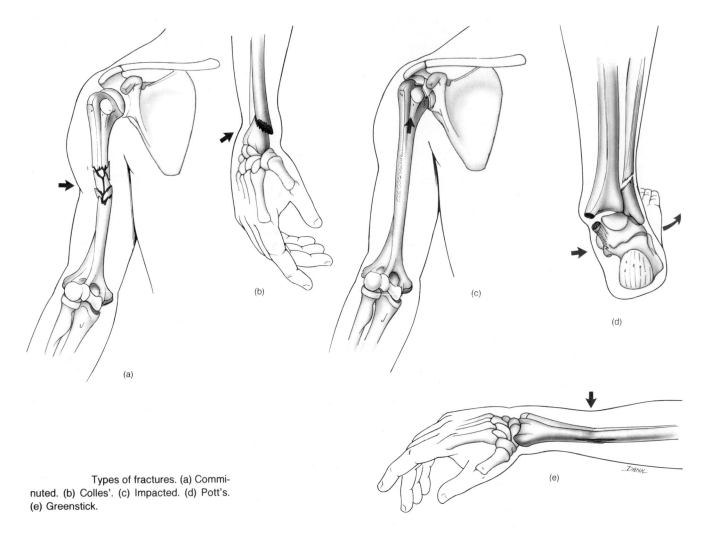

Types of fractures. (a) Comminuted. (b) Colles'. (c) Impacted. (d) Pott's. (e) Greenstick.

of vitamin D produces rickets in children and osteomalacia in adults.

Rickets

In the condition called **rickets,** there is a deficiency of vitamin D that results in an inability of the body to transport calcium and phosphorus from the digestive tract into the blood for utilization by bones. As a result, epiphyseal cartilage cells cease to degenerate and new cartilage continues to be produced. Epiphyseal cartilage thus becomes wider than normal. At the same time, the soft matrix laid down by the osteoblasts in the diaphysis fails to calcify. As a result, the bones stay soft. When the child walks, the weight of the body causes the bones in the legs to bow. Malformations of the head, chest, and pelvis also occur.

The cure and prevention of rickets consist of adding generous amounts of calcium, phosphorus, and vitamin D to the diet. Exposing the skin to the ultraviolet rays of sunlight also aids the body in manufacturing additional vitamin D.

Osteomalacia

Demineralization caused by vitamin D deficiency is called **osteomalacia** (os'-tē-ō-ma-LA-shē-a; *malacia* = softness). A deficiency of vitamin D in the adult causes the bones to give up excessive amounts of calcium and phosphorus. This loss, called *demineralization,* is especially heavy in the bones of the pelvis, legs, and spine. After the bones demineralize, the weight of the body produces a bowing of the leg bones, a shortening of the backbone, and a flattening of the pelvic bones. Osteomalacia mainly affects women who live on poor cereal diets devoid of

milk, are seldom exposed to the sun, and have repeated pregnancies that deplete the body of calcium. The condition responds to the same treatment as rickets; if the disease is severe enough and threatens life, large doses of vitamin D are given.

Osteomalacia may also result from failure to absorb fat (steatorrhea) because vitamin D is soluble in fats and calcium combines with fats. As a result, vitamin D and calcium remain with the unabsorbed fat and are lost in the feces.

Osteoporosis (os'-tē-ō-pō-RŌ-sis) is a bone disorder affecting the middle-aged and elderly, especially white females. Between puberty and the middle years, the sex hormones maintain osseous tissue by stimulating the osteoblasts to form new bone. Women produce smaller amounts of sex hormones after menopause, and both men and women produce smaller amounts during old age. As a result, the osteoblasts become less active and there is a decrease in bone mass. Osteoporosis affects the entire skeletal system, especially the spine, legs, and feet. As the spine collapses and curves, the thorax drops, and the ribs fall on the pelvic rim. This condition leads to gastrointestinal distension and an overall decrease in muscle tone.

Among the factors implicated in bone loss are a decrease in estrogen, calcium deficiency and malabsorption, vitamin D deficiency, loss of musle mass, inactivity, and high-protein diets.

Paget's disease is characterized by irregular thickening and softening of the bones. It rarely occurs in individuals under 50. Although the cause, or **etiology** (ē'-tē-OL-ō-jē), of the disease is unknown, some evidence suggests that it might be a slow virus infection. Bone-producing osteoblasts and bone-destroying osteoclasts apparently become uncoordinated. This alters the balance between bone formation and bone destruction. Paget's disease affects the skull, the pelvis, and the bones of the extremities.

The term **osteomyelitis** (os'-tē-ō-mī-i-LĪ-tis) includes all the infectious diseases of bone. These diseases may be localized or widespread and may also involve the periosteum, marrow, and cartilage. Various microorganisms may give rise to bone infection, but most often it is the bacterium *Staphylococcus aureus,* commonly called "staph." These bacteria may reach the bone by various means: the bloodstream, an injury such as a fracture, or an infection such as a sinus infection or a tooth abscess. The infection may destroy extensive areas of bone, spread to nearby joints, and, in rare cases, lead to death by producing abscesses. Antibiotics have been effective in treating the disease and in preventing it from spreading through extensive areas of bone.

In simplest terms, a **fracture** is any break in a bone. Usually, the fracture is restored to normal position by manipulation without surgery. This procedure of setting a fracture is called *closed reduction.* In other cases, the fracture must be exposed by surgery before the break is rejoined. This procedure is known as *open reduction.*

Types

Although fractures of bones of the extremities may be classified in several different ways, the following scheme is useful (Figure 6-6).

1. **Partial.** A fracture in which the break across the bone is incomplete.

2. **Complete.** A fracture in which the break across the bone is complete, so that the bone is broken into two pieces.
3. **Closed** or **simple.** A fracture in which the bone does not break through the skin.
4. **Open** or **compound.** A fracture in which the broken ends of the bone protrude through the skin.
5. **Comminuted** (KOM-i-nyoo'-ted). A fracture in which the bone is splintered at the site of impact and smaller fragments of bone are found between the two main fragments.
6. **Greenstick.** A partial fracture in which one side of the bone is broken and the other side bends; occurs only in children.
7. **Spiral.** A fracture in which the bone is usually twisted apart.
8. **Transverse.** A fracture at right angles to the long axis of the bone.
9. **Impacted.** A fracture in which one fragment is firmly driven into the other.
10. **Pott's.** A fracture of the distal end of the fibula, with serious injury of the distal tibial articulation.
11. **Colles'** (KOL-ēz). A fracture of the distal end of the radius in which the distal fragment is displaced posteriorly.
12. **Displaced.** A fracture in which the anatomical alignment of the bone fragments is not preserved.
13. **Nondisplaced.** A fracture in which the anatomical alignment of the bone fragments is preserved.

Fracture Repair

Unlike the skin, which may repair itself within days, or muscle, which may mend in weeks, a bone sometimes requires months to heal. A fractured femur, for example, may take six months to heal. Sufficient calcium to strengthen and harden new bone is deposited only gradually. Also, bone cells grow and reproduce slowly. Moreover, the blood supply to bone is decreased, which helps to explain the difficulty in the healing of an infected bone.

The following steps occur in the repair of a fracture (Figure 6-7).

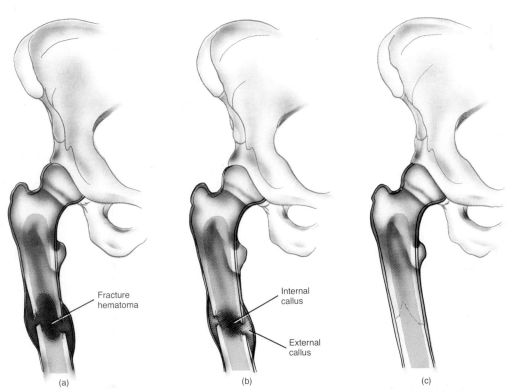

Fracture repair. (a) Formation of fracture hematoma. (b) Formation of external and internal calli. (c) Completely healed fracture.

1. As a result of the fracture, blood vessels crossing the fracture line are broken. These vessels are found in the periosteum, Haversian systems, and marrow cavity. As blood pours from the torn ends of the vessels, it coagulates and forms a clot in and about the site of the fracture. This clot, called a **fracture hematoma** (hē'-ma-TŌ-ma), usually occurs 6 to 8 hours after the injury. Since the circulation of blood ceases when the fracture hematoma forms, bone cells and periosteal cells at the fracture line die.

2. A growth of new bone tissue—a **callus**—develops in and around the fractured area. It forms a bridge between separated areas of bone. The part of the callus that forms from the osteogenic cells of the torn periosteum and develops around the outside of the fracture is called an *external callus.* The part of the callus that forms from the osteogenic cells of the endosteum and develops between the two ends of bone fragments and between the two marrow cavities is called the *internal callus.*

   Approximately 48 hours after a fracture occurs, the cells that ultimately repair the fracture become actively mitotic. These cells come from the osteogenic layer of the periosteum, the endosteum of the marrow cavity, and the bone marrow. As a result of their accelerated mitotic activity, the cells of the three regions grow toward the fracture. During the first week following the fracture, the cells of the endosteum and bone marrow form new trabeculae in the marrow cavity near the line of fracture. This is the internal callus. During the next few days, osteogenic cells of the periosteum form a collar around each bone fragment. The collar, or external callus, is replaced by trabeculae. The trabeculae of the calli are joined to living and dead portions of the original bone fragments.

3. The final phase of fracture repair is the **remodeling** of the calli. Dead portions of the original fragments are gradually resorbed by osteoclasts. Compact bone replaces spongy bone around the periphery of the fracture. In some cases, the heal-

ing is so complete that the fracture line is undetectable, even by x-ray. However, a thickened area on the surface of the bone usually remains as evidence of the fracture site.

CLINICAL APPLICATION

In the past when a fracture failed to unite, the patient could either wait, hoping that nature and time would solve the problem, or choose surgery. Today there is another alternative. The new procedure, called **pulsating electromagnetic fields (PEMFs),** involves electrotherapy to stimulate bone repair.

Treatment with PEMFs was first used clinically in 1974. Essentially, the fracture is exposed to a weak electrical current generated from coils that are fastened around the cast. Osteoblasts adjacent to the fracture site become more active metabolically in response to the stimulation of the alteration of electrical charges on the surfaces of their membranes. The increased activity apparently causes an acceleration of calcification, vascularization, and endochondral ossification and a subsequent acceleration of fracture repair. It was noted earlier that parathyroid hormone (PTH) increases osteoclastic activity and thus stimulates bone destruction. A recent hypothesis suggests that electricity also heals fractures by keeping PTH from acting on osteoclasts, thus increasing bone formation and repair.

Although the original application of PEMFs was to treat improperly healing fractures, research is now underway to determine if electrical stimulation can be used to regenerate limbs and to stop the growth of tumor cells in humans.

MEDICAL TERMINOLOGY

Achondroplasia (*a* = without; *chondro* = cartilage; *plasia* = growth)   Imperfect ossification within cartilage of long bones during fetal life; also called *fetal rickets.*

Brodie's abscess   Infection in the spongy tissue of a long bone, with a small inflammatory area.

Craniotomy (*cranium* = skull; *tome* = a cutting)   Any surgery that requires cutting through the bones surrounding the brain.

Necrosis (*necros* = death; *osis* = condition)   Death of tissues or organs; in the case of bone, results from deprivation of blood supply resulting from fracture, extensive removal of periosteum in surgery, exposure to radioactive substances, or other causes.

Osteitis (*osteo* = bone)   Inflammation or infection of bone.

Osteoarthritis (*arthro* = joint)   A degenerative condition of bone and also the joint.

Osteoblastoma (*oma* = tumor)   A benign tumor of the osteo blasts.

Osteochondroma (*chondro* = cartilage)   A benign tumor of the bone and cartilage.

Osteoma   A benign bone tumor.

Osteosarcoma (*sarcoma* = connective tissue tumor)   A malignant tumor composed of osseous tissue.

Pott's disease   Inflammation of the backbone, caused by the microorganism that produces tuberculosis.

STUDY OUTLINE

Functions (p. 124)

1. The skeletal system consists of all bones attached at joints and cartilage between joints.

2. The functions of the skeletal system include support, protection, leverage, mineral storage, and blood cell production.

Histology (p. 124)

1. Osseous tissue consists of widely separated cells surrounded by large amounts of intercellular substance. The intercellular substance contains collagenous fibers and abundant hydroxyapatites (mineral salts).

2. Parts of a typical long bone are the diaphysis (shaft), epiphyses (ends), metaphysis, articular cartilage, periosteum, medullary or marrow cavity, and endosteum.

3. Compact (dense) bone consists of Haversian systems with little space between them. Compact bone lies over spongy bone and composes most of the bone tissue of the diaphyses. Functionally, compact bone protects, supports, and resists stress.

4. Spongy (cancellous) bone consists of trabeculae surrounding many red marrow-filled spaces. It forms most of the structure of short, flat, and irregular bones, and the epiphyses of long bones. Functionally, spongy bone stores marrow and provides some support.

Ossification (p. 127)

1. Bone forms by a process called ossification or osteogenesis, which begins when mesenchymal cells become transformed into osteoblasts.
2. The process begins during the sixth or seventh week of embryonic life and continues throughout adulthood. The two types of ossification, intramembranous and endochondral, involve the replacement of a preexisting connective tissue with bone.
3. Intramembranous ossification occurs within fibrous membranes of the embryo and the adult.
4. Endochondral ossification occurs within a cartilage model. The primary ossification center of a long bone is in the diaphysis. Cartilage degenerates, leaving cavities that merge to form the marrow cavity. Osteoblasts lay down bone. Next, ossification occurs in the epiphyses, where bone replaces cartilage, except for the epiphyseal plate.
5. The anatomical zones of the epiphyseal plate are the zones of reserve cartilage, proliferating cartilage, hypertrophic cartilage, and calcified matrix.
6. Because of the activity of the epiphyseal plate, the diaphysis of a bone increases in length by appositional growth.
7. In both types of ossification, spongy bone is laid down first. Compact bone is later reconstructed from spongy bone.

Homeostasis (p. 131)

1. The homeostasis of bone growth and development depends on a balance between bone formation and resorption.
2. Old bone is constantly destroyed by osteoclasts, while new bone is constructed by osteoblasts. This process is called remodeling.
3. Normal growth depends on calcium, phosphorus, and vitamins (A, C, and D) and is controlled by hormones that are responsible for bone mineralization and resorption.

Disorders: Homeostatic Imbalances (p. 131)

1. Rickets is a vitamin D deficiency in children in which the body does not absorb calcium and phosphorus. The bones soften and bend under the body's weight.
2. Osteomalacia is a vitamin D deficiency in adults that leads to demineralization.
3. Osteoporosis is a decrease in the amount and strength of bone tissue due to decreases in hormone output.
4. Paget's disease is the irregular thickening and softening of bones, apparently related to an imbalance between osteoclast and osteoblast activities.
5. Osteomyelitis is a term for the infectious diseases of bones, marrow, and periosteum. It is frequently caused by "staph" bacteria.
6. A fracture is any break in a bone.
7. The types of fractures include: partial, complete, simple, compound, comminuted, greenstick, spiral, transverse, impacted, Pott's, Colles', displaced, and nondisplaced.
8. Fracture repair consists of forming a fracture hematoma, forming a callus, and remodeling.
9. Treatment by pulsating electromagnetic fields (PEMFs) has provided dramatic results in healing fractures that would otherwise not have mended properly. Its application for limb regeneration and stopping the growth of tumor cells is being investigated.

REVIEW QUESTIONS

1. Define the skeletal system. What are its five principal functions?
2. Why is osseous tissue considered a connective tissue? What is its composition?
3. Diagram the parts of a long bone, and list the functions of each part.
4. Distinguish between spongy and compact bone in terms of general appearance, location, and function.
5. Diagram the microscopic appearance of compact bone, and indicate the functions of the various components.
6. How does spongy bone differ histologically from compact bone?
7. What is meant by ossification? Describe the initial events of ossification.
8. Distinguish between the two principal kinds of ossification on the basis of the preexisting connective tissue present.
9. Outline the major events involved in intramembranous and endochondral ossification and explain the principal differences.
10. Describe the histology of the various zones of the epiphyseal plate. How does the plate grow?
11. What is the significance of the epiphyseal line?
12. Define remodeling.
13. How does osteoblast activity in balance with osteoclast activity demonstrate the homeostasis of bone?
14. List the primary factors involved in bone growth and replacement.
15. Distinguish between rickets and osteomalacia. What do the two diseases have in common?
16. What are the principal symptoms of osteoporosis, Paget's disease, and osteomyelitis? What is the etiology of each?
17. What is a fracture? Distinguish several principal kinds.
18. Outline the three basic steps involved in fracture repair.
19. Explain the principal of pulsating electromagnetic fields (PEMFs) in the repair of fractures.
20. Refer to the glossary of medical terminology associated with the skeletal system. Be sure that you can define each term.

STUDENT OBJECTIVES

- Define the four principal types of bones in the skeleton.

- Describe the various markings on the surfaces of bones.

- Relate the structure of the marking to its function.

- List the components of the axial and appendicular skeleton.

- Identify the bones of the skull and the major markings associated with each.

- Identify the principal sutures and fontanels of the skull.

- Identify the paranasal sinuses of the skull.

- Identify the principal foramina of the skull.

- Identify the bones of the vertebral column and their principal markings.

- List the defining characteristics and curves of each region of the vertebral column.

- Identify the bones of the thorax and their principal markings.

- Contrast herniated (slipped) disc, curvatures, spina bifida, and fractures of the vertebral column as disorders associated with the skeletal system.

7 The Skeletal System: The Axial Skeleton

The skeletal system forms the framework of the body. For this reason, a familiarity with the names, shapes, and positions of individual bones will help you to understand some of the other organ systems. For example, movements such as throwing a ball, typing, and walking require the coordinated use of bones and muscles. To understand how muscles produce different movements, you need to learn the parts of the bones to which the muscles attach. The respiratory system is also highly dependent on bone structure. The bones in the nasal cavity form a series of passageways that help clean, moisten, and warm inhaled air. Furthermore, the bones of the thorax are specially shaped and positioned so the chest can expand during inhalation. Many bones also serve as landmarks to students of anatomy as well as to surgeons. As you will see, bony landmarks can be used to locate the outlines of the lungs and heart, abdominal and pelvic viscera, and structures within the skull. Blood vessels and nerves often run parallel to bones. These structures can be located more easily if the bone is identified first.

We shall study bones by examining the various regions of the body. We shall look at the skull first and see how the bones of the skull relate to each other. We shall then move on to the vertebral column and the chest. This regional approach will allow you to see how all the many bones of the body relate to each other.

TYPES OF BONES

Almost all the bones of the body may be classified into four principal types on the basis of shape: long, short, flat, and irregular. **Long bones** have greater length than width and consist of a diaphysis and two epiphyses. They are slightly curved for strength. A curved bone is structurally designed to absorb the stress of the body weight at several different points so the stress is evenly distributed. If such bones were straight, the weight of the body would be unevenly distributed and the bone would easily fracture. Examples of long bones include bones of the thighs, legs, toes, arms, forearms, and fingers. Figure 6-1a shows the parts of a long bone.

Short bones are somewhat cube-shaped and nearly equal in length and width. Their texture is spongy except at the surface, where there is a thin layer of compact bone. Examples of short bones are the wrist and ankle bones.

Flat bones are generally thin and composed of two more or less parallel plates of compact bone enclosing a layer of spongy bone. The term *diploe* is applied to the spongy bone of the cranial bones. Flat bones afford considerable protection and provide extensive areas for muscle attachment. Examples of flat bones include the cranial bones, which protect the brain, the sternum and ribs, which protect organs in the thorax, and the scapulas.

Irregular bones have complex shapes and cannot be grouped into any of the three categories just described. They also vary in the amount of spongy and compact bone present. Such bones are the vertebrae and certain facial bones.

There are two additional types of bones which are not included in this classification by shape. **Wormian, or sutural** (SOO-chur-al), **bones** are small clusters of bones between the joints of certain cranial bones (see Figure 7-2e). Their number varies greatly from person to person. **Sesamoid bones** are small bones in tendons where considerable pressure develops, for instance, in the wrist. These, like the Wormian bones, are also variable in number. Two sesamoid bones, the patellas, or kneecaps, are present in all individuals.

SURFACE MARKINGS

The surfaces of bones reveal various structural features adapted to specific functions. These features are called **markings.** Long bones that bear a great deal of weight have large, rounded ends that can form sturdy joints. Other bones have depressions that receive the rounded ends. Rough areas serve as points of attachment for muscles, tendons, and ligaments. Grooves in the surfaces of bones provide for the passage of blood vessels. Openings occur where blood vessels and nerves pass through the bone. Exhibit 7-1 describes the different markings and their functions.

DIVISIONS OF THE SKELETAL SYSTEM

The adult human skeleton usually consists of 206 bones grouped in two principal divisions: the **axial** and the **appendicular.** The longitudinal *axis,* or center, of the human body is a straight line that runs vertically along the body's center of gravity. This imaginary line runs through the head and down to the space between the feet. The midsaggital section is drawn through this line. The axial division of the skeleton consists of the bones that lie around the axis: ribs, breastbone, bones of the skull, and backbone.

The appendicular division contains the bones of the free *appendages,* which are the upper and lower extremities, plus the bones called *girdles,* which connect the free appendages to the axial skeleton.

The 80 bones of the axial division and the 126 bones of the appendicular division are typically grouped as shown in Exhibit 7-2.

Now that you understand how the skeleton is organized into axial and appendicular divisions, refer to Figure 7-1 to see how the two divisions are joined to form the skeleton. The bones of the axial skeleton are shown in yellow. Be certain to locate the following regions of the skeleton: skull, cranium, face, hyoid bone, vertebral column, thorax, shoulder girdle, upper extremity, pelvic girdle, and lower extremity.

SKULL

The **skull,** which contains 22 bones, rests on the superior end of the vertebral column and is composed of two sets of bones: cranial bones and facial bones. The **cranial bones** enclose and protect the brain and the organs of sight, hearing, and balance. The 8 cranial bones are the frontal bone, parietal bones (2), temporal bones (2), occipital bone, sphenoid bone, and ethmoid bone. There are 14 **facial bones:** nasal bones (2), maxillae (2), zygomatic bones (2), mandible, lacrimal bones (2), palatine bones (2), inferior nasal conchae (2), and vomer. Be sure you can locate all the skull bones in the anterior, lateral, median, and posterior views of the skull (Figure 7-2).

EXHIBIT 7-1
BONE MARKINGS

MARKING	DESCRIPTION	EXAMPLE
DEPRESSIONS AND OPENINGS		
Fissure (FISH-ur)	A narrow, cleftlike opening between adjacent parts of bones through which blood vessels or nerves pass.	Superior orbital fissure of the sphenoid bone (Figure 7-2).
Foramen (fō-RĀ-men; *foramen* = hole)	An opening through which blood vessels, nerves, or ligaments pass.	Infraorbital foramen of the maxilla (Figure 7-2).
Meatus (mē-Ā-tus; *meatus* = canal)	A tubelike passageway running within a bone.	External auditory meatus of the temporal bone (Figure 7-2).
Paranasal sinus (*sin* = cavity)	An air-filled cavity within a bone connected to the nasal cavity.	Frontal sinus of the frontal bone (Figure 7-8).
Groove or sulcus (*sulcus* = ditchlike groove)	A furrow or groove that accommodates a soft structure such as a blood vessel, nerve, or tendon.	Intertubercular sulcus of the humerus (Figure 8-4).
Fossa (*fossa* = basinlike depression)	A depression in or on a bone.	Mandibular fossa of the temporal bone (Figure 7-4).
PROCESSES	Any prominent projection.	Mastoid process of the temporal bone (Figure 7-2).
Processes that form joints		
Condyle (KON-dīl; *condylus* = knucklelike process)	A large, articular prominence.	Medial condyle of the femur (Figure 8-10).
Head	A rounded, articular projection supported on the constricted portion (neck) of a bone.	Head of the femur (Figure 8-10).
Facet	A smooth, flat surface.	Articular facet for the tubercle of rib on a vertebra (Figure 7-18).
Processes to which tendons, ligaments, and other connective tissues attach		
Tubercle (TOO-ber-kul; *tube* = knob)	A small, rounded process.	Greater tubercle of the humerus (Figure 8-4).
Tuberosity	A large, rounded, usually roughened process.	Ischial tuberosity of the hipbone (Figure 8-8).
Trochanter (trō-KAN-ter)	A large, blunt projection found only on the femur.	Greater trochanter of the femur (Figure 8-10).
Crest	A prominent border or ridge on a bone.	Iliac crest of the hipbone (Figure 8-7).
Line	A less prominent ridge.	Linea aspera of the femur (Figure 8-10).
Spinous process or spine	A sharp, slender process.	Spinous process of a vertebra (Figure 7-12).
Epicondyle (*epi* = above)	A prominence above a condyle.	Medial epicondyle of the femur (Figure 8-10).

SUTURES

A **suture** (SOO-chur), meaning seam or stitch, is an immovable joint found only between skull bones. Very little connective tissue is found between the bones of the suture. Four prominent skull sutures are:

1. Coronal suture between the frontal bone and the two parietal bones.
2. Sagittal suture between the two parietal bones.
3. Lambdoidal (lam-DOY-dal) **suture** between the parietal bones and the occipital bone.
4. Squamosal (skwa-MŌ-sal) **suture** between the parietal bones and the temporal bones.

Refer to Figures 7-2 and 7-3 for the locations of these sutures. Several other sutures are also shown. Their names are descriptive of the bones they connect. For example, the frontonasal suture is between the frontal bone and the nasal bones. These sutures are indicated in Figures 7-2 to 7-5.

FONTANELS

The "skeleton" of a newly formed embryo consists of cartilage or fibrous membrane structures shaped like bones. Gradually the cartilage or fibrous membrane is replaced by bone. At birth, membrane-filled spaces called **fontanels** (fon'-ta-NELZ; = little fountains), are found between cranial bones (Figure 7-3). These "soft spots" are areas where the bone-making process is not yet complete. They allow the skull to be compressed during birth.

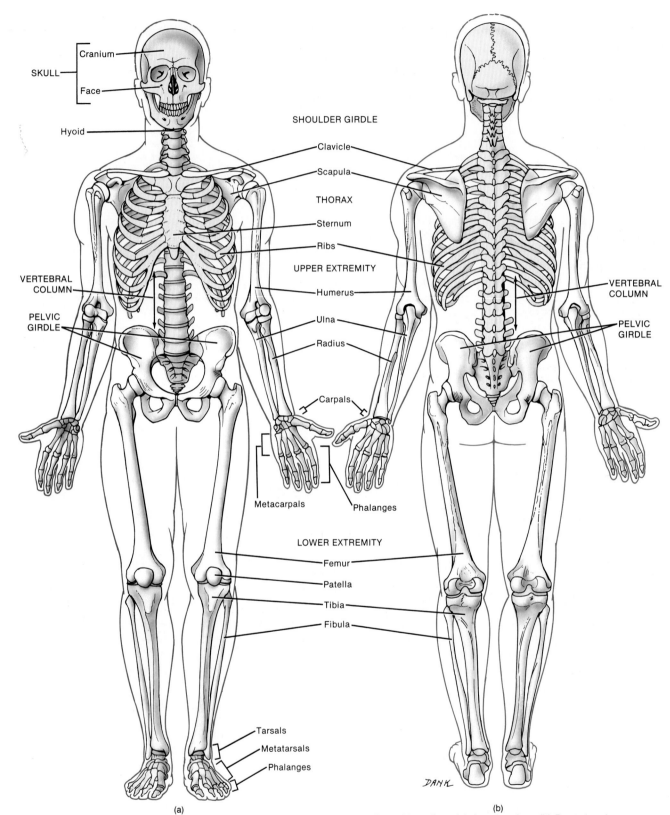

FIGURE 7-1 Divisions of the skeletal system. The axial skeleton is indicated in yellow. (a) Anterior view. (b) Posterior view.

Physicians find the fontanels helpful in determining the position of the infant's head prior to delivery. Although an infant may have many fontanels at birth, the form and location of six are fairly constant.

The **anterior (frontal) fontanel** is located between the angles of the two parietal bones and the two segments of the frontal bone. This fontanel is roughly diamond-shaped, and it is the largest of the six fontanels. It usually closes 18 to 24 months after birth.

The **posterior (occipital;** ok-SIP-i-tal) **fontanel** is situated between the two parietal bones and the occipital bone. This diamond-shaped fontanel is considerably

EXHIBIT 7-2

DIVISIONS OF THE SKELETAL SYSTEM

REGIONS OF THE SKELETON	NUMBER OF BONES	REGIONS OF THE SKELETON	NUMBER OF BONES
AXIAL SKELETON		**Upper extremities**	
Skull		Humerus	2
Cranium	8	Ulna	2
Face	14	Radius	2
Hyoid	1	Carpals	16
Auditory ossicles		Metacarpals	10
(3 in each ear)*	6	Phalanges	28
Vertebral column	26	**Pelvic girdle**	
Thorax		Coxal, hip, or pelvic bone	2
Sternum	1	**Lower extremities**	
Ribs	24	Femur	2
	80	Fibula	2
		Tibia	2
APPENDICULAR SKELETON		Patella	2
Pectoral (shoulder) girdles		Tarsals	14
Clavicle	2	Metatarsals	10
Scapula	2	Phalanges	28
			126

* Although the auditory ossicles are not considered part of the axial or appendicular skeleton, but rather as a separate group of bones, they are placed with the axial skeleton for convenience. They are discussed in detail in Chapter 17.

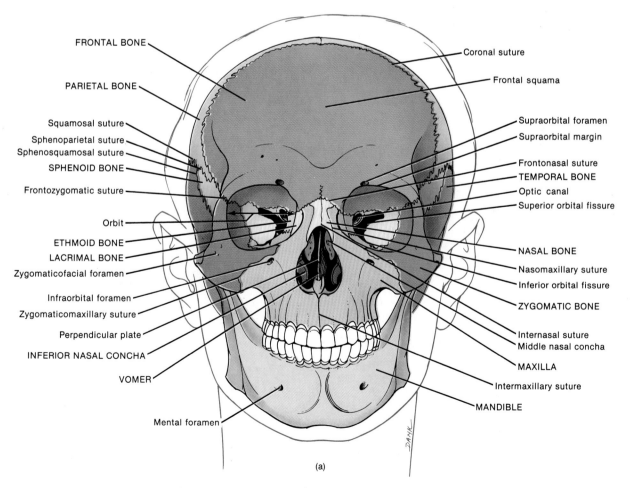

FRONTAL BONE — Coronal suture — Frontal squama

PARIETAL BONE — Supraorbital foramen — Supraorbital margin

Squamosal suture — Frontonasal suture — TEMPORAL BONE

Sphenoparietal suture — Optic canal — Superior orbital fissure

Sphenosquamosal suture — SPHENOID BONE

Frontozygomatic suture

Orbit — NASAL BONE

ETHMOID BONE — Nasomaxillary suture — Inferior orbital fissure

LACRIMAL BONE — ZYGOMATIC BONE

Zygomaticofacial foramen

Infraorbital foramen — Internasal suture — Middle nasal concha

Zygomaticomaxillary suture — MAXILLA

Perpendicular plate — Intermaxillary suture

INFERIOR NASAL CONCHA — MANDIBLE

VOMER

Mental foramen

(a)

FIGURE 7-2 Skull. (a) Anterior view.

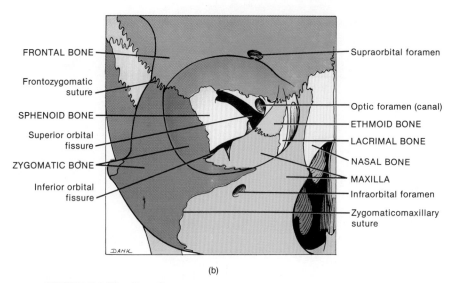

FRONTAL BONE

Frontozygomatic suture

SPHENOID BONE

Superior orbital fissure

ZYGOMATIC BONE

Inferior orbital fissure

Supraorbital foramen

Optic foramen (canal)

ETHMOID BONE

LACRIMAL BONE

NASAL BONE

MAXILLA

Infraorbital foramen

Zygomaticomaxillary suture

DANK

(b)

FIGURE 7-2 (*Continued*) Skull. (b) Details of the right orbit in anterior view.

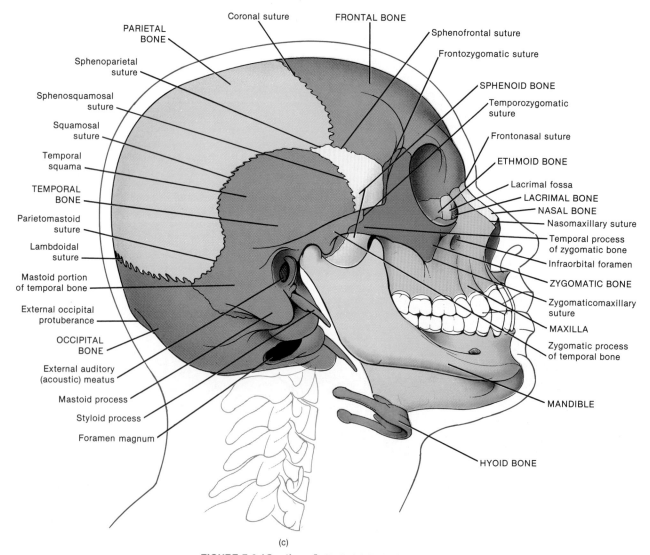

Coronal suture

FRONTAL BONE

PARIETAL BONE

Sphenofrontal suture

Frontozygomatic suture

Sphenoparietal suture

SPHENOID BONE

Sphenosquamosal suture

Temporozygomatic suture

Squamosal suture

Frontonasal suture

Temporal squama

ETHMOID BONE

TEMPORAL BONE

Lacrimal fossa

LACRIMAL BONE

Parietomastoid suture

NASAL BONE

Nasomaxillary suture

Lambdoidal suture

Temporal process of zygomatic bone

Mastoid portion of temporal bone

Infraorbital foramen

ZYGOMATIC BONE

External occipital protuberance

Zygomaticomaxillary suture

OCCIPITAL BONE

MAXILLA

External auditory (acoustic) meatus

Zygomatic process of temporal bone

Mastoid process

Styloid process

MANDIBLE

Foramen magnum

HYOID BONE

(c)

FIGURE 7-2 (*Continued*) Skull. (c) Right lateral view.

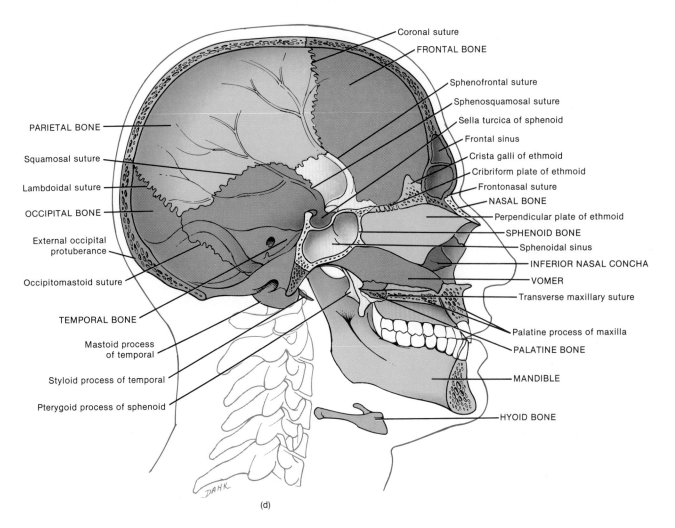

Coronal suture
FRONTAL BONE
Sphenofrontal suture
Sphenosquamosal suture
Sella turcica of sphenoid
Frontal sinus
Crista galli of ethmoid
Cribriform plate of ethmoid
Frontonasal suture
NASAL BONE
Perpendicular plate of ethmoid
SPHENOID BONE
Sphenoidal sinus
INFERIOR NASAL CONCHA
VOMER
Transverse maxillary suture
Palatine process of maxilla
PALATINE BONE
MANDIBLE
HYOID BONE

PARIETAL BONE
Squamosal suture
Lambdoidal suture
OCCIPITAL BONE
External occipital protuberance
Occipitomastoid suture
TEMPORAL BONE
Mastoid process of temporal
Styloid process of temporal
Pterygoid process of sphenoid

(d)

FIGURE 7-2 (Continued) Skull. (d) Median view.

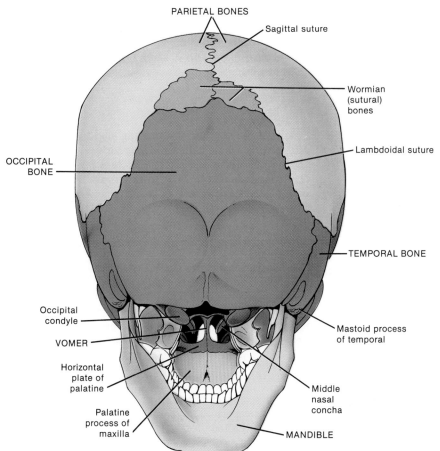

PARIETAL BONES
Sagittal suture
Wormian (sutural) bones
Lambdoidal suture
TEMPORAL BONE
Mastoid process of temporal
Middle nasal concha
MANDIBLE

OCCIPITAL BONE
Occipital condyle
VOMER
Horizontal plate of palatine
Palatine process of maxilla

(e)

FIGURE 7-2 (Continued) Skull. (e) Posterior view.

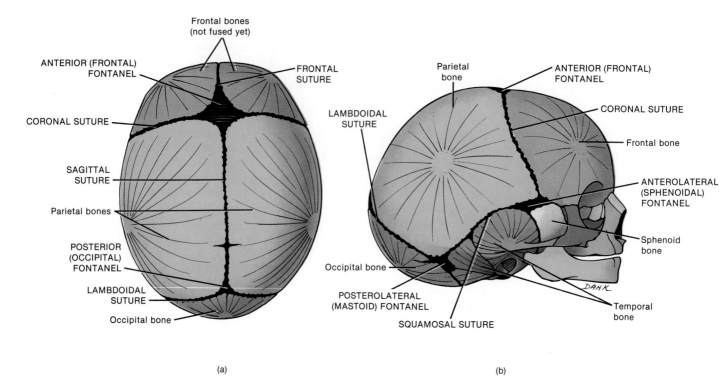

FIGURE 7-3 Fontanels of the skull at birth. (a) Superior view. (b) Right lateral view.

smaller than the anterior fontanel. It generally closes about 2 months after birth.

The **anterolateral (sphenoidal;** sfē-NOY-dal) **fontanels** are paired. One is located on each side of the skull at the junction of the frontal, parietal, temporal, and sphenoid bones. These fontanels are quite small and irregular in shape. They normally close 3 months after birth.

The **posterolateral (mastoid) fontanels** are also paired. One is situated on each side of the skull at the junction of the parietal, occipital, and temporal bones. These fontanels are irregularly shaped. They begin to close 1 or 2 months after birth, but closure is not generally complete until 12 months.

FRONTAL BONE

The **frontal bone** forms the forehead, the anterior part of the cranium; the roofs of the *orbits* (eye sockets); and most of the anterior part of the cranial floor. Soon after birth the left and right parts of the frontal bone are united by a suture. The suture usually disappears by age 6. If, however, the suture persists throughout life, it is referred to as the **metopic suture.**

If you examine the anterior and lateral views of the skull in Figure 7-2, you will note the **frontal squama** (SKWĀ-ma), or **vertical plate** (*squam* = scale). This scalelike plate, which corresponds to the forehead, gradually slopes down from the coronal suture, then turns abruptly downward.

A thickening of the frontal bone is called the **supraorbital margin.** From this margin the frontal bone extends posteriorly to form the roof of the orbit and part of the floor of the cranial cavity.

CLINICAL APPLICATION

Just above the supraorbital margin is a relatively sharp ridge that overlies the frontal sinus. A blow to the ridge frequently lacerates the skin over it, resulting in bleeding. Bruising of the skin over the ridge causes tissue fluid and blood to accumulate in the surrounding connective tissue and gravitate into the upper eyelid. The resulting swelling and discoloration is called a **"black eye."**

Within the supraorbital margin, slightly medial to its midpoint, is a hole called the **supraorbital foramen.** The supraorbital nerve and artery pass through this foramen. The **frontal sinuses** lie deep to the frontal squama. These mucus-lined cavities act as sound chambers which give the voice resonance.

PARIETAL BONES

The two **parietal bones** (pa-RĪ-i-tal; *paries* = wall) form the greater portion of the sides and roof of the cranial cavity (see Figure 7-2). The internal surfaces of the bones contain many eminences and depressions that accommodate the blood vessels supplying the outer meninx (covering) of the brain called the *dura mater.*

TEMPORAL BONES

The two **temporal bones** form the inferior sides of the cranium and part of the cranial floor. The term *tempora* pertains to the temples.

In the lateral view of the skull in Figure 7-2, notice the **squama** or **squamous portion**—a thin, large, expanded area that forms the anterior and superior part of the temple. Projecting from the inferior portion of the squama is the **zygomatic process,** which articulates with the temporal process of the zygomatic bone. The zygomatic process of the temporal bone and the temporal process of the zygomatic bone constitute the **zygomatic arch.**

At the floor of the cranial cavity, shown in Figure 7-5, is the **petrous portion** of the temporal bone. This portion is triangular and located at the base of the skull between the sphenoid and occipital bones. The petrous portion contains the internal ear, the essential part of the organ of hearing. It also contains the **carotid foramen (canal)** through which the internal carotid artery passes (see Figure 7-4). Posterior to the carotid foramen and anterior to the occipital bone is the **jugular foramen (fossa)** through which the internal jugular vein and the glossopharyngeal (IX) nerve, vagus (X) nerve, and accessory (XI) nerve pass. (As you will see later, the Roman numerals associated with cranial nerves indicate the order in which the nerves arise from the brain, from front to back.)

Between the squamous and petrous portions is a socket called the **mandibular fossa.** Anterior to the mandibular fossa is a rounded eminence, the **articular tubercle.** The mandibular fossa and articular tubercle articulate with the condylar process of the mandible (lower jawbone) to form the temporomandibular joint. The mandibular fossa and articular tubercle are seen best in Figure 7-4.

In the lateral view of the skull in Figure 7-2, you will see the **mastoid portion** of the temporal bone, located posterior and inferior to the external auditory meatus, or ear canal. In the adult, this portion of the bone contains a number of **mastoid air "cells."** These air spaces are separated from the brain only by thin bony partitions.

CLINICAL APPLICATION

If **mastoiditis,** the inflammation of these bony cells, occurs, the infection may spread to the brain or its outer covering. The mastoid air cells do not drain as do the paranasal sinuses.

The **mastoid process** is a rounded projection of the temporal bone posterior to the external auditory meatus. It serves as a point of attachment for several neck muscles. Near the posterior border of the mastoid process is the **mastoid foramen** through which a vein (emissary) to the transverse sinus and a small branch of the occipital artery to the dura mater pass. The **external auditory meatus** is the canal in the temporal bone that leads to the middle ear. The **internal acoustic meatus** is superior to the jugular foramen (see Figure 7-5a). It transmits the facial (VII) and vestibulocochlear (VIII) nerves and the internal auditory artery. The **styloid process** projects downward from the undersurface of the temporal bone and serves as a point of attachment for muscles and ligaments of the tongue and neck.

OCCIPITAL BONE

The **occipital** (ok-SIP-i-tal) **bone** forms the posterior part and a prominent portion of the base of the cranium (Figure 7-4).

The **foramen magnum** is a large hole in the inferior part of the bone through which the medulla oblongata and its membranes, the spinal portion of the accessory (XI) nerve, and the vertebral and spinal arteries pass.

The **occipital condyles** are oval processes with convex surfaces, one on either side of the foramen magnum, which articulate with depressions on the first cervical vertebra.

The **external occipital protuberance** is a prominent projection on the posterior surface of the bone just superior to the foramen magnum. You can feel this structure as a definite bump on the back of your head, just above your neck. The protuberance is also visible in Figure 7-2d.

SPHENOID BONE

The **sphenoid** (SFĒ-noyd) **bone** is situated at the middle part of the base of the skull (Figure 7-5). The combining form *spheno* means wedge. This bone is referred to as the keystone of the cranial floor because it articulates with all the other cranial bones. If you view the floor of the cranium from above, you will note that the sphenoid articulates with the temporal bones anteriorly and the occipital bone posteriorly. It lies posterior and slightly superior to the nasal cavities and forms part of the floor and sidewalls of the eye socket. The shape of the sphenoid is frequently described as a bat with outstretched wings.

The **body** of the sphenoid is the cubelike central portion between the ethmoid and occipital bones. It contains the **sphenoidal sinuses,** which drain into the nasal cavity (see Figure 7-8). On the superior surface of the sphenoid body is a depression called the **sella turcica** (SEL-a-TUR-si-ka = Turk's saddle). This depression houses the pituitary gland.

The **greater wings** of the sphenoid are lateral projections from the body and form the anterolateral floor of the cranium. The greater wings also form part of the lateral wall of the skull just anterior to the temporal bone. The **lesser wings** are anterior and superior to the greater wings. They form part of the floor of the cranium and the posterior part of the **orbit,** or eye socket.

Between the body and lesser wing, you can locate the **optic foramen** through which the optic (II) nerve and ophthalmic artery pass. Lateral to the body between the greater and lesser wings is a somewhat triangular slit called the **superior orbital fissure.** It is an opening for the oculomotor (III) nerve, trochlear (IV) nerve, ophthalmic branch of the trigeminal (V) nerve, and abducens (VI) nerve. This fissure may also be seen in the anterior view of the skull in Figure 7-2.

On the inferior part of the sphenoid bone you can see the **pterygoid** (TER-i-goyd) **processes.** These structures project inferiorly from the points where the body and greater wings unite. The pterygoid processes form part of the lateral walls of the nasal cavities.

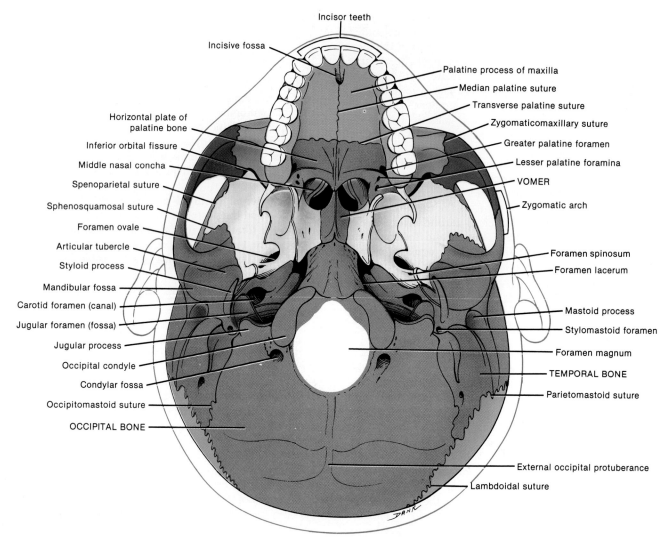

FIGURE 7-4 Skull in inferior view.

ETHMOID BONE

The **ethmoid bone** is a light, spongy bone located in the anterior part of the floor of the cranium between the orbits. It is anterior to the sphenoid and posterior to the nasal bones (Figure 7-6). This bone forms part of the anterior portion of the cranial floor, the medial wall of the orbits, the superior portions of the nasal septum, or partition, and most of the sidewalls of the nasal roof. The ethmoid is the principal supporting structure of the nasal cavities.

Its **lateral masses** or **labyrinths** compose most of the wall between the nasal cavities and the orbits. They contain several air spaces, or "cells," ranging in number from 3 to 18. It is from these "cells" that the bone derives its name (*ethmos* = sieve). The ethmoid "cells" together form the **ethmoidal sinuses.** The sinuses are shown in Figure 7-8. The **perpendicular plate** forms the superior portion of the nasal septum (see Figure 7-7). The **cribriform (horizontal) plate,** lies in the anterior floor of the cranium and forms the roof of the nasal cavity. Projecting upward from the horizontal plate is a triangular process

called the **crista galli** (= cock's comb). This structure serves as a point of attachment for the membranes that cover the brain.

The labyrinths contain two thin, scroll-shaped bones on either side of the nasal septum. These are called the **superior nasal concha** (KONG-ha; *concha* = shell) and the **middle nasal concha.** The conchae allow for the efficient circulation and filtration of inhaled air before it passes into the trachea, the bronchi, and the lungs.

NASAL BONES

The paired **nasal bones** are small, oblong bones that meet at the middle and superior part of the face (see Figures 7-2 and 7-7). Their fusion forms part of the bridge of the nose. The inferior portion of the nose, indeed the major portion, consists of cartilage.

MAXILLAE

The paired maxillary bones unite to form the upper jawbone (Figure 7-7). The **maxillae** (mak-SIL-ē) articulate

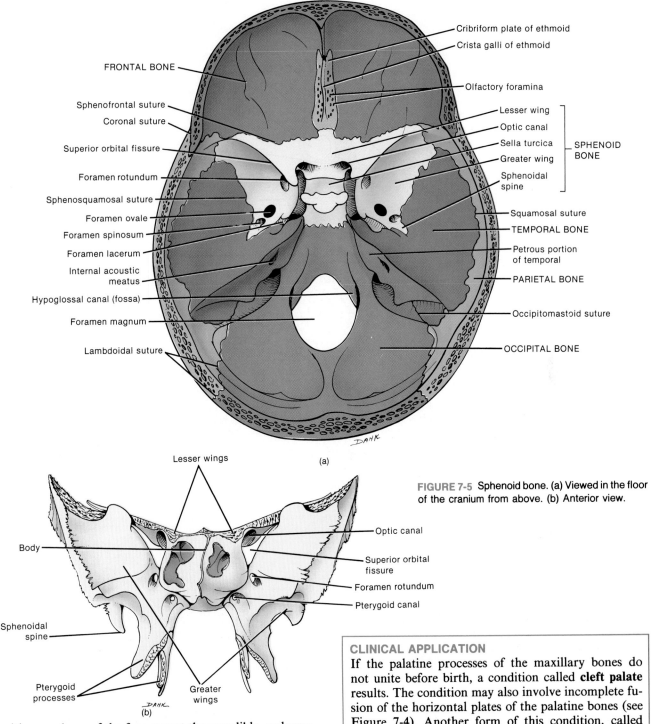

FRONTAL BONE

Sphenofrontal suture
Coronal suture
Superior orbital fissure
Foramen rotundum
Sphenosquamosal suture
Foramen ovale
Foramen spinosum
Foramen lacerum
Internal acoustic meatus
Hypoglossal canal (fossa)
Foramen magnum
Lambdoidal suture

Cribriform plate of ethmoid
Crista galli of ethmoid
Olfactory foramina
Lesser wing
Optic canal
Sella turcica
Greater wing
Sphenoidal spine

SPHENOID BONE

Squamosal suture
TEMPORAL BONE
Petrous portion of temporal
PARIETAL BONE
Occipitomastoid suture
OCCIPITAL BONE

(a)

Lesser wings
Body
Sphenoidal spine
Pterygoid processes
Greater wings

Optic canal
Superior orbital fissure
Foramen rotundum
Pterygoid canal

(b)

FIGURE 7-5 Sphenoid bone. (a) Viewed in the floor of the cranium from above. (b) Anterior view.

with every bone of the face except the mandible, or lower jawbone. They form part of the floors of the orbits, part of the roof of the mouth (most of the hard palate), and part of the lateral walls and floor of the nasal cavities.

Each maxillary bone contains a **maxillary sinus** that empties into the nasal cavity (see Figure 7-8). The **alveolar** (al-VĒ-ō-lar) **process** (*alveolus* = hollow) contains the bony sockets or **alveoli** into which the maxillary (upper) teeth are set. The **palatine process** is a horizontal projection of the maxilla that forms the anterior three-fourths of the hard palate, or anterior portion of the roof of the oral cavity. The two portions of the maxillary bones unite, and the fusion is normally completed before birth.

CLINICAL APPLICATION

If the palatine processes of the maxillary bones do not unite before birth, a condition called **cleft palate** results. The condition may also involve incomplete fusion of the horizontal plates of the palatine bones (see Figure 7-4). Another form of this condition, called **cleft lip,** involves a split in the upper lip. Cleft lip is often associated with cleft palate. Depending on the extent and position of the cleft, speech and swallowing may be affected. A surgical procedure can sometimes improve the condition.

A fissure associated with the maxilla and sphenoid bone is the **inferior orbital fissure.** It is located between the greater wing of the sphenoid and the maxilla (see Figure 7-4). It transmits the maxillary branch of the trigeminal (V) nerve, the infraorbital vessels, and the zygomatic nerve.

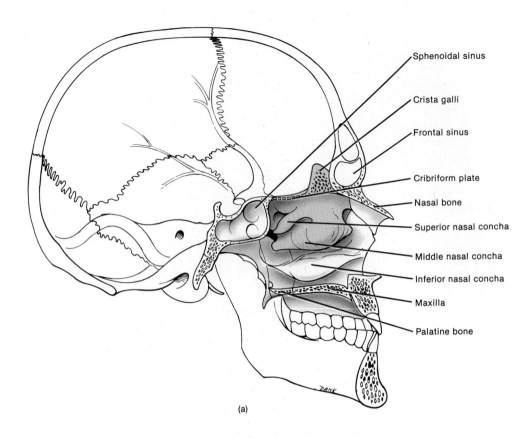

Sphenoidal sinus

Crista galli

Frontal sinus

Cribriform plate

Nasal bone

Superior nasal concha

Middle nasal concha

Inferior nasal concha

Maxilla

Palatine bone

(a)

FIGURE 7-6 Ethmoid bone. (a) Median view showing the ethmoid bone on the inner aspect of the left part of the skull.

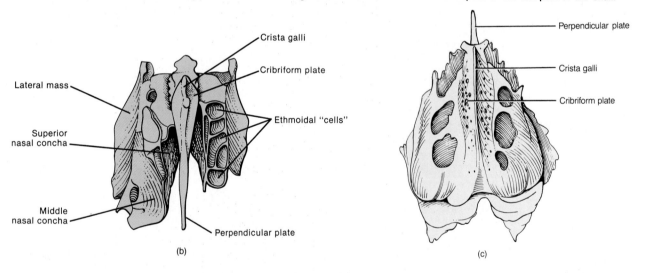

Crista galli

Cribriform plate

Lateral mass

Superior nasal concha

Ethmoidal "cells"

Middle nasal concha

Perpendicular plate

(b)

Perpendicular plate

Crista galli

Cribriform plate

(c)

FIGURE 7-6 (Continued) Ethmoid bone. (b) Anterior view. A frontal section has been made through the left side to expose the ethmoidal "cells." (c) Superior view. (d) Highly diagrammatic representation showing the approximate position of the ethmoid bone in the skull in anterior view.

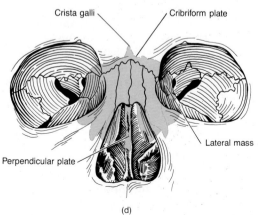

Crista galli

Cribriform plate

Perpendicular plate

Lateral mass

(d)

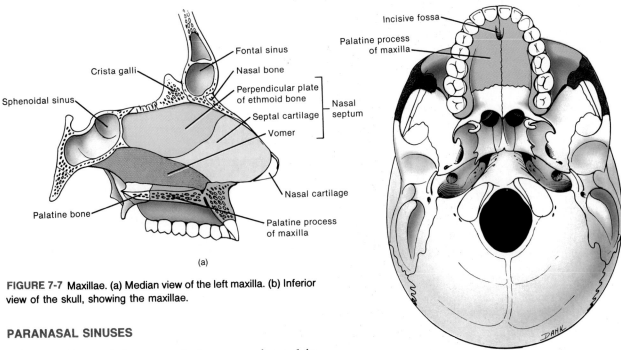

(a)

FIGURE 7-7 Maxillae. (a) Median view of the left maxilla. (b) Inferior view of the skull, showing the maxillae.

PARANASAL SINUSES

Paired cavities, called **paranasal sinuses,** are located in certain bones near the nasal cavity (Figure 7-8). The paranasal sinuses are lined with mucous membranes that are continuous with the lining of the nasal cavity. Cranial bones containing paranasal sinuses are the frontal bone, the sphenoid, the ethmoid, and the maxillae. (The sinuses were described in the discussion of each of these bones.) Besides producing mucus, the paranasal sinuses lighten the skull bones and serve as resonant chambers for sound.

CLINICAL APPLICATION

Secretions produced by the mucous membranes of the paranasal sinuses drain into the nasal cavity. An inflammation of the membranes due to an allergic reac-

tion or infection is called **sinusitis.** If the membranes swell enough to block drainage into the nasal cavity, fluid pressure builds up in the paranasal sinuses and a sinus headache results.

ZYGOMATIC BONES

The two **zygomatic bones (malars),** commonly referred to as the cheekbones, form the prominences of the cheeks and part of the outer wall and floor of the orbits (Figure 7-2b).

FIGURE 7-8 Paranasal sinuses. (a) Anterior view. (b) Median view.

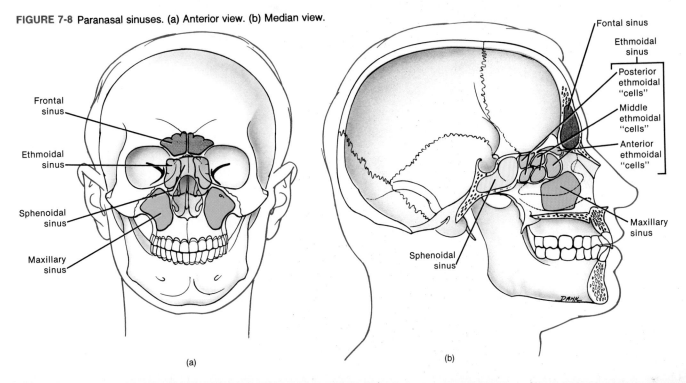

(a)

(b)

The **temporal process** of the zygomatic bone projects posteriorly and articulates with the zygomatic process of the temporal bone. These two processes form the **zygomatic arch** (Figure 7-4).

MANDIBLE

The **mandible** or lower jawbone is the largest, strongest facial bone (Figure 7-9). It is the only movable bone in the skull.

In the lateral view you can see that the mandible consists of a curved, horizontal portion called the **body** and two perpendicular portions called the **rami**. The **angle** of the mandible is the area where each ramus meets the body. Each ramus has a **condylar** (KON-di-lar) **process** that articulates with the mandibular fossa and articular tubercle of the temporal bone to form the temporomandibular (TM) joint. It also has a **coronoid** (KOR-ō-noyd) **process** to which the temporalis muscle attaches. The depression between the coronoid and condylar processes is called the **mandibular notch.** The **alveolar process** is an arch containing the sockets (**alveoli**) for the mandibular (lower) teeth.

CLINICAL APPLICATION

Opening of the mouth very wide, as in a very large yawn, may cause displacement of the condylar process of the mandible from the mandibular fossa of temporal bone, thereby producing a **dislocation of the jaw.** The displacement is usually bilateral and the patient is unable to close the mouth.

LACRIMAL BONES

The paired **lacrimal bones** (LAK-ri-mal; *lacrima* = tear) are thin bones roughly resembling a fingernail in size and shape. They are the smallest bones of the face. These bones are posterior and lateral to the nasal bones in the medial wall of the orbit. They can be seen in the anterior and lateral views of the skull in Figure 7-2. The lacrimal bones form a part of the medial wall of the orbit.

PALATINE BONES

The two **palatine** (PAL-a-tīn) **bones** are L-shaped and form the posterior portion of the hard palate, part of the floor and lateral wall of the nasal cavity, and a small portion of the floors of the orbits. The posterior portion of the hard palate, which separates the nasal cavity from the oral cavity, is formed by the **horizontal plates** of the palatine bones. These can be seen in Figure 7-4.

INFERIOR NASAL CONCHAE

Refer to the views of the skull in Figure 7-2a and 7-6a. The two **inferior nasal conchae** are scroll-like bones that form a part of the lateral wall of the nasal cavity and project into the nasal cavity inferior to the superior and middle nasal conchae of the ethmoid bone. They serve the same function as the superior and middle nasal conchae, that is, they allow for the circulation and filtration of air before it passes into the lungs. The inferior nasal conchae are separate bones and not part of the ethmoid.

VOMER

The **vomer** (= plowshare) is a roughly triangular bone that forms the inferior and posterior part of the nasal septum. It is clearly seen in the anterior view of the skull in Figure 7-2a and the inferior view in Figure 7-4.

The inferior border of the vomer articulates with the cartilage septum that divides the nose into a right and left nostril. Its superior border articulates with the perpendicular plate of the ethmoid bone. Thus the structures that form the **nasal septum,** or partition, are the perpendicular plate of the ethmoid, the septal cartilage, and the vomer (Figure 7-7a). If the vomer is deviated, that is, pushed to one side, the nasal chambers are of unequal size.

CLINICAL APPLICATION

A **deviated nasal septum** is deflected laterally from the midline of the nose. The deviation usually occurs

---

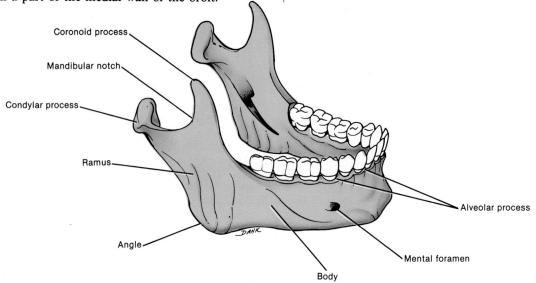

Coronoid process

Mandibular notch

Condylar process

Ramus

Angle

Alveolar process

Mental foramen

Body

FIGURE 7-9 Mandible in right lateral view.

at the junction of bone with the septal cartilage. If the deviation is severe, it may entirely block the nasal passageway. Even though the blockage may not be complete, infection and inflammation develop and cause nasal congestion, blockage of the paranasal sinus openings, and chronic sinusitis.

FORAMINA

Some foramina of the skull were mentioned along with the descriptions of the cranial and facial bones with which they are associated. As preparation for studying other systems of the body, especially the nervous and cardiovascular systems, these foramina, as well as additional ones, and the structures passing through them are listed in Exhibit 7-3. For your convenience and for future reference, the foramina are listed alphabetically.

HYOID BONE

The single **hyoid bone** (*hyoedes* = U-shaped) is a unique component of the axial skeleton because it does not articulate with any other bone. Rather, it is suspended from the styloid process of the temporal bone by ligaments and muscles. The hyoid is located in the neck between the mandible and larynx. It supports the tongue and provides attachment for some of its muscles. Refer to the anterior and lateral views of the skull in Figure 7-2 to see the position of the hyoid bone.

The hyoid consists of a horizontal body and paired projections called the **lesser cornu** (*cornu* = horns) and the **greater cornu** (Figure 7-10). Muscles and ligaments attach to these paired projections.

VERTEBRAL COLUMN

DIVISIONS

The **vertebral column,** or **spine,** together with the sternum and ribs, constitutes the skeleton of the **trunk** of the body. The vertebral column is composed of a series of bones called **vertebrae.** In the average adult, the column measures about 71 cm (28 inches) in length. In effect, the vertebral column is a strong, flexible rod that moves anteriorly, posteriorly, and laterally. It encloses and protects the spinal cord, supports the head, and serves as a point of attachment for the ribs and the muscles of the back.

Between the vertebrae are openings called **intervertebral foramina.** The nerves that connect the spinal cord to various parts of the body pass through these openings.

The adult vertebral column typically contains 26 vertebrae (Figure 7-11). These are distributed as follows: 7 **cervical vertebrae** (*cervix* = neck) in the neck region; 12 **thoracic vertebrae** posterior to the thoracic cavity; 5 **lumbar vertebrae** (*lumbus* = loin) supporting the lower back; 5 **sacral vertebrae** fused into one bone called the **sacrum;** and usually 4 **coccygeal** (kok-SIJ-ē-al) **vertebrae** fused into one or two bones called the **coccyx** (KOK-six). Prior to the fusion of the sacral and coccygeal vertebrae, the total number of vertebrae is 33.

Between adjacent vertebrae from the axis to the sacrum are fibrocartilaginous **intervertebral discs.** Each disc is composed of an outer fibrous ring consisting of fibrocartilage called the *annulus fibrosus* and an inner soft, pulpy, highly elastic structure called the *nucleus pulposus.* The discs form strong joints, permit various movements of the vertebral column, and absorb vertical shock. Under compression, they flatten, broaden, and bulge from their intervertebral spaces.

CURVES

When viewed from the side, the vertebral column shows four **curves** (Figure 7-11b). From the anterior view, these are alternately convex, meaning they curve toward the viewer, and concave, meaning they curve away from the viewer. The curves of the column, like the curves in a long bone, are important because they increase its strength, help maintain balance in the upright position, absorb shocks from walking, and help protect the column from fracture.

In the fetus, there is only a single anteriorly concave curve. At approximately the third postnatal month, when an infant begins to hold its head erect, the **cervical curve** develops. Later, when the child stands and walks, the **lumbar curve** develops. The cervical and lumbar curves are anteriorly convex. Because they are modifications of the fetal positions, they are called **secondary curves.** The other two curves, the **thoracic curve** and the **sacral curve,** are anteriorly concave. Since they retain the anterior concavity of the fetus, they are referred to as **primary curves.**

TYPICAL VERTEBRA

Although there are variations in size, shape, and detail in the vertebrae in different regions of the column, all the vertebrae are basically similar in structure (Figure 7-12). A typical vertebra consists of the following components.

1. The **body** is the thick, disc-shaped anterior portion that is the weight-bearing part of a vertebra. Its superior and inferior surfaces are roughened for the attachment of intervertebral discs. The anterior and lateral surfaces contain nutrient foramina for blood vessels.

2. The **vertebral arch (neural arch)** extends posteriorly from the body of the vertebra. With the body of the vertebra, it surrounds the spinal cord. It is formed by two short, thick processes, the **pedicles** (PED-i-kuls), which project posteriorly from the body to unite with the laminae. The **laminae** (LAM-

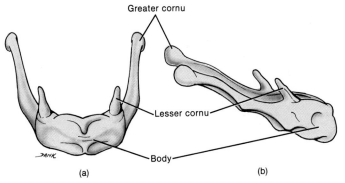

FIGURE 7-10 Hyoid bone. (a) Anterior view. (b) Right lateral view.

Greater cornu

Lesser cornu

Body

(a)   (b)

EXHIBIT 7-3
SUMMARY OF FORAMINA OF THE SKULL

FORAMEN	LOCATION	STRUCTURES PASSING THROUGH
Carotid (Figure 7-4)	Petrous portion of temporal.	Internal carotid artery.
Greater palatine (Figure 7-4)	Posterior angle of hard palate.	Greater palatine nerve and greater palatine vessels.
Hypoglossal (Figure 7-5)	Superior to base of occipital condyles.	Hypoglossal (XII) nerve and branch of ascending pharyngeal artery.
Incisive (Figure 7-7)	Posterior to incisor teeth.	Branches of descending palatine vessels and nasopalatine nerve.
Inferior orbital (Figure 7-4)	Between greater wing of sphenoid and maxilla.	Maxillary branch of trigeminal (V) nerve, zygomatic nerve, and infraorbital vessels.
Infraorbital (Figure 7-2a)	Inferior to orbit in maxilla.	Infraorbital nerve and artery.
Jugular (Figure 7-4)	Posterior to carotid canal between petrous portion of temporal and occipital.	Internal jugular vein, glossopharyngeal (IX) nerve, vagus (X) nerve, accessory (XI) nerve, and sigmoid sinus.
Lacerum (Figure 7-5)	Bounded anteriorly by sphenoid, posteriorly by petrous portion of temporal, and medially by the sphenoid and occipital.	Internal carotid artery and branch of ascending pharyngeal artery.
Lacrimal (Figure 7-2c)	Lacrimal bone.	Lacrimal (tear) duct.
Lesser palatine (Figure 7-4)	Posterior to greater palatine foramen.	Lesser palatine nerves and artery.
Magnum (Figure 7-4)	Occipital bone.	Medulla oblongata and its membranes, the accessory (XI) nerve, and the vertebral and spinal arteries and meninges.
Mandibular	Medial surface of ramus of mandible.	Inferior alveolar nerve and vessels.
Mastoid	Posterior border of mastoid process of temporal bone.	Emissary vein to transverse sinus and branch of occipital artery to dura mater.
Mental (Figure 7-9)	Inferior to second premolar tooth in mandible.	Mental nerve and vessels.
Olfactory (Figure 7-5)	Cribriform plate of ethmoid.	Olfactory (I) nerve.
Optic (Figure 7-5)	Between upper and lower portions of small wing of sphenoid.	Optic (II) nerve and ophthalmic artery.
Ovale (Figure 7-5)	Greater wing of sphenoid.	Mandibular branch of trigeminal (V) nerve.
Rotundum (Figure 7-5)	Junction of anterior and medial parts of sphenoid.	Maxillary branch of trigeminal (V) nerve.
Spinosum (Figure 7-5)	Posterior angle of sphenoid.	Middle meningeal vessels.
Stylomastoid (Figure 7-4)	Between styloid and mastoid processes of temporal.	Facial (VII) nerve and stylomastoid artery.
Superior orbital (Figure 7-5)	Between greater and lesser wings of sphenoid.	Oculomotor (III) nerve, trochlear (IV) nerve, ophthalmic branch of trigeminal (V) nerve, and abducens (VI) nerve.
Supraorbital (Figure 7-2a)	Supraorbital margin of orbit.	Supraorbital nerve and artery.
Zygomaticofacial (Figure 7-2a)	Zygomatic bone.	Zygomaticofacial nerve and vessels.

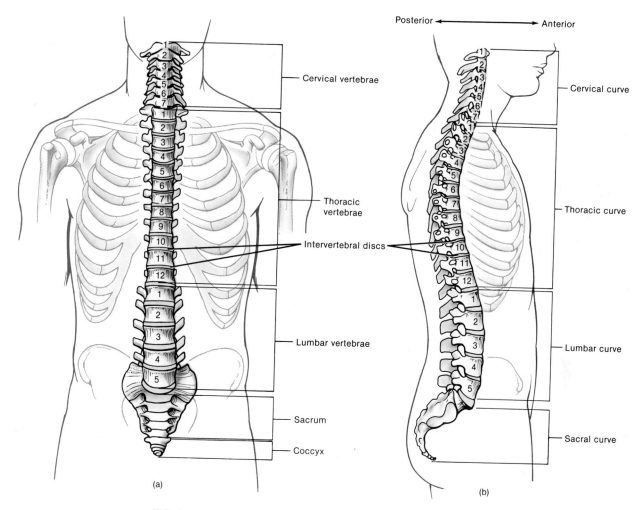

FIGURE 7-11 Vertebral column. (a) Anterior view. (b) Right lateral view.

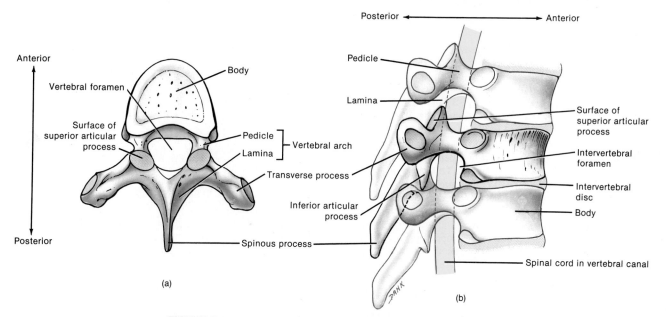

FIGURE 7-12 Typical vertebra. (a) Superior view. (b) Right lateral view.

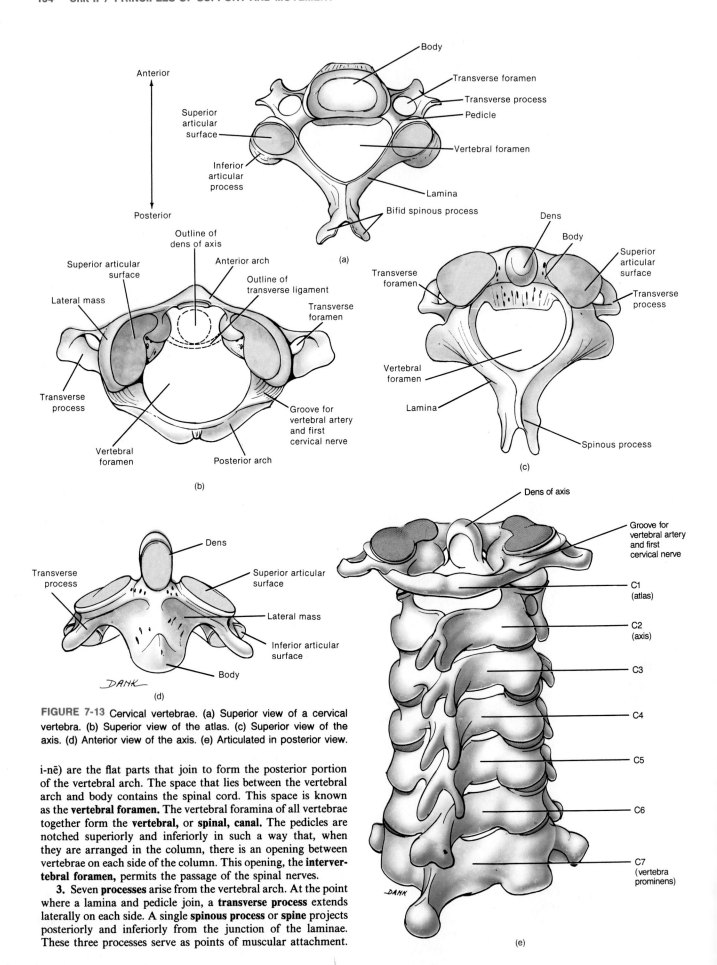

FIGURE 7-13 Cervical vertebrae. (a) Superior view of a cervical vertebra. (b) Superior view of the atlas. (c) Superior view of the axis. (d) Anterior view of the axis. (e) Articulated in posterior view.

i-nē) are the flat parts that join to form the posterior portion of the vertebral arch. The space that lies between the vertebral arch and body contains the spinal cord. This space is known as the **vertebral foramen.** The vertebral foramina of all vertebrae together form the **vertebral,** or spinal, **canal.** The pedicles are notched superiorly and inferiorly in such a way that, when they are arranged in the column, there is an opening between vertebrae on each side of the column. This opening, the **intervertebral foramen,** permits the passage of the spinal nerves.

3. Seven **processes** arise from the vertebral arch. At the point where a lamina and pedicle join, a **transverse process** extends laterally on each side. A single **spinous process** or **spine** projects posteriorly and inferiorly from the junction of the laminae. These three processes serve as points of muscular attachment.

The remaining four processes form joints with other vertebrae. The two **superior articular processes** of a vertebra articulate with the vertebra immediately superior to them. The two **inferior articular processes** of a vertebra articulate with the vertebra inferior to them.

CERVICAL REGION

When viewed from above, it can be seen that the bodies of **cervical vertebrae** are smaller than those of the thoracic vertebrae (Figure 7-13). The arches, however, are larger. The spinous processes of the second through sixth cervical vertebrae are often *bifid,* that is, with a cleft. All cervical vertebrae have three foramina: the vertebral foramen and two transverse foramina. Each cervical transverse process contains a **transverse foramen** through which the vertebral artery and its accompanying vein and nerve fibers pass.

The first two cervical vertebrae differ considerably from the others. The first cervical vertebra (C1), the **atlas,** is named for its support of the head. Essentially, the atlas is a ring of bone with **anterior** and **posterior arches** and large **lateral masses.** It lacks a body and a spinous process. The superior surfaces of the lateral masses, called **superior articular surfaces,** are concave and articulate with the occipital condyles of the occipital bone. This articulation permits the movement seen when nodding the head. The inferior surfaces of the lateral masses, the **inferior articular surfaces,** articulate with the second cervical vertebra. The transverse processes and transverse foramina of the atlas are quite large.

The second cervical vertebra (C2), the **axis,** does have a body. A peglike process called the **dens,** or **odontoid process,** projects up through the ring of the atlas. The dens makes a pivot on which the atlas and head rotate. This arrangement permits side-to-side rotation of the head.

CLINICAL APPLICATION

In various instances of trauma, the dens of the axis may be driven into the medulla oblongata of the brain, usually with instantly fatal results. This injury is the usual cause of fatality in **whiplash injuries.**

The third through sixth cervical vertebrae (C3–C6) correspond to the structural pattern of the typical cervical vertebra previously described.

The seventh cervical vertebra (C7), called the **vertebra prominens,** is somewhat different. It is marked by a large, nonbifid spinous process that may be seen and felt at the base of the neck (see Figure 7-11).

THORACIC REGION

Viewing a typical **thoracic vertebra** from above, you can see that it is considerably larger and stronger than a vertebra of the cervical region (Figure 7-14). In addition, the spinous process on each vertebra is long, pointed, and directed inferiorly. Thoracic vertebrae also have longer and heavier transverse processes than cervical vertebrae.

Except for the eleventh and twelfth thoracic vertebrae, the transverse processes have **facets** for articulating with the tubercles of the ribs. The bodies of thoracic vertebrae also have whole **facets** or half facets, called **demifacets,** for articulation with the heads of the ribs. The first thoracic vertebra (T1) has, on either side of its body, a superior whole facet and an inferior demifacet. The superior facet articulates with the first rib, and the inferior demifacet, together with the superior demifacet of the second thoracic vertebra (T2), forms a facet for articulation with the second rib. The second through eighth thoracic vertebrae (T2–T8) have two demifacets on each side, a larger superior demifacet and a smaller inferior demifacet. When the vertebrae are articulated, they form whole facets for the heads of the ribs. The ninth thoracic vertebra (T9) has a single superior demifacet on either side of its body. The tenth through twelfth thoracic vertebrae (T10–T12) have whole facets on either side of their bodies.

LUMBAR REGION

The **lumbar vertebrae** (L1–L5) are the largest and strongest in the column (Figure 7-15). Their various projections are short and thick. The superior articular processes are directed medially instead of superiorly. The inferior articular processes are directed laterally instead of inferiorly. The spinous process is quadrilateral in shape, thick and broad, and projects nearly straight posteriorly. The spinous process is well adapted for the attachment of the large back muscles.

SACRUM AND COCCYX

The **sacrum** is a triangular bone formed by the union of five sacral vertebrae. These are indicated in Figure 7-16 as S1 through S5. The sacrum serves as a strong foundation for the pelvic girdle. It is positioned at the posterior portion of the pelvic cavity between the two hipbones.

The concave anterior side of the sacrum faces the pelvic cavity. It is smooth and contains four **transverse lines** that mark the joining of the vertebral bodies. At the ends of these lines are four pairs of **anterior sacral (pelvic) foramina.**

The convex, posterior surface of the sacrum is irregular. It contains a **median sacral crest,** the fused spinous processes of the upper sacral vertebrae; a **lateral sacral crest,** the transverse processes of the sacral vertebrae; and four pairs of **posterior sacral (dorsal) foramina.** These foramina communicate with the anterior sacral foramina through which nerves and blood vessels pass. The **sacral canal** is a continuation of the vertebral canal. The laminae of the fifth sacral vertebra, and sometimes the fourth, fail to meet. This leaves an inferior entrance to the vertebral canal called the **sacral hiatus** (hi-Ā-tus).

The superior border of the sacrum exhibits an anteriorly projecting border, the **sacral promontory** (PROM-on-tō'-rē). It is an obstetrical landmark for measurements of the pelvis. An imaginary line running from the superior surface of the symphysis pubis to the sacral promontory separates the abdominal and pelvic cavities. Laterally, the sacrum has a large **auricular surface** for articulating with the ilium of the hipbone. Its **superior articular processes** articulate with the fifth lumbar vertebra.

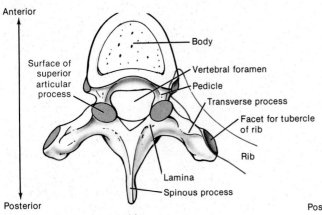

Anterior

Body

Surface of superior articular process

Vertebral foramen

Pedicle

Transverse process

Facet for tubercle of rib

Rib

Lamina

Spinous process

Posterior

(a)

FIGURE 7-14 Thoracic vertebrae. (a) Superior view. (b) Articulated in right lateral view.

The **coccyx** is also triangular in shape and is formed by the fusion of the coccygeal vertebrae, usually the last four. These are indicated in Figure 7-16 as Co1 through Co4. The coccyx articulates superiorly with the sacrum. The coccyx is the most rudimentary part of the column, representing the vestige of a tail.

THORAX

Anatomically, the term **thorax** refers to the chest. The skeletal portion of the thorax is a bony cage formed by the sternum, costal cartilage, ribs, and the bodies of the thoracic vertebrae (Figure 7-17).

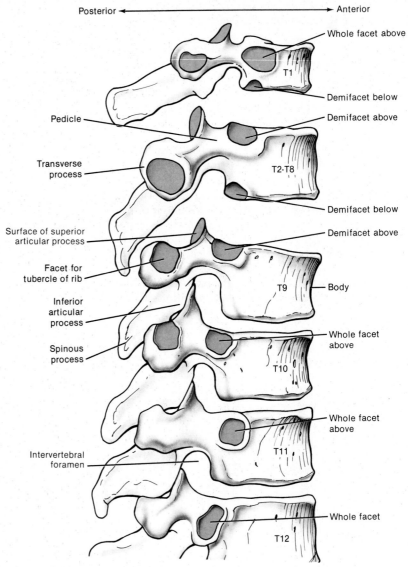

Posterior

Anterior

Whole facet above

T1

Demifacet below

Pedicle

Demifacet above

Transverse process

T2-T8

Demifacet below

Surface of superior articular process

Demifacet above

Facet for tubercle of rib

T9

Body

Inferior articular process

Whole facet above

Spinous process

T10

Whole facet above

T11

Intervertebral foramen

Whole facet

T12

(b)

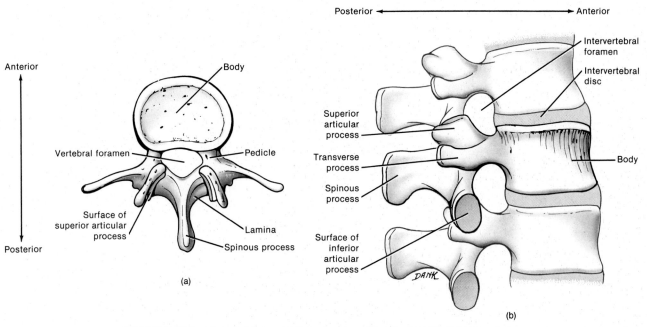

FIGURE 7-15 Lumbar vertebrae. (a) Superior view. (b) Articulated in right lateral view.

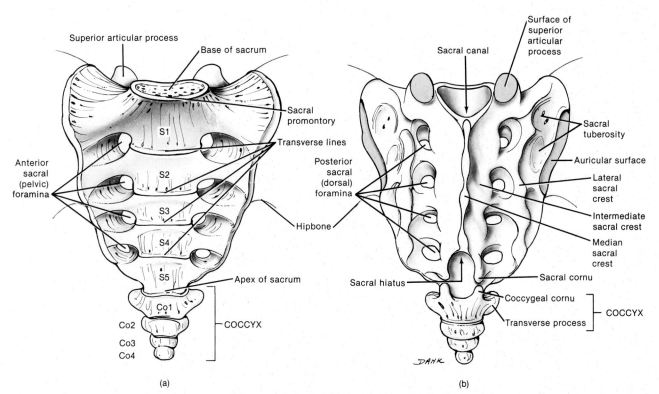

FIGURE 7-16 Sacrum and coccyx. (a) Anterior view. (b) Posterior view.

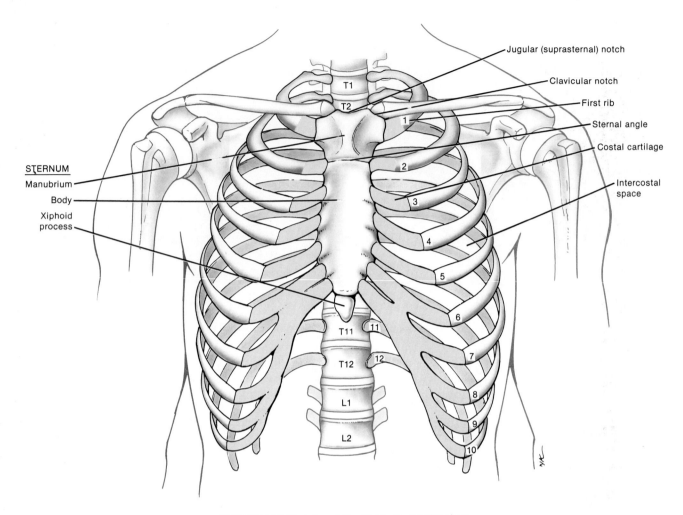

FIGURE 7-17 Skeleton of the thorax in anterior view.

The thoracic cage is roughly cone-shaped, the narrow portion being superior and the broad portion inferior. It is flattened from front to back. The thoracic cage encloses and protects the organs in the thoracic cavity. It also provides support for the bones of the shoulder girdle and upper extremities.

STERNUM

The **sternum,** or breastbone, is a flat, narrow bone measuring about 15 cm (6 inches) in length. It is located in the median line of the anterior thoracic wall.

The sternum (see Figure 7-17) consists of three basic portions: the **manubrium** (ma-NOO-brē-um), the triangular, superior portion; the **body,** the middle, largest portion; and the **xiphoid** (ZĪ-foyd) **process,** the inferior, smallest portion. The manubrium has a depression on its superior surface called the **jugular (suprasternal) notch.** On each side of the jugular notch are **clavicular notches** that articulate with the medial ends of the clavicles. The manubrium also articulates with the first and second ribs. The body of the sternum articulates directly or indirectly with the second through tenth ribs. The xiphoid process has no ribs attached to it but provides attachment for some abdominal muscles. The xiphoid process consists of hya-

line cartilage during infancy and childhood and does not ossify completely until about age 40. If the hands of a rescuer are mispositioned during cardiopulmonary resuscitation (CPR), there is the danger of fracturing the ossified xiphoid process, separating it from the body, and driving it into the liver.

CLINICAL APPLICATION

Since the sternum possesses red bone marrow throughout life and because it is readily accessible, it is a common site for **marrow biopsy.** Under a local anesthetic, a wide-bore needle is introduced into the marrow cavity of the sternum for aspiration of a sample of red bone marrow. This procedure is called a **sternal puncture.**

The sternum may also be split in the midsaggital plane to allow surgeons access to mediastinal structures such as the thymus gland, heart, and great vessels of the heart.

RIBS

Twelve pairs of **ribs** make up the sides of the thoracic cavity (see Figure 7-17). The ribs increase in length from the first through seventh. Then they decrease in length

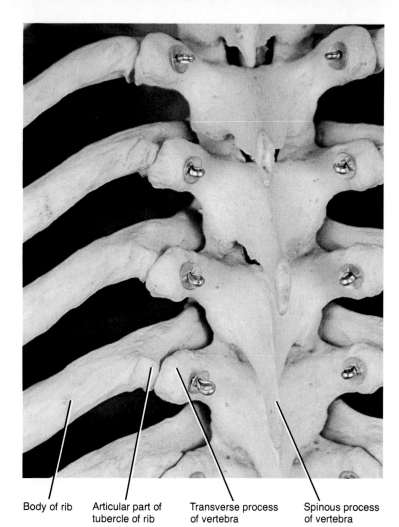

FIGURE 7-18 Typical rib. (a) A left rib viewed from below and behind. (b) Photograph of several ribs articulating with their respective vertebrae in posterior view. (Courtesy of Matt Iacobino.) (c) Photograph of several ribs articulating with their respective vertebrae in anterior view. (Courtesy of Matt Iacobino.)

Body of rib   Articular part of tubercle of rib   Transverse process of vertebra   Spinous process of vertebra

(b)

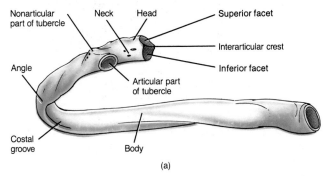

Nonarticular part of tubercle   Neck   Head   Superior facet

Interarticular crest

Inferior facet

Angle

Articular part of tubercle

Costal groove   Body

(a)

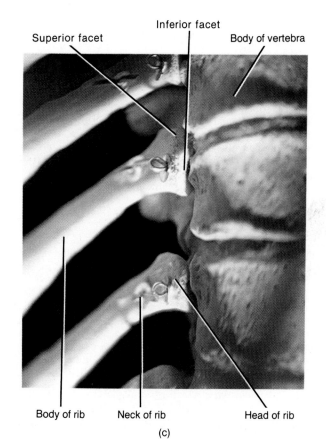

Superior facet   Inferior facet   Body of vertebra

Body of rib   Neck of rib   Head of rib

(c)

to the twelfth rib. Each rib articulates posteriorly with its corresponding thoracic vertebra.

The first through seventh ribs have a direct anterior attachment to the sternum by a strip of hyaline cartilage, called **costal cartilage** (*costa* = rib). These ribs are called **true ribs** or **vertebrosternal ribs.** The remaining five pairs of ribs are referred to as **false ribs** because their costal cartilages do not attach directly to the sternum. The cartilages of the eighth, ninth, and tenth ribs attach to each other and then to the cartilage of the seventh rib. These are called **vertebrochondral ribs.** The eleventh and twelfth ribs are designated as **floating ribs** because their anterior ends do not attach even indirectly to the sternum. They attach only posteriorly to the thoracic vertebrae.

Although there is some variation in rib structure, we will examine the parts of a typical (third through ninth) rib when viewed from the right side and from behind (Figure 7-18). The **head** of a typical rib is a projection at the posterior end of the rib. It is wedge-shaped and consists of one or two **facets** that articulate with facets on the bodies of adjacent thoracic vertebrae. The facets on the head of a rib are separated by a horizontal **interarticular crest.** The inferior facet on the head of a rib is larger than the superior facet. The **neck** is a constricted portion just lateral to the head. A knoblike structure on the posterior surface where the neck joins the body is called a **tubercle.** It consists of a **nonarticular part** which affords attachment to the ligament of the tubercle and an **articular part** which articulates with the facet of a transverse process of the inferior of the two vertebrae to which the head of the rib is connected. The **body** or **shaft** is the main part of the rib. A short distance beyond the tubercle, there is an abrupt change in the curvature of the shaft. This point is called the **angle.** The inner surface of the rib has a **costal groove** that protects blood vessels and a small nerve, artery, and vein.

The posterior portion of the rib is connected to a thoracic vertebra by its head and articular part of a tubercle.

The facet of the head fits into a facet on the body of a vertebra, and the articular part of the tubercle articulates with the facet of the transverse process of the vertebra. Each of the second through ninth ribs articulates with the bodies of two adjacent vertebrae. The first, tenth, eleventh, and twelfth ribs articulate with only one vertebra each. On the eleventh and twelfth ribs, there is no articulation between the tubercles and the transverse processes of their corresponding vertebrae.

Spaces between ribs, called **intercostal spaces,** are occupied by intercostal muscles, blood vessels, and nerves.

CLINICAL APPLICATION
Surgical access to the lungs or structures in the mediastinum is commonly undertaken through an intercostal space. Special rib retractors are used to create a wide separation between ribs. The costal cartilages are sufficiently elastic to permit considerable bending.

DISORDERS: HOMEOSTATIC IMBALANCES

Herniated (Slipped) Disc

In their function as shock absorbers, intervertebral discs are subject to compressional forces. The discs between the fourth and fifth lumbar vertebrae and between the fifth lumbar vertebra and sacrum usually are subject to more forces than other discs. If the anterior and posterior ligaments of the discs become injured or weakened, the pressure developed in the nucleus pulposus may be great enough to rupture the surrounding fibrocartilage. If this occurs, the nucleus pulposus may protrude posteriorly or into one of the adjacent vertebral bodies (herniate). This condition is called a **herniated (slipped) disc.**

Most often the nucleus pulposus slips posteriorly toward the spinal cord and spinal nerves. This movement exerts pressure on the spinal nerves, causing considerable, sometimes very acute, pain. If the roots of the sciatic nerve, which passes from the spinal cord to the foot, are pressured, the pain radiates down the back of the thigh, through the calf, and occasionally into the foot. If pressure is exerted on the spinal cord itself, nervous tissue may be destroyed. Traction, bed rest, and analgesia usually relieve the pain. If such treatment is ineffective, surgical decompression of the spinal nerves by laminectomy or removal of some of the nucleus pulposus may be necessary to relieve pain. Removal may involve conventional surgery or use of a proteolytic enzyme called chymopapain. This enzyme is extracted from a papaya plant and is injected into a herniated disc where it dissolves the nucleus pulposus, thus relieving pressure on spinal nerves and pain.

Curvatures

As a result of various conditions, the normal curves of the vertebral column may become exaggerated or the column may acquire a lateral bend. Such conditions are called **curvatures** of the spine.

Scoliosis (skō'-lē-Ō-sis; *scolio* = bent) is a lateral bending of the vertebral column, usually in the thoracic region. This is the most common of the curvatures. It may be congenital, due to an absence of the lateral half of a vertebra (hemivertebra). It may also be acquired from a persistent severe sciatica. Poliomyelitis may cause scoliosis by the paralysis of muscles on one side of the body, which produces a lateral deviation of the trunk toward the unaffected side. Poor posture is also a contributing factor.

Until recently, a child with progressive scoliosis faced either years of treatment with a brace or major corrective surgery. Several research teams are experimenting with using electrical stimulation of muscles to limit the progression of scoliosis and even to reduce curvatures in some cases. The skeletal muscles on the convex side of the spinal curve are electrically stimulated.

Kyphosis (kī-FŌ-sis; *kypho* = hunchback) is an exaggeration of the thoracic curve of the vertebral column. In tuberculosis of the spine, vertebral bodies may partially collapse, causing an acute angular bending of the vertebral column. In the elderly, degeneration of the intervertebral discs leads to kyphosis. Kyphosis may also be caused by rickets and poor posture. The term "round-shouldered" is an expression for mild kyphosis. Electrical muscle stimulation is being studied to assess its effects on kyphosis as well as on scoliosis.

Lordosis (lor-DŌ-sis; *lordo* = swayback) is an exaggeration of the lumbar curve of the vertebral column. It may result from increased weight of abdominal contents as in pregnancy or extreme obesity. Other causes include poor posture, rickets, and tuberculosis of the spine.

Spina Bifida

Spina bifida (SPĪ-na BIF-i-da) is a congenital defect of the vertebral column in which laminae fail to unite at the midline. The lumbar vertebrae are involved in about 50 percent of all cases. In less serious cases, the defect is small and the area is covered with skin. Only a dimple or tuft of hair may mark the site. Symptoms are mild, with perhaps intermittent urinary problems. An operation is not ordinarily required. Larger defects in the vertebral arches with protrusion of the membranes around the spinal cord or spinal cord tissue produce serious problems, such as partial or complete paralysis, partial or complete loss of urinary bladder control, and the absence of reflexes.

Fractures of the Vertebral Column

Fractures of the vertebral column most commonly involve T12, L1, and L2. They usually result from a flexion-compression type of injury such as might be sustained in landing on the feet or buttocks after a fall from a height or having a heavy weight fall on the shoulders. The forceful compression wedges the involved vertebrae. If, in addition to compression, there is forceful forward movement, one vertebra may displace forward on its adjacent vertebra below with either dislocation or fracture of the articular facets between the two (fracture dislocation) and with rupture of the interspinous ligaments.

The cervical vertebrae may be fractured or, more commonly, dislodged by a fall on the head with acute flexion of the neck, as might happen on diving into shallow water. Dislocation may even result from the sudden forward jerk which may occur

when an automobile or airplane crashes ("whiplash"). The relatively horizontal intervertebral facets of the cervical vertebrae allow dislocation to take place without their being fractured, whereas the relatively vertical thoracic and lumbar intervertebral facets nearly always fracture in forward dislocation of the thoracolumbar region. Spinal nerve damage may occur as a result of fractures of the vertebral column.

STUDY OUTLINE

Types of Bones (p. 138)
1. On the basis of shape, bones are classified as long, short, flat, or irregular.
2. Wormian or sutural bones are found between the sutures of certain cranial bones. Sesamoid bones develop in tendons or ligaments.

Surface Markings (p. 138)
1. Markings are areas on the surfaces of bones.
2. Each marking is structured for a specific function—joint formation, muscle attachment, or passage of nerves and blood vessels.
3. Terms that describe markings include fissure, foramen, meatus, fossa, process, condyle, head, facet, tuberosity, crest, and spine.

Divisions of the Skeletal System (p. 138)
1. The axial skeleton consists of bones arranged along the longitudinal axis. The parts of the axial skeleton are the skull, hyoid bone, auditory ossicles, vertebral column, sternum, and ribs.
2. The appendicular skeleton consists of the bones of the girdles and the upper and lower extremities. The parts of the appendicular skeleton are the shoulder girdles, the bones of the upper extremities, the pelvic girdle, and the bones of the lower extremities.

Skull (p. 138)
1. The skull consists of the cranium and the face. It is composed of 22 bones.
2. Sutures are immovable joints between bones of the skull. Examples are coronal, sagittal, lambdoidal, and squamosal sutures.
3. Fontanels are membrane-filled spaces between the cranial bones of fetuses and infants. The major fontanels are the anterior, posterior, anterolaterals, and posterolaterals.
4. The 8 cranial bones include the frontal, parietal (2), temporal (2), occipital, sphenoid, and ethmoid.
5. The 14 facial bones are the nasal (2), maxillae (2), zygomatic (2), mandible, lacrimal (2), palatine (2), inferior nasal conchae (2), and vomer.

6. Paranasal sinuses are cavities in bones of the skull that communicate with the nasal cavity. They are lined by mucous membranes. The cranial bones containing the paranasal sinuses are the frontal, sphenoid, ethmoid, and maxilla.
7. The foramina of the skull bones provide passages for nerves and blood vessels.

Hyoid Bone (p. 151)
1. The hyoid bone is a U-shaped bone that does not articulate with any other bone.
2. It supports the tongue and provides attachment for some of its muscles.

Vertebral Column (p. 151)
1. The vertebral column, the sternum, and the ribs constitute the skeleton of the trunk.
2. The bones of the adult vertebral column are the cervical vertebrae (7), thoracic vertebrae (12), lumbar vertebrae (5), the sacrum (5, fused) and the coccyx (4, fused).
3. The vertebral column contains primary curves (thoracic and sacral) and secondary curves (cervical and lumbar). These curves give strength, support, and balance.
4. The vertebra are similar in structure, each consisting of a body, vertebral arch, and seven processes. Vertebra in the different regions of the column vary in size, shape, and detail.

Thorax (p. 156)
1. The thoracic skeleton consists of the sternum, the ribs and costal cartilages, and the thoracic vertebrae.
2. The thorax protects vital organs in the chest area.

Disorders: Homeostatic Imbalances (p. 160)
1. Protrusion of the nucleus pulposus of an intervertebral disc posteriorly or into an adjacent vertebral body is called a herniated (slipped) disc.
2. Exaggeration of a normal curve of the vertebral column is called a curvature. Examples include scoliosis, kyphosis, and lordosis.
3. The imperfect union of the vertebral laminae at the midline, a congenital defect, is referred to as spina bifida.
4. Fractures of the vertebral column most often involve T12, L1, and L2.

REVIEW QUESTIONS

1. What are the four principal types of bones? Give an example of each. Distinguish between a sutural and a sesamoid bone.
2. What are surface markings? Describe and give an example of each.
3. Distinguish between the axial and appendicular skeletons. What subdivisions and bones are contained in each?
4. What are the bones that compose the skull? The cranium? The face?

5. Define a suture. What are the four prominent sutures of the skull? Where are they located?
6. What is a fontanel? Describe the location of the six fairly constant fontanels.
7. What is a paranasal sinus? What cranial bones contain paranasal sinuses?
8. Define the following: mastoiditis, cleft palate, cleft lip, sinusitis, dislocation of the jaw, and deviated nasal septum.

9. What is the hyoid bone? In what respect is it unique? What is its function?

10. What bones form the skeleton of the trunk? Distinguish between the number of nonfused vertebrae found in the adult vertebral column and that of a child.

11. What are the normal curves in the vertebral column? How are primary and secondary curves differentiated? What are the functions of the curves?

12. What are the principal distinguishing characteristics of the bones of the various regions of the vertebral column?

13. What bones form the skeleton of the thorax? What are the functions of the thoracic skeleton?

14. How are ribs classified on the basis of their attachment to the sternum?

15. What is a herniated (slipped) disc? Why does it cause pain? How is it treated?

16. What is a curvature? Describe the symptoms and causes of scoliosis, kyphosis, and lordosis.

17. Define spina bifida.

18. What are some common causes of fractures of the vertebral column?

8 The Skeletal System: The Appendicular Skeleton

This chapter discusses the bones of the appendicular skeleton, that is, the bones of the pectoral (shoulder) and pelvic girdles and extremities. The differences between male and female skeletons are also compared.

PECTORAL (SHOULDER) GIRDLES

The **pectoral** (PEK-tō-ral) or **shoulder girdles** attach the bones of the upper extremities to the axial skeleton (Figure 8-1). Each of the two pectoral girdles consists of two bones: a clavicle and a scapula. The clavicle is the anterior component and articulates with the sternum at the sternoclavicular joint. The posterior component, the scapula, which is positioned freely by complex muscle attachments, articulates with the clavicle and humerus. The pectoral girdles have no articulation with the vertebral column. Although the shoulder joints are not very stable, they are freely movable and thus allow movement in many directions.

CLAVICLE

The **clavicles** (KLAV-i-kuls), or collarbones, are long, slender bones with a double curvature (Figure 8-2). The two bones lie horizontally in the superior and anterior part of the thorax superior to the first rib.

The medial end of the clavicle, the **sternal extremity,** is rounded and articulates with the sternum. The broad, flat, lateral end, the **acromial** (a-KRŌ-mē-al) **extremity,** articulates with the acromion of the scapula. This joint is called the **acromioclavicular joint.** (Refer to Figure 8-1 for a view of these articulations.) The **conoid tubercle** on the inferior surface of the lateral end of the bone serves as a point of attachment for a ligament. The **costal tuberosity** on the inferior surface of the medial end also serves as a point of attachment for a ligament.

CLINICAL APPLICATION
Because of its position, the clavicle transmits forces from the upper extremity to the trunk. If such forces are excessive, as in falling on one's outstretched arm, a **fractured clavicle** may result. In fact, it is the most frequently broken bone in the body.
   Separation of the shoulder simply means dislocation of the acromioclavicular joint.

SCAPULA

The **scapulae** (SCAP-yoo-lē), or shoulder blades, are large, triangular, flat bones situated in the dorsal part of the thorax between the levels of the second and seventh ribs (Figure 8-3). Their medial borders are located about 5 cm (2 inches) from the vertebral column.

A sharp ridge, the **spine,** runs diagonally across the dorsal surface of the flattened, triangular **body.** The end of the spine projects as a flattened, expanded process called the **acromion** (a-KRŌ-mē-on). This process articulates with the clavicle. Inferior to the acromion is a depression called the **glenoid cavity.** This cavity articulates with the head of the humerus to form the shoulder joint.

The thin edge of the body near the vertebral column is the **medial** or **vertebral border.** The thick edge closer

to the arm is the **lateral** or **axillary border.** The medial and lateral borders join at the **inferior angle.** The superior edge of the scapular body, called the **superior border,** joins the vertebral border at the **superior angle.**

At the lateral end of the superior border is a projection of the anterior surface called the **coracoid** (KOR-a-koyd) **process** to which muscles attach. Above and below the spine are two fossae: the **supraspinous** (soo'-pra-SPĪ-nus) **fossa** and the **infraspinous fossa,** respectively. Both serve as surfaces of attachment for shoulder muscles. On the ventral (costal) surface is a lightly hollowed-out area called the **subscapular fossa,** also a surface of attachment for shoulder muscles.

UPPER EXTREMITIES

The **upper extremities** consist of 60 bones. The skeleton of the right upper extremity is shown in Figure 8-1. Each upper extremity includes the humerus in the arm, ulna and radius in the forearm, carpals (wrist bones), metacarpals (palm bones), and phalanges in the fingers of the hand.

HUMERUS

The **humerus** (HYOO-mer-us), or arm bone, is the longest and largest bone of the upper extremity (Figure 8-4). It articulates proximally with the scapula and distally at the elbow with both ulna and radius.

The proximal end of the humerus consists of a **head** that articulates with the glenoid cavity of the scapula. It also has an **anatomical neck,** which is an oblique groove just distal to the head. The **greater tubercle** is a lateral projection distal to the neck. The **lesser tubercle** is an anterior projection. Between these tubercles runs an **intertubercular sulcus (bicipital groove).** The **surgical neck** is a constricted portion just distal to the tubercles. It is so named because of its liability to fracture.

The **body** or shaft of the humerus is cylindrical at its proximal end. It gradually becomes triangular and is flattened and broad at its distal end. Along the middle portion of the shaft, there is a roughened, V-shaped area called the **deltoid tuberosity.** This area serves as a point of attachment for the deltoid muscle.

The following parts are found at the distal end of the humerus. The **capitulum** (ka-PIT-yoo-lum) is a rounded knob that articulates with the head of the radius. The **radial fossa** is a depression that receives the head of the radius when the forearm is flexed. The **trochlea** (TRŌK-lē-a) is a pulleylike surface that articulates with the ulna. The **coronoid fossa** is an anterior depression that receives part of the ulna when the forearm is flexed. The **olecranon** (ō-LEK-ra-non) **fossa** is a posterior depression that receives the olecranon of the ulna when the forearm is extended. The **medial epicondyle** and **lateral epicondyle** are rough projections on either side of the distal end.

ULNA AND RADIUS

The **ulna** is the medial bone of the forearm (Figure 8-5). In other words, it is located at the little finger side. The proximal end of the ulna presents an **olecranon (olec-**

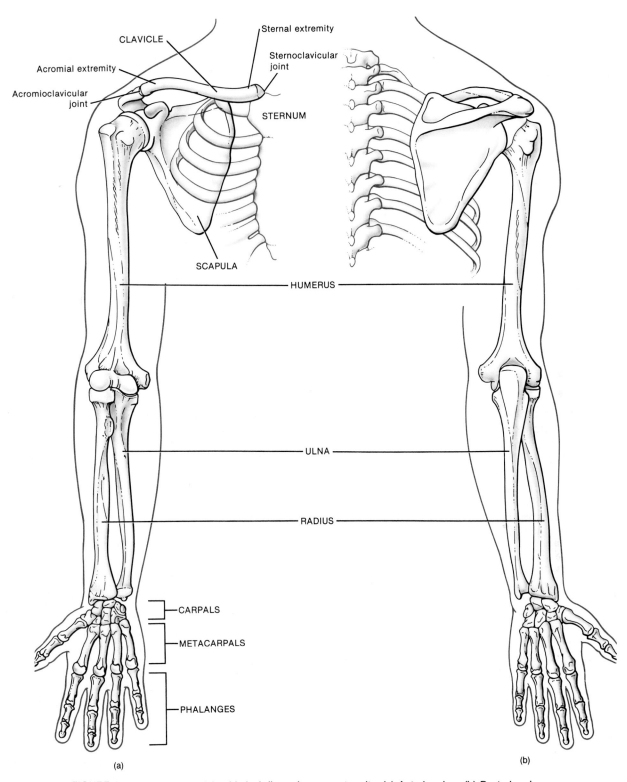

CLAVICLE

Sternal extremity

Sternoclavicular joint

Acromial extremity

STERNUM

Acromioclavicular joint

SCAPULA

HUMERUS

ULNA

RADIUS

CARPALS

METACARPALS

PHALANGES

(a)

(b)

FIGURE 8-1 Right pectoral (shoulder) girdle and upper extremity. (a) Anterior view. (b) Posterior view.

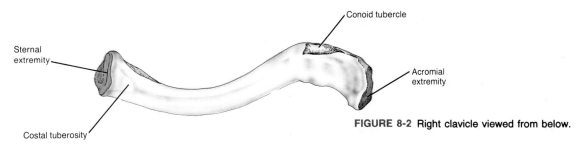

Conoid tubercle

Sternal extremity

Acromial extremity

Costal tuberosity

FIGURE 8-2 Right clavicle viewed from below.

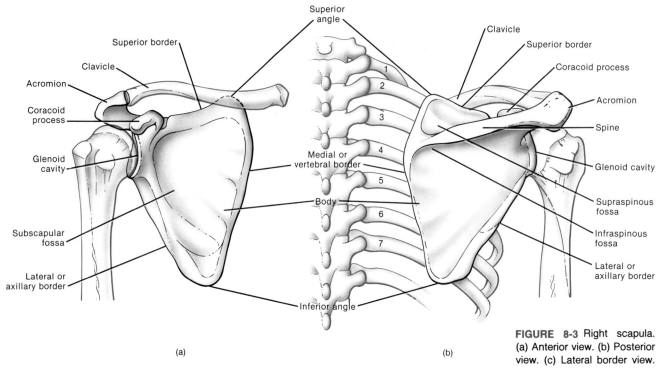

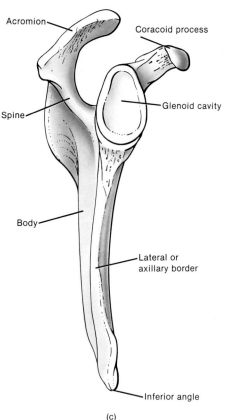

FIGURE 8-3 Right scapula. (a) Anterior view. (b) Posterior view. (c) Lateral border view.

to the trochlear notch. It receives the head of the radius. The distal end of the ulna consists of a **head** that is separated from the wrist by a fibrocartilage disc. A **styloid process** is on the posterior side of the distal end.

The **radius** is the lateral bone of the forearm, that is, it is situated on the thumb side. The proximal end of the radius has a disc-shaped **head** that articulates with the capitulum of the humerus and radial notch of the ulna. It also has a raised, roughened area on the medial side called the **radial tuberosity.** This is a point of attachment for the biceps muscle. The shaft of the radius widens distally to form a concave inferior surface that articulates with two bones of the wrist called the lunate and scaphoid bones. Also at the distal end is a **styloid process** on the lateral side and a medial, concave **ulnar notch** for articulation with the distal end of the ulna.

> **CLINICAL APPLICATION**
> When one falls on the outstretched arm, the radius bears the brunt of forces transmitted through the hand. If a fracture occurs in such a fall, it is usually a transverse break about 3 cm (1 inch) from the distal end of the bone. In this type of injury, called a **Colles' fracture,** the hand is displaced backward and upward.

CARPUS, METACARPUS, AND PHALANGES

The **carpus,** or wrist, consists of eight small bones, the **carpals,** united to each other by ligaments (Figure 8-6). The bones are arranged in two transverse rows, with four bones in each row. The proximal row of carpals, from the lateral to medial position, consists of the **scaphoid, lunate, triquetral,** and **pisiform.** In about 70 percent of cases involving carpal fractures, only the scaphoid is involved. The distal row of carpals, from the lateral to

ranon process), which forms the prominence of the elbow. The **coronoid process** is an anterior projection that, together with the olecranon, receives the trochlea of the humerus. The **trochlear (semilunar) notch** is a curved area between the olecranon and the coronoid process. The trochlea of the humerus fits into this notch. The **radial notch** is a depression located laterally and inferiorly

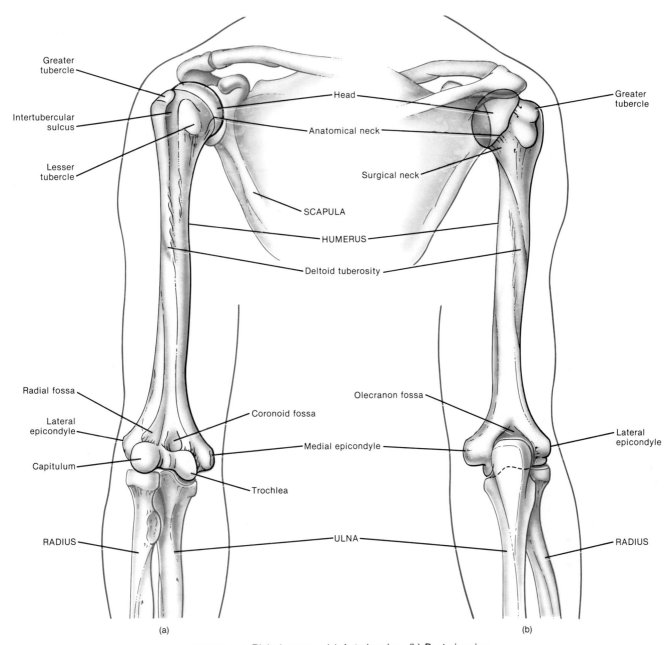

FIGURE 8-4 Right humerus. (a) Anterior view. (b) Posterior view.

medial position, consists of the **trapezium, trapezoid, capitate,** and **hamate.**

The five bones of the **metacarpus** constitute the palm of the hand. Each metacarpal bone consists of a proximal **base,** a **shaft,** and a distal **head.** The metacarpal bones are numbered I to V, starting with the lateral bone. The bases articulate with the distal row of carpal bones and with one another. The heads articulate with the proximal phalanges of the fingers. The heads of the metacarpals are commonly called the "knuckles" and are readily visible when the fist is clenched.

The **phalanges** (fa-LAN-jēz), or bones of the fingers, number 14 in each hand. Each consists of a proximal **base,** a **shaft,** and a distal **head.** There are two phalanges in the first digit, called the thumb or *pollex,* and three phalanges in each of the remaining four digits. These digits, moving medially from the thumb, are commonly

referred to as the index finger, middle finger, ring finger, and little finger. The first row of phalanges, the **proximal row,** articulates with the metacarpal bones and second row of phalanges. The second row of phalanges, the **middle row,** articulates with the proximal row and the third row. The third row of phalanges, the **distal row,** articulates with the middle row. A single finger bone is referred to as a **phalanx** (FĀ-lanks). The thumb has no middle phalanx.

PELVIC GIRDLE

The **pelvic girdle** consists of the two **coxal** (KOK-sal) **bones,** commonly called the pelvic, innominate, or hip-bones (Figure 8-7). The pelvic girdle provides a strong and stable support for the lower extremities on which the weight of the body is carried. The coxal bones are

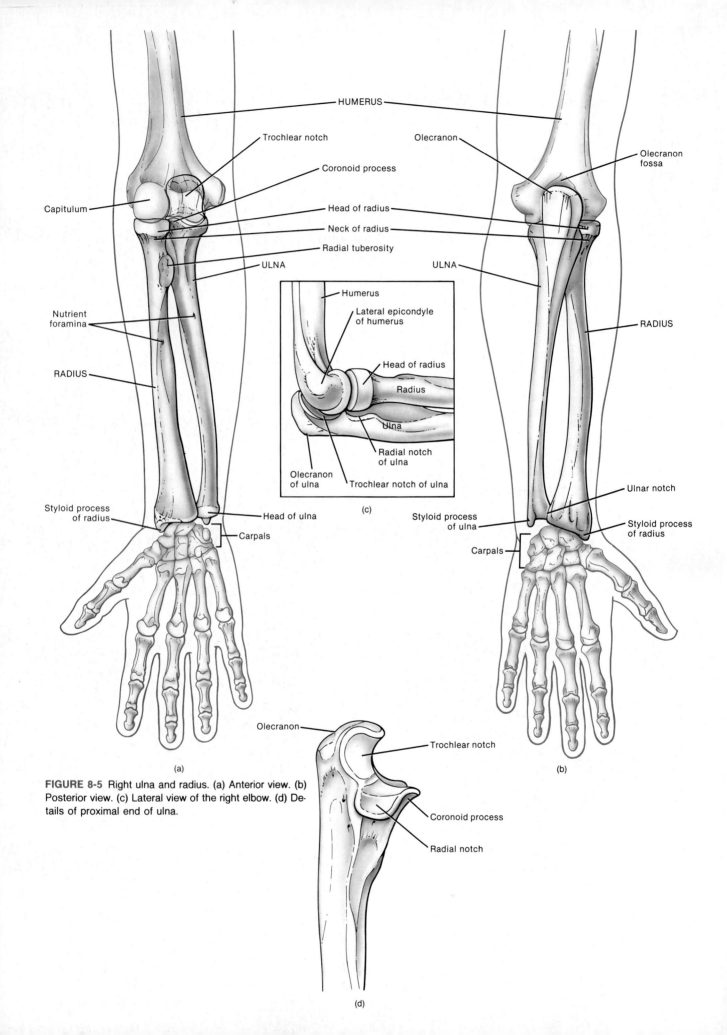

HUMERUS

Trochlear notch

Coronoid process

Capitulum

Head of radius

Neck of radius

Radial tuberosity

ULNA

Nutrient foramina

RADIUS

Styloid process of radius

Head of ulna

Carpals

Olecranon

Olecranon fossa

ULNA

RADIUS

Ulnar notch

Styloid process of ulna

Styloid process of radius

Carpals

Humerus

Lateral epicondyle of humerus

Head of radius

Radius

Ulna

Radial notch of ulna

Olecranon of ulna

Trochlear notch of ulna

(c)

(a)

(b)

Olecranon

Trochlear notch

Coronoid process

Radial notch

(d)

FIGURE 8-5 Right ulna and radius. (a) Anterior view. (b) Posterior view. (c) Lateral view of the right elbow. (d) Details of proximal end of ulna.

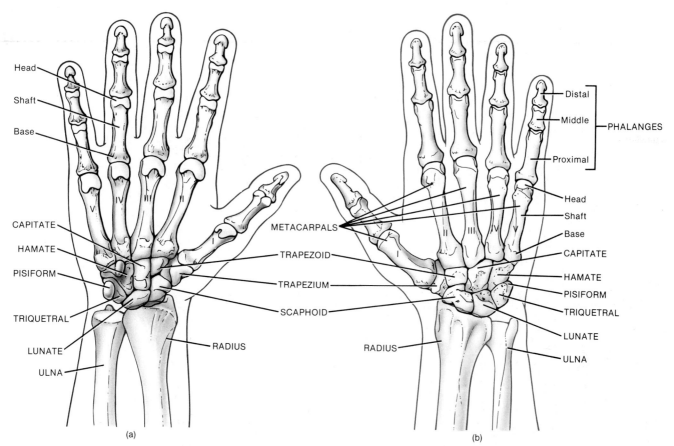

FIGURE 8-6 Right wrist and hand. (a) Anterior view. (b) Posterior view.

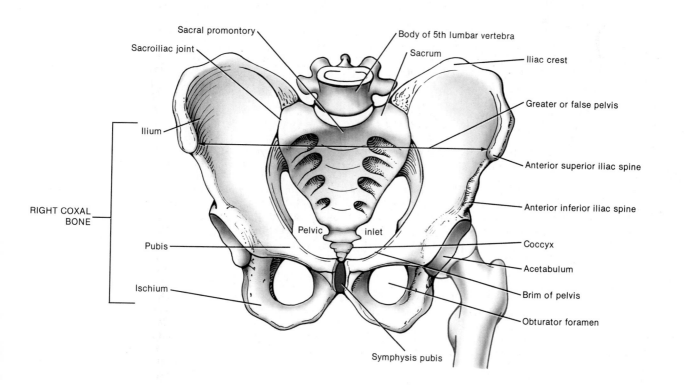

FIGURE 8-7 Pelvic girdle in anterior view.

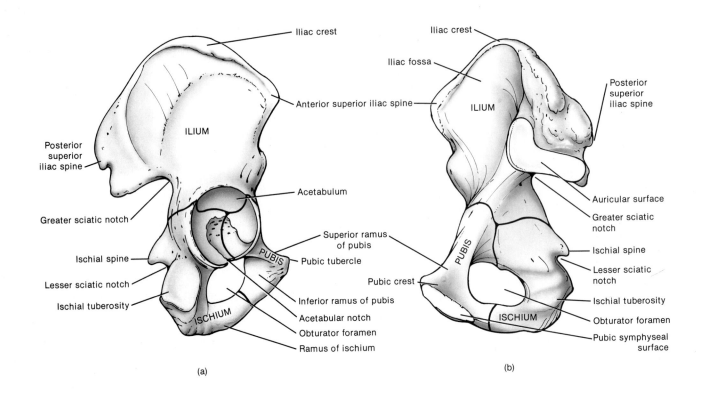

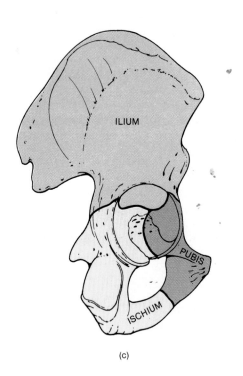

FIGURE 8-8 Right coxal bone. (a) Lateral view. (b) Medial view. The lines of fusion of the ilium, ischium, and pubis are not actually visible in an adult bone. (c) Three divisions of the coxal bone.

The **greater** or **false pelvis** represents the expanded portion situated superior to the brim of the pelvis. The greater pelvis consists laterally of the superior portions of the ilia and posteriorly of the superior portion of the sacrum. There is no bony component in the anterior aspect of the greater pelvis. Rather, the front is formed by the walls of the abdomen.

The **lesser** or **true pelvis** is inferior and posterior to the brim of the pelvis. It is formed by the inferior portions of the ilia and the sacrum, the coccyx, and the pubes. The lesser pelvis contains a superior opening called the **pelvic inlet** and an inferior opening called the **pelvic outlet.**

CLINICAL APPLICATION
Pelvimetry is the measurement of the size of the inlet and outlet of the birth canal. Measurement of the pelvic cavity is important to the physician, because the fetus must pass through the narrower opening of the lesser pelvis at birth.

Each of the two **coxal bones (os coxae)** of a newborn consists of three components: a superior **ilium,** an inferior and anterior **pubis,** and an inferior and posterior **ischium** (Figure 8-8). Eventually, the three separate bones fuse into one. The area of fusion is a deep, lateral fossa called the **acetabulum** (as'-e-TAB-yoo-lum). Although the adult coxae are both single bones, it is common to discuss the bones as if they still consisted of three portions.

united to each other anteriorly at the symphysis (SIM-fi-sis) pubis. They unite posteriorly to the sacrum.

Together with the sacrum and coccyx, the two bones of the pelvic girdle form the basinlike structure called the **pelvis.** The pelvis is divided into a greater pelvis and a lesser pelvis by an oblique plane that passes through sacral promontory (posterior), iliopectineal lines (laterally), and symphysis pubis (anteriorly). The circumference of this oblique plane is called the **brim of the pelvis.**

The ilium is the largest of the three subdivisions of the coxal bone. Its superior border, the **iliac crest,** ends anteriorly in the **anterior superior iliac spine.** Posteriorly, the iliac crest ends in the **posterior superior iliac spine.** The spines serve as points of attachment for muscles of the abdominal wall. Slightly inferior to the posterior inferior iliac spine is the **greater sciatic (sī-AT-ik) notch.** The internal surface of the ilium seen from the medial side is the **iliac fossa.** It is a concavity where the iliacus muscle attaches. Posterior to this fossa is the auricular surface, which articulates with the sacrum.

The ischium is the inferior, posterior portion of the coxal bone. It contains a prominent **ischial spine,** a **lesser sciatic notch** below the spine, and an **ischial tuberosity.** The rest of the ischium, the **ramus,** joins with the pubis and together they surround the **obturator (OB-too-rā'-ter) foramen.**

The pubis is the anterior and inferior part of the coxal bone. It consists of a **superior ramus,** an **inferior ramus,** and a **body** that contributes to the formation of the symphysis pubis.

The **symphysis pubis** is the joint between the two coxal bones (see Figure 8-7). It consists of fibrocartilage. The **acetabulum** is the fossa formed by the ilium, ischium, and pubis. It is the socket for the head of the femur. Two-fifths of the acetabulum is formed by the ilium, two-fifths by the ischium, and one-fifth by the pubis.

LOWER EXTREMITIES

The **lower extremities** are composed of 60 bones (Figure 8-9). Each extremity includes the femur in the thigh, patella (kneecap), fibula and tibia in the leg, tarsals (ankle bones), metasarsals, and phalanges in the toes.

FEMUR

The **femur,** or thighbone, is the longest and heaviest bone in the body (Figure 8-10). Its proximal end articulates with the coxal bone. Its distal end articulates with the tibia. The shaft of the femur bows medially so that it approaches the femur of the opposite thigh. As a result of this convergence, the knee joints are brought nearer to the body's line of gravity. The degree of convergence is greater in the female because the female pelvis is broader.

The proximal end of the femur consists of a rounded **head** that articulates with the acetabulum of the coxal bone. The **neck** of the femur is a constricted region distal to the head. A fairly common fracture in the elderly occurs at the neck of the femur. Apparently the neck becomes so weak that it fails to support the body. The **greater trochanter (trō-KAN-ter)** and **lesser trochanter** are projections that serve as points of attachment for some of the thigh and buttock muscles.

The shaft of the femur contains a rough vertical ridge on its posterior surface called the **linea aspera.** This ridge serves for the attachment of several thigh muscles.

The distal end of the femur is expanded and includes the **medial condyle** and **lateral condyle.** These articulate with the tibia. Superior to the condyles are the **medial epicondyle** and **lateral epicondyle.** A depressed area between the condyles on the posterior surface is called the **intercondylar (in'-ter-KON-di-lar) fossa.** The **patellar surface** is located between the condyles on the anterior surface.

PATELLA

The **patella,** or kneecap, is a small, triangular bone anterior to the knee joint (Figure 8-11). It is a sesamoid bone that develops in the tendon of the quadriceps femoris muscle. The broad superior end of the patella is called the **base.** The pointed inferior end is the **apex.** The posterior surface contains two **articular facets,** one for the medial condyle and the other for the lateral condyle of the femur.

TIBIA AND FIBULA

The **tibia,** or shinbone, is the larger, medial bone of the leg (Figure 8-12). It bears the major portion of the weight of the leg. The tibia articulates at its proximal end with the femur and fibula and at its distal end with the fibula of the leg and talus of the ankle.

The proximal end of the tibia is expanded into a **lateral condyle** and a **medial condyle.** These articulate with the condyles of the femur. The inferior surface of the lateral condyle articulates with the head of the fibula. The slightly concave condyles are separated by an upward projection called the **intercondylar eminence.** The **tibial tuberosity** on the anterior surface is a point of attachment for the patellar ligament.

The medial surface of the distal end of the tibia forms the **medial malleolus (mal-LĒ-ō-lus).** This structure articulates with the talus bone of the ankle and forms the prominence that can be felt on the medial surface of your ankle. The **fibular notch** articulates with the fibula.

The **fibula** is parallel and lateral to the tibia. It is considerably smaller than the tibia. The **head** of the fibula, the proximal end, articulates with the inferior surface of the lateral condyle of the tibia below the level of the knee joint. The distal end has a projection called the **lateral malleolus,** which articulates with the talus bone of the ankle. This forms the prominence on the lateral surface of the ankle. The inferior portion of the fibula also articulates with the tibia at the fibular notch. A fracture of the lower end of the fibula with injury to the tibial articulation is called a **Pott's fracture.**

TARSUS, METATARSUS, AND PHALANGES

The **tarsus** is a collective designation for the seven bones of the ankle called **tarsals** (Figure 8-13). The term *tarsos* pertains to a broad, flat surface. The **talus** and **calcaneus (kal-KĀ-nē-us)** are located on the posterior part of the foot. The anterior part contains the **cuboid, navicular,** and three **cuneiform bones** called the first (medial), second (intermediate), and third (lateral) cuneiform. The talus, the uppermost tarsal bone, is the only bone of the foot that articulates with the fibula and tibia. It is surrounded on one side by the medial malleolus of the tibia and on the other side by the lateral malleolus of the fibula. During walking, the talus initially bears the entire weight of the

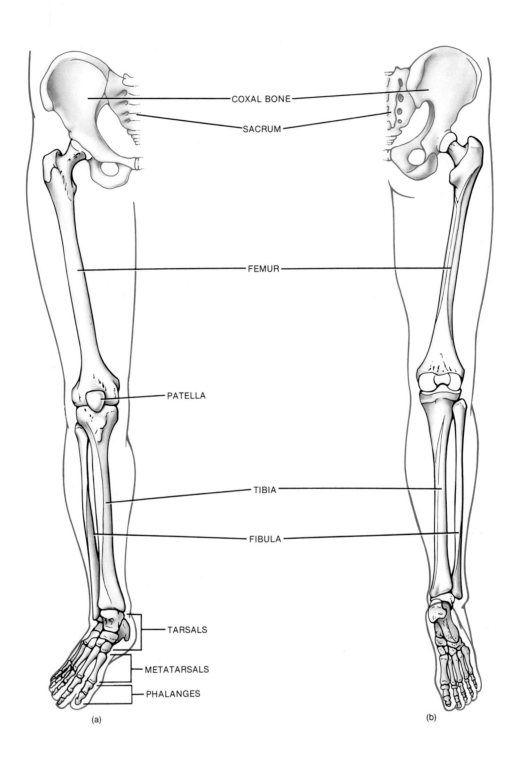

FIGURE 8-9 Right pelvic girdle and lower extremity. (a) Anterior view. (b) Posterior view.

extremity. About half the weight is then transmitted to the calcaneus. The remainder is transmitted to the other tarsal bones. The calcaneus, or heel bone, is the largest and strongest tarsal bone.

The **metatarsus** consists of five metatarsal bones numbered I to V from the medial to lateral position. Like the metacarpals of the palm of the hand, each metatarsal consists of a proximal **base,** a **shaft,** and a distal **head.**

The metatarsals articulate proximally with the first, second, and third cuneiform bones and with the cuboid. Distally, they articulate with the proximal row of phalanges. The first metatarsal is thicker than the others because it bears more weight.

The **phalanges** of the foot resemble those of the hand both in number and arrangement. Each also consists of a proximal **base,** a **shaft,** and a distal **head.** The great

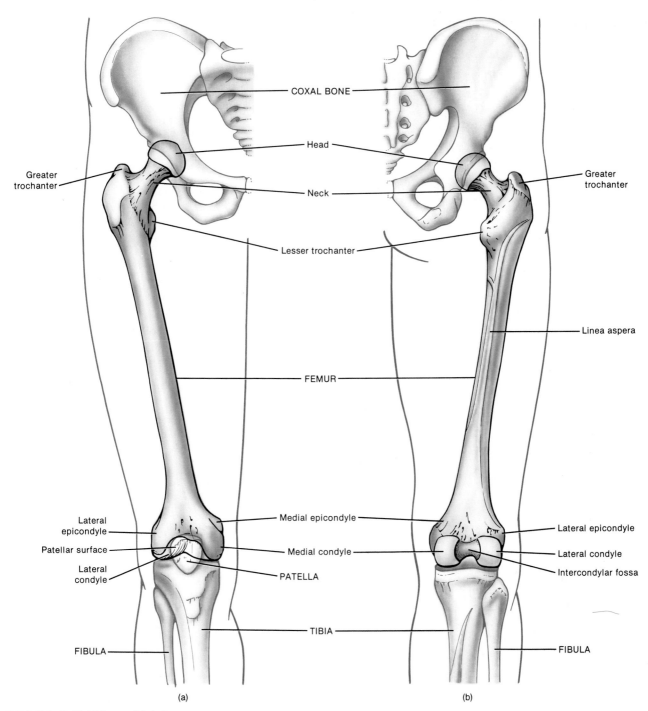

COXAL BONE

Head

Greater
trochanter

Neck

Greater
trochanter

Lesser trochanter

Linea aspera

FEMUR

Lateral
epicondyle

Medial epicondyle

Lateral epicondyle

Patellar surface

Medial condyle

Lateral condyle

Lateral
condyle

PATELLA

Intercondylar fossa

TIBIA

FIBULA

FIBULA

(a)

(b)

FIGURE 8-10 Right femur. (a) Anterior view. (b) Posterior view.

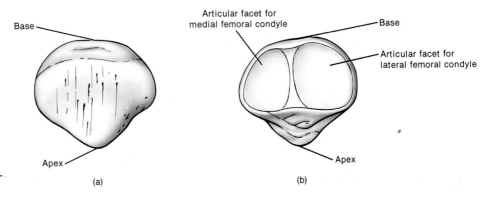

Base

Articular facet for
medial femoral condyle

Base

Articular facet for
lateral femoral condyle

Apex

Apex

(a)

(b)

FIGURE 8-11 Right patella. (a) Anterior view. (b) Posterior view.

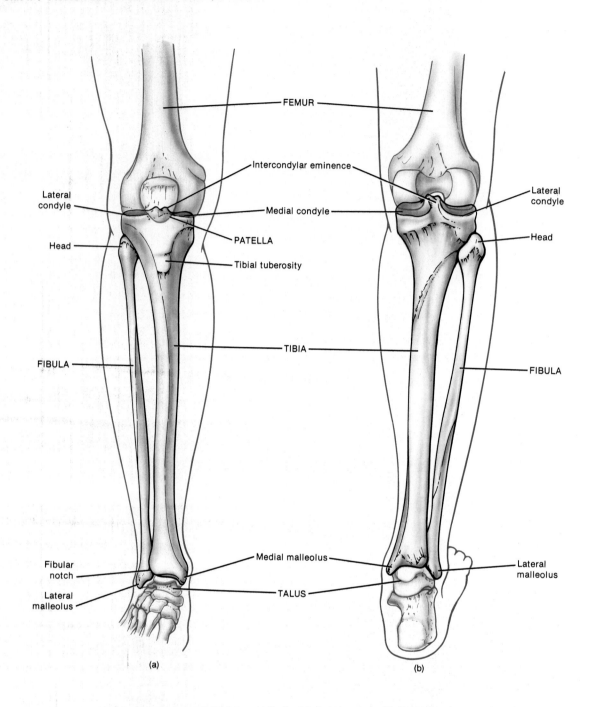

FIGURE 8-12 Right tibia and fibula. (a) Anterior view. (b) Posterior view.

(big) toe, or *hallux,* has two large, heavy phalanges called proximal and distal phalanges. The other four toes each have three phalanges—proximal, middle, and distal.

ARCHES OF THE FOOT

The bones of the foot are arranged in two **arches** (Figure 8-14). These arches enable the foot to support the weight of the body and provide leverage while walking. The arches are not rigid. They yield as weight is applied and spring back when the weight is lifted.

The **longitudinal arch** has two parts. Both consist of

tarsal and metatarsal bones arranged to form an arch from the anterior to the posterior part of the foot. The **medial,** or inner, part of the longitudinal arch originates at the calcaneus. It rises to the talus and descends through the navicular, the three cuneiforms, and the three medial metatarsals. The talus is the keystone of this arch. The **lateral,** or outer, part of the longitudinal arch also begins at the calcaneus. It rises at the cuboid and descends to the two lateral metatarsals. The cuboid is the keystone of the arch.

The **transverse arch** is formed by the calcaneus, navicular, cuboid, and the posterior parts of the five metatarsals.

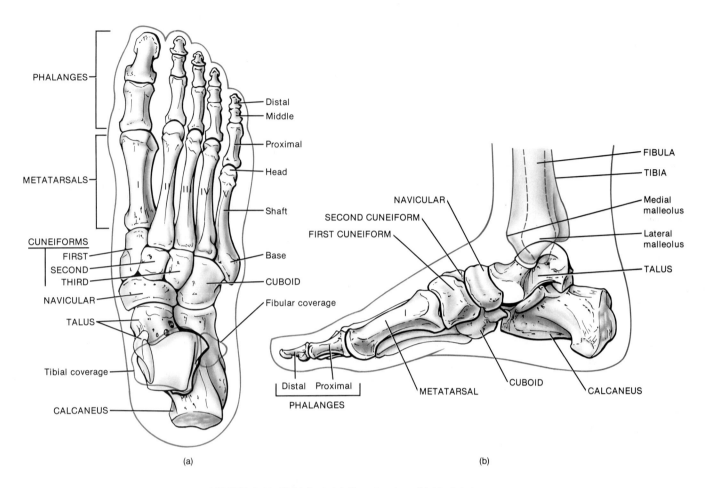

FIGURE 8-13 Right foot. (a) Superior view. (b) Medial view.

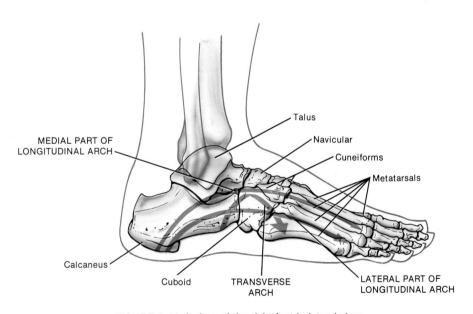

FIGURE 8-14 Arches of the right foot in lateral view.

CLINICAL APPLICATION

The bones composing the arches are held in position by ligaments and tendons. If these ligaments and tendons are weakened, the height of the medial longitudinal arch may decrease or "fall." The result is **flatfoot.**

Clawfoot is a condition in which the medial longitudinal arch is abnormally elevated. It is frequently caused by muscle imbalance, such as may result from poliomyelitis.

A **bunion** is an abnormal lateral displacement of the great toe from its natural position. This condition produces an inflammatory reaction of the bursae (a pouch of fluid located at a joint) that results in the formation of abnormal tissue.

MALE AND FEMALE SKELETONS

The bones of the male are generally larger and heavier than those of the female. The articular ends are thicker in relation to the shafts. In addition, since certain muscles of the male are larger than those of the female, the points of attachment—tuberosities, lines, ridges—are larger in the male skeleton.

Many significant structural differences between male and female skeletons are noted in the pelvis; most are related to pregnancy and childbirth. The typical differences are listed in Exhibit 8-1 and illustrated in Figure 8-15.

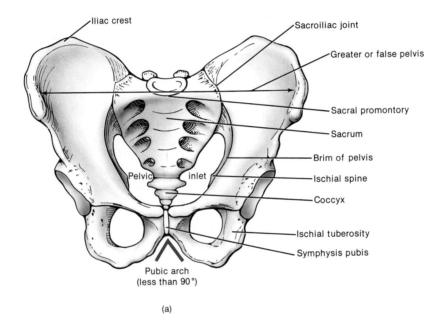

(a)

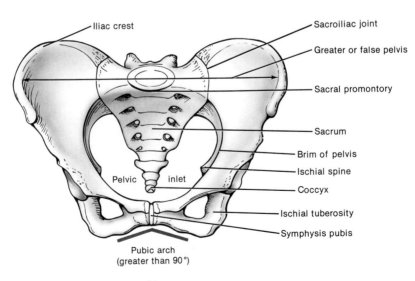

(b)

FIGURE 8-15 Pelvis. (a) Male pelvis in anterior view. (b) Female pelvis in anterior view.

EXHIBIT 8-1

COMPARISON OF TYPICAL MALE AND FEMALE PELVIS

POINT OF COMPARISON	MALE	FEMALE
General structure	Heavy and thick.	Light and thin.
Joint surfaces	Large.	Small.
Muscle attachments	Well marked.	Rather indistinct.
Greater pelvis	Deep.	Shallow.
Pelvic inlet	Heart-shaped.	Larger and more oval.
Pelvic outlet	Comparatively small.	Comparatively large.
First piece of sacrum	Superior surface of the body spans nearly half the width of sacrum.	Superior surface of the body spans about a third the width of sacrum.
Sacrum	Long, narrow, with smooth concavity.	Short, wide, flat, curving forward in lower part.
Auricular surface	Extends well down the third piece of the sacrum.	Extends only to upper border of third piece of the sacrum.
Pubic arch	Less than 90° angle.	Greater than 90° angle.
Inferior ramus of pubis	Presents strong everted surface for attachment of the crus of the penis.	Everted surface not present.
Pubic symphysis	More deep.	Less deep.
Ischial spine	Turned inward more.	Turned inward less.
Ischial tuberosity	Turned inward.	Turned outward.
Ilium	More vertical.	Less vertical.
Iliac fossa	Deep.	Shallow.
Iliac crest	More curved.	Less curved.
Anterior superior iliac spine	Closer.	Wider apart.
Acetabulum	Large.	Small.
Obturator foramen	Round.	Oval.
Greater sciatic notch	Narrow.	Wide.

STUDY OUTLINE

Pectoral (Shoulder) Girdles (p. 164)
1. Each pectoral or shoulder girdle consists of a clavicle and scapula.
2. Each attaches an upper extremity to the trunk.

Upper Extremities (p. 164)
The bones of each upper extremity include the humerus, ulna, radius, carpals, metacarpals, and phalanges.

Pelvic Girdle (p. 167)
1. The pelvic girdle consists of two coxal bones or hipbones.
2. It attaches the lower extremities to the trunk at the sacrum.
3. Each coxal bone consists of three fused components—ilium, pubis, and ischium.

Lower Extremities (p. 171)
1. The bones of each lower extremity include the femur, tibia, fibula, tarsals, metatarsals, and phalanges.
2. The bones of the foot are arranged in two arches, the longitudinal arch and the transverse arch, to provide support and leverage.

Male and Female Skeletons (p. 176)
1. Male bones are generally larger and heavier than female bones and have more prominent markings for muscle attachment.
2. The female pelvis is adapted for pregnancy and childbirth. Differences in pelvic structure are listed in Exhibit 8-1.

REVIEW QUESTIONS

1. What is the pectoral (shoulder) girdle? Why is it important?
2. What are the bones of the upper extremity? What is a Colles' fracture?
3. What is the pelvic girdle? Why is it important?
4. What are the bones of the lower extremity? What is a Pott's fracture?
5. What is pelvimetry? What is its clinical importance?
6. In what ways do the upper extremity and lower extremity differ structurally?
7. Describe the structure of the longitudinal and transverse arches of the foot. What is the function of an arch?
8. How do flatfeet, clawfeet, and bunions arise?
9. What are the principal structural differences between typical male and female skeletons? Use Exhibit 8-1 as a guide in formulating your response.

- Define an articulation and identify the factors that determine the degree of movement at a joint.

- Contrast the structure, kind of movement, and location of fibrous, cartilaginous, and synovial joints.

- Explain the principle of arthroscopy and its clinical importance.

- Discuss and compare the movements possible at various synovial joints.

- Describe selected articulations of the body with respect to the bones that enter into their formation, structural classification, and anatomical components.

- Describe the causes and symptoms of common joint disorders, including rheumatism, rheumatoid arthritis (RA), osteoarthritis, gouty arthritis, bursitis, dislocation and sprain, and tendinitis.

- Define medical terminology associated with articulations.

- Explain the actions of selected drugs associated with articulation.

9 Articulations

Bones are too rigid to bend without damage. Fortunately, the skeletal system consists of many separate bones, which are held together at joints by flexible connective tissue. All movements that change the positions of the bony parts of the body occur at joints. You can understand the importance of joints if you imagine how a cast over the knee joint prevents flexing the leg or how a splint on a finger limits the ability to manipulate small objects.

An **articulation,** or **joint,** is a point of contact between bones or between cartilage and bones. The joint's structure determines how it functions. Some joints permit no movement, others permit slight movement, and still others afford considerable movement. In general, the closer the fit at the point of contact, the stronger the joint. At tightly fitted joints, however, movement is restricted. The looser the fit, the greater the movement. Unfortunately, loosely fitted joints are prone to dislocation. Movement at joints is also determined by the flexibility of the connective tissue that binds the bones together and by the position of ligaments, muscles, and tendons.

CLASSIFICATION

FUNCTIONAL

The functional classification of joints takes into account the degree of movement they permit. Functionally, joints are classified as **synarthroses** (sin'-ar-THRŌ-sēz), which are immovable joints; **amphiarthroses** (am'-fē-ar-THRŌ-sēz), which are slightly movable joints; and **diarthroses** (dī'-ar-THRŌ-sēz), which are freely movable joints.

STRUCTURAL

The structural classification of joints is based on the presence or absence of a joint cavity (a space between the articulating bones) and the kind of connective tissue that binds the bones together. Structurally, joints are classified as **fibrous,** in which there is no joint cavity and the bones are held together by fibrous connective tissue; **cartilaginous,** in which there is no joint cavity and the bones are held together by cartilage; and **synovial,** in which there is a joint cavity and the bones forming the joint are united by a surrounding articular capsule and frequently by accessory ligaments (described in detail later). We will discuss the joints of the body based upon their structural classification, but with reference to their functional classification as well.

FIBROUS JOINTS

Fibrous joints lack a joint cavity, and the articulating bones are held very closely together by fibrous connective tissue. They permit little or no movement. The three types of fibrous joints are (1) sutures, (2) syndesmoses, and (3) gomphoses.

SUTURE

Sutures are found between bones of the skull. In a suture, the bones are united by a thin layer of dense fibrous connective tissue. Since sutures are immovable, they are functionally classified as synarthroses. Some sutures, present during growth, are replaced by bone in the adult. In this case they are called **synostoses** (sin'-os-TŌ-sēz), or bony joints—joints in which there is a complete fusion of bone across the suture line. An example is the frontal suture between the left and right sides of the frontal bone (see Figure 7-3a). Synostoses are also functionally classified as synarthroses.

SYNDESMOSIS

A **syndesmosis** (sin'-dez-MŌ-sis) is a fibrous joint in which the uniting fibrous connective tissue is present in a much greater amount than in a suture, but the fit between the bones is not quite as tight. The fibrous connective tissue forms an interosseous membrane or ligament. A syndesmosis is slightly movable because the bones are separated more than in a suture and some flexibility is permitted by the interosseous membrane or ligament. Thus, syndesmoses are functionally classified as amphiarthrotic. Examples of syndesmoses include the distal articulation of the tibia and fibula (see Figure 8-12) and the articulations between the shafts of the ulna and radius (see Figure 8-5).

GOMPHOSIS

A **gomphosis** (gom-FŌ-sis) is a type of fibrous joint in which a cone-shaped peg fits into a socket. The intervening substance is the periodontal ligament. A gomphosis is functionally classified as synarthrotic. Examples are the articulations of the roots of the teeth with the sockets (alveoli) of the maxillae and mandible.

CARTILAGINOUS JOINTS

Another joint that has no joint cavity is the **cartilaginous joint.** Here the articulating bones are tightly connected by cartilage. Like fibrous joints, they allow little or no movement. The two types of cartilaginous joints are (1) synchondroses and (2) symphyses.

SYNCHONDROSIS

A **synchondrosis** (sin'-kon-DRŌ-sis) is a cartilaginous joint in which the connecting material is hyaline cartilage. The most common type of synchondrosis is the epiphyseal plate (see Figure 6-5). Such a joint is found between the epiphysis and diaphysis of a growing bone and is immovable. Thus it is synarthrotic. Since the hyaline cartilage is eventually replaced by bone when growth ceases, the joint is temporary. It is replaced by a synostosis. Another example of a synchondrosis is the joint between the first rib and the sternum. The cartilage in this joint undergoes ossification during adult life.

SYMPHYSIS

A **symphysis** (SIM-fi-sis) is a cartilaginous joint in which the connecting material is a broad, flat disc of fibrocartilage. This joint is found between bodies of vertebrae (see

Figure 7-11). A portion of the intervertebral disc is cartilaginous material. The symphysis pubis between the anterior surfaces of the coxal bones is another example (see Figure 8-15). These joints are slightly movable, or amphiarthrotic.

SYNOVIAL JOINTS

STRUCTURE

A joint in which there is a space between articulating bones is called a **synovial** (si-NŌ-ve-al) **joint.** The space, or **joint cavity,** is also called a **synovial cavity** (Figure 9-1). Because of this cavity and because of the arrangement of the articular capsule and accessory ligaments, synovial joints are freely movable. Thus, synovial joints are functionally classified as diarthrotic.

Synovial joints are also characterized by the presence of **articular cartilage.** Articular cartilage covers the surfaces of the articulating bones, but does not bind the bones together. The articular cartilage of synovial joints is hyaline cartilage.

Synovial joints are surrounded by a sleevelike **articular capsule** that encloses the joint cavity and unites the articulating bones. The articular capsule is composed of two layers. The outer layer, the *fibrous capsule,* consists of dense connective (collagenous) tissue. It is attached to the periosteum of the articulating bones at a variable distance from the edge of the articular cartilage. The flexibility of the fibrous capsule permits movement at a joint, whereas its great tensile strength resists dislocation. The fibers of some fibrous capsules are arranged in parallel bundles and are therefore highly adapted to resist recurrent strain. Such fibers are called *ligaments* and are given

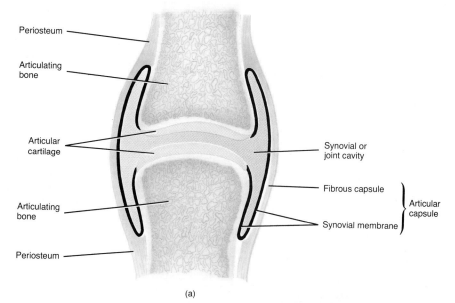

(a)

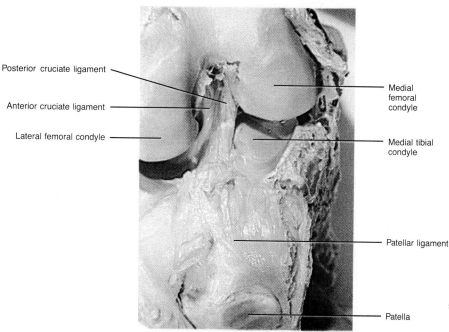

(b)

FIGURE 9-1 Synovial joint. (a) Diagram. (b) Photograph of the internal structure of the right knee joint. (Courtesy of John W. Eads.)

special names. The strength of the ligaments is one of the principal factors in holding bone to bone.

The inner layer of the articular capsule is formed by a *synovial membrane.* The synovial membrane is composed of loose connective tissue with elastic fibers and a variable amount of adipose tissue. It secretes *synovial fluid,* which lubricates the joint and provides nourishment for the articular cartilage. Synovial fluid also contains phagocytic cells that remove microbes and debris resulting from wear and tear in the joint. Synovial fluid consists of hyaluronic acid and an interstitial fluid formed from blood plasma and is similar in appearance and consistency to egg white. When there is no joint movement, the fluid is viscous, but as movement increases, the fluid becomes less viscous. The amount of synovial fluid varies in different joints of the body, ranging from a thin, viscous layer to about 3.5 ml (about 0.1 oz) of free fluid in a large joint such as the knee. The amount present in each joint is sufficient only to form a thin film over the surfaces within an articular capsule.

Many synovial joints also contain **accessory ligaments,** which are called extracapsular ligaments and intracapsular ligaments. *Extracapsular ligaments* are outside of the articular capsule. An example is the fibular collateral ligament of the knee joint (see Figure 9-8). *Intracapsular ligaments* occur within the articular capsule, but are excluded from the joint cavity by reflections of the synovial membrane. Examples are the cruciate ligaments of the knee joint (see Figure 9-8).

Inside some synovial joints, there are pads of fibrocartilage that lie between the articular surfaces of the bones and are attached by their margins to the fibrous capsule. These pads are called **articular discs (menisci).** The discs usually subdivide the joint cavity into two separate spaces. Articular discs allow two bones of different shapes to fit tightly; they modify the shape of the joint surfaces of the articulating bones. Articular discs also help to maintain the stability of the joint and direct the flow of synovial fluid to areas of greatest friction.

CLINICAL APPLICATION

A tearing of articular discs in the knee, commonly called **torn cartilage,** occurs frequently among athletes. Such damaged cartilage requires surgical removal (meniscectomy) or it will begin to wear and cause arthritis. Until recently, knee joint surgery for torn cartilage necessitated cutting through layers of healthy tissue and removing much, if not all, of the cartilage. This procedure is usually painful and expensive and does not always provide full recovery.

These problems have been overcome by **arthroscopy** or **arthroscopic surgery.** The arthroscope is a device, several inches long, that resembles a pencil and is designed like a telescope with a series of lenses aligned one above the other. Its optic fibers give off light, like a lamp on a miner's hard hat. It is inserted into the knee joint through an incision as small as a quarter-inch. A second small incision is made to insert a tube through which a salt solution is injected into the joint.

Another small incision is used for insertion of an instrument that shaves off and reshapes the damaged cartilage and then sucks out the shaved cartilage along with the salt solution. Some orthopedic surgeons attach a lightweight television camera to the arthroscope so that the image from inside the knee can be projected onto a screen. Since arthroscopy requires only small incisions, recovery is usually speedy and there is very little pain or discomfort. Although arthroscopy is used mostly to remove torn cartilage, it can also be used for other types of knee surgery and for surgery on other joints of the body.

The various movements of the body create friction between moving parts. To reduce this friction, saclike structures called **bursae** are situated in the body tissues. These sacs resemble joints in that their walls consist of connective tissue lined by a synovial membrane. They are also filled with a fluid similar to synovial fluid. Bursae are located between the skin and bone in places where skin rubs over bone. They are also found between tendons and bones, muscles and bones, and ligaments and bones. As fluid-filled sacs, they cushion the movement of one part of the body over another. An inflammation of a bursa is called **bursitis.**

The articular surfaces of synovial joints are kept in contact with each other by several factors. One factor is the fit of the articulating bones. This interlocking is very obvious at the hip joint, where the head of the femur articulates with the acetabulum of the coxal bone. Another factor is the strength of the joint ligaments. This is especially important in the hip joint. A third factor is the tension of the muscles around the joint. For example, the fibrous capsule of the knee joint is formed principally from tendinous expansions by muscles acting on the joint.

MOVEMENTS

The movement permitted at synovial joints is limited by several factors. One is the **apposition of soft parts.** For example, during bending of the elbow the anterior surface of the forearm is pressed against the anterior surface of the arm. This apposition limits movement. A second factor is the **tension of ligaments.** The different components of a fibrous capsule are tense only when the joint is in certain positions. Tense ligaments not only restrict the limitation of movement, but also direct the movement of the articulating bones with respect to each other. In the knee joint, for example, the major ligaments are lax when the knee is bent, but tense when the knee is straightened. Also, when the knee is straightened the surfaces of the articulating bones are in fullest contact with each other. A third factor that restricts movement at a synovial joint is **muscle tension.** This reinforces the restraint placed on a joint by ligaments. A good example of the effect of muscle tension on a joint is seen at the hip joint. When the thigh is raised with the knee straight, the movement is restricted by the tension of the hamstring muscles on the posterior surface of the thigh. But if the knee is bent, the tension on the hamstring muscles is lessened and the thigh can be raised further.

Following is a description of the specific movements that occur at synovial joints.

Gliding

A **gliding movement** is the simplest kind that can occur at a joint. One surface moves back and forth and from side to side over another surface without angular or rotary motion. Some joints that glide are those between the carpals and between the tarsals. The heads and tubercles of ribs glide on the bodies and transverse processes of vertebrae.

Angular

Angular movements increase or decrease the angle between bones. Among the angular movements are flexion, extension, abduction, and adduction (Figure 9-2). **Flexion** usually involves a decrease in the angle between the anterior surfaces of the articulating bones. An exception to this definition is flexion of the knee and the toe joints in which there is a decrease in the angle between the posterior surfaces of the articulating bones. Examples of flexion include bending the head forward (the joint is between the occipital bone and the atlas), bending the elbow, and bending the knee. Flexion of the foot at the ankle joint is called **dorsiflexion.**

Extension involves an increase in the angle between the anterior surfaces of the articulating bones, with the same exceptions of knee and toe joints. Extension restores a body part to its anatomical position after it has been flexed. Examples of extension are returning the head to the anatomical position after flexion, straightening the arm after flexion, and straightening the leg after flexion. Continuation of extension beyond the anatomical position, as in bending the head backward is called **hyperextension.** Extension of the foot at the ankle joint is **plantar flexion.**

Abduction usually means movement of a bone *away from* the midline of the body. An example of abduction is moving the arm upward and away from the body until it is held straight out at right angles to the chest. With the fingers and toes, however, the midline of the body is not used as the line of reference. Abduction of the fingers is a movement away from an imaginary line drawn through the middle finger; in other words, it is spreading the fingers. Abduction of the toes is relative to an imaginary line drawn through the second toe.

Adduction is usually movement of a part *toward* the midline of the body. An example of adduction is returning the arm to the side after abduction. As in abduction, adduction of the fingers is relative to the middle finger, and adduction of the toes is relative to the second toe.

Rotation

Rotation is the movement of a bone around its own longitudinal axis. During rotation, no other motion is permitted. In *medial rotation* the anterior surface of a bone or extremity moves toward the midline. In *lateral rotation,* the anterior surface moves away from the midline. We rotate the atlas around the odontoid process of the axis when we shake the head from side to side. Moving from the shoulder and turning the forearm up, then palm down, and then palm up again is an example of slight lateral and medial rotation of the humerus (Figure 9-3a).

Circumduction

Circumduction is a movement in which the distal end of a bone moves in a circle while the proximal end remains stable. The bone describes a cone in the air. Circumduction typically involves flexion, abduction, adduction, extension, and rotation. It involves a 360° rotation. An example is moving the outstretched arm in a circle to wind up to pitch a ball (Figure 9-3b).

Special

Special movements are those found only at the joints indicated in Figure 9-4. **Inversion** is the movement of the sole of the foot inward (medially) at the ankle joint. **Eversion** is the movement of the sole outward (laterally) at the ankle joint.

Protraction is the movement of the mandible or clavicle forward on a plane parallel to the ground. Thrusting the jaw outward is protraction of the mandible. Bringing your arms forward until the elbows touch requires protraction of the clavicle. **Retraction** is the movement of a protracted part of the body backward on a plane parallel to the ground. Pulling the lower jaw back in line with the upper jaw is retraction of the mandible.

Supination is a movement of the forearm in which the palm of the hand is turned anterior or superior. To demonstrate supination, flex your arm at the elbow to prevent rotation of the humerus in the shoulder joint. **Pronation** is a movement of the flexed forearm in which the palm is turned posterior or inferior.

Elevation is a movement in which a part of the body moves upward. You elevate your mandible when you close your mouth. **Depression** is a movement in which a part of the body moves downward. You depress your mandible when you open your mouth. The shoulders can also be elevated and depressed.

TYPES

Though all synovial joints are similar in structure, variations exist in the shape of the articulating surfaces. Accordingly, synovial joints are divided into six subtypes: gliding, hinge, pivot, ellipsoidal, saddle, and ball-and-socket joints.

Gliding

The articulating surfaces of bones in **gliding joints** or **arthrodia** (ar-THRŌ-dē-a) are usually flat. Only side-to-side and back-and-forth movements are permitted (Figure 9-5a). Since this joint allows movements in two planes, it is called *biaxial.* Twisting and rotation are inhibited

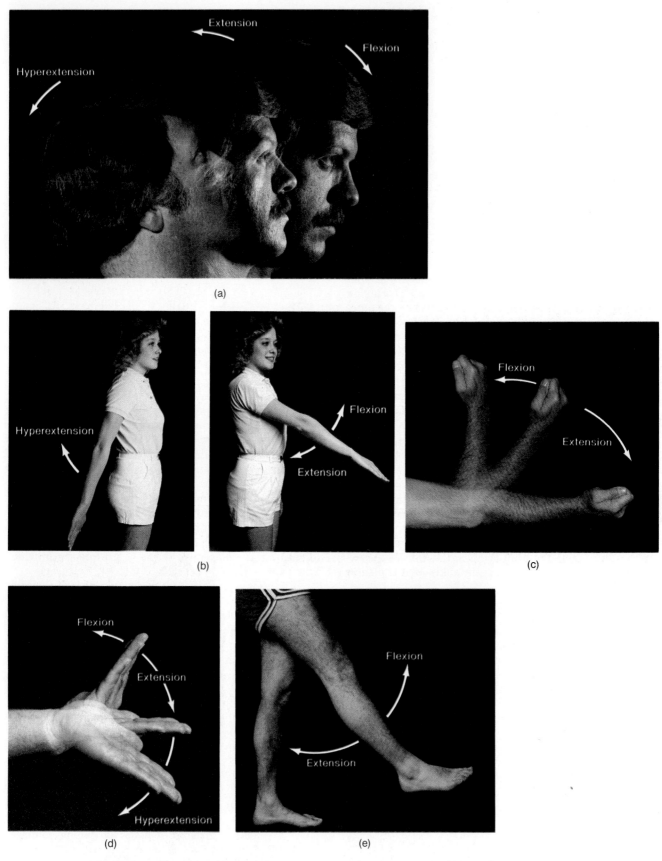

FIGURE 9-2 Angular movements at synovial joints. (© 1983 by Gerard J. Tortora. Courtesy of Lynne Tortora and James Borghesi.)

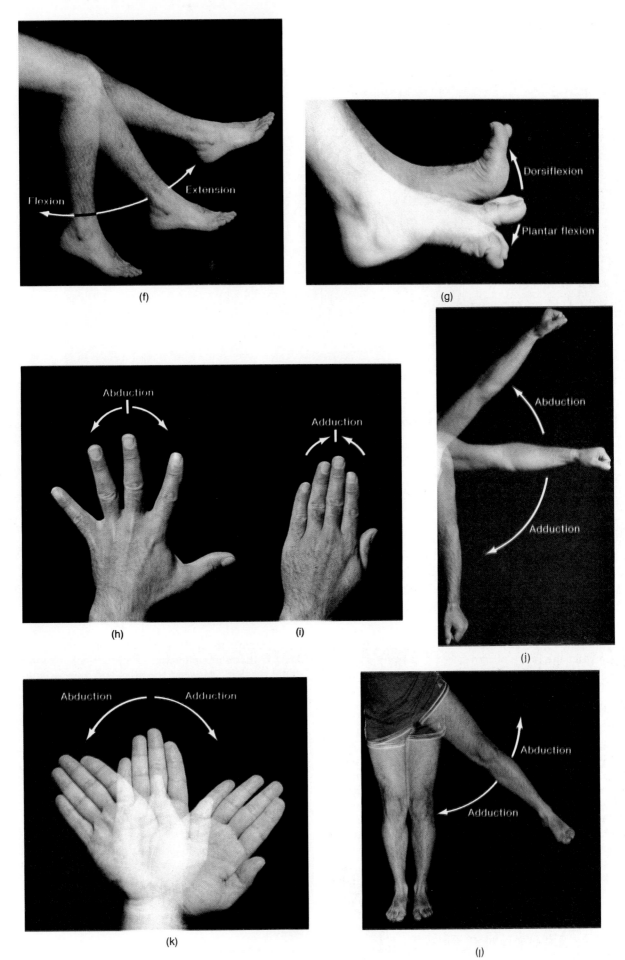

FIGURE 9-2 (*Continued*) Angular movements at synovial joints.

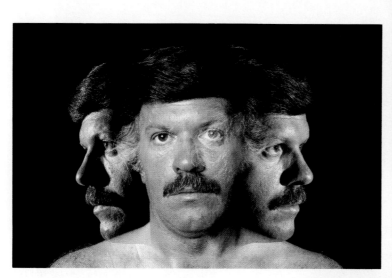

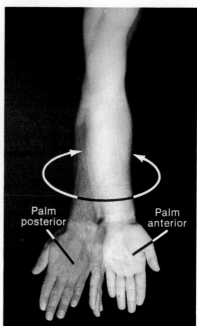

(a)

FIGURE 9-3 Rotation and circumduction. (a) Rotation at the atlantoaxial joint (left) and rotation of the humerus (right). (b) Circumduction of the humerus at the shoulder joint. (© 1983 by Gerard J. Tortora.)

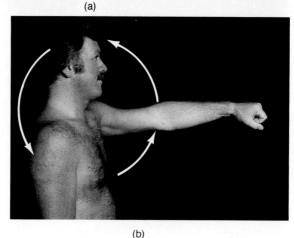

(b)

at gliding joints generally because ligaments or adjacent bones restrict the range of movement. Examples are the joints between carpal bones, tarsal bones, the sternum and clavicle, and the scapula and clavicle.

Hinge

A **hinge** or **ginglymus** (JIN-gli-mus) **joint** is one in which the convex surface of one bone fits into the concave surface of another bone. Movement is primarily in a single plane, and the joint is therefore known as *monaxial,* or *uniaxial* (Figure 9-5b). The motion is similar to that of a hinged door. Movement is usually flexion and extension. Examples of hinge joints are the elbow, ankle, and interphalangeal joints. The movement allowed by a hinge joint is illustrated by flexion and extension at the elbow (see Figure 9-2c).

Pivot

In a **pivot** or **trochoid** (TRŌ-koyd) **joint,** a rounded, pointed, or conical surface of one bone articulates within a ring formed partly by bone and partly by a ligament.

The primary movement permitted is rotation, and the joint is therefore *monaxial* (Figure 9-5c). Examples include the joint between the atlas and axis (atlantoaxial) and between the proximal ends of the radius and ulna. Movement at a pivot joint is illustrated by supination and pronation of the palms and rotation of the head from side to side (see Figure 9-3a).

Ellipsoidal

In an **ellipsoidal** or **condyloid** (KON-di-loyd) **joint,** an oval-shaped condyle of one bone fits into an elliptical cavity of another bone. Since the joint permits side-to-side and back-and-forth movements, it is *biaxial* (Figure 9-5d). The joint at the wrist between the radius and carpals is ellipsoidal. The movement permitted by such a joint is illustrated when you flex and extend and abduct and adduct the wrist (see Figure 9-3d,k).

Saddle

In a **saddle** or **sellaris** (sel-A-ris) **joint,** the articular surfaces of both bones are saddle-shaped, that is, concave

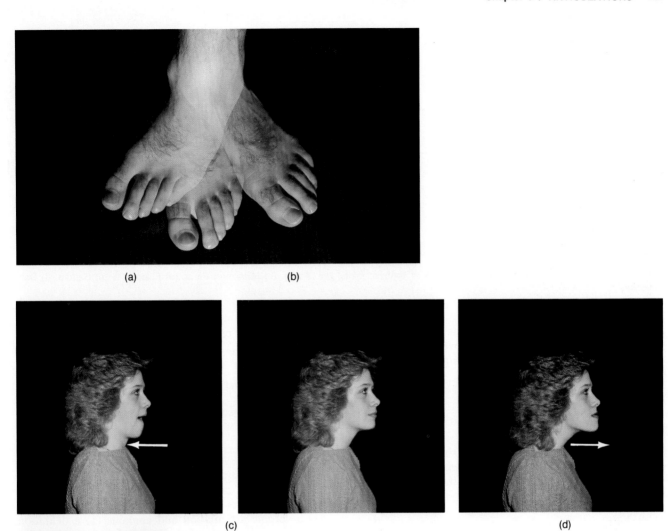

(a)   (b)

(c)   (d)

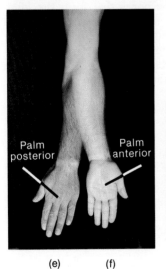

Palm posterior   Palm anterior

(e)   (f)

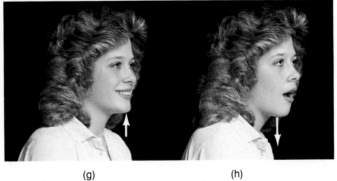

(g)   (h)

FIGURE 9-4 Special movements. (a) Inversion. (b) Eversion. (c) Retraction. (d) Protraction. (e) Pronation. (f) Supination. (g) Elevation. (h) Depression. (© 1983 by Gerard J. Tortora. Courtesy of Lynne Tortora and James Borghesi.)

in one direction and convex in the other. Essentially, the saddle joint is a modified ellipsoidal joint in which the movement is somewhat freer. Movements at a saddle joint are side to side and back and forth. Thus the joint is *biaxial* (Figure 9-5e). The joint between the trapezium and metacarpal of the thumb is an example of a saddle joint.

Ball-and-Socket

A **ball-and-socket** or **spheroid** (SFĒ-royd) **joint** consists of a ball-like surface of one bone fitted into a cuplike depression of another bone. Such a joint permits *triaxial* movement, or movement in three planes of motion: flexion-extension, abduction-adduction, and rotation (Figure

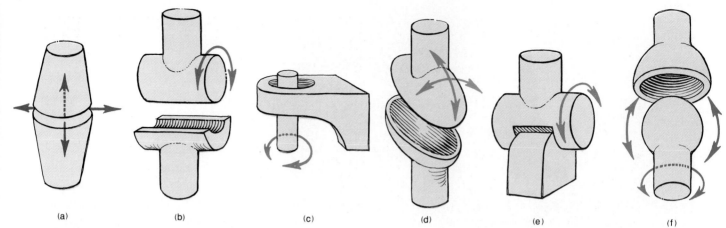

FIGURE 9-5 Simplified representations of subtypes of synovial joints. (a) Gliding. (b) Hinge. (c) Pivot. (d) Ellipsoidal. (e) Saddle. (f) Ball-and-socket.

9-5f). Examples of ball-and-socket joints are the shoulder joint and hip joint. The range of movement at a ball-and-socket joint is illustrated by circumduction of the arm (Figure 9-3b).

SUMMARY OF JOINTS

The summary of joints presented in Exhibit 9-1 is based on the anatomy of the joints. If we rearrange the types of joints into a classification based on movement, we arrive at the following:

Synarthroses: immovable joints
1. Suture
2. Synchondrosis
3. Gomphosis

Amphiarthroses: slightly movable joints
1. Syndesmosis
2. Symphysis

Diarthroses: freely movable joints
1. Gliding
2. Hinge
3. Pivot
4. Ellipsoidal
5. Saddle
6. Ball-and-socket

SELECTED ARTICULATIONS OF THE BODY

We will now examine in some detail selected articulations of the body. In order to simplify your learning efforts, a series of exhibits has been prepared. Each exhibit considers a specific articulation and contains (1) a definition, that is, a description of the bones that form the joint; (2) the type of joint (its structural classification); and (3) its anatomical components—a description of the major connecting ligaments, articular disc, articular capsule, and other distinguishing features of the joint. Each exhibit also refers you to an illustration of the joint.

EXHIBIT 9-1

JOINTS

TYPE	DESCRIPTION	MOVEMENT	EXAMPLES
FIBROUS	No joint cavity; bones held together by a thin layer of fibrous tissue or dense fibrous tissue.		
Suture	Found only between bones of the skull; articulating bones separated by a thin layer of fibrous tissue.	None (synarthrotic).	Lambdoidal suture between occipital and parietal bones.
Syndesmosis	Articulating bones united by dense fibrous tissue.	Slight (amphiarthrotic).	Distal ends of tibia and fibula.
Gomphosis	Cone-shaped peg fits into a socket; articulating bones separated by periodontal ligament.	None (synarthrotic).	Roots of teeth in alveolar processes.
CARTILAGINOUS	No joint cavity; articulating bones united by cartilage.		
Synchondrosis	Connecting material is hyaline cartilage.	None (synarthrotic).	Temporary joint between the diaphysis and epiphyses of a long bone and permanent joint between first rib and sternum.
Symphysis	Connecting material is a broad, flat disc of fibrocartilage.	Slight (amphiarthrotic).	Intervertebral joints and symphysis pubis.
SYNOVIAL	Joint cavity and articular cartilage present; articular capsule composed of an outer fibrous capsule and an inner synovial membrane; may contain accessory ligaments, articular discs (menisci), and bursae.	Freely movable (diarthrotic).	
Gliding	Articulating surfaces usually flat.	Biaxial (flexion-extension, abduction-adduction).	Intercarpal and intertarsal joints.
Hinge	Spoollike surface fits into a concave surface.	Monaxial (flexion-extension).	Elbow, ankle, and interphalangeal joints.
Pivot	Rounded, pointed, or concave surface fits into a ring formed partly by bone and partly by a ligament.	Monaxial (rotation).	Atlantoaxial and radioulnar joints.
Ellipsoidal	Oval-shaped condyle fits into an elliptical cavity.	Biaxial (flexion-extension, abduction-adduction).	Radiocarpal joint.
Saddle	Articular surfaces concave in one direction and convex in opposite direction.	Biaxial (flexion-extension, abduction-adduction).	Carpometacarpal joint of thumb.
Ball-and-socket	Ball-like surface fits into a cuplike depression.	Triaxial (flexion-extension, abduction-adduction, rotation).	Shoulder and hip joints.

EXHIBIT 9-2
HUMEROSCAPULAR (SHOULDER) JOINT (Figure 9-6)

DEFINITION	Joint formed by the head of the humerus and the glenoid cavity of the scapula.
TYPE OF JOINT	Synovial joint, ball-and-socket (spheroid) type.

ANATOMICAL COMPONENTS

1. **Articular capsule.** Loose sac that completely envelops the joint, extending from the circumference of the glenoid cavity to the anatomical neck of the humerus.
2. **Coracohumeral ligament.** Strong, broad ligament that extends from the coracoid process of the scapula to the greater tubercle of the humerus.
3. **Glenohumeral ligaments.** Three thickenings of the articular capsule over the ventral surface of the joint.
4. **Transverse humeral ligament.** Narrow sheet extending from the greater tubercle to the lesser tubercle of the humerus.
5. **Glenoid labrum.** Narrow rim of fibrocartilage around the edge of the glenoid cavity.
6. Among the bursae associated with the shoulder joint are:
   a. **Subscapular bursa** between the tendon of the subscapularis muscle and the underlying joint capsule.
   b. **Subdeltoid bursa** between the deltoid muscle and joint capsule.
   c. **Subacromial bursa** between the acromion and joint capsule.
   d. **Subcoracoid bursa** either lies between the coracoid process and joint capsule or appears as an extension from the subacromial bursa.

CLINICAL APPLICATION

The strength and stability of the shoulder joint are not provided by the shape of the articulating bones or its ligaments. Instead, deep muscles of the shoulder and their tendons—subscapularis, supraspinatus, infraspinatus, and teres minor—strengthen and stabilize the shoulder joint (see Exhibit 11-14 and Figure 11-15). The muscles and their tendons are so arranged as to form a nearly complete encirclement of the joint. This arrangement is referred to as the **rotator (musculotendinous) cuff** and is a common site of injury to baseball pitchers, especially tearing of the supraspinatus muscle.

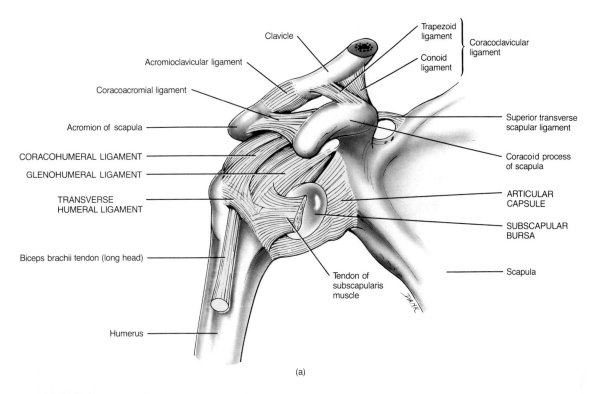

Clavicle

Trapezoid ligament

Conoid ligament

Coracoclavicular ligament

Acromioclavicular ligament

Coracoacromial ligament

Acromion of scapula

CORACOHUMERAL LIGAMENT

GLENOHUMERAL LIGAMENT

TRANSVERSE HUMERAL LIGAMENT

Biceps brachii tendon (long head)

Humerus

Superior transverse scapular ligament

Coracoid process of scapula

ARTICULAR CAPSULE

SUBSCAPULAR BURSA

Scapula

Tendon of subscapularis muscle

(a)

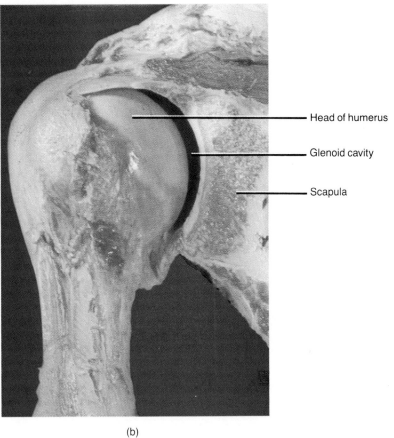

Head of humerus

Glenoid cavity

Scapula

(b)

FIGURE 9-6 Humeroscapular joint. (a) Diagram of anterior view. (b) Photograph of anterior view. (Courtesy of C. Yokochi and J. W. Rohen, *Photographic Anatomy of the Human Body,* 2d ed., 1979, IGAKU-SHOIN, Ltd., Tokyo, New York.)

EXHIBIT 9-3
COXAL (HIP) JOINT (Figure 9-7)

DEFINITION	Joint formed by the head of the femur and the acetabulum of the coxal bone.
TYPE OF JOINT	Synovial, ball-and-socket (spheroid) type.

ANATOMICAL
COMPONENTS

1. **Articular capsule.** Extends from the rim of the acetabulum to the neck of the femur. One of the strongest ligaments of the body, the capsule consists of circular and longitudinal fibers. The circular fibers, called the **zona orbicularis,** form a collar around the neck of the femur. The longitudinal fibers are reinforced by accessory ligaments known as the iliofemoral ligament, the pubofemoral ligament, and the ischiofemoral ligament.
2. **Iliofemoral ligament.** Thickened portion of the articular capsule that extends from the anterior inferior iliac spine of the coxal bone to the intertrochanteric line of the femur.
3. **Pubofemoral ligament.** Thickened portion of the articular capsule that extends from the pubic part of the rim of the acetabulum to the neck of the femur.
4. **Ischiofemoral ligament.** Thickened portion of the articular capsule that extends from the ischial wall of the acetabulum to the neck of the femur.
5. **Ligament of the head of the femur (capitate ligament).** Flat, triangular band that extends from the fossa of the acetabulum to the head of the femur.
6. **Acetabular labrum.** Fibrocartilage rim attached to the margin of the acetabulum.
7. **Transverse ligament of the acetabulum.** Strong ligament which crosses over the acetabular notch, converting it to a foramen. It supports part of the acetabular labrum and is connected with the ligament of the head of the femur and the articular capsule.

CLINICAL APPLICATION

Dislocation of the hip among adults is quite rare because of (1) the stability of the ball-and-socket joint, (2) the strong, tough, articular capsule, (3) the strength of the intracapsular ligaments, and (4) the extensive musculature over the joint.

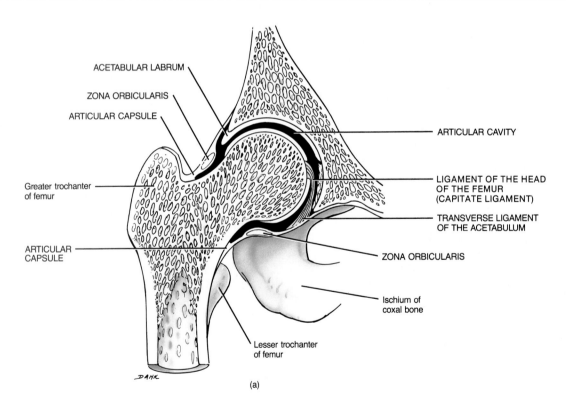

FIGURE 9-7 Coxal (hip) joint. (a) Frontal section.

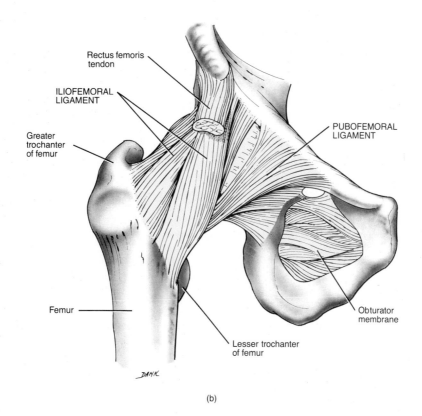

(b)

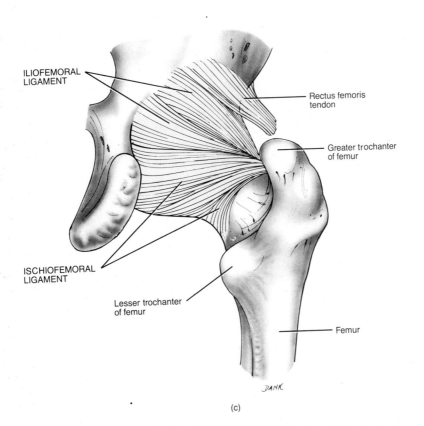

(c)

FIGURE 9-7 (*Continued*) Coxal (hip) joint. (b) Anterior view. (c) Posterior view.

EXHIBIT 9-4

TIBIOFEMORAL (KNEE) JOINT (Figure 9-8)

DEFINITION

The largest joint of the body, actually consisting of three joints: (1) an intermediate patellofemoral joint between the patella and the patellar surface of the femur, (2) a lateral tibiofemoral joint between the lateral condyle of the femur, lateral meniscus, and lateral condyle of the tibia, and (3) a medial tibiofemoral joint between the medial condyle of the femur, medial meniscus, and medial condyle of the tibia.

TYPE OF JOINT

Patellofemoral joint—partly synovial, gliding (arthrodial) type. Lateral and medial tibiofemoral joints—synovial, hinge (ginglymus) type.

ANATOMICAL COMPONENTS

1. **Articular capsule.** No complete, independent capsule uniting the bones. The ligamentous sheath surrounding the joint consists mostly of muscle tendons or expansions of them. There are, however, some capsular fibers connecting the articulating bones.
2. **Medial and lateral patellar retinacula.** Fused tendons of insertion of the quadriceps femoris muscle and the fascia lata that strengthen the anterior surface of the joint.
3. **Patellar ligament.** Central portion of the common tendon of insertion of the quadriceps femoris muscle that extends from the patella to the tibial tuberosity. This also strengthens the anterior surface of the joint. The posterior surface of the ligament is separated from the synovial membrane of the joint by an **infrapatellar fat pad.**
4. **Oblique popliteal ligament.** Broad, flat ligament that connects the intercondylar fossa of the femur to the head of the tibia. The tendon of the semimembranosus muscle is superficial to the ligament and passes from the medial condyle of the tibia to the lateral condyle of the femur. The ligament and tendon afford strength for the posterior surface of the joint.
5. **Arcuate popliteal ligament.** Extends from the lateral condyle of the femur to the styloid process of the head of the fibula. It strengthens the lower lateral part of the posterior surface of the joint.
6. **Tibial collateral ligament.** Broad, flat ligament on the medial surface of the joint that extends from the medial condyle of the femur to the medial condyle of the tibia. The ligament is crossed by tendons of the sartorius, gracilis, and semitendinosus muscles, all of which strengthen the medial aspect of the joint.
7. **Fibular collateral ligament.** Strong, rounded ligament on the lateral surface of the joint that extends from the lateral condyle of the femur to the lateral side of the head of the fibula. The ligament is covered by the tendon of the biceps femoris muscle. The tendon of the popliteus muscle is deep to the tendon.
8. **Intraarticular ligaments.** Ligaments within the capsule that connect the tibia and femur.
   a. **Anterior cruciate ligament.** Extends posteriorly and laterally from the area anterior to the intercondylar eminence of the tibia to the posterior part of the medial surface of the lateral condyle of the femur.

b. **Posterior cruciate ligament.** Extends anteriorly and medially from the posterior intercondylar fossa of the tibia and lateral meniscus to the anterior part of the medial surface of the medial condyle of the femur.

9. **Articular discs.** Fibrocartilage discs between the tibial and femoral condyles. They help to compensate for the incongruence of the articulating bones.

   a. **Medial meniscus.** Semicircular piece of fibrocartilage. Its anterior end is attached to the anterior intercondylar fossa of the tibia, in front of the anterior cruciate ligament. Its posterior end is attached to the posterior intercondylar fossa of the tibia between the attachments of the posterior cruciate ligament and lateral meniscus.

   b. **Lateral meniscus.** Circular piece of fibrocartilage. Its anterior end is attached anterior to the intercondylar eminence of the tibia and lateral and posterior to the anterior cruciate ligament. Its posterior end is attached posterior to the intercondylar eminence of the tibia and anterior to the posterior end of the medial meniscus. The medial and lateral menisci are connected to each other by the **transverse ligament** and to the margins of the head of the tibia by the **coronary ligaments.**

10. The principal **bursae** of the knee include:

    a. **Anterior bursae:** (1) between the patella and skin **(prepatellar bursa)**, (2) between upper part of tibia and patellar ligament **(infrapatellar bursa)**, (3) between lower part of tibial tuberosity and skin, and (4) between lower part of femur and deep surface of quadriceps femoris muscle **(suprapatellar bursa).**

    b. **Medial bursae:** (1) between medial head of gastrocnemius muscle and the articular capsule, (2) superficial to the tibial collateral ligament between the ligament and tendons of the sartorius, gracilis, and semitendinosus muscles, (3) deep to the tibial collateral ligament between the ligament and the tendon of the semimembranosus muscle, (4) between the tendon of the semimembranosus muscle and the head of the tibia, and (5) between the tendons of the semimembranosus and semitendinosus muscles.

    c. **Lateral bursae:** (1) between the lateral head of the gastrocnemius muscle and articular capsule, (2) between the tendon of the biceps femoris muscle and fibular collateral ligament, (3) between the tendon of the popliteus muscle and fibular collateral ligament, and (4) between the lateral condyle of the femur and the popliteus muscle.

CLINICAL APPLICATION

The most common type of **knee injury** in football is rupture of the tibial collateral ligament, often associated with tearing of the anterior cruciate ligament and medial meniscus (torn cartilage). It is caused by a blow to the lateral side of the knee. When a knee is examined for such an injury, the three C's are kept in mind: collateral ligament, cruciate ligament, and cartilage.

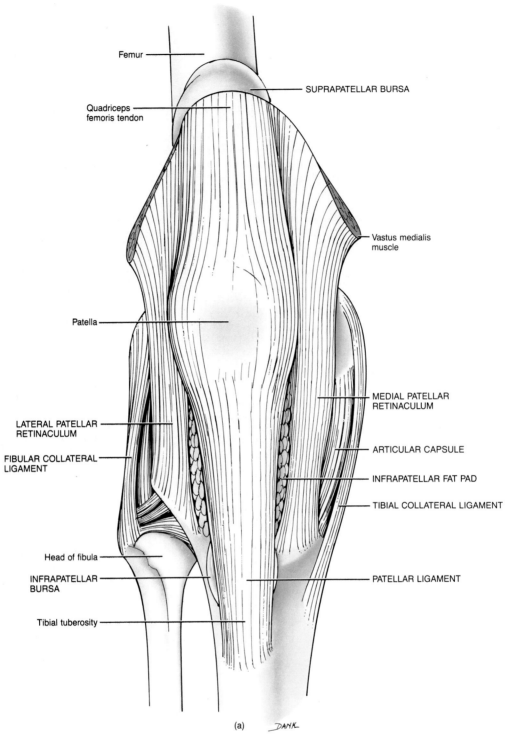

Femur

SUPRAPATELLAR BURSA

Quadriceps femoris tendon

Vastus medialis muscle

Patella

MEDIAL PATELLAR RETINACULUM

LATERAL PATELLAR RETINACULUM

ARTICULAR CAPSULE

FIBULAR COLLATERAL LIGAMENT

INFRAPATELLAR FAT PAD

TIBIAL COLLATERAL LIGAMENT

Head of fibula

INFRAPATELLAR BURSA

PATELLAR LIGAMENT

Tibial tuberosity

(a)   DANK

FIGURE 9-8 Tibiofemoral joint. (a) Diagram of anterior view.

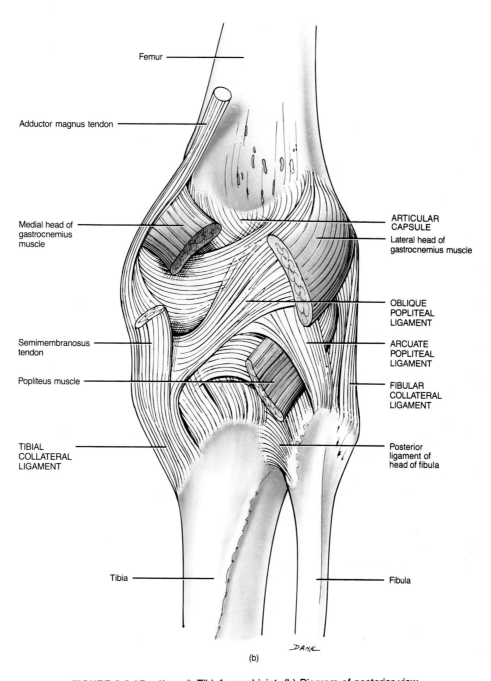

Femur

Adductor magnus tendon

Medial head of gastrocnemius muscle

ARTICULAR CAPSULE

Lateral head of gastrocnemius muscle

OBLIQUE POPLITEAL LIGAMENT

Semimembranosus tendon

ARCUATE POPLITEAL LIGAMENT

Popliteus muscle

FIBULAR COLLATERAL LIGAMENT

TIBIAL COLLATERAL LIGAMENT

Posterior ligament of head of fibula

Tibia

Fibula

DANK

(b)

FIGURE 9-8 (*Continued*) Tibiofemoral joint. (b) Diagram of posterior view.

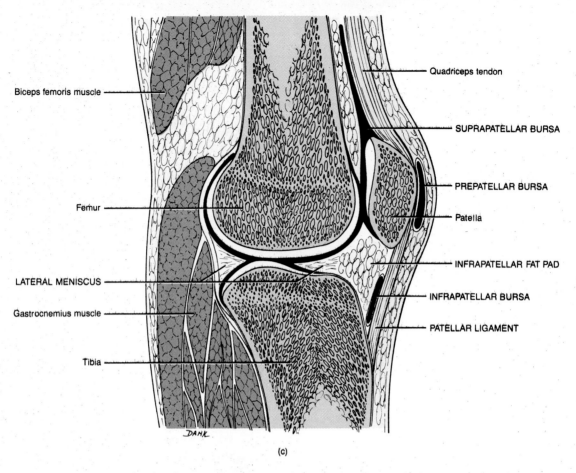

Biceps femoris muscle

Femur

LATERAL MENISCUS

Gastrocnemius muscle

Tibia

DANK

Quadriceps tendon

SUPRAPATELLAR BURSA

PREPATELLAR BURSA

Patella

INFRAPATELLAR FAT PAD

INFRAPATELLAR BURSA

PATELLAR LIGAMENT

(c)

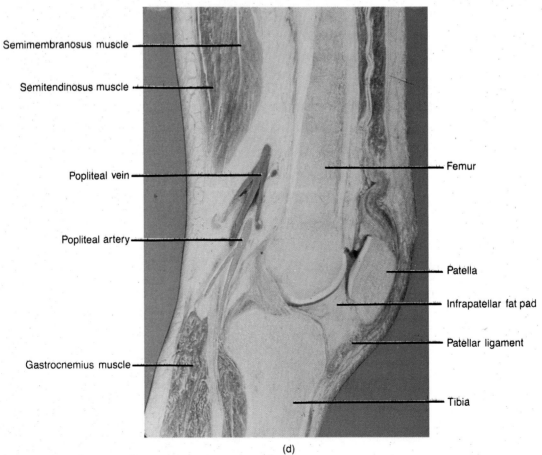

Semimembranosus muscle

Semitendinosus muscle

Popliteal vein

Popliteal artery

Gastrocnemius muscle

Femur

Patella

Infrapatellar fat pad

Patellar ligament

Tibia

(d)

FIGURE 9-8 (*Continued*) Tibiofemoral joint. (c) Diagram of sagittal section. (d) Photograph of sagittal section. (Courtesy of C. Yokochi and J. W. Rohen, *Photographic Anatomy of the Human Body,* 2d ed., 1979, IGAKU-SHOIN, Ltd., Tokyo, New York.)

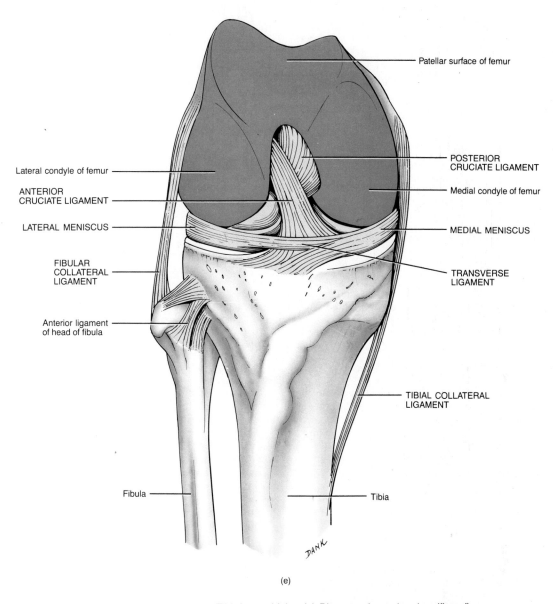

Patellar surface of femur

Lateral condyle of femur

ANTERIOR
CRUCIATE LIGAMENT

LATERAL MENISCUS

FIBULAR
COLLATERAL
LIGAMENT

Anterior ligament
of head of fibula

Fibula

POSTERIOR
CRUCIATE LIGAMENT

Medial condyle of femur

MEDIAL MENISCUS

TRANSVERSE
LIGAMENT

TIBIAL COLLATERAL
LIGAMENT

Tibia

(e)

FIGURE 9-8 (*Continued*) Tibiofemoral joint. (e) Diagram of anterior view (flexed).

DISORDERS: HOMEOSTATIC IMBALANCES

Rheumatism

Rheumatism (*rheumat* = subject to flux) refers to any painful state of the supporting structures of the body—its bones, ligaments, joints, tendons, or muscles. Arthritis is a form of rheumatism in which the joints have become inflamed.

Arthritis

The term **arthritis** refers to at least 25 different diseases, the most common of which are rheumatoid arthritis, osteoarthritis, and gouty arthritis. All these ailments are characterized by inflammation in one or more joints. Inflammation, pain, and stiffness may also be present in adjacent parts of the body, such as the muscles near the joint. Recent evidence suggests that the chronic pain that accompanies various arthritic conditions may be related to the patient's inability to produce endorphins. These chemicals are naturally produced painkillers (Chapter 14).

The causes of arthritis are unknown. In some cases, it follows the stress of sprains, infections, and joint injury. Some researchers think that the cause is a bacterium or virus, whereas others suspect an allergy. Some believe the nervous system or hormones are involved, whereas others suspect a metabolic disorder. Still others believe that certain types of prolonged psychological stress, such as inhibited hostility, can upset homeostatic balance and bring on arthritic attacks.

Rheumatoid Arthritis (RA)

Rheumatoid (ROO-ma-toyd) **arthritis** (RA) is the most common inflammatory form of arthritis. It involves inflammation of the joint, swelling, pain, and a loss of function. Usually this form occurs bilaterally—if your left knee is affected, your right knee may also be affected, although usually not to the same degree.

The primary symptom of rheumatoid arthritis is inflammation of the synovial membrane. If it is untreated, the following sequential pathology may occur. The membrane thickens and synovial fluid accumulates. The resulting pressure causes pain and tenderness. The membrane then produces an abnormal tissue called *pannus,* which adheres to the surface of the articular cartilage. The pannus formation sometimes erodes the cartilage completely. When the cartilage is destroyed, fibrous tissue joins

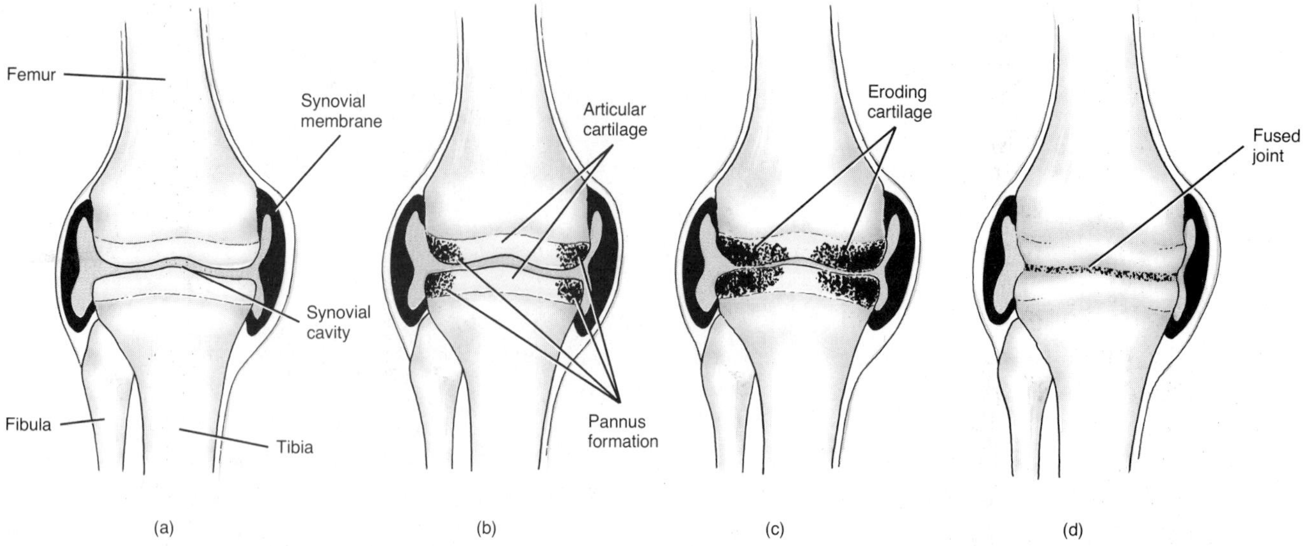

FIGURE 9-9 Rheumatoid arthritis. (a) Inflammation of the synovial membrane. (b) Early stage of pannus formation and erosion of articular cartilage. (c) Advanced stage of pannus formation and further erosion of articular cartilage. (d) Obliteration of joint cavity and fusion of articulating bones.

the exposed bone ends. The tissue ossifies and fuses the joint so that it is immovable—the ultimate crippling effect of rheumatoid arthritis (Figure 9-9). Most cases do not progress to this stage, but the range of motion of the joint is greatly inhibited by the severe inflammation and swelling.

Osteoarthritis

A degenerative joint disease far more common than rheumatoid arthritis, and usually less damaging, is **osteoarthritis** (os'-tē-ō-ar-THRĪ-tis). It apparently results from a combination of aging, irritation of the joints, and wear and abrasion.

Degenerative joint disease is a noninflammatory, progressive disorder of movable joints, particularly weight-bearing joints. It is characterized pathologically by the deterioration of articular cartilage and by formation of new bone in the subchondral areas and at the margins of the joint. The cartilage slowly degenerates, and as the bone ends become exposed, small bumps, or *spurs,* of new osseous tissue are deposited on them. These spurs decrease the space of the joint cavity and restrict joint movement. Unlike rheumatoid arthritis, osteoarthritis usually affects only the articular cartilage. The synovial membrane is rarely destroyed, and other tissues are unaffected.

Gouty Arthritis

Uric acid is a waste product produced during the metabolism of nucleic acids. Normally, all the acid is quickly excreted in the urine. In fact, it gives urine its name. The person who suffers from *gout* either produces excessive amounts of uric acid or is not able to excrete normal amounts. The result is a buildup of uric acid in the blood. This excess acid then reacts with sodium to form a salt called sodium urate. Crystals of this salt are deposited in soft tissues. Typical sites are the kidneys and the cartilage of the ears and joints.

In **gouty** (GOW-tē) **arthritis,** sodium urate crystals are deposited in the soft tissues of the joints. The crystals irritate the cartilage, causing inflammation, swelling, and acute pain. Eventually, the crystals destroy all the joint tissues. If the disorder is not treated, the ends of the articulating bones fuse and the joint becomes immovable.

Gouty arthritis occurs primarily in males of any age. It is believed to be the cause of 2 to 5 percent of all chronic joint diseases. Numerous studies indicate that gouty arthritis is sometimes caused by an abnormal gene. As a result of this gene, the body manufactures unusually large amounts of uric acid. Diet and environmental factors such as stress and climate are also suspected causes of gouty arthritis.

Although other forms of arthritis cannot be treated with complete success, the treatment of gouty arthritis with the use of various drugs has been quite effective. A chemical called colchicine has been utilized periodically since the sixth century to relieve the pain, swelling, and tissue destruction that occurs during attacks of gouty arthritis. This chemical is derived from the variety of crocus plant from which the spice saffron is obtained. Other drugs, which either inhibit uric acid production or assist in the elimination of excess uric acid by the kidneys, are used to prevent further attacks. The drug allopurinal is used for the treatment of gouty arthritis because it prevents the formation of uric acid without interfering with nucleic acid synthesis.

Bursitis

An acute chronic inflammation of a bursa is called **bursitis.** The condition may be caused by trauma, by an acute or chronic infection (including syphilis and tuberculosis), or by rheumatoid arthritis. Repeated excessive friction often results in a bursitis with local inflammation and the accumulation of fluid. Bunions are frequently associated with a friction bursitis over the head of the first metatarsal bone. Symptoms include pain, swelling, tenderness, and the limitation of motion involving the inflamed bursa. The prepatellar or subcutaneous infrapatellar bursa may become inflamed in individuals who spend a great deal of time kneeling. This bursitis is usually called **"housemaid's knee"** or **"carpet layer's knee."**

Aspiration of the knee joint may be necessary to relieve pressure, to evacuate blood, or to obtain a fluid sample for laboratory studies. In this procedure, the needle is inserted somewhat proximal and lateral to the patella through the tendinous part of the vastus lateralis muscle and is directed toward the middle of the joint. Through the same route, the joint cavity may be anesthetized or irrigated, and steroids may be administered in the treatment of knee pathologies.

Dislocation

A **dislocation** or **luxation** (luks-Ā-shun) is the displacement of a bone from a joint with tearing of ligaments, tendons, and articular capsules. A partial or incomplete dislocation is called a **subluxation.** The most common dislocations are those involving a finger, thumb, or shoulder. Those of the mandible, elbow, knee, or hip are less common. Symptoms include loss of motion, temporary paralysis of the involved joint, pain, swelling, and occasionally shock. A dislocation is usually caused by a blow or fall, although unusual physical effort may lead to this condition.

Sprain and Strain

A **sprain** is the forcible wrenching or twisting of a joint with partial rupture or other injury to its attachments without luxation. There may be damage to the associated blood vessels, muscles, tendons, ligaments, or nerves. A sprain is more serious than a **strain,** which is the overstretching of a muscle. Severe sprains may be so painful that the joint cannot be moved. There is considerable swelling, with reddish to blue discoloration due to hemorrhage from ruptured blood vessels. The ankle joint is most often sprained; the low back area is another frequent location for sprains.

MEDICAL TERMINOLOGY

Ankylosis (*ankyle* = stiff joint, *osis* = condition) Severe or complete loss of a movement at a joint.

Arthralgia (*arth* = joint; *algia* = pain) Pain in a joint.

Arthrosis Refers to an articulation; also a disease of a joint.

Bursectomy (*ectomy* = removal of) Removal of a bursa.

Chondritis (*chondro* = cartilage) Inflammation of cartilage.

Rheumatology (*rheumat* = subject to flux) The medical specialty devoted to arthritis.

Synovitis (*synov* = joint) Inflammation of a synovial membrane in a joint.

DRUGS ASSOCIATED WITH ARTICULATIONS

Nonsteroid Antiinflammatories

Without exception, rheumatologists (arthritis specialists) choose *aspirin* as the nonsteroid antiinflammatory drug to counteract joint inflammation and reduce the pain of rheumatoid arthritis. If aspirin is not well tolerated, other drugs include *phenylbutazone* (Butazolidin), *indomethacin* (Indocin), *ibuprofen* (Motrin), *fenoprofen* (Nalfon), *sulindac* (Clinoril), *naproxen* (Nafrosyn), and *tolmetin* (Tolectin).

Steroid Antiinflammatories

In severe cases of arthritis, especially rheumatoid arthritis, corticosteroids may be used for their antiinflammatory properties. The most commonly used of these drugs is *prednisone.*

Analgesics

Analgesics (pain relievers) can be used alone or in combination with antiinflammatories. Examples of analgesics are *acetaminophen* (Tylenol) and *propoxyphene* (Darvon).

Gold Compounds

Certain chemical compounds containing gold can, when given by injection, greatly reduce joint inflammation and slow down joint destruction in many people. An example is *gold sodium thiomalate* (Myochrysine).

Immunosuppressives

Immunosuppressive drugs, such as *azathioprine* (Imuran) and *cyclophosphamide* (Cytoxan), suppress the body's immune responses. They are used occasionally in cases of severe rheumatoid arthritis, but sparingly because of their serious side effects.

DMSO

DMSO stands for *dimethyl sulfoxide* (Rimso-50). It is a clear, colorless, and practically odorless liquid that has been used as an industrial solvent since the 1940s. Although the FDA has approved the use of a 50 percent solution of DMSO for the treatment of interstitial cystitis, a specific urinary bladder disorder, a 90% solution (approved for veterinary use) and a 99% solution (used as an industrial solvent) are also being used as an antiinflammatory, without FDA approval, to treat human disorders such as chronic arthritis, sprains, and strains.

The FDA points to the potential for side effects of DMSO since the drug is rapidly absorbed from the surface of the skin into the bloodstream. Among the side effects are allergic reactions (erythema and itching), headache, nausea, diarrhea, burning on urination, and disturbances in vision. Users also experience an oysterlike taste in their mouths and a garliclike body odor for up to 72 hours after administration. Lens opacities (cloudiness) have also been reported in laboratory animals. Studies of the effectiveness of DMSO for human disorders are now underway.

STUDY OUTLINE

Classification (p. 180)

1. A joint or articulation is a point of contact between two or more bones.
2. Functional classification of joints is based on the degree of movement permitted. Joints may be synarthroses, amphiarthroses, or diarthroses.
3. Structural classification is based on the presence of a joint cavity and type of connecting tissue. Structurally, joints are classified as fibrous, cartilaginous, or synovial.

Fibrous Joints (p. 180)

1. Bones held by fibrous connective tissue, with no joint cavity, are fibrous joints.

2. These joints include immovable sutures (found in the skull), slightly movable syndesmoses (such as the tibiofibular articulation), and immovable gomphoses (roots of teeth in alveoli of mandible and maxilla).

Cartilaginous Joints (p. 180)

1. Bones held together by cartilage, with no joint cavity, are cartilaginous joints.
2. These joints include immovable synchondroses united by hyaline cartilage (temporary cartilage between diaphysis and epiphyses) and partially movable symphyses united by fibrocartilage (the symphysis pubis).

Synovial Joints (p. 181)

1. Synovial joints contain a joint (synovial) cavity, articular cartilage, and a synovial membrane; some also contain ligaments, articular discs, and bursae.
2. All synovial joints are freely movable.
3. Movements at synovial joints are limited by the apposition of soft parts, tension of ligaments, and muscle tension.
4. Types of movements at synovial joints include gliding movements, angular movements, rotation, circumduction, inversion and eversion, protraction and retraction, supination and pronation, and elevation and depression.
5. Types of synovial joints include gliding joints (wrist bones), hinge joints (elbow), pivot joints (radioulnar), ellipsoidal joints (radiocarpal), saddle joints (carpometacarpal), and ball-and-socket joints (shoulder and hip).
6. A joint may be described according to the number of planes of movement it allows as nonaxial, biaxial, or triaxial.

Selected Articulations of the Body (p. 188)

1. The humeroscapular (shoulder joint) is formed by the humerus and scapula.
2. The coxal (hip) joint is formed by the femur and coxal bone.

3. The tibiofemoral (knee) joint is formed by the patella and femur and by the tibia and femur.

Disorders: Homeostatic Imbalances (p. 199)

1. Rheumatism is a painful state of supporting body structures such as bones, ligaments, tendons, joints, and muscles.
2. Arthritis refers to several disorders characterized by inflammation of joints, often accompanied by stiffness of adjacent structures.
3. Rheumatoid arthritis (RA) refers to inflammation of a joint accompanied by pain, swelling, and loss of function.
4. Osteoarthritis is a degenerative joint disease characterized by deterioration of articular cartilage and spur formation.
5. Gouty arthritis is a condition in which sodium urate crystals are deposited in the soft tissues of joints and eventually destroy the tissues.
6. Bursitis is an acute or chronic inflammation of bursae.
7. A dislocation, or luxation, is a displacement of a bone from its joint; a partial dislocation is called subluxation.
8. A sprain is the forcible wrenching or twisting of a joint with partial rupture to its attachments without dislocation, while a strain is the stretching of a muscle.

REVIEW QUESTIONS

1. Define an articulation. What factors determine the degree of movement at joints?
2. Distinguish among the three kinds of joints on the basis of structure and function. List the subtypes. Be sure to include degree of movement and specific examples.
3. Explain the components of a synovial joint. Indicate the relationship of ligaments and tendons to the strength of the joint and restrictions on movement.
4. Explain how the articulating bones in a synovial joint are held together.
5. What is an accessory ligament? Define the two principal types.
6. What is an articular disc? Why are they important?
7. Describe the principle and importance of arthroscopy.
8. What are bursae? What is their function?
9. Define the following principal movements: gliding, angular, rotation, circumduction, and special. Name a joint where each occurs.
10. Have another person assume the anatomical position and execute for you each of the movements at joints discussed in the text. Reverse roles, and see if you can execute the same movements.
11. Contrast monaxial, biaxial, and triaxial planes of movement. Give examples of each, and name a joint at which each occurs.
12. For each joint of the body discussed in Exhibits 9-2 through 9-4, be sure that you can name the bones that form the joint, identify the joint by type, and list the anatomical components of the joint.
13. Define arthritis. What are some suspected causes of arthritis?
14. What is rheumatism?
15. Distinguish between rheumatoid arthritis (RA), osteoarthritis, and gouty arthritis with respect to causes and symptoms.
16. Define bursitis. How is it caused?
17. What is tendinitis? What are some of its causes?
18. Define dislocation. What are the symptoms of dislocation? How is dislocation caused?
19. Distinguish between a sprain and a strain.
20. Refer to the glossary of medical terminology associated with articulations at the end of the chapter. Be sure that you can define each term.
21. Describe the actions of selected drugs associated with articulations.

- List the characteristics and functions of muscle tissue.

- Compare the location, microscopic appearance, nervous control, and functions of the three kinds of muscle tissue.

- Define fascia, epimysium, perimysium, endomysium, tendons, and aponeuroses and list their modes of attachment to muscles.

- Explain the relationship of blood vessels and nerves to skeletal muscles.

- Identify the histological characteristics of skeletal muscle tissue.

- List the principal events associated with the sliding-filament theory.

- Discuss the physiological importance of the motor unit.

- Identify the source of energy for muscular contraction.

- Define the all-or-none principle of muscular contraction.

- Describe different types of normal contractions performed by skeletal muscles.

- Describe the phases of contraction in a typical myogram of a twitch contraction.

- Compare fast (white) muscle with slow (red) muscle.

- Contrast cardiac muscle tissue with smooth muscle tissue.

- Compare oxygen debt, fatigue, and heat production as examples of muscle homeostasis.

- Define such common muscular disorders as fibrosis, fibrositis, "charleyhorse," muscular dystrophy, and myasthenia gravis.

- Compare spasms, cramps, convulsions, fibrillation, and tics as abnormal muscular contractions.

- Define medical terminology associated with the muscular system.

- Explain the actions of selected drugs associated with muscle tissue.

10 Muscle Tissue

Although bones and joints provide leverage and form the framework of the body, they are not capable of moving the body by themselves. Motion is an essential body function that results from the contraction and relaxation of muscles.

Muscle tissue constitutes about 40 to 50 percent of the total body weight and is composed of highly specialized cells.

CHARACTERISTICS

Muscle tissue has four principal characteristics that assume key roles in maintaining homeostasis.

1. **Excitability** is the ability of muscle tissue to receive and respond to stimuli. A stimulus is a change in the internal or external environment strong enough to initiate a nerve impulse (action potential).
2. **Contractility** is the ability to shorten and thicken, or contract, when a sufficient stimulus is received.
3. **Extensibility** is the ability of muscle tissue to stretch. Many skeletal muscles are arranged in opposing pairs. While one is contracting, the other is relaxed and is undergoing extension.
4. **Elasticity** is the ability of muscle to return to its original shape after contraction or extension.

FUNCTIONS

Through contraction, muscle performs three important functions:

1. Motion.
2. Maintenance of posture.
3. Heat production.

Motion is obvious in movements involving the whole body, such as walking and running, and in localized movements, such as grasping a pencil or nodding the head. All of these movements rely on the integrated functioning of the bones, joints, and muscles attached to the bones. Less noticeable kinds of motion produced by muscles are the beating of the heart, the churning of food in the stomach, the pushing of food through the intestines, the contraction of the gallbladder to release bile, and the contraction of the urinary bladder to expel urine.

In addition to the movement function, muscle tissue also enables the body to maintain posture. The contraction of skeletal muscles holds the body in stationary positions, such as standing and sitting.

The third function of muscle tissue is heat production. Skeletal muscle contractions produce heat and are thereby important in maintaining normal body temperature.

TYPES

Types of muscle tissue are categorized by location, microscopic structure, and nervous control.

Skeletal muscle tissue, which is named for its location, is attached to bones. It is **striated** muscle tissue because striations, or bandlike structures, are visible when the tissue is examined under a microscope. It is a **voluntary** muscle tissue because it can be made to contract by conscious control.

Visceral muscle tissue is located in the walls of hollow internal structures, such as blood vessels, the stomach, and the intestines. It is **smooth** or **nonstriated,** that is, it lacks the bandlike structures. It is **involuntary** muscle tissue because its contraction is usually not under conscious control.

Cardiac muscle tissue forms the walls of the heart. It is striated and involuntary.

Thus all muscle tissues are classified in the following way: (1) skeletal, striated, voluntary muscle, (2) visceral, smooth, involuntary muscle, and (3) cardiac, striated, involuntary muscle.

SKELETAL MUSCLE TISSUE

To understand the fundamental mechanisms of muscle movement, you will need some knowledge of its connective tissue components, its nerve and blood supply, and its histology.

FASCIA

The term **fascia** (FASH-ē-a) is applied to a sheet or broad band of fibrous connective tissue beneath the skin or around muscles and other organs of the body. Fasciae may be divided into three types: superficial, deep, and subserous.

The **superficial fascia,** or **subcutaneous layer,** is immediately deep to the skin. It covers the entire body and varies in thickness in different regions. On the back or dorsum of the hand it is quite thin, whereas over the inferior abdominal wall it is thick. The superficial fascia is composed of adipose tissue and loose connective tissue. The outer layer usually contains fat and varies considerably in thickness, while the inner layer is thin and elastic. Between the two layers are found arteries, veins, lymphatics, nerves, the mammary glands, and the facial muscles. Superficial fascia has a number of important functions. (1) It serves as a storehouse for water and particularly for fat. Much of the fat of an overweight person is in the superficial fascia. (2) It forms a layer of insulation protecting the body from loss of heat. (3) It provides mechanical protection from blows. (4) It provides a pathway for nerves and vessels.

The **deep fascia** is by far the most extensive of the three types. It is a dense connective tissue composed of a superficial layer (external investing layer) and a deep layer (internal investing layer). Unlike the superficial fascia, it does not contain fat. The deep fascia lines the body wall and extremities and holds muscles together, separating them into functioning groups. Functionally, deep fascia allows free movement of muscles, carries nerves and blood vessels, fills spaces between muscles, and sometimes provides the origin for muscles.

The **subserous (visceral) fascia** is located between the internal investing layer of deep fascia and a serous membrane. It is composed of loose connective tissue. It forms the fibrous layer of serous membranes, covers and supports the viscera, and attaches the parietal layer of serous membranes to the internal surface of the body wall.

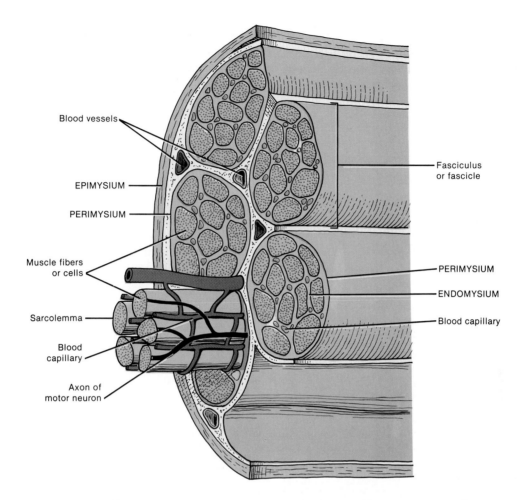

Blood vessels

EPIMYSIUM

PERIMYSIUM

Muscle fibers
or cells

Sarcolemma

Blood
capillary

Axon of
motor neuron

Fasciculus
or fascicle

PERIMYSIUM

ENDOMYSIUM

Blood capillary

FIGURE 10-1 Relationships of connective tissue to skeletal muscle. Shown is a cross section and longitudinal section of a skeletal muscle indicating the relative positions of the epimysium, perimysium and endomysium. Compare this figure with the photomicrograph in Figure 10-2b.

CONNECTIVE TISSUE COMPONENTS

Skeletal muscles are further protected, strengthened, and attached to other structures by several connective tissue components (Figure 10-1). The entire muscle is usually wrapped with a substantial quantity of fibrous connective tissue called the **epimysium** (ep'-i-MĪZ-ē-um). The epimysium is an extension of deep fascia. When the muscle is cut in cross section, invaginations of the epimysium are seen to divide the muscle into bundles of fibers (cells) called **fasciculi** (fa-SIK-yoo-lī) or **fascicles** (FAS-i-kuls). These invaginations of the epimysium are called the **perimysium** (per'-i-MĪZ-ē-um). Perimysium, like epimysium, is an extension of deep fascia. In turn, invaginations of the perimysium, called **endomysium** (en'-dō-MĪZ-ē-um), penetrate into the interior of each fascicle and separate the muscle cells. Endomysium is also an extension of deep fascia.

The epimysium, perimysium, and endomysium are all continuous with the connective tissue that attaches the muscle to another structure, such as bone or other muscle. All three elements may be extended beyond the muscle cells as a **tendon**—a cord of connective tissue that attaches a muscle to the periosteum of a bone. When the connective tissue elements extend as a broad, flat layer, the tendon

is called an **aponeurosis.** This structure also attaches to the coverings of a bone or another muscle. When a muscle contracts, the tendon and its corresponding bone or muscle are pulled toward the contracting muscle. In this way skeletal muscles produce movement. Certain tendons, especially those of the wrist and ankle, are enclosed by tubes of fibrous connective tissue called **tendon sheaths.** They are lined by a synovial membrane that permits the tendon to slide easily within the sheath. The sheaths also prevent the tendons from slipping out of place.

CLINICAL APPLICATION

Tendinitis or **tenosynovitis** (ten'-ō-sin-ō-VĪ-tis) frequently occurs as inflammation involving the tendon sheaths and synovial membrane surrounding certain joints. The wrists, shoulders, elbows (tennis elbow), finger joints (trigger finger), ankles, and associated tendons are most often affected. The affected sheaths may become visibly swollen because of fluid accumulation or they may remain dry. Local tenderness is variable, and there may be disabling pain with movement of the body part. The condition often follows some form of trauma, strain, or excessive exercise.

NERVE AND BLOOD SUPPLY

Skeletal muscles are well supplied with nerves and blood vessels. This innervation and vascularization is directly related to contraction, the chief characteristic of muscle. For a skeletal muscle cell to contract, it must first be stimulated by an impulse from a nerve cell. Muscle contraction also requires a good deal of energy and therefore large amounts of nutrients and oxygen. Moreover, the waste products of these energy-producing reactions must be eliminated. Thus muscle action depends on the blood supply.

Generally, an artery and one or two veins accompany each nerve that penetrates a skeletal muscle. The larger branches of the blood vessel accompany the nerve branches through the connective tissue of the muscle.

Microscopic blood vessels called capillaries are arranged in the endomysium. Each muscle cell is thus in close contact with one or more capillaries. Each skeletal muscle cell usually makes contact with a portion of a nerve cell.

HISTOLOGY

When a typical skeletal muscle is teased apart and viewed microscopically, it can be seen to consist of many elongated, cylindrical cells called **muscle fibers** (Figure 10-2a,b). These fibers lie parallel to one another and range from 10 to 100 μm in diameter. Some fibers may reach lengths of 30 cm (12 inches) or more. Each muscle fiber is enveloped by a plasma membrane called the **sarcolemma** (*sarco* = flesh; *lemma* = sheath). The sarcolemma surrounds a quantity of cytoplasm called **sarcoplasm.** Within the sacroplasm of a muscle fiber and lying close to the sarcolemma are many nuclei and a number of mitochondria. Skeletal muscle fibers are thus multinucleate. Also within a muscle fiber is the **sarcoplasmic reticulum** (sar'-kō-PLAZ-mik rē-TIK-yoo-lum), a network of membrane-enclosed tubules comparable to smooth endoplasmic reticulum (Figure 10-2c). Running transversely through the fiber and perpendicularly to the sarcoplasmic reticulum are **T tubules (transverse tubules).** The tubules are extensions of the sarcolemma that open to the outside of the fiber. A **triad** consists of a T tubule and the segments of sarcoplasmic reticulum on either side.

A highly magnified view of skeletal muscle fibers reveals threadlike structures, about 1 or 2 μm in diameter,

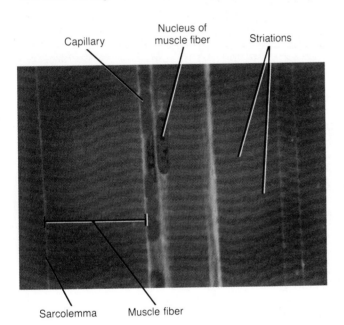

Capillary Nucleus of muscle fiber Striations

Sarcolemma Muscle fiber

(a)

FIGURE 10-2 Histology of skeletal muscle tissue. (a) Photomicrograph of several muscle fibers in longitudinal section at a magnification of 800×. (© 1983 by Michael H. Ross. Used by permission.) (b) Photomicrograph of several muscle fibers in cross section at a magnification of 500×. (© 1983 by Michael H. Ross. Used by permission.)

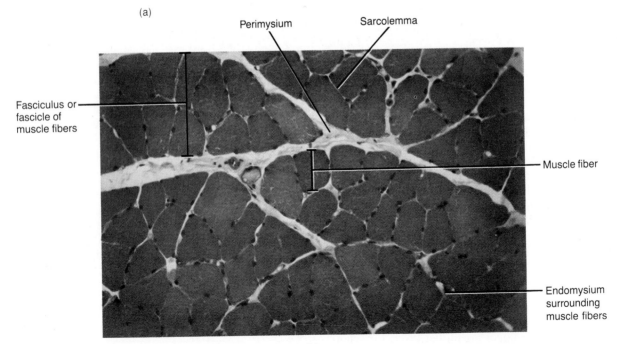

Perimysium Sarcolemma

Fasciculus or fascicle of muscle fibers

Muscle fiber

Endomysium surrounding muscle fibers

(b)

called **myofibrils** (Figure 10-2c,d,e). The prefix *myo* means muscle. The myofibrils, ranging in number from several hundred to several thousand, run longitudinally through the muscle fiber and consist of two kinds of even smaller structures called **myofilaments.** The **thin myofilaments** are about 6 nm in diameter. The **thick myofilaments** are about 16 nm in diameter.

The myofilaments of a myofibril do not extend the entire length of a muscle fiber—they are stacked in compartments called **sarcomeres.** Sarcomeres are separated from one another by narrow zones of dense material called **Z lines.** Each sarcomere is about 2.6 μm long. In a relaxed muscle fiber, that is, one that is not contracting, the thin and thick myofilaments overlap and form a dark, dense band called the **anisotropic band,** or **A band** (Figure 10-2d). Each A band is about 1.6 μm long. A light-colored, less dense area called the **isotropic band,** or **I band,** is composed of thin myofilaments only. The I bands are about 1 μm long. This combination of alternating dark and light bands gives the muscle fiber its striped appearance. A narrow **H zone** contains thick myofilaments only. The H zone is about 0.5 μm long. In the center of the H zone is the **M line,** a series of fine threads that appear to connect the middle parts of the adjacent thick filaments.

The thin myofilaments are composed mostly of the protein **actin.** The actin molecules are arranged in a double-stranded coil that gives the thin myofilaments their characteristic shape (Figure 10-3a). Besides actin, the thin myofilaments contain two other protein molecules, **tropomyosin** and **troponin.** Together they are referred to as a **tropomyosin-troponin complex.**

The thick myofilaments are composed mostly of the protein **myosin.** A myosin molecule is shaped like a rod with a round head. The rods form the long axis of the thick myofilaments, and the heads form projections called

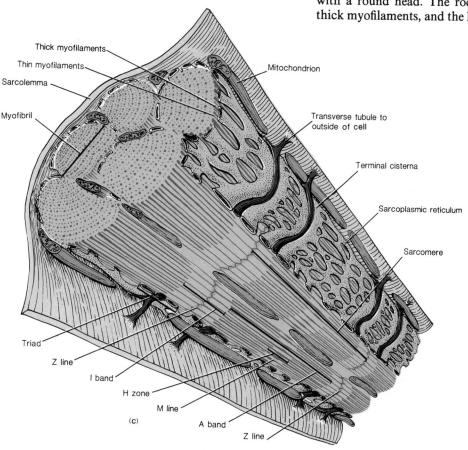

Thick myofilaments
Thin myofilaments
Sarcolemma
Myofibril
Mitochondrion
Transverse tubule to outside of cell
Terminal cisterna
Sarcoplasmic reticulum
Sarcomere
Triad
Z line
I band
H zone
M line
A band
Z line
(c)

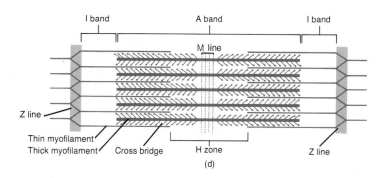

I band
A band
I band
M line
Z line
Thin myofilament
Thick myofilament
Cross bridge
H zone
Z line
(d)

FIGURE 10-2 (*Continued*) Histology of skeletal muscle tissue. (c) Enlarged aspect of several myofibrils of a muscle fiber based on an electron micrograph. (d) Enlarged aspect of a sarcomere showing thin and thick myofilaments.

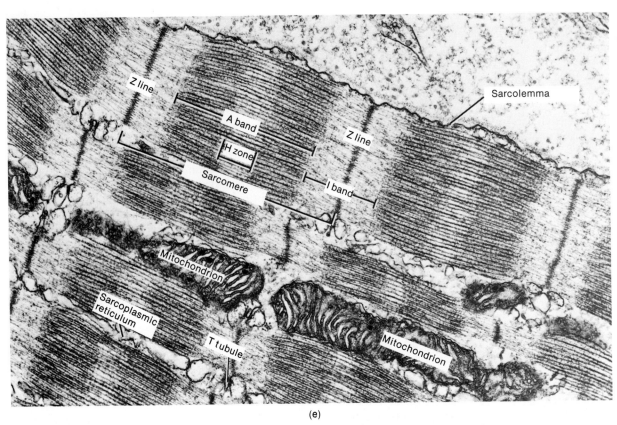

(e)

FIGURE 10-2 (*Continued*) Histology of skeletal muscle tissue. (e) Electron micrograph of several sarcomeres at a magnification of 35,000×. (Courtesy of D. E. Kelly, from *Introduction to the Musculoskeletal System,* by Cornelius Rosse and D. Kay Clawson, Harper & Row, Publishers, Inc., New York, 1970.)

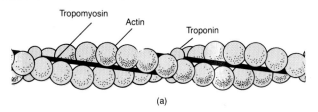

(a)

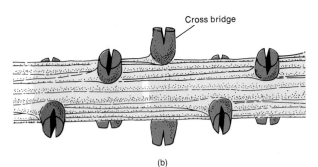

(b)

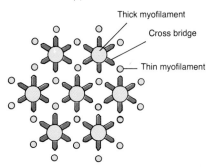

(c)

FIGURE 10-3 Detailed structure of myofilaments. (a) Thin myofilament. (b) Thick myofilament. (c) Cross section of several thin and thick myofilaments showing the arrangement of cross bridges. Note that each thick myofilament is surrounded by six thin myofilaments. (Parts (a) and (b) have been redrawn by permission from Arthur W. Ham, *Histology,* 7th ed., Philadelphia: J. B. Lippincott Company, 1974.)

cross bridges (Figure 10-3b). The cross bridges are arranged in pairs and seem to spiral around the main axis. The relationship between thin and thick myofilaments is shown in Figure 10-3c.

CONTRACTION

SLIDING-FILAMENT THEORY

During muscle contraction, the thin myofilaments slide inward toward the H zone. The sarcomere shortens, but the lengths of the thin and thick myofilaments do not change. The cross bridges of the thick myofilaments connect with portions of actin of the thin myofilaments. The myosin cross bridges move like the oars of a boat on the surface of the thin myofilaments, and the thin and thick myofilaments slide past each other (Figure 10-4). As the thin myofilaments move past the thick myofilaments, the H zone narrows and even disappears when the thin myofilaments meet at the center of the sarcomere (Figure 10-5). In fact, the cross bridges may pull the thin myofilaments of each sarcomere so far inward that

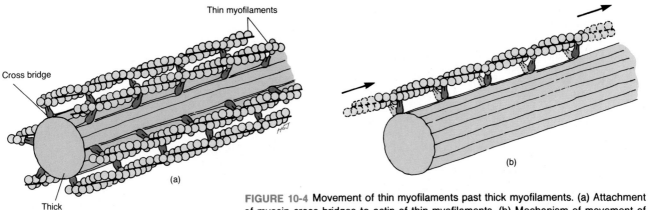

FIGURE 10-4 Movement of thin myofilaments past thick myofilaments. (a) Attachment of myosin cross bridges to actin of thin myofilaments. (b) Mechanism of movement of myosin cross bridges resulting in sliding of thin myofilaments toward H zone.

their ends overlap. As the thin myofilaments slide inward, the Z lines are drawn toward the A band and the sarcomere is shortened. The sliding of myofilaments and shortening of sarcomeres causes the shortening of the muscle fibers. All these events associated with the movement of myofilaments are known as the **sliding-filament theory** of muscle contraction.

MOTOR UNIT

For a skeletal muscle fiber to contract, a stimulus must be applied to it. Such a stimulus is normally transmitted by nerve cells, or **neurons.** A neuron has a threadlike process called a fiber, or axon, that may run 91 cm (3 ft) or more to a muscle. A bundle of such fibers from many different neurons composes a nerve. A neuron that transmits a stimulus to muscle tissue is called a **motor neuron.**

Upon entering a skeletal muscle, the axon of the motor neuron branches into fine endings that come into close approximation at grooves on the muscle membrane. The portion of the muscle cell membrane directly under the termination of the axon is called a **motor end plate** (Figure 10-6). The area of contact between neuron and muscular fiber is called a **neuromuscular junction,** or **myoneural junction.** When a nerve impulse reaches the terminal branches of the nerve fiber, small vesicles in the branches, called synaptic vesicles, release a chemical called **acetylcholine** (as'-ē-til-KŌ-lēn) or **ACh.** The ACh transmits the nerve impulse from the neuron, across the myoneural junction, to the motor end plate, thus initiating contraction. This mechanism is described in detail in Chapter 12.

A motor neuron, together with all the muscle cells it stimulates, is referred to as a **motor unit.** A single motor neuron may innervate about 150 muscle fibers, depending

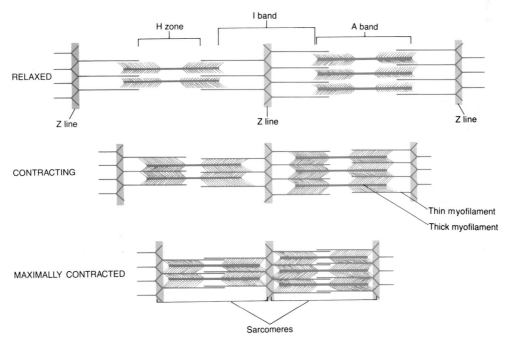

FIGURE 10-5 Sliding-filament theory of muscle contraction. Shown are the positions of the various parts of two sarcomeres in relaxed, contracting, and maximally contracted states. Note the movement of the thin myofilaments and the relative size of the H zone.

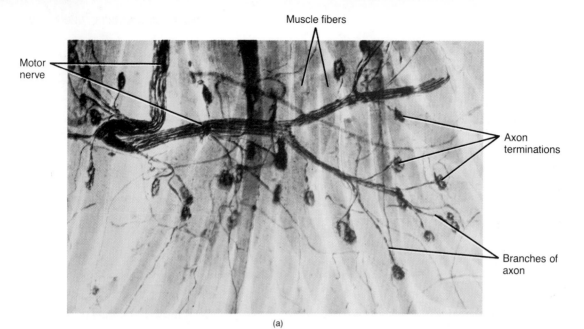

Muscle fibers

Motor nerve

Axon terminations

Branches of axon

(a)

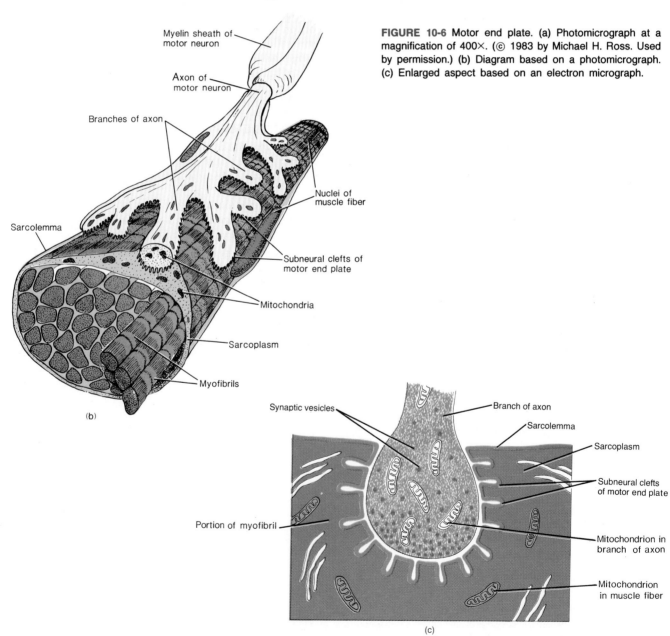

Myelin sheath of motor neuron

Axon of motor neuron

Branches of axon

Nuclei of muscle fiber

Subneural clefts of motor end plate

Sarcolemma

Mitochondria

Sarcoplasm

Myofibrils

(b)

FIGURE 10-6 Motor end plate. (a) Photomicrograph at a magnification of 400×. (© 1983 by Michael H. Ross. Used by permission.) (b) Diagram based on a photomicrograph. (c) Enlarged aspect based on an electron micrograph.

Synaptic vesicles

Branch of axon

Sarcolemma

Sarcoplasm

Subneural clefts of motor end plate

Portion of myofibril

Mitochondrion in branch of axon

Mitochondrion in muscle fiber

(c)

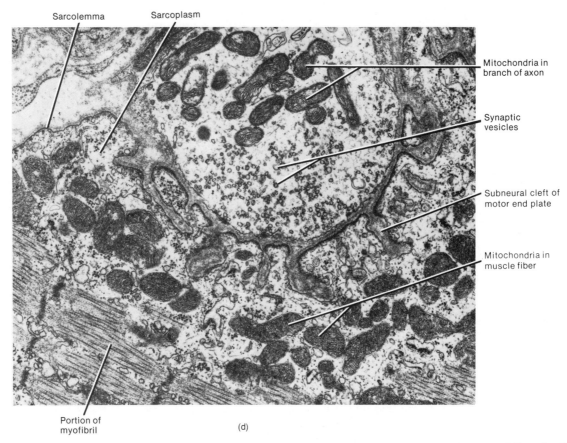

Sarcolemma   Sarcoplasm

Mitochondria in branch of axon

Synaptic vesicles

Subneural cleft of motor end plate

Mitochondria in muscle fiber

Portion of myofibril   (d)

FIGURE 10-6 (Continued) Motor end plate. (d) Electron micrograph at a magnification of 30,000×. (Courtesy of Cornelius Rosse and D. Kay Clawson, from *Introduction to the Musculoskeletal System,* Harper & Row, Publishers, New York, 1970.)

on the region of the body. This means that stimulation of one neuron will tend to cause the simultaneous contraction of about 150 muscle fibers. In addition, all the muscle fibers of a motor unit that are sufficiently stimulated will contract and relax together. Muscles that control precise movements, such as the extrinsic eye muscles, have fewer than 10 muscle fibers to each motor unit. Muscles of the body that are responsible for gross movements, such as the biceps and gastrocnemius, may have as many as 500 muscle fibers in each motor unit.

Stimulation of a motor neuron produces a contraction in all the muscle fibers in a particular motor unit. Accordingly, the total tension in a muscle can be varied by adjusting the number of motor units that are activated. The process of increasing the number of active motor units is called **recruitment** and is determined by the needs of the body at a given time. The various motor neurons to a given muscle fire asynchronously, that is, while some are excited, others are inhibited. This means that while some motor units are active, others are inactive; all the motor units are not contracting at the same time. Asynchronous firing of motor neurons prevents fatigue while maintaining contraction by allowing a brief rest for the inactive units. The alternating motor units relieve one another so smoothly that the contraction can be sustained for long periods. It also helps to maintain a nearly constant tension in a muscle and is one factor responsible for producing smooth movements during a muscle contraction, rather than a series of jerky movements.

PHYSIOLOGY

When a muscle fiber is relaxed, the concentration of calcium ions (Ca^{2+}) in the sarcoplasm is low; these ions are stored in the sarcoplasmic reticulum. Moreover, molecules of ATP are attached to the myosin cross bridges. The cross bridges are prevented from combining with actin of the thin myofilaments by the binding of the tropomyosin-troponin complex to actin and the binding of ATP to the myosin cross bridges. In other words, the muscle fiber remains relaxed as long as there are few calcium ions in the sarcoplasm, the tropomyosin-troponin complex is attached to actin, and ATP is attached to the myosin cross bridges.

When a nerve impulse reaches the nerve ending, a small amount of calcium enters the nerve ending, causing the synaptic vesicles to release acetylcholine. It diffuses across the myoneural junction and combines with receptor sites on the motor end plate. Acetylcholine alters the motor end plate and initiates an impulse that spreads from the motor end plate over the surface of the sarcolemma and into the T tubules. When the impulse is conveyed from the T tubules to the sarcoplasmic reticulum, the reticulum releases the calcium ions from storage into the sarcoplasm surrounding the myofilaments. The calcium ions move to the myosin cross bridges and activate the myosin. In the presence of calcium ions, myosin acts as an enzyme that catalyzes the breakdown of ATP into ADP + P. With the breakdown of ATP the myosin cross bridges are free to react with actin. Calcium ions also bind to the tropomyosin-troponin complex and permit it to split from the thin myofilament, leaving a free receptor site of actin that can attach to the myosin cross bridge. The energy released from the breakdown of ATP is used for the attachment and movement of the myosin cross bridges and thus the sliding of the myofilaments. As the thin myofilaments slide past the thick myofilaments, the Z lines are drawn toward each other and the sarcomere shortens—the muscle fibers contract and the muscle contracts.

What happens when a muscle fiber goes from a contracted state back to a relaxed state? Acetylcholine is rapidly destroyed by an enzyme called *acetylcholinesterase* (*AChE*) which is found on the membrane of muscle cells. The absence of ACh inhibits nerve impulse conduction from the axon terminals to the motor end plate. After the nerve impulse ends, the calcium ions are actively transported back into the sarcoplasmic reticulum for storage. This is accomplished by a transport protein called *calsequestrin* and involves an expenditure of some ATP. With the removal of calcium ions from the sarcoplasm, the enzymatic activity of myosin stops. The ADP is resynthesized into ATP, which again binds to the myosin cross bridges, and the tropomyosin-troponin complex is reattached to the actin of the thin myofilaments, so that the myosin cross bridges separate from the actin. Since the myosin cross bridges are broken, the thin myofilaments slip back to their relaxed position. The sarcomeres are thereby returned to their resting lengths and the muscle fiber resumes its resting state. ATP is also required to detach the cross bridges.

Exhibit 10-1 shows a summary of the events associated with the contraction and relaxation of a muscle fiber.

ENERGY

Contraction of a muscle requires energy. When a nerve impulse stimulates a muscle fiber, ATP, in the presence of ATPase (activated myosin), breaks down into ADP + P and energy is released. As far as we know, ATP is always the immediate source of energy for muscle contraction.

Like the other cells of the body, muscle cells synthesize ATP as follows:

$$ADP + P + E \text{ (energy)} \rightarrow ATP$$

The energy for replenishing ATP is derived from the breakdown of digested foods. However, unlike most other cells of the body, muscle fibers alternate between great activity and virtual inactivity. A resting muscle needs little energy and produces much more ATP than it can use. When a muscle is contracting, its energy requirements are high and the synthesis of ATP is accelerated. If the exercise is strenuous, ATP is used up even faster than it can be manufactured. Thus muscles must be able to build up a reserve supply of energy. They do this in two ways. First, the muscle fiber stores any excess ATP on the thick myofilaments. Second, when the fiber runs out of storage space for the ATP molecules, it combines the remainder of the ATP with a substance called *creatine*.

EXHIBIT 10-1

SUMMARY OF EVENTS INVOLVED IN CONTRACTION AND RELAXATION OF A SKELETAL MUSCLE FIBER

1. A nerve impulse causes synaptic vesicles in motor axon terminals to release acetylcholine.

2. Acetylcholine diffuses across the myoneural junction and initiates an impulse that spreads from the motor end plate over the surface of the sarcolemma.

3. The impulse enters the T tubules and sarcoplasmic reticulum and stimulates the sarcoplasmic reticulum to release calcium ions from storage into the sarcoplasm.

4. Calcium ions activate myosin, which breaks down ATP. Calcium ions also bind the tropomyosin-troponin complex to permit the complex to split from thin myofilaments.

5. Free receptor sites on actin of thin myofilaments attach to myosin cross bridges, using the energy from ATP breakdown, and thin myofilaments slide past thick myofilaments.

6. The sliding draws the Z lines toward each other, the sarcomere shortens, the muscle fibers contract, and the muscle contracts.

7. Acetylcholine is inactivated by acetylcholinesterase, thus inhibiting nerve impulse conduction from axon terminals to the motor end plate.

8. Once the nerve impulse is inhibited, calcium ions are actively transported back into the sarcoplasmic reticulum by a transport protein called calsequestrin, using energy from ATP breakdown.

9. The low calcium concentration in sarcoplasm stops the enzymatic activity of myosin, ADP is resynthesized to ATP, ATP binds to the myosin cross bridges, the tropomyosin-troponin complex is reattached to actin, myosin cross bridges separate from actin, and the thin myofilaments return to their relaxed position.

10. Sarcomeres return to their resting lengths, muscle fibers relax, and the muscle relaxes.

Creatine, which is produced in the liver, can accept a high-energy phosphate from ATP to become the high-energy compound *creatine phosphate*.

$$ATP + creatine \rightarrow creatine\ phosphate + ADP$$

This reaction is *anaerobic,* that is, it takes place in the absence of oxygen. (An *aerobic* reaction is one that requires oxygen.)

Creatine phosphate is produced when the muscle fibers are resting. During strenuous contraction, the reaction reverses itself. This reaction, also anaerobic, is shown in the following equation:

$$ADP + creatine\ phosphate \rightarrow ATP + creatine$$

CLINICAL APPLICATION

Following death, certain chemical changes occur in muscle tissue that affect the status of the muscles. Due to a lack of ATP, cross bridges of the thick myofilaments remain attached to the thin myofilaments, thus preventing relaxation. The resulting condition, in which muscles are in a state of partial contraction, is called **rigor mortis** (rigidity of death). The time elapsing between death and the onset of rigor mortis varies greatly among individuals. Those who have had long, wasting illnesses undergo rigor mortis more quickly.

ALL-OR-NONE PRINCIPLE

According to the **all-or-none principle,** individual muscle fibers of a motor unit will contract to their fullest extent or will not contract at all, providing conditions remain constant. In other words, *muscle fibers* do not partly contract. The principle does not mean that the entire muscle must be either fully relaxed or fully contracted because, of the many fibers that comprise the entire muscle, some are contracting and some are relaxing. Thus, the muscle as a whole can have graded contractions. The strength of contraction may be decreased by fatigue, lack of nutrients, or lack of oxygen. The weakest stimulus from a neuron that can initiate a contraction is called a **threshold** or **liminal stimulus.** A stimulus of lesser intensity, or one that cannot initiate contraction, is referred to as a **subthreshold** or **subliminal stimulus.**

KINDS

The various skeletal muscles are capable of producing different kinds of contractions, depending on the stimulation frequency.

Twitch

The **twitch contraction** is a rapid, jerky response to a single stimulus. Twitch contractions can be artifically brought about in muscles of an animal. A recording of a twitch contraction is the classic way to illustrate the different phases of one single contraction. Figure 10-7 is a graph of a twitch contraction. This kind of record of a contraction is called a **myogram.** Note that a brief period exists between application of the stimulus and the beginning of contraction: the **latent period.** In frog muscle, it lasts about 10 milliseconds (msec; 10 msec = 0.01 sec) (0.01 second). The second phase, the **contraction period,** lasts about 40 msec (0.04 sec) and is indicated by the

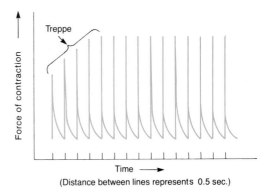

FIGURE 10-8 Myogram of treppe.

upward tracing. The third phase, the **relaxation period,** lasts about 50 msec (0.05 sec) and is indicated by the downward tracing. The duration of these periods depends on the muscle. The latent period, contraction period, and relaxation period for muscles that move the eyes are very short. Muscles that move the leg undergo longer periods.

If additional stimuli are applied to the muscle after the initial stimulus, other responses may be noted. For example, if two stimuli are applied one immediately after the other, the muscle will respond to the first stimulus but not to the second. When a muscle fiber receives enough stimulation to contract, it temporarily loses its irritability and cannot contract again until its responsiveness is regained. This period of lost irritability is the **refractory period.** Its duration also varies with the muscle involved. Skeletal muscle has a short refractory period of 5 msec (0.005 sec). Cardiac muscle has a long refractory period of 300 msec (0.30 sec).

Treppe

Treppe is the condition in which a skeletal muscle contracts more forcefully in response to the same strength of stimulus after it has contracted several times. It is demonstrated by stimulating an isolated muscle with a series of stimuli at the same frequency and voltage. Suppose a series of liminal stimuli are introduced into a muscle. Time must be allowed for the muscle to undergo its latent period, contract, and relax, but this takes only 0.1 sec in frog muscle. If stimuli are repeated at intervals of 0.5 sec, the first few tracings on the myogram will show an increasing height with each contraction. This is treppe—the staircase phenomenon (Figure 10-8). It is the principle athletes use when warming up. After the first few stimuli, the muscle reaches its peak of performance and undergoes its strongest contraction. Treppe is thought to result from an increase in the effectiveness of calcium ions in relieving tropomyosin-troponin inhibition after several contractions.

Tetanus

When two stimuli are applied and the second is delayed until the refractory period is over, the skeletal muscle will respond to both stimuli. In fact, if the second stimulus is applied after the refractory period, but before the muscle has finished relaxing, the second contraction will be

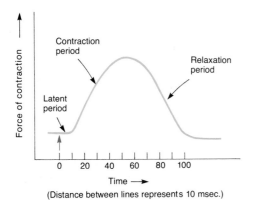

FIGURE 10-7 Myogram of a twitch contraction. The red arrow indicates the point at which the stimulus is applied.

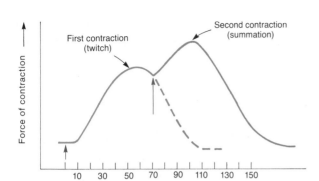

FIGURE 10-9 Myogram of summation of twitches. The second stimulus (long red arrow) is applied before the muscle has finished relaxing. The second contraction is even stronger than the first. The broken line represents the continuation of a twitch contraction.

stronger than the first. This phenomenon is **summation of twitches** (Figure 10-9).

If a frog muscle is stimulated at a rate of 20 to 30 stimuli per second, the muscle can only partly relax between stimuli. As a result, the muscle maintains a sustained contraction called **incomplete (unfused) tetanus** (Figure 10-10a). Stimulation at an increased rate (35 to 50 stimuli per second) results in **complete (fused) tetanus,** a sustained contraction that lacks even partial relaxation (Figure 10-10b). Essentially, both kinds of tetanus result from the addition of calcium ions released from the sarcoplasmic reticulum by the second stimulus to the calcium ions still in the sarcoplasm from the first stimulus. This causes the rapid succession of separate twitches. Relaxation is either partial or does not occur at all. Voluntary contractions, such as contraction of the biceps brachii muscle in order to flex the forearm, are tetanic contractions.

Isotonic and Isometric

Isotonic (*iso* = equal; *tonos* = tension) **contractions** are probably familiar to you. As the contraction occurs, the muscle shortens and pulls on another structure, such as a bone, to produce movement. During such a contraction, the tension remains constant and energy is expended.

In an **isometric contraction,** there is a minimal shortening of the muscle. It remains nearly the same length,

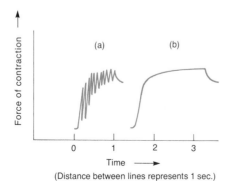

FIGURE 10-10 Myograms of (a) incomplete and (b) complete tetanus.

but the *tension* on the muscle increases greatly. Although isometric contractions do not result in body movement, energy is still expended. You can demonstrate such a contraction by carrying your books with your arm extended. The weight of the books pulls the arm downward, stretching the shoulder and arm muscles. The isometric contraction of the shoulder and arm muscles counteracts the stretch. The two forces—contraction and stretching—applied in opposite directions create the tension.

MUSCLE TONE

A muscle may be in a state of partial contraction even though contraction of the muscle fibers is always complete. At any given time, some cells in a muscle are contracted while others are relaxed. This contraction tightens a muscle, but there may not be enough fibers contracting at the time to produce movement. Asynchronous firing allows the contraction to be sustained for long periods.

A sustained partial contraction of portions of a skeletal muscle in response to activation of stretch receptors results in **muscle tone.** Tone is essential for maintaining posture. For example, when the muscles in the back of the neck are in tonic contraction, they keep the head in the anatomical position and prevent it from slumping forward onto the chest, but they do not apply enough force to pull the head back into hyperextension. The degree of tone in a skeletal muscle is monitored by receptors in the muscle called **muscle spindles.** They provide feedback information on tone to the brain so that adjustments can be made (Chapter 15).

CLINICAL APPLICATION

The term **flaccid** (FLAK-sid) is applied to muscles with less than normal tone. Such a loss of tone may be the result of damage or disease of the nerve that conducts a constant flow of impulses to the muscle. If the muscle does not receive impulses for an extended period of time, it may progress from flaccidity to **atrophy** (AT-rō-fē), which is a state of wasting away. Individual muscle cells decrease in size due to a progressive loss of myofibrils. Muscles may also become flaccid and atrophied if they are not used. Bedridden individuals and people with casts may experience atrophy because the flow of impulses to the inactive muscle is greatly reduced. If the nerve supply to a muscle is cut, it will undergo complete atrophy. In about six months to two years, the muscle will be one-quarter its original size and the muscle fibers will be replaced by fibrous tissue. The transition to fibrous tissue, when complete, cannot be reversed.

Muscular **hypertrophy** (hī-PER-trō-fē) is the reverse of atrophy. It refers to an increase in the diameters of muscle fibers due to the production of more myofibrils, mitochondria, sarcoplasmic reticulum, and nutrients. Hypertrophic muscles are capable of more forceful contractions. Weak muscular activity does not produce significant hypertrophy. It results from very forceful muscular activity or repetitive muscular activity at moderate levels.

FAST (WHITE) AND SLOW (RED) MUSCLE

The duration of contraction of various muscles of the body varies with the functions of the muscles in maintaining homeostasis. For example, the duration of contraction of eye muscles is less than $\frac{1}{100}$ sec, whereas the duration of contraction of the gastrocnemius muscle on the posterior aspect of the leg is about $\frac{1}{30}$ sec. Eye movements must be extremely rapid in order to fix the eyes on different objects very quickly. Movements of the legs during walking or running need not be as rapid as eye movements.

Muscles of the eye contain a predominance of **fast muscle fibers.** Such fibers have a more extensive sarcoplasmic reticulum for the rapid release and uptake of calcium ions needed for rapid contractions. Muscles like the gastrocnemius have a predominance of **slow muscle fibers.** These fibers are smaller and more aerobic; have more blood capillaries and more mitochondria; and have a large amount of myoglobin in the sarcoplasm. **Myoglobin** is a substance similar to the hemoglobin in red blood cells that can store oxygen until needed by the mitochondria. Since slow muscle has a reddish tint due to myoglobin and the large amount of red blood cells in capillaries, it is also known as **red muscle.** Fast muscle lacks myoglobin and has fewer capillaries and is thus also known as **white muscle.**

CARDIAC MUSCLE TISSUE

The principal constituent of the heart wall is **cardiac muscle tissue.** Although it is striated in appearance like skeletal muscle, it is involuntary. The fibers of cardiac muscle tissue are roughly quadrangular and usually have only a single centrally located nucleus (Figure 10-11). Skeletal muscle fibers contain several nuclei that are peripherally located. The thin sarcolemma of cardiac muscle fibers is similar to that of skeletal muscle, but the sarcoplasm is more abundant and the mitochondria are larger and more numerous. Cardiac muscle fibers have the same arrangement of actin and myosin and the same bands, zones, and lines as skeletal muscle fibers. Myofilaments in cardiac muscle fibers are not arranged in discrete myofibrils as in skeletal muscle. The T tubules of mammalian cardiac muscle are larger than those of skeletal muscle and are located at the Z lines rather than at the A-I band junctions as in skeletal muscle fibers. The sarcoplasmic reticulum of cardiac muscle is less well-developed than that in skeletal muscle.

While the groups of skeletal muscle fibers are arranged in a parallel fashion, those of cardiac muscle branch freely with other fibers to form two separate networks. The muscular walls and septum of the upper chambers of the heart (atria) compose one network. The muscular walls and septum of the lower chambers of the heart (ventricles) compose the other network. When a single fiber of either network is stimulated, all the fibers in the network become stimulated as well. Thus each network contracts as a functional unit. The fibers of each network were once thought to be fused together into a multinucleated mass called a syncytium. But it is now known that each fiber in a network is separated from the next fiber by an irregular transverse thickening of the sarco-

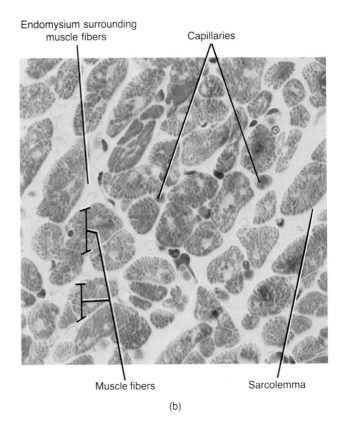

FIGURE 10-11 Histology of cardiac muscle tissue. (a) Photomicrograph of several muscle fibers in longitudinal section at a magnification of 400× (© 1983 by Michael H. Ross. Used by permission.) (b) Photomicrograph of several muscle fibers in cross section at a magnification of 640×. (© 1983 by Michael H. Ross. Used by permission.)

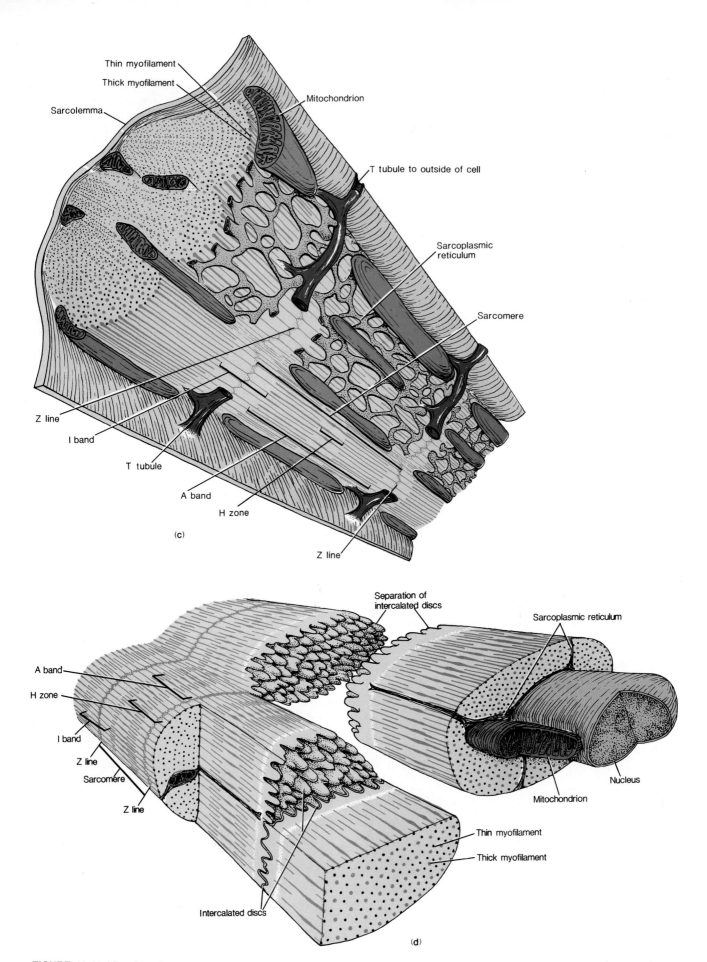

Thin myofilament

Thick myofilament

Sarcolemma

Mitochondrion

T tubule to outside of cell

Sarcoplasmic reticulum

Sarcomere

Z line

I band

T tubule

A band

H zone

Z line

(c)

Separation of intercalated discs

Sarcoplasmic reticulum

A band

H zone

I band

Z line

Sarcomere

Z line

Nucleus

Mitochondrion

Thin myofilament

Thick myofilament

Intercalated discs

(d)

FIGURE 10-11 (*Continued*) (c) Diagram based on an electron micrograph showing several myofibrils. (d) Diagram based on an electron micrograph showing intercalated discs and related structures.

lemma called an **intercalated** (in-TER-ka-lāt-ed) **disc.** These discs strengthen the cardiac muscle tissue and aid in impulse conduction.

Under normal conditions, cardiac muscle tissue contracts and relaxes rapidly, continuously, and rhythmically about 72 times a minute without stopping while a person is at rest. This is a major physiological difference between cardiac and skeletal muscle tissue. Another difference is the source of stimulation. Skeletal muscle tissue ordinarily contracts only when stimulated by a nerve impulse. In contrast, cardiac muscle tissue can contract without nerve stimulation. Its source of stimulation is a conducting tissue of specialized muscle within the heart. Nerve stimula-

tion merely causes the conducting tissue to increase or decrease its rate of discharge. Cardiac muscle tissue also has an extra long refractory period.

SMOOTH MUSCLE TISSUE

Like cardiac muscle tissue, **smooth muscle tissue** is usually involuntary. However, it is nonstriated. A single fiber of smooth muscle tissue is about 5 to 10 μm in diameter and 30 to 200 μm long. It is spindle-shaped, and within the fiber is a single, oval, centrally located nucleus (Figure 10-12). Smooth muscle cells contain actin and myosin filaments, but because the filaments are not as orderly as in skeletal and cardiac muscle tissue, the well-differentiated striations and sarcomeres do not occur.

Two kinds of smooth muscle tissue, visceral and multiunit, are recognized. The more common type is called **visceral muscle tissue.** It is found in wrap-around sheets that form part of the walls of the hollow viscera such as the stomach, intestines, uterus, and urinary bladder. The terms *smooth muscle tissue* and *visceral muscle tissue* are sometimes used interchangeably. The fibers in visceral muscle tissue are tightly bound together to form a continuous network. When a neuron stimulates one fiber, the impulse travels over the other fibers so that contraction

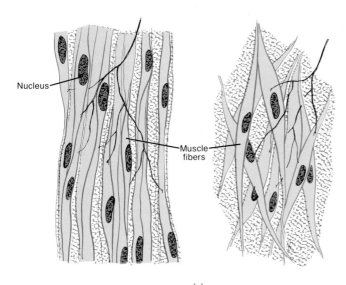

(a)

FIGURE 10-12 Histology of smooth muscle tissue. (a) Diagram of visceral smooth muscle tissue (left) and multiunit smooth muscle tissue (right). (b) Photomicrograph of several muscle fibers in longitudinal section at a magnification of 840×. (© 1983 by Michael H. Ross. Used by permission.) (c) Photomicrograph of several muscle fibers in cross section at a magnification of 840×. (© 1983 by Michael H. Ross. Used by permission.)

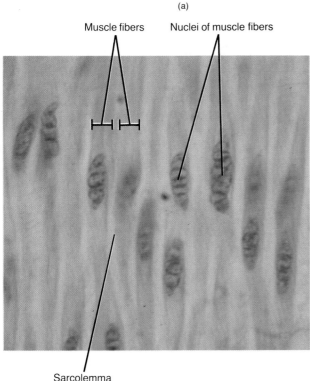

(b)

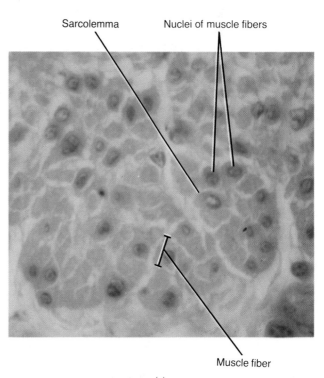

(c)

occurs in a wave over many adjacent fibers. Whereas skeletal muscle cells contract as individual units, visceral muscle cells contract in sequence as the impulse spreads from one cell to another.

The second kind of smooth muscle tissue, **multiunit smooth muscle tissue,** consists of individual fibers each with its own motor-nerve endings. Whereas stimulation of a single visceral muscle fiber causes contraction of many adjacent fibers, stimulation of a single multiunit fiber causes contraction of only that fiber. In this respect, multiunit muscle tissue is like skeletal muscle tissue. Multiunit smooth muscle tissue is found in the walls of blood vessels, in the arrector pili muscles that attach to hair follicles, and in the intrinsic muscles of the eye, such as the iris.

Both kinds of smooth muscle tissue contract and relax more slowly than skeletal muscle tissue. This characteristic is probably due to the arrangement of the thin and thick myofilaments of smooth muscle. Also, both kinds of smooth muscle can maintain a forceful contraction longer than skeletal muscle. Finally, unlike skeletal muscle fibers, smooth muscle fibers can stretch without developing tension. Thus, the smooth muscle in the wall of hollow organs such as the stomach, intestines, and urinary bladder can stretch as the viscera distend, while the pressure within them remains the same.

HOMEOSTASIS

Muscle tissue has a vital role in maintaining the body's homeostasis. Two examples are the relationship of muscle tissue to oxygen and to heat production.

OXYGEN DEBT

The energy required to convert ADP into ATP is produced by the breakdown of digested foods. The primary nutrient that usually supplies this energy is the sugar glucose. The reaction proceeds as follows:

Glucose $\rightarrow$ pyruvic acid + E (energy)

When the skeletal muscle is at rest, the breakdown of glucose is slow enough for the blood to supply sufficient oxygen to participate in the complete catabolism of pyruvic acid. The waste products of this aerobic reaction are carbon dioxide and water.

Pyruvic acid + $O_2 \rightarrow CO_2 + H_2O$ + E (energy)

When a skeletal muscle is contracting and ATP production increases, the breakdown of glucose occurs too rapidly for the blood to supply oxygen for the pyruvic acid to be completely catabolized. The pyruvic acid is only partly catabolized to lactic acid. The conversion of pyruvic acid to lactic acid proceeds *anaerobically*. About 80 percent of the lactic acid diffuses from the skeletal muscles and is transported to the liver. Some of the lactic acid accumulates in the muscle tissue, however. Physiologists suspect that this acid is responsible for the feeling of muscle fatigue. Ultimately, the lactic acid in the skeletal muscle, as well as that which has diffused into the blood, must be catabolized to CO_2 and H_2O. And for this conversion additional oxygen is needed—the **oxygen debt.** When vigorous activity is over, labored breathing continues in order to pay back the debt.

If a skeletal muscle or group of skeletal muscles is continuously stimulated for an extended period of time, the strength of contraction becomes progressively weaker until the muscle no longer responds. This condition is called **fatigue.** It results partly from the diminished availability of oxygen and partly from the toxic effects of lactic acid and carbon dioxide accumulated during exercise. The significant factors that contribute to muscle fatigue are:

1. Excessive activity, resulting in the accumulation of toxic products.

2. Malnutrition, resulting in insufficient supplies of glucose and, therefore, ATP.

3. Cardiovascular disturbances that impair the delivery of useful substances to muscles and the removal of waste products from muscles.

4. Respiratory disturbances that interfere with the oxygen supply and increase the oxygen debt.

HEAT PRODUCTION

The production of heat by skeletal muscle is an important homeostatic mechanism for maintaining normal body temperature. Of the total energy released during muscular contraction, only 20 to 30 percent is used for mechanical work (contraction). The rest is released as heat, which is utilized to help maintain a normal body temperature.

Heat production by muscles may be divided into two phases: (1) *initial heat,* which is produced by the contraction and relaxation of a muscle, and (2) *recovery heat,* which is produced after relaxation. Initial heat is independent of O_2 and is associated with ATP breakdown. Recovery heat is associated with ATP restoration. It includes the anaerobic breakdown of glucose to pyruvic acid and pyruvic acid to lactic acid. It also includes the aerobic breakdown of pyruvic acid to CO_2 and H_2O and the aerobic conversion of lactic acid to CO_2 and H_2O.

DISORDERS: HOMEOSTATIC IMBALANCES

Disorders of the muscular system are related to disruptions of homeostasis. The disorders may involve a lack of nutrients, the accumulation of toxic products, disease, injury, disuse, or faulty nervous connections (innervations).

Fibrosis

The formation of fibrous connective tissue in locations where it normally does not exist is called **fibrosis.** Skeletal and cardiac muscle fibers cannot undergo mitosis, and dead muscle fibers are normally replaced with fibrous connective tissue. Fibrosis, then, is often a consequence of muscle injury or degeneration.

Fibrositis

Fibrositis is an inflammation of fibrous tissue. If it occurs in the lumbar region, it is termed **lumbago** (lum-BĀ-gō). Fibrositis is a common condition characterized by pain, stiffness, or soreness of fibrous tissue, especially in the muscle coverings. It is not destructive or progressive. It may persist for years or spontaneously disappear. Attacks of fibrositis may follow an injury, repeated muscular strain, or prolonged muscular tension.

"Charleyhorse"

Fibromyositis refers to a group of symptoms that include pain, tenderness, and stiffness of the joints, muscles, or adjacent structures. These symptoms ordinarily occur in various combinations. When the thigh is involved, it is usually called **"charleyhorse."** It results from a contusion (bruising) and tearing of muscle fibers that produce a hematoma (collection of blood). "Charleyhorse" is characterized by the sudden onset of pain and is aggravated by motion. In some cases, local muscle spasms are noted. The condition is relieved with heat, massage, and rest and completely disappears, although occasionally it may become chronic or recur at frequent intervals.

Muscular Dystrophy

The term **muscular dystrophy** (*dystrophy* = degeneration) applies to a number of inherited myopathies, or muscle-destroying diseases. The disease is characterized by degeneration of the individual muscle cells, which leads to a progressive atrophy of the skeletal muscle. Usually the voluntary skeletal muscles are weakened equally on both sides of the body, whereas the internal muscles, such as the diaphragm, are not affected. Histologically, the changes that occur include the variation in muscle fiber size, degeneration of fibers, and deposition of fat.

The cause of muscular dystrophy has been variously attributed to a genetic defect, faulty metabolism of potassium, protein deficiency, and inability of the body to utilize creatine.

There is no specific drug therapy for this disorder. Treatment involves attempts to prolong ambulation by muscle-strengthening exercises, corrective surgical measures, and appropriate braces. Patients are encouraged to keep physically active.

A simple and reliable screening test for a wide variety of skeletal muscle disorders has been developed. Increased levels of an enzyme called *creatine phosphokinase* (*CPK*) in the blood are typical not only of the dystrophies but of other wasting diseases as well. The measurement of CPK, the most consistently elevated enzyme in cases of dystrophy, provides a diagnostic test even before clinical symptoms appear. Inflammatory muscle diseases can also be detected with this screening test.

Myasthenia Gravis

Myasthenia (mī'-as-THĒ-nē-a) **gravis** is a weakness of the skeletal muscles. It is caused by an abnormality at the neuromuscular junction that prevents the muscle fibers from contracting. Recall that motor neurons stimulate the skeletal muscle fibers to contract by releasing acetylcholine. Myasthenia gravis is believed to be due to antibodies directed against ACh receptors of the motor end plate. The antibodies bind to the receptors, leaving few available to ACh (see Chapter 12). As the disease progresses, more neuromuscular junctions become affected. The muscle becomes increasingly weaker and may eventually cease to function altogether.

Myasthenia gravis is more common in females, occurring most frequently between the ages of 20 and 50. The muscles of the face and neck are most apt to be involved. Initial symptoms include a weakness of the eye muscles and difficulty in swallowing. Later, the individual has difficulty chewing and talking. Eventually, the muscles of the limbs may become involved. Death may result from paralysis of the respiratory muscles, but usually the disorder does not progress to this stage.

Anticholinesterase drugs have been the primary treatment for the disease. More recently, steroid drugs, such as prednisone, have been used with great success. Immunosuppressant drugs are also used to decrease the production of antibodies that interfere with normal muscle contraction. Another recent treatment involves *plasmapheresis*, a procedure that separates blood cells from the plasma that contains the unwanted antibodies. The blood cells are then mixed with a plasma substitute and pumped back into the individual. Preliminary results indicate dramatic improvement in all treated individuals.

Abnormal Contractions

One kind of abnormal contraction of a muscle is **spasm:** a sudden, involuntary contraction of short duration. A **cramp** is a painful spasmodic contraction of a muscle. It is an involuntary, complete tetanic contraction. **Convulsions** are violent, involuntary tetanic contractions of an entire group of muscles. Convulsions occur when motor neurons are stimulated by fever, poisons, hysteria, or changes in body chemistry due to withdrawal of certain drugs. The stimulated neurons send many bursts of seemingly disordered impulses to the muscle fibers. **Fibrillation** is the uncoordinated contraction of individual muscle fibers preventing the smooth contraction of the muscle. A **tic** is a spasmodic twitching made involuntarily by muscles that are ordinarily under voluntary control. Twitching of the eyelid and face muscles are examples. In general, tics are of psychological origin.

MEDICAL TERMINOLOGY

Electromyography or **EMG** (*electro* = electricity; *myo* = muscle; *graph* = to write)   The recording and study of the electrical changes that occur in muscle tissue.

Gangrene (*gangraena* = an eating sore)   Death of a soft tissue, such as muscle, that results from interruption of its blood supply. It is caused by various species of *Clostridium*, bacteria that live anaerobically in the soil.

Myalgia (*algia* = painful condition)   Pain in or associated with muscles.

Myology (*logos* = study of)   Study of muscles.

Myoma (*oma* = tumor)   A tumor consisting of muscle tissue.

Myomalacia (*malaco* = soft)   Softening of a muscle.

Myopathy (*pathos* = disease)   Any disease of muscle tissue.

Myosclerosis (*scler* = hard)   Hardening of a muscle.

Myositis (*itis* = inflammation of)   Inflammation of muscle fibers (cells).

Myospasm   Spasm of a muscle.

Myotonia (*tonia* = tension)   Increased muscular excitability and contractility with decreased power of relaxation; tonic spasm of the muscle.

Paralysis (*para* = beyond; *lyein* = to loosen)   Loss or impairment of motor (muscular) function resulting from a lesion of nervous or muscular origin.

Shin splints   A soreness or pain along the shinbone (tibia) due to straining the flexor digitorum muscle. The condition is often caused by walking or running up and down hills or by vigorous activity of the legs following a period of relative inactivity.

Trichinosis   A myositis caused by the parasitic worm *Trichinella spiralis*, which may be found in the muscles of humans, rats, and pigs. People contract the disease by eating insufficiently cooked infected pork.

Volkmann's contracture (*contra* = against)   Permanent contraction of a muscle due to replacement of destroyed muscle cells with fibrous tissue that lacks ability to stretch. Destruction of muscle cells may occur from interference with circulation caused by a tight bandage, a piece of elastic, or a cast.

Wryneck or torticollis (*tortus* = twisted; *collam* = neck) Spasmodic contraction of several of the superficial and deep muscles of the neck; produces twisting of the neck and an unnatural position of the head.

DRUGS ASSOCIATED WITH THE MUSCULAR SYSTEM

Drugs associated with the muscular system are usually concerned with preventing or controlling abnormal muscular contractions. Specific drug therapy is lacking for the dystrophic diseases.

Skeletal Muscle Relaxants
Skeletal muscle spasms like "charleyhorse" or torticollis can be treated with analgesics for pain or with skeletal muscle relaxants like *methocarbamol* (Robaxin), *diazepam* (Valium), *chlorzoxazone* (Paraflex), *cyclobenzaprine HCl* (Flexeril), and *carisoprodol* (Soma). *Dantrolene* (Dantrium) is used to treat a rare, though deadly, muscle disorder known as malignant hyperpyrexia.

STUDY OUTLINE

Characteristics (p. 204)
1. Excitability is the property of receiving and responding to stimuli.
2. Contractility is the ability to shorten and thicken, contract.
3. Extensibility is the ability to be stretched or extended.
4. Elasticity is the ability to return to original shape after contraction or extension.

Functions (p. 204)
1. Through contraction, muscle tissue performs the three important functions of motion, maintenance of posture, and heat production.

Types (p. 204)
1. Skeletal muscle tissue is attached to bones. It is striated and voluntary.
2. Visceral muscle tissue is located in viscera. It is nonstriated (smooth) and involuntary.
3. Cardiac muscle tissue forms the walls of the heart. It is striated and involuntary.

Skeletal Muscle Tissue (p. 204)
1. The term fascia is applied to a sheet or broad band of fibrous connective tissue underneath the skin or around muscles and organs of the body. There are three types of fascia: superficial, deep, and subserous.
2. Connective tissue components are epimysium, covering the entire muscle; perimysium, covering fasciculi; and endomysium, covering fibers.
3. Tendons and aponeuroses are extensions of connective tissue beyond the muscle cells to attach the muscle to bone or other muscle.
4. Nerves convey impulses for muscular contraction.
5. Blood provides nutrients and oxygen for contraction.
6. Skeletal muscle consists of fibers covered by a sarcolemma. The fibers contain sarcoplasm, nuclei, sarcoplasmic reticulum, and T tubules.
7. Each fiber contains myofibrils that consist of thin and thick myofilaments. The myofilaments are compartmentalized into sarcomeres.
8. Thin myofilaments are composed of actin, tropomyosin, and troponin; thick myofilaments consist of myosin.

Contraction (p. 208)
Sliding-Filament Theory
1. A nerve impulse travels over the sarcolemma and enters the T tubules and sarcoplasmic reticulum.
2. The nerve impulse leads to the release of calcium ions from the sarcoplasmic reticulum, triggering the contractile process.
3. Actual contraction is brought about when the thin myofilaments of a sarcomere slide toward each other.

Motor Unit
1. A motor neuron transmits the stimulus to a skeletal muscle for contraction.
2. The region of the sarcolemma under the terminal branch of an axon of a motor neuron that is specialized to receive the nerve impulse is the motor end plate.
3. The area of contact between a motor neuron and muscle fiber is a neuromuscular, or myoneural, junction.
4. A motor neuron and the muscle fibers it stimulates form a motor unit.

Physiology
1. When a nerve impulse reaches the motor end plate, the neuron releases acetylcholine, which transmits the impulse to the motor end plate and then into the T tubules and sarcoplasmic reticulum.
2. This releases calcium ions that activate myosin, catalyzing the breakdown of ATP, and bind tropomyosin-troponin complex, so that actin of thin myofilaments can attach to myosin cross bridges of thick myofilaments.
3. The energy released from the breakdown of ATP causes the sliding of the myofilaments.

Energy
1. The only direct source of energy for muscle contraction is ATP.
2. A reserve supply of ATP may be built up by storage of excess in thick myofilaments or by anaerobic combination with creatine to form creatine phosphate, which breaks down to produce ATP when muscles contract strenuously.

All-or-None Principle
1. Muscle fibers of a motor unit contract to their fullest extent or not at all.
2. The weakest stimulus capable of causing contraction is a liminal, or threshold, stimulus.
3. A stimulus not capable of inducing contraction is a subliminal, or subthreshold, stimulus.

Kinds
1. The various kinds of contractions are twitch, treppe, tetanus, isotonic, and isometric.
2. A record of a contraction is called a myogram. The refractory period is the time when a muscle has temporarily lost excit-

ability. Skeletal muscles have a short refractory period. Cardiac muscle has a long refractory period.

3. Summation of twitches is the increased strength of a contraction resulting from the application of a second stimulus before the muscle has completely relaxed after a previous stimulus.

Muscle Tone

1. A sustained partial contraction of portions of a skeletal muscle results in muscle tone.
2. Tone is essential for maintaining posture.
3. Flaccidity is a condition of less than normal tone. Atrophy is a wasting away or decrease in size; hypertrophy is an enlargement or overgrowth.

Fast (White) and Slow (Red) Muscle

1. The duration of contraction of various muscles of the body varies with the functions of the muscles in maintaining homeostasis.
2. Fast or white muscles have an extensive sarcoplasmic reticulum.
3. Slow or red muscles have smaller fibers, more blood capillaries, and a large amount of myoglobin.

Cardiac Muscle Tissue (p. 215)

1. This muscle is found only in the heart. It is striated and involuntary.
2. The cells are quadrangular and usually contain a single centrally placed nucleus.
3. Compared to skeletal muscle tissue, cardiac muscle tissue has more sarcoplasm, more mitochondria, less well-developed sarcoplasmic reticulum, and larger T tubules located at Z lines rather than at A-I band junctions. Myofilaments are not arranged in discrete myofibrils.
4. The fibers branch freely to form two continuous networks, each of which contracts as a functional unit.

5. Intercalated discs provide strength and aid impulse conduction.

Smooth Muscle Tissue (p. 217)

1. Smooth muscle is nonstriated and involuntary.
2. Visceral smooth muscle is found in the walls of viscera. The fibers are arranged in a network.
3. Multiunit smooth muscle is found in blood vessels and the eye. The fibers operate singly rather than as a unit.

Homeostasis (p. 218)

1. Oxygen debt is the amount of O_2 needed to convert accumulated lactic acid into CO_2 and H_2O. It occurs during strenuous exercise and is paid back by continuing to breathe rapidly after exercising. Unit it is paid back, the homeostasis between muscular activity and oxygen requirements is not restored.
2. Muscle fatigue results from diminished availability of oxygen and toxic effects of carbon dioxide and lactic acid built up during exercise.
3. The heat given off during muscular contraction maintains the homeostasis of body temperature.

Disorders: Homeostatic Imbalances (p. 218)

1. Fibrosis is the formation of fibrous tissue where it normally does not exist; it frequently occurs in damaged muscle tissue.
2. Fibrositis is an inflammation of fibrous tissue. If it occurs in the lumbar region, it is called lumbago.
3. "Charleyhorse" refers to pain, tenderness, and stiffness of joints, muscles, and related structures in the thigh.
4. Muscular dystrophy is a hereditary disease of muscles characterized by degeneration of individual muscle cells.
5. Myasthenia gravis is a disease characterized by great muscular weakness and fatigability resulting from improper neuromuscular transmission.
6. Abnormal contractions include spasms, cramps, convulsions, fibrillations, and tics.

REVIEW QUESTIONS

1. How is the skeletal system related to the muscular system? What are the three basic functions of the muscular system?
2. What are the four characteristics of muscle tissue?
3. How can the three types of muscle tissue be distinguished?
4. What is fascia? What are the three different types of fascia and where are they found in the body?
5. Define epimysium, perimysium, endomysium, tendon, and aponeurosis. Describe the nerve and blood supply to a skeletal muscle.
6. Discuss the microscopic structure of skeletal muscle.
7. In considering the contraction of skeletal muscle, describe the following: motor unit, role of calcium, sources of energy, and sliding-filament theory.
8. What is the all-or-none principle? Relate it to a liminal and subliminal stimulus.
9. Define each of the following contractions and state the importance of each: twitch, treppe, tetanus, isotonic, and isometric.
10. What is a myogram?
11. Describe the latent period, contraction period, and relaxation period of muscle contraction. Construct a diagram to illustrate your answer.
12. Define the refractory period. How does it differ between skeletal and cardiac muscle? What is summation?
13. What is muscle tone? Why is it important? Distinguish between atrophy and hypertrophy.
14. Compare fast (white) and slow (red) muscle with respect to structure and function.
15. Compare cardiac and skeletal muscle with regard to microscopic structure, functions, and locations.
16. Compare cardiac and smooth muscle with regard to microscopic structure, functions, and locations.
17. Discuss each of the following as examples of muscle homeostasis: oxygen debt, fatigue, and heat production.
18. What do you think might be the relationship between shivering (uncontrolled muscular contractions) and body temperature? Can you relate sweating (cooling of the skin) after strenuous exercise to the homeostasis of body temperature?
19. Define fibrosis and fibrositis.
20. What is a "charleyhorse"?
21. What is muscular dystrophy?
22. What is myasthenia gravis? In this disease, why do the muscles not contract normally?
23. Define each of the following abnormal muscular contractions: spasm, cramp, convulsion, fibrillation, and tic.
24. Refer to the glossary of medical terminology associated with the muscular system. Be sure that you can define each term.
25. Describe the actions of selected drugs associated with muscle tissue.

- Describe the relationship between bones and skeletal muscles in producing body movements.

- Define a lever and fulcrum and compare the three classes of levers on the basis of placement of the fulcrum, effort, and resistance.

- Identify the various arrangements of muscle fibers in a skeletal muscle and relate the arrangements to the strength of contraction and range of movement.

- Discuss most body movements as activities of groups of muscles by explaining the roles of the prime mover, antagonist, and synergist.

- Define the criteria employed in naming skeletal muscles.

- Identify the principal skeletal muscles in different regions of the body by name, origin, insertion, action, and innervation.

- Discuss the administration of drugs by intramuscular injection.

- Compare the common sites of intramuscular injection.

11 The Muscular System

The term **muscle tissue** refers to all the contractile tissues of the body: skeletal, cardiac, and smooth muscle. The **muscular system,** however, refers to the *skeletal* muscle system: the skeletal muscle tissue and connective tissues that make up individual muscle organs, such as the biceps brachii muscle. Cardiac muscle tissue is located in the heart and is therefore considered part of the cardiovascular system. Smooth muscle tissue of the intestine is part of the digestive system, while smooth muscle tissue of the urinary bladder is part of the urinary system. In this chapter, we discuss only the muscular system. We will see how skeletal muscles produce movement and describe the principal skeletal muscles.

HOW SKELETAL MUSCLES PRODUCE MOVEMENT

ORIGIN AND INSERTION

Skeletal muscles produce movements by exerting force on tendons, which in turn pull on bones. Most muscles cross at least one joint and are attached to the articulating bones that form the joint (Figure 11-1). When such a muscle contracts, it draws one articulating bone toward the other. The two articulating bones usually do not move equally in response to the contraction. One is held nearly in its original position because other muscles contract to pull it in the opposite direction or because its structure makes it less movable. Ordinarily, the attachment of a muscle tendon to the stationary bone is called the **origin.** The attachment of the other muscle tendon to the movable bone is the **insertion.** A good analogy is a spring on a door. The part of the spring attached to the door represents the insertion; the part attached to the frame is the origin. The fleshy portion of the muscle between the tendons of the origin and insertion is called the **belly,** or **gaster.** The origin is usually proximal and the insertion distal, especially in the extremities. In addition, muscles that move a body part generally do not cover the moving part. Figure 11-1a shows that although contraction of the biceps brachii muscle moves the forearm, the belly of the muscle lies over the humerus.

LEVER SYSTEMS

In producing a body movement, bones act as levers and joints function as fulcrums of these levers. A **lever** may be defined as a rigid rod that moves about on some fixed point called a **fulcrum.** A fulcrum may be symbolized as ⚠. A lever is acted on at two different points by two different forces: the *resistance* Ⓡ and the *effort* (E). The resistance may be regarded as a force to be overcome, whereas the effort is the force exerted to overcome the resistance. The resistance may be the weight of the body part that is to be moved. The muscular effort (contraction) is applied to the bone at the insertion of the muscle and produces motion. Consider the biceps brachii flexing the forearm at the elbow as a weight is lifted (Figure 11-1b). When the forearm is raised, the elbow is the fulcrum. The weight of the forearm plus the weight in the hand is the resistance. The shortening of the biceps brachii pulling the forearm up is the effort.

Levers are categorized into three types according to the positions of the fulcrum, the effort, and the resistance.

1. In a **first-class lever,** the fulcrum is between the effort and resistance (Figure 11-2a). An example of a first-class lever is a seesaw. There are not many first-class levers in the body.

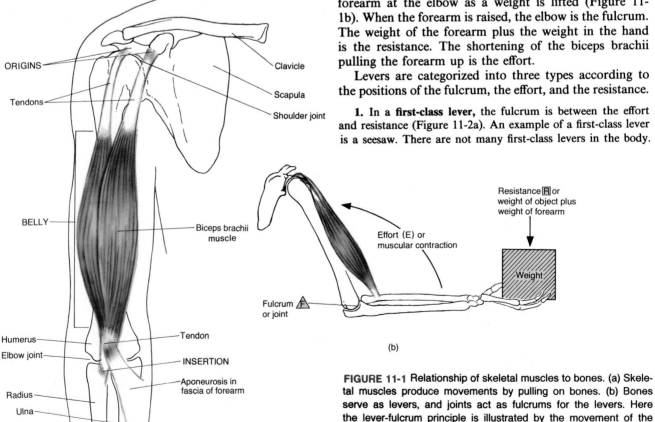

ORIGINS

Tendons

Clavicle

Scapula

Shoulder joint

BELLY

Biceps brachii muscle

Effort (E) or muscular contraction

Resistance Ⓡ or weight of object plus weight of forearm

Weight

Fulcrum ⚠ or joint

Humerus
Elbow joint

Tendon

INSERTION

Aponeurosis in fascia of forearm

Radius
Ulna

(a)

(b)

FIGURE 11-1 Relationship of skeletal muscles to bones. (a) Skeletal muscles produce movements by pulling on bones. (b) Bones serve as levers, and joints act as fulcrums for the levers. Here the lever-fulcrum principle is illustrated by the movement of the forearm lifting a weight. Note where the resistance and effort are applied in this example.

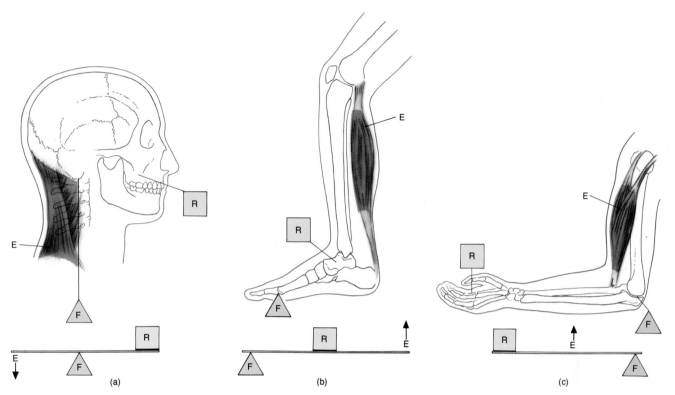

FIGURE 11-2 Classes of levers. Each is defined on the basis of the placement of the fulcrum, effort, and resistance. (a) First-class lever. (b) Second-class lever. (c) Third-class lever.

One example is the head resting on the vertebral column. When the head is raised, the facial portion of the skull is the resistance. The joint between the atlas and occipital bone (atlantooccipital joint) is the fulcrum. The contraction of the muscles of the back is the effort.

2. Second-class levers have the fulcrum at one end, the effort at the opposite end, and the resistance between them (Figure 11-2b). They operate like a wheelbarrow. Most authorities agree that there are very few examples of second-class levers in the body. One example is raising the body on the toes. The body is the resistance, the ball of the foot is the fulcrum, and the contraction of the calf muscles to pull the heel upward is the effort.

3. Third-class levers consist of the fulcrum at one end, the resistance at the opposite end, and the effort between them (Figure 11-2c). They are the most common levers in the body. An example is flexing the forearm at the elbow. As we have seen, the weight of the forearm is the resistance, the contraction of the biceps brachii is the effort, and the elbow joint is the fulcrum.

Leverage—the mechanical advantage gained by a lever—is largely responsible for a muscle's strength and range of movement. Consider strength first. Suppose we have two muscles of the same strength crossing and acting on a joint. Assume also that one is attached farther from the joint and one is closer. The muscle attached farther will produce the more powerful movement. Thus strength of movement depends on the placement of muscle attachments.

In considering the range of movement, again assume that we have two muscles of the same strength crossing and acting on a joint and that one is attached farther from the joint than the other. The muscle inserting closer to the joint will produce the greater range of movement.

Thus range of movement also depends on the placement of muscle attachments. Since strength increases with distance from the joint and range of movement decreases, maximal strength and maximal range are incompatible; strength and range vary inversely.

ARRANGEMENT OF FASCICULI

Recall from Chapter 10 that skeletal muscle fibers are arranged within the muscle in bundles called fasciculi or fascicles. The muscle fibers are arranged in a parallel fashion within each bundle, but the arrangement of the fasciculi with respect to the tendons may take one of four characteristic patterns.

The first pattern is called **parallel.** The fasciculi are parallel with the longitudinal axis and terminate at either end in flat tendons. The muscle is typically quadrilateral in shape. An example is the stylohyoid muscle (see Figure 11-7). In a modification of the parallel arrangement, called *fusiform,* the fasciculi are nearly parallel with the longitudinal axis and terminate at either end in flat tendons, but the muscle tapers toward the tendons, where the diameter is less than that of the belly. An example is the biceps brachii muscle (see Figure 11-16).

The second distinct pattern is called **convergent.** A broad origin of fasciculi converges to a narrow, restricted insertion. Such a pattern gives the muscle a triangular shape. An example is the deltoid muscle (see Figure 11-15).

The third distinct pattern is referred to as **pennate.** The fasciculi are short in relation to the entire length of the muscle and the tendon extends nearly the entire

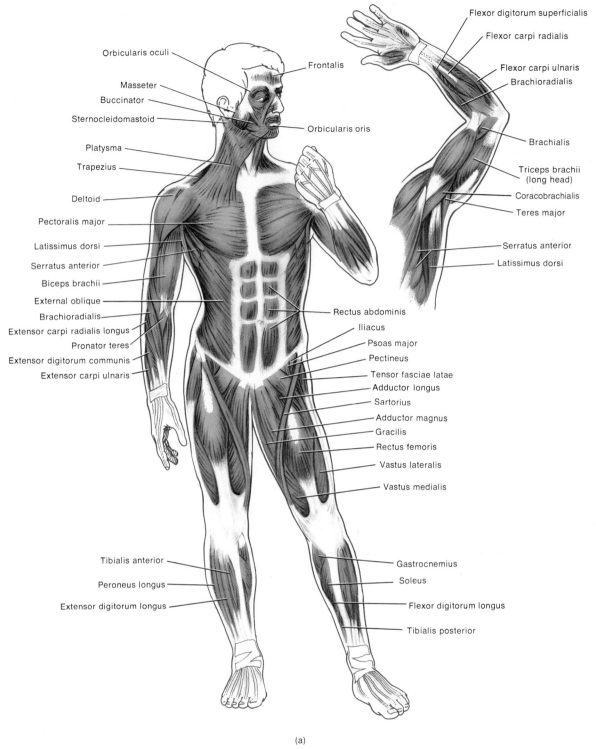

Orbicularis oculi
Masseter
Buccinator
Sternocleidomastoid
Platysma
Trapezius
Deltoid
Pectoralis major
Latissimus dorsi
Serratus anterior
Biceps brachii
External oblique
Brachioradialis
Extensor carpi radialis longus
Pronator teres
Extensor digitorum communis
Extensor carpi ulnaris

Frontalis
Orbicularis oris

Flexor digitorum superficialis
Flexor carpi radialis
Flexor carpi ulnaris
Brachioradialis

Brachialis
Triceps brachii (long head)
Coracobrachialis
Teres major

Serratus anterior
Latissimus dorsi

Rectus abdominis
Iliacus
Psoas major
Pectineus
Tensor fasciae latae
Adductor longus
Sartorius
Adductor magnus
Gracilis
Rectus femoris
Vastus lateralis
Vastus medialis

Tibialis anterior
Peroneus longus
Extensor digitorum longus

Gastrocnemius
Soleus
Flexor digitorum longus
Tibialis posterior

(a)

FIGURE 11-3 Principal superficial muscles. (a) Anterior view.

length of the muscle. The fasciculi are directed obliquely toward the tendon like the plumes of a feather. If the fasciculi are arranged on only one side of a tendon, as in the extensor digitorum muscle, the muscle is referred to as *unipennate* (see Figure 11-17). If the fasciculi are arranged on both sides of a centrally positioned tendon, as in the rectus femoris muscle, the muscle is referred to as *bipennate* (see Figure 11-20).

The final distinct pattern is referred to as **circular.** The fasciculi are arranged in a circular pattern and enclose an orifice. An example is the orbicularis oris muscle (see Figure 11-4).

Fascicular arrangement is correlated with the power of a muscle and range of movement. When a muscle fiber contracts, it shortens to a length just slightly greater than half of its resting length. Thus, the longer the fibers in a muscle, the greater the range of movement it can produce. By contrast, the strength of a muscle depends on the total number of fibers it contains, since a short fiber can contract as forcefully as a long one. Because a given muscle can either contain a small number of long fibers or a large number of short fibers, fascicular arrangement represents a compromise between power and range of movement. Pennate muscles, for example, have a large

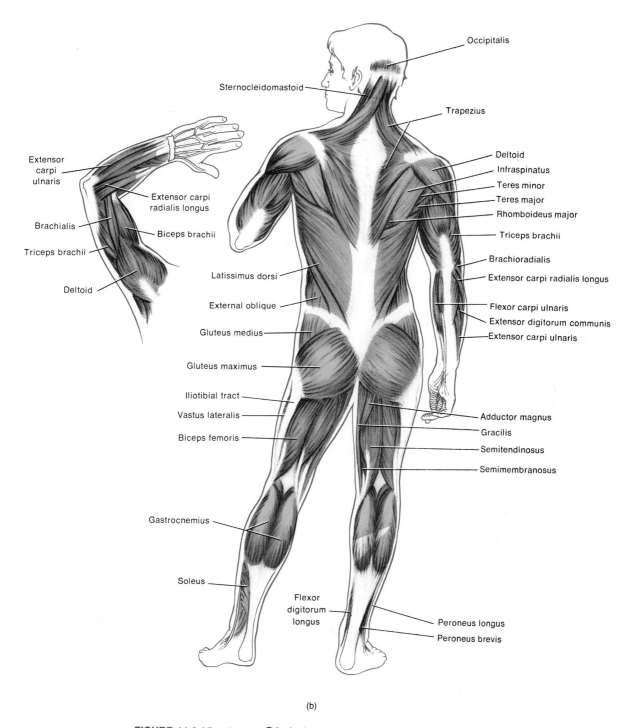

FIGURE 11-3 (*Continued*) Principal superficial muscles. (b) Posterior view.

number of fasciculi distributed over their tendons, giving them greater power, but a smaller range of movement. Parallel muscles, on the other hand, have comparatively few fasciculi that extend the length of the muscle. Thus, they have a greater range of movement, but less power.

GROUP ACTIONS

Most movements are coordinated by several skeletal muscles acting in groups rather than individually. A muscle that causes a desired action is referred to as the **agonist** or **prime mover.** Consider flexing the forearm at the elbow. In this instance, the biceps brachii is the agonist (see Figure 11-17). Simultaneously with the contraction of the agonist, another muscle, called the **antagonist,** is relaxing. In flexing the forearm, the triceps brachii serves as the antagonist (see Figure 11-17). The antagonist has an effect opposite to that of the agonist, that is, the antagonist relaxes and yields to the movement of the agonist. If we consider extending the forearm at the elbow, the triceps brachii is the agonist and the biceps brachii is the antagonist. Most joints are operated by antagonistic groups of muscles.

Other muscles called **synergists,** or **fixators,** assist the agonist by reducing undesired action or unnecessary movements in the less mobile articulating bone. While flexing the forearm, the synergists, in this case the deltoid and pectoralis major muscles, hold the arm and shoulder in a suitable position for the flexing action (see Figure 11-14). The deltoid abducts the humerus, and the pectoralis major adducts and medially rotates the humerus. Essentially, synergists contract at the same time as the prime mover and help the prime mover produce an effective movement. Many muscles are, at various times, prime movers, antagonists, or synergists, depending on the action.

NAMING SKELETAL MUSCLES

The names of most of the nearly 700 skeletal muscles are based on several types of characteristics. Learning the terms used to indicate specific characteristics will help you remember the names of muscles.

1. Muscle names may indicate the **direction of the muscle fibers.** *Rectus* fibers usually run parallel to the midline of the body. *Transverse* fibers run perpendicular to the midline. *Oblique* fibers are diagonal to the midline. Muscles named according to directions of fibers include the rectus abdominis, transversus abdominis, and external oblique.

2. A muscle may be named according to **location.** The temporalis is near the temporal bone. The tibialis anterior is near the tibia.

3. **Size** is another characteristic. The term *maximus* means largest, *minimus* smallest, *longus* long, and *brevis* short. Examples include the gluteus maximus, gluteus minimus, adductor longus, and peroneus brevis.

4. Some muscles are named for their **number of origins.** The biceps brachii has two origins, the triceps brachii three, and the quadriceps femoris four.

5. Other muscles are named on the basis of **shape.** Common examples include the deltoid (meaning triangular) and trapezius (meaning trapezoid).

6. Muscles may be named after their **origin** and **insertion.** The sternocleidomastoid originates on the sternum and clavicle and inserts at the mastoid process of the temporal bone; the stylohyoid originates on the styloid process of the temporal bone and inserts at the hyoid bone.

7. Still another characteristic of muscles used for naming is **action.** Exhibit 11-1 lists the principal actions of muscles, their definitions, and examples of muscles that perform the actions. For convenience, the actions are grouped as antagonistic pairs where possible.

PRINCIPAL SKELETAL MUSCLES

Exhibits 11-2 through 11-20 list the principal muscles of the body with their origins, insertions, actions, and innervations. (By no means have all the muscles of the body been included.) Refer to Chapters 7 and 8 to review bone markings, since they serve as points of origin and insertion for muscles. The muscles are divided into groups according to the part of the body on which they act. If you have mastered the naming of the muscles, their actions will have more meaning. Figure 11-3 shows general anterior and posterior views of the muscular system. Do not try to memorize all these muscles yet. As you study groups of muscles in the exhibits, refer to Figure 11-3 to see how each group is related to all others.

The figures that accompany the exhibits contain superficial and deep, anterior and posterior, or medial and lateral views to show the muscles' position as clearly as possible. An attempt has been made to show the relationship of the muscles under consideration to other muscles in the area you are studying.

EXHIBIT 11-1
PRINCIPAL ACTIONS OF MUSCLES

ACTION	DEFINITION	EXAMPLE
Flexor	Usually decreases the anterior angle at a joint; some decrease the posterior angle.	Flexor carpi radialis.
Extensor	Usually increases the anterior angle at a joint; some increase the posterior angle.	Extensor carpi ulnaris.
Abductor	Moves a bone away from the midline.	Abductor hallucis longus.
Adductor	Moves a bone closer to the midline.	Adductor longus.
Levator	Produces an upward movement.	Levator scapulae.
Depressor	Produces a downward movement.	Depressor labii inferioris.
Supinator	Turns the palm upward or anteriorly.	Supinator.
Pronator	Turns the palm downward or posteriorly.	Pronator teres.
Dorsiflexor	Flexes the foot at the ankle joint.	Tibialis anterior.
Plantar flexor	Extends the foot at the ankle joint.	Plantaris.
Invertor	Turns the sole of the foot inward.	Tibialis anterior.
Evertor	Turns the sole of the foot outward.	Peroneus tertius.
Sphincter	Decreases the size of an opening.	Orbicularis oculi.
Tensor	Makes a body part more rigid.	Tensor fasciae latae.
Rotator	Moves a bone around its longitudinal axis.	Obturator.

EXHIBIT 11-2
MUSCLES OF FACIAL EXPRESSION (Figure 11-4)

MUSCLE	ORIGIN	INSERTION	ACTION	INNERVATION
Epicranius (*epi* = over; *crani* = skull)	This muscle is divisible into two portions: the frontalis over the frontal bone and the occipitalis over the occipital bone. The two muscles are united by a strong aponeurosis, the galea aponeurotica, which covers the superior and lateral surfaces of the skull.			
Frontalis (*front* = forehead)	Galea aponeurotica.	Skin superior to supraorbital line.	Draws scalp forward, raises eyebrows, and wrinkles forehead horizontally.	Facial (VII) nerve.
Occipitalis (*occipito* = base of skull)	Occipital bone and mastoid process of temporal bone.	Galea aponeurotica.	Draws scalp backward.	Facial (VII) nerve.
Orbicularis oris (*orb* = circular; *or* = mouth)	Muscle fibers surrounding opening of mouth.	Skin at corner of mouth.	Closes lips, compresses lips against teeth, protrudes lips, and shapes lips during speech.	Facial (VII) nerve.
Zygomaticus major (*zygomatic* = cheek bone; *major* = greater)	Zygomatic bone.	Skin at angle of mouth and orbicularis oris.	Draws angle of mouth upward and outward as in smiling or laughing.	Facial (VII) nerve.
Levator labii superioris (*levator* = raises or elevates; *labii* = lip; *superioris* = upper)	Superior to infraorbital foramen of maxilla.	Skin at angle of mouth and orbicularis oris.	Elevates (raises) upper lip.	Facial (VII) nerve.
Depressor labii inferioris (*depressor* = depresses or lowers; *inferioris* = lower)	Mandible.	Skin of lower lip.	Depresses (lowers) lower lip.	Facial (VII) nerve.
Buccinator (*bucc* = cheek)	Alveolar processes of maxilla and mandible and pterygomandibular raphe (fibrous band extending from the pterygoid hamulus to the mandible).	Orbicularis oris.	Major cheek muscle; compresses cheek as in blowing air out of mouth and causes cheeks to cave in, producing the action of sucking.	Facial (VII) nerve.
Mentalis (*mentum* = chin)	Mandible.	Skin of chin.	Elevates and protrudes lower lip and pulls skin of chin up as in pouting.	Facial (VII) nerve.
Platysma (*platy* = flat, broad)	Fascia over deltoid and pectoralis major muscles.	Mandible, muscles around angle of mouth, and skin of lower face.	Draws outer part of lower lip downward and backward as in pouting; depresses mandible.	Facial (VII) nerve.
Risorius (*risor* = laughter)	Fascia over parotid (salivary) gland.	Skin at angle of mouth.	Draws angle of mouth laterally as in tenseness.	Facial (VII) nerve.
Orbicularis oculi (*ocul* = eye)	Medial wall of orbit.	Circular path around orbit.	Closes eye.	Facial (VII) nerve.
Corrugator supercilii (*corrugo* = to wrinkle; *supercilium* = eyebrow)	Medial end of superciliary arch of frontal bone.	Skin of eyebrow.	Draws eyebrow downward as in frowning.	Facial (VII) nerve.
Levator palpebrae superioris (*palpebrae* = eyelids) (see Figure 11-6b)	Roof of orbit (lesser wing of sphenoid bone).	Skin of upper eyelid.	Elevates upper eyelid.	Oculomotor (III) nerve.

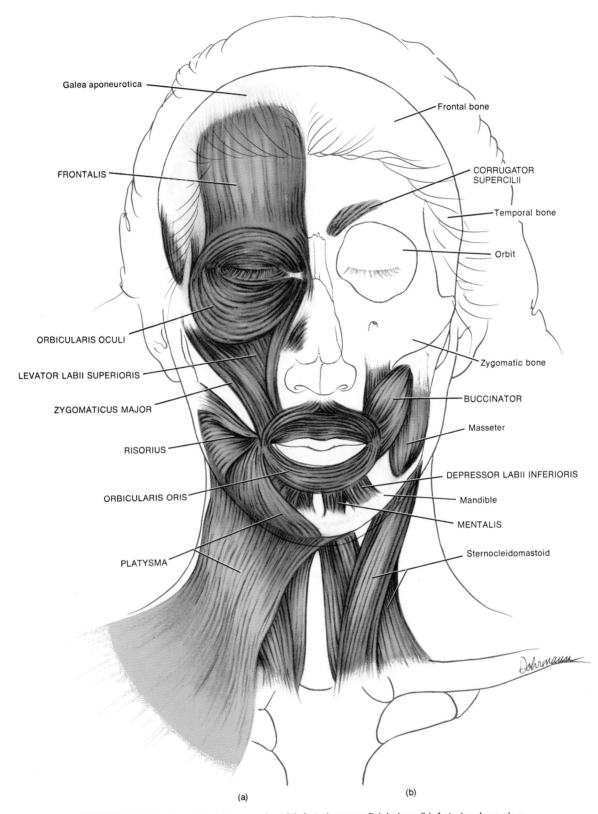

Galea aponeurotica

FRONTALIS

ORBICULARIS OCULI

LEVATOR LABII SUPERIORIS

ZYGOMATICUS MAJOR

RISORIUS

ORBICULARIS ORIS

PLATYSMA

Frontal bone

CORRUGATOR
SUPERCILII

Temporal bone

Orbit

Zygomatic bone

BUCCINATOR

Masseter

DEPRESSOR LABII INFERIORIS

Mandible

MENTALIS

Sternocleidomastoid

(a)

(b)

FIGURE 11-4 Muscles of facial expression. (a) Anterior superficial view. (b) Anterior deep view.

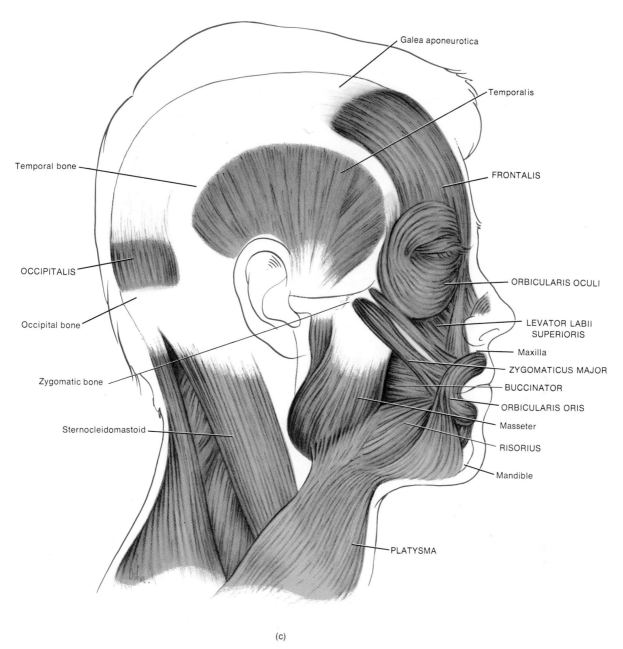

(c)

FIGURE 11-4 (*Continued*) Muscles of facial expression. (c) Right lateral superficial view.

EXHIBIT 11-3
MUSCLES THAT MOVE THE LOWER JAW (Figure 11-5)

MUSCLE	ORIGIN	INSERTION	ACTION	INNERVATION
Masseter (*maseter* = chewer)	Maxilla and zygomatic arch.	Angle and ramus of mandible.	Elevates mandible as in closing mouth and protracts (protrudes) mandible.	Mandibular branch of trigeminal (V) nerve.
Temporalis (*tempora* = temples)	Temporal bone.	Coronoid process of mandible.	Elevates and retracts mandible.	Temporal nerve from mandibular division of trigeminal (V) nerve.
Medial pterygoid (*medial* = closer to midline; *pterygoid* = like a wing; pterygoid plate of sphenoid bone)	Medial surface of lateral pterygoid plate of sphenoid; maxilla.	Angle and ramus of mandible.	Elevates and protracts mandible and moves mandible from side to side.	Mandibular branch of trigeminal (V) nerve.
Lateral pterygoid (*lateral* = farther from midline)	Greater wing and lateral surface of lateral pterygoid plate of sphenoid.	Condyle of mandible; temporomandibular articulation.	Protracts mandible, opens mouth, and moves mandible from side to side.	Mandibular branch of trigeminal (V) nerve.

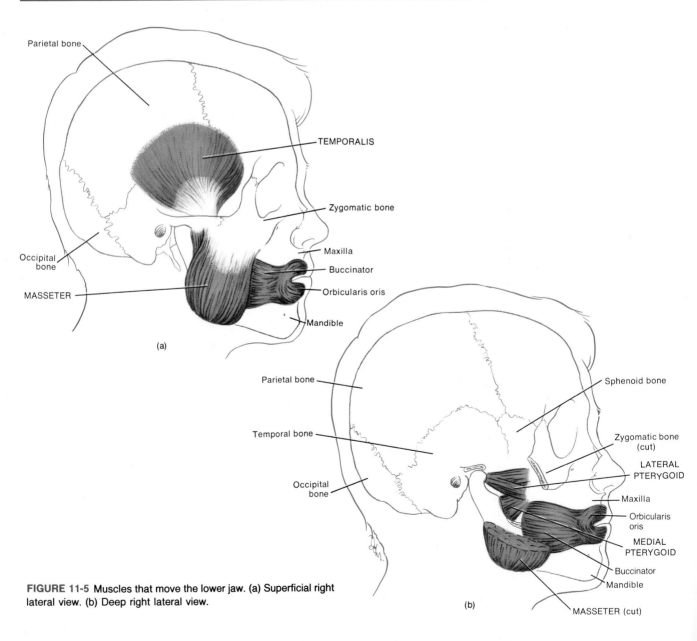

FIGURE 11-5 Muscles that move the lower jaw. (a) Superficial right lateral view. (b) Deep right lateral view.

EXHIBIT 11-4
MUSCLES THAT MOVE THE EYEBALLS—EXTRINSIC MUSCLES* (Figure 11-6)

MUSCLE	ORIGIN	INSERTION	ACTION	INNERVATION
Superior rectus (*superior* = above; *rectus* = in this case, muscle fibers running parallel to long axis of eyeball)	Tendinous ring attached to bony orbit around optic foramen.	Superior and central part of eyeball.	Rolls eyeball upward.	Oculomotor (III) nerve.
Inferior rectus (*inferior* = below)	Same as above.	Inferior and central part of eyeball.	Rolls eyeball downward.	Oculomotor (III) nerve.
Lateral rectus	Same as above.	Lateral side of eyeball.	Rolls eyeball laterally.	Abducens (VI) nerve.
Medial rectus	Same as above.	Medial side of eyeball.	Rolls eyeball medially.	Oculomotor (III) nerve.
Superior oblique (*oblique* = in this case, muscle fibers running diagonally to long axis of eyeball)	Same as above.	Eyeball between superior and lateral recti.	Rotates eyeball on its axis; directs cornea downward and laterally; note that it moves through a ring of fibrocartilaginous tissue called the trochlea (*trochlea* = pulley).	Trochlear (IV) nerve.
Inferior oblique	Maxilla (front of orbital cavity).	Eyeball between inferior and lateral recti.	Rotates eyeball on its axis; directs cornea upward and laterally.	Oculomotor (III) nerve.

* Muscles situated on the outside of the eyeball.

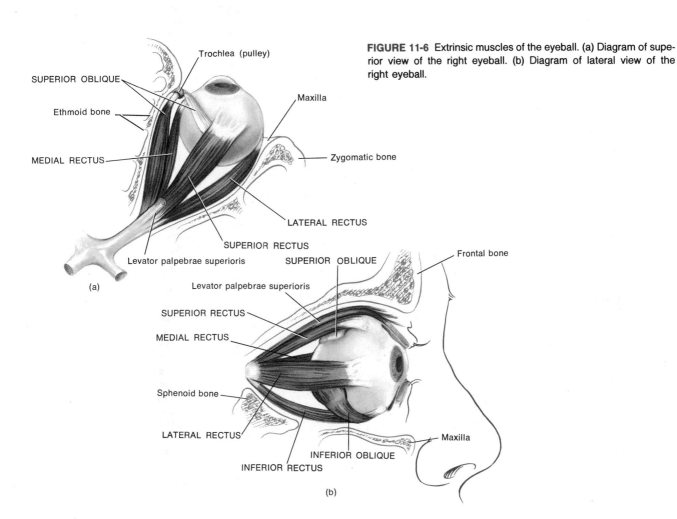

FIGURE 11-6 Extrinsic muscles of the eyeball. (a) Diagram of superior view of the right eyeball. (b) Diagram of lateral view of the right eyeball.

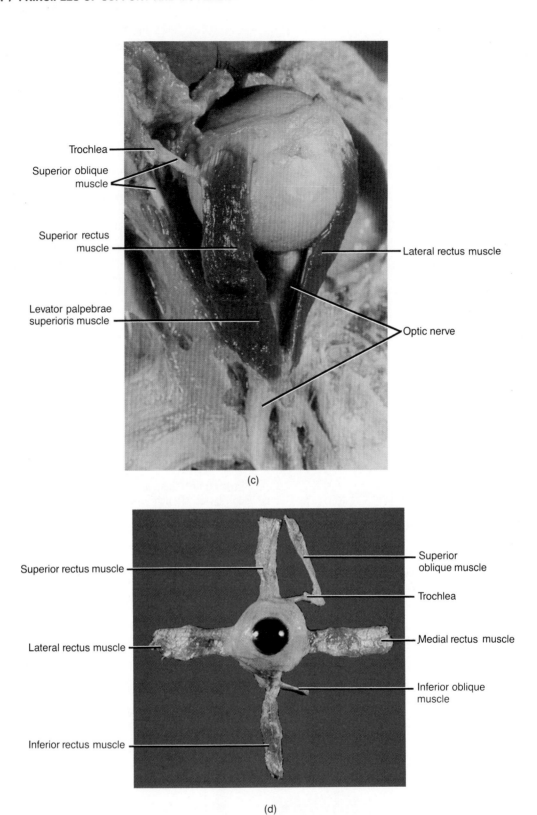

(c)

(d)

Figure 11-6 (*Continued*) Extrinsic muscles of the eyeball. (c) Photograph of superior view of the right eyeball. (d) Photograph of the right eye muscles extended. (Photographs courtesy of C. Yokochi and J. W. Rohen, *Photographic Anatomy of the Human Body,* 2d ed., 1979, IGAKU-SHOIN, Ltd., Tokyo, New York.)

EXHIBIT 11-5
MUSCLES THAT MOVE THE TONGUE (Figure 11-7)

MUSCLE	ORIGIN	INSERTION	ACTION	INNERVATION
Genioglossus (*geneion* = chin; *glossus* = tongue)	Mandible.	Undersurface of tongue and hyoid bone.	Depresses tongue and thrusts it forward (protraction).	Hypoglossal (XII) nerve.
Styloglossus (*stylo* = stake or pole; styloid process of temporal bone)	Styloid process of temporal bone.	Side and undersurface of tongue.	Elevates tongue and draws it backward (retraction).	Hypoglossal (XII) nerve.
Palatoglossus (*palato* = palate)	Anterior surface of soft palate.	Side of tongue.	Elevates tongue and draws soft palate down on tongue.	Hypoglossal (XII) nerve.
Hyoglossus	Body of hyoid bone.	Side of tongue.	Depresses tongue and draws down its sides.	Hypoglossal (XII) nerve.

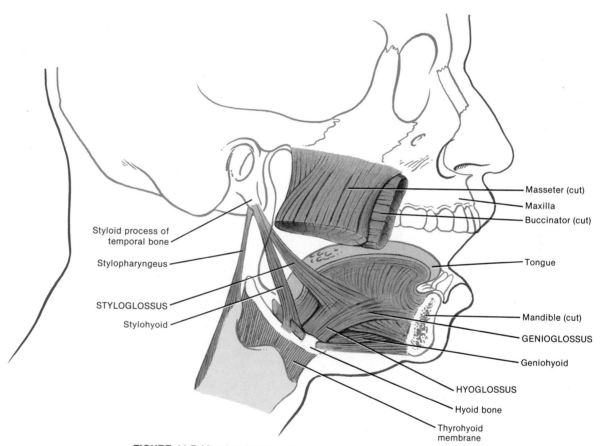

FIGURE 11-7 Muscles that move the tongue viewed from the right side.

EXHIBIT 11-6

MUSCLES OF THE PHARYNX (Figure 11-8)

MUSCLE	ORIGIN	INSERTION	ACTION	INNERVATION
Inferior constrictor (*inferior* = below; *constrictor* = decreases diameter of a lumen)	Cricoid and thyroid cartilages of larynx.	Posterior median raphe of pharynx.	Constricts inferior portion of pharynx to propel a bolus into esophagus.	Pharyngeal plexus.
Middle constrictor	Greater and lesser cornu of hyoid bone and stylohyoid ligament.	Posterior median raphe of pharynx.	Constricts middle portion of pharynx to propel a bolus into esophagus.	Pharyngeal plexus.
Superior constrictor (*superior* = above)	Pterygoid process, pterygomandibular raphe, and mylohyoid line of mandible.	Posterior median raphe of pharynx.	Constricts superior portion of pharynx to propel a bolus into esophagus.	Pharyngeal plexus.
Stylopharyngeus (*stylo* = stake or pole; styloid process of temporal bone; *pharyngo* = pharynx) (see also Figure 11-7)	Medial side of base of styloid process.	Lateral aspects of pharynx and thyroid cartilage.	Elevates larynx and dilates pharynx to help bolus descend.	Glossopharyngeal (IX) nerve.
Salpingopharyngeus (*salping* = pertaining to the auditory or uterine tube)	Inferior portion of auditory tube.	Posterior fibers of palatopharyngeus muscle.	Elevates superior portion of lateral wall of pharynx during swallowing and opens orifice of auditory tube.	Pharyngeal plexus.
Palatopharyngeus (*palato* = palate)	Soft palate.	Posterior border of thyroid cartilage and lateral and posterior wall of pharynx.	Elevates larynx and pharynx and helps close nasopharynx during swallowing.	Pharyngeal plexus.

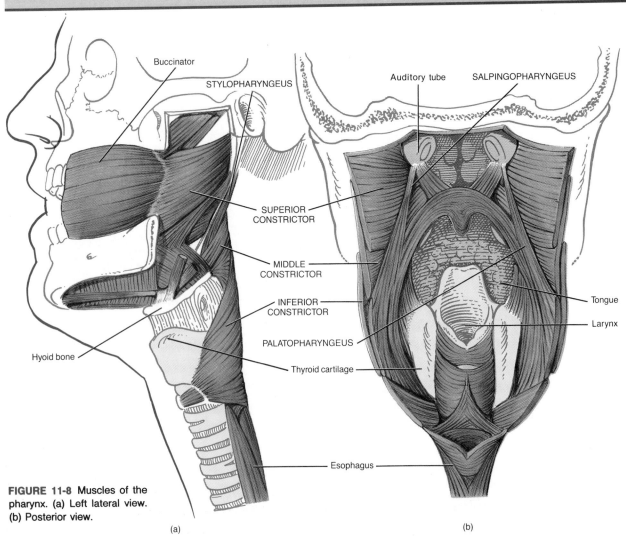

FIGURE 11-8 Muscles of the pharynx. (a) Left lateral view. (b) Posterior view.

(a)　　　　　(b)

EXHIBIT 11-7
MUSCLES OF THE LARYNX (Figure 11-9)

MUSCLE	ORIGIN	INSERTION	ACTION	INNERVATION
EXTRINSIC				
Omohyoid (*omo* = relationship to the shoulder; *hyoedes* = U-shaped; pertaining to hyoid bone)	Superior border of scapula and superior transverse ligament.	Body of hyoid bone.	Depresses hyoid bone.	Branches of ansa cervicalis nerve (C1–C3).
Sternohyoid (*sterno* = sternum)	Medial end of clavicle and manubrium of sternum.	Body of hyoid bone.	Depresses hyoid bone.	Branches of ansa cervicalis nerve (C1–C3).
Sternothyroid (*thyro* = thyroid gland)	Manubrium of sternum.	Thyroid cartilage of larynx.	Depresses thyroid cartilage.	Branches of ansa cervicalis nerve (C1–C3).
Thyrohyoid	Thyroid cartilage of larynx.	Greater cornu of hyoid bone.	Elevates thyroid cartilage and depresses hyoid bone.	Cervical nerves C1–C2 and descending hypoglossal (XII) nerve.
Stylopharyngeus	See Exhibit 11-6.			
Palatopharyngeus	See Exhibit 11-6.			
Inferior constrictor	See Exhibit 11-6.			
Middle constrictor	See Exhibit 11-6.			
INTRINSIC				
Cricothyroid (*crico* = cricoid cartilage of larynx)	Anterior and lateral portion of cricoid cartilage of larynx.	Anterior border of inferior cornu of thyroid cartilage of larynx and posterior part of inferior border of lamina of thyroid cartilage.	Produces tension and elongation of vocal folds.	External laryngeal branch of vagus (X) nerve.

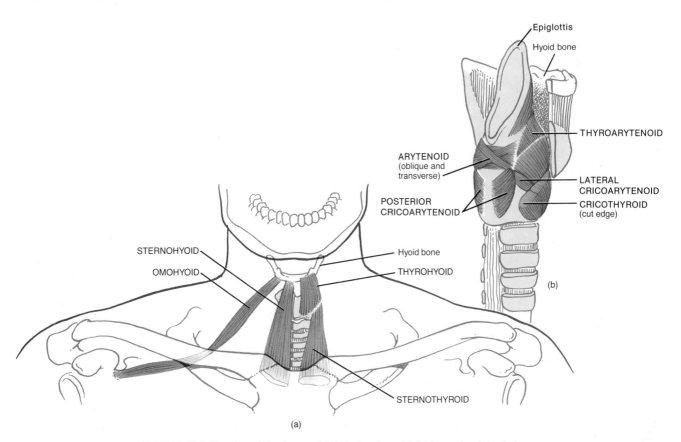

FIGURE 11-9 Muscles of the larynx. (a) Anterior view. (b) Right posterolateral view.

EXHIBIT 11-8

MUSCLES THAT MOVE THE HEAD

MUSCLE	ORIGIN	INSERTION	ACTION	INNERVATION
Sternocleidomastoid (*sternum* = breastbone; *cleido* = clavicle; *mastoid* = mastoid process of temporal bone) (see Figure 11-14)	Sternum and clavicle.	Mastoid process of temporal bone.	Contraction of both muscles flex the cervical part of the vertebral column, draw the head forward, and elevate chin; contraction of one muscle rotates face toward side opposite contracting muscle.	Accessory (XI) nerve; cervical nerves C2–C3.
Semispinalis capitis (*semi* = half; *spine* = spinous process; *caput* = head) (see Figure 11-18)	Articular process of seventh cervical vertebra and transverse processes of first six thoracic vertebrae.	Occipital bone.	Both muscles extend head; contraction of one muscle rotates face toward same side as contracting muscle.	Dorsal rami of spinal nerves.
Splenius capitis (*splenion* = bandage) (see Figure 11-18)	Ligamentum nuchae and spines of seventh cervical vertebra and first four thoracic vertebrae.	Occipital bone and mastoid process of temporal bone.	Both muscles extend head; contraction of one rotates it to same side as contracting muscle.	Dorsal rami of middle and lower cervical nerves.
Longissimus capitis (*longissimus* = longest) (see Figure 11-18)	Transverse processes of last four cervical vertebrae.	Mastoid process of temporal bone.	Extends head and rotates face toward side opposite contracting muscle.	Dorsal rami of middle and lower cervical nerves.

EXHIBIT 11-9

MUSCLES THAT ACT ON THE ANTERIOR ABDOMINAL WALL (Figure 11-10)

MUSCLE	ORIGIN	INSERTION	ACTION	INNERVATION
Rectus abdominis (*rectus* = fibers parallel to midline; *abdomino* = belly)	Pubic crest and symphysis pubis.	Cartilage of fifth to seventh ribs and xiphoid process.	Flexes vertebral column.	Branches of thoracic nerves T7–T12.
External oblique (*external* = closer to surface; *oblique* = fibers diagonal to midline)	Lower eight ribs.	Iliac crest and linea alba (midline aponeurosis).	Contraction of both compresses abdomen; contraction of one side alone bends vertebral column laterally.	Branches of thoracic nerves T7–T12 and iliohypogastric nerve.
Internal oblique (*internal* = farther from surface)	Iliac crest, inguinal ligament, and thoracolumbar fascia.	Cartilage of last three or four ribs.	Compresses abdomen; contraction of one side alone bends vertebral column laterally.	Branches of thoracic nerves T8–T12, iliohypogastric, and ilioinguinal nerves.
Transversus abdominis (*transverse* = fibers perpendicular to midline)	Iliac crest, inguinal ligament, lumbar fascia, and cartilages of last six ribs.	Xiphoid process, linea alba, and pubis.	Compresses abdomen.	Branches of thoracic nerves T8–T12, iliohypogastric, and ilioinguinal nerves.

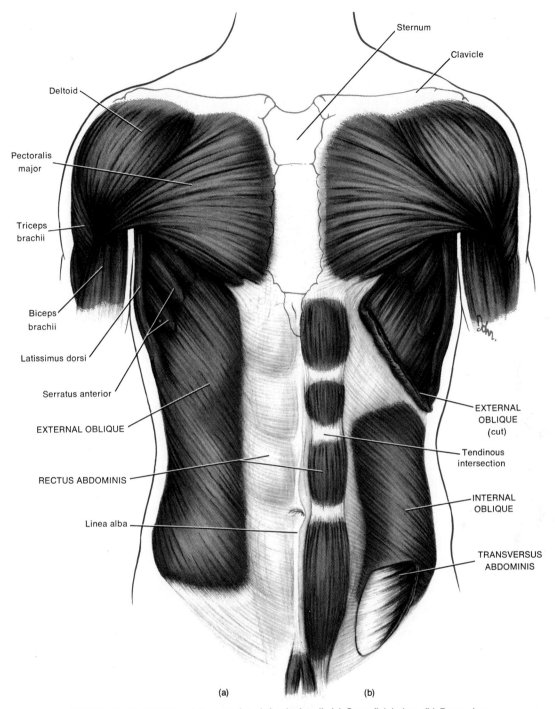

Sternum

Clavicle

Deltoid

Pectoralis
major

Triceps
brachii

Biceps
brachii

Latissimus dorsi

Serratus anterior

EXTERNAL OBLIQUE

RECTUS ABDOMINIS

Linea alba

EXTERNAL
OBLIQUE
(cut)

Tendinous
intersection

INTERNAL
OBLIQUE

TRANSVERSUS
ABDOMINIS

(a)   (b)

FIGURE 11-10 Muscles of the anterior abdominal wall. (a) Superficial view. (b) Deep view.

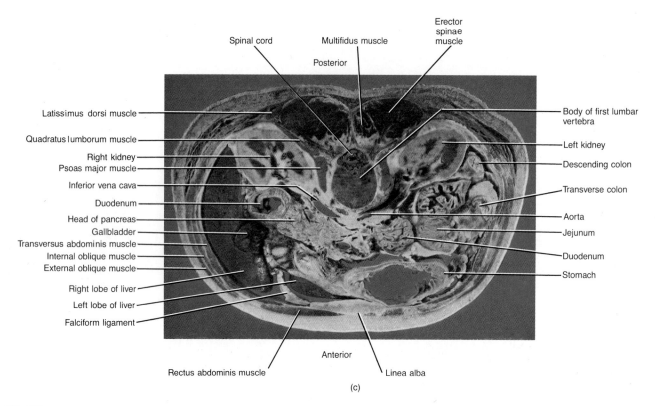

FIGURE 11-10 (*Continued*) Muscles of the anterior abdominal wall. (c) Photograph of a cross section through the abdomen showing the musculature and related viscera. (Courtesy of Stephen A. Kieffer and E. Robert Heitzman, *An Atlas of Cross-Sectional Anatomy,* Harper & Row, Publishers, Inc., New York, 1979.)

EXHIBIT 11-10

MUSCLES USED IN BREATHING (Figure 11-11)

MUSCLE	ORIGIN	INSERTION	ACTION	INNERVATION
Diaphragm (*dia* = across; *phragma* = wall)	Xiphoid process, costal cartilages of last six ribs, and lumbar vertebrae.	Central tendon.	Forms floor of thoracic cavity; pulls central tendon downward during inspiration and thus increases vertical length of thorax.	Phrenic nerve.
External intercostals (*inter* = between; *costa* = rib)	Inferior border of rib above.	Superior border of rib below.	Elevate ribs during inspiration and thus increase lateral and anteroposterior dimensions of thorax.	Intercostal nerves.
Internal intercostals	Superior border of rib below.	Inferior border of rib above.	Draw adjacent ribs together during forced expiration and thus decrease lateral and anteroposterior dimensions of thorax.	Intercostal nerves.

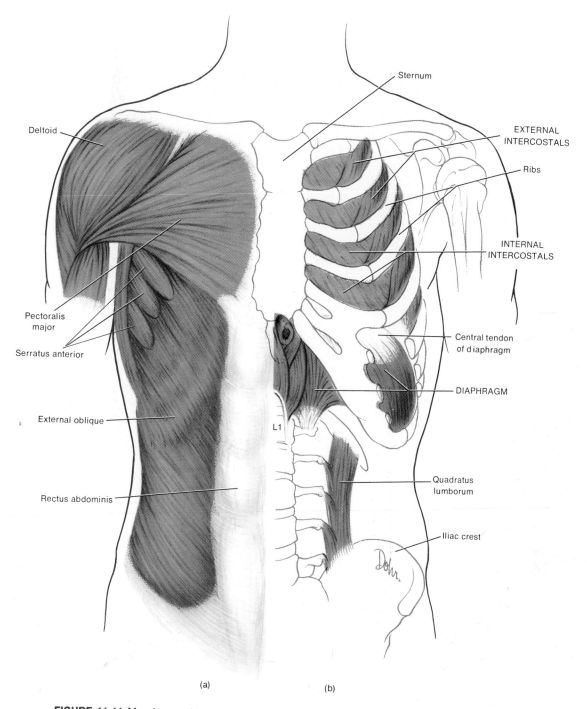

Deltoid

Pectoralis
major

Serratus anterior

External oblique

Rectus abdominis

Sternum

EXTERNAL
INTERCOSTALS

Ribs

INTERNAL
INTERCOSTALS

Central tendon
of diaphragm

DIAPHRAGM

Quadratus
lumborum

Iliac crest

L1

(a)   (b)

FIGURE 11-11 Muscles used in breathing. (a) Diagram of superficial view. (b) Diagram of deep view.

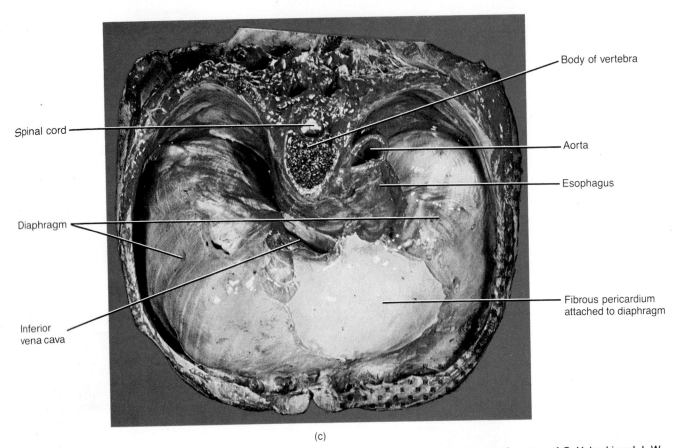

Spinal cord

Diaphragm

Inferior
vena cava

Body of vertebra

Aorta

Esophagus

Fibrous pericardium
attached to diaphragm

(c)

FIGURE 11-11 (*Continued*) Muscles used in breathing. (c) Photograph of superior view of diaphragm. (Courtesy of C. Yokochi and J. W. Rohen, *Photographic Anatomy of the Human Body,* 2d ed., 1979, IGAKU-SHOIN, Ltd., Tokyo, New York.)

EXHIBIT 11-11
MUSCLES OF THE PELVIC FLOOR* (Figure 11-12)

MUSCLE	ORIGIN	INSERTION	ACTION	INNERVATION
Levator ani (*levator* = raises; *ani* = anus)	This muscle is divisible into two parts, the pubococcygeus muscle and the iliococcygeus muscle. It forms the funnel-shaped floor of the pelvic cavity and supports the pelvic structures. It contains openings for the anal canal and urethra in both sexes and the vagina in the female.			
Pubococcygeus (*pubo* = pubis; *coccygeus* = coccyx)	Pubis.	Coccyx, urethra, anal canal, and central tendon of perineum.	Supports and slightly raises pelvic floor, resists increased intraabdominal pressure, and draws anus toward pubis and constricts it.	Sacral nerves S3–S4 or S4 and perineal branch of pudendal nerve.
Iliococcygeus (*ilio* = ilium)	Ischial spine.	Coccyx.	Supports and slightly raises pelvic floor, resists increased intraabdominal pressure, and draws anus toward pubis and constricts it.	Sacral nerves S3–S4 or S4 and perineal branch of pudendal nerve.
Coccygeus	Ischial spine.	Lower sacrum and upper coccyx.	Supports and slightly raises pelvic floor, resists intraabdominal pressure, and pulls coccyx forward following defecation or parturition.	Sacral nerve S3 or S4.

* The muscles of the pelvic floor, together with the fasciae covering their internal and external surfaces, are collectively referred to as the **pelvic diaphragm**.

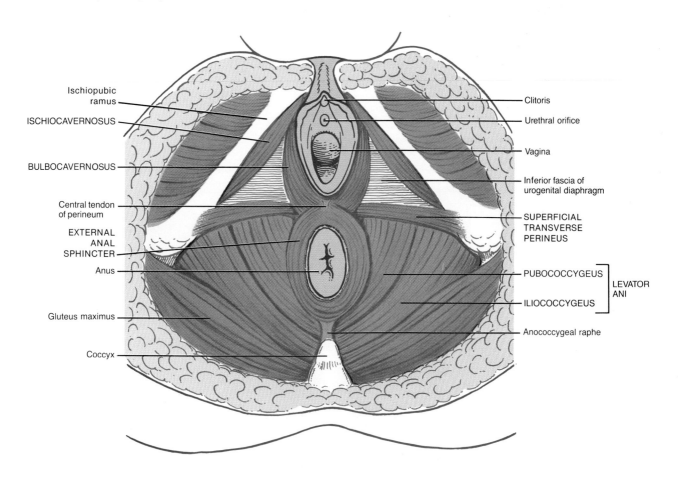

FIGURE 11-12 Muscles of the pelvic floor seen in the female perineum.

EXHIBIT 11-12

MUSCLES OF THE PERINEUM* (Figure 11-13)

MUSCLE	ORIGIN	INSERTION	ACTION	INNERVATION
Superficial transverse perineus (*superficial* = closer to surface; *transverse* = across; *perineus* = perineum)	Ischial tuberosity.	Central tendon of perineum.	Helps to stabilize the central tendon of the perineum.	Perineal branch of pudendal nerve.
Bulbocavernosus (*bulbus* = bulb; *caverna* = hollow place)	Central tendon of perineum.	Inferior fascia of urogenital diaphragm, corpus spongiosum of penis, and deep fascia on dorsum of penis in male; pubic arch and root and dorsum of clitoris in female.	Helps expel last drops of urine during micturition, helps propel semen along urethra, and assists in erection of the penis in male; decreases vaginal orifice and assists in erection of clitoris in female.	Perineal branch of pudendal nerve.
Ischiocavernosus (*ischion* = hip)	Ischial tuberosity and ischial and pubic rami.	Corpus cavernosum of penis in male and clitoris in female.	Maintains erection of penis in male and clitoris in female.	Perineal branch of pudendal nerve.
Deep transverse perineus † (*deep* = farther from surface)	Ischial rami.	Central tendon of perineum.	Helps eject last drops of urine and semen in male and urine in female.	Perineal branch of pudendal nerve.
Urethral sphincter † (*sphincter* = circular muscle that decreases size of an opening; *urethrae* = urethra)	Ischial and pubic rami.	Median raphe in male and vaginal wall in female.	Helps eject last drops of urine and semen in male and urine in female.	Perineal branch of pudendal nerve.
External anal sphincter	Anococcygeal raphe.	Central tendon of perineum.	Keeps anal canal and orifice closed.	Sacral nerve S4 and inferior rectal branch of pudendal nerve.

* The **perineum** is the entire outlet of the pelvis. It is a diamond-shaped area at the lower end of the trunk between the thighs and buttocks. It is bordered anteriorly by the symphysis pubis, laterally by the ischial tuberosities, and posteriorly by the coccyx. A transverse line drawn between the ischial tuberosities divides the perineum into an anterior **urogenital triangle** that contains the external genitals and a posterior **anal triangle** that contains the anus.

† The deep transverse perineus, the urethral sphincter, and a fibrous membrane constitute the **urogenital diaphragm.** It surrounds the urogenital ducts and helps to strengthen the pelvic floor.

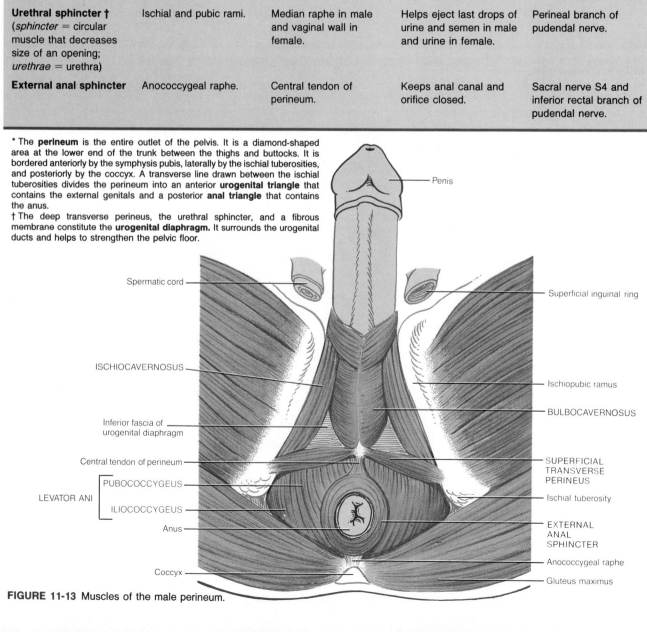

FIGURE 11-13 Muscles of the male perineum.

EXHIBIT 11-13
MUSCLES THAT MOVE THE PECTORAL (SHOULDER) GIRDLE (Figure 11-14)

MUSCLE	ORIGIN	INSERTION	ACTION	INNERVATION
Subclavius (*sub* = under; *clavius* = clavicle)	First rib.	Clavicle.	Depresses clavicle.	Nerve to subclavius.
Pectoralis minor (*pectus* = breast, chest, thorax; *minor* = lesser)	Third through fifth ribs.	Coracoid process of scapula.	Depresses scapula, rotates shoulder joint anteriorly, and elevates third through fifth ribs during forced inspiration when scapula is fixed.	Medial pectoral nerve.
Serratus anterior (*serratus* = saw-toothed; *anterior* = front)	Upper eight or nine ribs.	Vertebral border and inferior angle of scapula.	Rotates scapula laterally and elevates ribs when scapula is fixed.	Long thoracic nerve.
Trapezius (*trapezoides* = trapezoid-shaped)	Occipital bone, ligamentum nuchae, and spines of seventh cervical and all thoracic vertebrae.	Clavicle and acromion and spine of scapula.	Elevates clavicle, adducts scapula, elevates or depresses scapula, and extends head.	Accessory (XI) nerve and cervical nerves C3–C4.
Levator scapulae (*levator* = raises; *scapulae* = scapula)	Upper four or five cervical vertebrae.	Vertebral border of scapula.	Elevates scapula.	Dorsal scapular nerve and cervical nerves C3–C5.
Rhomboideus major (*rhomboides* = rhomboid or diamond-shaped)	Spines of second to fifth thoracic vertebrae.	Vertebral border of scapula.	Adducts scapula and slightly rotates it upward.	Dorsal scapular nerve.
Rhomboideus minor	Spines of seventh cervical and first thoracic vertebrae.	Superior angle of scapula.	Adducts scapula.	Dorsal scapular nerve.

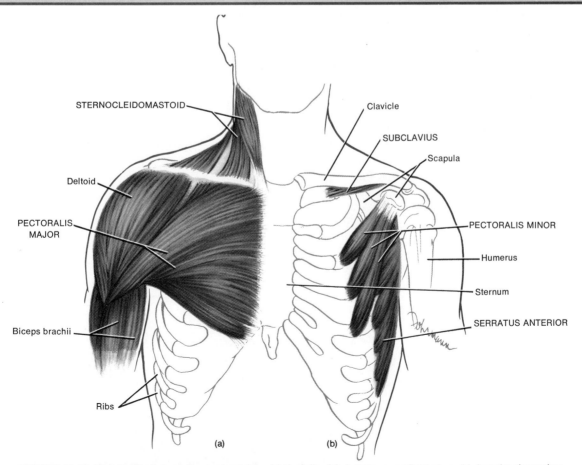

FIGURE 11-14 Muscles that move the pectoral (shoulder) girdle. (a) Anterior superficial view. (b) Anterior deep view.

EXHIBIT 11-14
MUSCLES THAT MOVE THE ARM (Figure 11-15)

MUSCLE	ORIGIN	INSERTION	ACTION	INNERVATION
Pectoralis major (see Figure 11-14a)	Clavicle, sternum, cartilages of second to sixth ribs.	Greater tubercle of humerus.	Flexes, adducts, and rotates arm medially.	Medial and lateral pectoral nerve.
Deltoid (*delta* = triangular)	Clavicle and acromion and spine of scapula.	Deltoid tuberosity of humerus.	Abducts arm.	Axillary nerve.
Subscapularis (*sub* = below; *scapularis* = scapula)	Subscapular fossa of scapula.	Lesser tubercle of humerus.	Rotates arm medially.	Upper and lower subscapular nerves.
Supraspinatus (*supra* = above; *spinatus* = spine of scapula)	Fossa superior to spine of scapula.	Greater tubercle of humerus.	Assists deltoid muscle in abducting arm.	Suprascapular nerve.

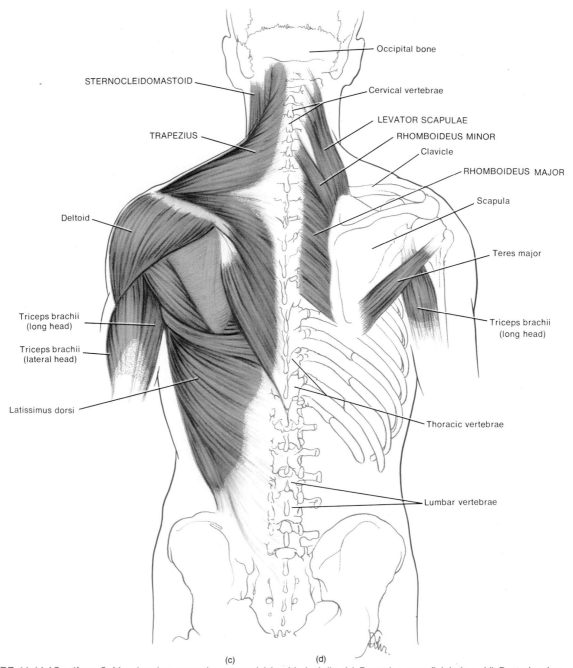

FIGURE 11-14 (*Continued*) Muscles that move the pectoral (shoulder) girdle. (c) Posterior superficial view. (d) Posterior deep view.

MUSCLE	ORIGIN	INSERTION	ACTION	INNERVATION
Infraspinatus (*infra* = below)	Fossa inferior to spine of scapula.	Greater tubercle of humerus.	Rotates arm laterally.	Suprascapular nerve.
Latissimus dorsi (*latissimus* = widest; *dorsum* = back)	Spines of lower six thoracic vertebrae, lumbar vertebrae, crests of sacrum and ilium, lower four ribs.	Intertubercular groove of humerus.	Extends, adducts, and rotates arm medially; draws shoulder downward and backward.	Thoracodorsal nerve.
Teres major (*teres* = long and round)	Inferior angle of scapula.	Distal to lesser tubercle of humerus.	Extends arm and draws it down; assists in adduction and medial rotation of arm.	Lower subscapular nerve.
Teres minor	Lateral border of scapula.	Greater tubercle of humerus.	Rotates arm laterally.	Axillary nerve.

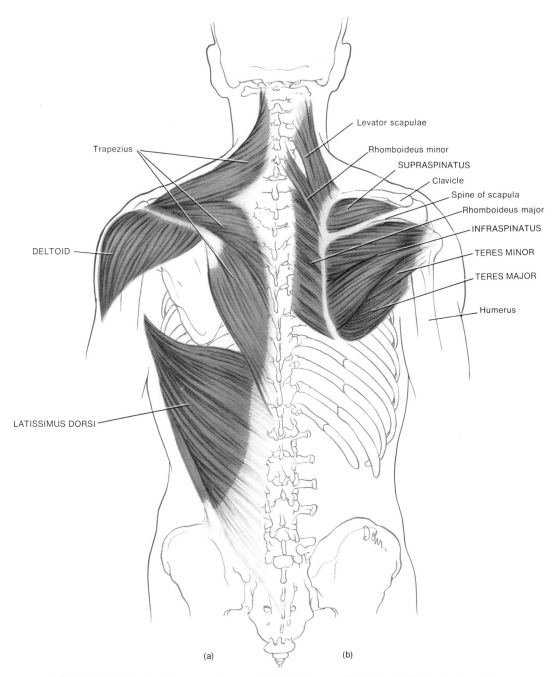

FIGURE 11-15 Muscles that move the arm. (a) Posterior superficial view. (b) Posterior deep view.

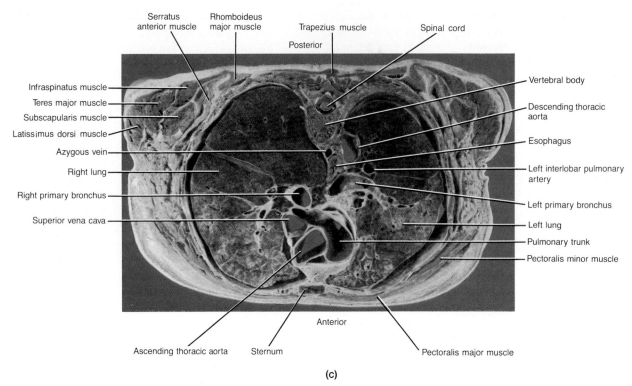

(c)

FIGURE 11-15 (*Continued*) Muscles that move the arm. (c) Photograph of a cross section of the thorax showing the musculature and related viscera. (Courtesy of Stephen A. Kieffer and E. Robert Heitzman, *An Atlas of Cross-Sectional Anatomy,* Harper & Row, Publishers, Inc., New York, 1979.)

EXHIBIT 11-15
MUSCLES THAT MOVE THE FOREARM (Figure 11-16)

MUSCLE	ORIGIN	INSERTION	ACTION	INNERVATION
Biceps brachii (*biceps* = two heads of origin; *brachion* = arm)	Long head originates from tubercle above glenoid cavity; short head originates from coracoid process of scapula.	Radial tuberosity and bicipital aponeurosis.	Flexes and supinates forearm.	Musculocutaneous nerve.
Brachialis	Anterior surface of humerus.	Tuberosity and coronoid process of ulna.	Flexes forearm.	Musculocutaneous, radial, and median nerves.
Brachioradialis (*radialis* = radius) (see also Figure 11-17)	Supracondyloid ridge of humerus.	Superior to styloid process of radius.	Flexes forearm.	Radial nerve.
Triceps brachii (*triceps* = three heads of origin)	Long head originates from infraglenoid tuberosity of scapula; lateral head originates from lateral and posterior surface of humerus superior to radial groove; medial head originates from posterior surface of humerus inferior to radial groove.	Olecranon of ulna.	Extends forearm.	Radial nerve.
Coracobrachialis (*coraco* = coracoid process)	Coracoid process of scapula.	Middle of medial surface of shaft of humerus.	Flexes and adducts forearm.	Musculocutaneous nerve.
Anconeus (*anconeal* = pertaining to elbow) (see Figure 11-17)	Lateral epicondyle of humerus.	Olecranon and superior portion of shaft of ulna.	Extends forearm.	Radial nerve.
Supinator (*supination* = turning palm upward or anteriorly)	Lateral epicondyle of humerus and ridge on ulna.	Oblique line of radius.	Supinates forearm.	Deep radial nerve.
Pronator teres (*pronation* = turning palm downward or posteriorly)	Medial epicondyle of humerus and coronoid process of ulna.	Midlateral surface of radius.	Pronates forearm.	Median nerve.
Pronator quadratus (*quadratus* = squared, four-sided)	Distal portion of shaft of ulna.	Inferior portion of shaft of radius.	Pronates and rotates forearm.	Median nerve.

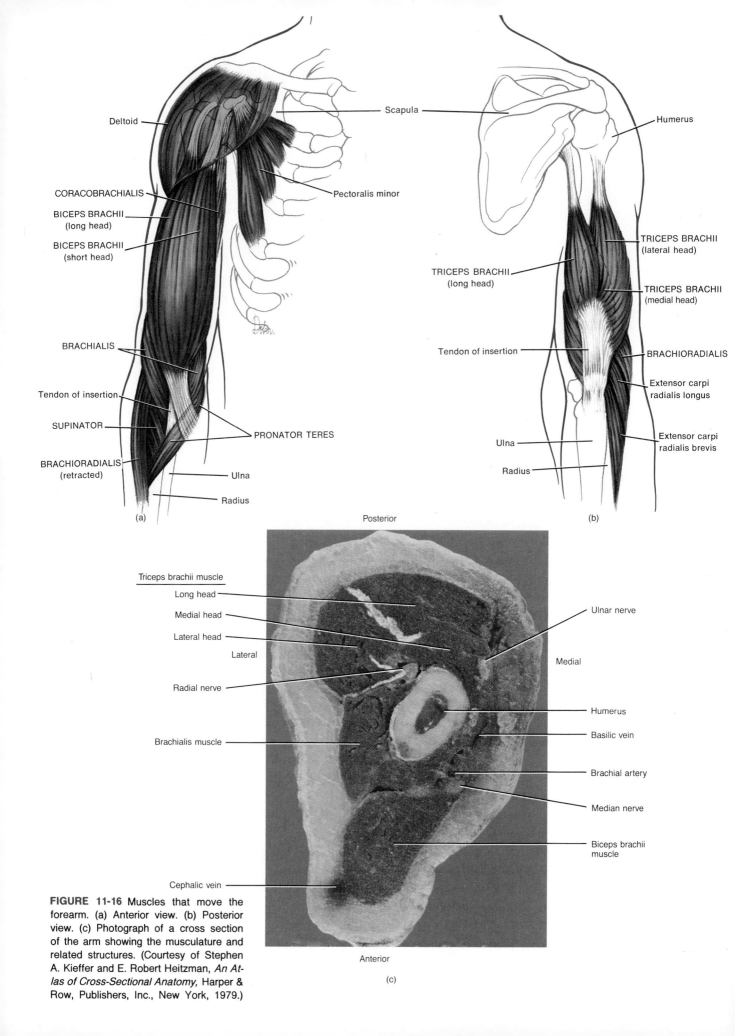

Deltoid

Scapula

Humerus

CORACOBRACHIALIS

BICEPS BRACHII
(long head)

BICEPS BRACHII
(short head)

Pectoralis minor

TRICEPS BRACHII
(long head)

TRICEPS BRACHII
(lateral head)

TRICEPS BRACHII
(medial head)

BRACHIALIS

Tendon of insertion

SUPINATOR

Tendon of insertion

PRONATOR TERES

BRACHIORADIALIS

Extensor carpi
radialis longus

BRACHIORADIALIS
(retracted)

Ulna

Ulna

Extensor carpi
radialis brevis

Radius

Radius

(a)

(b)

Posterior

Triceps brachii muscle

Long head

Medial head

Lateral head

Lateral

Radial nerve

Brachialis muscle

Ulnar nerve

Medial

Humerus

Basilic vein

Brachial artery

Median nerve

Biceps brachii
muscle

Cephalic vein

Anterior

(c)

FIGURE 11-16 Muscles that move the forearm. (a) Anterior view. (b) Posterior view. (c) Photograph of a cross section of the arm showing the musculature and related structures. (Courtesy of Stephen A. Kieffer and E. Robert Heitzman, *An Atlas of Cross-Sectional Anatomy,* Harper & Row, Publishers, Inc., New York, 1979.)

EXHIBIT 11-16

MUSCLES THAT MOVE THE WRIST AND FINGERS (Figure 11-17)

MUSCLE	ORIGIN	INSERTION	ACTION	INNERVATION
Flexor carpi radialis (*flexor* = decreases angle at joint; *carpus* = wrist; *radialis* = radius)	Medial epicondyle of humerus.	Second and third metacarpals.	Flexes and abducts wrist.	Median nerve.
Flexor carpi ulnaris (*ulnaris* = ulna)	Medial epicondyle of humerus and upper dorsal border of ulna.	Pisiform, hamate, and fifth metacarpal.	Flexes and adducts wrist.	Ulnar nerve.
Palmaris longus (*palma* = palm)	Medial epicondyle of humerus.	Transverse carpal ligament and palmar aponeurosis.	Flexes wrist and tenses palmar aponeurosis.	Median nerve.
Extensor carpi radialis longus (*extensor* = increases angle at joint; *longus* = long)	Lateral epicondyle of humerus.	Second metacarpal.	Extends and abducts wrist.	Radial nerve.
Extensor carpi ulnaris	Lateral epicondyle of humerus and dorsal border of ulna.	Fifth metacarpal.	Extends and adducts wrist.	Deep radial nerve.
Flexor digitorum profundus (*digit* = finger or toe; *profundus* = deep)	Anterior medial surface of body of ulna.	Bases of distal phalanges.	Flexes distal phalanges of each finger.	Median and ulnar nerves.
Flexor digitorum superficialis (*superficialis* = superficial)	Medial epicondyle of humerus, coronoid process of ulna, and oblique line of radius.	Middle phalanges.	Flexes middle phalanges of each finger.	Median nerve.
Extensor digitorum	Lateral epicondyle of humerus.	Middle and distal phalanges of each finger.	Extends phalanges.	Deep radial nerve.
Extensor indicis (*indicis* = index)	Dorsal surface of ulna.	Tendon of extensor digitorum of index finger.	Extends index finger.	Deep radial nerve.

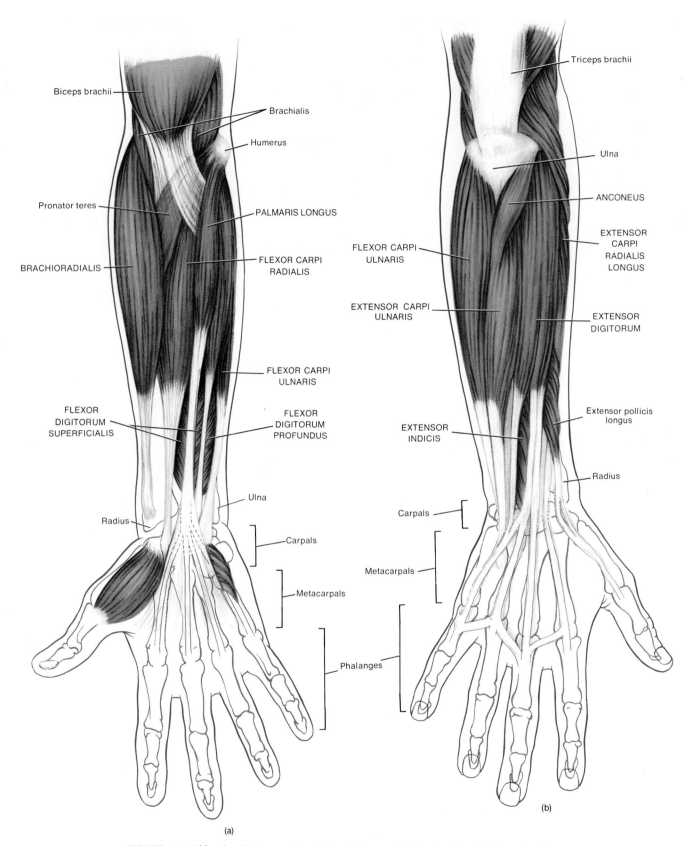

FIGURE 11-17 Muscles that move the wrist and fingers. (a) Anterior view. (b) Posterior view.

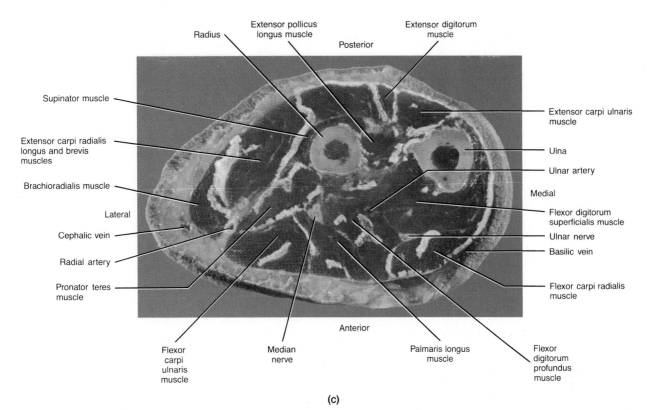

(c)

FIGURE 11-17 (*Continued*) Muscles that move the wrist and fingers. (c) Photograph of a cross section of the forearm showing the musculature and related structures. (Courtesy of Stephen A. Kieffer and E. Robert Heitzman, *An Atlas of Cross-Sectional Anatomy,* Harper & Row, Publishers, Inc., New York, 1979.)

EXHIBIT 11-17
MUSCLES THAT MOVE THE VERTEBRAL COLUMN (Figure 11-18)

MUSCLE	ORIGIN	INSERTION	ACTION	INNERVATION
Rectus abdominis (see Figure 11-10)	Pubic crest and symphisis pubis.	Cartilages of fifth through seventh ribs.	Flexes vertebral column at lumbar spine and compresses abdomen.	Intercostal (thoracic) nerves T7–T12.
Quadratus lumborum (*quadratus* = squared, four-sided; *lumb* = lumbar region)	Iliac crest.	Twelfth rib and upper four lumbar vertebrae.	Flexes vertebral column laterally.	Intercostal nerve T12 and lumbar nerve L1.
Sacrospinalis (erector spinae)	This posterior muscle consists of three groupings: iliocostalis, longissimus, and spinalis. These groups, in turn, consist of a series of overlapping muscles. The iliocostalis group is laterally placed, the longissimus group is intermediate in placement, and the spinalis is medially placed.			
LATERAL				
Iliocostalis lumborum (*ilium* = flank; *costa* = rib)	Iliac crest.	Lower six ribs.	Extends lumbar region of vertebral column.	Dorsal rami of lumbar nerves.
Iliocostalis thoracis (*thorax* = chest)	Lower six ribs.	Upper six ribs.	Maintains erect position of spine.	Dorsal rami of thoracic (intercostal) nerves.
Iliocostalis cervicis (*cervix* = neck)	First six ribs.	Transverse processes of fourth to sixth cervical vertebrae.	Extends cervical region of vertebral column.	Dorsal rami of cervical nerves.
INTERMEDIATE				
Longissimus thoracis	Transverse processes of lumbar vertebrae.	Transverse processes of all thoracic and upper lumbar vertebrae and ninth and tenth ribs.	Extends thoracic region of vertebral column.	Dorsal rami of spinal nerves.
Longissimus cervicis	Transverse processes of fourth and fifth thoracic vertebrae.	Transverse processes of second to sixth cervical vertebrae.	Extends cervical region of vertebral column.	Dorsal rami of spinal nerves.
Longissimus capitis (*caput* = head)	Transverse processes of upper four thoracic vertebrae.	Mastoid process of temporal bone.	Extends head and rotates it to opposite side.	Dorsal rami of middle and lower cervical nerves.
MEDIAL				
Spinalis thoracis	Spines of upper lumbar and lower thoracic vertebrae.	Spines of upper thoracic vertebrae.	Extends vertebral column.	Dorsal rami of spinal nerves.

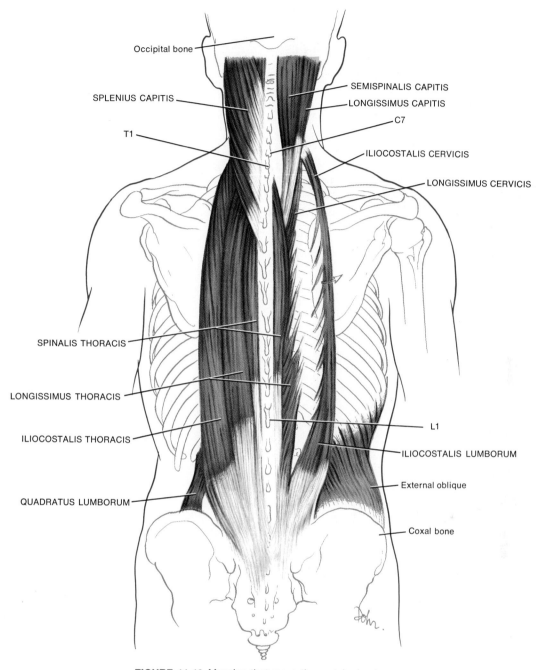

FIGURE 11-18 Muscles that move the vertebral column.

EXHIBIT 11-18
MUSCLES THAT MOVE THE THIGH (Figure 11-19)

MUSCLE	ORIGIN	INSERTION	ACTION	INNERVATION
Psoas major* (*psoa* = muscle of loin)	Transverse processes and bodies of lumbar vertebrae.	Lesser trochanter of femur.	Flexes and rotates thigh laterally; flexes vertebral column.	Lumbar nerves L2–L3.
Iliacus* (*iliac* = ilium)	Iliac fossa.	Tendon of psoas major.	Flexes and rotates thigh laterally; flexes vertebral column slightly.	Femoral nerve.
Gluteus maximus (*glutos* = buttock; *maximus* = largest)	Iliac crest, sacrum, coccyx, and aponeurosis of sacrospinalis.	Iliotibial tract of fascia lata and gluteal tuberosity of femur.	Extends and rotates thigh laterally.	Inferior gluteal nerve.
Gluteus medius (*media* = middle)	Ilium.	Greater trochanter of femur.	Abducts and rotates thigh medially.	Superior gluteal nerve.
Gluteus minimus (*minimus* = smallest)	Ilium.	Greater trochanter of femur.	Abducts and rotates thigh laterally.	Superior gluteal nerve.
Tensor fasciae latae (*tensor* = makes tense; *fascia* = band; *latus* = wide)	Iliac crest.	Tibia by way of the iliotibial tract.	Flexes and abducts thigh.	Superior gluteal nerve.
Adductor longus (*adductor* = moves part closer to midline; *longus* = long)	Pubic crest and symphysis pubis.	Linea aspera of femur.	Adducts, rotates, and flexes thigh.	Obturator nerve.
Adductor brevis (*brevis* = short)	Inferior ramus of pubis.	Linea aspera of femur.	Adducts, rotates, and flexes thigh.	Obturator nerve.
Adductor magnus (*magnus* = large)	Inferior ramus of pubis and ischium to ischial tuberosity.	Linea aspera of femur.	Adducts, flexes, and extends thigh (anterior part flexes, posterior part extends).	Obturator and sciatic nerves.
Piriformis (*pirum* = pear; *forma* = shape)	Sacrum.	Greater trochanter of femur.	Rotates thigh laterally and abducts it.	Sacral nerves S2 or S1–S2.
Obturator internus (*obturator* = obturator foramen; *internus* = inside)	Margin of obturator foramen, pubis, and ischium.	Greater trochanter of femur.	Rotates thigh laterally and abducts it.	Nerve to obturator internus.
Pectineus (*pecten* = comb-shaped)	Fascia of pubis.	Pectineal line of femur.	Flexes, adducts, and rotates thigh laterally.	Femoral nerve.

CLINICAL APPLICATION

A common sports injury is a **"pulled groin."** Very simply this means there is a strain, stretching, and probably some tearing away of the tendinous origins of the adductor muscles that move the thigh.

---

* Together the psoas major and iliacus are sometimes termed the **iliopsoas muscle.**

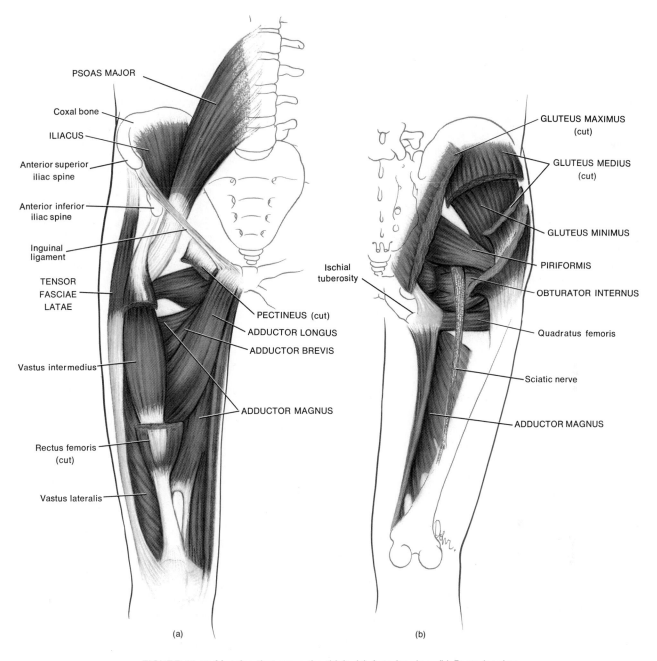

FIGURE 11-19 Muscles that move the thigh. (a) Anterior view. (b) Posterior view.

EXHIBIT 11-19
MUSCLES THAT ACT ON THE LEG (Figure 11-20)

MUSCLE	ORIGIN	INSERTION	ACTION	INNERVATION
Quadriceps femoris (*quadriceps* = four heads of origin; *femoris* = femur)	This composite muscle includes four distinct parts, usually described as four separate muscles. The common tendon that includes the patella and attaches to the tibial tuberosity is known as the patellar ligament.			
Rectus femoris (*rectus* = fibers parallel to midline)	Anterior inferior iliac spine.	Upper border of patella.		Femoral nerve.
Vastus lateralis (*vastus* = large; *lateralis* = lateral)	Greater trochanter and linea aspera of femur.		All four heads extend leg; rectus portion alone also flexes thigh.	Femoral nerve.
Vastus medialis (*medialis* = medial)	Linea aspera of femur.	Upper border and sides of patella; tibial tuberosity through patellar ligament (tendon of quadriceps).		Femoral nerve.
Vastus intermedius (*intermedius* = middle)	Anterior and lateral surfaces of body of femur.			Femoral nerve.
Hamstrings	A collective designation for three separate muscles.			
Biceps femoris (*biceps* = two heads of origin)	Long head arises from ischial tuberosity; short head arises from linea aspera of femur.	Head of fibula and lateral condyle of tibia.	Flexes leg and extends thigh.	Tibial nerve from sciatic nerve.
Semitendinosus (*semi* = half; *tendo* = tendon)	Ischial tuberosity.	Proximal part of medial surface of body of tibia.	Flexes leg and extends thigh.	Tibial nerve from sciatic nerve.
Semimembranosus (*membran* = membrane)	Ischial tuberosity.	Medial condyle of tibia.	Flexes leg and extends thigh.	Tibial nerve from sciatic nerve.
Gracilis (*gracilis* = slender)	Symphysis pubis and pubic arch.	Medial surface of body of tibia.	Flexes leg and adducts thigh.	Obturator nerve.
Sartorius (*sartor* = tailor; refers to cross-legged position of tailors)	Anterior superior spine of ilium.	Medial surface of body of tibia.	Flexes leg; flexes thigh and rotates it laterally, thus crossing leg.	Femoral nerve.

CLINICAL APPLICATION

"Pulled hamstrings" are common sports injuries in individuals who run very hard. Sometimes the violent muscular exertion required to perform a feat tears off part of the tendinous origins of the hamstrings from the ischial tuberosity. This is usually accompanied by a contusion (bruising) and tearing of some of the muscle fibers and rupture of blood vessels producing a hematoma (collection of blood).

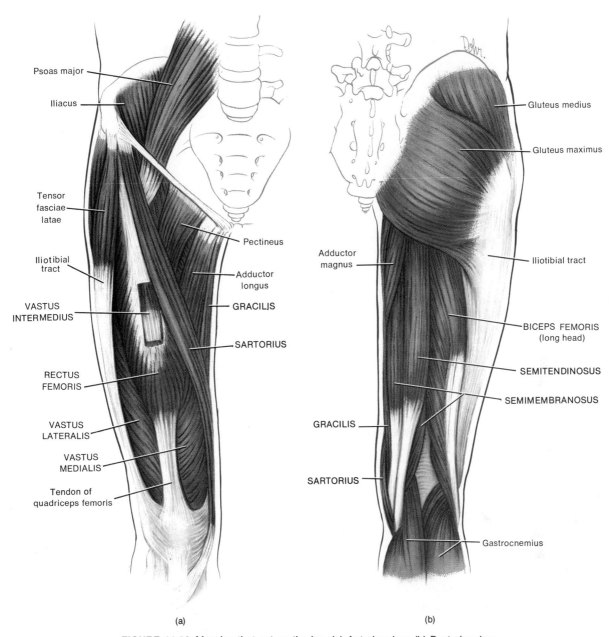

FIGURE 11-20 Muscles that act on the leg. (a) Anterior view. (b) Posterior view.

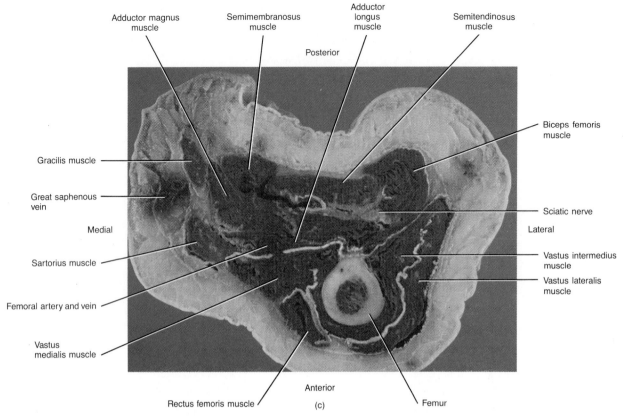

Adductor magnus muscle
Semimembranosus muscle
Adductor longus muscle
Semitendinosus muscle
Posterior
Biceps femoris muscle
Gracilis muscle
Great saphenous vein
Sciatic nerve
Medial
Lateral
Sartorius muscle
Vastus intermedius muscle
Vastus lateralis muscle
Femoral artery and vein
Vastus medialis muscle
Rectus femoris muscle
Anterior
(c)
Femur

FIGURE 11-20 (*Continued*) Muscles that act on the leg. (c) Cross section of the thigh showing the musculature and related structures. (Courtesy of Stephen A. Kieffer and E. Robert Heitzman, *An Atlas of Cross-Sectional Anatomy,* Harper & Row, Publishers, Inc., New York, 1979.)

EXHIBIT 11-20
MUSCLES THAT MOVE THE FOOT AND TOES (Figure 11-21)

MUSCLE	ORIGIN	INSERTION	ACTION	INNERVATION
Gastrocnemius (*gaster* = belly; *kneme* = leg)	Lateral and medial condyles of femur and capsule of knee.	Calcaneus by way of calcaneal ("Achilles") tendon.	Plantar flexes foot.	Tibial nerve.
Soleus (*soleus* = sole of foot)	Head of fibula and medial border of tibia.	Calcaneus by way of calcaneal ("Achilles") tendon.	Plantar flexes foot.	Tibial nerve.
Peroneus longus (*perone* = fibula; *longus* = long)	Head and body of fibula and lateral condyle of tibia.	First metatarsal and first cuneiform bone.	Plantar flexes and everts foot.	Superficial peroneal nerve.
Peroneus brevis (*brevis* = short)	Body of fibula.	Fifth metatarsal.	Plantar flexes and everts foot.	Superficial peroneal nerve.
Peroneus tertius (*tertius* = third)	Distal third of fibula.	Fifth metatarsal.	Dorsiflexes and everts foot.	Deep peroneal nerve.
Tibialis anterior (*tibialis* = tibia; *anterior* = front)	Lateral condyle and body of tibia.	First metatarsal and first cuneiform.	Dorsiflexes and inverts foot.	Deep peroneal nerve.
Tibialis posterior (*posterior* = back)	Interosseus membrane between tibia and fibula.	Second, third, and fourth metatarsals; navicular; third cuneiform; and cuboid.	Plantar flexes and inverts foot.	Tibial nerve.
Flexor digitorum longus (*flexor* = decreases angle at joint; *digitorum* = finger or toe)	Tibia.	Distal phalanges of four outer toes.	Flexes toes and plantar flexes and inverts foot.	Tibial nerve.
Extensor digitorum longus (*extensor* = increases angle at joint)	Lateral condyle of tibia and anterior surface of fibula.	Middle and distal phalanges of four outer toes.	Extends toes and dorsiflexes and everts foot.	Deep peroneal nerve.

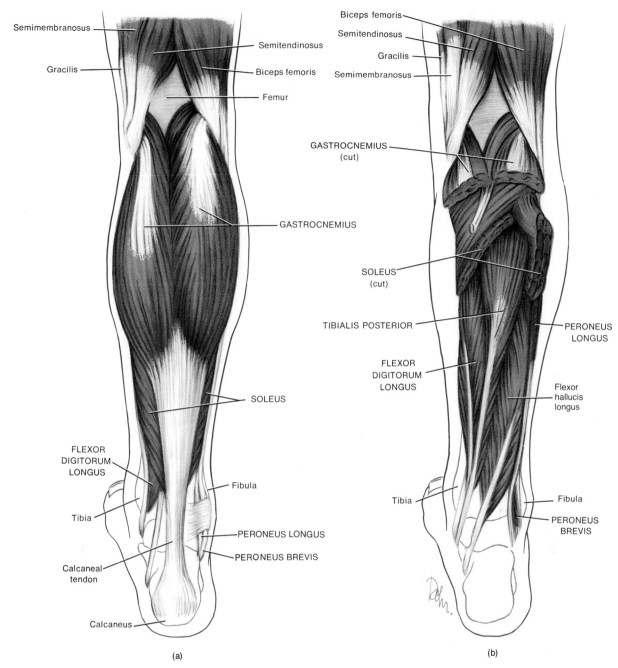

Semimembranosus

Gracilis

Semitendinosus

Biceps femoris

Femur

GASTROCNEMIUS

SOLEUS

FLEXOR
DIGITORUM
LONGUS

Tibia

Fibula

Calcaneal
tendon

PERONEUS LONGUS

PERONEUS BREVIS

Calcaneus

(a)

Biceps femoris

Semitendinosus

Gracilis

Semimembranosus

GASTROCNEMIUS
(cut)

SOLEUS
(cut)

TIBIALIS POSTERIOR

FLEXOR
DIGITORUM
LONGUS

Tibia

PERONEUS
LONGUS

Flexor
hallucis
longus

Fibula

PERONEUS
BREVIS

(b)

FIGURE 11-21 Muscles that move the foot and toes. (a) Superficial posterior view. (b) Deep posterior view.

INTRAMUSCULAR INJECTIONS

An intramuscular injection penetrates the skin and subcutaneous tissue to enter the muscle itself. Intramuscular injections are preferred when prompt absorption is desired, when larger doses than can be given cutaneously are indicated, or when the drug is too irritating to give subcutaneously. The common sites for intramuscular injections include the buttock, lateral side of the thigh, and the deltoid region of the arm. Muscles in these areas, especially the gluteal muscles in the buttock, are fairly thick. Because of the large number of muscle fibers and extensive fascia, the drug has a large surface area for absorption. Absorption is further promoted by the extensive blood supply to muscles. Ideally, intramuscular injec-

tions should be given deep within the muscle and away from major nerves and blood vessels.

For many intramuscular injections, the preferred site is the gluteus medius muscle of the buttock (Figure 11-22a). The buttock is divided into four quadrants and the upper outer quadrant used as the injection site. The iliac crest serves as a landmark for this quadrant. The spot for injection is usually about 5 to 7½ cm (2 to 3 inches) below the iliac crest. The upper outer quadrant is chosen because in this area the muscle is quite thick with few nerves. Thus there is less chance of injury to the sciatic nerve. Injury to the nerve can cause paralysis of the lower extremity. The probability of injecting the drug into a blood vessel is also remote in this area. After the needle is inserted into the gluteus medius muscle, the plunger

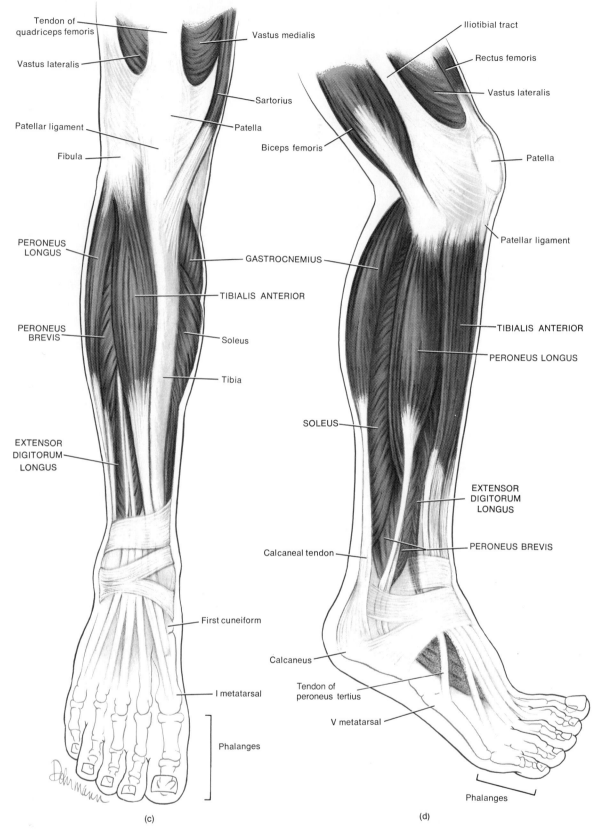

FIGURE 11-21 (*Continued*) Muscles that move the foot and toes. (c) Superficial anterior view. (d) Superficial lateral view.

is pulled up for a few seconds. If the syringe fills with blood, the needle is in a blood vessel and a different injection site on the opposite buttock is chosen.

Injections given in the lateral side of the thigh are inserted into the midportion of the vastus lateralis muscle (Figure 11-22b). This site is determined by using the knee and greater trochanter of the femur as landmarks. The

midportion of the muscle is located by measuring a handbreadth above the knee and a handbreadth below the greater trochanter.

The deltoid injection is given in the midportion of the muscle about two to three fingerbreadths below the acromion of the scapula and lateral to the axilla (Figure 11-22c).

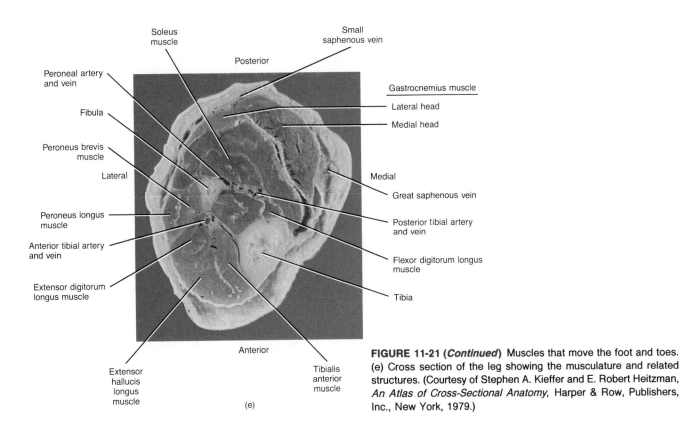

FIGURE 11-21 (*Continued*) Muscles that move the foot and toes. (e) Cross section of the leg showing the musculature and related structures. (Courtesy of Stephen A. Kieffer and E. Robert Heitzman, *An Atlas of Cross-Sectional Anatomy,* Harper & Row, Publishers, Inc., New York, 1979.)

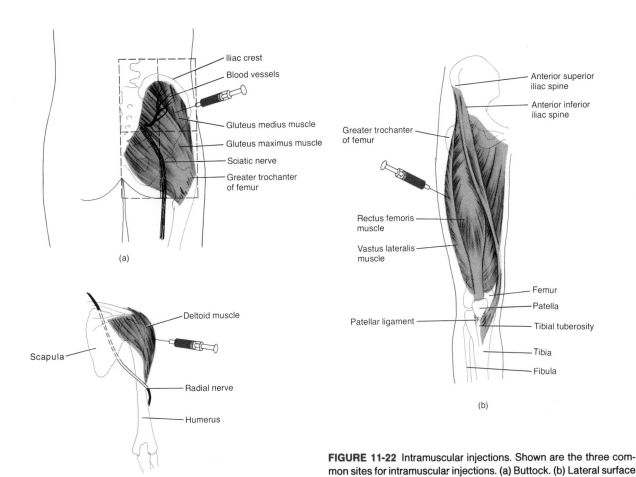

FIGURE 11-22 Intramuscular injections. Shown are the three common sites for intramuscular injections. (a) Buttock. (b) Lateral surface of the thigh. (c) Deltoid region of the arm.

STUDY OUTLINE

How Skeletal Muscles Produce Movement (p. 224)

1. Skeletal muscles produce movement by pulling on bones.
2. The attachment to the stationary bone is the origin. The attachment to the movable bone is the insertion.
3. Bones serve as levers and joints as fulcrums. The lever is acted on by two different forces: resistance and effort.
4. Levers are categorized into three types—first-class, second-class, and third-class—according to the position of the fulcrum, effort, and resistance on the lever.
5. Fascicular arrangements include parallel, convergent, pennate, and circular. Fascicular arrangement is correlated with the power of a muscle and the range of movement.
6. The agonist or prime mover produces the desired action. The antagonist produces an opposite action. The synergist assists the agonist by reducing unnecessary movement.

Naming Skeletal Muscles (p. 228)

Skeletal muscles are named on the basis of distinctive criteria: direction of fibers, location, size, number of origins (or heads), shape, origin and insertion, and action.

Principal Skeletal Muscles (p. 228)

The principal skeletal muscles of the body are grouped according to region in Exhibits 11-2 through 11-20.

Intramuscular Injections (p. 261)

1. Advantages of intramuscular injections are prompt absorption, use of larger doses than can be given cutaneously, and minimal irritation.
2. Common sites for intramuscular injections are the buttock, lateral side of the thigh, and deltoid region of the arm.

REVIEW QUESTIONS

1. What is meant by the muscular system? Explain fully.
2. Using the terms origin, insertion, and belly in your discussion, describe how skeletal muscles produce body movements by pulling on bones.
3. What is a lever? Fulcrum? Apply these terms to the body, and indicate the nature of the forces that act on levers. Describe the three classes of levers, and provide one example for each in the body.
4. Describe the various arrangements of fasciculi. How is fascicular arrangement correlated with the strength of a muscle and its range of movement?
5. Define the role of the agonist, antagonist, and synergist in producing body movements.
6. Select at random several muscles presented in Exhibits 11-2 through 11-20 and see if you can determine the criterion or criteria employed for naming each. In addition, refer to the prefixes, suffixes, roots, and definitions in each exhibit as a guide. Select as many muscles as you wish, as long as you feel you understand the concept involved.
7. Discuss the muscles and their actions involved in facial expression.
8. What muscles would you use to do the following : (a) frown, (b) pout, (c) show surprise, (d) show your upper teeth, (e) pucker your lips, (f) squint, (g) blow up a balloon, (h) smile?
9. What are the principal muscles that move the mandible? Give the function of each.
10. What would happen if you lost tone in the masseter and temporalis muscles?
11. What muscles move the eyeball? In which direction does each muscle move the eyeball?
12. Describe the action of each of the muscles acting on the tongue.
13. What tongue, facial, and mandibular muscles would you use when chewing a piece of gum?
14. What muscles constrict the pharynx? What muscle dilates the pharynx?
15. Describe the actions of the extrinsic and intrinsic muscles of the larynx.
16. What muscles are responsible for moving the head, and how do they move the head?
17. What muscles would you use to signify "yes" and "no" by moving your head?
18. What muscles accomplish compression of the anterior abdominal wall?
19. What are the principal muscles involved in breathing? What are their actions?
20. Describe the actions of the muscles of the pelvic floor.
21. Describe the actions of the muscles of the perineum.
22. In what directions is the pectoral (shoulder) girdle drawn? What muscles accomplish these movements?
23. What muscles are used to raise your shoulders, lower your shoulders, join your hands behind your back, and join your hands in front of your chest?
24. What movements are possible at the shoulder joint? What muscles accomplish these movements?
25. What muscles move the arm? In which directions do these movements occur?
26. What muscles move the forearm and what actions are used when striking a match?
27. Discuss the various movements possible at the wrist and fingers. What muscles accomplish these movements?
28. How many muscles and actions of the wrist and fingers used when writing can you list?
29. Discuss the various muscles and movements of the vertebral column.
30. Can you perform an exercise that would involve the use of each of the muscles listed in Exhibit 11-17?
31. What muscles accomplish movements of the femur? What actions are produced by these muscles?
32. Review the various movements involved in your favorite kind of dancing. What muscles listed in Exhibit 11-18 would you be using and what actions would you be performing?
33. What muscles act at the knee joint? What kinds of movements do these muscles perform?
34. Determine the muscles and their actions listed in Exhibit 11-19 that you would use to climb a ladder to a diving board, dive into the water, swim the length of a pool, and then sit at pool side.
35. Name the muscles that plantar flex, evert, pronate, dorsiflex, and supinate the foot.
36. In which directions are the toes moved? What muscles bring about these movements?
37. What are the advantages of intramuscular injections?
38. Describe how you would locate the sites for an intramuscular injection in the buttock, lateral side of the thigh, and deltoid region of the arms.

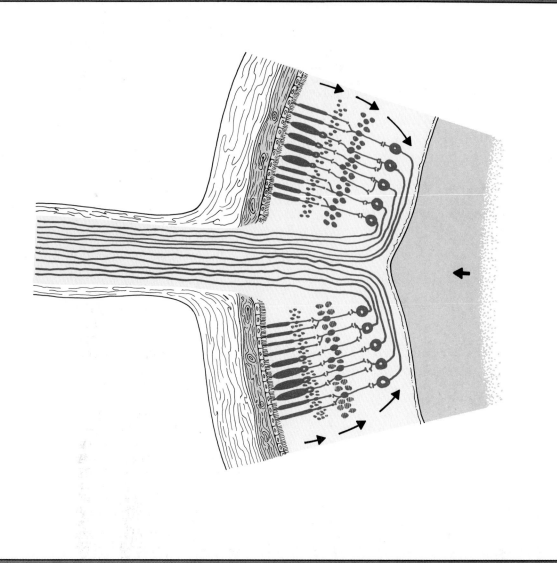

UNIT III

Control Systems of the Human Body

This unit will show you the significance of the nerve impulse in making rapid adjustments for maintaining homeostasis. You will learn how the nervous system detects changes in the environment, decides on a course of action, and responds to the change. We will also investigate the role of hormones in maintaining long-term homeostasis.

12 Nervous Tissue

The **nervous system** is the body's control center and communications network. In humans, the nervous system serves three broad functions. First, it senses changes within the body and in the outside environment. Second, it interprets the changes. Third, it responds to the interpretation by initiating action in the form of muscular contractions or glandular secretions.

Through sensation, integration, and response the nervous system represents the body's most rapid means of maintaining homeostasis. Its split-second reactions, carried out by nerve impulses, can normally make the adjustments necessary to keep the body functioning efficiently. As you will see later, the nervous system shares the maintenance of homeostasis with the endocrine system. Although the adjustments made by hormones secreted by endocrine glands are slower than those made by nerve impulses, they are no less effective.

ORGANIZATION

The nervous system may be divided into two principal divisions, the central nervous system and the peripheral nervous system, and several subdivisions (Figure 12-1).

The **central nervous system (CNS)** is the control center for the entire system and consists of the brain and spinal cord. All body sensations must be relayed from receptors to the central nervous system if they are to be interpreted and acted on. All the nerve impulses that stimulate muscles to contract and glands to secrete must also pass from the central nervous system.

The various nerve processes that connect the brain and spinal cord with receptors, muscles, and glands constitute the **peripheral (pe-RIF-er-al) nervous system (PNS).** The peripheral nervous system may be divided into an afferent system and an efferent system. The **affer-**ent (AF-er-ent) **system** consists of nerve cells that convey information from receptors in the periphery of the body to the central nervous system. These nerve cells, called *afferent (sensory) neurons,* are the first cells to pick up incoming information. The **efferent (EF-er-ent) system** consists of nerve cells that convey information from the central nervous system to muscles and glands. These nerve cells are called *efferent (motor) neurons.*

The efferent system is subdivided into a somatic nervous system and an autonomic nervous system. The **somatic nervous system** or **SNS** (*soma* = body) consists of efferent neurons that conduct impulses from the central nervous system to skeletal muscle tissue. Since the somatic nervous system produces movement only in skeletal muscle tissue, it is under conscious control and therefore voluntary. The **autonomic nervous system** or **ANS** (*auto* = self; *nomos* = law), by contrast, contains efferent neurons that convey impulses from the central nervous system to smooth muscle tissue, cardiac muscle tissue, and glands. Since it produces responses only in involuntary muscles and glands, it is usually considered to be involuntary.

With few exceptions, the viscera receive nerve fibers from the two divisions of the autonomic nervous system: the **sympathetic division** and the **parasympathetic division.** In general, the fibers of one division stimulate or increase an organ's activity, while the fibers from the other inhibit or decrease activity (see Chapter 16).

HISTOLOGY

Despite the organizational complexity of the nervous system, it consists of only two principal kinds of cells: neurons and neuroglia. Neurons make up the nervous tissue that forms the structural and functional portion of the

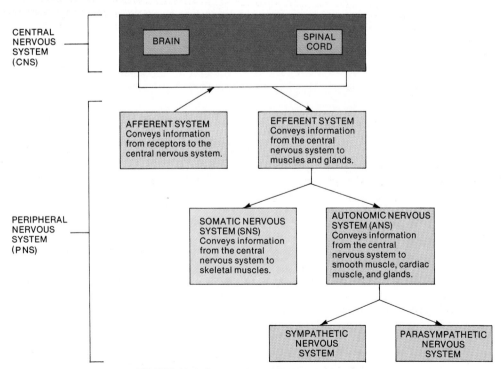

FIGURE 12-1 Organization of the nervous system.

EXHIBIT 12-1

NEUROGLIA OF CENTRAL NERVOUS SYSTEM

TYPE	DESCRIPTION	MICROSCOPIC APPEARANCE	FUNCTION
Astrocytes (*astro* = star; *cyte* = cell)	Star-shaped cells with numerous processes. *Protoplasmic astrocytes* are found in the gray matter of the CNS, and *fibrous astrocytes* are found in the white matter of the CNS.	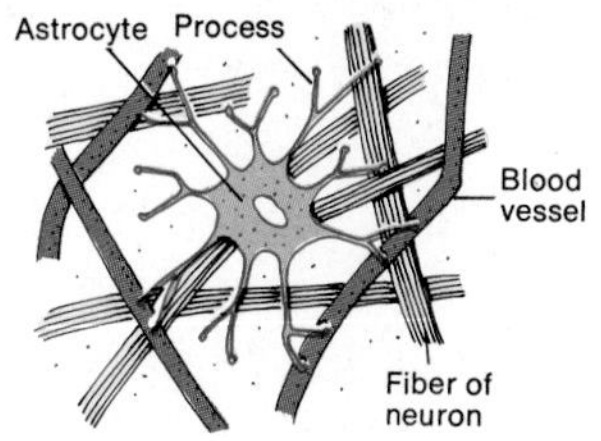	Twine around nerve cells to form supporting network in brain and spinal cord; attach neurons to their blood vessels.
Oligodendrocytes (*oligo* = few; *dendro* = tree)	Resemble astrocytes in some ways, but processes are fewer and shorter.	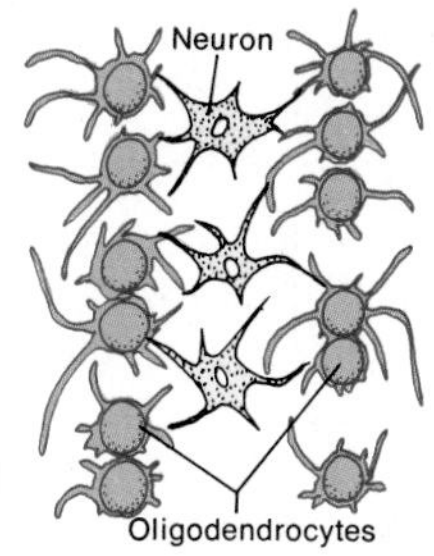	Give support by forming semi-rigid connective tissue rows between neurons in brain and spinal cord; produce a myelin sheath around axons of neurons of central nervous system.
Microglia (*micro* = small; *glia* = glue)	Small cells with few processes; derived from monocytes; normally stationary, but may migrate to site of injury; also called brain macrophages.	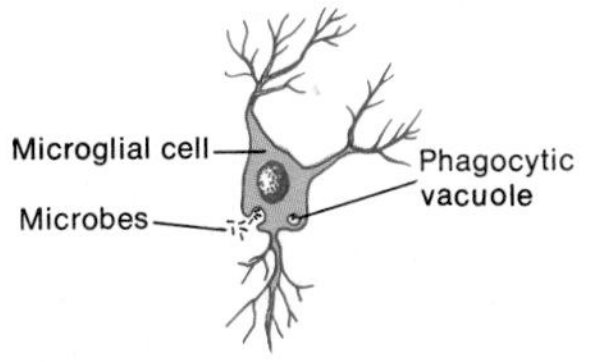	Engulf and destroy microbes and cellular debris; may migrate to area of injured nervous tissue and function as small macrophages.
Ependyma (*ependyma* = upper garment)	Epithelial cells arranged in a single layer and ranging in shape from squamous to columnar; many are ciliated.	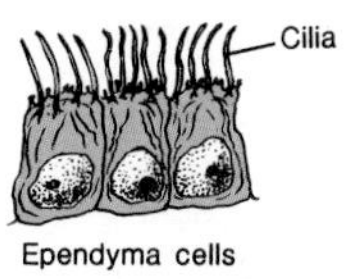	Form a continuous epithelial lining for the ventricles of the brain (spaces that form and circulate cerebrospinal fluid) and the central canal of the spinal cord.

system. Neurons are highly specialized for impulse conduction and for all special functions attributed to the nervous system: thinking, controlling muscle activity, regulating glands. Neuroglia serve as a special supporting and protective component of the nervous system.

NEUROGLIA

The cells of the nervous system that perform the functions of support and protection are called **neuroglia** (noo-ROG-lē-a; *neuro* = nerve; *glia* = glue) or **glial cells.** About 50 percent of all brain cells are neuroglial cells. Many of the glial cells form a supporting network by twining around nerve cells or lining certain structures in the brain and spinal cord. Others bind nervous tissue to supporting structures and attach the neurons to their blood vessels. A few types of glial cells also serve specialized protective functions. For example, many nerve fibers are coated with a thick, fatty sheath produced by one type of neuroglia.

Certain small glial cells are phagocytic; they protect the central nervous system from disease by engulfing invading microbes and clearing away debris. Neuroglia are of clinical interest because they are a common source of tumors of the nervous system. Exhibit 12-1 lists the neuroglial cells and summarizes their functions.

NEURONS

Nerve cells, called **neurons,** are responsible for conducting impulses from one part of the body to another. They are the structural and functional units of the nervous system.

Structure

A neuron consists of three distinct portions: (1) cell body, (2) dendrites, and (3) axon (Figure 12-2a). The **cell body, soma,** or **perikaryon** (per'-i-KAR-ē-on), contains a well-

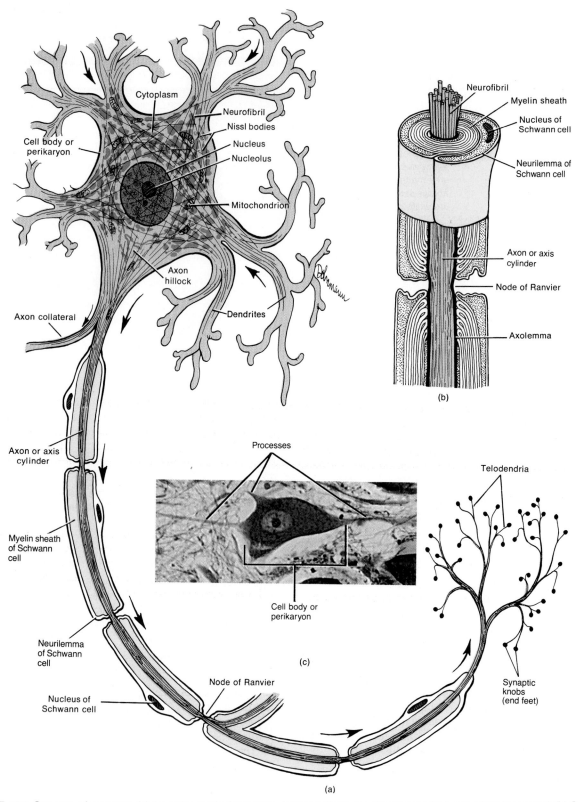

FIGURE 12-2 Structure of a neuron. (a) An entire multipolar neuron. Arrows indicate the direction in which nerve impulses travel. (b) Sections through a myelinated fiber. (c) Photomicrograph of a multipolar neuron at a magnification of 640×. (Courtesy of Biophoto Associates.)

defined nucleus and nucleolus surrounded by a granular cytoplasm. Within the cytoplasm are typical organelles such as lysosomes, mitochondria, and Golgi complexes. Many neurons also contain cytoplasmic inclusions such as *lipofuscin* pigment that occurs as clumps of yellowish-brown granules. Lipofuscin may be a by-product of lysosomal activity. Although its significance is unknown, lipo-

fuscin is related to aging; the amount of pigment increases with age. Also located in the cytoplasm are structures characteristic of neurons: Nissl bodies and neurofibrils. *Nissl bodies* are orderly arrangements of granular (rough) endoplasmic reticulum whose function is protein synthesis. Newly synthesized proteins pass from the perikaryon into the neuronal processes, mainly the axon, at the rate of about 1 mm (0.04 inch) per day. These proteins replace those lost during metabolism and are used for growth of neurons and regeneration of peripheral nerve fibers. *Neurofibrils* are long, thin fibrils composed of microtubules. They may assume a function in support and the transportation of nutrients. Mature neurons do not contain a mitotic apparatus. The significance of this is noted shortly.

The cytoplasmic processes of neurons generally depend on the direction in which they conduct impulses. There are two kinds: dendrites and axons. **Dendrites** (*dendro* = tree) are highly branched, thick extensions of the cytoplasm of the cell body. They typically contain Nissl bodies, mitochondria, and other cytoplasmic organelles. A neuron usually has several main dendrites. Their function is to conduct an impulse toward the cell body.

The second type of cytoplasmic process, called an **axon (axis cylinder),** is a single, highly specialized, long, thin process that conducts impulses away from the cell body to another neuron or tissue. It usually originates from the cell body as a small conical elevation called the *axon hillock.* An axon contains mitochondria and neurofibrils but no Nissl bodies; thus it does not carry on protein synthesis. Its cytoplasm, called *axoplasm,* is surrounded by a plasma membrane known as the *axolemma* (*lemma* = sheath or husk). Axons vary in length from a few millimeters (1 mm = 0.04 inch) in the brain to a meter (3.28 ft) or more between the spinal cord and toes. Along the course of an axon, there may be side branches called *axon collaterals.* The axon and its collaterals terminate by branching into many fine filaments called *telodendria.* The distal ends of telodendria are expanded into bulblike structures called *synaptic knobs* (*end feet*), which are important in nerve impulse conduction. They contain membrane-enclosed sacs called *synaptic vesicles* that store chemicals that determine whether impulse conduction occurs or not.

The cell body of a neuron is essential for the synthesis of many substances that sustain the life of the nerve cell. Neurons have two types of intracellular systems for transporting synthesized materials from the cell body. The slower one, called **axoplasmic flow,** conveys axoplasm in one direction only—from the cell body toward nerve fiber terminals. This mechanism may occur by protoplasmic streaming and supplies new axoplasm for developing or regenerating axons and renews axoplasm in growing and mature axons. The faster type of intracellular transport is called **axonal transport.** It conveys materials in both directions—away from the cell body and toward the cell body—possibly along tracks formed by microtubules and filaments. Axonal transport moves various organelles and materials that form the membranes of the axolemma, synaptic knobs, and synaptic vesicles. Materials returning to the cell body are degraded or recycled.

CLINICAL APPLICATION
The route taken by materials back to the cell body by axonal transport is the route by which the **herpes virus** and **rabies virus** make their way back to nerve cell bodies where they multiply and cause their damage. The toxin produced by the **tetanus bacterium** uses

---

the same route to reach the central nervous system. In fact, the time delay between the release of the toxin and the first appearance of symptoms is in part due to the time required for movement of the toxin by axonal transport.

The term **nerve fiber** is applied to an axon and its sheaths. Figure 12-2b shows two sections of a nerve fiber of the peripheral nervous system. Many axons, especially large, peripheral axons, are surrounded by a multilayered white, phospholipid, segmented covering called the **myelin sheath.** Axons containing such a covering are *myelinated,* while those without it are *unmyelinated.* The function of the myelin sheath is to increase the speed of nerve impulse conduction and to insulate and maintain the axon. Myelin is responsible for the color of the white matter in the nerves, brain, and spinal cord.

The myelin sheath of axons of the peripheral nervous system is produced by flattened cells, called **Schwann cells,** located along the axons. In the formation of a sheath, a developing Schwann cell encircles the axon until its ends meet and overlap (Figure 12-3). The cell then winds around the axon many times and, as it does so, the cytoplasm and nucleus are pushed to the outside layer. The inner portion, consisting of several layers of Schwann cell membrane, is the myelin sheath. The peripheral nucleated cytoplasmic layer of the Schwann cell (the outer layer that encloses the sheath) is called the **neurilemma (sheath of Schwann).**

The neurilemma is found only around fibers of the peripheral nervous system. Its function is to assist in the regeneration of injured axons. Between the segments of the myelin sheath are unmyelinated gaps called **nodes of Ranvier** (ron-VĒ-ā). Unmyelinated fibers are also enclosed by Schwann cells, but without multiple wrappings.

Nerve fibers of the central nervous system may also be myelinated or unmyelinated. Myelination of central nervous system axons is accomplished by oligodendrocytes in somewhat the same manner that Schwann cells myelinate peripheral nervous system axons (see Exhibit 12-1). Myelinated axons of the central nervous system also contain nodes of Ranvier, but they are not so numerous.

Myelin sheaths are first laid down during the later part of the fetal development and during the first year postnatally. The amount of myelin increases from birth to maturity. Since myelination is still in progress during infancy, an infant's responses to stimuli are not as rapid or coordinated as those of an older child or an adult.

Classification

The different neurons in the body may be classified by structure and function.

The structural classification is based on the number of processes extending from the cell body. **Multipolar neurons** have several dendrites and one axon. Most neurons in the brain and spinal cord are of this type. **Bipolar neurons** have one dendrite and one axon and are found in the retina of the eye, the inner ear, and the olfactory area. **Unipolar neurons** have only one process extending from the cell body. The single process divides into a cen-

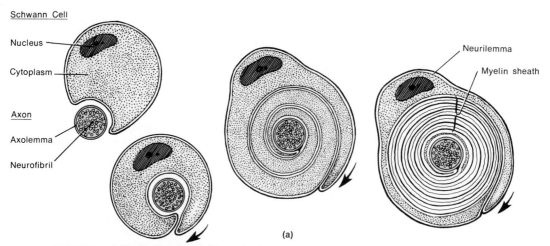

FIGURE 12-3 Myelin sheath. (a) Stages in the formation of a myelin sheath by a Schwann cell.

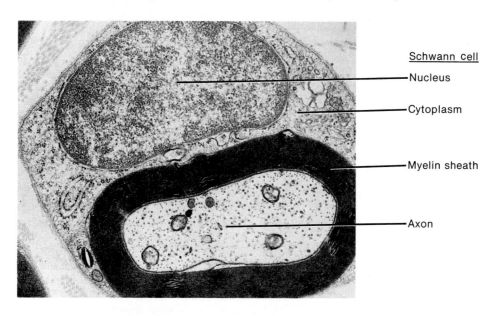

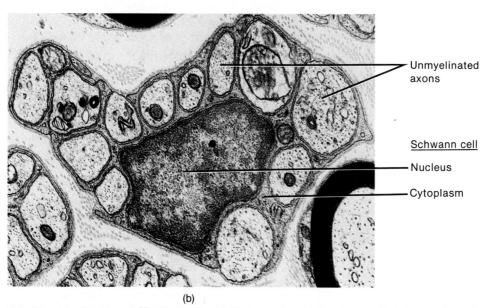

(b)

FIGURE 12-3 (*Continued*) Myelin sheath. (b) Electron micrographs of a myelinated axon (above) and several unmyelinated axons (below). (© Biology Media, Schulz, 1980, Photo Researchers.)

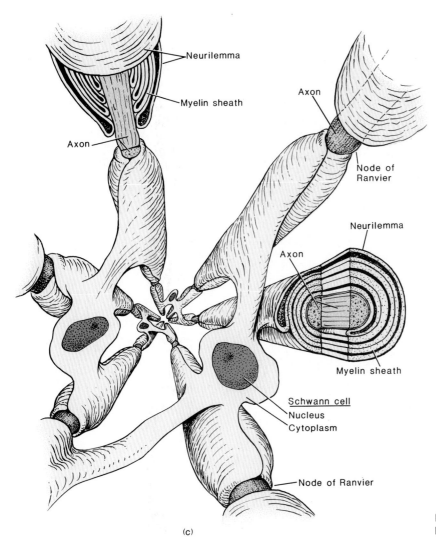

Neurilemma

Myelin sheath

Axon

Axon

Node of Ranvier

Neurilemma

Axon

Myelin sheath

Schwann cell
Nucleus
Cytoplasm

Node of Ranvier

(c)

FIGURE 12-3 (*Continued*) Myelin sheath. (c) Relationship of Schwann cells to axons.

tral branch, which functions as an axon, and a peripheral branch, which functions as a dendrite. Unipolar neurons originate in the embryo as bipolar neurons. During development, the axon and dendrite fuse into a single process and become a functional axon. Unipolar neurons are found in posterior (sensory) root ganglia of spinal nerves.

The functional classification of neurons is based on the direction in which they transmit impulses. **Sensory neurons,** called **afferent neurons,** transmit impulses from receptors in the skin, sense organs, and viscera to the brain and spinal cord. They are usually unipolar. **Motor neurons,** called **efferent neurons,** convey impulses from the brain and spinal cord to effectors, which may be either muscles or glands. Other neurons, called **association (connecting** or **internuncial) neurons,** carry impulses from sensory neurons to motor neurons and are located in the brain and spinal cord.

The processes of afferent and efferent neurons are arranged into bundles called *nerves.* Since nerves lie outside the central nervous system, they belong to the peripheral nervous system. The functional components of nerves are the nerve fibers, which may be grouped according to the following scheme.

1. General somatic afferent fibers conduct impulses from the skin, skeletal muscles, and joints to the central nervous system.

2. General somatic efferent fibers conduct impulses from the central nervous system to skeletal muscles. Impulses over these fibers cause the contraction of skeletal muscles.

3. General visceral afferent fibers convey impulses from the viscera and blood vessels to the central nervous system.

4. General visceral efferent fibers belong to the autonomic nervous system and are also called *autonomic fibers.* They convey impulses from the central nervous system to cause contractions of smooth and cardiac muscle and secretion by glands.

In addition to being grouped as nerves, neural tissue is also organized into other structures such as ganglia, tracts, nuclei, and horns. These are described in Chapter 13.

Structural Variation

Although all neurons conform to the general plan described, there are considerable differences in structure. For example, cell bodies range in diameter from 5 μm for the smallest cells to 135 μm for large motor neurons. The pattern of dendritic branching is varied and distinc-

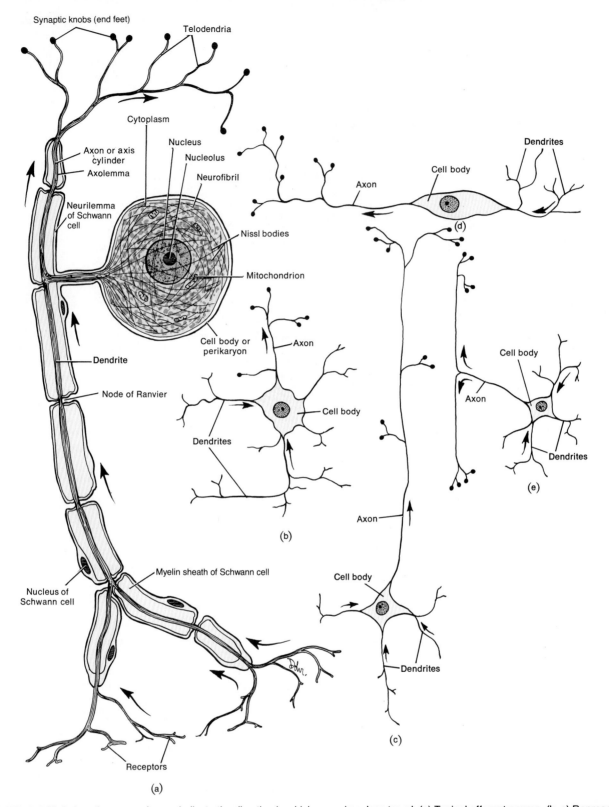

FIGURE 12-4 Varieties of neurons. Arrows indicate the direction in which nerve impulses travel. (a) Typical afferent neuron. (b–e) Representative association neurons. (b) Stellate cell. (c) Cell of Martinotti. (d) Horizontal cell of Cajal. (e) Granule cell.

tive for neurons in different parts of the body. The axons of very small neurons are only a fraction of a millimeter in length and lack a myelin sheath, while axons of large neurons are over a meter long and are usually enclosed in a myelin sheath.

A few patterns of diversity are shown in Figure 12-4. Compare the typical afferent (sensory) neuron to the typical efferent (motor) neuron shown in Figure 12-2a. What structural differences do you observe? Examples of association neurons shown are a *stellate cell,* a *cell of Martinotti,* and a *horizontal cell of Cajal.* All are found in the cerebral cortex, the outer layer of the cerebrum. The *granule cell* is an association neuron in the cortex of the cerebellum.

PHYSIOLOGY

Two striking features of nervous tissue are its limited ability to regenerate and its highly developed ability to produce and transmit electrical messages called nerve impulses.

REGENERATION

Unlike the cells of epithelial tissue, neurons have only limited powers for regeneration. Around the time of birth, the cell bodies of most developing nerve cells lose their mitotic apparatus (centrioles and mitotic spindles) and their ability to reproduce. Thus, when a neuron is damaged or destroyed, it cannot be replaced by the daughter cells of other neurons. A neuron destroyed is permanently lost and only some types of damage may be repaired.

Damage to some types of myelinated axons often can be repaired if the cell body remains intact and if the cell that performs the myelination remains active. Axons in the peripheral nervous system are myelinated by Schwann cells. Schwann cells proliferate following axonal damage and their neurilemmas form a tube that assists in regeneration (Chapter 13). Axons in the brain and spinal cord (central nervous system) are myelinated by oligodendroglial cells. These cells do not form neurilemmas to assist in regeneration and do not survive following axonal damage. An added complication in the central nervous system is that following axonal damage, the af-

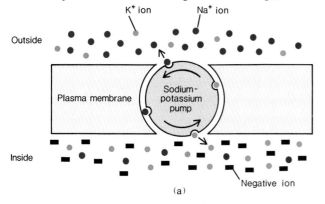

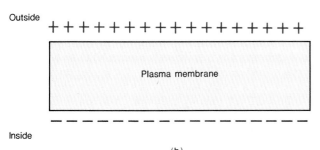

FIGURE 12-5 Development of the membrane (resting) potential. (a) Schematic representation of the sodium-potassium pump and distribution of ions. Note that the large number of Na$^+$ ions outside the membrane results in an external positive charge. Although there are more K$^+$ ions inside the cell membrane than outside, there are many more negative ions inside the membrane. This results in a net internal negative charge. (b) Simplified representation of a polarized membrane.

fected region is rapidly converted into a special form of scar tissue by astroglial proliferation. The scar tissue forms a barrier to regeneration. Thus, an injury to the brain or spinal cord is also permanent because axonal regeneration is blocked by rapid scar tissue formation. An injury to a nerve in the arm (peripheral nervous system) may repair itself before scar tissue forms and so some nerve function may be restored.

NERVE IMPULSE

At this point, we will consider the nerve impulse—your body's quickest way of controlling and maintaining homeostasis.

Membrane Potentials

Studies of cell membranes, especially in nerve and muscle cells, indicate that when a cell is at rest there is a considerable difference between the ion concentration outside and inside the plasma membrane. In a resting neuron (one that is not conducting an impulse), there is a difference in electrical charges on either side of the membrane. This difference, called a *potential difference,* is partly the result of an unequal distribution of potassium (K$^+$) ions and sodium (Na$^+$) ions on either side of the membrane. In resting neurons, the K$^+$ ion concentration inside the cell is about 28 to 30 times greater than it is outside. The Na$^+$ ion concentration is about 14 times greater outside than inside. Another significant factor is the presence of large, nondiffusible negatively charged ions trapped in the cell. Most of them are proteins. What causes the outside of the nerve cell membrane to differ from the inside?

Even when a nerve cell is not conducting an impulse, it is actively transporting ions across its membrane. (The mechanism of active transport may be reviewed in Figure 3-5a.) Na$^+$ ions are actively transported out and K$^+$ ions are actively transported in. The cellular system by which Na$^+$ and K$^+$ ions are actively transported simultaneously is called the **sodium-potassium pump** (Figure 12-5a). The operation of the pump requires the expenditure of ATP.

Neurons also contain a large number of negative ions, mostly protein anions, on the inside that cannot diffuse outside or diffuse very poorly. Since Na$^+$ ions are positive and are actively transported outside the cell by the sodium-potassium pump, a positive charge develops outside the membrane. Even though K$^+$ ions are positive and are actively transported to the inside of the cell by the sodium-potassium pump, there are insufficient K$^+$ ions to equalize the even larger number of nondiffusible negative ions trapped in the cell.

In addition, as a result of the operation of the sodium-potassium pump, there is a concentration gradient for Na$^+$ and K$^+$ ions. As a result, K$^+$ ions tend to diffuse (leak) out of the cell and Na$^+$ ions tend to diffuse into the cell. This diffusion occurs through membrane proteins called *potassium* and *sodium channels.* But, since membrane permeability to K$^+$ ions is 100 times greater than that to Na$^+$ ions, the number of K$^+$ ions that leave the cell is 100 times greater than the number of Na$^+$ ions that enter. The sodium-potassium pump not only actively

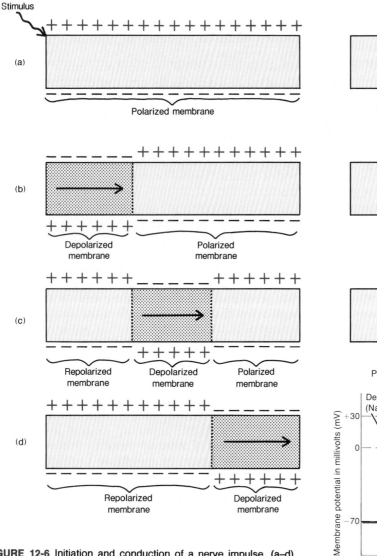

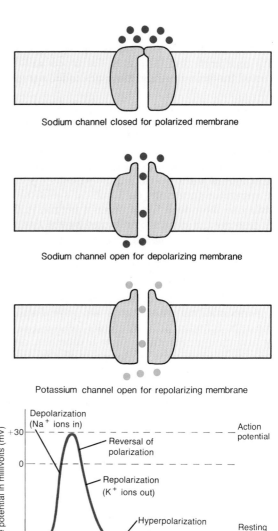

FIGURE 12-6 Initiation and conduction of a nerve impulse. (a–d) The stippled area containing the arrow represents the region of the membrane that has initiated and is conducting the nerve impulse. (e) Record of potential changes of a nerve impulse.

transports Na⁺ and K⁺ ions, it also establishes concentration gradients for the ions. The result is that there is a difference in charge on either side of the membrane— positive outside and negative inside. This difference in charge on either side of the membrane of a resting neuron is the **membrane (resting) potential.** Such a membrane is said to be **polarized** (Figure 12-5b).

Electrical measurements of a polarized membrane indicate a voltage of about 70 millivolts (mV). This means that the inside of the membrane is 70 mV less than the outside, that is, the membrane potential is −70 mV. In subsequent discussions of membrane potentials, we will use the mV value that refers to the inside of the membrane, −70 mV.

Excitability

The events associated with the generation of nerve impulses will now be examined. The ability of nerve cells to respond to stimuli and convert them into nerve im-

pulses is called **excitability.** A **stimulus** is a change in the environment of sufficient strength to initiate an impulse.

If a stimulus of adequate strength is applied to a polarized membrane, the membrane's permeability to Na⁺ ions is greatly increased at the point of stimulation (Figure 12-6a–d). As a result, sodium channels open and permit the influx of Na⁺ ions by diffusion. And since there are more Na⁺ ions entering than leaving, the electrical potential of the membrane begins to change. At first, the potential inside the membrane changes from −70 mV toward zero. At 0 mV the membrane is said to be **depolarized.** (Depolarization begins at the **threshold level:** about −60 mV.) Throughout depolarization, the Na⁺ ions continue to rush inside until the membrane potential is *reversed*— the inside of the membrane becomes positive and the outside negative. Electrical measurements indicate that the inside of the membrane is now +30 mV with respect to the outside. Thus the potential inside the membrane changes from −70 mV to 0 mV to +30 mV.

Once the events of depolarization have occurred, we say that an **action potential (nerve impulse)** is initiated. It lasts about 1 msec. The stimulated, negatively charged point on the outside of the membrane sends out an electrical current to the positive point (still polarized) adjacent to it. This local current causes the adjacent inner part of the membrane to reverse its potential from -70 mV to $+30$ mV. The reversal repeats itself over and over until the nerve impulse is conducted the length of the neuron. The nerve impulse is essentially a wave of negativity that self-propagates along the outside of a neuron cell membrane. Depolarization and reversal of potential require only about 0.5 msec. Of all the cells of the body, only muscle and nerve cells produce action potentials. Their ability to do this is called excitability.

By the time the impulse has traveled from one point on the membrane to the next, the previous point becomes **repolarized**—its resting potential is restored. Repolarization results from a new series of changes in membrane permeability. The membrane now becomes more permeable to K^+ ions than it was at its resting potential level and is relatively impermeable again to Na^+ ions. The outward movement of K^+ ions occurs through potassium channels. As the K^+ ions move through the open potassium channels, the outer surface of the membrane becomes electrically positive. The heavy loss of positive ions leaves the inner surface of the membrane negative again. Eventually any ions that have moved into or out of the nerve cell are restored to their original sites. Thus Na^+ ions are actively transported outside and K^+ ions are moved back into the cell. The repolarization period returns the cell to its resting potential, from $+30$ mV to -70 mV. The neuron is now prepared to receive another stimulus and conduct it in the same manner. In fact, until repolarization occurs, the neuron cannot conduct another impulse. The period of time during which the membrane recovers is called the **refractory period.** A record of the electrical changes associated with a nerve impulse is illustrated in Figure 12-6e.

A sensory neuron is generally stimulated at its distal end by a receptor, a structure sensitive to changes in the environment. Association and motor neurons are usually stimulated at their dendrites or cell bodies by another neuron. However, if a neuron's plasma membrane is depolarized at some point other than the usual one, the impulse will travel in both directions over the cell membrane of the entire neuron.

All-or-None Principle

Any stimulus strong enough to initiate an impulse is referred to as a **threshold,** or **liminal, stimulus.** When a stimulus is of threshold strength, we say the neuron has reached its threshold of stimulation. A nerve cell transmits an impulse according to the **all-or-none principle:** If a stimulus is strong enough to generate an action potential, the impulse is conducted along the entire neuron at a constant and maximum strength for the existing conditions. The conduction is independent of any further intensity of the stimulus. However, conduction may be altered by conditions such as toxic materials in cells, fatigue, and malaise. Any stimulus weaker than a threshold

stimulus is termed a **subthreshold,** or **subliminal, stimulus.** Such a stimulus is incapable of initiating a nerve impulse. If, however, a second stimulus or a series of subthreshold stimuli is quickly applied to the neuron, the cumulative effect may be sufficient to initiate an impulse. This phenomenon is called **summation.** If the second stimulus follows the original stimulus too closely, however, no response will occur because the nerve fiber needs sufficient time to recover from the passage of the first stimulus. This period of time, the **absolute refractory period,** depends on the fiber's diameter. Large fibers repolarize in about $\frac{1}{2,500}$ second. Their absolute refractory period is 0.4 msec. Thus a second nerve impulse can be transmitted $\frac{1}{2,500}$ second after the first—a total of up to 2,500 impulses per second. Small fibers, on the other hand, require as much as $\frac{1}{250}$ second to repolarize. Their absolute refractory period is 4 msec. Thus they can transmit only 250 impulses per second. Under normal body conditions, the frequency of conduction may range between 10 and 500 impulses per second.

Immediately following the absolute refractory period, there is a **relative refractory period.** It is the period of time during which a neuron will not respond to a normal-strength stimulus, but will respond to a greater-than-normal stimulus.

Saltatory Conduction

Thus far we have considered nerve impulse conduction via unmyelinated fibers. The step-by-step depolarization described is called **continuous conduction.** In myelinated fibers, conduction is somewhat different. The myelin sheath surrounding a fiber contains a lipoprotein substance that does not conduct an electric current. It thus forms an insulating layer around the fiber. The myelin sheath is interrupted at various intervals called nodes of Ranvier. At these nodes, membrane depolarization can occur. But beneath the myelin sheath depolarization is impossible. When an impulse is conducted along a myelinated fiber, it moves from one node to another through the surrounding extracellular fluids and through the axoplasm. Thus the impulse jumps from node to node. This type of impulse conduction, characteristic of myelinated fibers, is called **saltatory conduction** (*saltare* = leaping).

Saltatory conduction is a valuable asset to your homeostasis. Since the impulse jumps long intervals as it moves from one node to the next, the speed of conduction is greatly increased. The impulse travels much faster than in the step-by-step depolarization process involved in an unmyelinated fiber of equal diameter.

Speed of Nerve Impulses

The speed of a nerve impulse is independent of stimulus strength. Once a neuron reaches its threshold of stimulation, the speed of the nerve impulse is normally determined by the size, type, and physiological condition of the fiber. Fibers with large diameters conduct impulses faster than those with small ones. Fibers with the greatest diameter are called **A fibers** and are all myelinated. The A fibers have a brief absolute refractory period and are capable of saltatory conduction. They transmit impulses at speeds up to 130 m/sec. The A fibers are located in the axons of large sensory nerves that relay impulses associated with touch, pressure, position of joints, heat, and cold. They are also found in all motor nerves that convey impulses to the skeletal muscles. Sensory A fibers generally connect the brain and spinal cord with sensors that detect danger in the outside environment. Motor A fibers innervate the muscles that can do something about the situation. If you touch a hot object, information about the heat passes over sensory A fibers to the spinal cord. There it is relayed to motor A fibers that stimulate the muscles of the hand to withdraw instantaneously. The A fibers are located where split-second reaction may mean survival.

Other fibers, called B and C fibers, conduct impulses more slowly and are generally found where instantaneous response is not a life-and-death matter. **B fibers** have a middle-sized diameter and a somewhat longer absolute refractory period than A fibers. They are also myelinated and therefore capable of saltatory conduction. They conduct impulses at speeds of about 10 m/sec. B fibers are found in nerves that transmit impulses from the skin and viscera to the brain and spinal cord. They also constitute all the axons of the visceral efferent neurons located in the motor nerves that leave the lower part of the brain and spinal cord and terminate in relay stations called *ganglia* (GANG-glē-a). The ganglia ultimately link with other fibers that stimulate the smooth muscle and glands of the viscera.

C fibers have the smallest diameter and the longest absolute refractory periods. They conduct impulses at the rate of about 0.5 m/sec. C fibers are unmyelinated and incapable of saltatory conduction. They are located in nerves that conduct impulses from the skin and in visceral nerves. These fibers conduct impulses for pain and perhaps impulses for touch, pressure, heat, and cold from the skin and pain receptors from the viscera. C fibers are located in all motor nerves that lead from the ganglia and stimulate the smooth muscle and glands of the viscera. Examples of the motor functions of B and C fibers are constricting and dilating the pupils, increasing and decreasing the heart rate, and contracting and relaxing the urinary bladder—functions of the autonomic nervous system.

CONDUCTION ACROSS SYNAPSES

An impulse is conducted not only along the length of a neuron, but also from one neuron to another or to an effector such as a muscle or gland.

Impulses are conducted from one neuron to another across a **synapse**—a junction between two neurons. The term *synapsis* means connection. The synapse is essential for homeostasis because of its ability to transmit certain impulses and inhibit others. Much of an organism's ability to learn will probably be explained in terms of synapses. Moreover, most diseases of the brain and many psychiatric disorders result from a disruption of synaptic commu-

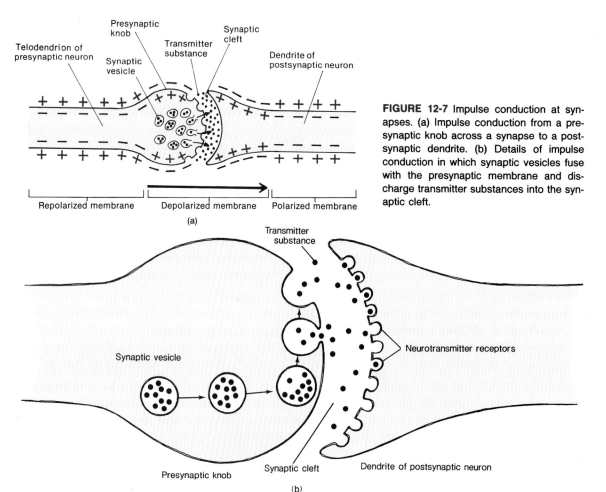

FIGURE 12-7 Impulse conduction at synapses. (a) Impulse conduction from a presynaptic knob across a synapse to a postsynaptic dendrite. (b) Details of impulse conduction in which synaptic vesicles fuse with the presynaptic membrane and discharge transmitter substances into the synaptic cleft.

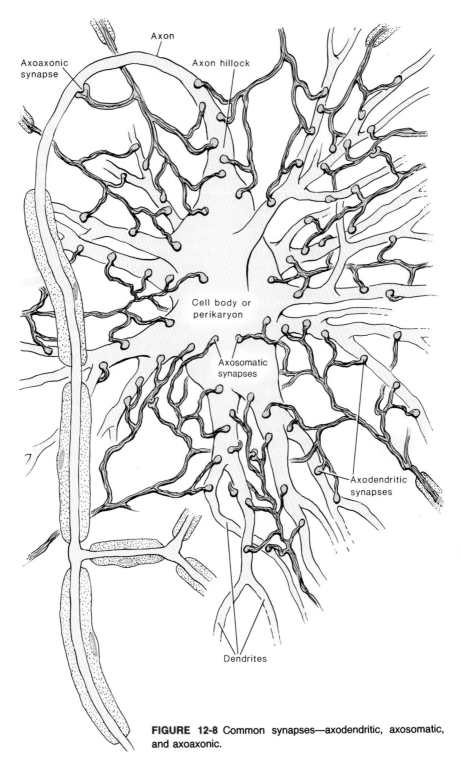

FIGURE 12-8 Common synapses—axodendritic, axosomatic, and axoaxonic.

nication. And synapses are the sites of action for most drugs that affect the brain, including therapeutic and addictive substances. Figure 12-7 shows that within a synapse is a minute gap, about 200 Å across, called the **synaptic cleft. A presynaptic neuron** is a neuron located before a synapse. A **postsynaptic neuron** is located after a synapse.

Impulses are conducted from a neuron to a muscle cell across an area of contact called a **neuromuscular (myoneural) junction.** The area of contact between a neuron and glandular cells is known as a **neuroglandular**

junction. Together, neuromuscular and neuroglandular junctions are known as **neuroeffector junctions.**

The telodendrium of an axon terminates in **synaptic knobs (end feet).** The synaptic knobs of a presynaptic neuron commonly synapse with the dendrites, cell body, or axon hillock of a postsynaptic neuron. Accordingly, synapses may be classified as *axodendritic, axosomatic,* and *axoaxonic* (Figure 12-8). The synaptic knobs from a single presynaptic neuron may synapse with several postsynaptic neurons. Such an arrangement, called **divergence,** permits a single presynaptic neuron to influence several postsynaptic neurons or several muscle or gland cells at the same time (see Figure 12-11a). In another arrangement, called **convergence,** the synaptic knobs of several presynaptic neurons synapse with a single postsynaptic neuron (see Figure 12-11b). This arrangement permits stimulation or inhibition of the postsynaptic neuron.

At a synapse there is only **one-way impulse conduction**—from a presynaptic axon to a postsynaptic dendrite, cell body, or axon hillock. Impulses must move forward over their pathway. They cannot back up into another presynaptic neuron. Such a mechanism is crucial in preventing impulse conduction along improper pathways.

Whether an impulse is conducted across a synapse, neuromuscular junction, or neuroglandular junction depends on the presence of chemicals called **transmitter substances (neurotransmitters).** These chemicals are made by the neuron, usually from amino acids. Following its production and transportation to the synaptic knobs, the transmitter substance is stored in the knobs in small membrane-enclosed sacs called **synaptic vesicles** (Figure 12-7). Each of the thousands of synaptic vesicles present may contain between 10,000 and 100,000 transmitter molecules.

When a nerve impulse arrives at a synaptic knob of a presynaptic neuron, it is believed that a small amount of calcium ions leaks into the knob, attracts synaptic vesicles to the plasma membrane, and helps to liberate the transmitter molecules from the vesicles. Some scientists believe that the synaptic vesicles fuse with the plasma membrane of the presynaptic neuron, form openings, and release the transmitter through the openings into the synaptic cleft. Other scientists think that the transmitter leaves through small channels to enter the synaptic cleft. In either case, the transmitter enters the synaptic cleft and, depending on the chemical nature of the transmitter and the interaction of the transmitter with receptors of the postsynaptic plasma membrane, several things can happen.

Excitatory Transmission

An **excitatory transmitter-receptor interaction** is one that can lower (make less negative) the postsynaptic neuron's membrane potential so that a new impulse can be generated across the synapse. If the potential is lowered enough, the membrane becomes depolarized, the potential inside the membrane becomes positive and the potential outside becomes negative, and a nerve impulse is initiated. Generally, the release of a transmitter by a single presynaptic knob is not sufficient to develop an action potential in a postsynaptic neuron. However, its release does bring the resting membrane potential closer to threshold level as Na⁺ ions move into the cell. This change from the resting membrane potential level in the direction of the threshold level is called the **excitatory postsynaptic potential (EPSP)** (Figure 12-9a). The EPSP is always less negative than the resting membrane potential of the neuron but more negative than its threshold level. Even though the release of a transmitter by a single presynaptic knob is not sufficient to initiate an action potential in a postsynaptic neuron, the postsynaptic neuron does become

more excitable to impulses from presynaptic neurons, so that the possibility for generating an impulse is greater. This effect is called **facilitation.** In other words, the postsynaptic neuron is prepared for subsequent stimuli that can trigger an impulse.

The EPSP lasts only a few milliseconds. If several presynaptic knobs release their transmitter at the same time, however, the combined effect may initiate a nerve impulse—this effect is **summation.** If hundreds of presynaptic knobs release their transmitter simultaneously, they increase the chance of initiating a nerve impulse in the postsynaptic neuron. The greater the summation, the greater the probability an impulse will be initiated. When the summation is the result of the accumulation of transmitter substance from several presynaptic knobs, it is called **spatial summation.** When summation is the result of the accumulation of transmitter substance from a single presynaptic knob firing two or more times in rapid succession, it is called **temporal summation.** Since the EPSP lasts only about 15 msec, the second firing must follow quickly if temporal summation is to occur. The time required for the impulse to cross a synapse—the **synaptic delay**—is about 0.5 msec. This delay is caused by the liberation of the transmitter substance, its passage across the synapse, stimulation of the postsynaptic neuron to become more permeable to Na⁺ ions, and the inward movement of Na⁺ ions that initiates the nerve impulse in the postsynaptic neuron.

Excitatory transmitter-receptor interactions are thought to occur by two different mechanisms. In the first mechanism, chemical transmitters bind to proteins

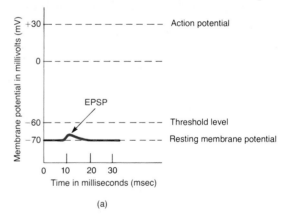

(a)

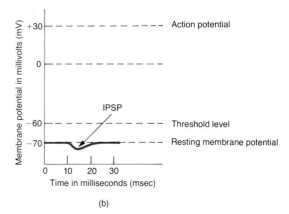

(b)

FIGURE 12-9 Comparison between (a) EPSP and (b) IPSP.

on the postsynaptic plasma membrane, called **neurotransmitter receptors** (see Figure 12-7b). This precise fit causes sodium channels in the plasma membrane to become more permeable to Na^+ ions. The inward movement of Na^+ results in depolarization and initiation of a nerve impulse. Other chemical transmitters use a different mechanism. They bind to neurotransmitter receptors and activate an enzyme in the postsynaptic membrane called **adenyl cyclase** (AD-en-nil SĪ-klās). Once activated, the enzyme converts ATP into **cyclic AMP**. Cyclic AMP then activates enzymes that increase the permeability of the postsynaptic membrane to Na^+ by opening channels through which the Na^+ passes. This initiates the impulse in the postsynaptic neuron. In this mechanism, the chemical transmitter is considered to be the first messenger and cyclic AMP the second messenger. The second messenger role of cyclic AMP in mediating certain hormonal responses is considered in detail in Chapter 18.

As you will see shortly, once a chemical transmitter attaches to a receptor site, it must be rapidly inactivated or it will stimulate the postsynaptic neuron, muscle, or gland indefinitely.

Inhibitory Transmission

An **inhibitory transmitter-receptor interaction** is one that can inhibit impulse generation at a synapse. Whereas excitatory transmitter-receptor interactions make the postsynaptic neuron's resting membrane potential less negative, inhibitory transmitter-receptor interactions make the postsynaptic neuron's resting membrane potential more negative. This is referred to as *hyperpolarization*. The cell interior becomes even more negative than the outside when it is at rest, making it even more difficult for the neuron to generate an action potential. The alteration of the postsynaptic membrane in which the resting membrane potential is made more negative is called the **inhibitory postsynaptic potential (IPSP)** (Figure 12-9b). When the transmitter attaches to the receptor site, the membrane becomes less permeable to Na^+ ions or more permeable to K^+ ions. As a result, there is an increase in internal negativity, or hyperpolarization. Just as the EPSP is less negative than the resting membrane potential of a neuron, the IPSP is more negative than the resting membrane potential.

Integration at Synapses

A single postsynaptic neuron synapses with many presynaptic neurons. Some presynaptic knobs produce excitation and some produce inhibition. The sum of all the effects, excitatory and inhibitory, determines the effect on the postsynaptic neuron. Thus the postsynaptic neuron is an **integrator**: It receives signals, integrates them, and then responds accordingly. The postsynaptic neuron may respond in the following ways:

1. If the excitatory effect is greater than the inhibitory effect, but equal to or less than the threshold level of stimulation, the result is facilitation.

2. If the excitatory effect is greater than the inhibitory effect, but equal to or higher than the threshold level of stimulation, the result is initiation of an impulse.

3. If the inhibitory effect is greater than the excitatory effect, the result is inhibition of the impulse.

TRANSMITTER SUBSTANCES

Perhaps the best studied transmitter substance is **acetylcholine** (as'-ē-til-KŌ-lēn) or **ACh.** It is a transmitter released by many neurons outside the brain and spinal cord and by some neurons inside the brain and cord. This is the same transmitter we discussed in Chapter 10 in relation to impulse conduction from a motor neuron to a muscle fiber across the neuromuscular junction. At neuromuscular junctions, ACh binds to receptor sites on the muscle fiber membrane and increases its permeability to Na^+ and K^+ ions. This depolarizes the membrane and a nerve impulse is generated, causing the muscle fiber to contract. As long as ACh is present in the neuromuscular junction or synapse, it can stimulate a muscle fiber almost indefinitely. The transmission of a continuous succession of impulses by ACh is normally prevented by an enzyme called **acetylcholinesterase (AChE)** or simply **cholinesterase** (kō'-lin-ES-ter-ās). AChE is found on the outside surfaces of the membranous folds of the muscle fiber membrane. Within $\frac{1}{500}$ sec, AChE inactivates ACh. This action permits the membrane of the muscle fiber to repolarize almost immediately so that another impulse may be generated. When the next impulse comes through, the synaptic vesicles release more ACh, the impulse is conducted, and AChE again inactivates ACh. This cycle is repeated over and over again.

ACh is released at some neuromuscular junctions that have cardiac and smooth muscle, as well as those that have skeletal muscle. It is also released at some neuroglandular junctions. In the brain and spinal cord, it appears that the effects of ACh are mediated through cyclic AMP. Although ACh leads to excitation in many parts of the body, it is inhibitory with respect to the heart (vagus nerve).

Another transmitter found in certain parts of the brain, some neuromuscular junctions (smooth and cardiac), and some neuroglandular junctions is **norepinephrine** (nor'-ep-i-NEF-rin) **(NE)**, also known as noradrenaline. It leads to excitation. Its inactivation is different from that of ACh. After NE is released from its synaptic vesicles, it is rapidly pumped back into the synaptic knob. Here it is either destroyed by the enzymes **catechol-O-methyltransferase** (kat'-e-kōl-ō-meth-il-TRANS-fer-ās) **(COMT)** and **monoamine oxidase** (mōn-ō-AM-ēn OK-si-dās) **(MAO)** or recycled back into the synaptic vesicles.

Other transmitters that probably lead to excitation include **serotonin** (ser'-ō-TŌ-nin) **(5-HT)**, which is found in high concentrations in the brain; **dopamine (DA); histamine; glutamic acid;** and **aspartic acid,** also found in the brain.

There is considerable evidence that a substance called **gamma aminobutyric** (GAM-ma am-i-nō-byoo-TĒR-ik) **acid (GABA)** leads to inhibition in the central nervous system. It probably exerts its effect by hyperpolarizing the postsynaptic membrane according to the mechanism previously described. The amino acid **glycine** leads to inhibition in the spinal cord. Transmitter substances in the central nervous system will be discussed in more detail in Chapter 14.

ORGANIZATION OF NEURONAL SYNAPSES

In subsequent chapters, we will be concerned with the structure and physiology of the nervous system. To give you a better understanding of how your nervous system helps maintain homeostasis, we now examine how synapses are organized into functional units.

The central nervous system contains millions of neurons. Their arrangement is not haphazard. They are organized into definite patterns called *neuronal pools.* Each pool differs from all others and has its own role in regulating homeostasis.

A neuronal pool may contain thousands or even millions of neurons. To illustrate the composition of a neuronal pool, a simplified version is given in Figure 12-10. This example contains only five postsynaptic neurons and two incoming presynaptic neurons. The postsynaptic neurons are subject to stimulation by the incoming presynaptic neurons. The postsynaptic neurons in the pool may be stimulated by one or several presynaptic knobs. Moreover, the incoming presynaptic knobs may produce facilitation, excitation, or inhibition. A principal feature of a neuronal pool is the location of the presynaptic neurons in relation to the postsynaptic neurons. Compare the location of presynaptic axon 1 with that of postsynaptic neuron B. Since they are aligned, more presynaptic knobs of axon number 1 synapse with postsynaptic neuron B than with postsynaptic neurons A or C. We say that postsynaptic neuron B is in the center of the field of presynaptic axon 1. Consequently, postsynaptic neuron B usually receives sufficient presynaptic knobs from axon 1 to generate an impulse. This region where the neuron in the pool fires is called the **discharge zone.**

Now look outside the field of presynaptic axon 1. Note its relation to postsynaptic neuron A. Here postsynaptic neuron A in the pool is receiving few presynaptic knobs from the axon supplying postsynaptic neuron B. Thus there are insufficient presynaptic knobs to fire postsynaptic neuron A, but enough to cause facilitation. We therefore call this area the **facilitated zone.** Presynaptic axon 1, when stimulated, will cause excitation of postsynaptic neuron B and facilitation of postsynaptic neuron A.

Neuronal pools in the central nervous system are arranged in patterns over which the impulses are conducted. These are termed **circuits. Simple series circuits** are arranged so that a presynaptic neuron stimulates a single neuron in a pool. The single neuron then stimulates another and so on. In other words, the impulse is relayed from one neuron to another in succession as a new impulse is generated at each synapse.

Most circuits, however, are more complex. In a **diverging circuit,** the impulse from a single presynaptic neuron causes the stimulation of increasing numbers of cells along the circuit (Figure 12-11a). An example of such a circuit is a single motor neuron in the brain stimulating numerous other motor neurons in the spinal cord that, in turn, leave the spinal cord where each stimulates many skeletal muscle fibers. Thus a single impulse may result in the contraction of several skeletal muscle fibers. In another kind of diverging circuit, impulses from one pathway are relayed to other pathways so the same information travels in various directions at the same time. This circuit is common along sensory pathways of the nervous system.

Another kind of circuit is called a **converging circuit** (Figure 12-11b). In one pattern of convergence, the postsynaptic neuron receives impulses from several fibers of the same source. Here there is the possibility of strong excitation or inhibition. In a second pattern, the postsynaptic neuron receives impulses from several different sources. Here there is a possibility of reacting the same way to different stimuli. Suppose your reaction to vomit is distinctly unpleasant. The smell of vomit (one kind of stimulus), the sight of vomit (another kind), or just reading about vomit (still another kind) might all have the same effect on you—an unpleasant one.

Some circuits in your body are constructed so that once the presynaptic cell is stimulated, it will cause the postsynaptic cell to transmit a series of impulses. One such circuit is called a **reverberating (oscillatory) circuit** (Figure 12-12a). In this pattern, the incoming impulse stimulates the first neuron, which stimulates the second,

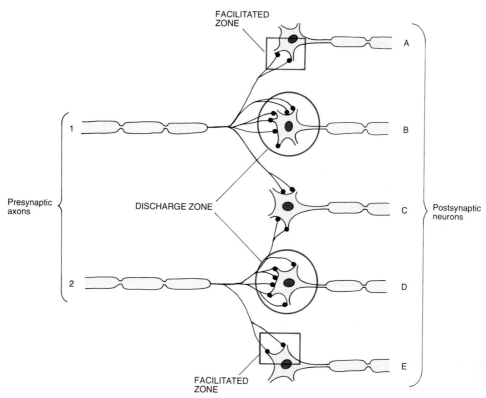

FIGURE 12-10 Relative positions of discharge and facilitated zones in a very simplified version of a neuronal pool.

which stimulates the third, and so on. Branches from the second and third neurons synapse with the first, however, sending the impulse back through the circuit again and again. A central feature of the reverberating circuit is that once fired, the output signal may last from a few seconds to many hours. The duration depends on the number and arrangement of neurons in the circuit. Among the body responses thought to be the result of output signals from reverberating circuits are the rate of breathing, coordinated muscular activities, waking up,

and sleeping (when reverberation stops). Some scientists think reverberating circuits are related to short-term memory.

A final circuit worth consideration is the **parallel after-discharge circuit** (Figure 12-12b). Like a reverberating circuit, a parallel after-discharge circuit is constructed so the postsynaptic cell transmits a series of impulses. In a parallel after-discharge circuit, a single presynaptic cell stimulates a group of neurons, each of which synapses with a common postsynaptic cell. The advantage of this

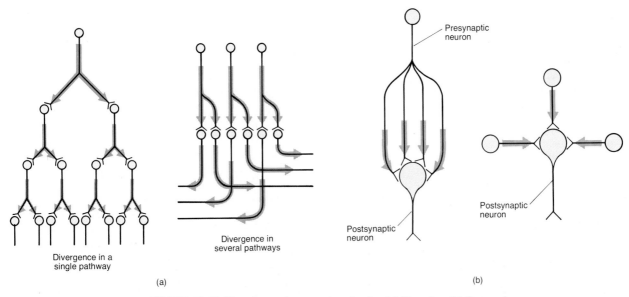

FIGURE 12-11 Diverging and converging circuits. (a) Diverging. (b) Converging.

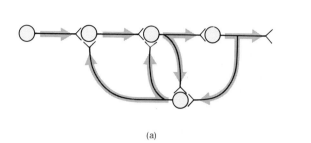 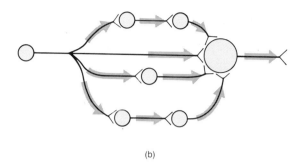

(a)   (b)

FIGURE 12-12 Reverberating and parallel after-discharge circuits. (a) Reverberating. (b) Parallel after-discharge.

circuit is that the postsynaptic neuron can send out a stream of impulses in succession as they are received. The impulses leave the postsynaptic neuron once every $\frac{1}{2,000}$ second. This circuit has no feedback system. Once all the neurons in the circuit have transmitted their im-

pulses to the postsynaptic neuron, the circuit is broken. It is thought that the parallel after-discharge circuit is employed for precise activities like mathematical calculations.

STUDY OUTLINE

Organization (p. 268)

1. The nervous system controls and integrates all body activities by sensing changes, interpreting them, and reacting to them.
2. The central nervous system consists of the brain and spinal cord.
3. The peripheral nervous system is classified into an afferent system and an efferent system.
4. The efferent system is subdivided into a somatic nervous system and an autonomic nervous system.
5. The somatic nervous system consists of efferent neurons that conduct impulses from the central nervous system to skeletal muscle tissue.
6. The autonomic nervous system contains efferent neurons that convey impulses from the central nervous system to smooth muscle tissue, cardiac muscle tissue, and glands.

Histology (p. 268)

Neuroglia

1. Neuroglia are specialized tissue cells that support neurons, attach neurons to blood vessels, produce the myelin sheath, and carry out phagocytosis.
2. Neuroglial cells include astrocytes, oligodendrocytes, microglia, and ependyma.

Neurons

1. Neurons, or nerve cells, consist of a perikaryon or cell body, dendrites that pick up stimuli and convey impulses to the cell body, and usually a single axon. The axon transmits impulses from the neuron to the dendrites or cell body of another neuron or to an effector organ of the body.
2. On the basis of structure, neurons are multipolar, bipolar, and unipolar.
3. On the basis of function, sensory (afferent) neurons transmit impulses to the central nervous system; association neurons transmit impulses to other neurons, including motor neurons; and motor (efferent) neurons transmit impulses to effectors.

Physiology (p. 275)

Regeneration

1. Around the time of birth, the cell body loses its mitotic apparatus and is no longer able to divide.
2. Nerve fibers (axis cylinders) that have a neurilemma are capable of regeneration.

Nerve Impulse

1. The nerve impulse is the body's quickest way of controlling and maintaining homeostasis.
2. The membrane of a nonconducting neuron is positive outside and negative inside due to the operation of the sodium-potassium pump. This difference in charge is called a resting potential, and the membrane is said to be polarized.
3. When a stimulus causes the inside of the cell membrane to become positive and the outside negative, the membrane is said to have an action potential, which travels from point to point along the membrane. The traveling action potential is a nerve impulse. The ability of a neuron to respond to a stimulus and convert it into a nerve impulse is called excitability.
4. Restoration of the resting potential is called repolarization. The period of time during which the membrane recovers is called the refractory period.
5. According to the all-or-none principle, if a stimulus is strong enough to generate an action potential, the impulse travels at a constant and maximum strength for the existing conditions.
6. Nerve impulse conduction in which the impulse jumps from node to node is called saltatory conduction.
7. Fibers with larger diameters conduct impulses faster than those with smaller diameters.

Conduction Across Synapses

1. Impulse conduction can occur from one neuron to another or from a neuron to an effector.
2. The junction between neurons is called a synapse.
3. At a synapse there is only one-way impulse conduction from a presynaptic axon to a postsynaptic dendrite, cell body, or axon hillock.
4. An excitatory transmitter-receptor interaction is one that can lower (make less negative) the postsynaptic neuron's membrane potential so that a new impulse can be generated across the synapse.
5. An inhibitory transmitter-receptor interaction is one that can raise (make more negative) the postsynaptic neuron's membrane potential and thus inhibit an impulse at a synapse.
6. It is thought that the transmitter that causes excitation in a major portion of the central nervous system is acetylcholine. An enzyme called acetylcholinesterase inactivates acetylcho-

line. Other probable transmitters that lead to excitation are norepinephrine, serotonin, dopamine, histamine, and glutamate. Transmitters that are probably inhibitory are gamma aminobutyric acid and glycine.

7. The postsynaptic neuron is an integrator. It receives signals, integrates them, and then responds accordingly.

Organization of Neuronal Synapses

1. Neurons in the central nervous system are organized into definite patterns called neuronal pools. Each pool differs from all others and has its own role in regulating homeostasis.

2. Neuronal pools are organized into circuits. These include simple series, diverging, converging, reverberating, and parallel after-discharge circuits.

REVIEW QUESTIONS

1. How does the nervous system maintain homeostasis? Distinguish between the central and peripheral nervous systems. Relate the terms *voluntary* and *involuntary* to the nervous system.
2. What are neuroglia? List the principal types and their functions.
3. Define a neuron. Diagram and label a neuron. Next to each part list its function.
4. What is a myelin sheath? What is its function?
5. How is the myelin sheath formed?
6. Define the neurilemma. Why is it important?
7. Discuss the structural classification of neurons. Give an example of each.
8. Describe the functional classification of neurons.
9. Distinguish among the following kinds of fibers: general somatic afferent, general somatic efferent, general visceral afferent, and general visceral efferent.
10. What are the structural differences between a typical afferent and a typical efferent neuron?
11. What determines neuron regeneration?
12. Define excitability.
13. Outline the principal steps in the origin and conduction of a nerve impulse.
14. Define the following: resting potential, polarized membrane, action potential, depolarized membrane, and repolarized membrane.
15. What is the all-or-none principle? Relate it to threshold stimulus, subthreshold stimulus, and summation.
16. What is saltatory conduction? Why is it important?
17. What determines the speed of nerve impulses? Give several examples.
18. What events are involved in the conduction of a nerve impulse across a synapse?
19. Distinguish between excitatory and inhibitory transmission.
20. Why does one-way impulse conduction occur?
21. How are nerve impulses inhibited? Of what advantage is this to the body?
22. Why is the postsynaptic neuron called an integrator?
23. What is a transmitter substance? List several probable transmitters, indicate their locations in the nervous system, and whether or not they lead to excitation or inhibition.
24. How do disease, drugs, and pressure affect synaptic transmission?
25. Distinguish between the discharge zone and facilitated zone and neuronal pool.
26. What is a neuron circuit? Distinguish among simple series, diverging, converging, reverberating, and parallel after-discharge circuits.

- Define white matter, gray matter, nerve, ganglion, tract, nucleus, and horn.
- Describe the gross anatomical features of the spinal cord.
- Explain how the spinal cord is protected.
- Describe the structure and location of the spinal meninges.
- Discuss the location, general technique, purpose, and significance of a spinal puncture.
- Describe the structure of the spinal cord in cross section.
- Explain the functions of the spinal cord as a conduction pathway and a reflex center.
- List the location, origin, termination, and function of the principal ascending and descending tracts of the spinal cord.
- Describe the components of a reflex arc and its relationship to homeostasis.
- Compare the functional anatomy of a stretch reflex, flexor reflex, and crossed extensor reflex.
- Define a reflex.

- Identify the relationship between reflexes and the maintenance of homeostasis.
- Classify reflexes on the basis of organs stimulated and location of receptors.
- List several clinically important reflexes.
- Describe the composition and coverings of a spinal nerve.
- Name the 31 pairs of spinal nerves.
- Explain how a spinal nerve branches upon leaving the intervertebral foramen.
- Define an intercostal nerve.
- Define a dermatome and its clinical importance.
- Describe spinal cord injury and list the immediate and long-range effects.
- Identify the effects of peripheral nerve damage and conditions necessary for its regeneration.
- Explain the causes and symptoms of sciatica, neuritis, and shingles.

13 The Spinal Cord and the Spinal Nerves

In this chapter, our main concern will be a study of the structure and function of the spinal cord and the nerves that originate from it. Keep in mind, however, that the spinal cord is continuous with the brain and together they constitute the central nervous system.

GROUPING OF NEURAL TISSUE

The term **white matter** refers to aggregations of myelinated axons from many neurons supported by neuroglia. The lipid substance myelin has a whitish color that gives white matter its name. The **gray matter** of the nervous system contains either nerve cell bodies and dendrites or bundles of unmyelinated axons and neuroglia. The absence of myelin in these areas accounts for their gray color.

A **nerve** is a bundle of fibers located outside the central nervous system. Since the dendrites of somatic afferent neurons and axons of somatic efferent neurons of the peripheral nervous system are myelinated, most nerves are white matter. Nerve cell bodies that lie outside the central nervous system are generally grouped with other nerve cell bodies to form **ganglia** (GANG-lē-a; *ganglion* = knot). Ganglia, since they are made up principally of nerve cell bodies, are masses of gray matter.

A **tract** is a bundle of fibers in the central nervous system. Tracts may run long distances up and down the spinal cord. Tracts also exist in the brain and connect parts of the brain with each other and with the spinal cord. The chief spinal tracts that conduct impulses up the cord are concerned with sensory impulses and are called *ascending tracts*. Spinal tracts that carry impulses down the cord are motor tracts and are called *descending tracts*. The major tracts consist of myelinated fibers and are therefore white matter. A **nucleus** is a mass of nerve cell bodies and dendrites in the central nervous system, all of which have similar functions. It forms gray matter. **Horns** or **columns** are the chief areas of gray matter in the spinal cord. The term *horn* describes the two-dimensional appearance of the organization of gray matter in the spinal cord as seen in cross section. The term *column* describes the three-dimensional appearance of the gray matter in longitudinal columns. Since the white matter of the spinal cord is also arranged in columns, we will refer to the gray matter as being arranged in horns.

SPINAL CORD

GENERAL FEATURES

The **spinal cord** is a cylindrical structure that is slightly flattened anteriorly and posteriorly. It begins as a continuation of the medulla oblongata, the inferior part of the brain stem, and extends from the foramen magnum of the occipital bone to the level of the second lumbar vertebra (Figure 13-1). The length of the adult spinal cord ranges from 42 to 45 cm (16 to 18 inches). The diameter of the cord varies at different levels.

When the cord is viewed externally, two conspicuous enlargements can be seen. The superior enlargement, the **cervical enlargement,** extends from the fourth cervical to the first thoracic vertebra. Nerves that supply the upper

extremities arise from the cervical enlargement. The inferior enlargement, called the **lumbar enlargement,** extends from the ninth to the twelfth thoracic vertebra. Nerves that supply the lower extremities arise from the lumbar enlargement.

Below the lumbar enlargement, the spinal cord tapers to a conical portion known as the **conus medullaris** (KŌ-nus med-yoo-LAR-is). The conus medullaris ends at the level of the intervertebral disc between the first and second lumbar vertebra. Arising from the conus medullaris is the **filum terminale** (FĪ-lum ter-mi-NAL-ē), a nonnervous fibrous tissue of the spinal cord that extends inferiorly to attach to the coccyx. The filum terminale consists mostly of pia mater, the innermost of three membranes that cover and protect the spinal cord and brain. Some nerves that arise from the lower portion of the cord do not leave the vertebral column immediately. They angle inferiorly in the vertebral canal like wisps of coarse hair flowing from the end of the cord. They are appropriately named the **cauda equina** (KAW-da ē-KWĪ-na), meaning horse's tail.

The spinal cord is a series of 31 segments, each giving rise to a pair of spinal nerves. **Spinal segment** refers to a region of the spinal cord from which a pair of spinal nerves arises. The cord is divided into right and left sides by two grooves (see Figure 13-3). The **anterior median fissure** is a deep, wide groove on the anterior (ventral) surface, and the **posterior median sulcus** is a shallower, narrow groove on the posterior (dorsal) surface.

PROTECTION AND COVERINGS

Vertebral Canal

The spinal cord is located in the vertebral canal of the vertebral column. The canal is formed by the vertebral foramina of all the vertebrae arranged on top of each other. Since the wall of the vertebral canal is essentially a ring of bone surrounding the spinal cord, the cord is well protected. A certain degree of protection is also provided by the meninges, cerebrospinal fluid, and the vertebral ligaments.

Meninges

The **meninges** (me-NIN-jēz) are coverings that run continuously around the spinal cord and brain. Those associated specifically with the cord are known as **spinal meninges** (Figure 13-2). The outer spinal meninx is called the **dura mater** (DYOO-ra MĀ-ter), meaning tough mother. It forms a tube from the level of the second sacral vertebra, where it is fused with the filum terminale, to the foramen magnum, where it is continuous with the dura mater of the brain. It is composed of dense, fibrous connective tissue. Between the dura mater and the wall of the vertebral canal is the *epidural space,* which is filled with fat, connective tissue, and blood vessels. It serves as padding around the cord. The epidural space inferior to the second lumbar vertebra is the site for the injection of anesthetics, such as a saddle-block for childbirth.

The middle spinal meninx is called the **arachnoid** (a-RAK-noyd), or spider layer. It is a delicate connective

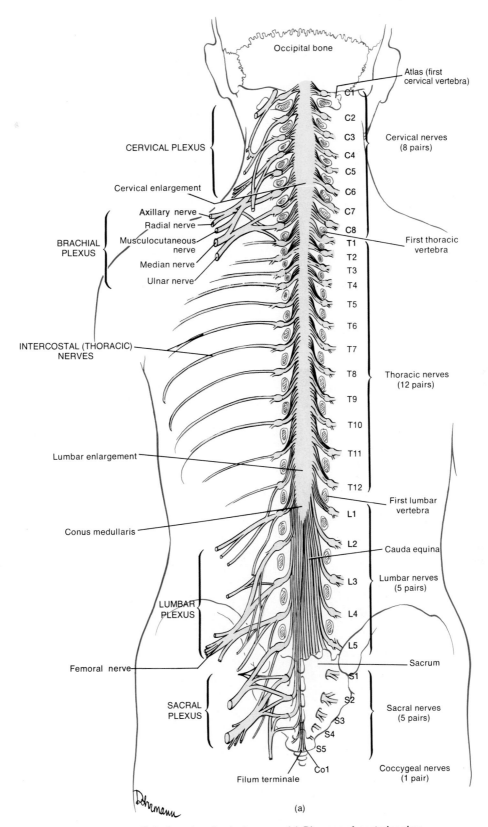

FIGURE 13-1 Spinal cord and spinal nerves. (a) Diagram of posterior view.

tissue membrane that forms a tube inside the dura mater. It is also continuous with the arachnoid of the brain. Between the dura mater and the arachnoid is the *subdural space,* which contains serous fluid.

The inner meninx is known as the **pia mater** (PĒ-a MĀ-ter), or delicate mother. It is a transparent fibrous membrane that forms a tube around and adheres to the surface of the spinal cord and brain. It contains numerous

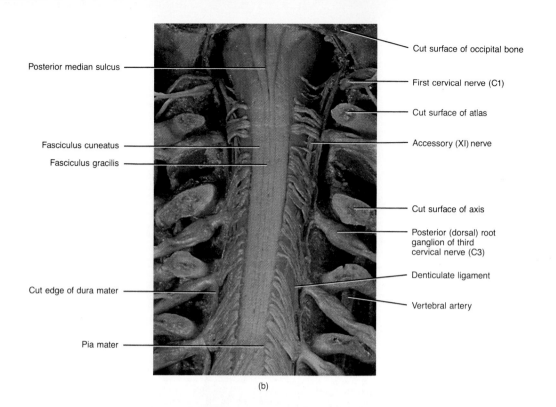

Posterior median sulcus

Fasciculus cuneatus

Fasciculus gracilis

Cut edge of dura mater

Pia mater

Cut surface of occipital bone

First cervical nerve (C1)

Cut surface of atlas

Accessory (XI) nerve

Cut surface of axis

Posterior (dorsal) root ganglion of third cervical nerve (C3)

Denticulate ligament

Vertebral artery

(b)

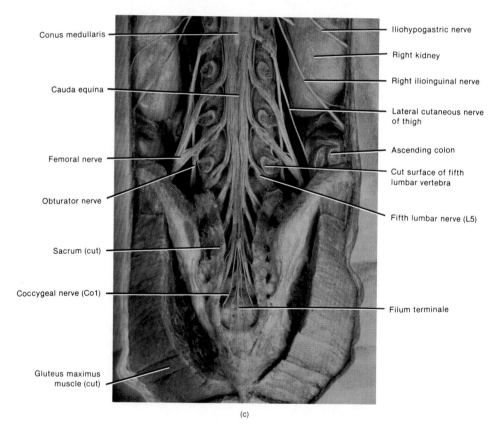

Conus medullaris

Cauda equina

Femoral nerve

Obturator nerve

Sacrum (cut)

Coccygeal nerve (Co1)

Gluteus maximus muscle (cut)

Iliohypogastric nerve

Right kidney

Right ilioinguinal nerve

Lateral cutaneous nerve of thigh

Ascending colon

Cut surface of fifth lumbar vertebra

Fifth lumbar nerve (L5)

Filum terminale

(c)

FIGURE 13-1 (*Continued*) Spinal cord and spinal nerves. (b) Photograph of the posterior aspect of the inferior portion of the medulla and the superior six segments of the cervical portion of the spinal cord. (c) Photograph of the posterior aspect of the conus medullaris and cauda equina. (Courtesy of N. Gluhbegovic and T. H. Williams, *The Human Brain: A Photographic Guide,* Harper & Row, New York, 1980.)

blood vessels. Between the arachnoid and the pia mater is the *subarachnoid space,* where the cerebrospinal fluid circulates.

All three spinal meninges cover the spinal nerves up to the point of exit from the spinal column through the intervertebral foramina. The spinal cord is suspended in the middle of its dural sheath by membranous extensions of the pia mater. These extensions, called the *denticulate* (den-TIK-yoo-lāt) *ligaments,* are attached laterally to the dura mater along the length of the cord between the ven-

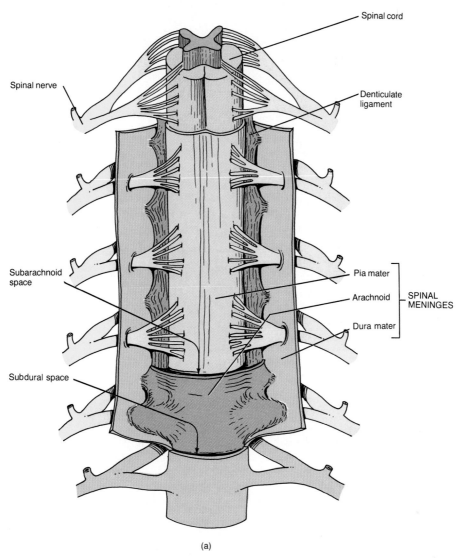

FIGURE 13-2 Spinal meninges. (a) Location of the spinal meninges as seen in a cross section of the spinal cord.

tral and dorsal nerve roots on either side. The ligaments protect the spinal cord against shock and sudden displacement.

CLINICAL APPLICATION

Inflammation of the meninges is known as **meningitis.** If only the dura mater becomes inflamed, the condition is *pachymeningitis.* Inflammation of the arachnoid and pia mater is *leptomeningitis,* the most common form of meningitis.

Cerebrospinal fluid is removed from the subarachnoid space in the inferior lumbar region of the spinal cord by a **spinal (lumbar) puncture** or **tap.** The procedure is normally performed between the third and fourth or fourth and fifth lumbar vertebrae. The spinous process of the fourth lumbar vertebra is easily located. A line drawn across the highest points of the iliac crests will pass through the spinous process. A lumbar puncture is below the spinal cord and thus poses little danger to it. If the patient lies on one side,

drawing the knees and chest together, the vertebrae separate slightly so that a needle can be conveniently inserted. In its course the needle pierces the skin, superficial fascia, supraspinous ligament, interspinous ligament, epidural space, and arachnoid, to enter the subarachnoid space. Lumbar punctures are used to withdraw fluid for diagnostic purposes and to introduce antibiotics (as in the case of meningitis) and radiopaque dyes.

STRUCTURE IN CROSS SECTION

The spinal cord consists of both gray and white matter. Figure 13-3 shows that the gray matter forms an H-shaped area within the white matter. The gray matter consists primarily of nerve cell bodies and unmyelinated axons and dendrites of association and motor neurons. The white matter consists of bundles of myelinated axons of motor and sensory neurons.

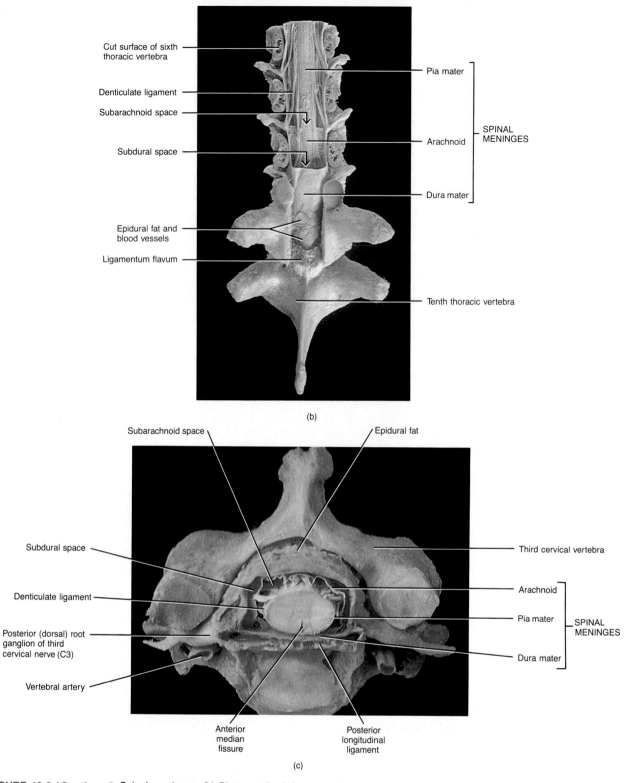

Cut surface of sixth
thoracic vertebra

Denticulate ligament

Subarachnoid space

Subdural space

Epidural fat and
blood vessels

Ligamentum flavum

Pia mater

Arachnoid

SPINAL
MENINGES

Dura mater

Tenth thoracic vertebra

(b)

Subarachnoid space

Epidural fat

Subdural space

Denticulate ligament

Posterior (dorsal) root
ganglion of third
cervical nerve (C3)

Vertebral artery

Third cervical vertebra

Arachnoid

Pia mater

SPINAL
MENINGES

Dura mater

Anterior
median
fissure

Posterior
longitudinal
ligament

(c)

FIGURE 13-2 (*Continued*) Spinal meninges. (b) Photograph of the posterior aspect of the spinal cord. (c) Photograph of a cross section through the spinal cord between the second and third thoracic vertebrae. (Courtesy of N. Gluhbegovic and T. H. Williams, *The Human Brain: A Photographic Guide,* Harper & Row, New York, 1980.)

The cross bar of the H is formed by the **gray commissure** (KOM-mi-shur). In the center of the gray commissure is a small space called the **central canal.** This canal runs the length of the spinal cord and is continuous with the fourth ventricle of the medulla. It contains cerebrospinal fluid. Anterior to the gray commissure is the **anterior (ventral) white commissure,** which connects the white matter of the right and left sides of the spinal cord.

The upright portions of the H are further subdivided into regions. Those closer to the front of the cord are called **anterior (ventral) gray horns.** They represent the motor part of the gray matter. The regions closer to the back of the cord are referred to as **posterior (dorsal) gray horns.** They represent the sensory part of the gray matter. The regions between the anterior and posterior gray horns are intermediate **lateral gray horns.** The lateral gray horns

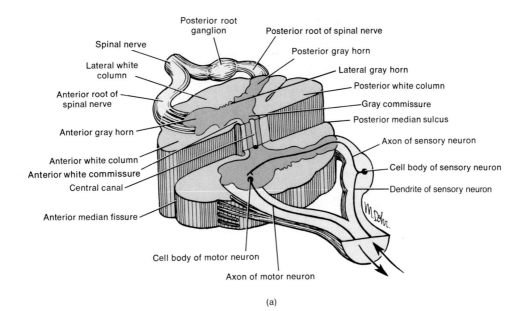

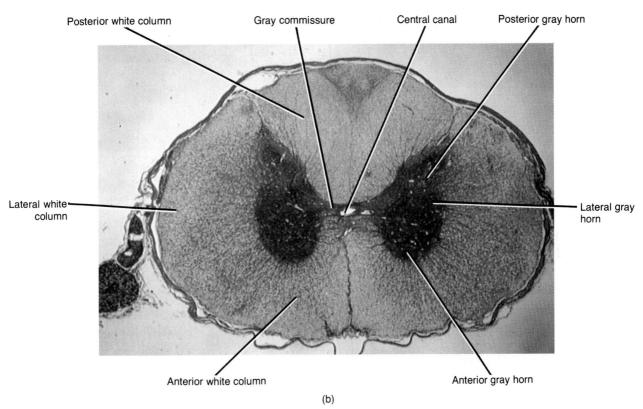

FIGURE 13-3 Spinal cord. (a) The organization of gray and white matter in the spinal cord as seen in cross section. The front of the figure has been sectioned at a lower level than the back so that you can see what is inside the posterior root ganglion, posterior root of the spinal nerve, anterior root of the spinal nerve, and the spinal nerve. (b) Photograph of the spinal cord at the seventh cervical segment. (Courtesy of Victor B. Eichler, Wichita State University.)

are present in the thoracic, upper lumbar, and sacral segments of the cord.

The gray matter of the cord also contains several nuclei that serve as relay stations for impulses and origins for certain nerves. Nuclei are clusters of nerve cell bodies and dendrites in the spinal cord and brain.

The white matter, like the gray matter, is organized into regions. The anterior and posterior gray horns divide the white matter on each side into three broad areas:

anterior (ventral) white columns, posterior (dorsal) white columns, and lateral white columns. Each column (or *funiculus*) in turn consists of distinct bundles of myelinated fibers that run the length of the cord. These bundles are called **tracts** or **fasciculi** (fa-SIK-yoo-lī). The longer **ascending tracts** consist of sensory axons that conduct impulses which enter the spinal cord upward to the brain. The longer **descending tracts** consist of motor axons that conduct impulses from the brain downward into the spinal

EXHIBIT 13-1

SELECTED ASCENDING AND DESCENDING TRACTS OF SPINAL CORD

TRACT	LOCATION (WHITE COLUMN)	ORIGIN	TERMINATION	FUNCTION
ASCENDING TRACTS				
Anterior (ventral) spinothalamic	Anterior (ventral) column.	Posterior (dorsal) gray horn on one side of cord, but crosses to opposite side of brain.	Thalamus; impulses eventually conveyed to cerebral cortex.	Conveys sensations for touch and pressure from one side of body to opposite side of thalamus. Eventually sensations reach cerebral cortex.
Lateral spinothalamic	Lateral column.	Posterior (dorsal) gray horn on one side of cord, but crosses to opposite side of brain.	Thalamus; impulses eventually conveyed to cerebral cortex.	Conveys sensations for pain and temperature from one side of body to opposite side of thalamus. Eventually sensations reach cerebral cortex.
Fasciculus gracilis and **fasciculus cuneatus**	Posterior (dorsal) column.	Axons of afferent neurons from periphery that enter posterior (dorsal) column on one side of cord and rise to same side of brain.	Nucleus gracilis and nucleus cuneatus of medulla; impulses eventually conveyed to cerebral cortex.	Convey sensations from one side of body to same side of medulla for touch; two-point discrimination (ability to distinguish that two points on skin are touched even though close together); proprioception (awareness of precise position of body parts and their direction of movement); stereognosis (ability to recognize size, shape, and texture of object); weight discrimination (ability to assess weight of an object); and vibration. Eventually sensations may reach cerebral cortex.
Posterior (dorsal) spinocerebellar	Posterior (dorsal) portion of lateral column.	Posterior (dorsal) gray horn on one side of cord and rises to same side of brain.	Cerebellum.	Conveys sensations from one side of body to same side of cerebellum for subconscious proprioception.
Anterior (ventral) spinocerebellar	Anterior (ventral) portion of lateral column.	Posterior (dorsal) gray horn on one side of cord; contains both crossed and uncrossed fibers.	Cerebellum.	Conveys sensations from both sides of body to cerebellum for subconscious proprioception.

cord where they synapse with other neurons whose axons pass out to muscles and glands. Thus the ascending tracts are sensory tracts and the descending tracts are motor tracts. Still other short tracts contain ascending or descending axons that convey impulses from one level of the cord to another.

FUNCTIONS

A major function of the spinal cord is to convey sensory impulses from the periphery to the brain and to conduct motor impulses from the brain to the periphery. This is accomplished within spinal tracts. A second principal function is to provide a means of integrating reflexes. Both functions are essential to maintaining homeostasis.

Spinal Tracts

The vital function of conveying sensory and motor information to and from the brain is carried out by the ascending and descending tracts of the cord. The names of the tracts indicate the white column (funiculus) in which the tract travels, where the cell bodies of the tract originate, and where the axons of the tract terminate. Since the origin and termination are specified, the direction of impulse conduction is also indicated by the name. For example, the anterior spinothalamic tract is located in the *anterior* white column, it originates in the *spinal* cord, and it terminates in the *thalamus* (a region of the brain). It is an ascending (sensory) tract since it conveys impulses from the cord upward to the brain.

The principal ascending and descending tracts are listed in Exhibit 13-1 and shown in Figure 13-4.

Reflex Center

The second principal function of the spinal cord is to provide reflexes. Spinal nerves are the paths of communication between the spinal cord tracts and the periphery. Figure 13-3 reveals that each pair of spinal nerves is connected to a segment of the cord by two points of attachment called roots. The **posterior** or **dorsal (sensory) root**

TRACT	LOCATION (WHITE COLUMN)	ORIGIN	TERMINATION	FUNCTION
DESCENDING TRACTS **Lateral corticospinal**	Lateral column.	Cerebral cortex on one side of brain, but crosses in base of medulla to opposite side of cord.	Anterior (ventral) gray horn.	Conveys motor impulses from one side of cortex to anterior gray horn of opposite side. Eventually impulses reach skeletal muscles on opposite side of body that coordinate precise, discrete movements.
Anterior (ventral) corticospinal	Anterior (ventral) column.	Cerebral cortex on one side of brain, uncrossed in medulla, but crosses to opposite side of cord.	Anterior (ventral) gray horn.	Conveys motor impulses from one side of cortex to anterior gray horn of opposite side. Eventually impulses reach skeletal muscles on opposite side of body that coordinate precise, discrete movements.
Rubrospinal	Lateral column.	Midbrain (red nucleus) on one side of brain, but crosses to opposite side of cord.	Anterior (ventral) gray horn.	Conveys motor impulses from one side of midbrain to skeletal muscles on opposite side of body that are concerned with muscle tone and posture.
Tectospinal	Anterior (ventral) column.	Midbrain on one side of brain, but crosses to opposite side of cord.	Anterior (ventral) gray horn.	Conveys motor impulses from one side of midbrain to skeletal muscles on opposite side of body that control movements of head in response to auditory, visual, and cutaneous stimuli.
Vestibulospinal	Anterior (ventral) column.	Medulla on one side of brain and descends to same side of cord.	Anterior (ventral) gray horn.	Conveys motor impulses from one side of medulla to skeletal muscles on same side of body that regulate body tone in response to movements of head (equilibrium).

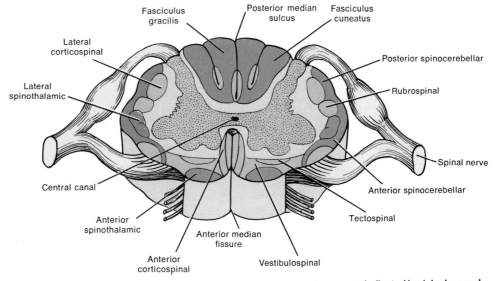

FIGURE 13-4 Selected tracts of the spinal cord. Ascending (sensory) tracts are indicated in pink; descending (motor) tracts are shown in blue.

contains sensory nerve fibers only and conducts impulses from the periphery to the spinal cord. These fibers extend into the posterior (dorsal) gray horn. Each dorsal root also has a swelling, the **posterior** or **dorsal (sensory) root ganglion,** which contains the cell bodies of the sensory neurons from the periphery. The other point of attachment of a spinal nerve to the cord is the **anterior** or **ventral (motor) root.** It contains motor nerve fibers only and conducts impulses from the spinal cord to the periphery.

The cell bodies of the motor neurons are located in the gray matter of the cord. If the motor impulse supplies a skeletal muscle, the cell bodies are located in the anterior (ventral) gray horn. If, however, the impulse supplies smooth muscle, cardiac muscle, or a gland through the autonomic nervous system, the cell bodies are located in the lateral gray horn.

Reflex Arc and Homeostasis

The path an impulse follows from its origin in the dendrites or cell body of a neuron in one part of the body to its termination elsewhere in the body is called a **conduction pathway.** All conduction pathways consist of circuits of neurons. One pathway is known as a **reflex arc,** the functional unit of the nervous system. A reflex arc contains two or more neurons over which impulses are conducted from a receptor to the brain or spinal cord and then to an effector. The basic components of a reflex arc are as follows (see Figure 13-5).

1. Receptor. The distal end of a dendrite or a sensory structure associated with the distal end of a dendrite. Its role in the reflex arc is to respond to a change in the internal or external environment by initiating a nerve impulse in a sensory neuron.

2. Sensory neuron. Passes the impulse from the receptor to its axonal termination in the central nervous system.

3. Center. A region, usually in the central nervous system, where an incoming sensory impulse generates an outgoing motor impulse. In the center, the impulse may be inhibited, transmitted, or rerouted. In the center of some reflex arcs, the sensory

neuron directly generates the impulse in the motor neuron. The center may also contain an association neuron between the sensory neuron and the motor neuron leading to a muscle or a gland.

4. Motor neuron. Transmits the impulse generated by the sensory or association neuron in the center to the organ of the body that will respond.

5. Effector. The organ of the body, either muscle or gland, that responds to the motor impulse. This response is called a **reflex action** or **reflex.**

Reflexes are fast responses to changes in the internal or external environment that allow the body to maintain homeostasis. Reflexes are associated not only with skeletal muscle contraction, but also with body functions such as heart rate, respiration, digestion, urination, and defecation. Reflexes carried out by the spinal cord alone are called **spinal reflexes.** Reflexes that result in the contraction of skeletal muscles are known as **somatic reflexes.** Those that cause the contraction of smooth or cardiac muscle or secretion by glands are **visceral (autonomic) reflexes.** Our concern at this point is to examine a few somatic spinal reflexes: the stretch reflex, the flexor reflex, and the crossed extensor reflex.

● *Stretch Reflex*   The **stretch reflex** is based on a two-neuron or *monosynaptic reflex arc.* Only two neurons are involved, and there is only one synapse in the pathway (Figure 13-5). This reflex results in the contraction of a muscle when it is stretched. Slight stretching of a muscle stimulates receptors in the muscle called *muscle spindles.* The spindles monitor changes in the length of the muscle. Once the spindle is stimulated, an impulse is sent along a sensory neuron to the spinal cord. The sensory neuron lies in the posterior root of a spinal nerve and synapses with a motor neuron in the anterior gray horn. The sensory neuron generates an impulse at the synapse that is transmitted along the motor neuron. The motor neuron lies in the anterior root of the spinal nerve and terminates in a skeletal muscle. Once the impulse reaches the stretched muscle, it contracts. Thus the stretch is counteracted by contraction.

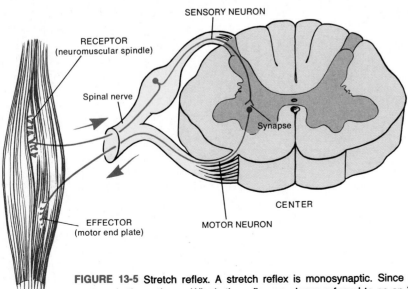

FIGURE 13-5 Stretch reflex. A stretch reflex is monosynaptic. Since only two neurons are involved, there is only one synapse in the pathway. Why is the reflex arc shown referred to as an ipsilateral reflex arc?

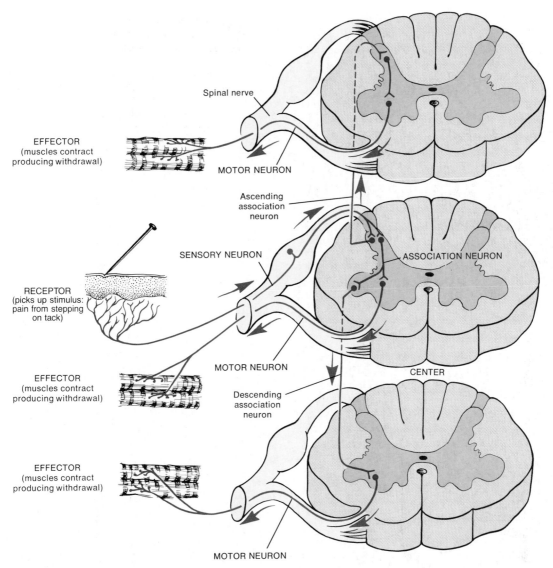

FIGURE 13-6 Flexor reflex. This reflex arc is polysynaptic and ipsilateral. It involves more than one synapse; it contains association neurons as well as sensory and motor neurons. The impulses leave and enter the spinal cord on the same side. Why is the reflex arc shown also an intersegmental reflex arc?

Since the sensory impulse enters the spinal cord on the same side that the motor impulse leaves the spinal cord, the reflex arc is called an *ipsilateral* (ip'-si-LAT-er-al) *reflex arc.* All monosynaptic reflex arcs are ipsilateral and are deep tendon reflexes.

The stretch reflex is essential in maintaining muscle tone and is important for muscle functions during exercise. Moreover, it is the basis for several tests used in neurological examinations. One such reflex is the *knee jerk,* or *patellar reflex.* This reflex is tested by tapping the patellar ligament (stimulus). Muscle spindles in the quadriceps femoris muscle attached to the ligament send the sensory impulse to the spinal cord, and the returning motor impulse causes contraction of the muscle. The response is extension of the leg at the knee, or a knee jerk.

● *Flexor Reflex and Crossed Extensor Reflex*   Reflexes other than stretch reflexes involve association neurons in addition to the sensory and motor neurons. Since more

than two neurons are involved, there is more than one synapse, and so these reflexes are *polysynaptic reflex arcs.* One example of a reflex based on a polysynaptic reflex arc is the **flexor reflex,** or **withdrawal reflex** (Figure 13-6). Suppose you step on a tack. As a result of the painful stimulus, you immediately withdraw your foot. What has happened? A sensory neuron transmits an impulse from the receptor to the spinal cord. A second impulse is generated in an association neuron, which generates a third impulse in a motor neuron. The motor neuron stimulates the muscles of your foot and you withdraw it. Thus a flexor reflex is protective in that it results in the movement of an extremity to avoid pain.

The flexor reflex, like the stretch reflex, is also ipsilateral. The incoming and outgoing impulses are on the same side of the spinal cord. The flexor reflex also illustrates another feature of polysynaptic reflex arcs. In the monosynaptic stretch reflex, the returning motor impulse affects only the quadriceps muscle of the thigh. When

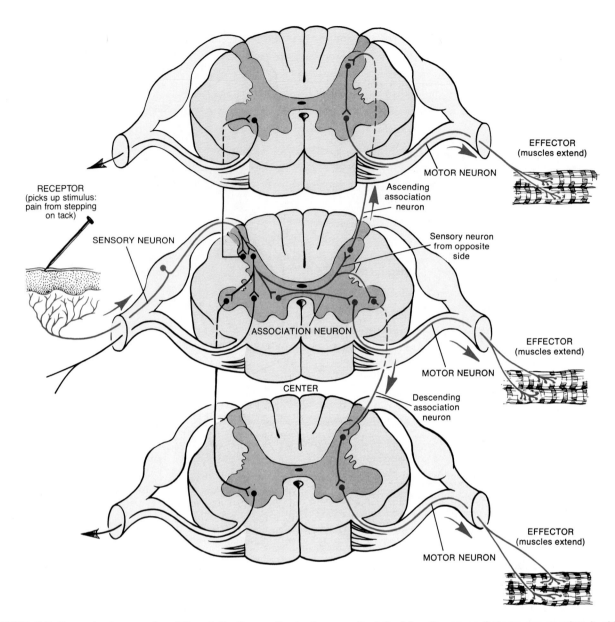

FIGURE 13-7 Crossed extensor reflex. Although the flexor reflex is shown on the left of the diagram so that you can correlate it with the crossed extensor reflex on the right, concentrate your attention on the crossed extensor reflex. Why is the crossed extensor reflex classified as a contralateral reflex arc?

you withdraw your entire lower or upper extremity from a noxious stimulus, more than one muscle is involved, and several motor neurons are simultaneously returning impulses to several upper and lower extremity muscles at the same time. Thus a single sensory impulse causes several motor responses. This kind of reflex arc, in which a single sensory neuron splits into ascending and descending branches, each forming a synapse with association neurons at different segments of the cord, is called an *intersegmental reflex arc.* Because of intersegmental reflex arcs, a single sensory neuron can activate several motor neurons and thereby cause stimulation of more than one effector.

Something else may happen when you step on a tack. You may lose your balance as your body weight shifts to the other foot. Then you do whatever you can to regain your balance so you do not fall. This means motor im-

pulses are also sent to your unstimulated foot and both upper extremities. The motor impulses that travel to your unaffected foot cause extension at the knee so you can place your entire body weight on the foot. These impulses cross the spinal cord as shown in Figure 13-7. The incoming sensory impulse not only initiates the flexor reflex that causes you to withdraw, it also initiates an extensor reflex. The incoming sensory impulse crosses to the opposite side of the spinal cord through association neurons at that level and several levels above and below the point of sensory stimulation. From these levels, the motor neurons cause extension of the knee, thus maintaining balance. Unlike the flexor reflex, which passes over an ipsilateral reflex arc, the extensor reflex passes over a *contralateral* (kon'-tra-LAT-er-al) *reflex arc*—the impulse enters one side of the spinal cord and exits on the opposite side. The reflex just described, in which extension of the muscles in one limb occurs as a result of contraction of the muscles of the opposite limb, is called a **crossed extensor reflex.**

The flexor reflex and crossed extensor reflex also illustrate *reciprocal innervation,* another feature of many reflexes. Reciprocal innervation occurs when a reflex excites a muscle to cause its contraction and also inhibits another muscle to allow its extension. Thus, in this reflex, excitation and inhibition occur simultaneously.

In the flexor reflex, when the flexor muscles of your lower extremity are contracting, the extensor muscles of the same extremity are being extended. If both sets of muscles contracted at the same time, you would not be able to flex your limb because both sets of muscles would pull on the limb bones. But because of reciprocal innervation, one set of muscles contracts while the other is being extended.

In the crossed extensor reflex, reciprocal innervation also occurs. While you are flexing the muscles of the limb that has been stimulated by the tack, the muscles of your other limb are producing extension to help maintain balance. Reciprocal innervation is vital in coordinating body movements. You may recall from Chapter 10 that skeletal muscles act in groups rather than alone and that each muscle in the group has a specific role in bringing about the movement. In flexing the forearm at the elbow, there is a prime mover, an antagonist, and a synergist. The prime mover (biceps) contracts to cause flexion, the antagonist (triceps) extends to yield to the action of the prime mover, and the synergist (deltoid) helps the prime mover perform its role efficiently.

Reflexes and Diagnosis

Reflexes are often used for diagnosing disorders of the nervous system and locating injured tissue. If a reflex ceases to function or functions abnormally, the physician may suspect that the damage lies somewhere along a particular conduction pathway. Visceral reflexes, however, are usually not practical tools for diagnosis. It is difficult to stimulate visceral receptors, since they are deep in the body. In contrast, many somatic reflexes can be tested simply by tapping or stroking the body.

Superficial reflexes are elicited by stroking the skin with a hard object, such as an applicator stick. The object is quickly passed over the skin once, depressing the skin but not producing a scratch.

Any skeletal muscle can normally be stimulated to contract by a slight sudden stretch of its tendon, which can be created by administering a light tap. Many muscle tendons are deeply buried, however, and cannot be readily tapped through the skin. Reflexes that involve a stretch stimulus to a tendon are called **deep tendon reflexes.** Deep tendon reflexes provide information about the integrity and function of the reflex arcs and spinal cord segments without involving the higher centers.

To obtain a substantial response when testing deep tendon reflexes, the muscle must be slightly stretched before the tap is administered. If it is stretched an appropriate amount, tapping the tendon elicits a muscle contraction.

If reflexes are weak or absent, *reinforcement* can be used. In this method, muscle groups other than those being tested are tensed voluntarily with isometric contractions to increase reflex activity in other parts of the body. For example, the person can be asked to hook the fingers together and then try to pull them apart. This action may increase the strength of reflexes involving other muscles. If the reflex can be demonstrated, it is certain that the sensory and motor nerve connections are intact between muscle and spinal cord.

Muscle reflexes can help determine the spinal cord's excitability. When a large number of facilitatory impulses are transmitted from the brain to the spinal cord, the muscle reflexes become so sensitive that simply tapping the knee tendon with the tip of one's finger may cause the leg to jump a considerable distance. On the other hand, the cord may be so intensely inhibited by other impulses from the brain that almost no degree of pounding on the muscles or tendons can elicit a response.

Neurological impairment can be evaluated by using a stopwatch to time the reflex response. Sensitivity of sensory end organs in a muscle is demonstrated by stretching it by as little as 0.05 mm and for as short a duration as $\frac{1}{20}$ sec.

Among the reflexes of clinical significance are:

1. Patellar reflex (knee jerk). This reflex involves extension of the leg by contraction of the quadriceps femoris muscle in response to tapping the patellar ligament (see Figure 13-5). The reflex is blocked by damaged afferent or efferent nerves to the muscle or reflex centers in the second, third, or fourth lumbar segments of the spinal cord. This reflex is also absent in people with chronic diabetes and neurosyphilis. The reflex is exaggerated in disease or injury involving the corticospinal tracts descending from cortex to spinal cord. This reflex may also be exaggerated by applying a second stimulus (a sudden loud noise) while tapping the patellar tendon.

2. Achilles reflex (ankle jerk). This reflex involves extension (plantar flexion) of the foot by contraction of the gastrocnemius and soleus muscles in response to tapping the Achilles tendon (Figure 13-8a). Blockage of the ankle jerk indicates damage to the nerves supplying the posterior leg muscles or to the nerve cells in the lumbosacral region of the spinal cord. This reflex is also absent in people with chronic diabetes, neurosyphilis, alcoholism, and subarachnoid hemorrhages. An exaggerated Achilles reflex indicates cervical cord compression or a lesion of the motor tracts of the first or second sacral segments of the cord.

3. Babinski sign. This reflex results from light stimulation to the outer margin of the sole of the foot (Figure 13-8b). The great toe is extended, with or without fanning of the other toes. This phenomenon occurs in normal children under 1½ years of age and is due to incomplete development of the nervous system. The myelination of fibers in the corticospinal tract has not reached completion. A positive Babinski sign after age 1½ is considered abnormal and indicates an interruption of the corticospinal tract as the result of a lesion of the tract, usually in the upper portion. The normal response after 1½ years of age is the **plantar reflex** (Figure 13-8c), or negative Babinski— a curling under of all the toes accompanied by a slight turning in and flexion of the anterior part of the foot.

4. Abdominal reflex. This reflex compresses the abdominal wall in response to stroking the side of the abdomen (Figure 13-8d). Two separate reflexes, the upper abdominal reflex and the lower abdominal reflex, are involved. The patient should

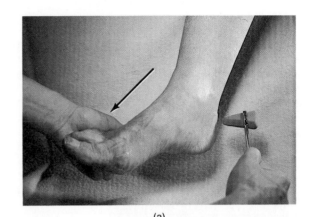

FIGURE 13-8 Clinically significant reflexes. (a) Achilles reflex. (b) Babinski sign. (c) Plantar reflex. (d) Abdominal reflex. ((a)–(c) Courtesy of Beckwith Studios. (d) © Joel Gordon.)

(a)

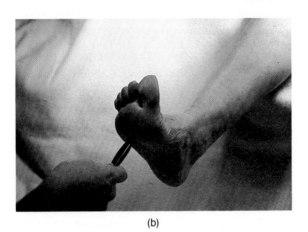

(b)

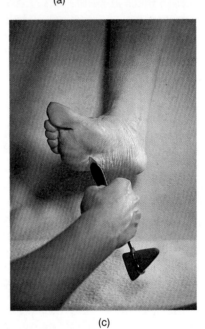

(c)

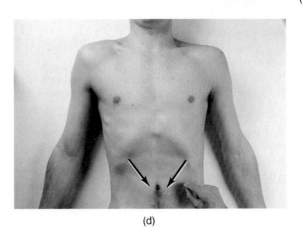

(d)

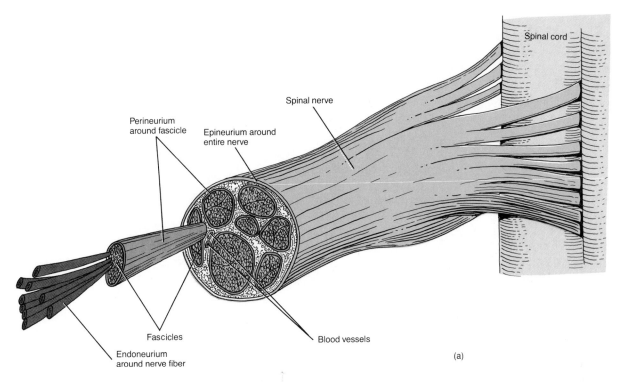

(a)

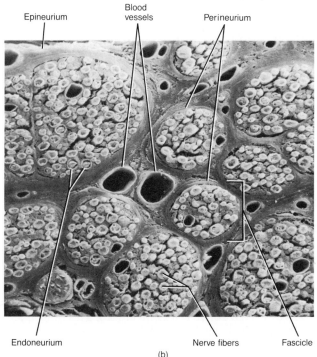

(b)

FIGURE 13-9 Coverings of a spinal nerve. (a) Diagram. (b) Scanning electron micrograph at a magnification of 900×. (Courtesy of Richard G. Kessel and Randy H. Kardon, from *Tissues and Organs: A Text-Atlas of Scanning Electron Microscopy.* Copyright © 1979 by Scientific American, Inc.)

be lying down and relaxed with arms at the sides and knees slightly flexed. The response is an abdominal muscle contraction that results in a lateral deviation of the umbilicus to the side opposite the stimulus. Absence of this reflex is associated with lesions of the corticospinal system. It may also be absent because of lesions of the peripheral nerves, lesions of reflex centers in the thoracic part of the cord, and multiple sclerosis.

SPINAL NERVES

NAMES

The 31 pairs of spinal nerves are named and numbered according to the region and level of the spinal cord from which they emerge (see Figure 13-1). The first cervical pair emerges between the atlas and the occipital bone. All other spinal nerves leave the vertebral column from the intervertebral foramina between adjoining vertebrae. There are 8 pairs of cervical nerves, 12 pairs of thoracic nerves, 5 pairs of lumbar nerves, 5 pairs of sacral nerves, and 1 pair of coccygeal nerves.

During fetal life, the spinal cord and vertebral column grow at different rates, the cord growing more slowly. Thus not all the spinal cord segments are in line with their corresponding vertebrae. Remember that the spinal cord terminates near the level of the first or second lumbar vertebra. Thus the lower lumbar, sacral, and coccygeal nerves must descend more and more to reach their foramina before emerging from the vertebral column. This arrangement constitutes the cauda equina.

COMPOSITION AND COVERINGS

A **spinal nerve** has two points of attachment to the cord: a posterior root and an anterior root. The posterior and anterior roots unite to form a spinal nerve at the intervertebral foramen. Since the posterior root contains sensory fibers and the anterior root contains motor fibers, a spinal nerve is a *mixed nerve.* The posterior (dorsal) root ganglion contains cell bodies of sensory neurons.

In Figure 13-9, you can see that the nerve contains many fibers surrounded by different coverings. The individual fibers, whether myelinated or unmyelinated, are wrapped in a connective tissue called the **endoneurium**

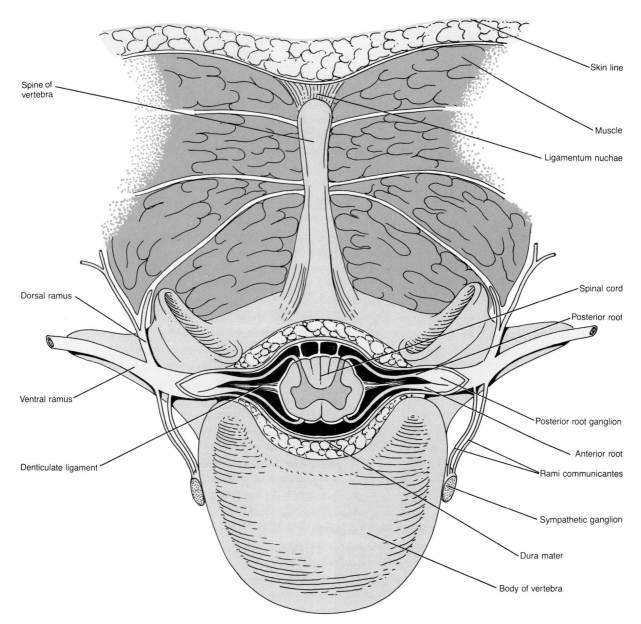

FIGURE 13-10 Branches of a typical spinal nerve.

(en'-dō-NYOO-rē-um). Groups of fibers with their endoneurium are arranged in bundles called fascicles, and each bundle is wrapped in connective tissue called the **perineurium** (per'-i-NYOO-rē-um). The outermost covering around the entire nerve is the **epineurium** (ep'-i-NYOO-rē-um). The spinal meninges fuse with the epineurium as the nerve exits from the vertebral canal.

DISTRIBUTION

Branches

Shortly after a spinal nerve leaves its intervertebral foramen, it divides into several branches (Figure 13-10). These branches are known as **rami** (RĀ-mē). The **dorsal ramus** (RĀ-mus) innervates the deep muscles and skin of the dorsal surface of the back. The **ventral ramus** of a spinal nerve innervates the superficial back muscles and all the

structures of the extremities and the lateral and ventral trunk. In addition to dorsal and ventral rami, spinal nerves also give off a **meningeal branch.** This branch reenters the spinal canal through the intervertebral foramen and supplies the vertebrae, vertebral ligaments, blood vessels of the spinal cord, and the meninges. Other branches of a spinal nerve are the **rami communicantes** (ko-myoo-nē-KAN-tēz), components of the autonomic nervous system whose structure and function are discussed in Chapter 16.

Plexuses

The ventral rami of spinal nerves, except for thoracic nerves T2–T11, do not go directly to the structures of the body they supply. Instead, they form networks by joining with adjacent nerves on either side of the body. Such a network is called a **plexus** (*plexus* = braid). The

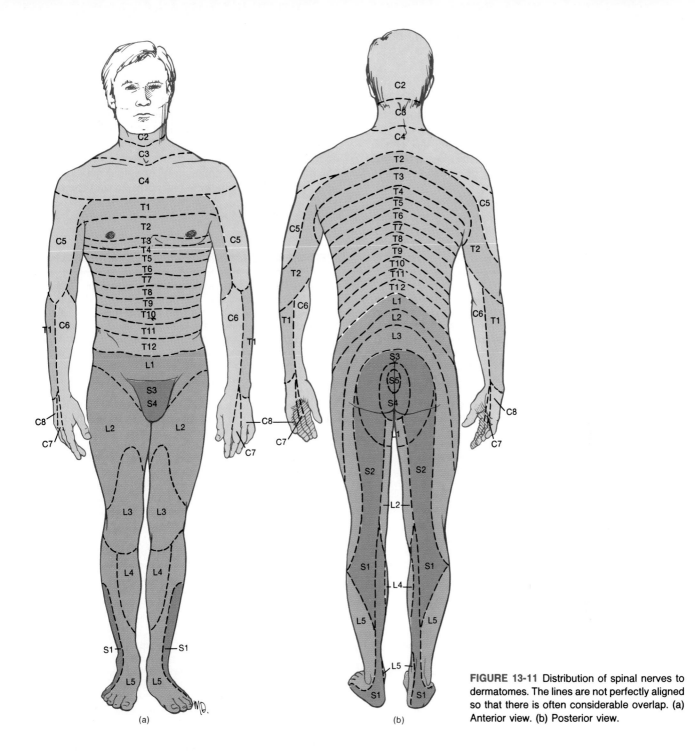

FIGURE 13-11 Distribution of spinal nerves to dermatomes. The lines are not perfectly aligned so that there is often considerable overlap. (a) Anterior view. (b) Posterior view.

principal plexuses are the cervical plexus, the brachial plexus, the lumbar plexus, and the sacral plexus (see Figure 13-1a). Emerging from the plexuses are nerves bearing names that are often descriptive of the general regions they supply or the course they take. Each of the nerves, in turn, may have several branches named for the specific structures they innervate.

The **cervical plexus** supplies the skin and muscles of the head, neck and upper part of the shoulders, connects with some cranial nerves, and supplies the diaphragm. The **brachial plexus** constitutes the nerve supply for the upper extremities and a number of neck and shoulder muscles. The **lumbar plexus** supplies the anterolateral abdominal wall, external genitals, and part of the lower extremities. The **sacral plexus** supplies the buttocks, perineum, and lower extremities.

Intercostal (Thoracic) Nerves

Spinal nerves T2–T11 do not enter into the formation of plexuses. Nerves T2–T11 are known as **intercostal (thoracic) nerves** and are distributed directly to the structures they supply in intercostal spaces (see Figure 13-1). After leaving its intervertebral foramen, the ventral ramus of nerve T2 supplies the intercostal muscles of the second intercostal space and the skin of the axilla and posteromedial aspect of the arm. Nerves T3 and T6 pass in the costal grooves of the ribs and are distributed to the intercostal muscles and skin of the anterior and lateral chest wall. Nerves T7–T11 supply the intercostal muscles and the abdominal muscles and overlying skin. The dorsal rami of the intercostal nerves supply the deep back muscles and skin of the dorsal aspect of the thorax.

DERMATOMES

The skin over the entire body is supplied segmentally by spinal nerves. This means that the spinal nerves innervate specific, constant segments of the skin. All spinal nerves except C1 supply branches to the skin. The skin segment supplied by the dorsal root of a spinal nerve is a **dermatome** (Figure 13-11).

In the neck and trunk, the dermatomes form consecutive bands of skin. In the trunk, there is an overlap of adjacent dermatome nerve supply. Thus there is little loss of sensation if only a single nerve supply to a dermatome is interrupted. Most of the skin of the face and scalp is supplied by the trigeminal (V) cranial nerve.

Since physicians know which spinal nerves are associated with each dermatome, it is possible to determine which segment of the spinal cord or spinal nerve is malfunctioning. If a dermatome is stimulated and the sensation is not perceived, it can be assumed that the nerves supplying the dermatome are involved.

DISORDERS: HOMEOSTATIC IMBALANCES

Spinal Cord Injury

The spinal cord may be damaged by fracture or dislocation of the vertebrae enclosing it or by wounds. All can result in **transection**—partial or complete severing of the spinal cord. Complete transection means that all ascending and descending pathways are cut. It results in loss of all sensation and voluntary muscular movement below the level of transection. A complete cervical transection close to the base of the skull usually leads to death by asphyxiation before treatment can be administered because impulses from the phrenic nerves to the breathing muscles are interrupted. If the upper cervical cord is partially transected, both the upper and lower extremities are paralyzed and the patient is classified as **quadriplegic** (*plege* = stroke). Partial transection between the cervical and lumbar enlargements results in paralysis of the lower extremities only, and the patient is classified as **paraplegic.**

In the case of partial transection, **spinal shock** lasts from a few days to several weeks. During this period, all reflex activity is abolished, a condition called *areflexia* (a'-rē-FLEK-sē-a). In time, however, there is a return of reflex activity. The first reflex to return is the knee jerk. Its reappearance may take several days. Next the flexion reflexes return. This may take up to several months. Then the crossed extensor reflexes return. Visceral reflexes such as erection and ejaculation are also affected by transection. Moreover, urinary bladder and bowel functions are no longer under voluntary control.

Until recently, severe damage resulting from transection was thought to be irreversible. However, a team of researchers has developed a technique for regenerating severed spinal cords in animals. The technique, called *delayed nerve grafting,* involves cutting away the crushed or injured section of the spinal cord and bridging the gap with nerve segments from the arm or leg. The original severed axons in the cord can then grow through the bridge. If the animal research is successful, the procedure may be attempted on humans. Delayed nerve grafting offers hope for paraplegics who have lost voluntary movements of the lower extremities, as well as urinary, bowel, and sexual functions.

Peripheral Nerve Damage and Repair

As we have seen, axons that have a neurilemma can be repaired as long as the cell body is intact, fibers are in association with Schwann cells, and scar tissue formation does not occur too rapidly. Most nerves that lie outside the brain and spinal cord consist of axons that are covered with a neurilemma. A person who injures a nerve in the upper extremity, for example, has a good chance of regaining nerve function. Axons in the brain and spinal cord do not have a neurilemma. Injury there is permanent.

When there is damage to an axon (or to dendrites of somatic afferent neurons), there are usually changes in the cell body and always changes in the portions of the nerve processes distal to the site of damage. The changes associated with the cell body are referred to as the **axon reaction** or **retrograde degeneration.** Those associated with the distal portion of the cut fiber are called **Wallerian** (wal-LE-rē-an) **degeneration.** The axon reaction occurs in essentially the same way, whether the damaged fiber is in the central or peripheral nervous system. The Wallerian reaction, however, depends on whether the fiber is central or peripheral.

Axon Reaction

When there is damage to an axon of a central or peripheral neuron, certain structural changes occur in the cell body. One of the most significant features of the axon reaction occurs 24 to 48 hours after damage. The Nissl bodies, arranged in an orderly fashion in an uninjured cell body, break down into finely granular masses. This alteration is called *chromatolysis* (krō'-ma-TOL-i-sis; *chromo* = color, *lysis* = dissolution). It begins between the axon hillock and nucleus, but spreads throughout the cell body. As a result of chromatolysis, the cell body swells and the swelling reaches its maximum between 10 and 20 days after injury (Figure 13-12b). Chromatolysis results in a loss of ribosomes by the rough endoplasmic reticulum and an increase in the number of free ribosomes. Another sign of the axon reaction is the off-center position of the nucleus in the cell body. This change makes it possible to identify the cell bodies of damaged fibers through a microscope.

Wallerian Degeneration

The part of the axon distal to the damage becomes slightly swollen and then breaks up into fragments by the third to fifth day. The myelin sheath around the axon also undergoes degeneration (Figure 13-12c). Degeneration of the distal portion of the axon and myelin sheath is called Wallerian degeneration. Following degeneration, there is phagocytosis of the remains by macrophages.

Even though there is degeneration of the axon and myelin sheath, the neurilemma of the Schwann cells remains. The Schwann cells on either side of the site of injury multiply by mitosis and grow toward each other until they form a tube across the injured area. The tube provides a means for new axons to grow from the proximal area across the injured area into the distal area previously occupied by the original nerve fiber (Figure 13-12d). The growth of new axons will not occur if the gap at the site of injury is too large or if the gap becomes filled with dense collagenous fibers.

Regeneration

Following chromatolysis, there are signs of recovery in the cell body. There is an acceleration of RNA and protein synthesis, which favors regeneration of the axon. Recovery often takes several months and involves the restoration of normal levels of RNA and proteins and the Nissl bodies to their usual, uninjured patterns.

Accelerated protein synthesis is required for repair of the

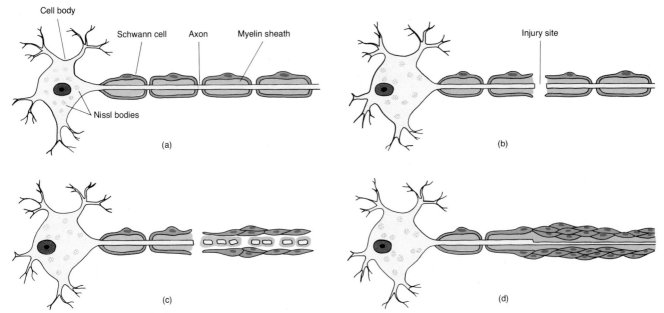

FIGURE 13-12 Peripheral nerve damage and repair. (a) Normal neuron. (b) Chromatolysis. (c) Wallerian degeneration. (d) Regeneration.

damaged axon. The proteins synthesized in the cell body pass into the axon by axoplasmic flow at about the rate of 1 mm (0.04 inch)/day. The proteins assist in regenerating the damaged axon. During the first few days following damage, regenerating axons begin to invade the tube formed by the Schwann cells. Axons from the proximal area grow at the rate of about 1.5 mm (0.06 inch)/day across the area of damage, find their way into the distal neurilemmal tubes, and grow toward the distally located receptors and effectors. Thus sensory and motor connections are reestablished. In time, a new myelin sheath is also produced by the Schwann cells. However, function is never completely restored after a nerve is severed.

Neuritis

Neuritis is inflammation of a single nerve, two or more nerves in separate areas, or many nerves simultaneously. It may result from irritation to the nerve produced by direct blows, bone fractures, contusions, or penetrating injuries. Additional causes include vitamin deficiency (usually thiamine) and poisons such as carbon monoxide, carbon tetrachloride, heavy metals, and some drugs.

Sciatica

Sciatica (sī-AT-i-ka) is a type of neuritis characterized by severe pain along the path of the sciatic nerve or its branches. The term is commonly applied to a number of disorders affecting this nerve. Because of its length and size, the sciatic nerve is exposed to many kinds of injury. Inflammation of or injury to the nerve causes pain that passes from the back or thigh down its length into the leg, foot, and toes.

Probably the most common cause of sciatica is a herniated (slipped) disc. Other causes include irritation from osteoarthritis, back injuries, or pressure on the nerve from certain types of exertion. Sciatica may be associated with diabetes mellitus, gout, or vitamin deficiencies. Other cases are idiopathic.

Shingles

Shingles is an acute infection of the peripheral nervous system. It is caused by a virus called herpes zoster (HER-pēz ZOS-ter), the chickenpox virus. It attacks the cell bodies in dorsal root ganglia, and resides in the ganglia until reactivated later in life. As the virus then spreads down sensory neurons, the inflammation spreads peripherally along the spinal nerves and infiltrates the epidermis and dermis over the nerves, frequently producing intense pain, a characteristic line of skin blisters, and discoloration of the skin along the line. The intercostal nerves and thoracic spinal nerves in the waist area are most commonly affected. Occasionally, the infection involves the first branch of the trigeminal (V) nerve, causing eye pain and corneal ulcerations and opacities.

STUDY OUTLINE

Grouping of Neural Tissue (p. 288)
1. White matter is an aggregation of myelinated axons and associated neuroglia.
2. Gray matter is a collection of nerve cell bodies and dendrites or unmyelinated axons along with associated neuroglia.
3. A nerve is a bundle of nerve fibers outside the central nervous system.
4. A ganglion is a collection of cell bodies outside the central nervous system.
5. A tract is a bundle of fibers of similar function in the central nervous system.
6. A nucleus is a mass of nerve cell bodies and dendrites in the gray matter of the brain and spinal cord.

7. A horn or column is an area of gray matter in the spinal cord.

Spinal Cord (p. 288)
General Features
1. The spinal cord begins as a continuation of the medulla oblongata and terminates at about the second lumbar vertebra.
2. It contains cervical lumbar enlargements which serve as points of origin for nerves to the extremities.
3. The tapered portion of the spinal cord is the conus medullaris, from which arise the filum terminale and cauda equina.
4. The spinal cord is partially divided into right and left sides by the anterior median fissure and posterior median sulcus.

5. The gray matter in the spinal cord is divided into horns and the white matter into funiculi or columns.
6. In the center of the spinal cord is the central canal, which runs the length of the spinal cord and contains cerebrospinal fluid.
7. There are ascending (sensory) tracts and descending (motor) tracts.

Protection and Coverings

1. The spinal cord is protected by the vertebral canal, meninges, cerebrospinal fluid, and vertebral ligaments.
2. The meninges are three coverings that run continuously around the spinal cord and brain: dura mater, arachnoid, and pia mater.
3. Removal of cerebrospinal fluid from the subarachnoid space or ventricle is called a spinal (lumbar) puncture. The procedure is used to diagnose pathologies and to introduce antibiotics.

Structure in Cross Section

1. Parts of the spinal cord observed in cross section are the gray commissure; central canal; anterior, posterior, and lateral gray horns; anterior, posterior, and lateral white columns; and ascending and descending tracts.
2. The spinal cord conveys sensory and motor information by way of the ascending and descending tracts, respectively.

Functions

1. A major function of the spinal cord is to convey sensory impulses from the periphery to the brain and to conduct motor impulses from the brain to the periphery.
2. Another function is to serve as a reflex center. The posterior root, posterior root ganglion, and anterior root are involved in conveying an impulse.
3. A reflex arc is the shortest route that can be taken by an impulse from a receptor to an effector. Its basic components are a receptor, a sensory neuron, a center, a motor neuron, and an effector.
4. A reflex is a quick, involuntary response to a stimulus that passes along a reflex arc. Reflexes represent the body's principal mechanisms for responding to changes in the internal and external environment.
5. Somatic spinal reflexes include the stretch reflex, flexor reflex, and crossed extensor reflex.
6. A two-neuron or monosynaptic reflex arc contains one sensory and one motor neuron. A stretch reflex, such as the patellar reflex, is an example.
7. A polysynaptic reflex arc contains a sensory, association, and motor neuron. A withdrawal or flexor reflex and a crossed extensor reflex are examples.
8. Stretch and flexor reflexes are ipsilateral. The crossed extensor reflex is contralateral.

9. The flexor and crossed extensor reflexes illustrate reciprocal innervation of muscles.
10. Among clinically important somatic reflexes are the patellar reflex, the Achilles reflex, the Babinski sign, and the abdominal reflex.

Spinal Nerves (p. 301)

Names

1. The 31 pairs of spinal nerves are named and numbered according to the region and level of the spinal cord from which they emerge.

Composition and Coverings

1. Spinal nerves are attached to the spinal cord by means of a posterior root and an anterior root. All spinal nerves are mixed.
2. Spinal nerves are covered by endoneurium, perineurium, and epineurium.

Distribution

1. Branches of a spinal nerve include the dorsal ramus, ventral ramus, meningeal branch, and rami communicantes.
2. The ventral rami of spinal nerves, except for T2–T11, form networks of nerves called plexuses.
3. Emerging from the plexuses are nerves bearing names that are often descriptive of the general regions they supply or the course they take.
4. The principal plexuses are called the cervical, brachial, lumbar, and sacral plexuses.
5. Nerves T2–T11 do not form plexuses and are called intercostal nerves. They are distributed directly to the structures they supply in intercostal spaces.

Dermatomes

1. All spinal nerves except C1 innervate specific, constant segments of the skin. The skin segments are called dermatomes.
2. Knowledge of dermatomes helps a physician to determine which segment of the spinal cord or spinal nerve is malfunctioning.

Disorders: Homeostatic Imbalances (p. 304)

1. Complete or partial severing of the spinal cord is called transection. It may result in quadriplegia or paraplegia. Partial transection is followed by a period of loss of reflex activity called areflexia.
2. Following peripheral nerve damage, repair is accomplished by an axon reaction, Wallerian degeneration, and regeneration.
3. Inflammation of nerves is known as neuritis.
4. Neuritis of the sciatic nerve and its branches is called sciatica.
5. Shingles is acute infection of peripheral nerves.

REVIEW QUESTIONS

1. Define the following terms: white matter, gray matter, nerve, ganglion, tract, nucleus, and horn.
2. Describe the location of the spinal cord. What are the cervical and lumbar enlargements?
3. Define conus medullaris, filum terminale, and cauda equina. What is a spinal segment? How is the spinal cord partially divided into a right and left side?
4. Describe the bony covering of the spinal cord.
5. Explain the location and composition of the spinal meninges. Describe the location of the epidural, subdural, and subarachnoid spaces. What are the denticulate ligaments?

6. Define meningitis.
7. What is a spinal puncture? Give several purposes served by a spinal puncture.
8. Based upon your knowledge of the structure of the spinal cord in cross section, define the following: gray commissure, central canal, anterior gray horn, lateral gray horn, posterior gray horn, anterior white column, lateral white column, posterior white column, ascending tract, and descending tract.
9. Describe the function of the spinal cord as a conduction pathway.

10. Using Exhibit 13-1 as a guide, be sure that you can list the location, origin, termination, and function of the principal ascending and descending tracts.
11. Describe how the spinal cord serves as a reflex center.
12. What is a reflex arc? List and define the components of a reflex arc.
13. Define a reflex. How are reflexes related to the maintenance of homeostasis?
14. Describe the mechanism of a stretch reflex, a flexor reflex, and a crossed extensor reflex.
15. Define the following terms relating to reflex arcs: monosynaptic, ipsilateral, polysynaptic, intersegmental, contralateral, and reciprocal innervation.
16. Why are reflexes important in diagnosis?
17. Indicate the clinical importance of the following reflexes: patellar, Achilles, Babinski sign, and abdominal.
18. Define a spinal nerve. Why are all spinal nerves classified as mixed nerves?
19. Describe how a spinal nerve is attached to the spinal cord.
20. Explain how a spinal nerve is enveloped by its several different coverings.
21. How are spinal nerves named and numbered?
22. Describe the branches and innervations of a typical spinal nerve.
23. What is a plexus?
24. What are intercostal nerves?
25. Define a dermatome. Why is a knowledge of dermatomes important?
26. What is transection? Distinguish between quadriplegia and paraplegia.
27. What is meant by spinal shock?
28. Outline the principal events that occur as part of the axon reaction, Wallerian degeneration, and regeneration following peripheral nerve damage.
29. Distinguish between sciatica and neuritis.
30. What is shingles?

- Identify the principal parts of the brain.

- Describe the location of the cranial meninges.

- Explain the formation and circulation of cerebrospinal fluid.

- Describe the blood supply to the brain and the concept of the blood–brain barrier (BBB).

- Compare the components of the brain stem with regard to structure and function.

- Identify the structure and functions of the diencephalon.

- Identify the structural features of the cerebrum.

- Describe the lobes, tracts, and basal ganglia of the cerebrum.

- Describe the structure and functions of the limbic system.

- Compare the motor, sensory, and association areas of the cerebrum.

- Describe the principle of the electroencephalograph and its significance in the diagnosis of certain disorders.

- Explain brain lateralization and the split-brain concept.

- Describe the anatomical characteristics and functions of the cerebellum.

- Discuss the various chemical transmitter substances found in the brain, as well as the different types of peptides and their functions.

- Define a cranial nerve.

- Identify the 12 pairs of cranial nerves by name, number, type, location, and function.

- Explain the effects of injury on cranial nerves.

- List the clinical symptoms of these disorders of the nervous system: poliomyelitis, cerebral palsy, Parkinsonism, epilepsy, multiple sclerosis (MS), cerebrovascular accidents (CVAs), dyslexia, Tay-Sachs disease, headache, trigeminal neuralgia, rabies, Reyes syndrome (RS), and senility.

- Define medical terminology associated with the central nervous system.

- Explain the actions of selected drugs on the central nervous system.

14 The Brain and the Cranial Nerves

Now we will consider the principal parts of the brain, how the brain is protected, how it is related to the spinal cord, and how it is related to the 12 pairs of cranial nerves.

BRAIN

EMBRYOLOGICAL DEVELOPMENT

At the end of the fourth week of the embryonic period, the brain develops from three rudimentary regions of the embryo called primary brain vesicles. These are the **prosencephalon** (forebrain), **mesencephalon** (midbrain), and **rhombencephalon** (hindbrain). During the fifth week of development, the prosencephalon develops into two secondary brain vesicles called the **diencephalon** and **telencephalon;** the rhombencephalon also develops into two secondary brain vesicles called the **myelencephalon** and **metencephalon.** Since the mesencephalon remains unchanged, there are five secondary brain vesicles. Ultimately, the diencephalon develops into the thalamus and hypothalamus of the brain, the telencephalon forms the cerebrum, the mesencephalon becomes the midbrain, the myelencephalon develops into the medulla oblongata, and the metencephalon becomes the pons varolii and cerebellum.

These relationships are summarized in Exhibit 14-1.

PRINCIPAL PARTS

The **brain** of an average adult is one of the largest organs of the body, weighing about 1,300 g (3 lb). Figure 14-1 shows that the brain is mushroom-shaped. It is divided into four principal parts: brain stem, diencephalon, cerebrum, and cerebellum. In some cases, embryological names are retained. The **brain stem,** the stalk of the mushroom, consists of the medulla oblongata, pons varolii, and midbrain or mesencephalon (mes-en-SEF-a-lon). The lower end of the brain stem is a continuation of the spinal cord. Above the brain stem is the **diencephalon** (dī-en-SEF-a-lon) consisting primarily of the thalamus and hypothalamus. The **cerebrum** spreads over the diencephalon. The cerebrum constitutes about seven-eighths of the total weight of the brain and occupies most of the cranium. Inferior to the cerebrum and posterior to the brain stem is the **cerebellum.**

PROTECTION AND COVERINGS

The brain is protected by the cranial bones (see Figure 7-2). Like the spinal cord, the brain is also protected by meninges. The **cranial meninges** surround the brain, are continuous with the spinal meninges, and have the same basic structure and bear the same names as the spinal meninges: the outermost **dura mater,** middle **arachnoid,** and innermost **pia mater** (Figure 14-2a).

The cranial dura mater consists of two layers. The thicker, outer layer (periosteal layer) tightly adheres to the cranial bones and serves as periosteum. The thinner, inner layer (meningeal layer) includes a mesothelial layer on its smooth surface. The spinal dura mater corresponds to the meningeal layer of the cranial dura mater.

> **CLINICAL APPLICATION**
> The middle meningeal artery, running on the inner surface of the temporal bone between the dura mater and the skull, is closely adherent to the bones of the skull. It and the meningeal veins may be ruptured by a blow on the temple, especially if the bone is fractured. This rupture produces an **extradural hemorrhage,** which results in gradually increasing cranial pressure, drowsiness, unconsciousness, and death unless there is surgical intervention.

CEREBROSPINAL FLUID (CSF)

The brain, as well as the rest of the central nervous system, is further protected against injury by **cerebrospinal fluid (CSF).** This fluid circulates through the subarachnoid space around the brain and spinal cord and through the ventricles of the brain. The subarachnoid space is the area between the arachnoid and pia mater.

The **ventricles** (VEN-tri-kuls) are cavities in the brain that communicate with each other, with the central canal of the spinal cord, and with the subarachnoid space (Figure 14-2a). Each of the two **lateral ventricles** is located in a hemisphere (side) of the cerebrum under the corpus callosum. The **third ventricle** is a slit between and inferior to the right and left halves of the thalamus and between the lateral ventricles. Each lateral ventricle communicates with the third ventricle by a narrow, oval opening, the **interventricular foramen.** The **fourth ventricle** lies between the inferior brain stem and the cerebellum. It communicates with the third ventricle via the **cerebral aqueduct,** which passes through the midbrain. The roof of the fourth ventricle has three openings: a **median aperture** and two **lateral apertures.** Through these openings, the fourth ventricle also communicates with the subarachnoid space of the brain and cord.

EXHIBIT 14-1 ■
DEVELOPMENT OF THE BRAIN FROM BRAIN VESICLES

PRIMARY BRAIN VESICLE	SECONDARY BRAIN VESICLE	PART OF BRAIN FORMED
Prosencephalon	Diencephalon	Thalamus and hypothalamus
	Telencephalon	Cerebrum
Mesencephalon	Mesencephalon	Midbrain
Rhombencephalon	Myelencephalon	Medulla oblongata
	Metencephalon	Pons varolii and cerebellum

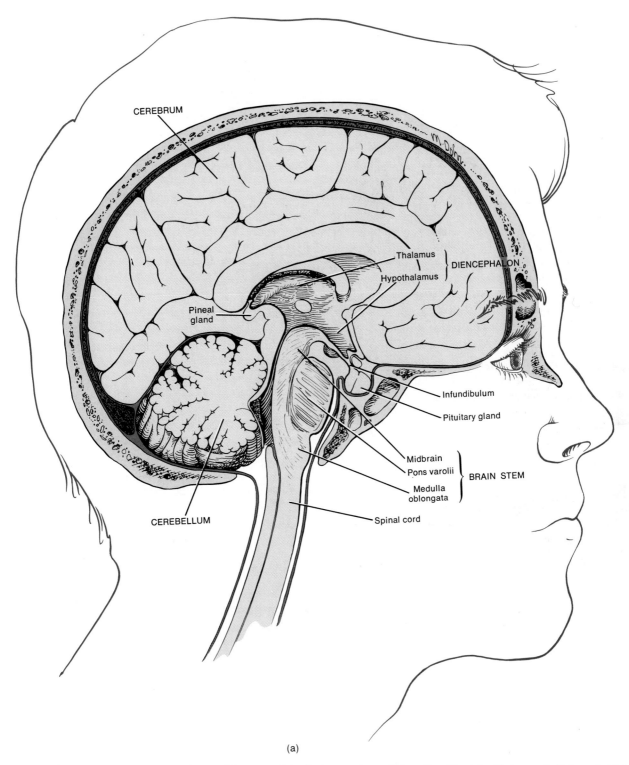

CEREBRUM

Thalamus

Hypothalamus

DIENCEPHALON

Pineal
gland

Infundibulum

Pituitary gland

Midbrain

Pons varolii

BRAIN STEM

Medulla
oblongata

CEREBELLUM

Spinal cord

M. Dohrn

(a)

FIGURE 14-1 Brain. (a) Principal parts of the medial aspect of the brain seen in sagittal section. The infundibulum and pituitary gland are discussed in conjunction with the endocrine system in Chapter 18.

The entire central nervous system contains about 125 ml (4 oz) of cerebrospinal fluid. It is a clear, colorless fluid of watery consistency. Chemically, it contains proteins, glucose, urea, and salts. It also contains some white blood cells. The fluid serves as a shock absorber for the central nervous system. It also circulates nutritive substances filtered from the blood.

Cerebrospinal fluid is formed primarily by filtration and secretion from networks of capillaries, called **choroid** (KŌ-royd; *chorion* = delicate) **plexuses,** located in the ventricles (Figure 14-2a). The fluid formed in the choroid plexuses of the lateral ventricles circulates through the interventricular foramen to the third ventricle, where more fluid is added by the choroid plexus of the third

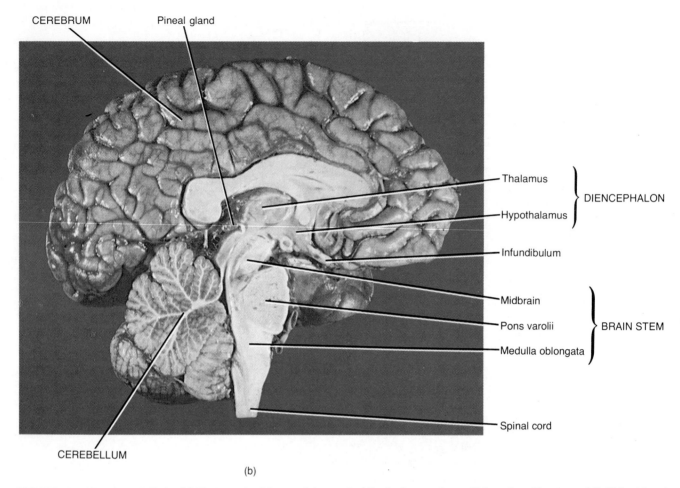

CEREBRUM   Pineal gland

Thalamus
Hypothalamus
} DIENCEPHALON

Infundibulum

Midbrain
Pons varolii
Medulla oblongata
} BRAIN STEM

Spinal cord

CEREBELLUM

(b)

FIGURE 14-1 (*Continued*) Brain. (b) Photograph of the medial aspect of the brain seen in sagittal section. (Courtesy of C. Yokochi and J. W. Rohen, *Photographic Anatomy of the Human Body*, 2d ed., 1979, IGAKU-SHOIN, Ltd., Tokyo, New York.)

ventricle. It then flows through the cerebral aqueduct into the fourth ventricle. Here there are contributions from the choroid plexus of the fourth ventricle. The fluid then circulates through the apertures of the fourth ventricle into the subarachnoid space around the back of the brain. It also passes downward to the subarachnoid space around the posterior surface of the spinal cord, up the anterior surface of the spinal cord, and around the anterior part of the brain. From here it is gradually reabsorbed into veins. Some cerebrospinal fluid may be formed by ependymal (neuroglial) cells lining the central canal of the spinal cord. This small quantity of fluid ascends to reach the fourth ventricle. Most of the fluid is absorbed into the superior sagittal sinus (Figure 14-2a–c). The absorption actually occurs through **arachnoid villi**—fingerlike projections of the arachnoid that push into the superior sagittal sinus. Normally, cerebrospinal fluid is absorbed as rapidly as it is formed.

CLINICAL APPLICATION

If an obstruction, such as a tumor, arises in the brain and interferes with the drainage of fluid from the ventricles into the subarachnoid space, large amounts of fluid accumulate in the ventricles. Fluid pressure inside the brain increases, and, if the fontanels have not yet closed, the head bulges to relieve the pressure. This condition is called **internal hydrocephalus** (*hydro* = water; *enkephalos* = brain). If an obstruction interferes with drainage somewhere in the subarachnoid space and cerebrospinal fluid accumulates inside the space, the condition is termed **external hydrocephalus.**

BLOOD SUPPLY

The brain is well supplied with blood vessels, which supply oxygen and nutrients. (Cerebral circulation is outlined in Exhibits 21-4 and 21-9 and Figures 21-14c and 21-18.)

Although the brain composes only about 2 percent of the body weight, it utilizes about 20 percent of the oxygen used by the entire body. The brain is one of the most metabolically active organs of the body and the amount of oxygen used varies with the degree of mental activity. If the blood flow to the brain is interrupted even briefly, unconsciousness may result. A one- or two-minute interruption may weaken the brain cells by starving them of oxygen. If the cells are totally deprived of oxygen for four minutes, many are permanently injured. Lysosomes of brain cells are sensitive to decreased oxygen concentration. If the condition persists long enough, lysosomes

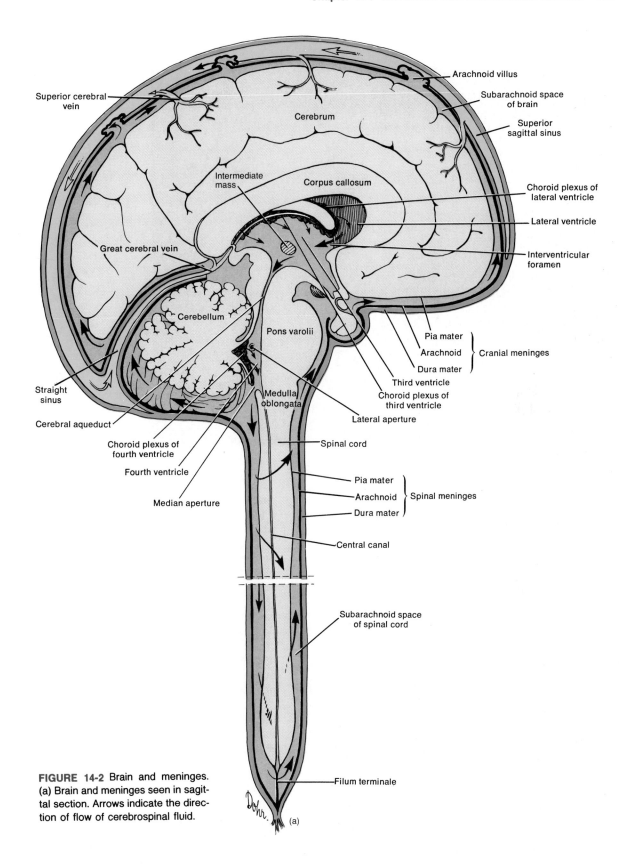

FIGURE 14-2 Brain and meninges. (a) Brain and meninges seen in sagittal section. Arrows indicate the direction of flow of cerebrospinal fluid.

break open and the released enzymes bring about self-destruction of brain cells. Occasionally during childbirth the oxygen supply from the mother's blood is interrupted before the baby leaves the birth canal and can breathe. Often such babies are stillborn or suffer permanent brain damage that may result in mental retardation, epilepsy, and paralysis.

Blood supplying the brain also contains glucose, the principal source of energy for brain cells. Because carbohydrate storage in the brain is limited, the supply of glu-

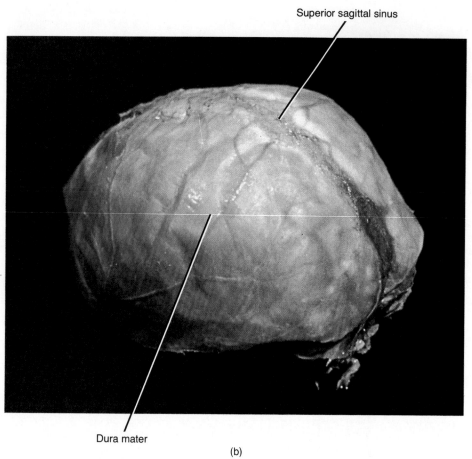

Superior sagittal sinus

Dura mater

(b)

FIGURE 14-2 (*Continued*) Brain and meninges. (b) Photograph of a brain with dura mater. Note the superior sagittal sinus. (Courtesy of J. W. Eads.) (c) Frontal section through the brain showing the relationship of the superior sagittal sinus to the arachnoid villi.

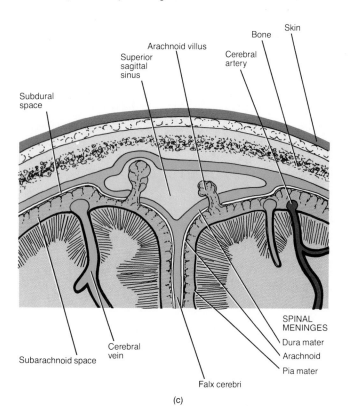

Arachnoid villus

Superior sagittal sinus

Subdural space

Bone

Skin

Cerebral artery

Cerebral vein

Subarachnoid space

Falx cerebri

SPINAL MENINGES

Dura mater

Arachnoid

Pia mater

(c)

cose must be continuous. If blood entering the brain has a low glucose level, mental confusion, dizziness, convulsions, and loss of consciousness may occur.

Both carbon dioxide and oxygen have potent effects on cerebral blood flow. Carbon dioxide increases cerebral blood flow by combining with water to form carbonic acid (H_2CO_3), which breaks down into hydrogen ions (H^+) and bicarbonate ions (HCO_3^-). The H^+ ions then cause vasodilation of cerebral vessels and increased blood flow. Dilation is almost directly proportional to an increase in H^+ ion concentration. A decrease in oxygen in the blood also causes vasodilation and increased cerebral blood flow. Other factors related to blood flow and pressure are discussed in detail in Chapter 21.

Glucose, oxygen, and certain ions pass rapidly from the circulating blood into brain cells. Other substances, such as creatinine, urea, chloride, insulin, and sucrose, enter quite slowly. Still other substances—proteins and most antibiotics—do not pass at all from the blood into brain cells. The differential rates of passage of certain materials from the blood into the brain suggest a concept called the **blood–brain barrier (BBB)**. Electron micrograph studies of the capillaries of the brain reveal that they differ structurally from other capillaries. Brain capillaries are constructed of more densely packed cells and are surrounded by large numbers of neuroglial cells and a continuous basement membrane. These features form a barrier to the passage of certain materials. Substances that cross the barrier are either very small molecules or require the assistance of a carrier molecule to cross by active transport. The blood–brain barrier functions as a selective barrier to protect brain cells from harmful substances. An injury to the brain due to trauma, inflammation, or toxins causes a breakdown of the blood–brain barrier, permitting the passage of normally restricted substances into brain tissue.

CLINICAL APPLICATION

Various **drugs** differ with respect to their passage through the blood–brain barrier. The antibiotics *chloramphenicol* (Chloromycetin), *tetracycline* (Achromycin V), and *sulfonamides* (Sonilyn) cross easily. *Penicillin* (Bicillin) crosses in only very small amounts.

Theopental sodium (Pentothal), a general anesthetic, rapidly crosses the blood–brain barrier following intravenous injection. *Atropine* (Atropisol), an antispasmodic, also quickly enters the brain. *Anisotropine methylbromide* (Valpin), another antispasmodic, and *phenylbutazone* (Azolid), an antiinflammatory drug, cannot cross the blood–brain barrier.

BRAIN STEM

Medulla Oblongata

The **medulla oblongata** (me-DULL-la ob'-long-GA-ta), or simply **medulla,** is a continuation of the upper portion of the spinal cord and forms the inferior part of the brain stem (Figure 14-3). Its position in relation to the other parts of the brain may be noted in Figure 14-1. It lies just superior to the level of the foramen magnum and extends upward to the inferior portion of the pons varolii. The medulla measures 3 cm (about 1 inch) in length.

The medulla contains all ascending and descending tracts that communicate between the spinal cord and various parts of the brain. These tracts constitute the white matter of the medulla. Some tracts cross as they pass through the medulla. Let us see how this crossing occurs and what it means.

On the ventral side of the medulla are two roughly triangular structures called **pyramids** (Figures 14-3 and 14-4). The pyramids are composed of the largest motor tracts that pass from the outer region of the cerebrum

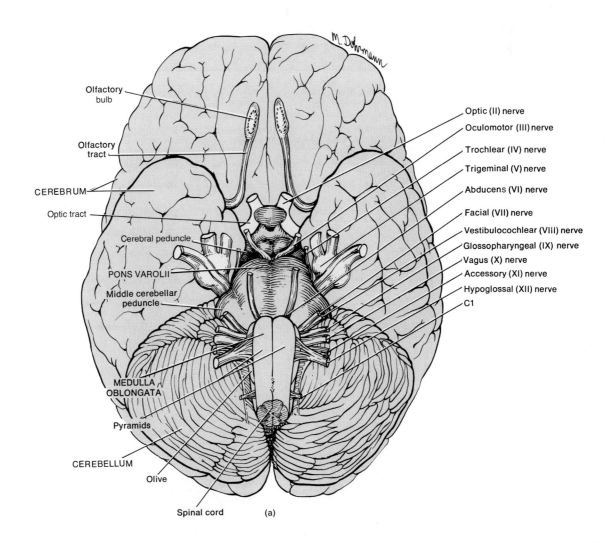

FIGURE 14-3 Brain stem. (a) Diagram of the ventral surface of the brain showing the structure of the brain stem in relation to the cranial nerves and associated structures.

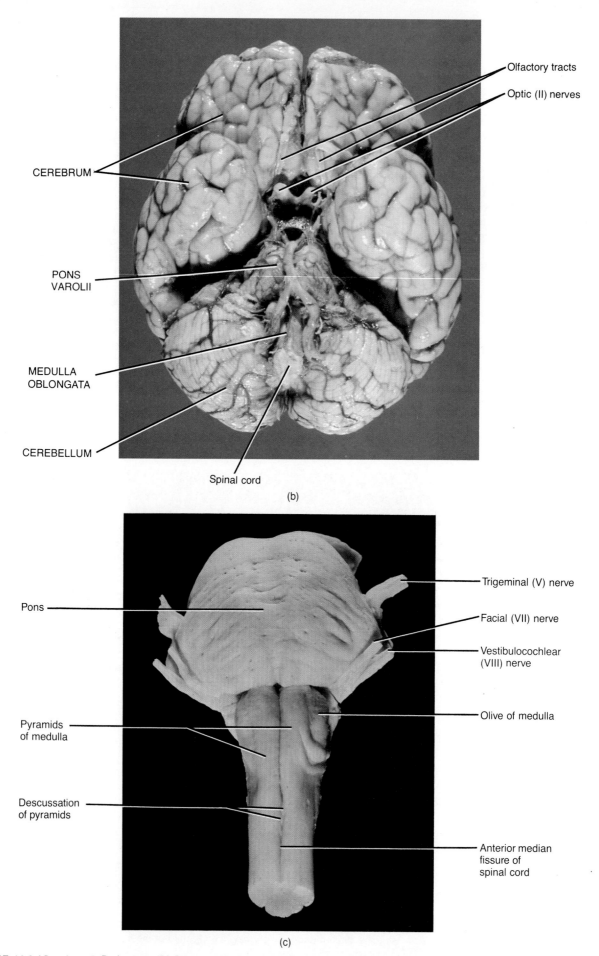

Olfactory tracts

Optic (II) nerves

CEREBRUM

PONS VAROLII

MEDULLA OBLONGATA

CEREBELLUM

Spinal cord

(b)

Pons

Trigeminal (V) nerve

Facial (VII) nerve

Vestibulocochlear (VIII) nerve

Pyramids of medulla

Olive of medulla

Descussation of pyramids

Anterior median fissure of spinal cord

(c)

FIGURE 14-3 (*Continued*) Brain stem. (b) Photograph of the ventral surface of the brain showing the brain stem in relation to associated structures. (Courtesy of Martin Rotker, Taurus Photos.) (c) Photograph of the ventral surface of the brain stem. (Courtesy of N. Gluhbegovic and T. H. Williams, *The Human Brain: A Photographic Guide,* Harper & Row, New York, 1980.)

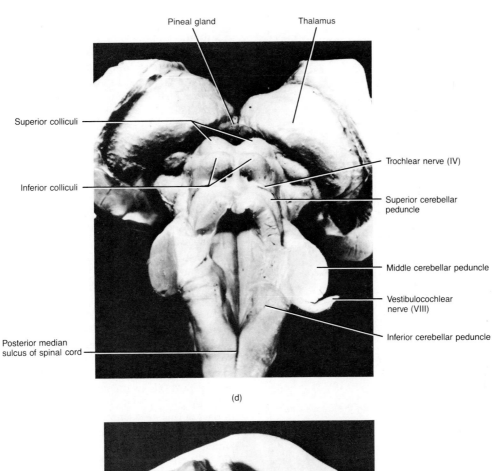

Pineal gland

Thalamus

Superior colliculi

Inferior colliculi

Posterior median
sulcus of spinal cord

Trochlear nerve (IV)

Superior cerebellar
peduncle

Middle cerebellar peduncle

Vestibulocochlear
nerve (VIII)

Inferior cerebellar peduncle

(d)

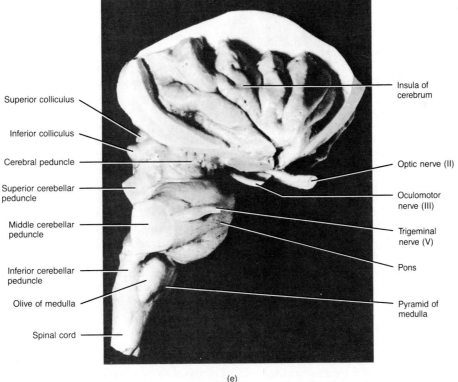

Superior colliculus

Inferior colliculus

Cerebral peduncle

Superior cerebellar
peduncle

Middle cerebellar
peduncle

Inferior cerebellar
peduncle

Olive of medulla

Spinal cord

Insula of
cerebrum

Optic nerve (II)

Oculomotor
nerve (III)

Trigeminal
nerve (V)

Pons

Pyramid of
medulla

(e)

FIGURE 14-3 (*Continued*) Brain stem. (d) Photograph of the posterior surface of the brain stem. (Courtesy of N. Gluhbegovic and T. H. Williams, *The Human Brain: A Photographic Guide,* Harper & Row, New York, 1980.) (e) Photograph of the brain stem in right lateral view. (Courtesy of N. Gluhbegovic and T. H. Williams, *The Human Brain: A Photographic Guide,* Harper & Row, New York, 1980.)

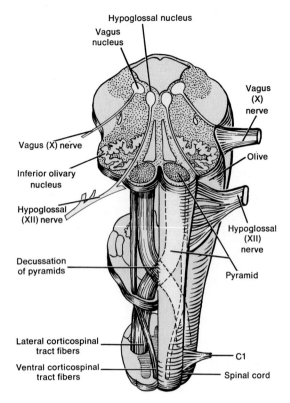

Hypoglossal nucleus

Vagus
nucleus

Vagus (X) nerve

Inferior olivary
nucleus

Hypoglossal
(XII) nerve

Decussation
of pyramids

Lateral corticospinal
tract fibers

Ventral corticospinal
tract fibers

Vagus
(X)
nerve

Olive

Hypoglossal
(XII)
nerve

Pyramid

C1

Spinal cord

FIGURE 14-4 Details of the medulla showing the decussation of pyramids.

(cerebral cortex) to the spinal cord. Just above the junction of the medulla with the spinal cord, most of the fibers in the left pyramid cross to the right side, and most of the fibers in the right pyramid cross to the left. This crossing is called the **decussation** (dē'-ku-SĀ-shun) **of pyramids.** The adaptive value, if any, of this phenomenon is unknown. The principal motor fibers that undergo decussation belong to the lateral corticospinal tracts. These tracts originate in the cerebral cortex and pass inferiorly to the medulla. The fibers cross in the pyramids and descend in the lateral columns of the spinal cord, terminating in the anterior gray horns. Here synapses occur with motor neurons that terminate in skeletal muscles. As a result of the crossing, fibers that originate in the left cerebral cortex activate muscles on the right side of the body, and fibers that originate in the right cerebral cortex activate muscles on the left side. Decussation explains why motor areas of one side of the cerebral cortex control muscular movements on the opposite side of the body.

The dorsal side of the medulla contains two pairs of prominent nuclei: the right and left **nucleus gracilis** (gras-I-lis; *gracilis* = slender) and **nucleus cuneatus** (kyoo-nē-Ā-tus; *cuneus* = wedge). These nuclei receive sensory fibers from ascending tracts (right and left fasciculus gracilis and fasciculus cuneatus) of the spinal cord and relay the sensory information to the opposite side of the medulla. This information is conveyed to the thalamus and then to the sensory areas of the cerebral cortex. Nearly all sensory impulses received on one side of the body cross in the medulla or spinal cord and are perceived in the opposite side of the cerebral cortex.

In addition to its function as a conduction pathway for motor and sensory impulses between the brain and spinal cord, the medulla also contains an area of dispersed gray matter containing some white fibers. This region is called the **reticular formation.** Actually, portions of the reticular formation are also located in the spinal cord, pons, midbrain, and diencephalon. The reticular formation functions in consciousness and arousal.

Within the medulla are three vital reflex centers of the reticular system. The **cardiac center** regulates heartbeat and force of contraction, the **medullary rhythmicity area** adjusts the basic rhythm of breathing, and the **vasomotor (vasoconstrictor) center** regulates the diameter of blood vessels. Other centers in the medulla are considered nonvital and coordinate swallowing, vomiting, coughing, sneezing, and hiccuping.

The medulla also contains the nuclei of origin for several pairs of cranial nerves (Figures 14-3a and 14-4). These are the cochlear and vestibular branches of the vestibulocochlear (VIII) nerves, which are concerned with hearing and equilibrium (there is also a nucleus for the vestibular branches in the pons); the glossopharyngeal (IX) nerves, which relay impulses related to swallowing, salivation, and taste; the vagus (X) nerves, which relay impulses to and from many thoracic and abdominal viscera; the cranial portions of the accessory (XI) nerves, which convey impulses related to head and shoulder movements (a part of this nerve also arises from the first five segments of the spinal cord); and the hypoglossal (XII) nerves, which convey impulses that involve tongue movements.

On each lateral surface of the medulla is an oval projection called the **olive** (Figure 14-3c,e). The olive contains an inferior olivary nucleus and two accessory olivary nuclei. The nuclei are connected to the cerebellum by fibers.

Also associated with the medulla is the greater part of the **vestibular nuclear complex.** This nuclear group consists of the *lateral, medial,* and *inferior vestibular nuclei* in the medulla and the *superior vestibular nucleus* in the pons. As you will see later (Chapter 17), the vestibular nuclei assume an important role in helping the body to maintain its sense of equilibrium.

In view of the many vital activities controlled by the medulla, it is not surprising that a hard blow to the base of the skull can be fatal. Nonfatal medullary injury may be indicated by cranial nerve malfunctions on the same side of the body as the area of medullary injury, paralysis and loss of sensation on the opposite side of the body, and irregularities in respiratory control.

Pons Varolii

The relationship of the **pons varolii** (va-RŌ-lē-ī) or **pons** to other parts of the brain can be seen in Figures 14-1 and 14-3. The pons, which means bridge, lies directly above the medulla and anterior to the cerebellum. It measures about 2.5 cm (1 inch) in length. Like the medulla, the pons consists of white fibers scattered throughout with nuclei. As the name implies, the pons is a bridge connecting the spinal cord with the brain and parts of the brain with each other. These connections are provided by fibers that run in two principal directions. The transverse fibers connect with the cerebellum through the *middle cerebellar peduncles*. The longitudinal fibers of the pons belong to the motor and sensory tracts that connect the spinal cord or medulla with the upper parts of the brain stem.

The nuclei for certain paired cranial nerves are also contained in the pons (Figure 14-3a). These include the trigeminal (V) nerves, which relay impulses for chewing and for sensations of the head and face; the abducens (VI) nerves, which regulate certain eyeball movements; the facial (VII) nerves, which conduct impulses related to taste, salivation, and facial expression; and the vestibular branches of the vestibulocochlear (VIII) nerves, which are concerned with equilibrium.

Other important nuclei in the reticular formation of the pons are the **pneumotaxic** (noo-mō-TAK-sik) **area** and the **apneustic** (ap-NOO-stik) **area.** Together with the medullary rhythmicity area in the medulla, they help control respiration.

Midbrain

The **midbrain,** or **mesencephalon** (*meso* = middle; *enkephalos* = brain), extends from the pons to the lower portion of the diencephalon (Figure 14-1 and 14-3). It is about 2.5 cm (1 inch) in length. The cerebral aqueduct passes through the midbrain and connects the third ventricle above with the fourth ventricle below.

The ventral portion of the midbrain contains a pair of fiber bundles referred to as **cerebral peduncles** (pe-DUNG-kulz). The cerebral peduncles contain many motor fibers that convey impulses from the cerebral cortex to the pons and spinal cord. They also contain sensory fibers that pass from the spinal cord to the thalamus. The cerebral peduncles constitute the main connection for tracts between upper parts of the brain and lower parts of the brain and the spinal cord.

The dorsal portion of the midbrain is called the **tectum** (*tectum* = roof) and contains four rounded eminences: the **corpora quadrigemina** (KOR-po-ra kwad-ri-JEM-in-a). Two of the eminences are known as the **superior colliculi** (ko-LIK-yoo-lī). These serve as reflex centers for movements of the eyeballs and head in response to visual and other stimuli. The other two eminences, the **inferior colliculi,** serve as reflex centers for movements of the head and trunk in response to auditory stimuli. The midbrain also contains the **substantia nigra** (sub-STAN-shē-a NĪ-gra), a large, heavily pigmented nucleus near the cerebral peduncles.

A major nucleus in the reticular formation of the midbrain is the **red nucleus.** Fibers from the cerebellum and cerebral cortex terminate in the red nucleus. The red nucleus is also the origin of cell bodies of the descending rubrospinal tract. Other nuclei in the midbrain are associated with cranial nerves (see Figure 14-3a). These include the oculomotor (III) nerves, which mediate some movements of the eyeballs and changes in pupil size and lens shape, and the trochlear (IV) nerves, which conduct impulses that move the eyeballs.

A structure called the **medial lemniscus** (*lemniskos* = ribbon or band) is common to the medulla, pons, and midbrain. The medial lemniscus is a band of white fibers containing axons that convey impulses for fine touch, proprioception, and vibrations from the medulla to the thalamus.

DIENCEPHALON

The **diencephalon** (*dia* = through; *enkephalos* = brain) consists principally of the thalamus and hypothalamus. The relationship of these structures to the rest of the brain is shown in Figure 14-1.

Thalamus

The **thalamus** (THAL-a-mus; *thalamos* = inner chamber) is an oval structure above the midbrain that measures about 3 cm (1 inch) in length and constitutes four-fifths of the diencephalon. It consists of two oval masses of mostly gray matter organized into nuclei that form the lateral walls of the third ventricle (Figure 14-5). The masses are joined by a bridge of gray matter called the **intermediate mass.** Each mass is deeply embedded in a cerebral hemisphere and is bounded laterally by the **internal capsule.**

Although the thalamic masses are primarily gray matter, some portions are white matter. Among the white matter portions are the **stratum zonale,** which covers the dorsal surface; the **external medullary lamina,** covering the lateral surface; and the **internal medullary lamina,** which divides the gray matter masses into an anterior nuclear group, a medial nuclear group, and a lateral nuclear group.

Within each group are nuclei that assume various roles. Some nuclei in the thalamus serve as relay stations for all sensory impulses, except smell, to the cerebral cortex. These include the **medial geniculate** (je-NIK-yoo-lāt) **nuclei** (hearing), the **lateral geniculate nuclei** (vision), and the **ventral posterior nuclei** (general sensations and taste). Other nuclei are centers for synapses in the somatic motor system. These include the **ventral lateral nuclei** (voluntary motor actions) and **ventral anterior nuclei** (voluntary motor actions and arousal). The thalamus is the principal relay station for sensory impulses that reach the cerebral cortex from the spinal cord, brain stem, cerebellum, and parts of the cerebrum.

The thalamus also functions as an interpretation center for some sensory impulses, such as pain, temperature, crude touch, and pressure. The thalamus also contains a **reticular nucleus** in its reticular formation, which in some way seems to modify neuronal activity in the thalamus, and an **anterior nucleus** in the floor of the lateral ventricle, which is concerned with certain emotions and memory.

Hypothalamus

The **hypothalamus** (*hypo* = under) is a small portion of the diencephalon. Its relationship to other parts of the brain is shown in Figures 14-1 and 14-5a. The hypothalamus forms the floor and part of the lateral walls of the third ventricle. It is partially protected by the sella turcica of the sphenoid bone.

Despite its small size, nuclei in the hypothalamus control many body activities, most of them related to homeo-

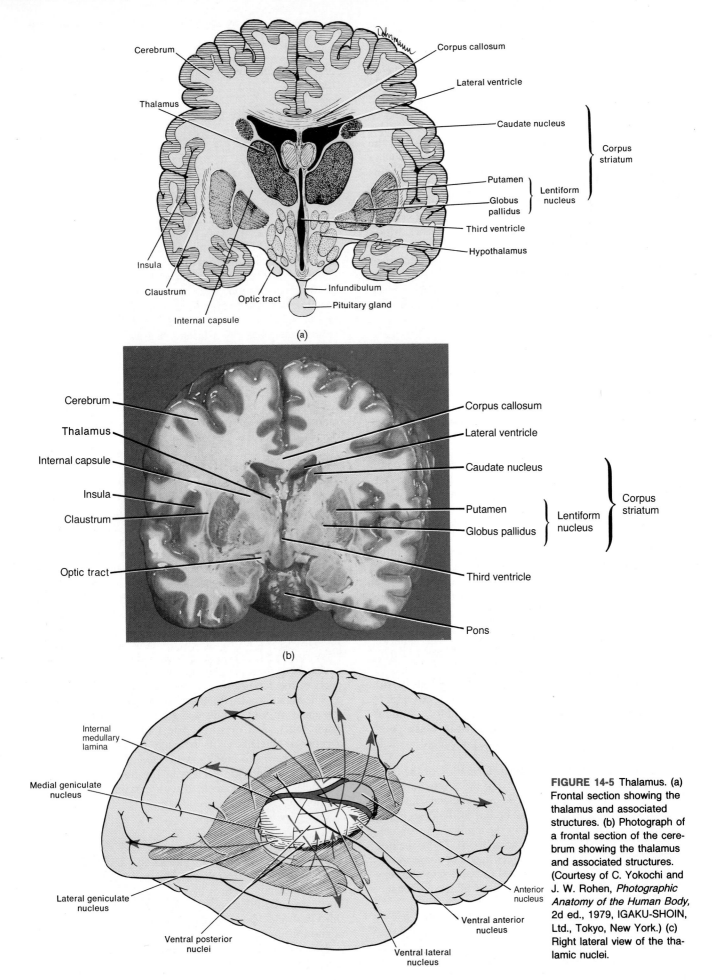

FIGURE 14-5 Thalamus. (a) Frontal section showing the thalamus and associated structures. (b) Photograph of a frontal section of the cerebrum showing the thalamus and associated structures. (Courtesy of C. Yokochi and J. W. Rohen, *Photographic Anatomy of the Human Body,* 2d ed., 1979, IGAKU-SHOIN, Ltd., Tokyo, New York.) (c) Right lateral view of the thalamic nuclei.

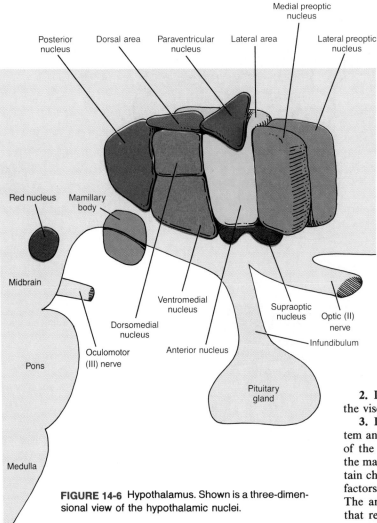

FIGURE 14-6 Hypothalamus. Shown is a three-dimensional view of the hypothalamic nuclei.

stasis. Although differentiation of the hypothalamic nuclei is far from precise, it is possible to identify certain nuclei. Some of these nuclei are more readily identified in lower animals and are more distinct in fetuses than adults. Also, within a given nucleus there may be several kinds of cells that can be differentiated histologically. The localization of function, with a few exceptions, is not specific to the individual nuclei; certain functions tend to overlap nuclear boundaries. For this reason, functions are attributed to regions rather than specific nuclei. Since the hypothalamic nuclei are useful landmarks in understanding subsequent discussion of functions, a three-dimensional view of the hypothalamic nuclei is shown in Figure 14-6.

The chief functions of the hypothalamus are as follows.

1. It controls and integrates the autonomic nervous system, which stimulates smooth muscle, regulates the rate of contraction of cardiac muscle, and controls the secretions of many glands. This is accomplished by axons of neurons whose dendrites and cell bodies are in hypothalamic nuclei. The axons form tracts from the hypothalamus to sympathetic and parasympathetic centers in the brain stem and spinal cord. Through the autonomic nervous system, the hypothalamus is the main regulator of visceral activities. It regulates heart rate, movement of food through the digestive tract, and contraction of the urinary bladder.

2. It is involved in the reception of sensory impulses from the viscera.

3. It is the principal intermediary between the nervous system and the endocrine system—the two major control systems of the body. The hypothalamus lies just above the pituitary, the main endocrine gland. When the hypothalamus detects certain changes in the body, it releases chemicals called regulating factors that stimulate or inhibit the anterior pituitary gland. The anterior pituitary then releases or holds back hormones that regulate various physiological activities of the body. The hypothalamus also produces two hormones, antidiuretic hormone (ADH) and oxytocin, which are transported to and stored in the posterior pituitary gland. The hormones are released from storage when needed by the body.

4. It is the center for the mind-over-body phenomenon. When the cerebral cortex interprets strong emotions, it often sends impulses along the tracts that connect the cortex with the hypothalamus. The hypothalamus then directs impulses via the autonomic nervous system and also releases chemicals that stimulate the anterior pituitary gland. The result can be a wide range of changes in body activities. For instance, when you panic, impulses leave the hypothalamus to stimulate your heart to beat faster. Likewise, continued psychological stress can produce long-term abnormalities in body function that result in serious illness. These so-called psychosomatic disorders are definitely real.

5. It is associated with feelings of rage and aggression.

6. It controls normal body temperature. Certain cells of the hypothalamus serve as a thermostat. If blood flowing through the hypothalamus is above normal temperature, the hypothalamus directs impulses along the autonomic nervous system to stimulate activities that promote heat loss. Heat can be lost through relaxation of the smooth muscle in the blood vessels, causing vasodilation of cutaneous vessels and increased heat loss from the skin. Heat loss also occurs by sweating. Conversely, if the temperature of the blood is below normal, the hypothalamus generates impulses that promote heat retention. Heat can be retained through the constriction of cutaneous blood vessels, cessation of sweating, and shivering.

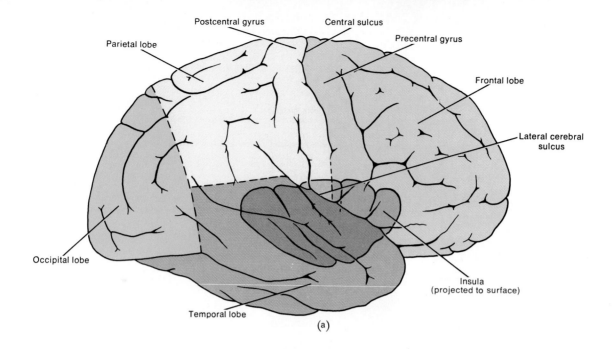

(a)

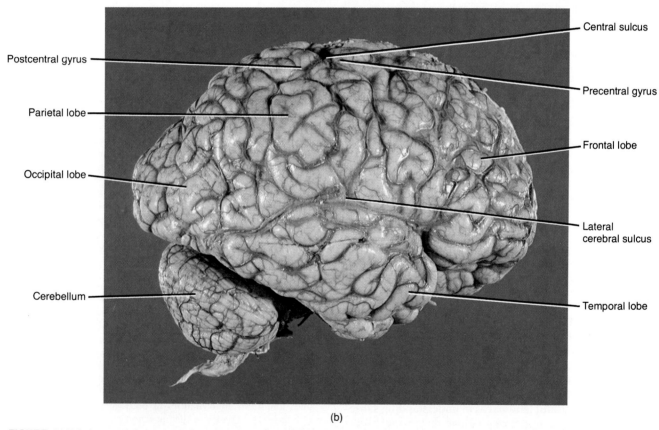

(b)

FIGURE 14-7 Lobes and fissures of the cerebrum. (a) Diagram of right lateral view. Since the insula cannot be seen externally, it has been projected to the surface. It can be seen in Figure 14-5a, b. (b) Photograph of right lateral view. (Courtesy of Martin Rotker, Taurus Photos.)

7. It regulates food intake through two centers. The **feeding center** is stimulated by hunger sensations from an empty stomach. When sufficient food has been ingested, the **satiety** (sa-TĪ-e-tē) **center** is stimulated and sends out impulses that inhibit the feeding center.

8. It contains a **thirst center.** Certain cells in the hypothalamus are stimulated when the extracellular fluid volume is reduced. The stimulated cells produce the sensation of thirst.

9. It is one of the centers that maintains the waking state and sleep patterns.

10. It exhibits properties of a self-sustained oscillator and, as such, acts as a pacemaker to drive many biological rhythms.

CEREBRUM

Supported on the brain stem and forming the bulk of the brain is the **cerebrum** (see Figure 14-1). The surface of the cerebrum is composed of gray matter 2 to 4 mm (0.08 to 0.16 inch) thick and is referred to as the **cerebral**

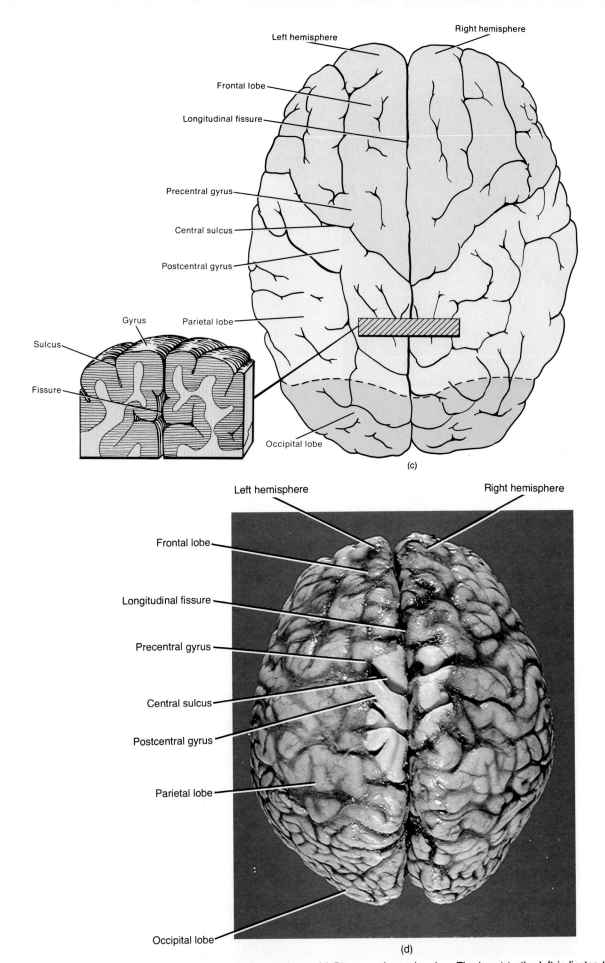

FIGURE 14-7 (*Continued*) Lobes and fissures of the cerebrum. (c) Diagram of superior view. The insert to the left indicates the relative differences among a gyrus, sulcus, and fissure. (d) Photograph of superior view. (Courtesy of Martin Rotker, Taurus Photos.)

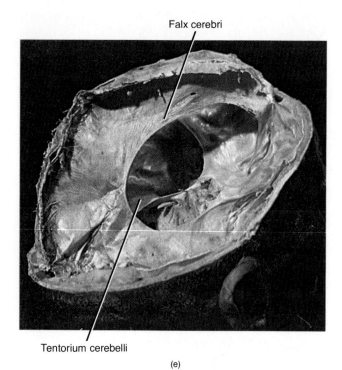

Falx cerebri

Tentorium cerebelli

(e)

FIGURE 14-7 (*Continued*) Lobes and fissures of the cerebrum. (e) Photograph of the skull showing the tentorium cerebelli and falx cerebri. (Courtesy of J. W. Eads.)

cortex (*cortex* = rind or bark). The cortex, containing roughly 15 billion cells, consists of six layers of nerve cell bodies in most areas. Beneath the cortex lies the cerebral white matter.

During embryonic development, when there is a rapid increase in brain size, the gray matter of the cortex enlarges out of proportion to the underlying white matter. As a result, the cortical region rolls and folds upon itself. The folds are called **gyri** (JĪ-rī) or **convolutions** (Figure 14-7a–d). The deep grooves between folds are referred to as **fissures;** the shallow grooves between folds are **sulci** (SUL-sī). The most prominent fissure, the **longitudinal fissure,** nearly separates the cerebrum into right and left

halves, or **hemispheres** (Figure 14-7c,d). The hemispheres, however, are connected internally by a large bundle of transverse fibers composed of white matter called the **corpus callosum** (kal-LŌ-sum; *corpus* = body; *callosus* = hard). Between the hemispheres is an extension of the cranial dura mater called the **falx (FALKS) cerebri** (Figure 14-7e).

Lobes

Each cerebral hemisphere is further subdivided into four lobes by deep sulci or fissures (Figure 14-7a–d). The **central sulcus** separates the **frontal lobe** from the **parietal lobe.** A major gyrus, the **precentral gyrus,** is located immediately anterior to the central sulcus. The gyrus is a landmark for the primary motor area of the cerebral cortex. Another major gyrus, the **postcentral gyrus,** is located immediately posterior to the central sulcus. This gyrus is a landmark for the general sensory area of the cerebral cortex. The **lateral cerebral sulcus** separates the **frontal lobe** from the **temporal lobe.** The **parietooccipital sulcus** separates the **parietal lobe** from the **occipital lobe.** Another prominent fissure, the **transverse fissure,** separates the cerebrum from the cerebellum. The frontal lobe, parietal lobe, temporal lobe, and occipital lobe are named after the bones that cover them. A fifth part of the cerebrum, the **insula,** lies deep within the lateral cerebral fissure, under the parietal, frontal, and temporal lobes. It cannot be seen in an external view of the brain (Figure 14-7a).

White Matter

The white matter underlying the cortex consists of myelinated axons running in three principal directions (Figure 14-8).

1. Association fibers connect and transmit impulses between gyri in the same hemisphere.

2. Commissural fibers transmit impulses from the gyri in one cerebral hemisphere to the corresponding gyri in the opposite cerebral hemisphere. Three important groups of commis-

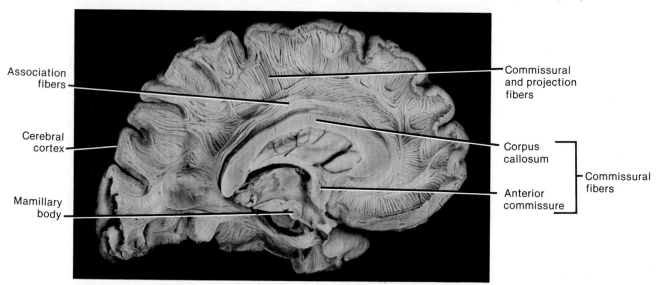

Association fibers

Cerebral cortex

Mamillary body

Commissural and projection fibers

Corpus callosum

Anterior commissure

Commissural fibers

FIGURE 14-8 White matter tracts of the left cerebral hemisphere seen in sagittal section. (Courtesy of N. Gluhbegovic and T. H. Williams, *The Human Brain: A Photographic Guide,* Harper & Row, New York, 1980.)

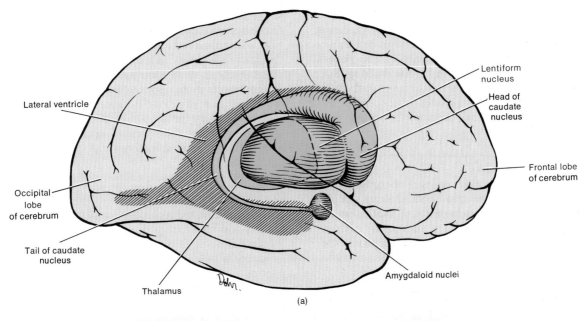

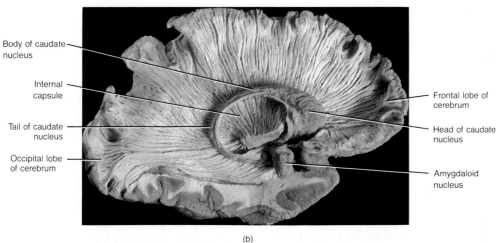

FIGURE 14-9 Basal ganglia. (a) In this diagram of the right lateral view of the cerebrum, the basal ganglia have been projected to the surface. Refer to Figure 14-5a, b to note the positions of the basal ganglia in the frontal section of the cerebrum. (b) Photograph of the medial surface of the left cerebral hemisphere showing portions of the basal ganglia. (Courtesy of N. Gluhbegovic and T. H. Williams, *The Human Brain: A Photographic Guide,* Harper & Row, New York, 1980.)

sural fibers are the *corpus callosum, anterior commissure,* and *posterior commissure.*

3. Projection fibers form ascending and descending tracts that transmit impulses from the cerebrum to other parts of the brain and spinal cord.

Basal Ganglia (Cerebral Nuclei)

The **basal ganglia,** or **cerebral nuclei,** are paired masses of gray matter in each cerebral hemisphere (Figures 14-5 and 14-9). The largest of the basal ganglia of each hemisphere is the **corpus striatum** (strī-Ā-tum; *corpus* = body; *striatus* = striped). It consists of the **caudate** (*cauda* = tail) **nucleus** and the **lentiform** (*lenticula* = shaped like a lentil or lens) **nucleus.** The lentiform nucleus, in turn, is subdivided into a lateral portion called the **putamen** (pu-TĀ-men; *putamen* = shell) and a medial portion called the **globus pallidus** (*globus* = ball; *pallid* = pale).

The portion of the *internal capsule* passing between the lentiform nucleus and the caudate nucleus and between the lentiform nucleus and thalamus is sometimes considered part of the corpus striatum. The internal capsule is made up of a group of sensory and motor white matter tracts that connect the cerebral cortex with the brain stem and spinal cord.

Other structures frequently considered part of the basal ganglia are the claustrum and amygdaloid nucleus. The **claustrum** (KLAWS-trum) is a thin sheet of gray matter lateral to the putamen. The **amygdaloid** (a-MIG-da-loyd; *amygda* = almond) **nucleus** is located at the tail end of the caudate nucleus. Some authorities also consider the **substantia nigra,** the **subthalamic nucleus,** and the **red nucleus** to be part of the basal ganglia. The substantia nigra is a large nucleus in the midbrain. The subthalamic nucleus lies against the internal capsule. Its major connection is with the globus pallidus.

The basal ganglia are interconnected by many fibers. They are also connected to the cerebral cortex, thalamus, and hypothalamus. The caudate nucleus and the putamen control large subconscious movements of the skeletal muscles, such as swinging the arms while walking. Such gross movements are also consciously controlled by the cerebral cortex. The globus pallidus is concerned with the regulation of muscle tone required for specific body movements.

CLINICAL APPLICATION

Damage to the basal ganglia results in abnormal body movements, such as uncontrollable shaking, called **tremor,** and **involuntary movements of skeletal muscles.** Moreover, destruction of a substantial portion of the caudate nucleus almost totally **paralyzes** the side of the body opposite to the damage. The caudate nucleus is an area often affected by a stroke.

A lesion in the subthalamic nucleus results in a motor disturbance on the opposite side of the body called **hemiballismus** (*hemi* = half; *ballismos* = jumping), which is characterized by involuntary movements occurring suddenly with great force and rapidity. The movements are purposeless and generally of the withdrawal type, although they may be jerky. The spontaneous movements affect the proximal portions of the extremities most severely, especially the arms.

Limbic System

Certain components of the cerebral hemispheres and diencephalon constitute the **limbic** (*limbus* = border) **system.** Among its components are the following regions of gray matter.

1. Limbic lobe. Formed by two gyri of the cerebral hemisphere: the cingulate gyrus and the hippocampal gyrus.

2. Hippocampus. An extension of the hippocampal gyrus that extends into the floor of the lateral ventricle.

3. Amygdaloid nucleus. Located at the tail end of the caudate nucleus.

4. Mammillary bodies of the hypothalamus. Two round masses close to the midline near the cerebral peduncles.

5. Anterior nucleus of the thalamus. Located in the floor of the lateral ventricle.

The limbic system is a wishbone-shaped group of structures that encircles the brain stem and functions in the emotional aspects of behavior related to survival. The hippocampus, together with portions of the cerebrum, also functions in memory. Although behavior is a function of the entire nervous system, the limbic system controls most of its involuntary aspects. Experiments on the limbic system of monkeys and other animals indicate that the amygdaloid nucleus assumes a major role in controlling the overall pattern of behavior.

Other experiments have shown that the limbic system is associated with pleasure and pain. When certain areas of the limbic system of the hypothalamus, thalamus, and midbrain are stimulated, the reactions of experimental animals indicate they are experiencing intense punishment. When other areas are stimulated, the animals' reactions indicate they are experiencing extreme pleasure. In still other studies, stimulation of the perifornical nuclei of the hypothalamus results in a behavioral pattern called *rage.* The animal assumes a defensive posture—extending its claws, raising its tail, hissing, spitting, growling, and opening its eyes wide. Stimulating other areas of the limbic system results in an opposite behavioral pattern: docility, tameness, and affection. Because the limbic system assumes a primary function in emotions such as pain, pleasure, anger, rage, fear, sorrow, sexual feelings, docility, and affection, it is sometimes called the "visceral" or "emotional" brain.

CLINICAL APPLICATION

Brain injuries are commonly associated with head injuries and result from displacement and distortion of neuronal tissue at the moment of impact. The various degrees of brain injury are described by the following terms.

1. **Concussion.** An abrupt but temporary loss of consciousness following a blow to the head or a sudden stopping of a moving head. A concussion produces no visible bruising of the brain.

2. **Contusion.** A visible bruising of the brain due to trauma and blood leaking from microscopic vessels. The pia mater is stripped from the brain over the injured area and may be torn, allowing blood to enter the subarachnoid space. A contusion usually results in an extended loss of consciousness, ranging from several minutes to many hours.

3. **Laceration.** Tearing of the brain, usually from a skull fracture or gunshot wound. A laceration results in rupture of large blood vessels with bleeding into the brain and subarachnoid space. Consequences include cerebral hematoma, edema, and increased intracranial pressure.

Functional Areas of Cerebral Cortex

The functions of the cerebrum are numerous and complex. In a general way, the cerebral cortex is divided into sensory, motor, and association areas. The **sensory areas** interpret sensory impulses, the **motor areas** control muscular movement, and the **association areas** are concerned with emotional and intellectual processes.

● *Sensory Areas* The **general sensory area** or **somesthetic** (sō'-mes-THET-ik; *soma* = body; *aisthesis* = perception) **area** is located directly posterior to the central sulcus of the cerebrum in the postcentral gyrus of the parietal lobe. It extends from the longitudinal fissure on the top of the cerebrum to the lateral cerebral sulcus. In Figure 14-10 the general sensory area is designated by the areas numbered 1, 2, and 3.*

The general sensory area receives sensations from cutaneous, muscular, and visceral receptors in various parts of the body. Each point of the area receives sensations from specific parts of the body, and essentially the entire

---

* These numbers, as well as most of the others shown, are based on K. Brodmann's cytoarchitectural map of the cerebral cortex. His map, first published in 1909, is an attempt to correlate structure and function.

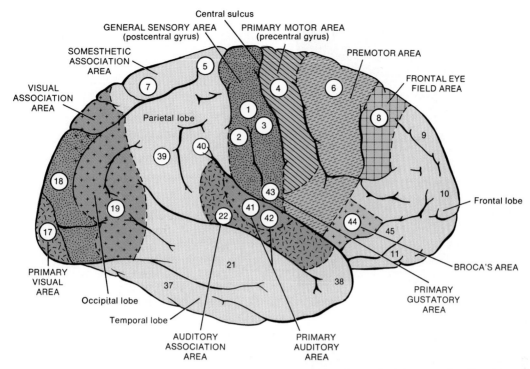

FIGURE 14-10 Functional areas of the cerebrum. This lateral view indicates the sensory and motor areas of the right hemisphere. Although Broca's area is in the left hemisphere of most people, it is shown here to indicate its location.

body is spatially represented in it. The size of the portion of the sensory area receiving stimuli from body parts is not dependent on the size of the part, but on the number of receptors the part contains. For example, a larger portion of the sensory area receives impulses from the lips than from the thorax (see Figure 15-3). The major function of the general sensory area is to localize exactly the points of the body where the sensations originate. The thalamus is capable of localizing sensations in a general way, that is, it receives sensations from large areas of the body, but cannot distinguish between specific areas of stimulation. This ability is reserved to the general sensory area of the cortex.

Posterior to the general sensory area is the **somesthetic association area.** It corresponds to the areas numbered 5 and 7 in Figure 14-10. The somesthetic association area receives input from the thalamus, other lower portions of the brain, and the general sensory area. Its role is to integrate and interpret sensations. This area permits you to determine the exact shape and texture of an object without looking at it, to determine the orientation of one object to another as they are felt, and to sense the relationship of one body part to another. Another role of the somesthetic association area is the storage of memories of past sensory experiences. Thus you can compare sensations with previous experiences.

Other sensory areas of the cortex include:

1. Primary visual area (area 17). Located on the medial surface of the occipital lobe and occasionally extends around to the lateral surface. It receives sensory impulses from the eyes and interprets shape and color.

2. Visual association area (areas 18 and 19). Located in the occipital lobe. It receives sensory signals from the primary visual area and the thalamus. It relates present to past visual experiences with recognition and evaluation of what is seen.

3. Primary auditory area (areas 41 and 42). Located in the superior part of the temporal lobe near the lateral cerebral sulcus. It interprets the basic characteristics of sound such as pitch and rhythm.

4. Auditory association area (area 22). Inferior to the primary auditory area in the temporal cortex. It determines if a sound is speech, music, or noise. It also interprets the meaning of speech by translating words into thoughts.

5. Primary gustatory area (area 43). Located at the base of the postcentral gyrus above the lateral cerebral sulcus in the parietal cortex. It interprets sensations related to taste.

6. Primary olfactory area. Located in the temporal lobe on the medial aspect. It interprets sensations related to smell.

7. Gnostic (NOS-tik; *gnosis* = knowledge) **area** (areas 5, 7, 39, and 40). This *common integrative area* is located between the somesthetic, visual, and auditory association areas. The gnostic area receives impulses from these areas, as well as from the taste and smell areas, the thalamus, and lower portions of the brain stem. It integrates sensory interpretations from the association areas and impulses from other areas so that a common thought can be formed from the various sensory inputs. It then transmits signals to other parts of the brain to cause the appropriate response to the sensory signal.

CLINICAL APPLICATION

In Chapter 2, we discussed the principle and clinical applications of a recently developed type of radioisotope scanning called **positron emission tomography (PET).** At that point it was indicated that PET is being used in the diagnosis of an assortment of diseases. It is also being used by scientists to probe the healthy brain. By detecting and recording changes in glucose

Brain activity:

—— Highest

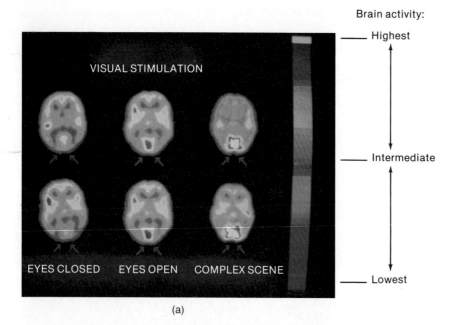

—— Intermediate

—— Lowest

(a)

Orientation of brain

Front

Left   Right

Back

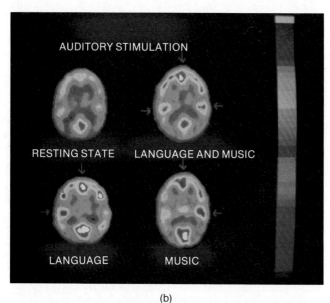

(b)

FIGURE 14-11 Positron emission tomography (PET) scans showing brain activity in response to visual and auditory stimulation. (a) Response of the visual areas of the cerebral cortex (areas between arrows) when the eyes are closed, open to white light, and open to a complex scene. (b) Response of auditory areas of the cerebral cortex (areas at horizontal arrows) and the frontal lobe (areas at vertical arrow) in response to language and music, language alone, and music alone. Note that when the auditory stimulation consists of both language and music, the auditory cortex on both sides of the brain is active; when the auditory stimulation is language only, there is a predominant left-sided activation of the auditory cortex; and when the auditory stimulation is music only, there is a predominant right-sided activation of the auditory cortex. Frontal lobe activation probably results from a high level of mentation and the promise that all subjects participating in the experiment would be paid in proportion to how much they could remember about the auditory stimulation for a subsequent test. (Courtesy of Dr. Michael E. Phelps, Division of Biophysics, Department of Radiological Sciences, UCLA School of Medicine, Los Angeles, California.)

metabolism (consumption), it is possible to identify which specific areas of the brain are involved in specific sensory and motor activities. The PET scans in Figure 14-11 indicate brain activity in response to certain visual and auditory stimulation.

● *Motor Areas*   The **primary motor area** (area 4) is located in the precentral gyrus of the frontal lobe (see Figure 14-10). Like the general sensory area, the primary motor area consists of regions that control specific muscles or groups of muscles (see Figure 15-7). Stimulation of a specific point of the primary motor area results in a muscular contraction, usually on the opposite side of the body.

The **premotor area** (area 6) is anterior to the primary motor area. It is concerned with learned motor activities of a complex and sequential nature. It generates impulses that cause a specific group of muscles to contract in a

specific sequence. An example of this is writing. Thus the premotor area controls skilled movements.

The **frontal eye field area** (area 8) in the frontal cortex is sometimes included in the premotor area. This area controls voluntary scanning movements of the eyes—searching for a word in a dictionary, for instance.

The **language areas** are also significant parts of the motor cortex. The translation of speech or written words into thoughts involves sensory areas—primary auditory, auditory association, primary visual, visual association, and gnostic—as we just described. The translation of thoughts into speech involves **Broca's** (BRŌ-kaz) **area** or the **motor speech area** (area 44), located in the frontal lobe just superior to the lateral cerebral sulcus. From this area, a sequence of signals is sent to the premotor regions that control the muscles of the larynx, pharynx, and mouth. The impulses from the premotor area to the muscles result in specific, coordinated contractions that enable you to speak. Simultaneously, impulses are sent from Broca's area to the primary motor area. From here, impulses reach your breathing muscles to regulate the proper flow of air past the vocal cords. The coordinated contractions of your speech and breathing muscles enable you to translate your thoughts into speech.

CLINICAL APPLICATION

Broca's area and other language areas are located in the left cerebral hemisphere of most individuals regardless of whether they are left-handed or right-handed. Injury to the sensory or motor speech areas results in **aphasia** (a-FĀ-zē-a; *a* = without; *phasis* = speech), an inability to speak; **agraphia** (*a* = without; *graph* = write), an inability to write; **word deafness,** an inability to understand spoken words; or **word blindness,** an inability to understand written words.

● *Association Areas*   The **association areas** of the cerebrum are made up of association tracts that connect motor and sensory areas (see Figure 14-8). The association region of the cortex occupies the greater portion of the lateral surfaces of the occipital, parietal, and temporal lobes and the frontal lobes anterior to the motor areas. The association areas are concerned with memory, emotions, reasoning, will, judgment, personality traits, and intelligence.

Brain Waves

Brain cells can generate electrical activity as a result of literally millions of action potentials of individual neurons. These electrical potentials are called **brain waves** and indicate activity of the cerebral cortex. Brain waves pass through the skull easily and can be detected by sensors called electrodes. A record of such waves is called

an **electroencephalogram (EEG).** An EEG is obtained by placing electrodes on the head and amplifying the waves with an electroencephalograph. As indicated in Figure 14-12, four kinds of waves are produced by normal individuals.

1. Alpha waves. These rhythmic waves occur at a frequency of about 10 to 12 cycles per second. (The unit commonly used to express frequency is the Hertz; 1 Hz = 1 cycle per second.) They are found in the EEGs of nearly all normal individuals when awake and in the resting state. These waves disappear entirely during sleep.

2. Beta waves. The frequency of these waves is between 15 and 60 Hz. Beta waves generally appear when the nervous system is active, that is, during periods of sensory input and mental activity.

3. Theta waves. These waves have frequencies of 5 to 8 Hz. Theta waves normally occur in children and in adults experiencing emotional stress.

4. Delta waves. The frequency of these waves is 1 to 5 Hz. Delta waves occur during sleep. They are normal in an awake infant. When produced by an awake adult, they indicate brain damage.

CLINICAL APPLICATION

Distinct **EEG patterns** appear in certain abnormalities. In fact, the EEG is used clinically in the diagnosis of epilepsy, infectious diseases, tumors, trauma, and hematomas. Electroencephalograms also furnish information regarding sleep and wakefulness.

BRAIN LATERALIZATION (SPLIT-BRAIN CONCEPT)

A distinctive characteristic of the human brain is the allocation of functions to the two cerebral hemispheres. At a glance the brain appears to have perfect bilateral symmetry, like most other organs of the body. It might therefore be expected that the two hemispheres of the brain would also be functionally equivalent. Actually many of the more specialized functions are found in only one hemisphere or the other. In fact, careful examination indicates that even the apparent anatomical symmetry is an illusion, since brain asymmetries have been detected by CT scans. In these images a peculiar departure from bilateral symmetry is observed. In right-handed people the right frontal lobe is usually wider than the left, but the left occipital and parietal lobes are wider than the right. The inner surfaces of the skull itself bulge at the right front and the left rear to accommodate the protuberances.

The distribution of the more specialized functions is profoundly asymmetrical. For example, the left hemisphere is more important for right-hand control, spoken and written language, numerical and scientific skills, and reasoning. The right hemisphere is more important for left-hand control, musical and artistic awareness, space and pattern perception, insight, imagination, and generating mental images of sight, sound, touch, taste, and smell in order to compare relationships.

In everyday life this lateralization of function can seldom be detected because information is readily passed between the hemispheres through several commissures, including the corpus callosum.

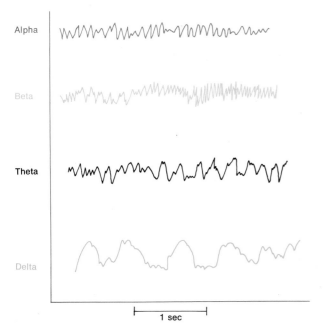

FIGURE 14-12 Kinds of waves recorded in an electroencephalogram (EEG).

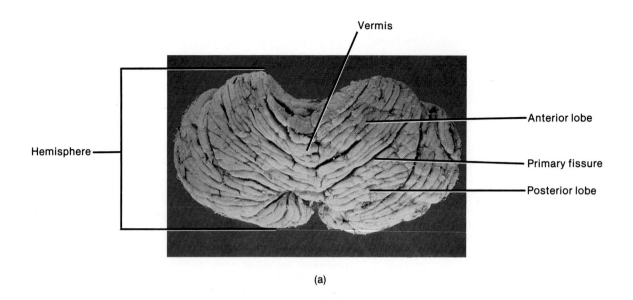

(a)

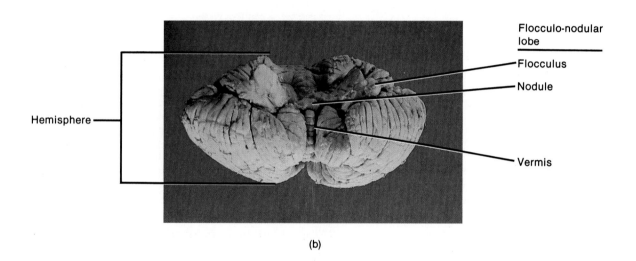

(b)

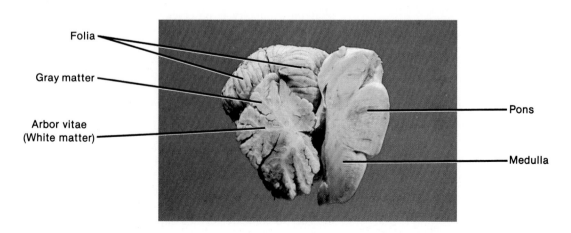

(c)

FIGURE 14-13 Cerebellum. (a) Superior view. (b) Posteroinferior view. (c) Viewed in sagittal section. (Courtesy of M. L. Barr, *The Human Nervous System,* 2nd ed., Harper & Row, New York, 1974.)

One of the most recent findings is that different emotional reactions result from damage to the right and left sides of the brain. Lesions in most areas on the left side are accompanied by the feelings of loss that might be expected as a result of any serious injury, and the patient is disturbed and depressed. Damage in much of the right hemisphere sometimes leaves the patient unconcerned with his condition.

A common cerebral dominance and also one of the most puzzling is the phenomenon of handedness. In the human population, no more than 9 percent are left-handed. This considerable bias toward right-handedness may represent a unique specialization of the human brain. Studies indicate that the distribution of brain asymmetries in left-handed people is different from that in right-handed people.

Scientists are testing the possibility that faulty connections between the hemispheres may be a cause of dyslexia and that other peculiarities in communication between the hemispheres may sometimes be clues to psychiatric symptoms and may also be involved in creativity and invention.

CEREBELLUM

The **cerebellum** is the second-largest portion of the brain (almost ⅛ of the brain's mass) and occupies the inferior and posterior aspects of the cranial cavity. Specifically, it is posterior to the medulla and pons and below the occipital lobes of the cerebrum. It is separated from the cerebrum by the **transverse fissure** (see Figure 14-1) and by an extension of the cranial dura mater called the **tentorium** (*tentorium* = tent) **cerebelli** (see Figure 14-7e).

The cerebellum is shaped somewhat like a butterfly. The central constricted area is the **vermis,** which means worm-shaped, and the lateral "wings" or lobes are referred to as **hemispheres** (Figure 14-13). Each hemisphere consists of lobes that are separated by deep and distinct fissures. The **anterior lobe** and **posterior lobe** are concerned with subconscious movements of skeletal muscles. The **flocculonodular lobe** is concerned with the sense of equilibrium (Chapter 17). Between the hemispheres is another extension of the cranial dura mater: the **falx cerebelli.** It passes only a short distance between the cerebellar hemispheres.

The surface of the cerebellum, called the **cortex,** consists of gray matter in a series of slender, parallel ridges called **folia.** They are less prominent than the gyri of the cerebral cortex. Beneath the gray matter are **white matter tracts (arbor vitae)** that resemble branches of a tree. Deep within the white matter are masses of gray matter, the **cerebellar nuclei.**

The cerebellum is attached to the brain stem by three paired bundles of fibers called **cerebellar peduncles** (see Figure 14-3d,e). **Inferior cerebellar peduncles** connect the cerebellum with the medulla at the base of the brain stem and with the spinal cord. **Middle cerebellar peduncles** connect the cerebellum with the pons. **Superior cerebellar peduncles** connect the cerebellum with the midbrain.

The cerebellum is a motor area of the brain concerned with certain subconscious movements in the skeletal muscles. These movements are required for coordination, maintenance of posture, and balance. The cerebellar peduncles are the fiber tracts that allow the cerebellum to perform its functions.

Let us now see how the cerebellum produces coordinated movement. Motor areas of the cerebral cortex voluntarily initiate muscle contraction. Once the movement has begun, the sensory areas of the cortex receive impulses from nerves in the joints. The impulses provide information about the extent of muscle contraction and the amount of joint movement. The term *proprioception* is applied to this sense of the position of one body part relative to another. The cerebral cortex uses the proprioceptive sensations to determine which muscles are required to contract next and with what strength they are to contract in order to continue moving in the desired direction. Then a pattern of impulses is generated by the cerebral cortex along tracts to the pons and midbrain, which relay the impulses over the middle and superior cerebellar peduncles to the cerebellum. The cerebellum then generates subconscious motor impulses along the inferior cerebellar peduncles to the medulla and spinal cord. The impulses pass downward along the spinal cord and out the nerves that stimulate the prime movers and synergists to contract and that inhibit the contraction of the antagonists. The result is a smooth, coordinated movement. A well-functioning cerebellum is essential for delicate movements such as playing the piano.

The cerebellum also transmits impulses that control postural muscles, that is, it maintains normal muscle tone. The cerebellum also maintains body equilibrium. The inner ear contains structures that sense balance. Information such as whether the body is leaning to the left or right is transmitted from the inner ear to the cerebellum. The cerebellum then discharges impulses that cause the contraction of the muscles necessary for maintaining equilibrium.

There is some evidence that the cerebellum may play a role in a person's emotional development, modulating sensations of anger and pleasure.

CLINICAL APPLICATION
Damage to the cerebellum through trauma or disease is characterized by certain symptoms involving skeletal muscles on the same side of the body as the damage. The effects are ipsilateral because of a double-crossing of tracts within the cerebellum. There may be lack of muscle coordination, called *ataxia* (*a* = without; *taxis* = order). Blindfolded people with ataxia cannot touch the tip of their nose with a finger because they cannot coordinate movement with their sense of where a body part is located. Another sign of ataxia is a change in the speech pattern due to a lack of coordination of speech muscles. Cerebellar damage may also result in *disturbances of gait,* in which the subject staggers or cannot coordinate normal walking movements, and *severe dizziness.*

TRANSMITTER SUBSTANCES IN THE BRAIN

There are over 40 different substances that are either known or suspected transmitter substances in the brain. These substances can facilitate, excite, or inhibit post-synaptic neurons. They establish the lines of communication between brain cells and are localized in certain parts of the brain. In Chapter 12, we noted that **acetylcholine (ACh)** is released at some neuromuscular junctions, at some neuroglandular junctions, and at synapses between certain brain and spinal cord cells. In most parts of the body, ACh leads to excitation. **Norepinephrine (NE),** another transmitter that leads to excitation, is released at some neuromuscular junctions and some neuroglandular junctions. It is also concentrated in a group of neurons in the brain stem near the fourth ventricle called the *locus coeruleus* (LŌ-kus sē-ROO-lē-us). This area, which literally means blue place, projects axons into the hypothalamus, cerebellum, and cerebral cortex. In these places, NE has been implicated in maintaining arousal (awakening from deep sleep), dreaming, and the regulation of mood. Neurons containing the transmitter **dopamine (DA)** are clustered in the midbrain in the substantia nigra. DA leads to excitation. Some axons projecting from the substantia nigra terminate in the cerebral cortex, where DA is thought to be involved in emotional responses. Other axons project to the corpus striatum (basal ganglia), where the DA is involved in gross subconscious movements of skeletal muscles. The degeneration of axons in the basal ganglia results in Parkinson's disease, which is discussed at the end of this chapter. **Serotonin (5-HT)** is a transmitter that leads to excitation and is concentrated in the neurons in a part of the brain stem called the *raphe nucleus.* Axons projecting from the nucleus terminate in the hypothalamus, thalamus, and other regions of the brain. Serotonin is thought to be involved in inducing sleep, sensory perception, and temperature regulation. The amino acids **glutamic acid** and **aspartic acid** are also believed to lead to excitation in the brain.

The most common transmitter in the brain leading to inhibition is **gamma aminobutyric acid (GABA).** It has been implicated as a likely target for antianxiety drugs such as diazempan (Valium). These drugs enhance the action of GABA. Another transmitter that leads to inhibition in the spinal cord is **glycine.**

In recent years, another group of chemical messengers in the brain has been identified. These are known as **peptides** and consist of chains of amino acids that occur naturally in the brain. In 1975 the first peptides, referred to as **enkephalins** (en-KEF-a-lins; *en* = without; *keph* = head), were discovered. These chemicals are similar in structure to morphine, the painkiller derived from the opium poppy. Enkephalins are concentrated in the thalamus, in parts of the limbic system, and in those spinal cord pathways that relay impulses for pain. It has been suggested that enkephalins are the body's natural painkillers. They do this by inhibiting impulses in the pain pathway and by binding to the same receptors in the brain as morphine.

Other naturally occurring peptides, called **endorphins** (en-DOR-fins), were subsequently isolated from the pituitary gland. Like the enkephalins, they have morphinelike properties that suppress pain. Endorphins have also been linked to memory and learning; sexual activity; control of body temperature; regulation of hormones that affect the onset of puberty, sexual drive, and reproduction; and mental illnesses such as depression and schizophrenia. One of the better studied endorphins is *beta-endorphin,* or *β-endorphin.* A peptide that is believed to work with endorphins is known as *substance P.* It is found in sensory nerves, spinal cord pathways, and parts of the brain associated with pain. When substance P is released by neurons, it conducts pain-related nerve impulses from peripheral pain receptors into the central nervous system. It is now suspected that endorphins may exert their analgesic effects by suppressing the release of substance P. Substance P has also been shown to counter the effects of certain nerve-damaging chemicals, prompting speculation that it might prove useful as a treatment for nerve degeneration.

A brain peptide called **dynorphin** (*dynamis* = power), discovered in 1979, is 200 times more powerful in action than morphine and 50 times more powerful than β-endorphin. Its exact functions have yet to be determined. However, the chemical might be a factor in controlling pain and registering emotions.

As research into brain peptides continues, a long list of substances is being drawn up. Interestingly enough, many of these brain peptides are also found in other parts of the body, where they serve as hormones or other regulators of physiological responses. Although we will be discussing many of them in subsequent chapters, a few will be mentioned here to give you some idea of the diversity of the peptides.

   1. **Angiotensin.** This is a substance that can raise blood pressure and is found in the bloodstream as well. It is produced by a kidney enzyme called renin. Some evidence suggests that angiotensin in the brain may be part of the brain's mechanism for regulating its own blood pressure.
   2. **Cholecystokinin** (ko'-lē-sis'-tō-KĪN-in). This peptide hormone (also called pancreozymin) is produced by the lining of the small intestine and causes the pancreas to release digestive juice and the gallbladder to release bile. Although the significance of cholecystokinin in the brain is unclear, there are some suggestions that it may be related to the control of feeding.
   3. **Neurotensin.** This substance helps to regulate blood glucose level by its action on glucose and glucagon. It is also possible that neurotensin may be involved in pain pathways.
   4. **Regulating factors.** These are chemicals produced by the hypothalamus that regulate the release of hormones by the pituitary gland.

CRANIAL NERVES

Of the 12 pairs of **cranial nerves,** 10 originate from the brain stem, but all leave the skull through foramina of the skull (see Figure 14-3a). The cranial nerves are designated with Roman numerals and with names. The Roman numerals indicate the order in which the nerves arise from the brain (front to back). The names indicate the distribution or function. Cranial nerves that contain both sensory and motor fibers are termed *mixed nerves.* Other cranial nerves contain sensory fibers only. The cell bodies of sensory fibers are located in ganglia outside the brain. The cell bodies of motor fibers lie in nuclei within the brain.

Some motor fibers control subconscious movements, yet the somatic nervous system has been defined as a *conscious* system. The reason for this apparent contradiction is that some fibers of the autonomic nervous system leave the brain bundled together with somatic fibers of the cranial nerves, as is the case for spinal nerves. Therefore, subconscious functions transmitted by the autonomic fibers are described along with the conscious functions of the somatic fibers of the cranial nerves.

Although the cranial nerves are mentioned singly in the following description of their type, location, and function, remember that they are paired structures.

OLFACTORY (I)

The **olfactory nerve** is entirely sensory and conveys impulses related to smell. It arises as bipolar neurons from the olfactory mucosa of the nasal cavity (see Figure 17-1b). The dendrites and cell bodies of these neurons are generally limited to the mucosa covering the superior nasal conchae and the adjacent nasal septum. Axons from the neurons pass through the cribriform plate of the ethmoid bone and synapse with other olfactory neurons in the *olfactory bulb,* an extension of the brain lying above the cribriform plate. The axons of these neurons make up the *olfactory tract.* The fibers from the tract terminate in the primary olfactory area in the cerebral cortex.

OPTIC (II)

The **optic nerve** is entirely sensory and conveys impulses related to vision. Impulses initiated by rods and cones of the retina are relayed by bipolar neurons to ganglion cells (see Figures 17-4 and 17-9). Axons of the ganglion cells, the optic nerve fibers, exit the optic foramen, after which the two optic nerves unite to form the *optic chiasma* (kī-AZ-ma). Within the chiasma, fibers from the medial half of each retina cross to the opposite side; those from the lateral half remain on the same side. From the chiasma, the regrouped fibers pass posteriorly to the *optic tracts.* From the optic tracts, the majority of fibers terminate in a nucleus (lateral geniculate) in the thalamus. They then synapse with neurons that pass to the visual areas of the cerebral cortex. Some fibers from the optic chiasma terminate in the superior colliculi of the midbrain. They synapse with neurons whose fibers terminate in the nuclei that convey impulses to the oculomotor (III), trochlear (IV), and abducens (VI) nerves—nerves that control the extrinsic (external) and intrinsic (internal) eye muscles. Through this relay, there are widespread motor responses to light stimuli.

OCULOMOTOR (III)

The **oculomotor nerve** is a mixed cranial nerve. It originates from neurons in a nucleus in the ventral portion of the midbrain. It runs forward, divides into superior and inferior divisions, both of which pass through the superior orbital fissure into the orbit. The superior branch is distributed to the superior rectus (an extrinsic eyeball muscle) and the levator palpebrae superioris (the muscle of the upper eyelid). The inferior branch is distributed to the medial rectus, inferior rectus, and inferior oblique muscles—all extrinsic eyeball muscles. These distributions to the levator palpebrae superioris and extrinsic eyeball muscles constitute the motor portion of the oculomotor nerve. Through these distributions, impulses are sent that control movements of the eyeball and upper eyelid.

The inferior branch of the oculomotor nerve also sends a branch to the *ciliary ganglion,* a relay center of the autonomic nervous system that connects a nucleus in the midbrain with the intrinsic eyeball muscles. These intrinsic muscles include the ciliary muscle of the eyeball and the sphincter muscle of the iris. Through the ciliary ganglion, the oculomotor nerve controls the smooth muscle (ciliary muscle) responsible for accommodation of the lens for near vision and the smooth muscle (sphincter muscle of iris) responsible for constriction of the pupil.

The sensory portion of the oculomotor nerve consists of afferent fibers from proprioceptors in the eyeball muscles supplied by the nerve to the midbrain. These fibers convey impulses related to muscle sense (proprioception).

Although the oculomotor (III), trochlear (IV), accessory (XI), and hypoglossal (XII) cranial nerves are referred to as mixed nerves because they contain fibers from proprioceptors, they are primarily motor, serving to stimulate skeletal muscle contractions.

TROCHLEAR (IV)

The **trochlear** (TROK-lē-ar) **nerve** is a mixed cranial nerve. It is the smallest of the 12 cranial nerves. The motor portion originates in a nucleus in the midbrain, and axons from the nucleus pass through the superior orbital fissure of the orbit. The motor fibers innervate the superior oblique muscle of the eyeball, another extrinsic eyeball muscle. It controls movement of the eyeball.

The sensory portion of the trochlear nerve consists of afferent fibers that run from proprioceptors in the superior oblique muscle to the nucleus of the nerve in the midbrain. The sensory portion is responsible for muscle sense.

TRIGEMINAL (V)

The **trigeminal nerve** is a mixed cranial nerve and the largest of the cranial nerves. As indicated by its name, the trigeminal nerve has three branches: ophthalmic (of-THAL-mik), maxillary, and mandibular. The trigeminal nerve contains two roots on the ventrolateral surface of the pons. The large sensory root has a swelling called the *semilunar (gasserian) ganglion* located in a fossa on the inner surface of the petrous portion of the temporal bone. From this ganglion, the **ophthalmic branch** enters the orbit via the superior orbital fissure, the **maxillary branch** enters the foramen rotundum, and the **mandibular branch** exits through the foramen ovale. The smaller motor root originates in a nucleus in the pons. The motor fibers join the mandibular branch and supply the muscles of mastication. These motor fibers, which control chewing movements, constitute the motor portion of the trigeminal nerve.

The sensory portion of the trigeminal nerve delivers impulses related to touch, pain, and temperature and consists of the ophthalmic, maxillary, and mandibular branches. The ophthalmic branch receives sensory fibers from the skin over the upper eyelid, eyeball, lacrimal glands, upper part of the nasal cavity, side of the nose, forehead, and anterior half of the scalp. The maxillary branch receives sensory fibers from the mucosa of the nose, palate, parts of the pharynx, upper teeth, upper lip, cheek, and lower eyelid. The mandibular branch transmits sensory fibers from the anterior two-thirds of the tongue (not taste), lower teeth, skin over the mandible and side of the head in front of the ear, and mucosa of the floor of the mouth. Sensory fibers from the three branches of the trigeminal nerve enter the semilunar ganglion and terminate in a nucleus in the pons. There are also sensory fibers from proprioceptors in the muscles of mastication.

CLINICAL APPLICATION

The inferior alveolar nerve, a branch of the mandibular nerve, supplies all the teeth in one half of the mandible and is frequently **anesthetized in dental procedures.** The same procedure will anesthetize the lower lip because the mental nerve is a branch of the inferior alveolar nerve. Since the lingual nerve runs very close to the inferior alveolar near the mental foramen, it too is often anesthetized at the same time. For anesthesia to the upper teeth, the superior alveolar nerve endings, branches of the maxillary branch, are blocked by inserting the needle beneath the mucous membrane; the anesthetic solution is then infiltrated slowly throughout the area of the roots of the teeth to be treated.

ABDUCENS (VI)

The **abducens** (ab-DOO-sens) **nerve** is a mixed cranial nerve that originates from a nucleus in the pons. The motor fibers extend from the nucleus to the lateral rectus muscle of the eyeball, an extrinsic eyeball muscle. Impulses over the fibers bring about movement of the eyeball. The sensory fibers run from proprioceptors in the lateral rectus muscle to the pons and mediate muscle sense. The abducens nerve reaches the lateral rectus muscle through the superior orbital fissure of the orbit.

FACIAL (VII)

The **facial nerve** is a mixed cranial nerve. Its motor fibers originate from a nucleus in the pons, enter the petrous portion of the temporal bone, and are distributed to facial, scalp, and neck muscles. Impulses along these fibers cause contraction of the muscles of facial expression. Some motor fibers are also distributed to the lacrimal, sublingual, submandibular, nasal, and palatine glands.

The sensory fibers extend from the taste buds of the anterior two-thirds of the tongue to the *geniculate ganglion,* a swelling of the facial nerve. From here, the fibers pass to a nucleus in the pons, which sends fibers to the thalamus for relay to the gustatory area of the cerebral cortex. The sensory portion of the facial nerve also conveys deep general sensations from the face. There are

also sensory fibers from proprioceptors in the muscles of the face and scalp.

VESTIBULOCOCHLEAR (VIII)

The **vestibulocochlear** (ves-tib'-yoo-lō-KŌK-lē-ar) **nerve,** formerly known as the **acoustic nerve,** is another sensory cranial nerve. It consists of two branches: the cochlear (auditory) branch and the vestibular branch. The **cochlear branch,** which conveys impulses associated with hearing, arises in the spiral organ in the cochlea of the internal ear. The cell bodies of the cochlear branch are located in the *spiral ganglion* of the cochlea. From here the axons pass through a nucleus in the medulla and terminate in the medial geniculate nucleus in the thalamus. Ultimately, the fibers synapse with neurons that relay the impulses to the auditory areas of the cerebral cortex.

The **vestibular branch** arises in the semicircular canals, the saccule, and the utricle of the inner ear. Fibers from the semicircular canals, saccule, and utricle extend to the *vestibular ganglion,* where the cell bodies are contained. The cell bodies of the fibers synapse in the ganglion with fibers that extend to a nucleus in the medulla and pons and terminate in the thalamus. Some fibers also enter the cerebellum. The vestibular branch transmits impulses related to equilibrium.

GLOSSOPHARYNGEAL (IX)

The **glossopharyngeal** (glos'-ō-fa-RIN-jē-al) **nerve** is a mixed cranial nerve. Its motor fibers originate in a nucleus in the medulla. The nerve exits the skull through the jugular foramen. The motor fibers are distributed to the swallowing muscles of the pharynx and the parotid gland to mediate swallowing movements and the secretion of saliva.

The sensory fibers of the glossopharyngeal nerve supply the pharynx and taste buds of the posterior third of the tongue. Some sensory fibers also originate from receptors in the carotid sinus, which assumes a major role in blood pressure regulation. The sensory fibers terminate in a nucleus in the thalamus. There are also sensory fibers from proprioceptors in the muscles innervated by this nerve.

VAGUS (X)

The **vagus nerve** is a mixed cranial nerve that is widely distributed from the head and neck into the thorax and abdomen. Its motor fibers originate in a nucleus of the medulla and terminate in the muscles of the pharynx, larynx, respiratory passageways, lungs, heart, esophagus, stomach, small intestine, most of the large intestine, and gallbladder. Impulses along the motor fibers generate visceral, cardiac, and skeletal muscle movement.

Sensory fibers of the vagus nerve supply essentially the same structures as the motor fibers. They convey impulses for various sensations from the larynx, the viscera, and the ear. The fibers terminate in the medulla and pons. There are also sensory fibers from proprioceptors in the muscles supplied by this nerve.

ACCESSORY (XI)

The **accessory nerve** (formerly the **spinal accessory nerve**) is a mixed cranial nerve. It differs from all other cranial nerves in that it originates from both the brain stem and the spinal cord. The **bulbar (medullary) portion** originates from nuclei in the medulla, passes through the jugular foramen, and supplies the voluntary muscles of the pharynx, larynx, and soft palate that are used in swallowing. The **spinal portion** originates in the anterior gray horn of the first five segments of the cervical portion of the spinal cord. The fibers from the segments join, enter the foramen magnum, and exit through the jugular foramen along with the bulbar portion.

The spinal portion conveys motor impulses to the sternocleidomastoid and trapezius muscles to coordinate head movements. The sensory fibers originate from propriocep-tors in the muscles supplied by its motor neurons and terminate in upper cervical posterior root ganglia.

HYPOGLOSSAL (XII)

The **hypoglossal nerve** is a mixed cranial nerve. The motor fibers originate in a nucleus in the medulla, pass through the hypoglossal canal, and supply the muscles of the tongue. These fibers conduct impulses related to speech and swallowing.

The sensory portion of the hypoglossal nerve consists of fibers originating from proprioceptors in the tongue muscles and terminating in the medulla. The sensory fibers conduct impulses for muscle sense.

A summary of cranial nerves is presented in Exhibit 14-2.

EXHIBIT 14-2
SUMMARY OF CRANIAL NERVES

NERVE	LOCATION	FUNCTION AND CLINICAL APPLICATION
Olfactory (I)	Arises in olfactory mucosa, passes through olfactory bulb and olfactory tract, and terminates in primary olfactory areas of cerebral cortex.	Smell: Loss of the sense of smell, called *anosmia,* may result from head injuries in which the cribriform plate of the ethmoid bone is fractured and from lesions along the olfactory pathway.
Optic (II)	Arises in retina of the eye, forms optic chiasma, passes through optic tracts, lateral geniculate nucleus in thalamus, and terminates in visual areas of cerebral cortex.	Vision: Fractures in the orbit, lesions along the visual pathway, and diseases of the nervous system may result in visual field defects and loss of visual acuity. A defect of vision is called *anopsia.*
Oculomotor (III)	Motor portion: originates in midbrain and is distributed to levator palpebrae superioris of upper eyelid, four extrinsic eyeball muscles (superior rectus, medial rectus, inferior rectus, and inferior oblique), ciliary muscle of eyeball, and sphincter muscle of iris. Sensory portion: consists of afferent fibers from proprioceptors in eyeball muscles and terminates in midbrain.	Motor: movement of eyelid and eyeball, accommodation of lens for near vision, and constriction of pupil. Sensory: muscle sense (proprioception). A lesion in the nerve causes *strabismus* (squinting), *ptosis* (drooping) of the upper eyelid, pupil dilation, the movement of the eyeball downward and outward on the damaged side, a loss of accommodation for near vision, and double vision (*diplopia*).
Trochlear (IV)	Motor portion: originates in midbrain and is distributed to superior oblique muscle, an extrinsic eyeball muscle. Sensory portion: consists of afferent fibers from proprioceptors in superior oblique muscles and terminates in midbrain.	Motor: movement of eyeball. Sensory: muscle sense (proprioception). In trochlear nerve paralysis, the head is tilted to the affected side and diplopia and strabismus occur.
Trigeminal (V)	Motor portion: originates in pons and terminates in muscles of mastication. Sensory portion: consists of three branches: *ophthalmic*—contains sensory fibers from skin over upper eyelid, eyeball, lacrimal glands, nasal cavity, side of nose, forehead, and anterior half of scalp; *maxillary*—contains sensory fibers from mucosa of nose, palate, parts of pharynx, upper teeth, upper lip, cheek, and lower eyelid; *mandibular*—contains sensory fibers from anterior two-thirds of tongue, lower teeth, skin over mandible, and side of	Motor: chewing. Sensory: conveys sensations for touch, pain, and temperature from structures supplied; muscle sense (proprioception). Injury results in paralysis of the muscles of mastication and a loss of sensation of touch and temperature. *Neuralgia* (pain) of one or more branches of trigeminal nerve is called *trigeminal neuralgia* (*tic douloureux*).

EXHIBIT 14-2
(*Continued*)

NERVE	LOCATION	FUNCTION AND CLINICAL APPLICATION
	head in front of ear. The three branches terminate in pons. Sensory portion also consists of afferent fibers from proprioceptors in muscles of mastication.	
Abducens (VI)	Motor portion: originates in pons and is distributed to lateral rectus muscle, an extrinsic eyeball muscle. Sensory portion: consists of afferent fibers from proprioceptors in lateral rectus muscle and terminates in pons.	Motor: movement of eyeball. Sensory: muscle sense (proprioception). With damage to this nerve, the affected eyeball cannot move laterally beyond the midpoint and the eye is usually directed medially.
Facial (VII)	Motor portion: originates in pons and is distributed to facial, scalp, and neck muscles and to lacrimal, sublingual, submandibular, nasal, and palatine glands. Sensory portion: arises from taste buds on anterior two-thirds of tongue, passes through geniculate ganglion, a nucleus in pons that sends fibers to thalamus for relay to gustatory areas of cerebral cortex. Also consists of afferent fibers from proprioceptors in muscles of face and scalp.	Motor: facial expression and secretion of saliva and tears. Sensory: taste; muscle sense (proprioception). Injury produces paralysis of the facial muscles, called *Bell's palsy,* loss of taste, and the eyes remain open, even during sleep.
Vestibulocochlear (VIII)	Cochlear branch: arises in spiral organ, forms spiral ganglion, passes through a nucleus in the medulla, and terminates in thalamus. Fibers synapse with neurons that relay impulses to auditory areas of cerebral cortex. Vestibular branch: arises in semicircular canals, saccule, and utricle and forms vestibular ganglion; fibers pass through medulla and pons and terminate in thalamus.	Cochlear branch: conveys impulses associated with hearing. Vestibular branch: conveys impulses associated with equilibrium. Injury to the cochlear branch may cause *tinnitus* (ringing) or deafness. Injury to the vestibular branch may cause *vertigo* (a subjective feeling of rotation), *ataxia,* and *nystagmus* (involuntary rapid movement of the eyeball).
Glossopharyngeal (IX)	Motor portion: originates in medulla and is distributed to swallowing muscles of pharynx and to parotid gland. Sensory portion: arises from taste buds on posterior one-third of tongue and from carotid sinus and terminates in thalamus. Also consists of afferent fibers from proprioceptors in swallowing muscles supplied.	Motor: swallowing movements and secretion of saliva. Sensory: taste and regulation of blood pressure; muscle sense (proprioception). Injury results in pain during swallowing, reduced secretion of saliva, loss of sensation in the throat, and loss of taste.
Vagus (X)	Motor portion: originates in medulla and terminates in muscles of pharynx, larynx, respiratory passageways, lungs, esophagus, heart, stomach, small intestine, most of large intestine, and gallbladder. Sensory portion: arises from essentially same structures supplied by motor fibers and terminates in medulla and pons. Also consists of afferent fibers from proprioceptors in muscles supplied.	Motor: visceral muscle movement and swallowing movements. Sensory: sensations from organs supplied; muscle sense (proprioception). Severing of both nerves in the upper body interferes with swallowing, paralyzes vocal cords, and interrupts sensations from many organs. Injury to both nerves in the abdominal area has little effect, since the abdominal organs are also supplied by autonomic fibers from the spinal cord.
Accessory (XI)	Motor portion: consists of a bulbar portion and a spinal portion. Bulbar portion originates from medulla and supplies voluntary muscles of pharynx, larynx, and soft palate. Spinal portion originates from anterior gray horn of first five cervical segments of spinal cord and supplies sternocleidomastoid and trapezius muscles. Sensory portion: consists of afferent fibers from proprioceptors in muscles supplied.	Motor: bulbar portion mediates swallowing movements; spinal portion mediates movements of head. Sensory: muscle sense (proprioception). If damaged, the sternocleidomastoid and trapezius muscles become paralyzed, with resulting inability to turn the head or raise the shoulders.

NERVE	LOCATION	FUNCTION AND CLINICAL APPLICATION
Hypoglossal (XII)	Motor portion: originates in medulla and supplies muscles of tongue. Sensory portion: consists of fibers from proprioceptors in tongue muscles that terminate in medulla.	Motor: movement of tongue during speech and swallowing. Sensory: muscle sense (proprioception). Injury results in difficulty in chewing, speaking, and swallowing. The tongue, when protruded, curls toward the affected side and the affected side becomes atrophied, shrunken, and deeply furrowed.

DISORDERS: HOMEOSTATIC IMBALANCES

Many disorders can affect the central nervous system. Some are caused by viruses or bacteria. Others are caused by damage to the nervous system during birth. The origins of many conditions, however, are unknown. Here we discuss the origins and symptoms of some common central nervous system disorders.

Poliomyelitis

Poliomyelitis, also known as **infantile paralysis** or simply **polio,** is a viral infection that is most common during childhood. The causative agent is a virus called poliovirus. The onset of the disease is marked by fever, severe headache, a stiff neck and back, deep muscle pain and weakness, and loss of certain somatic reflexes. The virus can be ingested in drinking water contaminated with feces containing the virus. It may affect almost all parts of the body. In its most serious form, called **bulbar polio,** the virus spreads via blood to the central nervous system, where it destroys the motor nerve cell bodies, specifically those in the anterior horns of the spinal cord and in the nuclei of the cranial nerves. Injury to the spinal gray matter is the basis for the name of this disease (*polio* = gray matter; *myel* = spinal cord). Destruction of the anterior horns produces paralysis. The first sign of bulbar polio is difficulty in swallowing, breathing, and speaking. Poliomyelitis can cause death from respiratory or heart failure if the virus invades the brain cells of the vital medullary centers. The incidence of polio in the United States has decreased markedly since the availability of polio vaccines (Salk vaccine and more recently Sabin vaccine).

Cerebral Palsy

The term **cerebral palsy** refers to a group of motor disorders caused by damage to the motor areas of the brain during fetal life, birth, or infancy. One cause is infection of the mother with German measles during the first three months of pregnancy. During early pregnancy, certain cells in the fetus are dividing and differentiating in order to lay down the basic structures of the brain. These cells can be abnormally changed by toxin from the measles virus. Radiation during fetal life, temporary oxygen starvation during birth, and hydrocephalus during infancy may also damage brain cells.

Cases of cerebral palsy are categorized into three groups depending on whether the cortex, the basal ganglia of the cerebrum, or the cerebellum is affected most severely. Most cerebral palsy victims have at least some damage in all three areas. The location and extent of motor damage determine the symptoms. The victim may be deaf or partially blind. About 70 percent of cerebral palsy victims appear to be mentally retarded. The apparent mental slowness, however, is often due to the person's inability to speak or hear well. Such individuals are often more mentally acute than they appear.

Cerebral palsy is not a progressive disease; it does not worsen as time elapses. Once the damage is done, however, it is irreversible.

Parkinsonism

This disorder, also called **Parkinson's disease,** is a progressive degeneration of the basal ganglia of the cerebrum. The basal ganglia regulate subconscious contractions of skeletal muscles that aid activities also consciously controlled by the motor areas of the cerebral cortex—swinging the arms when walking, for example. In Parkinsonism, neurons that release the excitatory transmitter dopamine (DA) degenerate. Diminished levels of DA cause unnecessary skeletal muscle movements that often interfere with voluntary movement. For instance, the muscles of the upper extremities may alternately contract and relax, causing the hands to shake. This shaking is called *tremor.* Other muscles may contract continuously, causing *rigidity* of the involved body part. Rigidity of the facial muscles gives the face a masklike appearance. The expression is characterized by a wide-eyed, unblinking stare and a slightly open mouth with uncontrolled drooling. Decreased dopamine production also results in *akinesia* (*a* = without; *kinesis* = motion) and *bradykinesia* (*brady* = slow). Vision, hearing, and intelligence are unaffected by the disorder, indicating that Parkinsonism does not attack the cerebral cortex.

Although people with Parkinsonism do not manufacture enough DA, injections of it are useless; the blood–brain barrier stops it. However, symptoms are somewhat relieved by a drug developed in the 1960s called levodopa (L-dopa). Administered by itself, levodopa may elevate brain levels of dopamine, causing undesirable side effects such as low blood pressure, nausea, mental changes, and liver dysfunction. In recent years, levodopa has been combined with carbidopa which inhibits the formation of dopamine outside the brain. The combined drugs diminish the undesirable side effects.

Multiple Sclerosis (MS)

Multiple sclerosis (MS) is the progressive destruction of the myelin sheaths of neurons in the central nervous system accompanied by disappearance of oligodendrocytes and the proliferation of astrocytes. The sheaths deteriorate to *scleroses,* which are hardened scars or plaques, in multiple regions. The destruction of myelin sheaths interferes with the transmission of impulses from one neuron to another, literally short-circuiting conduction pathways. Usually the first symptoms occur between the ages of 20 and 40. Early symptoms are generally produced by the formation of a few plaques and are, consequently, mild. Plaque formation in the cerebellum may produce lack of coordination in one hand. The patient's handwriting becomes strained and irregular. A short-circuiting of pathways in the corticospinal tract may partially paralyze the leg muscles so that the patient drags a foot when walking. Other early symptoms include double vision and urinary tract infections. Following a period of remission during which the symptoms temporarily disappear, a new series of plaques develop and the victim suffers a second attack. One attack follows another over the years. Each time the plaques form, some neurons are damaged by the hardening of their sheaths, while others are uninjured by their plaques. The result is a progressive loss of function interspersed with remission periods during which the undamaged neurons regain their ability to transmit impulses.

The symptoms of MS depend on the areas of the central nervous system most heavily laden with plaques. Sclerosis of the white matter of the spinal cord is common. As the sheaths of the neurons in the corticospinal tract deteriorate, the patient loses the ability to contract skeletal muscles. Damage to the ascending tracts produces numbness and short-circuits impulses related to position of body parts and flexion of joints. Damage to either set of tracts also destroys spinal cord reflexes.

As the disease progresses, most voluntary motor control is eventually lost and the patient becomes bedridden. Death occurs anywhere from 7 to 30 years after the first symptoms appear. The usual cause of death is a severe infection resulting from the loss of motor activity. Without the constricting action of the urinary bladder wall, for example, the bladder never totally empties and stagnant urine provides an environment for bacterial growth. Bladder infection may then spread to the kidney, damaging kidney cells.

Although the etiology of MS is unclear, there is increasing evidence that it might result from a viral infection that precipitates an autoimmune response. Viruses may trigger the destruction of oligodendrocytes by the antibodies and killer cells of the body's immune system. Like other demyelinating diseases, MS is incurable. However, in view of the evidence that it might be an autoimmune disease, immunosuppressive therapy is widely used (glucocorticoids such as prednisone). Treatment is also directed at management of complications such as spasticity, facial neuralgia and twitching, urinary bladder problems, and constipation. Electrical stimulation of the spinal cord can also improve function in certain patients.

Epilepsy

Epilepsy is the second most common neurological disorder after stroke. It is characterized by short, recurrent, periodic attacks of motor, sensory, or psychological malfunction. The attacks, called *epileptic seizures,* are initiated by abnormal and irregular discharges of electricity from millions of neurons in the brain. The discharges stimulate many of the neurons to send impulses over their conduction pathways. As a result, a person undergoing an attack may contract skeletal muscles involuntarily. Lights, noise, or smells may be sensed when the eyes, ears, and nose actually have not been stimulated. The electrical discharges may also inhibit certain brain centers. For instance, the waking center in the brain may be depressed so that the person loses consciousness.

Many different types of epileptic seizures exist. The particular type of seizure depends on the area of the brain that is electrically stimulated and whether the stimulation is restricted to a small area or spreads throughout the brain. *Grand mal* seizures are initiated by a burst of electrical discharges that travel throughout the motor areas and spread to the areas of consciousness in the brain. The person loses consciousness, has spasms of the voluntary muscles, and may also lose urinary and bowel control. Sensory and intellectual areas may also be involved. For instance, just as the attack begins, the person may sense a peculiar taste in the mouth or see flashes of light or have olfactory hallucinations. This is called an aura and is a warning that may allow the person to lie down and avoid injury. The unconsciousness and motor activity last a few minutes. Then the muscles relax, and the person awakens. Afterward, the individual may be mentally confused for a short period of time. Studies with EEGs show that grand mal attacks are characterized by a rapid rate of 25 to 30 brain waves per second. The normal adult rate is 10 waves per second.

Many epileptics suffer from electrical discharges that are restricted to one or several relatively small areas of the brain. An example is the *petit mal* form, which apparently involves the thalamus and hypothalamus. Petit mal seizures are characterized by an abnormally slow brain wave pattern of about 3 waves per second. The person may lose contact with the environment for about 5 to 30 seconds, but does not undergo the loss of motor control that is typical of a grand mal seizure. The victim merely seems to be daydreaming. A few people experience several hundred petit mal seizures each day. For them, the chief problems are a loss of productivity in school or work and periodic inattentiveness while driving a car.

A form of epilepsy that is sometimes confused with mental illness is *psychomotor epilepsy.* The electrical outburst occurs in the temporal lobe, where it causes the person to lose contact with reality.

The causes of epilepsy are varied. Many conditions can cause nerve cells to produce periodic bursts of impulses. These causes include head injuries, tumors and abscesses of the brain, and childhood infections, such as mumps, whooping cough, and measles. Epilepsy may also be *idiopathic,* that is, have no demonstrable cause. It should be noted that epilepsy almost never affects intelligence. If frequent severe seizures are allowed to occur over a long period of time, however, some cerebral damage may occasionally result. Damage can be prevented by controlling the seizures with drug therapy.

Epileptic seizures can be eliminated or alleviated by drugs that make neurons more difficult to stimulate. Many of these drugs change the permeability of the neuron cell membrane so that it does not depolarize as easily. One such drug is valproic acid, which increases the quantity of the inhibitory transmitter GABA.

Cerebrovascular Accidents (CVAs)

The most common brain disorder is a **cerebrovascular accident (CVA),** also called a **stroke** or **cerebral apoplexy.** A CVA is the destruction of brain tissue (infarction) resulting from disorders in the vessels that supply the brain. Common causes of CVAs are intracerebral hemorrhage from aneurysms, embolism, and atherosclerosis of the cerebral arteries. An *intracerebral hemorrhage* is a rupture of a vessel in the pia mater or brain. Blood seeps into the brain and damages neurons by increasing intracranial fluid pressure. An *embolus* is a blood clot, air bubble, or bit of foreign material, most often debris from an inflammation, that becomes lodged in an artery and blocks circulation. *Atherosclerosis* is the formation of plaques in the artery walls. The plaques may slow down circulation by constricting the vessel. Both emboli and atherosclerosis cause brain damage by reducing the supply of oxygen and glucose needed by brain cells.

Dyslexia

Dyslexia (dis-LEK-sē-a; *dys* = difficulty; *lexis* = words) is unre-

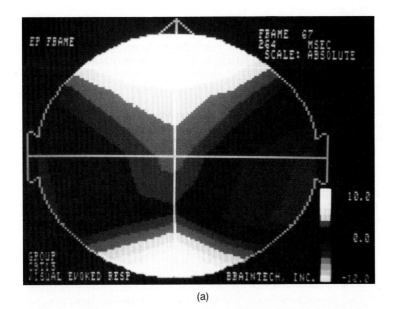

(a)

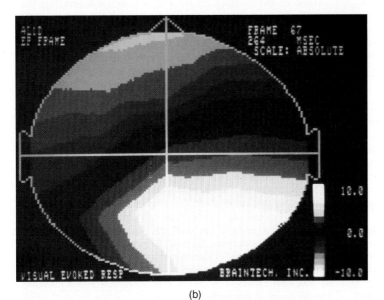

(b)

FIGURE 14-14 Brain electrical activity mapping (BEAM). (a) Image of a visual response (flash stimulus) produced by a normal child. Note the symmetry of the response in the anterior and posterior portions of the brain. (b) Image of a visual response (flash stimulus) produced by a dyslexic child. Note the asymmetry of the response. (Courtesy of Frank H. Duffy, M.D., and Gloria B. McAnulty, Ph.D., The Children's Hospital Medical Center, Boston, Massachusetts.)

lated to basic intellectual capacity, but it causes a mysterious difficulty in handling words and symbols. Apparently some peculiarity in the brain's organizational pattern distorts the ability to read, write, and count. Letters in words seem transposed, reversed, or upside-down—*dog* becomes *god; b* changes identity with *d;* a sign saying "OIL" inverts into "710." Many dyslexics cannot orient themselves in the three dimensions of space and may show bodily awkwardness.

The exact cause of dyslexia is unknown, since it is unaccompanied by outward scars of detectable neurological damage and its symptoms vary from victim to victim. It occurs three times as often among boys as among girls. It has been variously attributed to defective vision, brain damage, lead in the air, physical trauma, or oxygen deprivation during birth. A recent theory holds that it might be related to dysfunction of the vestibular apparatus and semicircular canals of the ear (Chapter 17).

CLINICAL APPLICATION

Now available to physicians is a new technique called **brain electrical activity mapping (BEAM).** It is a noninvasive procedure that measures and displays the electrical activity of the brain on a color television screen and compares the image produced with a normal image (Figure 14-14). BEAM is used primarily to diagnose dyslexia, but may have other diagnostic applications for learning disabilities, schizophrenia, depression, dementia, epilepsy, and early tumor occurrence and recurrence.

Tay-Sachs Disease

Tay-Sachs disease is a central nervous system affliction that brings death before age 5. The Tay-Sachs gene is carried mostly by individuals descended from the Ashkenazi Jews of Eastern Europe. Approximately 1 in 3,600 of their offspring will be afflicted with Tay-Sachs disease. The disease involves the neuronal degeneration of the central nervous system because of excessive amounts of a substance known as ganglioside G_{m2} in the nerve cells of the brain. The substance accumulates because of a deficient lysosomal enzyme. The afflicted child develops normally until the age of 4 to 8 months. Then the symptoms follow a course of progressive degeneration: paralysis, blindness, inability to eat, decubitus, and death from infection. There is no known cure.

Headache

One of the most common human afflictions is **headache** or **cephalgia** (*enkephalos* = brain, *algia* = painful condition). Based upon origin, two general types are distinguished: intracranial and extracranial. Serious headaches of intracranial origin are caused by brain tumors, blood vessel abnormalities, inflammation of the brain or meninges, decrease in oxygen supply to the brain, and damage to brain cells. Extracranial headaches are related to infections of the eyes, ears, nose, and sinuses and are commonly felt as headaches because of the location of these structures. The sinuses are pain sensitive in only one or two areas and do not usually cause pain except in acute sinusitis or with some obvious chronic sinus problem as a tumor or cyst formation or pressure on the conchae. Headaches that accompany sore throats and influenza may arise when the infection produces chemical changes in the blood that irritate pain sensors in the brain. Other extracranial causes include oral complications, occular disease such as glaucoma, and orthopedic disorders such as cervical spine disease.

Trigeminal Neuralgia (Tic Douloureux)

As noted earlier, pain arising from irritation of the trigeminal (V) nerve is known as **trigeminal neuralgia** or **tic douloureux** (doo-loo-ROO). The disorder is characterized by brief but extreme pain in the face and forehead on the affected side. Many patients describe sensitive regions around the mouth and nose that can cause an attack when touched. Eating, drinking, washing the face, and exposure to cold may also bring on an attack.

Rabies

Rabies is an acute infection that may result in fatal encephalitis, an inflammation of the brain. The disease is caused by a virus called rabiesvirus, usually transmitted by the bite of an animal that harbors the virus in its saliva. As the virus multiplies in skeletal muscle and connective tissues, it produces spasms of the muscles of the mouth and pharynx when swallowing liquids. The mere sight or thought of water can trigger the spasms. Encephalitis occurs when the virus migrates along peripheral nerves by axonal transport.

Until recently, antirabies treatment was the Pasteur treatment which consists of 14 to 21 injections of vaccine under the skin of the abdomen over a period of 2 to 3 weeks. This treatment sometimes causes an allergic reaction since the vaccine is prepared from rabbit brain. Today, antirabies treatment consists of administration of a vaccine prepared from human serum called *rabies immune globulin* (*RIG*) followed by vaccination with *human diploid cell vaccine* (*HDCV*). HDCV is prepared from human cell culture and is administered in five intramuscular (deltoid) injections over a 4-week period, with a sixth dose about two months later. Allergic reactions with the new vaccines are minimal.

Reye's Syndrome (RS)

Reye's syndrome (RS), first described in 1963 by the Australian pathologist R. Douglas Reye, seems to occur following a viral infection, particularly chickenpox or influenza. Aspirin at normal doses is believed to be a risk factor in the development of RS. The majority of persons affected are children or teenagers. The disease is characterized by vomiting and brain dysfunction (disorientation, lethargy, and personality changes) and may progress to coma. Also, the liver becomes infiltrated with small lipid droplets and loses some of its ability to detoxify ammonia. The disease runs its course in just a few days.

Brain dysfunction and death are typically caused by swelling of brain cells. The pressure not only kills the cells directly, but also results in hypoxia that kills them indirectly. The mortality rate is about 40 percent. Irreversible brain damage, including mental retardation, can also occur in children who survive, as a result of the swelling. Therapy is directed at controlling the swelling.

Senility

Senility or **dementia** (*de* = without; *mens* = mind) is typically a disorder of the elderly and refers to the development of widespread intellectual impairment, including memory loss, shortened attention span, decreased ability to perform mathematical calculations, and loss of orientation in time and space. In addition, personality changes such as irritability and moodiness often accompany the intellectual deficits. Some elderly people also experience delirium, the abrupt development of dramatic changes involving confusion, hallucinations, and sudden fluctuations in alertness.

Two outstanding hallmarks of senility have been known for years. They are cores of abnormal protein (senile plaques) between nerve cells and twisted fibers (neurofibrillary tangles) in the cell bodies of neurons. Both are prominent in the cerebral cortex and hippocampus, areas of the brain involved in memory and learning. There is evidence that a deficiency in acetylcholine (ACh) might contribute to memory and learning deficits of senility and that excessive amounts of aluminum are implicated in several types of dementia.

MEDICAL TERMINOLOGY

Agnosia (*a* = without: *gnosis* = knowledge) Inability to recognize the significance of sensory stimuli such as auditory, visual, olfactory, gustatory, and tactile.

Analgesia (*an* = without; *algia* = painful condition) Insensibility to pain.

Anesthesia (*esthesia* = feeling) Loss of feeling.

Apraxia (*pratto* = to do) Inability to carry out purposeful movements in the absence of paralysis.

Coma Abnormally deep unconsciousness with an absence of voluntary response to stimuli and with varying degrees of reflex activity. It may be due to illness or to an injury.

Huntington's chorea (*choreia* = dance) A rare hereditary disease characterized by involuntary jerky movements and mental deterioration that terminates in dementia.

Lethargy A condition of functional torpor or sluggishness.

Nerve block Loss of sensation in a region, such as in local dental anesthesia.

Neuralgia (*neur* = nerve) Attacks of pain along the entire course or branch of a peripheral sensory nerve.

Paralysis Diminished or total loss of motor function resulting from damage to nervous tissue or a muscle.

Spastic (*spas* = draw or pull) Resembling spasms or convulsions.

Stupor Condition of unconsciousness, torpor, or lethargy with suppression of sense or feeling.

Torpor Abnormal inactivity or lack of response to normal stimuli.

Viral encephalitis An acute inflammation of the brain caused by a direct attack by various viruses or by an allergic reaction to any of the many viruses that are normally harmless to the central nervous system. If the virus affects the spinal cord as well, it is called **encephalomyelitis.**

DRUGS ASSOCIATED WITH THE CENTRAL NERVOUS SYSTEM

Many disorders of the central nervous system, for example, poliomyelitis, neurosyphilis, cerebral palsy, and multiple sclerosis, cannot be treated and controlled with drugs. However, other disorders, such as headache, epilepsy, and Parkinson's disease, do respond to certain drugs.

Sedative-Hypnotics

Sedatives are drugs that calm nervousness, irritability, and excitement. Sedatives depress the central nervous system by depressing the higher cortical centers first and then extending down through the brain to the medullary centers and spinal cord. Hypnotics are agents that induce sleep. These drugs are indicated for treatment of anxiety or for sedation. The most widely used and known drug in this class is *diazepam* (Valium). Benzodiazepines generally—*diazepam, clonazepam, lorazepam,* and *chlordiazepoxide* (Librium)—belong to this class. Other examples are *flurazepam* (Dalmane), *methaqualone* (Quaalude), and barbiturates like *pentobarbital* (Nembutal).

Antidepressants

Antipressants are drugs that are effective against depressive psychological illness. The antidepressants cause elevations of biogenic amines in the central nervous system, especially norepinephrine and serotonin. Examples are the monoamine oxidase inhibitors (MAOIs) and the tricyclics such as *amitriptyline HCl* (Elavil, Endap), *imipramine* (Tofranil), and *doxepin HCl* (Sinequan, Adapin).

Psychotropics

Psychotropics are drugs that exert an effect on the mind. The phenothiazines like *chlorpromazine* (Thorazine) decrease uptake of norepinephrine, serotonin, and dopamine by the brain. Another example is *haloperidol* (Haldol).

Analgesics

Analgesics act by inhibiting impulses in the pain pathway, preventing them from reaching the limbic system. The nonnarcotic drugs are *aspirin* and *acetaminophen* (Tylenol). Intermediate analgesics are *codeine, proproxyphene* (Darvon), *pentazocine* (Talwin) and *oxycodone.* The more potent narcotics with high addiction potential include *morphine* and *meperidine* (Demerol, Pethidine).

Anticonvulsants

Anticonvulsants suppress convulsions. Most of these drugs make neurons more difficult to stimulate by changing the permeability of the neuron cell membrane so that it does not depolarize as easily. These drugs are also useful in treating epilepsy and trigeminal neuralgia. *Phenytoin* (Dilantin) is one of the principal drugs used to treat a number of types of epilepsy, especially grand mal. Also useful in grand mal epilepsy are *phenobarbital* (Luminol) and *primidone* (Mysoline). Petit mal epilepsy responds to *ethosuximide* (Zarontin) and *valproic acid* (Depakene). Psychomotor epilepsy is treated with *carbamazepine* (Tegretol) in addition to *phenytoin.* Parkinson's disease responds to dopamine-receptor agonists like *levodopa* (Sinemet).

STUDY OUTLINE

Brain (p. 310)

Embryological Development
1. During embryological development, brain vesicles are formed and serve as forerunners of various parts of the brain.
2. The diencephalon develops into the thalamus and hypothalamus, the telencephalon forms the cerebrum, the mesencephalon develops into the midbrain, the myelencephalon forms the medulla, and the metencephalon develops into the pons and cerebellum.

Principal Parts; Protection and Coverings
1. The principal parts of the brain are the brain stem, diencephalon, cerebrum, and cerebellum.
2. The brain is protected by the cranial bones, cranial meninges, and cerebrospinal fluid.

Cerebrospinal Fluid
1. Cerebrospinal fluid is formed in the choroid plexuses and circulates through the subarachnoid space, ventricles, and central canal. Most of the fluid is absorbed by the arachnoid villi of the superior sagittal sinus.
2. Cerebrospinal fluid protects by serving as a shock absorber. It also circulates nutritive substances from the blood.
3. The accumulation of cerebrospinal fluid in the head is called hydrocephalus. If the fluid accumulates in the ventricles, it is called internal hydrocephalus. If it accumulates in the subarachnoid space, it is called external hydrocephalus.

Blood Supply
1. The blood supply to the brain is via the circle of Willis.
2. Any interruption of the oxygen supply to the brain can result in weakening, permanent damage, or death of brain cells. Interruption of the mother's blood supply to a child during

childbirth before it can breathe may result in paralysis, mental retardation, epilepsy, or death.
3. Glucose deficiency may produce dizziness, convulsions, and unconsciousness.
4. The blood–brain barrier (BBB) is a concept that explains the differential rates of passage of certain materials from the blood into the brain.

Brain Stem
1. The medulla oblongata is continuous with the upper part of the spinal cord. It contains nuclei that are reflex centers for regulation of heart rate, respiratory rate, vasoconstriction, swallowing, coughing, vomiting, sneezing, and hiccuping. It also contains the nuclei of origin for cranial nerves VIII (cochlear and vestibular branches) through XII.
2. The pons is superior to the medulla. It connects the spinal cord with the brain and links parts of the brain with one another. It relays impulses from the cerebral cortex to the cerebellum related to voluntary skeletal movements. It contains the nuclei for cranial nerves V through VII and the vestibular branch of VIII. The reticular formation of the pons contains the pneumotaxic center, which helps control respiration.
3. The midbrain connects the pons and diencephalon. It conveys motor impulses from the cerebrum to the cerebellum and cord, sensory impulses from cord to thalamus, and regulates auditory and visual reflexes. It also contains the nuclei of origin for cranial nerves III and IV.

Diencephalon
1. The diencephalon consists of the thalamus and hypothalamus.
2. The thalamus is superior to the midbrain and contains nuclei that serve as relay stations for all sensory impulses, except

smell, to the cerebral cortex. It also registers conscious recognition of pain and temperature and some awareness of crude touch and pressure.

3. The hypothalamus is inferior to the thalamus. It controls the autonomic nervous system, connects the nervous and endocrine systems, controls body temperature, regulates food and fluid intake, and maintains the waking state and sleep patterns.

Cerebrum

1. The cerebrum is the largest part of the brain. Its cortex contains convolutions, fissures, and sulci.
2. The cerebral lobes are named the frontal, parietal, temporal, and occipital.
3. The white matter is under the cortex and consists of myelinated axons running in three principal directions.
4. The basal ganglia are paired masses of gray matter in the cerebral hemispheres. They help to control muscular movements.
5. The limbic system is found in the cerebral hemispheres and diencephalon. It functions in emotional aspects of behavior and memory.
6. The motor areas of the cerebral cortex are the regions that govern muscular movement. The sensory areas are concerned with the interpretation of sensory impulses. The association areas are concerned with emotional and intellectual processes.
7. Brain waves generated by the cerebral cortex are recorded as an EEG. They may be used to diagnose epilepsy, infections, and tumors.

Brain Lateralization (Split-Brain Concept)

1. Recent research indicates that the two hemispheres of the brain are not bilaterally symmetrical, either anatomically or functionally.
2. The left hemisphere is more important for right-handed control, spoken and written language, numerical and scientific skills, and reasoning.
3. The right hemisphere is more important for left-handed control, musical and artistic awareness, space and pattern perception, insight, imagination, and generating mental images of sight, sound, touch, taste, and smell.

Cerebellum

1. The cerebellum occupies the inferior and posterior aspects of the cranial cavity. It consists of two hemispheres and a central, constricted vermis.
2. It is attached to the brain stem by three pairs of cerebellar peduncles.
3. The cerebellum functions in the coordination of skeletal muscles and the maintenance of normal muscle tone and body equilibrium.

Transmitter Substances in the Brain (p. 332)

1. Over 40 different substances are known or suspected trans-
mitter substances in the brain that can facilitate, excite, or inhibit postsynaptic neurons.

2. Examples of transmitter substances include acetylcholine (ACh), norepinephrine (NE), dopamine (DA), serotonin (5-HT), glutamic acid, aspartic acid, gamma aminobutyric acid (GABA), and glycine.
3. Peptide chemical messengers that act as natural pain killers in the body are enkephalins, endorphins, and dynorphin.
4. Other peptides serve as hormones or other regulators of physiological responses. Examples include angiotensin, cholecystokinin, neurotensin, and regulating factors produced by the hypothalamus.

Cranial Nerves (p. 332)

1. Twelve pairs of cranial nerves originate from the brain.
2. The pairs are named primarily on the basis of distribution and numbered by order of attachment to the brain. (See Exhibit 14-2 for summary of cranial nerves.)

Disorders: Homeostatic Imbalances (p. 337)

1. Poliomyelitis is a viral infection that results in paralysis.
2. Cerebral palsy refers to a group of motor disorders caused by damage to motor centers of the cerebral cortex, cerebellum, or basal ganglia during fetal development, childbirth, or early infancy.
3. Parkinsonism is a progressive degeneration of the basal ganglia of the cerebrum resulting in insufficient dopamine.
4. Multiple sclerosis (MS) is the destruction of myelin sheaths of the neurons of the central nervous system. Impulse transmission is interrupted.
5. Epilepsy results from irregular electrical discharges of brain cells and may be diagnosed by an EEG. Depending on the form of the disease, the victim experiences degrees of motor, sensory, or psychological malfunction.
6. Cerebrovascular accidents (CVAs), also called strokes, are brain tissue destruction due to hemorrhage, thrombosis, or atherosclerosis.
7. Dyslexia involves an inability of an individual to comprehend written language.
8. Tay-Sachs disease is an inherited disorder that involves neurological degeneration of the CNS because of excessive amounts of ganglioside G_{m2}.
9. Headaches are of two types: intracranial and extracranial.
10. Irritation of the trigeminal nerve is known as trigeminal neuralgia.
11. Rabies is an acute viral infection that produces muscle spasms and encephalitis.
12. Reye's syndrome (RS) is characterized by vomiting, brain dysfunction, and liver damage.
13. Senility (dementia) is a disorder of the elderly that involves widespread intellectual impairment, personality changes, and sometimes delirium.

REVIEW QUESTIONS

1. Identify the four principal parts of the brain and the components of each, where applicable.
2. Describe the location of the cranial meninges.
3. Where is cerebrospinal fluid formed? Describe its circulation. Where is cerebrospinal fluid absorbed?
4. Distinguish between internal and external hydrocephalus.
5. Explain the importance of oxygen and glucose to brain cells.
6. What is the blood–brain barrier (BBB)? Is it of any advantage?

7. Describe the location and structure of the medulla. Define decussation of pyramids. Why is it important?
8. List the principal functions of the medulla.
9. Describe the location and structure of the pons. What are its functions?
10. Describe the location and structure of the midbrain. What are some of its functions?
11. Describe the location and structure of the thalamus. List some of its functions.

12. Where is the hypothalamus located? Explain some of its major functions.
13. Where is the cerebrum located? Describe the cortex, convolutions, fissures, and sulci of the cerebrum.
14. List and locate the lobes of the cerebrum. How are they separated from one another? What is the insula?
15. Describe the organization of cerebral white matter. Be sure to indicate the function of each group of fibers.
16. What are basal ganglia? Name the important basal ganglia and list the function of each.
17. Describe the effects of damage on the basal ganglia.
18. Define the limbic system. Explain several of its functions.
19. What is meant by a sensory area of the cerebral cortex? List, locate, and give the function of each sensory area.
20. What is meant by a motor area of the cerebral cortex? List, locate, and give the function of each motor area.
21. What is an association area of the cerebral cortex? What are its functions?
22. Define an electroencephalogram. What is the diagnostic value of an EEG?
23. Describe brain lateralization and the split-brain concept.
24. What evidence is there that lateralization exists in the human brain?
25. Discuss the phenomenon of handedness in the human population.
26. Describe the location of the cerebellum. List the principal parts of the cerebellum.
27. Describe the relationship of the dural extensions to the cerebellum.
28. What are cerebellar peduncles? List and explain the function of each.
29. Explain the functions of the cerebellum. What is ataxia?
30. What are transmitter substances? How do they affect post-synaptic neurons?
31. Give some examples of transmitter substances and indicate their functions.
32. Describe the peptide chemical messengers that act as natural painkillers in the body. Give their particular locations in the body and their functions.
33. Identify the following peptides as to location and function: angiotensin, cholecystokinin, and neurotensin.
34. Define a cranial nerve. How are cranial nerves named and numbered? Distinguish between a mixed and a sensory cranial nerve.
35. For each of the 12 pairs of cranial nerves, list (a) its name, number, and type; (b) its location; and (c) its function. In addition, list the effects of damage, where applicable.
36. Define each of the following: poliomyelitis, cerebral palsy, Parkinsonism, multiple sclerosis (MS), epilepsy, cerebrovascular accidents (CVAs), dyslexia, Tay-Sachs disease, headache, trigeminal neuralgia, rabies, Reye's syndrome (RS), and senility.
37. Refer to the glossary of medical terminology associated with the nervous system. Be sure that you can define each term.
38. Describe the actions of selected drugs on the central nervous system.

- Define a sensation and list the four prerequisites necessary for its transmission.

- Define projection, adaptation, afterimage, and modality as characteristics of sensations.

- Classify receptors on the basis of location, stimulus detected, and simplicity or complexity.

- Describe the distribution of cutaneous receptors.

- List the location and function of the receptors for tactile sensations (touch, pressure, vibration), thermoreceptive sensations (heat and cold), pain, and proprioception.

- Distinguish somatic, visceral, referred, and phantom pain.

- Describe how acupuncture is believed to relieve pain.

- Discuss the origin, neuronal components, and destination of the posterior column and spinothalamic pathways.

- Explain the neural pathways for pain and temperature; crude touch and pressure; and fine touch, proprioception, and vibration.

- Contrast the roles of the cerebellar tracts in conveying sensations.

- Compare the course of the pyramidal and extrapyramidal motor pathways.

- List the functions of the lateral corticospinal, anterior corticospinal, corticobulbar, rubrospinal, tectospinal, and vestibulospinal tracts.

- Compare integrative functions such as memory, sleep, and wakefulness.

15 The Sensory, Motor, and Integrative Systems

Now that you have examined the structure of the nervous system and its activities, we will see how its different parts cooperate in performing its three essential functions: (1) receiving sensory information; (2) transmitting motor impulses that result in movement or secretion; and (3) integration, an activity that deals with memory, sleep, and emotions.

SENSATIONS

Your ability to sense stimuli is vital to your survival. If pain could not be sensed, burns would be common. An inflamed appendix or stomach ulcer would progress unnoticed. A lack of sight would increase the risk of injury from unseen obstacles, a loss of smell would allow a harmful gas to be inhaled, a loss of hearing would prevent recognition of automobile horns, and a lack of taste would allow toxic substances to be ingested. In short, if you could not "sense" your environment and make the necessary homeostatic adjustments, you could not survive on your own.

DEFINITION

In its broadest context, **sensation** refers to a state of awareness of external or internal conditions of the body. **Perception** refers to the conscious registration of a sensory stimulus. For a sensation to occur, four prerequisites must be fulfilled.

1. A **stimulus,** or change in the environment, capable of initiating a response by the nervous system must be present.

2. A **receptor** or **sense organ** must pick up the stimulus and convert it to a nerve impulse. A sense receptor or sense organ may be viewed as specialized nervous tissue that is extremely sensitive to internal or external conditions.

3. The impulse must be **conducted** along a nervous pathway from the receptor or sense organ to the brain.

4. A region of the brain must **translate** the impulse into a sensation.

Receptors are capable of converting a specific stimulus into a nerve impulse. The stimulus may be light, heat, pressure, mechanical energy, or chemical energy. Each stimulus is capable of causing the membrane of the receptor to depolarize. This depolarization is called a **generator (receptor) potential.**

The generator potential is a graded response; within limits, the magnitude increases with stimulus strength and frequency. When the generator potential reaches the threshold level, it initiates an action potential (nerve impulse). Once initiated, the action potential is propagated along the nerve fiber. Whereas a generator potential is a local, graded response, an action potential obeys the all-or-none principle. The function of a generator potential is to initiate an action potential by transducing (converting) a stimulus into a nerve impulse.

A receptor may be quite simple. It may consist of the dendrites of a single neuron in the skin that are sensitive to pain stimuli. Or it may be contained in a complex organ such as the eye. Regardless of complexity, all sense receptors contain the dendrites of sensory neurons. The dendrites occur either alone or in close association with specialized cells of other tissues.

Receptors are at the same time very excitable and very specialized. Except for pain receptors, each has a low threshold of response to its specific stimulus and a high threshold to all others.

Once a stimulus is received by a receptor and converted into an impulse, the impulse is conducted along an afferent pathway that enters either the spinal cord or the brain. Many sensory impulses are conducted to the sensory areas of the cerebral cortex. It is in this region that stimuli produce conscious sensations. Sensory impulses that terminate in the spinal cord or brain stem can initiate motor activities, but typically do not produce conscious sensations. The thalamus detects pain sensations but cannot distinguish the intensity or location from which they arise. This is a function of the cerebrum.

CHARACTERISTICS

Most conscious sensations or perceptions occur in the cortical regions of the brain. In other words, you see, hear, and feel in the brain. You seem to see with your eyes, hear with your ears, and feel pain in an injured part of your body only because the cortex interprets the sensation as coming from the stimulated sense receptor. The term **projection** describes this process by which the brain refers sensations to their point of stimulation.

A second characteristic of many sensations is **adaptation.** The perception of a sensation may disappear even though a stimulus is still being applied. When you get into a tub of hot water, you might feel a burning sensation. But soon the sensation decreases to one of comfortable warmth, even though the stimulus (hot water) is still present. Other examples of adaptation include placing a ring on your finger, putting on your shoes or hat, and sitting on a chair. Initially, you are conscious of the sensations involved, but they are lost soon thereafter.

Sensations may also be characterized by **afterimages,** that is, some sensations persist even though the stimulus has been removed. This phenomenon is the reverse of adaptation. One common example of afterimage occurs when you look at a bright light and then look away or close your eyes. You still see the light for several seconds or minutes afterward.

Another characteristic of sensations is **modality:** the specific sensation felt. The sensation may be one of pain, pressure, touch, body position, equilibrium, hearing, vision, smell, or taste. In other words, the distinct property by which one sensation may be distinguished from another is its modality.

CLASSIFICATION OF RECEPTORS

Location

One convenient method of classifying receptors is by their location. **Exteroceptors** (eks'-ter-ō-SEP-tors) provide information about the external environment. They are sensitive to stimuli outside the body and transmit sensations of hearing, sight, smell, taste, touch, pressure, temperature, and pain. Exteroceptors are located near the surface of the body.

Visceroceptors (vis'-er-ō-SEP-tors), or **enteroceptors,** provide information about the internal environment. These sensations arise from within the body and may be felt as pain, pressure, fatigue, hunger, thirst, and nausea. Visceroceptors are located in blood vessels and viscera.

Proprioceptors (prō'-prē-ō-SEP-tors) provide information about body position and movement. Such sensations give us information about muscle tension, the position and tension of our joints, and equilibrium. These receptors are located in muscles, tendons, joints, and the internal ear.

Stimulus Detected

Another method of classifying receptors is by the type of stimuli they detect. **Mechanoreceptors** detect mechanical deformation of the receptor itself or in adjacent cells. Stimuli so detected include those related to touch, pressure, vibration, proprioception, hearing, equilibrium, and blood pressure. **Thermoreceptors** detect changes in temperature. **Nociceptors** detect pain, usually as a result of physical or chemical damage to tissues. **Electromagnetic (photo) receptors** detect light on the retina of the eye. **Chemoreceptors** detect taste in the mouth, smell in the nose, and chemicals in body fluids, such as oxygen, carbon dioxide, water, and glucose.

Simplicity or Complexity

As will be described shortly, receptors may also be classified according to the simplicity or complexity of their structure and the neural pathway involved. **Simple receptors** and neural pathways are associated with **general senses.** The receptors for general sensations are numerous and widespread. Examples include cutaneous sensations such as touch, pressure, vibration, heat, cold, and pain. **Complex receptors** and neural pathways are associated with **special senses.** The receptors for each special sense are found in only one or two specific areas of the body. Among the special senses are smell, taste, sight, and hearing.

GENERAL SENSES

CUTANEOUS SENSATIONS

Cutaneous sensations include tactile sensations (touch, pressure, vibration), thermoreceptive sensations (cold and heat), and pain. The receptors for these sensations are in the skin, connective tissue, and the ends of the gastrointestinal tract.

The cutaneous receptors are distributed over the body surface in such a way that certain parts of the body are densely populated with receptors and other parts contain only a few. Areas of the body that have few cutaneous receptors are insensitive; those containing many are very sensitive.

This type of distribution can be demonstrated in the skin by using the *two-point discrimination test* for touch. A compass is applied to the skin, and the distance in millimeters between the two points of the compass is var-

ied. The subject then indicates when two points are felt and when only one is felt. The compass may be placed on the tip of the tongue, an area where receptors are very densely packed. At a distance of 1.4 mm (0.06 inch), the points are able to stimulate two different receptors, and the subject feels touched by two objects. If the distance is less than 1.4 mm, the subject feels only one point, even though both points are touching the tongue, because the points are so close together that they reach only one receptor. If the compass is placed on the back of the neck, the subject feels two distinctly different points only if the distance between them is 36.2 mm (1.43 inch) or greater, because the receptors are few in number and far apart. The results of this test indicate that the more sensitive the area, the closer the compass points may be placed and still be felt separately. The following order for these receptors, from greatest sensitivity to least, has been established: tip of tongue, tip of finger, side of nose, back of hand, and back of neck.

Cutaneous receptors have simple structures. They consist of the dendrites of sensory neurons that may or may not be enclosed in a capsule of epithelial or connective tissue. Impulses generated by cutaneous receptors pass along somatic afferent neurons in spinal and cranial nerves, through the thalamus, to the general sensory area of the parietal lobe of the cortex.

Tactile Sensations

Even though the **tactile sensations** are divided into separate sensations of touch, pressure, and vibration, they are all detected by the same types of receptors, that is, receptors subject to deformation.

● *Touch*   Touch sensations generally result from stimulation of tactile receptors in the skin or tissues immediately beneath the skin. The term *light touch* refers to the ability to recognize exactly what point of the body is touched. *Crude touch* refers to the ability to perceive that something has touched the skin, although its exact location, shape, size, or texture cannot be determined.

Tactile receptors for touch include root hair plexuses, free nerve endings, Merkel's discs, Meissner's corpuscles, and end organs of Ruffini (Figure 15-1). **Root hair plexuses** are dendrites arranged in networks around the roots of hairs. They are not surrounded by supportive or protective structures. If a hair shaft is moved, the dendrites are stimulated. Root hair plexuses detect movements mainly on the surface of the body.

Other receptors that are not surrounded by supportive or protective structures are called **free (naked) nerve endings.** Free nerve endings are found everywhere in the skin and many other tissues.

Merkel's (MER-kelz) **discs** are receptors for touch that consist of disclike formations of dendrites attached to deeper layers of epidermal cells. They are distributed in many of the same locations as Meissner's corpuscles.

Meissner's (MĪS-nerz) **corpuscles** are egg-shaped receptors containing a mass of dendrites enclosed by connective tissue. They are located in the dermal papillae of the skin. Meissner's corpuscles are most numerous in the fingertips, palms of the hand, and soles of the feet.

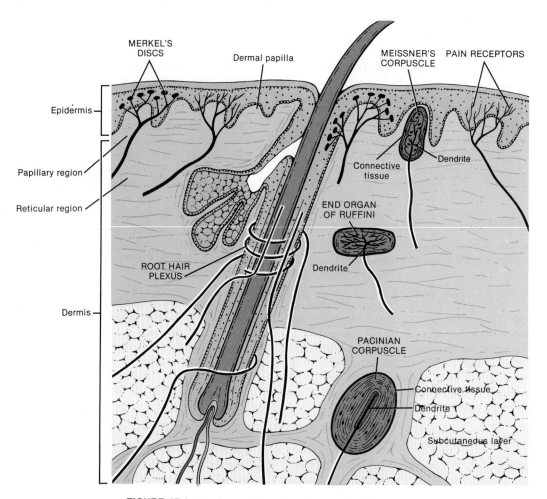

FIGURE 15-1 Structure and location of cutaneous receptors.

They are also abundant in the eyelids, tip of the tongue, lips, nipples, clitoris, and tip of penis.

End organs of Ruffini are embedded deeply in the dermis and in deeper tissues of the body. They detect heavy and continuous touch sensations.

● *Pressure*   **Pressure sensations** generally result from stimulation of tactile receptors in deeper tissues and are longer lasting and have less variation in intensity than touch sensations. Moreover, pressure is felt over a larger area than touch.

Pressure receptors are free nerve endings, end organs of Ruffini, and Pacinian corpuscles. **Pacinian** (pa-SIN-ē-an) **corpuscles** (Figure 15-1) are oval structures composed of a capsule resembling an onion and consist of connective tissue layers enclosing dendrites. Pacinian corpuscles are located in the subcutaneous tissue under the skin, the deep subcutaneous tissues that lie under mucous membranes, in serous membranes, around joints and tendons, in the perimysium of muscles, in the mammary glands, in the external genitalia of both sexes, and in certain viscera.

● *Vibration*   **Vibration sensations** result from rapidly repetitive sensory signals from tactile receptors.

The receptors for vibration sensations are Meissner's corpuscles and Pacinian corpuscles. Whereas Meissner's corpuscles detect low frequency vibration, Pacinian corpuscles detect higher frequency vibration.

Thermoreceptive Sensations

The **thermoreceptive sensations** are heat and cold. The **thermoreceptive receptors** are not known, but might be free nerve endings.

Pain Sensations

The receptors for **pain** are simply the branching ends of the dendrites of certain sensory neurons (Figure 15-1). Pain receptors are found in practically every tissue of the body. They may be stimulated by any type of stimulus. When stimuli for other sensations, such as touch, pressure, heat, and cold, reach a certain threshold, they stimulate the sensation of pain as well. Excessive stimulation of a sense organ causes pain. Additional stimuli for pain receptors include excessive distension or dilation of a structure, prolonged muscular contractions, muscle spasms, inadequate blood flow to an organ, or the presence of certain chemical substances. Pain receptors, because of their sensitivity to all stimuli, perform a protective function by identifying changes that may endanger the body. Pain receptors adapt only slightly or not at all.

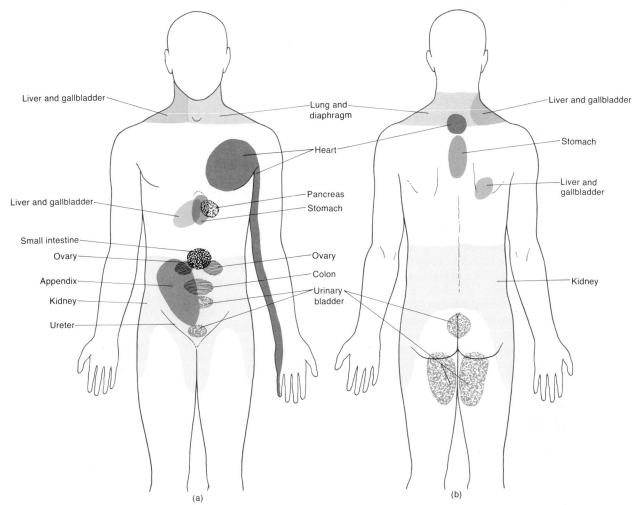

FIGURE 15-2 Referred pain. The colored parts of the diagrams indicate cutaneous areas to which visceral pain is referred. (a) Anterior view. (b) Posterior view.

Adaptation is the decrease or disappearance of the perception of a sensation even though the stimulus is still present. If there were adaptation to pain, it would cease to be sensed and irreparable damage could result.

Sensory impulses for pain are conducted to the central nervous system along spinal and cranial nerves. The lateral spinothalamic tracts of the spinal cord relay impulses to the thalamus. From here the impulses may be relayed to the postcentral gyrus of the parietal lobe. Recognition of the kind and intensity of most pain is ultimately localized in the cerebral cortex. Some awareness of pain occurs at subcortical levels.

Pain may be divided into two types: somatic and visceral. **Somatic pain** arises from stimulation of receptors in the skin, in which case it is called *superficial somatic pain,* or from stimulation of receptors in skeletal muscles, joints, tendons, and fascia, then called *deep somatic pain.* **Visceral pain** results from stimulation of receptors in the viscera.

The ability of the cerebral cortex to locate the origin of pain is related to past experience. In most instances of somatic pain and in some instances of visceral pain, the cortex accurately projects the pain back to the stimulated area. If you burn your finger, you feel the pain in your finger. If the lining of your pleural cavity is inflamed, you experience pain there. In most instances of visceral pain, however, the sensation is not projected back to the

point of stimulation. Rather, the pain may be felt in or just under the skin that overlies the stimulated organ. The pain may also be felt in a surface area far from the stimulated organ. This phenomenon is called **referred pain.** In general, the area to which the pain is referred and the visceral organ involved receive their innervation from the same segment of the spinal cord. Consider the following example. Afferent fibers from the heart as well as from the skin over the heart and along the medial aspect of the left upper extremity enter spinal cord segments T1 to T4. Thus the pain of a heart attack is typically felt in the skin over the heart and along the left arm. Figure 15-2 illustrates cutaneous regions to which visceral pain may be referred.

CLINICAL APPLICATION

A kind of pain frequently experienced by patients who have had a limb amputated is called **phantom pain.** They still experience pain or other sensations in the extremity as if the limb were still there. An explanation for this phenomenon is that the remaining proximal portions of the sensory nerves that previously received impulses from the limb are being stimulated by the trauma of the amputation. Stimuli from these nerves are interpreted by the brain as coming from the nonexistent (phantom) limb.

Inhibition of Pain

Pain sensations may be controlled by interrupting the pain impulse between the receptors and the interpretation centers of the brain. This may be done chemically, surgically, or by other means. Most pain sensations respond to pain-reducing drugs, which, in general, act to inhibit nerve impulse conduction at synapses.

Occasionally, however, pain may be controlled only by surgery. The purpose of surgical treatment is to interrupt the pain impulse somewhere between the receptors and the interpretation centers of the brain by severing the sensory nerve, its spinal root, or certain tracts in the spinal cord or brain. *Sympathectomy* is excision of portions of the neural tissue from the autonomic nervous system; *cordotomy* is severing of a spinal cord tract, usually the lateral spinothalamic; *rhizotomy* is the cutting of sensory nerve roots; *prefrontal lobotomy* is the destruction of the tracts that connect the thalamus with the prefrontal and frontal lobes of the cerebral cortex. In each instance, the pathway for pain is severed so that pain impulses are no longer conducted to the cortex.

Another method of inhibiting pain impulses is **acupuncture** (*acus* = needle; *pungere* = sting). Needles are inserted through selected areas of the skin and then twirled by the acupuncturist or by a mechanical device. After 20 to 30 minutes, pain is deadened for 6 to 8 hours. The location of needle insertion depends on the part of the body the acupuncturist wishes to anesthetize. To pull a tooth, a needle is inserted in the web between thumb and index finger. For a tonsillectomy, a needle is inserted approximately 5 cm (2 inches) above the wrist. For removal of a lung, a needle is placed in the forearm midway between wrist and elbow.

There is no satisfactory explanation of how acupuncture works. According to the "gate control" theory, twirling the acupuncture needle stimulates two sets of nerves that eventually enter the spinal cord and synapse with the same association neurons. One nerve contains nerve fibers of small diameter (C fibers) for pain, and the other nerve contains nerve fibers of larger diameter (A fibers) for touch. The impulse passing along the touch-conducting neurons is faster than that passing along the pain-conducting neurons—fibers with large diameters conduct impulses faster than those with small diameters. Because the touch impulse reaches the posterior gray horn of the cord first, it has priority over the pain impulse. It thus "closes the gate" to the brain before the pain impulse reaches the cord. Since the pain impulse does not pass to the brain, no pain is felt. Currently, acupuncture is used in the United States mostly for childbirth, tic douloureux, arthritis, and other nonsurgical conditions.

PROPRIOCEPTIVE SENSATIONS

An awareness of the activities of muscles, tendons, and joints is provided by the **proprioceptive** or **kinesthetic** (kin'-es-THET-ik) **sense.** It informs us of the degree to which muscles are contracted and the amount of tension created in the tendons. The proprioceptive sense enables us to recognize the location and rate of movement of one body part in relation to others. It also allows us to estimate weight and determine the muscular work necessary to perform a task. With the proprioceptive sense, we can judge the position and movements of our limbs without using our eyes when we walk, type, or dress in the dark.

Proprioceptive receptors are located in muscles, tendons, joints, and the internal ear. The **joint kinesthetic receptors** are located in the capsules of joints and ligaments about joints. These receptors provide feedback information on the degree and rate of angulation (change of position) of a joint.

Muscle spindles consist of the endings of sensory neurons that are wrapped around specialized muscle fibers. They are located in nearly all skeletal muscles and are most numerous in the muscles of the extremities. Muscle spindles provide feedback information on the degree of muscle stretch. This information is relayed to the central nervous system to assist in the coordination and efficiency of muscle contraction. Muscle spindles are involved in the stretch and extensor reflexes.

Tendon organs are also proprioceptive receptors that provide information about skeletal muscles. These organs are located at the junction of muscle and tendon. They function by sensing the tension applied to a tendon. The degree of tension is related to the degree of muscle contraction and is translated by the central nervous system.

The proprioceptors in the internal ear are the macula of the saccule and the utricle and cristae in the semicircular ducts. Their function in equilibrium is discussed in Chapter 17.

Proprioceptors adapt only slightly. This feature is advantageous since the brain must be apprised of the status of different parts of the body at all times so that adjustments can be made to ensure coordination.

The afferent pathway for muscle sense consists of impulses generated by proprioceptors via cranial and spinal nerves to the central nervous system. Impulses for conscious proprioception pass along ascending tracts in the cord, where they are relayed to the thalamus and cerebral cortex. The sensation is registered in the general sensory area in the parietal lobe of the cerebral cortex posterior to the central sulcus. Proprioceptive impulses that have resulted in reflex action pass to the cerebellum along spinocerebellar tracts.

LEVELS OF SENSATION

As we have said, a receptor converts a stimulus into a nerve impulse, and only after that impulse has been conducted to a region of the spinal cord or brain can it be translated into a sensation. The nature of the sensation and the type of reaction generated vary with the level of the central nervous system at which the sensation is translated.

Sensory fibers terminating in the spinal cord can generate spinal reflexes without immediate action by the brain. Sensory fibers terminating in the lower brain stem bring about far more complex motor reactions than simple spinal reflexes. When sensory impulses reach the lower brain stem, they cause subconscious motor reactions. Sensory impulses that reach the thalamus can be localized crudely in the body. At the thalamic levels sensations are sorted by modality, that is, identified as the *specific* sensation of touch, pressure, pain, position, hearing, vision, smell, or taste. When sensory information reaches the cerebral cortex, we experience precise localization. It is at this level that memories of previous sensory information are stored and the perception of sensation occurs on the basis of past experience.

SENSORY PATHWAYS

SOMATOSENSORY CORTEX

Sensory information from receptors on one side of the body crosses over to the opposite side in the spinal cord or brain stem and then to the **somatosensory cortex (general sensory** or **somesthetic area)** of the cerebral cortex where conscious sensations are produced (see Figure 14-10). Areas of the somatosensory cortex have been mapped out which represent the termination of sensory information from all parts of the body. Figure 15-3 shows the location and areas of representation of the somatosensory cortex of the right cerebral hemisphere. The left cerebral hemisphere has a duplicate somatosensory cortex.

Note that some parts of the body are represented by large areas in the somatosensory cortex. These include the lips, face, and thumb. Other parts of the body, such as the trunk and lower extremities, are represented by relatively small areas. The relative sizes of the somatosensory cortex are directly proportional to the number of specialized sensory receptors in each respective part of the body. Thus, there are numerous receptors in the skin

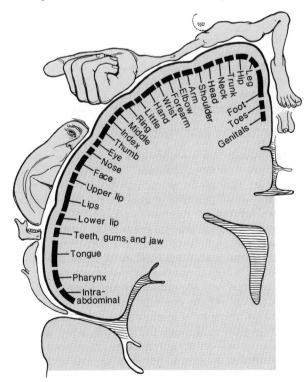

FIGURE 15-3 Somatosensory cortex of the right cerebral hemisphere.

of the lips, but relatively few in the skin of the trunk. Essentially, the size of the area for a particular part of the body is determined by the functional importance of the part and its need for sensitivity.

Let us now examine how sensory information is transmitted from receptors to the central nervous system. You will find it helpful to review the principal ascending and descending tracts of the spinal cord (see Exhibit 13-1 and Figure 13-4). Sensory information transmitted from the spinal cord to the brain is conducted along two general pathways: the posterior column pathway and the spinothalamic pathway.

POSTERIOR COLUMN PATHWAY

In the **posterior column pathway** to the cerebral cortex there are three separate sensory neurons. The **first-order neuron** connects the receptor with the spinal cord and medulla on the same side of the body. The cell body of the first-order neuron is in the posterior root ganglion of a spinal or cranial nerve. The first-order neuron synapses with a **second-order neuron.** The second-order neuron passes from the medulla upward to the thalamus. The cell body of the second-order neuron is located in the nuclei cuneatus or gracilis of the medulla. Before passing into the thalamus, the second-order neuron crosses to the opposite side of the medulla and enters the medial lemniscus, a projection tract that terminates at the thalamus. In the thalamus, the second-order neuron synapses with a **third-order neuron.** The third-order neuron terminates in the somesthetic sensory area of the cerebral cortex.

The posterior column pathway conducts impulses related to proprioception, light touch, two-point discrimination, and vibrations.

SPINOTHALAMIC PATHWAY

The **spinothalamic pathway** is also composed of three orders of sensory neurons. The first-order neuron connects a receptor of the neck, trunk, and extremities with the spinal cord. The cell body of the first-order neuron is in the posterior root ganglion also. The first-order neuron synapses with the second-order neuron, which has its cell body in the posterior gray horn of the spinal cord. The fiber of the second-order neuron crosses to the opposite side of the spinal cord and passes upward to the brain stem in the lateral spinothalamic tract or ventral spinothalamic tract. The fibers from the second-order neuron terminate in the thalamus. There the second-order neuron synapses with a third-order neuron. The third-order neuron terminates in the somesthetic sensory area of the cerebral cortex. The spinothalamic pathway conveys sensory impulses for pain and temperature as well as crude touch and pressure.

The second-order neurons of the spinothalamic pathway enter the medulla, pons, and midbrain. Thus the spinothalamic pathway conducts sensory signals that result in subconscious motor reactions. By contrast, second-order neurons of the posterior column pathway have a direct connection with the thalamus and cerebral cortex.

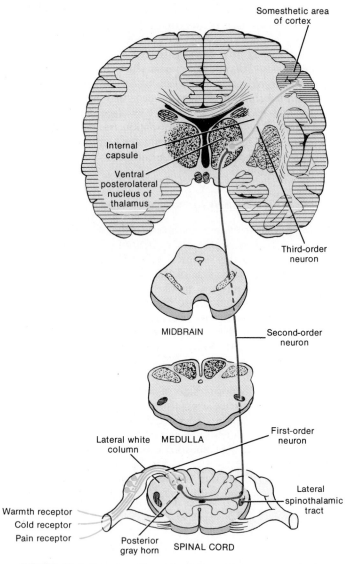

FIGURE 15-4 Sensory pathway for pain and temperature—the lateral spinothalamic pathway.

Thus the posterior column pathway conducts sensory information primarily into the conscious area of the brain.

Now we can examine the anatomy of specific sensory pathways—for pain and temperature; for crude touch and pressure; and for light touch, proprioception, and vibration.

Pain and Temperature

The sensory pathway for pain and temperature is called the **lateral spinothalamic pathway** (Figure 15-4). The first-order neuron conveys the impulse for pain or temperature from the appropriate receptor to the posterior gray horn on the same side of the spinal cord. In the horn, the first-order neuron synapses with a second-order neuron. The axon of the second-order neuron crosses to the opposite side of the cord. Here it becomes a component of the *lateral spinothalamic tract* in the lateral white column. The second-order neuron passes upward in the tract through the brain stem to a nucleus in the thalamus called

the ventral posterolateral nucleus. In the thalamus, conscious recognition of pain and temperature occurs. The sensory impulse is then conveyed from the thalamus through the internal capsule to the somesthetic area of the cerebral cortex by a third-order neuron. The cortex analyzes the sensory information for the precise source, severity, and quality of the pain and heat stimuli.

Crude Touch and Pressure

The neural pathway that conducts impulses for crude touch and pressure is the **anterior (ventral) spinothalamic pathway** (Figure 15-5). The first-order neuron conveys the impulse from a crude touch or pressure receptor to the posterior gray horn on the same side of the spinal cord. In the horn, the first-order neuron synapses with a second-order neuron. The axon of the second-order neuron crosses to the opposite side of the cord and becomes a component of the **anterior spinothalamic tract** in the anterior white column. The second-order neuron passes upward in the tract through the brain stem to the ventral posterolateral nucleus of the thalamus. The sensory impulse is then relayed from the thalamus through the internal capsule to the somesthetic area of the cerebral cortex by a third-order neuron. Although there is some awareness of crude touch and pressure at the thalamic level, it is not fully perceived until the impulses reach the cortex.

Light Touch, Proprioception, and Vibration

The neural pathway for light touch, proprioception, and vibration is called the **posterior column pathway** (Figure 15-6). This pathway conducts impulses that give rise to several discriminating senses.

1. Light touch: the ability to recognize the exact location of stimulation and to make two-point discriminations.
2. Stereognosis: the ability to recognize by "feel" the size, shape, and texture of an object.
3. Proprioception: the awareness of the precise position of body parts and directions of movement.
4. Weight discrimination: the ability to assess the weight of an object.
5. The ability to sense vibrations.

First-order neurons for the discriminating senses just noted follow a pathway different from those for pain and temperature and crude touch and pressure. Instead of terminating in the posterior gray horn, the first-order neurons from appropriate receptors pass upward in the fasciculus gracilis or fasciculus cuneatus in the posterior white column of the cord. From here the first-order neurons enter either the nucleus gracilis or nucleus cuneatus in the medulla, where they synapse with second-order neurons. The axons of the second-order neurons cross to the opposite side of the medulla and ascend to the thalamus through the medial lemniscus, a projection tract of white fibers passing through the medulla, pons, and midbrain. The second-order neuron axons synapse with third-order neurons in the ventral posterior nucleus in the thalamus. In the thalamus, there is no conscious awareness of the discriminating senses, except for a possible crude awareness of vibrations. The third-order neurons convey the sensory impulses to the somesthetic area of the cere-

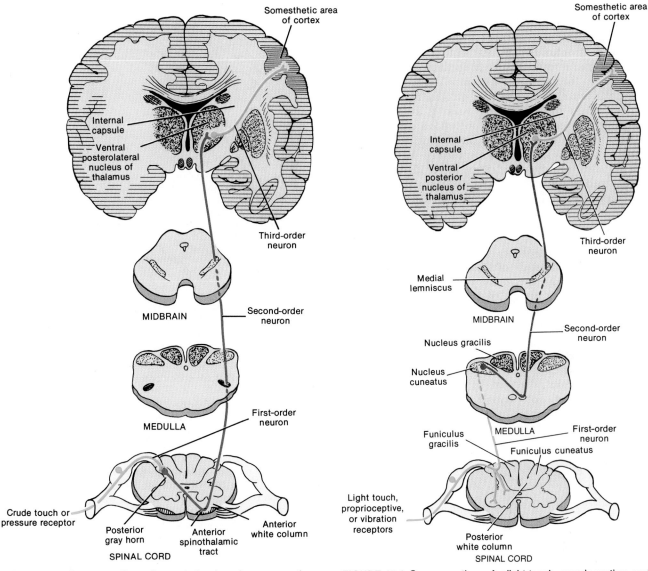

FIGURE 15-5 Sensory pathway for crude touch and pressure—the anterior spinothalamic pathway.

FIGURE 15-6 Sensory pathway for light touch, proprioception, and vibration—the posterior column pathway.

bral cortex. It is here that you perceive your sense of position and movement and light touch.

CEREBELLAR TRACTS

The *posterior spinocerebellar tract* is an uncrossed tract that conveys impulses concerned with subconscious muscle sense and thus assumes a role in reflex adjustments for posture and muscle tone. The nerve impulses originate in neurons that run between proprioceptors in muscles, tendons, and joints and the posterior gray horn of the spinal cord. Here the neurons synapse with afferent neurons that pass to the ipsilateral lateral white column of the cord to enter the posterior cerebellar tract. The tract enters the inferior cerebellar peduncles from the medulla and ends at the cerebellar cortex. In the cerebellum, synapses are made that ultimately result in the transmission of impulses back to the spinal cord to the anterior gray horn to synapse with the lower motor neurons leading to skeletal muscles.

The *anterior spinocerebellar tract* also conveys impulses for subconscious muscle sense. It, however, is made up of both crossed and uncrossed nerve fibers. Sensory neurons deliver impulses from proprioceptors to the posterior gray horn of the spinal cord. Here a synapse occurs with neurons that make up the anterior spinocerebellar tracts. Some fibers cross to the opposite side of the spinal cord in the anterior white commissure. Others pass laterally to the ipsilateral anterior spinocerebellar tract and move upward, through the brain stem, to the pons to enter the cerebellum through the superior cerebellar peduncles. Here again the impulses for subconscious muscle sense are registered.

MOTOR PATHWAYS

After receiving and interpreting sensations, the central nervous system generates impulses to direct responses to that sensory input. We have already considered somatic reflex arcs and visceral efferent pathways. Now we will look at the transmission of motor impulses that result in the movement of skeletal muscles.

The principal parts of the brain concerned with skeletal muscle control are the cerebral motor cortex, basal ganglia, reticular formation, and cerebellum. The motor cortex assumes the major role for controlling precise, discrete muscular movements. The basal ganglia largely integrate semivoluntary movements like walking, swimming, and laughing. The cerebellum, although not a control center, assists the motor cortex and basal ganglia by making body movements smooth and coordinated. Voluntary motor impulses are conveyed from the brain through the spinal cord by way of two major pathways: the pyramidal pathways and the extrapyramidal pathways.

MOTOR CORTEX

Just as the somatosensory cortex has been mapped to indicate the termination of sensory information from all parts of the body, the **motor cortex** has been mapped to indicate which groups of muscles are controlled by its specific areas (Figure 15-7). The right cerebral hemisphere has a duplicate of the motor cortex in the left cerebral hemisphere. Note that the different muscle groups are not represented equally in the motor cortex. In general, the degree of representation is proportional to the preciseness of movement required of a particular part of the body. For example, the thumb, fingers, lips, tongue, and vocal cords have large representations. The trunk has a relatively small representation. By comparing Figures 15-3 and 15-7, you will see that somatosensory and motor representations are not identical for the same part of the body.

PYRAMIDAL PATHWAYS

Voluntary motor impulses are conveyed from the motor areas of the brain to somatic efferent neurons leading to skeletal muscles via the **pyramidal (pi-RAM-i-dal) pathways.** Most pyramidal fibers originate from cell bodies in the precentral gyrus. They descend through the internal capsule of the cerebrum and cross to the opposite side of the brain. They terminate in nuclei of cranial nerves that innervate voluntary muscles or in the anterior gray horn of the spinal cord. A short connecting neuron probably completes the connection of the pyramidal fibers with the motor neurons that activate voluntary muscles.

The pathways over which the impulses travel from the motor cortex to skeletal muscles have two components: **upper motor neurons (pyramidal fibers)** in the brain and **lower motor neurons (peripheral fibers)** in the spinal cord. Here we consider three tracts of the pyramidal system.

1. Lateral corticospinal tract (pyramidal tract proper). This tract begins in the motor cortex and descends through the internal capsule of the cerebrum, the cerebral peduncle of the midbrain, and then the pons on the same side as the point of origin (Figure 15-8). In the medulla, the fibers decussate and descend through the spinal cord in the lateral white column in the lateral corticospinal tract. Thus the motor cortex of the right side of the brain controls muscles on the left side of the body and vice versa. Most upper motor neuron fibers of the lateral corticospinal tract synapse with short association neurons in the

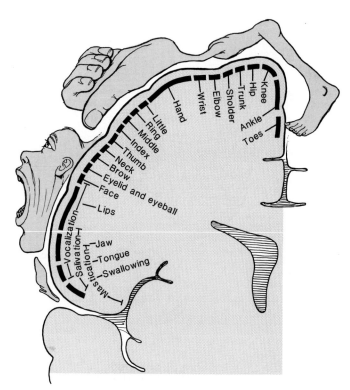

FIGURE 15-7 Motor cortex of the right cerebral hemisphere.

anterior gray horn of the cord. These then synapse in the anterior gray horn with lower motor neurons that exit all levels of the cord via the ventral roots of spinal nerves. The lower motor neurons terminate in skeletal muscles.

2. Anterior corticospinal tract. About 15 percent of the upper motor neurons from the motor cortex do not cross in the medulla. These pass through the medulla and continue to descend on the same side to the anterior white column to become part of the anterior (straight or uncrossed) corticospinal tract. The fibers of these upper motor neurons decussate and synapse with association neurons in the anterior gray horn of the spinal cord on the side opposite the origin of the anterior corticospinal tract. The association neurons in the horn synapse with lower motor neurons that exit the cervical and upper thoracic segments of the cord via the ventral roots of spinal nerves. The lower motor neurons terminate in skeletal muscles that control muscles of the neck and part of the trunk.

3. Corticobulbar tract. The fibers of this tract arise from upper motor neurons in the motor cortex. They accompany the corticospinal tracts through the internal capsule to the brain stem, where they decussate and terminate in the nuclei of cranial nerves in the pons and medulla. These cranial nerves include the oculomotor (III), trochlear (IV), trigeminal (V), abducens (VI), facial (VII), glossopharyngeal (IX), vagus (X), accessory (XI), and hypoglossal (XII). The corticobulbar tract conveys impulses that largely control voluntary movements of the head and neck.

The various tracts of the pyramidal system convey impulses from the cortex that result in precise muscular movements.

EXTRAPYRAMIDAL PATHWAYS

The **extrapyramidal pathways** include all descending tracts other than the pyramidal tracts. Generally, these include tracts that begin in the basal ganglia and reticular formation. The main extrapyramidal tracts are as follows.

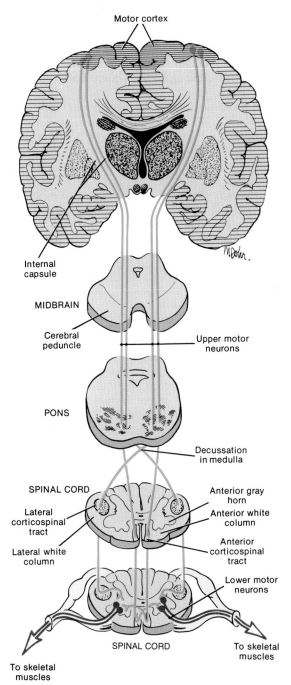

Motor cortex

Internal
capsule

MIDBRAIN

Cerebral
peduncle

Upper motor
neurons

PONS

Decussation
in medulla

SPINAL CORD

Lateral
corticospinal
tract

Anterior gray
horn

Anterior white
column

Lateral white
column

Anterior
corticospinal
tract

Lower motor
neurons

SPINAL CORD

To skeletal
muscles

To skeletal
muscles

FIGURE 15-8 Pyramidal pathways.

gray horns, mostly in the cervical and lumbosacral segments of the cord. It conveys impulses that regulate muscle tone in response to movements of the head. This tract, therefore, plays a major role in equilibrium.

Only one motor neuron carries the impulse from the cerebral cortex to the cranial nerve nuclei or spinal cord: an **upper motor neuron.** Only one motor neuron in the pathway actually terminates in a skeletal muscle: the **lower motor neuron.** This neuron, a somatic efferent neuron, always extends from the central nervous system to the skeletal muscle. Since it is the final transmitting neuron in the pathway, it is also called the **final common pathway.**

CLINICAL APPLICATION

The lower motor and upper motor neurons are important clinically. If the lower motor neuron is damaged or diseased, there is neither voluntary nor reflex action of the muscle it innervates, and the muscle remains in a relaxed state, a condition called **flaccid paralysis.** Injury or disease of upper motor neurons in a motor pathway is characterized by varying degrees of continued contraction of the muscle, a condition called **spasticity,** and exaggerated reflexes. Another characteristic is the **Babinski sign,** which is dorsiflexion of the great toe accompanied by fanning of the lateral toes in response to stroking the plantar surface along the outer border of the foot. As we have seen, this response is normal only in infants; the normal adult response is plantar flexion of the toes.

Lower motor neurons are subjected to stimulation by many other presynaptic neurons. Some signals are excitatory; others are inhibitory. The algebraic sum of the opposing signals determines the final response of the lower motor neuron. It is not just a simple matter of the brain sending an impulse and the muscle always contracting.

Association neurons are of considerable importance in the motor pathways. Impulses from the brain are conveyed to association neurons before being received by lower motor neurons. These association neurons integrate the pattern of muscle contraction.

The basal ganglia have many connections with other parts of the brain. Through these connections, they help to control subconscious movements. The caudate nucleus controls gross intentional movements. The caudate nucleus and putamen, together with the cortex, control patterns of movement. The globus pallidus controls positioning of the body for performing a complex movement. The subthalamic nucleus is thought to control walking and possibly rhythmic movements. Many potential functions of the basal ganglia are held in check by the cerebrum. Thus if the cerebral cortex is damaged early in life, a person can still perform many gross muscular movements.

The role of the cerebellum is significant also. The cerebellum is connected to other parts of the brain that are concerned with movement. The vestibulocerebellar tract transmits impulses from the equilibrium apparatus in the ear to the cerebellum. The olivocerebellar tract transmits impulses from the basal ganglia to the cerebellum. The

1. Rubrospinal tract. This tract originates in the red nucleus of the midbrain (after receiving fibers from the cerebellum), crosses over to descend in the lateral white column of the opposite side, and extends through the entire length of the spinal cord. The tract transmits impulses to skeletal muscles concerned with tone and posture.

2. Tectospinal tract. This tract originates in the superior colliculus of the midbrain, crosses to the opposite side, descends in the anterior white column, and enters the anterior gray horns in the cervical segments of the cord. Its function is to transmit impulses that control movements of the head in response to visual stimuli.

3. Vestibulospinal tract. This tract originates in the vestibular nucleus of the medulla, descends on the same side of the cord in the anterior white column, and terminates in the anterior

corticopontocerebellar tract conveys impulses from the cerebrum to the cerebellum. The spinocerebellar tracts relay proprioceptive information to the cerebellum. Thus the cerebellum receives considerable information regarding the overall physical status of the body. Using this information, the cerebellum generates impulses that integrate body responses.

Take tennis, for example. To make a good serve, you must bring your racket forward just far enough to make solid contact. How do you stop at the exact point without swinging too far? This is where the cerebellum comes in. It receives information about your body status while you are serving. Before you even hit the ball, the cerebellum has already sent information to the cerebral cortex and basal ganglia informing them that your swing must stop at an exact point. In response to cerebellar stimulation, the cortex and basal ganglia transmit motor impulses to your opposing body muscles to stop the swing. The cerebellar function of stopping overshoot when you want to zero in on a target is called its *damping function*. The cerebellum also helps you to coordinate different body parts while walking, running, and swimming. Finally, the cerebellum helps you maintain equilibrium.

INTEGRATIVE FUNCTIONS

We turn now to a fascinating, though poorly understood, function of the cerebrum: integration. The **integrative functions** include cerebral activities such as memory, sleep and wakefulness, and emotional response. The role of the limbic system in emotional behavior was discussed in Chapter 14.

MEMORY

Memory may be defined as the ability to recall thoughts. Memory may be generally classified into two kinds: activated and long-term.

Activated memory is the ability to recall bits of information. One example is finding a number in the phone book and then dialing it. If the number has no special significance, it is usually forgotten in a few seconds. One theory of activated memory claims that memories may be caused by reverberating neuronal circuits—an incoming impulse stimulates the first neuron, which stimulates the second, which stimulates the third, and so on (see Figure 12-12a). Branches from the second and third neurons synapse with the first, sending the impulse back through the circuit again and again. Thus the output neuron generates continuous impulses. Once fired, the output signal may last from a few seconds to many hours, depending on the arrangement of neurons in the circuit. If this pattern is applied to activated memory, an incoming thought—the phone number—continues in the brain even after the initial stimulus is gone. Thus you can recall the thought only for as long as the reverberation continues.

The concept of a reverberating circuit does not, however, explain long-term memory. **Long-term memory** is the persistence of an incoming impulse for years. One theory explains long-term memory on the basis of the principle of facilitation at synapses. When an incoming

thought enters a neuronal circuit, the synapses in the circuit become facilitated for the passage of a similar signal later on. Thus an incoming signal facilitates the synapses in the circuit used for that signal over and over and you recall the thought. Such a neuronal circuit is called an *engram*.

The first incoming thought leading to long-term memory lasts only a brief time. How then does an activated memory result in a long-term memory? One explanation is that the reverberating circuit of an activated memory may persist for up to an hour after the initial thought. This reverberation establishes the engram. Later another incoming thought can cause facilitation of the neurons in the engram and the result is long-term memory.

The portions of the brain thought to be associated with memory include the association cortex of the frontal, parietal, occipital, and temporal lobes and parts of the limbic system, especially the hippocampus.

WAKEFULNESS AND SLEEP

Humans sleep and awaken in a fairly constant 24-hour rhythm called a *circadian* (ser-KĀ-dē-an) *rhythm*. Since neuronal fatigue precedes sleep and the signs of fatigue disappear after sleep, fatigue is apparently one cause of sleep. Moreover, EEG recordings indicate that during wakefulness the cerebral cortex is very active, sending impulses continuously through the body. During sleep, however, fewer impulses are transmitted by the cerebral cortex. The activity of the cerebral cortex is thought to be related to the reticular formation.

The reticular formation has numerous connections with the cerebral cortex (Figure 15-9). Stimulation of portions of the reticular formation results in increased cortical activity. Thus the reticular formation is also known as the **reticular activating system (RAS).** One part of the system, the mesencephalic part, is composed of areas of gray matter of the pons and midbrain. When this area is stimulated, many impulses pass upward into the thalamus and disperse to widespread areas of the cerebral cortex. The effect is a generalized increase in cortical activity. The other part of the RAS, the thalamic part, consists of gray matter in the thalamus. When the thalamic part is stimulated, signals from specific parts of the thalamus cause activity in specific parts of the cerebral cortex. Apparently the mesencephalic part of the RAS causes general wakefulness (consciousness) and the thalamic part causes *arousal,* that is, awakening from deep sleep.

For arousal to occur, the RAS must be stimulated by input signals. Almost any sensory input can activate the RAS: pain stimuli, proprioceptive signals, bright light, an alarm clock. Once the RAS is activated, the cerebral cortex is also activated and you experience arousal. Signals from the cerebral cortex can also stimulate the RAS. Such signals may originate in the somesthetic cortex, the motor cortex, or the limbic system. When the signals activate the RAS, the RAS activates the cerebral cortex and arousal occurs.

Following arousal, the RAS and cerebral cortex continue to activate each other through a feedback system consisting of many circuits. The RAS also has a feedback

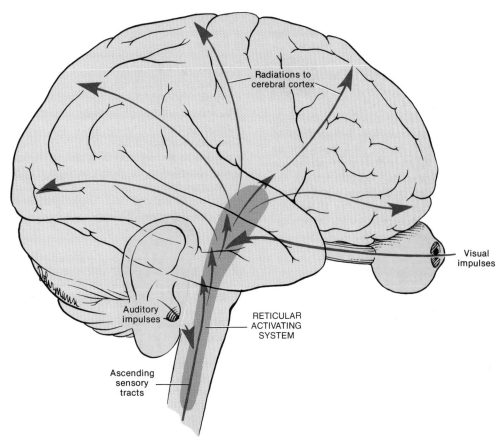

FIGURE 15-9 Reticular formation.

system with the spinal cord that is composed of many circuits. Impulses from the activated RAS are transmitted down the spinal cord and then to skeletal muscles. Muscle activation causes proprioceptors to return impulses that activate the RAS. The two feedback systems maintain activation of the RAS, which in turn maintains activation of the cerebral cortex. The result is a state of wakefulness called *consciousness.* The RAS is the physical basis of consciousness, the brain's chief watchguard. It continuously sifts and selects, forwarding only the essential, unusual, or dangerous to the conscious mind. Since humans experience different levels of consciousness (alertness, attentiveness, relaxation, nonattentiveness), it is assumed that the level of consciousness depends on the number of feedback circuits operating at the time.

CLINICAL APPLICATION

Consciousness may be altered by various factors. Amphetamines probably activate the RAS to produce a state of wakefulness and alertness. Meditation produces a lack of consciousness. Anesthetics produce a state of consciousness called anesthesia. Damage to the nervous system, as well as disease, can produce a lack of consciousness called coma. And drugs such as LSD can alter consciousness also.

If this theory of wakefulness is accepted, how then does sleep occur if the activating feedback systems are in continual operation? One explanation is that the feed-

back system slows tremendously or is inhibited. Inactivation of the RAS produces a state known as *sleep.*

Just as there are different levels of consciousness, there are different levels of sleep. Normal sleep consists of two types: nonrapid eye movement sleep (NREM) and rapid eye movement sleep (REM).

NREM sleep consists of four stages. Each has been identified by EEG recordings:

Stage 1. The person is relaxing with eyes closed. During this time, respirations are regular, pulse is even, and the person has fleeting thoughts. If awakened, the person will frequently say he has not been sleeping. Alpha waves appear on the EEG.

Stage 2. It is a little harder to awaken the person. Fragments of dreams may be experienced, and the eyes may slowly roll from side to side. The EEG shows *sleep spindles*—sudden, short bursts of sharply pointed alpha waves that occur at 14 to 16 Hz (cycles) per second.

Stage 3. The person is very relaxed. Body temperature begins to fall and blood pressure decreases. It is difficult to awaken the person and the *EEG* shows a mixture of sleep spindles and delta waves. This stage occurs about 20 minutes after falling asleep.

Stage 4. Deep sleep occurs. The person is very relaxed and responds slowly if awakened. Bed-wetting and sleep-walking may occur during this stage. The EEG is dominated by delta waves.

In a typical 7- or 8-hour sleep period, a person goes from stages 1 to 4 of NREM sleep. Then the person ascends to stages 3 and 2 and then to REM sleep within 50 to 90 minutes.

In **REM sleep,** the EEG readings are similar to those of stage 1 of NREM sleep. There are significant physiological differences, however. During REM sleep, respirations and pulse rate increase and are irregular. Blood pressure also fluctuates considerably. It is during REM sleep that most dreaming occurs. Following REM sleep, the person descends again to stages 3 and 4 of NREM sleep.

REM and NREM sleep alternate throughout the night with approximately 90-minute intervals between REM periods. This cycle repeats itself from 3 to 5 times during the entire sleep period. The REM periods start out lasting from 5 to 10 minutes and gradually lengthen until the final one lasts about 50 minutes. In a normal night's sleep, REM totals 90 to 120 minutes. As much as 50 percent of an infant's sleep is REM as contrasted with 20 percent for adults. Most sedatives significantly decrease REM sleep.

CLINICAL APPLICATION

A **polysomograph** (*poly* = many; *somnus* = sleep; *graph* = to write) is an instrument that uses electrodes to record several physiological variables during sleep. Among these variables are brain activity recorded as an electroencephalogram (EEG), eye movements recorded as an electrooculogram (EOG), and muscle tonus recorded as an electromyogram (EMG). These recordings indicate precisely when patients fall asleep, how many wake periods they experience, and the quality and duration of sleep.

STUDY OUTLINE

Sensations (p. 346)
Definition
1. Sensation is a state of awareness of external and internal conditions of the body.
2. The prerequisites for sensation are reception of a stimulus, conversion of the stimulus into a nerve impulse by a receptor, conduction of the impulse to the brain, and translation of the impulse into a sensation by a region of the brain.

Characteristics
1. Projection occurs when the brain refers a sensation to the point of stimulation.
2. Adaptation is the loss of sensation even though the stimulus is still applied.
3. An afterimage is the persistence of a sensation even though the stimulus is removed.
4. Modality is the property by which one sensation is distinguished from another.

Classification of Receptors
1. According to location, receptors are classified as exteroceptors, visceroceptors, and proprioceptors.
2. On the basis of type of stimulus detected, receptors are classified as mechanoreceptors, thermoreceptors, nociceptors, electromagnetic receptors, and chemoreceptors.
3. In terms of simplicity or complexity, simple receptors are associated with general senses and complex receptors are associated with special senses.

General Senses (p. 347)
Cutaneous Sensations
1. Cutaneous sensations include tactile sensations (touch, pressure, vibration), thermoreceptive sensations (heat and cold), and pain. Receptors for these sensations are located in the skin, connective tissues, and the ends of the gastrointestinal tract.
2. Receptors for touch are root hair plexuses, free nerve endings, Merkel's discs, Meissner's corpuscles, and end organs of Ruffini. Receptors for pressure are free nerve endings, end organs of Ruffini, and Pacinian corpuscles. Receptors for vibration are Meissner's corpuscles and Pacinian corpuscles.
3. Pain receptors are located in nearly every body tissue.
4. Two kinds of pain recognized in the parietal lobe of the cortex are somatic and visceral.
5. Referred pain is felt in the skin near or away from the organ sending pain impulses.
6. Phantom pain is the sensation of pain in a limb that has been amputated.
7. Pain impulses may be inhibited by drugs, surgery, and acupuncture.

Proprioceptive Sensations
1. Receptors located in muscles, tendons, and joints convey impulses related to muscle tone, movement of body parts, and body position.
2. The receptors include joint kinesthetic receptors, muscle spindles, and tendon organs.

Levels of Sensation
1. Sensory fibers terminating in the lower brain stem bring about far more complex motor reactions than simple spinal reflexes.
2. When sensory impulses reach the lower brain stem, they cause subconscious motor reactions.
3. Sensory impulses that reach the thalamus can be localized crudely in the body.
4. When sensory impulses reach the cerebral cortex, we experience precise localization.

Sensory Pathways (p. 351)
1. Sensory information from all parts of the body terminates in a specific area of the somatosensory cortex.
2. In the posterior column pathway and the spinothalamic pathway there are first-order, second-order, and third-order neurons.
3. The neural pathway for pain and temperature is the lateral spinothalamic pathway.
4. The neural pathway for crude touch and pressure is the anterior spinothalamic pathway.
5. The neural pathway for light touch, proprioception, and vibration is the posterior column pathway.
6. The pathways to the cerebellum are the anterior and posterior spinocerebellar tracts.

Motor Pathways (p. 353)
1. The muscles of all parts of the body are controlled by a specific area of the motor cortex.
2. Voluntary motor impulses are conveyed from the brain through the spinal cord along the pyramidal pathways and the extrapyramidal pathways.
3. Pyramidal pathways include the lateral corticospinal, anterior corticospinal, and corticobulbar tracts.
4. Major extrapyramidal tracts are the rubrospinal, tectospinal, and vestibulospinal tracts.

Integrative Functions (p. 356)

1. Memory is defined as the ability to recall thoughts; it consists of activated and long-term components.
2. Sleep and wakefulness are integrative functions that are controlled by the reticular activating system (RAS).
3. Nonrapid eye movement (NREM) sleep consists of four stages identified by EEG recordings.
4. Most dreaming occurs during rapid eye movement (REM) sleep.

REVIEW QUESTIONS

1. Define a sensation and a sense receptor. What prerequisites are necessary for the perception of a sensation?
2. Describe the following characteristics of a sensation: projection, adaptation, afterimage, modality.
3. Name some examples of adaptation not discussed in the text.
4. Classify receptors on the basis of location, stimulus detected, and simplicity or complexity.
5. Distinguish between a general sense and a special sense.
6. What is a cutaneous sensation? Distinguish tactile, thermoreceptive, and pain sensations.
7. How are cutaneous receptors distributed over the body? Relate your response to the two-point discrimination test.
8. For each of the following cutaneous sensations, describe the receptor involved in terms of structure, function, and location: touch, pressure, vibration, and pain.
9. How do cutaneous sensations help maintain homeostasis?
10. Why are pain receptors important? Differentiate somatic pain, visceral pain, referred pain, and phantom pain.
11. Why is the concept of referred pain useful to the physician in diagnosing internal disorders?
12. What is acupuncture? Describe the "gate control" theory of acupuncture.
13. How is acupuncture currently being used in the United States?
14. What is the proprioceptive sense? Where are the receptors for this sense located?
15. Relate proprioception to the maintenance of homeostasis.
16. Describe how various parts of the body are represented in the somatosensory cortex.
17. Distinguish between the posterior column and spinothalamic pathways.
18. What is the sensory pathway for pain and temperature? How does it function?
19. Which pathway controls crude touch and pressure? How does it function?
20. Which pathway is responsible for light touch, proprioception, and vibration? How does it function?
21. Describe how various parts of the body are represented in the motor cortex.
22. Which pathways control voluntary motor impulses from the brain through the spinal cord? How are they distinguished?
23. Define memory. What are the two kinds of memory?
24. Describe how sleep and wakefulness are related to the reticular activating system.
25. What are the four stages of nonrapid eye movement (NREM) sleep? How is NREM sleep distinguished from rapid eye movement (REM) sleep?

- Compare the structural and functional differences between the somatic efferent and autonomic portions of the nervous system.

- Identify the principal structural features of the autonomic nervous system.

- Compare the sympathetic and parasympathetic divisions of the autonomic nervous system in terms of structure, physiology, and chemical transmitters released.

- Describe a visceral autonomic reflex and its components.

- Explain the role of the hypothalamus and its relationship to the sympathetic and parasympathetic division.

- Explain the relationship between biofeedback and the autonomic nervous system.

- Describe the relationship between meditation and the autonomic nervous system.

16 The Autonomic Nervous System

The portion of the nervous system that regulates the activities of smooth muscle, cardiac muscle, and glands is the **autonomic nervous system (ANS).** Structurally, the system consists of visceral efferent neurons organized into nerves, ganglia, and plexuses. Functionally, it usually operates without conscious control. The system was originally named *autonomic* because physiologists thought it functioned with no control from the central nervous system, that it was autonomous or self-governing. It is now known that the autonomic system is neither structurally nor functionally independent of the central nervous system. It is regulated by centers in the brain, in particular by the cerebral cortex, hypothalamus, and medulla oblongata. However, the old terminology has been retained, and since the autonomic nervous system does differ from the somatic nervous system in some ways, the two are separated for convenience of study.

SOMATIC EFFERENT AND AUTONOMIC NERVOUS SYSTEMS

Whereas the somatic efferent nervous system produces conscious movement in skeletal muscles, the autonomic nervous system (visceral efferent nervous system) regulates visceral activities, and it generally does so involuntarily and automatically. Examples of visceral activities regulated by the autonomic nervous system are changes in the size of the pupil, accommodation for near vision, dilation and constriction of blood vessels, adjustment of the rate and force of the heartbeat, movements of the gastrointestinal tract, and secretion by most glands. These activities usually lie beyond conscious control. They are automatic.

The autonomic nervous system is entirely motor. All its axons are efferent fibers, which transmit impulses from the central nervous system to visceral effectors. Autonomic fibers are called **visceral efferent fibers. Visceral effectors** include cardiac muscle, smooth muscle, and glandular epithelium. This does not mean there are no afferent (sensory) impulses from visceral effectors, however. Impulses that give rise to visceral sensations pass over visceral afferent neurons that have cell bodies located in the posterior (dorsal) root ganglia of spinal nerves.

Some functions of these afferent neurons were described with the cranial and spinal nerves. The hypothalamus, which largely controls the autonomic nervous system, also receives impulses from the visceral sensory fibers.

The autonomic nervous system consists of two principal divisions: the **sympathetic** and the **parasympathetic.** Many organs innervated by the autonomic nervous system receive visceral efferent neurons from both components of the autonomic system—one set from the sympathetic division, another from the parasympathetic division. In general, impulses transmitted by the fibers of one division stimulate the organ to start or increase activity, whereas impulses from the other division decrease the organ's activity. Organs that receive impulses from both sympathetic and parasympathetic fibers are said to have *dual innervation.* In the somatic efferent nervous system, only one kind of motor neuron innervates an organ, which is always a skeletal muscle. When the somatic neurons stimulate the cells of the skeletal muscle, the muscle becomes active. When the neuron ceases to stimulate the muscle, contraction stops altogether.

STRUCTURE OF THE AUTONOMIC NERVOUS SYSTEM

VISCERAL EFFERENT PATHWAYS

Autonomic visceral efferent pathways always consist of two neurons. One extends from the central nervous system to a ganglion. The other extends directly from the ganglion to the effector (muscle or gland).

The first of the visceral efferent neurons in an autonomic pathway is called a **preganglionic neuron** (Figure 16-1). Its cell body is in the brain or spinal cord. Its myelinated axon, called a **preganglionic fiber,** passes out of the central nervous system as part of a cranial or spinal nerve. At some point, the fiber separates from the nerve and courses to an autonomic ganglion, where it synapses with the dendrites or cell body of the postganglionic neuron, the second neuron in the visceral efferent pathway.

The **postganglionic neuron** lies entirely outside the central nervous system. Its cell body and dendrites (if it has dendrites) are located in the autonomic ganglion,

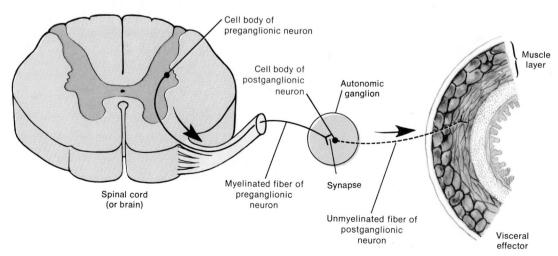

FIGURE 16-1 Relationship between preganglionic and postganglionic neurons.

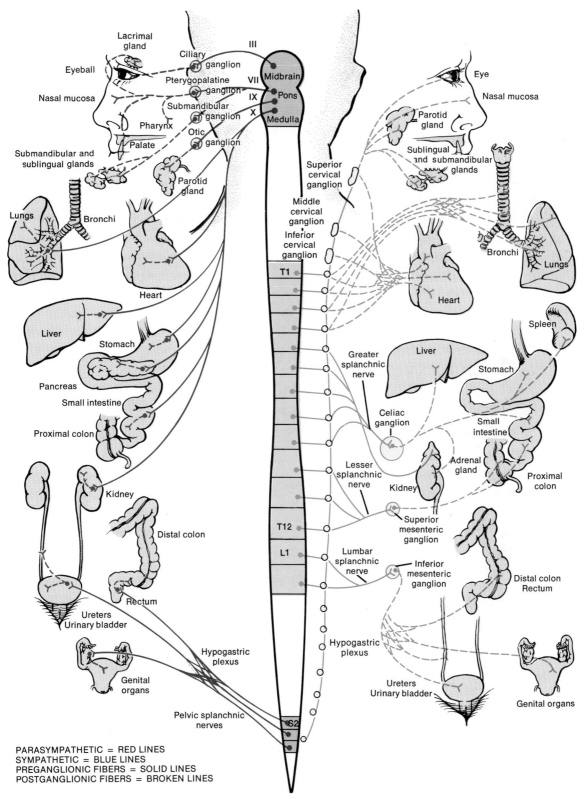

PARASYMPATHETIC = RED LINES
SYMPATHETIC = BLUE LINES
PREGANGLIONIC FIBERS = SOLID LINES
POSTGANGLIONIC FIBERS = BROKEN LINES

where the synapse with the preganglionic fibers occurs. The axon of a postganglionic neuron, called a **postganglionic fiber,** is unmyelinated and terminates in a visceral effector.

Thus preganglionic neurons convey efferent impulses from the central nervous system to autonomic ganglia. Postganglionic neurons relay the impulses from autonomic ganglia to visceral effectors.

FIGURE 16-2 Structure of the autonomic nervous system. Although the parasympathetic division is shown only on the left side of the figure and the sympathetic division is shown only on the right side, keep in mind that each division is actually on both sides (bilateral symmetry).

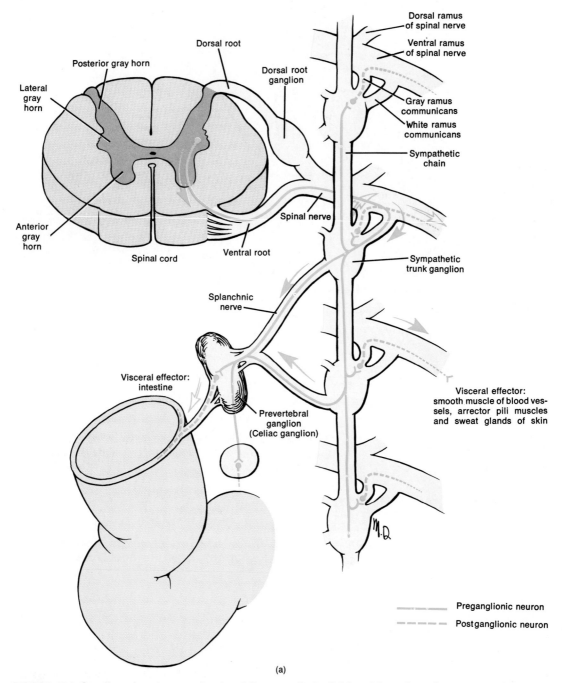

FIGURE 16-3 Ganglia and rami communicantes of the sympathetic division of the autonomic nervous system. (a) Diagram.

Preganglionic Neurons

In the sympathetic division, the preganglionic neurons have their cell bodies in the lateral gray horns of the twelve thoracic segments and first two lumbar segments of the spinal cord (Figure 16-2). It is for this reason that the sympathetic division is also called the **thoracolumbar** (thō'-ra-kō-LUM-bar) **division** and the fibers of the sympathetic preganglionic neurons are known as the **thoracolumbar outflow.**

The cell bodies of the preganglionic neurons of the parasympathetic division are located in the nuclei of cranial nerves III, VII, IX, and X in the brain stem and in the lateral gray horns of the second through fourth sacral segments of the spinal cord. Hence the parasympathetic division is also known as the **craniosacral division,** and the fibers of the parasympathetic preganglionic neurons are referred to as the **craniosacral outflow.**

Autonomic Ganglia

Autonomic pathways include **autonomic ganglia,** where synapses between visceral efferent neurons occur. Autonomic ganglia differ from posterior root ganglia. The latter contain cell bodies of sensory neurons and no synapses occur in them.

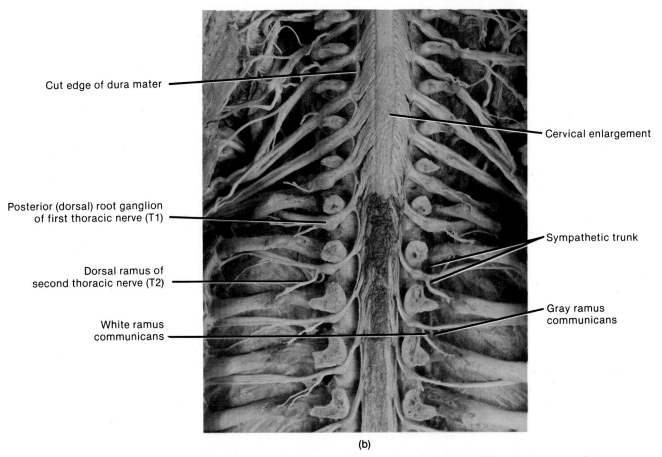

Cut edge of dura mater

Cervical enlargement

Posterior (dorsal) root ganglion
of first thoracic nerve (T1)

Sympathetic trunk

Dorsal ramus of
second thoracic nerve (T2)

Gray ramus
communicans

White ramus
communicans

(b)

FIGURE 16-3 (*Continued*) Ganglia and rami communicantes of the sympathetic division of the autonomic nervous system. (b) Photograph of the posterior aspect of the lower cervical and upper thoracic segments of the spinal cord. (Courtesy of N. Gluhbegovic and T. H. Williams, *The Human Brain: A Photographic Guide*, Harper & Row, New York, 1980.)

The autonomic ganglia may be divided into three general groups. The **sympathetic trunk** or **vertebral chain ganglia** are a series of ganglia that lie in a vertical row on either side of the vertebral column, extending from the base of the skull to the coccyx (Figure 16-3). They are also known as **paravertebral** or **lateral ganglia.** They receive preganglionic fibers only from the thoracolumbar (sympathetic) division (Figure 16-2).

The second kind of autonomic ganglion also belongs to the sympathetic division. It is called a **prevertebral** or **collateral ganglion** (Figure 16-3). The ganglia of this group lie anterior to the spinal column and close to the large abdominal arteries from which their names are derived. Examples of prevertebral ganglia so named are the celiac ganglion, on either side of the celiac artery just below the diaphragm; the superior mesenteric ganglion, near the beginning of the superior mesenteric artery in the upper abdomen; and the inferior mesenteric ganglion, located near the beginning of the inferior mesenteric artery in the middle of the abdomen (Figure 16-2). Prevertebral ganglia receive preganglionic fibers from the thoracolumbar (sympathetic) division.

The third kind of autonomic ganglion belongs to the parasympathetic division and is called a **terminal** or **intramural ganglion.** The ganglia of this group are located at the end of a visceral efferent pathway very close to visceral effectors or within the walls of visceral effectors. Terminal ganglia receive preganglionic fibers from the craniosacral (parasympathetic) division. The pregan-

glionic fibers do not pass through sympathetic trunk ganglia (Figure 16-2).

In addition to autonomic ganglia, the autonomic nervous system contains **autonomic plexuses.** Slender nerve fibers from ganglia containing postganglionic nerve cell bodies arranged in a branching network constitute an autonomic plexus.

Postganglionic Neurons

Axons from preganglionic neurons of the sympathetic division pass to ganglia of the sympathetic trunk. They can either synapse in the sympathetic chain ganglia with postganglionic sympathetics or they can continue, without synapsing, through the chain ganglia to end at a prevertebral ganglion where synapses with the postganglionic sympathetics can take place. Each sympathetic preganglionic fiber synapses with several postganglionic fibers in the ganglion, and the postganglionic fibers pass to several visceral effectors. Upon exiting their ganglia, the postsynaptic fibers innervate their visceral effectors.

Axons from preganglionic neurons of the parasympathetic division pass to terminal ganglia near or within a visceral effector. In the ganglion, the presynaptic neuron usually synapses with only four or five postsynaptic neurons to a single visceral effector. Upon exiting their ganglia, the postsynaptic fibers supply their visceral effectors.

With this background in mind, we can now examine some specific structural features of the sympathetic and parasympathetic divisions of the autonomic nervous system.

SYMPATHETIC DIVISION

The preganglionic fibers of the sympathetic division have their cell bodies located in the lateral gray horn of the spinal cord in the thoracic and first two lumbar segments (Figure 16-2). The preganglionic fibers are myelinated and leave the spinal cord through the ventral root of a spinal nerve along with the somatic efferent fibers at the same segmental levels. After exiting through the intervertebral foramina, the preganglionic sympathetic fibers enter a white ramus to pass to the nearest sympathetic trunk ganglion on the same side. Collectively, the white rami are called the **white rami communicantes** (kō-myoo-ne-KAN-tēz). Their name indicates that they contain myelinated fibers. Only thoracic and upper lumbar nerves have white rami communicantes. The white rami communicantes connect the ventral ramus of the spinal nerve with the ganglia of the sympathetic trunk.

The paired sympathetic trunks are situated anterolaterally to the spinal cord, one on either side. Each consists of a series of ganglia arranged more or less segmentally. The divisions of the sympathetic trunk are named on the basis of location. Typically, there are 22 ganglia in each chain: 3 cervical, 11 thoracic, 4 lumbar, and 4 sacral. Although the trunk extends downward from the neck, thorax, and abdomen to the coccyx, it receives preganglionic fibers only from the thoracic and lumbar segments of the spinal cord (Figure 16-2).

The cervical portion of each sympathetic trunk is located in the neck anterior to the prevertebral muscles. It is subdivided into a superior, middle, and inferior ganglion (Figure 16-2). The *superior cervical ganglion* is posterior to the internal carotid artery and anterior to the transverse processes of the second cervical vertebra. Postganglionic fibers leaving the ganglion serve the head, where they are distributed to the sweat glands, the smooth muscle of the eye and blood vessels of the face, the nasal mucosa, and the submandibular, sublingual, and parotid salivary glands. Gray rami communicantes from the ganglion also pass to the upper two to four cervical spinal nerves. The *middle cervical ganglion* is situated near the sixth cervical vertebra at the level of the cricoid cartilage. Postganglionic fibers from it innervate the heart. The *inferior cervical ganglion* is located near the first rib, anterior to the transverse processes of the seventh cervical vertebra. Its postganglionic fibers also supply the heart.

The thoracic portion of each sympathetic trunk usually consists of 11 segmentally arranged ganglia, lying ventral to the necks of the corresponding ribs. This portion of the sympathetic trunk receives most of the sympathetic preganglionic fibers. Postganglionic fibers from the thoracic sympathetic trunk innervate the heart, lungs, bronchi, and other thoracic viscera.

The lumbar portion of each sympathetic trunk is found on either side of the corresponding lumbar vertebrae. The sacral portion of the sympathetic trunk lies in the pelvic cavity on the medial side of the sacral foramina. Postganglionic fibers from the lumbar and sacral sympathetic chain ganglia are distributed with the respective spinal nerves via gray rami, or they may join the hypogastric plexus via direct visceral branches.

When a preganglionic fiber of a white ramus communicans enters the sympathetic trunk, it may terminate (synapse) in several ways. Some fibers synapse in the first ganglion at the level of entry. Others pass up or down the sympathetic trunk for a variable distance to form the fibers on which the ganglia are strung. These fibers, known as **sympathetic chains** (Figure 16-3), may not synapse until they reach a ganglion in the cervical or sacral area. Most rejoin the spinal nerves before supplying peripheral visceral effectors such as sweat glands and the smooth muscle in blood vessels and around hair follicles in the extremities. The **gray ramus communicans** (kō-MYOO-ne-kanz) is the structure containing the postganglionic fibers that connect the ganglion of the sympathetic trunk to the spinal nerve (Figure 16-3). The fibers are unmyelinated. All spinal nerves have gray rami communicantes. Gray rami communicantes outnumber the white rami, since there is a gray ramus leading to each of the 31 pairs of spinal nerves.

In most cases, a sympathetic preganglionic fiber terminates by synapsing with a large number of postganglionic cell bodies in a ganglion, usually 20 or more. Often the postganglionic fibers then terminate in widely separated organs of the body. Thus an impulse that starts in a single preganglionic neuron may reach several visceral effectors. For this reason, most sympathetic responses have widespread effects on the body.

Some preganglionic fibers pass through the sympathetic trunk without terminating in the trunk. Beyond the trunk, they form nerves known as **splanchnic** (SPLANK-nik) **nerves** (Figure 16-2). After passing through the trunk of ganglia, the splanchnic nerves terminate in the *celiac* (SĒ-lē-ak) or *solar plexus*. In the plexus, the preganglionic fibers synapse in ganglia with postganglionic cell bodies. These ganglia are prevertebral ganglia. The greater splanchnic nerve passes to the celiac ganglion of the celiac plexus. From here, postganglionic fibers are distributed to the stomach, spleen, liver, kidney, and small intestine. The lesser splanchnic nerve passes through the celiac plexus to the superior mesenteric ganglion of the superior mesenteric plexus. Postganglionic fibers from this ganglion innervate the small intestine and colon. The lowest splanchnic nerve, not always present, enters the renal plexus. Postganglionics supply the renal artery and ureter. The lumbar splanchnic nerve enters the inferior mesenteric plexus. In the plexus, the preganglionic fibers synapse with postganglionic fibers in the inferior mesenteric ganglion. These fibers pass through the hypogastric plexus and supply the distal colon and rectum, urinary bladder, and genital organs. As noted earlier, the postganglionic fibers leaving the prevertebral ganglia follow the course of various arteries to abdominal and pelvic visceral effectors.

PARASYMPATHETIC DIVISION

The preganglionic cell bodies of the parasympathetic division are found in nuclei in the brain stem and the lateral gray horn of the second through fourth sacral segments of the spinal cord (Figure 16-2). Their fibers emerge as part of a cranial nerve or as part of the ventral root of a spinal nerve. The **cranial parasympathetic outflow** consists of preganglionic fibers that leave the brain stem by way of the oculomotor (III) nerves, facial (VII) nerves, glossopharyngeal (IX) nerves, and vagus (X) nerves. The **sacral parasympathetic outflow** consists of preganglionic fibers that leave the ventral roots of the second through fourth sacral nerves. The preganglionic fibers of both the cranial and sacral outflows end in terminal ganglia, where they synapse with postganglionic neurons. We will first look at the cranial outflow.

The cranial outflow has five components: four pairs of ganglia and the plexuses associated with the vagus nerve. The four pairs of cranial parasympathetic ganglia innervate structures in the head and are located close to the organs they innervate. The *ciliary ganglion* is near the back of an orbit lateral to each optic nerve. Preganglionic fibers pass with the oculomotor (III) nerve to the ciliary ganglion. Postganglionic fibers from the ganglion innervate smooth muscle cells in the eyeball. Each *pterygopalatine* (ter'-i-gō-PAL-a-tin) *ganglion* is situated lateral to a sphenopalatine foramen. It receives preganglionic fibers from the facial (VII) nerve and transmits postganglionic fibers to the nasal mucosa, palate, pharynx, and lacrimal gland. Each *submandibular ganglion* is found near the duct of a submandibular salivary gland. It receives preganglionic fibers from the facial (VII) nerve and transmits postganglionic fibers that innervate the submandibular and sublingual salivary glands. The *otic ganglia* are situated just below each foramen ovale. The otic ganglion receives preganglionic fibers from the glossopharyngeal (IX) nerve and transmits postganglionic fibers that innervate the parotid salivary gland. Ganglia associated with the cranial outflow are classified as terminal ganglia. Since the terminal ganglia are close to their visceral effectors, postganglionic parasympathetic fibers are short. Postganglionic sympathetic fibers are relatively long.

The last component of the cranial outflow consists of the preganglionic fibers that leave the brain via the vagus (X) nerves. This component has the most extensive distribution of the parasympathetic fibers. It provides about 80 percent of the craniosacral outflow. Each vagus (X) nerve enters into the formation of several plexuses in the thorax and abdomen. As it passes through the thorax, it sends fibers to the *superficial cardiac plexus* in the arch of the aorta and the *deep cardiac plexus* anterior to the branching of the trachea. These plexuses contain terminal ganglia, and the postganglionic parasympathetic fibers emerging from them supply the heart. Also in the thorax is the *pulmonary plexus,* in front of and behind the roots of the lungs and within the lungs themselves. It receives preganglionic fibers from the vagus and transmits postganglionic parasympathetic fibers to the lungs and bronchi. Other plexuses associated with the vagus (X) nerve are described in later chapters in conjunction with the appropriate thoracic, abdominal, and pelvic viscera. Postganglionic fibers from these plexuses innervate viscera such as the liver, pancreas, stomach, kidneys, small intestine, and part of the colon.

The sacral parasympathetic outflow consists of pregan-

EXHIBIT 16-1

STRUCTURAL FEATURES OF SYMPATHETIC AND PARASYMPATHETIC DIVISIONS

SYMPATHETIC	PARASYMPATHETIC
Forms thoracolumbar outflow.	Forms craniosacral outflow.
Contains sympathetic trunk and prevertebral ganglia.	Contains terminal ganglia.
Ganglia are close to the CNS and distant from visceral effectors.	Ganglia are near or within visceral effectors.
Each preganglionic fiber synapses with many postganglionic neurons that pass to many visceral effectors.	Each preganglionic fiber usually synapses with four or five postganglionic neurons that pass to a single visceral effector.
Distributed throughout the body, including the skin.	Distribution limited primarily to head and viscera of thorax, abdomen, and pelvis.

glionic fibers from the ventral roots of the second through fourth sacral nerves. Collectively, they form the *pelvic splanchnic nerves.* They pass into the hypogastric plexus. From ganglia in the plexus, parasympathetic postganglionic fibers are distributed to the colon, ureters, urinary bladder, and reproductive organs.

The salient structural features of the sympathetic and parasympathetic divisions are compared in Exhibit 16-1.

PHYSIOLOGY

TRANSMITTER SUBSTANCES

Autonomic fibers, like other axons of the nervous system, release transmitter substances at synapses as well as at points of contact with visceral effectors. These latter points are called **neuroeffector junctions.** Neuroeffector junctions may be either neuromuscular or neuroglandular junctions. On the basis of the transmitter substance produced, autonomic fibers may be classified as either cholinergic or adrenergic.

Cholinergic (kō'-lin-ER-jik) **fibers** release **acetylcholine (ACh)** and include the following: (1) all sympathetic and parasympathetic preganglionic axons, (2) all parasympathetic postganglionic axons, and (3) some sympathetic postganglionic axons. The cholinergic sympathetic postganglionic axons include those to sweat glands and blood vessels in skeletal muscles, to the skin, and to the external genitalia. Since acetylcholine is quickly inactivated by the enzyme **acetylcholinesterase (AChE),** the effects of cholinergic fibers are short-lived and local.

Adrenergic (ad'-ren-ER-jik) **fibers** produce **norepinephrine (NE).** Most sympathetic postganglionic axons are adrenergic. Since norepinephrine is inactivated much more slowly by **catechol-o-methyltransferase (COMT)** or **monoamine oxidase (MAO)** than acetylcholine is by acetylcholinesterase, and since norepinephrine may enter the bloodstream, the effects of sympathetic stimulation are longer lasting and more widespread than parasympathetic stimulation.

EXHIBIT 16-2

ACTIVITIES OF AUTONOMIC NERVOUS SYSTEM

VISCERAL EFFECTOR	EFFECT OF SYMPATHETIC STIMULATION	EFFECT OF PARASYMPATHETIC STIMULATION
Eye		
Iris	Contraction of dilator muscle that results in dilation of pupil.	Contraction of sphincter muscle that results in constriction of pupil.
Ciliary muscle	No innervation.	Contraction that results in lens accommodation for near vision.
Glands		
Sweat	Stimulates secretion.	No innervation.
Lacrimal (tear)	Vasoconstriction, which inhibits secretion.	Normal or excessive secretion.
Salivary	Vasoconstriction, which decreases secretion.	Stimulates secretion and vasodilation.
Gastric	Vasoconstriction, which inhibits secretion.	Stimulates secretion.
Intestinal	Vasoconstriction, which inhibits secretion.	Stimulates secretion.
Adrenal medulla	Promotes epinephrine and norepinephrine secretion.	No innervation.
Adrenal cortex	Promotes glucocorticoid secretion.	No innervation.
Lungs (bronchial tubes)	Dilation.	Constriction.
Heart	Increases rate and strength of contraction; dilates coronary vessels that supply blood to heart muscle cells.	Decreases rate and strength of contraction; constricts coronary vessels.
Blood vessels		
Skin	Constriction.	No innervation for most.
Skeletal muscle	Dilation.	No innervation.
Visceral organs (except heart and lungs)	Constriction.	No innervation for most.
Liver	Promotes glycogenolysis; decreases bile secretion.	Promotes glycogenesis; increases bile secretion.
Gallbladder	Relaxation.	Contraction.
Stomach	Decreases motility.	Increases motility.
Intestines	Decreases motility.	Increases motility.
Gastrointestinal sphincters	Increases tone.	Relaxation.
Kidney	Constriction of blood vessels that results in decreased urine volume.	No innervation.
Pancreas	Inhibits secretion.	Promotes secretion.
Spleen	Contraction and discharge of stored blood into general circulation.	No innervation.
Urinary bladder	Relaxation of muscular wall; increases tone in internal sphincter.	Contraction of muscular wall; relaxation of internal sphincter.
Arrector pili of hair follicles	Contraction that results in erection of hairs.	No innervation.
Uterus	Inhibits contraction if nonpregnant; stimulates contraction if pregnant.	Minimal effect.
Sex organs	In male, vasoconstriction of ductus deferens, seminal vesicle, prostate; results in ejaculation. In female, reverse uterine peristalsis.	Vasodilation and erection in both sexes; secretion in female.

ACTIVITIES

Most visceral effectors have dual innervation, that is, they receive fibers from both the sympathetic and the parasympathetic divisions. In these cases, impulses from one division stimulate the organ's activities, whereas impulses from the other division inhibit the organ's activities. The stimulating division may be either the sympathetic or the parasympathetic, depending on the organ. For example, sympathetic impulses increase heart activity, whereas

parasympathetic impulses decrease it. On the other hand, parasympathetic impulses increase digestive activities, whereas sympathetic impulses inhibit them. The actions of the two systems are carefully integrated to help maintain homeostasis. A summary of the activities of the autonomic nervous system is presented in Exhibit 16-2.

The parasympathetic division is primarily concerned with activities that restore and conserve body energy. It is a **rest-repose system.** Under normal body conditions, for instance, parasympathetic impulses to the digestive glands and the smooth muscle of the digestive system dominate over sympathetic impulses. Thus energy-supplying food can be digested and absorbed by the body.

The sympathetic division, by contrast, is primarily concerned with processes involving the expenditure of energy. When the body is in homeostasis, the main function of the sympathetic division is to counteract the parasympathetic effects just enough to carry out normal processes requiring energy. During extreme stress, however, the sympathetic dominates the parasympathetic. When people are confronted with a stress condition, for example, their bodies become alert and they sometimes perform feats of unusual strength. Fear stimulates the sympathetic division.

Activation of the sympathetic division sets into operation a series of physiological responses collectively called the **fight-or-flight response.** It produces the following effects.

  1. The pupils of the eyes dilate.

  2. The heart rate increases.

  3. The blood vessels of the skin and viscera constrict.

  4. The remainder of the blood vessels dilate. This reaction causes a rise in blood pressure and a faster flow of blood into the dilated blood vessels of skeletal muscles, cardiac muscle, lungs, and brain—organs involved in fighting off danger.

  5. Rapid breathing occurs as the bronchioles dilate to allow faster movement of air in and out of the lungs.

  6. Blood sugar level rises as liver glycogen is converted to glucose to supply the body's additional energy needs.

  7. The medulla of the adrenal gland is stimulated to produce epinephrine and norepinephrine, hormones that intensify and prolong the sympathetic effects noted above.

  8. Processes that are not essential for meeting the stress situation are inhibited. For example, muscular movements of the gastrointestinal tract and digestive secretions are slowed down or even stopped.

CLINICAL APPLICATION

If the sympathetic trunk is cut on one side, the sympathetic supply to that side of the head is removed and the result is **Horner's syndrome,** in which the patient exhibits (on the affected side): ptosis (drooping of the upper eyelid), slight elevation of the lower eyelid, narrowing of the palpebral fissure, enophthalmos (the eye appears sunken), a constricted pupil, anhydrosis (lack of sweating), and flushing of the skin.

VISCERAL AUTONOMIC REFLEXES

A **visceral autonomic reflex** adjusts the activity of a visceral effector. In other words, it results in the contraction of smooth or cardiac muscle or secretion by a gland. Such reflexes assume a key role in activities such as regulating heart action, blood pressure, respiration, digestion, defecation, and urinary bladder functions.

A visceral autonomic reflex arc consists of the following components.

  1. Receptor. The receptor is the distal end of an afferent neuron in an exteroceptor or enteroceptor.

  2. Afferent neuron. This neuron, either a somatic afferent or visceral afferent neuron, conducts the sensory impulse to the spinal cord or brain.

  3. Association neurons. These neurons are found in the central nervous system.

  4. Visceral efferent preganglionic neuron. In the thoracic and abdominal regions, this neuron is in the lateral gray horn of the spinal cord. The axon passes through the ventral root of the spinal nerve, the spinal nerve, and the white ramus communicans. It then enters a sympathetic trunk or prevertebral ganglion, where it synapses with a postganglionic neuron. In the cranial and sacral regions, the visceral efferent preganglionic axon leaves the central nervous system and passes to a terminal ganglion, where it synapses with a postganglionic neuron. The role of the visceral efferent preganglionic neuron is to convey a motor impulse from the brain or spinal cord to an autonomic ganglion.

  5. Visceral efferent postganglionic neuron. This neuron conducts a motor impulse from a visceral efferent preganglionic neuron to the visceral effector.

  6. Visceral effector. A visceral effector is smooth muscle, cardiac muscle, or a gland.

The basic difference between a somatic reflex arc and a visceral autonomic reflex arc is that in a somatic reflex arc, only one efferent neuron is involved. In a visceral autonomic reflex arc, two efferent neurons are involved.

Visceral sensations do not always reach the cerebral cortex. Most remain at subconscious levels. Under normal conditions, you are not aware of muscular contractions of the digestive organs, heartbeat, changes in the diameter of blood vessels, and pupil dilation and constriction. When your body is making adjustments in such visceral activities, they are handled by visceral reflex arcs whose centers are in the spinal cord or lower regions of the brain. Among such centers are the cardiac, respiratory, vasomotor, swallowing, and vomiting centers in the medulla and the temperature control center in the hypothalamus. Stimuli delivered by somatic or visceral afferent neurons synapse in these centers, and the returning motor impulses conducted by visceral efferent neurons bring about an adjustment in the visceral effector without conscious recognition. The impulses are interpreted and acted on subconsciously. Some visceral sensations do give rise to conscious recognition: hunger, nausea, and fullness of the urinary bladder and rectum.

CONTROL BY HIGHER CENTERS

The autonomic nervous system is not a separate nervous system. Axons from many parts of the central nervous system are connected to both the sympathetic and the parasympathetic divisions of the autonomic nervous system and thus exert considerable control over it. Autonomic centers in the cerebral cortex are connected to autonomic centers of the thalamus, for example. These, in turn, are connected to the hypothalamus. In this hierarchy of command, the thalamus sorts incoming impulses before they reach the cerebral cortex. The cerebral cortex then turns over control and integration of visceral activities to the hypothalamus. It is at the level of the hypothalamus that the major control and integration of the autonomic nervous system is exerted.

The hypothalamus is connected to both the sympathetic and the parasympathetic divisions of the autonomic nervous system. The posterior and lateral portions of the hypothalamus appear to control the sympathetic division.

When these areas are stimulated, there is an increase in visceral activities—an increase in heart rate, a rise in blood pressure due to vasoconstriction of blood vessels, an increase in the rate and depth of respiration, dilation of the pupils, and inhibition of the digestive tract. On the other hand, the anterior and medial portions of the hypothalamus seem to control the parasympathetic division. Stimulation of these areas results in a decrease in heart rate, lowering of blood pressure, constriction of the pupils, and increased motility of the digestive tract.

Control of the autonomic nervous system by the cerebral cortex occurs primarily during emotional stress. In extreme anxiety, the cerebral cortex can stimulate the hypothalamus. This stimulation, in turn, increases heart rate and blood pressure. If the cortex is stimulated by an extremely unpleasant sight, the stimulation causes vasodilation of blood vessels, a lowering of blood pressure, and fainting.

Evidence of even more direct control of visceral responses is provided by data gathered from studies of biofeedback and meditation.

BIOFEEDBACK

In the simplest terms, **biofeedback** is a process in which people get constant signals, or feedback, about visceral body functions such as blood pressure, heart rate, and muscle tension. By using special monitoring devices, they can control these visceral functions consciously.

Suppose you are connected to a monitor that informs you of your heartbeat by means of lights. A red light indicates a fast heart beat, an amber light a normal rate, and a green light a slow rate. When you see the red light flash, you know your heart is beating too fast. You have been informed of a visceral response. This is the biofeedback. According to some researchers, you can be taught to slow down the heart rate by thinking of something pleasant and thus relaxing the body. The green light flashes and your reward is a slower heart rate. In similar experiments, some individuals have learned to control heart rhythm.

Researchers estimate that between 6 and 8 percent of the American population suffers from migraine headaches. Moreover, no effective treatment has been developed that does not have significant side effects and serious risks. One approach to alleviating migraine headaches is biofeedback.* The first clue that biofeedback could be used in this way came when a patient in the voluntary control laboratory of the Menninger Foundation demonstrated that with a 5°C (10°F) rise in hand skin temperature (as a result of vasodilation and increased blood flow) she could spontaneously recover from a migraine headache.

In a study conducted at the Menninger Foundation, subjects suffering from migraine headaches received instructions in the use of a monitor that registers the skin temperature of the right index finger. Subjects were also

* Much of the following discussion of the use of biofeedback for the treatment of migraine headaches is based upon the information provided by Dr. Joseph D. Sargent of the Menninger Foundation, Topeka, Kansas.

given a typewritten sheet containing two sets of phrases. The first set was designed to help them relax the entire body. The second set was designed to bring about an increased flow of blood in the hands. The subjects practiced raising their skin temperature at home for 5 to 15 minutes a day. When skin temperature increased, the monitor emitted a high-pitched sound. In time, the monitor was abandoned.

Once the subjects learned how to vasodilate their blood vessels, the migraine headaches lessened. Since migraine headaches are believed to involve a distension of blood vessels in the head, the shunting of blood from head to hands relieved the distension and thus the pain.

Other experiments have shown that biofeedback can be applied to childbirth. Women were given monitors hooked up to their fingers and arms to measure electrical conductivity of the skin and skeletal muscle tension. Both conductivity and tension increase with nervousness and make labor difficult. Muscle tension was recorded as a sirenlike sound that became louder with nervousness. Skin conductivity was recorded as a crackling noise that also increased with nervousness. The monitors kept the women informed of their nervousness. This was the biofeedback. Having pleasant thoughts reduced the sound levels. The reward was less nervousness. The results of the study indicate that the women needed less medication during labor and labor time itself was shortened.

There is no way to determine where biofeedback will lead. Perhaps the outstanding contribution of biofeedback research has been to demonstrate that the autonomic nervous system is not autonomous. Visceral responses can be controlled. Strong supporters point out that possible applications of biofeedback in medicine are legion. They envision its use in lowering blood pressure in patients with hypertension, altering heart rates and rhythms, relieving pain from migraine headaches, making delivery easier, and controlling anxiety related to a host of illnesses that may be linked to stress. One researcher concludes that "an important trend is beginning to take place in the areas of psychosomatic disorders and medicine. This is the increasing involvement of the patient in his own treatment. The traditional doctor-patient relationship is giving way slowly to a shared responsibility."

MEDITATION

Yoga, which literally means union, is defined as a higher consciousness achieved through a fully rested and relaxed body and a fully awake and relaxed mind. One widely practiced technique for achieving higher consciousness is called **transcendental meditation (TM).** One sits in a comfortable position with the eyes closed and concentrates on a suitable sound or thought.

Research indicates that transcendental meditation can alter physiological responses. Oxygen consumption decreases drastically along with carbon dioxide elimination. Subjects have experienced a reduction in metabolic rate and blood pressure. Researchers have also observed a decrease in heart rate, an increase in the intensity of alpha brain waves, a sharp decrease in the amount of lactic acid in the blood, and an increase in the skin's electrical resistance. These last four responses are characteristic of a highly relaxed state of mind. Alpha waves are found in the EEGs of almost all individuals in a resting, but awake state; they disappear during sleep.

These responses have been called an **integrated response**—essentially, a hypometabolic state due to inactivation of the sympathetic division of the autonomic nervous system. The response is the exact opposite of the fight-or-flight response, which is a hyperactive state of the sympathetic division. The existence of the integrated response suggests that the central nervous system does exert some control over the autonomic nervous system.

STUDY OUTLINE

Somatic Efferent and Autonomic Nervous Systems (p. 362)

1. The autonomic nervous system, or visceral efferent nervous system, regulates visceral activities, that is, activities of smooth muscle, cardiac muscle, and glands.
2. It usually operates without conscious control.
3. It is regulated by centers in the brain, in particular by the cerebral cortex, the hypothalamus, and the medulla oblongata.
4. The somatic efferent nervous system produces conscious movement in skeletal muscles.

Structure of the Autonomic Nervous System (p. 362)

1. The autonomic nervous system consists of visceral efferent neurons organized into nerves, ganglia, and plexuses.
2. It is entirely motor. All autonomic axons are efferent fibers.
3. Efferent neurons are preganglionic (with myelinated axons) and postganglionic (with unmyelinated axons).
4. The autonomic system consists of two principal divisions: sympathetic (thoracolumbar) and parasympathetic (craniosacral).
5. Autonomic ganglia are classified as sympathetic trunk ganglia (on sides of spinal column), prevertebral ganglia (anterior to spinal column), and terminal ganglia (near or inside visceral effectors).

Physiology (p. 367)

1. Autonomic fibers release chemical transmitters at synapses. On the basis of the transmitter produced, these fibers may be classified as cholinergic or adrenergic.
2. Cholinergic fibers release acetylcholine (ACh). Adrenergic fibers produce norepinephrine (NE).
3. Sympathetic responses are widespread and, in general, concerned with energy expenditure. Parasympathetic responses are restricted and are typically concerned with energy restoration and conservation.

Visceral Autonomic Reflexes (p. 369)

1. A visceral autonomic reflex adjusts the activity of a visceral effector.
2. A visceral autonomic reflex arc consists of a receptor, afferent neuron, association neuron, visceral efferent preganglionic neuron, visceral efferent postganglionic neuron, and visceral effector.

Control by Higher Centers (p. 369)

1. The hypothalamus controls and integrates the autonomic nervous system. It is connected to both the sympathetic and the parasympathetic divisions.
2. Biofeedback is a process in which people learn to monitor visceral functions and to control them consciously. It has been used to control heart rate, to alleviate migraine headaches, and to make childbirth easier.
3. Yoga is a higher consciousness achieved through a fully rested and relaxed body and a fully awake and relaxed mind.
4. Transcendental meditation (TM) produces the following physiological responses: decreased oxygen consumption and carbon dioxide elimination, reduced metabolic rate, decrease in heart rate, increase in the intensity of alpha brain waves, a sharp decrease in the amount of lactic acid in the blood, and an increase in the skin's electrical resistance.

REVIEW QUESTIONS

1. What are the principal components of the autonomic nervous system? What is its general function? Why is it called involuntary?
2. What is the principal anatomical difference between the voluntary nervous system and the autonomic nervous system?
3. Relate the role of visceral efferent fibers and visceral effectors to the autonomic nervous system.
4. Distinguish between preganglionic neurons and postganglionic neurons with respect to location and function.
5. What is an autonomic ganglion? Describe the location and function of the three types of autonomic ganglia. Define white and gray rami communicantes.
6. On what basis are the sympathetic and parasympathetic divisions of the autonomic nervous system differentiated anatomically and functionally?
7. Discuss the distinction between cholinergic and adrenergic fibers of the autonomic nervous system.
8. Give examples of the antagonistic effects of the sympathetic and parasympathetic divisions of the autonomic nervous system.
9. Summarize the principal functional differences between the voluntary nervous system and the autonomic nervous system.
10. Give the *sympathetic response* in a fear situation for each of the following body parts: hair follicles, sweat glands, iris of eye, lungs, spleen, adrenal glands, kidneys, urinary bladder, liver, heart, and blood vessels of the viscera, skeletal muscles, and skin.
11. Define a visceral autonomic reflex and give three examples.
12. Describe a complete visceral autonomic reflex in proper sequence.
13. Describe how the hypothalamus controls and integrates the autonomic nervous system.
14. Define biofeedback. Explain how it could be useful.
15. What is transcendental meditation (TM)? How is the integrated response related to the autonomic nervous system?

- Locate the receptors for olfaction and describe the neural pathway for smell.

- Identify the gustatory receptors and describe the neural pathway for taste.

- Explain the structure and physiology of the accessory visual structures.

- List the structural divisions of the eye.

- Discuss retinal image formation by describing refraction, accommodation, constriction of the pupil, convergence, and inverted image formation.

- Define emmetropia, myopia, hypermetropia, and astigmatism.

- Diagram and discuss the rhodopsin cycle responsible for light sensitivity of rods.

- Identify the afferent pathway of light impulses to the brain.

- Define the anatomical subdivisions of the ear.

- List the principal events in the physiology of hearing.

- Identify the receptor organs for equilibrium.

- Discuss the receptor organs' roles in maintaining static and dynamic equilibrium.

- Contrast the causes and symptoms of cataracts, glaucoma, conjunctivitis, trachoma, deafness, labyrinthine disease, Ménière's syndrome, impacted cerumen, otitis media, and motion sickness.

- Define medical terminology associated with the sense organs.

- Explain the actions of selected drugs on the sense organs.

17 The Special Senses

The special senses, like the general senses, allow us to detect changes in our environment. However, the special senses of smell, taste, sight, hearing, and equilibrium have receptor organs that are structurally more complex.

OLFACTORY SENSATIONS

STRUCTURE OF RECEPTORS

The receptors for the **olfactory** (ol-FAK-tō-rē) **sense** are located in the nasal epithelium in the superior portion of the nasal cavity on either side of the nasal septum (Figure 17-1). The nasal epithelium consists of two principal kinds of cells: supporting and olfactory. The **supporting cells** are columnar epithelial cells of the mucous membrane lining the nose. Olfactory glands in the mucosa keep the mucous membrane moist. The **olfactory cells** are bipolar neurons whose cell bodies lie between the supporting cells. The distal (free) end of each olfactory cell contains six to eight dendrites, called **olfactory hairs.** The hairs are believed to react to odors in the air and then to stimulate the olfactory cells, thus initiating the olfactory pathway.

STIMULATION OF RECEPTORS

In order for a substance to be smelled it must be volatile, that is, capable of entering into a gaseous state so that the gaseous particles can enter the nostrils. You have probably noticed that smell occurs in cycles each time inspiration occurs and that sensitivity to smell is greatly increased by sniffing. The substance to be smelled must be water-soluble so that it can dissolve in the mucus to make contact with olfactory cells. The substance must

also be lipid-soluble. Since the plasma membranes of olfactory hairs are largely lipid, the substance to be smelled must dissolve in the lipid covering to make contact with olfactory hairs in order to initiate an impulse.

Many attempts have been made to distinguish and classify the primary sensations of smell. One classification includes seven classes of primary sensations: camphoraceous, musky, floral, pepperminty, ethereal, pungent, and putrid. Recent data suggest there may be as many as fifty or more primary sensations of smell.

It is believed that olfactory cells react to olfactory stimuli in the same way that most sensory receptors react to their specific stimuli: first a generator potential is developed followed by initiation of a nerve impulse. Two principal theories have been formulated to explain how olfactory cells respond to each of the various primary sensations. The **chemical theory** assumes that there are different receptor chemicals in the membranes of olfactory hairs, each capable of reacting with a particular olfactory substance (stimulus). The interaction between chemical receptor and substance alters the permeability of the plasma membrane so that a generator potential is developed, followed by initiation of a nerve impulse. The **physical theory** holds that there are physical receptor sites on the plasma membranes of olfactory hairs that react with olfactory substances. This interaction then causes a change in membrane permeability, development of a generator potential, and initiation of a nerve impulse.

ADAPTATION AND ODOR THRESHOLDS

The sensation of smell happens quickly. Olfactory cells respond in milliseconds. Adaptation to odors also occurs rapidly and appears to involve the central nervous system.

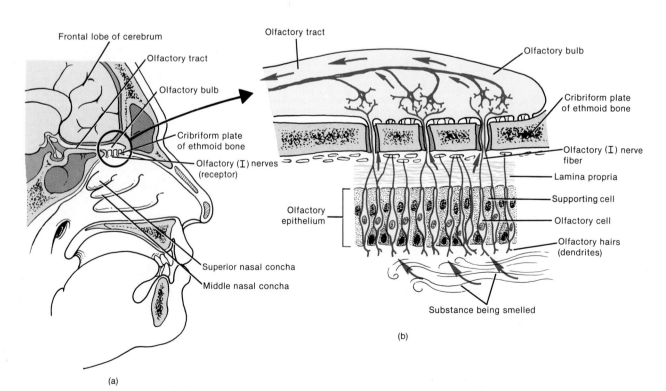

FIGURE 17-1 Olfactory receptors. (a) Location of receptors in nasal cavity. (b) Enlarged aspect of olfactory receptors.

Olfactory receptors themselves adapt about 50 percent in the first second or so after stimulation, but adapt very slowly after. Yet we know from common experience that complete adaptation to odors occurs in about a minute after exposure to a strong atmosphere, a rate far faster than can be explained on the basis of receptor adaptation alone. It has been suggested that most adaptation to odor involves a psychological component in the central nervous system. The mechanism and site of psychological adaptation are unknown.

One of the major characteristics of smell is its low threshold. Only a minute quantity of a substance need be present in air in order for it to be smelled (minimal identifiable odor or MIO). A good example is the substance methyl mercaptan which can be detected in concentrations as low as 1/25,000,000,000 mg per milliliter of air.

OLFACTORY PATHWAY

The unmyelinated axons of the olfactory cells unite to form the **olfactory (I) nerves,** which pass through foramina in the cribriform plate of the ethmoid bone. The olfactory (I) nerves terminate in paired masses of gray matter called the **olfactory bulbs.** The olfactory bulbs lie beneath the frontal lobes of the cerebrum on either side of the crista galli of the ethmoid bone. The first synapse of the olfactory neural pathway occurs in the olfactory bulbs between the axons of the olfactory (I) nerves and the dendrites of neurons inside the olfactory bulbs. Axons of these neurons run posteriorly to form the **olfactory tract.** From here, impulses are conveyed to the primary olfactory area of the cerebral cortex. In the cerebral cortex, the impulses are interpreted as odor and give rise to the sensation of smell. Impulses related to olfaction do not pass through the thalamus.

Both the supporting cells of the nasal epithelium and tear glands are innervated by branches of the trigeminal (V) nerve. The nerve receives stimuli of pain, cold, heat, tickling, and pressure. Olfactory stimuli such as pepper, ammonia, and chloroform are irritating and may cause tearing because they stimulate the lacrimal and nasal mucosal receptors of the trigeminal (V) nerve as well as the olfactory neurons.

GUSTATORY SENSATIONS

STRUCTURE OF RECEPTORS

The receptors for **gustatory** (GUS-ta-tō'-rē) **sensations,** or sensations of taste, are located in the taste buds (Figure 17-2). Taste buds are most numerous on the tongue, but they are also found on the soft palate and in the throat. The **taste buds** are oval bodies consisting of two kinds of cells. The **supporting cells** are a specialized epithelium that forms a capsule. Inside each capsule are 4 to 20 **gustatory cells.** Each gustatory cell contains a hairlike process **(gustatory hair)** that projects to the external surface through an opening in the taste bud called the **taste pore.** Gustatory cells make contact with taste stimuli through the taste pore.

Taste buds are found in some connective tissue elevations on the tongue called **papillae** (see Figure 24-5). The papillae give the upper surface of the tongue its rough appearance. **Circumvallate** (ser-kum-VAL-āt) or **vallate papillae,** the largest type, are circular and form an inverted V-shaped row at the posterior portion of the tongue. **Fungiform** (FUN-ji-form; meaning mushroom-shaped) **papillae** are knoblike elevations found primarily on the tip and sides of the tongue. All circumvallate and most fungiform papillae contain taste buds. **Filiform** (FIL-i-form) **papillae** are pointed threadlike structures that cover the anterior two-thirds of the tongue.

STIMULATION OF RECEPTORS

For gustatory cells to be stimulated, the substances we taste must be in solution in saliva so they can enter taste pores. Once the taste substance makes contact with plasma membranes of the gustatory hairs, the generator potential is developed. This is presumed to occur as a result of a membrane receptor–taste substance interaction. Then the generator potential initates a nerve impulse.

Despite the many substances we seem to taste, there are basically only four primary taste sensations: sour, salt, bitter, and sweet. All other "tastes," such as chocolate, pepper, and coffee, are combinations of these four modified by accompanying olfactory sensations.

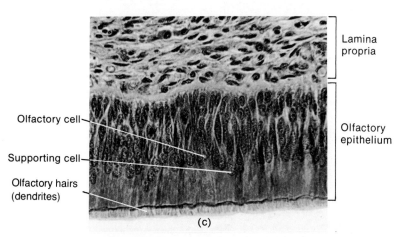

Lamina propria

Olfactory epithelium

Olfactory cell

Supporting cell

Olfactory hairs (dendrites)

(c)

FIGURE 17-1 (*Continued*) Olfactory receptors. (c) Photomicrograph of the olfactory mucosa at a magnification of 400×. (Courtesy of Donald I. Patt and Gail R. Patt, Harper & Row, Publishers, Inc., 1969.)

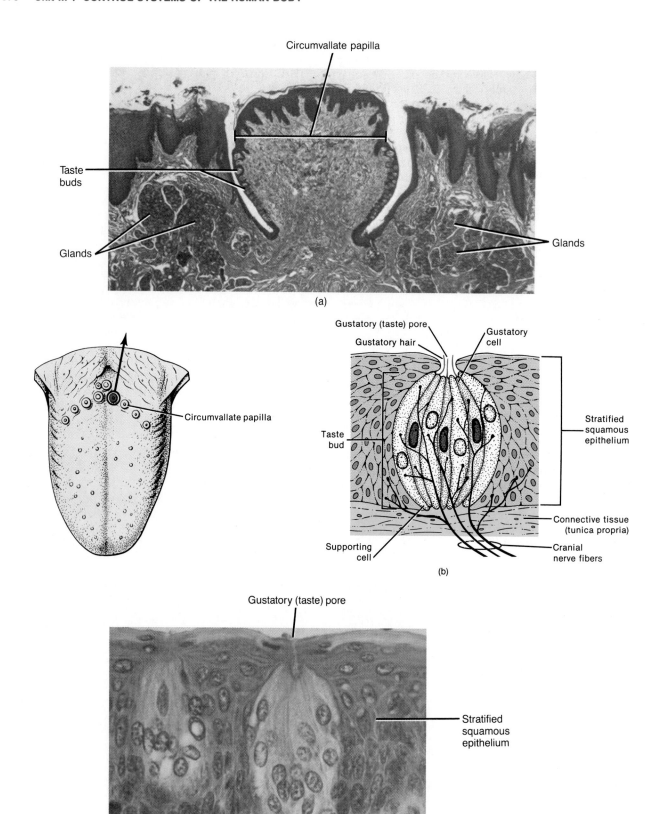

FIGURE 17-2 Gustatory receptors. (a) Photomicrograph of a circumvallate papilla containing taste buds at a magnification of 50×. (© 1983 by Michael H. Ross. Used by permission.) (b) Diagram of the structure of a taste bud. (c) Photomicrograph of a taste bud at a magnification of 600×. (© 1983 by Michael H. Ross. Used by permission.)

CLINICAL APPLICATION

Persons with colds or allergies sometimes complain that they cannot taste their food. Although their taste sensations may be operating normally, their olfactory sensations are not. This illustrates that much of **what we think of as taste is actually smell.** Odors from foods pass upward into the nasopharynx and stimulate the olfactory system. In fact, a given concentration of a substance will stimulate the olfactory system thousands of times more than it stimulates the gustatory system.

Each of the four primary tastes is caused by a different response to different chemicals. Certain regions of the tongue react more strongly than others to certain taste sensations. Although the tip of the tongue reacts to all four primary taste sensations, it is highly sensitive to sweet and salty substances. The posterior portion of the tongue is highly sensitive to bitter substances. The lateral edges of the tongue are more sensitive to sour substances (see Figure 24-5).

ADAPTATION AND TASTE THRESHOLDS

Adaptation to taste occurs rapidly. Complete adaptation can occur in 1 to 5 minutes of continuous stimulation. As with odors, receptor adaptation alone cannot account for the speed of complete adaptation. Although taste receptors have a period of rapid adaptation during the first 2 to 3 seconds after contact with food, adaptation is slow thereafter. Adaptation of receptors to smell contributes to taste adaptation, but still does not account for its speed. As with adaptation to odors, adaptation to taste involves a psychological adaptation in the central nervous system.

The threshold for taste varies for each of the primary tastes. The threshold for bitter substances, as measured by quinine, is lowest. This may have a protective function. The threshold for sour substances, as measured by hydrochloric acid, is somewhat higher. The thresholds for salt substances, as measured by sodium chloride, and sweet substances, as measured by sucrose, are about the same and are higher than both bitter and sour substances.

GUSTATORY PATHWAY

The cranial nerves that supply afferent fibers to taste buds are the facial (VII) nerve and trigeminal (V) nerve which supply the anterior two-thirds of the tongue; the glossopharyngeal (IX), which supplies the posterior one-third of the tongue; and the vagus (X), which supplies the epiglottis. Taste impulses are conveyed from the gustatory cells in taste buds along the nerves to the medulla and then to the thalamus. They terminate in the primary gustatory area in the parietal lobe of the cerebral cortex.

VISUAL SENSATIONS

The structures related to vision are the eyeball, the optic (II) nerve, the brain, and a number of accessory structures.

ACCESSORY STRUCTURES OF EYE

Among the **accessory structures** are the eyebrows, eyelids, eyelashes, and the lacrimal apparatus (Figure 17-3). The **eyebrows** form a transverse arch at the junction of the upper eyelid and forehead. Structurally, they resemble the hairy scalp. The skin of the eyebrows is richly supplied with sebaceous glands. The hairs are generally coarse and directed laterally. Deep to the skin of the eyebrows are the fibers of the orbicularis oculi muscles. The eyebrows help to protect the eyeballs from foreign objects, perspiration, and the direct rays of the sun.

The upper and lower **eyelids,** or **palpebrae** (PAL-pe-brē), have several important roles. They shade the eyes during sleep, protect the eyes from excessive light and foreign objects, and spread lubricating secretions over the eyeballs. The upper eyelid is more movable than the lower and contains in its superior region a special levator muscle known as the **levator palpebrae superioris.** The space between the upper and lower eyelids that exposes the eyeball is called the **palpebral fissure.** Its angles are known as the **lateral canthus** (KAN-thus), which is narrower and closer to the temporal bone, and the **medial canthus,** which is broader and nearer the nasal bone. In the medial canthus there is a small reddish elevation, the **lacrimal caruncle** (KAR-ung-kul), containing sebaceous and sudoriferous glands. A whitish material secreted by the caruncle collects in the medial canthus.

From superficial to deep, each eyelid consists of epidermis, dermis, subcutaneous areolar connective tissue, fibers of the orbicularis oculi muscle, a tarsal plate, tarsal glands, and a conjunctiva. The **tarsal plate** is a thick fold of connective tissue that forms much of the inner wall of each eyelid and gives form and support to the eyelids. Embedded in grooves on the deep surface of each tarsal plate is a row of elongated tarsal glands known as **Meibomian** (mī-BŌ-mē-an) **glands.** These are modified sebaceous glands, and their oily secretion helps keep the eyelids from adhering to each other. Infection of the Meibomian glands produces a tumor or cyst on the eyelid called a **chalazion** (ka-LĀ-zē-on). The **conjunctiva** (kon'-junk-TĪ-va) is a thin mucous membrane. It is called the **palpebral conjunctiva** when it lines the inner aspect of the eyelids. It is called the **bulbar** or **ocular conjunctiva** when it is reflected from the eyelids onto the anterior surface of the eyeball to the periphery of the cornea.

Projecting from the border of each eyelid, anterior to the Meibomian glands, is a row of short, thick hairs, the **eyelashes.** In the upper lid, they are long and turn upward; in the lower lid, they are short and turn downward. Sebaceous glands at the base of the hair follicles of the eyelashes called **glands of Zeis** (ZĪS) pour a lubricating fluid into the follicles. Infection of these glands is called a **sty.**

The **lacrimal apparatus** is a term used for a group of structures that manufactures and drains away tears. These structures are the lacrimal glands, the excretory lacrimal ducts, the lacrimal canals, the lacrimal sacs, and the nasolacrimal ducts. A **lacrimal gland** is a compound tubuloacinar gland located at the superior lateral portion of each orbit. Each is about the size and shape of an almond. Leading from the lacrimal glands are 6 to 12 **excretory**

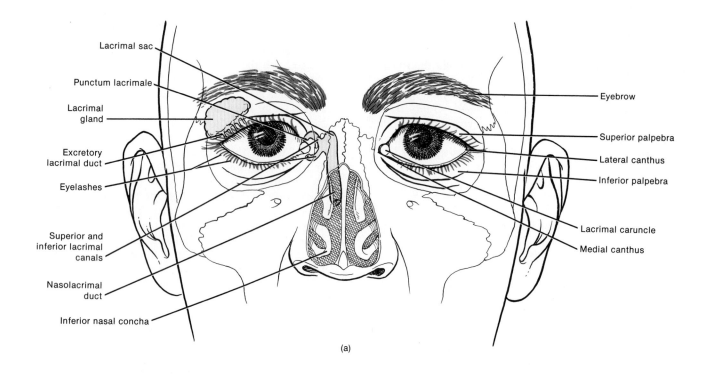

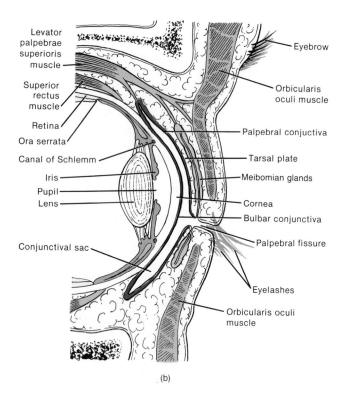

FIGURE 17-3 Accessory structures of the eye. (a) Anterior view. (b) Sagittal section of the eyelids and anterior portion of the eyeball.

lacrimal ducts that empty lacrimal fluid, or tears, onto the surface of the conjunctiva of the upper lid. From here the lacrimal fluid passes medially and enters two small openings called **puncta lacrimalia** that appear as two small pores, one in each papilla of the eyelid, at the medial canthus of the eye. The lacrimal secretion then passes into two ducts, the **lacrimal canals,** and is next conveyed into the lacrimal sac. The lacrimal canals are located in the lacrimal grooves of the lacrimal bones. The **lacrimal sac** is the superior expanded portion of the

nasolacrimal duct, a canal that transports the lacrimal secretion into the inferior meatus of the nose.

The **lacrimal secretion** is a watery solution containing salts, some mucus, and a bactericidal enzyme called **lysozyme.** It cleans, lubricates, and moistens the eyeball. After being secreted by the lacrimal glands, it is spread over the surface of the eyeball by the blinking of the eyelids. Usually, 1 ml per day is produced. Normally, the secretion is carried away by evaporation or by passing into the lacrimal canals and then into the nasal cavities as fast as it is produced. If, however, an irritating substance makes contact with the conjunctiva, the lacrimal glands are stimulated to oversecrete. Tears then accumulate more rapidly than they can be carried away. This is a protective mechanism since the tears dilute and wash away the irritating substance. "Watery" eyes also occur when an inflammation of the nasal mucosa, such as a cold, obstructs the nasolacrimal ducts so that drainage of tears is blocked.

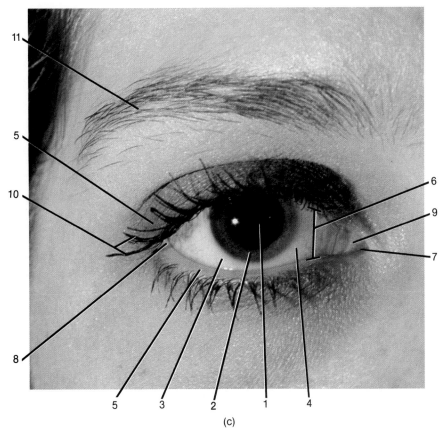

FIGURE 17-3 (*Continued*) Accessory structures of the eye. (c) Surface anatomy of the eye. (© 1982 by Gerard J. Tortora. Courtesy of Lynne Tortora.)

1. **Pupil.** Opening of center of iris of eyeball for light transmission.
2. **Iris.** Circular, pigmented muscular membrane behind cornea.
3. **Sclera.** "White" of eye, a coat of fibrous tissue that covers entire eyeball except for cornea.
4. **Conjunctiva.** Membrane that covers exposed surface of eyeball and lines eyelids.
5. **Palpebrae (eyelids).** Folds of skin and muscle lined by conjunctiva.
6. **Palpebral fissure.** Space between eyelids when they are open.
7. **Medial canthus.** Site of union of upper and lower eyelids near nose.
8. **Lateral canthus.** Site of union of upper and lower eyelids away from nose.
9. **Lacrimal caruncle.** Fleshy, yellowish projection of medial canthus that contains modified sweat and sebaceous glands.
10. **Eyelashes.** Hairs on margins of eyelids, usually arranged in two or three rows.
11. **Eyebrows.** Several rows of hairs superior to upper eyelids.

STRUCTURE OF EYEBALL

The adult **eyeball** measures about 2.5 cm (1 inch) in diameter. Of its total surface area, only the anterior one-sixth is exposed. The remainder is recessed and protected by the orbit into which it fits. Anatomically, the eyeball can be divided into three layers: fibrous tunic, vascular tunic, and retina or nervous tunic (Figure 17-4).

Fibrous Tunic

The **fibrous tunic** is the outer coat of the eyeball. It can be divided into two regions: the posterior portion is the sclera and the anterior portion is the cornea. The **sclera** (SKLE-ra), the "white of the eye," is a white coat of fibrous tissue that covers all the eyeball except the anterior colored portion. The sclera gives shape to the eyeball and protects its inner parts. Its posterior surface is pierced by the optic (II) nerve. The **cornea** (KOR-nē-a) is a nonvascular, nervous, transparent fibrous coat that covers the iris, the colored part of the eye. The cornea's outer surface is covered by an epithelial layer continuous with the epithelium of the bulbar conjunctiva. At the junction of the sclera and cornea is a venous sinus known as the **canal of Schlemm.** (Surface features of the eye are shown in Figure 17-3c.)

CLINICAL APPLICATION

Each year, about 10,000 **corneal transplants (keratoplasty)** are performed in the United States. Corneal transplants are considered to be the most successful type of transplantation. The surgical procedure is performed under an operating microscope at magnifications of between 6× and 40×. A general or local anesthesia is used. The defective cornea, usually 7 to 8 mm (about 0.3 inch) in diameter, is excised with microcorneal scissors and replaced with a donor cornea of similar diameter. The transplanted cornea is sewn into place with nylon sutures. Patients usually remain hospitalized for 3 to 5 days and are later fitted with hard or soft contact lenses or eyeglasses.

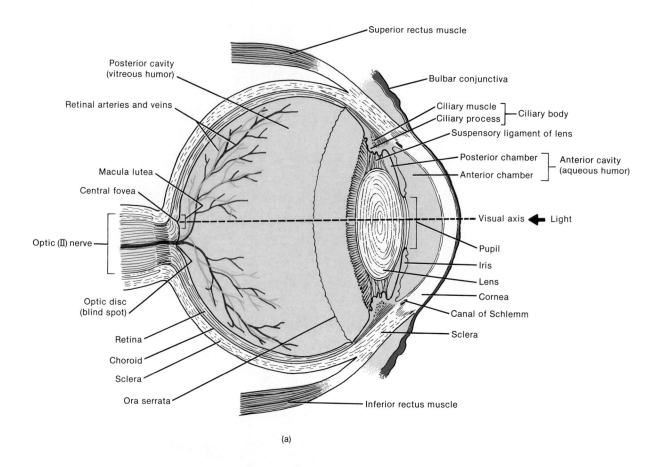

(a)

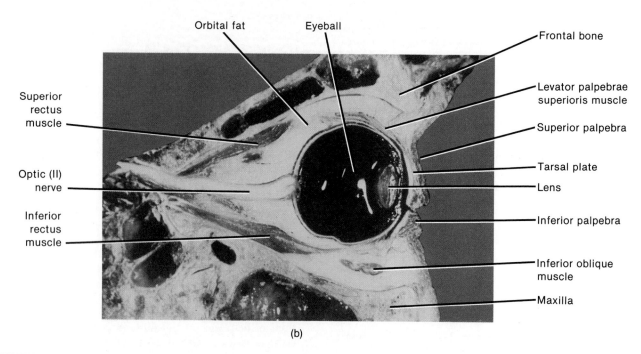

(b)

FIGURE 17-4 Structure of the eyeball. (a) Diagram of gross structure in transverse section. (b) Photograph of midsagittal section. (Courtesy of C. Yokochi and J. W. Rohen, *Photographic Anatomy of the Human Body,* 2nd ed., 1979, IGAKU-SHOIN, Tokyo, New York.)

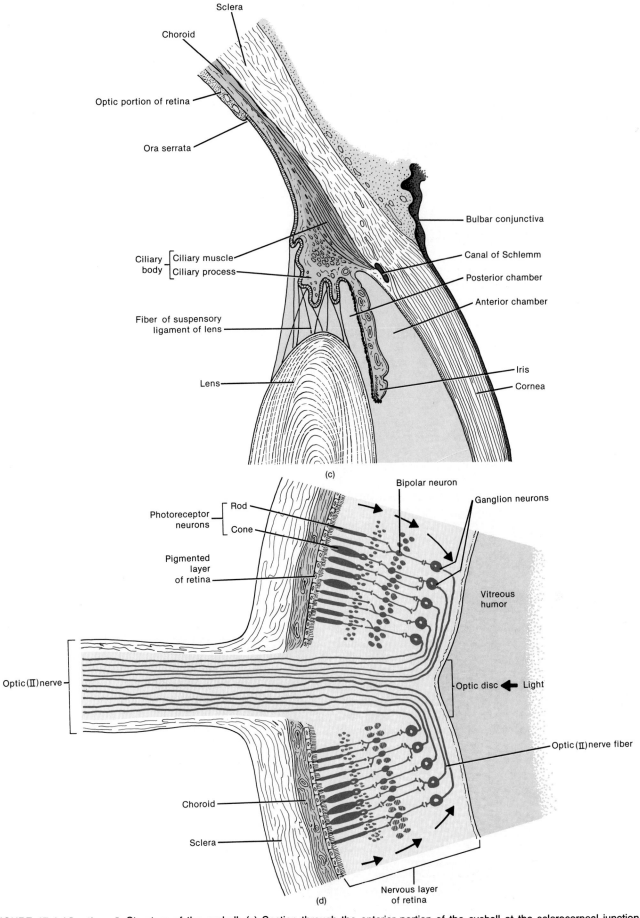

FIGURE 17-4 (*Continued*) Structure of the eyeball. (c) Section through the anterior portion of the eyeball at the sclerocorneal junction. (d) Diagram of the microscopic structure of the retina, exaggerated for emphasis.

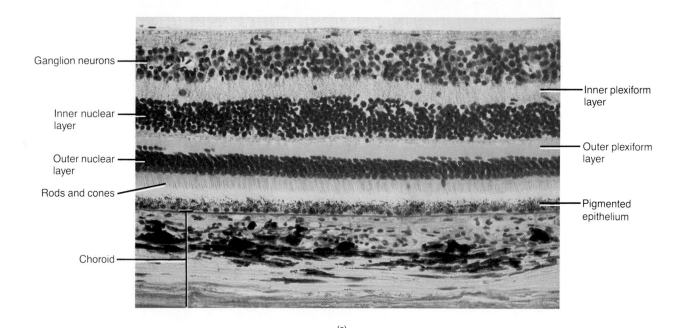

Ganglion neurons

Inner nuclear layer

Outer nuclear layer

Rods and cones

Choroid

Inner plexiform layer

Outer plexiform layer

Pigmented epithelium

(e)

FIGURE 17-4 (*Continued*) Structure of the eyeball. (e) Photomicrograph of a portion of the retina at a magnification of 300×. The outer nuclear layer contains nuclei of photoreceptor neurons; the outer plexiform layer is where photoreceptor and bipolar neurons synapse; the inner nuclear layer contains the nuclei of bipolar neurons; and the inner plexiform layer is where bipolar and ganglion neurons synapse. (© 1983 by Michael H. Ross. Used by permission.)

Vascular Tunic

The **vascular tunic** is the middle layer of the eyeball and is composed of three portions: the choroid, the ciliary body, and the iris. Collectively, these three structures are called the **uvea** (YOO-vē-a). The **choroid** (KŌ-royd), the posterior portion of the vascular tunic, is a thin, dark brown membrane that lines most of the internal surface of the sclera. It contains numerous blood vessels and a large amount of pigment. The choroid absorbs light rays so they are not reflected back out of the eyeball. Through its blood supply, it nourishes the retina. Like the sclera, it is pierced by the optic (II) nerve at the back of the eyeball.

In the anterior portion of the vascular tunic, the choroid becomes the **ciliary** (SIL-ē-ar'-ē) **body.** It is the thickest portion of the vascular tunic. It extends from the **ora serrata** (Ō-ra ser-RĀ-ta) of the retina (inner tunic) to a point just behind the sclerocorneal junction. The ora serrata is simply the jagged margin of the retina. The ciliary body consists of the ciliary processes and ciliary muscle. The **ciliary processes** consist of protrusions or folds on the internal surface of the ciliary body that secrete aqueous humor. The **ciliary muscle** is a smooth muscle that alters the shape of the lens for near or far vision.

The **iris** is the third portion of the vascular tunic. It consists of circular and radial smooth muscle fibers arranged to form a doughnut-shaped structure. The black hole in the center of the iris is the **pupil,** the area through which light enters the eyeball. The iris is suspended between the cornea and the lens and is attached at its outer margin to the ciliary process. A principal function of the iris is to regulate the amount of light entering the eyeball. When the eye is stimulated by bright light, the circular muscles of the iris contract and decrease the size of the pupil. When the eye must adjust to dim light, the radial muscles of the iris contract and increase the pupil's size.

Retina (Nervous Tunic)

The third and inner coat of the eye, the **retina (nervous tunic),** lies only in the posterior portion of the eye. Its primary function is image formation. It consists of an inner nervous tissue layer and an outer pigmented layer. The retina covers the choroid. At the edge of the ciliary body, it ends in a scalloped border called the ora serrata. This is where the nervous layer or visual portion of the retina ends. The pigmented layer extends anteriorly over the back of the ciliary body and the iris as the nonvisual portion of the retina.

CLINICAL APPLICATION

Detachment of the retina may occur in trauma, such as a blow to the head. The tear in the retina causes blindness in the corresponding field of vision. The actual detachment occurs between the sensory part of the retina and the underlying pigmented layer. Fluid accumulates between these layers, forcing the thin, pliable retina to billow out toward the vitreous humor. The retina may be reattached by photocoagulation by laser beam, cryosurgery, or scleral resection.

The nervous layer contains three zones of neurons. These three zones, named in the order in which they conduct impulses, are the **photoreceptor neurons,** the **bipolar neurons,** and the **ganglion neurons.** The dendrites of the photoreceptor neurons are called rods and cones because of their shapes. They are visual receptors highly specialized for stimulation by light rays. Functionally, rods and cones develop generator potentials. **Rods** are specialized for vision in dim light. They also allow us to discriminate between different shades of dark and light and permit us to see shapes and movement. **Cones** are specialized for color vision and sharpness of vision (*visual acuity*). They are stimulated only by bright light. This is why we cannot see color by moonlight. It is estimated that there are 7 million cones and somewhere between 10 and 20 times as many rods. Cones are most densely concentrated in the **central fovea,** a small depression in the center of the macula lutea. The **macula lutea** (MAK-yoo-la LOO-tē-a), or yellow spot, is in the exact center of the posterior portion of the retina, corresponding to the visual axis of the eye. The fovea is the area of sharpest vision because of the high concentration of cones. Rods are absent from the fovea and macula and increase in density toward the periphery of the retina.

CLINICAL APPLICATION

In a disease called **senile macular degeneration (SMD),** new blood vessels grow over the macula lutea. The effect ranges from distorted vision to blindness. SMD accounts for nearly all new cases of blindness in people over 65. Its cause is unknown. Laser beam treatment has been used effectively in arresting blood vessel proliferation and restoring normal vision in over 80 percent of cases. Treatment, however, must be obtained soon after the onset of symptoms. In order to determine if SMD exists, a simple test can be performed without any assistance or special instruments. Stare, one eye at a time (covering the other with your hand), at any long straight line, such as a door frame. If the line appears bent or twisted, or if a black spot appears, to either eye, a physician should be informed immediately.

When information has passed through the photoreceptor neurons, it is conducted across synapses to the bipolar neurons in the intermediate zone of the nervous layer of the retina. From here it is passed to the ganglion neurons. These cells transmit their signals through optic (II) nerve fibers to the brain in the form of nerve impulses.

The axons of the ganglion neurons extend posteriorly to a small area of the retina called the **optic disc (blind spot).** This region contains openings through which the axons of the ganglion neurons exit as the optic (II) nerve. Since it contains no rods or cones, an image striking it cannot be perceived. Thus it is called the blind spot.

Lens

In addition to the fibrous tunic, vascular tunic, and retina, the eyeball itself contains the lens, just behind the pupil and iris. The **lens** is constructed of numerous layers of protein fibers arranged like the layers of an onion. Normally, the lens is perfectly transparent. It is enclosed by a clear connective tissue capsule and held in position by the **suspensory ligament.** A loss of transparency of the lens is known as a **cataract.**

Interior

The interior of the eyeball is a large cavity divided into two smaller ones by the lens. The **anterior cavity,** the division anterior to the lens, is further divided into the **anterior chamber,** which lies behind the cornea and in front of the iris, and the **posterior chamber,** which lies behind the iris and in front of the suspensory ligament

and lens. The anterior cavity is filled with a watery fluid, similar to cerebrospinal fluid, called the **aqueous humor.** The fluid is believed to be secreted into the posterior chamber by choroid plexuses of the ciliary processes of the ciliary bodies behind the iris. Once the fluid is formed, it permeates the posterior chamber and then passes forward between the iris and the lens, through the pupil, into the anterior chamber. From the anterior chamber, the aqueous humor, which is continually produced, is drained off into the **canal of Schlemm** and then into the blood. The anterior chamber thus serves a function similar to the subarachnoid space around the brain and spinal cord. The canal of Schlemm is analogous to a venous sinus of the dura mater. The pressure in the eye, called **intraocular pressure,** is produced mainly by the aqueous humor. The intraocular pressure keeps the retina smoothly applied to the choroid so the retina will form clear images. Normal intraocular pressure (about 24 mm Hg) is maintained by drainage of the aqueous humor through the canal of Schlemm. Excessive intraocular pressure, called **glaucoma** (glow-KŌ-ma), results in degeneration of the retina and blindness. Besides maintaining intraocular pressure, the aqueous humor is also the principal link between the circulatory system and the lens and cornea. Neither the lens nor the cornea has blood vessels.

The second, and larger, cavity of the eyeball is the **posterior cavity.** It lies between the lens and the retina and contains a jellylike substance called the **vitreous humor.** This substance contributes to intraocular pressure, helps to prevent the eyeball from collapsing, and holds the retina flush against the internal portions of the eyeball. The vitreous humor, unlike the aqueous humor, does not undergo constant replacement. It is formed during embryonic life and is not replaced thereafter.

PHYSIOLOGY OF VISION

Before light can reach the rods and cones of the retina, it must pass through the cornea, aqueous humor, pupil, lens, and vitreous humor. For vision to occur, light reaching the rods and cones must form an image on the retina. The resulting nerve impulses must then be conducted to the visual areas of the cerebral cortex. In discussing the physiology of vision, let us first consider retinal image formation.

Retinal Image Formation

The formation of an image on the retina requires four basic processes, all concerned with focusing light rays: (1) refraction of light rays, (2) accommodation of the lens, (3) constriction of the pupil, and (4) convergence of the eyes. Accommodation and pupil size are functions of the smooth muscle cells of the ciliary muscle and the dilator and sphincter muscles of the iris. They are termed *intrinsic eye muscles* since they are inside the eyeball. Convergence is a function of the voluntary muscles attached to the outside of the eyeball called the *extrinsic eye muscles.*

● *Refraction of Light Rays*   When light rays traveling through a transparent medium (such as air) pass into a

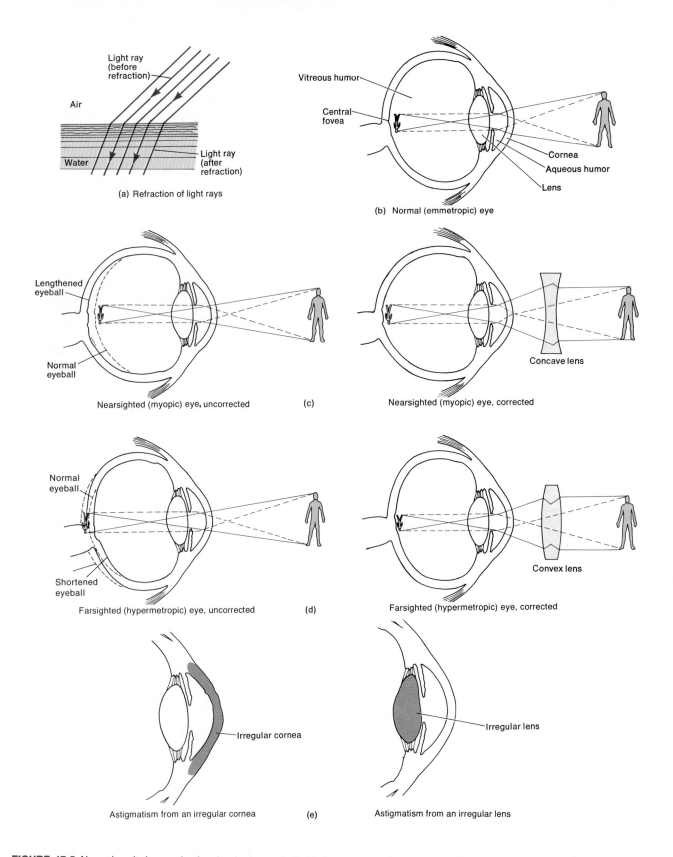

FIGURE 17-5 Normal and abnormal refraction in the eyeball. (a) Refraction of light rays passing from air into water. (b) In the normal or emmetropic eye, light rays from an object are bent sufficiently by the four refracting media and converged on the central fovea. A clear image is formed. (c) In the nearsighted or myopic eye, the image is focused in front of the retina. The condition may result from an elongated eyeball or thickened lens. Correction is by use of a concave lens which diverges entering light rays so that they have to travel further through the eyeball and are focused exactly on the retina. (d) In the farsighted or hypermetropic eye, the image is focused behind the retina. The condition results from a shortened eyeball or a thin lens. Correction is by a convex lens which converges entering light rays so that they focus exactly on the retina. (e) An astigmatism is an irregular curvature of the cornea (shown on left) or lens (shown on right). As a result, horizontal and vertical rays are focused at two different points on the retina, and the image is not focused on the area of sharpest vision of the retina. This results in blurred or distorted vision. Suitable glasses correct the refraction of an astigmatic eye.

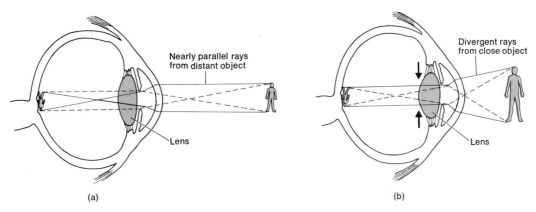

FIGURE 17-6 Accommodation. (a) For objects 6 m (20 ft) or more away. (b) For objects nearer than 6 m.

second transparent medium with a different density (such as water), they bend at the surface of the two media. This is **refraction** (Figure 17-5a). The eye has four such media of refraction: cornea, aqueous humor, lens, and vitreous humor. Light rays entering the eye from the air are refracted at the following points: (1) the anterior surface of the cornea as they pass from the lighter air into the denser cornea, (2) the posterior surface of the cornea as they pass into the less dense aqueous humor, (3) the anterior surface of the lens as they pass from the aqueous humor into the denser lens, and (4) the posterior surface of the lens as they pass from the lens into the less dense vitreous humor.

The degree of refraction that takes place at each surface in the eye is very precise. When an object is 6 m (20 ft) or more away from the viewer, the light rays reflected from the object are nearly parallel to one another. The parallel rays must be bent sufficiently to fall exactly on the central fovea, where vision is sharpest. Light rays that are reflected from near objects are divergent rather than parallel. As a result, they must be refracted toward each other to a greater extent. This change in refraction is brought about by the lens of the eye (Figure 17-5b and 17-6).

● *Accommodation of the Lens*   If the surface of a lens curves outward, as in a convex lens, the lens will refract incoming rays toward each other so they eventually intersect. The greater the curve, the more acutely it bends the rays toward each other. Conversely, when the surface of a lens curves inward, as in a concave lens, the rays bend away from each other. The lens of the eye is biconvex. Furthermore, it has the unique ability to change the focusing power of the eye by becoming moderately curved at one moment and greatly curved the next. When the eye is focusing on a close object, the lens curves greatly in order to bend the rays toward the central fovea. This increase in the curvature of the lens is called **accommodation** (Figure 17-6). In near vision, the ciliary muscle contracts, pulling the ciliary process and choroid forward toward the lens. This action releases the tension on the lens and suspensory ligament. Because of its elasticity, the lens shortens, thickens, and bulges. In far vision, the ciliary muscle is relaxed and the lens is flatter. With aging, the lens loses elasticity and, therefore, its ability to accommodate.

CLINICAL APPLICATION

The normal eye, known as an **emmetropic** (em'-e-TROP-ik) **eye,** can sufficiently refract light rays from an object 6 m (20 ft) away to focus a clear image on the retina. Many individuals, however, do not have this ability because of abnormalities related to improper refraction. Among these abnormalities are **myopia** (mī-Ō-pē-a) or nearsightedness, **hypermetropia** (hī'-per-mē-TRO-pē-a) or farsightedness, and **astigmatism** (a-STIG-ma-tizm) or irregularities in the surface of the lens or cornea. The conditions are illustrated and explained in Figure 17-5c–e. The inability to focus on nearby objects due to loss of elasticity of the lens with age is called **presbyopia** (prez-bē-OP-ē-a).

● *Constriction of Pupil*   The circular muscle fibers of the iris also assume a function in the formation of clear retinal images. Part of the accommodation mechanism consists of the contraction of the dilator and sphincter muscles of the iris to constrict the pupil. **Constriction of the pupil** means narrowing the diameter of the hole through which light enters the eye. This action occurs simultaneously with accommodation of the lens and prevents light rays from entering the eye through the periphery of the lens. Light rays entering at the periphery would not be brought to focus on the retina and would result in blurred vision. The pupil, as noted earlier, also constricts in bright light to protect the retina from sudden or intense stimulation.

● *Convergence*   Because of the alignment of the orbits in the skull, many animals see a set of objects off to the left through one eye and an entirely different set off to the right through the other. This characteristic doubles their field of vision and allows them to detect predators behind them. In humans, both eyes focus on only one set of objects—a characteristic called **single binocular vision.**

Single binocular vision occurs when light rays from an object are directed toward corresponding points on the two retinas. When we stare straight ahead at a distant object, the incoming light rays are aimed directly at both pupils and are refracted to identical spots on the retinas of both eyes. But as we move closer to the object, our eyes must rotate medially for the light rays from the object to hit the same points on both retinas. The term **convergence** refers to this medial movement of the two eyeballs so they are both directed toward the object being viewed. The nearer the object, the greater the degree of convergence necessary to maintain single binocular vision. Convergence is brought about by the coordinated action of the extrinsic eye muscles.

● *Inverted Image*   Images are focused upside down on the retina. They also undergo mirror reversal, that is, light reflected from the right side of an object hits the left side of the retina and vice versa. Note in Figure 17-5b that reflected light from the top of the object crosses light from the bottom of the object and strikes the retina below the central fovea. Reflected light from the bottom of the object crosses light from the top of the object and strikes the retina above the central fovea. The reason we do not see an inverted world is that the brain learns early in life to coordinate visual images with the exact locations of objects. The brain stores memories of reaching and touching objects and automatically turns visual images right-side-up and right-side-around.

Stimulation of Photoreceptors

● *Excitation of Rods*   After an image is formed on the retina by refraction, accommodation, constriction of the pupil, and convergence, light impulses must be converted into nerve impulses. The initial step is the development of generator potentials by rods and cones. Rods contain a reddish-purple photosensitive pigment called **rhodopsin (visual purple).** This substance consists of the protein **scotopsin** plus **retinene (visual yellow),** a derivative of vitamin A. When light rays strike a rod, rhodopsin rapidly breaks down into its components, neither of which can absorb light. The chemical breakdown stimulates generator potential development by the rods (Figure 17-7).

Rhodopsin is highly light sensitive—even the light rays from the moon or a candle will break down some of it and thereby allow us to see. The rods, then, are specialized for night vision. However, they are of only limited help for daylight vision. In bright light, the rhodopsin is broken down faster than it can be manufactured. In dim light, production is able to keep pace with a slower rate of breakdown. These characteristics of rhodopsin are responsible for the experience of having to adjust to a dark room after walking in from the sunshine. The normal period of adjustment is the time it takes for the completely dissociated rhodopsin to reform. **Night blindness,** which is also referred to as **nyctalopia** (nik'-ta-LŌ-pē-a), is the lack of normal night vision following the adjustment period. It is most often caused by vitamin A deficiency.

● *Excitation of Cones*   Cones are the receptors for bright light and color. The photosensitive pigments in cones are almost the same as those in rods. Both contain retinene, but the protein portion in cones is **photopsin,** which is slightly different from the scotopsin in rods. Unlike rhodopsin, the photosensitive pigments of the cones require bright light for their breakdown and they reform quickly. It is believed that there are three types of cones and that each contains a different combination of retinene

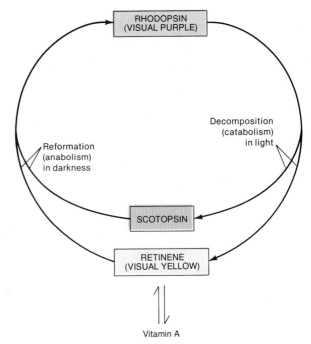

FIGURE 17-7 Rhodopsin cycle.

and photopsin. Each combination has a different maximum absorption of light of a different wavelength so that each responds best to light of a given color. One type of cone responds best to red light, the second to green light, and the third to blue light. Just as an artist can obtain almost any color by mixing colors, it is believed that cones can perceive any color by differential stimulation (Figure 17-8). Stimulation of a cone by two or more colors may produce any combination of colors.

CLINICAL APPLICATION

If a single group of color receptive cones is missing from the retina, an individual cannot distinguish some colors from others and is said to be **color-blind.** The most common type is **red-green color blindness** in which the cones best receptive to red light and to green light are missing. As a result, the person cannot distinguish between red and green. Color blindness is an inherited condition that affects males far more frequently than females. The inheritance of the condition is discussed in Chapter 29 and illustrated in Figure 29-22.

● *Light and Dark Adaptation*   If an individual is exposed to bright light for an extended period of time, a considerable amount of the photosensitive pigment in both rods and cones is broken down into retinene and opsin (scotopsin or photopsin). In addition, most of the retinene in both rods and cones is converted to vitamin A. As a result, the concentration of the photosensitive pigment in both is reduced considerably and the sensitivity of the eye to light is greatly reduced. This is called **light adaptation.**

Conversely, if an individual is exposed to darkness for an extended period of time, a considerable amount of retinene and opsins in the rods and cones is converted

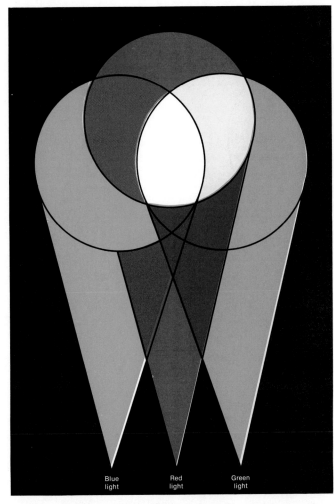

FIGURE 17-8 Differential stimulation of cones. Stimulation of a cone by two or more colors can produce any combination of colors.

Blue light   Red light   Green light

into the photosensitive pigments. Also, large amounts of vitamin A are converted into retinene, which is then converted into photosensitive pigments upon combination with opsins. As a result, the concentration of photosensitive pigments in rods and cones is increased considerably and the sensitivity of the eye to light is greatly increased. This is called **dark adaptation.**

Visual Pathway

From the rods and cones, information is transmitted through bipolar neurons to ganglion cells. Ganglion cells initiate nerve impulses. The cell bodies of the ganglion cells lie in the retina, and their axons leave the eyeball via the **optic (II) nerve** (Figure 17-9). The axons pass through the **optic chiasma** (kī-AZ-ma), a crossing point of the optic (II) nerves. Some fibers cross to the opposite side. Others remain uncrossed. On passing through the optic chiasma, the fibers, now part of the **optic tract,** enter the brain and terminate in the lateral geniculate nucleus of the thalamus. Here the fibers synapse with third-order neurons whose axons pass to the visual areas located in the occipital lobes of the cerebral cortex.

Analysis of the afferent pathway to the brain reveals that the visual field of each eye is divided into two regions:

the **nasal (medial) half** and the **temporal (lateral) half.** For each eye, light rays from an object in the nasal half of the visual field fall on the temporal half of the retina. Light rays from an object in the temporal half of the visual field fall on the nasal half of the retina (Figure 17-9). Also, light rays from objects at the top of the visual field of each eye fall on the inferior portion of the retina, and light rays from objects at the bottom of the visual field fall on the superior portion of the retina. In the optic chiasma, nerve fibers from the nasal halves of the retinas cross and continue on to the lateral geniculate nuclei of the thalamus; nerve fibers from the temporal halves of the retinas do not cross, but continue directly on to the lateral geniculate nuclei. As a result, the primary visual area of the cerebral cortex of the right occipital lobe interprets visual sensations from the left side of an object via impulses from the temporal half of the retina of the right eye and the nasal half of the retina of the left eye. The primary visual area of the cerebral cortex of the left occipital lobe interprets visual sensations from the right side of an object via impulses from the nasal half of the right eye and the temporal half of the left eye.

> **CLINICAL APPLICATION**
>
> A **scotoma** (skō-TŌ-ma) or **blind spot** in the field of vision, other than the normal blind spot (optic disc), may indicate a brain tumor along one of the afferent pathways. For instance, a symptom of a tumor in the right optic tract might be an inability to see the left side of a normal field of vision without moving the eyeball.

AUDITORY SENSATIONS AND EQUILIBRIUM

In addition to containing receptors for sound waves, the ear also contains receptors for equilibrium. Anatomically, the ear is divided into three principal regions: the external (outer) ear, the middle ear, and the internal (inner) ear.

EXTERNAL OR OUTER EAR

The **external (outer) ear** is structurally designed to collect sound waves and direct them inward (Figure 17-10a,b). It consists of the pinna, the external auditory canal, and the tympanic membrane, also called the eardrum.

The **pinna (auricle)** is a trumpet-shaped flap of elastic cartilage covered by thick skin. The rim of the pinna is called the **helix;** the inferior portion is the **lobule.** The pinna is attached to the head by ligaments and muscles. Additional surface anatomy features of the external ear are shown in Figure 17-10a.

The **external auditory canal** or **meatus** is a tube about 2.5 cm (1 inch) in length that lies in the external auditory meatus of the temporal bone. It leads from the pinna to the eardrum. The walls of the canal consist of bone lined with cartilage that is continuous with the cartilage of the pinna. The cartilage in the external auditory canal is covered with thin, highly sensitive skin. Near the exterior opening, the canal contains a few hairs and specialized sebaceous glands called **ceruminous** (se-ROO-mi-nus)

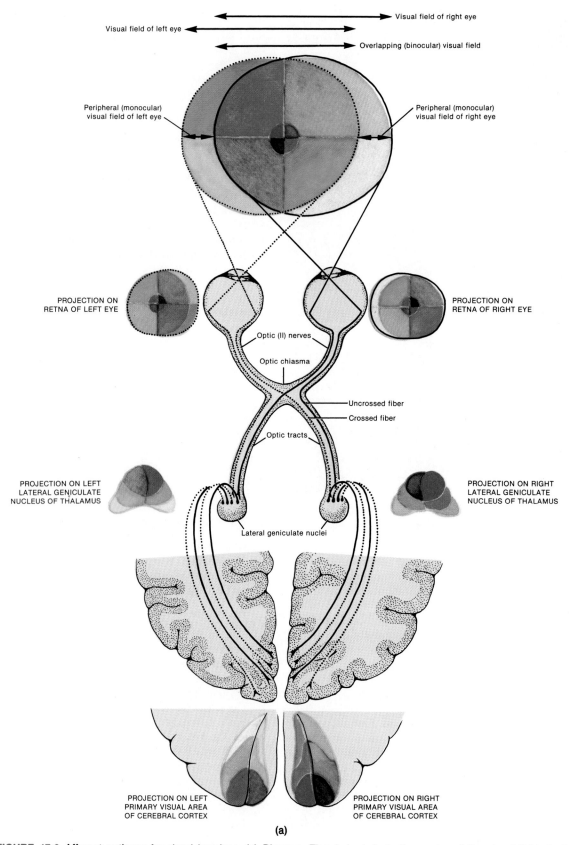

FIGURE 17-9 Afferent pathway for visual impulses. (a) Diagram. The dark circle in the center of the visual fields is the macula lutea. The center of the macula lutea is the central fovea, the area of sharpest vision.

glands, which secrete **cerumen** (earwax). The combination of hairs and cerumen (se-ROO-min) helps to prevent foreign objects from entering the ear.

The **tympanic** (tim-PAN-ik) **membrane,** or **eardrum,** is a thin, semitransparent partition of fibrous connective tissue between the external auditory canal and the middle ear. Its external surface is concave and covered with skin. Its internal surface is convex and covered with a mucous membrane.

CLINICAL APPLICATION

Rupture of the tympanic membrane may result from foreign bodies, pressure, or infection. Severe bleeding and escape of cerebrospinal fluid through a ruptured tympanic membrane may occur following a severe blow to the head and indicate skull fracture.

MIDDLE EAR

The **middle ear,** also called the **tympanic cavity,** is a small, epithelial-lined, air-filled cavity hollowed out of the temporal bone (Figure 17-10b–d). The cavity is separated from the external ear by the eardrum and from the internal ear by a thin bony partition that contains two small openings: the oval window and the round window.

The posterior wall of the cavity communicates with the mastoid air cells of the temporal bone through a chamber called the **tympanic antrum.** This anatomical fact explains why a middle ear infection may spread to the temporal bone, causing mastoiditis, or even to the brain.

The anterior wall of the cavity contains an opening that leads directly into the **auditory tube.** The auditory tube connects the middle ear with the nasopharynx of the throat. Through this passageway, infections may travel from the throat and nose to the ear. The function of the tube is to equalize air pressure on both sides of the tympanic membrane. Abrupt changes in external or internal air pressure might otherwise cause the eardrum to rupture. During swallowing and yawning, the tube opens to allow atmospheric air to enter or leave the middle ear until the internal pressure equals the external pressure. Any sudden pressure changes against the eardrum may be equalized by deliberately swallowing.

Extending across the middle ear are three exceedingly small bones called **auditory ossicles** (OS-si-kuls). The bones, named for their shape, are the malleus, incus, and stapes, commonly called the hammer, anvil, and stirrup, respectively. They are connected by synovial joints. The "handle" of the **malleus** is attached to the internal surface of the tympanic membrane. Its head articulates with the body of the incus. The **incus** is the intermediate bone in the series and articulates with the head of the stapes. The base or footplate of the **stapes** fits into a small opening in the thin bony partition between the middle and inner ear. The opening is called the **fenestra vestibuli** (fe-NES-tra ves-TIB-yoo-lī), or **oval window.** Directly below the oval window is another opening, the **fenestra cochlea** (fe-NES-tra KŌK-lē-a), or **round window.** This opening is enclosed by a membrane called the secondary tympanic membrane. The auditory ossicles are attached to the tympanic cavity by means of ligaments.

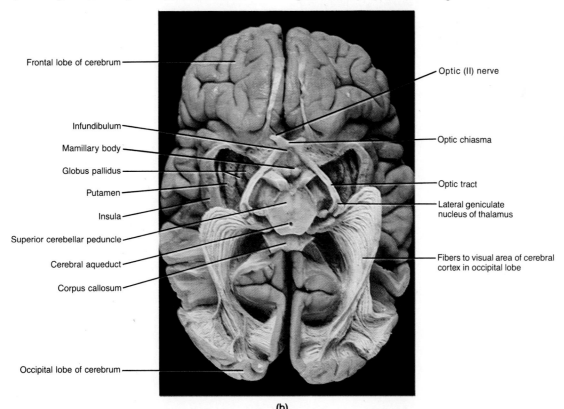

(b)

FIGURE 17-9 *(Continued)* Afferent pathway for visual impulses. (b) Photograph as seen from the ventral aspect of the brain. (Courtesy of N. Gluhbegovic and T. H. Williams, *The Human Brain: A Photographic Guide,* Harper & Row, New York, 1980.)

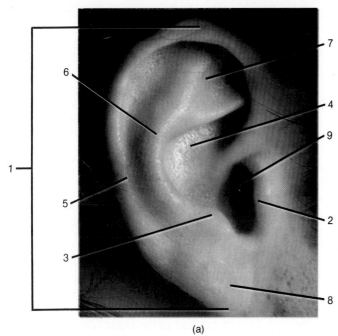

FIGURE 17-10 Structure of the auditory apparatus. (a) Surface anatomy of the right ear seen in lateral view. (© 1982 by Gerard J. Tortora. Courtesy of James Borghesi.) (b) Divisions of the right ear into external, middle, and internal portions seen in a frontal section through the right side of the skull. (c) Details of the middle ear and bony labyrinth of the internal ear. (d) Auditory ossicles of the middle ear.

1. **Pinna.** Portion of external ear not contained in head, also called auricle or trumpet.
2. **Tragus.** Cartilaginous projection anterior to external opening to ear.
3. **Antitragus.** Cartilaginous projection opposite tragus.
4. **Concha.** Hollow of auricle.
5. **Helix.** Superior and posterior free margin of auricle.
6. **Antihelix.** Semicircular ridge posterior and superior to concha.
7. **Triangular fossa.** Depression in superior portion of antihelix.
8. **Lobule.** Inferior portion of pinna devoid of cartilage.
9. **External auditory meatus.** Canal extending from external ear to eardrum.

(a)

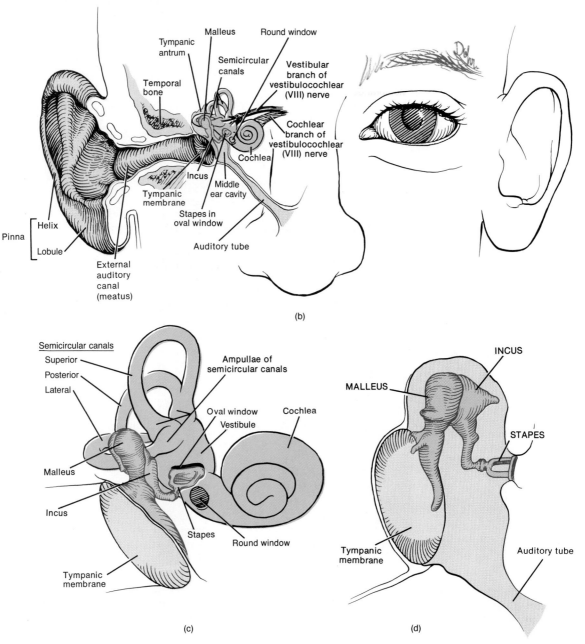

(b)

(c)

(d)

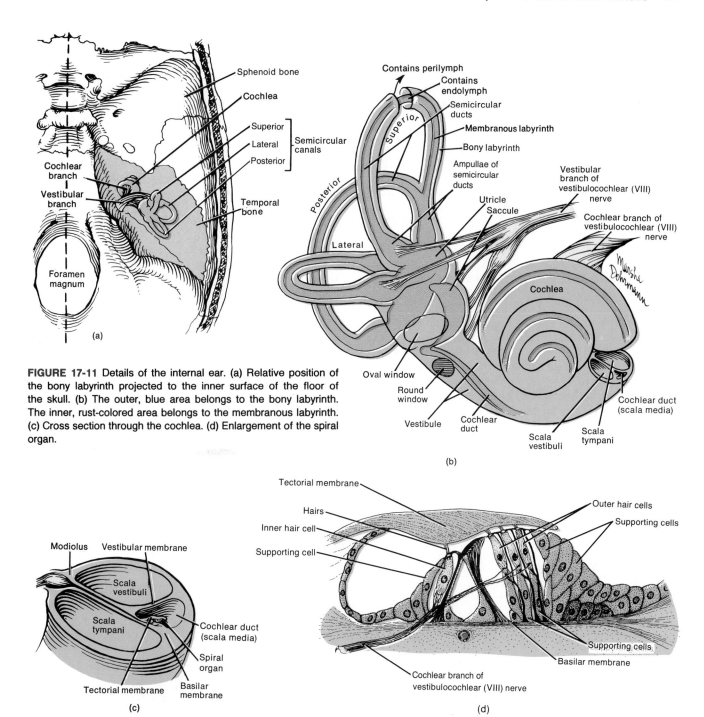

FIGURE 17-11 Details of the internal ear. (a) Relative position of the bony labyrinth projected to the inner surface of the floor of the skull. (b) The outer, blue area belongs to the bony labyrinth. The inner, rust-colored area belongs to the membranous labyrinth. (c) Cross section through the cochlea. (d) Enlargement of the spiral organ.

INTERNAL OR INNER EAR

The **internal (inner) ear** is also called the **labyrinth** (LAB-i-rinth) because of its complicated series of canals (Figure 17-11). Structurally, it consists of two main divisions: a bony labyrinth and a membranous labyrinth that fits in the bony labyrinth. The **bony labyrinth** is a series of cavities in the petrous portion of the temporal bone. It can be divided into three areas named on the basis of shape: the vestibule, cochlea, and semicircular canals. The bony labyrinth is lined with periosteum and contains a fluid called **perilymph.** This fluid surrounds the **membranous labyrinth,** a series of sacs and tubes lying inside and having the same general form as the bony labyrinth. The membra-nous labyrinth is lined with epithelium and contains a fluid called **endolymph.**

The **vestibule** constitutes the oval central portion of the bony labyrinth. The membranous labyrinth in the vestibule consists of two sacs called the **utricle** and **saccule.** These sacs are connected to each other by a small duct.

Projecting upward and posteriorly from the vestibule are the three bony **semicircular canals.** Each is arranged at approximately right angles to the other two. On the basis of their positions, they are called the superior, posterior, and lateral canals. One end of each canal enlarges into a swelling called the **ampulla** (am-POOL-la). The portions of the membranous labyrinth that lie inside the bony semicircular canals are called the **semicircular ducts (membranous semicircular canals)**. These structures are almost identical in shape to the semicircular canals and communicate with the utricle of the vestibule.

Lying in front of the vestibule is the **cochlea** (KŌK-lē-a), so designated because of its resemblance to a snail's shell. The cochlea consists of a bony spiral canal that makes about 2¾ turns around a central bony core called the **modiolus**. A cross section through the cochlea shows the canal is divided into three separate channels by partitions that together have the shape of the letter Y. The stem of the Y is a bony shelf that protrudes into the canal; the wings of the Y are composed mainly of membranous labyrinth. The channel above the bony partition is the **scala vestibuli**; the channel below is the **scala tympani**. The cochlea adjoins the wall of the vestibule, into which the scala vestibuli opens. The scala tympani terminates at the round window. The perilymph of the vestibule is continuous with that of the scala vestibuli. The third channel (between the wings of the Y) is the membranous labyrinth: the **cochlear duct (scala media)**. The cochlear duct is separated from the scala vestibuli by the **vestibular membrane**. It is separated from the scala tympani by the **basilar membrane**.

Resting on the basilar membrane is the **spiral organ**, the organ of hearing. The spiral organ is a series of epithelial cells on the inner surface of the basilar membrane. It consists of a number of supporting cells and hair cells, which are the receptors for auditory sensations. The inner hair cells are medially placed in a single row and extend the entire length of the cochlea. The outer hair cells are arranged in several rows throughout the cochlea. The hair cells have long hairlike processes at their free ends that extend into the endolymph of the cochlear duct. The basal ends of the hair cells are in contact with fibers of the cochlear branch of the vestibulocochlear (VIII) nerve. Projecting over and in contact with the hair cells of the spiral organ is the **tectorial membrane,** a delicate and flexible gelatinous membrane.

SOUND WAVES

Sound waves result from the alternate compression and decompression of air molecules. They originate from a vibrating object, much the same way that waves travel over the surface of water. The sounds heard most acutely by human ears are those from sources that vibrate at frequencies between 1,000 and 4,000 cycles per second (Hz). The entire range extends from 20 to 20,000 Hz.

The frequency of vibration of the source of the sound is directly related to pitch. The greater the vibration, the higher the pitch. Also, the greater the force of the vibration, the louder the sound. Sound intensities (loudness) are usually measured in units called **decibels (dB)**. Decibels measure the relative intensity of sounds; 1 dB represents an increase in intensity of 1.26 times. In the

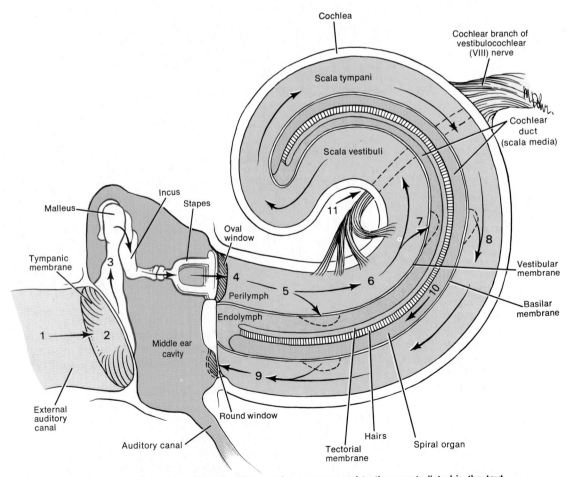

FIGURE 17-12 Physiology of hearing. The numbers correspond to the events listed in the text.

sound intensity range for normal communication, the human ear can detect approximately a 1-dB change in sound intensity. The hearing threshold, that is, the point at which an average young adult can just detect sound from silence is 0 dB. Between 115 and 120 dB is the threshold of pain. Just to give you some idea of how decibels relate to common sounds in the range between 0 and 120 dB, rustling leaves have a decibel rating of 15, normal conversation 45, crowd noise 60, a vacuum cleaner 75, and a pneumatic drill 90.

PHYSIOLOGY OF HEARING

The events involved in the physiology of hearing sound waves are as follows (Figure 17-12):

1. Sound waves that reach the ear are directed by the pinna into the external auditory canal.

2. When the waves strike the tympanic membrane, the alternate compression and decompression of the air causes the membrane to vibrate. The distance the membrane moves is always very small and is relative to the force and velocity with which air molecules strike it. It vibrates slowly in response to low-frequency sounds and rapidly in response to high-frequency sounds.

3. The central area of the tympanic membrane is connected to the malleus, which also starts to vibrate. The vibration is then picked up by the incus, which transmits the vibration to the stapes.

4. As the stapes moves back and forth, it pushes the oval window in and out.

If sound waves passed directly to the oval window without passing through the tympanic membrane and auditory bones, hearing would be inadequate. A minimal amount of sound energy is required to transmit sound waves through the perilymph of the cochlea. Since the tympanic membrane has a surface area about 22 times larger than that of the oval window, it can collect about 22 times more sound energy. This energy is sufficient to transmit sound waves through the perilymph.

5. The movement of the oval window sets up waves in the perilymph.

6. As the window bulges inward, it pushes the perilymph of the scala vestibuli and waves are propagated through the scala vestibuli into the fluid of the scala tympani.

7. This pressure pushes the vestibular membrane inward and increases the pressure of the endolymph inside the cochlear duct.

8. The basilar membrane gives under the pressure and bulges out into the scala tympani. Due to differences in structure of various areas of the basilar membrane, high-frequency resonance of the membrane occurs near its base and low-frequency resonance of the membrane occurs near its apex.

9. The sudden pressure in the scala tympani pushes the perilymph toward the round window, causing it to bulge back into the middle ear. Conversely, as the sound waves subside, the stapes moves backward and the procedure is reversed. That is, the fluid moves in the opposite direction along the same pathway and the basilar membrane bulges into the cochlear duct.

10. When the basilar membrane vibrates, the hair cells of the spiral organ are moved against the tectorial membrane. In some unknown manner, the movement of the hairs develops generator potentials that ultimately lead to the generation of nerve impulses.

11. The impulses are then passed on to the cochlear branch of the vestibulocochlear (VIII) nerve (Figure 17-13) and cochlear nuclei in the medulla. Here most impulses cross to the opposite side and then travel to the inferior colliculus of the midbrain, medial geniculate body of the thalamus, and finally to the auditory area of the temporal lobe of the cerebral cortex.

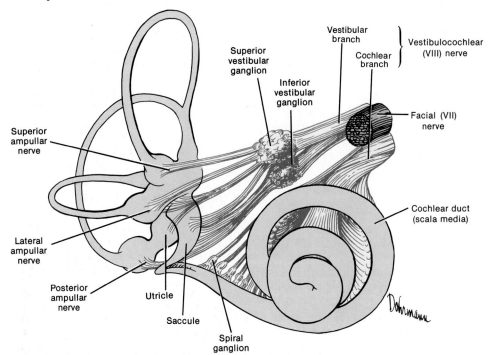

FIGURE 17-13 Principal nerves of the internal ear. Shown here is the right membranous labyrinth with the bony labyrinth removed. Nerve branches from each of the ampullae of the semicircular ducts and from the utricle and saccule form the vestibular branch of the vestibulocochlear (VIII) nerve. The vestibular ganglia contain the cell bodies of the vestibular branch. The cochlear branch of the vestibulocochlear (VIII) nerve arises in the spiral organ. The spiral ganglia contain the cell bodies of the cochlear branch.

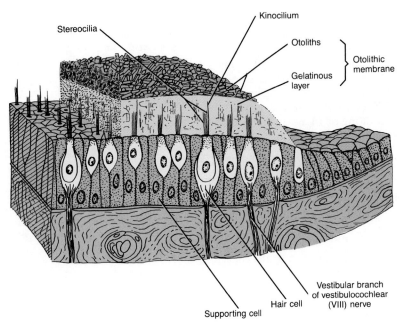

FIGURE 17-14 Structure of the macula.

PHYSIOLOGY OF EQUILIBRIUM

There are two kinds of **equilibrium.** One, called **static equilibrium,** refers to the orientation of the body (mainly the head) relative to the ground (gravity). The second kind, **dynamic equilibrium,** is the maintenance of body position (mainly the head) in response to sudden movements such as rotation, acceleration, and deceleration. The receptor organs for equilibrium are the maculae of the saccule and the utricle and cristae in the semicircular ducts.

Static Equilibrium

The walls of both the utricle and saccule contain a small, flat, plaquelike region called a **macula** (Figure 17-14). The maculae are the receptors that are concerned with static equilibrium. They provide sensory information regarding the orientation of the head in space and are essential for maintaining posture.

Microscopically, the two maculae resemble the spiral organ. They consist of differentiated neuroepithelial cells that are innervated by the vestibular branch of the vestibulocochlear (VIII) nerve. The maculae are anatomically located in planes perpendicular to one another and possess two kinds of cells: **hair (receptor) cells** and **supporting cells.** Two shapes of hair cells have been identified: one is more flask-shaped and the other is more cylindrical.

Both show long extensions of the cell membrane consisting of many *stereocilia* (they are actually microvilli) and one *kinocilium* (a conventional cilium) anchored firmly to its basal body and extending beyond the longest microvilli. Some of the microvilli reach lengths of over 100 µm, and some receptor cells have over 80 such projections.

The columnar supporting cells of the maculae are scattered between the hair cells. Floating directly over the hair cells is a thick, gelatinous, glycoprotein layer, probably secreted by the supporting cells, called the **otolithic membrane.** A layer of calcium carbonate crystals, called **otoliths** (*oto* = ear; *lithos* = stone), extends over the entire surface of the otolithic membrane. The specific gravity of these otoliths is about 3, which makes them much denser than the endolymph fluid that fills the rest of the utricle.

The otolithic membrane sits on top of the macula like a discus on a greased cookie sheet. If you tilt your head forward, the otolithic membrane (the discus in our analogy) slides downhill over the hair cells in the direction determined by the tilt of your head. In the same manner, if you are sitting upright in a car that suddenly jerks forward, the otolithic membrane, due to its inertia, slides backward and stimulates the hair cells. As the otoliths move, they pull on the gelatinous layer which pulls on the stereocilia and makes them bend. The movement of the stereocilia initiates a nerve impulse that is then transmitted to the vestibular branch of the vestibulocochlear (VIII) nerve (see Figure 17-13).

Most of the vestibular branch fibers enter the brain stem and terminate in the vestibular nuclear complex in the medulla. The remaining fibers enter the flocculonodular lobe of the cerebellum through the inferior cerebellar peduncle. Bidirectional pathways connect the vestibular nuclei and cerebellum. Fibers from all the vestibular nuclei form the medial longitudinal fasciculus which extends from the brain stem into the cervical portion of the spinal

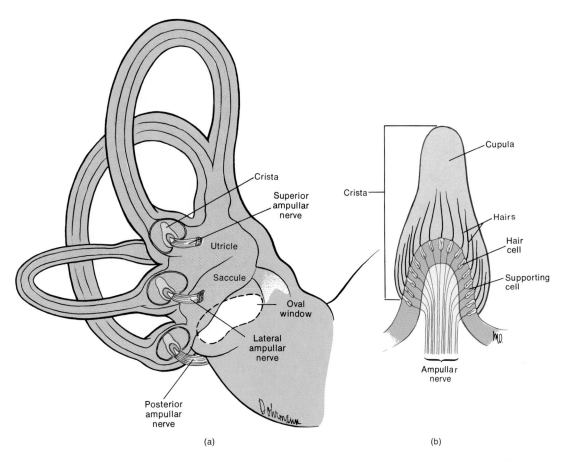

FIGURE 17-15 Semicircular ducts and dynamic equilibrium. (a) Position of the cristae relative to the membranous ampullae. (b) Enlarged aspect of a crista.

cord. The fasciculus sends impulses to the nuclei of cranial nerves that control eye movements [oculomotor (III), trochlear (IV), and abducens (VI)] and to the accessory (XI) nerve nucleus that helps control movements of the head and neck. In addition, fibers from the lateral vestibular nucleus form the vestibulospinal tract which conveys impulses to skeletal muscles that regulate body tone in response to head movements. Various pathways between the vestibular nuclei, cerebellum, and cerebrum enable the cerebellum to assume a key role in helping the body to maintain static equilibrium. The cerebellum continuously receives updated sensory information from the utricle and saccule concerning static equilibrium. Using this information, the cerebellum monitors and makes corrective adjustments in the motor activities that originate in the cerebral cortex. Essentially, the cerebellum sends continuous impulses to the motor areas of the cerebrum, in response to input from the utricle and saccule, causing the motor system to increase or decrease its impulses to specific skeletal muscles in order to maintain static equilibrium.

Dynamic Equilibrium

The cristae in the three semicircular ducts maintain dynamic equilibrium (Figure 17-15). The ducts are positioned at right angles to one another in three planes: frontal (the superior duct), sagittal (the posterior duct), and lateral (the lateral duct). This positioning permits detection of an imbalance in three planes. In the ampulla, the dilated portion of each duct, there is a small elevation called the **crista.** Each crista is composed of a group of **hair (receptor) cells** and **supporting cells** covered by a mass of gelatinous material called the **cupula.** When the head moves, the endolymph in the semicircular ducts flows over the hairs and bends them. The movement of the hairs stimulates sensory neurons, and the impulses pass over the vestibular branch of the vestibulocochlear (VIII) nerve. The impulses follow the same pathways as those involved in static equilibrium and are eventually sent to the muscles that must contract to maintain body balance in the new position.

DISORDERS: HOMEOSTATIC IMBALANCES

The special sense organs can be altered or damaged by numerous disorders. The causes of disorder range from congenital origins to the effects of old age. Here we discuss a few common disorders of the eyes and ears.

Cataract

The most prevalent disorder resulting in blindness is cataract formation. A **cataract** is a clouding of the lens or its capsule so that it becomes opaque or milk-white. As a result, the light from an object, which normally passes directly through the lens to produce a sharp image, produces only a degraded image. If the cataract is severe enough, no image at all is produced.

Cataracts can occur at any age, but we will discuss the type that develops with old age. As a person gets older, the cells in the lenses may degenerate and be replaced with nontransparent fibrous protein, or the lens may start to manufacture nontransparent protein. The cataract formation results in a progressive, painless loss of vision. The degree of loss depends on the location and extent of the opacity. If vision loss is gradual, frequent changes in glasses may help maintain useful vision for a while. However, when the changes become so extensive that light rays are blocked out altogether, surgery is indicated to remove the opaque lens. An artificial lens is substituted by means of eyeglasses or by surgical implantation of a plastic lens inside the eyeball.

Glaucoma

The second most common cause of blindness, especially in the elderly, is **glaucoma.** This disorder is characterized by an abnormally high pressure of fluid inside the eyeball. The aqueous humor does not return into the bloodstream through the canal of Schlemm as quickly as it is formed. The fluid accumulates and, by compressing the lens into the vitreous humor, puts pressure on the neurons of the retina. If the pressure continues over a long period of time, it destroys the neurons and brings about blindness. It can affect a person of any age, but 95 percent of the victims are over 40. Glaucoma affects the eyesight of more than 1 million people in the United States.

Treatment is through drugs or surgery. Drug treatment involves reducing the pressure in the eye by giving cholinergic compounds that cause the sphincter muscles of the iris and ciliary body to contract. These contractions prevent the obstruction of the anterior chamber and the canal of Schlemm, resulting in improved fluid outflow from the eye. Surgery consists of removing a small piece of the iris (iridectomy) to allow drainage and lessen the interior pressure.

Conjunctivitis

Many different eye inflammations exist, but the most common type is **conjunctivitis (pinkeye),** an inflammation of the membrane that lines the insides of the eyelids and covers the cornea. Conjunctivitis can be caused by microorganisms, most often the pneumococci or staphylococci bacteria. In such cases, the inflammation is very contagious. This epidemic type, common in children, is rarely serious. Conjunctivitis may also be caused by a number of irritants, in which case the inflammation is not contagious. Irritants include dust, smoke, wind, pollutants in the air, and excessive glare. The condition may be acute or chronic.

Trachoma

A form of chronic contagious conjunctivitis that is serious is **trachoma** (tra-KŌ-ma). It is caused by an organism, called the TRIC agent, that has characteristics of both viruses and bacteria. Trachoma is characterized by many granulations or fleshy projections on the eyelids. If untreated, these projections can irritate and inflame the cornea and reduce vision. The disease produces an excessive growth of subconjunctival tissue and the invasion of blood vessels into the upper half of the front of the cornea. The disease progresses until it covers the entire cornea, bringing about a loss of vision because of corneal opacity.

Antibiotics, such as tetracycline and the sulfa drugs, kill the organisms that cause trachoma and have reduced the seriousness of this infection.

Deafness

Deafness is the lack of the sense of hearing or significant hearing loss. It is generally divided into two principal types: nerve deafness and conduction deafness. *Nerve deafness* is caused by impairment of the cochlea or cochlear branch of the vestibulocochlear (VIII) nerve. *Conduction deafness* is caused by impairment of the middle ear mechanisms for transmitting sounds into the cochlea.

Labyrinthine Disease

Labyrinthine (lab'-i-RIN-thēn) **disease** refers to a malfunction of the inner ear that is characterized by deafness, tinnitus (ringing in the ears), vertigo (hallucination of movement), nausea, and vomiting. There may also be a blurring of vision, nystagmus (rapid, involuntary movement of the eyeballs), and a tendency to fall in a certain direction.

Among the causes of labyrinthine disease are: (1) infection of the middle ear, (2) trauma from brain concussion producing hemorrhage or splitting of the labyrinth, (3) cardiovascular diseases such as arteriosclerosis and blood vessel disturbances, (4) congenital malformation of the labyrinth, (5) excessive formation of endolymph, (6) allergy, (7) blood abnormalities, and (8) aging.

In severe cases of labyrinthine disease, surgical treatment is considered. Treatment with ultrasound, which consists of beaming ultrasound waves into the labyrinth, destroys the semicircular ducts (equilibrium) while preserving the cochlear duct (hearing). This treatment has produced excellent results. Another successful procedure involves electrical coagulation of the labyrinth, which changes the fluid to a jelly or a solid.

Ménière's Syndrome

Ménière's (men-YAIRZ) **syndrome** is one type of labyrinthine disorder that is characterized by fluctuating hearing loss, attacks of vertigo, and roaring tinnitus. Etiology of Ménière's syndrome is unknown. It is now thought that there is either an overproduction or underabsorption of endolymph in the cochlear duct. The hearing loss is caused by distortions in the basilar membrane of the cochlea. The classic type of Ménière's syndrome involves both the semicircular canals and the cochlea. The natural course of Ménière's syndrome may stretch out over a period of years, with the end result being almost total destruction of the patient's hearing.

The treatment for Ménière's syndrome may be either medical or surgical. Medical treatment is aimed at reducing the excess amount of endolymph in order to lessen the amount of distention and distortion of the basilar membrane. To accomplish this, the overall extracellular volume is diminished by means of salt restriction and the administration of diuretics. A popular treatment for the acute cases is the inhalation of 5 percent CO_2 with 95 percent O_2 for 10 minutes, four times a day. High levels of CO_2 in the cerebral circulation are powerful dilators of the cerebral vessels. For the 20 to 30 percent of Ménière's patients who do not respond to medical treatment, some type of vestibular surgery is indicated, for relief from vertigo. There are two basic procedures available to the otologic surgeon: those that destroy residual hearing and those that preserve it. As a rule, labyrinthectomy is reserved for advanced cases in which the hearing is poor.

Impacted Cerumen

Some people produce an abnormal amount of cerumen, or earwax, in the external auditory canal. It becomes impacted and prevents sound waves from reaching the tympanic membrane. The treatment for **impacted cerumen** is usually periodic ear irrigation or removal of wax with a blunt instrument.

Otitis Media

Otitis media is an acute infection of the middle ear cavity with a reddening and outward bulging of the eardrum, which may rupture. Second only to the common cold as a disease of childhood, otitis media is the most frequent diagnosis made by physicians who care for children. Infants and young children are at highest risk for otitis media; the peak prevalence is between 6 and 24 months of age.

Abnormal function of the auditory tube appears to be the most important mechanism in the pathogenesis of middle ear disease. This allows bacteria from the nasopharynx, which are a primary cause of middle ear infection, to enter the middle ear. Otitis media can be acute or chronic, and the primary etiology of both forms is bacterial infection. *Streptococcus pneumoniae* is the most common agent found in all age groups, followed by *Hemophilus influenzae*. Hearing loss is by far the most prevalent complication of otitis media.

Antimicrobial therapy has practically eliminated such complications of middle ear infections as mastoiditis and brain abscess, but perforation of the tympanic membrane and labyrinthitis still occur.

Motion Sickness

Motion sickness is a functional disorder brought on by repetitive angular, linear, or vertical motion and characterized by various symptoms, primarily nausea and vomiting. Warning symptoms include yawning, salivation, pallor, hyperventilation, profuse cold sweating, and prolonged drowsiness. Specific kinds of motion sickness are seasickness or air-, car-, space-, train-, or swing sickness.

Etiology is excessive stimulation of the vestibular apparatus by motion. The pathways taken by the afferent impulses from the labyrinth to the vomiting center in the medulla have not been defined. Visual stimuli and emotional factors like fear and anxiety can also contribute to motion sickness. Ideally, treatment with drugs of susceptible individuals should be instituted prior to their entering into conditions that would produce motion sickness, since prevention is more successful than treatment of symptoms once they have developed.

MEDICAL TERMINOLOGY

Achromatopsia (*a* = without; *chrom* = color) Complete color blindness.

Ametropia (*ametro* = disproportionate; *ops* = eye) Refractive defect of the eye resulting in an inability to focus images properly on the retina.

Anopsia (*opsia* = vision) A defect of vision.

Blepharitis (*blepharo* = eyelid; *itis* = inflammation of) An inflammation of the eyelid.

Eustachitis An inflammation or infection of the auditory tube.

Keratitis (*kerato* = cornea) An inflammation or infection of the cornea.

Labyrinthitis An inflammation of the labyrinth (inner ear).

Myringitis (*myringa* = eardrum) An inflammation of the eardrum; also called tympanitis.

Otalgia (*oto* = ear; *algia* = pain) Earache.

Otosclerosis (*sclerosis* = hardening) Pathological process in which new bone is deposited around the oval window. The result may be immobilization of the stapes, leading to deafness.

Ptosis (*ptosis* = fall) Falling or drooping of the eyelid. (This term is also used for the slipping of any organ below its normal position.)

Retinoblastoma (*blast* = bud; *oma* = tumor) A common tumor arising from immature retinal cells and accounting for 2 percent of childhood malignancies.

Strabismus An eye muscle disorder, commonly called "crossed eyes," in which the eyeballs do not move in unison. It may be caused by lack of coordination of the extrinsic eye muscles.

DRUGS ASSOCIATED WITH THE SPECIAL SENSES

Cholinergic drugs are used to produce *miosis* (constriction of the pupil) in the treatment of glaucoma. They reduce intraocular pressure by increasing the outflow of fluid from the anterior chamber of the eye. They can be divided into three classes: (1) direct-acting cholinergics; (2) reversible inhibitors of cholinesterase; and (3) irreversible inhibitors of cholinesterase.

Examples of direct-acting cholinergics are *pilocarpine* and *carbachol* (Doryl). Two reversible cholinesterase inhibitors are *physostigmine* and *neostigmine*. Irreversible cholinesterase inhibitors include *isoflurophate* (Floropryl) and *echothiophate* (Phospholine Iodide).

Carbonic Anhydrase Inhibitors

Carbonic anhydrase inhibitors are drugs that lower interocular pressure by inhibiting the formation of fluid. They are used to treat glaucoma. Examples are *acetazolamide* (Diamox) and *dichlorphenamide* (Daranide, Oratrol).

Adrenergic Blocking Agents

Adrenergic blocking agents are drugs that block the action of epinephrine and norepinephrine at the postganglionic nerve endings. *Timolol* (Timoptic) is used in the treatment of glaucoma because it lowers intraocular pressure by reducing aqueous humor production.

Antihistamines

Antihistamines are drugs that counteract the effects of histamine, a normal body chemical that is believed to cause the symptoms of persons who are hypersensitive to various allergic substances. There are some antihistamines that have an antinauseant action that is useful in the treatment of motion sickness, vertigo, Ménière's syndrome, and labyrinthitis. Examples are *cyproheptadine* (Periactin), *azatadine,* and *meclizine HCl* (Antivert).

Antibiotics

Antibiotics are drugs that are used to destroy pathogenic or noxious (injurious) organisms. The treatment for nearly all forms of otitis media is the antibiotic *ampicillin.*

STUDY OUTLINE

Olfactory Sensations (p. 374)

1. The receptors for olfaction are in the nasal epithelium.
2. Substances to be smelled must be volatile, water-soluble, and lipid-soluble.
3. Two theories that explain how olfactory cells respond to primary sensations are the chemical theory and physical theory.
4. Adaptation to odors occurs quickly, and the threshold of smell is low.
5. Olfactory cells convey impulses to olfactory (I) nerves, olfactory bulbs, olfactory tracts, and cerebral cortex.

Gustatory Sensations (p. 375)

1. The receptors for gustation are located in taste buds.
2. Substances to be tasted must be in solution in saliva.
3. The four primary tastes are salt, sweet, sour, and bitter.
4. Adaptation to taste occurs quickly, and the threshold varies with the taste involved.
5. Gustatory cells convey impulses to cranial nerves V, VII, IX, and X, medulla, thalamus, and cerebral cortex.

Visual Sensations (p. 377)

1. Accessory structures of the eyes include the eyebrows, eyelids, eyelashes, and the lacrimal apparatus.
2. The eye is constructed of three coats: (a) fibrous tunic (sclera and cornea), (b) vascular tunic (choroid, ciliary body, and iris), and (c) retina, which contains rods and cones.
3. The anterior cavity contains aqueous humor; the posterior cavity contains vitreous humor.
4. The refractive media of the eye are the cornea, aqueous humor, lens, and vitreous humor.
5. Retinal image formation involves refraction of light, accommodation of the lens, constriction of the pupil, convergence, and inverted image formation.
6. Improper refraction may result from myopia (nearsightedness), hypermetropia (farsightedness), and astigmatism (corneal or lens abnormalities).
7. Rods and cones develop generator potentials and ganglion cells initiate nerve impulses.
8. Impulses from ganglion cells are conveyed through the retina to the optic (II) nerve, the optic chiasma, the optic tract, the thalamus, and the cortex.

Auditory Sensations and Equilibrium (p. 387)

1. The ear consists of three anatomical subdivisions: (a) the external or outer ear (pinna, external auditory canal, and tympanic membrane), (b) the middle ear (auditory tube, ossicles, oval window, and round window), and (c) the internal or inner ear (bony labyrinth and membranous labyrinth). The internal ear contains the spiral organ, the organ of hearing.
2. Sound waves enter the external auditory canal, strike the tympanic membrane, pass through the ossicles, strike the oval window, set up waves in the perilymph, strike the vestibular membrane and scala tympani, increase pressure in the endolymph, strike the basilar membrane, and stimulate hairs on the spiral organ. A sound impulse is then initiated.
3. Static equilibrium is the orientation of the body relative to the pull of gravity. The maculae of the utricle and saccule are the sense organs of static equilibrium.
4. Dynamic equilibrium is the maintenance of body position in response to movement. The cristae in the semicircular ducts are the sense organs of dynamic equilibrium.

Disorders: Homeostatic Imbalances (p. 395)

1. Cataract is the loss of transparency of the lens or capsule.
2. Glaucoma is abnormally high intraocular pressure, which destroys neurons of the retina.
3. Conjunctivitis is an inflammation of the conjunctiva.
4. Trachoma is a chronic, contagious inflammation of the conjunctiva.
5. Deafness is the lack of the sense of hearing or significant hearing loss.
6. Labyrinthine disease is basically a malfunction of the inner ear that has a variety of causes.
7. Ménière's syndrome is the malfunction of the inner ear that may cause deafness and loss of equilibrium.
8. Impacted cerumen is an abnormal amount of earwax in the external auditory canal.
9. Otitis media is an acute infection of the middle ear cavity.
10. Motion sickness is a functional disorder precipitated by repetitive angular, linear, or vertical motion.

REVIEW QUESTIONS

1. What are the necessary conditions for substances to be smelled?
2. Outline the two principal theories of olfaction.
3. Discuss the origin and path of an impulse that results in smelling.
4. How are papillae related to taste buds? Describe the structure and location of the papillae.
5. How are gustatory receptors stimulated?
6. Describe the thresholds for the four primary tastes.
7. Discuss how an impulse for taste travels from a taste bud to the brain.
8. Describe the structure and importance of the following accessory structures of the eye: eyelids, eyelashes, and eyebrows.
9. What is the function of the lacrimal apparatus? Explain how it operates.
10. By means of a labeled diagram, indicate the principal anatomical structures of the eye.
11. How do extrinsic and intrinsic eye muscles differ?
12. Describe the location and contents of the chambers of the eye. What is intraocular pressure? How is the canal of Schlemm related to this pressure?
13. Explain how each of the following events is related to the physiology of vision: (a) refraction of light, (b) accommodation of the lens, (c) constriction of the pupil, (d) convergence, and (e) inverted image formation.
14. Distinguish emmetropia, myopia, hypermetropia, and astigmatism by means of a diagram.
15. How are rods and cones excited? Relate your discussion to the rhodopsin cycle by means of a diagram.
16. What is night blindness? What causes it?
17. Distinguish between light adaptation and dark adaptation.
18. Describe the path of a visual impulse from the optic (II) nerve to the brain.
19. Define visual field. Relate the visual field to image formation on the retina.

20. Diagram the principal parts of the outer, middle, and inner ear. Describe the function of each part labeled.
21. What are sound waves? How are sound intensities measured?
22. Explain the events involved in the transmission of sound from the pinna to the spiral organ.
23. What is the afferent pathway for sound impulses from the cochlear branch of the vestibulocochlear (VIII) nerve to the brain?
24. Compare the function of the maculae in the saccule and utricle in maintaining static equilibrium with the role of the cristae in the semicircular ducts in maintaining dynamic equilibrium.
25. Describe the path of an impulse that results in static and dynamic equilibrium.
26. Define each of the following: cataract, glaucoma, conjunctivitis, trachoma, deafness, labyrinthine disease, Ménière's syndrome, impacted cerumen, otitis media, and motion sickness.
27. Refer to the medical terminology associated with the sense organs. Be sure that you can define each term listed.
28. Describe the actions of selected drugs on the sense organs.

- Describe the relationship between the endocrine system and the nervous system in maintaining homeostasis.

- Discuss the functions of the endocrine system in maintaining homeostasis.

- Define an endocrine gland and list the endocrine glands of the body.

- Distinguish between water-soluble and lipid-soluble hormones.

- Explain the mechanism of hormonal action.

- Identify the role of prostaglandins in hormonal action.

- Describe the control of hormonal secretions via feedback cycles and explain several examples.

- Describe the structural and functional division of the pituitary gland into the adenohypophysis and the neurohypophysis.

- Discuss how the pituitary gland and hypothalamus are structurally and functionally related.

- List the hormones of the adenohypophysis, their principal actions, and their associated hypothalamic regulating factors.

- Describe the release of hormones stored in the neurohypophysis and their principal actions.

- Discuss the symptoms of pituitary dwarfism, giantism, acromegaly, and diabetes insipidus as pituitary gland disorders.

- Describe the location and histology of the thyroid gland.

- Explain the synthesis, storage, and release of thyroid hormones.

- Discuss the principal actions and control of thyroid gland hormones.

- Discuss the symptoms of cretinism, myxedema, exophthalmic goiter, and simple goiter as thyroid gland disorders.

- Describe the location and histology of the parathyroid glands.

- Explain the principal actions and control of the parathyroid hormone.

- Discuss the symptoms of tetany and osteitis fibrosa cystica as parathyroid gland disorders.

- Describe the location and histology of the adrenal (suprarenal) glands and explain their subdivision into cortical and medullary portions.

- Distinguish the effects of adrenal cortical mineralocorticoids, glucocorticoids, and gonadocorticoids on physiological activities, and explain how the hormones are controlled.

- Discuss the symptoms of aldosteronism, Addison's disease, Cushing's syndrome, and adrenogenital syndrome as adrenal cortical disorders.

- Identify the function of the adrenal medullary secretions as supplements of sympathetic responses.

- Discuss the symptoms of pheochromocytomas as an adrenal medullary disorder.

- Describe the location and histology of the pancreas.

- Compare the principal actions of the pancreatic hormones and describe how they are controlled.

- Discuss the symptoms of diabetes mellitus and hyperinsulinism as endocrine disorders of the pancreas.

- Explain why the ovaries and testes are classified as endocrine glands.

- Describe the location and histology of the pineal gland and the possible functions of its hormones.

- Describe the location and histology of the thymus gland and the function of its hormones in immunity.

- Explain why portions of the gastrointestinal tract, placenta, kidneys, and skin are considered as endocrine structures.

- Define the general adaptation syndrome and compare homeostatic responses and stress responses.

- Identify the body reactions during the alarm, resistance, and exhaustion stages of stress.

- Define medical terminology associated with the endocrine system.

18 The Endocrine System

The nervous system controls homeostasis through electrical impulses delivered over neurons. The body's other control system, the **endocrine system,** affects bodily activities by releasing chemical messengers, called hormones, into the bloodstream. Whereas the nervous system causes muscles to contract and glands to secrete, the endocrine system brings about changes in the metabolic activities of body tissues. The release of nerve impulses is much more rapid than that of hormones. Also, the effects of nervous system stimulation are generally brief compared to the effects of endocrine stimulation.

Obviously, the body could not function if the two great control systems were to pull in opposite directions. The nervous and endocrine systems coordinate their activities like an interlocking supersystem. Certain parts of the nervous system stimulate or inhibit the release of hormones, and hormones, in turn, are quite capable of stimulating or inhibiting the flow of nerve impulses.

Although the effects of hormones are many and varied, their actions can be categorized into four broad areas:

1. They help to control the internal environment by regulating its chemical composition and volume.

2. They respond to marked changes in environmental conditions to help the body cope with emergency demands such as infection, trauma, emotional stress, dehydration, starvation, hemorrhage, and temperature extremes.

3. They assume a role in the smooth, sequential integration of growth and development.

4. They contribute to the basic processes of reproduction, including gamete production, fertilization, nourishment of the embryo and fetus, delivery, and nourishment of the newborn.

ENDOCRINE GLANDS

The endocrine glands make up the **endocrine system.** The body contains two kinds of glands: exocrine and endocrine. **Exocrine glands** secrete their products onto a free surface or into ducts. The ducts carry the secretions into body cavities, into the lumens of various organs, or to the body's surface. Exocrine glands include sweat, sebaceous, mucous, and digestive glands. **Endocrine** (*endo* = within) **glands** by contrast, secrete their products (hormones) into the extracellular space around the secretory cells, rather than into ducts. The secretion then passes into capillaries to be transported in the blood. The endocrine glands of the body include the pituitary (hypophysis), thyroid, parathyroids, adrenals (suprarenals), pancreas, ovaries, testes, pineal (epiphysis cerebri), thymus, kidneys, stomach, small intestine, and placenta. (As you will see, some of these glands are mixed, that is, both endocrine and exocrine. The pancreas is an example). The location of many organs of the endocrine system is illustrated in Figure 18-1.

CHEMISTRY OF HORMONES

The secretions of endocrine glands are called **hormones** (*hormone* = set in motion). Although hormones are chemically diverse, they can be grouped into two principal classes on the basis of their solubility: water-soluble and lipid-soluble. **Water-soluble** hormones include protein hormones or derivatives of proteins (glycoproteins) or amino acids (catecholamines). Examples are hormones of the pituitary, parathyroids, pancreas, and adrenal medulla. These hormones are synthesized in the manner described for protein synthesis in Chapter 3. **Lipid-soluble** hormones include steroid hormones, lipids synthesized from cholesterol, and thyroid hormones. Examples are hormones of the thyroid, adrenal cortex, ovaries, and testes. These hormones are produced in mitochondria and agranular endoplasmic reticulum.

As you will learn in Chapter 29, all tissues and organs of the body are derived from three embryonic tissues called the endoderm, mesoderm, and ectoderm. Endocrine glands that originate from the endoderm (e.g., parathyroids and pancreas) or ectoderm (e.g., adrenal medulla and pituitary) generally secrete water-soluble hormones. Endocrine glands of mesodermal origin (e.g., adrenal cortex, ovaries, testes) generally secrete lipid-soluble hormones.

The one thing all hormones have in common—whether water-soluble or lipid-soluble—is the function of maintaining homeostasis by changing the physiological activities of cells.

MECHANISM OF HORMONAL ACTION

The general effects of most hormones are fairly well known. However, the manner in which specific hormones affect different cells of the body is just starting to become clear. The cells that respond to the effects of a hormone are called **target cells.** It appears as though hormones exert their effects on target cells in two quite different ways—one involves interaction with plasma membrane receptors and the other involves activation of genes.

INTERACTION WITH PLASMA MEMBRANE RECEPTORS

One mechanism of hormone action involves an interaction at the cell surface between the hormone and a receptor site on the plasma membrane. A hormone released from an endocrine gland circulates in the blood, reaches a target cell, and brings a specific message to that cell. The hormone is called the **first messenger.** To give the cell its message, the hormone must attach to a specific receptor site (integral protein) on the plasma membrane. This attachment can alter cell function by two different pathways.

In one pathway there is an increase in the synthesis of **cyclic adenosine-3′, 5′-monophosphate (cyclic AMP)** (see Figure 2-16). Cyclic AMP is synthesized from ATP, the main energy-storing chemical in cells. This synthesis requires an enzyme present in the plasma membrane called *adenyl cyclase.* When the first messenger attaches to its receptors, there is an activation of adenyl cyclase, resulting in the conversion of ATP into cyclic AMP in the cell (Figure 18-2a). Cyclic AMP then diffuses throughout the cell, binds to an intracellular receptor, and acts as a **second messenger,** altering cell function according to the message indicated by the hormone. Cyclic AMP can activate the appropriate cellular enzymes to get a specific biochemical response. Calcium ions may sometimes be involved as second messengers along with cyclic

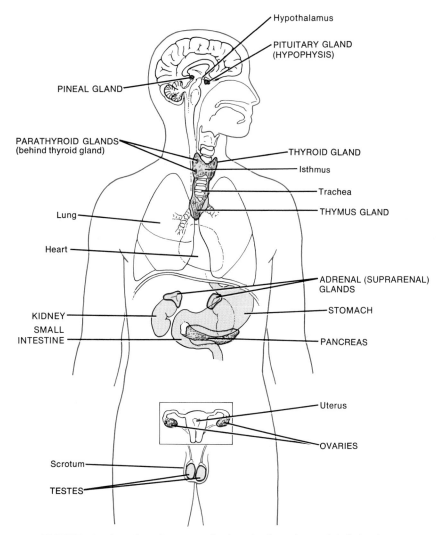

FIGURE 18-1 Location of many endocrine glands and associated structures.

AMP. Calcium ions bind to an intracellular protein called **calmodulin,** thus activating it, and activated calmodulin can, in turn, activate certain enzymes. Cyclic AMP can also stimulate protein synthesis, induce secretion, and alter membrane permeability. High levels of cyclic AMP persist only briefly because it is rapidly degraded by *cyclic AMP phosphodiesterase.* Since prostaglandins can influence the formation of cyclic AMP, they may assume a role in regulating hormonal actions. This will be discussed shortly. Most water-soluble hormones exert their effects by the pathway of increasing the synthesis of cyclic AMP. These include antidiuretic hormone (ADH), oxytocin (OT), follicle-stimulating hormone (FSH), luteinizing hormone (LH), thyroid-stimulating hormone (TSH), adrenocorticotropic hormone (ACTH), calcitonin (CT), parathyroid hormone (PTH), glucagon, epinephrine, and norepinephrine (NE).

The binding of a few water-soluble hormones—insulin, growth hormone (GH), and prolactin (PRL)—to their plasma membrane receptors does not lead to an increase in the synthesis of cyclic AMP. In an alternate pathway, the levels of cyclic AMP remain unchanged or decrease. Some investigators believe that calcium ions and other second messengers as yet unidentified may alter cell func-

tion. Some evidence even suggests that insulin, growth hormone, and prolactin may actually enter target cells and attach to various intracellular structures. Since the hormones might attach to various internal recognition sites that could alter cell function, second messengers would not be needed.

ACTIVATION OF GENES

Steroid hormones and thyroid hormones both alter cell function by activation of genes, but by somewhat different mechanisms. Since steroid hormones are lipid-soluble, they easily pass through the plasma membrane of the target cell. Upon entering the cell, a steroid hormone binds to a protein receptor site in the cytoplasm, and the hormonal-receptor complex is translocated into the nucleus of the cell (Figure 18-2b). The complex interacts with specific genes of the nuclear DNA and activates them to form the enzymes necessary to alter cell function in a specific way.

Thyroid hormones also enter target cells and interact with genes to alter cell function. However, thyroid hormones use binding sites in cell nuclei as well as protein receptors for translocation into the nucleus.

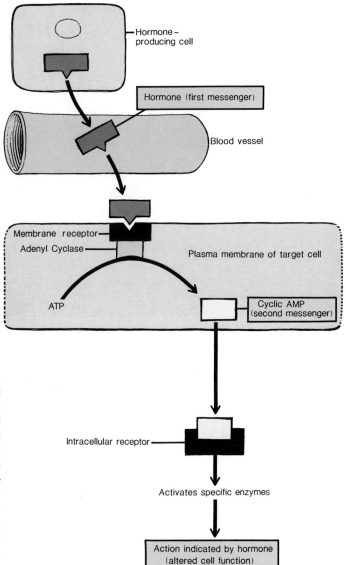

FIGURE 18-2 Proposed mechanisms of hormonal action. (a) Interaction with plasma membrane receptors in which there is an increase in the synthesis of cyclic AMP.

(a)

PROSTAGLANDINS (PGs) AND HORMONES

As we have noted earlier **prostaglandins** (pros'-ta-GLAN-dins) or **PGs** are membrane-associated, biologically active lipids that assume a number of important functions in the body. They are secreted into the blood in minute quantities and are potent in their action. Prostaglandins are also called *local* or *tissue hormones* because their site of action is the immediate area in which they are produced. This differentiates them from *circulating hormones,* which act on distant targets. In addition, prostaglandins are synthesized not by specialized endocrine organs, as are circulating hormones, but by nearly every mammalian cell and tissue. Chemical and mechanical stimuli, as well as anaphylaxis, leads to prostaglandin release.

Chemically, prostaglandins are composed of 20-carbon fatty acids containing 5 carbon atoms joined to form a cyclopentane ring. They are classified into four groups, E, F, A, and B. Prostaglandins are believed to be regulators or modulators of cell metabolism, although their precise actions remain to be determined. Their actions are closely linked to those of cyclic AMP. Varying with the tissue and species, prostaglandins either increase or decrease cyclic AMP formation. In this way, prostaglandins can alter the responses of cells to a hormone whose action involves cyclic AMP. In this respect, prostaglandins might be modulators of cyclic AMP–induced responses. Prostaglandins are rapidly inactivated, especially in the lungs, liver, and kidneys. The broad range of prostaglandin biological activity in relation to smooth muscle, secretion, blood flow, reproduction, platelet function, respiration, nerve impulse transmission, fat metabolism, immune response, and other life processes, as well as their involvement in inflammation, neoplasia, and other conditions, indicates their importance in both normal physiology and pathology.

What has intrigued investigators even more than the physiological role of prostaglandins has been their phar-macological effects and their implications for potential therapy. Prostaglandins are extraordinarily potent; one-billionth of a gram produces measurable effects. In vitro studies in animals including humans have revealed a rich diversity of pharmacological actions, varying with the type of prostaglandin and sometimes with the species. These effects include lowering or raising blood pressure; reducing gastric secretion; bronchodilation or broncho-constriction; stimulating or inhibiting platelet aggregation; contracting or relaxing intestinal and uterine smooth muscle; shrinking nasal passages; blocking or augmenting norepinephrine release; mediating inflammation; increasing intraocular pressure; causing sedation, stupor, and fever; inducing labor in pregnancy; stimulating steroid production; promoting natriuresis (excretion of sodium in the urine), potentiating the pain-producing effect of kinins; and many others.

Each year scientists pinpoint the specific roles of more and more prostaglandins. Some researchers feel that these

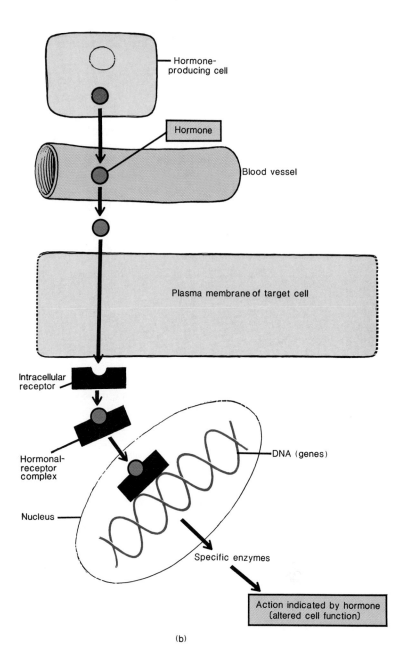

FIGURE 18-2 (*Continued*) Proposed mechanisms of hormonal action. (b) Activation of genes by a steroid hormone.

(b)

compounds will revolutionize the management of many common disorders, including immunological disorders, cancer, congestive heart failure, diarrhea, cholera, inflammations, fever, arthritis, kidney disease, glaucoma, trauma, and septicemia.

CONTROL OF HORMONAL SECRETIONS: FEEDBACK CONTROL

The amount of hormone released by an endocrine gland or tissue is determined by the body's need for the hormone at any given time. Secretion is normally regulated so that there is no overproduction or underproduction of a particular hormone. This regulation is one of the very important ways that the body attempts to maintain homeostasis. Unfortunately, there are times when the regulating mechanism does not operate properly and hormonal levels are excessive or deficient. When this happens, disorders result.

The typical way in which hormonal secretions are regulated is by **negative feedback control** (Chapter 1). As applied to hormones, information regarding the hormone level or its effect is fed back to the gland, which then responds accordingly. In one type of negative feedback system, the regulation of hormonal secretion does not involve direct participation by the nervous system. For example, blood calcium level is controlled by parathyroid hormone (PTH), produced by the parathyroid glands. If, for some reason, blood calcium level is low, this serves as a stimulus for the parathyroids to release more PTH (see Figure 18-10). PTH then exerts its effects in various parts of the body until the blood calcium level is raised to normal. A high blood calcium level serves as a stimulus for the parathyroids to cease their production of PTH. In the absence of the hormone, other mechanisms take over until blood calcium level is lowered to normal. Note that in negative feedback control the body's response (increased or decreased calcium level) is opposite (negative) to the stimulus (low or high calcium level). Other hormones that are regulated without direct involvement of the nervous system include calcitonin (CT), produced by the thyroid gland; insulin by the pancreas; and aldosterone by the adrenal cortex.

In other negative feedback systems, the hormone is released as a direct result of nerve impulses that stimulate the endocrine gland. Epinephrine and norepinephrine (NE) are released from the adrenal medulla in response to sympathetic nerve impulses. Antidiuretic hormone (ADH) is released from the posterior pituitary in response to nerve impulses from the hypothalamus (see Figure 26-10).

One of the few exceptions to the rule of negative feedback control is oxytocin (OT). The regulating system for the release of OT from the pituitary in response to nerve impulses from the hypothalamus is a positive feedback cycle, that is, the output intensifies the input (see Figure 18-6).

There are also negative feedback systems that involve the nervous system through chemical secretions from the hypothalamus, called **regulating factors** (see Figure 18-4). Some regulating factors, called **releasing factors,** stimulate the release of the hormone into the blood, so that it can exert its influence. Other regulating factors, called **inhibiting factors,** prevent the release of the hormone.

As we discuss the effects of various hormones in this chapter, we will also describe how the secretions of the hormones are controlled. At that time you will be able to see which type of negative feedback system is operating.

PITUITARY (HYPOPHYSIS)

The hormones of the **pituitary gland,** also called the **hypophysis** (hī-POF-i-sis), regulate so many body activities that the pituitary has been nicknamed the "master gland." It is a round structure and surprisingly small, measuring about 1.3 cm (0.5 inch) in diameter. The pituitary lies in the sella turcica of the sphenoid bone and is attached to the hypothalamus of the brain via a stalklike structure, the **infundibulum** (see Figure 14-1a).

The pituitary is divided structurally and functionally into an anterior lobe and a posterior lobe. Both are connected to the hypothalamus. Between the lobes is a small, relatively avascular zone, the **pars intermedia.** Although much larger and more clearly defined in structure and function in some lower animals, its role in humans is obscure. The **anterior lobe** constitutes about 75 percent of the total weight of the gland. It is derived from the ectoderm from an embryological invagination of pharyngeal epithelium. Accordingly, the anterior lobe contains many glandular epithelial cells and forms the glandular

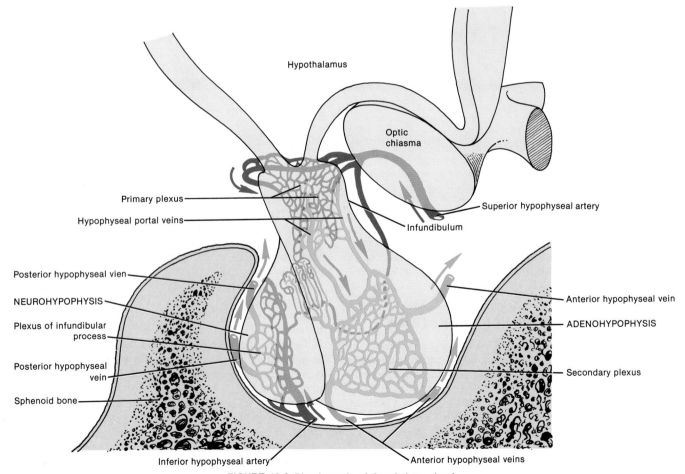

FIGURE 18-3 Blood supply of the pituitary gland.

part of the pituitary. A system of blood vessels connects the anterior lobe with the hypothalamus.

The **posterior lobe** is also derived from the ectoderm, but from an outgrowth of the hypothalamus of the brain. Accordingly, the posterior lobe contains axonic ends of neurons whose cell bodies are located in the hypothalamus. The nerve fibers that terminate in the posterior lobe are supported by cells called pituicytes. Other nerve fibers connect the posterior lobe directly with the hypothalamus.

ADENOHYPOPHYSIS

The anterior lobe of the pituitary is also called the **adenohypophysis** (ad'-i-nō-hī-POF-i-sis). It releases hormones that regulate a whole range of bodily activities from growth to reproduction. The release of these hormones is either stimulated or inhibited by chemical secretions from the hypothalamus called **regulating factors.** Regulating factors will be considered along with each of the anterior pituitary hormones. These factors constitute an important link between the nervous system and the endocrine system.

The hypothalamic regulating factors are delivered to the adenohypophysis in the following way. The blood supply to the adenohypophysis and infundibulum is derived principally from several *superior hypophyseal* (hī'-po-FIZ-ē-al) *arteries.* These arteries are branches of the internal carotid and posterior communicating arteries (Figure 18-3). The superior hypophyseal arteries form a network or plexus of capillaries, the *primary plexus,* in the infundibulum near the inferior portion of the hypothalamus. Regulating factors from the hypothalamus diffuse into this plexus. This plexus drains into the *hypophyseal portal veins,* that pass down the infundibulum. At the inferior portion of the infundibulum, the veins form a *secondary plexus* in the adenohypophysis. From this plexus, hormones of the adenohypophysis pass into the anterior hypophyseal veins for distribution to tissue cells. Such a delivery system permits regulating factors to act quickly on the adenohypophysis without first circulating through the heart. The short route prevents dilution or destruction of the regulating factors.

When the adenohypophysis receives proper stimulation from the hypothalamus via regulating factors, its glandular cells secrete any one of seven hormones. Recently, special staining techniques have established the division of glandular cells into five principal types.

1. Growth hormone cells produce **growth hormone (GH),** which controls general body growth.

2. Prolactin cells synthesize **prolactin (PRL),** which initiates milk production by the mammary glands.

3. TSH cells manufacture **thyroid-stimulating hormone (TSH)** which controls the thyroid gland.

4. Gonadotroph cells produce **follicle-stimulating hormone (FSH),** which stimulates the production of eggs and sperm in the ovaries and testes, respectively, and **luteinizing hormone (LH),** which stimulates other sexual and reproductive activities.

5. Corticotroph-lipotroph cells synthesize **adrenocorticotropic hormone (ACTH),** which stimulates the adrenal cortex to secrete its hormones, and **melanocyte-stimulating hormone (MSH),** which is related to skin pigmentation.

Except for the growth hormone (GH), melanocyte-stimulating hormone (MSH), and prolactin (PRL), all

the secretions are referred to as **tropic hormones** (*trop* = turn on), which means that they stimulate other endocrine glands. Follicle-stimulating hormone (FSH) and luteinizing hormone (LH) are also called **gonadotropic** (gō-nad-ō-TRŌ-pik) **hormones** because they regulate the functions of the gonads. The gonads (ovaries and testes) are the endocrine glands that produce sex steroid hormones.

Growth Hormone (GH)

Growth hormone (GH) is also known as **somatotropin** (sō'-ma-tō-TRŌ-pin) and **somatotropic hormone (STH),** because it turns on body cells to grow. Its principal function is to act on the skeleton and skeletal muscles, in particular, to increase their rate of growth and maintain their size once growth is attained. GH causes cells to grow and multiply by directly increasing the rate at which amino acids enter cells and are built up into proteins. GH is considered to be a hormone of protein anabolism since it increases the rate of protein synthesis. GH also promotes fat catabolism, that is, it causes cells to switch from burning carbohydrates to burning fats for energy. For example, it stimulates adipose tissue to release fat. And it stimulates other cells to break down the released fat molecules. At the same time, GH accelerates the rate at which glycogen stored in the liver is converted into glucose and released into the blood. Since the cells are using fats for energy, however, they do not consume as much glucose. The result is an increase in blood sugar level, a condition called **hyperglycemia** (hī'-per-glī-SĒ-mē-a). This process is called the **diabetogenic** (dī'-a-bet'-ō-JEN-ik) **effect** because it mimics the elevated blood glucose level of diabetes mellitus.

GH seems to produce many of its effects by converting other factors into growth-promoting substances called **somatomedins** and **insulinlike growth factors (IGF).** These are small peptides made in the liver under the influence of GH. Both mediate most of the effects of GH and are structurally and functionally similar to insulin. However, their growth-promoting effects are much more potent than those of insulin.

The control of GH secretion is not yet clearly understood. Its release from the anterior pituitary is apparently controlled by at least two regulating factors from the hypothalamus: **growth hormone releasing factor (GHRF)** and **growth hormone inhibiting factor (GHIF),** or **somatostatin.** When GHRF is released by the hypothalamus into the bloodstream, it circulates to the anterior pituitary and stimulates the release of GH. On the other hand, GHIF inhibits the release of GH.

Among the stimuli that promote GH secretion is **hypoglycemia,** that is, low blood sugar level. Other promoting stimuli are listed in Exhibit 18-1. When blood sugar level is low, the hypothalamus is stimulated to secrete GHRF (Figure 18-4). Upon reaching the anterior pituitary, GHRF causes the anterior pituitary to release GH. Somatomedins, under the influence of GH, raise blood sugar level by converting glycogen into glucose and releasing it into the blood. As soon as blood sugar level returns to normal, GHRF secretion shuts off.

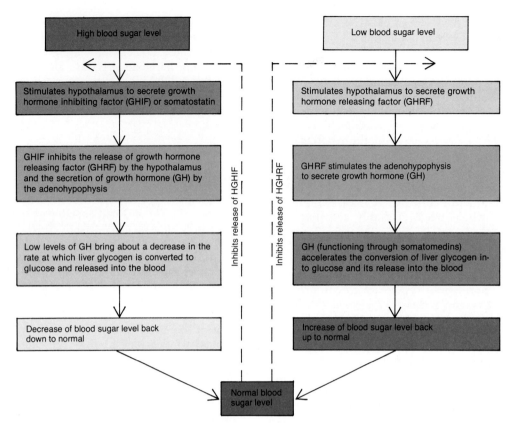

FIGURE 18-4 Regulation of the secretion of growth hormone (GH). Like other hormones of the adenohypophysis, the secretion of GH is controlled by regulating factors. Like most hormones of the body, GH secretion and inhibition involve negative feedback systems.

One stimulus that inhibits GH secretion is hyperglycemia, high blood sugar level (Figure 18-4). Other inhibiting stimuli are listed in Exhibit 18-1. An abnormally high blood sugar level stimulates the hypothalamus to secrete the regulating factor GHIF (somatostatin). GHIF inhibits the release of GHRF and thus the secretion of GH. As a result, blood sugar level decreases. You will see later that certain endocrine cells of the pancreas, called delta cells, also secrete GHIF (somatostatin) and it can inhibit the secretion of pancreatic hormones (insulin and glucagon) just as it inhibits secretion of GH.

The regulation of secretion of GH illustrates two phenomena that are typical of secretions of the adenohypophysis. First, each hormone is believed to be controlled by its own regulating factor from the hypothalamus. In some cases the regulating factor stimulates hormone secretion; in other cases, it inhibits secretion. Second, secretion is generally regulated through negative feedback systems. Since hormones are chemical regulators of homeostasis, these feedback systems are hardly surprising. Continued heavy secretion of a hormone would overshoot the goal and send the body out of balance in the opposite direction.

EXHIBIT 18-1 ▌▌▌▌▌▌
STIMULI THAT INFLUENCE NORMAL GROWTH HORMONE (GH) SECRETION

PROMOTES	INHIBITS
Hypoglycemia	Hyperglycemia
Decreased fatty acids	Increased fatty acids
Increased amino acids	Decreased amino acids
Low levels of GH	High levels of GH
Somatomedins	Growth hormone inhibiting factor (GHIF)
Stages 3 and 4 of NREM sleep	REM sleep
Stress	Emotional deprivation
Vigorous physical exercise	Obesity
Estrogens	Hypothyroidism
Glucagon	
Insulin	
Glucocorticoids	
Dopamine (DA)	
Acetylcholine (ACh)	

CLINICAL APPLICATION

Disorders of the endocrine system, in general, involve **hyposecretion** (underproduction) of hormones or **hypersecretion** (overproduction).

Among the clinically interesting disorders related to the adenohypophysis are those involving GH. If GH is hyposecreted during the growth years, bone growth is slow and the epiphyseal plates close before normal height is reached. This condition is called **pituitary dwarfism.** Other organs of the body also fail to grow, and the pituitary dwarf is childlike in many physical respects. Treatment requires administration of GH during childhood before the epiphyseal plates close.

Hypersecretion of GH during childhood results in **giantism,** an abnormal increase in the length of long bones. Hypersecretion during adulthood is called **acromegaly** (ak'-rō-MEG-a-lē). Acromegaly cannot produce further lengthening of the long bones because the epiphyseal plates are already closed. Instead, the bones of the hands, feet, cheeks, and jaws thicken. Other tissues also grow. The eyelids, lips, tongue, and nose enlarge and the skin thickens and furrows, especially on the forehead and soles of the feet.

Thyroid-Stimulating Hormone (TSH)

Thyroid-stimulating hormone (TSH) is also called **thyrotropin** (thī-rō-TRŌ-pin). It stimulates the synthesis and secretion of the hormones produced by the thyroid gland. Secretion is controlled by a regulating factor produced by the hypothalamus called **thyrotropin releasing factor (TRF).** Release of TRF depends on blood levels of thyroxine and the body's metabolic rate among other factors, and operates according to a negative feedback system.

Adrenocorticotropic Hormone (ACTH)

Adrenocorticotropic hormone (ACTH) is also called **adrenocorticotropin** (ad-rē'-nō-kor'-ti-kō-TRŌ-pin). Its tropic function is to control the production and secretion of certain adrenal cortex hormones. Secretion of ACTH is governed by a regulating factor produced by the hypothalamus called **corticotropin releasing factor (CRF).** Release of CRF depends on a number of stimuli and hormones and operates as a negative feedback system.

Follicle-Stimulating Hormone (FSH)

In the female, **follicle-stimulating hormone (FSH)** is transported from the adenohypophysis by the blood to the ovaries, where it initiates the development of ova each month. FSH also stimulates cells in the ovaries to secrete estrogens, or female sex hormones. In the male, FSH stimulates the testes to initiate sperm production. Secretion of FSH is subject to a regulating factor produced by the hypothalamus called **gonadotropin releasing factor (GnRF).** GnRF is released in response to estrogens and possibly progesterone in the female and testosterone in the male and involves a negative feedback system.

Luteinizing Hormone (LH)

In the female, **luteinizing** (LOO-tē-in'-īz-ing) **hormone (LH),** together with estrogens, stimulates the ovary to release an ovum (ovulation) and prepares the uterus for implantation of a fertilized ovum. It also stimulates formation of the corpus luteum in the ovary, which secretes progesterone (another female sex hormone) and readies the mammary glands for milk secretion. In the male, LH stimulates the interstitial endocrinocytes in the testes to develop and secrete large amounts of testosterone. Secretion of LH, like that of FSH, is controlled by GnRF. Release of GnRF is governed by a negative feedback system involving estrogens, progesterone, and testosterone. Recent evidence suggests that the placenta, formed during pregnancy, may be an extrahypothalamic source of GnRF. The GnRF it secretes has been designated as **placental luteotropic releasing factor (pLRF).** Its role in pregnancy will be discussed in Chapter 29.

Prolactin (PRL)

Prolactin (PRL) or the **lactogenic hormone,** together with other hormones, initiates and maintains milk secretion by the mammary glands. The actual ejection of milk by the mammary glands is controlled by a hormone stored in the posterior lobe called oxytocin. Together, milk secretion and ejection are referred to as lactation. PRL acts directly on tissues. By itself, it has little effect; it requires preparation by estrogens, progesterone, corticosteroids, growth hormone, thyroxine, and insulin. When the mammary glands have been primed by these hormones, PRL brings about milk secretion.

PRL has both an inhibitory and an excitatory negative control system. During menstrual cycles, **prolactin inhibiting factor (PIF),** a regulating factor from the hypothalamus, inhibits the release of PRL from the anterior pituitary. As the levels of estrogens and progesterone fall during the late secretory phase of the menstrual cycle, the secretion of PIF diminishes and the blood level of PRL rises. However, its rising level does not last long enough to have much effect on the breasts, which may be tender because of the presence of PRL just before menstruation. As the menstrual cycle starts up again and the level of estrogens again rises, PIF is again secreted and the PRL level drops.

PRL levels rise during pregnancy. Apparently a regulating factor from the hypothalamus, called **prolactin releasing factor (PRF),** stimulates PRL secretion after long periods of inhibition. PRL levels fall after delivery and rise again during breast feeding. Sucking reduces hypothalamic secretion of PIF. Mechanical stimulation of the nonlactating female breast brings about increased secretion of PRL; stimulation of the male breast does not.

Melanocyte-Stimulating Hormone (MSH)

The **melanocyte-stimulating hormone (MSH)** increases skin pigmentation by stimulating the dispersion of melanin granules in melanocytes. In the absence of the hormone, the skin may be pallid. An excess of MSH may cause darkening of the skin. Secretion of MSH is stimulated by a hypothalamic regulating factor called **melanocyte-stimulating hormone releasing factor (MRF).** It is inhibited by a **melanocyte-stimulating hormone inhibiting factor (MIF).**

A summary of anterior pituitary hormones, their principal actions, and associated hypothalamic regulating factors is presented in Exhibit 18-2.

EXHIBIT 18-2

ANTERIOR PITUITARY HORMONES, PRINCIPAL ACTIONS, AND ASSOCIATED HYPOTHALAMIC
REGULATING FACTORS

HORMONE	PRINCIPAL ACTIONS	ASSOCIATED HYPOTHALAMIC REGULATING FACTORS
Growth hormone (GH)	Growth of body cells; protein anabolism.	Growth hormone releasing factor (GHRF); Growth hormone inhibiting factor (GHIF).
Thyroid-stimulating hormone (TSH)	Controls secretion of hormones by thyroid gland.	Thyrotropin releasing factor (TRF).
Adrenocorticotropic hormone (ACTH)	Controls secretion of some hormones by adrenal cortex.	Corticotropin releasing factor (CRF).
Follicle-stimulating hormone (FSH)	In female, initiates development of ova and induces ovarian secretion of estrogens. In male, stimulates testes to produce sperm.	Gonadotropin releasing factor (GnRF).
Luteinizing hormone (LH)	In female, together with estrogens, stimulates ovulation and formation of progesterone-producing corpus luteum, prepares uterus for implantation, and readies mammary glands to secrete milk. In male, stimulates interstitial cells in testes to develop and produce testosterone.	Gonadotropin releasing factor (GnRF).
Prolactin (PRL)	Together with other hormones, initiates and maintains milk secretion by the mammary glands.	Prolactin inhibiting factor (PIF); Prolactin releasing factor (PRF).
Melanocyte-stimulating hormone (MSH)	Stimulates dispersion of melanin granules in melanocytes.	Melanocyte-stimulating hormone releasing factor (MRF); Melanocyte-stimulating hormone inhibiting factor (MIF).

NEUROHYPOPHYSIS

In a strict sense, the posterior lobe, or **neurohypophysis,** is not an endocrine gland, since it does not synthesize hormones. The posterior lobe consists of cells called **pituicytes** (pi-TOO-i-sītz), which are similar in appearance to the neuroglia of the nervous system. It also contains axon terminations of secretory neurons of the hypothalamus (Figure 18-5). Such neurons are called **neurosecretory cells.** The cell bodies of the neurons originate in nuclei in the hypothalamus. The fibers project from the hypothalamus, form the **hypothalamic-hypophyseal tract,** and terminate on blood capillaries in the neurohypophysis. The cell bodies of the neurosecretory cells produce two hormones: **oxytocin (OT)** and **antidiuretic hormone (ADH).** OT is produced primarily in the paraventricular nucleus and ADH is synthesized primarily in the supraoptic nucleus.

Following their production, the hormones are transported in the neuron fibers by a carrier protein called *neurophysin* into the neurohypophysis and stored in the axon terminals. Later, when the hypothalamus is properly stimulated, it sends impulses over the neurosecretory cells. The impulses cause the release of the hormones from the axon terminals into the blood.

The blood supply to the neurohypophysis is from the *inferior hypophyseal arteries,* derived from the internal carotid arteries. In the neurohypophysis, the inferior hypophyseal arteries form a plexus of capillaries called the *plexus of the infundibular process.* From this plexus hormones stored in the neurohypophysis pass into the *posterior hypophyseal veins* for distribution to tissue cells.

Oxytocin (OT)

Oxytocin (ok'-sē-TŌ-sin) or **OT** stimulates the contraction of the smooth muscle cells in the pregnant uterus and the contractile cells around the ducts of the mammary glands. It is released in large quantities just prior to giving birth (Figure 18-6). When labor begins, the cervix of the uterus is distended. This distension initiates afferent impulses to the neurosecretory cells of the paraventricular nucleus in the hypothalamus that stimulate the secretion of OT. The OT is transported by neurophysin to the neurohypophysis. The impulses also cause the neurohypophysis to release OT into the blood. It is then carried by the blood to the uterus to reinforce uterine contractions. As the contractions become more forceful, the resulting afferent impulses stimulate the synthesis of more OT. Then, a positive feedback cycle is established. Note also that the afferent part of the cycle is neural, whereas the efferent part is hormonal.

OT affects milk ejection. Milk formed by the glandular cells of the breasts is stored until the baby begins active sucking. For about 30 seconds to 1 minute after nursing begins, the baby receives no milk. During this latent period, nerve impulses from the nipple are transmitted to the hypothalamus. According to a mechanism similar to that involved in forming and releasing OT for uterine muscle contractions, OT flows from the neurohypophysis

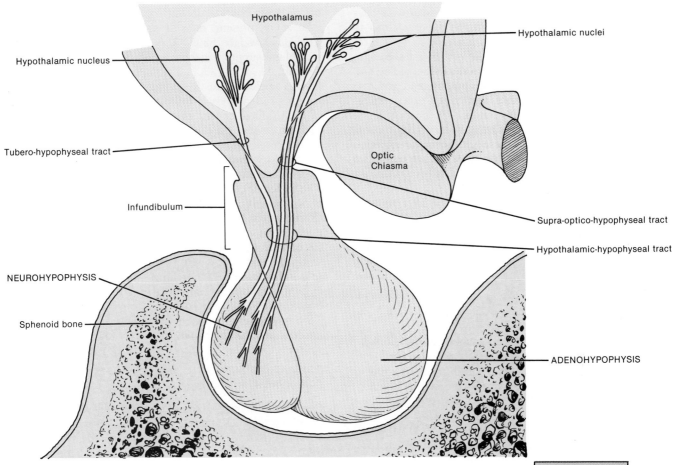

FIGURE 18-5 Hypothalamic-hypophyseal tract.

via the blood to the mammary glands, where it stimulates smooth muscle cells to contract and eject milk.

OT is inhibited by progesterone, but works together with estrogens. Estrogens and progesterone inhibit, via PIF, the release of PRL. PRL accumulates in the pituitary gland during gestation. The sucking stimulation that produces the release of OT also inhibits the release of PIF.

Antidiuretic Hormone (ADH)

An **antidiuretic** is any chemical substance that prevents excessive urine production. The principal physiological activity of **antidiuretic hormone (ADH)** is its effect on urine volume. ADH causes the kidneys to remove water from newly forming urine and return it to the bloodstream, thus decreasing urine volume. This involves an increase in the permeability of plasma membranes of kidney water-reabsorbing cells so that more water passes from urine into kidney cells. In the absence of ADH, urine output may be increased tenfold.

ADH can also raise blood pressure by bringing about constriction of arterioles. This effect is noted if there is a severe loss of blood volume due to hemorrhage.

The amount of ADH normally secreted varies with the body's needs. When the body is dehydrated, the concentration of water in the blood falls below normal limits as the salt-to-water ratio changes. Receptors in the hypothalamus called osmoreceptors detect the low water con-

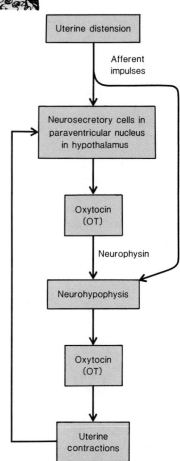

FIGURE 18-6 Regulation of the secretion of oxytocin (OT) during labor.

centration in the plasma and stimulate the neurosecretory cells of the supraoptic nucleus in the hypothalamus to produce ADH. The hormone is transported by neurophysin to the neurohypophysis. It is then released into the bloodstream and transported to the kidneys. The kidneys respond by decreasing urine output, and water is conserved. ADH also decreases the rate at which perspiration is produced during dehydration. By contrast, if the blood contains a higher than normal water concentration, the receptors detect the increase and hormone secretion is stopped. The kidneys can then release large quantities of urine, and the volume of body fluid is brought down to normal.

Secretion of ADH can also be altered by a number of other conditions. Pain, stress, trauma, anxiety, acetylcholine (ACh), nicotine, and drugs, such as morphine, tranquilizers, and some anesthetics stimulate secretion of the hormone. Alcohol inhibits secretion and thereby increases urine output. This may be why thirst is one symptom of a hangover.

CLINICAL APPLICATION

The principal abnormality associated with dysfunction of the neurohypophysis is **diabetes insipidus** (in-SIP-i-dus). Diabetes means overflow and insipidus means tasteless. This disorder should not be confused with diabetes mellitus (*meli* = honey), a disorder of the pancreas characterized by sugar in the urine. Diabetes insipidus is the result of a hyposecretion of ADH, usually caused by damage to the neurohypophysis or the supraoptic nucleus in the hypothalamus. Symptoms include excretion of large amounts of urine and subsequent thirst. Diabetes insipidus is treated by administering ADH.

A summary of posterior pituitary hormones, their principal actions, and control of secretion is presented in Exhibit 18-3.

THYROID

The **thyroid gland** is located just below the larynx. The right and left **lateral lobes** lie one on either side of the trachea. The lobes are connected by a mass of tissue called an **isthmus** (IS-mus) that lies in front of the trachea, just below the cricoid cartilage (Figure 18-7). The **pyramidal lobe,** when present, extends upward from the isthmus. The gland weighs about 25 g (almost 1 oz) and has a rich blood supply, receiving about 80 to 120 ml of blood per minute.

Histologically, the thyroid gland is composed of spherical sacs called **thyroid follicles** (Figure 18-8). The walls of each follicle consist of cells that reach the surface of the lumen of the follicle (**follicular cells**) and cells that do not reach the lumen (**parafollicular** or **C cells**). When the cells are inactive, they tend to be cuboidal, but when actively secreting hormones, they become more columnar. The follicular cells manufacture **thyroxine** (thī-ROK-sēn) or **T$_4$,** since it contains four atoms of iodine, and **triiodothyronine** (trī-ī'-od-ō-THĪ-rō-nēn) or **T$_3$,** since it contains three atoms of iodine. Together these hormones are referred to as the **thyroid hormones.** Thyroxine is normally secreted in greater quantity than triiodothyronine, but triiodothyronine is 3 to 4 times more potent. Moreover, in peripheral tissues, especially the liver and lungs, about one-third of triiodothyronine is converted to thyroxine. Both hormones are similar functionally. The parafollicular cells produce **calcitonin** (kal-si-TŌ-nin) or **CT.**

FORMATION, STORAGE, AND RELEASE OF THYROID HORMONES

One of the thyroid gland's unique features is its ability to store hormones and release them in a steady flow over a long period of time. The first stage in the synthesis of thyroid hormones is the active transport of iodine ions from the blood into the follicle cells. Under normal conditions, the iodide concentration in the cells is about 40 times that of the blood; during periods of maximal activity, the concentration can increase to over 300 times that of the blood. In the follicle cells, the iodide is oxidized to iodine. Through a series of enzymatically controlled reactions, the iodine combines with the amino acid tyrosine to form the thyroid hormones. This combination occurs within a large glycoprotein molecule, called **thyroglobulin (TGB),** which is secreted by the follicle cells into the follicle.

The thyroid hormones are an integral part of TGB and, as such, are stored (sometimes for months) until needed. The entire complex in the follicle, consisting of TGB and the stored hormones, constitutes the **thyroid**

EXHIBIT 18-3

SUMMARY OF POSTERIOR PITUITARY HORMONES, PRINCIPAL ACTIONS, AND CONTROL OF SECRETION

HORMONE	PRINCIPAL ACTIONS	CONTROL OF SECRETION
Oxytocin (OT)	Stimulates contraction of smooth muscle cells of pregnant uterus during labor and stimulates contraction of contractile cells of mammary glands for milk ejection.	Neurosecretory cells of hypothalamus secrete OT in response to uterine distension and stimulation of nipples.
Antidiuretic hormone (ADH)	Principal effect is to decrease urine volume; also raise blood pressure by constricting arteries during severe hemorrhage.	Neurosecretory cells of hypothalamus secrete ADH in response to low water concentration of the blood, pain, stress, trauma, anxiety, acetylcholine (ACh), nicotine, morphine, and tranquilizers; alcohol inhibits secretion.

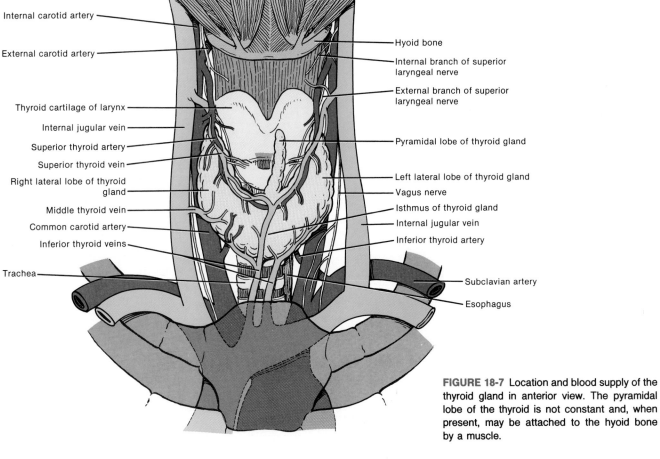

Internal carotid artery

External carotid artery

Thyroid cartilage of larynx

Internal jugular vein

Superior thyroid artery

Superior thyroid vein

Right lateral lobe of thyroid gland

Middle thyroid vein

Common carotid artery

Inferior thyroid veins

Trachea

Hyoid bone

Internal branch of superior laryngeal nerve

External branch of superior laryngeal nerve

Pyramidal lobe of thyroid gland

Left lateral lobe of thyroid gland

Vagus nerve

Isthmus of thyroid gland

Internal jugular vein

Inferior thyroid artery

Subclavian artery

Esophagus

FIGURE 18-7 Location and blood supply of the thyroid gland in anterior view. The pyramidal lobe of the thyroid is not constant and, when present, may be attached to the hyoid bone by a muscle.

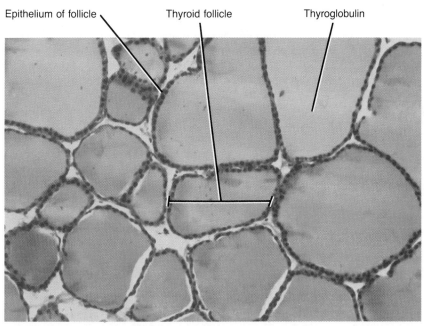

Epithelium of follicle

Thyroid follicle

Thyroglobulin

FIGURE 18-8 Histology of the thyroid gland. Photomicrograph at a magnification of 230×. (© 1983 by Michael H. Ross. Used by permission.)

colloid. Before the thyroid hormones can be released into the blood, droplets of colloid are absorbed into the follicular cells by pinocytosis and the thyroid hormones are split from TGB. Following their diffusion into the blood, most of the thyroid hormones combine with plasma proteins, mainly **thyroxine-binding globulin (TBG).** Thyroid hormones combined with plasma proteins are referred to as **protein-bound iodine (PBI).**

FUNCTION AND CONTROL OF THYROID HORMONES

The thyroid hormones have three principal effects on the body: (1) regulation of metabolism, (2) regulation of growth and development, and (3) regulation of the activity of the nervous system. With respect to the regulation of metabolism, the thyroid hormones stimulate virtually all aspects of carbohydrate and lipid catabolism in most

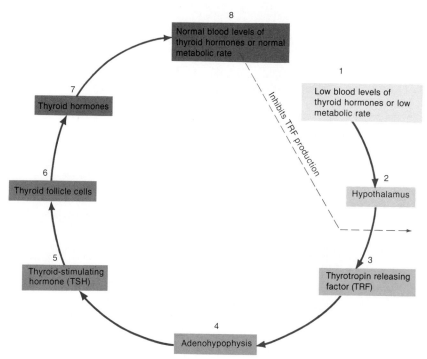

FIGURE 18-9 Regulation of the secretion of thyroid hormones.

cells of the body. They also increase the rate of protein synthesis. Since their overall effect is to increase catabolism, they increase the basal metabolic rate. The energy produced raises body temperature as heat is given off and this phenomenon is called the **calorigenic effect.**

The thyroid hormones help to regulate tissue growth and development, especially in children. They work with GH to accelerate body growth, particularly the growth of nervous tissue. Deficiency of the hormones during fetal development can result in fewer and smaller neurons, defective myelination of axons, and mental retardation. During the early years of life, deficiency of the hormones results in small stature and poor development of certain organs.

Finally, the thyroid hormones increase the reactivity of the nervous system. This results in increased blood flow, increased and more forceful heartbeats, increased blood pressure, increased motility of the gastrointestinal tract, and increased nervousness.

The secretion of thyroid hormones is stimulated by several factors (Figure 18-9). If thyroid hormone levels in the blood fall below normal or the metabolic rate decreases, chemical sensors in the hypothalamus detect the change in blood chemistry and stimulate the hypothalamus to secrete a regulating factor called thyrotropin releasing factor (TRF). TRF stimulates the adenohypophysis to secrete thyroid-stimulating hormone (TSH). Then TSH stimulates the thyroid to release thyroid hormones until the metabolic rate returns to normal. Conditions that increase the body's need for energy—a cold environment, high altitude, pregnancy—also trigger this feedback system and increase the secretion of thyroid hormones.

Thyroid activity can be inhibited by a number of other factors. When large amounts of certain sex hormones (estrogens and androgens) are circulating in the blood, for example, TSH secretion diminishes. Aging slows down

the activities of most glands, and thyroid production may decrease.

CLINICAL APPLICATION

Hyposecretion of thyroid hormones during the growth years results in **cretinism** (KRĒ-tin-izm). Two outstanding clinical symptoms of the cretin are dwarfism and mental retardation. The first is caused by failure of the skeleton to grow and mature. The second is caused by failure of the brain to develop fully. Recall that one function of thyroid hormones is to control tissue growth and development. Cretins also exhibit retarded sexual development and a yellowish skin color. Flat pads of fat develop, giving the cretin a characteristic round face and thick nose; a large, thick, protruding tongue; and protruding abdomen. Because the energy-producing metabolic reactions are slow, the cretin has a low body temperature and general lethargy. Carbohydrates are stored rather than utilized. Heart rate is also slow. If the condition is diagnosed early, the symptoms can be eliminated by administering thyroid hormones.

Hypothyroidism during the adult years produces **myxedema** (mix-e-DĒ-ma). A hallmark of this disorder is an edema that causes the facial tissues to swell and look puffy. Like the cretin, the person with myxedema suffers from slow heart rate, low body temperature, muscular weakness, general lethargy, and a tendency to gain weight easily. The long-term effect of a slow heart rate may overwork the heart muscle, causing the heart to enlarge. Because the brain has already reached maturity, the person with myxedema does not experience mental retardation. However, in moderately severe cases, nerve reactivity may be dulled so that the person lacks mental alertness. Myxedema occurs eight times more frequently in females than in males. Its symptoms are abolished by the administration of thyroxine.

Hypersecretion of thyroid hormones gives rise to **exophthalmic** (ek'-sof-THAL-mik) **goiter** (GOY-ter). This disease, like myxedema, is also more frequent in females. One of its primary symptoms is an enlarged thyroid, called a **goiter,** which may be two to three times its original size. Two other symptoms are an edema behind the eye, which causes the eye to protrude **(exophthalmos),** and an abnormally high metabolic rate. The high metabolic rate produces a range of effects that are generally opposite to those of myxedema—increased pulse, high body temperature, and moist, flushed skin. The person loses weight and is usually full of "nervous" energy. The thyroid hormones also increase the responsiveness of the nervous system, causing the person to become irritable and exhibit tremors of the extended fingers. Hyperthyroidism is usually treated by administering drugs that suppress thyroid hormone synthesis or by surgically removing part of the gland.

Goiter is a symptom of many thyroid disorders. It may also occur if the gland does not receive enough iodine to produce sufficient thyroxine for the body's needs. The follicular cells then enlarge in a futile attempt to produce more thyroid hormones and they secrete large quantities of thyroglobulin (TGB). This condition is called **simple goiter.** Simple goiter is most often caused by a lower than average amount of iodine in the diet. It may also develop if iodine intake is not increased during certain conditions that put a high demand on the body for thyroxine, such as frequent exposure to cold and high fat and protein diets.

CALCITONIN (CT)

The hormone produced by the parafollicular cells of the thyroid gland is **calcitonin** (kal-si-TŌ-nin) or **CT.** It is involved in the homeostasis of blood calcium level. CT lowers the amount of calcium and phosphate in the blood by inhibiting bone breakdown and accelerating the absorption of calcium by the bones. It appears to exert its effect in lowering calcium and phosphate blood levels by inhibiting osteoclastic action and the parathyroid hormone (PTH). If CT is administered to a person with a normal level of blood calcium, it causes **hypocalcemia** (low blood calcium level). Hypocalcemia is also a complication of magnesium deficiency. If CT is given to a person with **hypercalcemia** (high blood calcium level), the level returns to normal. It is suspected that the blood calcium level directly controls the secretion of CT according to a negative feedback system that does not involve the pituitary gland (Figure 18-10).

A summary of hormones produced by the thyroid gland, their principal actions, and control of secretion is presented in Exhibit 18-4.

PARATHYROIDS

Typically embedded on the posterior surfaces of the lateral lobes of the thyroid are small, round masses of tissue called the **parathyroid glands.** Usually, two parathyroids, superior and inferior, are attached to each lateral thyroid lobe (Figure 18-11). They measure about 3 to 6 mm (0.1 to 0.3 inch) in length, 2 to 5 mm (0.07 to 0.2 inch) in width, and 0.5 to 2 mm (0.02 to 0.07 inch) in thickness.

Histologically, the parathyroids contain two kinds of epithelial cells (Figure 18-12). The more numerous cells, called **principal** or **chief cells,** are believed to be the major synthesizer of **parathyroid hormone (PTH).** Some researchers believe that the other kind of cell, called an **oxyphil cell,** synthesizes a reserve capacity of hormone.

PARATHYROID HORMONE (PTH)

Parathyroid hormone (PTH) or **parathormone** controls the homeostasis of ions in the blood, especially calcium and phosphate ions. If adequate amounts of vitamin D are present, PTH increases the rate of calcium, some magnesium, and phosphate absorption from the gastrointestinal tract into the blood. It also leads to the activation of vitamin D. Moreover, PTH increases the number and activity of osteoclasts, or bone-destroying cells. As a result, bone tissue is broken down and calcium and phosphate are released into the blood. PTH produces two

---

EXHIBIT 18-4

SUMMARY OF THYROID GLAND HORMONES, PRINCIPAL ACTIONS, AND CONTROL OF SECRETION

HORMONE	PRINCIPAL ACTIONS	CONTROL OF SECRETION
Thyroid hormones **Thyroxine (T4)**	Regulates metabolism, growth and development, and activity of nervous system.	Thyrotropin releasing factor (TRF) is released from hypothalamus in response to low thyroid hormone levels, low metabolic rate, cold, pregnancy, and high altitudes; TRF secretion is inhibited in response to high thyroid hormone levels, high metabolic rate, high levels of estrogens and androgens, and aging.
Triiodothyronine (T3)	Same as above.	Same as above.
Calcitonin (CT)	Lowers blood levels of calcium by accelerating calcium absorption by bones.	High blood calcium levels stimulate secretion; low levels inhibit secretion.

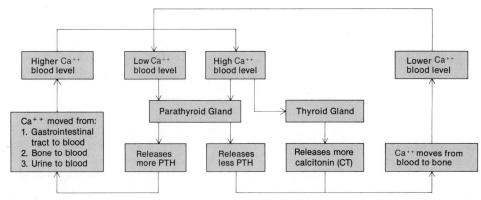

FIGURE 18-10 Regulation of the secretion of the parathyroid hormone (PTH) and calcitonin (CT).

changes in the kidneys. It increases the rate at which the kidneys remove calcium and magnesium from urine that is being formed and returns them to the blood. It accelerates the transportation of phosphate from the blood into the urine for elimination. More phosphate is lost through the urine than is gained from the bones.

The overall effect of PTH with respect to ions, then, is to decrease blood phosphate level and increase blood calcium level. As far as blood calcium level is concerned, PTH and CT are antagonists.

PTH secretion is not controlled by the pituitary gland. When the calcium level of the blood falls, more PTH is released (see Figure 18-10). Conversely, when the calcium level of the blood rises, less PTH (and more CT) is secreted. This is another example of a negative feedback control system that does not involve the pituitary gland.

CLINICAL APPLICATION
A normal amount of calcium in the extracellular fluid is necessary to maintain the resting state of neurons. A deficiency of calcium caused by *hypoparathyroidism*

causes neurons to depolarize without the usual stimulus. As a result, nervous impulses increase and result in muscle twitches, spasms, and convulsions. This condition is called **tetany.** The effects of hypocalcemic tetany are observed in the *Trousseau* (troo-SŌ) and *Chvostek* (VOS-tek) *signs.* Trousseau sign is observed when the binding of a blood pressure cuff around the arm produces contraction of the fingers and inability to open the hand. The Chvostek sign is a contracture of the facial muscles elicited by tapping the facial nerves at the angle of the jaw. Hypoparathyroidism results from surgical removal of the parathyroids or from parathyroid damage caused by parathyroid disease, infection, hemorrhage, or mechanical injury.

Hyperparathyroidism causes demineralization of bone. This condition is called **osteitis fibrosa cystica** because the areas of destroyed bone tissue are replaced by cavities that fill with fibrous tissue. The bones thus become deformed and are highly susceptible to fracture. Hyperparathyroidism is usually caused by a tumor in the parathyroids.

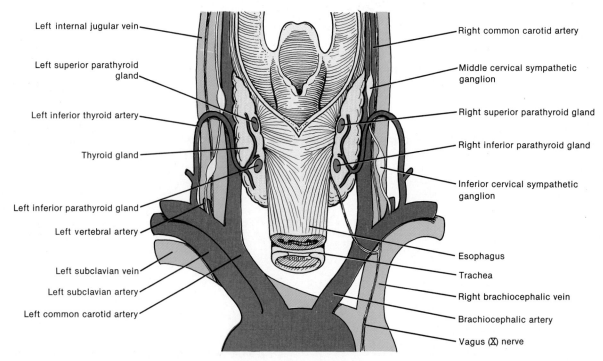

FIGURE 18-11 Location and blood supply of the parathyroid glands in posterior view.

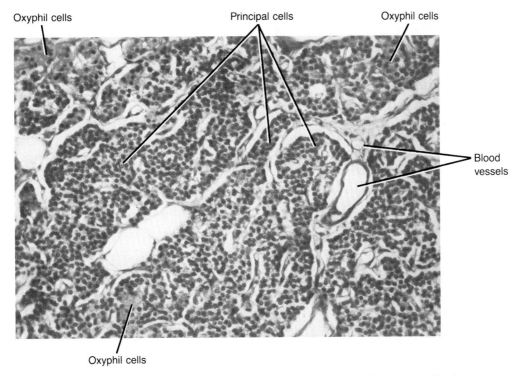

Oxyphil cells   Principal cells   Oxyphil cells

Blood vessels

Oxyphil cells

FIGURE 18-12 Histology of the parathyroid glands. Photomicrograph at a magnification of 180×. (© 1983 by Michael H. Ross. Used by permission.)

A summary of the principal actions and control of secretion of parathyroid hormone (PTH) is presented in Exhibit 18-5.

ADRENALS (SUPRARENALS)

The body has two **adrenal (suprarenal) glands,** one of which is located superior to each kidney (Figure 18-13). Each adrenal gland is structurally and functionally differentiated into two sections: the outer **adrenal cortex,** which makes up the bulk of the gland, and the inner **adrenal medulla** (Figure 18-14). Whereas the adrenal cortex is derived from mesoderm, the adrenal medulla is derived from the ectoderm. Covering the gland are an inner, thick layer of fatty connective tissue and an outer, thin fibrous capsule.

The average dimensions of the adult adrenal gland are about 50 mm (2 inches) in length, 30 mm (1.1 inch) in width, and 10 mm (0.4 inch) in thickness. The adrenals, like the thyroid, are among the most vascular organs of the body.

ADRENAL CORTEX

Histologically, the cortex is subdivided into three zones (Figure 18-14). Each zone has a different cellular arrangement and secretes different groups of steroid hormones. The outer zone, directly underneath the connective tissue capsule, is referred to as the **zona glomerulosa.** It comprises about 15 percent of the total cortical volume. Its cells are arranged in arched loops or round balls. Its primary secretions are a group of hormones called mineralocorticoids (min'-er-al-ō-KOR-ti-koyds).

The middle zone, or **zona fasciculata,** is the widest of the three zones and consists of cells arranged in long, straight cords. The zona fasciculata secretes mainly glucocorticoids (gloo'-kō-KŌR-ti-koyds).

The inner zone, the **zona reticularis,** contains cords of cells that branch freely. This zone synthesizes minute amounts of hormones, mainly the sex hormones called gonadocorticoids (gō-na-dō-KŌR-ti-koyds) and of these chiefly male hormones called androgens.

Mineralocorticoids

Mineralocorticoids help control water and electrolyte homeostasis, particularly the concentrations of sodium and potassium. Although the adrenal cortex secretes at least

EXHIBIT 18-5

PRINCIPAL ACTIONS AND CONTROL OF SECRETION OF PARATHYROID HORMONE (PTH)

PRINCIPAL ACTIONS	CONTROL OF SECRETION
Increases blood calcium level and decreases blood phosphate level by increasing rate of calcium absorption from gastrointestinal tract into blood; increases number and activity of osteoclasts; increases calcium absorption by kidneys; increases phosphate excretion by kidneys; and activates vitamin D.	Low blood calcium levels stimulate secretion; high levels inhibit secretion.

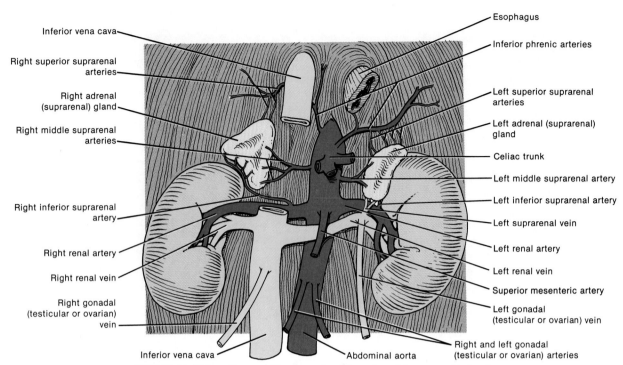

FIGURE 18-13 Location and blood supply of the adrenal (suprarenal) glands.

three different hormones classified as mineralocorticoids, one of these hormones is responsible for about 95 percent of the mineralocorticoid activity—**aldosterone** (al-dō-STĒR-ōn). Aldosterone acts on the tubule cells in the kidneys and causes them to increase their reabsorption of sodium and water. As a result, sodium ions are removed from the urine and returned to the blood. In this manner, aldosterone prevents rapid depletion of sodium and water

from the body. On the other hand, aldosterone decreases reabsorption of potassium, so large amounts of potassium are lost in the urine.

These two basic functions—conservation of sodium and water and elimination of potassium—cause a number of secondary effects. For example, a large proportion of the sodium reabsorption occurs through an exchange reaction whereby positive hydrogen ions pass into the urine

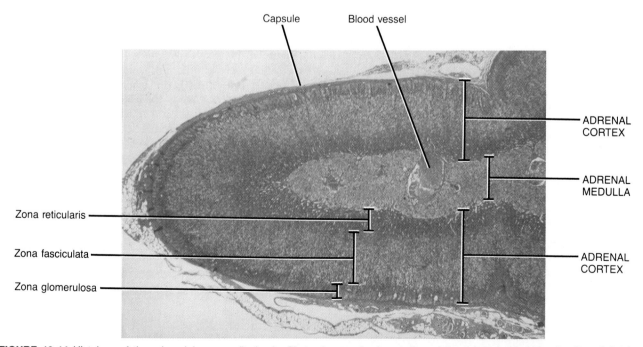

FIGURE 18-14 Histology of the adrenal (suprarenal) glands. Photomicrograph of a section of the adrenal gland showing its subdivisions and zones at a magnification of 25×. (© 1983 by Michael H. Ross. Used by permission.)

to replace the positive sodium ions. Since this mechanism removes hydrogen ions, it makes the blood less acidic and prevents acidosis. The movement of sodium ions also sets up a positively charged field in the blood vessels around the kidney tubules. As a result, negatively charged chloride and bicarbonate ions are drawn out of the urine and back into the blood. Finally, the increase in sodium ion concentration in the blood vessels causes water to move by osmosis from the urine into the blood. In summary, aldosterone causes potassium excretion and sodium reabsorption. The sodium reabsorption leads to the elimination of H^+ ions; the retention of Na^+, Cl^-, and HCO_3^- ions; and the retention of water.

The control of aldosterone secretion is complex. Apparently, several mechanisms operate. One of these is the **renin-angiotensin** (an'-jē-ō-TEN-sin) **pathway** (Figure 18-15). A decrease in blood volume from dehydration, Na^+ deficiency, or hemorrhage brings about a drop in blood pressure. The low blood pressure stimulates certain kidney cells, called juxtaglomerular cells, to secrete into the blood an enzyme called *renin* (RĒ-nin) (see Figure 26-9). In this pathway, renin converts *angiotensinogen*, a plasma protein produced by the liver, into *angiotensin I*, which is then converted into *angiotensin II* by a plasma enzyme in the lungs. Angiotensin II stimulates the adrenal cortex to produce more aldosterone. Aldosterone brings about increased Na^+ and water reabsorption. This reabsorption leads to an increase in extracellular fluid volume and a restoration of blood pressure to normal. Since an-

giotensin II is a powerful vasoconstrictor, this action also helps to elevate blood pressure.

A second mechanism for the control of aldosterone involves potassium ion concentration. An increased K^+ concentration in extracellular fluid directly stimulates aldosterone secretion by the adrenal cortex and causes the elimination of excess K^+ by the kidneys. A decreased K^+ concentration in extracellular fluid decreases aldosterone production, and thus less K^+ than usual is eliminated by the kidneys.

Adrenocorticotropic hormone (ACTH) from the anterior pituitary has a minor effect on the secretion of aldosterone. Only in the complete absence of ACTH is there a mild to moderate deficiency of aldosterone. The major controlling effect of ACTH is on glucocorticoids, as will be discussed shortly.

CLINICAL APPLICATION
Hypersecretion of the mineralocorticoid aldosterone results in **aldosteronism,** characterized by a decrease in the body's potassium concentration. If potassium depletion is great, neurons cannot depolarize and muscular paralysis results. Hypersecretion also brings about excessive retention of sodium and water. The water increases the volume of the blood and causes high blood pressure. It also increases the volume of the interstitial fluid, producing edema.

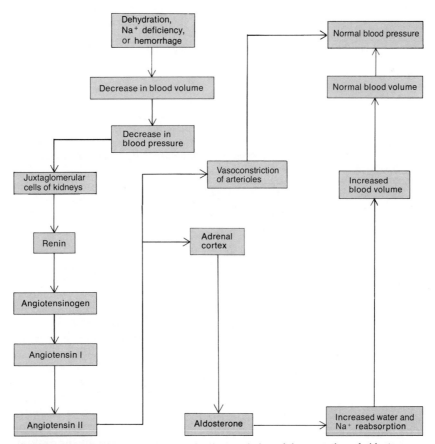

FIGURE 18-15 Proposed mechanism for the regulation of the secretion of aldosterone by the renin-angiotensin pathway.

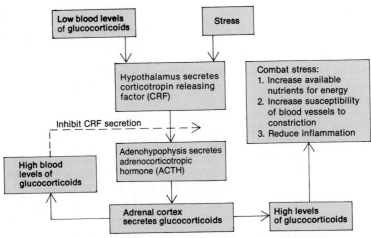

FIGURE 18-16 Regulation of the secretion of glucocorticoids.

Glucocorticoids

The **glucocorticoids** are a group of hormones concerned with normal metabolism and resistance to stress. Three glucocorticoids are **cortisol (hydrocortisone), corticosterone,** and **cortisone.** Of the three, cortisol is the most abundant. The glucocorticoids have the following effects on the body.

1. Glucocorticoids work with other hormones in promoting normal metabolism. Their role is to make sure enough energy is provided. They increase the rate at which proteins are catabolized and amino acids are removed from cells, primarily muscle cells, and transported to the liver. The amino acids may be synthesized into new proteins, such as the enzymes needed for metabolic reactions. If the body's reserves of glycogen and fat are low, the liver may convert the amino acids to glucose. This conversion of a substance other than carbohydrate into glucose is called **gluconeogenesis** (gloo'-kō-nē'-ō-JEN-e-sis). Glucocorticoids also release fatty acids from adipose tissue for conversion to glucose.

2. Glucocorticoids work in many ways to provide resistance to stress. A sudden increase in available glucose by way of gluconeogenesis from amino acids makes the body more alert. Additional glucose gives the body energy for combating a range of stresses: fright, temperature extremes, high altitude, bleeding, infection, surgery, trauma, and almost any debilitating disease. Glucocorticoids also make the blood vessels more sensitive to vessel-constricting chemicals. They thereby raise blood pressure. This effect is advantageous if the stress happens to be blood loss, which causes a drop in blood pressure.

3. Glucocorticoids are antiinflammatory compounds. They stabilize lysosomal membranes to inhibit the release of inflammatory substances, decrease blood capillary permeability, and depress phagocytosis. Unfortunately, they also retard connective tissue regeneration and are thereby responsible for slow wound healing. High doses of glucocorticoids cause atrophy of the thymus gland, spleen, and lymph nodes, thus depressing immune responses. However, high doses may be useful in the treatment of rheumatism.

The control of glucocorticoid secretion is a typical negative feedback mechanism (Figure 18-16). The two principal stimuli are stress and low blood level of glucocorticoids. Either condition stimulates the hypothalamus to secrete a regulating factor called **corticotropin releasing factor (CRF).** This secretion initiates the release of ACTH from the anterior lobe of the pituitary. ACTH is carried through the blood to the adrenal cortex, where it then stimulates glucocorticoid secretion.

CLINICAL APPLICATION

Hyposecretion of glucocorticoids results in the condition called **Addison's disease.** Clinical symptoms include hypoglycemia, which leads to muscular weakness, mental lethargy, and weight loss. Increased potassium and decreased sodium lead to low blood pressure and dehydration.

Cushing's syndrome is a hypersecretion of glucocorticoids, especially cortisol and cortisone. The condition is characterized by the redistribution of fat. The result is spindly legs accompanied by a "moon face," "buffalo hump" on the back, and pendulous abdomen. Facial skin is flushed, and the skin covering the abdomen develops stretch marks. The individual also bruises easily, and wound healing is poor.

Gonadocorticoids

The adrenal cortex secretes both male and female **gonadocorticoids** or **sex hormones.** These are estrogens and androgens. Estrogens are several closely related female sex hormones that are also produced by the ovaries and placenta. Androgens are male sex hormones. An important androgen, called testosterone, is produced by the testes. The concentration of sex hormones secreted by normal adult adrenals is usually so low that their effects are insignificant.

CLINICAL APPLICATION

The **adrenogenital syndrome** usually refers to a group of enzyme deficiencies that block the synthesis of glucocorticoids. In an attempt to compensate, the anterior pituitary secretes more ACTH. As a result, excess androgenic (male) hormones are produced, causing *virilism,* or masculinization. For instance, the female develops extremely virile characteristics such as growth of a beard, development of a much deeper voice, occasionally development of baldness, development of a masculine distribution of hair on the body and on the pubis, growth of the clitoris that resembles a penis, and deposition of proteins in the skin and muscles producing typical masculine characteristics. Such virilism may also result from tumors of the adrenal gland called *virilizing adenomas* (*aden* = gland; *oma* = tumor).

In the prepubertal male, the syndrome causes the same characteristics as in the female, plus rapid development of the male sexual organs and creation of male sexual desires. In the adult male, the virilizing characteristics of the adrenogenital syndrome are usually completely obscured by the normal virilizing characteristics of the testosterone secreted by the testes. As a result, it is often difficult to make a diagnosis of adrenogenital syndrome in the male adult. However, an occasional adrenal tumor secretes sufficient quantities of feminizing hormones that the male patient develops **gynecomastia** (*gyneca* = woman, *mast* = breast), which means excessive growth of the male mammary glands. Such a tumor is called a *feminizing adenoma*.

ADRENAL MEDULLA

The adrenal medulla consists of hormone-producing cells, called **chromaffin** (krō-MAF-in) **cells,** which surround large blood-containing sinuses. Chromaffin cells develop from the same source as the postganglionic cells of the sympathetic division of the nervous system. They are directly innervated by preganglionic cells of the sympathetic division of the autonomic nervous system and may be regarded as postganglionic cells that are specialized to secrete. In all other visceral effectors, preganglionic sympathetic fibers first synapse with postganglionic neurons before innervating the effector. In the adrenal medulla, however, the preganglionic fibers pass directly into the chromaffin cells of the gland. The secretion of hormones from the chromaffin cells is directly controlled by the autonomic nervous system, and innervation by the preganglionic fibers allows the gland to respond rapidly to a stimulus.

Epinephrine and Norepinephrine (NE)

The two principal hormones synthesized by the adrenal medulla are **epinephrine** and **norepinephrine (NE),** also called adrenaline and noradrenaline, respectively. Epinephrine constitutes about 80 percent of the total secretion of the gland and is more potent in its action than norepinephrine. Both hormones are **sympathomimetic** (sim'-pa-thō-mi-MET-ik), that is, they produce effects that mimic those brought about by the sympathetic division of the autonomic nervous system. To a large extent, they are responsible for the fight-or-flight response. Like the glucocorticoids of the adrenal cortices, these hormones help the body resist stress. However, unlike the cortical hormones, the medullary hormones are not essential for life.

Under stress, impulses received by the hypothalamus are conveyed to sympathetic preganglionic neurons, which cause the chromaffin cells to increase their output of epinephrine and norepinephrine. Epinephrine increases blood pressure by increasing heart rate and constricting the blood vessels. It accelerates the rate of respiration, dilates respiratory passageways, decreases the rate of digestion, increases the efficiency of muscular contractions, increases blood sugar level, and stimulates cellular metab-

olism. Hypoglycemia may also stimulate medullary secretion of epinephrine and norepinephrone.

CLINICAL APPLICATION

Tumors of the chromaffin cells of the adrenal medulla, called **pheochromocytomas** (fē-ō-krō'-mō-sī-TŌ-mas), cause hypersecretion of the medullary hormones. The oversecretion causes high blood pressure, high levels of sugar in the blood and urine, an elevated basal metabolic rate, nervousness, and sweating. Since the medullary hormones create the same effects as does sympathetic nervous stimulation, hypersecretion puts the individual into a prolonged version of the fight-or-flight response. This condition ultimately wears out the body, and the individual eventually suffers from general weakness.

A summary of the hormones produced by the adrenal glands, their principal actions, and control of secretion is presented in Exhibit 18-6.

PANCREAS

The **pancreas** can be classified as both an endocrine and an exocrine gland. We shall treat its endocrine functions at this point; its exocrine functions are discussed in the chapter on the digestive system (Chapter 24). The pancreas is a flattened organ located posterior and slightly inferior to the stomach (Figure 18-17). The adult pancreas consists of a head, body, and tail. Its average length is about 12.5 cm (6 inches), and its average weight is about 85 g (3 oz).

The endocrine portion of the pancreas consists of clusters of cells called **islets of Langerhans** (LAHNG-er-hanz) (Figure 18-18). Three kinds of cells are found in these clusters: (1) **alpha cells,** which secrete the hormone glucagon; (2) **beta cells,** which secrete the hormone insulin, and (3) **delta cells,** which secrete growth hormone inhibiting factor (GHIF) or somatostatin. The islets are surrounded by blood capillaries and by cells (acini) that form the exocrine part of the gland. Glucagon and insulin are the endocrine secretions of the pancreas and are concerned with regulation of blood sugar level.

GLUCAGON

The product of the alpha cells is **glucagon** (GLOO-ka-gon), a hormone whose principal physiological activity is to increase the blood glucose level (Figure 18-19). Glucagon does this by accelerating the conversion of glycogen in the liver into glucose (glycogenolysis) and the conversion in the liver of other nutrients, such as amino acids, glycerol, and lactic acid, into glucose (gluconeogenesis). The liver then releases the glucose into the blood, and the blood sugar level rises. Secretion of glucagon is directly controlled by the level of blood sugar via a negative feedback system. When the blood sugar level falls below normal, chemical sensors in the alpha cells of the islets stimulate the cells to secrete glucagon. When blood sugar rises, the cells are no longer stimulated and production

EXHIBIT 18-6

SUMMARY OF HORMONES PRODUCED BY THE ADRENAL GLANDS, PRINCIPAL ACTIONS, AND CONTROL OF SECRETION

HORMONE	PRINCIPAL ACTIONS	CONTROL OF SECRETION
Adrenal Cortical Hormones **Mineralocorticoids** **(mainly aldosterone)**	Increase blood levels of sodium and water and decrease blood levels of potassium.	Decreased blood volume or sodium levels initiate renin-angiotensin pathway to stimulate aldosterone secretion; increased blood levels of potassium stimulate aldosterone secretion; ACTH has only a minor effect in promoting aldosterone secretion.
Glucocorticoids (mainly cortisol)	Help promote normal metabolism, resistance to stress, and counter inflammatory response.	ACTH is released in response to stress and low blood levels of glucocorticoids.
Gonadocorticoids	Concentrations secreted by adults are so low that their effects are insignificant.	Discussed in detail in Chapter 28.
Adrenal Medullary Hormones **Epinephrine**	Sympathomimetic, that is, produces effects that mimic those of ANS during stress.	
Norepinephrine (NE)	Same as above.	

slackens. If for some reason the self-regulating device fails and the alpha cells secrete glucagon continuously, hyperglycemia may result. Exercise and largely or entirely protein meals that raise the amino acid level of the blood also cause an increase in glucagon secretion. Glucagon secretion is inhibited by GHIF (somatostatin).

INSULIN

The beta cells of the islets produce the hormone **insulin.** Insulin also increases the build-up of proteins in cells. Its chief physiological action is opposite that of glucagon. Insulin decreases blood sugar level in several ways (Figure

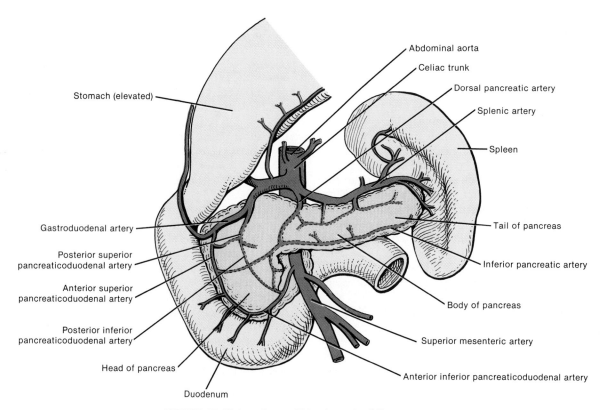

FIGURE 18-17 Location and blood supply of the pancreas.

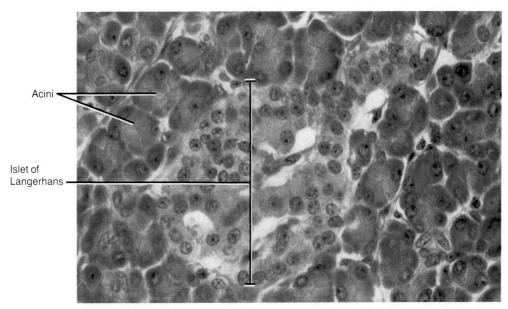

Acini

Islet of
Langerhans

FIGURE 18-18 Histology of the pancreas. Photomicrograph at a magnification of 600×. (© 1983 by Michael H. Ross. Used by permission.)

18-19). It accelerates the transport of glucose from the blood into cells, especially skeletal muscle cells. It also accelerates the conversion of glucose into glycogen (glycogenesis). Insulin also decreases glycogenolysis and gluconeogenesis, stimulates the conversion of glucose or other nutrients into fatty acids (lipogenesis), and helps stimulate protein synthesis.

The regulation of insulin secretion, like that of glucagon secretion, is directly determined by the level of sugar in the blood and is based on a negative feedback system.

Other hormones can indirectly affect insulin production, however. For instance, GH raises blood glucose level, and the rise in glucose level triggers insulin secretion. ACTH, by stimulating the secretion of glucocorticoids, brings about hyperglycemia and also indirectly stimulates the release of insulin. Gastrointestinal hormones like stomach and intestinal gastrin, secretin, cholecystokinin (CCK), and gastric inhibitory peptide (GIP) also stimulate insulin secretion. GHIF (somatostatin) inhibits the secretion of insulin.

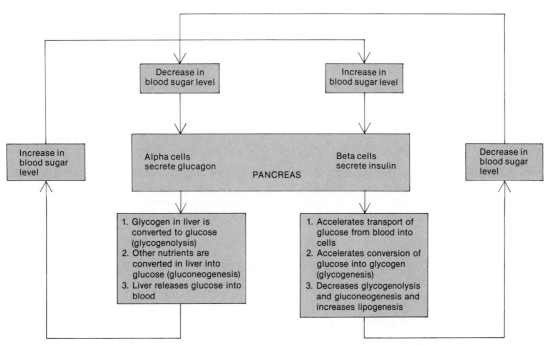

FIGURE 18-19 Regulation of the secretion of glucagon and insulin.

CLINICAL APPLICATION

An absolute or relative deficiency of insulin results in a number of clinical symptoms referred to as **diabetes mellitus** (MEL-it-us). Research has led many to conclude that diabetes mellitus is not a single hereditary disease, but rather a heterogeneous group of hereditary diseases, all of which ultimately lead to an elevation of glucose in the blood (hyperglycemia) and excretion of glucose in the urine as hyperglycemia increases. Diabetes mellitus is also characterized by the three "polys": an inability to reabsorb water, resulting in increased urine production (*polyuria*); excessive thirst (*polydipsia*); and excessive eating (*polyphagia*).

Two major types of diabetes have been distinguished: the maturity-onset type and the juvenile-onset type. *Maturity-onset diabetes* is much more common, representing more than 90 percent of all cases. It most often occurs in people who are over 40 and overweight. Clinical symptoms are mild, and the high glucose levels in the blood can usually be controlled by diet alone. Many maturity-onset diabetics have a sufficiency or even a surplus of insulin in the blood. For these individuals, diabetes arises not from a shortage of insulin, but probably from defects in the molecular machinery that mediates the action of insulin on its target cells. Maturity-onset diabetes is therefore called *non-insulin-dependent diabetes.*

Juvenile-onset diabetes develops in people younger than age 20. Its onset is more abrupt and it is generally more severe than maturity-onset diabetes. It is usually preceded by a virus infection like measles or mumps. The disease is characterized by a marked decline in the number of beta cells in the pancreas leading to insufficient production of insulin and an elevation of glucose in the blood. It is known as *insulin-dependent diabetes.* The deficiency of insulin accelerates the breakdown of the body's reserve of fat resulting in the production of organic acids called ketones. This causes a form of acidosis called *ketosis,* which lowers the pH of the blood and can result in death. The catabolism of stored fats and proteins also causes weight loss. As lipids are transported by the blood from storage depots to hungry cells, lipid particles are deposited on the walls of blood vessels. The deposition leads to atherosclerosis and a multitude of cardiovascular problems.

Some cases of juvenile-onset diabetes may be explained primarily on a genetic basis and others on an environmental basis. Still other cases appear to rise from a complex interaction between genetic background and environment.

Hyperinsulinism is much rarer than hyposecretion and is generally the result of a malignant tumor in an islet. The principal symptom is a decreased blood glucose level, which stimulates the secretion of epinephrine, glucagon, and GH. As a consequence, anxiety, sweating, tremor, increased heart rate, and weakness occur. Moreover, brain cells do not have enough glucose to function efficiently. This condition leads to mental disorientation, convulsions, unconsciousness, shock, and eventual death as the vital centers in the medulla are affected.

A summary of the hormones produced by the pancreas, their principal actions, and control of secretion is presented in Exhibit 18-7.

OVARIES AND TESTES

The female gonads, called the **ovaries,** are paired oval bodies located in the pelvic cavity. The ovaries produce female sex hormones called **estrogens** and **progesterone.** These hormones are responsible for the development and maintenance of the female sexual characteristics. Along with the gonadotropic hormones of the pituitary, the sex hormones also regulate the menstrual cycle, maintain pregnancy, and prepare the mammary glands for lactation. The ovaries (and placenta) also produce a hormone called **relaxin** which relaxes the symphysis pubis and helps dilate the uterine cervix toward the end of pregnancy.

The male has two oval glands, called **testes,** that lie in the scrotum. The testes produce **testosterone,** the primary male sex hormone, that stimulates the development and maintenance of the male sexual characteristics. The testes also produce the hormone **inhibin** that inhibits secretion of FSH. The detailed structure of the ovaries and testes and the specific roles of gonadotropic hormones and sex hormones will be discussed in Chapter 28.

EXHIBIT 18-7

SUMMARY OF HORMONES PRODUCED BY THE PANCREAS, PRINCIPAL ACTIONS, AND CONTROL OF SECRETION

HORMONE	PRINCIPAL ACTIONS	CONTROL OF SECRETION
Glucagon	Raises blood sugar level by accelerating conversion of glucogen into glucose in liver (glycogenolysis) and conversion of other nutrients into glucose in the liver (gluconeogenesis) and releasing glucose into blood.	Decreased blood level of glucose, exercise, and largely protein meals stimulate glucagon secretion; GHIF (somatostatin) inhibits glucagon secretion.
Insulin	Lowers blood sugar level by accelerating transport of glucose into cells, converting glucose into glycogen (glycogenesis), and decreasing glycogenolysis and gluconeogenesis; also increases lipogenesis and stimulates protein synthesis.	Increased blood level of glucose, GH, ACTH, and gastrointestinal hormones stimulate insulin secretion; GHIF (somatostatin) inhibits insulin secretion.

PINEAL (EPIPHYSIS CEREBRI)

The endocrine gland attached to the roof of the third ventricle is known as the **pineal gland** (because of its resemblance to a pine cone) or **epiphysis cerebri** (see Figure 18-1). The gland is about 5 to 8 mm (0.2 to 0.3 inch) long and 5 mm wide. It weighs about 0.2 g. It is covered by a capsule formed by the pia mater and consists of masses of **neuroglial cells** and parenchymal secretory cells called **pinealocytes.** Around the cells are scattered preganglionic sympathetic fibers. The pineal gland starts to calcify at about the time of puberty. Such calcium deposits are referred to as **brain sand.** Contrary to a once widely held belief, there is no evidence that the pineal atrophies with age and that the presence of brain sand is an indication of atrophy. In fact, the presence of brain sand may indicate increased secretory activity.

Although many anatomical facts concerning the pineal gland have been known for years, its physiology is still somewhat obscure. One hormone secreted by the pineal gland is **melatonin,** which appears to inhibit reproductive activities by inhibiting gonadotropic hormones. Some evidence also exists that the pineal secretes a second hormone called **adrenoglomerulotropin** (a-drē'-nō-glō-mer'-yoo-lō-TRŌ-pin). This hormone may stimulate the adrenal cortex to secrete aldosterone. Other substances found in the pineal gland include norepinephrine (NE), serotonin, histamine, GnRF, and GABA.

THYMUS

Usually a bilobed lymphatic organ, the **thymus gland** is located in the superior mediastinum, posterior to the sternum and between the lungs (Figure 18-20). The two **thymic lobes** are held in close proximity by an enveloping layer of connective tissue. Each lobe is enclosed by a fibrous connective tissue **capsule.** The capsule gives off extensions into the lobes called **trabeculae** (tra-BEK-yoo-lē), which divide the lobes into **lobules** (Figure 18-21a).

Each lobule consists of a deeply staining peripheral **cortex** and a lighter staining central **medulla.** The cortex is composed almost entirely of small, medium, and large tightly packed lymphocytes held in place by reticular tissue fibers (Figure 18-21b). Since the reticular tissue differs in origin and structure from that usually found in other lymphatic organs, it is referred to as **epithelioreticular** supporting tissue. The medulla consists mostly of epithelial cells and more widely scattered lymphocytes, and its reticulum is more cellular than fibrous. In addition, the medulla contains characteristic **thymic corpuscles,** concentric layers of epithelial cells. Their significance is unknown.

The thymus gland is conspicuous in the infant, and it reaches its maximum size of about 40 grams during puberty. After puberty, most of the thymic tissue is replaced by fat and connective tissue. By the time the person reaches maturity, the gland has atrophied substantially.

The function of the thymus gland is related to immunity. Lymphoid tissue of the body consists primarily of lymphocytes that may be distinguished into two kinds: B cells and T cells. Both are derived originally in the embryo from lymphocytic stem cells in bone marrow. Before migrating to their positions in lymphoid tissue, the descendants of the stem cells follow two distinct pathways. About half of them migrate to the thymus gland, where they are processed to become thymus-dependent lymphocytes, or **T cells.** The thymus gland confers on some of them the ability to destroy antigens (foreign microbes and substances) directly. The remaining stem cells are processed in some as yet undetermined area of the body, possibly bone marrow, the fetal liver and spleen, or gut-associated lymphoid tissue, and are known as **B cells.** These cells, under the influence of hormones produced by the thymus gland, called **thymosin, thymic humoral factor (THF), thymic factor (TF),** and **thymopoietin,** differentiate into plasma cells. Plasma cells, in turn, produce antibodies against antigens. Thymosin is also secreted by macrophages. The details of the roles of B cells and T cells in immunity are presented in Chapter 22.

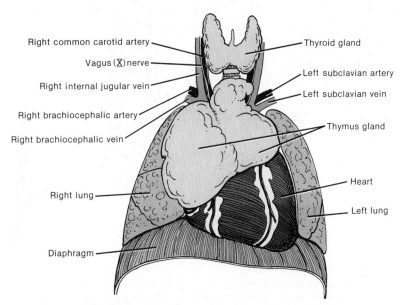

Right common carotid artery — Thyroid gland
Vagus (X) nerve — Left subclavian artery
Right internal jugular vein — Left subclavian vein
Right brachiocephalic artery — Thymus gland
Right brachiocephalic vein
Right lung — Heart
— Left lung
Diaphragm

FIGURE 18-20 Location of the thymus gland in a young child.

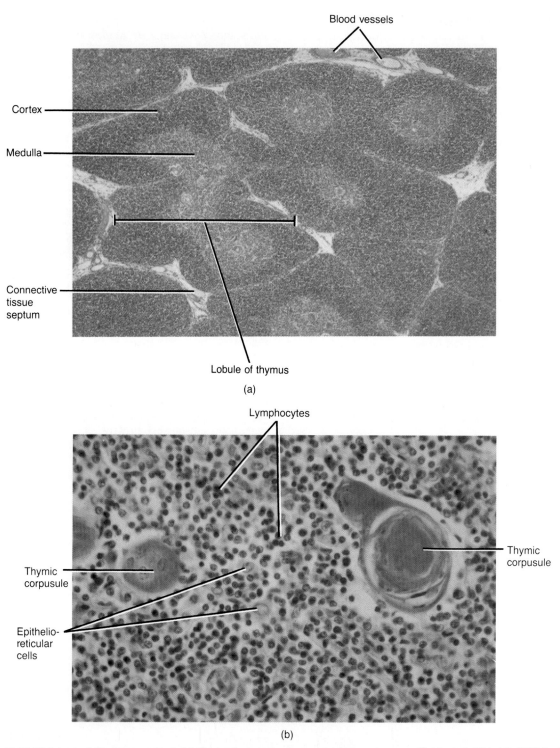

FIGURE 18-21 Histology of the thymus gland. (a) Photomicrograph of several lobules at a magnification of 40×. (© 1983 by Michael H. Ross. Used by permission.) (b) Photomicrograph of an enlarged aspect of thymic corpuscles at a magnification of 600×. (© 1983 by Michael H. Ross. Used by permission.)

OTHER ENDOCRINE TISSUES

Before leaving our discussion of hormones, it should be noted that body tissues other than endocrine glands also secrete hormones. The gastrointestinal tract synthesizes several hormones that regulate digestion in the stomach and small intestine. Among these hormones are **stomach gastrin, enteric gastrin, secretin, cholecystokinin (CCK),** **enterocrinin,** and **gastric inhibitory peptide (GIP).** The actions of these hormones and their control of secretion are summarized in Exhibit 24-1.

The placenta produces **human chorionic gonadotropin (HCG), estrogens, progesterone, relaxin,** and **human chorionic somatomammotropin (HCS),** all of which are related to pregnancy (Chapter 29).

When the kidneys (and liver to a lesser extent) become hypoxic, it is believed that they release an enzyme called **renal erythropoietic factor.** It is secreted into the blood where it acts on a plasma protein to bring about the production of a hormone called **erythropoietin** (ē-rith'-rō-POY-ē-tin), which stimulates red blood cell production (Chapter 19). The kidneys also help bring about the activation of the hormone vitamin D.

Finally, the skin produces vitamin D in the presence of sunlight.

STRESS AND THE GENERAL ADAPTATION SYNDROME

Homeostatic mechanisms are geared toward counteracting the everyday stresses of living. If they are successful, the internal environment maintains a uniform chemistry, temperature, and pressure. If a stress is extreme, unusual, or long-lasting, however, the normal mechanisms may not be sufficient. In this case, the stress triggers a wide-ranging set of bodily changes called the **general adaptation syndrome.** Hans Selye, who is the world's acknowledged authority on stress, was responsible for the concept of the general adaptation syndrome. Unlike the homeostatic mechanisms, the general adaptation syndrome does not maintain a normal internal environment. In fact, it does just the opposite. For instance, blood pressure and blood sugar level are raised above normal. The purpose of these changes in the internal environment is to gear up the body to meet an emergency.

STRESSORS

The hypothalamus can be called the body's watchdog. It has sensors that detect changes in the chemistry, temperature, and pressure of the blood. It is informed of emotions through tracts that connect it with the emotional centers of the cerebral cortex. When the hypothalamus senses stress, it initiates a chain of reactions that produce the general adaptation syndrome. The stresses that produce the syndrome are called **stressors.** A stressor may be almost any disturbance—heat or cold, environmental poisons, poisons given off by bacteria during a raging infection, heavy bleeding from a wound or surgery, or a strong emotional reaction.

When a stressor appears, it stimulates the hypothalamus to initiate the syndrome through two pathways. The first pathway is stimulation of the sympathetic nervous system and adrenal medulla. This stimulation produces an immediate set of responses called the alarm reaction. The second pathway, called the resistance reaction, involves the anterior pituitary gland and adrenal cortex. The resistance reaction is slower to start, but its effects last longer.

ALARM REACTION

The **alarm reaction,** or **fight-or-flight response,** is the body's initial reaction to a stressor (Figure 18-22a). It is actually a complex of reactions initiated by the hypothalamic stimulation of the sympathetic nervous system

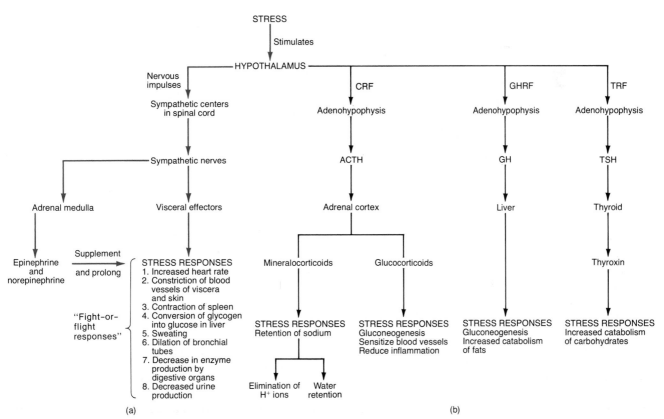

FIGURE 18-22 Responses to stressors during the general adaptation syndrome. (a) During the alarm stage. (b) During the resistance stage. Colored arrows indicate immediate reactions. Black arrows indicate long-term reactions.

and the adrenal medulla. The responses of the visceral effectors are immediate and short-lived. They are designed to counteract a danger by mobilizing the body's resources for immediate physical activity. In essence, the alarm reaction brings tremendous amounts of glucose and oxygen to the organs that are most active in warding off danger. These are the brain, which must become highly alert; the skeletal muscles, which may have to fight off an attacker; and the heart, which must work furiously to pump enough materials to the brain and muscles. The hyperglycemia associated with sympathetic activity is produced by liver glycogenolysis, stimulated by epinephrine and norepinephrine from the adrenal medulla and liver gluconeogenesis, stimulated by fat mobilization and glucose-sparing by GH and protein mobilization by glucocorticoids.

Among the stress responses that characterize the alarm stage are the following.

1. The heart rate and the strength of cardiac muscle contraction increase. This response circulates substances in the blood very quickly to areas where they are needed to combat the stress.

2. Blood vessels supplying the skin and viscera, except the heart and lungs, undergo constriction. At the same time, blood vessels supplying the skeletal muscles and brain undergo dilation. These responses route more blood to organs active in the stress responses while decreasing blood supply to organs that do not assume an immediate, active role.

3. The spleen contracts and discharges stored blood into the general circulation to provide additional blood. Moreover, red blood cell production is accelerated and the ability of the blood to clot is increased. These preparations are made to combat bleeding.

4. The liver transforms large amounts of stored glycogen into glucose and releases it into the bloodstream. The glucose is broken down by the active cells to provide the energy needed to meet the stressor.

5. Sweat production increases. This response helps lower body temperature, which is elevated as circulation increases and body catabolism increases. Profuse sweating also helps to eliminate wastes produced as a result of accelerated catabolism.

6. The rate of breathing increases, and the respiratory passageways widen to accommodate more air. This response enables the body to acquire more oxygen, which is needed in the decomposition reactions of catabolism. It also allows the body to eliminate more carbon dioxide, which is produced as a side product during catabolism.

7. Production of saliva, stomach enzymes, and intestinal enzymes decreases. This reaction takes place since digestive activity is not essential for counteracting the stress.

8. Sympathetic impulses to the adrenal medulla increase its secretion of epinephrine and norepinephrine. These hormones supplement and prolong many sympathetic responses—increasing heart rate and strength, constricting blood vessels, accelerating the rate of breathing, widening respiratory passageways, increasing the rate of catabolism, decreasing the rate of digestion, and increasing blood sugar level.

If you group the stress responses of the alarm stage by function, you will note that they are designed to increase circulation rapidly, promote catabolism for energy production, and decrease nonessential activities. If the stress is great enough, the body mechanisms may not be able to cope and death can result.

RESISTANCE REACTION

The second stage in the stress responses is the **resistance reaction** (Figure 18-22b). Unlike the short-lived alarm reaction that is initiated by nervous impulses from the hypothalamus, the resistance reaction is initiated by regulating factors secreted by the hypothalamus and is a long-term reaction. The regulating factors are CRF, GHRF, and TRF.

CRF stimulates the adenohypophysis to increase its secretion of ACTH. ACTH stimulates the adrenal cortex to secrete more of its hormones. The adrenal cortex is also indirectly stimulated by the alarm reaction. During the alarm reaction, kidney activity is cut back because of decreased blood circulation to the kidneys, since it is not essential for meeting sudden danger. The resultant decrease in urine production stimulates the secretion of the mineralocorticoids.

The mineralocorticoids secreted by the adrenal cortex bring about the conservation of sodium ions by the body. A secondary effect of sodium conservation is the elimination of hydrogen ions. The hydrogen ions build up in high concentrations as a result of increased catabolism and tend to make the blood more acidic. Thus, during stress, a lowering of body pH is prevented. Sodium retention also leads to water retention, thus maintaining the high blood pressure that is typical of the alarm reaction. It also helps make up for fluid lost through severe bleeding.

The glucocorticoids, which are produced in high concentrations during stress, bring about the following reactions.

1. The glucocorticoids accelerate protein catabolism and the conversion of amino acids into glucose so that the body has a large supply of energy long after the immediate stores of glucose have been used up. The glucocorticoids also stimulate the removal of proteins from cell structures and stimulate the liver to break them down into amino acids. The amino acids can then be rebuilt into enzymes that are needed to catalyze the increased chemical activities of the cells or be converted to glucose.

2. The glucocorticoids make blood vessels more sensitive to stimuli that bring about their constriction. This response counteracts a drop in blood pressure caused by bleeding.

3. The glucocorticoids also inhibit the production of fibroblasts, which develop into connective tissue cells. Injured fibroblasts release chemicals that play a role in stimulating the inflammatory response. Thus the glucocorticoids reduce inflammation and prevent it from becoming disruptive rather than protective. Unfortunately, through their effect on fibroblasts, the glucocorticoids also discourage connective tissue formation. Wound healing is therefore slow during a prolonged resistance stage.

Two other regulating factors are secreted by the hypothalamus in response to a stressor: TRF and GHRF. TRF causes the adenohypophysis to secrete TSH. GHRF causes it to secrete GH. TSH stimulates the thyroid to secrete thyroxine, which increases the catabolism of carbohydrates. HGH stimulates the catabolism of fats and the conversion of glycogen to glucose. The combined actions of TSH and GH increase catabolism and thereby supply additional energy for the body.

The resistance stage of the general adaptation syndrome allows the body to continue fighting a stressor long after the effects of the alarm reaction have dissipated. It increases the rate at which life processes occur. It also provides the energy, functional proteins, and circulatory changes required for meeting emotional crises, performing strenuous tasks, fighting infection, or resisting the threat of bleeding to death. During the resistance stage, blood chemistry returns to nearly normal. The cells use glucose at the same rate it enters the bloodstream. Thus blood sugar level returns to normal. Blood pH is brought under control by the kidneys as they excrete more hydrogen ions. However, blood pressure remains abnormally high because the retention of water increases the volume of blood.

All of us are confronted by stressors from time to time, and we have all experienced the resistance stage. Generally, this stage is successful in seeing us through a stressful situation, and our bodies then return to normal. Occasionally, the resistance stage fails to combat the stressor, however, and the body "gives up." In this case, the general adaptation syndrome moves into the stage of exhaustion.

EXHAUSTION

A major cause of exhaustion is loss of potassium ions. When the mineralocorticoids stimulate the kidney to retain sodium ions, potassium and hydrogen ions are traded off for sodium ions and secreted in the urine. As the chief positive ion in cells, potassium is partly responsible for controlling the water concentration of the cytoplasm. As the cells lose more and more potassium, they function less and less effectively. Finally they start to die. This condition is called the **stage of exhaustion.** Unless it is rapidly reversed, vital organs cease functioning and the person dies. Another cause of exhaustion is depletion of the adrenal glucocorticoids. In this case, blood glucose level suddenly falls and the cells do not receive enough nutrients. A final cause of exhaustion is weakening organs. A long-term or strong resistance reaction puts heavy demands on the body, particularly on the heart, blood vessels, and adrenal cortex. They may not be up to handling the demands, or they may suddenly fail under the strain. In this respect, ability to handle stressors is determined to a large degree by general health.

MEDICAL TERMINOLOGY

Feminizing adenoma (*aden* = gland; *oma* = tumor) Malignant tumor of the adrenal gland that secretes abnormally high amounts of female sex hormones and produces female secondary sexual characteristics in the male.

Hyperplasia (*hyper* = over, *plas* = grow) Increase in the number of cells due to an increase in the frequency of cell division.

Hypoplasia (*hypo* = under) Defective development of tissue.

Neuroblastoma (*neuro* = nerve) Malignant tumor arising from the adrenal medulla associated with metastases to bones.

Thyroid storm An aggravation of all symptoms of hyperthyroidism characterized by unregulated hypermetabolism with fever and rapid heart rate; results from trauma, surgery, and unusual emotional stress or labor.

Virilizing adenoma Malignant tumor of the adrenal gland that secretes high amounts of male sex hormones and produces male secondary sexual characteristics in the female.

STUDY OUTLINE

Endocrine Glands (p. 402)
1. Both the endocrine and nervous systems assume a role in maintaining homeostasis.
2. Hormones help regulate the internal environment, respond to stress, help regulate growth and development, and contribute to reproductive processes.
3. Exocrine glands (sweat, sebaceous, digestive) secrete their products through ducts into body cavities or onto body surfaces.
4. Endocrine glands secrete hormones into the blood.

Chemistry of Hormones (p. 402)
1. On the basis of solubility, hormones are classified as water-soluble and lipid-soluble.
2. Cells that respond to the effects of hormones are called target cells.

Mechanism of Hormonal Action (p. 402)
1. Water-soluble hormones exert their effects by interacting with plasma membrane receptors; some utilize cyclic AMP as a second messenger.
2. Lipid-soluble hormones exert their effects by interacting with genes.
3. Prostaglandins (PG) can increase or decrease cyclic AMP formation and thus modulate hormone responses that use cyclic AMP.

Control of Hormonal Secretions: Feedback Control (p. 405)
1. A negative feedback control mechanism prevents overproduction or underproduction of a hormone.
2. Hormone secretions are controlled by levels of circulating hormone itself, nerve impulses, and regulating factors.

Pituitary (Hypophysis) (p. 406)
1. The pituitary is located in the sella turcica and is differentiated into the adenohypophysis (the anterior lobe and glandular portion) and the neurohypophysis (the posterior lobe and nervous portion).
2. Hormones of the adenohypophysis are released or inhibited by regulating factors produced by the hypothalamus.
3. The blood supply to the adenohypophysis is from the superior hypophyseal arteries.
4. Histologically, the adenohypophysis consists of growth hormone cells that produce growth hormone (GH); prolactin cells that produce prolactin (PRL); TSH cells that secrete thyroid-stimulating hormone (TSH); gonadotroph cells that synthesize follicle-stimulating hormone (FSH) and luteinizing hormone (LH); and corticotroph-lipotroph cells that

secrete adrenocorticotropin hormone (ACTH) and melano-cyte-stimulating hormone (MSH).

5. GH stimulates body growth through somatomedins and insulinlike growth factors (IGF) and is controlled by GHIF (growth hormone inhibiting factor or somatostatin) and GHRF (growth hormone releasing factor).

6. Disorders associated with improper levels of GH are pituitary dwarfism, giantism, and acromegaly.

7. TSH regulates thyroid gland activities and is controlled by TRF (thyrotropin releasing factor).

8. ACTH regulates the activities of the adrenal cortex and is controlled by CRF (corticotropin releasing factor).

9. FSH regulates the activities of the ovaries and testes and is controlled by GnRF (gonadotropin releasing factor).

10. LH regulates female and male reproductive activities and is controlled by GnRF.

11. PRL helps initiate milk secretion and is controlled by PIF (prolactin inhibiting factor) and PRF (prolactin releasing factor).

12. MSH increases skin pigmentation and is controlled by MRF (melanocyte-stimulating hormone releasing factor) and MIF (melanocyte-stimulating hormone inhibiting factor).

13. The neural connection between the hypothalamus and neurohypophysis is via the hypothalamic-hypophyseal tract.

14. Hormones made by the hypothalamus and stored in the neurohypophysis are oxytocin or OT (stimulates contraction of uterus and ejection of milk) and antidiuretic hormone or ADH (stimulates water reabsorption by the kidneys and arteriole constriction).

15. OT secretion is controlled by uterine distension and sucking during nursing; ADH is controlled primarily by water concentration.

16. A disorder associated with dysfunction of the neurohypophysis is diabetes insipidus.

Thyroid (p. 412)

1. The thyroid gland is located below the larynx.

2. Histologically, the thyroid consists of thyroid follicles composed of follicular cells, which secrete the thyroid hormones thyroxine (T_4) and triiodothyronine (T_3), and parafollicular cells, which secrete calcitonin (CT).

3. Thyroid hormones are synthesized from iodine and tyrosine within thyroglobulin (TBG) and carried in the blood with plasma proteins, mostly thyroxine-binding globulin (TBG).

4. Thyroid hormones regulate the rate of metabolism, growth and development, and the reactivity of the nervous system. Secretion is controlled by TRF.

5. Cretinism, myxedema, exophthalmic goiter, and simple goiter are disorders associated with dysfunction of the thyroid gland.

6. Calcitonin (CT) lowers the blood level of calcium. Secretion is controlled by its own level in blood.

Parathyroids (p. 415)

1. The parathyroids are embedded on the posterior surfaces of the lateral lobes of the thyroid.

2. Histologically, the parathyroids consist of principal and oxyphil cells.

3. Parathyroid hormone (PTH) regulates the homeostasis of calcium and phosphate by increasing blood calcium level and decreasing blood phosphate level. Secretion is controlled by its own level in blood.

4. Tetany and osteitis fibrosa cystica are disorders associated with the parathyroid glands.

Adrenals (Suprarenals) (p. 417)

1. The adrenal glands are located superior to the kidneys. They consist of an outer cortex and inner medulla.

2. Histologically, the cortex is divided into a zona glomerulosa, zona fasciculata, and zona reticularis; the medulla consists of chromaffin cells.

3. Cortical secretions are mineralocorticoids, glucocorticoids, and gonadocorticoids.

4. Mineralocorticoids (e.g., aldosterone) increase sodium and water reabsorption and decrease potassium reabsorption. Secretion is controlled by the renin-angiotensin pathway and blood level of potassium.

5. A dysfunction related to aldosterone secretion is aldosteronism.

6. Glucocorticoids (e.g., cortisol) promote normal metabolism, help resist stress, and serve as antiinflammatories. Secretion is controlled by CRF.

7. Disorders associated with glucocorticoid secretion are Addison's disease and Cushing's syndrome.

8. Gonadocorticoids secreted by the adrenal cortex have minimal effects. Excessive production results in adrenogenital syndrome.

9. Medullary secretions are epinephrine and norepinephrine (NE) which produce effects similar to sympathetic responses. They are released under stress.

10. Tumors of medullary chromaffin cells are called pheochromocytomas.

Pancreas (p. 421)

1. The pancreas is posterior and slightly inferior to the stomach.

2. Histologically, it consists of islets of Langerhans (endocrine cells) and acini (enzyme-producing cells). Three types of cells in the endocrine portion are alpha cells, beta cells, and delta cells.

3. Alpha cells secrete glucagon, beta cells secrete insulin, and delta cells secrete growth hormone inhibiting factor (GHIF) or somatostatin.

4. Glucagon increases blood sugar level. Secretion is controlled by its own level in the blood.

5. Insulin decreases blood sugar level. Secretion is controlled by its own level in the blood.

6. Disorders associated with insulin production are diabetes mellitus and hyperinsulinism.

Ovaries and Testes (p. 424)

1. Ovaries are located in the pelvic cavity and produce sex hormones related to development and maintenance of female sexual characteristics, menstrual cycle, pregnancy, and lactation.

2. Testes lie inside the scrotum and produce sex hormones related to the development and maintenance of male sexual characteristics.

Pineal (Epiphysis Cerebri) (p. 425)

1. The pineal is attached to the roof of the third ventricle.

2. Histologically, it consists of secretory parenchymal cells called pinealocytes, neuroglial cells, and scattered preganglionic sympathetic fibers. Calcified deposits are referred to as brain sand.

3. It secretes melatonin (possibly regulates reproductive activities by inhibiting gonadotropic hormones) and adrenoglomerulotropin (may stimulate adrenal cortex to secrete aldosterone).

Thymus (p. 425)

1. The thymus is a bilobed lymphatic gland located in the superior mediastinum posterior to the sternum and between the lungs.

2. Histologically, it consists primarily of various sizes of lymphocytes.

3. The gland is necessary for the maturation of the thymus-

dependent lymphocytes (T cells) of the immune system; its hormones (thymosin, thymic humoral factor, thymic factor, and thymopoietin) may cause B cells to differentiate into antibody-producing plasma cells.

Other Endocrine Tissues (p. 426)

1. The gastrointestinal tract synthesizes stomach and intestinal gastrin, secretin, cholecystokinin (CCK), enterocrinin, and gastric inhibitory peptide (GIP).
2. The placenta produces human chorionic gonadotropin (HCG), estrogens, progesterone, relaxin, and human chorionic somatomammotropin (HCS).
3. The kidneys release an enzyme that produces erythropoietin.
4. The skin synthesizes vitamin D.

Stress and the General Adaptation Syndrome (p. 427)

General Adaptation Syndrome

1. If the stress is extreme or unusual, it triggers a wide-ranging set of bodily changes called the general adaptation syndrome.
2. Unlike the homeostatic mechanisms, this syndrome does not maintain a constant internal environment.

Stressors

1. The stresses that produce the general adaptation syndrome are called stressors.
2. Stressors include surgical operations, poisons, infections, fever, and strong emotional responses.

Alarm Reaction

1. The alarm reaction is initiated by nerve impulses from the hypothalamus to the sympathetic division of the autonomic nervous system and adrenal medulla. Responses are the immediate and short-lived fight-or-flight responses that increase circulation, promote catabolism for energy production, and decrease nonessential activities.

Resistance Reaction

1. The resistance reaction is initiated by regulating factors secreted by the hypothalamus.
2. The regulating factors are CRF, GHRF, and TRF.
3. CRF stimulates the adenohypophysis to increase its secretion of ACTH, which in turn stimulates the adrenal cortex to secrete hormones.
4. Resistance reactions are long-term and accelerate catabolism to provide energy to counteract stress.
5. Glucocorticoids are produced in high concentrations during stress. They create many distinct physiological effects.

Exhaustion

1. The stage of exhaustion results from dramatic changes during alarm and resistance reactions.
2. Exhaustion is caused mainly by loss of potassium, depletion of adrenal glucocorticoids, and weakened organs, and if stress is too great, it may lead to death.

REVIEW QUESTIONS

1. Contrast endocrine and nervous system control of homeostasis.
2. Distinguish between an endocrine gland and an exocrine gland.
3. What is a hormone? Distinguish between tropic and gonadotropic hormones.
4. How do water-soluble and lipid-soluble hormones differ?
5. Explain the mechanism of hormonal action involving (a) interaction with plasma membrane receptors and (b) activation of genes.
6. Define a prostaglandin. Give several functions of prostaglandins.
7. How are prostaglandins related to hormonal action?
8. How are negative feedback systems related to hormonal control? Discuss the three models of operation.
9. In what respect is the pituitary gland actually two glands?
10. Describe the histology of the adenohypophysis. Why does the anterior lobe of the gland have such an abundant blood supply?
11. What hormones are produced by the adenohypophysis? What are their functions? How are they controlled?
12. Relate the importance of regulating factors to secretions of the adenohypophysis.
13. Describe the clinical symptoms of pituitary dwarfism, giantism, and acromegaly.
14. Discuss the histology of the neurohypophysis and the function and regulation of its hormones.
15. What are the clinical symptoms of diabetes insipidus?
16. Describe the location and histology of the thyroid gland.
17. How are the thyroid hormones made, stored, and secreted?
18. Discuss the physiological effects of the thyroid hormones. How are these hormones regulated?
19. Discuss the clinical symptoms of cretinism, myxedema, exophthalmic goiter, and simple goiter.
20. Describe the function and control of calcitonin.
21. Where are the parathyroids located? What is their histology?
22. What are the functions of the parathyroid hormone?
23. Discuss the clinical symptoms of tetany and osteitis fibrosa cystica.
24. Compare the adrenal cortex and adrenal medulla with regard to location and histology.
25. Describe the hormones produced by the adrenal cortex in terms of type, normal function, and control.
26. Describe the clinical symptoms of aldosteronism, Addison's disease, Cushing's syndrome, and adrenogenital syndrome.
27. What relationship does the adrenal medulla have to the autonomic nervous system? What is the action of adrenal medullary hormones?
28. What is a pheochromocytoma?
29. Describe the location of the pancreas and the histology of the islets of Langerhans.
30. What are the actions of glucagon and insulin? How are the hormones controlled?
31. Describe the clinical symptoms of diabetes mellitus and hyperinsulinism.
32. Why are the ovaries and testes considered to be endocrine glands?
33. Where is the pineal gland located? What are its assumed functions?
34. Describe the location and histology of the thymus gland. What is its proposed function?
35. List the hormones secreted by the gastrointestinal tract, placenta, blood, and skin.
36. Define the general adaptation syndrome. What is a stressor?
37. How do homeostatic responses differ from stress responses?
38. Outline the reactions of the body during the alarm stage, resistance stage, and stage of exhaustion when placed under stress. What is the central role of the hypothalamus during stress?
39. Refer to the glossary of medical terminology associated with the endocrine system. Be sure that you can define each term.

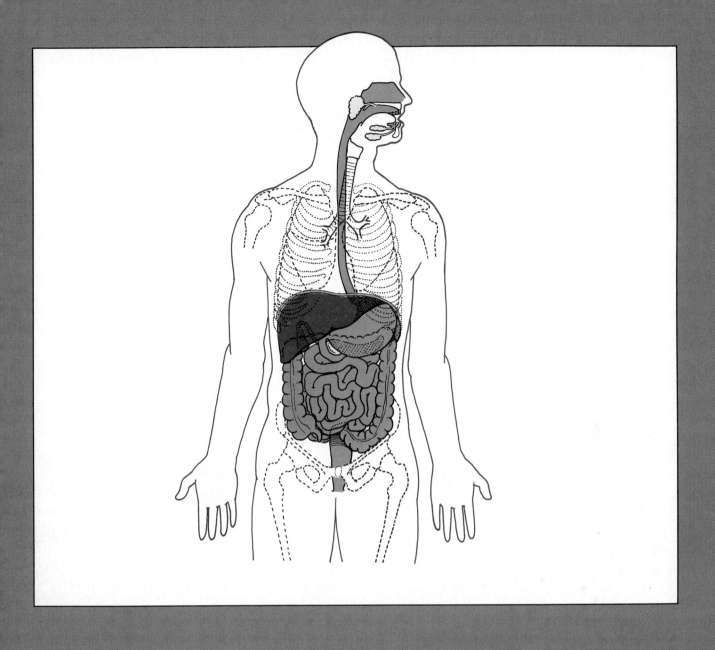

UNIT IV

Maintenance of the Human Body

This unit explains how the body maintains homeostasis on a day-to-day basis. In these chapters you will be studying the interrelations among the cardiovascular, lymphatic, respiratory, digestive, and urinary systems. You will also learn about metabolism, fluid and electrolyte balance, and acid–base dynamics.

- Contrast the general roles of blood, lymph, and interstitial fluid in maintaining homeostasis.

- Define the principal physical characteristics of blood and its functions in the body.

- Compare the origins of the formed elements in blood.

- Discuss the structure of erythrocytes and their function in the transport of oxygen and carbon dioxide.

- Define erythropoiesis and describe erythrocyte production and destruction.

- Explain the importance of a reticulocyte count and hematocrit.

- List the structural features and types of leucocytes.

- Explain the significance of a differential count.

- Discuss the role of leucocytes in phagocytosis and antibody production.

- Discuss the structure of thrombocytes and explain their role in blood clotting.

- List the components of plasma and explain their importance.

- Explain how the body attempts to prevent blood loss.

- Identify the stages involved in blood clotting.

- Explain the various factors that promote and inhibit blood clotting.

- Define clotting time, bleeding time, and prothrombin time.

- Explain ABO and Rh blood grouping.

- Define the antigen-antibody reaction as the basis for ABO blood grouping.

- Define the antigen-antibody reaction of the Rh blood grouping system.

- Define erythroblastosis fetalis as a harmful antigen-antibody reaction.

- Compare the location, composition, and function of interstitial fluid and lymph.

- Contract the causes of nutritional, pernicious, hemorrhagic, hemolytic, aplastic, and sickle cell anemia.

- Define polycythemia and describe the importance of hematocrit in its diagnosis.

- Identify the clinical symptoms of infectious mononucleosis and leukemia.

- Define medical terminology associated with blood.

19 The Cardiovascular System: The Blood

The more specialized a cell becomes, the less capable it is of carrying on an independent existence. For instance, a specialized cell is less able to protect itself from extreme temperatures, toxic chemicals, and changes in pH. Often it cannot go looking for food or devour whole bits of food. And, if it is firmly implanted in a tissue, it cannot move away from its own wastes. The substance that bathes the cell and carries out these vital functions for it is called **interstitial fluid** (also known as **intercellular** or **tissue fluid**).

The interstitial fluid, in turn, must be serviced by blood and lymph. The blood picks up oxygen from the lungs, nutrients from the digestive tract, hormones from the endocrine glands, and enzymes from still other parts of the body. It transports these substances to all the tissues where they diffuse from the capillaries into the interstitial fluid. In the interstitial fluid, the substances are passed on to the cells and exchanged for wastes.

Since the blood services all the tissues of the body, it can be an important medium for the transport of disease-causing organisms. To protect itself from the spread of disease, the body has a lymphatic system—a collection of vessels containing a fluid called lymph. The lymph picks up materials, including wastes, from the interstitial fluid, cleanses them of bacteria, and returns them to the blood. The blood then carries the wastes to the lungs, kidneys, and sweat glands, where they are eliminated from the body. The blood also takes wastes to the liver, where they are detoxified.

Blood inside blood vessels, interstitial fluid around body cells, and lymph inside lymph vessels constitute the *internal environment* of the human organism. Because the body cells are too specialized to adjust to more than very limited changes in their environment, the internal environment must be kept constant. This condition we have called homeostasis. In preceding chapters, we have discussed how the internal environment is kept in homeostasis. Now we will look at that environment itself.

The blood, heart, and blood vessels constitute the **cardiovascular system.** The lymph, lymph vessels, and lymph glands make up the **lymphatic system.** Let us first take a look at the substance known as blood.

PHYSICAL CHARACTERISTICS

The red body fluid that flows through all the vessels except the lymph vessels is called **blood.** Blood is a viscous fluid—it is thicker and more adhesive than water. Water is considered to have a viscosity of 1.0. The viscosity of blood, by comparison, ranges from 4.5 to 5.5. It flows more slowly than water, at least in part because of its viscosity. The adhesive quality of blood, or its stickiness, may be felt by touching it. Blood is also slightly heavier than water.

Other physical characteristics of blood include a temperature of about 38°C (100.4°F), a pH range of 7.35 to 7.45 (slightly alkaline), and a 0.85 to 0.90 percent concentration of salt (NaCl).

Blood constitutes about 8 percent of the total body weight. The blood volume of an average-sized male is 5 to 6 liters (5 to 6 qt). An average-sized female has 4 to 5 liters.

FUNCTIONS

Blood is a complex liquid that performs a number of critical functions.

1. It transports oxygen from the lungs to all cells of the body.

2. It transports carbon dioxide from the cells to the lungs.

3. It transports nutrients from the digestive organs to the cells.

4. It transports waste products from the cells to the kidneys, lungs, and sweat glands.

5. It transports hormones from endocrine glands to the cells.

6. It transports enzymes to various cells.

7. It regulates body pH through buffers and amino acids.

8. It plays a role in the regulation of normal body temperature because it contains a large volume of water (an excellent heat absorber and coolant).

9. It regulates the water content of cells, principally through dissolved sodium ions.

10. It prevents body fluid loss through the clotting mechanism.

11. It protects against toxins and foreign microbes through special combat-unit cells.

COMPONENTS

Microscopically, blood is composed of two portions: formed elements (cells and cell-like structures) and plasma (liquid containing dissolved substances). The formed elements compose about 45 percent of the volume of blood; plasma constitutes about 55 percent (Figure 19-1).

FORMED ELEMENTS

In clinical practice, the most common classification of the **formed elements** of the blood is the following.

Erythrocytes (red blood cells)

Leucocytes (white blood cells)
    Granular leucocytes (granulocytes)
        Neutrophils
        Eosinophils
        Basophils
    Agranular leucocytes (agranulocytes)
        Lymphocytes
        Monocytes

Thrombocytes (platelets)

Origin

The process by which blood cells are formed is called **hemopoiesis** (hē-mō-poy-Ē-sis) or **hematopoiesis.** During embryonic and fetal life, there are no clear-cut centers for blood cell production. The yolk sac, liver, spleen, thymus gland, lymph nodes, and bone marrow all participate at various times in producing the formed elements. In the adult, however, we can pinpoint the production process to the red bone marrow in the sternum, ribs, vertebrae, and pelvis and to lymphoid tissue. Red blood cells, granular leucocytes, and platelets are produced in red bone marrow (myeloid tissue). Agranular leucocytes arise from both myeloid tissue and from lymphoid tissue—spleen, tonsils, lymph nodes. Undifferentiated mes-

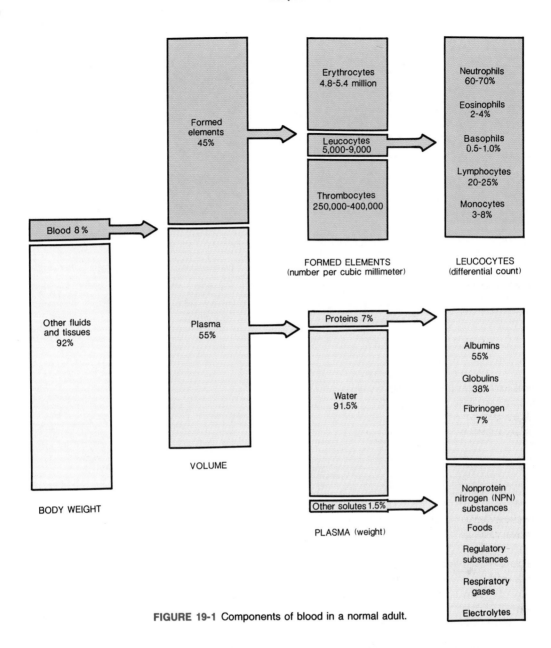

FIGURE 19-1 Components of blood in a normal adult.

enchymal cells in red bone marrow are transformed into **hemocytoblasts** (hē'-mō-SĪ-tō-blasts), immature cells that are eventually capable of developing into mature blood cells (Figure 19-2a). The hemocytoblasts undergo differentiation into five types of cells from which the major types of blood cells develop.

1. Rubriblasts (proerythroblasts) form mature red blood cells.

2. Myeloblasts form mature neutrophils, eosinophils, and basophils.

3. Megakaryoblasts form mature thrombocytes (platelets).

4. Lymphoblasts form lymphocytes.

5. Monoblasts form monocytes.

Erythrocytes

● *Structure* Microscopically, **red blood cells (rbc's)** or **erythrocytes** (e-RITH-rō-sīts), appear as biconcave discs averaging about 7.7 μm in diameter (Figure 19-2b). Mature red blood cells are quite simple in structure. They

lack a nucleus and can neither reproduce nor carry on extensive metabolic activities. The plasma membrane is selectively permeable and consists of protein (stromatin) and lipids (lecithin and cholesterol). The membrane encloses cytoplasm and a red pigment called *hemoglobin*. Hemoglobin, which constitutes about 33 percent of the cell weight, is responsible for the red color of blood. Normal values for hemoglobin are 14 to 20 gm/100 ml in infants, 12 to 15 gm/100 ml in adult females, and 14 to 16.5 gm/100 ml in adult males.

● *Functions* The hemoglobin in erythrocytes combines with oxygen to form oxyhemoglobin and carbon dioxide to form carbaminohemoglobin and transports them through the blood vessels. The hemoglobin molecule consists of a protein called globin and a pigment called heme, which contains iron (Figure 19-3). As the erythrocytes pass through the lungs, each of the four iron atoms in the hemoglobin molecules combines with a molecule of oxygen. The oxygen is transported in this state to other

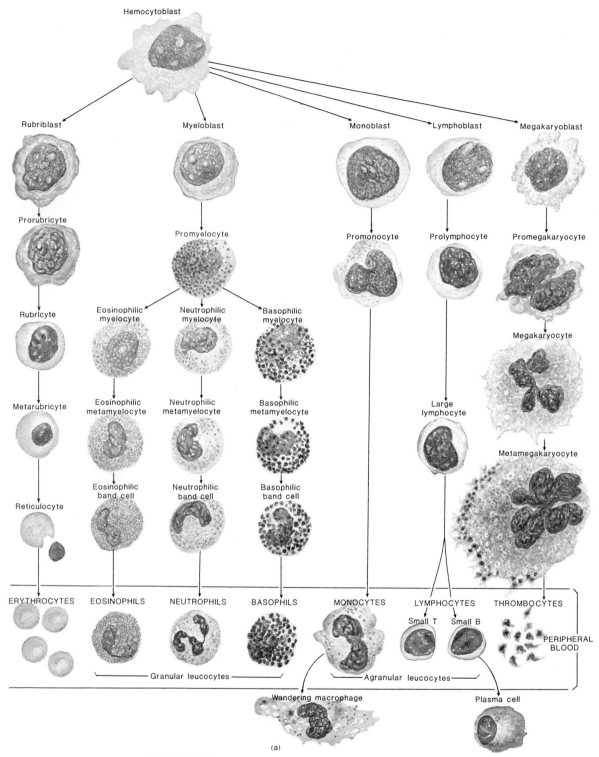

FIGURE 19-2 Blood cells. (a) Origin, development, and structure.

tissues of the body. In the tissues, the iron-oxygen reaction reverses, and the oxygen is released to diffuse into the interstitial fluid. On the return trip, the globin portion combines with a molecule of carbon dioxide from the interstitial fluid to form carbaminohemoglobin. This complex is transported to the lungs, where the carbon dioxide is released and then exhaled. Although about 23 percent of carbon dioxide is transported by hemoglobin in this

manner, the greater portion, about 70 percent, is transported in blood plasma as the bicarbonate ion, HCO_3^- (Chapter 23).

Red blood cells are highly specialized for their transport function. They contain a large number of hemoglobin molecules in order to increase their oxygen-carrying capacity. One estimate is 280 million molecules of hemoglobin per erythrocyte. Hemoglobin is contained inside red

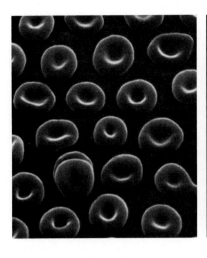
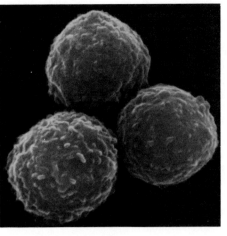

(b)

FIGURE 19-2 (*Continued*) Blood cells. (b) Scanning electron micrographs of erythrocytes (left), leucocytes (center), and a platelet (right) at a magnification of 10,000×. (Courtesy of Fisher Scientific Company and S.T.E.M. Laboratories, Inc., Copyright 1975.)

blood cells because hemoglobin molecules are small and, if they were free in plasma, they would leak through the endothelial membranes of blood vessels and be lost in the urine. The biconcave shape of a red blood cell has a much greater surface area than, say, a sphere or a cube. The erythrocyte thus presents the maximum surface area for the diffusion of gas molecules that pass through the membrane to combine with hemoglobin.

CLINICAL APPLICATION
Clinical trials have been performed to test the effectiveness of a **blood substitute** called *Fluosol-DA*. It is a

---

FIGURE 19-3 Diagram of a hemoglobin molecule. The globin (protein) portions of the molecule are indicated in green and blue and the four heme (iron-containing) portions are in the center of each globin molecule.

slippery, white liquid that has a very high solubility of oxygen. Since its only function is to transport oxygen, it is actually a hemoglobin substitute. It may eventually turn out to be useful in providing oxygen to tissues which are oxygen-deficient as a result of carbon monoxide poisoning, sickle cell anemia, strokes, heart attacks, and burns. Its use is limited by the fact that it must be kept frozen until just before use and recipients must receive oxygen by mask. Nevertheless, patients would be free of the risks of transfusion reactions and posttransfusion hepatitis.

● *Life Span and Number*   The cell membrane of a red blood cell becomes fragile and the cell is nonfunctional in about 120 days. A healthy male has about 5.4 million red blood cells per cubic millimeter (mm³) of blood, and a healthy female has about 4.8 million. The higher value in the male is because of his higher rate of metabolism. To maintain normal quantities of erythrocytes, the body must produce new mature cells at the astonishing rate of 2 million per second. In the adult, production takes place in the red bone marrow in the spongy bone of the cranium, ribs, sternum, bodies of vertebrae, and proximal epiphyses of the humerus and femur.

● *Production*   The process by which erythrocytes are formed is called **erythropoiesis** (e-rith'-rō-poy-Ē-sis). Erythropoiesis starts with the transformation of a hemocytoblast into a rubriblast (Figure 19-2a). The *rubriblast* (proerythroblast) gives rise to a *prorubricyte* (early erythroblast), which then develops into a *rubricyte* (intermediate erythroblast), the first cell in the sequence that begins to synthesize hemoglobin. The rubricyte next develops into a *metarubricyte* (late erythroblast). In the metarubricyte, hemoglobin synthesis is at a maximum. In the next stage, the metarubricyte develops into a *reticulocyte* and the nucleus is lost by extrusion. The reticulocyte in turn becomes an *erythrocyte,* or mature red blood cell. Once the erythrocyte is formed, it leaves the marrow and enters the bloodstream. Aged erythrocytes are destroyed by reticuloendothelial cells in the liver and spleen. The hemoglobin molecules are split apart, the iron is reused, and the rest of the molecule is converted into other substances for reuse or elimination.

Normally erythropoiesis and red cell destruction proceed at the same pace. But if the body suddenly needs more erythrocytes or if erythropoiesis is not keeping up with red blood cell destruction, a homeostatic mechanism steps up erythrocyte production. The mechanism is triggered by the reduced supply of oxygen for body cells that results from a reduced number of erythrocytes. If certain kidney cells become oxygen-deficient, they release an enzyme called **renal erythropoietic factor** that converts a plasma protein into the hormone **erythropoietin.** This hormone circulates through the blood to the red bone marrow, where it stimulates more hemocytoblasts to develop into red blood cells. Erythropoietin is also produced by other body tissues, particularly the liver.

Cellular oxygen deficiency, called **hypoxia** (hī-POKS-ē-a), may occur if you do not breathe in enough oxygen. This deficiency commonly occurs at high altitudes, where the air contains less oxygen. Oxygen deficiency may also occur because of anemia. The term **anemia** means that the number of functional red blood cells or their hemoglobin content is below normal. Consequently, the erythrocytes are unable to transport enough oxygen from the lungs to the cells. Anemia has many causes; lack of iron, lack of certain amino acids, and lack of vitamin B_{12} are but a few. Iron is needed for the oxygen-carrying part of the hemoglobin molecule. The amino acids are needed for the protein, or globin, part. Vitamin B_{12} helps the red bone marrow to produce erythrocytes. This vitamin is obtained from meat, especially liver, but it cannot be absorbed by the lining of the small intestine without the help of another substance—**intrinsic factor** produced by the mucosal cells of the stomach. Intrinsic factor facilitates the absorption of vitamin B_{12}. Once absorbed, it is stored in the liver. (Several types of anemia are described later in the chapter.)

CLINICAL APPLICATION

The rate of erythropoiesis is measured by a procedure called a **reticulocyte** (re-TIK-yoo-lō-sīt) **count.** Some reticulocytes are normally released into the bloodstream before they become mature red blood cells. If the number of reticulocytes in a sample of blood is less than 0.5 percent of the number of mature red blood cells in the sample, erythropoiesis is occurring too slowly. A low reticulocyte count might confirm a diagnosis of nutritional or pernicious anemia. Or it might indicate a kidney disease that prevents the kidney cells from producing erythropoietin. If the reticulocytes number more than 1.5 percent of the mature red blood cells, erythropoiesis is abnormally rapid. Any number of problems may be responsible for a high reticulocyte count: anemia, oxygen deficiency, and uncontrolled red blood cell production caused by a cancer in the bone marrow. A high count may also indicate that treatment for a condition causing anemia has been effective and the bone marrow is making up for lost time.

Another test with important clinical applications is the hematocrit. **Hematocrit** is the percentage of blood that is made up of red blood cells. It is determined by centrifuging blood and noting the ratio of red blood cells to whole blood. The average hematocrit for males is 47 percent. This means that in 100 ml of blood there are 47 ml of cells and 53 ml of plasma. The average hematocrit for females is 42 percent. Anemic blood may have a hematocrit of 15 percent; polycythemic blood (an abnormal increase in the number of functional erythrocytes) may have a hematocrit of 65 percent. Athletes, however, may have higher than normal hematocrits, reflecting constant physical activity rather than a pathological condition.

Leucocytes

● *Structure and Types*   Unlike red blood cells, **leucocytes** (LOO-kō-sīts), or **white blood cells** (**wbc's**) have nuclei and do not contain hemoglobin (see Figure 19-2). Leucocytes fall into two major groups. The first group is the **granular leucocytes.** They develop from red bone marrow, have granules in the cytoplasm, and possess lobed nuclei. The three kinds of granular leucocytes are **neutrophils** or **polymorphs** (10 to 12 μm in diameter), **eosinophils** (10 to 12 μm in diameter), and **basophils** (8 to 10 μm in diameter).

The second principal group of leucocytes is the **agranular leucocytes.** They develop from lymphoid and myeloid tissue, no cytoplasmic granules can be seen under a light microscope, and their nuclei are usually spherical. The two kinds of agranular leucocytes are **lymphocytes** (7 to 15 μm in diameter) and **monocytes** (14 to 19 μm in diameter).

● *Functions*   The skin and mucous membranes of the body are continuously exposed to microbes and their toxins. Some of these microbes are capable of invading deeper tissues to cause disease. The general function of leucocytes is to combat these microbes once they enter the body by phagocytosis or antibody production. Neutrophils and monocytes are actively **phagocytotic**—they can ingest bacteria and dispose of dead matter (see Figure 4-4). Neutrophils (NOO-trō-fils) are the most active leucocytes in response to tissue destruction by bacteria. In addition to carrying on phagocytosis, they release the enzyme lysozyme, which destroys certain bacteria. Apparently monocytes (MON-ō-sīts) take longer to reach the site of infection than do neutrophils, but once they arrive, they do so in larger numbers and destroy more microbes. Monocytes that have migrated to infected tissues are called **wandering macrophages.** They clean up cellular debris following an infection.

A number of different chemicals in inflamed tissue cause phagocytes to migrate toward the tissue. This phenomenon is called **chemotaxis.** Among the substances that provide stimuli for chemotaxis are toxins and degenerative products of damaged tissues.

Most leucocytes possess, to some degree, the ability to crawl through the minute spaces between the cells that form the walls of capillaries and through connective and epithelial tissue. This movement, like that of amoebas, is called **diapedesis** (dī'-a-pe-DĒ-sis). First, part of the cell membrane stretches out like an arm. Then the cytoplasm and nucleus flow into the projection. Finally, the rest of the membrane snaps up into place. Another projection is made, and so on, until the cell has crawled to its destination (see Figure 4-3b).

Eosinophils (ē'-ō-SIN-ō-fils) are believed to combat the irritants that cause allergies. Eosinophils leave the capillaries, enter the tissue fluid, and produce antihistamines that destroy antigen-antibody complexes. Eosinophils might also be phagocytic.

Basophils (BĀ-sō-fils) are also believed to be involved in allergic reactions. Basophils leave the capillaries, enter the tissues, and become the mast cells of the tissues, liberating heparin, histamine, and serotonin.

Lymphocytes (LIM-fō-sīts) are involved in the production of antibodies. **Antibodies** (AN-ti-bod'-ēz) are special proteins that inactivate antigens. An **antigen** (AN-ti-jen) is any substance that will stimulate the production of antibodies and is capable of reacting specifically with the antibody. Most antigens are proteins, and most are not synthesized by the body. Many of the proteins that make up the cell structures and enzymes of bacteria are antigens. The toxins released by bacteria are also antigens. When antigens enter the body, they react chemically with substances in the lymphocytes and stimulate some lymphocytes, called B cells, to become **plasma cells** (Figure 19-4). The plasma cells then produce antibodies, globulin-type proteins that attach to antigens much as enzymes attach to substrates. Like enzymes, a specific antibody will generally attach only to a certain antigen. However, unlike enzymes, which enhance the reactivity of the substrate, antibodies "cover" their antigens so the antigens cannot come in contact with other chemicals in the body. In this way, bacterial poisons can be sealed up and rendered harmless. The bacteria themselves are destroyed by the antibodies. This process is called the **antigen-antibody response.** Phagocytes in tissues destroy the antigen-antibody complexes. The antigen-antibody response helps us to combat infection and gives us immunity to some diseases. It is also responsible for blood types, allergies, and the body's rejection of organs transplanted from an individual with a different genetic makeup.

CLINICAL APPLICATION

An increase in the number of white cells present in the blood indicates a state of inflammation or infection. Because each type of white cell plays a different role, determining which types have increased aids in diagnosis of the condition. A **differential count** is the number of each kind of white cell in 100 white blood cells. A normal differential count falls within the following percentages.

Neutrophils	60–70%
Eosinophils	2–4%
Basophils	0.5–1%
Lymphocytes	20–25%
Monocytes	3–8%
	100%

Particular attention is paid to the neutrophils in a differential count. More often than not, a high neutrophil count indicates damage by invading bacteria. An increase in the number of monocytes generally indicates a chronic infection. Eosinophils and basophils are elevated during allergic reactions. High lymphocyte counts indicate antigen-antibody reactions.

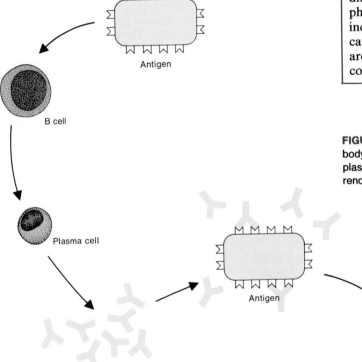

FIGURE 19-4 Antigen-antibody response. An antigen entering the body stimulates a B cell to develop into an antibody-producing plasma cell. The antibodies attach to the antigen, cover it, and render it harmless.

Antigen

B cell

Plasma cell

Antibodies

Antigen

Covered antigen

● *Life Span and Number*   Foreign bacteria exist everywhere in the environment and have continuous access to the body through the mouth, nose, and pores of the skin. Furthermore, many cells, especially those of epithelial tissue, age and die, and their remains must be disposed of daily. Even when the body is healthy, the leucocytes actively ingest bacteria and debris. However, a leucocyte can phagocytose only a certain number of substances before they interfere with the leucocyte's normal metabolic activities and bring on its death. Consequently, the life span of most leucocytes is very short. In a healthy body, some white blood cells will live only a few days. During a period of infection they may live only a few hours.

Leucocytes are far less numerous than red blood cells, averaging from 5,000 to 9,000 cells per cubic millimeter of blood. Red blood cells, therefore, outnumber white blood cells about 700 to 1. The term **leucocytosis** (loo'-kō-sī-TŌ-sis) refers to an increase in the number of white blood cells. If the increase exceeds 10,000, a pathological condition is usually indicated. An abnormally low level of white blood cells (below 5,000/mm³) is termed **leucopenia** (loo-kō-PĒ-nē-a).

● *Production*   Granular leucocytes are produced in red bone marrow (myeloid tissue); agranular leucocytes are produced in both myeloid and lymphoid tissue. The developmental sequences for the five types of leucocytes are shown in Figure 19-2a.

Thrombocytes

● *Structure*   In addition to the immature cell types that develop into erythrocytes and leucocytes, hemocytoblasts differentiate into still another kind of cell, called a megakaryoblast (see Figure 19-2). Megakaryoblasts are transformed into megakaryocytes, large cells that shed fragments of cytoplasm. Each fragment becomes enclosed by a piece of the cell membrane and is called a **thrombocyte** (THROM-bō-sīt) or **platelet.** Platelets are disc-shaped

cells without a nucleus. They average from 2 to 4 μm in diameter.

● *Function*   Platelets prevent fluid loss by initiating a chain of reactions that results in blood clotting. This mechanism is described shortly.

● *Life Span and Number*   Like the other formed elements of the blood, platelets have a short life, probably only 5 to 9 days. Between 250,000 and 400,000 platelets appear in each cubic millimeter of blood.

● *Production*   Platelets are produced in red bone marrow according to the developmental sequence shown in Figure 19-2a.

A summary of the formed elements in blood is presented in Exhibit 19-1.

CLINICAL APPLICATION
The most commonly ordered hematology test is the **complete blood count (CBC)**. It generally includes determination of hemoglobin, hematocrit, red blood cell count, white blood cell count, differential count, and comments about red blood cell, white blood cell, and platelet morphology.

PLASMA

When the formed elements are removed from blood, a straw-colored liquid called **plasma** is left. Exhibit 19-2 outlines the chemical composition of plasma. Note that about 7 percent of the solutes are proteins. Some of these proteins are also found elsewhere in the body, but in blood they are called *plasma proteins*. Albumins, which constitute 55 percent of plasma proteins, are largely responsible for blood's viscosity. The concentration of albumins is about four times higher in plasma than in interstitial fluid. Along with the electrolytes, albumins also help

EXHIBIT 19-1 ▮▮▮▮
SUMMARY OF THE FORMED ELEMENTS IN BLOOD

FORMED ELEMENT	NUMBER	DIAMETER (in μm)	LIFE SPAN	FUNCTION
Erythrocyte (red blood cell)	4.8 million/mm³ in females 5.4 million/mm³ in males	7.7	120 days	Transport oxygen and carbon dioxide.
Leucocyte (white blood cell)	5,000–9,000/mm³		Few hours to a few days	
Granular				
Neutrophil	60–70% of total	10–12		Phagocytosis.
Eosinophil	2–4% of total	10–12		Combat allergens.
Basophil	0.5–1% of total	8–10		Combat allergens.
Agranular				
Lymphocyte	20–25% of total	7–15		Immunity (antigen-antibody reactions).
Monocyte	3–8% of total	14–19		Phagocytosis.
Thrombocyte (platelet)	250,000–400,000/mm³	2–4	5–9 days	Blood clotting.

EXHIBIT 19-2

CHEMICAL COMPOSITION AND DESCRIPTION OF SUBSTANCES IN PLASMA

CONSTITUENT	DESCRIPTION
WATER	Liquid portion of blood; constitutes about 91.5 percent of plasma. Ninety percent of water derived from absorption from digestive tract; 10 percent from cellular respiration. Acts as solvent and suspending medium for solid components of blood and absorbs heat.
SOLUTES	Constitute about 8.5 percent of plasma.
Proteins Albumins	Smallest plasma proteins. Produced by liver and provide blood with viscosity, a factor related to maintenance and regulation of blood pressure. Also exert considerable osmotic pressure to maintain water balance between blood and tissues and regulate blood volume.
Globulins	Protein group to which antibodies belong. Gamma globulins attack measles, hepatitis, and polio viruses and tetanus bacterium.
Fibrinogen	Produced by liver. Plays essential role in clotting.
Nonprotein nitrogen (NPN) substances	Contain nitrogen but are not proteins. Include urea, uric acid, creatine, creatinine, and ammonium salts. Represent breakdown products of protein metabolism and are carried by blood to organs of excretion.
Food substances	Products of digestion passed into blood for distribution to all body cells. Include amino acids (from proteins), glucose (from carbohydrates), and fatty acids and glycerol (from fats).
Regulatory substances	Enzymes, produced by body cells, to catalyze chemical reactions. Hormones, produced by endocrine glands, to regulate growth and development in body.
Respiratory gases	Oxygen and carbon dioxide. These gases are more closely associated with hemoglobin or red blood cells than plasma itself.
Electrolytes	Inorganic salts of plasma. Cations include Na^+, K^+, Ca^{2+}, Mg^{2+}; anions include Cl^-, PO_4^{3-}, SO_4^{2-}, HCO_3^-. Help maintain osmotic pressure, normal pH, physiological balance between tissues and blood.

to regulate blood volume by preventing the water in the blood from diffusing into the interstitial fluid. Recall that water moves by osmosis from an area of high water (low solute) concentration to an area of low water (high solute) concentration. Globulins, which comprise 38 percent of plasma proteins, are antibody proteins released by plasma cells. Gamma globulin is especially well known because it is able to form an antigen-antibody complex with the proteins of the hepatitis and measles viruses and the tetanus bacterium, among others. Fibrinogen makes up about 7 percent of plasma proteins and takes part in the blood-clotting mechanism along with the platelets.

HEMOSTASIS

Hemostasis refers to the stopping of bleeding. When blood vessels are damaged or ruptured, three basic mechanisms operate to prevent blood loss: (1) vascular spasm, (2) platelet plug formation, and (3) blood coagulation (clotting). Although these mechanisms are useful for preventing hemorrhage in smaller blood vessels, extensive hemorrhage usually requires intervention of some type.

VASCULAR SPASM

When a blood vessel is damaged, the smooth muscle in its wall contracts immediately. Such a **vascular spasm** reduces blood loss for up to 30 minutes, during which time the other hemostatic mechanisms can go into operation. The spasm is probably caused by damage to the vessel wall and from reflexes initiated by pain receptors.

PLATELET PLUG FORMATION

When platelets come into contact with parts of a damaged blood vessel, such as collagen or endothelium, their characteristics change drastically. They begin to enlarge and their shapes become even more irregular. They also become sticky and begin to adhere to the collagen fibers. Their synthesis of ADP and enzymes causes the formation of substances that activate more platelets, causing them to stick to the original platelets. The accumulation and attachment of large numbers of platelets form a **platelet plug.** The plug is very effective in preventing blood loss in a small vessel. Although the platelet plug is initially loose, it becomes quite tight when reinforced by fibrin threads formed during coagulation.

COAGULATION

Normally, blood maintains its liquid state as long as it remains in the vessels. If it is drawn from the body, however, it thickens and forms a gel. Eventually, the gel separates from the liquid. The straw-colored liquid, called **serum,** is simply plasma minus its clotting proteins. The gel is called a **clot** and consists of a network of insoluble fibers in which the cellular components of blood are trapped.

The process of clotting is called **coagulation.** If the blood clots too easily, the result can be thrombosis—clotting in an unbroken blood vessel. If the blood takes too long to clot, a hemorrhage can result.

Clotting involves various chemicals known as **coagulation factors.** In plasma, these factors are called *plasma coagulation factors.* A few *platelet coagulation factors* are released by platelets. One coagulation factor is released by damaged body tissues. Coagulation factors are listed in Exhibit 19-3.

Clotting is a complex process, but it can be described basically as a sequence of three stages.

Stage 1. Formation of a substance called thromboplastin.
Stage 2. Conversion of prothrombin, a plasma protein, into thrombin, an enzyme. This stage requires the presence of thromboplastin and several other plasma coagulation factors.
Stage 3. Conversion of fibrinogen, another plasma protein, into insoluble fibrin. The reaction is catalyzed by thrombin.

EXHIBIT 19-3

COAGULATION FACTORS

FACTOR	COMMENTS
PLASMA COAGULATION FACTORS	
I: Fibrinogen	Synthesized in liver. Important factor in stage 3 of clotting, in which it is converted to fibrin. Plasma minus fibrinogen is called serum.
II: Prothrombin	Synthesized in liver; formation requires vitamin K. Important in stage 2 of clotting, in which it is converted to thrombin.
III: Thromboplastin	In extrinsic pathway it is known as extrinsic thromboplastin and is formed from tissue thromboplastin. In intrinsic pathway it is called intrinsic thromboplastin and is formed from platelet disintegration. Formation of thromboplastin signifies end of stage 1 of clotting.
IV: Calcium ions	Apparently involved in all three stages of clotting. Removal of calcium or its binding in plasma prevents coagulation.
V: Proaccelerin or labile factor	Synthesized in liver. Required for stages 1 and 2 of both extrinsic and intrinsic pathways.
VI	No longer used in coagulation theory. Number has not been reassigned.
VII: Serum prothrombin conversion accelerator (SPCA) or stable factor	Synthesized in liver; formation requires vitamin K. Required in stage 1 of extrinsic pathway.
VIII: Antihemophilic factor	Synthesized by liver. Required for stage 1 of intrinsic pathway. Deficiency causes classic hemophilia A.
IX: Christmas factor or plasma thromboplastin component (PTC)	Synthesized by liver; formation requires vitamin K. Required for stage 1 of intrinsic pathway. Deficiency causes hemophilia B.
X: Stuart factor or Stuart-Prower factor	Synthesized by liver; formation requires vitamin K. Required for stages 1 and 2 of extrinsic and intrinsic pathways. Deficiency results in nosebleeds, bleeding into joints, or bleeding into soft tissues.
XI: Plasma thromboplastin antecedent (PTA)	Synthesized by liver. Required for stage 1 of intrinsic pathway. Deficiency results in hemophilia C.
XII: Hageman factor	Required for stage 1 of intrinsic pathway. Known to be activated by contact with glass and may assume role in initiating coagulation outside body.
XIII: Fibrin stabilizing factor (FSF)	Required for stage 3 of clotting.
PLATELET COAGULATION FACTORS	
Pf_1: Platelet factor 1 or platelet accelerator	Essentially same as plasma coagulation factor V.
Pf_2: Platelet factor 2 or thrombin accelerator	Accelerates formation of thrombin in stage 1 of intrinsic pathway and conversion of fibrinogen to fibrin.
Pf_3: Platelet factor 3 or platelet thromboplastic factor	Required for stage 1 of intrinsic pathway.
Pf_4: Platelet factor 4	Binds heparin, an anticoagulant, during clotting.

Fibrin forms the threads of the clot. (Cigarette smoke contains at least two substances that interfere with fibrin formation.)

Depending on whether thromboplastin is released by damaged tissues or formed by platelet disintegration, blood clotting may proceed along one of two routes: the extrinsic pathway or the intrinsic pathway.

Extrinsic Pathway

The **extrinsic pathway** of blood clotting begins when a blood vessel is ruptured (Figure 19-5). The damaged tissues surrounding the blood vessel or the ruptured area of the blood vessel itself release a lipoprotein called *tissue thromboplastin*. Tissue thromboplastin, in reaction with plasma coagulation factors IV, V, VII, and X, forms *extrinsic thromboplastin*. This process is stage 1 of the extrinsic pathway. In stage 2 prothrombin is converted to thrombin. This conversion requires extrinsic thromboplastin and several plasma coagulating factors such as IV, V, VII, and X. In stage 3 fibrinogen, which is soluble, is converted to fibrin, which is insoluble, by the thrombin formed in stage 2. This reaction requires plasma coagulation factors IV and XIII. Except for a few initial steps of the extrinsic pathway, all other stages require calcium ions (Ca^{2+}).

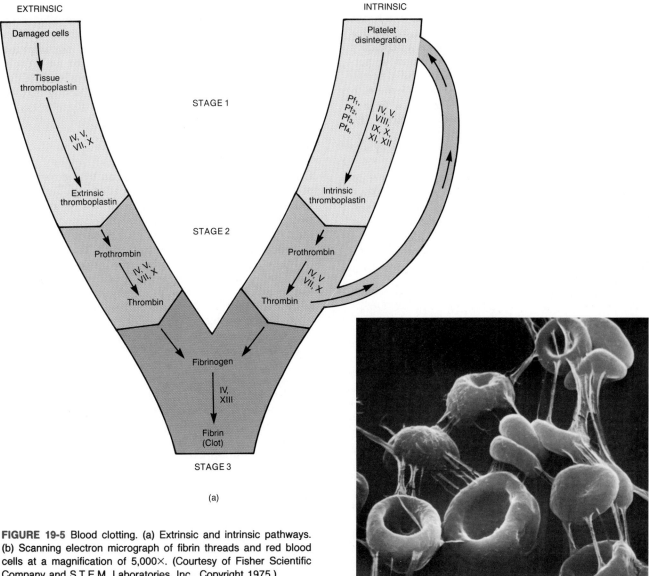

FIGURE 19-5 Blood clotting. (a) Extrinsic and intrinsic pathways. (b) Scanning electron micrograph of fibrin threads and red blood cells at a magnification of 5,000×. (Courtesy of Fisher Scientific Company and S.T.E.M. Laboratories, Inc., Copyright 1975.)

Intrinsic Pathway

The **intrinsic pathway** of clotting begins with the rough surface created by a ruptured vessel (Figure 19-5). Under normal conditions the negatively charged platelet membrane and negatively charged endothelial lining of a blood vessel repel each other. Thus platelets do not adhere to the endothelial lining. However, a ruptured blood vessel results in the loss of the negative charge by the endothelial lining. Thus the platelets adhere to the ruptured area. This clumping together of platelets causes them to disintegrate and release their platelet coagulation factors into the plasma. The clumping reaction can also plug up small injuries without initiating the clotting sequence.

In stage 1 of the intrinsic pathway, four platelet coagulation factors (Pf_1, Pf_2, Pf_3, Pf_4), in reaction with seven plasma coagulation factors (IV, V, VIII, IX, X, XI, XII), form *intrinsic thromboplastin*. In stage 2 prothrombin is converted to thrombin by intrinsic thromboplastin and several plasma coagulating factors, including IV, V, VII, and X. Stage 3 of the intrinsic pathway is the same as

that for the extrinsic pathway. It involves the conversion of fibrinogen to fibrin in the presence of thrombin and plasma coagulation factors IV and XIII. All stages of the intrinsic pathway require Ca^{2+} ions.

Besides helping to convert fibrinogen to fibrin, thrombin causes more platelets to adhere to one another, disintegrate, and release even more platelet coagulation factors. This cyclic feature of the intrinsic pathway ensures continual platelet disintegration until the clot is formed. Once the clot is formed, it plugs the ruptured area of the blood vessel and thus prevents hemorrhage (bleeding). Permanent repair of the blood vessel can then take place. In time, fibroblasts form connective tissue in the ruptured area and new endothelial cells repair the lining.

CLINICAL APPLICATION

Hemophilia (*hemo* = blood; *philein* = to love) refers to several different hereditary deficiencies of coagulation in which bleeding may occur spontaneously or after only minor trauma. The effects of all the forms of the disorder are so similar that they are hardly distinguishable from one another, but each is a deficiency of a different blood clotting factor. For example, persons with the most common type of hemophilia, hemophilia A (classic hemophilia), lack factor VIII, the antihemophilic factor. Persons with hemophilia B lack factor IX. Those with hemophilia C lack factor XI. Hemophilia A and B occur primarily among males; hemophilia C is a mild form affecting both males and females.

Hemophilia is characterized by spontaneous or traumatic subcutaneous and intramuscular hemorrhaging, nosebleeds, blood in the urine, and joint pain and damage due to joint hemorrhaging. Treatment involves the application of pressure to accessible bleeding sites and transfusions of fresh plasma or the appropriate deficient clotting factor to relieve the bleeding tendency for several days.

Retraction and Fibrinolysis

Normal coagulation involves two additional events after clot formation: clot retraction and fibrinolysis. **Clot retraction** or **syneresis** (si-NER-e-sis) is the consolidation or tightening of the fibrin clot. The fibrin threads attached to the damaged surfaces of the blood vessel gradually contract. As the clot retracts, the ruptured area of the blood vessel gets smaller. Thus the risk of hemorrhage is further decreased. During retraction, some serum escapes between the fibrin threads, but the formed elements in blood remain trapped in the fibrin threads. Normal clot retraction depends on an adequate number of platelets. Apparently, platelets in the clot disintegrate and, through the intrinsic pathway, form more fibrin.

The second event following clot formation—**fibrinolysis** (fī-brin-OL-i-sis)—involves dissolution of the blood clot. Once the blood vessel is repaired, the clot dissolves. At the beginning of the coagulation pathway, certain enzymes are released by damaged tissues. These enzymes activate an inactive plasma enzyme called *plasminogen* into an active form called *plasmin*. Plasmin dissolves the fibrin clot.

CLINICAL APPLICATION

Physicians have now developed a **technique for dissolving clots** by administering fibrinolytic enzymes (streptokinase and urokinase) via catheterization. These enzymes are introduced by means of a catheter at the site of the blood clots and exert their effect by converting plasminogen into plasmin, thus dissolving the clot. Such fibrinolytic therapy has been shown to be effective for blood clots in the lungs and deep veins and seems promising for dissolution of blood clots in coronary arteries.

Clot formation is a vital mechanism that prevents excessive loss of blood from the body. To form clots, the body needs calcium and vitamin K. Vitamin K is not involved in the actual clot formation, but it is required for the synthesis of prothrombin (factor II) and factors VII, IX, and X. The vitamin is normally produced by bacteria that live in the intestine. Because it is fat-soluble, it can be absorbed through the mucosa of the intestine and into the blood only if it is attached to fat. People suffering from disorders that prevent absorption of fat often experience uncontrolled bleeding. Clotting may be encouraged by applying a thrombin or fibrin spray, a rough surface such as gauze, or heat.

Prevention

Unwanted clotting may be brought on by the formation of cholesterol-containing masses called *plaques* in the walls of the blood vessels. They result in a rough surface that is perfect for the adhesion of platelets and is often the site of clotting. Clotting in an unbroken blood vessel is called **thrombosis.** The clot itself is a **thrombus** (*thrombo* = clot). A thrombus may dissolve spontaneously, but if it remains intact, it may damage tissues by cutting off the oxygen supply. Equally serious is the possibility that the thrombus will become dislodged and be carried with the blood to a smaller vessel. In a smaller vessel, the clot may block the circulation to a vital organ. A blood clot, bubble of air, fat from broken bones, or a piece of debris transported by the bloodstream is called an **embolus** (*em* = in; *bolus* = a mass). When an embolus becomes lodged in a vessel and cuts off circulation, the condition is called an **embolism.**

No matter how healthy the body is, occasional rough spots appear on uncut vessel walls. In fact, it is believed that blood clotting is a continuous process inside blood vessels that is continually combated by clot-preventing and clot-dissolving mechanisms. Blood contains antithrombic substances—substances that prevent thrombin formation.

In general, any chemical substance that prevents clotting is an **anticoagulant.** *Heparin* is a quick-acting anticoagulant that blocks the clotting mechanism by inhibiting the conversion of prothrombin to thrombin and prevents most thrombus formation. It is used in open-heart surgery to prevent clotting. It is extracted from donated bovine lung tissue plus bovine and porcine intestinal mucosa. The pharmaceutical preparation *dicumarol* may be given to patients who are thrombosis-prone. Dicumarol acts as an antagonist to vitamin K and thus lowers the level of prothrombin. Dicumarol is slower acting than heparin and is used primarily as a preventative. *CPD* (*citrate phosphate dextrose*), *ACD* (*acid citrate dextrose*), and *EDTA* (*ethylenediamine tetracetic acid*) are used by laboratories and blood banks to prevent blood samples from clotting. These substances react with Ca^{2+} to form insoluble compounds. In this way, the blood Ca^{2+} is tied up and is no longer free to catalyze the conversion of prothrombin to thrombin.

Tests

● *Clotting Time* The time required for blood to coagulate, usually from 5 to 15 minutes, is known as **clotting time.** This time is used as an index of a person's blood-clotting properties. One method for determining clotting time involves taking a sample of blood from a vein and

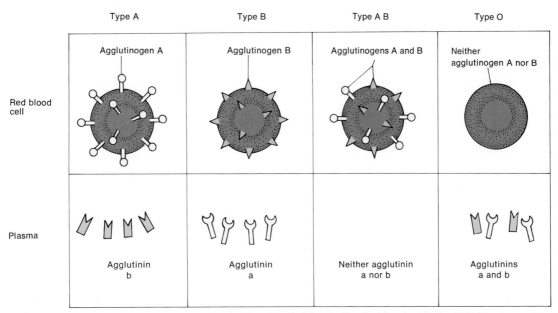

FIGURE 19-6 Agglutinogens (antigens) and agglutinins (antibodies) involved in the ABO blood grouping system.

placing 1 ml into each of three Pyrex tubes. The tubes are then submerged in a water bath at 37°C (98.6°F) and examined every 30 seconds for clot formation. Clotting is initiated when the platelets break up upon coming into contact with the glass. When the clot adheres to the tube, the end point is reached and the time is recorded. Blood taken from individuals with hemophilia clots very slowly or not at all.

● *Bleeding Time*   **Bleeding time** is the time required for the cessation of bleeding from a small skin puncture. This time is usually measured by puncturing the ear lobe. As the droplets of blood escape, they are blotted by gently touching the wound with filter paper. When the paper is no longer stained, the bleeding has stopped. Normally, bleeding time varies from 1 to 4 minutes. Unlike clotting time, which involves only the breakdown of platelets, bleeding time also involves constriction of injured blood vessels and all the stages of clot formation.

● *Prothrombin Time*   **Prothrombin time** is a test used to determine the amount of prothrombin in the blood. A blood sample is treated with CPD or a similar compound to tie up the Ca^{2+} and make the blood incoagulable. Then Ca^{2+}, thromboplastin, and plasma coagulation factors V and VII are all mixed with the blood sample. The length of time required for the blood to clot is the prothrombin time and depends on the amount of prothrombin in the sample. Normal prothrombin time is about 12 seconds.

GROUPING (TYPING)

The surfaces of erythrocytes contain genetically determined antigens called **agglutinogens** (ag'-loo-TIN-ō-jens) or **isoantigens.** Two major blood group classifications—ABO grouping system and the Rh system—are based on the presence or absence of these proteins.

ABO

The *ABO blood grouping* is based on two agglutinogens symbolized as A and B (Figure 19-6). Individuals whose erythrocytes manufacture only agglutinogen A are said to have blood type A. Those who manufacture only agglutinogen B are type B. Individuals who manufacture both A and B are type AB. Those who manufacture neither are type O.

These four blood types are not equally distributed. The incidence in the white population in the United States: type A, 41 percent; type B, 10 percent; type AB, 4 percent; type O, 45 percent. Among blacks, the frequencies are: type A 27 percent; type B, 20 percent; type AB, 7 percent; and type O, 46 percent.

The blood plasma of many people contains genetically determined antibodies referred to as **agglutinins** (a-GLOO-ti-nins) or **isoantibodies.** These are agglutinin a (anti-A), which attacks agglutinogen A, and agglutinin b (anti-B), which attacks B. The agglutinins formed by each of the four blood types are shown in Figure 19-6. You do not have agglutinins that attack the agglutinogens of your own erythrocytes, but you do have an agglutinin against any agglutinogen you do not synthesize. For example, the plasma of type B blood contains agglutinin a. Suppose type A blood is accidentally given to a person who does not have A agglutinogens (type B or type O). The person's body recognizes that the A protein is foreign and therefore treats it as an antigen. Agglutinin a's attack the foreign erythrocytes and cause them to *agglutinate* (clump). The degree of agglutination depends on the titer (strength or amount) of agglutinin in the blood. This reaction is another example of an antigen-antibody response (Figure 19-7).

When blood is given to a patient, care must be taken to ensure that the individual's agglutinins will not agglutinate the donated erythrocytes and cause clumping. Destruction of the donated cells will not only undo the work

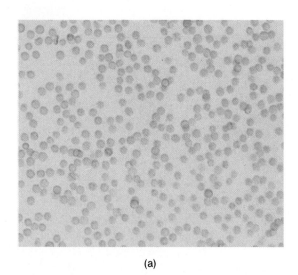

(a)

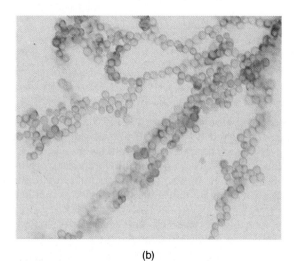

(b)

FIGURE 19-7 Compatibility of blood types as viewed through a microscope. (a) Since the blood type shown has no agglutinins to attack the agglutinogens, no agglutination occurs. (b) But if a blood type has incompatible agglutinins, they will attack the agglutinogens and agglutination occurs. (Courtesy of Lester V. Bergman & Associates.)

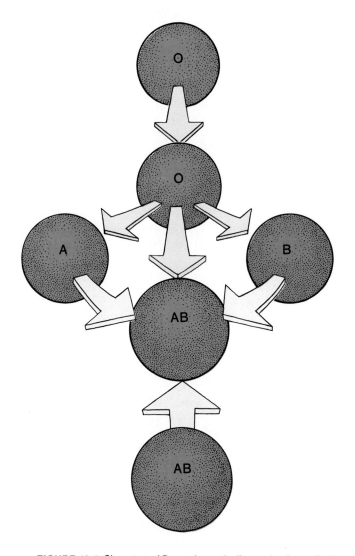

FIGURE 19-8 Since type AB can theoretically receive from all other blood types, it is called the universal recipient. Type O is called the universal donor since it can theoretically give to all other blood types.

of the transfusion, but the clumps can block vessels and may lead to kidney damage and death.

Type A blood will agglutinate when mixed with types B and O. It may be given to individuals whose blood type is A or AB, but never to those whose blood type is B or O. Type B blood will agglutinate when mixed with types A and O. Thus it may be given to individuals whose blood type is B or AB, but never to individuals whose blood type is either A or O. Type AB blood will agglutinate when mixed with blood types A, B, or O. It may be given to individuals whose blood type is AB only. No other blood types may receive it. Type AB individuals requiring blood may be given types AB, A, B, and O—type AB blood has neither a nor b agglutinins.

EXHIBIT 19-4

INTERACTIONS OF CELLS AND PLASMA OF ABO SYSTEM

BLOOD TYPE	AGGLUTINOGEN (ANTIGEN)	AGGLUTININ (ANTIBODY)	PLASMA CAUSES AGGLUTINATION OF	CELLS AGGLUTINATED BY PLASMA OF
A	A	b	B, AB	B, O
B	B	a	A, AB	A, O
AB	A, B	Neither a nor b	None	A, B, O
O	Neither A nor B	a, b	A, B, AB	None

Since type AB blood can theoretically receive all other blood types, it is called the **universal recipient** (Figure 19-8). Type O blood will not agglutinate with any of the other types. Theoretically, it may be given to individuals whose blood type is O, A, B, and AB. Since type O blood has no agglutinogens to serve as antigens in another person's body, it is known as the **universal donor.** Type O persons requiring blood may receive only type O blood. In practice, only matching blood types are used for transfusions.

> **CLINICAL APPLICATION**
> Scientists recently converted type B blood into type O blood by using an enzyme called alpha galactosidase. The enzyme is isolated from green coffee beans and has the ability to remove a sugar from the surface of red blood cells, thus **changing type B into type O blood.** Preliminary tests on animals and a few human volunteers showed no adverse reactions. It is hoped that further testing will confirm the safety of the converted blood and that other conversions can be made so that blood banks could juggle their supplies as needed.

Exhibit 19-4 summarizes the interreactions of the four blood types.

Rh

The **Rh system** of blood classification is so named because it was first worked out in the blood of the *Rhesus* monkey. Like the ABO grouping, the Rh system is based on agglutinogens that lie on the surfaces of erythrocytes. Individuals whose erythrocytes have the Rh agglutinogens are designated Rh^+. Those who lack Rh agglutinogens are designated Rh^-. It is estimated that 85 percent of whites and 88 percent of blacks in the United States are Rh^+, whereas 15 percent of whites and 12 percent of blacks are Rh^-.

Under normal circumstances, human plasma does not contain anti-Rh agglutinins. However, if an Rh^- person receives Rh^+ blood, the body starts to make anti-Rh agglutinins that will remain in the blood. If a second transfusion of Rh^+ blood is given later, the previously formed anti-Rh agglutinins will react against the donated blood and a severe reaction may occur.

> **CLINICAL APPLICATION**
> One of the most common problems with Rh incompatibility arises from pregnancy (Figure 19-9). During pregnancy, some of the fetus's blood may leak from

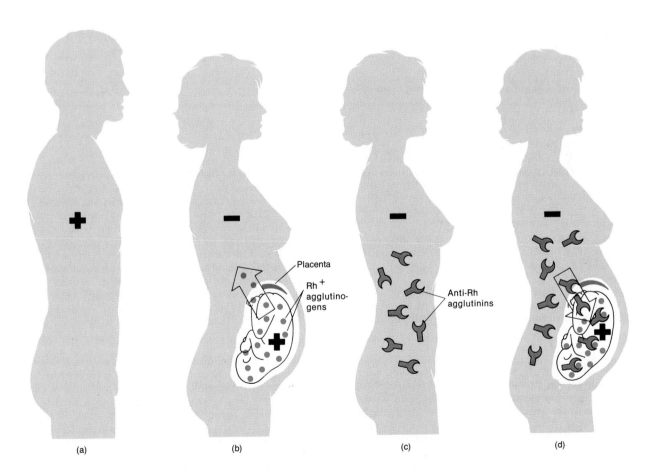

FIGURE 19-9 Development of erythroblastosis fetalis. (a) Rh^+ father. (b) Rh^- mother and Rh^+ fetus. If an Rh^- female is impregnated by an Rh^+ male and the fetus is Rh^+, fetal Rh^+ agglutinogens may enter the maternal blood via the placenta during delivery. (c) Upon exposure to the fetal Rh^+ agglutinogens, the mother will make anti-Rh agglutinins. (d) If the female becomes pregnant again, her anti-Rh agglutinins will cross the placenta into the fetal blood. If the fetus is Rh^+, erythroblastosis fetalis will result.

the placenta (afterbirth) into the mother's bloodstream. If the fetus is Rh$^+$ and the woman is Rh$^-$, she, upon exposure to the Rh$^+$ fetal cells, will make anti-Rh agglutinins. If the woman becomes pregnant again, her anti-Rh agglutinins will cross the placenta and make their way into the bloodstream of the baby. If the fetus is Rh$^-$, no problem will occur, since Rh$^-$ blood does not have the Rh agglutinogen. If the fetus is Rh$^+$, an antigen-antibody response called **hemolysis** (*lysis =* dissolve) may occur in the fetal blood. Hemolysis means a swelling and subsequent rupture of erythrocytes resulting in the liberation of hemoglobin. The hemolysis brought on by fetal-maternal incompatibility is called **erythroblastosis fetalis** or **hemolytic disease of newborn.** When a baby is born with this condition, all the blood is slowly removed and replaced with agglutinin-free blood. It is even possible to transfuse blood into the unborn child if the disease is suspected. More important, though, is the fact that the disorder can be prevented with an injection of an anti-Rh gamma$_2$-globulin agglutinin preparation, administered to Rh$^-$ mothers right after delivery or abortion. These agglutinins tie up the fetal agglutinogens so the mother cannot respond to the foreign agglutinogens by producing agglutinins. Thus the fetus of the next pregnancy is protected. In the case of an Rh$^+$ mother and an Rh$^-$ child, there are no complications, since the fetus cannot make agglutinins.

INTERSTITIAL FLUID AND LYMPH

Whole blood does not flow into the tissue spaces; it remains in closed vessels. However, certain constituents of the plasma do move through the capillary walls, and once they move out of the blood, they are called interstitial fluid. Interstitial fluid and lymph are basically the same. The major difference between the two is location. When the fluid bathes the cells, it is called **interstitial fluid, intercellular fluid,** or **tissue fluid.** When it flows through the lymphatic vessels, it is called **lymph.**

Both fluids are similar in composition to plasma. The principal chemical difference is that they contain less protein than plasma because the larger protein molecules are not easily filtered through the cells that form the walls of the capillaries. The transfer of materials between blood and interstitial fluid occurs by osmosis, diffusion, and filtration across the cells that make up the capillary walls.

Interstitial fluid and lymph also differ from plasma in that they contain variable numbers of leucocytes. Leucocytes can enter the tissue fluid by diapedesis, and the lymphoid tissue itself is one of the sites of agranular leucocyte production. Like plasma, interstitial fluid and lymph lack erythrocytes and platelets.

Other substances, especially organic molecules, in interstitial fluid and lymph vary in relation to the location of the sample analyzed. The lymph vessels that drain the organs of the digestive tract, for example, contain a great deal of lipid absorbed from food.

DISORDERS: HOMEOSTATIC IMBALANCES

Anemia

Anemia is a sign, not a diagnosis. Many kinds of anemia exist, all characterized by insufficient erythrocytes or hemoglobin. These conditions lead to fatigue and intolerance to cold, both of which are related to lack of oxygen needed for energy and heat production, and to paleness, which is due to low hemoglobin content.

Nutritional Anemia

Nutritional anemia arises from an inadequate diet, one that provides insufficient amounts of iron, the necessary amino acids, or vitamin B$_{12}$.

Pernicious Anemia

Pernicious anemia is the insufficient production of erythrocytes resulting from an inability of the body to produce intrinsic factor.

Hemorrhagic Anemia

An excessive loss of erythrocytes through bleeding is called **hemorrhagic anemia.** Common causes are large wounds, stomach ulcers, and heavy menstrual bleeding. If bleeding is extraordinarily heavy, the anemia is termed acute. Excessive blood loss can be fatal. Slow, prolonged bleeding is apt to produce a chronic anemia; the chief symptom is fatigue.

Hemolytic Anemia

If erythrocyte cell membranes rupture prematurely, the cells remain as "ghosts" and their hemoglobin pours out into the plasma. A characteristic sign of this condition, called **hemolytic anemia,** is distortion in the shape of erythrocytes that are pro-

gressing toward hemolysis. There may also be a sharp increase in the number of reticulocytes, since the destruction of red blood cells stimulates erythropoiesis.

The premature destruction of red blood cells may result from inherent defects, such as hemoglobin defects, abnormal red blood cell enzymes, or defects of the red blood cell membrane. Agents that may cause hemolytic anemia are parasites, toxins, and antibodies from incompatible blood (Rh$^-$ mother and Rh$^+$ fetus, for instance). *Erythroblastosis fetalis* is an example of a hemolytic anemia.

The term **thalassemia** (thal'-a-SĒ-mē-a) represents a group of hereditary hemolytic anemias, resulting from a defect in the synthesis of hemoglobin, which produces extremely thin and fragile erythrocytes. It occurs primarily in populations from countries bordering the Mediterranean Sea. Treatment generally consists of blood transfusions.

Aplastic Anemia

Destruction or inhibition of the red bone marrow results in **aplastic anemia.** Typically, the marrow is replaced by fatty tissue, fibrous tissue, or tumor cells. Toxins, gamma radiation, and certain medications are causes. Many of the medications inhibit the enzymes involved in hemopoiesis. Bone marrow transplants can now be done with a reasonable hope of success in patients with aplastic anemia. They are done early, before the victim is sensitized by transfusions. Immunosuppressive drugs are given for 4 days before the transplant and with decreasing frequency for 100 days afterward.

Sickle Cell Anemia

The erythrocytes of a person with **sickle cell anemia** manufac-

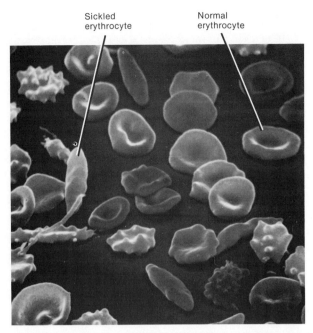

Sickled   Normal
erythrocyte   erythrocyte

FIGURE 19-10 Scanning electron micrograph of erythrocytes in sickle cell anemia at a magnification of 5,000×. (Courtesy of Fisher Scientific Company and S.T.E.M. Laboratories, Inc., Copyright 1975.)

ture an abnormal kind of hemoglobin. When an erythrocyte gives up its oxygen to the interstitial fluid, its hemoglobin tends to lose its integrity in places of low oxygen tension and forms long, stiff, rodlike structures that bend the erythrocyte into a sickle shape (Figure 19-10). The sickled cells rupture easily. Even though erythropoiesis is stimulated by the loss of the cells, it cannot keep pace with the hemolysis. The individual consequently suffers from a hemolytic anemia that reduces the amount of oxygen that can be supplied to the tissues. Prolonged oxygen reduction may eventually cause extensive tissue damage. Furthermore, because of the shape of the sickled cells, they tend to get stuck in blood vessels and can cut off blood supply to an organ altogether.

Sickle cell anemia is inherited. The gene responsible for the tendency of the erythrocytes to sickle during hypoxia also seems to prevent erythrocytes from rupturing during a malarial crisis. The gene also alters the permeability of the plasma membranes of sickled cells, causing potassium to leak out. Low levels of potassium kill the malarial parasites that infect sickled cells. Sickle cell genes are found primarily among populations, or descendants of populations, that live in the malaria belt around the world, including parts of Mediterranean Europe and sub-

tropical Africa and Asia. A person with only one of the sickling genes is said to have sickle cell trait. Such an individual has a high resistance to malaria—a factor that may have tremendous survival value—but does not develop the anemia. Only people who inherit a sickling gene from both parents get sickle cell anemia.

Polycythemia

The term **polycythemia** (pol'-ē-sī-THĒ-mē-a) refers to an abnormal increase in the number of red blood cells. The hematocrit is an important procedure in diagnosing the condition. Increases of 2 to 3 million cells per cubic millimeter above normal are considered to be polycythemic. The blood's viscosity is greatly increased because of the extra red blood cells. The increased viscosity causes a rise in blood pressure, and contributes to thrombosis and hemorrhage. The thrombosis results from too many red blood cells piling up as they try to enter smaller vessels. The hemorrhage is due to widespread hyperemia (unusually large amount of blood in an organ part).

Infectious Mononucleosis

Infectious mononucleosis is a contagious disease caused by a virus. It occurs mainly in children and young adults. Its trademark is an elevated white blood cell count with an abnormally high percentage of lymphocytes and mononucleocytes. An increase in the number of monocytes usually indicates a chronic infection. Symptoms include slight fever, sore throat, brilliant red throat and soft palate, stiff neck, cough, and malaise. The spleen may enlarge. Secondary complications involving the liver, heart, kidneys, and nervous system may develop. There is no cure for mononucleosis, and treatment consists of watching for and treating complications. Usually the disease runs its course in a few weeks, and the individual generally suffers no permanent ill effects.

Leukemia

Also called "cancer of the blood," **leukemia** is an uncontrolled, greatly accelerated production of white blood cells. Many of the cells fail to reach maturity. As with most cancers, the symptoms result not so much from the cancer cells themselves as from their interference with normal body processes. The anemia and bleeding problems commonly seen in leukemia result from the crowding out of normal bone marrow cells, preventing normal production of red blood cells and platelets. The most common cause of death from leukemia is internal hemorrhaging, especially cerebral hemorrhage that destroys the vital centers in the brain. Another frequent cause of death is uncontrolled infection owing to lack of mature or normal white blood cells. The abnormal accumulation of leucocytes may be reduced by using x-rays and antileukemic drugs. Partial or complete remissions may be induced, with some lasting as long as 15 years.

MEDICAL TERMINOLOGY

Blood plasma substitute   This substance mimics the characteristics of plasma. It is used to maintain blood volume during emergency conditions (such as hemorrhage) until blood can be matched or to prevent dehydration if a patient cannot swallow liquids. It is also used to replace fluid and electrolytes after loss of blood during surgery.

Citrated whole blood   Whole blood protected from coagulation by CPD or a similar compound.

Corpuscles (*corpus* = body)   Cellular elements in the blood such as red and white blood cells.

Direct (immediate) transfusion (*trans* = through)   Transfer of blood directly from one person to another without exposing the blood to air.

Exchange transfusion   Removing blood from the recipient while simultaneously replacing it with donor blood. This method is used for erythroblastosis fetalis and poisoning.

Fibrinogen   A plasma protein; also, a sterile, freeze-dried preparation of this component of normal plasma. In solution, the preparation can be converted to insoluble fibrin by adding thrombin. It can be applied to wounds to stop bleeding and hemorrhagic disorders caused by fibrinogen deficiency.

Fractionated blood (*fract* = break)   Blood that has been separated into its components. Only the part needed by the patient is given.

Gamma globulin (immune serum globulin)   Solution of globulins from nonhuman blood consisting of antibodies that react with specific pathogens, such as measles, epidemic hepatitis, tetanus, and possibly poliomyelitis viruses. It is prepared by injecting the specific virus into animals, removing blood from the animals after antibodies have accumulated, isolating antibodies, and injecting them into a human for short-term immunity.

Hem, Hemo, Hema, Hemato (*heme* = iron)   Various combining forms meaning blood.

Hemorrhage (*rrhage* = bursting forth)   Bleeding, either internal (from blood vessels into tissues) or external (from blood vessels directly to the surface of the body).

Heparinized whole blood   Whole blood in a heparin solution, which prevents coagulation.

Indirect (mediate) transfusion   Transfer of blood from a donor to a container and then to the recipient, permitting blood to be stored for an emergency. The blood may be separated into its components so that a patient will receive only a needed part.

Normal plasma   Cell-free plasma containing normal concentrations of all solutes. It is used to bring blood volume up to normal when excessive numbers of blood cells have not been lost.

Platelet concentrates   A preparation of platelets obtained from freshly drawn whole blood and used for transfusions in platelet-deficiency disorders such as hemophilia.

Reciprocal transfusion   Transfer of blood from a person who has recovered from a contagious infection into the vessels of a patient suffering with the same infection. An equal amount of blood is returned from the patient to the well person. This method allows the patient to receive antibody-bearing lymphocytes from the recovered person.

Septicemia (*sep* = decay; *emia* = condition of blood)   Toxins or disease-causing bacteria in the blood. Also called "blood poisoning."

Thrombocytopenia (*thrombo* = clot; *penia* = poverty)   Very low platelet count that results in a tendency to bleed from capillaries.

Transfusion   Transfer of whole blood, blood components (red blood cells only or plasma only), or bone marrow directly into the bloodstream.

Venesection (*veno* = vein)   Opening of a vein for withdrawal of blood.

Whole blood   Blood containing all formed elements, plasma, and plasma solutes in natural concentration.

STUDY OUTLINE

Physical Characteristics (p. 436)

1. The cardiovascular system consists of blood, the heart, and blood vessels. The lymphatic system consists of lymph, lymph vessels, and lymph glands.
2. Physical characteristics of blood include viscosity, 4.5 to 5.5; temperature, 38°C (100.4°F); pH, 7.35 to 7.45; and salinity, 0.85 to 0.90 NaCl. Blood constitutes about 8 percent of body weight.

Functions (p. 436)

1. Blood transports oxygen, carbon dioxide, nutrients, wastes, hormones, and enzymes.
2. It helps to regulate pH, body temperature, and water content of cells.
3. It prevents excessive fluid loss through clotting.
4. It protects against toxins and microbes.

Components (p. 436)

1. The formed elements in blood include erythrocytes (red blood cells), leucocytes (white blood cells), and thrombocytes (platelets).
2. Blood cells are formed by a process called hemopoiesis.
3. Red bone marrow (myeloid tissue) is responsible for producing red blood cells, granular leucocytes, and platelets; lymphoid tissue and myeloid tissue produce agranular leucocytes.

Erythrocytes

1. Erythrocytes are biconcave discs without nuclei and containing hemoglobin.
2. The function of red blood cells is to transport oxygen and carbon dioxide.
3. Red blood cells live about 120 days. A healthy male has about 5.4 million/mm³ of blood; a healthy female, about 4.8 million/mm³.
4. Erythrocyte formation, called erythropoiesis, occurs in adult red marrow of certain bones.
5. A reticulocyte count is a diagnostic test that indicates the rate of erythropoiesis.
6. A hematocrit measures the percentage of red blood cells in whole blood.

Leucocytes

1. Leucocytes are nucleated cells. Two principal types are granular (neutrophils, eosinophils, basophils) and agranular (lymphocytes and monocytes).
2. The general function of leucocytes is to combat inflammation and infection. Neutrophils and monocytes (wandering macrophages) do so through phagocytosis.
3. Eosinophils and basophils are believed to be involved in combating allergic reactions.
4. Lymphocytes, in response to the presence of foreign substances called antigens, differentiate into tissue plasma cells which produce antibodies. Antibodies attach to the antigens and render them harmless. This antigen-antibody response combats infection and provides immunity.
5. A differential count is a diagnostic test in which white blood cells are enumerated.
6. White blood cells usually live for only a few hours or a few days. Normal blood contains 5,000 to 9,000/mm³.

Thrombocytes

1. Thrombocytes are disc-shaped structures without nuclei.
2. They are formed from megakaryocytes and are involved in clotting.
3. Normal blood contains 250,000 to 400,000/mm³.

Plasma

1. The liquid portion of blood, called plasma, consists of 91.5 percent water and 8.5 percent solutes.
2. Principal solutes include proteins (albumins, globulins, fibrinogen), nonprotein nitrogen (NPN) substances, foods, enzymes and hormones, respiratory gases, and electrolytes.

Hemostasis (p. 443)

1. Hemostasis refers to the prevention of blood loss.
2. It involves vascular spasm, platelet plug formation, and blood coagulation.
3. In vascular spasm, the smooth muscle of a blood vessel wall contracts to stop bleeding.
4. Platelet plug formation involves the clumping of platelets to stop bleeding.

5. A clot is a network of insoluble protein (fibrin) in which formed elements of blood are trapped.
6. The chemicals involved in clotting are known as coagulation factors. There are two kinds: plasma and platelet coagulation factors.
7. Blood clotting involves two pathways: the intrinsic and the extrinsic.
8. Normal coagulation also involves clot retraction (tightening of the clot) and fibrinolysis (dissolution of the clot).
9. Clotting in an unbroken blood vessel is called thrombosis. A thrombus that moves from its site of origin is called an embolus.
10. Anticoagulatants (e.g., heparin) prevent clogging.
11. Clinically important clotting tests are clotting time (time required for blood to coagulate), bleeding time (time required for the cessation of bleeding from a small skin puncture), and prothrombin time (time required for the blood to coagulate, which depends on the amount of prothrombin in the blood sample).

Grouping (Typing) (p. 447)

1. ABO and Rh systems are based on antigen-antibody responses.
2. In the ABO system, agglutinogens (antigens) A and B determine blood type. Plasma contains agglutinins (antibodies) that clump agglutinogens which are foreign to the individual.

3. In the Rh system, individuals whose erythrocytes have Rh agglutinogens are classified as Rh^+. Those who lack the antigen are Rh^-.
4. A disorder due to Rh incompatibility between mother and fetus is called erythroblastosis fetalis.

1. Interstitial fluid bathes body cells, whereas lymph is found in lymphatic vessels.
2. These fluids are similar in chemical composition. They differ chemically from plasma in that both contain less protein and a variable number of leucocytes. Like plasma, they contain no platelets or erythrocytes.

Disorders: Homeostatic Imbalances (p. 450)

1. Anemia is a decreased erythrocyte count or hemoglobin deficiency. Kinds of anemia include nutritional, pernicious, hemorrhagic, hemolytic, aplastic, and sickle cell anemia.
2. Polycythemia is an abnormal increase in the number of erythrocytes.
3. Infectious mononucleosis is characterized by an elevated white cell count, especially lymphocytes and mononucleocytes. The cause is a virus.
4. Leukemia is the uncontrolled production of white blood cells that interferes with normal clotting and vital body activities.

REVIEW QUESTIONS

1. How are blood, interstitial fluid, and lymph related to the maintenance of homeostasis?
2. Distinguish between the cardiovascular system and lymphatic system.
3. List the principal physical characteristics of blood.
4. List the functions of blood and their relationship to other systems of the body.
5. Distinguish between plasma and formed elements.
6. Describe the origin of blood cells.
7. Describe the microscopic appearance of erythrocytes. What is the essential function of erythrocytes?
8. Define erythropoiesis. Relate erythropoiesis to red blood cell count. What factors accelerate and decelerate erythropoiesis?
9. What is a reticulocyte count? What is its diagnostic significance?
10. Define hematocrit. Compare the hematocrits of anemic and polycythemic blood.
11. Describe the classification of leucocytes. What are their functions?
12. What is the importance of diapedesis, chemotaxis, and phagocytosis in fighting bacterial invasion?
13. What is a differential count? What is its significance?
14. Distinguish between leucocytosis and leucopenia.
15. Describe the antigen-antibody response. How is it protective?
16. Describe the structure and function of thrombocytes.
17. Compare erythrocytes, leucocytes, and thrombocytes with

respect to: size, number per mm^3, and life span.
18. What are the major constituents in plasma? What do they do?
19. What is the difference between plasma and serum?
20. Define hemostasis. Explain the mechanism involved in vascular spasm and platelet plug formation.
21. Briefly describe the process of clot formation. What is fibrinolysis? Why does blood usually not remain clotted in vessels?
22. List the various coagulation factors.
23. What are the pathways involved in blood clotting?
24. Define the following: thrombus, embolus, anticoagulant, clotting time, bleeding time, prothrombin time.
25. What is the basis for ABO blood grouping? What are agglutinogens and agglutinins?
26. What is the basis for the Rh system? How does erythroblastosis fetalis occur? How may it be prevented?
27. Compare interstitial fluid and lymph with regard to location, chemical composition, and function.
28. Define anemia. Contrast the causes of nutritional, pernicious, hemorrhagic, hemolytic, aplastic, and sickle cell anemias.
29. What is infectious mononucleosis?
30. What is leukemia, and what is the cause of some of its symptoms?
31. Refer to the glossary of medical terminology associated with blood. Be sure that you can define each term.

STUDENT OBJECTIVES

- Identify the location of the heart in the mediastinum.
- Distinguish between the structure and location of the fibrous and serous pericardium.
- Contrast the structure of the epicardium, myocardium, and endocardium.
- Identify the chambers, great vessels, and valves of the heart.
- Explain the initiation and conduction of nerve impulses through the electrical conduction system of the heart.
- Discuss the route of blood in coronary (cardiac) circulation.
- Compare angina pectoris and myocardial infarction as abnormalities of coronary (cardiac) circulation.
- Explain the pressure changes associated with blood flow through the heart.
- Define systole and diastole as the two principal events of the cardiac cycle.
- Relate the events of the cardiac cycle to time.
- Contrast the sounds of the heart and their clinical significance.
- Discuss the surface anatomy features of the heart.

- Define cardiac output and explain what determines it.
- Explain what is meant by cardiac reserve.
- Define Starling's law of the heart.
- Contrast the effects of sympathetic and parasympathetic stimulation of the heart.
- Define the role of pressoreceptors in reflex pathways in controlling heart rate.
- Explain how chemicals, temperature, emotions, and age affect heart rate.
- Define circulatory shock and explain the homeostatic mechanisms that compensate for it.
- List the risk factors involved in heart disease.
- Explain why inadequate blood supply, anatomical disorders, and malfunctions of conduction are primary reasons for heart disease.
- Define congestive heart failure (CHF).
- Explain the principle and diagnostic importance of cardiac catheterization.
- Define medical terminology associated with the heart.

20 The Cardiovascular System: The Heart

The heart is the center of the cardiovascular system. It is a hollow, muscular organ that weighs about 342 grams (11 oz) and beats over 100,000 times a day to pump blood through over 60,000 miles of blood vessels. The blood vessels form a network of tubes that carry blood from the heart to the tissues of the body and then return it to the heart. The anatomical features of the heart that we will consider are its location, covering, wall and chambers, great vessels, valves, and conduction system. Functional aspects of the heart to be considered are the electrocardiogram, cardiac cycle (heartbeat), cardiac output, and circulatory shock.

LOCATION

The **heart** is situated obliquely between the lungs in the mediastinum. About two-thirds of its mass lies to the left of the body's midline (Figures 20-1a and 20-10). The heart is shaped like a blunt cone about the size of your closed fist—12 cm (5 inches) long, 9 cm (3½ inches) wide at its broadest point, and 6 cm (2½ inches) thick.

Its pointed end, the *apex*, is formed by the tip of the left ventricle, projects inferiorly, anteriorly, and to the left and lies superior to the central depression of the diaphragm. Anteriorly, the apex is in the fifth intercostal space—about 7.5 to 8 cm (3 inches) from the midline of the body.

The *left border* is formed almost entirely by the left ventricle, although the left atrium forms part of the upper end of the border. The left border of the heart is indicated approximately as a curved line drawn from the left fifth costochondral junction to the left second costochondral junction.

The *superior border,* where the great vessels enter and leave the heart, is formed by both atria. It is represented by a line joining the second left intercostal space to the third right costal cartilage.

The *base* of the heart projects superiorly, posteriorly, and to the right. It is formed by the atria, mostly the left atrium. It lies opposite the fifth to ninth thoracic vertebrae. Anteriorly, it lies just inferior to the second rib.

The *right border* is formed by the right atrium and corresponds to a curved line drawn from the xiphisternal articulation to about the middle of the right third costal cartilage.

The *inferior border* is formed by the right ventricle and slightly by the left ventricle. It is represented by a line passing from the inferior end of the right border through the xiphisternal joint to the apex of the heart.

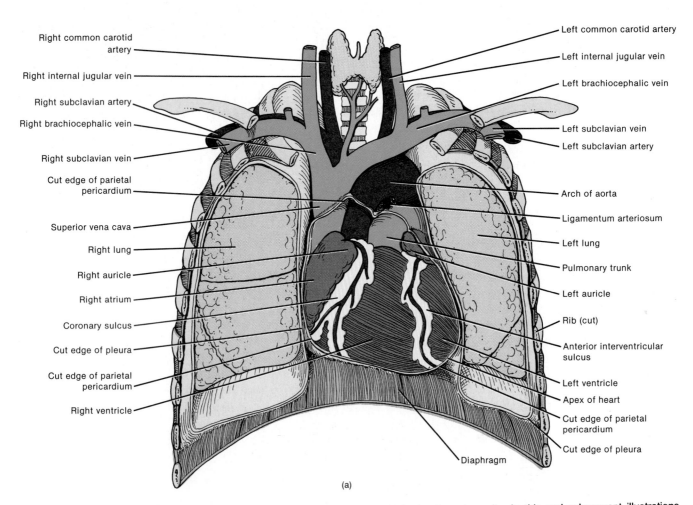

FIGURE 20-1 Heart. (a) Position of the heart and associated blood vessels in the thoracic cavity. In this and subsequent illustrations, vessels that carry oxygenated blood are colored red; vessels that carry deoxygenated blood are blue.

The *sternocostal* (*anterior*) *surface* is formed mainly by the right ventricle and right atrium, whereas the *diaphragmatic* (*inferior*) *surface* is formed by the left and right ventricles, mostly the left.

PARIETAL PERICARDIUM (PERICARDIAL SAC)

The heart is enclosed in a loose-fitting serous membrane called the **parietal pericardium** or **pericardial sac** (Figure 20-1a, b). It consists of two layers: the fibrous layer and the serous layer. The *fibrous layer* (*fibrous pericardium*) is the outer layer and consists of a tough, fibrous connective tissue. It is attached to the large blood vessels entering and leaving the heart, to the diaphragm, and to the inside of the sternal wall of the thorax. It also adheres to the parietal pleurae. The fibrous pericardium prevents overdistension of the heart, surrounds it with a tough protective membrane, and anchors it in the mediastinum.

The inner layer of the parietal pericardium, known as the *serous layer* (*serous pericardium*), is thinner and more delicate. It is continuous with the visceral pericardium (the outer layer of the wall of the heart) at the base of the heart and around the large blood vessels.

An inflammation of the parietal pericardium is known as **pericarditis.**

HEART WALL

The wall of the heart (Figure 20-1b) is divided into three portions: the epicardium (external layer), the myocardium (middle layer), and the endocardium (inner layer). The **epicardium (visceral pericardium)** is the thin, transparent outer layer of the wall. It is composed of serous tissue and mesothelium. Between the serous pericardium and the epicardium is a potential space called the *pericardial cavity*. The cavity contains a watery fluid, known as *pericardial fluid,* which prevents friction between the membranes as the heart moves.

The **myocardium,** which is cardiac muscle tissue, constitutes the bulk of the heart. Cardiac muscle fibers are involuntary, striated, and branched, and the tissue is arranged in interlacing bundles of fibers. The myocardium is responsible for the contraction of the heart.

The **endocardium** is a thin layer of endothelium overlying a thin layer of connective tissue pierced by tiny blood vessels and bundles of smooth muscle. It lines the inside of the myocardium and covers the valves of the heart and the tendons that hold them open. It is continuous with the endothelial lining of the large blood vessels of the heart.

Inflammation of the epicardium, myocardium, and endocardium are referred to as **epicarditis, myocarditis,** and **endocarditis,** respectively.

CHAMBERS OF THE HEART

The interior of the heart is divided into four cavities, or **chambers,** which receive the circulating blood (Figure 20-2d–f). The two upper chambers are called the right and left **atria.** Each atrium has an appendage called an *auricle* (AWR-i-kul; *auris* = ear), so named because its shape resembles a dog's ear. The auricle increases the atrium's surface area. The lining of the atria is smooth, except for the anterior atrial walls and the lining of the auricles, which contain projecting muscle bundles that are parallel to one another and resemble the teeth of a comb: the *musculi pectinati* (MUS-kyoo-lī pek-ti-NA-tē). These bundles give the lining of the auricles a ridged appearance.

The atria are separated by a partition called the *interatrial septum.* A prominent feature of this septum is an oval depression, the *fossa ovalis,* which corresponds to the site of the foramen ovale, an opening in the interatrial septum of the fetal heart. The fossa ovalis faces the opening of the inferior vena cava and is located in the septal wall of the right atrium.

The two lower chambers are the right and left **ventricles.** They are separated by an *interventricular septum.*

The muscle tissue of the atria and ventricles is separated by connective tissue that also forms the valves. This "cardiac skeleton" effectively divides the myocardium

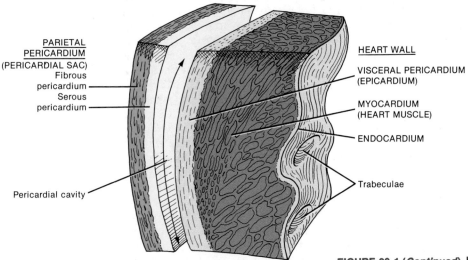

PARIETAL
PERICARDIUM
(PERICARDIAL SAC)
Fibrous
pericardium
Serous
pericardium

Pericardial cavity

HEART WALL

VISCERAL PERICARDIUM
(EPICARDIUM)

MYOCARDIUM
(HEART MUSCLE)

ENDOCARDIUM

Trabeculae

(b)

FIGURE 20-1 (Continued) Heart. (b) Structure of the parietal pericardium and heart wall.

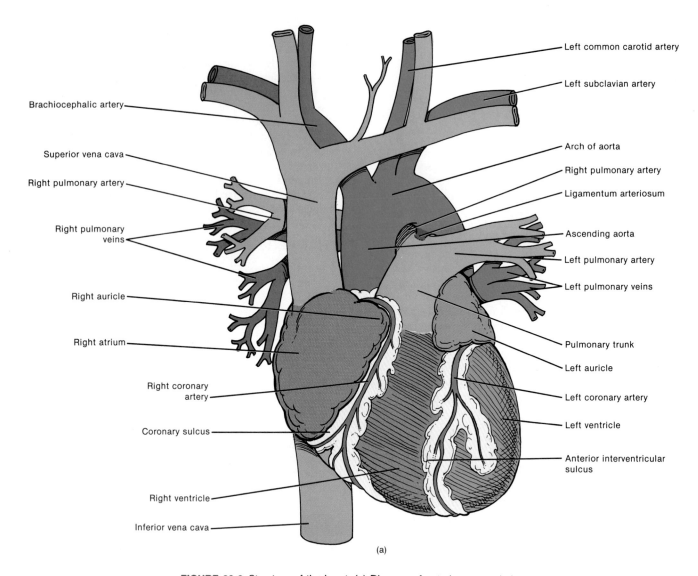

Brachiocephalic artery

Superior vena cava

Right pulmonary artery

Right pulmonary veins

Right auricle

Right atrium

Right coronary artery

Coronary sulcus

Right ventricle

Inferior vena cava

Left common carotid artery

Left subclavian artery

Arch of aorta

Right pulmonary artery

Ligamentum arteriosum

Ascending aorta

Left pulmonary artery

Left pulmonary veins

Pulmonary trunk

Left auricle

Left coronary artery

Left ventricle

Anterior interventricular sulcus

(a)

FIGURE 20-2 Structure of the heart. (a) Diagram of anterior external view.

into two separate muscle masses. Externally, a groove known as the *coronary sulcus* (SUL-kus) separates the atria from the ventricles. It encircles the heart and houses the coronary sinus and circumflex branch of the left coronary artery. The *anterior interventricular sulcus* and *posterior interventricular sulcus* separate the right and left ventricles externally. The sulci contain coronary blood vessels and a variable amount of fat (Figure 20-2a–c).

GREAT VESSELS OF THE HEART

The right atrium receives blood from all parts of the body except the lungs. It receives the blood through three veins. The **superior vena cava** brings blood from parts of the body superior to the heart; the **inferior vena cava,** brings blood from parts of the body inferior to the heart; and the **coronary sinus** drains blood from most of the vessels supplying the walls of the heart (Figure 20-2c). The right atrium then delivers the blood into the right ventricle, which pumps it into the **pulmonary trunk.** The pulmonary trunk divides into a **right** and **left pulmonary artery,** each of which carries blood to the lungs. In the

lungs, the blood releases its carbon dioxide and takes on oxygen. It returns to the heart via four **pulmonary veins** that empty into the left atrium. The blood then passes into the left ventricle, which pumps the blood into the **ascending aorta.** From here the blood is passed into the **coronary arteries, arch of the aorta, thoracic aorta, and abdominal aorta.** These blood vessels transport the blood to all body parts except the lungs.

The sizes of the four chambers vary according to function (Figure 20-2d–f). The right atrium, which must collect blood coming from almost all parts of the body, is slightly larger than the left atrium, which receives blood only from the lungs. The thickness of the chamber walls varies too. The atria are thin-walled because they need only enough cardiac muscle tissue to deliver the blood into the ventricles with the aid of a reduced pressure created by the expanding ventricles. The right ventricle has a thicker layer of myocardium than the atria, since it must send blood to the lungs and around back to the left atrium. The left ventricle has the thickest walls, since it must pump blood at high pressure through literally thousands of miles of vessels in the head, trunk, and extremities.

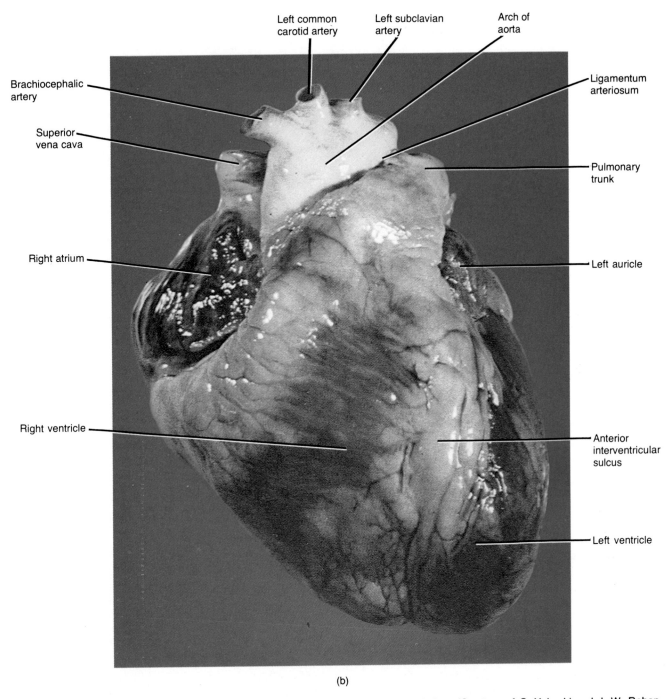

Left common carotid artery

Left subclavian artery

Arch of aorta

Brachiocephalic artery

Ligamentum arteriosum

Superior vena cava

Pulmonary trunk

Right atrium

Left auricle

Right ventricle

Anterior interventricular sulcus

Left ventricle

(b)

FIGURE 20-2 (*Continued*) Structure of the heart. (b) Photograph of anterior external view. (Courtesy of C. Yokochi and J. W. Rohen, *Photographic Anatomy of the Human Body,* 2nd ed., 1979, IGAKU-SHOIN, Ltd., Tokyo, New York.)

VALVES OF THE HEART

As each chamber of the heart contracts, it pushes a portion of blood into a ventricle or out of the heart through an artery. In order to keep the blood from flowing backward, the heart has **valves.**

ATRIOVENTRICULAR (AV) VALVES

Atrioventricular (AV) valves lie between the atria and ventricles (Figure 20-2d). The right atrioventricular valve

between the right atrium and right ventricle, is also called the **tricuspid valve** because it consists of three flaps, or cusps. These flaps are fibrous tissues that grow out of the walls of the heart and are covered with endocardium. The pointed ends of the cusps project into the ventricle. Cords called *chordae tendineae* (KOR-dē TEN-di-nē) connect the pointed ends to small conical projections— the *papillary muscles* (muscular columns)—located on the inner surface of the ventricles. The irregular surface of ridges and folds of the myocardium in the ventricles is known as the *trabeculae carneae* (tra-BEK-yoo-lē

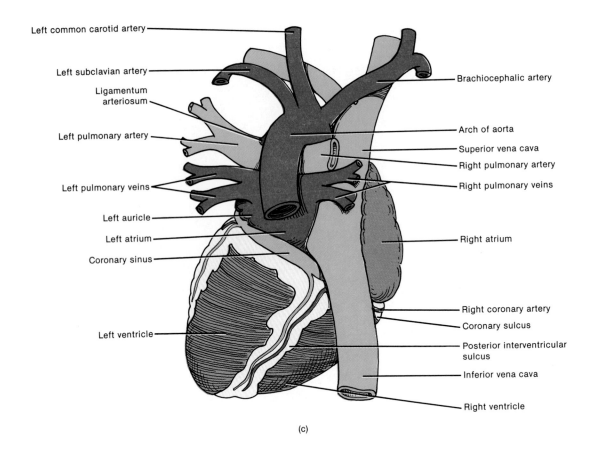

(c)

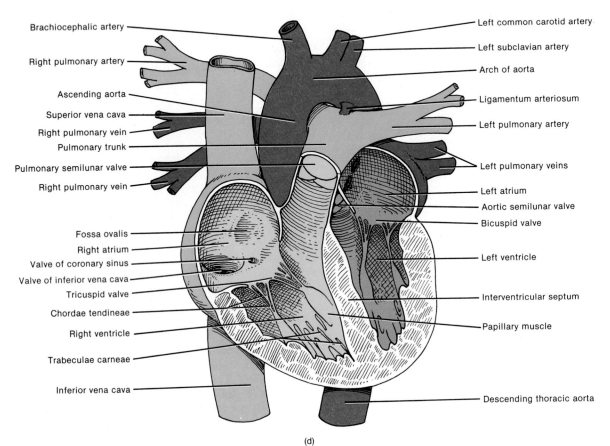

(d)

FIGURE 20-2 (*Continued*) Structure of the heart. (c) Diagram of posterior external view. (d) Diagram of anterior internal view.

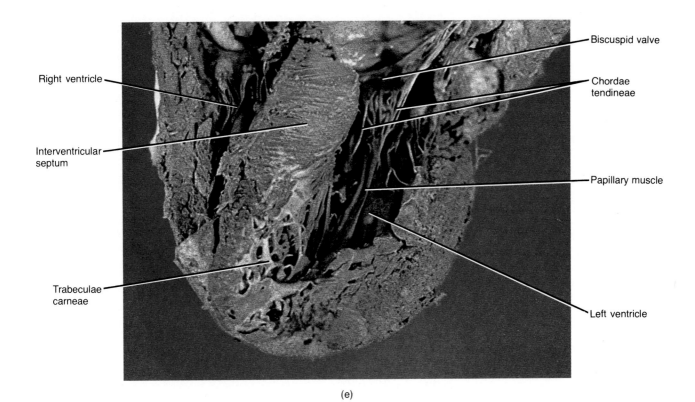

Right ventricle

Interventricular septum

Trabeculae carneae

Biscuspid valve

Chordae tendineae

Papillary muscle

Left ventricle

(e)

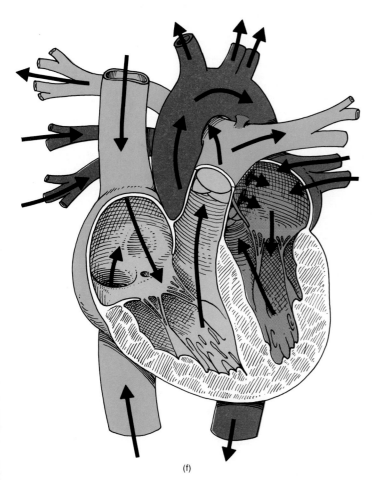

(f)

FIGURE 20-2 (*Continued*) Structure of the heart. (e) Photograph of a portion of anterior internal view. (Courtesy of J. W. Eads.) (f) Path of blood through the heart.

KAR-enē). The chordae tendineae and their muscles keep the flaps pointing in the direction of the blood flow. As the atrium relaxes and the ventricle pumps the blood out of the heart, any blood driven back toward the atrium is pushed between the flaps and the ventricle walls (Figure 20-3). This action drives the cusps upward until their edges meet and close the opening. At the same time, contraction of the papillary muscles helps prevent the valve from swinging upward into the atrium.

The atrioventricular valve between the left atrium and left ventricle is called the **bicuspid (mitral) valve.** It is also known as the left atrioventricular (left AV) valve. It has two cusps that work in the same way as the cusps of the tricuspid valve. Its cusps are also attached by way of the chordae tendineae to papillary muscles.

SEMILUNAR VALVES

Both arteries that leave the heart have a valve that prevents blood from flowing back into the heart. These are the **semilunar valves** (Figure 20-3). The **pulmonary semilunar valve** lies in the opening where the pulmonary trunk leaves the right ventricle. The **aortic semilunar valve** is situated at the opening between the left ventricle and the aorta.

Both valves consist of three semilunar (half-moon or crescent-shaped) cusps. Each cusp is attached by its convex margin to the artery wall. The free borders of the cusps curve outward and project into the opening inside the blood vessel. Like the atrioventricular valves, the semilunar valves permit blood to flow in only one direction—in this case, from the ventricles into the arteries.

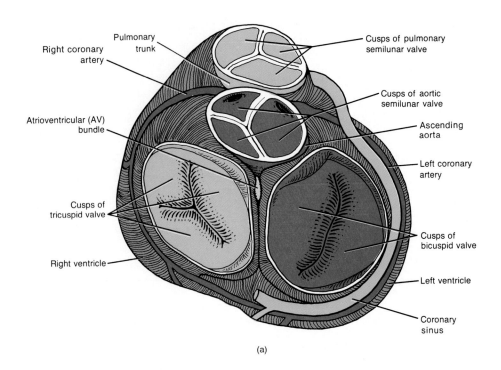

(a)

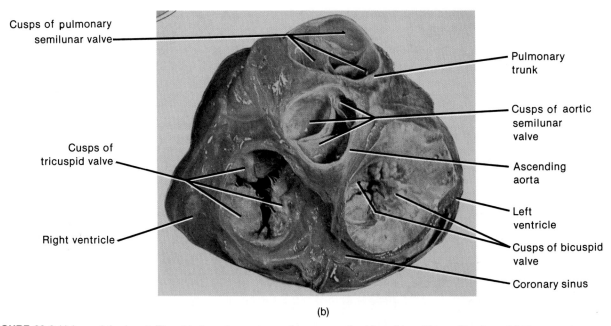

(b)

FIGURE 20-3 Valves of the heart. The atria have been removed to expose the tricuspid and bicuspid valves. (a) Diagram. (b) Photograph. (Courtesy of C. Yokochi and J. W. Rohen, *Photographic Anatomy of the Human Body,* 2nd ed., 1979, IGAKU-SHOIN, Ltd., Tokyo, New York.)

CLINICAL APPLICATION

The development of **artificial heart valves** made from synthetic materials has made it possible for cardiac surgeons to repair defective natural valves to alleviate many valvular problems (Figure 20-4). In some valvular disorders, a heart valve does not close properly, causing blood to backflow.

Even more amazing than artificial heart valves is the development of a permanently implanted **artificial**

heart. On December 2, 1982, Dr. Barney B. Clark, a 61-year-old retired dentist, made medical history by becoming the first human to receive such an artificial heart. The 7½-hour operation was performed by Dr. William DeVries, a surgeon at the University of Utah Medical Center.

The artificial heart, known as Jarvik-7 after its inventor, Dr. Robert Jarvik, consists of an aluminum base and a pair of rigid plastic chambers that serve as ventricles (Figure 20-5). Each chamber contains a flexible diaphragm driven by compressed air that forces blood past mechanical valves into the major blood vessels. The surgery essentially consisted of removing the ventricles from Dr. Clark's heart, but leaving the atria intact. Then Dacron connectors were sutured onto the atria, aorta, and pulmonary trunk. The artificial heart was then snapped into position via the connectors.

The external power source for the artificial heart is a 375-pound unit that includes two compressors and a back-up compressor that force compressed air through six-foot-long air tubes into the artificial heart. These tubes enter through the abdomen. The unit also contains a three-hour supply of pressurized air should there be a power failure, a dehumidifier that removes moisture from the compressed air, and mechanisms that control air pressure and heart rate. The unit is always within six feet of the patient, the length of the air tubes. The artificial heart is designed to pump sufficient blood to sustain only moderate activity.

On the 112th day after the implantation, Dr. Clark died from massive circulatory collapse, shock, and failure of all of his natural organs. This was probably due to complications of antibiotic therapy to treat pneumonia and the damage that his body suffered just prior to implantation of the artificial heart. While in his body, the artificial heart beat almost 13 million times for 2,688 hours. Except for a defective biscuspid valve that had to be replaced, the heart showed no evidence of clots, infection, or physical deterioration. Dr. Clark's courage and determination have made him one of the outstanding pioneers in the advancement of medical research; his efforts will help scientists to develop even more successful artificial hearts.

CONDUCTION SYSTEM

The heart is innervated by the autonomic nervous system (Chapter 16), but the autonomic neurons only increase or decrease the time it takes to complete a cardiac cycle, that is, they do not initiate contraction. The chamber walls can go on contracting and relaxing, contracting and relaxing, without any direct stimulus from the nervous system. This action is possible because the heart has an intrinsic regulating system called the **conduction system.** The conduction system is composed of specialized muscle tissue that generates and distributes the electrical impulses which stimulate the cardiac muscle fibers to contract. These tissues are the sinoatrial (sinuatrial) node, the atrioventricular (AV) node, the atrioventricular (AV) bundle, the bundle branches, and the Purkinje fibers. The cells of the conduction system develop during embryological life from certain cardiac muscle cells. These cells lose their ability to contract and become specialized for impulse transmission.

A **node** of the conducting system is a compact mass of conducting cells. The **sinoatrial (sinuatrial) node,** also known as the **SA node** or **pacemaker,** is located in the right atrial wall inferior to the opening of the superior vena cava (Figure 20-6a). The SA node initiates each cardiac cycle, and thereby sets the basic pace for the heart rate—hence its common name, pacemaker. The rate set by the SA node may be altered by nervous impulses from the autonomic nervous system or by certain blood-borne chemicals such as thyroid hormones and epinephrine.

Once an action potential is initiated by the SA node, the impulse spreads out over both atria, causing them to contract and at the same time depolarizing the **atrioventricular (AV) node.** Because of its location near the inferior portion of the interatrial septum, the AV node is one of the last portions of the atria to be depolarized.

Artificial atrioventricular valve

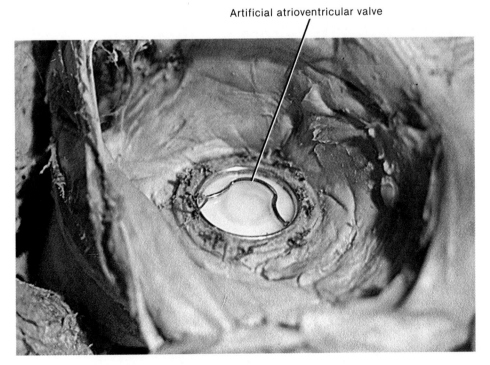

FIGURE 20-4 Photograph of a superior view of an artificial atrioventricular valve. (Courtesy of John W. Eads.)

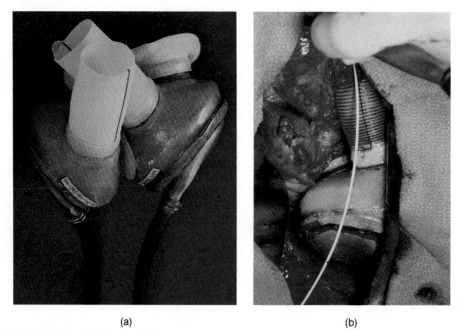

(a) (b)

FIGURE 20-5 Artificial heart. (a) Photograph of an artificial heart prior to implantation. (b) Photograph of an implanted artificial heart. (Courtesy Nelson, University of Utah.)

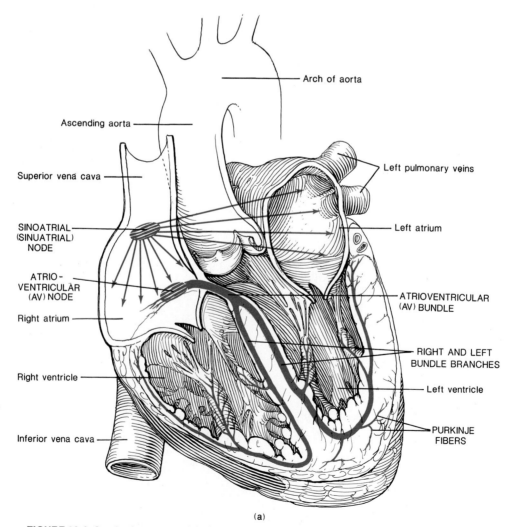

(a)

FIGURE 20-6 Conduction system of the heart. (a) Location of the nodes and bundles of the conduction system.

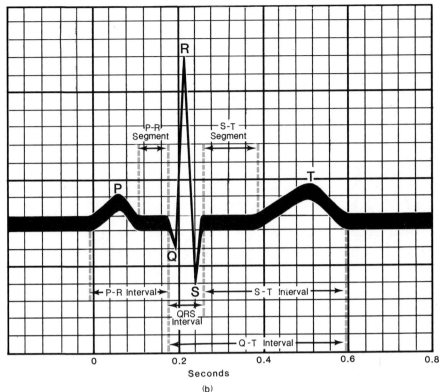

(b)

FIGURE 20-6 (*Continued*) Conduction system of the heart. (b) Normal electrocardiogram of a single heartbeat, enlarged for emphasis. (c) Normal electrocardiogram of several heartbeats.

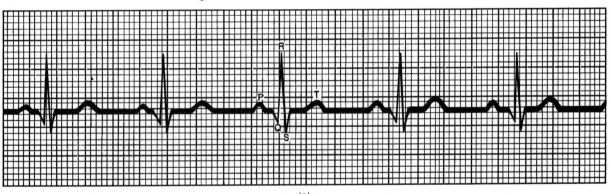

(c)

From the AV node, a tract of conducting fibers called the **atrioventricular (AV) bundle** runs through the cardiac skeleton to the top of the interventricular septum. It then continues down both sides of the septum as the **right** and **left bundle branches.** The atrioventricular bundle distributes the action potential over the medial surfaces of the ventricles. Actual contraction of the ventricles is stimulated by the **Purkinje (pur-KIN-jē) fibers** that emerge from the bundle branches and pass into the cells of the myocardium.

ELECTROCARDIOGRAM

Impulse transmission through the conduction system generates electrical currents that may be detected on the body's surface. A recording of the electrical changes that accompany the cardiac cycle is called an **electrocardiogram (ECG** or **EKG).** The instrument used to record the changes is an *electrocardiograph.*

Each portion of the cardiac cycle produces a different electrical impulse. These impulses are transmitted from the electrodes to a recording pen that graphs the impulses as a series of up-and-down waves called *deflection waves.* In a typical record (Figure 20-6b, c), three clearly recognizable waves accompany each cardiac cycle. The first, called the **P wave,** is a small upward wave. It indicates atrial depolarization—the spread of an impulse from the SA node through the muscle of the two atria. A fraction of a second after the P wave begins, the atria contract. The second wave, called the **QRS wave (complex),** begins as a downward deflection, continues as a large, upright, triangular wave, and ends as a downward wave at its base. This deflection represents ventricular depolarization, that is, the spread of the electrical impulse through the ventricles. The third recognizable deflection is a dome-shaped **T wave.** This wave indicates ventricular repolarization. There is no deflection to show atrial repolarization because the stronger QRS wave masks this event.

In reading an electrocardiogram, it is important to note the size of the deflection waves and certain time intervals. Enlargement of the P wave, for example, indicates enlargement of the atrium, as in mitral stenosis.

(a)

In this condition, the mitral valve narrows, blood backs up into the atrium, and there is expansion of the atrial wall.

The *P-R interval* is measured from the beginning of the P wave to the beginning of the R wave. It represents the conduction time from the beginning of atrial excitation to the beginning of ventricular excitation. The P-R interval is the time required for an impulse to travel through the atria and atrioventricular node to the remaining conducting tissues. The lengthening of this interval, as in atherosclerotic heart disease and rheumatic fever, occurs because the heart tissue covered by the P-R interval, namely the atria and atrioventricular node, is scarred or inflamed. Thus the impulse must travel at a slower rate and the interval is lengthened. The normal P-R interval covers no more than 0.2 sec.

An enlarged Q wave may indicate a myocardial infarction (heart attack). An enlarged R wave generally indicates enlarged ventricles.

The *S-T segment* begins at the end of the S wave and terminates at the beginning of the T wave. It represents the time between the end of the spread of the impulse through the ventricles and repolarization of the ventricles. The S-T segment is elevated in acute myocardial infarction and depressed when the heart muscle receives insufficient oxygen.

The T wave represents ventricular repolarization. It is flat when the heart muscle is receiving insufficient oxygen, as in atherosclerotic heart disease. It may be elevated when the body's potassium level is increased.

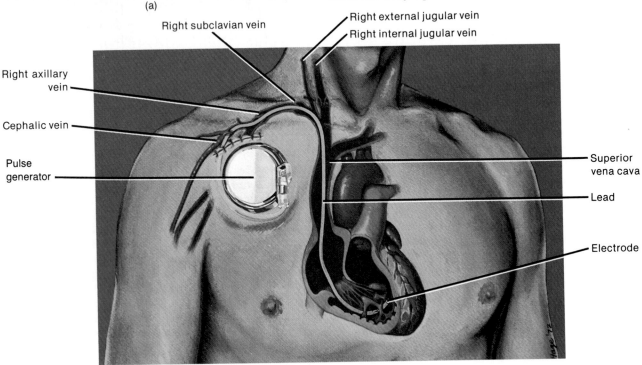

(b)

FIGURE 20-7 Artificial pacemaker. (a) Photograph of a pacemaker prior to implantation. (Courtesy of Medtronic, Inc.) (b) Diagram of an implanted artificial pacemaker. In this method of implantation, the lead is inserted into the cephalic vein under fluoroscopic guidance, following an incision inferior to the clavicle. The lead is then guided into the axillary and subclavian veins, superior vena cava, right atrium, and finally into the right ventricle to make contact with the ventricular myocardium. The pulse generator is placed in a subcutaneous pocket. (Courtesy of Medtronic, Inc.)

The ECG is useful in diagnosing abnormal cardiac rhythms and conduction patterns and following the course of recovery from a heart attack. It can also detect the presence of fetal life or determine the presence of several fetuses. For some people, ECG monitoring may be done by using a *Holter monitor.* This is an ambulatory system in which the monitoring hardware is worn by an individual while going about everyday routines. The Holter monitor is used especially to detect rhythm disorders in the conduction system. It is also used to correlate rhythm disorders and symptoms and to follow the effectiveness of drugs.

CLINICAL APPLICATION

When major elements of the conduction system are disrupted, heart block of varying degrees may result. Complete heart block results most commonly from a disturbance in AV conduction. The ventricles fail to receive atrial impulses, causing the ventricles and atria to beat independently of each other. In patients with heart block, normal heart rate can be restored and maintained with an **artificial pacemaker** (Figure 20-7). A pacemaker is a device that sends out small electrical charges that stimulate the heart. It consists of three basic parts: a *pulse generator,* which contains the battery cells; a *lead,* which is a flexible wire connected to the pulse generator; and an *electrode,* which delivers the charge to the heart.

Pulse generators are of two types. Demand (ventricular-inhibited) generators sense when the heart is sending out its own electrical charges and therefore will not send an electrical charge to the heart. But when the heart fails to beat, the demand generator senses the lack of activity and immediately sends an electrical charge to the heart. Demand generators are used for patients who still have some natural heart activity. The other pulse generator is called a fixed-rate (asynchronous) generator. This unit sends a series of electrical charges to the heart at a steady rate, usually 80 beats per minute. Unlike demand generators, fixed-rate generators supply all the stimulation, since the natural activity of the heart is too slow.

Pacemaker leads are referred to as endocardial and myocardial. Endocardial leads are passed through a vein so the tip, called the electrode, makes contact with the endocardium. This is the type shown in Figure 20-7b. Myocardial leads are designed so that the electrode attaches directly to the outside of the heart.

BLOOD SUPPLY

The wall of the heart, like any other tissue, including large blood vessels, has its own blood vessels. Nutrients could not possibly diffuse through all the layers of cells that make up the heart tissue. The flow of blood through the numerous vessels that pierce the myocardium is called the **coronary (cardiac) circulation** (Figure 20-8).

The vessels that serve the myocardium include the *left coronary artery,* which originates as a branch of the ascending aorta. This artery runs under the left atrium and divides into the anterior interventricular and circumflex branches. The *anterior interventricular branch* follows the anterior interventricular sulcus and supplies oxygenated blood to the walls of both ventricles. The *circumflex branch* distributes oxygenated blood to the walls of the left ventricle and left atrium.

The *right coronary artery* also originates as a branch of the ascending aorta. It runs under the right atrium and divides into the posterior interventricular and marginal branches. The *posterior interventricular branch* follows the posterior interventricular sulcus and supplies the walls of the two ventricles with oxygenated blood. The *marginal branch* transports oxygenated blood to the myocardium of the right ventricle and right atrium. The left ventricle receives the most abundant blood supply because of the enormous work it must do.

As blood passes through the atrial system of the heart, it delivers oxygen and nutrients and collects carbon dioxide and wastes. Most of the deoxygenated blood, which carries the carbon dioxide and wastes, is collected by a large vein, the *coronary sinus,* which empties into the right atrium. A vascular sinus is a vein with a thin wall that has no smooth muscle to alter its diameter. The principal tributaries of the coronary sinus are the *great cardiac vein,* which drains the anterior aspect of the heart, and the *middle cardiac vein,* which drains the posterior aspect of the heart.

CLINICAL APPLICATION

Most heart problems result from faulty coronary circulation. If a reduced oxygen supply weakens the cells but does not actually kill them, the condition is called **ischemia** (is-KĒ-mē-a). **Angina pectoris** (an-JĪ-na PEK-tō-ris), meaning "chest pain," is ischemia of the myocardium. (Remember that pain impulses originating from most visceral muscles are referred to an area on the surface of the body.) Angina pectoris occurs when coronary circulation is somewhat reduced for some reason. Stress, which produces constriction of vessel walls, is a common cause. Equally common is strenous exercise after a heavy meal. When any quantity of food enters the stomach, the body increases blood flow to the digestive tract. As a consequence, some blood is diverted away from other organs, including the heart. Exercise, however, increases heart muscle activity and thus the heart's need for oxygen. Doing heavy work while food is in the stomach can therefore lead to oxygen deficiency in the myocardium. Agina pectoris weakens the heart muscle, but it does not produce a full-scale heart attack. The simple remedy of taking nitroglycerin, a drug that dilates coronary vessels and thereby increased blood flow to the area, brings coronary circulation back to normal and stops the pain of angina.

A much more serious problem is **myocardial infarction** (in-FARK-shun), commonly called a "coronary" or "heart attack." **Infarction** means the death of an area of tissue because of an interrupted blood supply. Myocardial infarction may result from a thrombus or embolus in one of the coronary arteries. The tissue distal to the obstruction dies and is replaced by noncon-

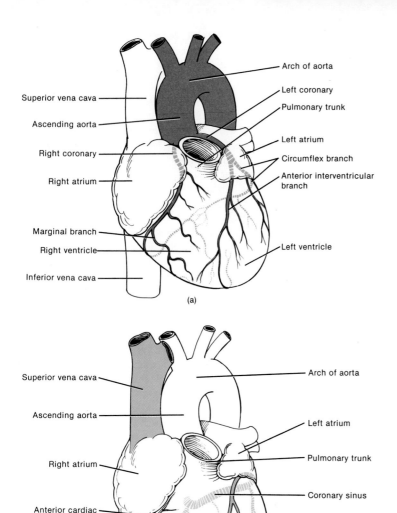

(a)

Arch of aorta
Superior vena cava
Left coronary
Pulmonary trunk
Ascending aorta
Left atrium
Right coronary
Circumflex branch
Right atrium
Anterior interventricular branch
Marginal branch
Right ventricle
Left ventricle
Inferior vena cava

(b)

Arch of aorta
Superior vena cava
Left atrium
Ascending aorta
Pulmonary trunk
Right atrium
Coronary sinus
Great cardiac
Anterior cardiac
Right ventricle
Small cardiac
Left ventricle
Inferior vena cava
Middle cardiac

FIGURE 20-8 Coronary (cardiac) circulation. (a) Anterior view of arterial distribution. (b) Anterior view of venous drainage.

tractile scar tissue. Thus, the heart muscle loses at least some of its strength. The aftereffects depend partly on the size and location of the infarcted, or dead, area.

Diagnostic tests have been developed that enable physicians to diagnose heart attacks within hours. During a heart attack, certain enzymes are released into the bloodstream and are the strongest indicators that a heart attack has occurred. One test detects creatine kinase MB isoenzyme (CK-MB), which is found only in heart cells. The test detects nearly all the CK-MB released after a heart attack and gives physicians a more accurate measure of the extent of the heart damage. Another test measures lactic dehydrogenase-1 (LDH-1), an enzyme liberated during a myocardial infarction. This test will yield results in an hour. Other enzymes that are always elevated after a myocardial infarction are serum glutamic oxaloacetic transaminase (SGOT) and creatine phosphokinase (CPK).

CARDIAC CYCLE

Before we examine the events of the cardiac cycle, it will be useful to first look at how blood flows through the heart.

BLOOD FLOW THROUGH THE HEART

Two phenomena control the movement of blood through the heart: the opening and closing of the valves and the contraction and relaxation of the myocardium. Both these activities occur without direct stimulation from the nervous system. The valves are controlled by pressure changes in each heart chamber. The contraction of the cardiac muscle is stimulated by its conduction system.

Blood flows from an area of higher pressure to an area of lower pressure. The pressure developed in a heart chamber is related primarily to the chamber's size and the volume of blood it contains. The greater the volume of blood, the higher the pressure. Figure 20-9b indicates pressure changes associated with the left side of the heart during the cardiac cycle. Although the pressures in the right side of the heart are somewhat lower, the same results are achieved.

The pressure in the atria is called **atrial pressure.** When the atria are relaxed, the pressure in them steadily increases because blood flows into them continuously from their vessels. When the ventricles contract, ventricular pressure drops below atrial pressure and the atrioventricular valves open. Atrial blood drains into the ventricles, which are empty and consequently have a lower pressure. When atrial pressure builds up, the atria

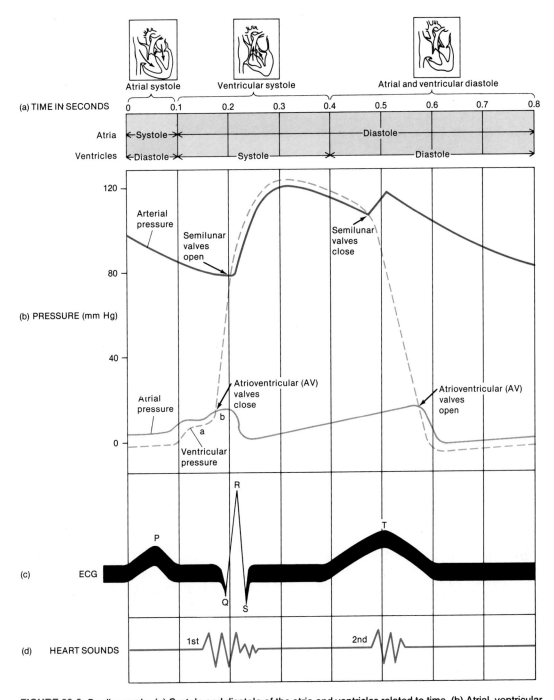

FIGURE 20-9 Cardiac cycle. (a) Systole and diastole of the atria and ventricles related to time. (b) Atrial, ventricular, and arterial pressure changes along with the opening and closing of valves during the cardiac cycle. (c) ECG related to the cardiac cycle. (d) Heart sounds related to the cardiac cycle.

contract and send the remaining blood rushing into the ventricles.

As the ventricles fill with blood, their pressure, called **ventricular pressure,** becomes greater than atrial pressure. At this point, a backflow of blood closes the atrioventricular valves. The ventricles start to contract, but the semilunar valves remain closed from the previous cardiac cycle. No blood moves out into the aorta or pulmonary artery. The ventricles are closed.

Ventricular pressure continues to rise until it is greater than the **arterial pressure,** the pressure in the aorta and pulmonary trunk. When ventricular pressure rises above

arterial pressure, the semilunar valves are forced open and blood is ejected from the ventricles into the great vessels.

Once the ventricles eject their blood, ventricular pressure falls below that in the great vessels. As a result, the semilunar valves are pushed closed by the backflow of blood created by the recoil of the elastic walls of the arteries. Once again, the ventricles are closed chambers. As the ventricles relax, ventricular pressure decreases until it becomes less than atrial pressure. At that point, atrial pressure forces the atrioventricular valves open, blood fills the ventricles, and another cycle begins.

ATRIAL AND VENTRICULAR SYSTOLE AND DIASTOLE

In a normal heartbeat, the two atria contract simultaneously while the two ventricles relax. Then, when the two ventricles contract, the two atria relax. The term **systole** (SIS-tō-lē) refers to the phase of contraction; **diastole** (dī-AS-tō-lē) is the phase of relaxation. A **cardiac cycle**, or complete heartbeat, consists of a systole and diastole of both atria plus the systole and diastole of both ventricles.

    1. Atrial diastole. During **atrial diastole** deoxygenated blood from the various parts of the body enters the right atrium through the superior and inferior venae cavae and coronary sinus. Simultaneously, oxygenated blood from the lungs enters the left atrium via the pulmonary veins. Blood normally flows continually from the veins into the atria. During the first part of atrial diastole, the atrioventricular valves are closed, since the ventricles are in systole (contraction). During the latter part of atrial diastole, the atrioventricular valves open. During this time, before the atria contracts, about 70 percent of the blood that will enter the ventricles flows directly through the atria into the ventricles. Atrial contraction, and increased atrial pressure, forces the additional 30 percent into the ventricles.

    2. Ventricular systole. As the atria go into diastole, the ventricles go into systole. **Ventricular systole** is initiated by the spread of the action potential through the ventricles as indicated by the QRS wave. During most of ventricular systole, the atrioventricular valves are closed, the ventricles contract, and, as ventricular pressure increases, they force blood into their respective vessels. The right ventricle pumps deoxygenated blood to the lungs through the open semilunar valve of the pulmonary trunk. The left ventricle pumps oxygenated blood through the open semilunar valve of the aorta. At the end of the ventricular systole, the semilunar valves close and both atria and ventricles relax.

    3. Atrial systole. When atrial diastole and ventricular systole are completed, the events are reversed—the atria go into systole and the ventricles go into diastole. When the SA node fires, the atria depolarize and contract. This action produces the P wave on the ECG and signals the start of **atrial systole**. Contraction of the atria causes an increase in atrial pressure and forces about 30 percent of the blood into the ventricles. Deoxygenated blood from the right atrium passes into the right ventricle through the open tricuspid valve. Oxygenated blood passes from the left atrium into the left ventricle through the open bicuspid valve.

    4. Ventricular diastole. While the atria are contracting, the ventricles are in diastole. During **ventricular diastole,** the atrioventricular valves are open, the ventricles are filling with blood, and the semilunar valves in the aorta and pulmonary trunk are closed.

TIMING

If we assume that the average heart beats 75 times per minute, then each cardiac cycle requires about 0.8 sec (Figure 20-9a). During the first 0.1 sec, the atria contract and the ventricles relax. The atrioventricular valves are open, and the semilunar valves are closed. For the next 0.3 sec, the atria are relaxing and the ventricles are contracting. During the first part of this period, all valves are closed; during the second part, the semilunar valves are open. The last 0.4 sec of the cycle is the relaxation, or quiescent, period and all chambers are in diastole. In a complete cycle, then, the atria are in systole 0.1 sec and in diastole 0.7 sec; the ventricles are in systole

0.3 sec and in diastole 0.5 sec. For the first part of the quiescent period, all valves are closed; during the latter part, the atrioventricular valves open and blood starts draining into the ventricles. When the heart beats faster than normal, the quiescent period is shortened accordingly.

SOUNDS

The sound of the heartbeat comes primarily from turbulence in blood flow created by the closure of the valves, not from the contraction of the heart muscle. The first sound, which can be described as a **lubb** ($\overline{oo}$) sound, is a long, booming sound. The lubb is the sound created by the closure of the atrioventricular valves soon after ventricular systole begins (Figure 20-9d). The second sound, which is heard as a short, sharp sound, can be described as a **dupp** ($\breve{u}$) sound. Dupp is the sound created as the semilunar valves close toward the end of ventricular systole. A pause between the second sound and the first sound of the next cycle is about two times longer than the pause between the first and second sound of each cycle. Thus the cardiac cycle can be heard as a lubb, dupp, pause; lubb, dupp, pause; lubb, dupp, pause.

CLINICAL APPLICATION

Heart sounds provide valuable information about the valves. Unusual sounds are called **murmurs.** Some murmurs are caused by the noise made by a little blood bubbling back up into an atrium because of improper closure of an atrioventricular valve.

    Frequently a portion of a normal mitral valve is pushed back too far (prolapsed) during contraction. Such a condition is called **mitral valve prolapse (MVP).** Or a murmur might be caused by a diseased mitral valve due to rheumatic fever (**valvular heart disease**).

    In the absence of any complications, many murmurs have no clinical significance.

    The act of listening to sounds within the body is called **auscultation** (*auscultare* = to listen; aws-kul-TĀ-shun) and it is usually done with a stethoscope.

The valves of the heart may be identified by surface projection (Figure 20-10). The pulmonary and aortic semilunar valves are represented on the surface by a line about 2.5 cm (1 inch) in length. The pulmonary semilunar valve lies horizontally behind the inner end of the left third costal cartilage and the adjoining part of the sternum. The aortic semilunar valve is placed obliquely behind the left side of the sternum, opposite the third intercostal space. The tricuspid valve lies behind the sternum, extending from the midline at the level of the fourth costal cartilage down toward the right sixth chondrosternal junction. The bicuspid lies behind the left side of the sternum obliquely at the level of the fourth costal cartilage. It is represented by a line about 3 cm in length.

    Although heart sounds are produced in part by the closure of valves, they are not necessarily heard best over these valves. Each sound tends to be clearest in a slightly different location closest to the surface of the body (Figure 20-10).

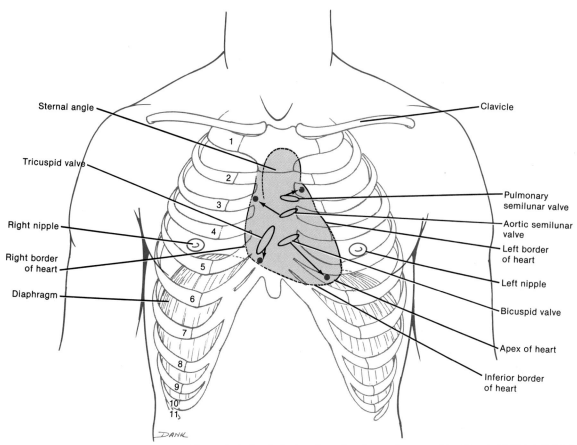

FIGURE 20-10 Surface projection of the heart. The red circles indicate where heart sounds caused by the respective valves are best heard.

CARDIAC OUTPUT (CO)

The amount of blood ejected from the left ventricle into the aorta per minute is called the **cardiac output (CO)** or **cardiac minute output.** Cardiac output is determined by (1) the amount of blood pumped by the left ventricle during each beat and (2) the number of heartbeats per minute. The amount of blood ejected by a ventricle during each systole is called the **stroke volume (SV).** In a resting adult, stroke volume averages 70 ml and heart rate is about 75 beats per minute. The average cardiac output, then, in a resting adult is

Cardiac output = stroke volume × beats per minute
= 70 ml × 75/min
= 5250 ml/min   or   5.25 liters/min

Factors that increase stroke volume or heart rate tend to increase cardiac output. Factors that decrease stroke volume or heart rate tend to decrease cardiac output.

STROKE VOLUME

The actual amount of blood ejected by a ventricle with each heartbeat depends on how much blood enters the ventricle during diastole and how much blood is left in the ventricle following its systole.

End-Diastolic Volume (EDV)

The term **end-diastolic volume (EDV)** refers to the volume of blood that enters a ventricle during diastole. As noted earlier about 70 percent of the blood flows from the atria into the ventricles before atrial systole; atrial contraction causes the additional 30 percent to enter the ventricles. During diastole, the ventricles normally increase their volume to 120 to 130 ml, the end-diastolic volume. This volume is determined principally by the length of ventricular diastole and venous pressure. When heart rate increases, the duration of diastole is shorter, the ventricles may contract before they are adequately filled, and the EDV is reduced. When venous pressure increases, a greater volume of blood is forced into the ventricles, and the EDV is increased. As you will see shortly, this results in a more forceful ventricular contraction.

End-Systolic Volume (ESV)

End-systolic volume (ESV) refers to the volume of blood still left in a ventricle following its systole. Since end-diastolic volume is about 120 to 130 ml and stroke volume is about 70 ml, end-systolic volume is about 50 to 60 ml (end-diastolic volume minus stroke volume).

End-systolic volume is determined principally by arterial pressure and the force of ventricular contraction. Arterial pressure in the aorta and pulmonary trunk just prior to ventricular systole is the pressure that must be overcome in order for opening of the semilunar valves. If arterial pressure is significantly elevated, the resistance prevents the ventricles from pumping out as much as they should and stroke volume decreases.

Within limits, contraction is more forceful when muscle fibers are stretched. During exercise, for example, a large amount of blood enters the heart and the increased diastolic filling stretches the fibers of the right ventricle. This increased length of the cardiac muscle fibers intensifies the force of the ventricular contraction, that is, the force of the beat. The increased incoming volume of blood is handled by an increased output through a more forceful ventricular contraction. As the increased amount of blood returns from the lungs to the left side of the heart, left ventricular stroke volume also increases. Thus, during exercise, cardiac output is increased. This phenomenon, by which the length of the cardiac muscle fiber determines the force of contraction, is referred to as **Starling's law of the heart.** The force of ventricular contraction is also determined by factors such as nervous control, chemicals, temperature, emotions, sex, and age. These factors are considered as part of the discussion dealing with the regulation of heart rate.

CLINICAL APPLICATION

The maximum percentage that the cardiac output can increase above normal is referred to as **cardiac reserve.** For example, during strenous exercise the cardiac output of a normal adult may increase to about four times normal. Since this is an increase of 400 percent above normal, we say that the cardiac reserve is 400 percent. Well-trained athletes may have a cardiac reserve as high as 600 percent, an increase in cardiac output of about six times normal. Although cardiac reserve depends on many factors, it is markedly influenced by heart conditions such as ischemic heart disease, valvular disorders, and myocardial damage.

HEART RATE

During certain pathological conditions, stroke volume may fall dangerously low. If the ventricular myocardium is weak or damaged by an infarction, it cannot contract strongly. Or blood volume may be reduced by excessive bleeding. Stroke volume then falls because the cardiac fibers are not sufficiently stretched. In these cases, the body attempts to maintain a safe cardiac output by increasing the rate and strength of contraction. The heart rate is regulated by several factors.

Autonomic Control

The most important control of heart rate is the effect of the autonomic nervous system. Within the medulla of the brain is a group of neurons called the **cardioacceleratory center (CAC).** Arising from this center are sympathetic fibers that travel down a tract in the spinal cord and then pass outward in the *cardiac (accelerator) nerves* and innervate the SA node, AV node, and portions of the myocardium (Figure 20-11). When the cardioacceleratory center is stimulated, nerve impulses travel along the sympathetic fibers. This causes them to release norepinephrine, which increases the rate of heartbeat and the strength of contraction.

The medulla also contains a group of neurons that form the **cardioinhibitory center (CIC).** Arising from this center are parasympathetic fibers that reach the heart via the *vagus (X) nerve.* These fibers innervate the SA node and AV node. When this center is stimulated, nerve impulses transmitted along the parasympathetic fibers cause the release of acetylcholine. This decreases the rate of heartbeat and strength of contraction.

The autonomic control of the heart is therefore the result of opposing sympathetic (stimulatory) and parasympathetic (inhibitory) influences. Sensory impulses from receptors in different parts of the cardiovascular system act upon the centers so that a balance between stimulation and inhibition is maintained. Nerve cells capable of responding to changes in blood pressure are called **pressoreceptors (baroreceptors).** These receptors affect the rate of heartbeat and are involved in three reflex pathways: the carotid sinus reflex, the aortic reflex, and the right heart (atrial) reflex.

● *Carotid Sinus Reflex*   The **carotid sinus reflex** is concerned with maintaining normal blood pressure in the brain. The *carotid sinus* is a small widening of the internal carotid artery just above the point where it branches off from the common carotid artery (Figure 20-11). In the wall of the carotid sinus lie pressoreceptors. Any increase in blood pressure stretches the wall of the sinus, and the stretching stimulates the pressoreceptors. The impulses then travel from the pressoreceptors over sensory neurons in the glossopharyngeal (IX) nerves. Within the medulla, the impulses stimulate the cardioinhibitory center and inhibit the cardioacceleratory center. Consequently, more parasympathetic impulses pass from the cardioinhibitory center via the vagus (X) nerves to the heart and fewer sympathetic impulses pass from the cardioacceleratory center via cardiac (accelerator) nerves to the heart. The result is a decrease in heart rate and force of contraction. There is a subsequent decrease in cardiac output, a decrease in arterial blood pressure, and restoration of blood pressure to normal.

If blood pressure falls, reflex acceleration of the heart takes place. The pressoreceptors in the carotid sinus do not stimulate the cardioinhibitory center, and the cardioacceleratory center is free to dominate. The heart then beats faster and more forcefully to restore normal blood pressure. This inverse relationship between blood pressure and heart rate is referred to as **Marey's law of the heart** (Figure 20-12).

The ability of the carotid sinus reflex (and aortic reflex described shortly) to maintain a relatively constant blood pressure is very important when a person sits or stands after having been lying down. Immediately upon moving from a prone to erect position, blood pressure in the head and upper part of the body falls. The decrease in pressure,

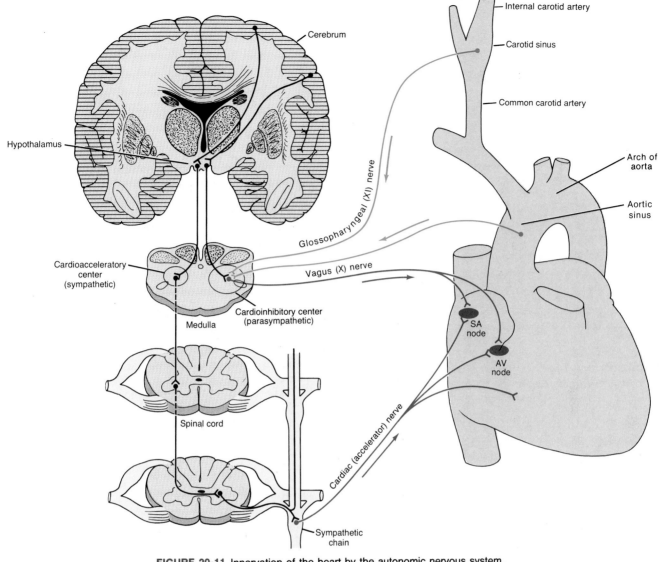

FIGURE 20-11 Innervation of the heart by the autonomic nervous system.

however, is counteracted by the reflexes. If the pressure was to fall markedly, unconsciousness could occur.

● **Aortic Reflex** The **aortic reflex** is concerned with general systemic blood pressure. It is initiated by pressoreceptors in the wall of the arch of the aorta (see Figure 20-11), and it operates like the carotid sinus reflex.

● **Right Heart (Atrial) Reflex** The **right heart (atrial) reflex** responds to venous blood pressure. It is initiated by pressoreceptors in the superior and inferior venae cavae and in the right atrium. When venous pressure increases, the pressoreceptors send impulses that stimulate the cardioacceleratory center. This action causes the heart rate to increase. This mechanism is called the **Bainbridge reflex.**

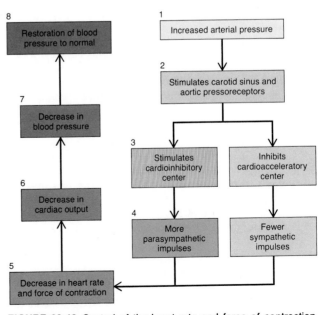

FIGURE 20-12 Control of the heart rate and force of contraction by pressoreceptors in the carotid sinus reflex and aortic reflex.

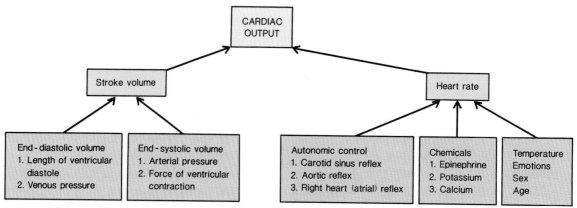

FIGURE 20-13 Summary of factors that influence cardiac output.

Chemicals

Certain chemicals present in the body also have an effect on heart rate. For example, epinephrine, produced by the adrenal medulla in response to sympathetic stimulation, increases the excitability of the SA node. This, in turn, increases the rate and strength of contraction. Elevated levels of potassium or sodium decrease the heart rate and strength of contraction. It appears that excess potassium interferes with the generation of nerve impulses and excess sodium interferes with calcium participation in muscular contraction. An excess of calcium increases heart rate and strength of contraction.

Temperature

Increased body temperature, such as occurs during strenuous exercise, causes the AV node to discharge impulses faster and thereby increases heart rate. Decreased body temperature resulting from exposure to cold or deliberately cooling the body prior to surgery decreases heart rate and strength of contraction.

Emotions

Strong emotions such as fear, anger, and anxiety, along with a multitude of physiological stressors, increase heart rate through the general adaptation syndrome. Mental states such as depression and grief tend to stimulate the cardioinhibitory center and decrease heart rate.

Sex and Age

Sex is another factor—the heartbeat is somewhat faster in females. Age is yet another factor—the heartbeat is fastest at birth, moderately fast in youth, average in adulthood, and below average in old age.

A summary of the factors that influence cardiac output is presented in Figure 20-13.

CIRCULATORY SHOCK AND HOMEOSTASIS

When cardiac output or blood volume is reduced to the point where body tissues do not receive an adequate blood supply, **circulatory shock** results. Circulatory shock is caused by loss of blood volume through hemorrhage or through the release of histamine due to damage to body tissues (trauma). The characteristic symptoms of circulatory shock are a pale, clammy skin, cyanosis of ears and fingers, a feeble though rapid pulse, shallow and rapid breathing, lowered body temperature, and mental confusion or unconsciousness.

If the shock is mild, certain homeostatic mechanisms of the circulatory system compensate so that no serious damage results. Lowered blood pressure is compensated by constriction of blood vessels and water retention. Renin is secreted by the kidneys, aldosterone by the adrenal cortex, epinephrine by the adrenal medulla, and ADH by the posterior pituitary. Even though some blood is lost from circulation, blood return to the heart is normal and cardiac output remains essentially unchanged. Veins and many arterioles are constricted during compensation, but there is no constriction of arterioles supplying the heart and brain. As a consequence, blood flow to the heart and brain is normal or nearly so. Compensation is an effective homeostatic mechanism until about 900 ml of blood is lost.

If the shock is severe, death may occur. For instance, if the return of venous blood is greatly diminished by excessive blood loss, the compensatory mechanisms are

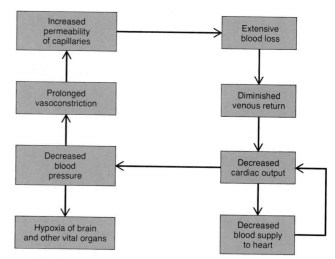

FIGURE 20-14 Circulatory shock cycle. Note how the cycle perpetuates itself until death results.

insufficient. When the cardiac output decreases, the heart fails to pump enough blood to supply its own coronary vessels and the heart muscle weakens. In addition, prolonged vasoconstriction ultimately leads to tissue hypoxia, and vital organs such as the kidneys and liver are damaged. Essentially, the initial shock promotes more shock and a **circulatory shock cycle** is established (Figure 20-14). Once the shock reaches a certain level of severity, damage to the circulatory organs is so extensive that death ensues.

DISORDERS: HOMEOSTATIC IMBALANCES

Risk Factors in Heart Disease

It is estimated that one in every five persons who reaches 60 will have a **heart attack.** One in every four persons between 30 and 60 has the potential to be stricken. Heart disease is epidemic in this country, despite the fact that some of the causes can be foreseen and prevented. The results of research indicate that people who develop combinations of certain risk factors eventually have heart attacks. These factors are:

1. High blood cholesterol level.
2. High blood pressure.
3. Cigarette smoking.
4. Obesity.
5. Lack of exercise.
6. Diabetes mellitus.
7. Genetic predisposition.

The first five risk factors all contribute to increasing the heart's work load. High blood cholesterol and hypertension are discussed later in the next chapter. Cigarette smoking, through the effects of nicotine, stimulates the adrenal gland to oversecrete aldosterone, epinephrine, and norepinephrine—powerful vasoconstrictors. Overweight people develop miles of extra capillaries to nourish fat tissue. The heart has to work harder to pump the blood through more vessels. Without exercise, venous return gets less help from contracting skeletal muscles. In addition, regular exercise strengthens the smooth muscle of blood vessels and enables them to assist general circulation. Exercise also increases cardiac efficiency and output. In diabetes mellitus, fat metabolism dominates glucose metabolism. As a result, cholesterol levels get progressively higher and result in plaque formation, a situation that may lead to high blood pressure. High blood pressure drives fat into the vessel wall, encouraging atherosclerosis.

There are significant differences between males and females in terms of coronary heart disease. For example, white males have more severe coronary artery atherosclerosis and more frequent myocardial infarction and sudden death than white females. By contrast, white females have a greater incidence of angina pectoris than white males. These differences are not as great in nonwhites.

Research indicates that muscle cells of the atria and ventricles contain receptors for androgens, steroid hormones produced by the adrenal cortex. Although the significance of this is unknown, it does suggest that the sex steroid hormones may affect heart function directly and may explain the differences between males and females in terms of coronary heart disease.

Causes of Heart Disease

Generally, the immediate cause of heart trouble is one of the following: inadequate coronary blood supply, anatomical disorders, or faulty electrical conduction in the heart.

Inadequate Coronary Blood Supply

Angina pectoris and myocardial infarction result from insufficient oxygen supply to the myocardium. One cause is a thrombus or an embolus in a coronary artery. Another is the buildup of fatty deposits in arterial walls (Chapter 21).

Anatomical Disorders

Less than 1 percent of all new babies have a **congenital,** or **inborn, heart defect.** Even so, the total number in this country each year is estimated to be 30,000 to 40,000. Some of these infants may live quite healthy and long lives without any need for repairing their hearts. But sometimes an inborn heart defect is so severe that an infant lives only a few hours. A common anatomical defect is **patent ductus arteriosus.** The connection between the aorta and the pulmonary artery remains open instead of closing completely after birth. As a result, aortic blood flows into the lower-pressure pulmonary trunk, thus increasing the pulmonary trunk blood pressure and overworking both ventricles and the heart.

A **septal defect** is an opening in the septum that separates the interior of the heart into a left and right side. **Interatrial septal defect** is failure of the fetal foramen ovale between the two atria to close after birth. Because pressure in the right atrium is low, interatrial septal defect generally allows a good deal of blood to flow from the left atrium to the right without going through systemic circulation. This defect, an example of left-to-right shunt, overloads the pulmonary circulation, produces fatigue, and increases respiratory infections. If it occurs early in life, it inhibits growth because the systemic circulation may be deprived of a considerable portion of the blood destined for the organs and tissues of the body.

Interventricular septal defect is caused by an incomplete closure of the interventricular septum. Owing to the higher pressure in the left ventricle, blood is shunted directly from the left ventricle into the right ventricle. This is another example of a left-to-right shunt. Septal openings can now be sewn shut or covered with synthetic patches.

Valvular stenosis is a narrowing, or *stenosis,* of one of the valves regulating blood flow in the heart. Stenosis may occur in the valve itself, most commonly in the mitral valve from rheumatic heart disease or in the aortic valve from sclerosis or rheumatic fever. Or it may occur near a valve. All stenoses are serious because they place a severe work load on the heart by making it work harder to push the blood through the abnormally narrow valve openings. As a result of mitral stenosis, blood pressure is increased. Angina pectoris and heart failure may accompany this disorder. Most stenosed valves are totally replaced with artificial valves (see Figure 20-4).

Tetralogy of Fallot (tet-RAL-ō-jē fal-Ō) is a combination of four defects: an interventricular septal opening, an aorta that emerges from both ventricles instead of from the left ventricle, a stenosed pulmonary semilunar valve, and an enlarged right ventricle. The condition is an example of a right-to-left shunt, in which blood is shunted from the right ventricle to the left ventricle without going through pulmonary circulation. Because there is stenosis of the pulmonary semilunar valve, the increased right ventricular pressure forces deoxygenated blood from the right ventricle to enter the left ventricle through the interventricular septum. As a result, deoxygenated blood gets mixed with the oxygenated blood that is pumped into systemic circulation. Also, because the aorta emerges from the right ventricle and the pulmonary artery is stenosed, very little blood ever gets to the lungs and pulmonary circulation is bypassed almost completely. The insufficient amount of oxygenated blood in systemic circulation results in *cyanosis* (sī-a-NŌ-sis), a blue or dark purple discoloration of skin. For this reason the tetralogy of Fallot is one of the conditions that causes a "blue baby." Lung disorders and suffocation also result in cyanosis.

Today it is possible to correct cases of tetralogy of Fallot when the patient is of proper age and condition. Open-heart operations are performed in which the narrowed pulmonary valve is cut open and the interventricular septal defect is sealed with a Dacron patch.

Faulty Conduction: Arrhythmias

At least half the deaths from myocardial infarction occur before the patient reaches the hospital. These early deaths could result from an irregular heart rhythm—an **arrhythmia** (a-RITH-mē-a). Sometimes this condition progresses to the stage where the heart stops functioning. An arrhythmia is caused by disturbances in the conduction system. This abnormal rhythm of the heartbeat can result in cardiac arrest if the heart cannot supply its own oxygen demands, as well as those of the rest of the body. Serious arrhythmias can be controlled, and the normal heart rhythm can be reestablished, if they are detected and treated early enough.

Arrhythmia arises when electrical impulses through the heart are blocked at critical points in the conduction system. One such arrhythmia is called a **heart block.** Perhaps the most common blockage is in the atrioventricular node, which conducts impulses from the atria to the ventricles. This disturbance is called *atrioventricular (AV) block.* It usually indicates a myocardial infarction, atherosclerosis, rheumatic heart disease, diphtheria, or syphilis. In a *first-degree AV block,* which can be detected only with an electrocardiograph, the conduction of impulses from the atria to the ventricles through the AV node is delayed. As a result, the P-R interval is longer than normal.

Occasionally, the delay at the AV node will progressively increase from beat to beat until conduction to the ventricles is blocked. This causes the P-R interval to become longer and longer until finally a P wave occurs without a following QRS complex and T wave. By the time the sinoatrial node fires again, AV conduction has had time to recover and the sequence starts over. This condition represents one variation of a *second-degree AV block.*

In another variation of a second-degree AV block, the delay at the AV node is sufficiently long that occasionally an atrial impulse fails to reach the ventricles. This usually occurs in a regular sequence such as 2:1, 3:1, or 4:1, the ratios being that of atrial to ventricular beats. When ventricular contraction does not occur (dropped beat), oxygenated blood is not pumped efficiently to all parts of the body. The patient may feel faint or may collapse if there are many dropped ventricular beats.

In a *third-degree* or *complete AV block,* there is no conduction of impulses between the atria and ventricles. Atrial and ventricular rates get out of synchronization (Figure 20-15a). The ventricles may go into systole at any time. This condition could occur when the atria are in systole or just before. Or the ventricles may rest for a few cardiac cycles. With complete AV block, patients may have vertigo, unconsciousness, or convulsions. These symptoms result from a decreased cardiac output with diminished cerebral blood flow and cerebral hypoxia or lack of sufficient oxygen.

Among the causes of AV block are excessive stimulation by the vagus (X) nerves that depresses conductivity of the junctional fibers, destruction of the AV bundle as a result of coronary infarct, atherosclerosis, myocarditis, or depression caused by various drugs. Other heart blocks include *intraatrial (IA) block,* *interventricular (IV) block,* and *bundle branch block (BBB).*

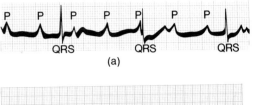

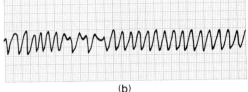

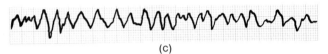

FIGURE 20-15 Abnormal electrocardiograms. (a) Third-degree (complete) AV block. There is no fixed ratio between atrial depolarizations (P waves) and ventricular depolarizations (QRS complexes). (b) Atrial fibrillation. Atrial depolarization is very rapid (400–600/min). This gives rise to very small electrical changes replacing P waves. Those that are conducted through the AV node are done so at very irregular time intervals, hence the QRS complexes are totally irregular in spacing. (c) Ventricular fibrillation. Total disorganization of the electrical pattern occurs. There are only very rapid and irregular waves with no suggestion of normal complexes.

In the latter condition, the ventricles do not contract together because of the delayed impulse in the blocked branch.

Two other abnormal rhythms are **flutter** and **fibrillation.** In **atrial flutter** the atrial rhythm averages between 240 and 360 beats per minute. The condition is essentially rapid atrial contractions accompanied by a second-degree AV block. It generally indicates severe damage to heart muscle. Atrial flutter usually becomes fibrillation after a few hours, days, or weeks. **Atrial fibrillation** is synchronous contraction of the atrial muscles that causes the atria to contract irregularly and still faster (Figure 20-15b). Atrial flutter and fibrillation occur in myocardial infarction, acute and chronic rheumatic heart disease, and hyperthyroidism. Atrial fibrillation results in complete uncoordination of atrial contraction so that atrial pumping ceases altogether. When the muscle fibrillates, the muscle fibers of the atrium quiver individually instead of contracting together. The quivering cancels out the pumping of the atrium. In a strong heart, atrial fibrillation reduces the pumping effectiveness of the heart by only 25 to 30 percent.

Ventricular fibrillation is another abnormality that almost always indicates imminent cardiac arrest and death unless corrected quickly. It is characterized by asynchronous, haphazard, ventricular muscle contractions. The rate may be rapid or slow. The impulse travels to the different parts of the ventricles at different rates. Thus part of the ventricle may be contracting while other parts are still unstimulated. Ventricular contraction becomes ineffective and circulatory failure and death occur (Figure 20-15c). Ventricular fibrillation may be caused by coronary occlusion. It sometimes occurs during surgical procedures on the heart or pericardium. It may be the cause of death in electrocution.

In order to stop ventricular fibrillation, a very strong electrical current can be applied for a short period of time. This process is called **defibrillation** and the instrument used to deliver the current is called a **defibrillator.** Through two metal paddles pressed to the chest, a strong electrical current is delivered and it causes all the cardiac muscle fibers to become depolarized simultaneously. All impulses stop and the heart remains quiescent for 3 to 5 seconds. It is hoped that the SA node will again begin functioning to restore a normal heart rhythm. Clinical trials are now underway to test the feasibility of a defibrillator implanted in the abdomen called the *Automatic Implantable Defibrillator (AID)*. It automatically monitors heart activity and can deliver up to four electrical currents to correct the fibrillation.

Another form of arrhythmia arises when a small region of the heart becomes more excitable than normal, causing an occasional abnormal impulse to be generated between normal impulses. The region from which the abnormal impulse is generated is called an **ectopic focus.** As a wave of depolarization spreads outward from the ectopic focus, it causes a *premature contraction (extrasystole)*. The contraction occurs early in diastole before the SA node is normally scheduled to discharge its impulses. A person with such a ventricular premature contraction might feel a thump in the chest. Ectopic foci may be caused by emotional stress, excessive intake of stimulants such as caffeine or nicotine, lack of sleep, and local ischemic areas of cardiac muscle.

Congestive Heart Failure (CHF)

Congestive heart failure (CHF) may be defined as a chronic or acute state that results when the heart is not capable of supplying the oxygen demands of the body. Symptoms and signs of CHF are produced by diminished blood flow to the various tissues of the body and by accumulation of excess blood in the various organs because the heart is unable to pump out the blood returned to it by the great veins. Both diminished blood flow and accumulation of excess blood in the organs occur together, but certain symptoms result from congestion, while others are produced by poor tissue nutrition. CHF is usually caused by coronary artery disease in which atherosclerosis leads to myocardial ischemia.

Although the treatment of CHF should be directed toward correcting the cause, its medical management is basically the same in all cases. CHF occurs when the heart pumps less blood than the body requires. Consequently, the aim of therapy is to increase the output of the heart, improve tissue oxygenation, remove excess fluid, and prevent fluid reaccumulation. Medical management of CHF requires digitalis (a drug that strengthens the contraction of heart muscles), diuretics, dietary salt restriction, possibly oxygen, and antiarrhythmic medications (drugs that help maintain normal cardiac rhythm).

Cardiac Catheterization

In **cardiac catheterization,** the tip of a long plastic *catheter,* or tube, is introduced into a vein in the arm or leg. The catheter is radiopaque so that it can be seen with a fluoroscope. With the help of the fluoroscope, it is then threaded through the vena cava and into the right atrium, right ventricle, or pulmonary trunk. The catheter can also be inserted into an artery of the arm or leg and worked up through the aorta to the left atrium and ventricle.

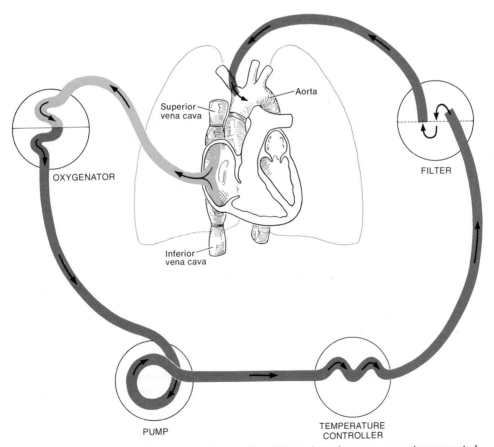

FIGURE 20-16 Principle of the heart-lung bypass. Blood drawn from the venae cavae is oxygenated, rewarmed to body temperature or cooled, filtered to remove air and emboli, and returned to the aorta and coronary arteries at the proper pressure. If necessary, drugs, anesthetics, and transfusions may be added to the circuit.

A recently developed technique using an electrode-type catheter allows the physician to obtain an intracardiac electrocardiogram in order to evaluate the status of the conduction system. Even newer is the development of a balloon catheter to correct a cardiac defect. In one common infantile defect, there is transposition of the great vessels, and the infant frequently dies of hypoxia because of inadequate mixing of arterial and venous blood. To correct this defect without a thoracotomy (chest incision), a catheter with an uninflated balloon on the tip is inserted into the right atrium through the foramen ovale into the left atrium. Inflating the balloon and rapidly withdrawing it with a sudden vigorous pull ruptures the atrial septum establishing a significant interatrial communication. It produces adequate pulmonary-systemic venous mixing at the atrial level, overcoming the hypoxia.

Balloon catheters are also being used to dilate partially or completely obstructed coronary arteries as well as other blood vessels. The catheter is guided to the obstructed artery, blown up briefly, then withdrawn. This procedure, called **percutaneous transluminal angioplasty (PTA),** compresses the obstructing material against the vessel wall allowing better blood flow. In some instances, this procedure has prevented limb amputation and coronary bypass surgery.

A number of clinical uses are now made of cardiac catheterization. One allows the physician to investigate congenital or acquired heart disease to determine the type and severity of defects and thereby determine if there is a need for surgery and the type of surgery required. Another is to find out the status of the coronary arteries and ventricular function in coronary artery disease. Visualization of the coronary arteries from various projections enables the cardiologist to determine the need for coronary bypass surgery. Also, pressures may be recorded to provide information about the functioning of valves, and the oxygen content of blood near atrial or ventricular septal defects may be analyzed.

The development of cardiac catheterization equipment and techniques has now made it possible to place catheters into the vascular system of almost every organ. This procedure is not only an aid to diagnosis, but can also be used to apply chemotherapy locally.

Open-Heart Surgery

Before surgeons can perform **open-heart surgery** to correct even the simplest heart defect, they have to be able to open up the heart and expose the chambers. Techniques have been developed to capture the blood spurting out of the open chamber and pump it back into the vessels. One such life-support technique is the heart-lung bypass. Coupled with hypothermia, it has now made possible both heart surgery and heart transplants. **Hypothermia** (hī'-pō-THER-mē-a) refers to a low body temperature. In various surgical procedures, it refers to a deliberate cooling of the body in order to slow metabolism and reduce the oxygen needs of the tissues. Thus the heart and brain can withstand short periods of interrupted or reduced blood flow. Lost blood is then replaced by transfusion during and after the operation.

In the **heart-lung bypass,** blood bypasses the heart and lungs completely (Figure 20-16). It is pumped and oxygenated by a heart-lung machine outside the body. Modern heart-lung machines may also chill the blood to produce hypothermia as well.

MEDICAL TERMINOLOGY

Aortic insufficiency   An improper closure of the aortic semilunar valve that permits a backflow of blood.

Cardiac arrest   Complete stoppage of the heartbeat.

Cardiomegaly (*mega* = large)   Heart enlargement. Long-distance runners and weight lifters frequently develop heart enlargement as a natural adaptation to increased work load produced by regular exercise.

Commissurotomy   An operation that is performed to widen the opening in a heart valve which has become narrowed by scar tissue.

Compensation   A change in the circulatory system made to compensate for some abnormality; an adjustment of size of heart or rate of heartbeat made to counterbalance a defect in structure or function; often used specifically to describe the maintenance of adequate circulation in spite of the presence of heart disease.

Compliance   The passive or diastolic stiffness properties of the left ventricle. A hypertrophied or fibrosed heart with a stiff wall, for example, has decreased compliance. Also, the stiffness properties of the lung.

Constrictive pericarditis   A shrinking and thickening of the pericardium which prevents the heart muscle from expanding and contracting normally.

Incompetent valve   Any valve that does not close properly thus permitting a backflow of blood; also called **valvular insufficiency.**

Palpitation   A fluttering of the heart or abnormal rate or rhythm of the heart.

Pancarditis (*pan* = all)   Inflammation of the whole heart including inner layer (endocardium), heart muscle (myocardium), and outer sac (pericardium).

Paroxysmal tachycardia   A period of rapid heartbeats which begins and ends suddenly.

Percussion   Tapping a part of the body as an aid in diagnosing the condition of parts of the body by the sound obtained.

Stokes-Adams syndrome   Sudden attacks of unconsciousness, sometimes with convulsions, which may accompany heart block.

Syncope   A temporary cessation of consciousness; a faint. One cause might be insufficient blood supply to the brain.

STUDY OUTLINE

Location (p. 456)
1. The heart is situated obliquely between the lungs in the mediastinum.
2. About two-thirds of its mass is to the left of the midline.

Parietal Pericardium (Pericardial Sac) (p. 457)
1. The parietal pericardium, consisting of an outer fibrous layer and an inner serous layer, encloses the heart.
2. Between the serous pericardium and the epicardium is the pericardial cavity, a space filled with pericardial fluid that prevents friction between the two membranes.

Wall; Chambers; Vessels; and Valves (pp. 457–459)
1. The wall of the heart has three layers: epicardium, myocardium, and endocardium.
2. The chambers include two upper atria and two lower ventricles.
3. The blood flows through the heart from the superior and

inferior venae cavae and the coronary sinus to the right atrium, through the tricuspid valve to the right ventricle, through the pulmonary trunk to the lungs, through the pulmonary veins into the left atrium, through the bicuspid valve to the left ventricle, and out through the aorta.

4. Valves prevent backflow of blood in the heart.
5. Atrioventricular (AV) valves, between the atria and their ventricles, are the tricuspid valve on the right side of the heart and the bicuspid (mitral) valve on the left.
6. The chordae tendineae and their muscles keep the flaps of the valves pointing in the direction of blood flow.
7. The two arteries that leave the heart both have a semilunar valve.

Conduction System (p. 463)

1. The conduction system consists of tissue specialized for impulse conduction.
2. Components of this system are the sinoatrial node (pacemaker), atrioventricular (AV) node, atrioventricular (AV) bundle, bundle branches, and Purkinje fibers.

Electrocardiogram (p. 465)

1. The record of electrical changes during each cardiac cycle is referred to as an electrocardiogram (ECG).
2. A normal ECG consists of a P wave (spread of impulse from SA node over atria), QRS wave (spread of impulse through ventricles), and T wave (ventricular repolarization). The P-R interval represents the conduction time from the beginning of atrial excitation to the beginning of ventricular excitation. The S-T segment represents the time between the end of the spread of the impulse through the ventricles and repolarization of the ventricles.
3. The ECG is invaluable in diagnosing abnormal cardiac rhythms and conduction patterns, detecting the presence of fetal life, determining the presence of several fetuses, and following the course of recovery from a heart attack.
4. An artificial pacemaker may be used to restore an abnormal cardiac rhythm.

Blood Supply (p. 467)

1. The coronary (cardiac) circulation takes oxygenated blood through the arterial system of the myocardium.
2. Deoxygenated blood returns to the right atrium via the coronary sinus.
3. Complications of this system are angina pectoris and myocardial infarction.

Cardiac Cycle (p. 468)

1. Blood flows through the heart from an area of higher to lower pressure.
2. The pressure developed is related to the size and volume of a chamber.
3. A cardiac cycle consists of the systole (contraction) and diastole (relaxation) of both atria plus the systole and diastole of both ventricles followed by a short pause.
4. The movement of blood through the heart is controlled by the opening and closing of the valves and the contraction and relaxation of the myocardium.

5. With an average heartbeat of 75/min, a complete cardiac cycle requires 0.8 sec.
6. The first sound (lubb) represents the closing of the atrioventricular valves. The second sound (dupp) represents the closing of semilunar valves.
7. A peculiar sound is called a murmur.

Cardiac Output (CO) (p. 471)

1. Cardiac output (CO) is the amount of blood ejected by the left ventricle into the aorta per minute. It is calculated as follows: CO = stroke volume × beats per minute.
2. Stroke volume (SV) is the amount of blood ejected by a ventricle during each systole.
3. Stroke volume (SV) depends on how much blood enters a ventricle during diastole (end-diastolic volume) and how much blood is left in a ventricle following its systole (end-systolic volume).
4. The maximum percentage that cardiac output can be increased above normal is cardiac reserve.
5. Heart rate and strength of contraction may be increased by sympathetic stimulation from the cardioacceleratory center in the medulla and decreased by parasympathetic stimulation from the cardioinhibitory center in the medulla.
6. Pressoreceptors are nerve cells that respond to changes in blood pressure. They act on the cardiac centers in the medulla through three reflex pathways: carotid sinus reflex, aortic reflex, and right heart (atrial) reflex.
7. Other influences on heart rate include chemicals (epinephrine, sodium, potassium), temperature, emotion, sex, and age.

Circulatory Shock and Homeostasis (p. 474)

1. Shock results when cardiac output is reduced or blood volume decreases to the point where body tissues become hypoxic.
2. Mild shock is compensated by vasoconstriction and water retention.
3. In severe shock, venous return is diminished and cardiac output decreases. The heart becomes hypoxic, prolonged vasoconstriction leads to hypoxia of other organs, and the shock cycle is intensified.

Disorders: Homeostatic Imbalances (p. 475)

1. Risk factors in heart disease include high blood cholesterol, high blood pressure, cigarette smoking, obesity, lack of exercise, diabetes mellitus, and genetic disposition.
2. The immediate causes of heart disease are inadequate coronary blood supply, anatomical disorders (patent ductus arteriosus, septal defects, valvular stenosis, and tetralogy of Fallot), and arrhythmias (heart block, flutter, fibrillation, and premature contractions).
3. Congestive heart failure (CHF) results when the heart cannot supply the oxygen demands of the body.
4. Cardiac catheterization permits physicians to determine heart disorders and pressures, to correct some defects, and to apply chemotherapy locally.
5. Hypothermia (deliberate body cooling) and the heart-lung bypass permit open-heart surgery.

REVIEW QUESTIONS

1. Describe the location of the heart in the mediastinum. Distinguish the subdivisions of the pericardium. What is the purpose of this structure?
2. Compare the three portions of the heart wall. Define atria and ventricles. What vessels enter or exit the atria and ventricles?
3. Describe the principal valves in the heart and how they operate.

4. Describe the path of a nerve impulse through the heart's conducting system.
5. Define and label the deflection waves of a normal electrocardiogram. Explain why the ECG is an important diagnostic tool.
6. What is a Holter monitor? Why is it used?
7. Describe the types of artificial pacemakers.
8. Describe the route of blood in the coronary (cardiac) circulation. Distinguish between angina pectoris and myocardial infarction.
9. Discuss the pressure changes associated with the cardiac cycle.
10. List the principal events of atrial diastole, ventricular systole, atrial systole, and ventricular diastole.
11. By means of a labeled diagram, relate the events of the cardiac cycle to time. What is the quiescent period?
12. Describe the significance of the heart sounds. What is a heart murmur?
13. What is cardiac output (CO)? How is it calculated?
14. Define stroke volume (SV). Explain how end-diastolic volume (EDV) and end-systolic volume (ESV) are related to stroke volume.
15. What is Starling's law of the heart? What is its significance?
16. Define cardiac reserve. Why is it important?
17. Distinguish between the cardioacceleratory and cardioinhibitory centers with respect to the regulation of heart rate.
18. What is a pressoreceptor? Outline the operation of the carotid sinus reflex, aortic reflex, and right heart (atrial) reflex.
19. Explain how each of the following affects heart rate: chemicals, temperature, emotions, sex, and age.
20. Define circulatory shock. What are its symptoms?
21. How is homeostasis restored after mild shock?
22. Describe the effects of severe shock by drawing a shock cycle.
23. Describe the risk factors involved in heart disease.
24. Describe patent ductus arteriosus and the various types of septal defects.
25. What are the common causes of valvular stenosis? What are the consequences of this disorder?
26. What is tetralogy of Fallot?
27. What is heart block? Describe the various types of atrioventricular (AV) block.
28. Distinguish between atrial flutter, atrial fibrillation, ventricular fibrillation, and premature contractions as types of arrhythmias.
29. What is defibrillation?
30. Describe the symptoms of congestive heart failure (CHF).
31. What is cardiac catheterization? List some of its clinical applications.
32. Explain the use of hypothermia and the heart-lung bypass in open-heart surgery.
33. Refer to the glossary of medical terminology associated with the heart. Be sure that you can define each term.

- Contrast the structure and function of arteries, arterioles, capillaries, venules, and veins.
- Explain why blood circulates between different regions of the cardiovascular system.
- Relate the importance of cardiac output, blood volume, and peripheral resistance to blood pressure.
- Explain the role of the vasomotor center in controlling blood pressure.
- Contrast the roles of pressoreceptors and chemoreceptors in regulating blood pressure.
- Describe the effects of epinephrine, norepinephrine (NE), antidiuretic hormone (ADH), angiotensin II, histamine, and kinins on blood pressure.
- Define autoregulation and explain its importance.
- Discuss the movement of fluids through capillaries.
- Explain how velocity of blood flow, skeletal muscle contractions, valves in veins, and breathing assist in the return of venous blood to the heart.
- Define a blood reservoir and explain its importance.
- Explain the effects of exercise on the cardiovascular system.
- Define pulse and identify the arteries where pulse may be felt.
- Define blood pressure (BP).

- Explain one clinical method for recording systolic and diastolic pressure.
- Contrast the clinical significance of systolic, diastolic, and pulse pressures.
- Compare systemic, hepatic portal, pulmonary, fetal, and cerebral circulation.
- Identify the principal arteries and veins of systemic circulation.
- Trace the route of blood involved in hepatic portal circulation and explain its importance.
- Identify the major blood vessels of pulmonary circulation.
- Contrast fetal and adult circulation.
- Describe the importance of cerebral circulation.
- Explain the fate of fetal circulation structures once postnatal circulation is established.
- List the causes and symptoms of aneurysms, coronary artery disease, atherosclerosis, coronary artery spasm, and hypertension.
- Define medical terminology associated with the cardiovascular system.
- Explain the actions of selected drugs on the cardiovascular system.

21 The Cardiovascular System: Vessels and Routes

The blood vessels form a network of tubes that carry blood away from the heart, transport it to the tissues of the body, and then return it to the heart. **Arteries** are the vessels that carry blood from the heart to the tissues. Large, elastic arteries leave the heart and divide into medium-sized, muscular vessels that head toward the various regions of the body. The medium-sized arteries divide into small arteries, which, in turn, divide into still smaller arteries called **arterioles.** As the arterioles enter a tissue, they branch into countless microscopic vessels called **capillaries.** Through the walls of the capillaries, substances are exchanged between the blood and body tissues. Before leaving the tissue, groups of capillaries reunite to form small veins called **venules.** These, in turn, merge to form progressively larger tubes called veins. **Veins,** then, are blood vessels that convey blood from the tissues back to the heart. Since blood vessels require oxygen and nutrients just like other tissues of the body, they also have blood vessels in their own walls called **vasa vasorum.**

ARTERIES

Arteries have walls constructed of three coats or tunics and a hollow core, called a *lumen,* through which the blood flows (Figures 21-1 and 21-4). The inner coat of an arterial wall is the **tunica interna (intima).** It is composed of a lining of endothelium (simple squamous epithelium) that is in contact with the blood and a layer of elastic tissue called the internal elastic membrane. The middle coat, or **tunica media,** is usually the thickest layer. It consists of elastic fibers and smooth muscle. The outer coat, the **tunica externa (adventitia),** is composed principally of elastic and collagenous fibers. An external elastic membrane may separate the tunica media from the tunica externa.

As a result of the structure of the middle coat especially, arteries have two major properties: elasticity and contractility. When the ventricles of the heart contract and eject blood into the large arteries, the arteries expand to contain the extra blood. Then, as the ventricles relax, the elastic recoil of the arteries forces the blood onward. The contractility of an artery comes from its smooth muscle. The smooth muscle is arranged longitudinally and in rings around the lumen somewhat like a doughnut and is innervated by sympathetic branches of the autonomic nervous system. When there is sympathetic stimulation, the smooth muscle contracts, squeezes the wall around the lumen, and narrows the vessel. Such a decrease in the size of the lumen is called **vasoconstriction.** Conversely, when sympathetic stimulation is removed, the smooth muscle fibers relax and the size of the arterial lumen increases. This increase is called **vasodilation** and is often due to the inhibition of vasoconstriction.

The contractility of arteries also serves a function in stopping bleeding. This is called vascular spasm, one of the three mechanisms involved in hemostasis (Chapter 19). The blood flowing through an artery is under a great deal of pressure. Thus great quantities of blood can be quickly lost from a broken artery. When an artery is cut, its wall constricts so that blood does not escape quite

so rapidly. However, there is a limit to how much vasoconstriction can help.

ELASTIC ARTERIES

Large arteries are referred to as **elastic** or **conducting arteries.** They include the aorta and brachiocephalic, common carotid, subclavian, vertebral, and common iliac arteries. The wall of elastic arteries is relatively thin in proportion to their diameter and their tunica media contains more elastic fibers and less smooth muscle. As the heart alternately contracts and relaxes, the rate of blood flow tends to be intermittent. When the heart contracts and forces blood into the aorta, the wall of the elastic arteries stretches. During relaxation of the heart, the wall of the elastic arteries recoils, moving the blood forward in a more continuous flow. Elastic arteries are called conducting arteries because they conduct blood from the heart to medium-sized arteries.

MUSCULAR ARTERIES

Medium-sized arteries are called **muscular** or **distributing arteries.** They include the axillary, brachial, radial, intercostal, splenic, mesenteric, femoral, popliteal, and tibial arteries. Their tunica media contains more smooth muscle than elastic fibers and they are capable of vasoconstriction and vasodilation to adjust the volume of blood to suit the needs of the structure supplied. The wall of muscular arteries is relatively thick, mainly due to the large amounts of smooth muscle. Muscular arteries are called distributing arteries because they distribute blood to various parts of the body.

ANASTOMOSIS

Most parts of the body receive branches from more than one artery. In such areas the distal ends of the vessels unite. The junction of two or more vessels supplying the same body region is called an **anastomosis** (a-nas-tō-MŌ-sis). Anastomoses may also occur between the origins of veins and between arterioles and venules. Anastomoses between arteries provide alternate routes by which blood can reach a tissue or organ. Thus if a vessel is occluded by disease, injury, or surgery, circulation to a part of the body is not necessarily stopped. The alternate route of blood to a body part through an anastomosis is known as **collateral circulation.** An alternate blood route may also be from nonanastomosing vessels that supply the same region of the body.

Arteries that do not anastomose are known as **end arteries.** Occlusion of an end artery interrupts the blood supply to a whole segment of an organ, producing necrosis (death) of that segment.

CLINICAL APPLICATION
A procedure that provides physicians with a much clearer look at diseased arteries is called **digital subtraction angiography (DSA).** The procedure employs a computer technique that compares an x-ray image

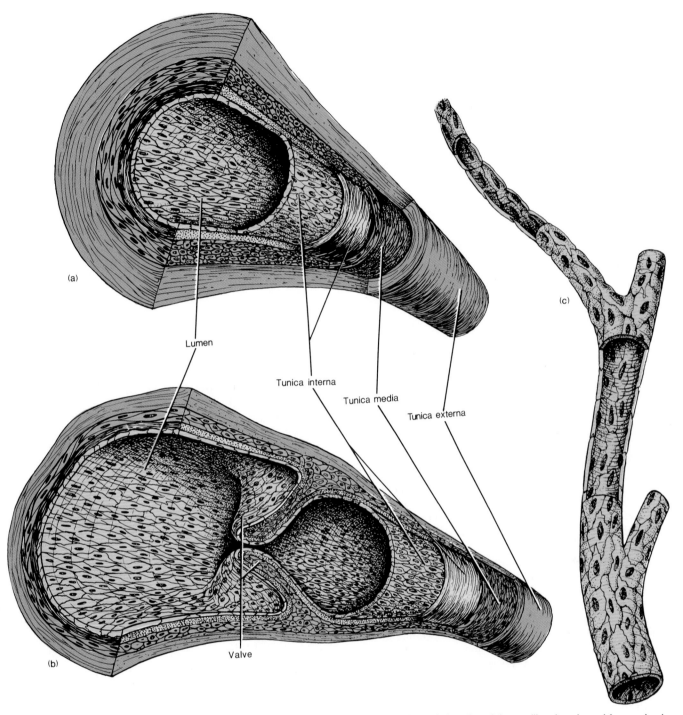

Lumen

Tunica interna

Tunica media

Tunica externa

Valve

(a)

(b)

(c)

FIGURE 21-1 Comparative structure of (a) an artery, (b) a vein, and (c) a capillary. The relative size of the capillary is enlarged for emphasis.

of the same region of the body before and after a contrast substance has been introduced intravenously. Any tissue or blood vessels that show up in the first image can be subtracted (erased) from the second image, leaving an unobstructed view of an artery. DSA figuratively lifts an artery out of the body so that it can be studied in isolation. DSA helps in diagnosing lesions in the carotid arteries leading to the brain, a potential cause of strokes, and in evaluating patients before surgery and after coronary bypass and certain transplant operations.

ARTERIOLES

An **arteriole** is a small artery that delivers blood to capillaries. Arterioles closer to the arteries from which they branch have a tunica interna like that of arteries, a tunica media composed of smooth muscle and very few elastic fibers, and a tunica externa composed mostly of elastic and collagenous fibers (see Figure 21-4c). As arterioles get smaller in size, the tunics change character so that arterioles closest to capillaries consist of little more than a layer of endothelium surrounded by a few scattered smooth muscle cells (Figure 21-2).

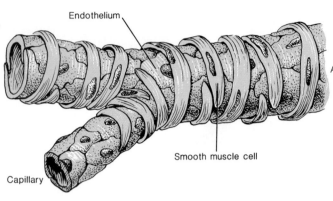

FIGURE 21-2 Structure of an arteriole.

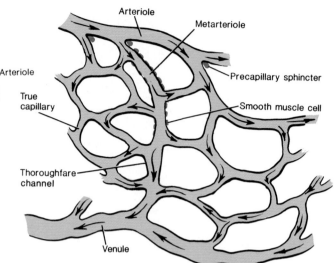

FIGURE 21-3 Details of a capillary network.

Arterioles play a key role in regulating blood flow from arteries into capillaries. The smooth muscle of arterioles, like that of arteries, is subject to vasoconstriction and vasodilation. During vasoconstriction, blood flow into capillaries is restricted; during vasodilation, the flow is significantly increased. The relationship of arterioles to blood flow will be considered in detail later in the chapter.

CAPILLARIES

Capillaries are microscopic vessels that usually connect arterioles and venules (Figure 21-1). They are found near almost every cell in the body. The distribution of capillaries in the body varies with the activity of the tissue. For example, in places where activity is higher, such as muscles, liver, kidneys, lungs, and nervous system, there are rich capillary supplies. In areas where activity is lower, such as tendons and ligaments, the capillary supply is not as extensive. The epidermis, cornea of the eye, and cartilage are devoid of capillaries.

The primary function of capillaries is to permit the exchange of nutrients and wastes between the blood and tissue cells. The structure of the capillaries is admirably suited to this purpose. Capillary walls are composed of only a single layer of cells (endothelium). They have no tunica media or tunica externa. Thus a substance in the blood must pass through the plasma membrane of just one cell to reach tissue cells. This vital exchange of materials occurs only through capillary walls—the thick walls of arteries and veins present too great a barrier.

Although in some places in the body capillaries pass directly from arterioles to venules, in other places they form extensive branching networks. These networks increase the surface area for diffusion and thereby allow a rapid exchange of large quantities of materials. In most tissues, blood normally flows through only a small portion of the capillary network when metabolic needs are low. But, when a tissue becomes active, the entire capillary network fills with blood.

The flow of blood through capillaries is regulated by vessels with smooth muscle in their walls. A **metarteriole** (*met* = beyond) is a vessel that emerges from an arteriole, transverses the capillary network, and empties into a venule (Figure 21-3). The proximal portions of metarterioles are surrounded by scattered smooth muscle cells whose

contraction and relaxation help to regulate the amount and force of blood. The distal portion of a metarteriole has no smooth muscle cells and is called a *thoroughfare channel.* It serves as a low-resistance channel that increases blood flow. **True capillaries** emerge from arterioles or metarterioles and are not on the direct flow route from arteriole to venule. At their sites of origin, there is a ring of smooth muscle called a **precapillary sphincter** that controls the flow of blood entering a true capillary. The various factors that regulate the contraction of smooth muscle cells and precapillary sphincters are discussed later.

Some capillaries of the body, such as those found in muscle tissue and other locations, are referred to as **continuous capillaries.** These capillaries are so named because the cytoplasm of the endothelial cells is continuous when viewed in cross section through a microscope; the cytoplasm appears as an uninterrupted ring, except for the endothelial junction. Other capillaries of the body are referred to as **fenestrated capillaries.** They differ from continuous capillaries in that their endothelial cells have numerous fenestrae (pores) where the cytoplasm is absent. The fenestrae range from 700 to 1000 Å in diameter and are closed by a thin diaphragm, except in the capillaries in the kidneys where they are assumed to be open. Fenestrated capillaries are also found in the villi of the small intestine, choroid plexuses of the ventricles in the brain, ciliary processes of the eyes, and endocrine glands.

Microscopic blood vessels in certain parts of the body, such as the liver, are termed **sinusoids.** They are wider than capillaries and more tortuous. Also, instead of the usual endothelial lining, sinusoids are lined largely by phagocytic cells. In the liver, such cells are called *stellate reticuloendothelial cells.* Like capillaries, sinusoids convey blood from arterioles to venules. Other regions containing sinusoids include the spleen, adenohypophysis, parathyroid glands, and adrenal cortex.

VENULES

When several capillaries unite, they form small veins called **venules** (see Figure 21-4c). Venules collect blood

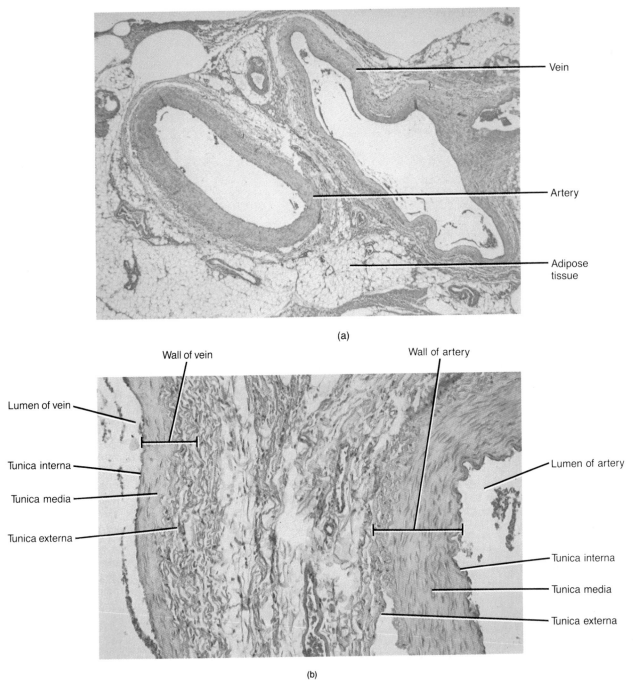

(a)

(b)

FIGURE 21-4 Histology of blood vessels. (a) Photomicrograph of an artery and its accompanying vein at a magnification of 50×. (b) Photomicrograph of a portion of the wall of an artery and its accompanying vein at a magnification of 250×.

from capillaries and drain it into veins. The venules closest to the capillaries consist of a tunica interna of endothelium and a tunica externa of connective tissue. As the venules approach the veins, they also contain the tunica media characteristic of veins.

VEINS

Veins are composed of essentially the same three coats as arteries, but they have considerably less elastic tissue and smooth muscle and contain more white fibrous tissue

(Figures 21-1 and 21-4). However, they are still distensible enough to adapt to variations in the volume and pressure of blood passing through them.

By the time the blood leaves the capillaries and moves into the veins, it has lost a great deal of pressure. The difference in pressure can be observed in the blood flow from a cut vessel; blood leaves a cut vein in an even flow rather than in the rapid spurts characteristic of arteries. Most of the structural differences between arteries and veins reflect this pressure difference. For example, veins do not need walls as strong as those of arteries.

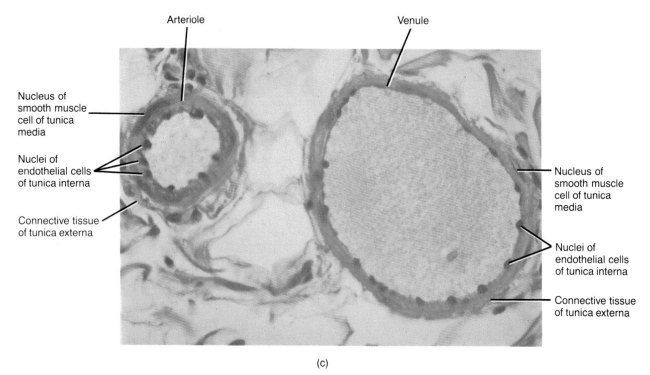

Arteriole   Venule

Nucleus of
smooth muscle
cell of tunica
media

Nuclei of
endothelial cells
of tunica interna

Connective tissue
of tunica externa

Nucleus of
smooth muscle
cell of tunica
media

Nuclei of
endothelial cells
of tunica interna

Connective tissue
of tunica externa

(c)

FIGURE 21-4 (*Continued*) Histology of blood vessels. (c) Photomicrograph of an arteriole and a venule at a magnification of 800×. (© 1983 by Michael H. Ross. Used by permission.)

The low pressure in veins, however, has its disadvantages. When you stand, the pressure pushing blood up the veins in your lower extremities is barely enough to balance the force of gravity pushing it back down. For this reason, many veins, especially those in the limbs, contain valves that prevent backflow. Normal valves ensure the flow of blood toward the heart.

CLINICAL APPLICATION
In people with weak venous valves, large quantities of blood are forced by gravity back down into distal parts of the vein. This pressure overloads the vein and pushes the walls outward. After repeated overloading, the walls lose their elasticity and become stretched and flabby. A vein damaged in this way is called a **varicose vein.** Varicose veins may be caused by prolonged standing and pregnancy. Because a varicosed wall is not able to exert a firm resistance against the blood, blood tends to accumulate in the pouched-out area of the vein, causing it to swell and forcing fluid into the surrounding tissue. Veins close to the surface of the legs are highly susceptible to varicosities. Veins that lie deeper are not as vulnerable because surrounding skeletal muscles prevent their walls from overstretching.

Varicosities are also common in the veins that lie in the wall of the anal canal. These varicosities are called **hemorrhoids** (HEM-o-royds). Hemorrhoids may be caused by constipation. Repeated straining during defecation forces blood down into the superior hemorrhoidal plexus, increasing pressure in these veins. Constipation is related to low-fiber diets, especially in

North America. One hypothesis links increased intra-abdominal pressure, caused by straining during evacuation of firm feces, directly to hemorrhoids or varicose veins. A new development in the treatment of hemorrhoids is *cryosurgery* (*cryo* = cold), in which external hemorrhoids are destroyed by freezing them with a solution of nitrous oxide or liquid nitrogen.

A **vascular (venous) sinus** is a vein with a thin endothelial wall that has no smooth muscle to alter its diameter. Surrounding tissue replaces the tunica media and tunica externa to provide support. Intracranial vascular sinuses, which are supported by the dura mater, return cerebrospinal fluid and deoxygenated blood from the brain to the heart. Another example of a vascular sinus is the coronary sinus of the heart.

PHYSIOLOGY OF CIRCULATION

BLOOD FLOW AND BLOOD PRESSURE

Blood flows through its system of closed vessels because of different pressures in various parts of the system. It always flows from regions of higher pressure to regions of lower pressure. The mean (average) pressure in the aorta is about 100 mm Hg (mercury). This pressure continually decreases rapidly through the arterial system and more slowly through the venous system (Figure 21-5). Because of the continuous drop in pressure, blood flows from the aorta (100 mm Hg) to the arteries (100–40 mm Hg) to arterioles (40–25 mm Hg) to capillaries (25–12 mm Hg) to venules (12–8 mm Hg) to veins (10–5 mm

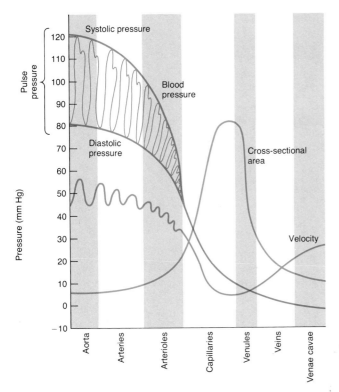

FIGURE 21-5 Relationship between blood pressure, velocity of blood flow, and total cross-sectional area in various blood vessels of the cardiovascular system. The velocity curve (green) and cross-sectional area curve (blue) are not related to the pressures shown in millimeters of mercury (mm Hg).

Hg) to the venae cavae (2 mm Hg). The pressure in the right atrium is 0 mm Hg.

Other mechanisms aid the flow of blood also. When blood leaves the capillaries, it enters the venules and veins, which are larger in diameter and thereby offer less resistance to flow. Contraction of skeletal muscles around the veins also helps drive blood toward the heart.

Factors That Affect Arterial Blood Pressure

We have defined blood pressure as the pressure exerted by blood on the wall of any blood vessel. In clinical use, however, the term refers to the pressure in arteries. Here we will identify several factors that influence arterial blood pressure.

● *Cardiac Output (CO)*   Cardiac output (CO), the amount of blood ejected by the left ventricle into the aorta each minute, is the principal determinant of blood pressure. As noted in Chapter 20, CO is calculated by multiplying stroke volume by heart rate. In a normal, resting adult, it is about 5.25 liters/min (70 ml × 75 beats/min). Blood pressure varies directly with CO. If CO is increased by any increase in stroke volume or heart rate, then blood pressure increases. A decrease in CO causes a decrease in blood pressure.

● *Blood Volume*   Blood pressure is directly proportional to the volume of blood in the cardiovascular system. The normal volume of blood in a human body is about 5

liters (5 qt). Any decrease in this volume, as from hemorrhage, decreases the amount of blood that is circulated through the arteries each minute. As a result, blood pressure drops. Conversely, anything that increases blood volume, such as high salt intake and therefore water retention, increases blood pressure.

● *Peripheral Resistance*   **Peripheral resistance** refers to the resistance (impedance) to blood flow by the force of friction between blood and the walls of blood vessels. It is related to the viscosity of blood and blood vessel diameter.

The viscosity of blood is a function of the number of red blood cells and amount of plasma proteins and also the ratio of red blood cells and solutes to fluid. Any condition that increases the viscosity of blood, such as dehydration or an unusually high number of red blood cells, increases blood pressure. A depletion of plasma proteins or red blood cells, as a result of anemia or hemorrhage, decreases blood viscosity and blood pressure.

The smaller the diameter of a vessel, the more resistance it offers the blood. A major function of arterioles is to control peripheral resistance and, therefore, blood pressure, by changing their diameters. The center for this regulation is the vasomotor center in the medulla, which will be described shortly.

The factors that affect arterial blood pressure are summarized in Figure 21-6.

Control of Blood Pressure

In Chapter 20 we considered the increase and decrease of heart rate and force of contraction by the cardioacceleratory center (CAC) and the cardioinhibitory center (CIC). We also took a look at how certain chemicals (epinephrine, potassium, sodium, and calcium), temperature, emotions, sex, and age affect heart rate. By their effect on heart rate and force of contraction, these factors also control blood pressure. Any increase in heart rate and force of contraction increases blood pressure. Conversely, any decrease will lower blood pressure.

At this point we will examine those factors that help to regulate blood pressure by acting on the blood vessels themselves.

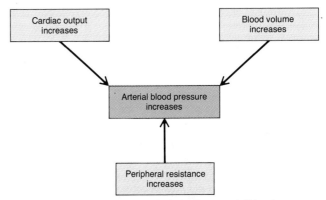

FIGURE 21-6 Summary of factors that affect arterial blood pressure.

● **Vasomotor Center**   Within the medulla is a cluster of neurons referred to as the **vasomotor** (*vas* = vessel, *motor* = movement) **center.** The function of this center is to control the diameter of blood vessels, especially arterioles. It continually sends impulses to the smooth muscle in arteriole walls that result in a moderate state of vasoconstriction at all times. This state of tonic contraction, called *vasomotor tone,* assumes a role in maintaining peripheral resistance and blood pressure. The vasomotor center brings about vasoconstriction by increasing the number of sympathetic impulses above normal. It causes vasodilation by decreasing the number of sympathetic impulses below normal. In other words, in this case, the sympathetic division of the ANS can bring about both vasoconstriction and vasodilation.

The vasomotor center is affected by several factors, all of which influence blood pressure.

● **Pressoreceptors**   It will be recalled that stimulated **pressoreceptors** in the carotid sinus and aorta send impulses to the cardiac center that result in increased or decreased cardiac output to help regulate blood pressure (see Figure 20-12). This reflex not only acts on the heart, but it also involves arterioles. For example, if there is an increase in blood pressure, the pressoreceptors stimulate the cardioinhibitory center and inhibit the cardioacceleratory center. The result is a decrease in cardiac output and a decrease in blood pressure. The pressoreceptors also send impulses to the vasomotor center. In response, the vasomotor center decreases sympathetic stimulation to the arterioles. The result is vasodilation and a decrease in blood pressure (Figure 21-7). If there is a decrease in blood pressure, the pressoreceptors inhibit the cardioinhibitory center and stimulate the cardioacceleratory center. The result is an increase in cardiac output and an increase in blood pressure. Also, the pressoreceptors send impulses to the vasomotor center that increase sympathetic stimulation to arterioles. The result is vasoconstriction and an increase in blood pressure.

● **Chemoreceptors**   Receptors sensitive to chemicals in the blood are called **chemoreceptors.** Chemoreceptors in the carotid sinus and aorta are called *carotid* and *aortic bodies,* respectively. They are sensitive to arterial blood levels of oxygen, carbon dioxide, and hydrogen ions. Given a deficiency of oxygen (hypoxia), a decrease in hydrogen ion concentration, or an excess of carbon dioxide (hypercapnia), the chemoreceptors are stimulated and send impulses to the vasomotor center. In response, the vasomotor center increases sympathetic stimulation to arterioles. This brings about vasoconstriction and an increase in blood pressure. As you will see in Chapter 23, chemoreceptors also stimulate the medullary rhythmicity area in response to these chemical stimuli to adjust the rate of respiration.

● **Control by Higher Brain Centers**   In response to strong emotions, **higher brain centers,** such as the cerebral cortex, can have a significant influence on blood pressure. For example, during periods of intense anger, the cerebral cortex stimulates the vasomotor center to fire sympathetic impulses to arterioles. This causes vasoconstriction and

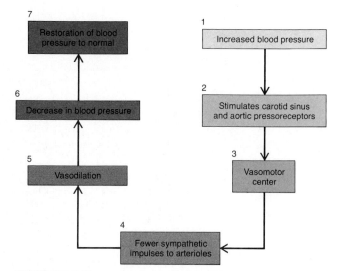

FIGURE 21-7 Vasomotor center control of blood pressure by stimulation of pressoreceptors.

an increase in blood pressure. When a person is emotionally upset, impulses from higher brain centers cause a decrease in sympathetic stimulation by the vasomotor center. This produces vasodilation and a decrease in blood pressure. A frequent result is fainting because blood flow to the brain is diminished.

● **Chemicals**   Several **chemicals** affect blood pressure by causing vasoconstriction. Epinephrine and norepinephrine (NE), produced by the adrenal medulla, increase the rate and force of heart contractions and bring about vasoconstriction of abdominal and cutaneous arterioles. They also bring about dilation of cardiac and skeletal arterioles. Antidiuretic hormone (ADH), produced by the hypothalamus and released from the neurohypophysis, causes vasoconstriction if there is a severe loss of blood due to hemorrhage. Angiotensin II helps to raise blood pressure by stimulating secretion of aldosterone (increases sodium ion concentration and water reabsorption) and causing vasoconstriction due to release of renin. Histamine produced by mast cells and kinins found in plasma are vasodilators that assume key functions during the inflammatory response.

● **Autoregulation (Local Control)**   Autoregulation refers to a local, automatic adjustment of blood flow in a given region of the body in response to the particular needs of the tissue. In most body tissues, oxygen is the principal stimulus for autoregulation. A suggested mechanism for autoregulation is as follows. In response to low oxygen supplies, the cells in the immediate area produce and release *vasodilator substances.* Such substances are thought to include potassium ions, hydrogen ions, carbon dioxide, lactic acid, and adenosine. Once released, the vasodilator substances produce a local dilation of arterioles and relaxation of precapillary sphincters. The result is an increased flow of blood into the tissue, which restores oxygen levels to normal. The autoregulation mechanism is important in meeting the nutritional demands of active tissues, such as muscle tissue, where the demand might increase as much as tenfold.

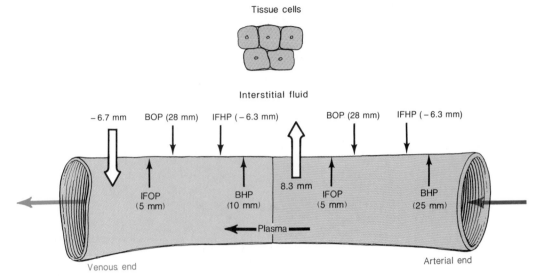

Key:

BHP = Blood hydrostatic pressure
IFHP = Interstitial fluid hydrostatic pressure
BOP = Blood osmotic pressure
IFOP = Interstitial fluid osmotic pressure

FIGURE 21-8 Hydrostatic and osmotic pressure forces involved in moving fluid out of plasma (filtration) and into plasma (reabsorption).

Capillary Exchange

For reasons to be discussed shortly, you will see that, although blood pressure decreases consistently from the aorta to the venae cavae, the velocity of blood decreases as it flows through the aorta, arterioles, and capillaries and then increases as it passes into venules and veins. The velocity of blood flow in capillaries is the slowest in the cardiovascular system, and this is important to allow for the exchange of materials between blood and body tissues.

Blood usually does not flow in a continuous manner through capillary networks. Rather, it flows intermittently because of contraction and relaxation of the smooth muscle cells of metarterioles and the precapillary sphincters of true capillaries. The intermittent contraction and relaxation, which may occur 5 to 10 times per minute, is called **vasomotion.** The most important factor in controlling vasomotion is the oxygen concentration in tissues, a mechanism similar to autoregulation in arterioles.

Diffusion is the principal mechanism by which substances are exchanged between capillary blood and body cells. Most substances are exchanged by this process through endothelial junctions of continuous capillaries and fenestrae of fenestrated capillaries. Some larger molecules are exchanged by pinocytosis.

The movement of water and dissolved substances, except proteins, through capillaries is greatly affected by hydrostatic pressure and osmotic pressure. Blood pressure in capillaries is called **blood hydrostatic pressure (BHP).** It tends to move fluid out of a capillary and averages 25 mm Hg at the arterial end of a capillary and 10 mm Hg at the venous end (Figure 21-8). **Interstitial fluid hydrostatic pressure (IFHP)** is the pressure of interstitial fluid against cells of a tissue and the endothelial cells of capillaries. It also tends to move fluid into capillaries and averages −6.3 mm Hg at both the arterial and venous ends of capillaries. **Blood osmotic pressure (BOP)** is due

principally to the presence of large amounts of plasma proteins and tends to move fluid into capillaries by osmosis. It averages 28 mm Hg at both ends of capillaries. **Interstitial fluid osmotic pressure (IFOP),** due to small amounts of proteins in interstitial fluid, tends to move fluid out of capillaries. It averages 5.0 mm Hg at both ends of capillaries.

Whether fluids leave or enter capillaries depends on how the pressures relate to each other. If the forces that tend to move fluid out of capillaries are greater than the forces that tend to move fluid into capillaries, the fluid will move from capillaries into tissue spaces. If, on the other hand, the inward force is greater, then fluid will move from tissue spaces into capillaries. The term **effective filtration pressure (Peff)** is used to show the direction of fluid movement. It is calculated as follows:

$$P\text{eff} = (BHP + IFOP) - (IFHP + BOP)$$

Substituting measured values at the arterial end of a capillary,

$$
\begin{aligned}
P\text{eff} &= (25 + 5) - (-6.3 + 28) \\
&= (30) - (21.7) \\
&= 8.3 \text{ mm Hg}
\end{aligned}
$$

Substituting measured values at the venous end of a capillary,

$$
\begin{aligned}
P\text{eff} &= (10 + 5) - (-6.3 + (28) \\
&= (15) - (21.7) \\
&= -6.7 \text{ mm Hg}
\end{aligned}
$$

Thus, at the arterial end of a capillary, there is a *net outward force* (8.3 mm Hg), and fluid moves out of the capillary (filtered) into tissue spaces. At the venous end of a capillary, the negative value (−6.7 mm Hg) represents a *net inward force,* and fluid moves into the capillary (reabsorbed) from tissue spaces.

Not all the fluid filtered at one end of the capillary is reabsorbed at the other end. When we discuss fluid dynamics in more detail in Chapter 27, you will see that some of the filtered fluid and any proteins that escape from blood into interstitial fluid are returned by the lymphatic system to the cardiovascular system. Under normal conditions there is a state of near equilibrium at the arterial and venous ends of a capillary in which filtered fluid and absorbed fluid plus that returned to the lymphatic system are nearly equal. This near equilibrium is known as **Starling's law of the capillaries.**

Factors That Aid Venous Return

The establishment of a pressure gradient is the primary reason why blood flows. But a number of other factors help the blood to return through the veins: an increased rate of flow, contractions of the skeletal muscles, valves, and breathing.

● *Velocity of Blood Flow* The velocity of blood flow is inversely related to the cross-sectional area of the blood vessels. Blood flows most rapidly where the cross-sectional area is least (see Figure 21-5). Each time an artery branches, the total cross-sectional area of all the branches is greater than that of the original vessel. When branches combine, the resulting cross-sectional area is less than that of the original branches. The cross-sectional area of the aorta is 2.5 cm², and the velocity of the blood there is 40 cm/sec. Capillaries have a cross-sectional area of 2,500 cm², and the velocity is less than 0.1 cm/sec. In the venae cavae, the cross-sectional area is 8 cm² and the velocity is 5 to 20 cm/sec. The area and velocity for the different types of blood vessels are listed in Exhibit 21-1.

Thus, the velocity of blood decreases as it flows from the aorta to arteries to arterioles to capillaries. The velocity of blood in the capillaries is the slowest in the cardiovascular system. Thus there is adequate exchange (diffusion) time between the capillaries and adjacent tissues. As blood vessels leave capillaries and approach the heart, their cross-sectional area decreases. Therefore, the velocity of blood increases as it flows from capillaries to venules to veins to the heart.

CLINICAL APPLICATION
Blood flow can be measured to diagnose certain circulatory disorders. **Circulation time** is the time required for blood to pass from the right atrium, through pulmonary circulation, back to the left ventricle, through systemic circulation down to the foot, and back again to the right atrium. Such a trip usually takes about 1 minute and requires about 28 heartbeats.

● *Skeletal Muscle Contractions and Valves* The combination of skeletal muscle contractions and valves in veins is important in returning venous blood to the heart. Many veins, especially those in the extremities, contain valves. When skeletal muscles contract, they tighten around the veins running through them and the valves open. This pressure drives the blood toward the heart—the action is called **milking** (Figure 21-9). When the muscles relax,

EXHIBIT 21-1

RELATIONSHIP BETWEEN BLOOD VELOCITY AND CROSS-SECTIONAL AREA

VESSEL	CROSS-SECTIONAL AREA (cm²)	VELOCITY (cm/sec)
Aorta	2.5	40
Arteries	20	10–40
Arterioles	40	0.1
Capillaries	2,500	less than 0.1
Venules	250	0.3
Veins	80	0.3–5.0
Venae cavae	8	5–20

the valves close to prevent the backflow of blood away from the heart. Individuals who are immobilized through injury or disease cannot take advantage of these contractions. As a result, the return of venous blood to the heart is slower and the heart has to work harder. For this reason, periodic massage is helpful.

● *Breathing* Another factor that is important in maintaining venous circulation is breathing. During inspiration, the diaphragm moves downward. This causes a decrease in pressure in the chest cavity and an increase in pressure in the abdominal cavity. Once the pressure difference is established, blood is squeezed from the abdominal veins into the thoracic veins. When the pressures reverse during expiration, blood in the veins is prevented from backflowing by the valves.

BLOOD RESERVOIRS

The volume of blood in various parts of the cardiovascular system varies considerably. Veins, venules, and venous sinuses contain about 59 percent of the blood in the system, arteries about 13 percent, pulmonary vessels about 12 percent, the heart about 9 percent, and arterioles and capillaries about 7 percent. Since systemic veins contain so much of the blood, they are referred to as **blood reservoirs.** They serve as storage depots for blood, which can be moved out of them quickly to other parts of the body if the need arises. When there is increased muscular activity, the vasomotor center sends increasing sympathetic impulses to veins that serve as blood reservoirs. The result is vasoconstriction. This permits the distribution of blood from venous reservoirs to skeletal muscles, where it is needed most. A similar mechanism operates in cases of hemorrhage when blood volume and pressure decrease. Vasoconstriction of veins in venous reservoirs helps to compensate for the blood loss. Among the principal blood reservoirs are the veins of the abdominal organs (especially the liver and spleen) and the veins of the skin.

CLINICAL APPLICATION
Of the various types of **exercise,** some are more effective than others for improving the health of the cardio-

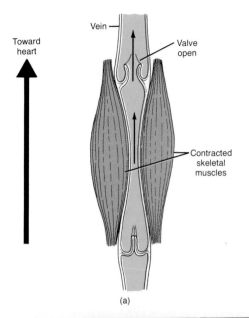

Toward heart

Vein

Valve open

Contracted skeletal muscles

(a)

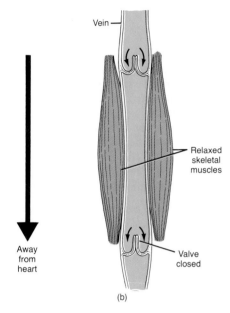

Vein

Away from heart

Relaxed skeletal muscles

Valve closed

(b)

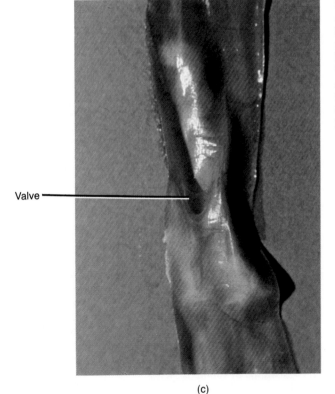

Valve

(c)

FIGURE 21-9 Role of skeletal muscle contractions and venous valves in returning blood to the heart. (a) When skeletal muscles contract, the valves open, and blood is forced toward the heart. (b) When skeletal muscles relax, the valves close to prevent the backflowing of blood from the heart. (c) Photograph of a one-way valve in a vein. (Courtesy of J. W. Eads.)

vascular system because they involve movements of large body muscles. The sustained movements of large muscles increases the flow of blood to the heart. Brisk walking, running, bicycling, cross-country skiing, and swimming are examples of sustained exercises.

Sustained exercise increases the oxygen demand of the muscles, and whether the demand is met depends primarily on the adequacy of cardiac output. After several weeks of training, the healthy individual increases cardiac output and thereby increases the rate of oxygen delivery to the tissues.

Another benefit of sustained exercise is the reduced oxygen requirement of skeletal muscles. After conditioning, an individual will have a decrease in skeletal muscle blood flow and lactic acid production. After conditioning, therefore, the heart is required to pump less blood to the muscles for a given amount of oxygen required by exercise.

Physical conditioning also causes an interesting effect upon systemic blood pressure. After a conditioning period, hypertensive individuals show a reduction in systolic pressure amounting to an average of 13 mm Hg. Blood pressure response to a given work load also is less after training, while lower pressure itself helps reduce myocardial oxygen requirements. So conditioning also may be useful in the management of systemic hypertension.

Additional benefits to be gained from physical conditioning are a reduction in high-density lipoprotein (HDL), a substance that seems to counter the impact of cholesterol in heart disease (discussed shortly). Exercise also helps to control weight and increases the body's ability to dissolve blood clots by increasing fibrinolytic activity. Intense exercise increases levels of endorphins, the body's natural painkillers. This may explain the psychological "high" that runners experience with strenuous training and the "low" they feel when they miss regular workouts.

CHECKING CIRCULATION

PULSE

The alternate expansion and elastic recoil of an artery with each systole of the left ventricle is called the **pulse.** Pulse is strongest in the arteries closest to the heart. It becomes weaker as it passes over the arterial system, and it disappears altogether in the capillaries. The pulse may be felt in any artery that lies near the surface of the body and over a bone or other firm tissue. The radial artery at the wrist is most commonly used. Other arteries that may be used for determining pulse are the:

1. Temporal artery, above and toward the outside of the eye.
2. Facial artery, at the lower jawbone on a line with the corners of the mouth.
3. Common carotid artery, on the side of the neck.
4. Brachial artery, along the inner side of the biceps brachii muscle.
5. Femoral artery, near the pelvic bone.
6. Popliteal artery, behind the knee.
7. Posterior tibial artery, behind the medial malleolus of the tibia.
8. Dorsalis pedis artery, over the instep of the foot.

The pulse rate is the same as the heart rate and averages between 70 and 90 beats per minute in the resting state. The term **tachycardia** (tak'-ē-KAR-dē-a; *tachy* = fast) is applied to a rapid heart or pulse rate (over 100/min). The term **bradycardia** (brad'-ē-KAR-dē-a; *brady* = slow) indicates a slow heart or pulse rate (under 50/min).

Other characteristics of the pulse may give additional information about circulation. For example, the intervals between beats should be equal in length. If a pulse is

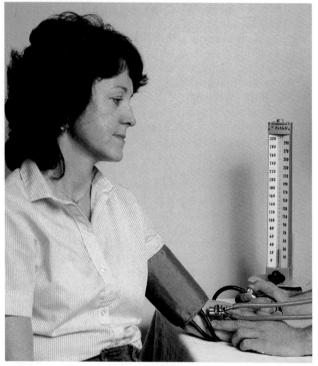

FIGURE 21-10 Measurement of blood pressure using a sphygmomanometer and a stethoscope. (© 1982 by Gerard J. Tortora. Courtesy of Geraldine C. Tortora.)

missed at intervals, the pulse is said to be irregular. Also, each pulse beat should be of equal strength. Irregularities in strength may indicate a lack of muscle tone in the heart or arteries.

MEASUREMENT OF BLOOD PRESSURE (BP)

In clinical use, the term **blood pressure (BP)** refers to the pressure in arteries exerted by the left ventricle when it undergoes systole and the pressure remaining in the arteries when the ventricle is in diastole. Blood pressure is usually taken in the left brachial artery, and it is measured by a *sphygmomanometer* (sfig'-mō-ma-NOM-e-ter; *sphygmo* = pulse). A commonly used sphygmomanometer consists of a rubber cuff attached by a rubber tube to a compressible hand pump or bulb (Figure 21-10). Another tube attaches to the cuff and to a column of mercury or pressure dial marked off in millimeters. This column measures the pressure. The cuff is wrapped around the arm over the brachial artery and inflated by squeezing the bulb. The inflation creates a pressure on the artery. The bulb is squeezed until the pressure in the cuff exceeds the pressure in the artery. At this point, the walls of the brachial artery are compressed tightly against each other, and no blood can flow through. Compression of the artery may be evidenced in two ways. First, if a stethoscope is placed over the artery below the cuff, no pulse can be heard. Second, no pulse can be felt by placing the fingers over the radial artery at the wrist.

Next the cuff is deflated gradually until the pressure in the cuff is slightly less than the maximal pressure in the brachial artery. At this point, the artery opens, a spurt of blood passes through, and the pulse may be heard through the stethoscope. As cuff pressure is further reduced, the sound suddenly becomes faint. Finally, the sound disappears altogether. When the first sound is heard, a reading on the mercury column is made. This sound corresponds to **systolic blood pressure**—the force with which blood is pushing against arterial walls during ventricular contraction. The pressure recorded on the mercury column when the sounds suddenly become faint is called **diastolic blood pressure.** It measures the force of blood in arteries during ventricular relaxation. Whereas systolic pressure indicates the force of the left ventricular contraction, diastolic pressure provides information about the resistance of blood vessels.

CLINICAL APPLICATION
The average blood pressure of a young adult male is about 120 mm Hg systolic and 80 mm Hg diastolic, expressed as 120/80. In young adult females, the pressures are 8 to 10 mm Hg less. The difference between systolic and diastolic pressure is called **pulse pressure.** This pressure, which averages 40 mm Hg, provides information about the condition of the arteries. For example, conditions such as atherosclerosis and patent ductus arteriosus greatly increase pulse pressure. The normal ratio of systolic pressure to diastolic pressure to pulse pressure is about 3:2:1.

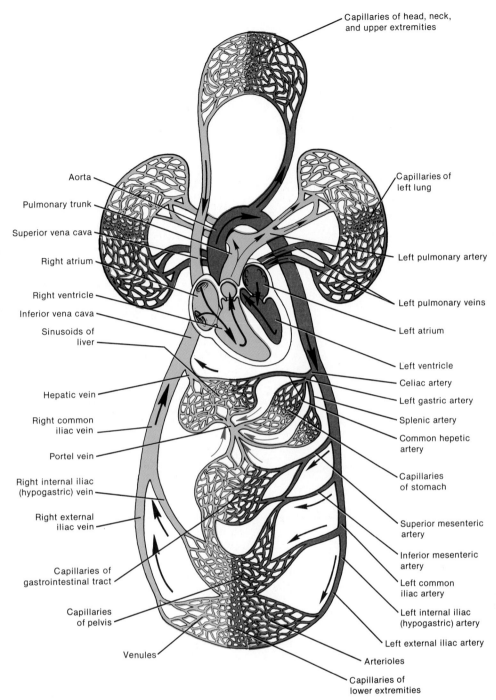

FIGURE 21-11 Circulatory routes. Systemic circulation is indicated by heavy black arrows, pulmonary circulation by thin black arrows, and hepatic portal circulation by thin colored arrows. Refer to Figure 20-7 for the details of coronary (cardiac) circulation and Figure 21-24 for the details of fetal circulation.

CIRCULATORY ROUTES

The arteries, arterioles, capillaries, venules, and veins are organized into definite routes in order to circulate the blood throughout the body. We can now look at the basic routes the blood takes as it is transported through its vessels.

Figure 21-11 shows a number of basic **circulatory routes** through which the blood travels. **Systemic circulation** includes all the oxygenated blood that leaves the left ventricle through the aorta and the deoxygenated blood that returns to the right atrium after traveling to all the organs including the nutrient arteries to the lungs. Two of the many subdivisions of the systemic circulation

are the **coronary (cardiac) circulation** (see Figure 20-7), which supplies the myocardium of the heart, and the **hepatic portal circulation,** which runs from the digestive tract to the liver. Blood leaving the aorta and traveling through the systemic arteries is a bright red color. As it moves through the capillaries, it loses its oxygen and takes on carbon dioxide, which gives the blood in the systemic veins its dark red color.

When blood returns to the heart from the systemic route, it goes out of the right ventricle through the **pulmonary circulation** to the lungs. In the lungs, it loses its carbon dioxide and takes on oxygen. It is now bright red again. It returns to the left atrium of the heart and reenters the systemic circulation.

EXHIBIT 21-2
AORTA AND ITS BRANCHES (Figure 21-12)

DIVISION OF AORTA	ARTERIAL BRANCH	REGION SUPPLIED
Ascending aorta	Right and left coronary	Heart.
Arch of aorta	Brachiocephalic — Right common carotid	Right side of head and neck.
	Brachiocephalic — Right subclavian	Right upper extremity.
	Left common carotid	Left side of head and neck.
	Left subclavian	Left upper extremity.
Thoracic aorta	Intercostals	Intercostal and chest muscles and pleurae.
	Superior phrenics	Posterior and superior surfaces of diaphragm.
	Bronchials	Bronchi of lungs.
	Esophageals	Esophagus.
Abdominal aorta	Inferior phrenics	Inferior surface of diaphragm.
	Celiac — Common hepatic	Liver.
	Celiac — Left gastric	Stomach and esophagus.
	Celiac — Splenic	Spleen, pancreas, and stomach.
	Superior mesenteric	Small intestine, cecum, and ascending and transverse colons.
	Suprarenals	Adrenal (suprarenal) glands.
	Renals	Kidneys.
	Gonadals — Testicular	Testes.
	Gonadals — Ovarians	Ovaries.
	Inferior mesenteric	Transverse, descending, and sigmoid colons and rectum.
	Common iliacs — External iliacs	Lower extremities.
	Common iliacs — Internal iliacs (hypogastrics)	Uterus, prostate, muscles of buttocks, and urinary bladder.

Another major route—the **fetal circulation**—exists only in the fetus and contains special structures that allow the developing fetus to exchange materials with its mother.

Cerebral circulation (circle of Willis) is discussed in Exhibit 21-4.

SYSTEMIC CIRCULATION

The flow of blood from the left ventricle to all parts of the body and back to the right atrium is called the **systemic circulation.** The purpose of systemic circulation is to carry oxygen and nutrients to body tissues and to remove carbon dioxide and other wastes from the tissues. All systemic arteries branch from the *aorta,* which arises from the left ventricle of the heart.

As the aorta emerges from the left ventricle, it passes upward and deep to the pulmonary artery. At this point, it is called the *ascending aorta.* The ascending aorta gives off two coronary branches to the heart muscle. Then it turns to the left, forming the *arch of the aorta* before descending to the level of the fourth thoracic vertebra as the *descending aorta.* The descending aorta lies close to the vertebral bodies, passes through the diaphragm, and terminates at the level of the fourth lumbar vertebra by dividing into two *common iliac arteries,* which carry blood to the lower extremities. The section of the descending aorta between the arch of aorta and the diaphragm is referred to as the *thoracic aorta.* The section between the diaphragm and the common iliac arteries is termed the *abdominal aorta.* Each section of the aorta gives off arteries that continue to branch into distributing arteries leading to organs and finally into the arterioles and capillaries that pierce the tissues.

Blood is returned to the heart through the systemic veins. All the veins of the systemic circulation flow into either the *superior* or *inferior venae cavae* or the *coronary sinus.* They in turn empty into the right atrium. The principal arteries and veins of systemic circulation are described and illustrated in Exhibits 21-2 to 21-13 and Figures 21-12 to 21-21.

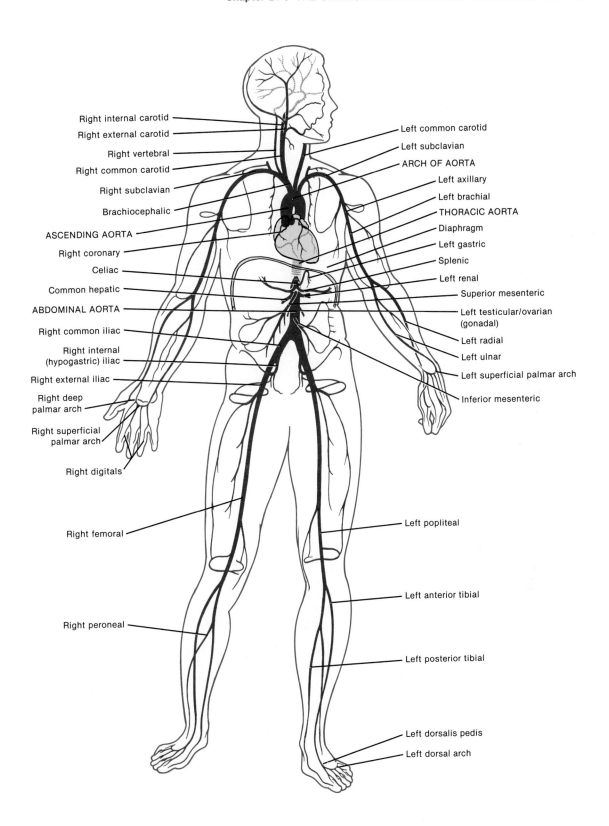

FIGURE 21-12 Aorta and its principal branches in anterior view.

EXHIBIT 21-3
ASCENDING AORTA (Figure 21-13)

BRANCH	DESCRIPTION AND REGION SUPPLIED
Coronary arteries	Right and left branches arise from ascending aorta just superior to aortic semilunar valve. They form crown around heart, giving off branches to atrial and ventricular myocardium.

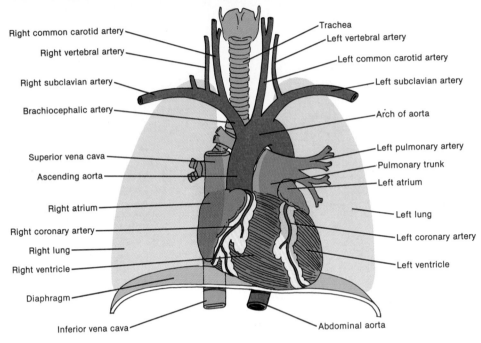

FIGURE 21-13 Ascending aorta and its branches in anterior view.

EXHIBIT 21-4
ARCH OF AORTA (Figure 21-14)

BRANCH	DESCRIPTION AND REGION SUPPLIED
Brachiocephalic	**Brachiocephalic artery** is first branch off arch of aorta. It divides to form right subclavian artery and right common carotid artery. **Right subclavian artery** extends from brachiocephalic to first rib and then passes into armpit (axilla) and supplies arm, forearm, and hand. Continuation of right subclavian into axilla is called **axillary artery.*** From here, it continues into arm as **brachial artery.** At bend of elbow, brachial artery divides into medial **ulnar** and lateral **radial arteries.** These vessels pass down to palm, one on each side of forearm. In palm, branches of two arteries anastomose to form two palmar arches—**superficial palmar arch** and **deep palmar arch.** From these arches arise **digital arteries,** which supply fingers and thumb. Before passing into axilla, right subclavian gives off major branch to brain called **vertebral artery.** Right vertebral artery passes through foramina of transverse processes of cervical vertebrae and enters skull through foramen magnum to reach undersurface of brain. Here it unites with left vertebral artery to form **basilar artery.**
Right common carotid	**Right common carotid artery** passes upward in neck. At upper level of larynx, it divides into **right external** and **right internal carotid arteries.** External carotid supplies right side of thyroid gland, tongue, throat, face, ear, scalp, and dura mater. Internal carotid supplies brain, right eye, and right sides of forehead and nose. Anastomoses of left and right internal carotids along with basilar artery form arterial circle at base of brain called **circle of Willis.** From this anastomosis arise arteries supplying brain. Essentially, circle of Willis is formed by union of **anterior cerebral arteries** (branches of internal carotids) and **posterior cerebral arteries** (branches of basilar artery). Posterior cerebral arteries are connected with internal carotids by **posterior communicating arteries.** Anterior cerebral arteries are connected by **anterior communicating arteries.** Circle of Willis equalizes blood pressure to brain and provides alternate routes for blood to brain should arteries become damaged.
Left common carotid	**Left common carotid** is second branch off arch of aorta (see Figure 21-13). Corresponding to right common carotid, it divides into basically same branches with same names, except that arteries are now labeled "left" instead of "right."
Left subclavian	**Left subclavian artery** is third branch off arch of aorta (see Figure 21-13). It distributes blood to left vertebral artery and vessels of left upper extremity. Arteries branching from left subclavian are named like those of right subclavian.

* The right subclavian artery is a good example of the practice of giving the same vessel different names as it passes through different regions.

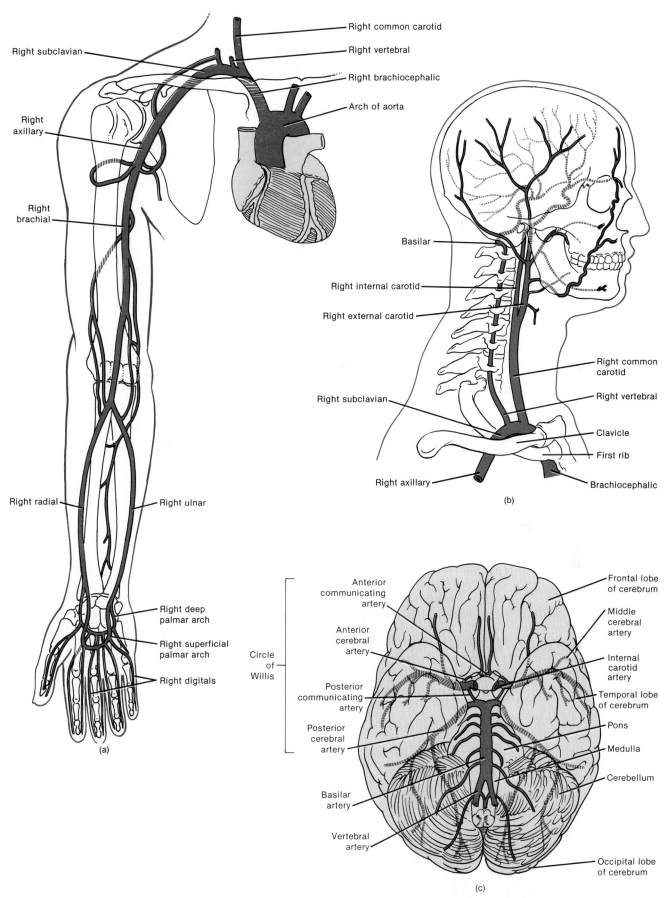

FIGURE 21-14 Arch of the aorta and its branches. (a) Anterior view of the arteries of the right upper extremity. (b) Right lateral view of the arteries of the neck and head. (c) Arteries of the base of the brain. Note the arteries that comprise the circle of Willis.

EXHIBIT 21-5
THORACIC AORTA (Figure 21-15)

BRANCH	DESCRIPTION AND REGION SUPPLIED
	Thoracic aorta runs from fourth to twelfth thoracic vertebrae. Along its course, it sends off numerous small arteries to viscera and skeletal muscles of the chest. Branches of an artery that supply viscera are called **visceral branches.** Those that supply body wall structures are **parietal branches.**
VISCERAL **Pericardial**	Several minute **pericardial arteries** supply blood to dorsal aspect of pericardium.
Bronchial	One **right** and two **left bronchial arteries** supply the bronchial tubes, areolar tissue of the lungs, bronchial lymph nodes, and esophagus.
Esophageal	Four or five **esophageal arteries** supply the esophagus.
Mediastinal	Numerous small **mediastinal arteries** supply blood to structures in the posterior mediastinum.
PARIETAL **Posterior intercostal**	Nine pairs of **posterior intercostal arteries** supply the intercostal, pectoral, and abdominal muscles; overlying subcutaneous tissue and skin; mammary glands; and vertebral canal and its contents.
Subcostal	The **left** and **right subcostal arteries** have a distribution similar to that of the posterior intercostals.
Superior phrenic	Small **superior phrenic arteries** supply the posterior surface of the diaphragm.

EXHIBIT 21-6
ABDOMINAL AORTA (Figure 21-15)

BRANCH	DESCRIPTION AND REGION SUPPLIED
VISCERAL **Celiac**	**Celiac artery (trunk)** is first visceral aortic branch below diaphragm. It has three branches: (1) **common hepatic artery,** which supplies tissues of liver, (2) **left gastric artery,** which supplies stomach, and (3) **splenic artery,** which supplies spleen, pancreas, and stomach.
Superior mesenteric	**Superior mesenteric artery** distributes blood to small intestine and part of large intestine.
Suprarenals	Right and left **suprarenal arteries** supply blood to adrenal (suprarenal) glands.
Renals	Right and left **renal arteries** carry blood to kidneys.
Gonadals (testiculars and ovarians)	Right and left **testicular arteries** extend into scrotum and terminate in testes; right and left **ovarian arteries** are distributed to ovaries.
Inferior mesenteric	**Inferior mesenteric artery** supplies major part of large intestine and rectum.
PARIETAL **Inferior phrenics**	**Inferior phrenic arteries** are distributed to undersurface of diaphragm.
Lumbars	**Lumbar arteries** supply spinal cord and its meninges and muscles and skin of lumbar region of back.
Middle sacral	**Middle sacral artery** supplies sacrum, coccyx, gluteus maximus muscles, and rectum.

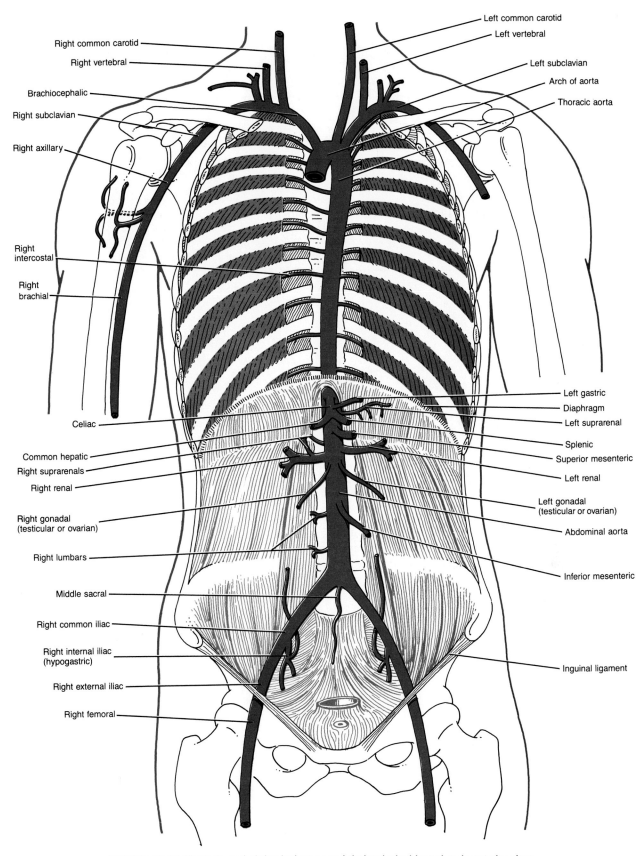

FIGURE 21-15 Thoracic and abdominal aorta and their principal branches in anterior view.

EXHIBIT 21-7

ARTERIES OF PELVIS AND LOWER EXTREMITIES (Figure 21-16)

BRANCH	DESCRIPTION AND REGION SUPPLIED
Common iliacs	At about level of fourth lumbar vertebra, abdominal aorta divides into right and left **common iliac arteries.** Each passes downward about 5 cm (2 inches) and gives rise to two branches: internal iliac and external iliac.
Internal iliacs	**Internal iliac** or **hypogastric arteries** form branches that supply psoas major, quadratus lumborum, and medial side of each thigh, urinary bladder, rectum, prostate gland, ductus deferens, uterus, and vagina.
External iliacs	**External iliac arteries** diverge through pelvis, enter thighs, and here become right and left **femoral arteries.** Both femorals send branches back up to genitals and wall of abdomen. Other branches run to muscles of thigh. Femoral continues down medial and posterior side of thigh at back of knee joint, where it becomes **popliteal artery.** Between knee and ankle, popliteal runs down back of leg and is called **posterior tibial artery.** Below knee, **peroneal artery** branches off posterior tibial to supply structures on medial side of fibula and calcaneus. In calf, **anterior tibial artery** branches off popliteal and runs along front of leg. At ankle, it becomes **dorsalis pedis artery.** At ankle, posterior tibial divides into **medial** and **lateral plantar arteries.** These arteries anastomose with dorsalis pedis and supply blood to foot.

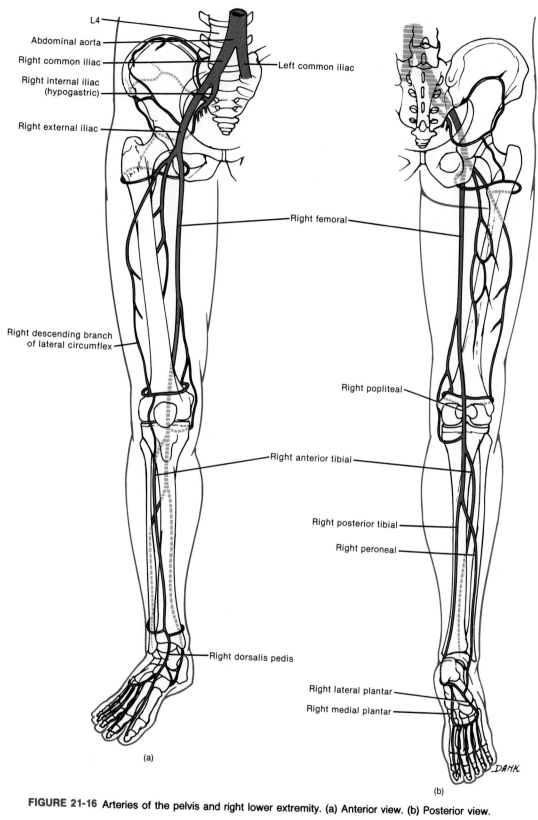

L4

Abdominal aorta

Right common iliac

Right internal iliac
(hypogastric)

Right external iliac

Left common iliac

Right femoral

Right descending branch
of lateral circumflex

Right popliteal

Right anterior tibial

Right posterior tibial

Right peroneal

Right dorsalis pedis

Right lateral plantar

Right medial plantar

(a)

(b)

DANK

FIGURE 21-16 Arteries of the pelvis and right lower extremity. (a) Anterior view. (b) Posterior view.

EXHIBIT 21-8
VEINS OF SYSTEMIC CIRCULATION (Figure 21-17)

VEIN	DESCRIPTION AND REGION DRAINED
	All systemic and cardiac veins return blood to the right atrium of the heart through one of three large vessels. Return flow in coronary circulation is taken up by **cardiac veins,** which empty into the large vein of the heart, the **coronary sinus.** From here, the blood empties into the right atrium of the heart (see Figure 20-7). Return flow in systemic circulation empties into the superior vena cava or inferior vena cava.
Superior vena cava	Veins that empty into the **superior vena cava** are veins of the head and neck, upper extremities, and some from the thorax.
Inferior vena cava	Veins that empty into the **inferior vena cava** are some from the thorax and veins of the abdomen, pelvis, and lower extremities.

CLINICAL APPLICATION

The inferior vena cava is commonly compressed during the later stages of pregnancy due to the enlargement of the uterus. This produces edema of the ankles and feet and temporary varicose veins.

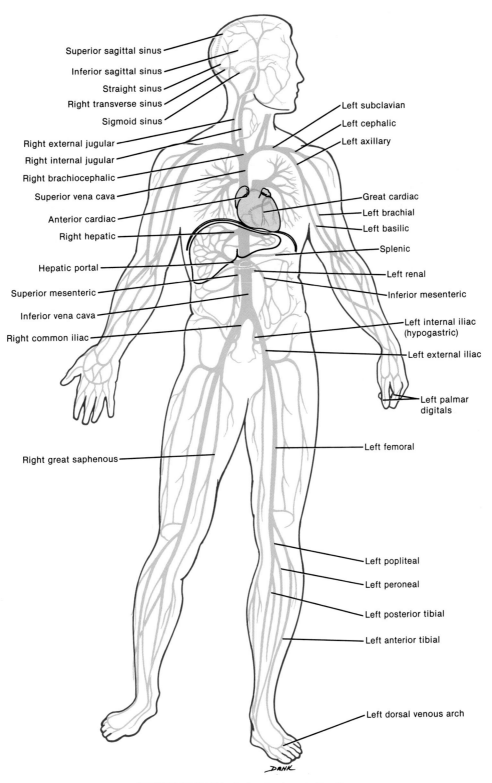

Superior sagittal sinus
Inferior sagittal sinus
Straight sinus
Right transverse sinus
Sigmoid sinus
Right external jugular
Right internal jugular
Right brachiocephalic
Superior vena cava
Anterior cardiac
Right hepatic
Hepatic portal
Superior mesenteric
Inferior vena cava
Right common iliac

Right great saphenous

Left subclavian
Left cephalic
Left axillary
Great cardiac
Left brachial
Left basilic
Splenic
Left renal
Inferior mesenteric
Left internal iliac (hypogastric)
Left external iliac
Left palmar digitals
Left femoral
Left popliteal
Left peroneal
Left posterior tibial
Left anterior tibial
Left dorsal venous arch

DANK

FIGURE 21-17 Principal veins in anterior view.

EXHIBIT 21-9
VEINS OF HEAD AND NECK (Figure 21-18)

VEIN	DESCRIPTION AND REGION DRAINED
Internal jugulars	Right and left **internal jugular veins** receive blood from face and neck. They arise as continuation of **sigmoid sinuses** at base of skull. Intracranial vascular sinuses are located between layers of dura mater and receive blood from brain. Other sinuses that drain into internal jugular include **superior sagittal sinus, inferior sagittal sinus, straight sinus,** and **transverse (lateral) sinuses.** Internal jugulars descend on either side of neck and pass behind clavicles, where they join with right and left subclavian veins. Unions of internal jugulars and subclavians form right and left brachiocephalic veins. From here blood flows into superior vena cava.
External jugulars	Right and left **external jugular veins** run down neck along outside of internal jugulars. They drain blood from parotid (salivary) glands, facial muscles, scalp, and other superficial structures into subclavian veins.

> **CLINICAL APPLICATION**
> In cases of heart failure, the venous pressure in the right atrium may rise. In such patients the pressure in the column of blood in the external jugular vein rises so that, even with the patient at rest and sitting in a chair, the external jugular vein will be visibly distended. Temporary distention of the vein is often seen in healthy adults when the intrathoracic pressure is raised in coughing and physical exertion

EXHIBIT 21-10
VEINS OF UPPER EXTREMITIES (Figure 21-19)

VEIN	DESCRIPTION AND REGION DRAINED
	Blood from each upper extremity is returned to the heart by deep and superficial veins. Both sets of veins contain valves. **Deep veins** are located deep in the body. They usually accompany arteries, and many have the same names as corresponding arteries. **Superficial veins** are located just below the skin and are often visible. They anastomose extensively with each other and deep veins.
SUPERFICIAL	
Cephalics	**Cephalic vein** of each upper extremity begins in the medial part of **dorsal arch** and winds upward around radial border of forearm. In front of elbow, it is connected to basilic vein by the **median cubital vein.** Just below elbow, cephalic vein unites with **accessory cephalic vein** to form cephalic vein of upper extremity. Ultimately, cephalic vein empties into axillary vein.
Basilics	**Basilic vein** of each upper extremity originates in the ulnar part of dorsal arch. It extends along posterior surface of ulna to point below elbow where it receives **median cubital vein.** If a vein must be punctured for an injection, transfusion, or removal of a blood sample, median cubitals are preferred. The median cubital vein joins the basilic vein to form the axillary vein.
Median antebrachials	**Median antebrachial veins** drain venous plexus on palmar surface of hand, ascend on ulnar side of anterior forearm, and end in median cubital veins.
DEEP	
Radials	**Radial veins** receive **dorsal metacarpal veins.**
Ulnars	**Ulnar veins** receive tributaries from **deep palmar arch.** Radial and ulnar veins unite in bend of elbow to form brachial vein.
Brachials	Located on either side of brachial artery, **brachial veins** join into axillary veins.
Axillaries	**Axillary veins** are a continuation of brachials and basilics. Axillaries end at first rib, where they become subclavians.
Subclavians	Right and left **subclavian veins** unite with internal jugulars to form brachiocephalic veins. Thoracic duct of lymphatic system delivers lymph into left subclavian vein at junction with internal jugular. Right lymphatic duct delivers lymph into right subclavian vein at corresponding junction.

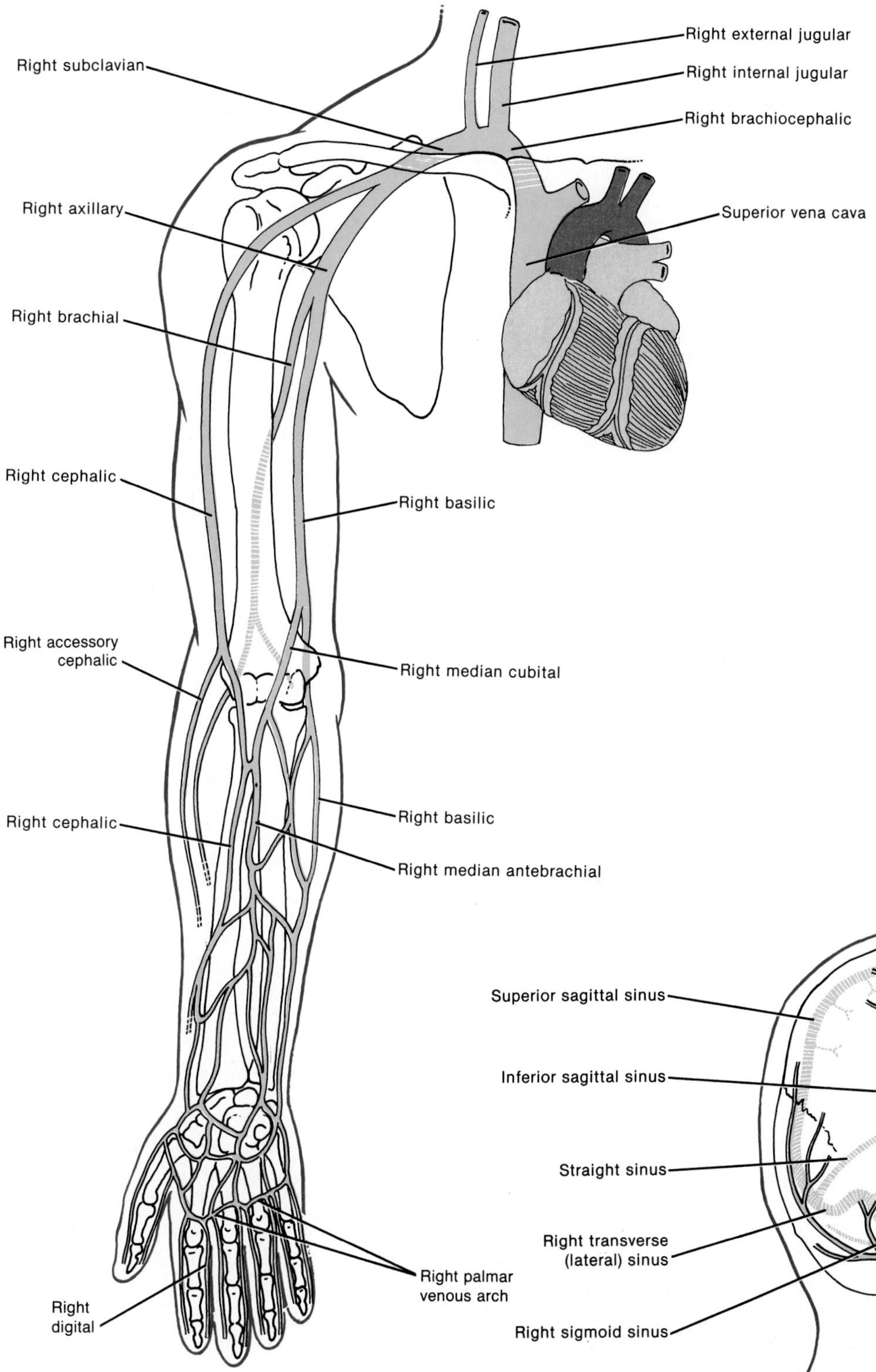

Right external jugular

Right internal jugular

Right brachiocephalic

FIGURE 21-19 Veins of the right upper extremity in anterior view.

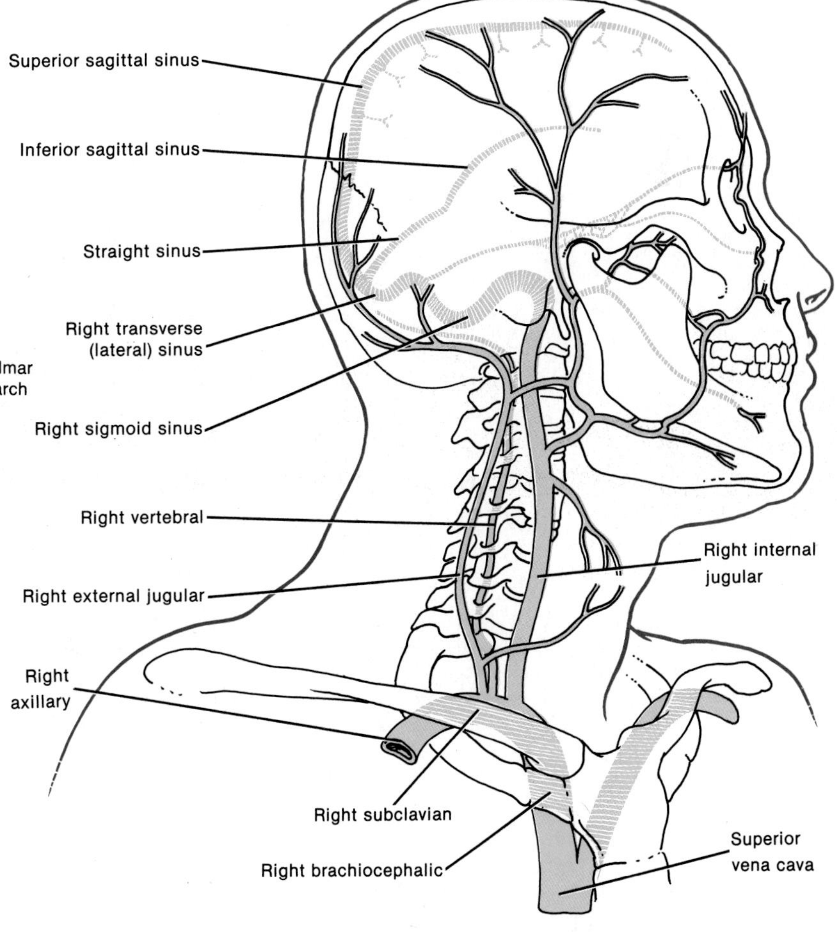

Right subclavian

Right axillary

Superior vena cava

Right brachial

Right cephalic

Right basilic

Right accessory cephalic

Right median cubital

Right cephalic

Right basilic

Right median antebrachial

Superior sagittal sinus

Inferior sagittal sinus

Straight sinus

Right transverse (lateral) sinus

Right sigmoid sinus

Right palmar venous arch

Right digital

Right vertebral

Right internal jugular

Right external jugular

FIGURE 21-18 Veins of the head and neck in right lateral view.

Right axillary

Right subclavian

Right brachiocephalic

Superior vena cava

EXHIBIT 21-11
VEINS OF THORAX (Figure 21-20)

VEIN	DESCRIPTION AND REGION DRAINED
Brachiocephalic	Right and left **brachiocephalic veins,** formed by union of subclavians and internal jugulars, drain blood from head, neck, upper extremities, mammary glands, and upper thorax. Brachiocephalics unite to form superior vena cava.
Azygos veins	**Azygos veins,** besides collecting blood from thorax, may serve as bypass for inferior vena cava that drains blood from lower body. Several small veins directly link azygos veins with inferior vena cava. And large veins that drain lower extremities and abdomen dump blood into azygos. If inferior vena cava or hepatic portal vein becomes obstructed, azygos veins can return blood from lower body to superior vena cava.
Azygos	**Azygos vein** lies in front of vertebral column, slightly right of midline. It begins as continuation of right ascending lumbar vein. It connects with inferior vena cava, right common iliac, and lumbar veins. Azygos receives blood from **right intercostal veins** that drain chest muscles; from hemiazygos and accessory hemiazygos veins; from several **esophageal, mediastinal,** and **pericardial veins;** and from right **bronchial vein.** Vein ascends to fourth thoracic vertebra, arches over right lung, and empties into superior vena cava.
Hemiazygos	**Hemiazygos vein** is in front of vertebral column and slightly left of midline. It begins as continuation of left ascending lumbar vein. It receives blood from lower four or five **intercostal veins** and some **esophageal** and **mediastinal veins.** At level of ninth thoracic vertebra, it joins azygos vein.
Accessory hemiazygos	**Accessory hemiazygos vein** is also in front and to left of vertebral column. It receives blood from three or four **intercostal veins** and left **bronchial vein.** It joins azygos at level of eighth thoracic vertebra.

EXHIBIT 21-12
VEINS OF THE ABDOMEN AND PELVIS (Figure 21-20)

VEIN	DESCRIPTION AND REGION DRAINED
Inferior vena cava	**Inferior vena cava** is the largest vein of the body. It is formed by union of two common iliac veins that drain lower extremities and abdomen. Inferior vena cava extends upward through abdomen and thorax to right atrium. Numerous small veins enter the inferior vena cava. Most carry return flow from branches of abdominal aorta and names correspond to names of arteries.
Common iliacs	**Common iliac veins** are formed by union of internal (hypogastric) and external iliac veins and represent distal continuation of inferior vena cava at its bifurcation.
Internal iliacs (hypogastrics)	Tributaries of **internal iliac (hypogastric) veins** basically correspond to branches of external iliac arteries. Internal iliacs drain gluteal muscles, medial side of thigh, urinary bladder, rectum, prostate gland, ductus deferens, uterus, and vagina.
External iliacs	**External iliac veins** are continuation of femoral veins and receive blood from lower extremities and inferior part of anterior abdominal wall.
Renals	**Renal veins** drain kidneys.
Gonadals (testiculars and ovarians)	**Testicular veins** drain testes (left testicular vein empties into left renal vein); **ovarian veins** drain ovaries (left ovarian vein empties into left renal vein).
Suprarenals	**Suprarenal veins** drain suprarenal glands (left suprarenal vein empties into left renal vein).
Inferior phrenics	**Inferior phrenic veins** drain diaphragm (left inferior phrenic vein sends tributary to left renal vein).
Hepatics	**Hepatic veins** drain liver.
Lumbars	A series of parallel **lumbar veins** drain blood from both sides of posterior abdominal wall. Lumbars connect at right angles with right and left **ascending lumbar veins,** which form origin of corresponding azygos or hemiazygos vein. Lumbars drain blood into ascending lumbars and then run to inferior vena cava, where they release remainder of flow.

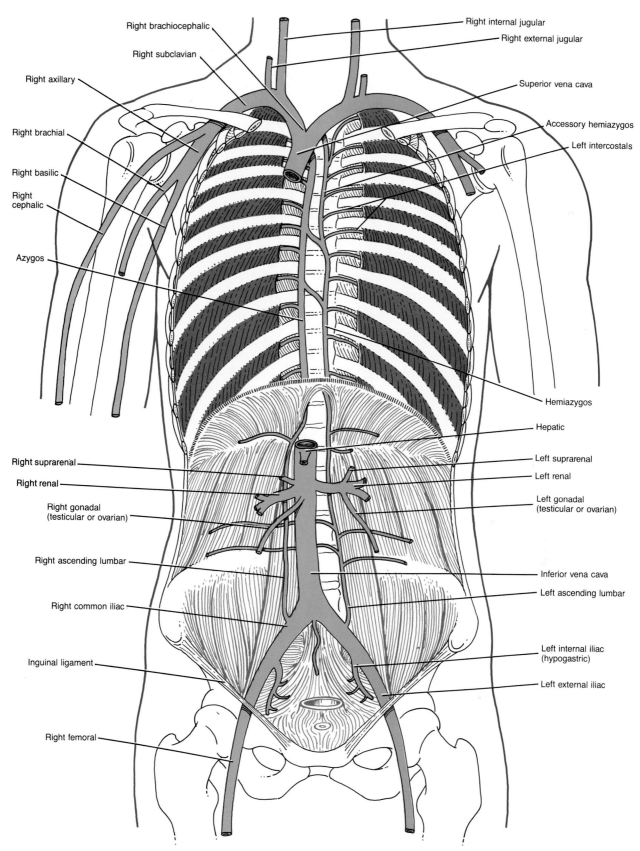

FIGURE 21-20 Veins of the thorax, abdomen, and pelvis in anterior view.

EXHIBIT 21-13

VEINS OF LOWER EXTREMITIES (Figure 21-21)

VEIN	DESCRIPTION AND REGION DRAINED
	Blood from each lower extremity is returned by superficial set and deep set of veins.
SUPERFICIAL VEINS	
Great saphenous	**Great saphenous vein,** longest vein in body, begins at medial end of **dorsal venous arch** of foot. It passes in front of medial malleolus and then upward along medial aspect of leg and thigh. It receives tributaries from superficial tissues and connects with deep veins as well. It empties into femoral vein in groin.

> **CLINICAL APPLICATION**
>
> The great saphenous vein is very constant in its position anterior to the medial malleolus. It is frequently used for prolonged administration of intravenous fluids. This is particularly important in very young babies and in patients of any age who are in shock and whose veins are collapsed. It and the small saphenous vein are subject to varicosity.

Small saphenous	**Small saphenous vein** begins at lateral end of dorsal venous arch of foot. It passes behind lateral malleolus and ascends under skin of back of leg. It receives blood from foot and posterior portion of leg. It empties into popliteal vein behind knee.
DEEP VEINS	
Posterior tibial	**Posterior tibial vein** is formed by union of **medial** and **lateral plantar veins** behind medial malleolus. It ascends deep in muscle at back of leg, receives blood from **peroneal vein,** and unites with anterior tibial vein just below knee.
Anterior tibial	**Anterior tibial vein** is upward continuation of **dorsalis pedis veins** in foot. It runs between tibia and fibula and unites with posterior tibial to form popliteal vein.
Popliteal	**Popliteal vein,** just behind knee, receives blood from anterior and posterior tibials and small saphenous vein.
Femoral	**Femoral vein** is upward continuation of popliteal just above knee. Femorals run up posterior of thighs and drain deep structures of thighs. After receiving great saphenous veins in groin, they continue as right and left external iliac veins.

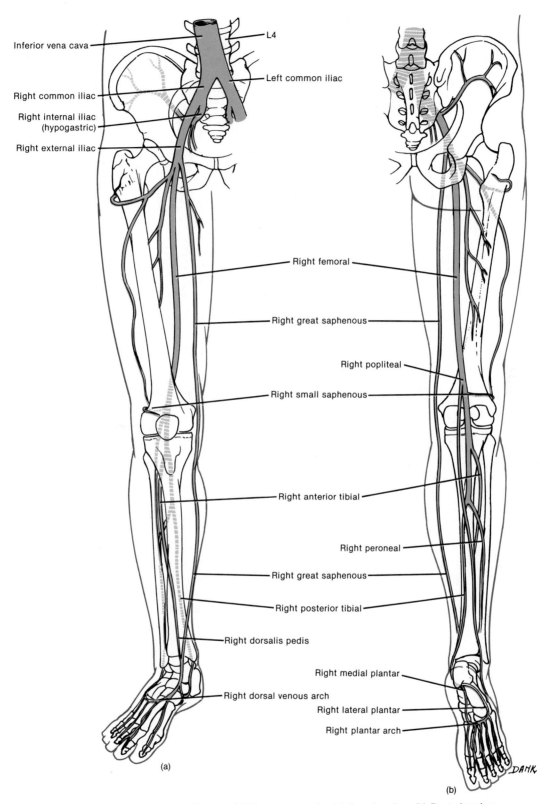

FIGURE 21-21 Veins of the pelvis and right lower extremity. (a) Anterior view. (b) Posterior view.

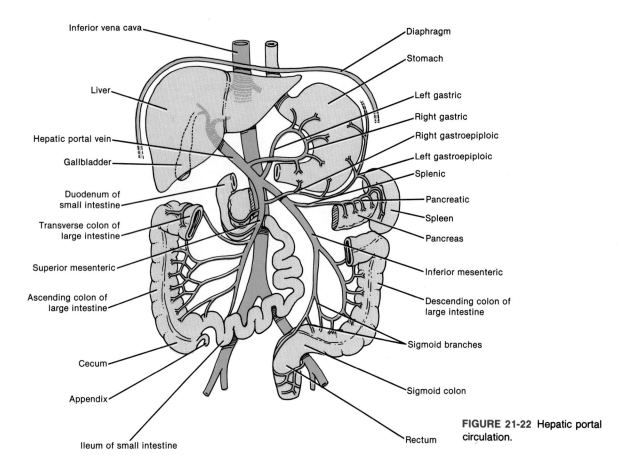

FIGURE 21-22 Hepatic portal circulation.

HEPATIC PORTAL CIRCULATION

Blood enters the liver from two sources. The hepatic artery delivers oxygenated blood from the systemic circulation; the hepatic portal vein delivers deoxygenated blood from the digestive organs. The term **hepatic portal circulation** refers to this flow of venous blood from the digestive organs to the liver before returning to the heart (Figure 21-22). Hepatic portal blood is rich with substances absorbed from the digestive tract. The liver monitors these substances before they pass into the general circulation. For example, the liver stores nutrients such as glucose. It modifies other digested substances so they may be used by cells. It detoxifies harmful substances that have been absorbed by the digestive tract and destroys bacteria by phagocytosis.

The hepatic portal system includes veins that drain blood from the pancreas, spleen, stomach, intestines, and gallbladder and transport it to the portal vein of the liver. The *hepatic portal vein* is formed by the union of the superior mesenteric and splenic veins. The *superior mesenteric vein* drains blood from the small intestine and portions of the large intestine and stomach. The *splenic vein* drains the spleen and receives tributaries from the stomach, pancreas, and portions of the colon. The tributaries from the stomach are the *gastric, pyloric,* and *gastroepiploic veins.* The *pancreatic veins* come from the pancreas, and the *inferior mesenteric veins* come from portions of the colon. Before the hepatic portal vein enters the liver, it receives the *cystic vein* from the gallbladder and other veins. Ultimately, blood leaves the liver through the *hepatic veins,* which enter the inferior vena cava.

PULMONARY CIRCULATION

The flow of deoxygenated blood from the right ventricle to the lungs and the return of oxygenated blood from the lungs to the left atrium is called the **pulmonary circulation** (Figure 21-23). The *pulmonary trunk* emerges from the right ventricle and passes upward, backward, and to the left. It then divides into two branches. The *right pulmonary artery* runs to the right lung; the *left pulmonary artery* goes to the left lung. On entering the lungs, the branches divide and subdivide until ultimately they form capillaries around the alveoli in the lungs. Carbon dioxide is passed from the blood into the alveoli to be breathed out of the lungs. Oxygen breathed in by the lungs is passed from the alveoli into the blood. The capillaries unite, venules and veins are formed, and eventually, two *pulmonary veins* exit from each lung and transport the oxygenated blood to the left atrium. The pulmonary veins are the only postnatal veins that carry oxygenated blood. Contractions of the left ventricle then sends the blood into the systemic circulation.

FETAL CIRCULATION

The circulatory system of a fetus, called **fetal circulation,** differs from an adult's because the lungs, kidneys, and digestive tract of a fetus are nonfunctional. The fetus derives its oxygen and nutrients from the maternal blood and eliminates its carbon dioxide and wastes into the maternal blood (Figure 21-24).

The exchange of materials between fetal and maternal circulation occurs through a structure called the *placenta*

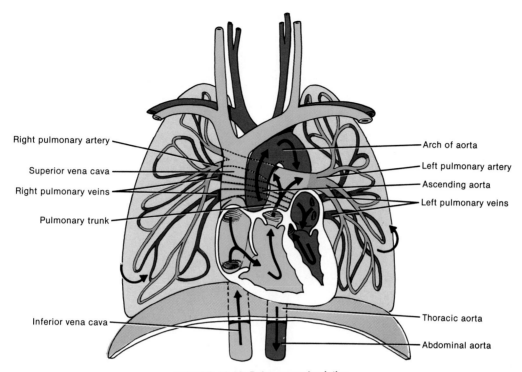

Right pulmonary artery

Superior vena cava

Right pulmonary veins

Pulmonary trunk

Arch of aorta

Left pulmonary artery

Ascending aorta

Left pulmonary veins

Inferior vena cava

Thoracic aorta

Abdominal aorta

FIGURE 21-23 Pulmonary circulation.

(pla-SEN-ta). It is attached to the umbilicus of the fetus by the umbilical (um-BIL-i-kal) cord, and it communicates with the mother through countless small blood vessels that emerge from the uterine wall. The umbilical cord contains blood vessels that branch into capillaries in the placenta. Wastes from the fetal blood diffuse out of the capillaries, into spaces containing maternal blood (intervillous spaces) in the placenta, and finally into the mother's uterine blood vessels. Nutrients travel the opposite route—from the maternal blood vessels to the intervillous spaces to the fetal capillaries. Normally there is no mixing of maternal and fetal blood since all exchanges occur through capillaries.

Blood passes from the fetus to the placenta via two *umbilical arteries.* These branches of the internal iliac (hypogastric) arteries are included in the umbilical cord. At the placenta, the blood picks up oxygen and nutrients and eliminates carbon dioxide and wastes. The oxygenated blood returns from the placenta via a single *umbilical vein.* This vein ascends to the liver of the fetus, where it divides into two branches. Some blood flows through the branch that joins the hepatic portal vein and enters the liver. Although the fetal liver manufactures red blood cells, it does not function in digestion. Therefore, most of the blood flows into the second branch, the *ductus venosus* (DUK-tus ve-NŌ-sus). The ductus venosus eventually passes its blood to the inferior vena cava, bypassing the liver.

In general, circulation through other portions of the fetus is not unlike postnatal circulation. Deoxygenated blood returning from the lower regions is mingled with oxygenated blood from the ductus venosus in the inferior vena cava. This mixed blood then enters the right atrium. The circulation of blood through the upper portion of the fetus is also similar to postnatal flow. Deoxygenated

blood returning from the upper regions of the fetus is collected by the superior vena cava, and it also passes into the right atrium.

Most of the blood does not pass through the right ventricle to the lungs, as it does in postnatal circulation, since the fetal lungs do not operate. In the fetus, an opening called the *foramen ovale* (fō-RĀ-men ō-VAL-ē) exists in the septum between the right and left atria. A valve in the inferior vena cava directs about a third of the blood through the foramen ovale so that it may be sent directly into the systemic circulation. The blood that does descend into the right ventricle is pumped into the pulmonary trunk, but little of this blood actually reaches the lungs. Most blood in the pulmonary trunk is sent through the *ductus arteriosus* (ar-tē-rē-Ō-sus). This small vessel connecting the pulmonary trunk with the aorta enables blood in excess of nutrient requirements to bypass the fetal lungs. The blood in the aorta is carried to all parts of the fetus through its systemic branches. When the common iliac arteries branch into the external and internal iliacs (hypogastric), part of the blood flows into the internal iliacs (hypogastrics). It then goes to the umbilical arteries and back to the placenta for another exchange of materials. The only vessel that carries fully oxygenated blood is the umbilical vein.

At birth, when lung, renal, digestive, and liver functions are established, the special structures of fetal circulation are no longer needed and the following changes occur.

1. The umbilical arteries atrophy to become the *lateral umbilical ligaments.*

2. The umbilical vein becomes the *round ligament* of the liver.

3. The placenta is delivered by the mother as the *"afterbirth."*

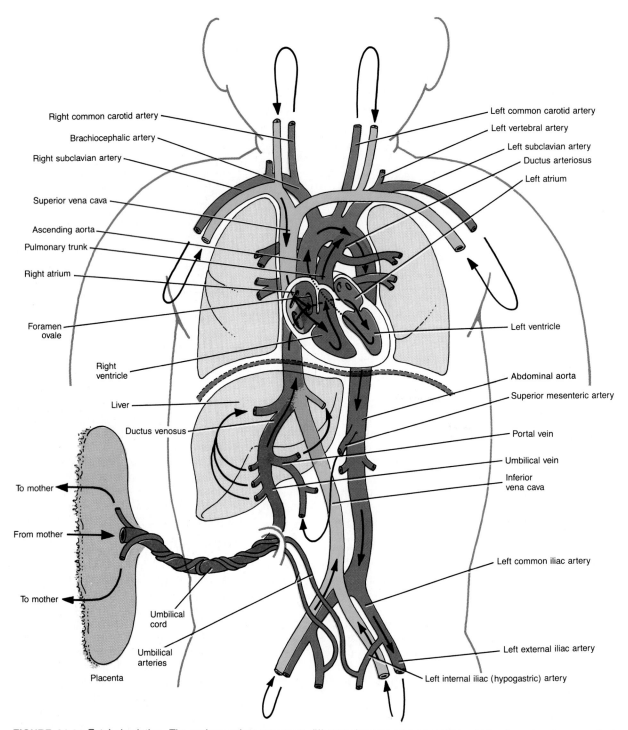

FIGURE 21-24 Fetal circulation. The various colors represent different degrees of oxygenation of blood ranging from greatest (red) to intermediate (two shades of purple) to least (blue).

4. The ductus venosus becomes the *ligamentum venosum*, a fibrous cord in the liver.

5. The foramen ovale normally closes shortly after birth to become the *fossa ovalis*, a depression in the interatrial septum.

6. The ductus arteriosus closes, atrophies, and becomes the *ligamentum arteriosum*.

Anatomical defects resulting from failure of these changes to occur are described in Chapter 20.

DISORDERS: HOMEOSTATIC IMBALANCES

Aneurysm

A blood-filled sac formed by an outpouching in an arterial or venous wall is called an **aneurysm** (AN-yoo-rizm). Aneurysms may occur in any major blood vessel. A **berry** aneurysm is a small aneurysm, frequently in the cerebral artery. If it ruptures, it may cause a hemorrhage below the dura mater. Hemorrhaging is one cause of a stroke. A **ventricular** aneurysm is a dilation of a ventricle of the heart. An **aortic** aneurysm is a dilation of the aorta. The aorta has a higher incidence of aneurysms than any other artery, probably because of its curved shape, large size, and high pressure. In the thorax, aneurysms usually occur in the ascending or descending aorta but seldom in the arch of the aorta. Symptoms of thoracic aortic aneurysm depend on pressure exerted on adjacent structures. For example, pressure on the inferior (recurrent) laryngeal nerve causes hoarseness and a brassy cough. Pressure on the esophagus may cause dysphagia (difficulty in swallowing). Dyspnea (difficulty in breathing) may follow pressure on the trachea, the root of the lung, or the phrenic nerve.

Coronary Artery Disease (CAD)

In Chapter 20 it was indicated that the common causes of heart disease are related to inadequate coronary blood supply, anatomical disorders, and arrhythmias. **Coronary artery disease (CAD)** is a condition in which the heart muscle receives inadequate blood because of an interruption of blood supply. Depending on the degree of interruption, symptoms can range from a mild chest pain to a full-scale heart attack. The underlying causes of CAD are many and varied. Two of the principal ones are atherosclerosis and coronary artery spasm.

Atherosclerosis

Atherosclerosis (ath'-er-ō-skle-RŌ-sis) is a process in which fatty substances, especially cholesterol and triglycerides (ingested fats), are deposited in the walls of medium-sized and large arteries in response to certain stimuli. It is believed that the first event in atherosclerosis is damage to the endothelial lining of the artery. Contributing factors include high blood pressure (hypertension), carbon monoxide in cigarettes, and a high cholesterol level. Following endothelial damage, platelets aggregate in the area of injury, adhere to the injured surface, and begin releasing their clot-forming chemicals. One of the chemicals increases permeability of the endothelial lining and modifies the cholesterol so that it is taken up by smooth muscle cells in the tunica media. The more cholesterol there is in the blood, the faster the process occurs. Platelets also release a chemical that stimulates the smooth muscle cells to proliferate. The result of enlarged, proliferating muscle cells is the development of an **atherosclerotic plaque** which deforms the arterial wall (Figure 21-25).

The plaque looks like a pearly gray or yellow mound of tissue. As it grows, it may obstruct blood flow in the affected artery and damage the tissues the artery supplies. An additional danger is that the plaque may provide a roughened surface for clot formation, and a thrombus may form. If the clot breaks off and forms an embolus, it may obstruct small arteries, capillaries, and veins quite a distance from the site of formation.

Cholesterol is a major building block of all plasma membranes and a key compound for the synthesis of some hormones and bile salts. Some cholesterol is present in foods (eggs, dairy products, organ meats, beef, pork, and processed luncheon meats); some is synthesized by the liver. As it turns out, foods containing cholesterol are not the main source of cholesterol in the blood. It is the type of fat eaten. Saturated fats (Chapter 2) have their carbon atoms filled with all the hydrogen atoms they can accommodate and are high in cholesterol. They occur mostly in animal foods and contribute significantly to a rise in blood cholesterol levels.

Neither cholesterol nor triglycerides dissolve in water and thus cannot travel in the blood in their unaltered forms. They are made water-soluble by combination with proteins produced by the liver and intestine and the complexes are called **lipoproteins.** The two major classes are called **low-density lipoproteins (LDL)** and **high-density lipoproteins (HDL).** There is an important difference between the two: LDL seems to pick up cholesterol and deposit it in body cells including, under abnormal conditions, smooth muscle cells in arteries. HDL seems to gather cholesterol from body cells and transports it to the liver for elimination. According to this proposed relationship, a person's susceptibility to CAD from atherosclerosis depends on the relative balance between LDL and HDL. Supposedly, the higher the ratio of HDL to total cholesterol (and to LDL), the less the risk of CAD. Several ways to improve the HDL-LDL ratio are regular sustained exercise, diet (reducing animal fats, especially), and giving up cigarettes.

Diagnosis of atherosclerosis is based upon a stress ECG, angiography, and nuclear cardiology. The *stress ECG* is done while a person is exercising on a treadmill or bicycle in order to determine the presence and extent of CAD that might be missed on a resting ECG. *Angiography* (an-jē-OG-ra-fē) involves injecting a dye directly into the blood and then taking x-rays of the arteries. The technique is also called *arteriography,* and the film is called an *arteriogram* (Figure 21-25c). Angiography of the blood vessels of the heart is called *coronary angiography. Nuclear cardiology* involves injection of a radioactive substance into a vein followed by analysis of its distribution in the heart. An example of such a test is a *thallium scan.* As the radioisotope thallium 201 circulates through the coronary arteries, a camera placed over the chest picks up the signals from the radioisotope, relays them to a computer for translation, and a picture of the heart is displayed. An area of poor circulation of thallium in the heart indicates blockage of a coronary artery.

The treatment of CAD varies with the nature and urgency of the symptoms. Among the possible treatments are drug therapy, coronary bypass surgery, and percutaneous transluminal angioplasty (PTA). The two most commonly used drugs are *nitroglycerin* (which dilates blood vessels to increase blood flow to the heart) and *beta blockers* (which block the heart's response to sympathetic stimulation and cut down on the work load of the heart). Antithrombotic drugs that suppress the activity of platelets are being tested on a trial basis. In *coronary bypass surgery,* a vein taken from a leg or a synthetic vein is used as a conduit to literally bypass an area (or areas) of blockage in one or more coronary arteries. *Percutaneous transluminal angioplasty (PTA)* is the balloon catheterization technique to dilate obstructed vessels, described in Chapter 20. Unfortunately, calcified plaques are not crushed by the balloon, and the balloon can damage the arterial wall. Despite these drawbacks, it is hoped that the procedure can be refined so that it will come into general use in the treatment of atherosclerosis.

Coronary Artery Spasm

Atherosclerosis results in a fixed obstruction to blood flow. Recently, it has been learned that obstruction can also be caused by **coronary artery spasm,** a condition in which the smooth muscle of a coronary artery undergoes a sudden contraction, resulting in vasoconstriction. Coronary artery spasm can occur in individuals with or without atherosclerosis and may result in chest pain during rest (variant angina), chest pain during exertion (typical angina), heart attacks, and sudden death. Although the causes of coronary artery spasm are unknown, sev-

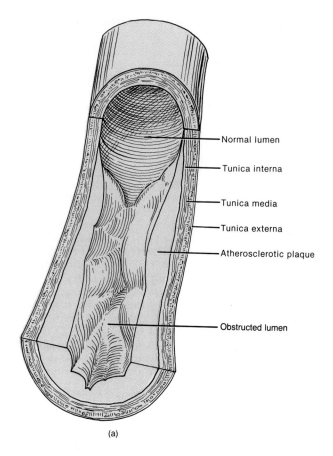

(a)

Normal lumen

Tunica interna

Tunica media

Tunica externa

Atherosclerotic plaque

Obstructed lumen

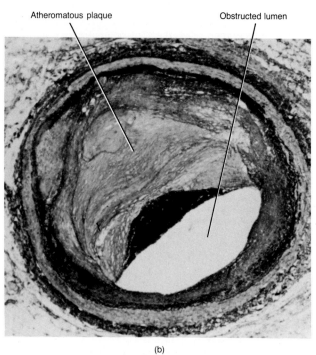

Atheromatous plaque

Obstructed lumen

(b)

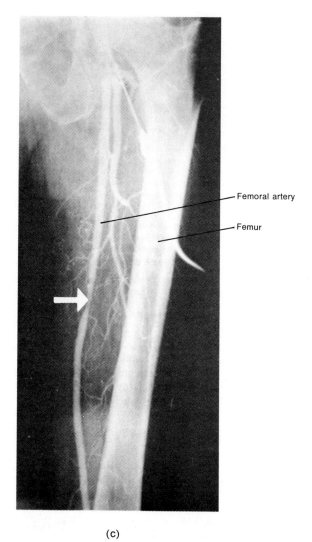

Femoral artery

Femur

(c)

FIGURE 21-25 Atherosclerosis. (a) Atherosclerotic plaque formation. Compare the size of the normal lumen and the obstructed lumen. (b) Photomicrograph of a cross section of an artery partially obstructed by an atherosclerotic plaque. (Courtesy of National Heart, Lung, and Blood Institute.) (c) Femoral arteriogram showing an atherosclerotic plaque (arrow) in the middle third of the thigh. (Courtesy of Lester W. Paul and John H. Juhl, *The Essentials of Roentgen Interpretation,* 3rd ed., Harper & Row, Publishers, Inc., New York, 1972.)

eral factors are receiving attention. These include smoking, stress, and a vasoconstrictor chemical released by platelets. There is considerable interest in aspirin and other drugs that might inhibit the vasoconstrictor chemical. Coronary artery spasm is treated with long-acting nitrates (nitroglycerin) and calcium-blocking agents (drugs that block calcium transport into muscle cells, a requirement for muscle contraction).

Hypertension

Hypertension, or high blood pressure, is the most common disease affecting the heart and blood vessels. Statistics indicate that hypertension afflicts one out of every five American adults. Although there is some disagreement as to what defines hypertension, a strong consensus has emerged suggesting that a blood pressure of 120/80 is normal and desirable in a healthy adult.

A reading of 140/90 is generally regarded as the threshold of hypertension, while higher values, especially those over 160/95, are classified as dangerously hypertensive.

Primary hypertension, or essential hypertension, is a persistently elevated blood pressure that cannot be attributed to any particular organic cause. Approximately 85 percent of all hypertension cases fit this definition. The other 15 percent are *secondary hypertension.* Secondary hypertension has an identifiable underlying cause such as atherosclerosis, kidney disease, and adrenal hypersecretion. Atherosclerosis increases blood pressure by reducing the elasticity of the arterial wall and narrowing the lumen through which the blood can flow. Kidney disease and obstruction of blood flow may cause the kidneys to release renin into the blood. This enzyme catalyzes the formation of angiotensin II from a plasma protein. Angiotensin II is a powerful blood-vessel constrictor—and the most potent agent known for raising blood pressure. It also stimulates aldosterone release. Aldosteronism, the hypersecretion of aldosterone, may also cause an increase in blood pressure. Aldosterone is the adrenal cortex hormone that promotes the retention of salt and water by the kidneys. It thus tends to increase plasma volume. Pheochromocytoma is a tumor of the adrenal medulla. It produces and releases into the blood large quantities of norepinephrine and epinephrine. These hormones also raise blood pressure. Epinephrine causes an increase in heart rate and norepinephrine causes vasoconstriction.

High blood pressure is of considerable concern because of the harm it can do to the heart, brain, and kidneys if it remains uncontrolled. The heart is most commonly affected by high blood pressure. When pressure is high, the heart uses more energy in pumping. Because of the increased effort, the heart muscle thickens and the heart becomes enlarged. The heart also needs more oxygen. If it cannot meet the demands put on it, angina pectoris or even myocardial infarction may develop. Hypertension is also a factor in the development of atherosclerosis. Continued high blood pressure may produce a cerebral vascular accident (CVA) or stroke. In this case, severe strain has been imposed on the cerebral arteries that supply the brain. These arteries are usually less protected by surrounding tissues than are the major arteries in other parts of the body. These weakened cerebral arteries may finally rupture, and a brain hemorrhage follows.

The kidneys are also prime targets of hypertension. The principal site of damage is in the arterioles that supply them. The continual high blood pressure pushing against the walls of the arterioles causes them to thicken, thus narrowing the lumen. The blood supply to the kidneys is thereby gradually reduced. In response, the kidneys may secrete renin, which raises the blood pressure even higher and complicates the problem. The reduced blood flow to the kidney cells may eventually lead to the death of the cells.

Medical science cannot cure hypertension. However, almost all cases of hypertension, whether mild or severe, can be controlled in a number of ways. The overweight person with hypertension will usually be placed on a reducing diet, because blood pressure often falls with weight loss. Treatment often involves the restriction of sodium intake. Sodium restriction curbs fluid retention by the body and also tends to reduce blood volume. Since nicotine is a vasoconstrictor, it elevates blood pressure. Stopping smoking may help to decrease blood pressure. As indicated earlier, exercise can help to reduce hypertension. In recent years, some physicians have advocated relaxation techniques (yoga, meditation, and biofeedback) to treat hypertension. For those individuals who require medication, a number of drugs are available. Many people can be treated with diuretics which eliminate large amounts of water and sodium, thus decreasing blood volume and reducing blood pressure. Vasodilators are often used in combination with diuretics. They relax the smooth muscle in arterial walls causing vasodilation and lowering blood pressure. Beta blockers are also used to lower blood pressure, often in combination with diuretics.

MEDICAL TERMINOLOGY

Angiocardiography (*angio* = vessel; *cardio* = heart; *graph* = writing) X-ray examination of the heart and great blood vessels after injection of a radiopaque dye into the blood stream.

Aortography X-ray examination of the aorta and its main branches after injection of a radiopaque dye.

Claudication Pain and lameness or limping caused by defective circulation of the blood in the vessels of the limbs.

Coronary endarterectomy The removal of the obstructing area within the lumen of the vessel.

Cor pulmonale Heart disease resulting from disease of the lungs or the blood vessels in the lungs. This is due to resistance to the passage of blood through the lungs.

Cyanosis (*cyano* = blue) Slightly bluish, dark purple skin coloration due to oxygen deficiency in systemic blood.

Hypercholesteremia (*hyper* = over; *heme* = blood) An excess of cholesterol in the blood.

Hypotension (*hypo* = below; *tension* = pressure) Low blood pressure; most commonly used to describe an acute drop in blood pressure, as occurs in circulatory shock.

Normotensive Characterized by normal blood pressure.

Occlusion The closure or obstruction of the lumen of a structure such as a blood vessel.

Phlebitis (*phleb* = vein) Inflammation of a vein often in a leg.

Shunt A passage between two blood vessels or between the two sides of the heart.

Thrombectomy (*thrombo* = clot) An operation to remove a blood clot from a blood vessel.

Thrombophlebitis Inflammation of a vein with clot formation.

DRUGS ASSOCIATED WITH THE CARDIOVASCULAR SYSTEM

Fibrinolytic Drugs

Fibrinolytic drugs trigger the body's own fibrinolytic system that dissolves fresh fibrin clots. The drugs convert the inactive enzyme plasminogen to the active fibrinolytic enzyme plasmin (fibrinolysin). They are used in acute pulmonary embolism, deep vein thrombosis, and occluded access shunts in renal dialysis patients. Examples are *streptokinase* (Streptase) and *urokinase* (Abbokinase).

Vasodilators

Vasodilators are used to dilate blood vessels. Drugs that are used in the treatment of angina pectoris include *nitroglycerin, isosorbide dinitrate* (Isordil), and *pentaerythritol tetranitrate* (PETN, Peritrate).

Inotropics

Inotropics are drugs that affect the force of muscular contractions. Positive inotropic effects increase contractility, while negative inotropic effects decrease contractility. In heart conditions, some inotropic drugs increase stroke volume. Examples of drugs with positive inotropic effects are *digoxin* (Lanoxin) and *digitoxin* (Crystodigin), both digitalis preparations; *dobutamine* (Dobutrex); *isoproterenol* (Isuprel); *dopamine* (Intropin); and *epinephrine*. Drugs with negative inotropic properties that are used to reduce myocardial oxygen consumption and dilate coronary arteries are *verapamil* (Calan, Isoptin), *diltiazen* (Cardizem), and *nifedipine* (Procardia). These negative inotropic drugs are also **calcium-blocking agents** because they literally block channels through which calcium ions move into smooth muscle cells in arterial walls, a requirement for muscle contraction. As a result, the arterial wall relaxes and blood flow is increased. This improves the functioning of the heart and relieves angina.

Antiarrhythmics

Antiarrhythmics are drugs that overcome heartbeat arrhythmias. Arrhythmia is any abnormality of cardiac impulse conduction or timing of beats. Arrhythmias are produced in many cases of myocardial infarction and heart failure. Examples of antiarrhythmics are *quinidine* (Cardioquin), *procainamide hydrochloride* (Pronestyl), *phenytoin* (Dilantin), *bretylium* (Bretylol), and local anesthetics such as *lidocaine* (Xylocaine).

Cholesterol-Lowering Agents

Since many coronary patients exhibit elevated blood cholesterol levels, and since this appears to be one of the coronary risk factors, cholesterol-lowering drugs are sometimes indicated in select patients. The drugs that lower blood cholesterol levels bind bile acids in the intestine and this leads to a greater movement of sterols from the liver to bile and out of the body.

Examples are *clofibrate* (Atromid-S) and *probucol* (Lorelco).

Antihypertensives

Antihypertensives are drugs that lower blood pressure either immediately, in cases of emergencies, or gradually. Diuretic agents (drugs that stimulate the flow of urine) are considered first line drugs in the management of hypertension. Examples are *ticrynafen* (Selacryn), *furosemide* (Lasix), and *chlorothiazide* (Diuril).

Beta-blockers are drugs that compete with epinephrine for available receptor sites, thus diminishing the stimulatory effect of epinephrine. Beta-blockers protect the heart from overresponding to physical strain and emotional stress by decreasing heart rate. By doing this and causing blood vessel dilation, they act as antihypertensives. Examples are *propranolol* (Inderal), *metoprolol* (Lopressor), *nadolol* (Corgard), *timolol* (Blocardren), *pindolol* (Visken), and *atenolol* (Tenormin). Drugs used in treatment of acute hypertension are *nitroprusside* and *diazoxide* (Hyperstat), while chronic hypertension drugs include *hydralazine* (Apresoline) and *minoxidil*. Miscellaneous antihypertensives which have sites of action in the central nervous system include *prazosin* (Minipress), *methyldopa* (Aldomet), and *clonidine* (Catapres). A drug called *captopril* prevents the body from producing angiotensin II, thus effectively lowering blood pressure.

Anticoagulants

Anticoagulants prevent blood clotting. These agents prevent chemical changes in blood proteins concerned with coagulation and thereby prevent clotting. They are used in transfusions and to prevent thrombosis and the formation of embolisms. Examples are *coumarin* (Dicumarol), *heparin,* and *warfarin* (Coumadin, Panwarfin).

STUDY OUTLINE

Arteries (p. 482)
1. Arteries carry blood away from the heart. Their wall consists of a tunica interna, tunica media (which maintains elasticity and contractility), and tunica externa.
2. Large arteries are referred to as elastic (conducting) arteries and medium-sized arteries are called muscular (distributing) arteries.
3. Many arteries anastomose—the distal ends of two or more vessels unite. An alternate blood route from an anastomosis is called collateral circulation. Arteries that do not anastomose are called end arteries.

Arterioles (p. 483)
1. Arterioles are small arteries that deliver blood to capillaries.
2. Through constriction and dilation they assume a key role in regulating blood flow from arteries into capillaries.

Capillaries (p. 484)
1. Capillaries are microscopic blood vessels through which materials are exchanged between blood and tissue cells; some capillaries are continuous, others are fenestrated.
2. Capillaries branch to form an extensive capillary network throughout the tissue. This network increases the surface area, allowing a rapid exchange of large quantities of materials.
3. Precapillary sphincters regulate blood flow through capillaries.
4. Microscopic blood vessels in the liver are called sinusoids.

Venules (p. 484)
1. Venules are small vessels that continue from capillaries and merge to form veins.
2. They drain blood from capillaries into veins.

Veins (p. 485)
1. Veins consist of the same three tunics as arteries, but have less elastic tissue and smooth muscle.
2. They contain valves to prevent backflow of blood.
3. Weak valves can lead to varicose veins or hemorrhoids.
4. Vascular (venous) sinuses are veins with very thin walls.

Physiology of Circulation (p. 486)
Blood Flow and Blood Pressure
1. Blood flows from regions of higher to lower pressure. The established pressure gradient is from aorta (100 mm Hg) to arteries (100–40 mm Hg) to arterioles 40–25 mm Hg) to capillaries (25–12 mm Hg) to venules (12–8 mm Hg) to veins (10–5 mm Hg) to venae cavae (2 mm Hg) to right atrium (0 mm Hg).
2. Any factor that increases cardiac output increases blood pressure.
3. As blood volume increases, blood pressure increases.
4. Peripheral resistance is determined by blood viscosity and blood vessel diameter. Increased viscosity and vasoconstriction increase peripheral resistance and thus increase blood pressure.
5. Factors that determine heart rate and force of contraction,

and therefore blood pressure, are the autonomic nervous system through the cardiac center, chemicals, temperature, emotions, sex, and age.
6. Factors that regulate blood pressure by acting on blood vessels include the vasomotor center in the medulla together with pressoreceptors, chemoreceptors, and higher brain centers; chemicals; and autoregulation.
7. The movement of water and dissolved substances (except proteins) through capillaries by diffusion is dependent on hydrostatic and osmotic pressures.
8. The near equilibrium at the arterial and venous ends of a capillary by which fluids exit and enter is called Starling's law of the capillaries.
9. Blood return to the heart is maintained by several factors including increasing velocity of blood in veins, skeletal muscular contractions, valves in veins (especially in the extremities), and breathing.

Blood Reservoirs
1. Systemic veins are collectively called blood reservoirs.
2. They store blood which through vasoconstriction can move to other parts of the body if the need arises.
3. The principal reservoirs are the veins of the abdominal organs (liver and spleen) and skin.

Checking Circulation (p. 492)
Pulse
1. Pulse is the alternate expansion and elastic recoil of an artery with each heartbeat. It may be felt in any artery that lies near the surface or over a hard tissue.
2. A normal rate is between 70 and 80 beats per minute.

Measurement of Blood Pressure
1. Blood pressure is the pressure exerted by blood on the wall of an artery when the left ventricle undergoes systole and then diastole. It is measured by the use of a sphygmomanometer.
2. Systolic blood pressure is the force of blood recorded during ventricular contraction. Diastolic blood pressure is the force of blood recorded during ventricular relaxation. The average blood pressure is 120/80 mm Hg.
3. Pulse pressure is the difference between systolic and diastolic pressure. It averages 40 mm Hg and provides information about the condition of arteries.

Circulatory Routes (p. 493)
1. The largest circulatory route is the systemic circulation.
2. Two of the many subdivisions of the systemic circulation are the coronary (cardiac) circulation and the hepatic portal circulation.

3. Other routes include the cerebral, pulmonary, and fetal circulation.

Systemic Circulation
1. The systemic circulation takes oxygenated blood from the left ventricle through the aorta to all parts of the body including lung tissue.
2. The aorta is divided into the ascending aorta, the arch of the aorta, and the descending aorta. Each section gives off arteries that branch to supply the whole body.
3. Blood is returned to the heart through the systemic veins. All the veins of the systemic circulation flow into either the superior or inferior venae cavae or the coronary sinus. They in turn empty into the right atrium.

Hepatic Portal Circulation
1. The hepatic portal circulation collects blood from the veins of the pancreas, spleen, stomach, intestines, and gallbladder and directs it into the hepatic portal vein of the liver.
2. This circulation enables the liver to utilize nutrients and detoxify harmful substances in the blood.

Pulmonary Circulation
1. The pulmonary circulation takes deoxygenated blood from the right ventricle to the lungs and returns oxygenated blood from the lungs to the left atrium.
2. It allows blood to be oxygenated for systemic circulation.

Fetal Circulation
1. The fetal circulation involves the exchange of materials between fetus and mother.
2. The fetus derives its oxygen and nutrients and eliminates its carbon dioxide and wastes through the maternal blood supply by means of a structure called the placenta.
3. At birth, when lung, digestive, and liver functions are established, the special structures of fetal circulation are no longer needed.

Disorders: Homeostatic Imbalances (p. 513)
1. An aneurysm is a sac formed by an outpocketing of a portion of an arterial or venous wall.
2. Coronary artery disease (CAD) refers to an inadequate blood supply to the heart muscle. Two principal causes are atherosclerosis and coronary artery spasm.
3. Atherosclerosis is a process in which fatty substances are deposited in the walls of arteries.
4. Coronary artery spasm is caused by a sudden contraction of the smooth muscle in an arterial wall that produces vasoconstriction.
5. Hypertension is high blood pressure and may damage the heart, brain, and kidneys.

REVIEW QUESTIONS

1. Describe the structural and functional differences among arteries, arterioles, capillaries, venules, and veins.
2. Discuss the importance of the elasticity and contractility of arteries.
3. Distinguish between elastic and muscular arteries in terms of location, histology, and function. What is an anastomosis? What is collateral circulation?
4. Describe how capillaries are structurally adapted for exchanging materials with body cells.
5. Define varicose veins and hemorrhoids.
6. Why does blood flow faster in arteries and veins than in capillaries?
7. Describe how each of the following affects blood pressure: cardiac output, blood volume, and peripheral resistance.
8. What is the vasomotor center? What are its functions?
9. Discuss how the vasomotor center operates with pressoreceptors, chemoreceptors, and higher brain centers to control blood pressures.
10. Explain the effects of different chemicals, such as epinephrine, norepinephrine (NE), antidiuretic hormone (ADH), angiotensin II, histamine, and kinins on blood pressure.
11. What is meant by autoregulation? Describe its probable mechanism.
12. Describe how hydrostatic and osmotic pressures determine

fluid movement through capillaries.

13. What is Starling's law of the capillaries?
14. Identify the factors that assist the return of venous blood to the heart.
15. What are blood reservoirs? Why are they important?
16. Explain the cardiovascular benefits of exercise.
17. Define pulse. Where may pulse be felt?
18. Contrast tachycardia, bradycardia, and irregular pulse.
19. What is blood pressure? Describe how systolic and diastolic blood pressure are recorded by means of a sphygmomanometer.
20. Compare the clinical significance of systolic and diastolic pressure. How are these pressures written?
21. Define pulse pressure. What does this pressure indicate?
22. What is meant by a circulatory route? Define systemic circulation.
23. Diagram the major divisions of the aorta, their principal arterial branches, and the regions supplied.
24. Trace a drop of blood from the arch of the aorta through its systemic circulatory route to the tip of the big toe on your left foot and back to the heart again. Remember that the major branches of the arch are the brachiocephalic artery, left common carotid artery, and left subclavian artery. Be sure to indicate which veins return the blood to the heart.
25. What is the circle of Willis? Why is it important?
26. What are visceral branches of an artery? Parietal branches?
27. What major organs are supplied by branches of the thoracic aorta? How is blood returned from these organs to the heart?
28. What organs are supplied by the celiac, superior mesenteric, renal, inferior mesenteric, inferior phrenic, and middle sacral arteries? How is blood returned to the heart?

29. Trace a drop of blood from the brachiocephalic artery into the digits of the right upper extremity and back again to the right atrium.
30. What is a deep vein? A superficial vein? Define a venous sinus in relation to blood vessels. What are the three major groups of systemic veins?
31. What is hepatic portal circulation? Describe the route by means of a diagram. Why is this route significant?
32. Define pulmonary circulation. Prepare a diagram to indicate the route. What is the purpose of the route?
33. Discuss in detail the anatomy and physiology of fetal circulation. Be sure to indicate the function of the umbilical arteries, umbilical vein, ductus venosus, foramen ovale, and ductus arteriosus.
34. What is the fate of the special structures involved in fetal circulation once postnatal circulation is established?
35. What is an aneurysm? Distinguish three types on the basis of location.
36. What is coronary artery disease (CAD)?
37. Describe how atherosclerosis develops? How is it diagnosed and treated?
38. Distinguish between low-density lipoproteins (LDL) and high-density lipoproteins (HDL).
39. What is coronary artery spasm? How is it treated?
40. Compare primary and secondary hypertension with regard to cause. How does hypertension affect the body? How is hypertension treated?
41. Refer to the glossary of medical terminology associated with blood vessels. Be sure that you can define each term.
42. Describe the actions of selected drugs on the cardiovascular system.

- Identify the components and functions of the lymphatic system.

- Describe the structure and origin of lymphatics and contrast them with veins.

- Describe the histological aspects of lymph nodes and explain their functions.

- Trace the general plan of lymph circulation from lymphatics into the thoracic duct or right lymphatic duct.

- Explain the forces responsible for maintaining the circulation of lymph.

- Give the locations and compare the functions of the tonsils, spleen, and thymus gland.

- Explain the difference between nonspecific and specific resistance.

- Discuss the roles of the skin and mucous membranes, antimicrobial substances, phagocytosis, inflammation, and fever as components of nonspecific resistance.

- Define immunity.

- Explain the relationship between an antigen and an antibody.

- Contrast the main characteristics of antigens and antibodies.

- Contrast the role of T cells in cellular immunity and the role of B cells in humoral immunity.

- Discuss the relationship of immunology to cancer.

- Define a monoclonal antibody and explain its importance.

- Describe hypersensitivity, tissue rejection, autoimmune diseases, and acquired immune deficiency syndrome (AIDS) as malfunctions of the immune system.

- Define medical terminology associated with the lymphatic system.

22 The Lymphatic System and Immunity

Lymph, lymphatic vessels, a series of small masses of lymphoid tissue called lymph nodes, and three organs—tonsils, thymus, and spleen—make up the **lymphatic** (lim-FAT-ik) **system.** The primary function of the lymphatic system is to drain from the tissue spaces protein-containing fluid that escapes from the blood capillaries. Such proteins cannot be directly reabsorbed by the cardiovascular system. Other functions of the lymphatic system are to transport fats from the digestive tract to the blood, to produce lymphocytes, and to develop antibodies.

LYMPHATIC VESSELS

Lymphatic vessels originate as blind-ended tubes that begin in spaces between cells (Figure 22-1). The tubes, called **lymph capillaries,** occur singly or in extensive plexuses. Lymph capillaries originate throughout the body, but not in avascular tissue, the central nervous system, splenic pulp, and bone marrow. They are slightly larger and more permeable than blood capillaries.

Just as blood capillaries converge to form venules and veins, lymph capillaries unite to form larger and larger lymph vessels called **lymphatics** (Figure 22-2). Lymphatics resemble veins in structure, but have thinner walls and more valves and contain lymph nodes at various intervals. Lymphatics of the skin travel in loose subcutaneous tissue and generally follow veins. Lymphatics of the viscera generally follow arteries, forming plexuses around them. Ultimately, lymphatics converge into two main channels—the thoracic duct and the right lymphatic duct. This will be described shortly.

CLINICAL APPLICATION

Lymphangiography (lim-fan'-jē-OG-ra-fē) is the x-ray examination of lymphatic vessels and lymph organs after they are filled with a radiopaque substance. Such

---

an x-ray is called a **lymphangiogram** (lim-FAN-jē-ō-gram). Lymphangiograms are useful in detecting edema and carcinomas and in locating lymph nodes for surgical or radiotherapeutic treatment. A normal lymphangiogram of lymphatic vessels and a few nodes in the upper thighs and pelvis is shown in Figure 22-3.

STRUCTURE OF LYMPH NODES

The oval or bean-shaped structures located along the length of lymphatics are called **lymph nodes.** They range from 1 to 25 mm (0.04 to 1 inch) in length. A lymph node contains a slight depression on one side called a *hilus* (HĪ-lus) or *hilum,* where blood vessels and an efferent lymphatic vessel leave the node (Figure 22-4). Each node is covered by a *capsule* of fibrous connective tissue that extends into the node. The capsular extensions are called *trabeculae* (tra-BEK-yoo-lē). The capsule, trabeculae, and hilus constitute the stroma (framework) of a lymph node. The parenchyma of a lymph node is specialized into two regions. The outer *cortex* contains densely packed lymphocytes arranged in masses called *lymph nodules.* The nodules often contain lighter-staining central areas, the *germinal centers,* where lymphocytes are produced. The inner region of a lymph node is called the *medulla.* In the medulla, the lymphocytes are arranged in strands called *medullary cords.*

The circulation of lymph through a node involves afferent (to convey toward a center) lymphatic vessels, sinuses in the node, and efferent (to convey away from a center) lymphatic vessels. *Afferent lymphatic vessels* enter the convex surface of the node at several points. They contain valves that open toward the node so that the lymph is directed inward. Once inside the node, the lymph enters the sinuses, which are a series of irregular channels. Lymph from the afferent lymphatic vessels enters the *cortical sinuses* under the capsule. From here it circulates to the *medullary sinuses* between the medullary cords. From these sinuses the lymph circulates into a usually single *efferent lymphatic vessel.* This vessel is located at the hilus of the lymph node. The efferent vessel is wider than the afferent vessels and contains valves that open away from the node to convey the lymph out of the node.

Lymph nodes are scattered throughout the body, usually in groups (see Figure 22-2b). Typically, these groups are arranged in two sets: *superficial* and *deep.*

As the lymph circulates through the nodes, it is processed by macrophages. The macrophages are fixed phagocytic cells of the reticuloendothelial system that line the sinuses. They remove bacteria, foreign material, and cell debris from the lymph. Lymph nodes also give rise to lymphocytes and plasma cells. The plasma cells produce antibodies.

Histological features of lymphatics and lymph nodes are shown in Figure 22-5.

CLINICAL APPLICATION

At times the number of microbes entering the nodes in the lymph is so great that the macrophages and

---

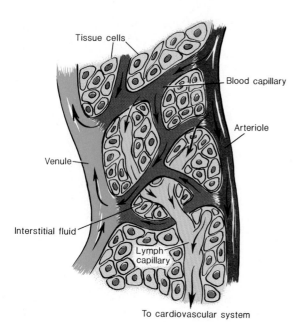

Tissue cells

Blood capillary

Arteriole

Venule

Interstitial fluid

Lymph capillary

To cardiovascular system

FIGURE 22-1 Relationship of lymph capillaries to tissue cells and blood capillaries.

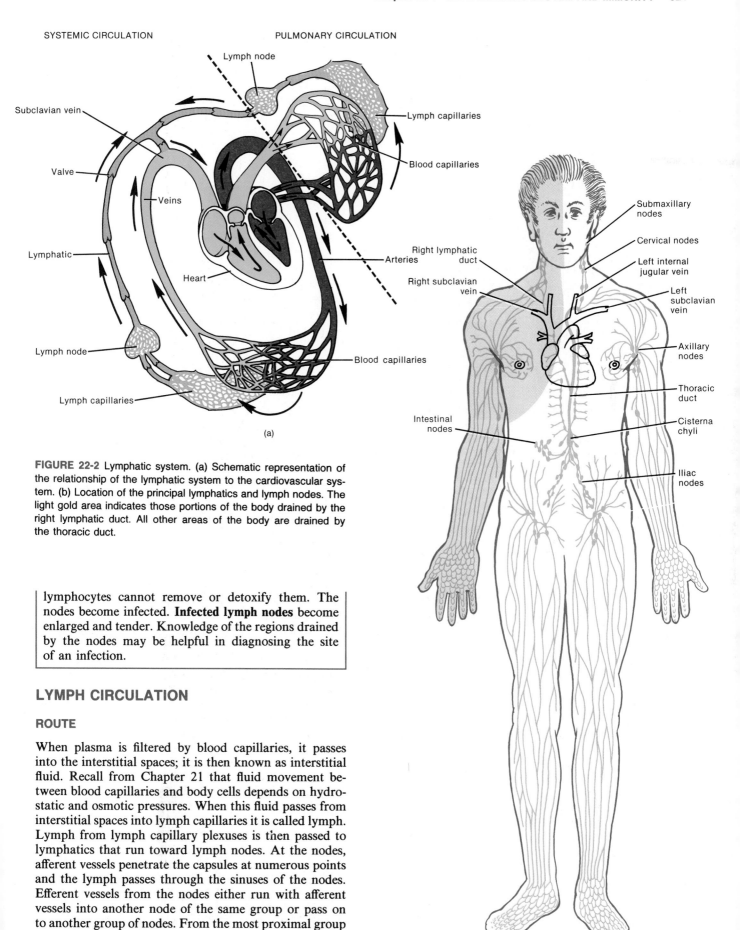

SYSTEMIC CIRCULATION

PULMONARY CIRCULATION

Lymph node

Subclavian vein

Valve

Veins

Lymphatic

Heart

Lymph node

Lymph capillaries

Lymph capillaries

Blood capillaries

Arteries

Blood capillaries

(a)

Right lymphatic duct

Right subclavian vein

Submaxillary nodes

Cervical nodes

Left internal jugular vein

Left subclavian vein

Axillary nodes

Thoracic duct

Intestinal nodes

Cisterna chyli

Iliac nodes

FIGURE 22-2 Lymphatic system. (a) Schematic representation of the relationship of the lymphatic system to the cardiovascular system. (b) Location of the principal lymphatics and lymph nodes. The light gold area indicates those portions of the body drained by the right lymphatic duct. All other areas of the body are drained by the thoracic duct.

lymphocytes cannot remove or detoxify them. The nodes become infected. **Infected lymph nodes** become enlarged and tender. Knowledge of the regions drained by the nodes may be helpful in diagnosing the site of an infection.

LYMPH CIRCULATION

ROUTE

When plasma is filtered by blood capillaries, it passes into the interstitial spaces; it is then known as interstitial fluid. Recall from Chapter 21 that fluid movement between blood capillaries and body cells depends on hydrostatic and osmotic pressures. When this fluid passes from interstitial spaces into lymph capillaries it is called lymph. Lymph from lymph capillary plexuses is then passed to lymphatics that run toward lymph nodes. At the nodes, afferent vessels penetrate the capsules at numerous points and the lymph passes through the sinuses of the nodes. Efferent vessels from the nodes either run with afferent vessels into another node of the same group or pass on to another group of nodes. From the most proximal group of each chain of nodes, the efferent vessels unite to form

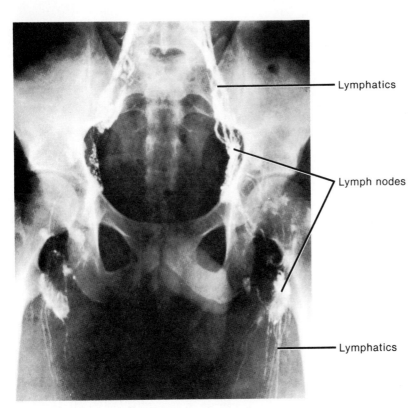

FIGURE 22-3 Normal lymphangiogram of the upper thighs and pelvis. (Courtesy of Lester W. Paul and John H. Juhl, *The Essentials of Roentgen Interpretation,* 3rd ed., Harper & Row, Publishers, Inc., New York, 1972.)

lymph trunks. The principal trunks are the *lumbar, intestinal, bronchomediastinal, subclavian,* and *jugular trunks* (Figure 22-6).

Thoracic (Left Lymphatic) Duct

The principal trunks pass their lymph into two main channels, the thoracic duct and the right lymphatic duct. The **thoracic (left lymphatic) duct** is about 38 to 45 cm (15 to 18 inches) in length and begins as a dilation in front of the second lumbar vertebra called the *cisterna chyli* (sis-TER-na KĪ-lē). It is the main collecting duct of the lymphatic system and receives lymph from the left side of the head, neck, and chest, the left upper extremity, and the entire body below the ribs.

The cisterna chyli receives lymph from the right and left lumbar trunks and from the intestinal trunk. The lumbar trunks drain lymph from the lower extremities, walls and viscera of the pelvis, kidneys, suprarenals, and the deep lymphatics from most of the abdominal wall. The intestinal trunk drains lymph from the stomach, intestines, pancreas, spleen, and visceral surface of the liver. In the neck, the thoracic duct also receives lymph from the left jugular, left subclavian, and left bronchomediastinal trunks. The left jugular trunk drains lymph from the left side of the head and neck; the left subclavian trunk drains lymph from the upper left extremity; and the left bronchomediastinal trunk drains lymph from the

left side of the deeper parts of the anterior thoracic wall, upper part of the anterior abdominal wall, anterior part of the diaphragm, left lung, and left side of the heart.

Right Lymphatic Duct

The **right lymphatic duct** is about 1.25 cm (0.5 inch) long and drains lymph from the upper right side of the body (see Figure 22-2b). The right lymphatic duct collects lymph from its trunks as follows (Figure 22-6). It receives lymph from the right jugular trunk, which drains the right side of the head and neck, from the right subclavian trunk, which drains the right upper extremity, and from the right bronchomediastinal trunk, which drains the right side of the thorax, right lung, right side of the heart, and part of the convex surface of the liver.

Ultimately, the thoracic duct empties all of its lymph into the junction of the left internal jugular vein and left subclavian vein and the right lymphatic duct empties all of its lymph into the junction of the right internal jugular vein and right subclavian vein. Thus, lymph is drained back into the blood and the cycle repeats itself continuously.

MAINTENANCE

The flow of lymph from tissue spaces to the large lymphatic ducts to the subclavian veins is maintained primarily by the milking action of muscle tissue. Skeletal muscle contractions compress lymph vessels and force lymph toward the subclavian veins. Lymph vessels, like veins, contain valves, and the valves ensure the movement of lymph toward the subclavian veins (see Figure 22-5a).

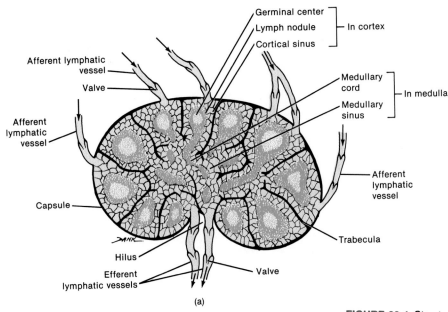

(a)

Inguinal lymph
node

(b)

FIGURE 22-4 Structure of a lymph node. (a) Diagram showing the path taken by circulating lymph. (b) Photograph of an inguinal lymph node. (Courtesy of C. Yokochi and J. W. Rohen, *Photographic Anatomy of the Human Body,* 2nd ed., 1979, IGAKU-SHOIN, Tokyo, New York.)

Another factor that maintains lymph flow is respiratory movements. These movements create a pressure gradient between the two ends of the lymphatic system. Lymph flows from the tissue spaces, where the pressure is higher, toward the thoracic region, where it is lower.

CLINICAL APPLICATION

Edema, an excessive accumulation of interstitial fluid in tissue spaces, may be caused by an obstruction, such as an infected node or a blockage of vessels, in the pathway between the lymphatic capillaries and the subclavian veins. Another cause is excessive lymph formation and increased permeability of blood capillary walls. A rise in capillary blood pressure, in which interstitial fluid is formed faster than it is passed into lymphatics, also may result in edema.

LYMPHATIC ORGANS

TONSILS

Tonsils are basically masses of lymphoid tissue embedded in mucous membrane. The *pharyngeal* (fa-RIN-jē-al) *tonsil* or *adenoid* is embedded in the posterior wall of the nasopharynx (see Figure 23-3a). The *palatine* (PAL-a-tīn) *tonsils* are situated in the tonsillar fossae between the pharyngopalatine and glossopalatine arches (see Figure 24-4). These are the ones commonly removed by a tonsillectomy. The *lingual* (LIN-gwal) *tonsil* is located at the base of the tongue and may also have to be removed by a tonsillectomy (see Figure 24-5a).

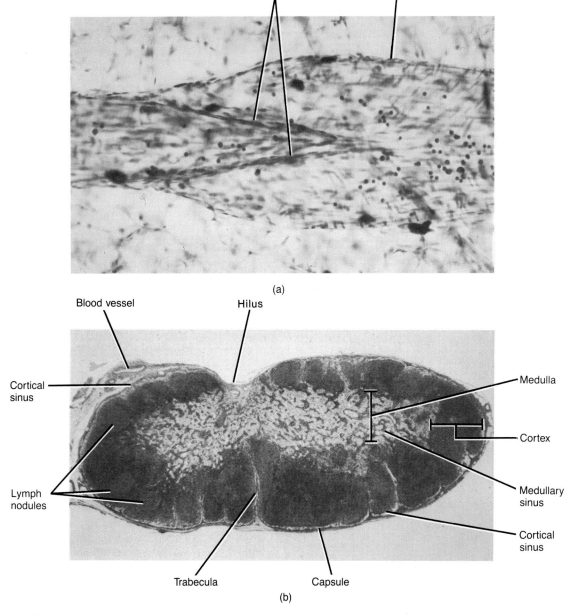

FIGURE 22-5 Histology of lymphatics and lymph nodes. (a) Photomicrograph of a lymphatic at a magnification of 180×. (b) Photomicrograph of a lymph node at a magnification of 25×. (© 1983 by Michael H. Ross. Used by permission.)

SPLEEN

The oval **spleen** is the largest mass of lymphatic tissue in the body, measuring about 12 cm (5 inches) in length. It is situated in the left hypochondriac region between the fundus of the stomach and diaphragm (see Figure 1-8d). Its *visceral surface* (Figure 22-7a) contains the contours of the organs adjacent to it—the gastric impression (stomach), renal impression (left kidney), and colic impression (left flexure of colon). The *diaphragmatic* (dī-a-fra-MAT-ik) *surface* (Figure 22-7b) is smooth and convex and conforms to the concave surface of the diaphragm to which it is adjacent.

The spleen is surrounded by a capsule of fibroelastic tissue and scattered smooth muscle (Figure 22-7c). The capsule, in turn, is covered by a serous membrane, the peritoneum. Like lymph nodes, the spleen contains trabeculae and a hilus. The capsule, trabeculae, and hilus constitute the stroma of the spleen.

The parenchyma of the spleen consists of two different kinds of tissue called white pulp and red pulp. *White pulp* is essentially lymphoid tissue arranged around arteries. The clusters of lymphocytes surrounding the arteries at intervals and the expansions are referred to as *splenic nodules*. The *red pulp* consists of *venous sinuses* filled with blood and cords of splenic tissue called *splenic cords*.

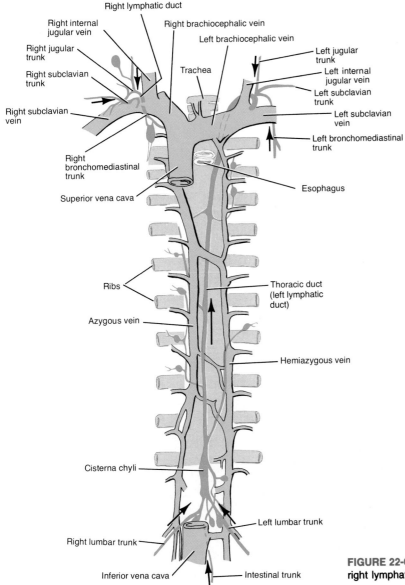

Right lymphatic duct

Right internal jugular vein

Right brachiocephalic vein

Right jugular trunk

Left brachiocephalic vein

Right subclavian trunk

Left jugular trunk

Trachea

Left internal jugular vein

Right subclavian vein

Left subclavian trunk

Left subclavian vein

Right bronchomediastinal trunk

Left bronchomediastinal trunk

Superior vena cava

Esophagus

Ribs

Thoracic duct (left lymphatic duct)

Azygous vein

Hemiazygous vein

Cisterna chyli

Left lumbar trunk

Right lumbar trunk

Inferior vena cava

Intestinal trunk

FIGURE 22-6 Relationship of lymph trunks to the thoracic duct and right lymphatic duct.

Veins are closely associated with the red pulp.

The splenic artery and vein and the efferent lymphatics pass through the hilus. Since the spleen has no afferent lymphatic vessels or lymph sinuses, it does not filter lymph. The spleen phagocytoses bacteria and worn-out red blood cells and platelets. It also produces lymphocytes and antibody-producing plasma cells. In addition, the spleen stores and releases blood in case of demand, such as during hemorrhage. The release seems to be purely sympathetic. The sympathetic impulses cause the smooth muscle of the spleen to contract.

CLINICAL APPLICATION

Although the spleen is protected from traumatic injuries, it is the most frequently damaged organ in cases of abdominal trauma, particularly those involving se-

vere blows over the lower left chest or upper abdomen that fracture the protecting ribs. Such a crushing injury may **rupture the spleen,** which causes severe intraperitoneal hemorrhage and shock. This requires prompt removal of the spleen, called a **splenectomy,** to prevent the patient from bleeding to death. The functions of the spleen are then assumed by other reticuloendothelial organs.

THYMUS GLAND

The **thymus gland** is a bilobed mass of lymphatic tissue in the upper thoracic cavity. It is found along the trachea behind the sternum (see Figure 18-20). The thymus is relatively large in children. It reaches its maximum size at puberty and then undergoes involution. Eventually, most of it is replaced by fat and connective tissue. Its

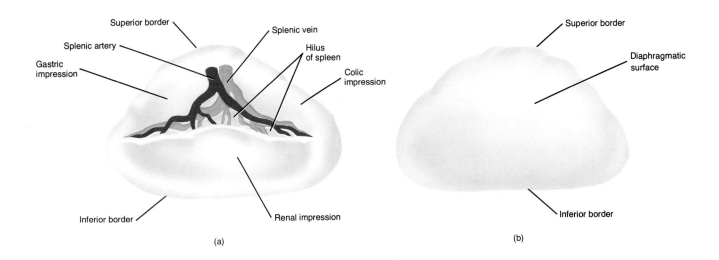

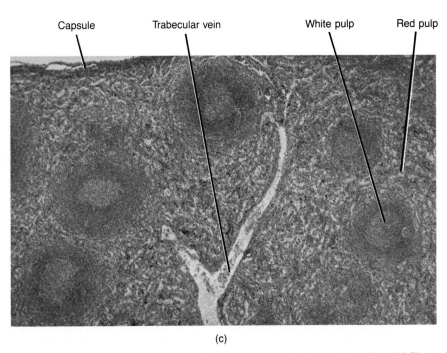

FIGURE 22-7 Spleen. (a) Gross structure of visceral surface. (b) Gross structure of diaphragmatic surface. (c) Photomicrograph of a portion of the spleen at a magnification of 60×. (Photomicrograph © 1983 by Michael H. Ross. Used by permission.)

role in immunity is to help produce cells that destroy invading microbes directly or cells that manufacture antibodies against invading microbes. This will be discussed in detail shortly.

NONSPECIFIC RESISTANCE TO DISEASE

The human body continually attempts to maintain homeostasis by counteracting harmful stimuli in the environment. Frequently, these stimuli are disease-producing organisms, called *pathogens* (PATH-ō-jens), or their toxins. The ability to ward off disease is called **resistance.** Lack of resistance is called **susceptibility.** Defenses against dis-

ease may be grouped into two broad areas: nonspecific resistance and specific resistance. Nonspecific resistance represents a wide variety of body reactions against a wide range of pathogens. Specific resistance or **immunity** involves the production of a specific antibody against a specific pathogen or its toxin. We will first consider mechanisms of nonspecific resistance.

SKIN AND MUCOUS MEMBRANES

The skin and mucous membranes of the body possess certain mechanical factors and chemical factors that are involved in combating the initial attempt of a microbe to cause disease.

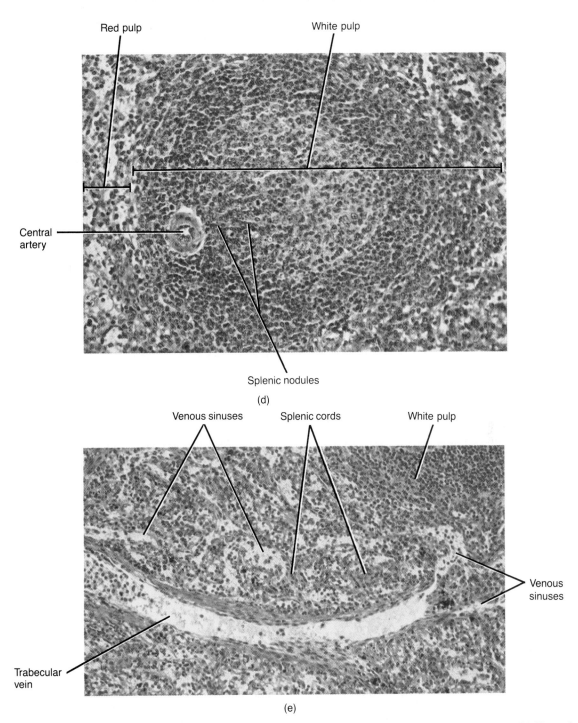

Red pulp

White pulp

Central artery

Splenic nodules

(d)

Venous sinuses   Splenic cords   White pulp

Venous sinuses

Trabecular vein

(e)

FIGURE 22-7 (*Continued*) Spleen. (d) Photomicrograph of an enlarged aspect of white pulp at a magnification of 150×. (e) Photomicrograph of an enlarged aspect of red pulp at a magnification of 150×. (Photomicrographs © 1983 by Michael H. Ross. Used by permission.)

Mechanical Factors

The **intact skin,** as noted in Chapter 5, consists of two distinct portions called the dermis and epidermis. If you consider the closely packed cells of the epidermis, its continuous layering, and the presence of keratin, you can see that the intact skin provides a formidable physical barrier to the entrance of microbes. The intact surface of healthy epidermis seems to be rarely penetrated by bacteria. But when the epithelial surface is broken, a subcutaneous infection often develops. The bacteria most likely to cause such an infection are staphylococci, which normally inhabit the hair follicles and sweat glands of the skin. Moreover, when the skin is moist, as in hot humid climates, dermal infections, especially fungus infections, such as athlete's foot, are quite common.

Mucous membranes, like the skin, also consist of an epithelial layer and an underlying connective tissue layer. But unlike the skin, a mucous membrane lines a body cavity that opens to the exterior. Examples include the membranes lining the entire digestive, respiratory, urinary, and reproductive tracts. The epithelial layer of a mucous membrane secretes a fluid called *mucus,* which prevents the cavities from drying out. Since mucus is slightly viscous, it traps many microbes that enter the respiratory and digestive tracts. The mucous membrane of the nose has mucus-coated **hairs** that trap and filter air containing microbes, dust, and pollutants. The mucous membrane of the upper respiratory tract contains **cilia,** microscopic hairlike projections of the epithelial cells (see Figure 3-13c). These cilia move in such a manner that they pass inhaled dust and microbes that have become trapped in mucus toward the throat. This so-called "ciliary escalator" keeps the "mucus blanket" moving toward the throat at a rate of 1 to 3 cm per hour. Coughing and sneezing speed up the "escalator."

Since some pathogens can thrive on the moist secretions of a mucous membrane, they are able to penetrate the membrane if present in sufficient numbers. This penetration may be related to toxic products produced by the microbes, prior injury by viral infections, or mucosal irritations. Although mucous membranes do inhibit the entrance of many microbes, they are less effective than the skin.

There are several other mechanical factors that help to protect epithelial surfaces of the skin and mucous membranes. One such mechanism that protects the eyes is the **lacrimal** (LAK-ri-mal) **apparatus** (see Figure 17-3a). It consists of a group of structures that manufactures and drains away tears, which are spread over the surface of the eyeball by blinking. Normally, the tears are carried away by evaporation or pass into the nose as fast as they are produced. The continual washing action of tears helps to keep microbes from settling on the surface of the eye. If an irritating substance or large numbers of microbes make contact with the eye, the lacrimal glands start to oversecrete. Tears then accumulate more rapidly than they can be carried away. This is also a protective mechanism to dilute and wash away the irritating substance or microbes.

Saliva, produced by the salivary glands, washes microbes from the surfaces of the teeth and the mucous membrane of the mouth, much as tears wash the eyes. This helps to prevent colonization by microbes.

Sebum, produced by the sebaceous (oil) glands of the skin, forms a protective film over the surface of the skin. **Perspiration** flushes some microbes from the skin. Microbes are additionally prevented from entering the lower respiratory tract by a small lid of cartilage called the **epiglottis** that covers the voice box during swallowing (see Figure 23-2b). The cleansing of the urethra by the **flow of urine** represents another mechanical factor that prevents microbial colonization in the urinary system.

Chemical Factors

Mechanical factors alone do not account for the high degree of resistance of skin and mucous membranes to microbial invasion. Certain chemical factors play an important role. **Gastric juice** is a collection of hydrochloric acid, enzymes, and mucus produced by the glands of the stomach. The very high acidity of gastric juice (pH 1.2–3.0) is sufficient to preserve the usual sterility of the stomach. The acidity of gastric juice destroys bacteria and almost all important bacterial toxins.

The **acidic pH** of the skin, between 3 and 5, is due in part to the acid products of bacterial metabolism. This acidity probably discourages the growth of many microbes that contact the skin. One of the components of sebum is **unsaturated fatty acids.** These fatty acids kill certain pathogenic bacteria, such as *Streptococcus pyogenes,* the causative agent of scarlet fever and streptococcal sore throat. **Lysozyme** is an enzyme capable of breaking down cell walls of various bacteria under certain conditions. Lysozyme is normally found in perspiration, tears, saliva, nasal secretions, and tissue fluids.

ANTIMICROBIAL SUBSTANCES

In addition to the mechanical and chemical barriers of the skin and mucous membranes, the body also produces certain antimicrobial substances. Among these are interferon, complement, and properdin.

Interferon (IFN)

Host cells infected with viruses produce a protein called **interferon** (in'-ter-FĒR-on) or **IFN.** In humans, it has been shown to be produced by leucocytes, fibroblasts, and T cells. Once released from virus-infected cells, it diffuses to uninfected neighboring cells and binds to surface receptors. This somehow induces uninfected cells to synthesize another antiviral protein that inhibits intracellular viral replication. Interferon appears to be the body's first line of defense against infection by many different viruses. It appears to decrease the virulence (disease-producing power) of viruses associated with chickenpox, genital herpes, rabies, rubella, chronic hepatitis, shingles, eye infections, encephalitis, and at least one type of common cold.

The suggested relationship between viruses and cancer has prompted scientists to study the effects of interferon on cancer patients. Among the human cancers recently shown to respond in some way to interferon are osteogenic carcinoma, multiple myeloma, melanoma (highly malignant form of skin cancer), breast cancer, and certain types of leukemia and lymphoma. Although the initial results are generally encouraging, much more research is needed before any definitive conclusions can be drawn. For example, scientists still have to determine dosage levels, frequency of administration, and usefulness alone or in combination with other cancer treatments, such as chemotherapy and radiation. Moreover, its side effects have yet to be fully analyzed. Presently observed side effects are chills, fever, and suppression of bone marrow function, all of which seem to be mild.

Complement

Another antimicrobial substance that is very important to nonspecific resistance (and immunity) is complement. **Complement** is actually a group of eleven proteins found in normal blood serum. The system is called complement because it "complements" certain immune and allergic reactions involving antibodies. These will be discussed shortly. The function of antibody is to recognize the microbe as a foreign organism, form an antigen-antibody complex, and activate complement for attack. The antigen-antibody complex also fixes (attaches) the complement to the surface of the invading microbe. Once complement is activated, it destroys microbes as follows:

1. Some complement proteins initiate a series of reactions leading to **cell lysis** caused by holes in the plasma membrane of the microbe.

2. Some complement proteins interact with receptors on phagocytes, promoting phagocytosis. This is called **opsonization** (op-sō-ni-ZĀ-shun) or **immune adherence.**

3. Some complement proteins contribute to the development of **inflammation** by causing the release of **histamine** from mast cells, leucocytes, and platelets. Histamine increases the permeability of blood capillaries, a process that enables leucocytes to penetrate tissues in order to combat infection or allergy.

4. Some complement proteins serve as **chemotactic** (kē-mō-TAK-tic) **agents,** attracting large numbers of leucocytes to the area.

Properdin

Properdin (prō-PER-din; *pro* = in behalf of, *perdere* = to destroy), like complement, is also a protein found in serum. It is a complex consisting of three proteins. Properdin, acting together with complement, leads to the destruction of several types of bacteria, the enhancement of phagocytosis, and triggering of inflammatory responses.

PHAGOCYTOSIS

When microbes penetrate the skin and mucous membranes or bypass the antimicrobial substances in blood, there is another nonspecific resistance of the body called phagocytosis. Very simply, **phagocytosis** (*phagein* = to eat; *cyto* = cell) means the ingestion and destruction of microbes or any foreign particulate matter by certain cells called phagocytes.

Kinds of Phagocytes

The kinds of phagocytes that participate in the process fall into two broad categories: microphages and macrophages. The granulocytes of blood are **microphages** (MĪ-krō-fā-jez). However, not all granulocytes exhibit the same phagocytic capabilities. Neutrophils have the most prominent phagocytic activity. Eosinophils are believed to have some phagocytic capability, and the role of basophils in phagocytosis is debatable.

When an infection occurs, both microphages (especially neutrophils) and monocytes migrate to the infected area. During this migration, the monocytes enlarge and develop into actively phagocytic cells called **macrophages** (MAK-rō-fā-jez). Since these cells leave the blood and

migrate to infected areas, they are called **wandering macrophages.** Some of the macrophages, called **fixed macrophages** or **histiocytes** enter certain tissues and organs of the body and remain there. Fixed macrophages are found in the liver (stellate reticuloendothelial cells), lungs (alveolar macrophages), brain (microglia), spleen, lymph nodes, and bone marrow.

Mechanism

For convenience of study, phagocytosis will be divided into two phases: adherence and ingestion. **Adherence** or **attachment** is the formation of firm contact between the cell membrane of the phagocyte and the microbe (or foreign material). In some instances, adherence occurs easily, and the microbe is readily phagocytozed (see Figure 4-4). In other cases, adherence is more difficult, but the particle can be phagocytozed if the phagocyte traps the particle to be ingested against a rough surface, like a blood vessel, blood clot, or connective tissue fibers, where it cannot slide away. This is sometimes called **nonimmune** or **surface phagocytosis.** Or bacteria can be phagocytozed if they are first coated with complement or antibody to promote the attachment of the microbe to the phagocyte. This is **opsonization** or **immune adherence.** A final factor that helps adherence is chemotaxis.

Following adherence, **ingestion** occurs. In the process of ingestion, projections of the cell membrane of the phagocyte, called pseudopodia, engulf the microbe (Figure 22-8). Once the microbe is surrounded, the membrane folds inward, forming a sac around the microbe called a **phagocytic vacuole.** The vacuole pinches off from the membrane and enters the cytoplasm. Within the cytoplasm, it collides with lysosomes that contain digestive enzymes and bactericidal substances. Upon contact, the vacuolar and lysosomal membranes fuse together to form a single, larger structure called a **phagolysosome,** or **digestive vacuole.** Within the phagolysosome, most bacteria are usually killed within 10 to 30 minutes. It is assumed that destruction occurs as a result of the contents of the lysosomes: lactic acid (which lowers the pH in the phagolysosome), the production of hydrogen peroxide, lysozyme, and the destructive capabilities of enzymes that degrade carbohydrates, proteins, lipids, and nucleic acids.

Some microbes, such as toxin-producing staphylococci, may be ingested, but are not necessarily killed. In fact,

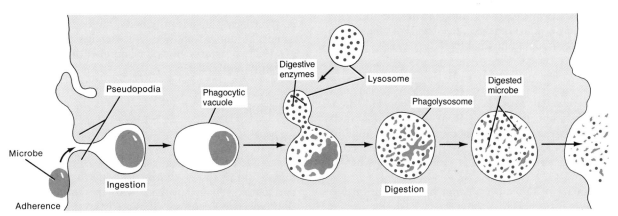

FIGURE 22-8 Phases of phagocytosis.

EXHIBIT 22-1

SUMMARY OF NONSPECIFIC RESISTANCE

COMPONENT	FUNCTIONS	COMPONENT	FUNCTIONS
Skin and mucous membranes **Mechanical factors**		**Acid pH of skin**	Discourages growth of many microbes.
Intact skin	Forms a physical barrier to the entrance of microbes.	**Unsaturated fatty acids**	Antibacterial substance in sebum.
Mucous membranes	Inhibit the entrance of many microbes, but not as effective as intact skin.	**Lysozyme**	Antimicrobial substance in perspiration, tears, saliva, nasal secretions, and tissue fluids.
Mucus	Traps microbes in respiratory and digestive tracts.		
Hairs	Filter microbes and dust in nose.	**Antimicrobial substances**	
Cilia	Together with mucus, trap and remove microbes and dust from upper respiratory tract.	**Interferon (IFN)**	Protects uninfected host cells from viral infection.
Lacrimal apparatus	Tears dilute and wash away irritating substances and microbes.	**Complement**	Causes lysis of microbes, promotes phagocytosis, contributes to inflammation, serves as chemotactic agent.
Saliva	Washes microbes from surfaces of teeth and mucous membranes of mouth.	**Properdin**	Works with complement to bring about same responses as complement.
Epiglottis	Prevents microbes and dust from entering trachea.	**Phagocytosis**	Ingestion and destruction of foreign particulate matter by microphages and macrophages.
Urine	Washes microbes from urethra.	**Inflammation**	Confines and destroys microbes and repairs tissues.
Chemical Factors		**Fever**	Inhibits microbial growth and speeds up body reactions that aid repair.
Gastric juice	Destroys bacteria and most toxins in stomach.		

their toxins can actually kill the phagocytes. Other microbes, such as the tubercle bacillus, may multiply within the phagolysosome and eventually destroy the phagocyte. Still other microbes, such as the causative agents of tularemia and brucellosis, may remain dormant in phagocytes for months or years at a time.

INFLAMMATION

When tissues of the body are damaged by microbes, physical agents, or chemical agents, **inflammation** occurs. This is another nonspecific resistance, the details of which have already been discussed in Chapter 4.

FEVER

The most frequent cause of **fever,** an abnormally high body temperature, is infection from bacteria (and their toxins) and viruses. The high body temperature inhibits some microbial growth and speeds up body reactions that aid repair. (Fever is discussed in more detail in Chapter 25.)

A summary of the various components of nonspecific resistance is presented in Exhibit 22-1.

IMMUNITY (SPECIFIC RESISTANCE TO DISEASE)

Despite the variety of mechanisms of nonspecific resistance, they all have one thing in common. They are designed to protect the body from any kind of pathogen. They are not specifically directed against a particular microbe. Specific resistance to disease, called **immunity,** involves the production of a specific type of cell or specific molecule (antibody) to destroy a particular antigen. If antigen 1 invades the body, antibody 1 is produced against it. If antigen 2 invades the body, antibody 2 is produced against it, and so on. We will examine the components of specific resistance by first examining the nature of antigens and antibodies.

ANTIGENS

Definition

An **antigen** is any chemical substance that, when introduced into the body, causes the body to produce specific antibodies, which can react with the antigen. Antigens thus have two important characteristics. The first is **immu-**

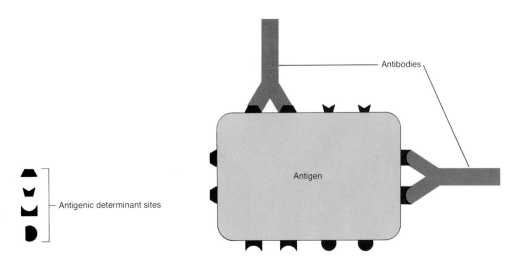

FIGURE 22-9 Relationship of an antigen to antibodies. Since most antigens contain more than one antigenic determinant site, they are referred to as multivalent. Most human antibodies are bivalent; they have two reaction sites that are complementary to the antigenic determinant sites for which they are specific. Each antibody has reaction sites for specific antigenic determinant sites only.

nogenicity (i'-myoo-nō-jen-IS-it-ē), the ability to stimulate the formation of specific antibodies. The second is **reactivity,** the ability of the antigen to react specifically with the produced antibodies. An antigen with both of these characterisitics is called a **complete antigen.**

Characteristics

Chemically, the vast majority of antigens are proteins, nucleoproteins (nucleic acid + protein), lipoproteins (lipid + protein), glycoproteins (carbohydrate + protein), and certain large polysaccharides. In general, they have molecular weights of 10,000 or greater.

The entire microbe, such as a bacterium or virus, or components of microbes may act as an antigen. For example, bacterial structures such as flagella, capsules, and cell walls are also antigenic; bacterial toxins are highly antigenic. Nonmicrobial examples of antigens include pollen, egg white, incompatible blood cells, and transplanted tissues and organs. The myriad of antigens in the environment provides many opportunities for the production of antibodies by the body.

Antibodies do not form against the whole antigen. At specific regions on the surface of the antigen, called **antigenic determinant sites** (Figure 22-9), specific chemical groups of the antigen combine with the antibody. This combination depends upon the size and shape of the determinant site and the manner in which it corresponds to the chemical structure of the antibody. The combination is very much like the lock and key analogy used to describe the combination of enzyme and substrate molecule.

The number of antigenic determinant sites on the surface of an antigen is called *valence.* Most antigens are *multivalent,* that is, they have several antigenic determinant sites. It is believed that in order to induce antibody formation, an antigen must have at least two determinant sites (*bivalent*). If an antigen is chemically broken down, it is possible to separate the determinant site. Its molecular weight, compared to that of the antigen as a whole, is

small. It might be only 200 to 1,000. An isolated determinant site still has the ability to react with an antibody in response to the original antigen (reactivity), but does not have the ability to stimulate the production of antibodies (immunogenicity) when injected into an animal. Its lower molecular weight prevents it from serving as a complete antigen. A determinant site that has reactivity but not immunogenicity is called a **partial antigen** or **hapten** (*haptein* = to grasp). A hapten can stimulate an immune response if it is attached to a larger carrier molecule so that the combined molecule has two determinant sites. Some drugs of low molecular weight, like penicillin, may combine with proteins of high molecular weight in the body. This combination forms two determinant sites, and the complex becomes antigenic. Antibodies formed against the complex are responsible for the allergic reactions to drugs and other chemicals.

As a rule, antigens are foreign substances. They are not usually part of the chemistry of the body. The body's own substances, recognized as "self," do not act as antigens, identified as "nonself," to stimulate antibody production. However, there are certain conditions in which the distinction between self and nonself breaks down, and antibodies which attack the body are produced. These conditions (to be described later) result in what are called autoimmune disease.

Let's now look at some of the characteristics of antibodies.

ANTIBODIES

Definition

An **antibody** is a protein produced by the body in response to the presence of an antigen and is capable of combining specifically with the antigen. This is essentially the complementary definition of an antigen.

The specific fit of the antibody with antigen depends not only on the size and shape of the antigenic determi-

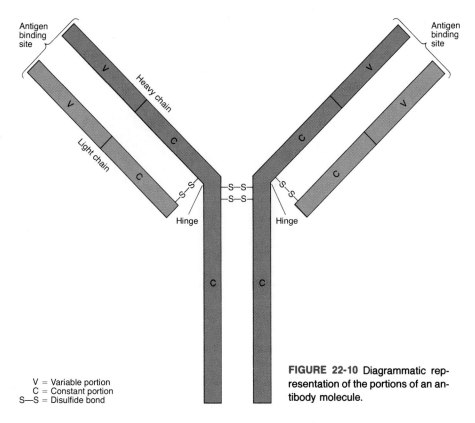

V = Variable portion
C = Constant portion
S—S = Disulfide bond

FIGURE 22-10 Diagrammatic representation of the portions of an antibody molecule.

nant site, but also on the corresponding antibody site, much like the lock and key analogy (see Figure 22-9). An antibody, like an antigen, also has a valence. Whereas most antigens are multivalent, antibodies are bivalent or multivalent. The majority of human antibodies are bivalent.

Antibodies belong to a group of proteins called globulins, and for this reason they are also known as **immunoglobulins** (i'-myoo-nō-GLOB-yoo-lins) or **Ig.** Five different classes of immunoglobulins are known to exist in humans. These are designated as IgG, IgA, IgM, IgD, and IgE. Each has a distinct chemical structure and a specific biological role. IgG antibodies enhance phagocytosis, neutralize toxins, and protect the fetus and newborn. IgA antibodies provide localized protection on mucosal surfaces; IgM antibodies are especially effective against microbes by causing agglutination and lysis. IgD antibodies may be involved in stimulating antibody-producing cells to manufacture antibodies. IgE antibodies are involved in allergic reactions.

Structure

Since antibodies are proteins, they consist of polypeptide chains. Most antibodies contain four pairs of polypeptide chains (Figure 22-10). Two of the chains are identical to each other and are called *heavy (H) chains.* Each consists of more than 400 amino acids. The other two chains, also identical to each other, are called *light (L) chains,* and each consists of 200 amino acids. The antibody consists of identical halves held together by disulfide bonds (S—S). Each half consists of a heavy and a light chain, also held together by a disulfide bond (S—S). Over-

all, the antibody molecule sometimes assumes the shape of a letter Y. At other times, it resembles a letter T. The reasons for these differences in shape will be explained shortly.

Within each H and L chain, there are two distinct regions. The tops of the H and L chains, called the *variable portions,* contain the antigen binding site. The variable portion is different for each kind of antibody, and it is this portion that allows the antibody to recognize and attach specifically to a particular type of antigen. Since most antibodies have two variable portions for attachment of antigens, they are bivalent. The remainder of each polypeptide chain is called the *constant portion.* The constant portion is the same in all antibodies of the same class and is responsible for the type of antigen-antibody reaction that occurs. However, the constant portion differs from one class of antibody to another, and its structure serves as a basis for distinguishing the classes.

In performing its role in immunity, the antibody is believed to behave as a switch. When antibodies are viewed in the electron microscope before combination with antigen, they resemble the letter T. But after combination with antigen, they appear to be smaller and resemble the letter Y. It is possible that the binding of antigen causes a rearrangement of the structure of antibody. If you examine Figure 22-10 again, you will note that there is a hinge area along the H chains; this hinge provides the antibody molecule with flexibility. When antigen and antibody are combined, the rearrangement of antibody might consist of a pivoting movement in which the H chain moves from a T shape into a Y shape. This pivoting could cause exposure of the constant portion, which then functions in a particular antigen-antibody reaction, such as fixing complement.

CELLULAR AND HUMORAL IMMUNITY

The ability of the body to defend itself against invading agents such as bacteria, toxins, viruses, and foreign tissues consists of two closely allied components. One component consists of the formation of specially sensitized lymphocytes that have the capacity to attach to the foreign agent and destroy it. This is called **cellular (cell-mediated) immunity** and is particularly effective against fungi, parasites, intracellular viral infections, cancer cells, and foreign tissue transplants. In the other component, the body produces circulating antibodies that are capable of attacking an invading agent. This is called **humoral immunity** and is particularly effective against bacterial and viral infections.

Cellular immunity and humoral immunity are the product of the body's lymphoid tissue. The bulk of lymphoid tissue is located in the lymph nodes, but it is also found in the spleen and gastrointestinal tract and, to a lesser extent, in bone marrow. The placement of lymphoid tissue is strategically designed to intercept an invading agent before it can spread too extensively into the general circulation.

Formation of T Cells and B Cells

Lymphoid tissue consists primarily of lymphocytes that may be distinguished into two kinds. **T cells** are responsible for cellular immunity. **B cells** develop into specialized cells (plasma cells) that produce antibodies and provide humoral immunity.

Both types of lymphocytes are derived originally in the embryo from lymphocytic stem cells in bone marrow (Figure 22-11). Before migrating to their positions in lymphoid tissues, the descendants of the stem cells follow two distinct pathways. About half of them first migrate

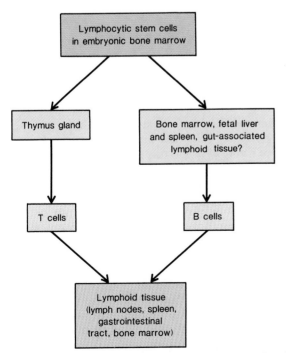

FIGURE 22-11 Origin and differentiation of T cells and B cells.

to the thymus gland, where they are processed to become T cells. Since the processing occurs in the thymus gland, they are called T cells. The thymus gland in some way confers what is called *immunologic competence* on the T cells. This means that they develop the ability to differentiate into cells that perform specific immune reactions. They then leave the thymus gland and become embedded in lymphoid tissue. Immunologic competence is conferred by the thymus gland shortly before birth and for a few months after birth. Removal of the thymus gland from an animal before it has processed the T cells results in a failure to develop cellular immunity. Removal after processing has little effect.

The remaining stem cells are processed in some unknown area of the body, possibly bone marrow, fetal liver and spleen, or gut-associated lymphoid tissue. They become the B cells. They are given this name because in birds they are processed in the bursa of Fabricius, a small pouch of lymphoid tissue attached to the intestine. Once processed, B cells migrate to lymphoid tissue and take up their positions there. Although both T cells and B cells occupy the same lymphoid tissues, they localize in separate areas of the tissues.

T Cells and Cellular Immunity

There are literally thousands of different types of T cells, and each is capable of responding to a specific antigen. At any given time, most T cells are inactive. When an antigen enters the body, only the particular T cell specifically programmed to react with the antigen becomes activated. Such an activated T cell is said to be *sensitized.* Activation occurs when macrophages phagocytose the antigen and present it to the T cell. Sensitized T cells increase in size and divide, each cell giving rise to a *clone,* or population of cells identical to itself (Figure 22-12). Four subpopulations of cells within the clone can be recognized: killer T cells, helper T cells, suppressor T cells, and memory T cells.

Killer T cells leave the lymphoid tissue and migrate to the site of invasion. Here, several things happen. They attach to the invading cell and secrete *cytotoxic substances* that destroy the antigen directly. The cytotoxic substances are probably lysosomal enzymes synthesized by the killer T cells. The cells can also release a protein substance called *transfer factor.* This substance reacts with nonsensitized lymphocytes at the site of invasion, causing them to take on the same characteristics as the sensitized killer T cells. In this way, additional lymphocytes are recruited, and the effect of the sensitized killer T cells is intensified. Killer T cells also secrete a substance called *macrophage chemotactic factor.* This substance attracts macrophages to the site of invasion where they destroy the antigens by phagocytosis. Another substance released by killer T cells, called *macrophage activating factor* (*MAF*), greatly increases the phagocytic activity of macrophages. In summary, then, killer T cells can destroy antigens directly by releasing cytotoxic substances and indirectly by recruiting additional lymphocytes, attracting macrophages, and intensifying phagocytosis by macrophages. Killer T cells are especially effective against slowly developing bac-

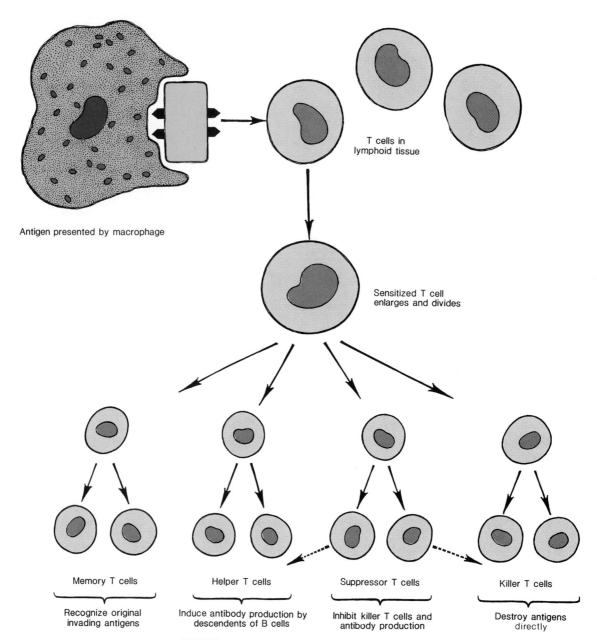

Antigen presented by macrophage

T cells in lymphoid tissue

Sensitized T cell enlarges and divides

Memory T cells

Recognize original invading antigens

Helper T cells

Induce antibody production by descendents of B cells

Suppressor T cells

Inhibit killer T cells and antibody production

Killer T cells

Destroy antigens directly

FIGURE 22-12 Role of T cells in cellular immunity.

terial diseases, such as tuberculosis and brucellosis, some viruses, fungi, transplanted cells, and cancer cells.

Helper T cells cooperate with B cells to help amplify antibody production. Although helper T cells do not themselves secrete antibody, antigens first interact with them before inducing antibody production by the descendants of B cells.

Suppressor T cells assume a role in regulating parts of the immune response. They can inhibit secretion of injurious substances by killer T cells and inhibit the development of B cells into antibody-producing plasma cells. The normal ratio of helper T cells to suppressor T cells is 2:1.

Memory T cells are programmed to recognize the original invading antigen. Should the pathogen invade the body at a later date, the memory cells initiate a far swifter

reaction than during the first invasion. In fact, the second response is so swift that the pathogens are usually destroyed before any signs or symptoms of the disease occur.

B Cells and Humoral Immunity

The body not only contains thousands of different T cells, it also contains thousands of different B cells, each capable of responding to a specific antigen. Whereas killer T cells leave their reservoirs of lymphoid tissue to meet a foreign antigen, B cells respond differently. They differentiate into cells that produce specific antibodies that then circulate in the lymph and blood to reach the site of invasion. When a foreign antigen has been prepared and presented by macrophages to B cells in lymph nodes, the spleen, or lymphoid tissue in the gastrointestinal tract, B cells

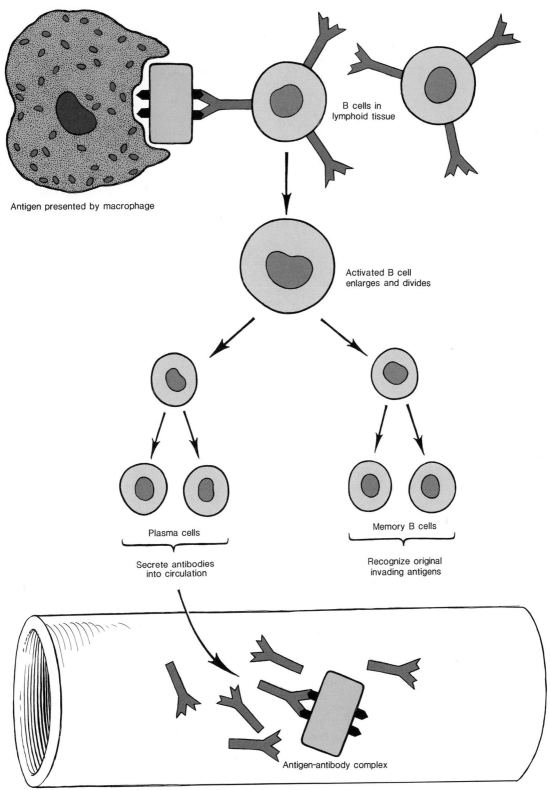

Antigen presented by macrophage

B cells in lymphoid tissue

Activated B cell enlarges and divides

Plasma cells

Secrete antibodies into circulation

Memory B cells

Recognize original invading antigens

Antigen-antibody complex

FIGURE 22-13 Role of B cells in humoral immunity.

specific for the antigen are activated. Some of them enlarge and divide and differentiate into a clone of **plasma cells** under the influence of thymic hormones (Figure 22-13). Plasma cells secrete the antibody. The phenomenal

rate of secretion is about 2,000 antibody molecules per second for each cell, and it occurs for several days until the plasma cell dies (in 4 to 5 days). The activated B cells that do not differentiate into plasma cells remain

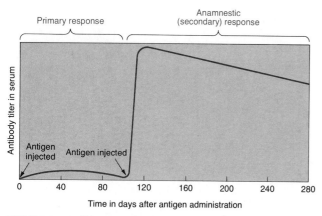

FIGURE 22-14 Primary and anamnestic (secondary) responses to the injection of an antigen.

as **memory B cells,** ready to respond more rapidly and forcefully should the same antigen appear at a future time.

Different antigens stimulate different B cells to develop into plasma cells and their accompanying memory B cells because the B cells of a particular clone are capable of secreting only one kind of antibody. Each specific antigen activates only those B cells already predetermined to secrete antibody specific to that antigen. The specific antigen selects a specific B cell because that particular B cell has on its surface the antibody molecules that it is capable of producing. The surface antibodies serve as receptor sites with which the antigen can combine.

Once the antigen-antibody complex is formed, the antibody activates complement enzymes for attack and fixes the complement to the surface of the antigen.

CLINICAL APPLICATION

The immune response of the body, whether cellular or humoral, is much more intense after a second or subsequent exposure to an antigen than after the initial exposure. This can be demonstrated by measuring the amount of antibody in serum, called the *antibody titer* (TĪ-ter). After an initial contact with an antigen, there is a period of several days during which no antibody is present. Then there is a slow rise in the antibody titer, followed by a gradual decline. Such a response of the body to the first contact is called the **primary response** (Figure 22-14). During the primary response, in which the body is said to be primed or sensitized, there is proliferation of immunocompetent lymphocytes. When the antigen is contacted again, whether for the second time or the two hundredth, there is an immediate proliferation of immunocompetent lymphocytes, and the antibody titer is far greater than that for the primary response. This accelerated, more intense response is called the **anamnestic** (an'-am-NES-tik; *anamne* = recall) or **secondary response.** The reason for the anamnestic response is that some of the immunocompetent lymphocytes formed during the primary response remain as memory cells. These memory cells not only add to the pool of cells that can respond to the antigen—their response is more intense as well.

Primary and anamnestic responses occur during microbial infection. When you recover from an infection without taking antibiotics, it is usually because of the primary response. If, at a later time, you contact the same microbe, the anamnestic response could be so swift that the microbes are quickly destroyed and you do not exhibit any signs or symptoms.

You have probably guessed by now that the anamnestic response provides the basis for immunization against certain diseases. When you receive the initial immunization, your body is sensitized. Should you encounter the pathogen again as an infecting microbe or "booster dose," your body experiences the anamnestic response. The booster dose literally boosts the antibody titer to a higher level. Since their effects are not permanent, boosters must be given periodically, and immunization schedules are designed to maintain high antibody titers.

IMMUNOLOGY AND CANCER

When a normal cell becomes transformed into a cancer cell, the tumor cell assumes cell surface components called **tumor-specific antigens.** It is believed that the immune system usually recognizes tumor-specific antigens as "nonself" and destroys the cancer cells carrying them. Such an immune response is called **immunologic surveillance.** Most investigators believe that cellular immunity is the basic mechanism involved in tumor destruction. In this regard, it is presumed that sensitized killer T cells react with tumor-specific antigens, initiating lysis of tumor cells. Sensitized macrophages may also be involved.

Despite the mechanism of immunologic surveillance, some cancer cells escape destruction, a phenomenon called **immunologic escape.** One possible explanation is that tumor cells may shed their tumor-specific antigens, thus evading recognition. Another theory is that antibodies produced by plasma cells bind to tumor-specific antigens, preventing recognition by killer T cells. It is reasonable to assume that immunologic techniques may be used in the future to prevent cancer.

MONOCLONAL ANTIBODIES

Scientists have known for many years how to induce laboratory animals (or humans) to produce antibodies. By injecting a particular antigen, antibodies are produced against the antigen by plasma cells (descendants of B cells). However, since the antibodies are produced by many different plasma cells, they vary physically and chemically; they are not pure. What scientists have learned to do is isolate a single B cell and fuse it with a tumor cell that is capable of proliferating endlessly. The resulting hybrid cell is called a **hybridoma cell.** These cells are a long-term source of substantial quantities of pure antibodies called **monoclonal antibodies.** Such antibodies will go after one and only one antigen.

One clinical use for monoclonal antibodies is the diagnosis of allergies and diseases such as hepatitis, rabies, and certain sexually transmitted diseases. Another application is the detection of cancer at an early stage and determining the extent of metastasis. One of the most exciting applications is in the area of treating diseases.

Monoclonal antibodies, either alone or in combination with drugs, are being used in trial studies to treat cancer. This approach has the advantage of destroying diseased tissues only, while sparing healthy tissues, thus overcoming some of the major adverse effects of chemotherapy and radiation treatment. Monoclonal antibodies may also be useful in preparing vaccines to counteract the rejection associated with transplants and to treat autoimmune diseases. One such disease that has been treated with some success is severe combined immunodeficiency (SCID), the most serious of all immune deficiency diseases. SCID victims are born lacking all immune defenses for fighting disease. This is the rare disease that affects the boy who is internationally known as the "boy in a bubble."

DISORDERS: HOMEOSTATIC IMBALANCES

The antigen-antibody response is essential to survival and typically provides a state of immunity. Under certain circumstances, however, it may create problems. Three such problems are hypersensitivity (allergy), tissue rejection, and autoimmunity.

Hypersensitivity (Allergy)

A person who is overly reactive to an antigen is said to be **hypersensitive** or **allergic.** Whenever an allergic reaction occurs, there is tissue injury. The antigens that induce an allergic reaction are called **allergens.** Almost any substance can be an allergen for some individual. Common allergens include certain foods (milk, eggs), antibiotics such as penicillin, cosmetics, chemicals in plants such as poison ivy, pollens, dust, molds, and even microbes.

One type of hypersensitivity is called **anaphylaxis** (an'-a-fi-LAK-sis), the interaction of humoral antibodies (IgE) with mast cells and basophils. Anaphylaxis literally means "against protection." Some anaphylactic reactions are localized and produce *allergic rhinitis* (hay fever), *bronchial asthma, atopic dermatitis* (eczema), and *urticaria* (hives). A severe anaphylactic reaction, called *acute anaphylaxis (anaphylactic shock),* may produce life-threatening systemic effects such as circulatory shock and asphyxia.

In response to certain allergens, some people produce IgE antibodies which bind to basophils in the blood and mast cells that are especially numerous in the skin, respiratory system, and endothelium of blood vessels. The antibodies may remain attached to the cells for weeks before a response occurs. This attachment, however, has the effect of sensitizing the cells, that is, they become subject to injury if the allergen enters the body at a later date. When the allergen enters the body again, tissue damage results. The allergen is believed to bind the cell-bound IgE antibodies together, forming a lattice. This stimulates the release of mediators of anaphylaxis stored in granules in the cytoplasm of basophils and mast cells. Among these mediators are histamine, eosinophil chemotactic factor of anaphylaxis (ECFA), platelet-activating factors (PAFs), slow-reacting substance of anaphylaxis (SRS-A), and serotonin. Collectively, these mediators of anaphylaxis increase blood capillary permeability, increase smooth muscle contraction in bronchial tubes, increase gastic secretion of hydrochloric acid, enhance complement reactions, increase mucus secretion, lyse platelets, and inhibit coagulation. Increased capillary permeability results in edema and redness and other symptoms of inflammation seen in anaphylaxis. Constriction of bronchial tubes results in difficulty in breathing. Excessive mucus secretion is responsible for the "runny" nose.

In acute anaphylaxis, large amounts of histamine are released into circulation. As a result, there is flushing of the skin; the respiratory tubes are continuously constricted, resulting in wheezing and shortness of breath; and edema is accelerated. This condition lowers the blood volume even more. If the effects of histamine are not counteracted, death may result, usually from respiratory failure. The effects of histamine may be reversed by administering epinephrine or antihistamines, drugs that inactivate histamine.

Tissue Rejection

Transplantation involves the replacement of an injured or diseased tissue or organ. Usually, the body recognizes the proteins in the transplanted tissue or organ as foreign and produces antibodies against them. This phenomenon is known as **tissue rejection.** Rejection can be somewhat reduced by matching donor and recipient and by administering drugs that inhibit the body's ability to form antibodies.

Types of Transplants

The closer the relationship of donor and recipient, the more successful the transplant. The most successful transplants are **isografts,** transplants in which the donor and recipient have identical genetic backgrounds. They include transplants between identical twins and from one part of the body to another.

An **allograft (homograft)** is a transplant between individuals of the same species, but with different genetic backgrounds. The success of this type of transplant has been moderate. Frequently, it is used as a temporary measure until the damaged or diseased tissue is able to repair itself. Skin transplants from other individuals and blood transfusions might properly be considered allografts. The one organ allograft that has been quite successful is the thymus. Children born without a thymus can now receive the gland from an aborted fetus. The thymus-deficient child cannot produce antibodies and thus cannot reject the transplant. Rejection later on indicates that the child is manufacturing antibodies and no longer needs the organ.

A **xenograft** is a transplant between animals of different species. This type of transplantation is used primarily as a physiological dressing over severe burns. Xenografts are presently restricted to laboratory animals.

Immunosuppressive Therapy

Until recently, **immunosuppressive drugs** suppressed not only the recipient's immune rejection of the donor organ, but the immune response to all antigens as well. This causes patients to become very susceptible to infectious diseases. A new drug called cyclosporin A, derived from a fungus, has largely overcome this problem. It is a selective immunosuppressive drug. It inhibits T cells which are responsible for tissue rejection, but has only a minimal effect on B cells. Thus, rejection is avoided and resistance against disease is still maintained.

Autoimmune Diseases

Under normal conditions, the body's immune mechanism is able to recognize its own tissues and chemicals. It normally does not produce T cells or B cells against its own substances. Such recognition of self is called **immunologic tolerance.** Although the mechanism of tolerance is not completely understood, it is believed that suppressor T cells may inhibit the differentiation of B cells into antibody-producing plasma cells or inhibit helper T cells which cooperate with B cells to amplify antibody production.

At times, however, immunologic tolerance breaks down and the body has difficulty in discriminating between its own antigens and foreign antigens. This loss of immunologic tolerance leads to an **autoimmune disease (autoimmunity).** Such diseases are immunologic responses mediated by antibodies against a person's own tissue antigens. Among human autoimmune diseases are rheumatoid arthritis (RA), systemic lupus erythematosus (SLE), thyroiditis, rheumatic fever, glomerulonephritis, encephalomyelitis, hemolytic and pernicious anemia, myasthenia gravis, and multiple sclerosis (MS).

Acquired Immune Deficiency Syndrome (AIDS)

Never before has science been confronted with an epidemic in which the primary disease only lowers the victim's immunity and then a second unrelated disease produces the symptoms that may result in death. This primary disease, first recognized in 1981, is called **acquired immune deficiency syndrome (AIDS)** and has a mortality rate of nearly 40 percent. One of the principal problems associated with AIDS is that its victims have too few T cells. In addition, the ratio of helper T cells to suppressor T cells, normally 2:1, is reversed. Symptoms of AIDS may develop for months or years and include malaise, a low-grade fever, coughing, shortness of breath, sore throat, muscle aches,

weight loss, and enlarged lymph nodes in the neck, axilla, and groin.

The two diseases that most often kill AIDS victims are Kaposi's sarcoma and *Pneumocystis carinii* pneumonia. Kaposi's sarcoma is a deadly form of skin cancer prevalent in equatorial Africa, but previously almost unknown in the United States. *Pneumocystis carinii* pneumonia is a rare form of pneumonia caused by the protozoan, *Pneumocystis carinii.* AIDS victims are also subject to a form of herpes that attacks the central nervous system and a bacterial infection that usually causes tuberculosis in chickens and pigs.

AIDS is found primarily among homosexual males, intravenous drug users, hemophiliacs, and Haitian immigrants. It is also found among infants and other patients who have received blood transfusions. The cause of AIDS is unknown. However, most experts feel that it is caused by an infectious organism. The pattern of AIDS closely resembles the occurrence of hepatitis B (serum hepatitis), a liver disease that commonly strikes homosexual drug addicts using contaminated needles, and sometimes patients getting blood transfusions. Like hepatitis B, AIDS might be transmitted by sexual contact or blood.

MEDICAL TERMINOLOGY

Adenitis (*adeno* = gland; *itis* = inflammation of) Enlarged, tender, and inflamed lymph nodes resulting from an infection.

Elephantiasis Great enlargement of a limb (especially lower limbs) and scrotum resulting from obstruction of lymph glands or vessels by a parasitic worm.

Hypersplenism (*hyper* = over) Abnormal splenic activity involving highly increased blood cell destruction.

Lymphadenectomy (*ectomy* = removal) Removal of a lymph node.

Lymphadenopathy (*patho* = disease) Enlarged, sometimes

tender lymph glands.

Lymphangioma (*angio* = vessel; *oma* = tumor). A benign tumor of the lymph vessels.

Lymphangitis Inflammation of the lymphatic vessels.

Lymphedema (*edema* = swelling) Accumulation of lymph fluid producing subcutaneous tissue swelling.

Lymphoma Any tumor composed of lymph tissue. Malignancy of reticuloendothelial cells of lymph nodes is called Hodgkin's disease.

Lymphostasis (*stasis* = halt) A lymph flow stoppage.

Splenomegaly (*mega* = large) Enlarged spleen.

STUDY OUTLINE

Lymphatic Vessels (p. 520)

1. The lymphatic system consists of lymph, lymphatic vessels, lymph nodes, and lymph organs.
2. Lymphatic vessels begin as blind-ended lymph capillaries in tissue spaces between cells.
3. Lymph capillaries merge to form larger vessels, called lymphatics, which ultimately converge into the thoracic duct or right lymphatic duct.
4. Lymphatics have thinner walls and more valves than veins.

Structure of Lymph Nodes (p. 520)

1. Lymph nodes are oval structures located along lymphatics.
2. Lymph enters nodes through afferent lymphatic vessels and exits through efferent lymphatic vessels.
3. Lymph passing through the nodes is processed by macrophages.

Lymph Circulation (p. 521)

1. The passage of lymph is from interstitial fluid, to lymph capillaries, to lymphatics, to lymph trunks, to the thoracic duct or right lymphatic trunk, to the subclavian veins.
2. Lymph flows as a result of skeletal muscle contractions and respiratory movements. It is also aided by valves in the lymphatics.

Lymphatic Organs (p. 523)

1. Tonsils are masses of lymphoid tissue embedded in mucous

membranes. They include the pharyngeal, palatine, and lingual tonsils.
2. The spleen functions as a lymphatic organ in phagocytosis of bacteria and worn-out cells and production of lymphocytes and plasma cells. It also acts as a reservoir for blood.
3. The thymus gland functions in immunity by processing T cells and stimulating B cells to develop into antibody-producing plasma cells.

Nonspecific Resistance to Disease (p. 526)

1. The ability to ward off disease using a number of defenses is called resistance. Lack of resistance is called susceptibility.
2. Nonspecific resistance refers to a wide variety of body responses against a wide range of pathogens.
3. Nonspecific resistance includes mechanical factors (skin, mucous membranes, lacrimal apparatus, saliva, mucus, cilia, epiglottis, and flow of urine), chemical factors (gastric juice, acid pH of skin, unsaturated fatty acids, and lysozyme), antimicrobial substances (interferon, complement, and properdin), phagocytosis, inflammation, and fever.

Immunity (Specific Resistance to Disease) (p. 530)

1. Specific resistance to disease involves the production of a specific lymphocyte or antibody against a specific antigen and is called immunity.
2. Antigens are chemical substances that, when introduced into the body, stimulate the production of antibodies that react with the antigen.

3. Examples of antigens are microbes, microbial structures, pollen, incompatible blood cells, and transplants.
4. Antigens are characterized by immunogenicity, reactivity, and multivalence.
5. Antibodies are proteins produced in response to antigens.
6. Based on chemistry and structure, antibodies are distinguished into five principal classes, each with specific biological roles (IgG, IgA, IgM, IgD, and IgE).
7. Antibodies consist of heavy and light chains and variable and constant portions.
8. Cellular immunity refers to destruction of antigens by T cells and humoral immunity refers to destruction of antigens by antibodies.
9. T cells are processed in the thymus gland; B cells may be processed in bone marrow, fetal liver and spleen, or gut-associated lymphoid tissue.
10. T cells consist of subpopulations: killer T cells destroy antigens directly; helper T cells help B cells to produce antibodies; suppressor T cells help to regulate the immune response; and memory T cells initiate response to subsequent invasions by the antigen.
11. B cells develop into antibody-producing plasma cells under the influence of thymic hormones; memory B cells recognize the original, invading antigen.
12. The anamnestic response provides the basis for immunization against certain diseases.
13. Cancer cells contain tumor-specific antigens and are frequently destroyed by the body's immune system (immuno-logic surveillance); some cancer cells escape detection and destruction, a phenomenon called immunologic escape.
14. Monoclonal antibodies are pure antibodies produced by fusing a B cell with a tumor cell; they are important in diagnosis, detection of disease, treatment, preparing vaccines, and countering rejection by transplants and autoimmune diseases.

Disorders: Homeostatic Imbalances (p. 537)

1. Hypersensitivity is overreactivity to an antigen. Localized anaphylactic reactions include hay fever, asthma, eczema, and hives; acute anaphylaxis is a severe reaction with systemic effects.
2. Tissue rejection of a transplanted tissue or organ involves antibody production against the proteins (antigens) in the transplant. It may be overcome with immunosuppressive drugs.
3. Autoimmune diseases result when the body does not recognize "self" antigens and produces antibodies against them. Several human autoimmune diseases are rheumatoid arthritis (RA), systemic lupus erythematosus (SLE), rheumatic fever, hemolytic and pernicious anemias, myasthenia gravis, and multiple sclerosis (MS).
4. Acquired immune deficiency syndrome (AIDS) lowers the body's immunity by decreasing the number of T cells and reversing the ratio of helper T cells to suppressor T cells. AIDS victims frequently develop Karposi's sarcoma and *Pneumocystis carinii* pneumonia.

REVIEW QUESTIONS

1. Identify the components and functions of the lymphatic system.
2. How do lymphatic vessels originate? Compare veins and lymphatics with regard to structure.
3. What is a lymphangiogram? What is its diagnostic value?
4. Describe the structure of a lymph node. What functions do lymph nodes serve?
5. Construct a diagram to indicate the route of lymph circulation.
6. List and explain the various factors involved in the maintenance of lymph circulation.
7. Define edema. What are some of its causes?
8. Identify the tonsils by location.
9. Describe the location, gross anatomy, histology, and functions of the spleen. What is a splenectomy?
10. Describe the role of the thymus gland in immunity.
11. Describe the operation of the various mechanical and chemical factors involved in nonspecific resistance.
12. Outline the role of the following antimicrobial substances: interferon (IFN), complement, and properdin.
13. What is phagocytosis? Describe the steps involved in adherence and ingestion.
14. Compare the roles of microphages and macrophages in phagocytosis.
15. Define immunity.
16. Distinguish between an antigen and an antibody.
17. List the various characteristics of antigens.
18. By means of a labeled diagram, describe the structure of an antibody.
19. What are the characteristics of antibodies?
20. Where are T cells and B cells processed? How do they differ in function?
21. Describe the role of T cells in cellular immunity. Be sure to cite the roles of each type of T cell.
22. Describe the role of B cells in humoral immunity.
23. Discuss the importance of the secondary (anamnestic) response of the body to an antigen.
24. Explain how immunology is related to cancer.
25. What are monoclonal antibodies? How are they produced? Why are they clinically important?
26. Define hypersensitivity, anaphylaxis, and anaphylactic shock.
27. Describe the mechanism of anaphylaxis.
28. Why does tissue rejection occur? How is this problem overcome?
29. How are transplants classified?
30. Define an autoimmune disease. Give several examples.
31. Describe the symptoms of acquired immune deficiency syndrome (AIDS). What are the complications of AIDS?
32. Refer to the glossary of medical terminology associated with the lymphatic system. Be sure that you define each term.

- Identify the organs of the respiratory system.
- Compare the structure and function of the external and internal nose.
- Differentiate the three regions of the pharynx and describe their roles in respiration.
- Describe the structure of the larynx and explain its function in respiration and voice production.
- Explain the structure and function of the trachea.
- Contrast tracheostomy and intubation as alternative methods for clearing obstructed air passageways.
- Discuss the composition of the bronchial tree.
- Identify the coverings of the lungs and the division of the lungs into lobes.
- Describe the composition of a lobule of the lung.
- Explain the structure of the alveolar-capillary membrane and its function in the diffusion of respiratory gases.
- List the events involved in inspiration and expiration.
- Explain how compliance and airway resistance relate to breathing.
- Define coughing, sneezing, sighing, yawning, sobbing, crying, laughing, and hiccuping as modified respiratory movements.
- Compare the volumes and capacities of air exchanged during respiration.
- Define the partial pressure of a gas, Boyle's law, Charles' law, Dalton's law, and Henry's law.

- Explain how external and internal respiration occur, basing your explanation on differences in the partial pressure of the respiratory gases.
- Explain how the respiratory gases are transported by the blood.
- Describe how the oxygen-carrying capacity of the blood is affected by pO_2, pCO_2, temperature, and DPG.
- Explain how the respiratory center functions in establishing the basic rhythm of respiration.
- Describe how various factors may modify the rate of respiration.
- List the basic steps involved in cardiopulmonary resuscitation (CPR).
- Explain how the abdominal thrust (Heimlich) maneuver is performed.
- Define bronchogenic carcinoma, nasal polyps, bronchial asthma, emphysema, pneumonia, tuberculosis, infant respiratory distress syndrome (RDS), sudden infant death syndrome (SIDS), coryza (common cold), and influenza (flu) as disorders of the respiratory system.
- Define medical terminology associated with the respiratory system.
- Explain the actions of selected drugs on the respiratory system.

23 The Respiratory System

Cells need a continuous supply of oxygen to carry out the activities that are vital to their survival. Many of these activities release quantities of carbon dioxide. Since an excessive amount of carbon dioxide produces acid conditions that are poisonous to cells, the gas must be eliminated quickly and efficiently. The two systems that supply oxygen and eliminate carbon dioxide are the cardiovascular system and the respiratory system. The **respiratory system** consists of organs that exchange gases between the atmosphere and blood. These organs are the nose, pharynx, larynx, trachea, bronchi, and lungs (Figure 23-1). The cardiovascular system transports the gases in the blood between the lungs and the cells. The term *upper respiratory system* refers to the nose and throat and associated structures. The *lower respiratory system* refers to the remainder of the system.

The overall exchange of gases between the atmosphere, the blood, and the cells is **respiration.** Three basic processes are involved. The first process, **pulmonary ventilation,** or breathing, is the inflow and outflow of air between the atmosphere and the lungs. The second and third processes involve the exchange of gases within the body. **External respiration** is the exchange of gases between the lungs and blood. **Internal respiration** is the exchange of gases between the blood and the cells.

The respiratory and cardiovascular systems participate equally in respiration. Failure of either system has the same effect on the body: disruption of homeostasis and rapid death of cells from oxygen starvation.

ORGANS

NOSE

Anatomy

The **nose** has an external portion and an internal portion inside the skull (Figure 23-2). The external portion consists of a supporting framework of bone and cartilage covered with skin and lined with mucous membrane. The bridge of the nose is formed by the nasal bones, which hold it in a fixed position. Because it has a framework of pliable cartilage, the rest of the external nose is quite flexible. On the undersurface of the external nose are two openings called the **nostrils** or **external nares** (NA-rēz; *sing.,* naris). The surface anatomy of the nose is shown in Figure 23-2a.

The internal portion of the nose is a large cavity in the skull that lies inferior to the cranium and superior to the mouth. Anteriorly, the internal nose merges with the external nose, and posteriorly it communicates with the throat (pharynx) through two openings called the **internal nares (choanae).** Four paranasal sinuses (frontal, sphenoidal, maxillary, and ethmoidal) and the nasolacrimal ducts also open into the internal nose. The lateral walls of the internal nose are formed by the ethmoid, maxillae, and inferior nasal conchae bones. The ethmoid also forms the roof. The floor is formed by the palatine bones and the palatine process of the maxilla, which together comprise the hard palate.

The inside of both the external and internal nose consists of a **nasal cavity,** divided into right and left sides

by a vertical partition called the **nasal septum.** The anterior portion of the septum is made primarily of cartilage. The remainder is formed by the vomer and the perpendicular plate of the ethmoid (see Figure 7-7a). The anterior portion of the nasal cavity, just inside the nostrils, is called the **vestibule.** It is surrounded by cartilage. The upper nasal cavity is surrounded by bone.

Physiology

The interior structures of the nose are specialized for three functions: incoming air is warmed, moistened, and filtered; olfactory stimuli are received; and large hollow resonating chambers are provided for speech sounds.

When air enters the nostrils, it passes first through the vestibule. The vestibule is lined by skin containing coarse hairs that filter out large dust particles. The air then passes into the upper nasal cavity. Three shelves formed by projections of the superior, middle, and inferior nasal conchae extend out of the lateral wall of the cavity. The conchae, almost reaching the septum, subdivide each side of the nasal cavity into a series of groovelike passageways—the **superior, middle,** and **inferior meatuses.** Mucous membrane lines the cavity and its shelves. The olfactory receptors lie in the membrane lining the area superior to the superior nasal conchae and is also called the **olfactory region.** Below the olfactory region, the membrane

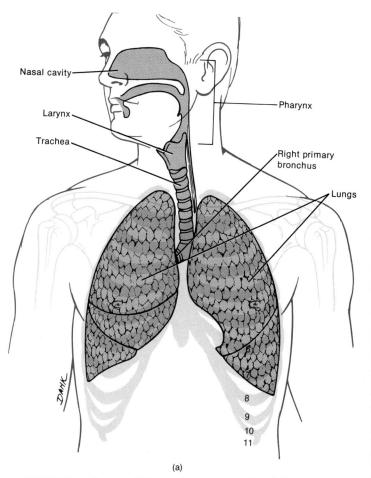

(a)

FIGURE 23-1 Organs of the respiratory system in relation to surrounding structures. (a) Diagram.

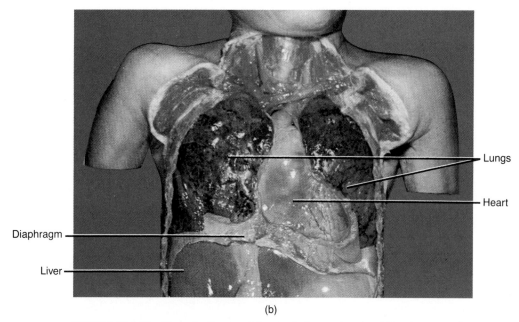

(b)

FIGURE 23-1 (*Continued*) Organs of the respiratory system in relation to surrounding structures. (b) Photograph. (Courtesy of C. Yokochi and J. W. Rohen, *Photographic Anatomy of the Human Body,* 2nd ed., 1979, IGAKU-SHOIN, Ltd., Tokyo, New York.)

contains pseudostratified ciliated columnar cells with many goblet cells and capillaries. As the air whirls around the conchae and meatuses, it is warmed by the capillaries. Mucus secreted by the goblet cells moistens the air and traps dust particles. Drainage from the lacrimal ducts, and perhaps secretions from the paranasal sinuses, also help moisten the air. The cilia move the mucus-dust packages along the pharynx so they can be eliminated from the body.

CLINICAL APPLICATION

Nosebleed, or **epistaxis** (ep'-i-STAK-sis), is common because of the exposure of the nose to trauma and the extensive blood supply of the nose. Bleeding, either

---

arterial or venous, usually occurs on the anterior part of the septum and can be arrested by firm packing of the external nares. If the point of bleeding is in the posterior region, plugging of both the external and internal nares may be necessary. In extreme emergency, the external carotid artery may have to be ligated (tied) in order to control the hemorrhage.

PHARYNX

The **pharynx** (FAR-inks), or throat, is a somewhat funnel-shaped tube about 13 cm (5 inches) long that starts at the internal nares and extends partway down the neck

1. **Root.** Superior attachment of nose at forehead located between eyes.
2. **Apex.** Tip of nose.
3. **Dorsum nasi.** Rounded anterior border connecting root and apex; in profile, may be straight, convex, concave, or wavy.
4. **Nasofacial angle.** Point at which side of nose blends with tissues of face.
5. **Ala.** Convex flared portio of inferior lateral surface; unites with upper lip.
6. **External naris.** External opening into nose.
7. **Bridge.** Superior portion of dorsum nasi, superficial to nasal bones.

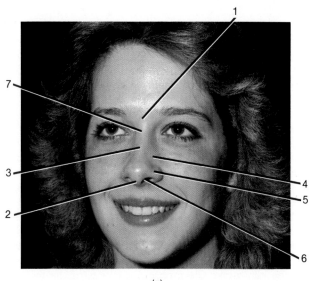

(a)

FIGURE 23-2 Nose. (a) Surface anatomy of the nose in anterior view. (© 1982 by Gerard J. Tortora. Courtesy of Lynne Tortora.)

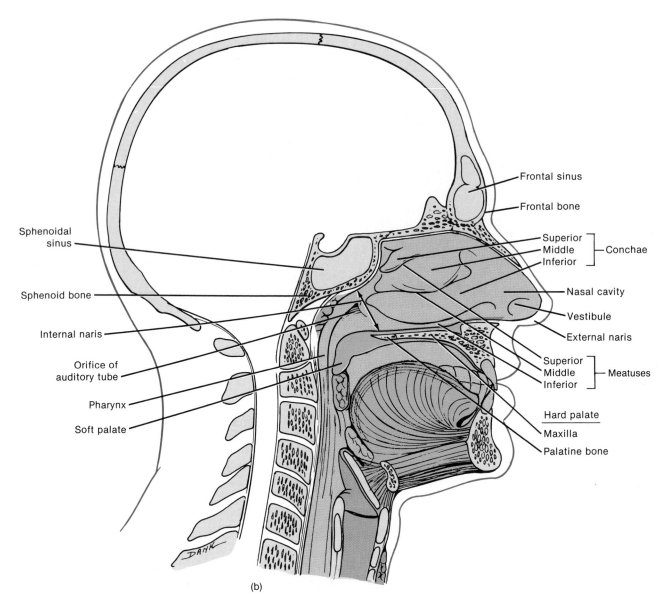

(b)

FIGURE 23-2 (*Continued*) Nose. (b) Diagram of the left side seen in sagittal section with the nasal septum removed.

(Figure 23-3). It lies just posterior to the nasal cavity and oral cavity and just anterior to the cervical vertebrae. Its wall is composed of skeletal muscles and lined with mucous membrane. The functions of the pharynx are to serve as a passageway for air and food and to provide a resonating chamber for speech sounds.

The uppermost portion of the pharynx, called the **nasopharynx,** lies posterior to the internal nasal cavity and extends to the plane of the soft palate. There are four openings in its wall: two internal nares and two openings that lead into the auditory tubes. The posterior wall also contains the pharyngeal tonsil, or adenoid. Through the internal nares the nasopharynx exchanges air with the nasal cavities and receives the packages of dust-laden mucus. It is lined with pseudostratified ciliated epithelium, and the cilia move the mucus down toward the mouth. The nasopharynx also exchanges small amounts of air

with the auditory tubes so that the air pressure inside the middle ear equals the pressure of the atmospheric air flowing through the nose and pharynx.

The middle portion of the pharynx, the **oropharynx,** lies posterior to the oral cavity and extends from the soft palate inferior to the level of the hyoid bone. It has only one opening, the *fauces* (FAW-sēz), or opening from the mouth. It is lined by stratified squamous epithelium. This portion of the pharynx is both respiratory and digestive in function, since it is a common passageway for both air and food. Two pairs of tonsils, the palatine and lingual tonsils, are found in the oropharynx. The lingual tonsil lies at the base of the tongue (see also Figure 24-5a).

The lowest portion of the pharynx, the **laryngopharynx** (la-rin'-gō-FAR-inks), extends downward from the hyoid bone and becomes continuous with the esophagus (food

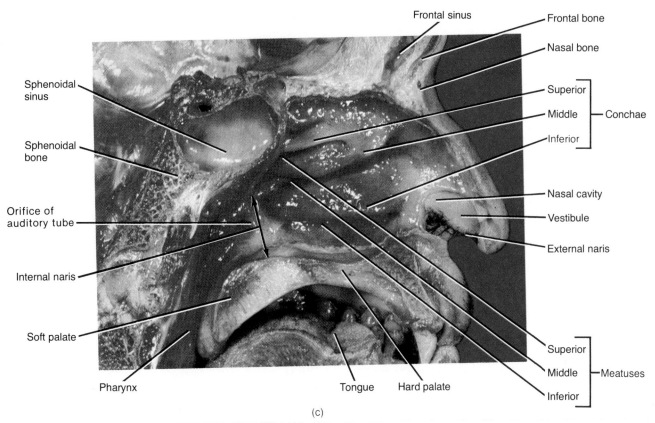

Frontal sinus — Frontal bone
— Nasal bone
Sphenoidal sinus
— Superior
— Middle — Conchae
— Inferior
Sphenoidal bone
— Nasal cavity
Orifice of auditory tube — Vestibule
— External naris
Internal naris
Soft palate — Superior
— Middle — Meatuses
Pharynx   Tongue   Hard palate   Inferior

(c)

FIGURE 23-2 (Continued) Nose. (c) Photograph of the left side seen in sagittal section with the nasal septum removed. (Courtesy of C. Yokochi and J. W. Rohen, *Photographic Anatomy of the Human Body,* 2nd ed., 1979, IGAKU-SHOIN, Ltd., Tokyo, New York.)

tube) posteriorly and the larynx (voice box) anteriorly. Like the oropharynx, the laryngopharynx is a respiratory and a digestive pathway and is lined by stratified squamous epithelium.

LARYNX

The **larynx,** or voice box, is a short passageway that connects the pharynx with the trachea. It lies in the midline of the neck anterior to the fourth through sixth cervical vertebrae.

Anatomy

The wall of the larynx is supported by nine pieces of cartilage (Figure 23-4). Three are single and three are paired. The three single pieces are the thyroid cartilage, epiglottic cartilage (epiglottis), and cricoid cartilage. Of the paired cartilages, the arytenoid cartilages are the most important. The paired corniculate and cuneiform cartilages are of lesser significance.

The **thyroid cartilage,** or Adam's apple, consists of two fused plates that form the anterior wall of the larynx and give it its triangular shape. It is larger in males than in females.

The **epiglottis** is a large, leaf-shaped piece of cartilage lying on top of the larynx (see also Figure 23-3). The

"stem" of the epiglottis is attached to the thyroid cartilage, but the "leaf" portion is unattached and free to move up and down like a trapdoor. During swallowing, there is elevation of the larynx. This causes the free edge of the epiglottis to form a lid over the glottis and closure of the glottis. The **glottis** is the space between the vocal folds (true vocal cords) in the larynx. In this way, the larynx is closed off and liquids and foods are routed into the esophagus and kept out of the trachea. If anything but air passes into the larynx, a cough reflex attempts to expel the material.

The **cricoid** (KRĪ-koyd) **cartilage** is a ring of cartilage forming the inferior wall of the larynx. It is attached to the first ring of cartilage of the trachea.

The paired **arytenoid** (ar'-i-TĒ-noyd) **cartilages** are pyramidal in shape and located at the superior border of the cricoid cartilage. They attach to the vocal folds and pharyngeal muscles and by their action can move the vocal cords.

The paired **corniculate** (kor-NIK-yoo-lāt) **cartilages** are cone-shaped. One is located at the apex of each arytenoid cartilage. The paired **cuneiform** (kyoo-NĒ-i-form) **cartilages** are rod-shaped cartilages that connect the epiglottis to the arytenoid cartilages.

The epithelium lining the larynx below the vocal folds is pseudostratified. It consists of ciliated columnar cells, goblet cells, and basal cells, and it helps to trap dust not removed in the upper passages.

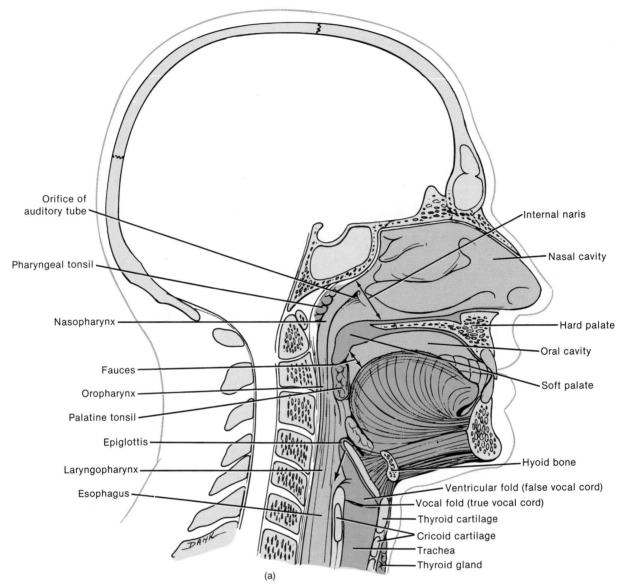

FIGURE 23-3 Head and neck seen in sagittal section. (a) Diagram.

Voice Production

The mucous membrane of the larynx is arranged into two pairs of folds—an upper pair called the **ventricular folds (false vocal cords)** and a lower pair called simply the **vocal folds (true vocal cords)** (Figure 23-4c). When the ventricular folds are brought together they function in holding the breath against pressure in the thoracic cavity such as might occur when a person exerts a strain while lifting a heavy weight. The mucous membrane of the vocal folds is lined by nonkeratinized stratified squamous epithelium. Under the membrane lie bands of elastic ligaments stretched between pieces of rigid cartilage like the strings on a guitar. Skeletal muscles of the larynx, called intrinsic muscles, are attached internally to the pieces of rigid cartilage and to the vocal folds themselves (see Figure 11-9). When the muscles contract, they pull the strings of elastic ligaments tight and stretch the vocal folds out into the air passageways so that the glottis is narrowed. If air is directed against the vocal folds, they

vibrate and set up sound waves in the column of air in the pharynx, nose, and mouth. The greater the pressure of air, the louder the sound.

Pitch is controlled by the tension on the vocal folds. If they are pulled taut by the muscles, they vibrate more rapidly and a higher pitch results. Lower sounds are produced by decreasing the muscular tension on the vocal folds. Vocal folds are usually thicker and longer in males than in females, and therefore they vibrate more slowly. Thus men generally have a lower range of pitch than women.

Sound originates from the vibration of the vocal folds, but other structures are necessary for converting the sound into recognizable speech. The pharynx, mouth, nasal cavity, and paranasal sinuses all act as resonating chambers that give the voice its human and individual quality. By constricting and relaxing the muscles in the walls of the pharynx, we produce the vowel sounds. Muscles of the face, tongue, and lips help us to enunciate words.

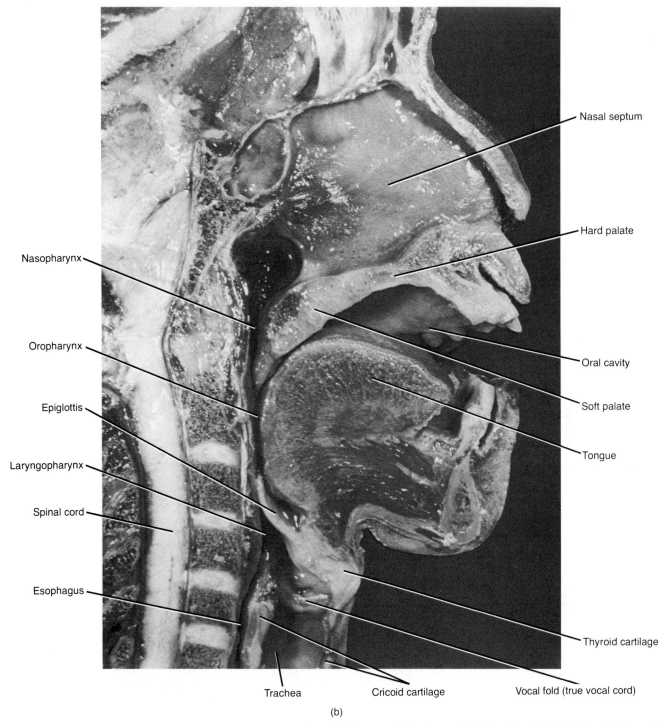

Nasal septum

Hard palate

Oral cavity

Soft palate

Tongue

Thyroid cartilage

Vocal fold (true vocal cord)

Nasopharynx

Oropharynx

Epiglottis

Laryngopharynx

Spinal cord

Esophagus

Trachea

Cricoid cartilage

(b)

FIGURE 23-3 (*Continued*) Head and neck seen in sagittal section. (b) Photograph. (Courtesy of C. Yokochi and J. W. Rohen, *Photographic Anatomy of the Human Body,* 2nd ed., 1979, IGAKU-SHOIN, Ltd., Tokyo, New York.)

CLINICAL APPLICATION

Laryngitis is an inflammation of the larynx that is most often caused by a respiratory infection or irritants such as cigarette smoke. Inflammation of the vocal folds themselves causes hoarseness or loss of voice by interfering with the contraction of the folds or by causing them to swell to the point where they cannot vibrate freely. Many long-term smokers acquire a permanent hoarseness from the damage done by chronic inflammation.

TRACHEA

The **trachea** (TRĀ-kē-a), or windpipe, is a tubular passageway for air about 12 cm (4½ inches) in length and 2.5 cm (1 inch) in diameter. It is located anterior to the esophagus and extends from the larynx to the fifth thoracic vertebra, where it divides into right and left primary bronchi (see Figure 23-6a).

The tracheal epithelium is pseudostratified. It consists of ciliated columnar cells that reach the lumenal surface, goblet cells, and basal cells that do not reach the lumenal

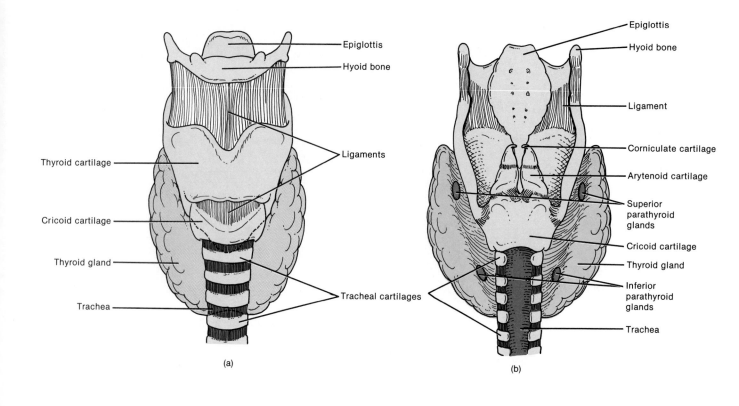

(a)

(b)

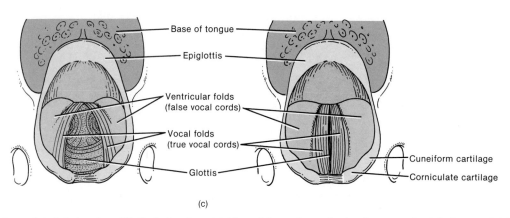

(c)

FIGURE 23-4 Larynx. (a) Anterior view. (b) Posterior view. (c) Viewed from above. In the figure on the left the vocal folds (true vocal cords) are relaxed and the glottis is open. In the figure on the right the vocal folds are pulled taut and the glottis is closed.

surface (Figure 23-5). The epithelium provides the same protection against dust as the membrane lining the larynx. The wall of the trachea is composed of smooth muscle and elastic connective tissue. It also consists of a series of 16 to 20 horizontal incomplete rings of hyaline cartilage that look like a series of letter Cs stacked one on top of another. The open parts of the Cs face the esophagus and permit it to expand into the trachea during swallowing. Transverse smooth muscle fibers, called the *trachealis muscle,* attach the open ends of the cartilage rings. The open ends of the rings of cartilage are also attached by elastic connective tissue. The solid parts of the Cs provide a rigid support so the tracheal wall does not collapse inward and obstruct the air passageway.

CLINICAL APPLICATION
At the point where the trachea bifurcates into right and left primary bronchi, there is an internal ridge called the **carina** (ka-RĪ-na). It is formed by a posterior and somewhat inferior projection of the last tracheal cartilage. Widening and distortion of the carina, which can be seen in an examination by bronchoscopy, is a serious prognostic sign, since it usually indicates a carcinoma of the lymph nodes around the bifurcation of the trachea. **Bronchoscopy** is the visual examination of the bronchi through a **bronchoscope,** an illuminated, tubular instrument which can be passed through the trachea into the bronchi.

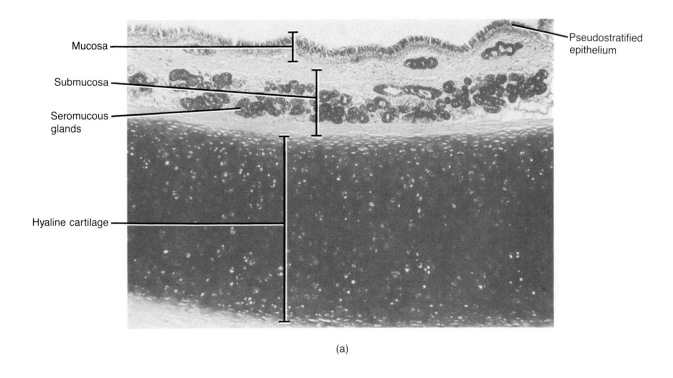

Mucosa

Submucosa

Seromucous
glands

Hyaline cartilage

Pseudostratified
epithelium

(a)

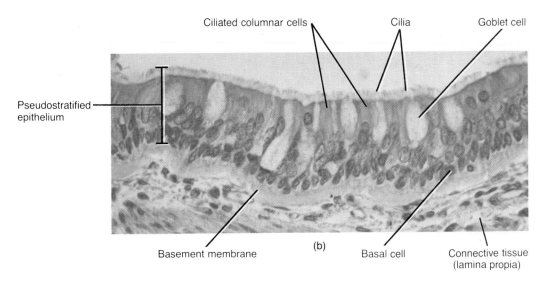

Ciliated columnar cells   Cilia   Goblet cell

Pseudostratified
epithelium

Basement membrane   (b)   Basal cell   Connective tissue
(lamina propia)

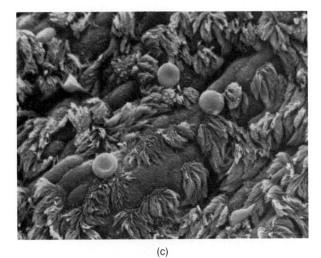

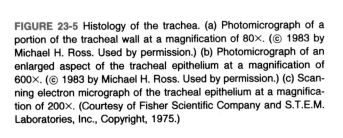

FIGURE 23-5 Histology of the trachea. (a) Photomicrograph of a
portion of the tracheal wall at a magnification of 80×. (© 1983 by
Michael H. Ross. Used by permission.) (b) Photomicrograph of an
enlarged aspect of the tracheal epithelium at a magnification of
600×. (© 1983 by Michael H. Ross. Used by permission.) (c) Scanning electron micrograph of the tracheal epithelium at a magnification of 200×. (Courtesy of Fisher Scientific Company and S.T.E.M.
Laboratories, Inc., Copyright, 1975.)

(c)

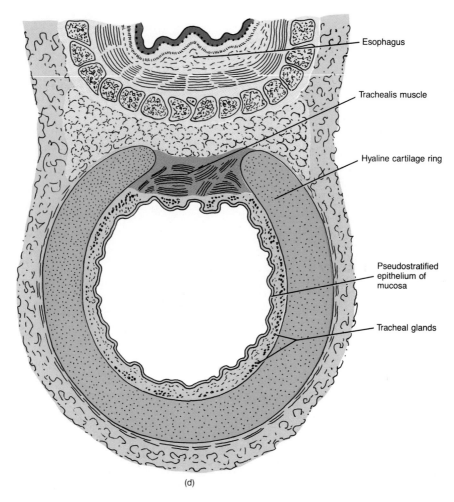

(d)

FIGURE 23-5 (Continued) Histology of the trachea. (d) Diagram of a cross section through the trachea and a portion of the esophagus.

Occasionally the respiratory passageways are unable to protect themselves from obstruction. The rings of cartilage may be accidentally crushed; the mucous membrane may become inflamed and swell so much that it closes off the air space; inflamed membranes secrete a great deal of mucus that may clog the lower respiratory passageways; or a large object may be breathed in (aspirated) while the glottis is open. The passageways must be cleared quickly. If the obstruction is above the level of the chest, a **tracheostomy** (trā-kē-OS-tō-mē) may be performed. An incision is made in the neck and into the part of the trachea below the obstructed area. The patient breathes through a tube inserted through the incision. Another method is **intubation.** A tube is inserted into the mouth or nose and passed down through the larynx and trachea. The firm wall of the tube pushes back any flexible obstruction, and the inside of the tube provides a passageway for air. If mucus is clogging the trachea, it can be suctioned out through the tube.

BRONCHI

The trachea terminates in the chest by dividing at the sternal angle into a **right primary bronchus** (BRON-kus),

which goes to the right lung, and a **left primary bronchus,** which goes to the left lung (Figure 23-6a, b). The right primary bronchus is more vertical, shorter, and wider than the left. As a result, foreign objects in the air passageways are more likely to enter it than the left and frequently lodge in it. Like the trachea, the primary bronchi (BRON-kē) contain incomplete rings of cartilage and are lined by a pseudostratified ciliated epithelium.

Upon entering the lungs, the primary bronchi divide to form smaller bronchi—the **secondary (lobar) bronchi,** one for each lobe of the lung (the right lung has three lobes; the left lung has two). The secondary bronchi continue to branch, forming still smaller bronchi, called **tertiary (segmental) bronchi** which divide into **bronchioles.** Bronchioles, in turn, branch into even smaller tubes called **terminal bronchioles.** This continuous branching from the trachea resembles a tree trunk with its branches and is commonly referred to as the **bronchial tree.**

As the branching becomes more extensive in the bronchial tree, several structural changes may be noted. First, rings of cartilage are replaced by plates of cartilage that finally disappear in the bronchioles. Second, as the cartilage decreases, the amount of smooth muscle increases. Third, the epithelium changes from pseudostratified ciliated to simple cuboidal in the terminal bronchioles.

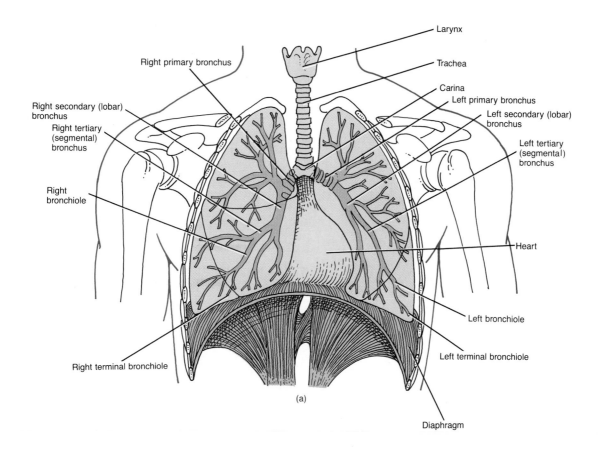

(a)

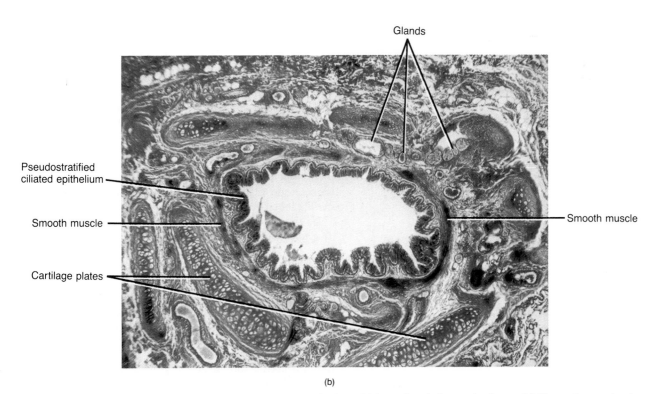

(b)

FIGURE 23-6 Air passageways to the lungs. (a) Diagram of the bronchial tree in relation to the lungs. (b) Photomicrograph of a cross section of a primary bronchus at a magnification of 100×. (© 1983 by Michael H. Ross. Used by permission.)

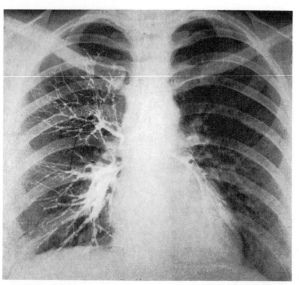

(c)

FIGURE 23-6 (Continued) Air passageways to the lungs. (c) Anteroposterior bronchogram. (Courtesy of Lester W. Paul and John H. Juhl, *The Essentials of Roentgen Interpretation,* 3rd ed., Harper & Row, Publishers, Inc., New York, 1972.)

CLINICAL APPLICATION

The fact that the walls of the bronchioles contain a great deal of smooth muscle but no cartilage is clinically significant. During an **asthma attack** the muscles go into spasm. Because there is no supporting cartilage, the spasms can close off the air passageways.

Bronchography (brong-KOG-ra-fē) is a technique for examining the bronchial tree. An intratracheal catheter is passed into the mouth or nose, through the glottis, and into the trachea. Then, an opaque contrast medium, usually containing iodine is introduced, by means of gravity, into the trachea and distributed through the bronchial branches. Roentgenograms of the chest in various positions are taken and the developed film, a **bronchogram** (BRONG-kō-gram), provides a picture of the tree (Figure 23-6c).

LUNGS

The **lungs** are paired, cone-shaped organs lying in the thoracic cavity. They are separated from each other by the heart and other structures in the mediastinum (see Figure 20-1a). Two layers of serous membrane, collectively called the **pleural membrane,** enclose and protect each lung. The outer layer is attached to the wall of the thoracic cavity and is called the **parietal pleura.** The inner layer, the **visceral pleura,** covers the lungs themselves. Between the visceral and parietal pleura is a small potential space, the **pleural cavity,** which contains a lubricating fluid secreted by the membranes (see Figure 1-7b). This fluid prevents friction between the membranes and allows them to move easily on one another during breathing.

CLINICAL APPLICATION

In certain conditions, the pleural cavity may fill with air (**pneumothorax**), blood (**hemothorax**), or pus. Air in the pleural cavity, mostly commonly introduced in a surgical opening of the chest, may cause the lung to collapse. Fluid can be drained from the pleural cavity by inserting a needle, usually posteriorly through the seventh intercostal space. The needle is passed along the superior border of the lower rib to avoid damage to the intercostal nerves and blood vessels. Below the seventh intercostal space there is danger of penetrating the diaphragm.

Inflammation of the pleural membrane, or **pleurisy,** causes friction during breathing that can be quite painful when the swollen membranes rub against each other.

Gross Anatomy

The lungs extend from the diaphragm to a point about 1.5 to 2.5 cm (¾ to 1 inch) superior to the clavicles and lie against the ribs anteriorly and posteriorly. The broad inferior portion of the lung, the **base,** is concave and fits over the convex area of the diaphragm (Figure 23-7). The narrow superior portion of the lung is termed the **apex (cupula).** The surface of the lung lying against the ribs, the **costal surface,** is rounded to match the curvature of the ribs. The **mediastinal (medial) surface** of each lung contains a vertical slit, the **hilus,** through which bronchi, pulmonary vessels, and nerves enter and exit. The blood vessels, bronchi, and nerves are held together by the pleura and connective tissue, and they constitute the **root** of the lung. Medially, the left lung also contains a concavity, the **cardiac notch,** in which the heart lies.

The right lung is thicker and broader than the left. It is also somewhat shorter than the left because the diaphragm is higher on the right side to accommodate the liver that lies below it.

Lobes and Fissures

Each lung is divided into lobes by one or more fissures. Both lungs have an **oblique fissure,** which extends downward and forward. The right lung also has a **horizontal fissure.** The oblique fissure in the left lung separates the **superior lobe** from the **inferior lobe.** The upper part of the oblique fissure of the right lung separates the superior lobe from the inferior lobe, whereas the lower part of the oblique fissure separates the inferior lobe from the **middle lobe.** The horizontal fissure of the right lung subdivides the superior lobe, thus forming a middle lobe.

Each lobe receives its own secondary (lobar) bronchus. Thus the right primary bronchus gives rise to three secondary (lobar) bronchi called the **superior, middle,** and **inferior secondary (lobar) bronchi.** The left primary bronchus gives rise to a **superior** and an **inferior secondary (lobar) bronchus.** Within the substance of the lung, the secondary bronchi give rise to the **tertiary (segmental) bronchi,** which are constant in both origin and distribution. The segment of lung tissue that each supplies is called a **bronchopulmonary segment.** Bronchial and pulmonary disorders, such as tumors or abscesses, may be localized in a bronchopulmonary segment and may be surgically removed without seriously disrupting surrounding lung tissue.

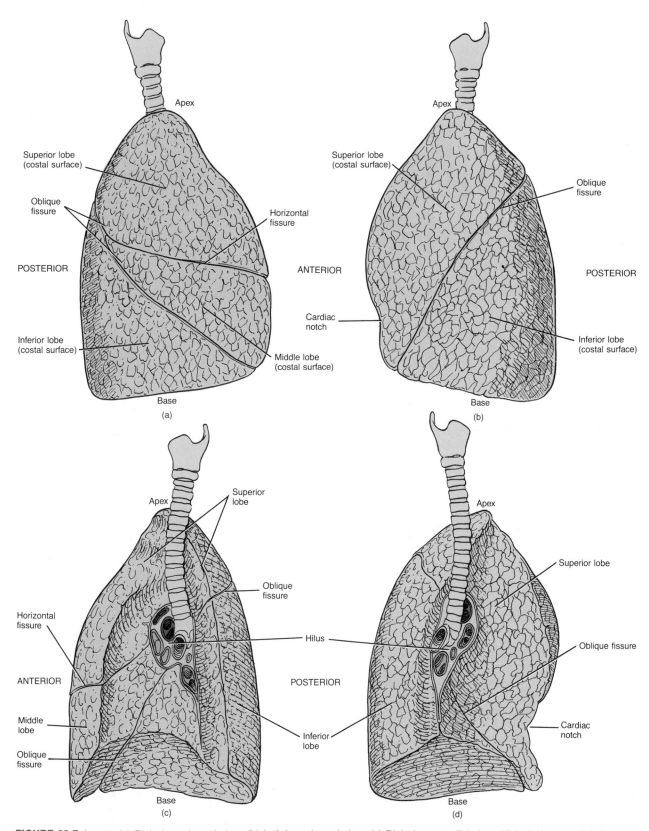

FIGURE 23-7 Lungs. (a) Right lung, lateral view. (b) Left lung, lateral view. (c) Right lung, medial view. (d) Left lung, medial view.

Lobules

Each bronchopulmonary segment of the lungs is broken up into many small compartments called **lobules** (Figure

23-8a). Each lobule is wrapped in elastic connective tissue and contains a lymphatic vessel, an arteriole, a venule, and a branch from a terminal bronchiole. Terminal bronchioles subdivide into microscopic branches called **respi-**

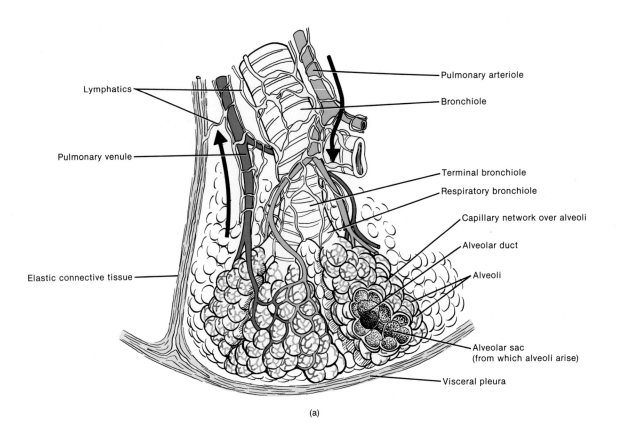

Lymphatics

Pulmonary venule

Elastic connective tissue

Pulmonary arteriole

Bronchiole

Terminal bronchiole

Respiratory bronchiole

Capillary network over alveoli

Alveolar duct

Alveoli

Alveolar sac
(from which alveoli arise)

Visceral pleura

(a)

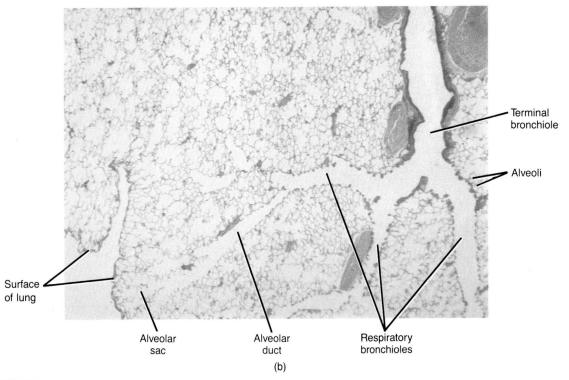

Terminal
bronchiole

Alveoli

Surface
of lung

Alveolar
sac

Alveolar
duct

Respiratory
bronchioles

(b)

FIGURE 23-8 Histology of the lungs. (a) Diagram of a lobule of the lung. (b) Photomicrograph of a terminal bronchiole, respiratory bronchioles, alveolar duct, alveolar sac, and alveoli at a magnification of 50×. (© 1983 by Michael H. Ross. Used by permission.)

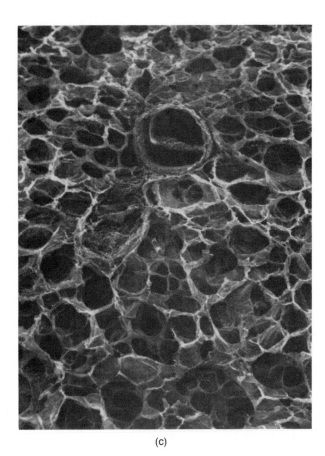

(c)

FIGURE 23-8 (*Continued*) Histology of the lungs. (c) Scanning electron micrograph of alveolar sacs at a magnification of 100×. (Courtesy of Fisher Scientific Company and S.T.E.M. Laboratories, Inc., Copyright, 1975). (d) Scanning electron micrograph of an alveolus showing squamous pulmonary alveolar cells and an alveolar macrophage at a magnification of 3,430×. (Courtesy of Richard K. Kessel and Randy H. Kardon, *Tissues and Organs: A Text-Atlas of Scanning Electron Microscopy*. Copyright © 1979 by Scientific American, Inc.) (e) Diagram of the structure of the alveolar-capillary membrane.

Alveolar
macrophage   Septal cells

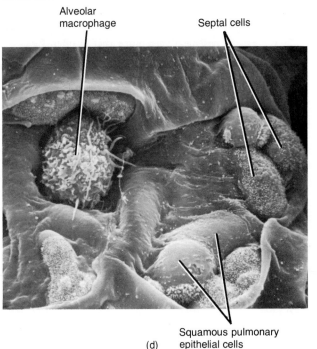

(d)   Squamous pulmonary
epithelial cells

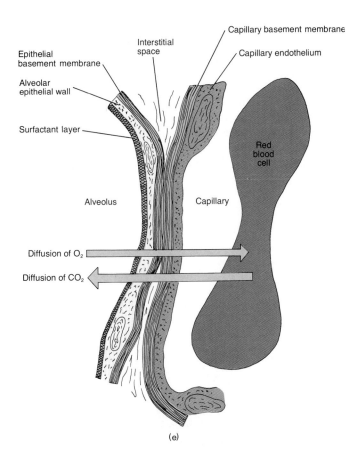

(e)

ratory bronchioles (Figure 23-8a, b). As the respiratory bronchioles become more distal, the epithelial lining changes from cuboidal to squamous. Respiratory bronchioles, in turn, subdivide into several (2 to 11) **alveolar ducts (atria).**

Around the alveolar ducts are numerous alveoli and alveolar sacs. An **alveolus** (al-VĒ-ō-lus) is a cup-shaped outpouching lined by epithelium and supported by a thin elastic basement membrane. **Alveolar sacs** are two or more alveoli that share a common opening (Figure 23-8a–c). The alveolar walls consist of two principal types of epithelial cells: *squamous pulmonary epithelial cells* and *septal cells.* The squamous pulmonary epithelial cells are the larger of the two types of cells and form a continuous lining of the alveolar wall, except for occasional septal cells. Septal cells are much smaller, cuboidal in shape, and are dispersed among the squamous pulmonary epithelial cells. Septal cells produce a phospholipid substance called *surfactant* (sur-FAK-tant), which lowers surface tension. Also found within the alveolar wall are free *alveolar macrophages (dust cells).* They are highly phagocytic and serve to remove dust particles or other debris that gain entrance to alveolar spaces. Deep to the layer of squamous pulmonary epithelial cells is an elastic basement membrane. Over the alveoli, the arteriole and venule disperse into a capillary network (Figure 23-8a). The blood capillaries consist of a single layer of endothelial cells and a basement membrane.

Alveolar-Capillary (Respiratory) Membrane

The exchange of respiratory gases between the lungs and blood takes place by diffusion across the alveoli and capillary walls. This membrane, through which the respiratory gases move, is collectively known as the **alveolar-capillary (respiratory) membrane** (Figure 23-8e). It consists of:

1. A layer of squamous pulmonary epithelial cells with septal cells and free alveolar macrophages that constitute the alveolar (epithelial) wall.
2. An epithelial basement membrane underneath the alveolar wall.
3. A capillary basement membrane that is often fused to the epithelial basement membrane.
4. The endothelial cells of the capillary.

Despite the large number of layers, the alveolar-capillary membrane averages only 0.5 μm in thickness. This is of considerable importance to the efficient diffusion of respiratory gases. Moreover, it has been estimated that the lungs contain 300 million alveoli, providing an immense surface area of 70 m² (753 ft²) for the exchange of gases.

Blood Supply

The arterial supply of the lungs is derived from the pulmonary trunk. It divides into a left pulmonary artery which enters the left lung and a right pulmonary artery which enters the right lung. The venous return of the oxygenated blood is by way of the pulmonary veins, typically two in number on each side—the right and left superior and inferior pulmonary veins. All four veins drain into the left atrium (see Figure 21-23).

CLINICAL APPLICATION

Pulmonary embolism is a common cause of sickness and death. An embolus may arise in a distant site, such as a leg vein, and pass through the right side of the heart to a lung via a pulmonary artery. The immediate effect of a pulmonary embolism is complete or partial obstruction of the pulmonary arterial blood flow to the lung resulting in dysfunction of the affected lung tissue. A large embolus can produce death in a few minutes.

RESPIRATION

The principal purpose of **respiration** is to supply the cells of the body with oxygen and remove the carbon dioxide produced by cellular activities. The three basic processes of respiration are pulmonary ventilation, external respiration, and internal respiration.

PULMONARY VENTILATION

Pulmonary ventilation (breathing) is the process by which gases are exchanged between the atmosphere and lung alveoli. Air flows between the atmosphere and lungs for the same reason that blood flows through the body—a pressure gradient exists. We breathe in when the pressure inside the lungs is less than the air pressure in the atmo-

sphere. We breathe out when the pressure inside the lungs is greater than the pressure in the atmosphere. Let us examine the mechanics of pulmonary ventilation by first looking at inspiration.

Inspiration

Breathing in is called **inspiration** or **inhalation.** Just before each inspiration, the air pressure inside the lungs equals the pressure of the atmosphere, which is about 760 mm Hg, or 1 atmosphere (atm), at sea level. For air to flow into the lungs, the pressure inside the lungs must become lower than the pressure in the atmosphere. This condition is achieved by increasing the volume of the lungs.

The pressure of a gas in a closed container is inversely proportional to the volume of the container. If the size of a closed container is increased, the pressure of the air inside the container decreases. If the size of the container is decreased, then the pressure inside it increases. This is referred to as **Boyle's law** and may be demonstrated as follows. Suppose we place a gas in a cylinder that has a movable piston and a pressure gauge, and the initial pressure is 1 atm (Figure 23-9). This pressure is created by the gas molecules striking the wall of the container. If the piston is pushed down, the gas is concentrated in a smaller volume. This means that the same number of gas molecules are striking less wall space. The gauge shows that the pressure doubles as the gas is compressed to half its volume. In other words, the same number of molecules in half the space produces twice the pressure. Conversely, if the piston is raised to increase the volume, the pressure decreases. Thus, the volume of a gas varies inversely with pressure (assuming that temperature is constant). Boyle's law applies to the operation of a bicycle pump and the blowing up of a balloon. Differences in pressure force air into our lungs when we inhale and force the air out when we exhale.

In order for inspiration to occur, the lungs must be expanded. This increases lung volume and thus decreases the pressure in the lungs. The first step toward increasing lung volume involves contraction of the principal inspiratory muscles—the diaphragm and external intercostals

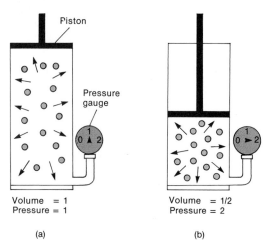

FIGURE 23-9 Boyle's law. The volume of gas varies inversely with the pressure. If the volume is decreased to ¼, what happens to the pressure?

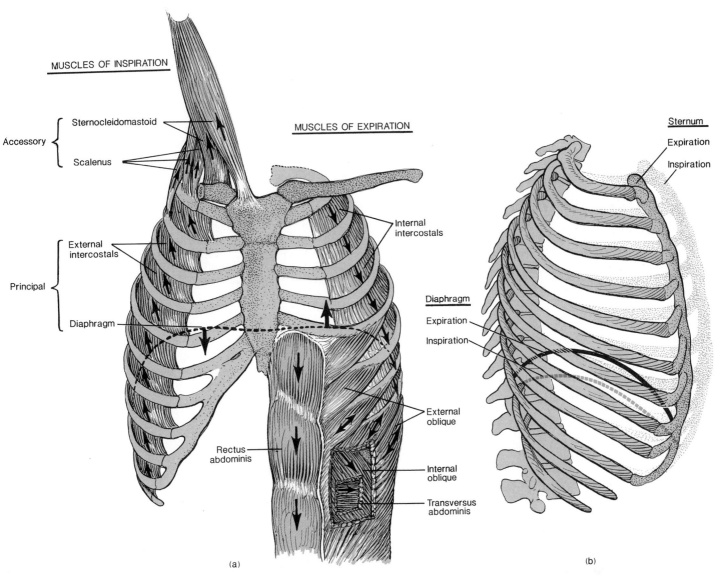

FIGURE 23-10 Pulmonary ventilation: muscles of inspiration and expiration. (a) Inspiratory muscles and their actions (left) and expiratory muscles and their actions (right). (b) Changes in size of thoracic cavity during inspiration (blue) and expiration (black).

(Figure 23-10; see also Figure 11-11). The diaphragm is a sheet of skeletal muscle that forms the floor of the thoracic cavity. Contraction of the diaphragm causes it to flatten, lowering its dome. This increases the vertical diameter of the thoracic cavity and accounts for the movement of more than two-thirds of the air that enters the lungs during inspiration. At the same time the diaphragm contracts, the external intercostals contract. As a result, the ribs are pulled upward and the sternum is pushed forward. This increases the anterior-posterior diameter of the thoracic cavity.

The term applied to normal quiet breathing is **eupnea** (yoop-NĒ-a; *eu* = normal). Eupnea involves shallow, deep, or combined shallow and deep breathing. Shallow (chest) breathing is called **costal breathing.** It consists of an upward and outward movement of the chest as a result of contraction of the external intercostal muscles. Deep (abdominal) breathing is called **diaphragmatic breathing.** It consists of the outward movement of the abdomen as a result of the contraction and descent of

the diaphragm. During *forced inspiration,* accessory muscles of inspiration also participate in increasing the size of the thoracic cavity. Contraction of the sternocleidomastoids elevates the sternum and contraction of the scalenes elevates the superior ribs. Inspiration is referred to as an active process because it is initiated by muscle contraction.

During normal breathing, the pressure between the two pleural layers, called **intrapleural (intrathoracic) pressure,** is always subatmospheric. (It may become temporarily positive only during some modified respiratory movements such as coughing or straining during defecation.) Just before inspiration, it is about 756 mm Hg (Figure 23-11). The overall increase in the size of the thoracic cavity causes intrapleural pressure to fall below the pressure of the air inside the lungs, to about 754 mm Hg. Consequently, the walls of the lungs are sucked outward by the partial vacuum. Expansion of the lungs is further aided by movement of the pleura. The parietal and visceral pleurae are normally strongly attached to each other due

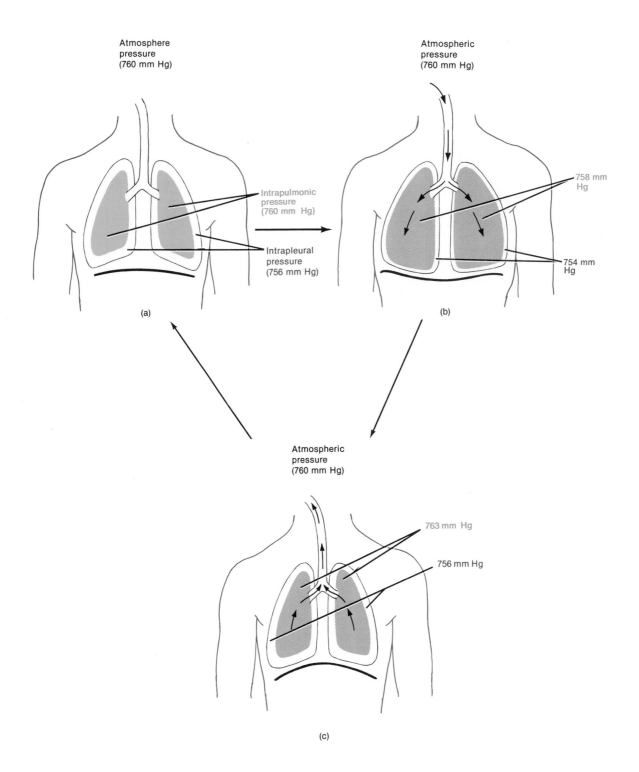

FIGURE 23-11 Pulmonary ventilation: pressure changes. (a) Lungs and pleural cavity just before inspiration. (b) Chest expanded and intrapleural pressure decreased; lungs pulled outward and intrapulmonic pressure decreased. (c) Chest relaxes, intrapleural pressure rises, and lungs snap inward. Intrapulmonic pressure raised, forcing air out until intrapulmonic pressure equals atmospheric pressure (a).

to surface tension created by their moist adjoining surfaces. As the thoracic cavity expands, the parietal pleura lining the cavity is pulled in all directions, and the visceral pleura is pulled along with it.

When the volume of the lungs increases, the pressure inside the lungs, called the **intrapulmonic (intraalveolar)** **pressure,** drops from 760 to 758 mm Hg. A pressure gradient is thus established between the atmosphere and the alveoli. Air rushes from the atmosphere into the lungs, and an inspiration takes place. Air continues to move into the lungs until intrapulmonic pressure equals atmospheric pressure.

Expiration

Breathing out, called **expiration** or **exhalation,** is also achieved by a pressure gradient, but in this case the gradient is reversed so that the pressure in the lungs is greater than the pressure of the atmosphere. Normal expiration, unlike inspiration, is a passive process since no muscular contractions are involved. Expiration starts when the inspiratory muscles relax. As the external intercostals relax, the ribs move downward, and as the diaphragm relaxes, its dome moves upward. These movements decrease the vertical and anterior-posterior diameters of the thoracic cavity, and it returns to its resting size (see Figure 23-10).

Expiration becomes active during higher levels of ventilation and when air movement out of the lungs is impeded. During these times, contraction of the internal intercostals moves the ribs downward and contraction of the abdominal muscles moves the inferior ribs downward and compresses the abdominal viscera, thus forcing the diaphragm upward.

As intrapleural pressure returns to its preinspiration level (756 mm Hg), the walls of the lungs are no longer sucked out. The elastic basement membranes of the alveoli and elastic fibers in bronchioles and alveolar ducts recoil into their relaxed shape, and lung volume decreases. Intrapulmonic pressure increases to 763 mm Hg, and air moves from the area of higher pressure in the alveoli to the area of lower pressure in the atmosphere (see Figure 23-11).

It was noted earlier that intrapleural pressure is normally subatmospheric. The pleural cavities are sealed off from the outside environment and cannot equalize their pressure with that of the atmosphere. Nor can the diaphragm and rib cage move inward enough to bring the intrapleural pressure up to atmospheric pressure. Maintenance of a low intrapleural pressure is vital to the functioning of the lungs. The alveoli are so elastic that at the end of an expiration they attempt to recoil inward and collapse on themselves like the walls of a deflated balloon. A collapsed lung or portion of a lung is called **atelectasis** (at'-ē-LEK-ta-sis; *ateles* = incomplete, *ektasis* = dilation). Such a collapse, which would obstruct the movement of air, is prevented by the slightly lower pressure in the pleural cavities that keeps the alveoli slightly inflated.

Another factor preventing the collapse of alveoli is the presence of **surfactant,** the phospholipid produced by the septal cells of the alveolar walls. Surfactant decreases surface tension in the lungs. That is, it forms a thin lining on the alveoli and prevents them from sticking together following expiration. Thus, as alveoli become smaller, for example following expiration, the tendency of alveoli to collapse is minimized because the surface tension does not increase. As you will see later, a deficiency of surfactant results in a disorder called infant respiratory distress syndrome (RDS).

Compliance

Compliance refers to the ease with which the lungs and thoracic wall can be expanded. High compliance means that the lungs and thoracic wall expand easily; low compliance means that they resist expansion. Compliance is related to two principal factors: elasticity and surface tension. The presence of elastic fibers in lung tissue results in high compliance. If surface tension within lung tissue were high, the tissues would resist expansion, but surfactant lowers surface tension and thus increases compliance. Any condition that destroys lung tissue, causes it to become fibrotic or edematous, causes a deficiency in surfactant, or in any way impedes lung expansion or contraction decreases compliance.

Airway Resistance

The walls of the respiratory passageways, especially the bronchi and bronchioles, offer some resistance to the normal flow of air into the lungs. The muscular contraction of normal inspiration not only expands the thoracic cavity, but also helps to overcome resistance to airflow. Any condition that obstructs the air passageways would increase resistance and require more pressure to force air through. During a forced expiration, as in coughing, straining, or playing a wind instrument, intrapleural pressure may increase from its normally subatmospheric (negative) value to a positive one. This greatly increases airway resistance because it results in compression of the airways. This is an important consideration for people with chronic obstructive pulmonary disease in which airway resistance is high even during rest.

MODIFIED RESPIRATORY MOVEMENTS

Respirations also provide humans with methods for expressing emotions such as laughing, yawning, sighing, and sobbing. Moreover, respiratory air can be used to expel foreign matter from the upper air passages through actions such as sneezing and coughing. Some of the modified respiratory movements that express emotion or clear the air passageways are listed in Exhibit 23-1. All these movements are reflexes, but some of them also can be initiated voluntarily.

PULMONARY AIR VOLUMES AND CAPACITIES

In clinical practice, the word **respiration** means one inspiration plus one expiration. The healthy adult averages about 12 respirations a minute while at rest. During each respiration, the lungs exchange various amounts of air with the atmosphere. A lower than normal amount of exchange is usually a sign of pulmonary malfunction.

Spirometry

The apparatus commonly used to measure the amount of air exchanged during breathing and rate of ventilation is a **spirometer** or **respirometer** (Figure 23-12). A spirometer consists of a weighted drum inverted over a chamber of water. The drum usually contains oxygen or air. A tube connects the air-filled chamber with the subject's mouth. During inspiration, air is removed from the chamber, the drum sinks, and an upward deflection is recorded by a stylus on graph paper on the rotating drum. During

EXHIBIT 23-1
MODIFIED RESPIRATORY MOVEMENTS

MOVEMENT	COMMENT
Coughing	A long-drawn and deep inspiration followed by a complete closure of the glottis, which results in a strong expiration that suddenly pushes the glottis open and sends a blast of air through the upper respiratory passages. Stimulus for this reflex act may be a foreign body lodged in the larynx, trachea, or epiglottis.
Sneezing	Spasmodic contraction of muscles of expiration which forcefully expels air through the nose and mouth. Stimulus may be an irritation of the nasal mucosa.
Sighing	A long-drawn and deep inspiration immediately followed by a shorter but forceful expiration.
Yawning	A deep inspiration through the widely opened mouth producing an exaggerated depression of the lower jaw. It may be stimulated by drowsiness or fatigue, but precise stimulus-receptor cause is unknown.
Sobbing	A series of convulsive inspirations followed by a single prolonged expiration. The glottis closes earlier than normal after each inspiration so only a little air enters the lungs with each inspiration.
Crying	An inspiration followed by many short convulsive expirations, during which the glottis remains open and the vocal cords vibrate; accompanied by characteristic facial expressions.
Laughing	The same basic movements as crying, but the rhythm of the movements and the facial expressions usually differ from those of crying. Laughing and crying are sometimes indistinguishable.
Hiccuping	Spasmodic contraction of the diaphragm followed by a spasmodic closure of the glottis to produce a sharp inspiratory sound. Stimulus is usually irritation of the sensory nerve endings of the digestive tract.

expiration, air is added, the drum rises, and a downward deflection is recorded. The record is called a **spirogram** (Figure 23-13).

Pulmonary Volumes

During the process of normal quiet breathing, about 500 ml of air moves into the respiratory passageways with each inspiration. The same amount moves out with each expiration. This volume of air inspired (or expired) is called **tidal volume** (Figure 23-13). Only about 350 ml of the tidal volume actually reaches the alveoli. The other 150 ml remains in air spaces of the nose, pharynx, larynx, trachea, and bronchi and is known as **dead air volume.** The total air taken in during 1 minute is called the **minute volume of respiration.** It is calculated by multiplying the tidal volume by the normal breathing rate per minute. An average volume would be 500 ml times 12 respirations per minute, or 6,000 ml/min.

By taking a very deep breath, we can inspire a good deal more than 500 ml. This excess inhaled air, called the **inspiratory reserve volume,** averages 3,100 ml above the 500 ml of tidal volume. Thus the respiratory system can pull in as much as 3,600 ml of air.

If we inhale normally and then exhale as forcibly as possible, we should be able to push out 1,200 ml of air in addition to the 500 ml tidal volume. This extra 1,200 ml is called the **expiratory reserve volume.**

Even after the expiratory reserve volume is expelled, a good deal of air remains in the lungs because the lower intrapleural pressure keeps the alveoli slightly inflated and some air also remains in the noncollapsible air passageways. This air, the **residual volume,** amounts to about 1,200 ml.

Opening the thoracic cavity allows the intrapleural pressure to equal the atmospheric pressure, forcing out some of the residual volume. The air remaining is called the **minimal volume.** Minimal volume provides a medical and legal tool for determining whether a baby was born dead or died after birth. The presence of minimal volume can be demonstrated by placing a piece of lung in water and watching it float. Fetal lungs contain no air, and so the lung of a stillborn will not float in water.

Pulmonary Capacities

Lung capacity can be calculated by combining various lung volumes. **Inspiratory capacity,** the total inspiratory ability of the lungs, is the sum of tidal volume plus inspiratory reserve volume (3,600 ml). **Functional residual capacity** is the sum of residual volume plus expiratory reserve volume (2,400 ml). **Vital capacity** is the sum of inspiratory reserve volume, tidal volume, and expiratory reserve volume (4,800 ml). Finally, **total lung capacity** is the sum of all volumes (6,000 ml).

CLINICAL APPLICATION

The **measurement of respiratory volumes and capacities** is an essential tool for determining how well the lungs are functioning. Spirometry is indicated for an individual exhibiting labored breathing and is used in the diagnosis of respiratory disorders such as bronchial asthma and emphysema. For instance, during the early stages of emphysema, many of the alveoli lose their elasticity. During expiration they fail to snap inward, and consequently they fail to force out a normal amount of air. Thus the residual volume is increased at the expense of the expiratory reserve volume. Pulmonary infections can cause inflammation and an accumulation of fluid in the air spaces of the lungs (pulmonary edema). The fluid reduces the amount of space available for air and consequently decreases the vital capacity.

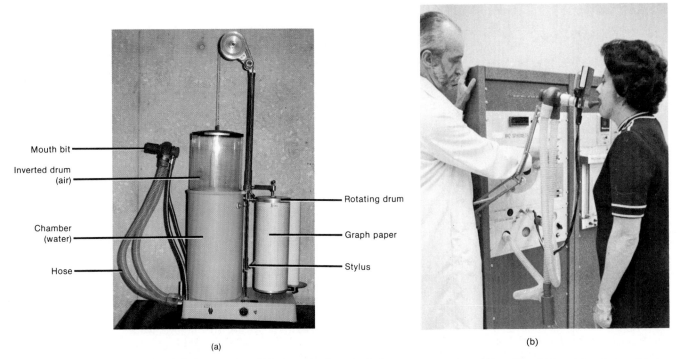

Mouth bit

Inverted drum
(air)

Chamber
(water)

Hose

Rotating drum

Graph paper

Stylus

(a)

(b)

FIGURE 23-12 Spirometers. (a) Collins spirometer. This type of spirometer is the one commonly used in college biology laboratories. (Photographed in a biology laboratory at Bergen Community College by Gerard J, Tortora.) (b) Ohio 842 spirometer. This instrument is a highly sophisticated spirometer that utilizes a computerized mechanism for recording results. (Courtesy of Lenny Patti.)

EXCHANGE OF RESPIRATORY GASES

As soon as the lungs fill with air, oxygen diffuses from the alveoli into the blood, through the interstitial fluid, and finally into the cells. Carbon dioxide diffuses in the opposite direction—from the cells, through interstitial fluid to the blood, and to the alveoli. To understand how respiratory gases are exchanged in the body, you need to know a few gas laws.

Charles' Law

According to **Charles' law,** the volume of a gas is directly proportional to its absolute temperature, assuming that the pressure remains constant. Recall the cylinder we used to demonstrate Boyle's law. Suppose now that the gas in the cylinder exerts an initial pressure of 1 atm when the piston is halfway down (Figure 23-14). When the gas is heated, the gas molecules move faster and the

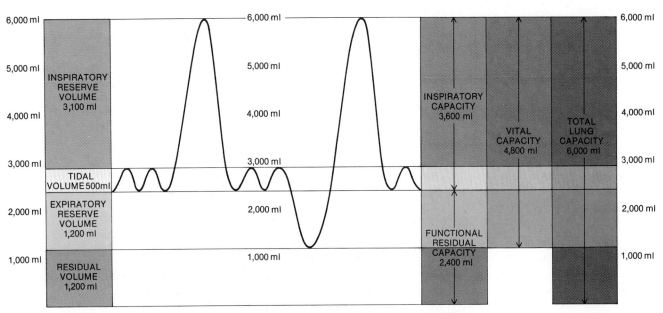

FIGURE 23-13 Spirogram of pulmonary volumes and capacities.

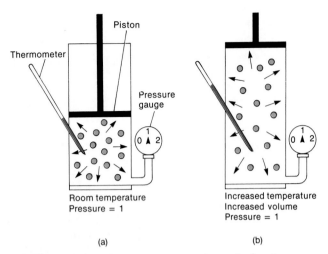

FIGURE 23-14 Charles' law. The volume of a gas is directly proportional to its absolute temperature, assuming that the pressure is constant.

number of collisions within the cylinder increases. Assuming that the piston moves freely (no external pressure is applied), the force of the molecules hitting it moves it upward. As the gas expands, the movement of the piston provides a measure of the increase in volume. As the space in the cylinder increases, the molecules have farther to travel, so the number of collisions decreases as the space increases. The pressure of 1 atm is maintained, and the volume increases in direct proportion to the temperature increase.

Dalton's Law

According to **Dalton's law,** each gas in a mixture of gases exerts its own pressure as if all the other gases were not present. This *partial pressure* is denoted as *p*. The total pressure of the mixture is calculated by simply adding all the partial pressures. Atmospheric air is a mixture of several gases—oxygen, carbon dioxide, nitrogen, water vapor, and a number of other gases that appear in such small quantities that we will ignore them. Atmospheric pressure is the sum of the pressures of all these gases:

Atmospheric pressure = $pO_2 + pCO_2 + pN_2 + pH_2O$
   (760 mm Hg)

We can determine the partial pressure exerted by each component in the mixture by multiplying the percentage of the mixture the particular gas constitutes by the total pressure of the mixture. For example, to find the partial pressure of oxygen in the atmosphere, multiply the percentage of atmospheric air composed of oxygen (21 percent) by the total atmospheric pressure (760 mm Hg):

Atmospheric pO_2 = 21% × 760 mm Hg
   = 159.60 or 160 mm Hg

Since the percentage of CO_2 in the atmosphere is 0.04,

Atmospheric pCO_2 = 0.04% × 760 mm Hg
   = 0.3040 or 0.3 mm Hg

The partial pressures of the respiratory gases and nitrogen in the atmosphere, alveoli, blood, and tissue cells are shown in Exhibit 23-2. These partial pressures are important in determining the movement of oxygen and carbon dioxide between the atmosphere and lungs, the lungs and blood, and the blood and body cells. When a mixture of gases diffuses across a semipermeable membrane, each gas diffuses from the area where its partial pressure is greater to the area where its partial pressure is less. Every gas is on its own and behaves as if the other gases in the mixture did not exist.

The amounts of respiratory gases vary in inspired (atmospheric), alveolar, and expired air (Exhibit 23-3). Inspired (atmospheric) air contains about 21 percent oxygen and 0.04 percent carbon dioxide. Expired air contains less oxygen (about 16 percent) and more carbon dioxide (about 4.5 percent) than inspired air. Compared to alveolar air, expired air contains more oxygen (about 16 percent versus 14 percent) and less carbon dioxide (about 4.5 percent versus 5.5 percent) because some of the expired air is dead air volume that has not participated in gaseous exchange. Expired air is actually a mixture of inspired and alveolar air.

Henry's Law

You have probably noticed that a bottle of soda makes a hissing sound when the top is removed, and bubbles rise to the surface for some time afterward. The gas dissolved in carbonated beverages is carbon dioxide. The ability of a gas to stay in solution depends on its partial pressure and solubility coefficient, that is, its physical or chemical attraction for water. The solubility coefficient of carbon dioxide is high (0.57), that of oxygen is lower (0.024), and that of nitrogen is still lower (0.012). The higher the partial pressure of a gas over a liquid and

EXHIBIT 23-2

PARTIAL PRESSURES (IN mm Hg) OF RESPIRATORY GASES AND NITROGEN IN ATMOSPHERIC AIR, ALVEOLAR AIR, BLOOD, AND TISSUE CELLS

	ATMOSPHERIC AIR (SEA LEVEL)	ALVEOLAR AIR	DEOXYGENATED BLOOD	OXYGENATED BLOOD	TISSUE CELLS
pO_2	160	105	40	105	40
pCO_2	0.3	40	45	40	45
pN_2	597	569	569	569	569

EXHIBIT 23-3

APPROXIMATE PERCENTAGE OF OXYGEN AND CARBON DIOXIDE IN INSPIRED AIR, ALVEOLAR AIR, AND EXPIRED AIR

	INSPIRED AIR	ALVEOLAR AIR	EXPIRED AIR
Oxygen	21	14	16
Carbon dioxide	0.04	5.50	4.5

the higher the solubility coefficient, the more gas will stay in solution. Since the soda is bottled under pressure and capped, the CO_2 remains dissolved as long as the bottle is unopened. Once you remove the cap, the pressure is released and the gas begins to bubble out. This phenomenon is explained by **Henry's law:** The quantity of a gas that will dissolve in a liquid is proportional to the partial pressure of the gas and its solubility coefficient, when the temperature remains constant.

Henry's law explains two conditions resulting from changes in the solubility of nitrogen in body fluids. Even though the air we breathe contains about 79 percent nitrogen, this gas has no known effect on bodily functions since very little of it dissolves in blood plasma because of its low solubility coefficient at sea level pressure. But, when a deep-sea diver, scuba **(self-contained underwater breathing apparatus)** diver, or caisson worker (person who builds tunnels under rivers) breathes air under high pressure, the nitrogen in the mixture can affect the body. Partial pressure is a function of total pressure, and therefore the partial pressure of all the components of a mixture increases as the total increases. Since the partial pressure of nitrogen is higher in a mixture of compressed air than in air at sea level pressure, a considerable amount of nitrogen goes into solution in plasma and interstitial fluid. Excessive amounts of dissolved nitrogen may produce giddiness and other symptoms similar to alcohol intoxication. The condition is called **nitrogen narcosis** or "rapture of the depths." The greater the depth, the more severe the condition.

If a diver is brought to the surface slowly, the dissolved nitrogen can be eliminated through the lungs. However, if a diver ascends too rapidly, the nitrogen comes out of solution too quickly to be eliminated by respiration. Instead it forms gas bubbles in the tissues which result in **decompression sickness (caisson disease** or **bends).** The effects of decompression sickness typically result from bubbles in nervous tissue and can be mild or severe, depending on the amount of bubbles formed. Symptoms include joint pain, especially in the arms and legs, dizziness, shortness of breath, extreme fatigue, paralysis, and unconsciousness. Decompression sickness can be prevented by a slow ascent or by the use of a special decompression tank within five minutes after arriving at the surface. The use of helium-oxygen mixtures instead of air containing nitrogen may reduce the dangers of decompression sickness since helium is only about 40 percent as soluble as nitrogen in blood.

CLINICAL APPLICATION

A major clinical application of Henry's law is **hyperbaric** (*hyper* = over; *baros* = pressure) **oxygenation (HBO).** Using pressure to cause more oxygen to dissolve in the blood is an effective technique in treating patients infected by anaerobic bacteria, such as those that cause tetanus and gangrene. (Anaerobic bacteria cannot live in the presence of free oxygen.) A person undergoing hyperbaric oxygenation is placed in a hyperbaric chamber, which contains oxygen at a pressure of 3 to 4 atm (2,280 to 3,040 mm Hg). The body tissues pick up the oxygen, and the bacteria are killed. Hyperbaric chambers may also be used for treating certain heart disorders, carbon monoxide poisoning, smoke inhalation, drowning, asphyxia, vascular insufficiencies, and burns.

EXTERNAL RESPIRATION

External respiration is the exchange of oxygen and carbon dioxide between the alveoli of the lungs and pulmonary blood capillaries (Figure 23-15a). It results in the conversion of *deoxygenated blood* (more CO_2 than O_2) coming from the heart to *oxygenated blood* (more O_2 than CO_2) returning to the heart. During inspiration, atmospheric air containing oxygen enters the alveoli. Deoxygenated blood is pumped from the right ventricle through the pulmonary arteries into the pulmonary capillaries overlying the alveoli. The pO_2 of alveolar air is 105 mm Hg. The pO_2 of the deoxygenated blood entering the pulmonary capillaries is only 40 mm Hg. As a result of this difference in pO_2, oxygen diffuses from the alveoli into the deoxygenated blood until equilibrium is reached and the pO_2 of the now oxygenated blood is 105 mm Hg. While oxygen diffuses from the alveoli into deoxygenated blood, carbon dioxide diffuses in the opposite direction. On arriving at the lungs, the pCO_2 of pulmonary deoxygenated blood is 45 mm Hg, while that of the alveoli is 40 mm Hg. Because of this difference in pCO_2, carbon dioxide diffuses from pulmonary deoxygenated blood into the alveoli until the pCO_2 of the blood decreases to 40 mm Hg, the pCO_2 of pulmonary oxygenated blood. Thus the pO_2 and pCO_2 of oxygenated blood leaving the lungs are the same as in alveolar air. The carbon dioxide that diffuses into the alveoli is eliminated from the lungs during expiration.

External respiration is aided by several anatomic adaptations. The total thickness of the alveolar-capillary (respiratory) membranes is only 0.5 μm. Thicker membranes would inhibit diffusion. The surface area over which diffusion may occur is large. The total surface area of the alveoli is about 70 m² (753 ft²), many more times the total surface area of the skin. Lying over the alveoli are countless capillaries—so many that 900 ml of blood is able to participate in gas exchange at any time. Finally, the capillaries are so narrow that the red blood cells must flow through them in single file. This feature gives each red blood cell maximum exposure to the available oxygen.

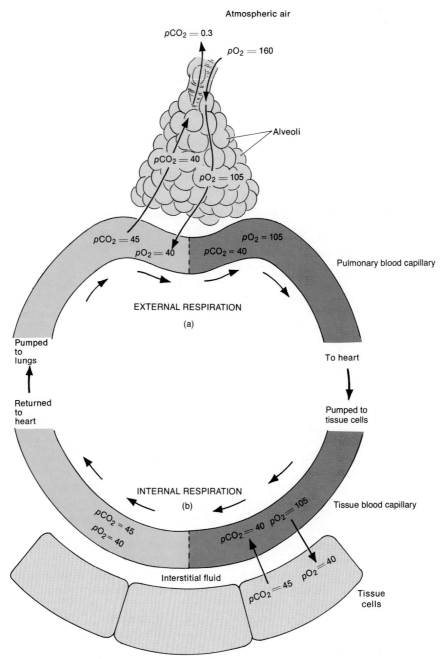

Atmospheric air

$pCO_2 = 0.3$

$pO_2 = 160$

Alveoli

$pCO_2 = 40$

$pO_2 = 105$

$pCO_2 = 45$ $pO_2 = 105$

$pO_2 = 40$ $pCO_2 = 40$ Pulmonary blood capillary

EXTERNAL RESPIRATION

(a)

Pumped
to
lungs   To heart

Returned
to
heart   Pumped to
                                                         tissue cells

INTERNAL RESPIRATION

(b)   Tissue blood capillary

$pCO_2 = 45$ $pCO_2 = 40$ $pO_2 = 105$

$pO_2 = 40$

Interstitial fluid $pCO_2 = 45$ $pO_2 = 40$ Tissue
                                                                      cells

FIGURE 23-15 Partial pressures involved in respiration. (a) External. (b) Internal. All pressures
are in mm Hg.

CLINICAL APPLICATION

The efficiency of external respiration depends on several factors. One of the most important is altitude. As long as alveolar pO_2 is higher than venous blood pO_2, oxygen diffuses from the alveoli into the blood. As a person ascends in altitude, the atmospheric pO_2 decreases, the alveolar pO_2 decreases correspondingly, and less oxygen diffuses into the blood. For example, at sea level, pO_2 is 160 mm Hg. At 10,000 ft, it decreases to 110 mm Hg, at 20,000 ft to 73 mm Hg, and at 50,000 ft to 18 mm Hg. The common symptoms of **altitude sickness**—shortness of breath, nausea, dizziness—are attributable to the low concentrations of oxygen in the blood.

Another factor that affects external respiration is the total surface area available for O_2–CO_2 exchange. Any **pulmonary disorder** that decreases the functional surface area formed by the alveolar-capillary membranes decreases the efficiency of external respiration.

A third factor that influences external respiration is the **minute volume of respiration.** Certain drugs, such as morphine, slow down the respiration rate, thereby decreasing the amount of oxygen and carbon dioxide that can be exchanged between the alveoli and the blood.

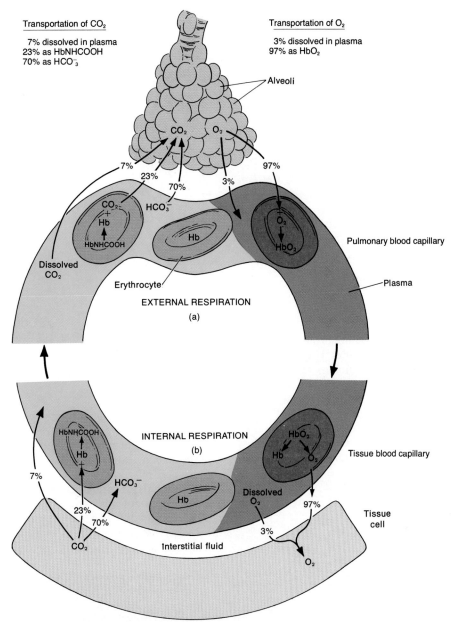

Transportation of CO_2

7% dissolved in plasma
23% as HbNHCOOH
70% as HCO_3^-

Transportation of O_2

3% dissolved in plasma
97% as HbO_2

Alveoli

CO_2 O_2

7%   23%   70%   3%   97%

CO_2^-
$+$
Hb

HbNHCOOH

HCO_3^-

O_2
$\downarrow$
HbO_2

Pulmonary blood capillary

Dissolved
CO_2

Hb

Plasma

Erythrocyte

EXTERNAL RESPIRATION
(a)

HbNHCOOH

Hb
$+$

INTERNAL RESPIRATION
(b)

HbO_2

Hb O_2

Tissue blood capillary

7%

HCO_3^-

Hb

23%   70%

CO_2

Dissolved
O_2

97%

Tissue
cell

Interstitial fluid

3%

O_2

FIGURE 23-16 Transportation of respiratory gases in respiration. (a) External. (b) Internal. The role of red blood cells in the transportation of carbon dioxide is detailed in Figure 23-20.

INTERNAL RESPIRATION

As soon as external respiration is completed, oxygenated blood leaves the lungs through the pulmonary veins and returns to the heart. From here it is pumped from the left ventricle into the aorta and through the systemic arteries to tissue cells. The exchange of oxygen and carbon dioxide between tissue blood capillaries and tissue cells is called **internal respiration** (Figure 23-15b). It results in the conversion of oxygenated blood into deoxygenated blood. Oxygenated blood in tissue capillaries has a pO_2 of 105 mm Hg, whereas tissue cells have a pO_2 of 40 mm Hg. Because of this difference in pO_2, oxygen diffuses from the oxygenated blood through interstitial fluid and into tissue cells until the pO_2 in blood decreases to 40 mm Hg, the pO_2 of tissue capillary deoxygenated blood. While oxygen diffuses from the tissue blood capillaries

into tissue cells, carbon dioxide diffuses in the opposite direction. The pCO_2 of tissue cells is 45 mm Hg, while that of tissue capillary oxygenated blood is 40 mm Hg. As a result, carbon dioxide diffuses from tissue cells through interstitial fluid into the oxygenated blood until the pCO_2 in the blood increases to 45 mm Hg, the pCO_2 of tissue capillary deoxygenated blood. The deoxygenated blood now returns to the heart. From here it is pumped to the lungs for another cycle of external respiration.

TRANSPORT OF RESPIRATORY GASES

The transportation of respiratory gases between the lungs and body tissues is a function of the blood. When oxygen and carbon dioxide enter the blood, certain physical and chemical changes occur that aid in gas transport and exchange.

Oxygen

Under normal resting conditions, each 100 ml of oxygenated blood contains 20 ml of oxygen. Oxygen does not dissolve easily in water and, therefore, very little oxygen is carried in the dissolved state in the water in blood plasma. In fact, 100 ml of oxygenated blood contains only about 3 percent oxygen dissolved in plasma. The remainder of the oxygen, about 97 percent, is carried in chemical combination with hemoglobin in red blood cells (Figure 23-16).

Hemoglobin consists of a protein portion called globin and a pigment portion called heme. The heme portion contains four atoms of iron, each capable of combining with a molecule of oxygen (see Figure 19-3). Oxygen and hemoglobin combine in an easily reversible reaction to form **oxyhemoglobin.**

$$Hb + O_2 \rightleftharpoons HbO_2$$

| Reduced hemoglobin (uncombined hemoglobin) | Oxygen | Oxyhemoglobin (combined hemoglobin) |

● *Hemoglobin and pO_2*   The most important factor that determines how much oxygen combines with hemoglobin is pO_2. When Hb (reduced hemoglobin) is completely converted to HbO_2, it is referred to as **fully saturated.** When hemoglobin consists of a mixture of Hb and HbO_2, it is **partially saturated.** The **percent saturation of hemoglobin** is the percent of HbO_2 in total hemoglobin. The degree of saturation of hemoglobin with oxygen is illustrated in Figure 23-17, the oxygen-hemoglobin dissociation curve. Note that when the pO_2 is high, hemoglobin binds with large amounts of oxygen and is almost fully saturated. When pO_2 is low, hemoglobin is only partially

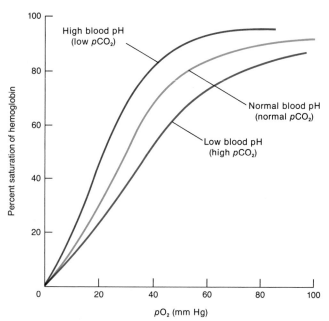

FIGURE 23-18 Oxygen-hemoglobin dissociation curve at normal body temperature showing the relationship between hemoglobin saturation and pH (pCO_2). As pH increases (decreasing pCO_2), more oxygen combines with hemoglobin.

saturated and oxygen is released from hemoglobin. In other words, the greater the pO_2, the more oxygen will combine with hemoglobin, until the available hemoglobin molecules are saturated. Therefore, in pulmonary capillaries, where pO_2 is high, a lot of oxygen binds with hemoglobin, but in tissue capillaries, where the pO_2 is lower, hemoglobin does not hold as much oxygen and the oxygen is released for diffusion into the tissue cells. The binding of oxygen to hemoglobin is an example of a positive feedback system.

● *Hemoglobin and pH*   The amount of oxygen released from hemoglobin is determined by several factors in addition to pO_2. For example, in an acid environment, oxygen splits more readily from hemoglobin (Figure 23-18). This is referred to as the **Bohr effect** and is based on the belief that when hydrogen ions bind to hemoglobin they alter the structure of hemoglobin and thereby decrease its oxygen-carrying capacity.

Low blood pH (acidic condition) results from the presence of lactic acid, a reaction product of muscle contraction, and high pCO_2. As carbon dioxide is taken up by the blood, much of it is temporarily converted to carbonic acid. This conversion is catalyzed by an enzyme in red blood cells called *carbonic anhydrase.*

$$CO_2 + H_2O \underset{\text{carbonic anhydrase}}{\rightleftharpoons} H_2CO_3 \rightleftharpoons H^+ + HCO_3^-$$

| Carbon dioxide | Water | Carbonic acid | Hydrogen ion | Bicarbonate ion |

The carbonic acid thus formed in red blood cells dissociates into hydrogen ions and bicarbonate ions. As the hydrogen ion concentration increases, pH decreases. Thus, an increased pCO_2 produces a more acid environment that helps to split oxygen from hemoglobin.

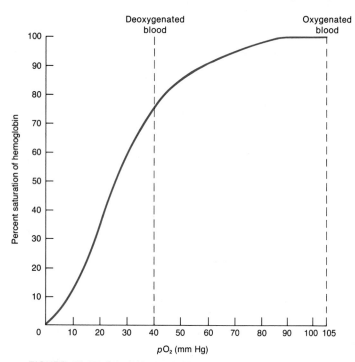

FIGURE 23-17 Oxygen-hemoglobin dissociation curve at normal body temperature showing the relationship between hemoglobin saturation and pO_2. As pO_2 increases, more oxygen combines with hemoglobin.

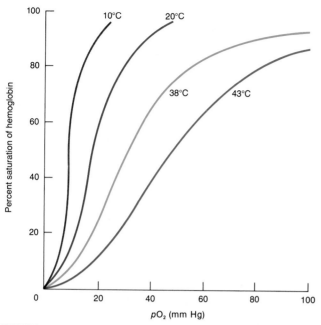

FIGURE 23-19 Oxygen-hemoglobin dissociation curve showing the relationship between hemoglobin saturation and temperature. As temperature increases, less oxygen combines with hemoglobin.

● *Hemoglobin and Temperature*   Within limits, as temperature increases, so does the amount of oxygen released from hemoglobin (Figure 23-19). Heat energy is a by-product of the metabolic reactions of all cells, and contracting muscle cells release an especially large amount of heat. Splitting the oxyhemoglobin molecule is another example of how homeostatic mechanisms adjust body activities to cellular needs. Active cells require more oxygen, and active cells liberate more acid and heat. The acid and heat, in turn, stimulate the oxyhemoglobin to release its oxygen.

● *Hemoglobin and DPG*   A final factor that helps to release oxygen from hemoglobin is a substance found in red blood cells called **2,3-diphosphoglycerate,** or simply **DPG.** It is an intermediate compound formed in red blood cells during glycolysis (Chapter 25). It has the ability to combine reversibly with hemoglobin and thus alter its structure to release oxygen. The production of DPG is greatest when there is decreased oxygen delivery to tissues. Thus, DPG enhances oxygen delivery to tissues and helps to maintain the release of oxygen by hemoglobin.

CLINICAL APPLICATION

Carbon monoxide (CO) is a colorless and odorless gas found in exhaust fumes from automobiles and in tobacco smoke. It is a by-product of burning carbon-containing materials such as coal and wood. One of its interesting features is that it combines with hemoglobin very much as oxygen does, except that the combination of carbon monoxide and hemoglobin is over 200 times as tenacious as the combination of oxygen and hemoglobin. In addition, in concentrations as small as 0.1 percent ($pCO = 0.5$ mm Hg), carbon monoxide will combine with half the hemoglobin molecules. Thus, the oxygen-carrying capacity of the blood is reduced by one half. Increased levels of carbon monoxide lead to hypoxia, and the result is **carbon monoxide poisoning.** The condition may be treated by administering pure oxygen ($pO_2 = 600$ mm Hg), which slowly replaces the carbon monoxide combined with the hemoglobin.

Carbon Dioxide

Under normal resting conditions, each 100 ml of deoxygenated blood contains 4 ml of carbon dioxide. The CO_2 is carried by the blood in several forms (see Figure 23-16). The smallest percentage, about 7 percent, is dissolved in plasma. Upon reaching the lungs, it diffuses into the alveoli. A somewhat higher percentage, about 23 percent, combines with the globin portion of hemoglobin to form **carbaminohemoglobin.**

$$\underset{\text{Hemoglobin}}{\text{Hb}} + \underset{\text{Carbon dioxide}}{\text{CO}_2} \rightleftharpoons \underset{\text{Carbaminohemoglobin}}{\text{HbNHCOOH}}$$

The formation of carbaminohemoglobin is greatly influenced by pCO_2. For example, in tissue capillaries pCO_2 is relatively high, and this encourages the formation of carbaminohemoglobin. But in pulmonary capillaries, pCO_2 is relatively low, and the CO_2 readily splits apart from globin and enters the alveoli by diffusion.

The greatest percentage of CO_2, about 70 percent, is transported in plasma as bicarbonate ions. The reaction that brings about this method of transportation of CO_2 is the same one noted earlier:

$$\underset{\substack{\text{Carbon} \\ \text{dioxide}}}{\text{CO}_2} + \underset{\text{Water}}{\text{H}_2\text{O}} \underset{}{\overset{\substack{\text{carbonic} \\ \text{anhydrase}}}{\rightleftharpoons}} \underset{\substack{\text{Carbonic} \\ \text{acid}}}{\text{H}_2\text{CO}_3} \rightleftharpoons \underset{\substack{\text{Hydrogen} \\ \text{ion}}}{\text{H}^+} + \underset{\substack{\text{Bicarbonate} \\ \text{ion}}}{\text{HCO}_3^-}$$

As CO_2 diffuses into tissue capillaries and enters the red blood cells, it reacts with water, in the presence of carbonic anhydrase, to form carbonic acid. The carbonic acid dissociates into H^+ ions and HCO_3^- ions. The H^+ ions combine mainly with hemoglobin. The HCO_3^- ions leave the red blood cells and enter the plasma. In exchange, chloride ions (Cl^-) diffuse from plasma into the red blood cells. This exchange of negative ions maintains the ionic balance between plasma and red blood cells and is known as the **chloride shift** (Figure 23-20a). The Cl^- ions that enter red blood cells combine with potassium ions (K^+) to form the salt potassium chloride (KCl). The HCO_3^- ions that enter plasma from the red blood cells combine with sodium (Na^+), the principal positive ion in extracellular fluid, to form sodium bicarbonate ($NaHCO_3$). The net effect of all of these reactions is that CO_2 is carried from tissue cells as bicarbonate ions in plasma.

Deoxygenated blood returning to the lungs contains CO_2 dissolved in plasma, CO_2 combined with globin as carbaminohemoglobin, and CO_2 incorporated in bicarbonate ions. In the pulmonary capillaries, the events are

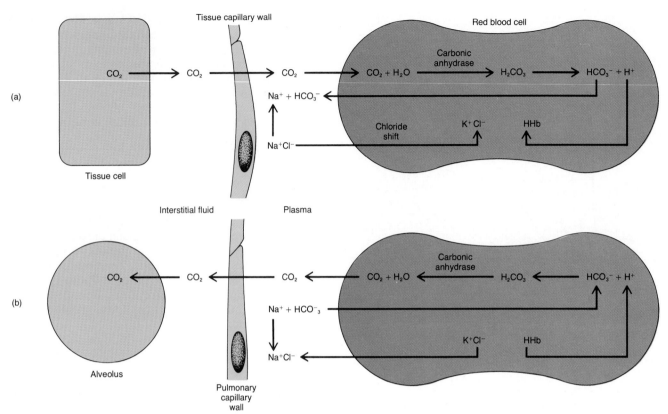

FIGURE 23-20 Transportation of carbon dioxide. (a) Between tissue cells and tissue capillaries. (b) Between pulmonary capillaries and alveoli.

reversed. The CO_2 dissolved in plasma diffuses into the alveoli. The CO_2 combined with globin splits from the globin and diffuses into the alveoli. The CO_2 carried as bicarbonate is released as follows (Figure 23-20b). As the hemoglobin in pulmonary blood picks up oxygen, the H^+ ions are released from hemoglobin. The Cl^- ions simultaneously split from K^+ ions, and HCO_3^- ions reenter the red blood cells after splitting from Na^+ ions. The H^+ and HCO_3^- ions recombine to form H_2CO_3, which, in the presence of carbonic anhydrase, splits into CO_2 and H_2O. The CO_2 leaves the red blood cells and diffuses into the alveoli. The direction of the carbonic acid reaction depends mostly on pCO_2. In tissue capillaries, where pCO_2 is high, bicarbonate is formed. In pulmonary capillaries, where pCO_2 is low, CO_2 and H_2O are formed.

CONTROL OF RESPIRATION

The basic rhythm of respiration is controlled by portions of the nervous system in the medulla and pons. In response to the demands of the body, this rhythm can be modified. We will first examine the principal mechanisms involved in the nervous control of the rhythm of respiration.

NERVOUS CONTROL

The size of the thorax is affected by the action of the respiratory muscles. These muscles contract and relax as a result of nerve impulses transmitted to them from centers in the brain. The area from which nerve impulses are sent to respiratory muscles is located bilaterally in the reticular formation of the brain stem; it is referred to as the **respiratory center.** The respiratory center consists of a widely dispersed group of neurons that is functionally divided into three areas: (1) the medullary rhythmicity area, in the medulla, (2) the pneumotaxic (noo-mō-TAK-sik) area, in the pons, and (3) the apneustic (ap-NOO-stik) area, in the pons (Figure 23-21).

Medullary Rhythmicity Area

The function of the **medullary rhythmicity area** is to control the basic rhythm of respiration. In the normal resting state, inspiration usually lasts for about 2 seconds and expiration for about 3 seconds. This is the basic rhythm of respiration. Within the medullary rhythmicity area are found both inspiratory and expiratory neurons that comprise inspiratory and expiratory areas, respectively. Let us first consider the proposed role of the inspiratory neurons in respiration.

The basic rhythm of respiration is determined by nerve impulses generated in the inspiratory area (Figure 23-22a). At the beginning of expiration, the inspiratory area is inactive, but, after 3 seconds, it suddenly and automatically becomes active. This activity seems to result from an intrinsic excitability of the inspiratory neurons. In fact, when all incoming nerve connections to the inspiratory area are cut or blocked, the area still rhythmically discharges impulses that result in inspiration. Nerve impulses

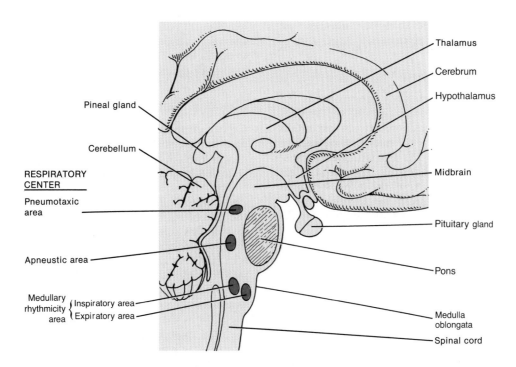

FIGURE 23-21 Approximate location of areas of the respiratory center.

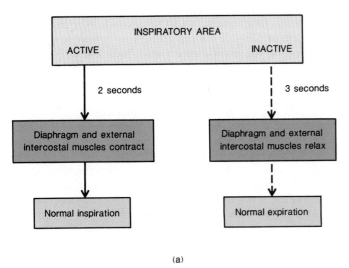

(a)

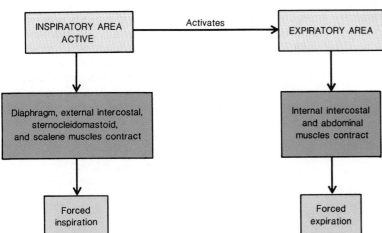

(b)

from the active inspiratory area last for about 2 seconds and travel to the muscles of inspiration. The impulses reach the diaphragm by the phrenic nerves and the external intercostal muscles by the intercostal nerves. When the impulses reach the inspiratory muscles, the muscles contract and inspiration occurs. At the end of 2 seconds, the inspiratory muscles become inactive again, and the cycle repeats itself over and over.

It is believed that the expiratory neurons remain inactive during most normal quiet respiration. During quiet respiration, inspiration is accomplished by active contraction of the inspiratory muscles and expiration results from passive recoil of the lungs and thoracic wall as the inspiratory muscles relax. However, during high levels of ventilation, it is believed that impulses from the inspiratory area activate the expiratory area (Figure 23-22b). Impulses discharged from the expiratory area cause contraction

FIGURE 23-22 Proposed role of the medullary rhythmicity area in controlling the basic rhythm of respiration. (a) During normal quiet breathing. (b) During high levels of respiration.

of the internal intercostals and abdominal muscles that decrease the size of the thoracic cavity.

Pneumotaxic Area

Although the medullary rhythmicity area controls the basic rhythm of respiration, other parts of the nervous system help coordinate the transition between inspiration and expiration. One of these is the **pneumotaxic area** in the upper pons (see Figure 23-21). It continuously transmits impulses to the inspiratory area. The major effect of these impulses is to help turn off the inspiratory area before the lungs become too full of air. In other words, the impulses limit inspiration and thus facilitate expiration.

Apneustic Area

Another part of the nervous system that coordinates the transition between inspiration and expiration is the **apneustic area** in the lower pons (see Figure 23-21). It sends impulses to the inspiratory area that activate it and prolong inspiration, thus inhibiting expiration. This occurs when the pneumotaxic area is inactive. When the pneumotaxic area is active, it overrides the apneustic area.

REGULATION OF RESPIRATORY CENTER ACTIVITY

Although the basic rhythm of respiration is set and coordinated by the respiratory center, the rhythm can be modified in response to the demands of the body by neural input to the center.

Cortical Influences

The respiratory center has connections with the cerebral cortex, which means we can voluntarily alter our pattern of breathing. We can even refuse to breathe at all for a short time. Voluntary control is protective because it enables us to prevent water or irritating gases from entering the lungs. The ability to stop breathing is limited by the buildup of CO_2 in the blood, however. When the pCO_2 increases to a certain level, the inspiratory area is stimulated, impulses are sent to inspiratory muscles, and breathing resumes whether or not the person wishes. It is impossible for people to kill themselves by holding their breath.

Inflation Reflex

Located in the walls of bronchi and bronchioles throughout the lungs are **stretch receptors.** When the receptors become overstretched, impulses are sent along the vagus nerves to the inspiratory area and apneustic area. In response, the inspiratory area is inhibited and the apneustic area is inhibited from activating the inspiratory area. The result is that expiration follows. As air leaves the lungs during expiration, the lungs deflate and the stretch receptors are no longer stimulated. Thus, the inspiratory and apneustic areas are no longer inhibited and a new inspiration begins. This reflex is referred to as the **inflation (Hering-Breuer) reflex.** Some evidence seems to suggest that the reflex is mainly a protective mechanism for preventing overinflation of the lungs rather than a key component in the regulation of respiration.

Chemical Stimuli

Certain chemical stimuli determine how fast we breathe. The ultimate goal of the respiratory system is to maintain proper levels of carbon dioxide and oxygen and the system is highly responsive to changes in the blood levels of either. Although it is convenient to speak of carbon dioxide as the most important chemical stimulus for regulating the rate of respiration, it is actually hydrogen ions that assume this role. For example, carbon dioxide (CO_2) in the blood combines with water (H_2O) to form carbonic acid (H_2CO_3). But the carbonic acid quickly breaks down into H^+ ions and bicarbonate (HCO_3^-) ions. Any increase in CO_2 will cause an increase in H^+ ions, and any decrease in CO_2 will cause a decrease in H^+ ions. In effect, it is the H^+ ions that alter the rate of respiration rather than the CO_2 molecules. Although the following discussion refers to the levels of CO_2 and their effect on respiration, keep in mind that it is really the H^+ ions that cause the effects.

Within the medulla is an area that is highly sensitive to blood concentrations of CO_2. It is called the **chemosensitive area.** Outside the central nervous system are **chemoreceptors,** the carotid and aortic bodies, located in the carotid and aortic sinuses, respectively. They are sensitive to changes in CO_2 and O_2 levels in the blood.

Under normal circumstances, arterial blood pCO_2 is 40 mm Hg. If there is even a slight increase in pCO_2— a condition called **hypercapnia**—the chemosensitive area in the medulla and chemoreceptors in the carotid and aortic sinuses are stimulated. Stimulation of the chemosensitive area and chemoreceptors causes the inspiratory area to become highly active, and the rate of respiration increases. This increased rate, **hyperventilation,** allows the body to expel more CO_2 until the pCO_2 is lowered to normal. If arterial pCO_2 is lower than 40 mm Hg, the chemosensitive area and chemoreceptors are not stimulated and stimulatory impulses are not sent to the inspiratory area. Consequently, the area sets its own moderate pace until CO_2 accumulates and the pCO_2 rises to 40 mm Hg. A slow rate of respiration is called **hypoventilation.**

The oxygen chemoreceptors are sensitive only to large decreases in the pO_2. If arterial pO_2 falls from a normal of 105 mm Hg to 70 mm Hg, the oxygen chemoreceptors become stimulated and send impulses to the inspiratory area. But if the pO_2 falls much below 70 mm Hg, the cells of the inspiratory area suffer oxygen starvation and do not respond well to any chemical receptors. They send fewer impulses to the inspiratory muscles, and the respiration rate decreases or breathing ceases altogether.

Other Influences

The carotid and aortic sinuses also contain pressoreceptors (baroreceptors) that are stimulated by a rise in blood pressure. Although these pressoreceptors are concerned mainly with the control of circulation (see Figure 20-11), they help control respiration. For example, a sudden rise in blood pressure decreases the rate of respiration, and a drop in blood pressure increases the respiratory rate.

Other factors that control respiration are:

1. An increase in body temperature, as during a fever or vigorous muscular exercise, increases the rate of respiration. A decrease in body temperature decreases respiratory rate. A sudden cold stimulus such as plunging into cold water causes a temporary cessation of breathing called **apnea** (ap-NĒ-a; *a* = without, *pnoia* = breath).

2. A sudden, severe pain brings about apnea, but a prolonged pain triggers the general adaptation syndrome and increases respiration rate.

3. Stretching the anal sphincter muscle increases the respiratory rate. This technique is sometimes employed to stimulate respiration during emergencies.

4. Irritation of the pharynx or larynx by touch or chemicals brings about an immediate cessation of breathing followed by coughing.

INTERVENTION IN RESPIRATORY CRISES

CARDIOPULMONARY RESUSCITATION (CPR)

A serious decrease in respiration or heart rate presents an urgent crisis because the body's cells cannot survive long if they are starved of oxygenated blood. If oxygen is withheld from the cells of the brain for 5 to 6 minutes, there is usually severe and permanent brain injury or death. **Cardiopulmonary resuscitation (CPR)** is the artificial reestablishment of normal or near normal respiration and circulation.

The **A, B, C**'s of cardiopulmonary resuscitation are **Airway, Breathing,** and **Circulation.** The rescuer must establish an airway, provide artificial ventilation if breathing has stopped, and reestablish circulation if there is inadequate cardiac action. Cardiopulmonary resuscitation can be administered at the site of emergency and can be used for any heart or respiratory failure, whether the cause be drowning, strangulation, carbon monoxide or insecticide poisoning, overdose of a drug or anesthesia, electrocution, or myocardial infarction. The success of cardiopulmonary resuscitation is directly related to speed and efficiency. Delay may be fatal. The sequential steps must be continued uniformly and without interruption until the patient recovers or is pronounced dead.

Airway

The first step in cardiopulmonary resuscitation is the immediate opening of the airway. One method of doing this is the **head-tilt maneuver.** With the victim on his or her back, the rescuer places one hand under the victim's neck and the other hand on the forehead. The rescuer simultaneously lifts the neck with one hand and tilts the head backward by pressure on the forehead with the other (Figure 23-23a). The tilted position opens the upper air passageways to their maximum size.

Breathing

If the patient does not resume spontaneous breathing after the head has been tilted backward, artificial ventilation must be given mouth-to-mouth or mouth-to-nose. This is called **external air ventilation** or exhaled air ventilation. In the more usual mouth-to-mouth method, the rescuer maintains the head tilt on the victim and the victim's nostrils are pinched together with the thumb and index finger (Figure 23-23b). The rescuer then opens his or her mouth widely, takes a deep breath, makes a tight seal with his or her mouth around the patient's mouth, and blows in about twice the amount the patient normally breathes. The rescuer then removes his or her mouth and allows the patient to exhale passively. This cycle is repeated approximately 12 times per minute for adults.

Even though the air supplied to the victim is air exhaled by the rescuer, it can still provide sufficient oxygen. Atmospheric air contains about 21 percent O_2 and 0.04 percent of CO_2. Exhaled air still contains about 16 percent O_2 and 4.5 percent CO_2. This amount is more than adequate to maintain a victim's blood O_2 and CO_2 at normal levels if air is given at the prescribed rate and amount.

If the rescuer observes the following three signs, ventilation is adequate: (1) The chest rises and falls with every breath. (2) The lungs can be felt to resist as they expand. (3) Air can be heard escaping during exhalation.

Circulation

One method of reestablishing circulation is called **external cardiac compression,** or **closed-chest cardiac compression** (CCCC). It consists of the application of rhythmic pressure over the sternum (Figure 23-23c). The rescuer places the heels of the hands on the lower half of the sternum and presses down firmly and smoothly at least 60 times a minute. This action compresses the heart and produces an artificial circulation because the heart lies almost in the middle of the chest between the lower portion of the sternum and the spine. When properly done, external cardiac compression can produce systolic blood pressure peaks of over 100 mm Hg. It can also bring carotid arterial blood flow up to 35 percent of normal.

Complications that can occur from the use of cardiac compression include fracture of the ribs and sternum, laceration of the liver, and the formation of fat emboli. They can be minimized by the following precautions.

1. Never compress over the xiphoid process at the tip of the sternum. It extends down over the abdomen, and pressure on it may cause laceration of the liver, which can be fatal.

2. Never let your fingers touch the patient's ribs when you compress. Keep your fingers off the patient, and place the heel of your hand in the middle of the patient's chest over the lower half of the sternum.

3. Never compress the abdomen and chest simultaneously since this action traps the liver and may rupture it.

4. Never use sudden or jerking movements to compress the chest. Compression should be smooth, regular, and uninterrupted, with 50 percent of the cycle compression and 50 percent relaxation.

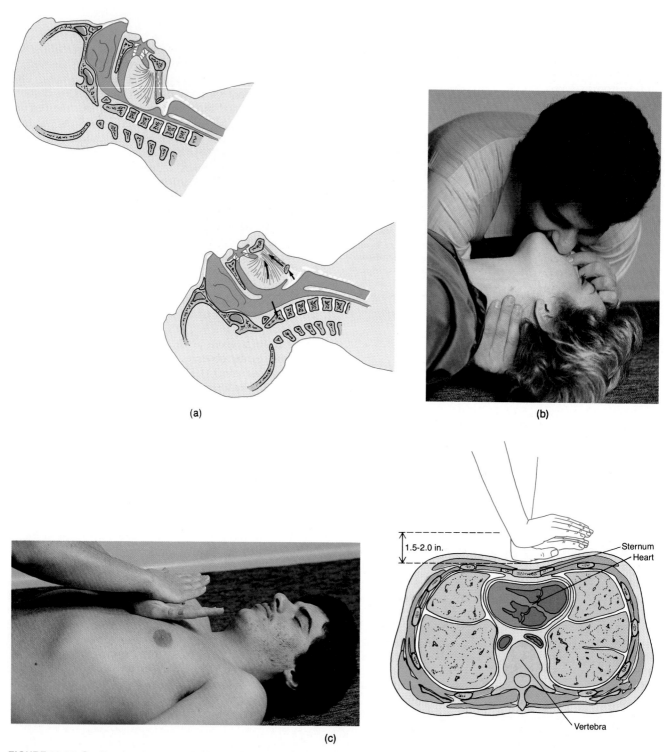

(a)

(b)

(c)

1.5-2.0 in.

Sternum

Heart

Vertebra

FIGURE 23-23 Cardiopulmonary resuscitation (CPR). (a) Head-tilt maneuver to open airway. (b) Exhaled air ventilation. Shown is the procedure for mouth-to-mouth resuscitation. (© 1982 by Gerard J. Tortora. Courtesy of Lynne Tortora and James Borghesi.) (c) External cardiac compression technique. On the left is shown the position of the hands. (© 1982 by Gerard J. Tortora. Courtesy of Geraldine C. Tortora and James Borghesi.) On the right is shown the effect of external cardiac compression on the heart.

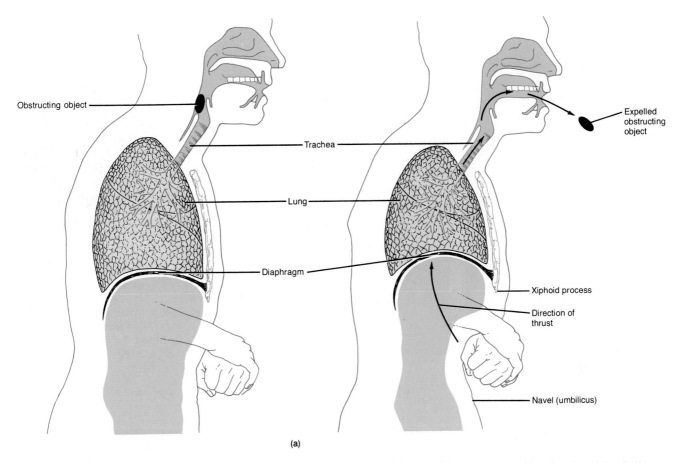

Obstructing object

Trachea

Lung

Diaphragm

Expelled
obstructing
object

Xiphoid process

Direction of
thrust

Navel (umbilicus)

(a)

FIGURE 23-24 Abdominal thrust (Heimlich) maneuver. (a) The application of a quick upward thrust causes sudden elevation of the diaphragm and forceful, rapid expulsion of air in the lungs. Air forced through the trachea and larynx expels the obstructing object.

Compression of the sternum produces some artificial ventilation but not enough for adequate oxygenation of the blood. Therefore exhaled air ventilation must always be used with it. When there are two rescuers, the best technique is to have one rescuer apply at least 60 cardiac compressions per minute. The other rescuer should exhale into the patient's mouth between every fifth and sixth compression.

ABDOMINAL THRUST (HEIMLICH) MANEUVER

Food choking (café coronary), the sixth leading cause of accidental death, is often mistaken for myocardial infarction. However, it can be recognized easily. The victim cannot speak or breathe and may become panic-stricken and run from the room. He or she becomes pale, then deeply cyanotic, and collapses. Without intervention, severe brain damage or death can occur in 5 to 6 minutes.

The food or other obstructing object causing asphyxiation may lodge in the back of the throat or enter the trachea to occlude the airway. Tracheostomy, even when a physician performs it, can be hazardous in a nonclinical setting. An instrument for removing food from the back of the throat has been developed but is seldom at hand in an emergency. However, there is a first aid procedure that does not require special instruments and can be performed by any informed layman. This procedure is called the **abdominal thrust (Heimlich) maneuver.**

Food choking probably occurs most frequenty during inspiration, which causes the bolus of food to be sucked against the opening into the larynx. At the time of the accident, the lungs are therefore expanded. Even during normal expiration, however, some tidal air (500 ml) and the entire expiratory reserve volume (1,200 ml) are in the lungs. The principle of the abdominal thrust maneuver is to expel the air forcibly in order to dislodge the obstruction.

Essentially, the abdominal thrust maneuver consists of pressing one's fist upward into the epigastric region of the abdomen to elevate the diaphragm suddenly. This thrust compresses the lungs in the rib cage and increases the air pressure in the bronchopulmonary tree. This forces air out through the trachea and will eject the food (or object) blocking the airway. The action can be simulated by inserting a cork in a compressible plastic bottle and then squeezing it suddenly—the cork flies out because of the increased pressure. Figure 23-24 shows how to administer the abdominal thrust maneuver.

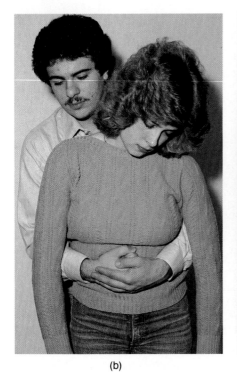

(b)

(c)

FIGURE 23-24 (*Continued*) Abdominal thrust (Heimlich) maneuver. (b) If the victim is upright, stand behind the victim and place both arms around the waist just above the belt line. Then make a fist with one hand and place the thumb side of the fist against the victim's abdomen, slightly above the navel and below the rib cage. Next, grasp your fist with your other hand and press into the victim's abdomen with a quick upward thrust. This should be done by a sharp flexion at the elbows rather than a "bear hug" that compresses the rib cage. The thrust is repeated several times if necessary. (© 1982 by Gerard J. Tortora. Courtesy of Lynne Tortora and James Borghesi.) (c) If the victim is recumbent, kneel astride the victim's hips. Place the hands one on top of the other with the heel of the bottom hand on the victim's abdomen slightly above the navel and below the rib cage. Then apply the quick upward thrust. (© 1982 by Gerard J. Tortora. Courtesy of Lynne Tortora and James Borghesi.)

DISORDERS: HOMEOSTATIC IMBALANCES

Bronchogenic Carcinoma (Lung Cancer)

As part of ordinary breathing, many irritating substances are inhaled. Almost all pollutants, including inhaled smoke, have an irritating effect on the bronchial tubes and lungs and may be regarded as stresses or irritating stimuli. The effects of irritation on the bronchial epithelium are important clinically because a common lung cancer, **bronchogenic carcinoma,** starts in the walls of the bronchi.

Microscopic examination of the pseudostratified epithelium of a bronchial tube reveals three kinds of cells, all of which are in contact with the basement membrane. The columnar cells contain cilia and extend upward from the basement membrane and reach the lumenal surface. At intervals between the ciliated columnar cells are the mucus-secreting goblet cells. The basal cells do not reach the lumenal surface. The basal cells divide continuously, replacing the columnar cells as they wear down and are sloughed off.

The constant irritation by inhaled smoke and pollutants causes an enlargement of the goblet cells of the bronchial epithelium. They respond by secreting excessive mucus. The basal cells also respond to the stress by undergoing cell division so fast that they push into the area occupied by the goblet and columnar cells. As many as 20 rows of basal cells may be produced. Many researchers believe that if the stress is removed at this point, the epithelium can return to normal.

If the stress persists, more and more mucus is secreted and the cilia become less effective. As a result, mucus is not carried toward the throat but remains trapped in the bronchial tubes. The individual then develops a "smoker's cough." Moreover, the constant irritation from the pollutant slowly destroys the alveoli, which are replaced with thick, inelastic connective tissue. Mucus that has accumulated becomes trapped in the air sacs. Millions of sacs rupture, reducing the diffusion surface for the exchange of oxygen and carbon dioxide. The individual has

now developed emphysema. If the stress is removed at this point, there is little chance for improvement. Alveolar tissue that has been destroyed cannot be repaired. But removal of the stress can stop further destruction of lung tissue.

If the stress continues, the emphysema gets progressively worse, and the basal cells of the bronchial tubes continue to divide and break through the basement membrane. At this point the stage is set for bronchogenic carcinoma. Columnar and goblet cells disappear and may be replaced with squamous cancer cells. If this happens, the malignant growth spreads throughout the lung and may block a bronchial tube. If the obstruction occurs in a large bronchial tube, very little oxygen enters the lung, and disease-producing bacteria thrive on the mucoid secretions. In the end, the patient may develop emphysema, carcinoma, and a host of infectious diseases. Treatment involves surgical removal of the diseased lung. However, metastasis (spreading) of the growth through the lymphatic or blood system may result in new growths in other parts of the body such as the brain and liver.

Other factors may be associated with lung cancer. For instance, breast, stomach, and prostate malignancies can metastasize to the lungs. People who apparently have not been exposed to pollutants do occasionally develop bronchogenic carcinoma. However, the occurrence of bronchogenic carcinoma is probably over 20 times higher in heavy cigarette smokers than it is in nonsmokers.

Nasal Polyps

Nasal polyps are protruding growths of the mucous membrane that usually hang down from the posterior wall of the nasal septum. The polyps appear as bluish-white tumors and may fill the nasopharynx as they become larger. Nasal polyps usually undergo atrophy if untreated, but they are easily removed by a physician with a nasal snare and cautery.

Bronchial Asthma

Bronchial asthma is a reaction, usually allergic, characterized by attacks of wheezing and difficult breathing. Attacks are brought on by spasms of the smooth muscles that lie in the walls of the smaller bronchi and bronchioles, causing the passageways to close partially. The patient has trouble exhaling, and the alveoli may remain inflated during expiration. Usually the mucous membranes that line the respiratory passageways become irritated and secrete excessive amounts of mucus that may clog the bronchi and bronchioles and worsen the attack. About three out of four asthma victims are allergic to edible or air-borne substances as common as wheat or dust. Others are sensitive to the proteins of harmless bacteria that inhabit the paranasal sinuses, nose, and throat. Asthma might also have a psychosomatic origin. (The anaphylactic mechanism involved in bronchial asthma may be reviewed in Chapter 22.)

Bronchitis

Bronchitis is inflammation of the bronchi. The typical symptom is a productive cough in which a thick greenish-yellow sputum is raised. This signifies an underlying infection that is causing excessive secretion of mucus. Cigarette smoking remains the most important cause of chronic bronchitis, that is, bronchitis that lasts for at least three months of the year for two successive years.

Emphysema

In **emphysema** (em'-fi-SĒ-ma), the alveolar walls lose their elasticity and remain filled with air during expiration. The name means "blown up" or "full of air." Reduced forced expiratory volume is the first symptom. Later, alveoli in other areas of the lungs are damaged. Many alveoli may merge to form larger air sacs with a reduced overall volume. The lungs become permanently inflated because they have lost elasticity. To adjust to the increased lung size, the size of the chest cage increases. The patient has to work voluntarily to exhale. Oxygen diffusion does not occur as easily across the damaged alveolar-capillary membrane, blood O_2 is somewhat lowered, and any mild exercise that raises the oxygen requirements of the cells leaves the patient breathless. As the disease progresses, the alveoli are replaced with thick fibrous connective tissue. Even carbon dioxide does not diffuse easily through this fibrous tissue. If the blood cannot buffer all the hydrogen ions that accumulate, the blood pH drops or unusually high amounts of carbon dioxide may dissolve in the plasma. High carbon dioxide levels produce acid conditions that are toxic to brain cells. Consequently, the inspiratory area becomes less active and the respiration rate slows down, further aggravating the problem. The compressed and damaged capillaries around the deteriorating alveoli may no longer be able to receive blood. As a result, resistance to blood flow increases in the pulmonary trunk and the right ventricle overworks as it attempts to force blood through the remaining capillaries.

Emphysema is generally caused by a long-term irritation. Air pollution, occupational exposure to industrial dust, and cigarette smoke are the most common irritants. Chronic bronchial asthma also may produce alveolar damage. Cases of emphysema are becoming more and more frequent in the United States. The irony is that the disease can be prevented and the progressive deterioration can be stopped by eliminating the harmful stimuli.

Diseases, such as bronchial asthma, bronchitis, and emphysema have in common some degree of obstruction of the air passageways. The term **chronic obstructive pulmonary disease (COPD)** is used to refer to these disorders.

Pneumonia

The term **pneumonia** means an acute infection or inflammation of the alveoli. In this disease, the alveolar sacs fill up with fluid and dead white blood cells, reducing the amount of air space in the lungs. (Remember that one of the cardinal signs of inflammation is edema.) Oxygen has difficulty diffusing through the inflamed alveoli, and the blood O_2 may be drastically reduced. Blood CO_2 usually remains normal because carbon dioxide diffuses through the alveoli more easily than oxygen does. If all the alveoli of a lobe are inflamed, the pneumonia is called *lobar pneumonia*. If only parts of the lobe are involved, it is called *lobular (segmental) pneumonia*. If both the alveoli and the bronchial tubes are included, it is called *bronchopneumonia*.

The most common cause of pneumonia is the pneumococcus bacterium, but other bacteria or a fungus may be the source of trouble. Viral pneumonia is caused by several viruses, including the influenza virus.

Tuberculosis

The bacterium called *Mycobacterium tuberculosis* produces an inflammation called **tuberculosis.** This disease still ranks as the number-one killer in the communicable disease category. Tuberculosis most often affects the lungs and the pleurae. The bacteria destroy parts of the lung tissue, and the tissue is replaced by fibrous connective tissue. Because the connective tissue is inelastic and thick, the affected areas of the lungs do not snap back during expiration and larger amounts of air are retained. Gases no longer diffuse easily through the fibrous tissue.

Tuberculosis bacteria are spread by inhalation. Although they can withstand exposure to many disinfectants, they die quickly in sunlight. Thus, tuberculosis is sometimes associated with crowded, poorly lit housing. Many drugs are successful in treating tuberculosis. Rest, sunlight, and good diet are vital parts of treatment.

Infant Respiratory Distress Syndrome (RDS)

Sometimes called glassy-lung disease or hyaline membrane disease (HMD), **infant respiratory distress syndrome (RDS)** is responsible for approximately 20,000 newborn infant deaths per year. Before birth, the respiratory passages are filled with fluid. Part of this fluid is amniotic fluid inhaled during respiratory movements in utero. The remainder is produced by the submucosal glands and the goblet cells of the respiratory epithelium.

At birth, this fluid-filled passageway must become an air-filled passageway, and the collapsed primitive alveoli (terminal sacs) must expand and function in gas exchange. The success of this transition depends largely on surfactant, the phospholipid produced by alveolar cells that lowers surface tension in the fluid layer lining the primitive alveoli once air enters the lungs. Surfactant is present in the fetus's lungs as early as 23 weeks and by 28 to 32 weeks the amount present is sufficient to prevent alveolar collapse during breathing. Surfactant is produced continuously by septal alveolar cells. The presence of surfactant can be detected by amniocentesis.

Although in a normal, full-term infant the second and subsequent breaths require less respiratory effort than the first, breathing is not completely normal until about 40 minutes after birth. The entire lung is not inflated fully with the first one or two breaths. In fact, for the first 7 to 10 days, small areas of the lungs may remain uninflated.

In the newborn whose lungs are deficient in surfactant, the effort required for the first breath is essentially the same as that required in normal newborns. However, the surface tension of the alveolar fluid is 7 to 14 times higher than the surface tension of alveolar fluid with a monomolecular layer of surfactant. Consequently, during expiration after the first inspiration, the surface tension of the alveoli increases as the alveoli deflate. The alveoli collapse almost to their original uninflated state.

Idiopathic RDS usually appears within a few hours after

birth. (An idiopathic condition is one that occurs spontaneously in an individual from an unknown or obscure cause.) Affected infants show difficult and labored breathing with withdrawal of the intercostal and subcostal spaces. Death may occur soon after onset of respiratory difficulty or may be delayed for a few days, although many infants survive. At autopsy, the lungs are underinflated and areas of atelectasis (collapse of lung) are prominent. If the infant survives for at least a few hours after developing respiratory distress, the alveoli are often filled with a fluid of high-protein content that resembles a hyaline (or glassy) membrane. RDS occurs frequently in premature infants and also in infants of diabetic mothers, particularly if the diabetes is untreated or poorly controlled.

A treatment called PEEP (positive end expiratory pressure) is now being used to treat RDS. It consists of passing a tube through the air passages to the top of the lungs to provide needed oxygen-rich air at continuous pressures of up to 14 mm Hg. Continuous pressure keeps the baby's alveoli open and available for gas exchange.

Sudden Infant Death Syndrome (SIDS)

Sudden infant death syndrome (SIDS), also called crib death or cot death, kills approximately 10,000 American babies every year. It kills more infants between the ages of 1 week and 12 months than any other disease. SIDS occurs without warning. Although about half of its victims had an upper respiratory infection within two weeks of death, the babies tended to be remarkably healthy. Two other findings are important: the apparently *silent* nature of death and the *disarray* frequently found at the scene.

Given this death scene, it is not surprising that for centuries "crib death" was mistakenly thought to be due to suffocation. Most medical people, however, believe that a healthy child *cannot* smother in its bedclothes. The diaper of a SIDS victim is usually full of stool and urine, and the urinary bladder is almost always empty. External examination reveals only a blood-tinged froth that often exudes from the nostrils. These signs all point to a sudden lethal episode associated with tonic motor activity, not to a gradually deepening coma leading to death.

The similarity of findings in SIDS cases strengthens the theory that there is one underlying cause. The seasonal distribution of SIDS incidences, being lowest during the summer months and highest in the late fall, strongly suggests an infectious agent, viruses being the most likely possibility. Most cases occur at a time in life when gamma globulin levels are low and the infant is at a critical period of susceptibility.

At least thirteen different hypotheses have been proposed as explanations for SIDS. Although the precise pathogenic mechanisms have yet to be clarified, several valuable observations have emerged as a result of investigations during the past decade. Most, if not all, such cases appear to share a common mechanism ("final common pathway"). Thus, SIDS is probably a single disease entity. According to one study, a likely cause of death is laryngospasm, possibly triggered by a previous, mild, nonspecific virus infection of the upper respiratory tract. Another possible factor includes periods of prolonged apnea as a result of malfunction of the respiratory center. Hypoxia may also be involved. Other suspected causes are an allergic reaction or bacterial poisoning.

Coryza (Common Cold) and Influenza (Flu)

Several viruses are responsible for **coryza (common cold).** Typical symptoms include sneezing, excessive nasal secretion, and congestion. The uncomplicated common cold is not usually accompanied by a fever. Complications include sinusitis, ear infections, and laryngitis. Since common colds are caused by viruses, antibiotics are of no use in treatment. Although over-the-counter drugs may lessen the severity of certain symptoms of the common cold, they do not lessen recovery time.

Influenza (flu) is also caused by a virus. Its symptoms include chills, fever (usually higher than 101°F), headache, and muscular aches. Coldlike symptoms appear as the fever subsides. Diarrhea is not a symptom of influenza.

MEDICAL TERMINOLOGY

Asphyxia (*sphyxis* = pulse) Oxygen starvation due to low atmospheric oxygen or interference with ventilation, external respiration, or internal respiration.

Aspiration (*spirare* = breathe) The act of breathing, especially breathing in; drawing of a substance in or out by suction.

Bronchiectasis (*ektasis* = dilation) A chronic dilation of the bronchi or bronchioles.

Cheyne-Stokes respiration A repeated cycle of irregular breathing beginning with shallow breaths that increase in depth and rapidity, then decrease and cease altogether for 15 to 20 seconds. Cheyne-Stokes is normal in infants. It is also often seen just before death from pulmonary, cerebral, cardiac, and kidney disease and is referred to as the "death rattle."

Diphtheria (*diphthera* = membrane) An acute bacterial infection that causes the mucous membranes of the oropharynx, nasopharynx, and larynx to enlarge and become leathery. Enlarged membranes may obstruct airways and cause death from asphyxiation.

Dyspnea (*dys* = painful, difficult; *pnoia* = breath) Labored breathing (short-windedness).

Hypoxia (*hypo* = below, under) Reduction in oxygen supply to cells.

Orthopnea (*ortho* = straight) Inability to breathe in a horizontal position.

Pneumonectomy (*pneumo* = lung; *tome* = cutting) Surgical removal of a lung.

Pulmonary (*pulmo* = lung) **edema** Excess amounts of interstitial fluid in the lungs producing cough and dyspnea. Common in failure of the left side of the heart.

Rales Sounds sometimes heard in the lungs that resemble bubbling or rattling. May be caused by air or an abnormal secretion in the lungs.

Respirator A metal chamber that entombs the chest; also called an "iron lung." Used to produce inspiration and expiration in patient with paralyzed respiratory muscles. Pressure inside the chamber is rhythmically alternated to suck out and push in chest walls.

Rhinitis (*rhino* = nose) Chronic or acute inflammation of the mucous membrane of the nose.

DRUGS ASSOCIATED WITH THE RESPIRATORY SYSTEM

Many respiratory disorders are treated by means of **nebulization** (neb'-yoo-li-ZĀ-shun). This procedure is the administering of medication in the form of droplets that are suspended in air to selected areas of the respiratory tract. The patient inhales the medication as a fine mist. The number of droplets suspended in the mist and the area of the respiratory tract that the medication will reach both depend on droplet size. Smaller droplets (approximately 2 µm in diameter) can be suspended in greater numbers than can large droplets and will reach the alveolar ducts and sacs. Larger droplets (approximately 7 to 16 µm in diameter) will be deposited mostly in the bronchi and bronchioles. Droplets of 40 µm and larger will be deposited in the upper respiratory tract—the mouth, pharynx, trachea, and main bronchi. Nebulization therapy can be used with many different types of drugs, such as chemicals that relax the smooth muscle of the respiratory passageways, chemicals that reduce the thickness of mucus, and antibiotics.

Bronchodilators
Bronchodilators reverse the effects of bronchospasm. Bronchospasm occurs when the smooth muscles of the walls of the bronchioles go into spasm, narrowing or closing off the air passageway and causing labored breathing. Bronchodilators are used to treat asthma. Examples are *oxtriphylline* (Choledyl), *aminophylline* (Aminophyllin), *terbutaline sulfate* (Bricanyl, Brethine), and *ephedrine sulfate*. Bronchodilator aerosols, the most reliable and uniformly effective agents for this purpose, must be given by a nebulizer which produces a fine mist. Examples are *metaproterenol* (Alupent), *isoetharine* (Bronkosol, Bronkometer), and *isoproterenol* (Isuprel).

Antihistamines
Antihistamines are drugs that counteract the effects of histamine, a normal body chemical that is believed to cause the symptoms of persons who are hypersensitive to various allergens. Antihistamines are used to relieve the symptoms of allergic reactions, especially hay fever and other allergic disorders of the nasal passages. Examples are *diphenhydramine* (Benadryl) and *chlorpheniramine* (Chlor-Trimeton).

Corticosteroids
Corticosteroids reduce the inflammation of chronic bronchitis, thereby relieving airway obstruction, and produce bronchial relaxation. An example is *prednisone*.

Expectorants
Expectorants are drugs that promote expectoration, which is the ejection of sputum and other materials from the air passages. Liquefying expectorants are drugs that promote the ejection of mucus from the respiratory tract by decreasing the viscosity of that already present. Examples are *potassium iodide* and *guaifenesin*.

Mucolytics
Mucolytics destroy or dissolve mucus. Examples are *propylene glycol* and *acetylcysteine* (Mucomyst). These drugs may be administered as nebulized aerosol mists.

Antitussives
Antitussives combat and prevent excessive coughing. In chronic bronchitis, excessive coughing may weaken the walls of the bronchioles. Examples of antitussives are *codeine* and *dextromethorphan*.

STUDY OUTLINE

Organs (p. 542)
1. Respiratory organs include the nose, pharynx, larynx, trachea, bronchi, and lungs.
2. They act with the cardiovascular system to supply oxygen and remove carbon dioxide from the blood.

Nose
1. The external portion is made of cartilage and skin and is lined with mucous membrane. Openings to the exterior are the external nares.
2. The internal portion communicates with the nasopharynx through the internal nares and the paranasal sinuses.
3. The nasal cavity is divided by a septum. The anterior portion of the cavity is called the vestibule.
4. The nose is adapted for warming, moistening, and filtering air, for olfaction, and for speech.

Pharynx
1. The pharynx, or throat, is a muscular tube lined by a mucous membrane.
2. The anatomic regions are nasopharynx, oropharynx, and laryngopharynx.
3. The nasopharynx functions in respiration. The oropharynx and laryngopharynx function both in digestion and in respiration.

Larynx
1. The larynx is a passageway that connects the pharynx with the trachea.
2. It contains the thyroid cartilage (Adam's apple); the epiglottis, which prevents food from entering the larynx; the cricoid cartilage, which connects the larynx and trachea; and the paired arytenoid corniculate, and cuneiform cartilages.

3. The larynx contains vocal folds, which produce sound. Taut folds produce high pitches, and relaxed ones produce low pitches.

Trachea
1. The trachea extends from the larynx to the primary bronchi.
2. It is composed of smooth muscle and C-shaped rings of cartilage and is lined with pseudostratified epithelium.
3. Two methods of removing obstructions from the respiratory passageways are tracheostomy and intubation.

Bronchi
1. The bronchial tree consists of the trachea, primary bronchi, secondary bronchi, tertiary bronchi, bronchioles, and terminal bronchioles. Walls of bronchi contain rings of cartilage; walls of bronchioles do not.
2. A bronchogram is a roentgenogram of the tree after introduction of an opaque contrast medium usually containing iodine.

Lungs
1. Lungs are paired organs in the thoracic cavity. They are enclosed by the pleural membrane. The parietal pleura is the outer layer; the visceral pleura is the inner layer.
2. The right lung has three lobes separated by two fissures; the left lung has two lobes separated by one fissure and a depression, the cardiac notch.
3. The secondary bronchi give rise to branches called segmental bronchi which supply segments of lung tissue called bronchopulmonary segments.
4. Each bronchopulmonary segment consists of lobules, which contain lymphatics, arterioles, venules, terminal bronchioles, respiratory bronchioles, alveolar ducts, alveolar sacs, and alveoli.
5. Gas exchange occurs across the alveolar-capillary (respiratory) membranes.

Respiration (p. 556)

Pulmonary Ventilation

1. Pulmonary ventilation or breathing consists of inspiration and expiration.
2. The movement of air into and out of the lungs depends on pressure changes governed in part by Boyle's law, which states that the volume of a gas varies inversely with pressure, assuming that temperature is constant.
3. Inspiration occurs when intrapulmonic pressure falls below atmospheric pressure. Contraction of the diaphragm and external intercostal muscles increases the size of the thorax, thus decreasing the intrapleural pressure so that the lungs expand. Expansion of the lungs decreases intrapulmonic pressure, so that air moves along the pressure gradient from the atmosphere into the lungs.
4. Expiration occurs when intrapulmonic pressure is higher than atmospheric pressure. Relaxation of the diaphragm and external intercostal muscles increases intrapleural pressure, lung volume decreases, and intrapulmonic pressure increases so that air moves from the lungs to the atmosphere.
5. During forced inspiration, accessory muscles of inspiration (sternocleidomastoids and scalenes) are also used.
6. Forced expiration employs contraction of the internal intercostals and abdominal muscles.
7. Compliance is the ease with which the lungs and thoracic wall expand.
8. The walls of the respiratory passageways offer some resistance to breathing.

Modified Respiratory Movements

1. Modified respiratory movements are used to express emotions and to clear air passageways.
2. Coughing, sneezing, sighing, yawning, sobbing, crying, laughing, and hiccuping are types of modified respiratory movements.

Pulmonary Air Volumes and Capacities

1. Air volumes exchanged during breathing and rate of respiration are measured with a spirometer.
2. Among the pulmonary air volumes exchanged in ventilation are tidal volume, inspiratory reserve, expiratory reserve, residual volume, and minimal volumes.
3. Pulmonary lung capacities, the sum of two or more volumes, include inspiratory, functional residual, vital, and total.
4. The minute volume of respiration is the total air taken in during 1 minute (tidal volume times 12 respirations per minute).

Exchange of Respiratory Gases

1. The partial pressure of a gas is the pressure exerted by that gas in a mixture of gases. It is symbolized by p.
2. Charles' law indicates that the volume of a gas is directly proportional to its absolute temperature, assuming that the pressure remains constant.
3. According to Dalton's law, each gas in a mixture of gases exerts its own pressure as if all the other gases were not present.
4. Henry's law states that the quantity of a gas that will dissolve in a liquid is proportional to the partial pressure of the gas and its solubility coefficient, when the temperature remains constant.

External Respiration; Internal Respiration

1. In internal and external expiration O_2 and CO_2 move from areas of their higher partial pressure to areas of their lower partial pressure.
2. External respiration is the exchange of gases between alveoli and pulmonary blood capillaries. It is aided by a thin alveolar-capillary membrane, a large alveolar surface area, and a rich blood supply.
3. Internal respiration is the exchange of gases between tissue blood capillaries and tissue cells.

Transport of Respiratory Gases

1. In each 100 ml of oxygenated blood, 3 percent of the O_2 is dissolved in plasma and 97 percent is carried with hemoglobin as oxyhemoglobin (HbO_2).
2. The association of oxygen and hemoglobin is affected by pO_2, pCO_2, temperature, and DPG.
3. In each 100 ml of deoxygenated blood, 7 percent of CO_2 is dissolved in plasma, 23 percent combines with hemoglobin as carbaminohemoglobin ($HbNHCOOH$), and 70 percent is converted to the bicarbonate ion (HCO_3^-).

Control of Respiration (p. 568)

Nervous Control

1. The respiratory center consists of a medullary rhythmicity area (inspiratory and expiratory area), pneumotaxic area, and apneustic area.
2. The inspiratory area has an intrinsic excitability that sets the basic rhythm of respiration.
3. The pneumotaxic and apneustic areas coordinate the transition between inspiration and expiration.

Regulation of Respiratory Center Activity

1. Respirations may be modified by a number of factors, both in the brain and outside.
2. Among the modifying factors are cortical influences, the inflation reflex, chemical stimuli (O_2 and CO_2 levels), blood pressure, temperature, pain, and irritation to the respiratory mucosa.

Intervention in Respiratory Crises (p. 571)

1. Cardiopulmonary resuscitation (CPR) is the artificial reestablishment of respiration and circulation. The A, B, C's of CPR are Airway, Breathing, and Circulation.
2. The abdominal thrust (Heimlich) maneuver is a first aid procedure used in case of food choking. It consists of an abdominal thrust that elevates the diaphragm, compresses the lungs, and increases air pressure in the bronchial tree.

Disorders: Homeostatic Imbalances (p. 574)

1. In bronchogenic carcinoma, bronchial epithelial cells are replaced by cancer cells after constant irritation has disrupted the normal growth, division, and function of the epithelial cells.
2. Nasal polyps are growths of mucous membrane in the nasal cavity.
3. Bronchial asthma occurs when spasms of smooth muscle in bronchial tubes result in partial closure of air passageways, inflammation, inflated alveoli, and excess mucus production.
4. Emphysema is characterized by deterioration of alveoli leading to loss of their elasticity. Symptoms are reduced expiratory volume, inflated lungs, and enlarged chest.
5. Pneumonia is an acute inflammation or infection of alveoli.
6. Tuberculosis is an inflammation of pleura and lungs produced by the organism *Mycobacterium tuberculosis*.
7. Infant respiratory distress syndrome (RDS) is an infant disorder in which surfactant is lacking and alveolar ducts and alveoli have a glassy appearance.
8. Sudden infant death syndrome (SIDS) has recently been linked to laryngospasm, possibly triggered by a viral infection of the upper respiratory tract.
9. Coryza (common cold) is caused by viruses and is usually not accompanied by a fever, whereas influenza (flu) is usually accompanied by a fever greater than 101°F.

REVIEW QUESTIONS

1. What organs make up the respiratory system? What function do the respiratory and cardiovascular systems have in common?
2. Describe the structure of the external and internal nose and describe their functions in filtering, warming, and moistening air.
3. What is the pharynx? Differentiate the three regions of the pharynx and indicate their roles in respiration.
4. Describe the structure of the larynx and explain how it functions in respiration and voice production.
5. Describe the location and structure of the trachea. What is tracheostomy? Intubation?
6. What is the bronchial tree? Describe its structure. What is a bronchogram?
7. Where are the lungs located? Distinguish the parietal pleura from the visceral pleura. What is pleurisy?
8. Define each of the following parts of a lung: base, apex, costal surface, medial surface, hilus, root, cardiac notch, and lobe.
9. What is a lobule of the lung? Describe its composition and function in respiration.
10. What is a bronchopulmonary segment?
11. Describe the histology of the alveolar-capillary (respiratory) membrane.
12. What are the basic differences among pulmonary ventilation, external respiration, and internal respiration?
13. Discuss the basic steps involved in inspiration and expiration. Be sure to include values for all pressures involved.
14. Distinguish between quiet and forced inspiration and expiration.
15. Describe how compliance and airway resistance relate to pulmonary ventilation.
16. Define the various kinds of modified respiratory movements.
17. What is a spirometer? Define the various lung volumes and capacities. How is the minute volume of respiration calculated?
18. Define the partial pressure of a gas. How is it calculated?
19. Define Boyle's law, Charles' law, Dalton's law, and Henry's law.
20. Construct a diagram to illustrate how and why the respiratory gases move during external and internal respiration.
21. How are oxygen and carbon dioxide carried by the blood?
22. Describe the relationship between hemoglobin and pO_2, pCO_2, temperature, and DPG.
23. Explain how CO_2 is picked up by tissue capillary blood and then discharged into the alveoli.
24. How does the medullary rhythmicity area function in controlling respiration? How are the apneustic area and pneumotaxic area related to the control of respiration?
25. Explain how each of the following modifies respiration: cerebral cortex, inflation reflex, CO_2, O_2, blood pressure, temperature, pain, and irritations of the respiratory mucosa.
26. How does the control of respiration demonstrate the principle of homeostasis.
27. What is the objective of cardiopulmonary resuscitation (CPR)? What cautions must be observed in exhaled air ventilation and external cardiac compression? Why?
28. Describe the steps involved in the abdominal thrust (Heimlich) maneuver.
29. For each of the following disorders, list the principal clinical symptoms: bronchogenic carcinoma, nasal polyps, bronchial asthma, emphysema, pneumonia, tuberculosis, infant respiratory distress syndrome (RDS), sudden infant death syndrome (SIDS), coryza (common cold), and influenza (flu).
30. Refer to the glossary of medical terminology associated with the respiratory system. Be sure that you can define each term.
31. Describe the actions of selected drugs on the respiratory system.

24 The Digestive System

We all know that food is vital to life. It is required for the chemical reactions that occur in every cell—both those that synthesize new enzymes, cell structures, bone, and all the other components of the body and those that release the energy needed for the building processes. However, most of the foods we eat are simply too large to pass through the plasma membranes of the cells. The breaking down of food molecules for use by body cells is called **digestion** and the organs that collectively perform this function comprise the **digestive system.**

REGULATION OF FOOD INTAKE

Basically, the intake of food is regulated by two sensations—hunger and appetite. Whereas **hunger** implies a craving for food in general, **appetite** usually means desire for a specific food. Moreover, hunger is a stronger sensation that is accompanied by a greater degree of discomfort.

The control center for food intake is the hypothalamus. A cluster of nerve cells in the lateral hypothalamus called the **appetite center** discharges impulses that result in increased food intake. A cluster of nerve cells in the medial hypothalamus constitutes the **satiety center.** When this

center is stimulated, food intake is inhibited. According to one theory, a low blood glucose level stimulates the appetite center, and a high blood glucose level inhibits it. Decreased amounts of amino acids and fatty acids seem to have the same effect.

Once eating has begun, food intake must continue to be regulated to prevent overfilling the digestive tract. As food fills the upper digestive tract, distension stimulates receptors in its wall. These receptors convey visceral sensory impulses to the satiety center, thus inhibiting food intake until the contents of the tract have been digested.

DIGESTIVE PROCESSES

The digestive system prepares food for consumption by the cells through five basic activities.

1. Ingestion. Taking food into the body (eating).
2. Peristalsis. The movement of food along the digestive tract.
3. Digestion. The breakdown of food by both chemical and mechanical processes.
4. Absorption. The passage of digested food from the diges-

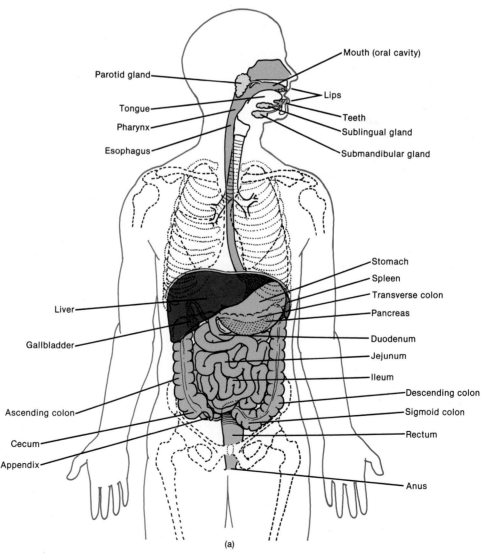

(a)

FIGURE 24-1 Organs of the digestive system and related structures. (a) Diagram.

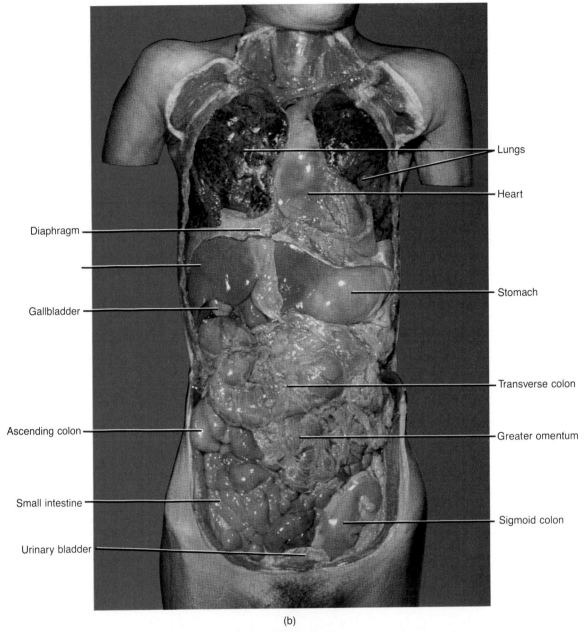

(b)

FIGURE 24-1 (*Continued*)   Organs of the digestive system and related structures. (b) Photograph. (Courtesy of C. Yokochi and J. W. Rohen, *Photographic Anatomy of the Human Body,* 2nd ed., 1979, IGAKU-SHOIN, Tokyo, New York.)

tive tract into the cardiovascular and lymphatic systems for distribution to cells.

5. Defecation. The elimination of indigestible substances from the body.

Chemical digestion is a series of catabolic reactions that break down the large carbohydrate, lipid, and protein molecules that we eat into molecules usable by body cells. These products of digestion are small enough to pass through the walls of the digestive organs, into the blood and lymph capillaries, and eventually into the body's cells. **Mechanical digestion** consists of various movements that aid chemical digestion. Food is prepared by the teeth before it can be swallowed. Then the smooth muscles of the stomach and small intestine churn the food so it is thoroughly mixed with the enzymes that catalyze the reactions.

ORGANIZATION

The organs of digestion are traditionally divided into two main groups. First is the **gastrointestinal (GI) tract** or **alimentary canal,** a continuous tube running through the ventral body cavity and extending from the mouth to the anus (Figure 24-1). The relationship of the digestive organs to the nine regions of the abdominopelvic cavity may be reviewed in Figure 1-8b. The length of a tract taken from a cadaver is about 9 m (30 ft). In a living person it is somewhat shorter because the muscles in its wall are in a state of tone. Organs composing the gastrointestinal tract include the mouth, pharynx, esophagus, stomach, small intestine, and large intestine. The GI tract contains the food from the time it is eaten until it is digested and prepared for elimination. Muscular contrac-

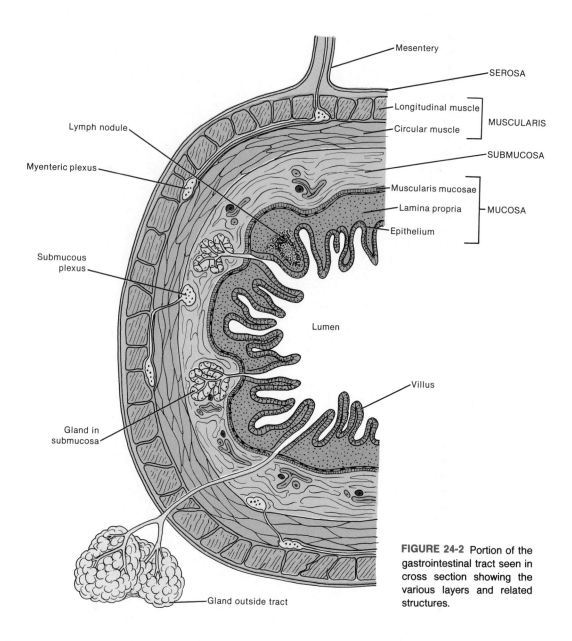

FIGURE 24-2 Portion of the gastrointestinal tract seen in cross section showing the various layers and related structures.

tions in the walls of the GI tract break down the food physically by churning it. Secretions produced by cells along the tract break down the food chemically.

The second group of organs composing the digestive system consists of the **accessory structures**—the teeth, tongue, salivary glands, liver, gallbladder, and pancreas. Teeth protrude into the GI tract and aid in the physical breakdown of food. The other accessory structures, except for the tongue, lie totally outside the tract and produce or store secretions that aid in the chemical breakdown of food. These secretions are released into the tract through ducts.

GENERAL HISTOLOGY

The wall of the GI tract, especially from the esophagus to the anal canal, has the same basic arrangement of tissues. The four coats or tunics of the tract from the inside out are the mucosa, submucosa, muscularis, and serosa or adventitia (Figure 24-2).

Mucosa

The **mucosa**, or inner lining of the tract, is a mucous membrane attached to a thin layer of visceral muscle. Two layers compose the membrane: a *lining epithelium,* which is in direct contact with the contents of the GI tract, and an underlying layer of loose connective tissue called the *lamina propria.* Under the lamina propria is visceral muscle called the *muscularis mucosae.*

The epithelial layer is composed of nonkeratinized cells that are stratified in the mouth and esophagus, but are simple throughout the rest of the tract. The functions of the stratified epithelium are protection and secretion. The functions of the simple epithelium are secretion and absorption. However, the lack of keratin allows some absorption to occur in all parts of the tract.

The lamina propria is made of loose connective tissue containing many blood and lymph vessels and scattered lymph nodules, masses of lymphatic tissue that are not encapsulated. This layer supports the epithelium, binds it to the muscularis mucosae, and provides it with a blood and lymph supply. The blood and lymph vessels are the avenues by which nutrients in the tract reach the other tissues of the body. The lymph tissue also protects against disease. Remember that the GI tract is in contact with the outside environment and contains food which often carries harmful bacteria. Unlike the skin, the mucous membrane of the tract is not protected from bacterial entry by keratin. The lamina propria also contains glandular epithelium that secretes products necessary for chemical digestion.

The muscularis mucosae contains smooth muscle fibers that throw the mucous membrane of the intestine into small folds which increase the digestive and absorptive area. With one exception, which will be described later, the other three coats of the intestine contain no glandular epithelium.

Submucosa

The **submucosa** consists of loose connective tissue that binds the mucosa to the third tunic, the muscularis. It is highly vascular and contains a portion of the *submucous plexus,* which is part of the autonomic nerve supply to the muscularis mucosae. This plexus is important in controlling secretion of the GI tract.

Muscularis

The **muscularis** of the mouth, pharynx, and esophagus consists in part of skeletal muscle that produces voluntary swallowing. Throughout the rest of the tract, the muscularis consists of smooth muscle that is generally found in two sheets: an inner ring of circular fibers and an outer sheet of longitudinal fibers. Contractions of the smooth muscles help to break down food physically, mix it with digestive secretions, and propel it through the tract. The muscularis also contains the major nerve supply to the alimentary tract—the *myenteric plexus,* which consists of fibers from both autonomic divisions. This plexus controls mostly GI motility.

Serosa

The **serosa** is the outermost layer of most portions of the canal. It is a serous membrane composed of connective tissue and epithelium. This tunic is also called the **visceral peritoneum** and forms a portion of the peritoneum, which we will describe in detail.

PERITONEUM

The **peritoneum** (per'-i-tō-NĒ-um) is the largest serous membrane of the body. Serous membranes are also associated with the heart (pericardium) and lungs (pleurae). Serous membranes consist of a layer of simple squamous epithelium (called mesothelium) and an underlying supporting layer of connective tissue. The **parietal peritoneum** lines the wall of the abdominal cavity. The **visceral peritoneum** covers some of the organs and constitutes their serosa. The potential space between the parietal and visceral portions of the peritoneum is called the **peritoneal**

cavity and contains serous fluid. In certain diseases, the peritoneal cavity may become distended by several liters of fluid so that it forms an actual space. Such an accumulation of serous fluid is called **ascites** (a-SĪ-tēz).

Unlike the pericardium and pleurae, the peritoneum contains large folds that weave in between the viscera. The folds bind the organs to each other and to the walls of the cavity and contain the blood and lymph vessels and the nerves that supply the abdominal organs. One extension of the peritoneum is called the **mesentery** (MEZ-en-ter'-ē). It is an outward fold of the serous coat of the small intestine (Figure 24-3). The tip of the fold is attached to the posterior abdominal wall. The mesentery binds the small intestine to the wall. A similar fold of parietal peritoneum, called the **mesocolon** (mez'-ō-KŌ-lon), binds the large intestine to the posterior body wall. It also carries blood vessels and lymphatics to the intestines.

Other important peritoneal folds are the falciform ligament, the lesser omentum, and the greater omentum. The **falciform** (FAL-si-form) **ligament** attaches the liver to the anterior abdominal wall and diaphragm. The **lesser omentum** (ō-MENT-um) arises as two folds in the serosa of the stomach and duodenum suspending the stomach and duodenum from the liver. The **greater omentum** is a four-layered fold in the serosa of the stomach that hangs down like an apron over the front of the intestines. It then passes up to part of the large intestine (the transverse colon), wraps itself around it, and finally attaches to the parietal peritoneum of the posterior wall of the abdominal cavity. Because the greater omentum contains large quantities of adipose tissue, it commonly is called the "fatty apron." The greater omentum contains numerous lymph nodes. If an infection occurs in the intestine, plasma cells formed in the lymph nodes combat the infection and help prevent it from spreading to the peritoneum.

MOUTH (ORAL CAVITY)

The **mouth,** also referred to as the **oral** or **buccal** (BUK-al) **cavity,** is formed by the cheeks, hard and soft palates, and tongue (Figure 24-4). Forming the lateral walls of the oral cavity are the **cheeks**—muscular structures covered on the outside by skin and lined by nonkeratinized stratified squamous epithelium. The anterior portions of the cheeks terminate in the superior and inferior lips.

The **lips (labia)** are fleshy folds surrounding the orifice of the mouth. They are covered on the outside by skin and on the inside by a mucous membrane. The transition zone where the two kinds of covering tissue meet is called the **vermilion** (ver-MIL-yon). This portion of the lips is nonkeratinized and the color of the blood in the underlying blood vessels is visible through the transparent surface layer of the vermilion. The inner surface of each lip is attached to its corresponding gum by a midline fold of mucous membrane called the **labial frenulum** (LĀ-bē-al FREN-yoo-lum).

The orbicularis oris muscle and connective tissue lie between the external integumentary covering and the internal mucosal lining. During chewing, the cheeks and lips help to keep food between the upper and lower teeth. They also assist in speech.

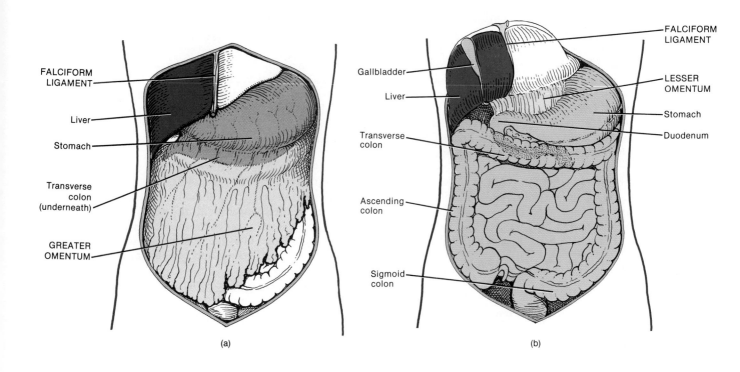

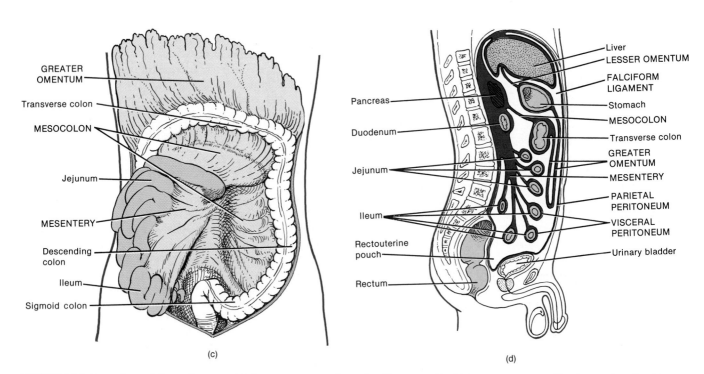

FIGURE 24-3 Extensions of the peritoneum. (a) Greater omentum (see also Figure 24-1b). (b) Lesser omentum. The liver and gallbladder have been lifted. (c) Mesentery. The greater omentum has been lifted. (d) Sagittal section through the abdomen and pelvis indicating the relationship of the peritoneal extensions to each other.

The **vestibule** of the oral cavity is bounded externally by the cheeks and lips and internally by the gums and teeth. The **oral cavity proper** extends from the vestibule to the **fauces** (FAW-sēs), the opening between the oral cavity and the pharynx or throat.

The **hard palate,** the anterior portion of the roof of

the mouth, is formed by the maxillae and palatine bones, covered by mucous membrane, and forms a bony partition between the oral and nasal cavities. The **soft palate** forms the posterior portion of the roof of the mouth. It is an arch-shaped muscular partition between the oropharynx and nasopharynx and is lined by mucous membrane.

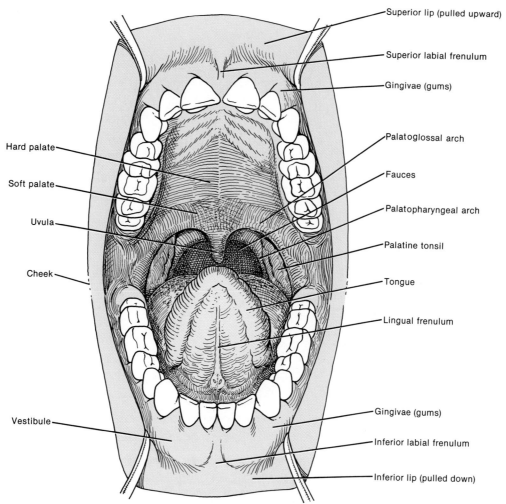

FIGURE 24-4 Mouth or oral cavity.

Hanging from the free border of the soft palate is a conical muscular process called the **uvula** (OO-vyoo-la). On either side of the base of the uvula are two muscular folds that run down the lateral side of the soft palate. Anteriorly, the **palatoglossal arch (anterior pillar)** extends inferiorly, laterally, and anteriorly to the side of the base of the tongue. Posteriorly, the **palatopharyngeal (PAL-a-tō-fa-rin'-jē-al) arch (posterior pillar)** projects inferiorly, laterally, and posteriorly to the side of the pharynx. The palatine tonsils are situated between the arches, and the lingual tonsil is situated at the base of the tongue. At the posterior border of the soft palate, the mouth opens into the oropharynx through the fauces.

TONGUE

The **tongue,** together with its associated muscles, forms the floor of the oral cavity. It is an accessory structure of the digestive system composed of skeletal muscle covered with mucous membrane (Figure 24-5). The tongue is divided into symmetrical lateral halves by a median septum that extends throughout its entire length and is attached inferiorly to the hyoid bone. Each half of the tongue consists of an identical complement of extrinsic and intrinsic muscles.

The **extrinsic muscles** of the tongue originate outside the tongue and insert into it. They include the hyoglossus, chondroglossus, genioglossus, styloglossus, and palatoglossus (see Figure 11-7). The extrinsic muscles move the tongue from side to side and in and out. These movements maneuver food for chewing, shape the food into a rounded mass, called a **bolus,** and force the food to the back of the mouth for swallowing. They also form the floor of the mouth and hold the tongue in position. The **intrinsic muscles** originate and insert within the tongue and alter the shape and size of the tongue for speech and swallowing. The intrinsic muscles include the longitudinalis superior, longitudinalis inferior, transversus linguae, and verticalis linguae (Figure 24-5d). The **lingual frenulum,** a fold of mucous membrane in the midline of the undersurface of the tongue, aids in limiting the movement of the tongue posteriorly.

CLINICAL APPLICATION

If the lingual frenulum is too short, tongue movements are restricted, speech is faulty, and the person is said to be "tongue-tied." This congenital problem is referred to as **ankyloglossia** (ang-ki-lō-GLOSS-ē-a). It can be corrected by cutting the lingual frenulum.

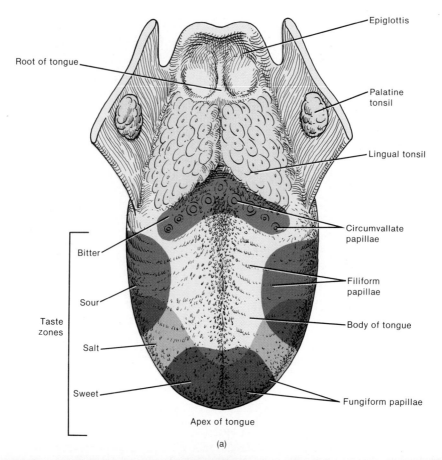

Epiglottis

Root of tongue

Palatine tonsil

Lingual tonsil

Circumvallate papillae

Bitter

Sour

Filiform papillae

Taste zones

Body of tongue

Salt

Sweet

Fungiform papillae

Apex of tongue

(a)

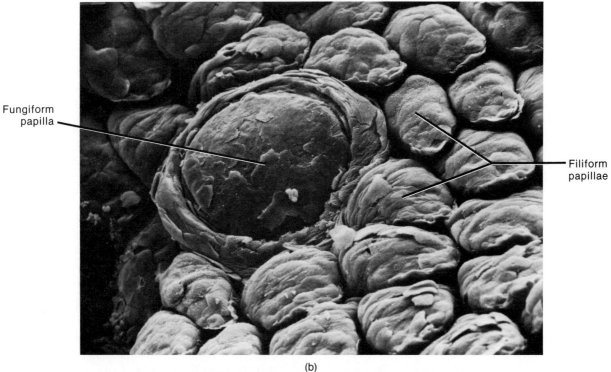

Fungiform papilla

Filiform papillae

(b)

FIGURE 24-5 Tongue. (a) Locations of the papillae and the four taste zones. (b) Scanning electron micrograph of a fungiform papilla surrounded by filiform papillae at a magnification of 245×. (Courtesy of Richard K. Kessel and Randy H. Kardon, *Tissues and Organs: A Text-Atlas of Scanning Electron Microscopy.* Copyright © 1979 by Scientific American, Inc.)

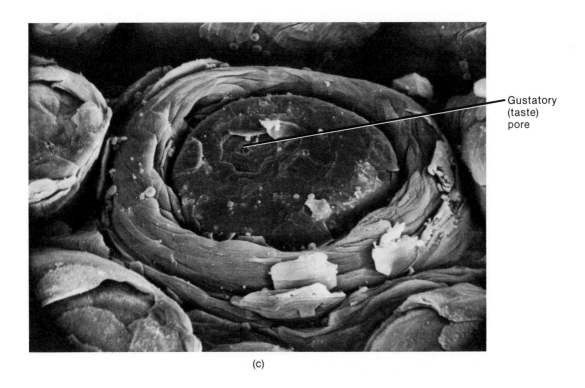

Gustatory
(taste)
pore

(c)

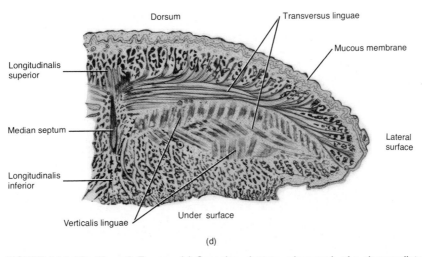

Dorsum

Transversus linguae

Mucous membrane

Longitudinalis
superior

Median septum

Lateral
surface

Longitudinalis
inferior

Verticalis linguae

Under surface

(d)

FIGURE 24-5 (*Continued*) Tongue. (c) Scanning electron micrograph of a circumvallate papilla at a magnification of 1020×. (d) Frontal section through the left side of the tongue showing its intrinsic muscles. (Courtesy of Richard K. Kessel and Randy H. Kardon, *Tissues and Organs: A Text-Atlas of Scanning Microscopy*. Copyright © 1979 by Scientific American, Inc.)

The upper surface and sides of the tongue are covered with **papillae** (pa-PIL-ē), projections of the lamina propria covered with epithelium (Figure 24-5a–c). **Filiform papillae** are conical projections distributed in parallel rows over the anterior two-thirds of the tongue. They contain no taste buds. **Fungiform papillae** are mushroomlike elevations distributed among the filiform papillae and more numerous near the tip of the tongue. They appear as red dots on the surface of the tongue, and most of them contain taste buds. **Circumvallate papillae** are arranged in the form of an inverted V on the posterior surface of the tongue, and all of them contain taste buds. Note the taste zones of the tongue in Figure 24-5a.

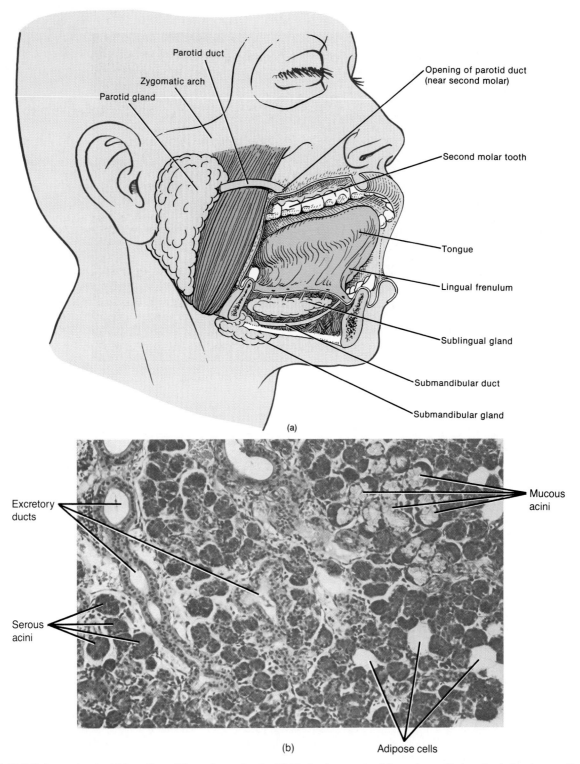

FIGURE 24-6 Salivary glands. (a) Location of the salivary glands. (b) Photomicrograph of the submandibular gland showing mostly serous acini and a few mucous acini at a magnification of 120×. The parotids consist of all serous acini and the sublinguals consist of mostly mucous acini and a few serous acini. (© by Michael H. Ross. Used by permission.)

SALIVARY GLANDS

Saliva is a fluid that is continuously secreted by glands in or near the mouth. Ordinarily, just enough saliva is secreted to keep the mucous membranes of the mouth moist, but when food enters the mouth, secretion increases so the saliva can lubricate, dissolve, and begin the chemical breakdown of the food. The mucous membrane lining the mouth contains many small glands, the **buccal glands,** that secrete small amounts of saliva. However, the major portion of saliva is secreted by the **salivary glands,** accessory structures that lie outside the mouth and pour their contents into ducts that empty into the oral cavity. There are three pairs of salivary glands: parotid, submandibular (submaxillary), and sublingual glands (Figure 24-6a).

The **parotid glands** are located under and in front of the ears between the skin and the masseter muscle. They are compound tubuloacinar glands. Each secretes into the oral cavity vestibule via a duct, called the parotid duct, that pierces the buccinator muscle to open into the vestibule opposite the upper second molar tooth. The **submandibular glands,** which are compound acinar glands, are found beneath the base of the tongue in the posterior part of the floor of the mouth (Figure 24-6b). Their ducts, the submandibular ducts, run superficially under the mucosa on either side of the midline of the floor of the mouth and enter the oral cavity proper just behind the central incisors. The **sublingual glands,** also compound acinar glands, are anterior to the submandibular glands, and their ducts, the lesser sublingual ducts, open into the floor of the mouth in the oral cavity proper.

CLINICAL APPLICATION

Although any of the salivary glands may become infected as a result of a nasopharyngeal infection, the parotids are typically the target of the mumps viruses. **Mumps** is an inflammation and enlargement of the parotid glands accompanied by fever and extreme pain during swallowing. In about 20 to 35 percent of males past puberty, the testes may also become inflamed, and although it rarely occurs, sterility is a possible consequence. In some people, the pancreas is also involved.

Composition of Saliva

The fluids secreted by the buccal glands and the three pairs of salivary glands constitute **saliva.** Amounts of saliva secreted daily vary considerably but range from 1,000 to 1,500 ml. Chemically, saliva is 99.5 percent water and 0.5 percent solutes. Among the solutes are salts—chlorides, bicarbonates, and phosphates of sodium and potassium. Some dissolved gases and various organic substances including urea and uric acid, serum albumin and globulin, mucin, the bacteriolytic enzyme lysozyme, and the digestive enzyme salivary amylase are also present.

Each saliva-producing gland supplies different ingredients to saliva. The parotids contain cells that secrete a watery serous liquid containing the enzyme salivary amylase. The submandibular glands contain cells similar to those found in the parotids plus some mucous cells. Therefore, they secrete a fluid that is thickened with mucus, but still contains quite a bit of enzyme. The sublingual glands contain mostly mucous cells, so they secrete a much thicker fluid that contributes only a small amount of enzyme to the saliva.

The water in saliva provides a medium for dissolving foods so they can be tasted and digestive reactions can take place. The chlorides in the saliva activate the salivary amylase. The bicarbonates and phosphates buffer chemicals that enter the mouth and keep the saliva at a slightly acidic pH of 6.35 to 6.85. Urea and uric acid are found in saliva because the saliva-producing glands (like the sweat glands of the skin) help the body to get rid of wastes. Mucin is a protein that forms mucus when dissolved in water. Mucus lubricates the food so it can be

easily turned in the mouth, formed into a ball, and swallowed. The enzyme lysozyme destroys bacteria, thereby protecting the mucous membrane from infection and thus the teeth from decay.

Secretion of Saliva

Salivation is entirely under nervous control. Normally, moderate amounts of saliva are continuously secreted in response to parasympathetic stimulation to keep the mucous membranes moist and to lubricate the movements of the tongue and lips during speech. The saliva is then swallowed and reabsorbed to prevent fluid loss. Dehydration causes the salivary and buccal glands to cease secreting saliva to conserve water. The subsequent feeling of dryness in the mouth promotes sensations of thirst. This phenomenon is also noted during fear or anxiety, when sympathetic stimulation dominates.

Food stimulates the glands to secrete heavily. When food is taken into the mouth, chemicals in the food stimulate receptors in taste buds on the tongue. Rolling a dry, indigestible object over the tongue produces friction, which also may stimulate the receptors. Impulses are conveyed from the receptors to two salivary nuclei in the brain stem called the **superior** and **inferior salivatory nuclei.** The nuclei are located at about the junction of the medulla and pons. Returning parasympathetic autonomic impulses from the nuclei activate the secretion of saliva (see Figure 16-2).

The smell, sight, touch, or sound of food preparation also stimulates increased saliva secretion. These stimuli constitute psychological activation and involve learned behavior. Memories stored in the cerebral cortex that associate the stimuli with food are stimulated. The cortex sends impulses to the nuclei in the brain stem via extrapyramidal pathways, and the salivary glands are activated. Psychological activation of the glands has some benefit to the body because it allows the mouth to start chemical digestion as soon as the food is ingested.

Salivation also occurs in response to swallowing irritating foods or during nausea. Reflexes originating in the stomach and upper small intestine stimulate salivation. This mechanism presumably helps to dilute or neutralize the irritating substance.

Saliva continues to be secreted heavily some time after food is swallowed. This flow of saliva washes out the mouth and dilutes and buffers the chemical remnants of irritating substances.

TEETH

The **teeth,** or **dentes,** are accessory structures of the digestive system located in sockets of the alveolar processes of the mandible and maxillae. The alveolar processes are covered by the **gingivae** (jin-JI-vē) or gums, which extend slightly into each socket forming the gingival sulcus (Figure 24-7). The sockets are lined by the **periodontal ligament,** which consists of dense fibrous connective tissue and is attached to the socket walls and the cemental surface of the roots. Thus it anchors the teeth in position and also acts as a shock absorber to dissipate the forces of chewing.

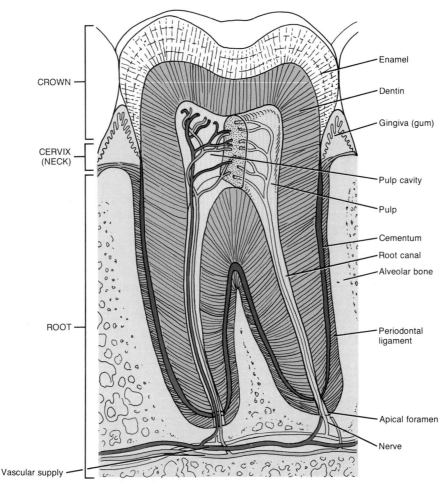

CROWN

CERVIX
(NECK)

ROOT

Vascular supply

Enamel

Dentin

Gingiva (gum)

Pulp cavity

Pulp

Cementum

Root canal

Alveolar bone

Periodontal
ligament

Apical foramen

Nerve

FIGURE 24-7 Parts of a typical tooth as seen in a section through a molar.

A typical tooth consists of three principal portions. The **crown** is the portion above the level of the gums. The **root** consists of one to three projections embedded in the socket. The **cervix** is the constricted junction line of the crown and the root.

Teeth are composed primarily of **dentin,** a bonelike substance that gives the tooth its basic shape and rigidity. The dentin encloses a cavity. The enlarged part of the cavity, the **pulp cavity,** lies in the crown and is filled with **pulp,** a connective tissue containing blood vessels, nerves, and lymphatics. Narrow extensions of the pulp cavity run through the root of the tooth and are called **root canals.** Each root canal has an opening at its base, the **apical foramen.** Through the foramen enter blood vessels bearing nourishment, lymphatics affording protection, and nerves providing sensation. The dentin of the crown is covered by **enamel** that consists primarily of calcium phosphate and calcium carbonate. Enamel is the hardest substance in the body and protects the tooth from the wear of chewing. It is also a barrier against acids that easily dissolve the dentin. The dentin of the root is covered by **cementum,** another bonelike substance, which attaches the root to the periodontal ligament.

Dental Terminology

Because of the curvature of the dental arches, it is necessary to use terms other than anterior, posterior, medial, and lateral in describing the surfaces of the teeth. Accordingly, the following directional terms are used. **Labial** refers to the surface of a tooth in contact with or directed toward the lips. **Buccal** refers to the surface in contact with or directed toward the cheeks. **Lingual** is restricted to the teeth of the lower jaw and refers to the surface directed toward the tongue. **Palatal,** on the other hand, is restricted to the teeth of the upper jaw and refers to the surface directed toward the palate. The term **mesial** designates the anterior or medial side of the tooth relative to its position in the dental arch. **Distal** refers to the posterior or lateral side of the tooth relative to its position in the dental arch. Essentially, mesial and distal refer to the sides of adjacent teeth that are in contact with each other. Finally, **occlusal** refers to the biting surface of a tooth.

Dentitions

Everyone has two **dentitions,** or sets of teeth. The first of these—the **deciduous teeth, milk teeth,** or **baby teeth**— begin to erupt at about 6 months of age, and one pair appears at about each month thereafter until all 20 are present. Figure 24-8a illustrates the deciduous teeth. The **incisors,** which are closest to the midline, are chisel-shaped and adapted for cutting into food. They are referred to as **central** or **lateral incisors** on the basis of their position. Next to the incisors, moving posteriorly,

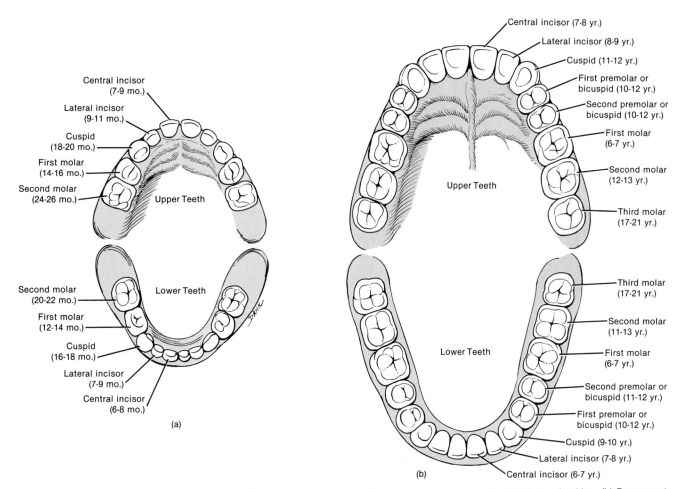

FIGURE 24-8 Dentitions and times of eruptions. The times of eruptions are indicated in parentheses. (a) Deciduous dentition. (b) Permanent dentition.

are the **cuspids (canines),** which have a pointed surface called a cusp. Cuspids are used to tear and shred food. The incisors and cuspids have only one root apiece. Behind them lie the **first** and **second molars,** which have four cusps. Upper molars have three roots; lower molars have two roots. The molars crush and grind food.

All the deciduous teeth are lost—generally between 6 and 12 years of age—and are replaced by the **permanent dentition** (Figure 24-8b). The permanent dentition contains 32 teeth that appear between the age of 6 and adulthood. It resembles the deciduous dentition with the following exceptions. The deciduous molars are replaced with the **first** and **second premolars (bicuspids)** that have two cusps and one root (upper first bicuspids have two roots) and are used for crushing and grinding. The permanent molars erupt into the mouth behind the bicuspids. They do not replace any deciduous teeth and erupt as the jaw grows to accommodate them—the **first molars** at age 6, the **second molars** at age 12, the **third molars (wisdom teeth)** after age 18. The human jaw has become smaller through time and often does not afford enough room behind the second molars for the eruption of the third molars. In this case, the third molars remain embedded in the alveolar bone and are said to be "impacted." Most often they cause pressure and pain and must be surgically removed. In some individuals, third molars may be dwarfed in size or may not develop at all.

DIGESTION IN THE MOUTH

Mechanical

Through chewing, or **mastication,** the tongue manipulates food, the teeth grind it, and the food is mixed with saliva. As a result, the food is reduced to a soft, flexible bolus that is easily swallowed.

Chemical

The enzyme **salivary amylase** (AM-i-lās) initiates the breakdown of starch. This is the only chemical digestion that occurs in the mouth. Carbohydrates are either monosaccharide and disaccharide sugars or polysaccharide starches (see Chapter 2). Most of the carbohydrates we eat are polysaccharides. Since only monosaccharides can be absorbed into the bloodstream, ingested disaccharides and polysaccharides must be broken down. The function of salivary amylase is to break the chemical bonds between some of the monosaccharides in the starches to reduce the long-chain polysaccharides to the disaccharide maltose. Food usually is swallowed too quickly for all of the starches to be reduced to disaccharides in the mouth. However, salivary amylase in the swallowed food continues to act on starches for another 15 to 30 minutes in

EXHIBIT 24-1

DIGESTION IN THE MOUTH

STRUCTURE	ACTIVITY	RESULT
Cheeks	Keep food between teeth during mastication.	Foods uniformly chewed.
Lips	Keep food between teeth during mastication.	Foods uniformly chewed.
Tongue Extrinsic muscles	Move tongue from side to side and in and out.	Food maneuvered for mastication, shaped into bolus, and maneuvered for deglutition.
Intrinsic muscles	Alter shape of tongue.	Deglutition and speech.
Taste buds	Serve as receptors for food stimulus.	Secretion of saliva stimulated by nerve impulses from taste buds to salivatory nuclei in brain stem to salivary glands.
Buccal glands	Secrete saliva.	Lining of mouth and pharynx moistened and lubricated.
Salivary glands	Secrete saliva.	Lining of mouth and pharynx moistened and lubricated. Saliva softens, moistens, and dissolves food, coats food with mucin, cleanses mouth and teeth. Salivary amylase reduces polysaccharides to the disaccharide maltose.
Teeth	Cut, tear, and pulverize food.	Solid foods reduced to smaller particles for swallowing.

the stomach before the stomach acids eventually inactivate it.

Exhibit 24-1 summarizes digestion in the mouth.

DEGLUTITION

Swallowing, or **deglutition** (dē-gloo-TISH-un), is a mechanism that moves food from the mouth to the stomach. It is facilitated by saliva and mucus and involves the mouth, pharynx, and esophagus. Swallowing is conveniently divided into three stages: (1) the voluntary stage, in which the bolus is moved into the oropharynx, (2) the pharyngeal stage, the involuntary passage of the bolus through the pharynx into the esophagus, and (3) the esophageal stage, the involuntary passage of the bolus through the esophagus into the stomach.

Swallowing starts when the bolus is forced to the back of the mouth cavity and into the oropharynx by the movement of the tongue upward and backward against the palate. This represents the **voluntary stage** of swallowing (Figure 24-9).

With the passage of the bolus into the oropharynx, the involuntary **pharyngeal stage** of swallowing begins. The respiratory passageways close and breathing is temporarily interrupted. The bolus stimulates receptors in the oropharynx, which send impulses to the **deglutition center** in the medulla and lower pons of the brain stem. The returning impulses cause the soft palate and uvula to move upward to close off the nasopharynx, and the larynx is pulled forward and upward under the tongue. As the larynx rises, it meets the epiglottis, which seals off the glottis. The movement of the larynx also pulls the vocal cords together, further sealing off the respiratory tract, and widens the opening between the laryngopharynx and esophagus. The bolus passes through the laryngopharynx and enters the esophagus in 1 to 2 seconds. The respiratory passageways then reopen and breathing resumes.

ESOPHAGUS

The **esophagus** (e-SOF-a-gus), the third organ involved in deglutition, is a muscular, collapsible tube that lies behind the trachea. It is about 23 to 25 cm (10 inches) long and begins at the end of the laryngopharynx, passes through the mediastinum anterior to the vertebral column, pierces the diaphragm through an opening called the *esophageal hiatus,* and terminates in the superior portion of the stomach.

The *mucosa* of the esophagus consists of nonkeratinized stratified squamous epithelium, lamina propria, and a muscularis mucosae (Figure 24-10a, b). The *submucosa* contains connective tissue and blood vessels. The *muscularis* of the upper third is striated, the middle third is striated and smooth, and the lower third is smooth. The outer layer is known as the *adventitia* (ad-ven-TISH-a) and not the serosa because the loose connective tissue of the layer is not covered by epithelium (mesothelium) and because the connective tissue merges with the connective tissue of surrounding structures.

The esophagus does not produce digestive enzymes and does not carry on absorption. It secretes mucus and transports food to the stomach. The passage of food from the laryngopharynx into the esophagus is regulated by a sphincter (thick circle of muscle around an opening) at the entrance to the esophagus called the **upper esophageal** (e-sof'-a-JĒ-al) **sphincter.** It consists of the cricopharyngeus muscle attached to the cricoid cartilage. The elevation of the larynx during the pharyngeal stage of swallowing causes the sphincter to relax and the bolus enters the esophagus. The sphincter also relaxes during expiration.

During the **esophageal phase** of swallowing, food is pushed through the esophagus by involuntary muscular movements called **peristalsis** (per'-i-STAL-sis) (Figure 24-10c, d). Peristalsis is a function of the muscularis and is controlled by the medulla. In the section of the esophagus lying just above and around the top of the bolus,

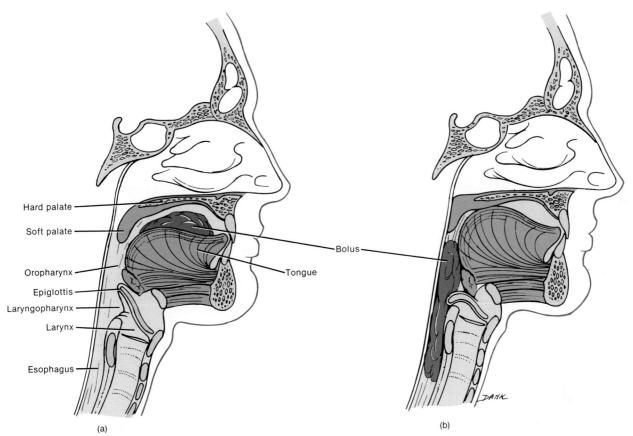

FIGURE 24-9 Deglutition. (a) Position of structures prior to deglutition. (b) During deglutition, the tongue rises against the palate, the nose is closed off, the larynx rises, the epiglottis seals off the larynx, and the bolus is passed into the esophagus.

the circular muscle fibers contract. The contraction constricts the esophageal wall and squeezes the bolus downward. Meanwhile, longitudinal fibers lying around the bottom of and just below the bolus also contract. Contraction of the longitudinal fibers shortens this lower section, pushing its walls outward so it can receive the bolus. The contractions are repeated in a wave that moves down the esophagus, pushing the food toward the stomach. Passage of the bolus is further facilitated by glands secreting mucus. The passage of solid or semisolid food from the mouth to the stomach takes 4 to 8 seconds. Very soft foods and liquids pass through in about 1 second.

Just above the level of the diaphragm, the esophagus is slightly narrowed. This narrowing has been attributed to a physiological sphincter in the inferior part of the esophagus known as the **lower esophageal (gastroesophageal) sphincter.** The lower esophageal sphincter relaxes during swallowing and thus aids the passage of the bolus from the esophagus into the stomach.

CLINICAL APPLICATION

If the lower esophageal sphincter fails to relax normally as food approaches, the condition is called **achalasia** (ak'-a-LĀ-zē-a; *a* = without; *chalasis* = relaxation). As a result, food passage from the esophagus into the stomach is greatly impeded. A whole meal may become lodged in the esophagus, entering the stomach very slowly. Distension of the esophagus results in chest pain that is often confused with pain originating from the heart. The condition is caused by malfunction of the myenteric plexus.

If, on the other hand, the lower esophageal sphincter fails to close adequately after food has entered the stomach, the stomach contents can enter the lower esophagus. Hydrochloric acid from the stomach contents can irritate the esophageal wall, resulting in a burning sensation. The sensation is known as **heartburn** because it is experienced in the region over the heart, but it is not, in fact, related to any cardiac problem. Heartburn can be treated by taking antacids (Tums, Gelusil, Rolaids, Maalox) that neutralize the hydrochloric acid and lessen the severity of the burning sensation and discomfort.

Exhibit 24-2 summarizes the digestion-related activities of the pharynx and esophagus.

STOMACH

The **stomach** is a J-shaped enlargement of the GI tract directly under the diaphragm in the epigastric, umbilical, and left hypochondriac regions of the abdomen (see Figure 1-8b). The superior portion of the stomach is a continuation of the esophagus. The inferior portion empties into the duodenum, the first part of the small intestine. Within each individual, the position and size of the stomach vary continually. For instance, the diaphragm pushes the stomach downward with each inspiration and pulls it upward with each expiration. Empty, it is about the size of a large sausage, but it can stretch to accommodate large amounts of food.

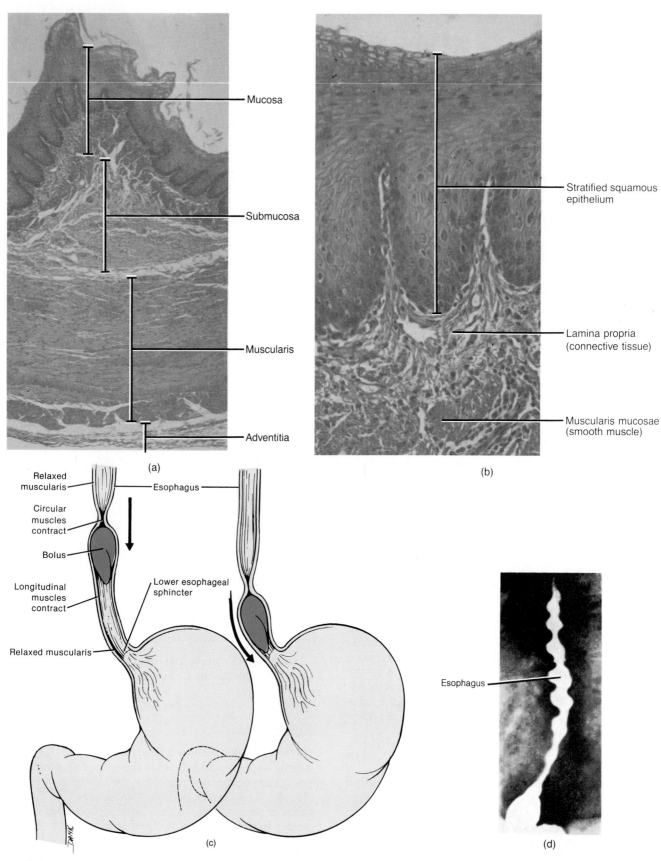

FIGURE 24-10 Esophagus. (a) Photomicrograph of a portion of the wall of the esophagus at a magnification of 60×. (b) Photomicrograph of an enlarged aspect of the mucosa of the esophagus at a magnification of 180×. (© 1983 by Michael H. Ross. Used by permission.) (c) Diagram of peristalsis. (d) Anteroposterior projection of peristalsis made during fluoroscopic examination while a patient was swallowing barium. (Courtesy of Lester W. Paul and John H. Juhl, *The Essentials of Roentgen Interpretation,* 3rd ed., Harper & Row, Publishers, Inc., New York, 1972.)

EXHIBIT 24-2

DIGESTIVE ACTIVITIES OF THE PHARYNX AND ESOPHAGUS

STRUCTURE	ACTIVITY	RESULT
Pharynx	Deglutition.	Moves bolus from oropharynx to laryngopharynx and into esophagus; closes air passageways.
	Relaxation of upper esophageal sphincter.	Moves bolus from laryngopharynx into esophagus.
Esophagus	Peristalsis.	Forces bolus down esophagus.
	Relaxation of lower esophageal sphincter.	Moves bolus into stomach.
	Secretion of mucus.	Lubricates esophagus for smooth passage of bolus.

ANATOMY

The stomach is divided into four areas: cardia, fundus, body, and pylorus (Figure 24-11). The **cardia** surrounds the lower esophageal sphincter. The rounded portion above and to the left of the cardia is the **fundus.** Below the fundus is the large central portion of the stomach, called the **body.** The narrow, inferior region is the **pylorus (antrum).** The concave medial border of the stomach is called the **lesser curvature,** and the convex lateral border is the **greater curvature.** The pylorus communicates with the duodenum of the small intestine via a sphincter called the **pyloric sphincter (valve).**

CLINICAL APPLICATION

Two abnormalities of the pyloric sphincter can occur in infants. **Pylorospasm** is characterized by failure of the muscle fibers encircling the opening to relax normally. It can be caused by hypertrophy or continuous spasm of the sphincter and usually occurs between the second and twelfth weeks of life. Ingested food does not pass easily from the stomach to the small intestine, the stomach becomes overly full, and the infant vomits frequently to relieve the pressure. Pylorospasm is treated by adrenergic drugs that relax the muscle fibers of the sphincter. **Pyloric stenosis** is a narrowing of the pyloric sphincter caused by a tumorlike mass that apparently is formed by enlargement of the circular muscle fibers. It must be surgically corrected.

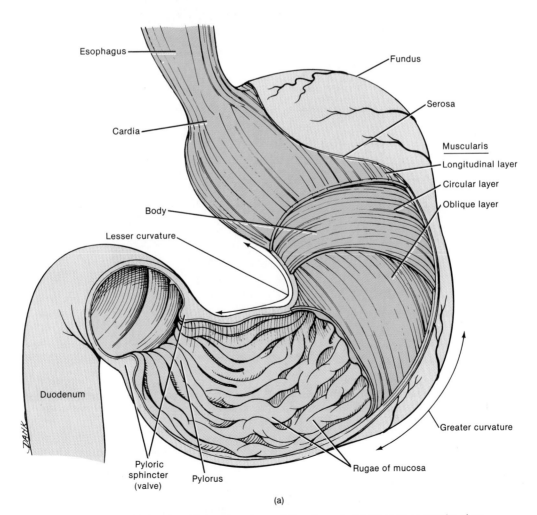

(a)

FIGURE 24-11 External and internal anatomy of the stomach. (a) Diagram in anterior view.

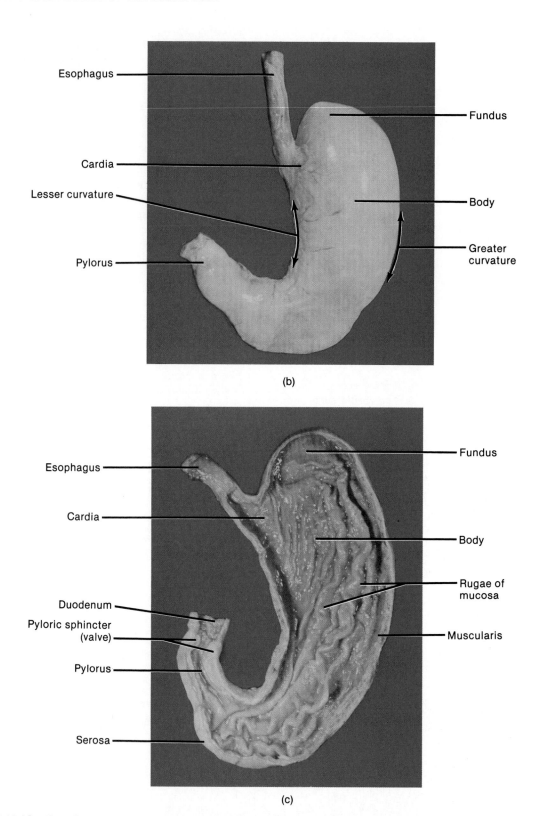

(b)

(c)

FIGURE 24-11 (*Continued*)   External and internal anatomy of the stomach. (b) Photograph of the external surface in anterior view. (c) Photograph of the internal surface in anterior view showing rugae. (Photographs courtesy of C. Yokochi and J. W. Rohen, *Photographic Anatomy of the Human Body,* 2nd ed., 1979, IGAKU-SHOIN, Ltd., Tokyo, New York.)

HISTOLOGY

The stomach wall is composed of the same four basic layers as the rest of the alimentary canal, with certain modifications. When the stomach is empty, the *mucosa* lies in large folds, called **rugae** (ROO-jē), that can be seen with the naked eye. Microscopic inspection of the mucosa reveals a layer of simple columnar epithelium containing many narrow openings that extend down into the lamina propria (Figure 24-12a–d). These pits—**gastric**

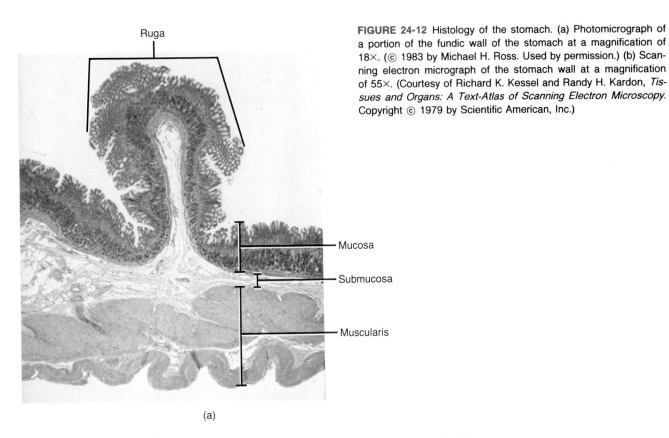

Ruga

Mucosa

Submucosa

Muscularis

(a)

FIGURE 24-12 Histology of the stomach. (a) Photomicrograph of a portion of the fundic wall of the stomach at a magnification of 18×. (© 1983 by Michael H. Ross. Used by permission.) (b) Scanning electron micrograph of the stomach wall at a magnification of 55×. (Courtesy of Richard K. Kessel and Randy H. Kardon, *Tissues and Organs: A Text-Atlas of Scanning Electron Microscopy.* Copyright © 1979 by Scientific American, Inc.)

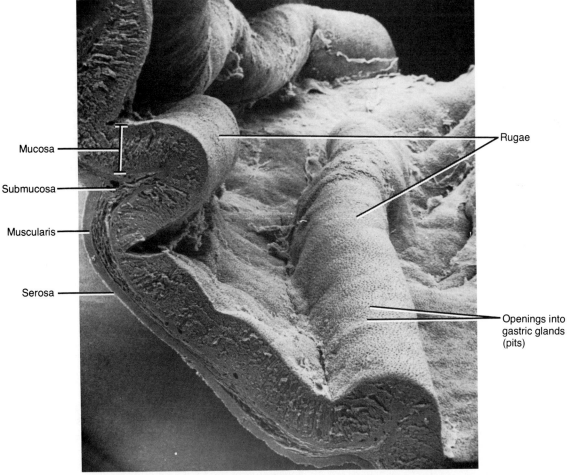

Mucosa

Submucosa

Muscularis

Serosa

Rugae

Openings into gastric glands (pits)

(b)

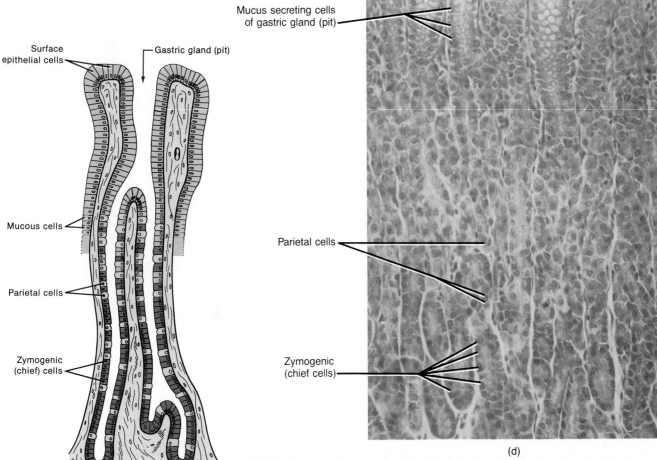

Mucus secreting cells of gastric gland (pit)

Surface epithelial cells

Gastric gland (pit)

Mucous cells

Parietal cells

Parietal cells

Zymogenic (chief) cells

Zymogenic (chief cells)

(c)

(d)

FIGURE 24-12 (*Continued*) Histology of the stomach. (c) Diagram of gastric glands from the fundic wall of the stomach. (d) Photomicrograph of an enlarged aspect of the mucosa of the fundic wall of the stomach at a magnification of 180×. (© 1983 by Michael H. Ross. Used by permission.)

glands—are lined with several kinds of secreting cells: zymogenic, parietal, mucous, and enteroendocrine. The **zymogenic (chief) cells** secrete the principal gastric enzyme precursor, pepsinogen. Hydrochloric acid, involved in the conversion of pepsinogen to the active enzyme pepsin, and intrinsic factor, involved in the absorption of vitamin B_{12} for red blood cell production, are produced by the **parietal cells.** You may recall from Chapter 19 that inability to produce intrinsic factor can result in pernicious anemia. The **mucous cells** secrete mucus. The **enteroendocrine (argentaffin) cells** secrete stomach gastrin, a hormone that stimulates secretion of hydrochloric acid and pepsinogen, contracts the lower esophageal sphincter, mildly increases motility of the GI tract, and relaxes the pyloric sphincter. Secretions of the gastric glands are together called **gastric juice.**

CLINICAL APPLICATION

Viewing of the mucosa of the stomach through an illuminated tube fitted with a lens system is called **gastroscopy.** During the procedure, the patient is anesthetized and the gastroscope is passed into the stomach, which is then inflated with air. A flexible fiberoptic instrument makes possible direct visualization of different parts of the gastric mucosa. It is also possible to perform a gastric mucosal biopsy through a gastroscope.

The *submucosa* of the stomach is composed of loose areolar connective tissue, which connects the mucosa to the muscularis.

The *muscularis,* unlike that in other areas of the alimentary canal, has three layers of smooth muscle: an outer longitudinal layer, a middle circular layer, and an inner oblique layer. This arrangement of fibers allows the stomach to contract in a variety of ways to churn food, break it into small particles, mix it with gastric juice, and pass it to the duodenum.

The *serosa* covering the stomach is part of the visceral peritoneum. At the lesser curvature the two layers of the visceral peritoneum come together and extend upward to the liver as the lesser omentum. At the greater curvature, the visceral peritoneum continues downward as the greater omentum hanging over the intestines.

DIGESTION IN THE STOMACH

Mechanical

Several minutes after food enters the stomach, gentle, rippling, peristaltic movements called **mixing waves** pass over the stomach every 15 to 25 seconds. These waves macerate food, mix it with the secretions of the gastric glands, and reduce it to a thin liquid called **chyme** (kīm). Few mixing waves are observed in the fundus, which is primarily a storage area. Foods may remain in the fundus for an hour or more without becoming mixed with gastric juice. During this time, salivary digestion continues.

As digestion proceeds in the stomach, more vigorous mixing waves begin at the body of the stomach and intensify as they reach the pylorus. The pyloric sphincter normally remains almost, but not completely, closed. As food reaches the pylorus, each mixing wave forces a small amount of the gastric contents into the duodenum through the pyloric sphincter. Most of the food is forced back into the body of the stomach where it is subjected to further mixing. The next wave pushes it forward again and forces a little more into the duodenum. The forward and backward movement of the gastric contents are responsible for almost all of the mixing in the stomach.

Chemical

The principal chemical activity of the stomach is to begin the digestion of proteins. In the adult, digestion is achieved primarily through the enzyme **pepsin.** Pepsin breaks certain peptide bonds between the amino acids making up proteins. Thus a protein chain of many amino acids is broken down into smaller fragments called *peptides*. Pepsin is most effective in the very acidic environment of the stomach (pH 2). It becomes inactive in an alkaline environment.

What keeps pepsin from digesting the protein in stomach cells along with the food? First, pepsin is secreted in an inactive form called **pepsinogen,** so it cannot digest the proteins in the zymogenic cells that produce it. It is not converted into active pepsin unil it comes in contact with the hydrochloric acid secreted by the parietal cells. Second, the stomach cells are protected by mucus, especially after pepsin has been activated. The mucus coats the mucosa to form a barrier between it and the gastric juices.

Another enzyme of the stomach is **gastric lipase.** Gastric lipase splits the butterfat molecules found in milk. This enzyme operates best at a pH of 5 to 6 and has a limited role in the adult stomach. Adults rely almost exclusively on an enzyme found in the small intestine to digest fats.

The infant stomach also secretes **rennin,** which is important in the digestion of milk. Rennin and calcium act on the casein of milk to produce a curd. The coagulation prevents too rapid a passage of milk from the stomach. Rennin is absent in the gastric secretions of adults.

The principal activities of gastric digestion are summarized in Exhibit 24-3.

REGULATION OF GASTRIC SECRETION

Stimulation

The secretion of gastric juice is regulated by both nervous and hormonal mechanisms (Figure 24-13). Parasympathetic impulses from nuclei in the medulla are transmitted via the vagus (X) nerves and stimulate the gastric glands to secrete pepsinogen, hydrochloric acid, mucus, and stomach gastrin. Stomach gastrin is also secreted by gastric glands in response to certain foods that enter the stomach.

● *Cephalic (Reflex) Phase*   The **cephalic phase** of gastric secretions occurs before food enters the stomach and prepares the stomach for digestion. The sight, smell, taste, or thought of food initiates this reflex. Nerve impulses from the cerebral cortex or feeding center in the hypothalamus send impulses to the medulla. The medulla relays impulses over the parasympathetic fibers in the vagus (X) nerve to stimulate the gastric glands to secrete.

● *Gastric Phase*   Once the food reaches the stomach, both nervous and hormonal mechanisms ensure that gastric secretion continues. This is the **gastric phase** of secretion. Food of any kind causes distension and stimulates receptors in the walls of the stomach. These receptors send impulses to the medulla and back to the gastric glands, and they may send messages directly to the glands as well. The impulses stimulate the flow of gastric juice.

EXHIBIT 24-3

SUMMARY OF GASTRIC DIGESTION

STRUCTURE	ACTIVITY	RESULT
Mucosa		
Zymogenic (chief) cells	Secrete pepsinogen.	Precursor of pepsin is produced.
Parietal cells	Secrete hydrochloric acid.	Converts pepsinogen into pepsin, which digests proteins into peptides.
	Secrete intrinsic factor.	Required for absorption of vitamin B_{12} and erythrocyte formation.
Mucous cells	Secrete mucus.	Prevents digestion of stomach wall.
Enteroendocrine cells	Secrete stomach gastrin.	Stimulates gastric secretion, contracts lower esophageal sphincter, increases motility of the stomach, and relaxes pyloric sphincter.
Muscularis	Mixing waves.	Macerate food, mix it with gastric juice, reduce food to chyme, and force chyme through pyloric sphincter.
Pyloric sphincter	Opens to permit passage of chyme into duodenum.	Prevents backflow of food from duodenum to stomach.

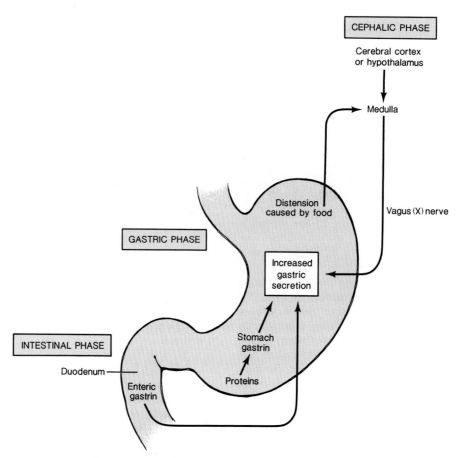

FIGURE 24-13 Summary of factors that stimulate gastric secretion.

Emotions such as anger, fear, and anxiety may slow down digestion in the stomach because they stimulate the sympathetic nervous system, which inhibits the impulses of the parasympathetic fibers.

Protein foods and alcohol stimulate the pyloric mucosa to secrete the hormone **stomach gastrin.** It is absorbed into the bloodstream, circulated through the body and finally reaches its target cells, the gastric glands, where it stimulates secretion of large amounts of gastric juice. It also contracts the lower esophageal sphincter, increases motility of the GI tract, and relaxes the pyloric sphincter and ileocecal sphincter.

● *Intestinal Phase*   Some investigators believe that when partially digested proteins leave the stomach and enter the duodenum, they stimulate the duodenal mucosa to release **enteric gastrin,** a hormone that stimulates the gastric glands to continue their secretion. This constitutes the **intestinal phase** of secretion. However, this mechanism produces relatively small amounts of gastric juice.

Inhibition

Even though chyme stimulates gastric secretion during the intestinal phase, it can inhibit secretion during the gastric phase. For example, the presence of food in the small intestine during the gastric phase initiates an **enterogastric reflex** in which nerve impulses carried to the medulla from the duodenum return to the stomach and inhibit gastric secretion. These impulses ultimately inhibit

parasympathetic stimulation and stimulate sympathetic activity. Stimuli that initiate this reflex are distension of the duodenum, the presence of acid or partially digested proteins in food in the duodenum, or irritation of the duodenal mucosa.

Several intestinal hormones also inhibit gastric secretion. In the presence of acid, partially digested proteins, fats, hypertonic or hypotonic fluids, or irritating substances in chyme, the intestinal mucosa releases **secretin, cholecystokinin (CCK),** and **gastric inhibiting peptide (GIP).** All three hormones inhibit gastric secretion and decrease motility of the GI tract (Exhibit 24-4). Secretin and cholecystokinin are also important in the control of pancreatic and intestinal secretion, and cholecystokinin also helps regulate secretion of bile from the gallbladder (see Exhibit 24-5).

GASTRIC EMPTYING

Gastric emptying is stimulated by two principal factors: nerve impulses in response to distension and stomach gastrin released in the presence of certain types of foods. As noted earlier, during the gastric phase of secretion, distension and the presence of partially digested proteins and alcohol stimulate secretion of gastric juice and stomach gastrin. In the presence of stomach gastrin, the lower esophageal sphincter contracts, the motility of the stomach increases, and the pyloric sphincter relaxes. The net effect of these actions is stomach emptying (Figure 24-14a).

EXHIBIT 24-4

HORMONAL CONTROL OF GASTRIC SECRETION, PANCREATIC SECRETION, AND SECRETION AND RELEASE OF BILE

HORMONE	WHERE PRODUCED	STIMULANT	ACTION
Stomach gastrin	Pyloric mucosa.	Partially digested proteins in stomach.	Stimulates secretion of gastric juice, constricts lower esophageal sphincter, increases motility of GI tract, and relaxes pyloric sphincter and ileocecal sphincter.
Enteric gastrin	Intestinal mucosa.	Partially digested proteins in chyme in small intestine.	Same as above.
Secretin	Intestinal mucosa.	Acid, partially digested proteins, fats, hypertonic or hypotonic fluids, or irritants in chyme in small intestine.	Inhibits secretion of gastric juice, decreases motility of the GI tract, stimulates secretion of pancreatic juice rich in sodium bicarbonate ions, stimulates secretion of bile by hepatic cells of liver, and stimulates secretion of intestinal juice.
Cholecystokinin (CCK)	Intestinal mucosa.	Same as for secretin.	Inhibits secretion of gastric juice, decreases motility of the GI tract, stimulates the secretion of pancreatic juice rich in digestive enzymes, causes ejection of bile from the gallbladder and opening of the sphincter of the hepatopancreatic ampulla, and stimulates secretion of intestinal juice.
Gastric inhibiting peptide (GIP)	Intestinal mucosa.	Same as for secretin.	Inhibits secretion of gastric juice and decreases motility of the GI tract.

The stomach empties all its contents into the duodenum 2 to 6 hours after ingestion. Food rich in carbohydrate leaves the stomach in a few hours. Protein foods are somewhat slower, and emptying is slowest after a meal containing large amounts of fat.

Stomach emptying is inhibited by the enterogastric reflex and hormones released in response to certain constituents in chyme. The enterogastric reflex not only inhibits gastric secretion, it also inhibits gastric motility. The hormones secretin, cholecystokinin (CCK) and gastric inhibiting peptide (GIP) also inhibit gastric secretion and inhibit gastric motility (Figure 24-14b). The rate of stomach emptying is limited to the amount of chyme that the small intestine can process.

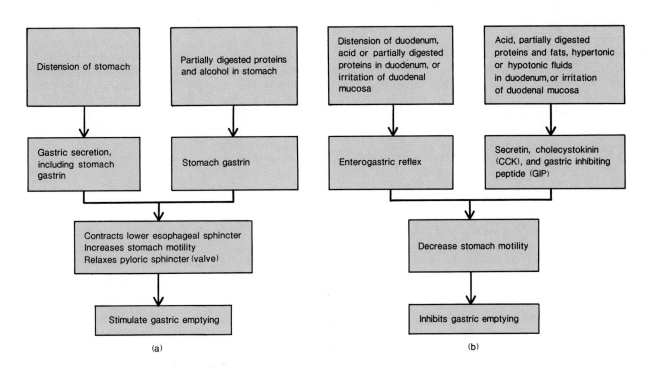

FIGURE 24-14 Factors that (a) stimulate gastric emptying and (b) inhibit gastric emptying.

CLINICAL APPLICATION

Excessive gastric emptying in the wrong direction sometimes occurs. **Vomiting** is the forcible expulsion of the contents of the upper GI tract (stomach and sometimes duodenum) through the mouth. The strongest stimuli for vomiting are irritation and distension of the stomach. Other stimuli include unpleasant sights and dizziness. Impulses are transmitted to the vomiting center in the medulla and returning impulses to the upper GI tract organs, diaphragm, and abdominal muscles bring about the vomiting act. The steps typically involved in the act are: (1) a deep breath inspiration, (2) opening of the upper esophageal sphincter, (3) closure of the glottis, (4) elevation of the soft palate to close off the internal nares, (5) contraction of the diaphragm and abdominal muscles, and (6) opening of the lower esophageal sphincter. Basically, vomiting involves squeezing the stomach between the diaphragm and abdominal muscles and expulsion of the contents through open esophageal sphincters.

Prolonged vomiting, especially in infants and elderly people, can be serious because the loss of gastric juice and fluids can lead to disturbances in fluid and acid–base balance.

ABSORPTION

The stomach wall is impermeable to the passage of most materials into the blood, so most substances are not absorbed until they reach the small intestine. However, the stomach does participate in the absorption of some water, electrolytes, certain drugs, and alcohol.

PANCREAS

The next organ of the GI tract involved in the breakdown of food is the small intestine. Chemical digestion in the small intestine depends not only on its own secretions, but on activities of three accessory structures of digestion outside the alimentary canal: the pancreas, liver, and gallbladder. We will first consider the activities of the accessory structures and then examine their contributions to digestion in the small intestine.

ANATOMY

The **pancreas** is a soft, oblong tubuloacinar gland about 12.5 cm (6 inches) long and 2.5 cm (1 inch) thick. It lies posterior to the greater curvature of the stomach and is connected by a duct (sometimes two) to the duodenum (Figure 24-15a). The pancreas is divided into a head, body, and tail. The **head** is the expanded portion near the C-shaped curve of the duodenum. Moving superiorly and to the left of the head are the centrally located **body** and the terminal tapering **tail.**

The pancreas is linked to the small intestine by a series of ducts. The products of its secreting cells are dumped into small ducts attached to the cells and eventually leave the pancreas through a large main tube called the **pancreatic duct.** In most people the pancreatic duct unites with the common bile duct from the liver and gallbladder and enters the duodenum in a common duct, called the **hepatopancreatic ampulla.** The ampulla opens on an elevation of the duodenal mucosa known as the **duodenal papilla,** about 10 cm (4 inches) below the pylorus of the stomach. An **accessory duct** may also lead from the pancreas and empty into the duodenum about 2.5 cm (1 inch) above the ampulla.

HISTOLOGY

The pancreas is made up of small clusters of glandular epithelial cells. About 1 percent of the cells, the **islets of Langerhans,** form the endocrine portion of the pancreas and consist of alpha, beta, and delta cells that secrete hormones (glucagon, insulin, and somatostatin, respectively). The functions of these hormones may be reviewed in Chapter 18. The remaining 99 percent of the cells, called **acini** (AS-i-nē), are the exocrine portions of the organ (see Figure 18-18). Secreting cells of the acini release a mixture of digestive enzymes called pancreatic juice.

PANCREATIC JUICE

Each day the pancreas produces 1,200 to 1,500 ml (about 1.2 to 1.5 qt) of **pancreatic juice,** a clear, colorless liquid. It consists mostly of water, some salts, sodium bicarbonate, and enzymes. The sodium bicarbonate gives pancreatic juice a slightly alkaline pH (7.1 to 8.2) that stops the action of pepsin from the stomach and creates the proper environment for the enzymes in the small intestine. The enzymes in pancreatic juice include a carbohydrate-digesting enzyme called *pancreatic amylase;* several protein-digesting enzymes called *trypsin* (TRIP-sin), *chymotrypsin* (kī'-mō-TRIP-sin) and *carboxypolypeptidase* (kar-bok'-sē-polē'-PEP-ti-dās); the principal fat-digesting enzyme in the adult body called *pancreatic lipase;* and nucleic acid–digesting enzymes called *ribonuclease* and *deoxyribonuclease.*

Just as pepsin is produced in the stomach in an inactive form (pepsinogen), so too are the protein-digesting enzymes of the pancreas. This prevents the enzymes from digesting cells of the pancreas. The active enzyme trypsin is secreted in an inactive form called *trypsinogen* (trip-SIN-ō-jen). Its activation to trypsin is accomplished in the small intestine by an enzyme secreted by the intestinal mucosa when chyme comes in contract with the mucosa. The activating enzyme is called *enterokinase* (en'-ter-ō-KĪ-nās). Chymotrypsin is activated in the small intestine by trypsin from its inactive form, *chymotrypsinogen.* Carboxypolypeptidase is also activated in the small intestine by trypsin. Its inactive form is called procarboxypolypeptidase.

REGULATION OF PANCREATIC SECRETIONS

Pancreatic secretion, like gastric secretion, is regulated by both nervous and hormonal mechanisms (Figure 24-16). When the cephalic and gastric phases of gastric secretion occur, parasympathetic impulses are simultaneously transmitted along the vagus (X) nerves to the pancreas that result in the secretion of pancreatic enzymes.

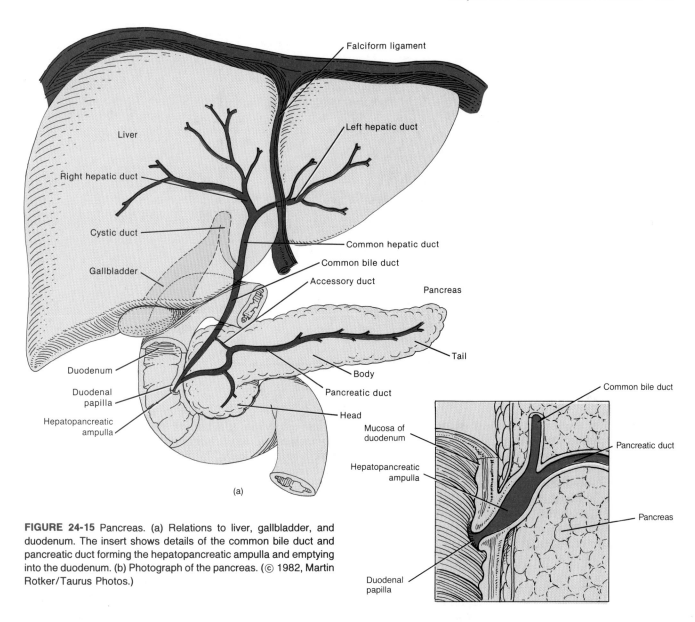

(a)

FIGURE 24-15 Pancreas. (a) Relations to liver, gallbladder, and duodenum. The insert shows details of the common bile duct and pancreatic duct forming the hepatopancreatic ampulla and emptying into the duodenum. (b) Photograph of the pancreas. (© 1982, Martin Rotker/Taurus Photos.)

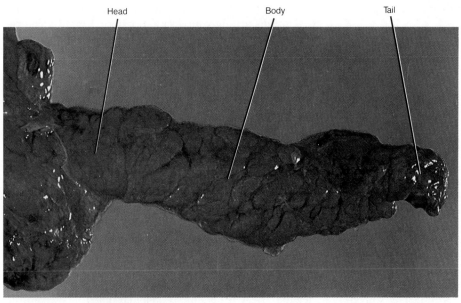

(b)

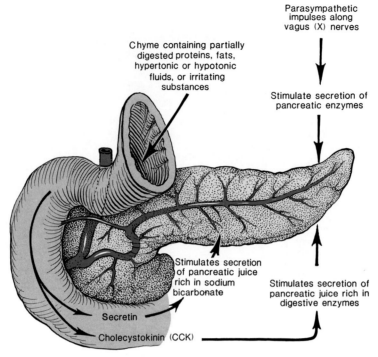

FIGURE 24-16 Control of pancreatic secretion.

In response to chyme in the small intestine, especially chyme that contains partially digested proteins, fats, hypertonic or hypotonic fluids, or irritating substances, the small intestinal mucosa secretes secretin and cholecystokinin (CCK), two hormones that affect pancreatic secretion. Secretin stimulates the pancreas to secrete pancreatic juice that is rich in sodium bicarbonate ions. Cholecystokinin (CCK) stimulates a pancreatic secretion rich in digestive enzymes (see Exhibit 24-4).

LIVER

The **liver** weighs about 1.4 kg (about 3 lb) in the average adult. It is located under the diaphragm and occupies most of the right hypochondrium and part of the epigastrium of the abdomen (see Figure 1-8c).

ANATOMY

The liver is almost completely covered by peritoneum and completely covered by a dense connective tissue layer that lies beneath the peritoneum. It is divided into two principal lobes—the **right lobe** and the **left lobe**—separated by the **falciform ligament** (Figure 24-17). Associated with the right lobe are the inferior **quadrate lobe** and posterior **caudate lobe.** The falciform ligament is a reflection of the parietal peritoneum, which extends from the undersurface of the diaphragm to the superior surface of the liver, between the two principal lobes of the liver. In the free border of the falciform ligament is the **ligamentum teres (round ligament).** It extends from the liver to the umbilicus. The ligamentum teres is a fibrous cord derived from the umbilical vein of the fetus.

The liver, like the pancreas is connected to the small intestine by a series of ducts. Bile, one of the liver's products, enters **bile capillaries** or **canaliculi** (kan'-a-LIK-yoo-

lī) that empty into small ducts. These small ducts eventually merge to form the larger **right** and **left hepatic ducts,** which unite to leave the liver as the **common hepatic duct** (Figures 24-15a, 24-17b, and 24-18). Further on, the common hepatic duct joins the **cystic duct** from the gallbladder. The two tubes become the **common bile duct.** The common bile duct and pancreatic duct enter the duodenum in a common duct called the **hepatopancreatic ampulla.**

HISTOLOGY

The lobes of the liver are made up of numerous functional units called **lobules,** which may be seen under a microscope (Figure 24-18). A lobule consists of cords of **hepatic (liver) cells** arranged in a radial pattern around a **central vein.** Between the cords are endothelial-lined spaces called **sinusoids,** through which blood passes. The sinusoids are also partly lined with phagocytic cells, termed **stellate reticuloendothelial cells,** that destroy worn-out white and red blood cells and bacteria.

> **CLINICAL APPLICATION**
> Liver tissue for diagnostic purposes may be obtained by **liver biopsy.** In the procedure, the needle is inserted through the seventh, eighth, or ninth intercostal space while the patient is holding his or her breath in full expiration. This lessens the possibility of damage to the lung and contamination of the pleural cavity.

BLOOD SUPPLY

The liver receives a double supply of blood. From the hepatic artery it obtains oxygenated blood, and from the hepatic portal vein it receives deoxygenated blood con-

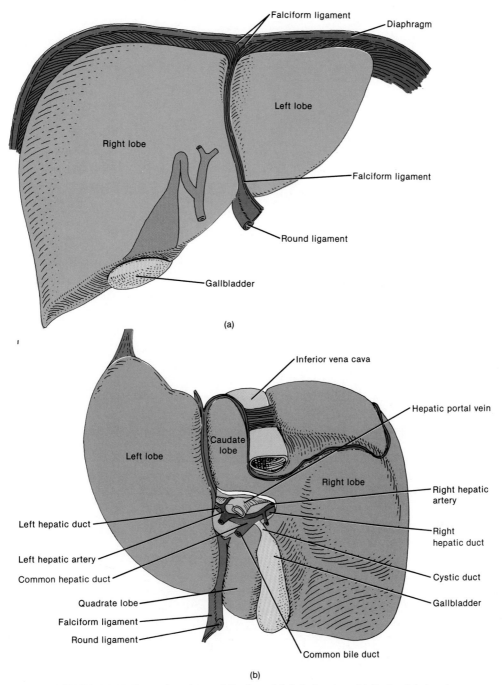

FIGURE 24-17 External anatomy of the liver. (a) Anterior view. (b) Posteroinferior view.

taining newly absorbed nutrients (see Figures 21-15, 21-22, and 24-18). Branches of both the hepatic artery and the hepatic portal vein carry the blood into the sinusoids of the lobules, where oxygen, most of the nutrients, and certain poisons are extracted by the hepatic cells. Nutrients are stored or used to make new materials. The poisons are stored or detoxified. Products manufactured by the hepatic cells and nutrients needed by other cells are secreted back into the blood. The blood then drains into the central vein and eventually passes into a hepatic vein. Unlike the other products of the liver, bile normally is not secreted into the bloodstream.

BILE

Each day the hepatic cells secrete 800 to 1,000 ml (about 1 qt) of bile, a yellow, brownish, or olive-green liquid. It has a pH of 7.6 to 8.6. Bile consists mostly of water and bile salts, cholesterol, a phospholipid called lecithin, bile pigments, and several ions.

Bile is partially an excretory product and partially a digestive secretion. Bile salts assume a role in **emulsification,** the breakdown of fat globules into a suspension of fat droplets about 1 μm in diameter, and absorption of fats following their digestion. Cholesterol is made soluble

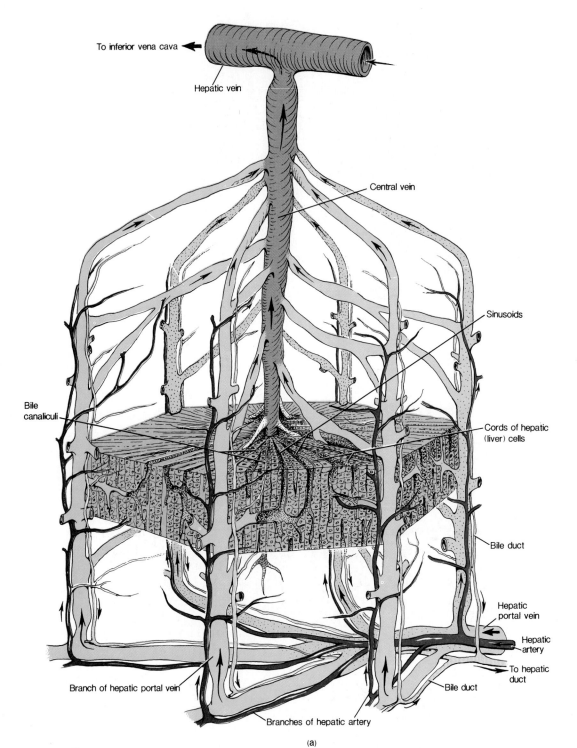

To inferior vena cava

Hepatic vein

Central vein

Sinusoids

Bile canaliculi

Cords of hepatic (liver) cells

Bile duct

Hepatic portal vein

Hepatic artery

To hepatic duct

Branch of hepatic portal vein

Bile duct

Branches of hepatic artery

(a)

FIGURE 24-18 Histology of the liver. (a) Diagram of the microscopic appearance of a liver lobule.

in bile by bile salts and lecithin. The principal bile pigment is **bilirubin.** When red blood cells are broken down, iron, globin, and bilirubin are released. The iron and globin are recycled, but some of the bilirubin is excreted into the bile ducts. Bilirubin eventually is broken down in the intestine, and one of its breakdown products (urobilinogen) give feces their color.

CLINICAL APPLICATION

If insufficient bile salts or lecithin are present in bile or if there is excessive cholesterol, the cholesterol precipitates out of solution and crystalizes to form **gallstones (biliary calculi).** The problems associated with

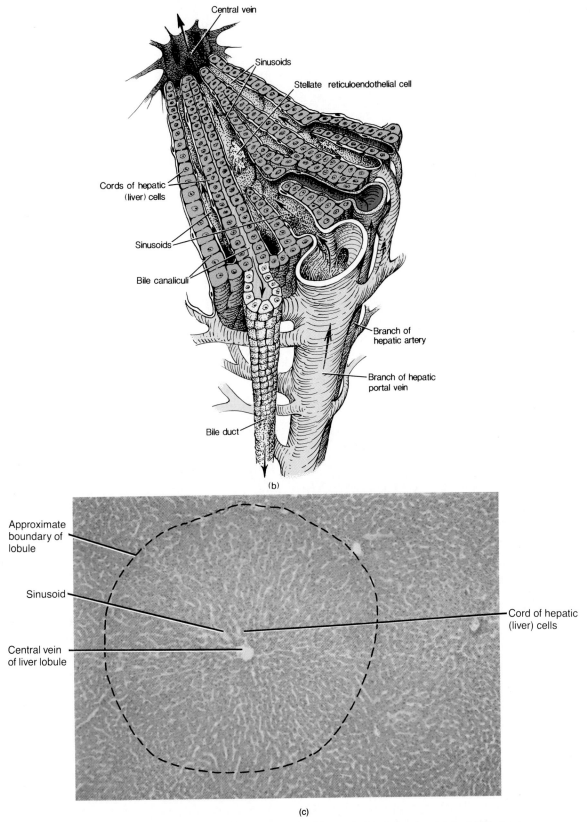

Central vein

Sinusoids

Stellate reticuloendothelial cell

Cords of hepatic (liver) cells

Sinusoids

Bile canaliculi

Branch of hepatic artery

Branch of hepatic portal vein

Bile duct

(b)

Approximate boundary of lobule

Sinusoid

Cord of hepatic (liver) cells

Central vein of liver lobule

(c)

FIGURE 24-18 (*Continued*) Histology of the liver. (b) Enlarged aspect showing details. (c) Photomicrograph of a liver lobule at a magnification of 65×. (© 1983 by Michael H. Ross. Used by permission.)

gallstone formation and the diagnosis and treatment of gallstones are discussed in detail at the end of the chapter.

If the liver is unable to remove bilirubin from the blood because of increased destruction of red blood cells or obstruction of bile ducts, large amounts of bilirubin circulate through the bloodstream and collect in other tissues, giving the skin and eyes a yellow color. This condition is called **jaundice.** If the jaundice is due to damaged red blood cells, it is called **hemolytic jaundice;** if it is due to obstruction in the biliary system, it is known as **obstructive jaundice.**

REGULATION OF BILE SECRETION

The rate at which bile is secreted is determined by several factors. Vagal stimulation can increase the production of bile to more than twice the normal rate. Secretin, the hormone that stimulates the synthesis of pancreatic juice rich in sodium bicarbonate, also stimulates the secretion of bile (see Exhibit 24-4). Within limits, as blood flow through the liver increases, so does the secretion of bile. Finally, the presence of large amounts of bile salts in the blood also increases the rate of bile production.

FUNCTIONS OF THE LIVER

The liver performs many vital functions. Among these are the following.

1. The liver manufactures bile salts which are used in the small intestine for the emulsification and absorption of fats.

2. The liver, together with mast cells, manufactures the anticoagulant heparin and most of the other plasma proteins, such as prothrombin, fibrinogen, and albumin.

3. The stellate reticuloendothelial cells of the liver phagocytose worn-out red and white blood cells and some bacteria.

4. Liver cells contain enzymes that either break down poisons or transform them into less harmful compounds. When amino acids are burned for energy, for example, they leave behind toxic nitrogenous wastes (such as ammonia), that are converted to urea by the liver cells. Moderate amounts of urea are harmless to the body and are easily excreted by the kidneys and sweat glands.

5. Newly absorbed nutrients are collected in the liver. Depending on the body's needs, it can change any excess monosaccharides into glycogen or fat, both of which can be stored, or it can transform glycogen, fat, and protein into glucose.

6. The liver stores glycogen, copper, iron, and vitamins A, D, E, and K. It also stores some poisons that cannot be broken down and excreted. (High levels of DDT are found in the livers of animals, including humans, who eat sprayed fruits and vegetables.)

7. The liver and kidneys participate in the activation of vitamin D.

GALLBLADDER

The **gallbladder** is a pear-shaped sac about 7 to 10 cm (3 to 4 inches) long. It is located in a fossa of the visceral surface of the liver (see Figures 24-15a and 24-17).

HISTOLOGY

The inner wall of the gallbladder consists of a mucous membrane arranged in rugae resembling those of the stomach. The middle, muscular coat of the wall consists of smooth muscle fibers. Contraction of these fibers by hormonal stimulation ejects the contents of the gallbladder into the cystic duct. The outer coat is the visceral peritoneum.

FUNCTION

The function of the gallbladder is to store and concentrate bile (up to 10-fold) until it is needed in the small intestine. In the concentration process, water and many ions are absorbed by the gallbladder mucosa. Bile from the liver enters the small intestine through the common bile duct. When the small intestine is empty, a valve in the duct, called the **sphincter of the hepatopancreatic ampulla,** closes, and the backed-up bile overflows into the cystic duct to the gallbladder.

EMPTYING OF THE GALLBLADDER

In order for the gallbladder to eject bile into the small intestine to participate in the digestive process, the muscularis must contract to force bile into the common bile duct and the sphincter of the hepatopancreatic ampulla must relax. Chyme entering the duodenum that contains particularly high concentrations of fats or partially digested proteins stimulates the intestinal mucosa to secrete cholecystokinin (CCK). This hormone brings about contraction of the muscularis coupled with relaxation of the sphincter of the hepatopancreatic ampulla, resulting in emptying of the gallbladder (see Exhibit 24-4).

SMALL INTESTINE

The major portions of digestion and absorption occur in a long tube called the **small intestine.** The small intestine begins at the pyloric sphincter of the stomach, coils through the central and lower part of the abdominal cavity, and eventually opens into the large intestine. It averages 2.5 cm (1 inch) in diameter and about 6.35 m (21 ft) in length.

ANATOMY

The small intestine is divided into three segments (see Figure 24-1). The **duodenum** (doo'-ō-DĒ-num), the shortest part, originates at the pyloric sphincter of the stomach and extends about 25 cm (10 inches) until it merges with the jejunum. The **jejunum** (jē-JOO-num) is about 2.5 m (8 ft) long and extends to the ileum. The final portion of the small intestine, the **ileum** (IL-ē-um), measures about 3.6 m (12 ft) and joins the large intestine at the **ileocecal** (il'-ē-ō-SĒ-kal) **valve.**

HISTOLOGY

The wall of the small intestine is composed of the same four tunics that make up most of the GI tract. However, both the mucosa and the submucosa are modified to allow the small intestine to complete the processes of digestion and absorption (Figure 24-19).

The mucosa contains many pits lined with glandular epithelium. These pits—the **intestinal glands**—secrete the intestinal digestive enzymes, collectively called intestinal juice. The submucosa of the duodenum contains **Brunner's glands,** which secrete an alkaline mucus to protect the wall of the small intestine from the action of the enzymes and to aid in neutralizing acid in the chyme. Some of the epithelial cells in the mucosa and submucosa have been transformed to goblet cells, which secrete additional mucus.

Since almost all the absorption of nutrients occurs in the small intestine, its structure is specially adapted for this function. Its length alone provides a large surface

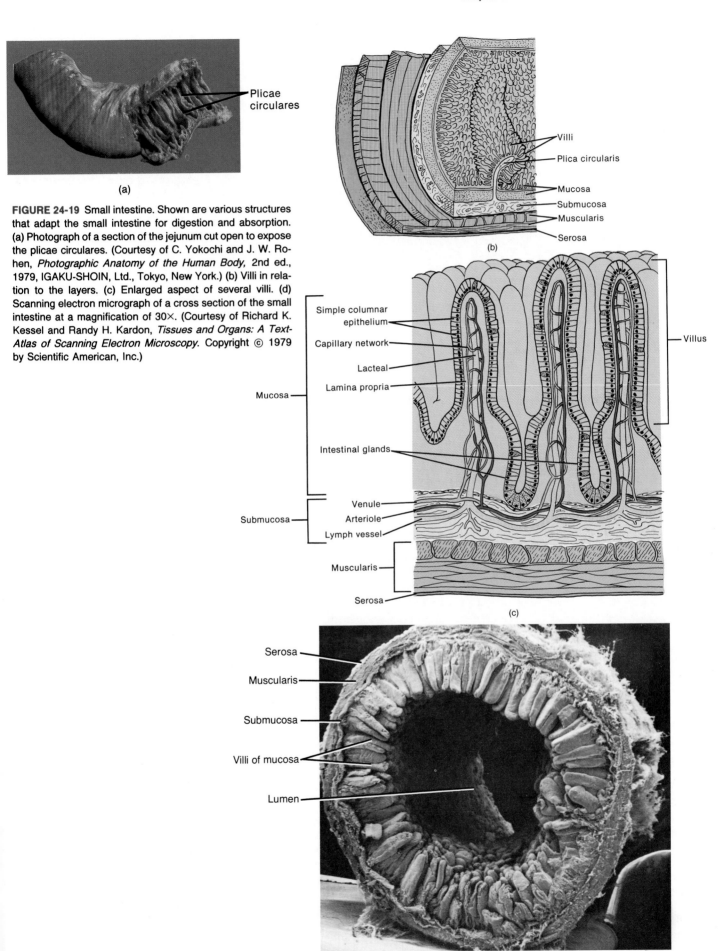

(a)

FIGURE 24-19 Small intestine. Shown are various structures that adapt the small intestine for digestion and absorption. (a) Photograph of a section of the jejunum cut open to expose the plicae circulares. (Courtesy of C. Yokochi and J. W. Rohen, *Photographic Anatomy of the Human Body*, 2nd ed., 1979, IGAKU-SHOIN, Ltd., Tokyo, New York.) (b) Villi in relation to the layers. (c) Enlarged aspect of several villi. (d) Scanning electron micrograph of a cross section of the small intestine at a magnification of 30×. (Courtesy of Richard K. Kessel and Randy H. Kardon, *Tissues and Organs: A Text-Atlas of Scanning Electron Microscopy.* Copyright © 1979 by Scientific American, Inc.)

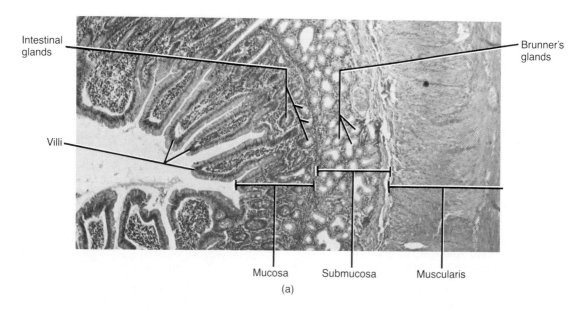

(a)

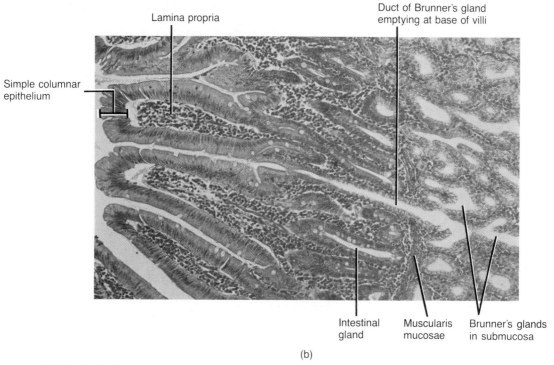

(b)

FIGURE 24-20 Histology of the small intestine. (a) Photomicrograph of a portion of the wall of the duodenum at a magnification of 40×. (b) Photomicrograph of an enlarged aspect of the duodenal mucosa at a magnification of 165×. (© 1983 by Michael H. Ross. Used by permission.)

area for absorption and that area is further increased by modifications in the structure of its wall. The epithelium covering and lining the mucosa consists of simple columnar epithelium. These epithelial cells, except those transformed into goblet cells, contain **microvilli,** fingerlike projections of the plasma membrane. Larger amounts of digested nutrients diffuse into the intestinal wall because the microvilli increase the surface area of the plasma membrane. They also increase the surface area for digestion.

The mucosa lies in a series of **villi,** projections 0.5 to 1 mm high, giving the intestinal mucosa its velvety appearance. The enormous number of villi (4 to 5 million)

vastly increases the surface area of the epithelium available for absorption and digestion. Each villus has a core of lamina propria, the connective tissue layer of the mucosa. Embedded in this connective tissue are an arteriole, a venule, a capillary network, and a **lacteal** (LAK-tē-al) or lymphatic vessel. Nutrients that diffuse through the epithelial cells which cover the villus are able to pass through the capillary walls and the lacteal and enter the cardiovascular and lymphatic systems.

In addition to the microvilli and villi, a third set of projections called **plicae circulares** (PLĪ-kē SER-kyoo-lar-es) further increases the surface area for absorption and digestion. The plicae are permanent deep folds in

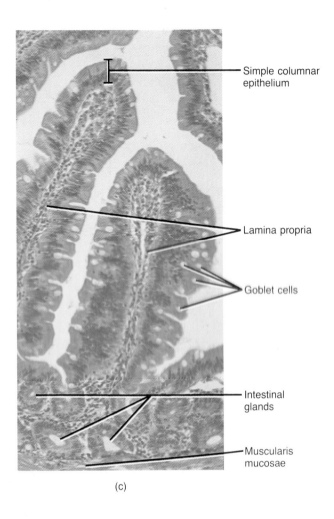

Simple columnar epithelium

Lamina propria

Goblet cells

Intestinal glands

Muscularis mucosae

(c)

FIGURE 24-20 (*Continued*) Histology of the small intestine. (c) Photomicrograph of an enlarged aspect of two villi from the ileum at a magnification of 120×. (© 1983 by Michael H. Ross. Used by permission.) (d) Scanning electron micrograph of several villi at a magnification of 120×. (Courtesy of Biophoto Associates.)

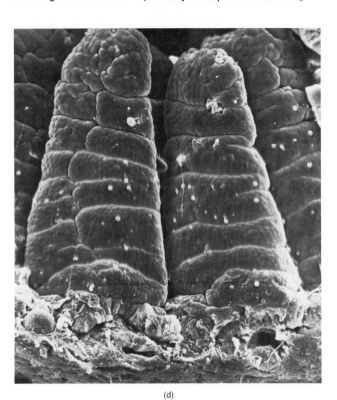

(d)

the mucosa and submucosa. Some of the folds extend all the way around the intestine, and others extend only part way around. The plicae circulares enhance absorption by causing the chyme to spiral, rather than moving in a straight line, as it passes through the small intestine.

The muscularis of the small intestine consists of two layers of smooth muscle. The outer, thinner layer contains longitudinally arranged fibers. The inner, thicker layer contains circularly arranged fibers. Except for a major portion of the duodenum, the serosa (or visceral peritoneum) completely covers the small intestine. Additional histological aspects of the small intestine are shown in Figure 24-20.

There is an abundance of lymphatic tissue in the form of lymph nodules, masses of lymphatic tissue not covered by a capsule wall. **Solitary lymph nodules** are most numerous in the lower part of the ileum. Aggregated lymph nodules, referred to as **Peyer's (PĪ-erz) patches,** are numerous in the ileum.

INTESTINAL JUICE

Intestinal juice is a clear yellow fluid secreted in amounts of about 2 to 3 liters (about 2 to 3 qt) a day. It has a pH of 7.6, which is slightly alkaline, and contains water, mucus, and several enzymes. The intestinal enzymes include three carbohydrate-digesting enzymes called *maltase, sucrase,* and *lactase;* several protein-digesting en-

zymes called *peptidases;* and two nucleic acid–digesting enzymes, *ribonuclease* and *deoxyribonuclease.* Much of the digestion of foods by enzymes produced by the small intestine actually occurs in epithelial cells lining the small intestine rather than in a fluid outside the cells in the lumen of the tube.

DIGESTION IN THE SMALL INTESTINE

Mechanical

The movements of the small intestine are arbitrarily divided into two types: segmentation and peristalsis. **Segmentation** is the major movement of the small intestine. It is strictly a localized contraction in areas containing food. It mixes chyme with the digestive juices and brings the particles of food into contact with the mucosa for absorption. It does not push the intestinal contents along the tract. Segmentation starts with the contractions of circular muscle fibers in a portion of the small intestine, an action that constricts the intestine into segments. Next, muscle fibers that encircle the middle of each segment also contract, dividing each segment again. Finally, the fibers that contracted first relax, and each small segment unites with an adjoining small segment so that large segments are reformed. This sequence of events is repeated 12 to 16 times a minute, sloshing the chyme back and forth (Figure 24-21). Segmentation depends mainly on

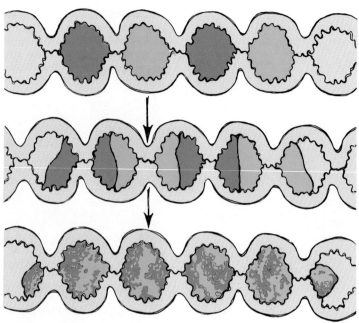

FIGURE 24-21 Segmentation. Localized contractions thoroughly mix the small intestinal contents.

intestinal distension which initiates impulses to the central nervous system. Returning parasympathetic impulses increase motility. Sympathetic impulses decrease intestinal motility.

Peristalsis propels the chyme onward through the intestinal tract. Peristaltic contractions in the small intestine are normally very weak compared to those in the esophagus or stomach. Chyme moves through the small intestine at a rate of about 1 cm/min. Thus, chyme remains in the small intestine for 3 to 5 hours. Peristalsis, like segmentation, is initiated by distension and controlled by the autonomic nervous system.

Chemical

In the mouth, salivary amylase converts starch (polysaccharide) to maltose (disaccharide). In the stomach, pepsin converts proteins to peptides (small proteins). Thus, chyme entering the small intestine contains partially digested carbohydrates, partially digested proteins, and essentially undigested lipids. The completion of the digestion of carbohydrates, proteins, and lipids is a collective effort of pancreatic juice, bile, and intestinal juice in the small intestine.

● *Carbohydrates* Even though the action of **salivary amylase** may continue in the stomach for some time, its activity is blocked by the acid pH of the stomach. Thus, few starches are reduced to maltose by the time chyme leaves the stomach. Any starches not already broken down into the disaccharide are converted by **pancreatic amylase,** an enzyme in pancreatic juice that acts in the small intestine.

Sucrose and lactose, two disaccharides, are ingested ~~ such and are not acted upon until they reach the small

intestine. Three enzymes in the intestinal juice digest the disaccharides into monosaccharides. **Maltase** splits maltose into two molecules of glucose. **Sucrase** breaks sucrose into a molecule of glucose and a molecule of fructose. **Lactase** digests lactose into a molecule of glucose and a molecule of galactose. This completes the digestion of carbohydrates.

CLINICAL APPLICATION
In some individuals the mucosal cells of the small intestine fail to produce lactase, which is essential for the digestion of lactose. This condition is called **lactose intolerance.** The deficiency can lead to an electrolyte imbalance. Its symptoms include diarrhea, gas and bloating, and abdominal cramps. The deficiency occurs in 30 to 70 percent of blacks and 15 percent of Caucasians. It sometimes occurs following gastric surgery.

● *Proteins* Protein digestion starts in the stomach, where proteins are fragmented by the action of **pepsin** into peptides. Enzymes found in pancreatic juice continue the digestion. **Trypsin** and **chymotrypsin** continue to break down proteins into peptides. Although pepsin, trypsin, and chymotrypsin all convert whole proteins into peptides, their actions differ somewhat since each splits peptide bonds between different amino acids. **Carboxypolypeptidase** acts on peptides and breaks the peptide bond that attaches the terminal amino acid to the carboxyl (acid) end of the peptide. Protein digestion is completed by the **peptidases. Aminopeptidase** acts on peptides and breaks the peptide bonds that attach amino acids to the amino end of the peptide. **Dipeptidase** splits dipeptides (two amino acids joined by a peptide bond) into amino acids that can be absorbed.

EXHIBIT 24-5

SUMMARY OF DIGESTIVE ENZYMES

ENZYME	SOURCE	SUBSTRATE	PRODUCT
Salivary amylase	Salivary glands.	Starches (polysaccharides).	Maltose (disaccharide).
Pepsin (activated from pepsinogen by hydrochloric acid)	Stomach (zymogenic cells).	Proteins.	Peptides.
Pancreatic amylase	Pancreas.	Starches (polysaccharides).	Maltose (disaccharide).
Trypsin (activated from trypsinogen by enterokinase)	Pancreas.	Proteins.	Peptides.
Chymotrypsin (activated from chymotrypsinogen by trypsin)	Pancreas.	Proteins.	Peptides.
Carboxypolypeptidase (activated from procarboxypolypeptidase by trypsin)	Pancreas.	Terminal amino acid at carboxyl (acid) end of peptides.	Peptides and amino acids.
Pancreatic lipase	Pancreas.	Neutral fats (triglycerides) that have been emulsified by bile salts.	Fatty acids and monoglycerides.
Maltase	Small intestine.	Maltose.	Glucose.
Sucrase	Small intestine.	Sucrose.	Glucose and fructose.
Lactase	Small intestine.	Lactose.	Glucose and galactose.
Peptidases			
Aminopeptidase	Small intestine.	Terminal amino acids at amino end of peptides.	Amino acids.
Dipeptidase	Small intestine.	Dipeptides.	Amino acids.
Ribonuclease	Pancreas and small intestine.	Ribonucleic acid nucleotides.	Pentoses and nitrogen bases.
Deoxyribonuclease	Pancreas and small intestine.	Deoxyribonucleic acid nucleotides.	Pentoses and nitrogen bases.

● *Lipids* In an adult, almost all lipid digestion occurs in the small intestine. The first step in the process involves the preparation of neutral fats (triglycerides) by bile salts. Neutral fats, or just simply fats, are the most abundant lipids in the diet. They are called triglycerides because they consist of a molecule of glycerol and three molecules of fatty acid (see Figure 2-11). Bile salts break the globules of fat into droplets about 1 μm in diameter. This process is called **emulsification.** It is necessary so that the fat-splitting enzyme can get at the lipid molecules. In the second step, **pancreatic lipase,** an enzyme found in pancreatic juice, hydrolyzes each fat molecule into fatty acids and monoglycerides, end products of fat digestion. Lipase removes two of the three fatty acids from glycerol; the third remains attached to the glycerol, thus forming monoglycerides.

● *Nucleic Acids* Both intestinal juice and pancreatic juice contain **nucleases** that digest nucleotides into their constituent pentoses and nitrogen bases. **Ribonuclease** acts on ribonucleic acid nucleotides and **deoxyribonuclease** acts on deoxyribonucleic acid nucleotides.

A summary of digestive enzymes in terms of source, substrate acted on, and product is presented in Exhibit 24-5.

REGULATION OF INTESTINAL SECRETION

The most important means for regulating small intestinal secretion is local reflexes in response to the presence of chyme. Also, secretin and cholecystokinin (CCK) stimulate the production of intestinal juice.

ABSORPTION

All the chemical and mechanical phases of digestion from the mouth down through the small intestine are directed toward changing food into forms that can pass through the epithelial cells lining the mucosa into the underlying blood and lymph vessels. These forms are monosaccharides (glucose, fructose, and galactose), amino acids, fatty acids, glycerol, and glycerides. Passage of these digested nutrients from the alimentary canal into the blood or lymph is called **absorption.**

About 90 percent of all absorption of nutrients takes place throughout the length of the small intestine. The other 10 percent occurs in the stomach and large intestine. Any undigested or unabsorbed material left in the small intestine is passed on to the large intestine. Absorption of materials in the small intestine occurs specifically through the villi and depends on diffusion, facilitated diffusion, osmosis, and active transport.

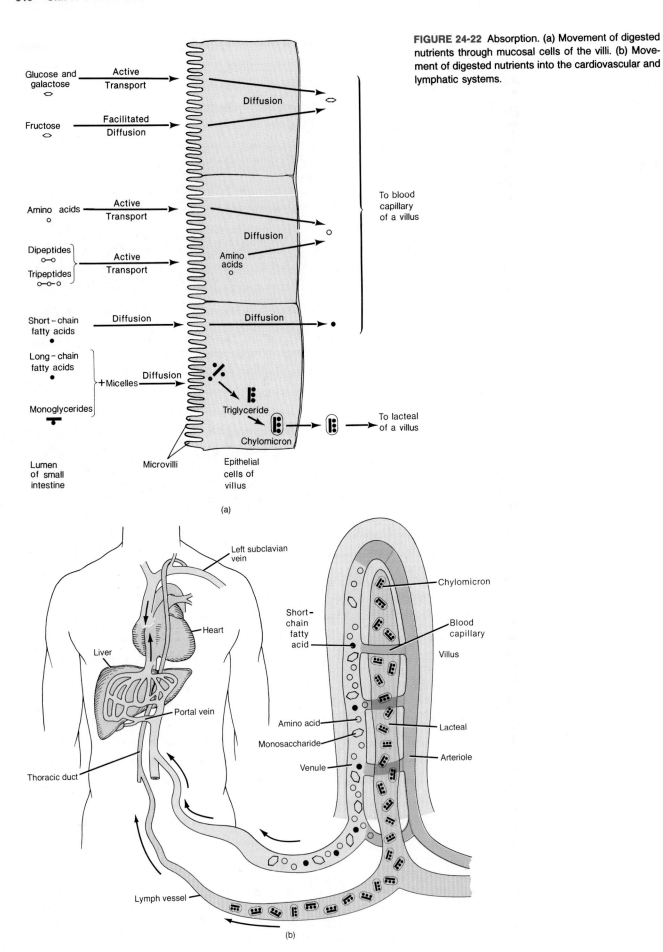

FIGURE 24-22 Absorption. (a) Movement of digested nutrients through mucosal cells of the villi. (b) Movement of digested nutrients into the cardiovascular and lymphatic systems.

Carbohydrates

Essentially all carbohydrates are absorbed as monosaccharides. Glucose and galactose are transported into epithelial cells of the villi by an active process that is somehow coupled with the active transport of sodium. Fructose is transported by facilitated diffusion. Transported monosaccharides then move out of the epithelial cells by diffusion and enter the capillaries of the villi. From here they are transported in the bloodstream to the liver via the hepatic portal system. After their passage through the liver, they move through the heart and then enter general circulation (Figure 24-22).

Proteins

Most proteins are absorbed as amino acids and the process occurs mostly in the duodenum and jejunum. Amino acid transport into epithelial cells of the villi is an active transport process coupled with active sodium transport. Amino acids move out of the epithelial cells by diffusion to enter the bloodstream. They follow the same route as that taken by monosaccharides. At times, dipeptides and tripeptides can be taken in by epithelial cells by active transport. Most of them are hydrolyzed to amino acids in the cells and then passed into the capillaries of the villi (Figure 24-22).

Lipids

As a result of emulsification and fat digestion, neutral fats (triglycerides) are broken down into monoglycerides and fatty acids. Lipase effectively removes two of the three fatty acids from glycerol during fat digestion; the other fatty acid remains attached to glycerol, thus forming monoglycerides. Short-chain fatty acids (those with less than 10 to 12 carbon atoms) pass into the epithelial cells by diffusion and follow the same route taken by monosaccharides and amino acids (Figure 24-22).

Most fatty acids are long-chain fatty acids. They and the monoglycerides are transported differently. Bile salts form spherical aggregates called **micelles** (mī-SELZ). They are about 25 Å in diameter and consist of 20 to 50 molecules of bile salt. Despite their large size, micelles have the ability to dissolve in water in the intestinal fluid. During fat digestion, fatty acids and monoglycerides dissolve in the center of the micelles, and it is in this form that they reach the epithelial cells of the villi. On coming into contact with the surfaces of the epithelial cells, fatty acids and monoglycerides diffuse into the cells, leaving the micelles behind in chyme. The micelles continually repeat this ferrying function. The majority of bile salts in the small intestine are ultimately reabsorbed in the ileum to be returned by the blood to the liver for resecretion. This cycle is called **enterohepatic circulation.**

Within the epithelial cells, many monoglycerides are further digested by lipase in the cells to glycerol and fatty acids. Then, the fatty acids and glycerol are recombined to form triglycerides in the smooth endoplasmic reticulum of the epithelial cell. The triglycerides, still in the endoplasmic reticulum, aggregate into globules along with phospholipids and cholesterol and become coated with proteins. These masses are called **chylomicrons.** The protein coat keeps the chylomicrons suspended and from sticking to each other. The chylomicrons leave the epithelial cell and enter the lacteal of a villus. From here, they are transported by way of lymphatic vessels to the thoracic duct and enter the cardiovascular system at the left subclavian vein. Finally, they arrive at the liver through the hepatic artery (Figure 24-22).

Water

The total volume of fluid that enters the small intestine each day is about 9 liters (about 9 qt). This fluid is derived from ingestion of liquids (about 1.5 liters) and from various gastrointestinal secretions (about 7.5 liters). Roughly 8 to 8.5 liters of the fluid in the small intestine is absorbed; the remainder, about 0.5 to 1.0 liter, passes into the large intestine. There most of it is also absorbed.

The absorption of water by the small intestine occurs by osmosis from the lumen of the small intestine through epithelial cells and into the blood capillaries in the villi. The normal rate of absorption is about 200 to 400 ml/h. Water can move across the intestinal mucosa in both directions. The absorption of water from the small intestine is associated with the absorption of electrolytes and digested foods in order to maintain an osmotic balance with the blood.

Electrolytes

The electrolytes absorbed by the small intestine are mostly constituents of gastrointestinal secretions. Some are also components of ingested foods and liquids. Sodium is able to move in and out of epithelial cells by diffusion. It can also move into mucosal cells by active transport for removal from the small intestine. Chloride, iodide, and nitrate ions can passively follow sodium ions or be actively transported. Calcium ions are also actively transported and their movement depends on parathyroid hormone (PTH) and vitamin D. Other electrolytes such as iron, potassium, magnesium, and phosphate also move by active transport.

Vitamins

Fat-soluble vitamins, such as A, D, E, and K, are absorbed along with ingested dietary fats in micelles. In fact, they cannot be absorbed unless they are ingested with some fat. Most water-soluble vitamins, such as the B vitamins and C, are absorbed by diffusion. Vitamin B_{12}, you may recall, requires combination with intrinsic factor produced by the stomach for its absorption.

A summary of the digestive and absorptive activities of the small intestine is presented in Exhibit 24-6.

LARGE INTESTINE

The overall functions of the large intestine are the completion of absorption, the manufacture of certain vitamins, the formation of feces, and the expulsion of feces from the body.

EXHIBIT 24-6

SUMMARY OF DIGESTION AND ABSORPTION IN THE SMALL INTESTINE

STRUCTURE	ACTIVITY
Pancreas	Delivers pancreatic juice into the duodenum via the pancreatic duct (see Exhibit 24-5 for pancreatic enzymes and their functions).
Liver	Produces bile which is necessary for emulsification of fats.
Gallbladder	Stores, concentrates, and delivers bile into the duodenum via the common bile duct.
Small Intestine **Mucosa and submucosa**	
Intestinal glands	Secrete intestinal juice (see Exhibit 24-5 for intestinal enzymes and their functions).
Brunner's glands	Secrete mucus for protection and lubrication.
Microvilli	Fingerlike projections of epithelial cells that increase surface area for absorption and digestion.
Villi	Projections of mucosa that are the site of absorption of digested food and also increase the surface area for absorption and digestion.
Plicae circulares	Circular folds of mucosa and submucosa that increase surface area for absorption and digestion.
Muscularis Segmentation	Consists of alternating contractions of circular fibers that produce segmentation and resegmentation of portions of the small intestine; mixes chyme with digestive juices and brings food into contact with the mucosa for absorption.
Peristalsis	Consists of mild waves of contraction and relaxation of circular and longitudinal muscle passing the length of the small intestine; moves chyme toward ileocecal valve.

ANATOMY

The **large intestine** is about 1.5 m (5 ft) in length and averages 6.5 cm (2.5 inches) in diameter. It extends from the ileum to the anus and is attached to the posterior abdominal wall by its **mesocolon** of visceral peritoneum. Structurally, the large intestine is divided into four principal regions: cecum, colon, rectum, and anal canal (Figure 24-23).

The opening from the ileum into the large intestine is guarded by a fold of mucous membrane called the **ileocecal valve.** This structure allows materials from the small intestine to pass into the large intestine. Hanging below the ileocecal valve is the **cecum,** a blind pouch about 6 cm (2.5 inches) long. Attached to the cecum is a twisted, coiled tube, measuring about 8 cm (3 inches) in length, called the **vermiform appendix** (*vermis* = worm). The visceral peritoneum of the appendix, called the **mesoappendix,** attaches the appendix to the inferior part of the ileum and adjacent part of the posterior abdominal wall.

The open end of the cecum merges with a long tube called the **colon.** The colon is divided into ascending, transverse, descending, and sigmoid portions. The **ascending colon** ascends on the right side of the abdomen, reaches the undersurface of the liver, and turns abruptly to the left. Here it forms the **right colic (hepatic) flexure.** The colon continues across the abdomen to the left side as the **transverse colon.** It curves beneath the lower end of the spleen on the left side as the **left colic (splenic) flexure** and passes downward to the level of the iliac crest as the **descending colon.** The **sigmoid colon** begins at the left iliac crest, projects inward to the midline, and terminates as the rectum at about the level of the third sacral vertebra.

The **rectum,** the last 20 cm (8 inches) of GI tract, lies anterior to the sacrum and coccyx. The terminal 2 to 3 cm (1 inch) of the rectum is called the **anal canal** (Figure 24-24). The mucous membrane of the anal canal is arranged in longitudinal folds called **anal columns** that contain a network of arteries and veins. The opening of the anal canal to the exterior is called the **anus.** It is guarded by an internal sphincter of smooth muscle and an external sphincter of skeletal muscle. Normally the anus is closed except during the elimination of the wastes of digestion.

CLINICAL APPLICATION

Inflammation and enlargement of the rectal veins (varicose veins) due to weakening of the valves is known as **hemorrhoids (piles).** Initially contained within the anus (first degree), they gradually enlarge until they prolapse or extend outward on defecation (second degree) and finally remain prolapsed through the anal orifice (third degree).

HISTOLOGY

The wall of the large intestine differs from that of the small intestine in several respects (Figure 24-25). No villi or permanent circular folds are found in the mucosa, which does, however, contain simple columnar epithelium with numerous goblet cells. These cells secrete mucus that lubricates the colonic contents as they pass through the colon. Solitary lymph nodules also are found in the mucosa. The submucosa of the large intestine is similar to that found in the rest of the alimentary canal. The muscularis consists of an external layer of longitudinal muscles

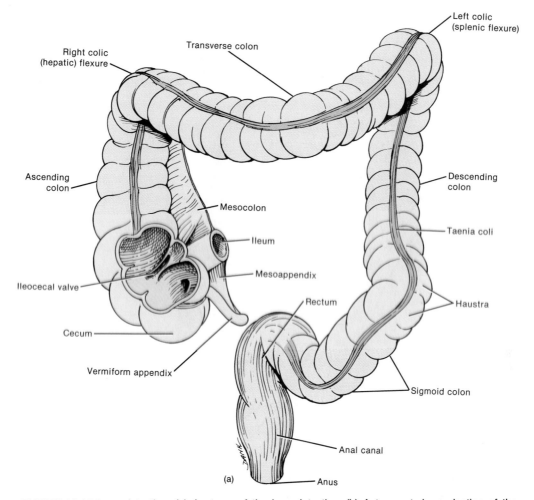

FIGURE 24-23 Large intestine. (a) Anatomy of the large intestine. (b) Anteroposterior projection of the large intestine in which several haustra are clearly visible. (Courtesy of Lester W. Paul and John H. Juhl, *The Essentials of Roentgen Interpretation,* 3rd ed., Harper & Row, Publishers, Inc., New York, 1972.)

and an internal layer of circular muscles. Unlike other parts of the digestive tract, the longitudinal muscles do not form a continuous sheet around the wall, but are broken up into three flat bands called **taeniae coli** (TĔ-ni-ē KŌ-lī). Each band runs the length of most of the large intestine. Tonic contractions of the bands gather the colon into a series of pouches called **haustra** (HAWS-tra), which give the colon its puckered appearance (see Figure 24-23). The serosa of the large intestine is part of the visceral peritoneum. Small pouches of visceral peritoneum filled with fat are attached to taeniae and are called **epiploic appendages.**

DIGESTION IN THE LARGE INTESTINE

Mechanical

The passage of chyme from the ileum into the cecum is regulated by the action of the ileocecal valve. The valve normally remains mildly contracted so that the passage of chyme into the cecum is usually a slow process. Immediately following a meal, there is a **gastroileal reflex** in which ileal peristalsis is intensified and any chyme in the ileum is forced into the cecum. The hormone stomach gastrin also relaxes the valve. Whenever the cecum is

Haustra of transverse colon

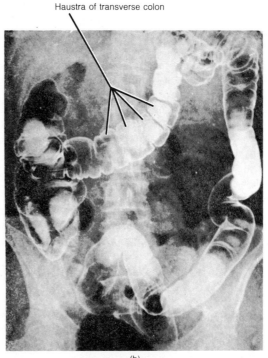

(b)

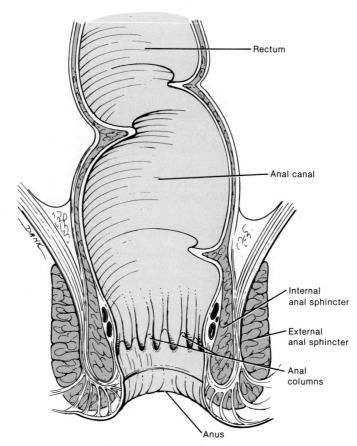

FIGURE 24-24 Anal canal seen in longitudinal section.

distended, the degree of contraction of the ileocecal valve is intensified.

Movements of the colon begin when substances enter through the ileocecal valve. Since chyme moves through the small intestine at a fairly constant rate, the time required for a meal to pass into the colon is determined by gastric evacuation time. As food passes through the ileocecal valve, it fills the cecum and accumulates in the ascending colon.

One movement characteristic of the large intestine is **haustral churning.** In this process, the haustra remain relaxed and distended while they fill up. When the distension reaches a certain point, the walls contract and squeeze the contents into the next haustrum. **Peristalsis** also occurs, although at a slower rate than in other portions of the tract (3 to 12 contractions per minute). A final type of movement is **mass peristalsis,** a strong peristaltic wave that begins in about the middle of the transverse colon and drives the colonic contents into the rectum. Food in the stomach initiates this reflex action in the colon. Thus mass peristalsis usually takes place three or four times a day, during a meal or immediately after.

Chemical

The last stage of digestion occurs through bacterial, not enzymatic action. Mucus is secreted by the glands of the large intestine, but no enzymes are secreted. Chyme is prepared for elimination by the action of bacteria. These bacteria ferment any remaining carbohydrates and release hydrogen, carbon dioxide, and methane gas. These gases contribute to flatus (gas) in the colon. They also convert remaining proteins to amino acids and break down the amino acids into simpler substances: indole, skatole, hydrogen sulfide, and fatty acids. Some of the indole and skatole is carried off in the feces and contributes to its odor. The rest is absorbed and transported to the liver, where they are converted to less toxic compounds and excreted in the urine. Bacteria also decompose bilirubin to simpler pigments (urobilinogen) which give feces their brown color. Several vitamins needed for normal metabolism, including some B vitamins and vitamin K, are synthesized by bacterial action and absorbed.

ABSORPTION AND FECES FORMATION

By the time the chyme has remained in the large intestine 3 to 10 hours, it has become solid or semisolid as a result of absorption and is now known as **feces.** Chemically, feces consist of water, inorganic salts, sloughed off epithelial cells from the mucosa of the alimentary canal, bacteria, products of bacterial decomposition, and undigested parts of food not attacked by bacteria.

Although most water absorption occurs in the small intestine, the large intestine absorbs enough to make it an important organ in maintaining the body's water balance. Of the 0.5 to 1.0 liter that enters the large intestine, all but about 100 ml is absorbed. The absorption is greatest

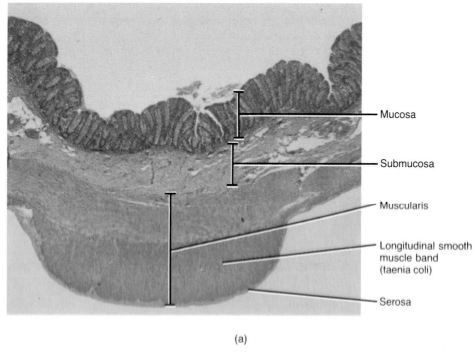

Mucosa

Submucosa

Muscularis

Longitudinal smooth
muscle band
(taenia coli)

Serosa

(a)

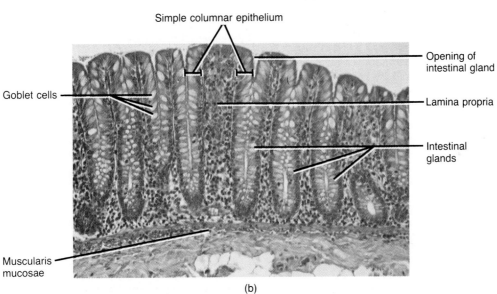

Simple columnar epithelium

Goblet cells

Muscularis
mucosae

Opening of
intestinal gland

Lamina propria

Intestinal
glands

(b)

FIGURE 24-25 Histology of the large intestine. (a) Photomicrograph of a portion of the wall of the large intestine at a magnification of 25×. (b) Photomicrograph of an enlarged aspect of the mucosa of the large intestine at a magnification of 140×. (© 1983 by Michael H. Ross. Used by permission.)

in the cecum and ascending colon. The large intestine also absorbs electrolytes, including sodium and chloride.

DEFECATION

Mass peristaltic movements push fecal material into the rectum. The resulting distension of the rectal wall stimulates pressure-sensitive receptors, initiating a reflex for **defecation,** which is emptying of the rectum. Contraction of the longitudinal rectal muscles shortens the rectum, thereby increasing the pressure inside it. The pressure along with voluntary contractions of the diaphragm and abdominal muscles forces the sphincters open, and the feces are expelled through the anus. Voluntary contrac-

tions of the diaphragm and abdominal muscles aid defecation by increasing the pressure inside the abdomen, which pushes the walls of the sigmoid colon and rectum inward. If defecation does not occur, the feces remain in the rectum until the next wave of mass peristalsis again stimulates the pressure-sensitive receptors, creating the desire to defecate.

In infants, the defecation reflex causes automatic emptying of the rectum without the voluntary control of the external anal sphincter. In certain instances of spinal cord injury, the reflex is abolished and defecation requires supportive measues, such as cathartics (laxatives).

Activities of the large intestine are summarized in Exhibit 24-7.

EXHIBIT 24-7

DIGESTIVE ACTIVITIES OF LARGE INTESTINE

STRUCTURE	ACTION	FUNCTION
Mucosa	Secretes mucus.	Lubricates colon and protects mucosa.
	Absorbs water and other soluble compounds.	Maintains water balance; solidifies feces. Vitamins and electrolytes absorbed and toxic substances sent to liver to be detoxified.
	Bacterial activity.	Breaks down undigested carbohydrates, proteins, and amino acids into products that can be expelled in feces or absorbed and detoxified by liver. Certain B vitamins and vitamin K synthesized.
Muscularis	Haustral churning.	Contents moved from haustrum to haustrum by muscular contractions.
	Peristalsis.	Contents moved along length of colon by contractions of circular and longitudinal muscles.
	Mass peristalsis.	Contents forced into sigmoid colon and rectum by strong peristaltic wave.
	Defecation.	Feces eliminated by contractions in sigmoid colon and rectum.

CLINICAL APPLICATION

Diarrhea refers to frequent defecation of liquid feces caused by increased motility of the intestines. Since chyme passes too quickly through the small intestine and feces passes too quickly through the large intestine, there is not enough time for absorption. Like vomiting, diarrhea can result in dehydration and electrolyte imbalances. Diarrhea may be caused by microbes that irritate the gastrointestinal mucosa and stress.

Constipation refers to infrequent or difficult defecation. It is caused by decreased motility of the intestines in which feces remain in the colon for prolonged periods of time. As it does so, there is considerable absorption and feces become dry and hard. Constipation may be caused by improper bowel habits, spasms of the colon, insufficient bulk in the diet, lack of exercise, and emotions. Usual treatment for constipation is a mild cathartic (laxative) that induces defecation.

DISORDERS: HOMEOSTATIC IMBALANCES

Dentral Caries

Dental caries, or tooth decay, involve a gradual demineralization (softening) of the enamel and dentin. If this condition remains untreated, various microorganisms may invade the pulp, causing inflammation and infection with subsequent death (necrosis) of the dental pulp and abscess of the alveolar bone surrounding the root's apex. Such teeth are treated by root canal therapy.

The process of dental caries is initiated when bacteria act on sugars, giving off acids which demineralize the enamel. Microbes that digest sugar into lactic acid are common in the mouth cavity. One that seems to be cariogenic (caries causing) is the bacterium *Streptococcus mutans.* **Dextran,** a sticky polysaccharide produced from sucrose, forms a capsule around the bacteria causing them to stick to the teeth. Masses of bacterial cells, dextran, and other debris adhering to teeth is collectively called **dental plaque.** Saliva cannot reach the tooth surface to buffer the acid because the plaque covers the teeth.

Preventive measures include prenatal diet supplements (chiefly vitamin D, calcium, and phosphorus) and fluoride treatments to protect against acids during the period when teeth are being calcified. Naturally occurring excessive fluoride may cause a light brown to brownish black discoloration of the enamel of the permanent teeth called mottling.

Brushing the teeth immediately after eating removes the plaque from flat surfaces before the bacteria have a chance to go to work. Dentists also suggest that the plaque between the teeth be removed every 24 hours with dental floss or by flushing with a water irrigation device.

Periodontal Disease

Periodontal disease is a collective term for a variety of conditions characterized by inflammation and degeneration of the gingivae, alveolar bone, periodontal ligament, and cementum. One such condition is called **pyorrhea.** The initial symptoms are enlargement and inflammation of the soft tissue and bleeding gums. Without treatment, the soft tissue may deteriorate and the alveolar bone may be resorbed, causing loosening of the teeth and recession of the gums.

Periodontal diseases are frequently caused by poor oral hygiene; local irritants, such as bacteria, impacted food, and cigarette smoke; or by a poor "bite." The latter may put a strain on the tissues supporting the teeth. Periodontal diseases may also be caused by allergies, vitamin deficiencies (especially vitamin C), and a number of systemic disorders, especially those that affect bone, connective tissue, or circulation.

Peritonitis

Peritonitis is an acute inflammation of the serous membrane lining the abdominal cavity and covering the abdominal viscera. One possible cause is contamination of the peritoneum by pathogenic bacteria from the external environment. This contamination could result from accidental or surgical wounds in the abdominal wall or from perforation or rupture of organs with consequent exposure to the outside environment. Another possible cause is perforation of the walls of organs that contain bacteria or chemicals beneficial to the organ but toxic to the peritoneum. For example, the large intestine contains colonies of bacteria that live on undigested nutrients and break them down so they can be eliminated. But if the bacteria enter the peritoneal cavity, they attack the cells of the peritoneum for food and produce acute infection. Moreover, the peritoneum has no natural barriers that keep it from being irritated or digested by

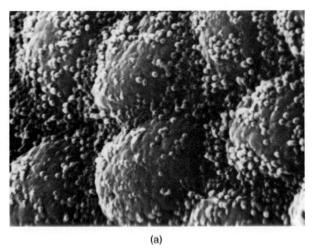

(a)

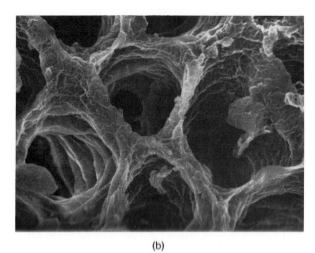

(b)

FIGURE 24-26 Peptic ulcers. (a) Scanning electron micrograph of the normal stomach mucosa at a magnification of 5,000×. (b) Scanning electron micrograph of ulceration of the stomach mucosa at a magnification of 1,000×. (Courtesy of Fisher Scientific Company and S.T.E.M. Laboratories, Inc., Copyright 1975.)

chemical substances such as bile and digestive enzymes.

Although it contains a great deal of lymphatic tissue and can combat infection fairly well, the peritoneum is in contact with most of the abdominal organs. If infection gets out of hand, it may destroy vital organs and bring on death. For these reasons, perforation of the alimentary canal from an ulcer is considered serious. A surgeon planning to do extensive surgery on the colon may give the patient high doses of antibiotics for several days prior to surgery to kill intestinal bacteria and reduce the risk of peritoneal contamination.

Peptic Ulcers

An **ulcer** is a craterlike lesion in a membrane. Ulcers that develop in areas of the alimentary canal exposed to acid gastric juice are called **peptic ulcers** (Figure 24-26). Peptic ulcers occasionally develop in the lower end of the esophagus, but most occur on the lesser curvature of the stomach, where they are called **gastric ulcers,** or in the first part of the duodenum, where they are called **duodenal ulcers.**

Hypersecretion of acid gastric juice seems to be the immediate cause of duodenal ulcers. In gastric ulcer patients, because the stomach wall is highly adapted to resist gastric juice through the secretion of mucus, the cause may be hyposecretion of mucus. Hypersecretion of pepsin also may contribute to ulcer formation.

Among the factors believed to stimulate an increase in acid secretion are emotions, certain foods or medications (alcohol, coffee, aspirin), and overstimulation of the vagus (X) nerve. Normally, the mucous membrane lining the stomach and duodenal walls resists the secretions of hydrochloric acid and pepsin. In some people, however, this resistance breaks down and an ulcer develops.

The danger inherent in ulcers is the erosion of the muscular portion of the wall of the stomach or duodenum. This erosion may damage blood vessels and produce fatal hemorrhaging. If an ulcer erodes all the way through the wall, the condition is called *perforation*. Perforation allows bacteria and partially digested food to pass into the peritoneal cavity, producing peritonitis.

Appendicitis

Appendicitis is an inflammation of the vermiform appendix. It is preceded by obstruction of the lumen of the appendix by fecal material, inflammation, a foreign body, carcinoma of the cecum, stenosis, or kinking of the organ. The infection that follows as a result of obstruction causes inflammation of all

layers of the appendix. This may result in edema, ischemia, gangrene, and perforation. Rupture of the appendix results in peritonitis. Loops of the intestines, the omentum, and the parietal peritoneum may become adherent and form an abscess, either at the site of the appendix or elsewhere in the abdominal cavity.

Typically, appendicitis begins with referred pain in the umbilical region of the abdomen followed by anorexia (lack or loss of appetite for food), nausea, and vomiting. After several hours, the pain localizes in the lower right quadrant and is continuous, dull or severe, and intensified by coughing, sneezing, or body movements.

Early appendectomy (removal of the appendix) is recommended in all suspected cases because it is safer to operate than to risk gangrene, rupture, and peritonitis. Appendectomy may be performed through a muscle-splitting incision in the right lower quadrant. The cecum is brought into the incision. The base of the appendix is tied, the appendix is excised, and the stump is usually cauterized and then invaginated into the cecum.

Tumors

Both benign and malignant **tumors** can occur in all parts of the gastrointestinal tract. The benign growths are much more common, but malignant tumors are responsible for 30 percent of all deaths from cancer in the United States.

For early diagnosis, complete routine examinations are necessary. Cancers of the mouth usually are detected through routine dental checkups. A regular physical checkup should include rectal examination. Fifty percent of all rectal carcinomas are within reach of the finger, and 75 percent of all colonic carcinomas can be seen with the sigmoidoscope.

Both the fiberoptic sigmoidoscope and the more recent fiberoptic endoscope are flexible tubular instruments composed of a light and many tiny glass fibers. They allow visualization, magnification, biopsy, electrosurgery, and even photography of the entire length of the gastrointestinal tract. Colonoscopy, the endoscopic examination of the colon, makes possible the identification and removal of malignant polyps of the colon (gastric polypectomy) before invasion of the intestinal wall or lymphatic metastasis. It has proved to be a safe and effective treatment that avoids the significant expense, risk, and discomfort of major surgery. Colonoscopy may be the greatest advance yet toward lowering the death rate from cancer of the colon. Unfortunately, this type of cancer has shown a considerable increase in incidence over the past 20 years.

Another test in a routine examination for intestinal disorders is the filling of the gastrointestinal tract with barium, which is either swallowed or given in an enema. Barium, a mineral, shows up on x-rays the same way that calcium appears in bones. Tumors as well as ulcers can be diagnosed this way. The only definitive treatment of gastrointestinal carcinomas, if they cannot be removed using the endoscope, is surgery.

Diverticulitis

Diverticula are saclike outpouchings of the wall of the colon when the muscularis becomes weak. The development of diverticula is called **diverticulosis.** Many people who develop diverticulosis are asymptomatic and experience no complications. About 15 percent of people with diverticulosis will eventually develop an inflammation within diverticula, a condition known as **diverticulitis.**

Research done by placing inflated balloons in the colon and measuring the pressure created by segmentation indicates that diverticula form because of lack of sufficient bulk in the colon during segmentation. The powerful contractions, working against insufficient bulk, create a pressure so high that it causes the colonic walls to blow out.

The increase in diverticular disease, from a rarity at the turn of the century to a disorder of an estimated 25 to 33 percent of middle-aged and older Americans today, has been attributed to a shift to a low-residue diet. A recent test result reported that 62 to 70 patients treated for diverticular disease with a high-residue diet containing unprocessed bran showed marked relief of symptoms.

Treatment consists of bed rest, cleansing enemas, and drugs to reduce infection. In severe cases, portions of the affected colon may require surgical removal and a temporary colostomy.

Cirrhosis

Cirrhosis refers to a distorted or scarred liver as a result of chronic inflammation. The parenchymal (functional) liver cells are replaced by fibrous or adipose connective tissue, a process called stromal repair. The liver has a high capacity for parenchymal regeneration, and stromal repair occurs whenever a parenchymal cell is killed or cells are damaged continuously for a long time. The symptoms of cirrhosis include jaundice, edema in the legs, uncontrolled bleeding, and increased sensitivity to drugs. Cirrhosis may be caused by hepatitis (inflammation of the liver), certain chemicals that destroy liver cells, parasites that infect the liver, and alcoholism.

Hepatitis

Hepatitis refers to inflammation of the liver and can be caused by viruses, drugs, and chemicals, including alcohol. Clinically several types are recognized.

Hepatitis A (infectious hepatitis) is caused by hepatitis A virus and is spread by fecal contamination of food, clothing, toys, eating utensils, and so forth (fecal-oral route). It is generally a mild disease of children and young adults characterized by anorexia, malaise, nausea, diarrhea, fever, and chills. Eventually jaundice appears. It does not cause lasting liver damage. Most people recover in 4 to 6 weeks.

Hepatitis B (serum hepatitis) is caused by hepatitis B virus and is spread primarily by contaminated syringes and transfusion equipment. It can also be spread by any secretion of fluid by the body (tears, saliva, semen). Hepatitis B can produce chronic liver inflammation and can persist for years or even a lifetime. Persons who harbor the active hepatitis B virus are at risk for cirrhosis and also become carriers.

Non-A, non-B (NANB) hepatitis is a form of hepatitis that cannot be traced to either hepatitis A or hepatitis B viruses. It is clinically similar to hepatitis B and is often spread by blood transfusions. It is believed to account for considerably more post-transfusion hepatitis than that related to hepatitis B.

Gallstones

Gallstones (biliary calculi) are a major health problem in the United States today, affecting more than 15 million people. **Cholecystectomy** (removal of the gallbladder) is the most frequently performed major operation. The ailment is four times more common in women that in men, and the number of persons suffering with gallstones is highest in the 55- to 65-year-old age group. The cholesterol in bile may crystallize at any point between bile canaliculi, where it is first apparent, and the hepatopancreatic ampulla, where the bile enters the duodenum. The fusion of single crystals is the beginning of 95 percent of all gallstones.

Cholesterol gallstones can cause obstruction to the outflow of bile from the gallbladder and irritate the mucosal surface. These two factors, combined with the presence of bacteria, may lead to precipitation of other substances upon the pure cholesterol core of the gallstone. Following their formation, gallstones gradually grow in size and number and may cause minimal, intermittent, or complete obstruction to the flow of bile from the gallbladder into the duct system. If obstruction of the outlet occurs, and the gallbladder cannot empty as it normally does after eating, the pressure within it increases and the individual may have intense pain or discomfort (**biliary colic**). The pain may radiate to the right shoulder or to the lower back. Bacterial content in the bile is great. With obstruction and stasis, bacteria increase in number. The products of bacterial action may produce a number of additional symptoms, including fever. The longer the calculi are in the gallbladder, the greater the incidence of calculi in the duct system. Small calculi may pass from the gallbladder to the cystic duct and into the common bile duct. Calculi may also ascend into the intrahepatic duct system and obstruct segments of the liver. Complete obstruction of the flow of bile into the duodenum may result in death.

Several gallstone-dissolving drugs are being studied. One of them, chenodeoxycholic acid (CDCA), appears to be effective and capable of desaturating bile and dissolving cholesterol gallstones. Unfortunately, it seems to work only in patients with well-functioning gallbladders and even then only about half the time. The second drug, ursodeoxycholic acid, seems to be more effective in lower doses. Another gallstone-dissolving drug being tested is monooctanoin. However, much more research is needed on these drugs to determine the long-term effects of interfering with cholesterol's metabolic pathways. At the present, it seems that surgery offers the only definitive treatment of gallstones.

Anorexia Nervosa

Anorexia nervosa is a disorder characterized by loss of appetite and bizarre patterns of eating. The subconsciously self-imposed starvation appears to be a response to emotional conflicts about self-identification and acceptance of a normal adult sex role. The disorder is found predominantly in young, single females. The physical consequence of the disorder is severe and progressive starvation. Amenorrhea (absence of menstruation) and a lowered basal metabolic rate reflect the depressant effects of the starvation. Individuals may become emaciated and may ultimately die of starvation or one of its complications. Also associated with the disorder are brain abnormalities coupled with impaired mental performance. Treatment consists of psychotherapy and dietary regulation.

Bulimia

A disorder that typically affects single, middle-class, young, white females is known as **bulimia** (*bous* = ox; *limos* = hunger) or **binge-purge syndrome.** It is characterized by uncontrollable overeating followed by forced vomiting or overdoses of laxatives. This binge-purge cycle occurs in response to fears of being overweight, stress, depression, and physiological disorders such as hypothalamic tumors.

Bulimia can upset the body's electrolyte balance and increase susceptibility to flu, salivary gland infections, dry skin, acne, muscle spasms, loss of hair, kidney and liver diseases, tooth decay, ulcers, hernias, constipation, and hormone imbalances. Treatment of bulimia combines nutrition counseling, psychotherapy, and medical treatment.

Dietary Fiber (Roughage) and GI Disorders

A deficiency in our diet that has received recent attention is lack of dietary fiber or roughage. The bulk of our intake consists of extensively purified carbohydrates—mainly starches and sugars—as well as substantial amounts of fats and oils. People who choose a fiber-rich, unrefined diet will greatly reduce their chances of developing diseases of overnutrition (such as obesity, diabetes, gallstones, and coronary heart disease), diseases of the underworked mouth (caries and periodontal disease), and diseases of an underfed large bowel (constipation, varicose veins, hemorrhoids, colon spasm, diverticulitis, appendicitis, and large intestinal cancer). Each of these conditions is directly related to the digestion and metabolism of food and the operation of the digestive system. The daily addition of a small amount of bran or other foods high in fiber would overcome the roughage deficiency.

MEDICAL TERMINOLOGY

Botulism (*botulus* = sausage)   A type of food poisoning caused by a toxin produced by *Clostridium botulinum.* The bacterium is ingested in improperly cooked or preserved foods. The toxin inhibits nerve impulse conduction at synapses by inhibiting the release of acetylcholine. Symptoms include paralysis, nausea, vomiting, blurred or double vision, difficulty in speech, difficulty in swallowing, dryness of the mouth, and general weakness.

Cholecystitis (*chole* = bile; *kystis* = bladder; *itis* = inflammation of)   Inflammation of the gallbladder that often leads to infection. Some cases are caused by obstruction of the cystic duct with bile stones. Stagnating bile salts irritate the mucosa. Dead mucosal cells provide a medium for the growth of bacteria.

Cholelithiasis (*lithos* = stone)   The presence of gallstones.

Colitis   Inflammation of the colon and rectum. Inflammation of the mucosa reduces absorption of water and salts, producing watery, bloody feces, and, in severe cases, dehydration and salt depletion. Irritated muscularis spasms produce cramps.

Colostomy (*stomoun* = provide an opening)   An incision of the colon to create an artificial opening or "stoma" to the exterior. This opening serves as a substitute anus through which feces are eliminated. A temporary colostomy may be done to allow a badly inflamed colon to rest and heal. If the rectum is removed for malignancy, the colostomy provides a permanent outlet for feces.

Dysphagia (*dys* = abnormal; *phagein* = to eat)   Difficulty in swallowing that may be caused by inflammation, paralysis, obstruction, or trauma.

Enteritis (*enteron* = intestine)   An inflammation of the intestine, particularly the small intestine.

Flatus   Excessive amounts of air (gas) in the stomach or intestine, usually expelled through the anus. If the gas is expelled through the mouth, it is called **eructation** or **belching** (burping). Flatus may result from gas released during the breakdown of foods in the stomach or from swallowing air or gas-containing substances such as carbonated drinks.

Gastrectomy (*gastro* = stomach; *tome* = excision)   Removal of a portion of or the entire stomach.

Hernia   Protrusion of an organ or part of an organ through a membrane or cavity wall, usually the abdominal cavity. *Diaphragmatic* (*hiatal*) *hernia* is the protrusion of the lower esophagus, stomach, or intestine into the thoracic cavity through the opening in the diaphragm (esophageal hiatus) that allows passage of the esophagus. *Umbilical hernia* is the protrusion of abdominal organs through the navel area of the abdominal wall. *Inguinal hernia* is the protrusion of the hernial sac containing the intestine into the inguinal opening. It may extend into the scrotal compartment, causing strangulation of the herniated part.

Inflammatory bowel disease (IBD)   Disorder that exists in two forms: (1) Crohn's disease (inflamed intestine) and (2) ulcerative colitis (inflammation of the colon consisting of ulcerations and usually accompanied by rectal bleeding).

Irritable bowel syndrome   Disease of the entire gastrointestinal tract characterized by abnormal muscular contractions, especially a spastic colon, excessive mucus in stools, and alternating diarrhea and constipation.

Nausea (*nausia* = seasickness)   Discomfort preceding vomiting. Possibly, it is caused by distension or irritation of the gastrointestinal tract, most commonly the stomach.

Pancreatitis   Inflammation of the pancreas. The pancreas secretes active trypsin instead of trypsinogen, and the trypsin digests the pancreatic cells and blood vessels.

DRUGS ASSOCIATED WITH THE DIGESTIVE SYSTEM

Antacids

Antacids act by chemical neutralization and by absorption of excessive stomach acid. Examples are *sodium bicarbonate* (also called *baking soda*), *calcium carbonate* (Titralac), *magnesium hydroxide* (Mylanta II, Maalox), and *aluminum hydroxide gels* (Amphojel, Creamalin). Antacids are frequently used in cases of heartburn and ulcers.

Antisecretories

Antisecretories decrease the secretion of hydrochloric acid and pepsin in the stomach. They decrease nocturnal, meal-stimulated, and abnormal acid secretion. They include anticholinergic drugs such as *clidinium* (Quarzan) and a newer blocker of certain histamine receptors, *cimetidine* (Tagamet). Antisecretories are used in many cases of ulcers, especially recurrent ulcers.

Antispasmodics

Antispasmodics calm stomach and intestinal muscles, dry up excess fluids produced during gastrointestinal spasms, and quiet hypermotility (excessive activity of the stomach and intestines). These compounds are useful in ulcers and colitis. Examples are *dicyclomine HCl* (Bentyl), *atropine,* and *belladonna.*

Digestants

Some diseases and parasites destroy digestive juices. It is necessary to replace these juices to allow normal nutrition. Digestant agents help to break down solid food and act as catalysts in the extraction of proteins, carbohydrates, and minerals. Examples are *bile salts, pancreatic enzymes* (Cotazym, Ilozyme, Viokase), *hydrochloric acid* (Acidulin), and *pepsin.*

Emetics

Emetics are preparations that induce vomiting. Examples are *ipecac, mustard,* and *salt water.* Emetics are sometimes used to empty the stomach after poisons have been swallowed.

Cathartics and Laxatives

Cathartics and laxatives are used to bring about emptying of the bowel. They do so by speeding the passage of the intestinal contents through the gastrointestinal tract. Cathartics that are rather severe in their action tend to dry out the natural intestinal mucosa; examples are *citrate of magnesia* and *magnesium sulfate* (also called *epsom salt*). Other cathartics, used when a thorough cleansing of the intestinal tract is necessary before surgery or x-rays, include *castor oil, cascara sagrada,* and *phenolphthalein* derivatives such as *bisacodyl* (Dulcolax). Milder laxatives such as *mineral oil, agar,* and *milk of magnesia,* as well as or along with stool softeners such as *dioctyl sodium sulfosuccinate* (Colace) and *psyllium seed* (Metamucil), are useful in preventing constipation in bedridden patients and to prevent straining (in patients with hemorrhoids or severe heart disease, for example).

STUDY OUTLINE

Regulation of Food Intake (p. 582)

1. Food intake is regulated by two sensations: hunger and appetite.
2. The control centers for food intake (appetite center and satiety center) are located in the hypothalamus.

Digestive Processes (p. 582)

1. Food is prepared for use by cells by five basic activities: ingestion, peristalsis, mechanical and chemical digestion, absorption, and defecation.
2. Chemical digestion is a series of catabolic reactions that break down the large carbohydrate, lipid, and protein molecules of food into molecules that are usable by body cells.
3. Mechanical digestion consists of movements that aid chemical digestion.
4. Absorption is the passage of end products of digestion from the digestive tract into blood or lymph for distribution to cells.

Organization (p. 583)

1. The organs of digestion are usually divided into two main groups: those composing the gastrointestinal (GI) tract, or alimentary canal, and accessory structures.
2. The GI tract is a continuous tube running through the ventral body cavity from the mouth to the anus.
3. The accessory structures include the teeth, tongue, salivary glands, liver, gallbladder, and pancreas.
4. The basic arrangement of tissues in the alimentary canal from the inside outward is the mucosa, submucosa, muscularis, and serosa (peritoneum).
5. Extensions of the peritoneum include the mesentery, mesocolon, falciform ligament, lesser omentum, and greater omentum.

Mouth (Oral Cavity) (p. 585)

1. The mouth is formed by the cheeks, palates, lips, and tongue, which aid mechanical digestion.
2. The vestibule is the space between the cheeks and lips and teeth and gums.
3. The oral cavity proper extends from the vestibule to the fauces.

Tongue

1. The tongue, together with its associated muscles, forms the floor of the oral cavity. It is composed of skeletal muscle covered with mucous membrane.
2. The upper surface and sides of the tongue are covered with papillae. Some papillae contain taste buds.

Salivary Glands

1. The major portion of saliva is secreted by the salivary glands, which lie outside the mouth and pour their contents into ducts that empty into the oral cavity.
2. There are three pairs of salivary glands: the parotid, submandibular (submaxillary), and sublingual glands.
3. Saliva lubricates food and starts the chemical digestion of carbohydrates.
4. Salivation is entirely under nervous control.

Teeth

1. The teeth, or dentes, project into the mouth and are adapted for mechanical digestion.
2. A typical tooth consists of three principal portions: crown, root, and cervix.
3. Teeth are composed primarily of dentin covered by enamel, the hardest substance in the body.
4. There are two dentitions—deciduous and permanent.

Digestion in the Mouth

1. Through mastication food is mixed with saliva and shaped into a bolus.
2. Salivary amylase converts polysaccharides (starches) to disaccharides (maltose).

Deglutition

1. Deglutition or swallowing moves a bolus from the mouth to the stomach.
2. It consists of a voluntary stage, pharyngeal stage (involuntary) and esophageal stage (involuntary).

Esophagus (p. 594)

1. The esophagus is a collapsible, muscular tube that connects the pharynx to the stomach.
2. It passes a bolus into the stomach by peristalsis.
3. It contains an upper and lower esophageal sphincter.

Stomach (p. 595)

Anatomy; Histology

1. The stomach begins at the bottom of the esophagus and ends at the pyloric sphincter.
2. Adaptations of the stomach for digestion include rugae; glands that produce mucus, hydrochloric acid, a protein-digesting enzyme, intrinsic factor, and stomach gastrin; and a three-layered muscularis for efficient mechanical movement.

Digestion in the Stomach

1. Mechanical digestion consists of mixing waves.

2. Chemical digestion consists of the conversion of proteins into peptides by pepsin.

Regulation of Gastric Secretion
1. Gastric secretion is regulated by nervous and hormonal mechanisms.
2. Stimulation occurs in three phases: cephalic (reflex), gastric, and intestinal.

Absorption
1. The stomach wall is impermeable to most substances.
2. Among the substances absorbed are some water, certain electrolytes and drugs, and alcohol.

Pancreas (p. 604)
1. The pancreas is connected to the duodenum via the pancreatic and accessory ducts.
2. Pancreatic juice contains enzymes that digest starch to maltose (pancreatic amylase), proteins to peptides (trypsin and chymotrypsin), terminal amino acids at the carboxyl ends of peptides (carboxypolypeptidase), neutral fats to fatty acids and monoglycerides (pancreatic lipase), and nucleotides to pentoses and nitrogen bases (nucleases).
3. Pancreatic secretion is regulated by nervous and hormonal mechanisms.

Liver (p. 606)
1. Hepatic cells of the liver produce bile that is transported by a duct system to the gallbladder for storage.
2. Bile's contribution to digestion is the emulsification of neutral fats.
3. Bile secretion is regulated by nervous and hormonal mechanisms.

Gallbladder (p. 610)
1. The gallbladder stores and concentrates bile.
2. Bile is ejected into the common bile duct under the influence of cholecystokinin (CCK).

Small Intestine (p. 610)
Anatomy; Histology
1. The small intestine extends from the pyloric sphincter to the ileocecal valve.
2. It is highly adapted for digestion and absorption. Its glands produce enzymes and mucus, and the microvilli, villi, and plicae circulares of its wall provide a large surface area for digestion and absorption.
3. Intestinal enzymes break down foods inside epithelial cells of the mucosa.

Intestinal Juice; Digestion
1. Intestinal enzymes break down maltose to glucose (maltase), sucrose to glucose and fructose (sucrase), lactose to glucose and galactose (lactase), terminal amino acids at the amino ends of peptides (aminopeptidase), dipeptides to amino acids (dipeptidase), and nucleotides to pentoses and nitrogen bases (nucleases).
2. Mechanical digestion in the small intestine involves segmentation and peristalsis.

Regulation of Intestinal Secretion
1. The most important mechanism is local reflexes.
2. Hormones also assume a role.

Absorption
1. Absorption is the passage of the end products of digestion from the alimentary canal into the blood or lymph.
2. Monosaccharides, amino acids, and short-chain fatty acids

pass into the blood capillaries.
3. Long-chain fatty acids and monoglycerides are absorbed as part of micelles, resynthesized to triglycerides, and transported as chylomicrons.
4. Chylomicrons are taken up by the lacteal of a villus.
5. The small intestine also absorbs water, electrolytes, and vitamins.

Large Intestine (p. 617)
Anatomy; Histology
1. The large intestine extends from the ileocecal valve to the anus.
2. Its subdivisions include the cecum, colon, rectum, and anal canal.
3. The mucosa contains numerous goblet cells and the muscularis consists of taeniae coli.

Digestion in the Large Intestine
1. Mechanical movements of the large intestine include haustral churning, peristalsis, and mass peristalsis.
2. The last stages of chemical digestion occur in the large intestine through bacterial, rather than enzymatic, action. Substances are further broken down and some vitamins are synthesized.

Absorption and Feces Formation
1. The large intestine absorbs water, electrolytes, and vitamins.
2. Feces consists of water, inorganic salts, epithelial cells, bacteria, and undigested foods.

Defecation
1. The elimination of feces from the large intestine is called defecation.
2. Defecation is a reflex action aided by voluntary contractions of the diaphragm and abdominal muscles.

Disorders: Homeostatic Imbalances (p. 622)
1. Dental caries are started by acid-producing bacteria that reside in dental plaque.
2. Periodontal diseases are characterized by inflammation and degeneration of gingivae, alveolar bone, periodontal membrane, and cementum.
3. Peritonitis is inflammation of the peritoneum.
4. Peptic ulcers are craterlike lesions that develop in the mucous membrane of the alimentary canal in areas exposed to gastric juice.
5. Appendicitis is an inflammation of the vermiform appendix resulting from obstruction of the lumen of the appendix by inflammation, a foreign body, carcinoma of the cecum, stenosis, or kinking of the organ.
6. Tumors of the gastrointestinal tract may be detected by sigmoidoscopy, colonoscopy, and barium x-ray.
7. Diverticulitis is inflammation of diverticula in the colon.
8. Cirrhosis is a condition in which parenchymal cells of the liver damaged by chronic inflammation are replaced by fibrous or adipose connective tissue.
9. Hepatitis is an inflammation of the liver. Types include hepatitis A; hepatitis B; and non-A, non-B (NANB) hepatitis.
10. The fusion of individual crystals of cholesterol is the beginning of 95 percent of all gallstones. Gallstones can cause obstruction to the outflow of bile in any portion of the duct system.
11. Anorexia nervosa is a disorder characterized by loss of appetite.
12. Bulimia is a binge-purge syndrome of behavior in which uncontrollable overeating is followed by forced vomiting or overdoses of laxatives.

REVIEW QUESTIONS

1. Describe how food intake is regulated.
2. Define digestion. Distinguish between chemical and mechanical digestion.
3. Identify the organs of the gastrointestinal (GI) tract in sequence. How does the gastrointestinal (GI) tract differ from the accessory structures of digestion?
4. Describe the structure of each of the four coats of the gastrointestinal (GI) tract.
5. What is the peritoneum? Describe the location and function of the mesentery, mesocolon, falciform ligament, lesser omentum, and greater omentum.
6. What structures form the oral cavity?
7. Make a simple diagram of the tongue. Indicate the location of the papillae and the four taste zones. What is ankyloglossia?
8. Describe the location of the salivary glands and their ducts. What are buccal glands?
9. How are salivary glands distinguished histologically?
10. Describe the composition of saliva and the role of each of its components in digestion. What is the pH of saliva?
11. How is salivary secretion regulated?
12. What are the principal portions of a typical tooth? What are the functions of each part?
13. Compare deciduous and permanent dentitions with regard to number of teeth and time of eruption.
14. Contrast the functions of incisors, cuspids, premolars, and molars.
15. What is a bolus? How is it formed?
16. Define deglutition. List the sequence of events involved in passing a bolus from the mouth to the stomach. Be sure to discuss the voluntary, pharyngeal, and esophageal stages of swallowing.
17. Describe the location and histology of the esophagus. What is its role in digestion?
18. Explain the operation of the upper and lower esophageal sphincters. How is achalasia related to heartburn?
19. Describe the location of the stomach. List and briefly explain the anatomical features of the stomach.
20. Distinguish between pyloric stenosis and pylorospasm.
21. What is the importance of rugae, zymogenic cells, parietal cells, mucous cells, and enteroendocrine cells in the stomach?
22. Describe mechanical digestion in the stomach.
23. What is the role of pepsin? Why is it secreted in an inactive form?
24. Outline the factors that stimulate and inhibit gastric secretion. Be sure to discuss the cephalic, gastric, and intestinal phases.
25. How is gastric emptying stimulated and inhibited?
26. Explain how vomiting occurs.
27. Describe the role of the stomach in absorption.
28. Where is the pancreas located? Describe the duct system connecting the pancreas to the duodenum.
29. What are pancreatic acini? Contrast their functions with those of the islets of Langerhans.
30. Describe the composition of pancreatic juice and the digestive functions of each component.
31. How is pancreatic juice secretion regulated?
32. Where is the liver located? What are its principal functions?
33. Describe the anatomy of the liver. Draw a labeled diagram of a liver lobule.
34. How is blood carried to and from the liver?
35. Once bile has been formed by the liver, how is it collected and transported to the gallbladder for storage?
36. What is the function of bile?
37. How is bile secretion regulated? Define jaundice and distinguish two principal types.
38. Where is the gallbladder located? How is it connected to the duodenum?
39. Describe the function of the gallbladder. How is emptying of the gallbladder regulated?
40. What are the subdivisions of the small intestine? How are the mucosa and submucosa of the small intestine adapted for digestion and absorption?
41. Describe the movements in the small intestine.
42. Explain the function of each enzyme in intestinal juice.
43. How is small intestinal secretion regulated?
44. Define absorption. How are the end products of carbohydrate and protein digestion absorbed? How are the end products of fat digestion absorbed?
45. What routes are taken by absorbed nutrients to reach the liver?
46. Describe the absorption of water, electrolytes, and vitamins by the small intestine.
47. What are the principal subdivisions of the large intestine? How does the muscularis of the large intestine differ from that of the rest of the digestive tract? What are haustra?
48. Describe the mechanical movements that occur in the large intestine.
49. How is the ileocecal valve regulated?
50. Explain the activities of the large intestine that change its contents into feces.
51. Define defecation. How does it occur?
52. Distinguish between diarrhea and constipation.
53. Describe the causes (where known) and clinical symptoms for: dental caries, periodontal disease, peritonitis, peptic ulcers, appendicitis, tumors, diverticulitis, cirrhosis, hepatitis, gallstones, anorexia nervosa, and bulimia.
54. Refer to the glossary of medical terminology associated with the digestive system. Be sure that you can define each term.
55. Describe the actions of selected drugs on the digestive system.

25 Metabolism

Nutrients are chemical substances in food that provide energy, form new body components, or assist in the functioning of various body processes. There are six principal classes of nutrients: carbohydrates, lipids, proteins, minerals, vitamins, and water. Carbohydrates, proteins, and lipids are digested by enzymes in the gastrointestinal tract. The end products of digestion that ultimately reach body cells are monosaccharides, amino acids, and fatty acids and monoglycerides. Some are used to synthesize new structural molecules in cells or to synthesize new regulatory molecules, such as hormones and enzymes. Most are used to produce energy to sustain life processes. This energy is used for processes such as active transport, DNA replication, synthesis of proteins and other molecules, muscle contraction, and nerve impulse conduction. Until the energy is needed, it is stored in ATP.

Some minerals and many vitamins are part of enzyme systems that catalyze the reactions undergone by carbohydrates, proteins, and lipids.

Water has five major functions. It is an excellent solvent and suspending medium, it participates in hydrolysis reactions, it acts as a coolant, it lubricates, and it helps to maintain a constant body temperature due to its ability to release and absorb heat slowly.

METABOLISM

Metabolism (me-TAB-ō-lizm) refers to all the chemical reactions of the body. The body's metabolism may be thought of as an energy-balancing act between catabolic reactions, which provide energy, and anabolic reactions, which require energy.

CATABOLISM

Catabolism (ka-TAB-ō-lizm) is the term for decomposition chemical reactions that provide energy (Figure

25-1). Digestion is a catabolic process in which the breaking of bonds of food molecules releases energy. Another example is a process called oxidation (cellular respiration). It involves the breakdown of absorbed nutrients resulting in a release of energy. **Oxidation** is the removal of electrons and hydrogen ions (hydrogen atoms) from a molecule or, less commonly, the addition of oxygen to a molecule. Most oxidations are actually dehydrogenations, that is, they involve the loss of hydrogen atoms. As you will see later, each time one substance is oxidized, another is almost simultaneously reduced. **Reduction** is the addition of electrons and hydrogen ions (hydrogen atoms) to a molecule or, less commonly, the removal of oxygen from a molecule. It is the opposite of oxidation. Glucose is the body's favorite nutrient for oxidation, but fats and proteins are also oxidized. Oxidation also involves the synthesis of ATP from ADP. As substances are oxidized, energy is produced. The energy is removed in a stepwise fashion and ultimately trapped in ATP for storage. When body cells require energy, ATP is broken down and the energy is released.

ANABOLISM

Anabolism (a-NAB-ō-lizm) is just the opposite of catabolism—a series of synthetic reactions whereby small molecules are built up into larger ones that form the body's structural and functional components. Anabolic reactions require energy and the energy is supplied by catabolic reactions of the body (Figure 25-1). One example of an anabolic process is the formation of peptide bonds between amino acids, thereby building up the amino acids into the protein portions of cytoplasm, enzymes, hormones, and antibodies. Fats also participate in the body's anabolism. For instance, fats can be built into the phospholipids that form the plasma membrane. They are also part of the steroid hormones.

We will now consider the metabolism of carbohydrates, lipids, and proteins in body cells.

CARBOHYDRATE METABOLISM

During the process of digestion, polysaccharides and disaccharides are hydrolyzed to become the monosaccharides—glucose, fructose, and galactose—which are absorbed into the capillaries of the villi of the small intestine. They are then carried through the hepatic portal vein to the liver, where fructose and galactose are converted to glucose. The liver is the only organ that has the necessary enzymes to make this conversion. Thus the story of carbohydrate metabolism is really the story of glucose metabolism.

FATE OF CARBOHYDRATES

Since glucose is the body's preferred source of energy, the fate of absorbed glucose depends on the body cells' energy needs. If the cells require immediate energy, the glucose is oxidized by the cells. Each gram of carbohydrate produces about 4.0 Calories (Cal). (The caloric content of a food is a measure of the heat it releases upon

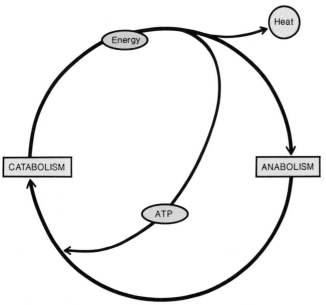

FIGURE 25-1 Relationship of ATP to catabolism and anabolism.

oxidation. Determination of caloric value is described later in the chapter.) The glucose not needed for immediate use is handled in several ways. First, the liver can convert excess glucose to glycogen and then store it. Skeletal muscle cells can also store glycogen. Second, if the glycogen storage areas are filled up, the liver cells can transform the glucose to fat that can be stored in adipose tissue. Later, when the cells need more energy, the glycogen and fat can be converted back to glucose, which is released into the bloodstream so it can be transported to cells for oxidation. Third, excess glucose can be excreted in the urine. Normally, this happens only when a meal containing mostly carbohydrates and no fats is eaten. Without the inhibiting effect of fats, the stomach empties its contents quickly and the carbohydrates are all digested at the same time. As a result, large numbers of monosaccharides suddenly flood into the bloodstream. Since the liver is unable to process all of them simultaneously, the blood glucose level rises and the condition of hyperglycemia may result in glucose in the urine (glycosuria).

GLUCOSE MOVEMENT INTO CELLS

Before glucose can be used by body cells, it must first pass through the plasma membrane and then enter the cytoplasm. The process by which this occurs is facilitated diffusion (Chapter 3) and the rate of glucose transport is greatly increased by insulin (Chapter 18). Immediately upon entry into cells, glucose combines with a phosphate group, produced by the breakdown of ATP. This addition of a phosphate group to a molecule is called *phosphorylation*. The product of this enzymatically catalyzed reaction is glucose-6-phosphate (see Figure 25-7). In most cells of the body, phosphorylation serves to capture glucose in the cell so that it cannot move back out. Liver cells, kidney tubule cells, and intestinal epithelial cells have the necessary enzyme (phosphatase) to remove the phosphate group and transport glucose out of the cell (see Figure 25-7).

GLUCOSE CATABOLISM

The **oxidation** of glucose is also known as **cellular respiration.** It occurs in every cell in the body and provides the cell's chief source of energy. The complete oxidation of glucose to carbon dioxide and water produces large amounts of energy. It occurs in three successive stages: glycolysis, the Krebs cycle, and the electron transport chain.

Glycolysis

The term **glycolysis** (glī-KOL-i-sis; *glyco* = sugar, *lysis* = breakdown) refers to a series of chemical reactions in the cytoplasm of a cell that convert a six-carbon molecule of glucose into two three-carbon molecules of pyruvic acid. The simplified overall reaction for glycolysis is:

$$C—C—C—C—C—C \longrightarrow 2\ C—C—C + 2\ ATP$$

Glucose → Pyruvic acid

The details of glycolysis are shown in Figure 25-2a. The essential features of the process are:

1. Each of the 10 reactions is catalyzed by a specific enzyme. (The names of the enzymes are indicated in the rust color.)

2. All of the reactions are reversible.

3. The first three reactions involve the addition of a phosphate group (phosphorylation) to glucose, its conversion to fructose, and the addition of another phosphate group to fructose to prime the pathway. This involves an input of energy in which two molecules of ATP are converted to ADP.

4. The double phosphorylated molecule of fructose splits into two three-carbon compounds, glyceraldehyde-3-phosphate and dihydroxyacetone phosphate. These compounds are interconvertible.

5. Each of the two three-carbon compounds is degraded to form a molecule of pyruvic acid. In the process, each three-carbon compound generates two molecules of ATP.

6. Since two molecules of ATP are needed to get glycolysis started and four molecules of ATP are generated to complete the process, there is a net gain of two molecules of ATP for each molecule of glucose that is oxidized (Figure 25-2b).

7. Most of the energy produced during glycolysis is used to generate ATP. The remainder is expended as heat energy, some of which helps to maintain body temperature.

The fate of pyruvic acid depends on the availability of oxygen. If anaerobic conditions exist in cells (as during strenuous exercise), pyruvic acid is reduced by the addition of two hydrogen atoms to form lactic acid (Figure 25-2a). The lactic acid may be transported to the liver, where it is converted back to pyruvic acid. Or, it may remain in the cells until aerobic conditions are restored (recall the mechanisms for the repayment of the oxygen debt described in Chapter 10) and it is then converted to pyruvic acid in the cells.

Under aerobic conditions, the process of the complete oxidation of glucose continues, and pyruvic acid is oxidized to form carbon dioxide and water in two sets of reactions: the Krebs cycle and the electron transport chain. However, before pyruvic acid can enter the Krebs cycle, it must be prepared. This process occurs in mitochondria.

Formation of Acetyl Coenzyme A

Although all enzymes are proteins, many enzymes are attached to nonprotein compounds. If a nonprotein group detaches from the enzyme and acts as a carrier molecule, it is called a **coenzyme.** Enzymes work together with their conezymes. The enzyme catalyzes the reaction, and the coenzyme attaches to the end product of the reaction and carries it to the next reaction. Each step in the oxidation of glucose requires a different enzyme and often a coenzyme as well. We are interested in only one coenzyme at this point: a substance called **coenzyme A (CoA).**

During the transitional step between glycolysis and the Krebs cycle, pyruvic acid is prepared for entrance into the cycle. Essentially, pyruvic acid is converted to a two-carbon compound by the loss of carbon dioxide. The loss of a molecule of CO_2 by a substance is called **decarboxylation** (dē-kar-bok'-si-LĀ-shun). (It is indicated in the diagram by the arrow labeled CO_2.) This two-carbon fragment, called an **acetyl group,** attaches itself to coenzyme A, and the whole complex is called **acetyl coenzyme A** (Figure 25-3). It is in this form that pyruvic acid enters the Krebs cycle.

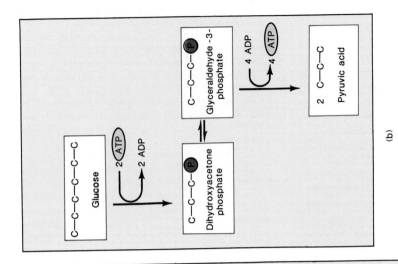

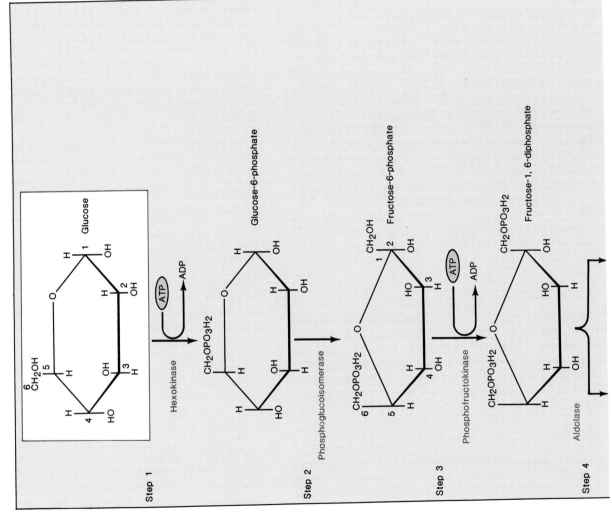

(b)

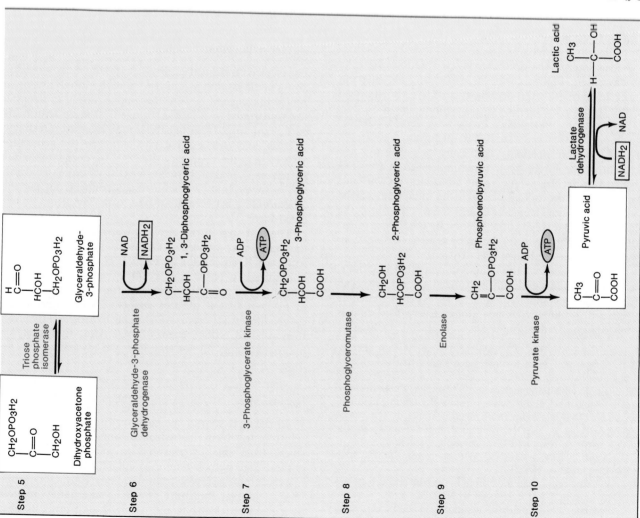

FIGURE 25-2 Glycolysis. As a result of glycolysis, there is a net gain of two molecules of ATP. (a) Detailed version. (b) Simplified version.

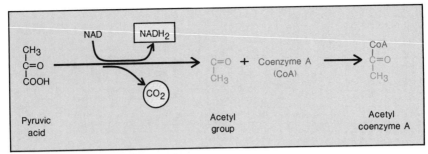

FIGURE 25-3 Formation of acetyl coenzyme A from pyruvic acid for entrance into the Krebs cycle.

Krebs Cycle

The **Krebs cycle** is also called the **citric acid cycle** or the **tricarboxylic acid (TCA) cycle.** It is a series of reactions that occur in the matrix of mitochondria of cells (Figure 25-4). Coenzyme A carries an acetyl group to a mitochondrion, detaches itself, and goes back into the cytoplasm to pick up another fragment. The acetyl group combines with oxaloacetic acid to form citric acid. From this point the Krebs cycle consists mainly of a series of decarboxylation and oxidation-reduction reactions, each controlled by a different enzyme.

Let us look at the decarboxylation reactions first. As we have seen (Figure 25-3), in preparation for entering the cycle, pyruvic acid is decarboxylated to form an acetyl group. In the cycle isocitric acid loses a molecule of CO_2 to form α-ketoglutaric acid. In the next step, α-ketoglutaric acid is decarboxylated (and picks up a molecule of CoA) to form succinyl CoA. Thus each time pyruvic acid enters the Krebs cycle three molecules of CO_2 are liberated by decarboxylation. The molecules of CO_2 leave the mitochondria, diffuse through the cytoplasm of the cell to the plasma membrane, and then diffuse into the blood by internal respiration. Eventually the CO_2 is transported by the blood to the lungs by external respiration. Once it reaches the lungs, it is exhaled.

Now let us look at the oxidation-reduction reactions. Remember, when a molecule is oxidized, it loses hydrogen atoms (electrons and hydrogen ions). When a molecule is reduced, it gains hydrogen atoms. Every oxidation is coupled with a reduction.

1. In the conversion of pyruvic acid to acetyl CoA (Figure 25-3), pyruvic acid loses two hydrogen atoms, that is, it is oxidized. The hydrogen atoms are picked up by a coenzyme called **nicotinamide adenine dinucleotide (NAD).** NAD contains two adenine nucleotides and is derived from a B vitamin called niacin. Since NAD picks up the hydrogen atoms, it is reduced and represented as $NADH_2$ (indicated in the diagram by the curved arrow entering and then leaving the reaction).

2. Isocitric acid is oxidized to form α-ketoglutaric acid. NAD is reduced to $NADH_2$.

3. α-Ketoglutaric acid is oxidized in the conversion to succinyl CoA. Again, NAD is reduced to $NADH_2$.

4. Succinic acid is oxidized to fumaric acid. In this case, the coenzyme that picks up the two hydrogen atoms is **flavin adenine dinucleotide (FAD).** FAD, like NAD, contains two adenine nucleotides, but it is derived from vitamin B_2 (riboflavin). It belongs to a class of compounds called **flavoproteins.** FAD is reduced to $FADH_2$.

5. Malic acid is oxidized to oxalocetic acid. NAD is reduced to $NADH_2$.

Thus, each time a molecule of pyruvic acid enters the Krebs cycle, four molecules of $NADH_2$ and one molecule of $FADH_2$ are produced by the various oxidation-reduction reactions. These reduced coenzymes are very important because they now contain the stored energy originally in glucose and then in pyruvic acid. In the next stage in the complete oxidation of glucose, the electron transport chain, the energy in the coenzymes is transferred to ATP for storage.

Electron Transport Chain

The **electron transport chain** is a series of oxidation-reduction reactions that occur on mitochondrial cristae in which the energy in $NADH_2$ and $FADH_2$ is liberated and transferred to ATP for storage. In the electron transport chain, three types of carrier molecules are alternately oxidized and reduced and participate in the generation of ATP: (1) the coenzyme **FAD,** (2) a coenzyme called **coenzyme Q,** and (3) **cytochromes** (SĪ-tō-krōms), red protein pigments that have an iron-containing group capable of alternating between a reduced form (Fe^{2+}) and an oxidized form (Fe^{3+}).

The first step in the electron transport chain involves the transfer of hydrogen atoms from $NADH_2$ (mostly from the Krebs cycle) to FAD (Figure 25-5). $NADH_2$ is oxidized to NAD and FAD is reduced to $FADH_2$. The importance of the hydrogen transfer from $NADH_2$ to FAD is that it releases energy. This energy is used to produce ATP from ADP. The hydrogen atoms are then used to reduce coenzyme Q. After that step, the hydrogen atoms do not stay intact. They ionize into hydrogen ions and electrons.

$$\underset{\text{Hydrogen atom}}{H} \quad \rightarrow \quad \underset{\text{Hydrogen ion}}{H^+} \quad + \quad \underset{\text{Electron}}{e^-}$$

In the next step of the electron transport system, the electrons (e^-) from the hydrogen atoms are passed successively from one cytochrome to another—from cytochrome b to cytochrome c to cytochrome a and finally to cytochrome a_3 (cytochrome oxidase). Each cytochrome in the electron transport system is alternately reduced as it picks up electrons and oxidized as it gives up electrons. Because of the involvement of cytochromes in the electron transport chain, it is also known as the **cytochrome system.**

In the process of electron transfer between cytochromes, more energy is liberated and then stored in ATP. The formation of ATP from the liberated energy occurs

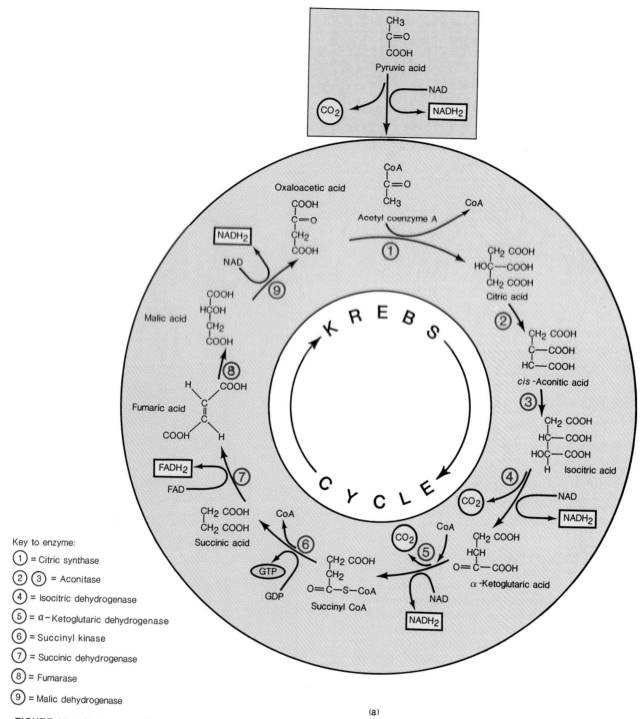

(a)

FIGURE 25-4 Krebs cycle. The net results of the Krebs cycle are the production of reduced coenzymes ($NADH_2$ and $FADH_2$), which contain stored energy; the generation of GTP, a high-energy compound that is used to produce ATP; and the formation of CO_2, which is transported to the lungs for expiration. (a) Detailed version.

between cytochrome b and cytochrome c and between cytochrome a and cytochrome a_3. Thus, ATP molecules are formed at three different places in the electron transport chain. At the end of the electron transport chain, the electrons are passed to oxygen, which becomes negatively charged. This oxygen, required for the complete oxidation of glucose, is supplied to your body cells through inspiration, external respiration, and internal res-

piration. The oxygen combines with the H^+ ions released from the breakdown of hydrogen atoms and forms water.

ATP Production

Let's add up the number of ATP molecules formed during the complete oxidation of glucose. During glycolysis there is a gain of two molecules of ATP (see Figure 25-2).

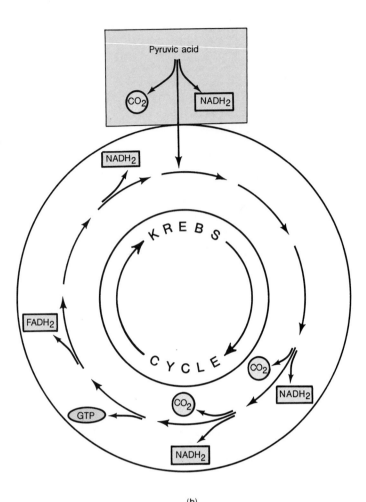

(b)

FIGURE 25-4 (*Continued*) Krebs cycle. (b) Simplified version.

Each molecule of glucose that is oxidized yields two molecules of pyruvic acid, both of which enter the Krebs cycle. Although the Krebs cycle does not produce any ATP, it does generate a substance called **guanosine triphosphate (GTP)**. GTP is produced from guanosine diphosphate (GDP), using the energy of succinyl coenzyme A (see Figure 25-4). GTP is a high-energy compound that may be used to convert ADP to ATP. Accordingly, we will assume that there is a production of two molecules of ATP in the Krebs cycle since the energy in GTP is incorporated into ATP. The bulk of the ATP is generated in the electron transport chain, where a total of 34 are produced. This ATP accounting can be summarized as follows:

Process	Net Gain of ATP
Glycolysis	2
Krebs cycle	2
Electron transport chain	34
Total	38

The complete oxidation of a molecule of glucose can be summarized as follows (Figure 25-6):

$$C_6H_{12}O_6 + 6O_2 \rightarrow 38ATP + 6CO_2 + 6H_2O$$

Glucose Oxygen Adenosine triphosphate Carbon dioxide Water

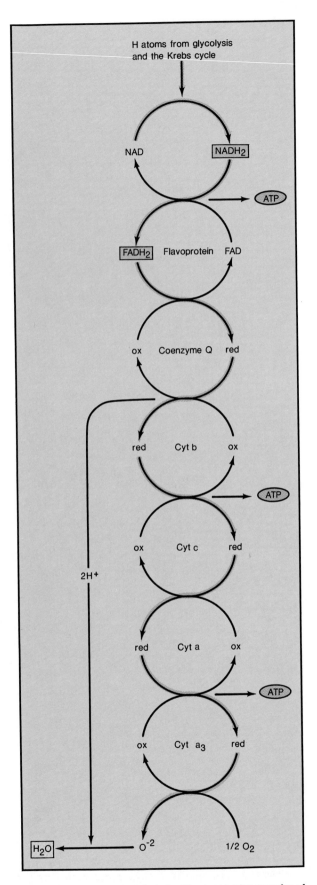

FIGURE 25-5 Electron transport chain. The successive transfer of hydrogen atoms and electrons results in the formation of ATP at three different places.

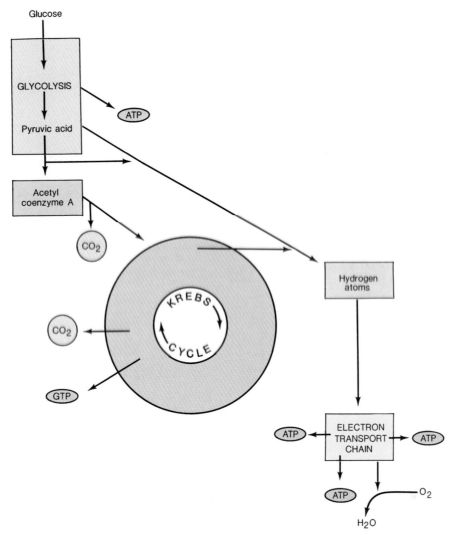

FIGURE 25-6 Summary of the complete oxidation of glucose.

About 43 percent of the energy originally in glucose is captured by ATP, the remainder is given off as heat.

Glycolysis, the Krebs cycle, and especially the electron transport chain provide all the ATP for cellular activities. And because the Krebs cycle and electron transport chain are aerobic processes, the cells cannot carry on their activities for long without sufficient oxygen.

GLUCOSE ANABOLISM

Most of the glucose in the body is catabolized to supply energy. However, some glucose participates in a number of anabolic reactions. One is the synthesis of glycogen from many glucose molecules. Another is the manufacture of glucose from the breakdown products of proteins and lipids.

Glucose Storage: Glycogenesis

If glucose is not needed immediately for energy, it is combined with many other molecules of glucose to form a long-chain molecule called glycogen. This process is called **glycogenesis** (*glyco* = sugar; *genesis* = origin). The body can store about 500 g (about 1.1 lb) of glycogen in the liver and skeletal muscle cells. Roughly 80 percent of the glycogen is stored in the skeletal muscles.

In the process of glycogenesis (Figure 25-7), glucose that enters cells is first phosphorylated to glucose-6-phosphate. This is then converted to glucose-1-phosphate, then to uridine diphosphate glucose, and finally to glycogen. Glycogenesis is stimulated by insulin from the pancreas.

Glucose Release: Glycogenolysis

When the body needs energy, the glycogen stored in the liver is broken down into glucose and released into the bloodstream to be transported to cells, where it will be catabolized. The process of converting glycogen back to glucose is called **glycogenolysis** (*lysis* = breakdown). Glycogenolysis usually occurs between meals.

Glycogenolysis does not occur exactly by the reverse processes involved in glycogenesis (Figure 25-7). The first step in glycogenolysis involves the splitting off of glucose molecules from the branched glycogen molecule by phosphorylation to form glucose-1-phosphate. Phosphorylase, the enzyme that catalyzes this reaction, is activated by

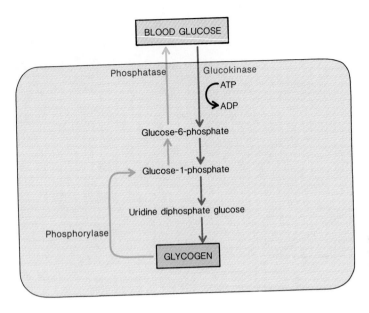

FIGURE 25-7 Glycogenesis and glycogenolysis. The glycogenesis pathway, the conversion of glucose into glycogen, is indicated by red arrows. The glycogenolysis pathway, the conversion of glycogen into glucose, is indicated by blue arrows. Some of the enzymes involved are indicated by parentheses.

the hormones glucagon from the pancreas and epinephrine from the adrenal medulla. Glucose-1-phosphate is then converted to glucose-6-phosphate and finally to glucose. Phosphatase, the enzyme that converts glucose-6-phosphate into glucose and releases glucose into the blood, is present in liver cells, but absent in skeletal muscle cells.

Formation of Glucose from Proteins and Fats: Gluconeogenesis

When your liver runs low on glycogen, it is time to eat. If you do not eat, your body starts catabolizing fats and proteins. Actually, the body normally catabolizes some of its fats and a few of its proteins. But large-scale fat and protein catabolism does not happen unless you are starving, eating meals that contain very few carbohydrates, or suffering from an endocrine disorder.

Both fat molecules and protein molecules may be converted in the liver to glucose. The process by which glucose is formed from noncarbohydrate sources is called **gluconeogenesis** (gloo'-kō-nē'-o-JEN-e-sis).

In the process of gluconeogenesis, moderate quantities of glucose can be formed from amino acids and the glycerol portion of fat molecules (Figure 25-8). About 60 percent of the amino acids in the body can undergo this conversion. Amino acids such as alanine, cysteine, glycine, serine, and threonine are converted to pyruvic acid. The pyruvic acid may be resynthesized into glucose or enter the Krebs cycle. Glycerol may be converted into glyceraldehyde-3-phosphate, which may also be resynthesized into glucose or form pyruvic acid. Figure 25-8 also shows how gluconeogenesis is related to other metabolic reactions.

Gluconeogenesis is stimulated by cortisol, one of the glucocorticoid hormones of the adrenal cortex, and thyroxine from the thyroid gland. Cortisol mobilizes proteins from body cells, making them available in the form of amino acids, thus supplying a pool of amino acids for gluconeogenesis. Thyroxine also mobilizes proteins and may mobilize fats from fat depots by making glycerol available for gluconeogenesis. Gluconeogenesis is also stimulated by epinephrine, glucagon, and growth hormone (GH).

LIPID METABOLISM

Lipids are second to carbohydrates as a source of energy. More frequently they are used as building blocks to form essential structures. When neutral fats (triglycerides) are eaten, they are ultimately digested into fatty acids and monoglycerides. Short-chain fatty acids enter the blood capillaries in villi, while long-chain fatty acids and monoglycerides are carried in micelles to epithelial cells of the villi for entrance. Once inside, they are further digested to glycerol and fatty acids, recombined to form triglycerides, and transported as chylomicrons through the lacteals of villi into the thoracic duct, subclavian vein, and to the liver.

FATE OF LIPIDS

Lipids, such as fatty acids and glycerol, like carbohydrates, may be oxidized to produce ATP. Each gram of fat produces about 9.0 Cal. If the body has no immediate need to utilize fats this way, they are stored in adipose tissue (fat depots) throughout the body and in the liver. Other lipids are used as structural molecules or to synthesize other essential substances. For example, phospholipids are constituents of plasma membranes and substrates for the synthesis of prostaglandins, lipoproteins are used to transport cholesterol throughout the body, thromboplastin is needed for blood clotting, and myelin sheaths speed up nerve impulse conduction. Cholesterol, another lipid, is used in the synthesis of bile salts and steroid hormones (adrenocortical hormones and sex hormones). The various functions of lipids in the body may be reviewed in Exhibit 2-5.

FAT STORAGE

The major function of adipose tissue is to store fats until they are needed for energy in other parts of the body.

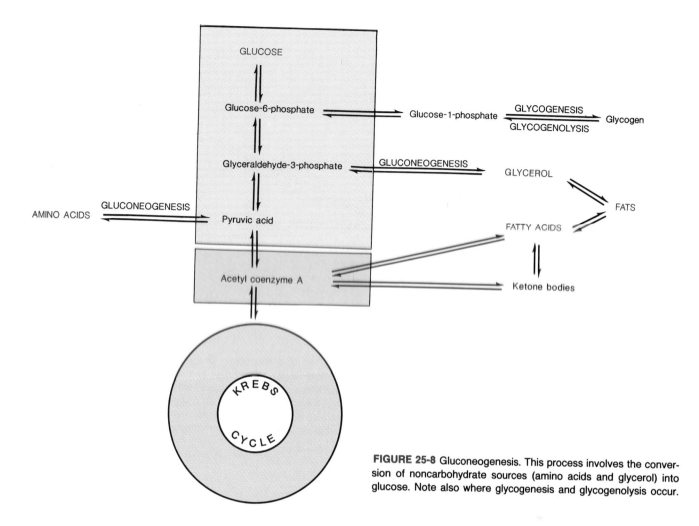

FIGURE 25-8 Gluconeogenesis. This process involves the conversion of noncarbohydrate sources (amino acids and glycerol) into glucose. Note also where glycogenesis and glycogenolysis occur.

It also insulates and protects. About 50 percent of stored fat is deposited in subcutaneous tissue—approximately 12 percent around the kidneys, 10 to 15 percent in the omenta, 20 percent in genital areas, and 5 to 8 percent between muscles. Fat is also stored behind the eyes, in the sulci of the heart, and in the folds of the large intestine.

Adipose cells contain lipases that catalyze the deposition of fats from chylomicrons and hydrolyze fats into fatty acids and glycerol. Because of the rapid exchange of fatty acids, fats are renewed about once every two to three weeks. Thus, the fat stored in your adipose tissue today is not the same fat that was present last month. Moreover, fat is continually released from storage, transported through the blood, and redeposited in other adipose tissue cells.

LIPID CATABOLISM

Fats stored in adipose tissue constitute the largest reserve of energy. The body can store much more fat than it can glycogen. Moreover, the energy yield of fats is more than twice that of carbohydrates. Nevertheless, fats are only the body's second-favorite source of energy because they are more difficult to catabolize than carbohydrates.

Glycerol

Before fat molecules can be metabolized as an energy source, they must first be released from fat depots and split into glycerol and fatty acids. Both these processes are stimulated by growth hormone (GH). The glycerol and fatty acids are then catabolized separately (Figure 25-9).

Glycerol is converted easily by many cells of the body to glyceraldehyde-3-phosphate, one of the compounds also formed during the catabolism of glucose. The cells then transform glyceraldehyde-3-phosphate into glucose

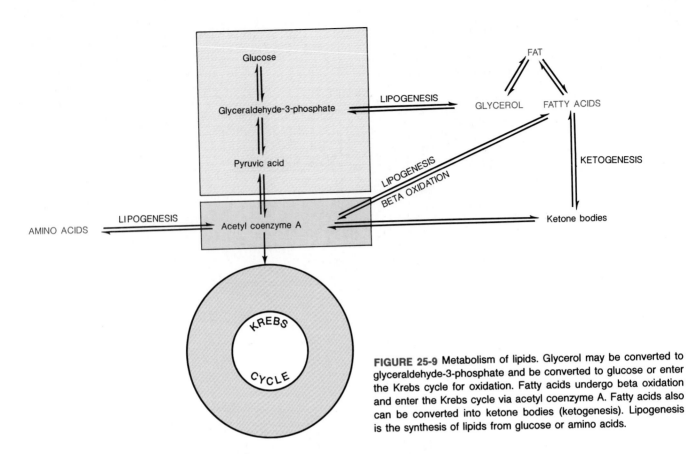

FIGURE 25-9 Metabolism of lipids. Glycerol may be converted to glyceraldehyde-3-phosphate and be converted to glucose or enter the Krebs cycle for oxidation. Fatty acids undergo beta oxidation and enter the Krebs cycle via acetyl coenzyme A. Fatty acids also can be converted into ketone bodies (ketogenesis). Lipogenesis is the synthesis of lipids from glucose or amino acids.

for catabolism. This is one example of gluconeogenesis. Since glyceraldehyde-3-phosphate is an intermediate product in the conversion of glycerol to glucose, glycerol thus enters the glycolytic pathway to be used for energy.

Fatty Acids

Fatty acids are catabolized differently and the process occurs in the matrix of mitochondria. The first step in fatty-acid catabolism involves a series of reactions called **beta oxidation.** Through a series of complex reactions involving dehydrogenation, hydration, and cleavage, enzymes remove pairs of carbon atoms at a time from the long chain of carbon atoms comprising a fatty acid. The result of beta oxidation is a series of two-carbon fragments, **acetyl coenzyme A (CoA).** Since the majority of fatty acids have even numbers of carbon atoms, the number of acetyl CoA molecules produced is easily calculated by dividing the number of carbon atoms in the fatty acid by two. Thus, palmitic acid, a fatty acid with 18 carbon atoms, will produce 9 acetyl CoA molecules upon beta oxidation.

In the second step of fatty acid catabolism, the acetyl CoA formed as a result of beta oxidation enters the Krebs cycle (Figure 25-9). In terms of energy yield, an 18-carbon fatty acid, such as palmitic acid, can yield a net of 129 ATPs upon its complete oxidation via the Krebs cycle and electron transport chain.

As part of normal fatty acid catabolism, the liver can take acetyl CoA molecules, two at a time, and condense

them to form a substance called **acetoacetic acid** which is converted mostly into β-hydroxybutyric acid and partially into **acetone.** These substances are collectively known as **ketone bodies,** and their formation is called **ketogenesis** (Figure 25-9). They then leave the liver, enter the bloodstream, and diffuse into other body cells where they are broken down into acetyl CoA, which enters the Krebs cycle for oxidation. During periods of excessive beta oxidation, large amounts of acetyl CoA are produced. This might occur following a meal rich in fats or during fasting or starvation because essentially no carbohydrates are available for catabolism. It may also occur in diabetes mellitus because insulin is not available to stimulate glycogenesis and glucose transport into cells. When a diabetic becomes seriously insulin deficient, one of the telltale signs is a sweet smell of acetone on the breath.

CLINICAL APPLICATION

Since the body prefers glucose as a source of energy, ketone bodies are generally produced in very small quantities (1.5 to 2.0 mg/100 ml blood). When the number of ketone bodies in the blood rises above normal—a condition called **ketosis**—the ketone bodies, most of which are acids, must be buffered by the body. If too many accumulate, they use up the body's buffers and the blood pH falls. Thus extreme or prolonged ketosis can lead to **acidosis,** or abnormally low blood pH.

LIPID ANABOLISM: LIPOGENESIS

Liver cells can synthesize lipids from glucose or amino acids through a process called **lipogenesis** (Figure 25-9). Lipogenesis occurs when a greater quantity of carbohydrate enters the body than can be used for energy or stored as glycogen. The excess carbohydrate is synthesized into fats. The steps in the conversion of glucose to lipids are complex and involve the formation of glyceraldehyde-3-phosphate, which can be converted to glycerol, and acetyl CoA, which can be converted to fatty acids (see Figure 25-8). The process is enhanced by insulin. The resulting glycerol and fatty acids can undergo anabolic reactions to become fat that can be stored or go through a series of anabolic reactions that produce other lipids such as lipoproteins, phospholipids, and cholesterol.

Many amino acids can be converted into acetyl CoA, which can be converted into fats (Figure 25-9). When people have more proteins in their diets than can be uti-

lized as such, much of the excess protein is converted to and stored as fats.

PROTEIN METABOLISM

During the process of digestion, proteins are broken down into their constituent amino acids. The amino acids are then absorbed by the blood capillaries in villi and transported to the liver via the hepatic portal vein.

FATE OF PROTEINS

Amino acids enter body cells by active transport. This process is stimulated by growth hormone (GH) and insulin. Almost immediately after entrance, they are synthesized into proteins. Generally, the body uses very little protein for energy, as long as it ingests or stores sufficient amounts of carbohydrates and fats. Each gram of protein

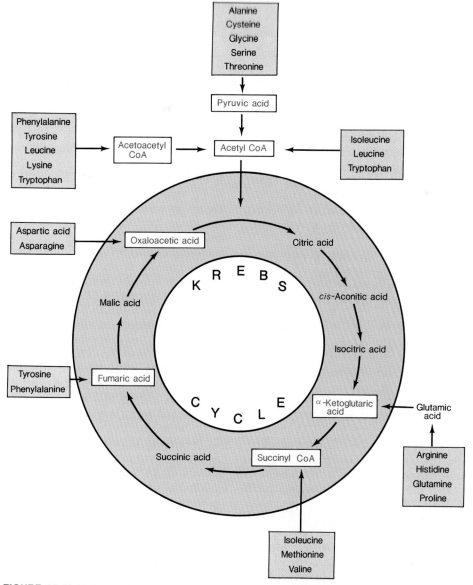

FIGURE 25-10 Various points at which amino acids, shown in gold boxes, enter the Krebs cycle for oxidation.

produces about 4.0 Cal. Most proteins function as enzymes. Other proteins are involved in transportation (hemoglobin). Others serve as antibodies, clotting chemicals (fibrinogen), hormones (insulin), and contractile elements in muscle cells (actin and myosin). Several proteins serve as structural components of the body (collagen, elastin, keratin, and nucleoproteins). The various functions of proteins in the body may be reviewed in Exhibit 2-6.

PROTEIN CATABOLISM

A certain amount of protein catabolism occurs in the body each day, although much of this is only partial catabolism. Proteins are extracted from worn-out cells, such as red blood cells, and broken down into free amino acids. Some amino acids are converted into other amino acids, peptide bonds are reformed, and new proteins are made.

If other energy sources are used up or if other sources are adequate and protein intake is high, the liver can convert protein to fat or glucose or oxidize it to carbon dioxide and water. However, before amino acids can be catabolized, they must first be converted to various substances that can enter the Krebs cycle. One such conversion consists of removing the amino group (NH_2) from the amino acid, a process called **deamination.** The liver cells then convert the NH_2 to ammonia (NH_3) and finally to urea, which is excreted in the urine. Other conversions are decarboxylation and dehydrogenation. The fate of the remaining part of the amino acid depends on what kind of amino acid it is. Figure 25-10 shows that amino acids enter the Krebs cycle at different points. They may be converted to pyruvic acid, acetyl CoA, α-ketoglutaric acid, succinyl CoA, fumaric acid, oxaloacetic acid, or acetoacetyl CoA. The point is that amino acids can be altered in various ways to enter the Krebs cycle at various locations.

The gluconeogenesis of amino acids into glucose may be reviewed in Figure 25-8. The conversion of amino acids into fatty acids or ketone bodies is also shown in Figure 25-8.

PROTEIN ANABOLISM

Protein anabolism involves the formation of peptide bonds between amino acids to produce new proteins. Protein anabolism, or synthesis, is carried out on the ribosomes of almost every cell in the body, directed by the cells' DNA and RNA (see Figures 3-16 and 3-17). Protein synthesis is stimulated by growth hormone (GH), thyroxine, and insulin. The synthesized proteins are the primary constituents of enzymes, antibodies, clotting chemicals, hormones, structural components of cells, and so forth. Because proteins are a primary ingredient of most cell structures, high-protein diets are essential during the growth years, during pregnancy, and when tissue has been damaged by disease or injury.

Of the naturally occurring amino acids, 10 are referred to as **essential amino acids.** These amino acids cannot be synthesized by the human body from molecules present in the body. They are synthesized by plants or bacteria and so foods containing these amino acids are "essential"

for human growth and must be part of the diet. **Nonessential amino acids** can be synthesized by body cells by a process called **transamination,** the transfer of an amino group from an amino acid to a substance such as pyruvic acid or an acid of the Krebs cycle. Once the appropriate essential and nonessential amino acids are present in cells, protein synthesis occurs rapidly.

A summary of carbohydrate, lipid, and protein metabolism is presented in Exhibit 25-1.

ABSORPTIVE AND POSTABSORPTIVE STATES

The body actually alternates between two metabolic states, referred to as absorptive and postabsorptive. During the **absorptive state** ingested nutrients are entering the cardiovascular and lymphatic systems from the gastrointestinal tract. During the **postabsorptive (fasting) state,** absorption is complete and the energy needs of the body must be satisfied by nutrients already in the body. By analyzing the major events of both states we can also gain a better understanding of how the metabolic pathways are interrelated.

ABSORPTIVE STATE

An average meal usually requires about 4 hours for complete absorption and, given 3 meals a day, the body spends about 12 hours in the absorptive state. The other 12 hours, during late morning, late afternoon, and most of the evening are spent in the postabsorptive state.

During the absorptive state, glucose transported to the liver is mostly converted to fats or glycogen; little is oxidized for energy in the liver. Some fat synthesized in the liver remains there, but most of it is released into the blood for storage in adipose tissue. Adipose tissue cells also take up glucose not picked up by the liver and convert it into fat for storage. Some blood glucose is stored as glycogen in skeletal muscles. The majority of blood glucose is used by body cells for oxidation to carbon dioxide, water, and ATP (Figure 25-11).

During the absorptive state, most fat is stored in adipose tissue; only a small portion is used for synthesis. Adipose cells derive the fat from chylomicrons, from synthetic activities of the liver, and from their own synthetic reactions (Figure 25-11).

Many absorbed amino acids that enter liver cells are converted to carbohydrates called keto acids. These are catabolized to carbon dioxide and water to provide energy. Some keto acids are also synthesized into fatty acids in liver cells. These fatty acids may be synthesized into fats and stored in adipose tissue. Some amino acids that enter liver cells are used to synthesize proteins, such as plasma proteins. Amino acids not taken up by liver cells enter other cells of the body, such as muscle cells, for synthesis into proteins.

POSTABSORPTIVE (FASTING) STATE

The principal concern of the body during the postabsorptive state is to maintain normal blood glucose level (80 to 100 mg/100 ml of blood) even though no absorption

EXHIBIT 25-1
SUMMARY OF METABOLISM

PROCESS	COMMENT
CARBOHYDRATE METABOLISM	
Glucose catabolism	Complete oxidation of glucose, also referred to as cellular respiration, is chief source of energy in cells. The process requires glycolysis, Krebs cycle, and electron transport chain. The complete oxidation of 1 molecule of glucose results in the net production of 36 molecules of ATP.
Glycolysis	Conversion of glucose into pyruvic acid results in the production of some ATP. Reactions do not require oxygen.
Krebs cycle	Cycle includes series of oxidation-reduction reactions in which coenzymes (NAD and FAD) pick up hydrogen atoms from oxidized organic acids and some ATP is produced. CO_2 and H_2O are by-products. Reactions are aerobic.
Electron transport chain	Third set of reactions in glucose catabolism is another series of oxidation-reduction reactions, in which electrons are passed between FAD, coenzyme Q, and cytochromes and most of the ATP is produced. Reactions are aerobic.
Glucose anabolism	Some glucose is converted into glycogen (glycogenesis) for storage if not needed immediately for energy. Glycogen can be converted to glucose (glycogenolysis) if needed for energy. The conversion of fats and proteins into glucose is called gluconeogenesis.
LIPID METABOLISM	
Catabolism	Glycerol may be converted into glucose (gluconeogenesis) or catabolized via glycolysis. Fatty acids are catabolized via beta oxidation into acetyl CoA that is catabolized in the Krebs cycle. Acetyl CoA can also be converted into ketone bodies.
Anabolism	The synthesis of lipids from glucose and amino acids is lipogenesis. Fats are stored in adipose tissue.
PROTEIN METABOLISM	
Catabolism	Amino acids are oxidized via the Krebs cycle after conversion by processes such as deamination. Ammonia resulting from the conversions is converted into urea in the liver and excreted in urine. Amino acids may be converted into glucose (gluconeogenesis) and fatty acids or ketone bodies.
Anabolism	Protein synthesis is directed by DNA and utilizes the cells' RNA and ribosomes.

is occurring. The maintenance of this level is especially important to the functioning of the nervous system, since it can't use any other nutrient for energy.

In general, glucose is made available during the postabsorptive state by utilization of all glucose sources and by glucose sparing and fat utilization (Figure 25-12). One source of blood glucose during fasting is liver glycogen. This, however, only provides enough glucose for about 4 hours. Another source is skeletal muscle glycogen, through a very indirect route in which the glycogen eventually reaches the liver for release. This mechanism supplies about as much glucose as the liver. A third source of blood glucose is the catabolism of fats, especially in adipose tissue. The glycerol thus produced can be converted into glucose in the liver. Another source of blood glucose is protein. In fact, during fasting it is the major source. Proteins in muscle tissue and, to a lesser extent, in other tissues can be converted to amino acids and the amino acids can be converted to keto acids and then to glucose in the liver.

Despite all of these mechanisms for supplying blood glucose, they cannot maintain blood glucose level for very long. Thus, a major body adjustment must be made during the postabsorptive state. Although the nervous system continues to utilize blood glucose normally, all other body tissues reduce their oxidation of glucose and switch over to fat as their energy source. Accordingly, fats in adipose tissue are broken down and fatty acids are released into the blood. The fatty acids are picked up by body cells, except nervous tissue, and oxidized to carbon dioxide, water, and ATP. The liver converts fatty acids to ketone bodies which enter most body cells and are then oxidized to carbon dioxide, water, and ATP. As a result of this glucose sparing and fat utilization, an individual can fast for several weeks, provided water is consumed.

REGULATION OF METABOLISM

Absorbed nutrients have several alternatives, based upon the needs of the body. They may be oxidized for energy, stored, or converted. The pathway taken by a particular nutrient is enzymatically controlled and is regulated by hormones. In the discussion of the metabolism of carbohydrates, lipids, and proteins, reference has been made to the hormonal regulation involved. The actions of the principal hormones related to metabolism are summarized in Exhibit 25-2.

Hormones are the primary regulators of metabolism. However, hormonal control is ineffective without the proper minerals and vitamins. Some minerals and many vitamins are components of the enzyme systems that catalyze the metabolic reactions.

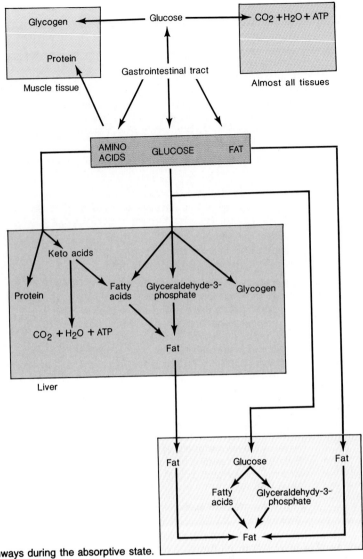

FIGURE 25-11 Principal pathways during the absorptive state.

FIGURE 25-12 Principal pathways during the postabsorptive (fasting) state.

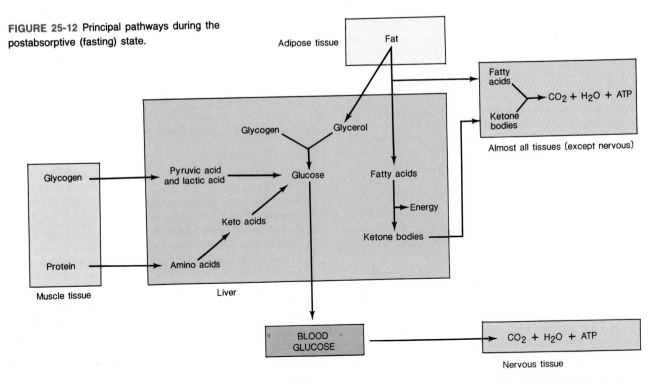

EXHIBIT 25-2

SUMMARY OF HORMONAL REGULATION OF THE METABOLISM OF CARBOHYDRATES, LIPIDS, AND PROTEINS

HORMONE	ACTIONS
Insulin	Increases glucose uptake by cells, especially muscle and adipose cells.
	Stimulates conversion of glucose into glycogen (glycogenesis).
	Stimulates fat synthesis and inhibits fat breakdown.
	Stimulates active transport of amino acids into cells, especially muscle cells.
	Stimulates protein synthesis.
Glucagon	Stimulates conversion of glycogen into glucose (glycogenolysis).
	Stimulates conversion of noncarbohydrates into glucose (gluconeogenesis).
	Stimulates breakdown of fats (fat mobilization).
Epinephrine	Stimulates conversion of glycogen into glucose (glycogenolysis).
	Stimulates conversion of noncarbohydrates into glucose (gluconeogenesis).
	Stimulates breakdown of fats (fat mobilization).
Growth hormone (GH)	Stimulates active transport of amino acids into body cells, especially muscle cells.
	Stimulates protein synthesis.
	Stimulates conversion of noncarbohydrates into glucose (gluconeogenesis).
	Stimulates breakdown of fats (fat mobilization).
Thyroxine	Stimulates protein synthesis.
	Stimulates conversion of noncarbohydrates into glucose (gluconeogenesis).
	Stimulates breakdown of fats (fat mobilization).
Cortisol	Stimulates conversion of noncarbohydrates into glucose (gluconeogenesis).
Testosterone	Increases protein deposition in body cells, especially muscle cells.

MINERALS

Minerals are inorganic substances. They may appear in combination with each other or in combination with organic compounds. Minerals constitute about 4 percent of the total body weight, and they are concentrated most heavily in the skeleton. Minerals known to perform functions essential to life include calcium, phosphorus, sodium, chlorine, potassium, magnesium, iron, sulfur, iodine, manganese, cobalt, copper, zinc, selenium, and chromium. Other minerals—aluminum, silicon, arsenic, nickel—are present in the body, but their functions have not yet been determined.

Calcium and phosphorus form part of the structure of bone. But since minerals do not form long-chain compounds, they are otherwise poor building materials. Their chief role is to help regulate body processes. Calcium, iron, magnesium, and manganese are constituents of some coenzymes. Magnesium also serves as a catalyst for the conversion of ADP to ATP. Without these minerals, metabolism would stop and the body would die. Minerals such as sodium and phorphorus work in buffer systems. Sodium helps regulate the osmosis of water and, along with other ions, is involved in the generation of nerve impulses. Exhibit 25-3 describes the functions of some minerals vital to the body. Note that the body generally uses the ions of the minerals rather than the nonionized form. Some minerals, such as chlorine, are toxic or even fatal to the body if ingested in the nonionized form.

VITAMINS

Organic nutrients required in minute amounts to maintain growth and normal metabolism are called **vitamins.** Un-like carbohydrates, fats, or proteins, vitamins do not provide energy or serve as building materials. The essential function of vitamins is the regulation of physiological processes. Of the vitamins whose functions are known, most serve as coenzymes.

Most vitamins cannot be synthesized by the body. They may be ingested in foods or pills. Other vitamins, such as vitamin K, are produced by bacteria in the gastrointestinal tract. The body can assemble some vitamins if the raw materials called *provitamins* are provided. Vitamin A is produced by the body from the provitamin carotene, a chemical present in spinach, carrots, liver, and milk. No single food contains all the required vitamins—one of the best reasons for eating a balanced diet.

CLINICAL APPLICATION

The term **avitaminosis** refers to a deficiency of any vitamin in the diet. **Hypervitaminosis** refers to an excess of one or more vitamins. An excess of vitamin A, called hypervitaminosis A, results in peeling and itching of the skin, hair loss, headache, and dizziness. An excess of vitamin D, called hypervitaminosis D, results in weakness, fatigue, weight loss, and nausea.

On the basis of solubility, vitamins are divided into two principal groups: fat-soluble and water-soluble. **Fat-soluble** vitamins are absorbed along with ingested dietary fats by the small intestine as micelles. In fact, they cannot be absorbed unless they are ingested with some fat. Fat-soluble vitamins are generally stored in cells, particularly liver cells, so reserves can be built up. Examples of fat-soluble vitamins are vitamins A, D, E, and K. **Water-**

soluble vitamins, by contrast, are absorbed along with water in the gastrointestinal tract and dissolve in the body fluids. Excess quantities of these vitamins are excreted in the urine. Thus the body does not store water-soluble vitamins well. Examples of water-soluble vitamins are the B vitamins and vitamin C.

Exhibit 25-4 lists the principal vitamins, their sources, functions, and related disorders.

METABOLISM AND BODY HEAT

We will now consider the relationship of foods to body heat, mechanisms of heat gain and loss, and the regulation of body temperature.

MEASURING HEAT

Heat is a form of energy that can be measured as **temperature** and expressed in units called calories. A **calorie (cal)** is the amount of heat energy required to raise the temperature of 1 g of water 1°C, from 14° to 15°C. Since the calorie is a small unit and large amounts of energy are stored in foods, the **Calorie** or **kilocalorie** is used instead. A Calorie is equal to 1,000 cal and is defined as the amount of heat required to raise the temperature of 1,000 g of water 1°C. The Calorie is the unit we use to express the heating value of foods and to measure the body's metabolic rate.

The apparatus used to determine the caloric value of foods is called a **calorimeter.** A weighed sample of a dehydrated food is burned completely in an insulated metal container. The energy released by the burning food is absorbed by the container and then transferred to a known volume of water that surrounds the container. The change in the water's temperature is directly related to the number of Calories released by the food. Knowing the caloric value of foods is important—if we know the amount of energy the body uses for various activities, we can adjust our food intake. In this way we can control body weight by taking in only enough Calories to sustain our activities.

PRODUCTION OF BODY HEAT

Most of the heat produced by the body comes from oxidation of the food we eat. The rate at which this heat is produced—the **metabolic rate**—is also measured in Calories. Among the factors that affect metabolic rate are the following:

EXHIBIT 25-3

MINERALS VITAL TO THE BODY

MINERAL	COMMENTS	IMPORTANCE
Calcium	Most abundant cation in body. Appears in combination with phosphorus in ratio of 2:1.5. About 99 percent is stored in bone and teeth. Remainder stored in muscle, other soft tissues, and blood plasma. Blood calcium level controlled by calcitonin (CT) and parathyroid hormone (PTH). Absorption occurs only in the presence of vitamin D. Most is excreted in feces and small amount in urine. Sources are milk, egg yolk, shellfish, green leafy vegetables.	Formation of bones and teeth, blood clotting, normal muscles and nerve activity, endocytosis and exocytosis, cellular motility, chromosome movement prior to cell division, glycogen metabolism, and synthesis and release of transmitter substances.
Phosphorus	About 80 percent found in bones and teeth. Remainder distributed in muscle, brain cells, blood. More functions than any other mineral. Blood phosphorus level controlled by calcitonin (CT) and parathyroid hormone (PTH). Most excreted in urine, small amount eliminated in feces. Sources are dairy products, meat, fish, poultry, nuts.	Formation of bones and teeth. Constitutes a major buffer system of blood. Plays important role in muscle contraction and nerve activity. Component of many enzymes. Involved in transfer and storage of energy (ATP). Component of DNA and RNA.
Iron	About 66 percent found in hemoglobin of blood. Remainder distributed in skeletal muscles, liver, spleen, enzymes. Normal losses of iron occur by shedding of hair, epithelial cells, and mucosal cells and in sweat, urine, feces, and bile. Sources are meat, liver, shellfish, egg yolk, beans, legumes, dried fruits, nuts, cereals.	As component of hemoglobin, carries O_2 to body cells. Component of cytochromes involved in formation of ATP from catabolism.
Iodine	Essential component of thyroid hormones. Excreted in urine. Sources are seafood, cod-liver oil, and vegetables grown in iodine-rich soils or iodized salt.	Required by thyroid gland to synthesize thyroid hormones, hormones that regulate metabolic rate.
Copper	Some stored in liver and spleen. Most excreted in feces. Sources include eggs, whole-wheat flour, beans, beets, liver, fish, spinach, asparagus.	Required with iron for synthesis of hemoglobin. Component of enzyme necessary for melanin pigment formation.
Sodium	Most found in extracellular fluids, some in bones. Excreted in urine and perspiration. Normal intake of NaCl (table salt) supplies required amounts.	As most abundant cation in extracellular fluid, strongly affects distribution of water through osmosis. Part of bicarbonate buffer system. Functions in nerve impulse conduction.

1. Exercise. During strenuous exercise, the metabolic rate may increase to as much as 50 times the normal rate.

2. Nervous system. In a stress situation, the sympathetic nervous system is stimulated and the nerves release norepinephrine (NE), which increases the metabolic rate of body cells. Strong sympathetic stimulation may increase the metabolic rate by 160 times, but only for a few minutes.

3. Hormones. In addition to norepinephrine (NE), two other hormones affect metabolic rate: epinephrine produced by the adrenal glands and thyroid hormones produced by the thyroid gland. Epinephrine is secreted in stress situations. Increased secretions of thyroid hormones increase the metabolic rate.

4. Body temperature. The higher the body temperature, the higher the metabolic rate. Each 1°C rise in temperature increases the rate of biochemical reactions by about 10 percent. The metabolic rate may be substantially increased during fever.

BASAL METABOLIC RATE

Since many factors affect metabolic rate, it is measured under standard conditions designed to reduce or eliminate those factors as much as possible. These conditions of the body are called the **basal state** and the measurement obtained is the **basal metabolic rate (BMR).** The person should not exercise for 30 to 60 minutes before the measurement is taken. The individual must be completely at rest, but awake. Air temperature should be comfortable. The person should fast for at least 12 hours. Body temperature should be normal. Basal metabolic rate is a measure of the rate at which the body breaks down foods (and therefore releases heat). It is expressed in Calories per square meter of body surface area per hour. BMR is also a measure of how much thyroxine the thyroid gland is producing, since thyroxine regulates the rate of food breakdown and is not a controllable factor under basal conditions.

The basal metabolic rate is most often measured indirectly by measuring oxygen consumption. If a given amount of food releases a given amount of heat energy when it is oxidized, it must combine with a given amount of oxygen. Thus, by measuring the amount of oxygen needed for the metabolism of foods, we can determine how many Calories are produced. The amount of heat energy released when 1 liter of oxygen combines with carbohydrates is 5.05 Calories; with fats, the heat released is 4.70 Calories; with proteins, the heat released is 4.60 Calories. The average of the three values is 4.825 Calories. Therefore, every time a liter of oxygen is consumed, 4.825 Calories are produced. For the normal adult, the average BMR is about 70 Calories per hour.

MINERAL	COMMENTS	IMPORTANCE
Potassium	Principal cation in intracellular fluid. Most is excreted in urine. Normal food intake supplies required amounts.	Functions in transmission of nerve impulses and muscle contraction.
Chlorine	Found in extracellular and intracellular fluids. Principal anion of extracellular fluid. Most excreted in urine. Normal intake of NaCl supplies required amounts.	Assumes role in acid–base balance of blood, water balance, and formation of HCl in stomach.
Magnesium	Component of soft tissues and bone. Excreted in urine and feces. Widespread in various foods.	Required for normal functioning of muscle and nervous tissue. Participates in bone formation. Constituent of many coenzymes.
Sulfur	Constituent of many proteins (such as insulin) and some vitamins (thiamine and biotin). Excreted in urine. Sources include beef, liver, lamb, fish, poultry, eggs, cheese, beans.	As component of hormones and vitamins, regulates various body activities.
Zinc	Important component of certain enzymes. Widespread in many foods, especially meats.	As a component of carbonic anhydrase, important in carbon dioxide metabolism. Necessary for normal growth, proper functioning of prostate gland, normal taste sensations and appetite, and normal sperm counts in males. As a component of peptidases, it is involved in protein digestion.
Fluorine	Component of bones, teeth, other tissues.	Appears to improve tooth structure and inhibit cariogenesis.
Manganese	Some stored in liver and spleen. Most excreted in feces.	Activates several enzymes. Needed for hemoglobin synthesis, urea formation, growth, reproduction, and lactation.
Cobalt	Constituent of vitamin B_{12}.	As part of B_{12}, required for maturation of erythropoeisis.
Chromium	Found in high concentrations in brewer's yeast. Also found in wine and some brands of beer.	Necessary for the proper utilization of dietary sugars and other carbohydrates by optimizing the production and effects of insulin. Helps increase blood levels of HDL, while decreasing levels of LDL.
Selenium		An antioxidant. Prevents chromosome breakage and may assume a role in preventing certain birth defects.

EXHIBIT 25-4
THE PRINCIPAL VITAMINS

VITAMIN	COMMENT AND SOURCE	FUNCTION	DEFICIENCY SYMPTOMS AND DISORDERS
FAT-SOLUBLE			
A	Formed from provitamin carotene (and other provitamins) in intestinal tract. Requires bile salts and fat for absorption. Stored in liver. Sources of carotene and other provitamins include yellow and green vegetables; sources of vitamin A include fish-liver oils, milk, butter.	Maintains general health and vigor of epithelial cells.	Deficiency results in atrophy and keratinization of epithelium, leading to dry skin and hair, increased incidence of ear, sinus, respiratory, urinary, and digestive infections, inability to gain weight, drying of cornea with ulceration (*xerophthalmia*), nervous disorders, and skin sores.
		Essential for formation of rhodopsin, light-sensitive chemical in rods of retina.	*Night blindness* or decreased ability for dark adaptation.
		Growth of bones and teeth by apparently helping to regulate activity of osteoblasts and osteoclasts.	Slow and faulty development of bones and teeth.
D	In presence of sunlight, provitamin D₃ (derivative of cholesterol) converted to vitamin D in the skin, liver, and kidneys. Dietary vitamin D requires moderate amounts of bile salts and fat for absorption. Stored in tissues to slight extent. Most excreted via bile. Sources include fish-liver oils, egg yolk, fortified milk.	Essential for absorption and utilization of calcium and phosphorus from gastrointestinal tract. May work with parathyroid hormone (PTH) that controls calcium metabolism.	Defective utilization of calcium by bones leads to *rickets* in children and *osteomalacia* in adults. Possible loss of muscle tone.
E (tocopherols)	Stored in liver, adipose tissue, and muscles. Requires bile salts and fat for absorption. Sources include fresh nuts and wheat germ, seed oils, green leafy vegetables.	Believed to inhibit catabolism of certain fatty acids that help form cell structures, especially membranes. Involved in formation of DNA, RNA, and red blood cells. May promote wound healing, contribute to the normal structure and functioning of the nervous system, prevent scarring and reduce the severity of visual loss associated with retrolental fibroplasia (an eye disease in premature infants caused by too much oxygen in incubators) by functioning as an antioxidant. Believed to help protect liver from toxic chemicals like carbon tetrachloride.	May cause the oxidation of unsaturated fats resulting in abnormal structure and function of mitochondria, lysosomes, and plasma membranes, a possible consequence being hemolytic anemia. Deficiency also causes muscular dystrophy in monkeys and sterility in rats.
K	Produced in considerable quantities by intestinal bacteria. Requires bile salts and fat for absorption. Stored in liver and spleen. Other sources include spinach, cauliflower, cabbage, liver.	Coenzyme believed essential for synthesis of prothrombin by liver and several clotting factors. Also known as antihemorrhagic vitamin.	Delayed clotting time results in excessive bleeding.

LOSS OF BODY HEAT

Body heat is produced by the oxidation of foods we eat. This heat must be removed continuously or body temperature would rise steadily. The principal routes of heat loss include radiation, conduction, convection, and evaporation.

Radiation

Radiation is the transfer of heat as infrared heat rays from one object to another without physical contact. Your body loses heat by the radiation of heat waves to cooler objects nearby such as ceilings, floors, and walls. If these objects are at a higher temperature, you absorb heat by

EXHIBIT 25-4 Continued ▬▬▬▬▬▬▬▬▬▬▬▬▬▬▬▬▬▬▬▬▬▬▬▬▬

VITAMIN	COMMENT AND SOURCE	FUNCTION	DEFICIENCY SYMPTOMS AND DISORDERS
WATER-SOLUBLE			
B₁ (thiamine)	Rapidly destroyed by heat. Not stored in body. Excessive intake eliminated in urine. Sources include whole-grain products, eggs, pork, nuts, liver, yeast.	Acts as coenzyme for many different enzymes involved in carbohydrate metabolism of pyruvic acid to CO_2 and H_2O. Essential for synthesis of acetylcholine.	Improper carbohydrate metabolism leads to buildup of pyruvic and lactic acids and insufficient energy for muscle and nerve cells. Deficiency leads to two syndromes: (1) *beriberi*—partial paralysis of smooth muscle of GI tract causing digestive disturbances, skeletal muscle paralysis, atrophy of limbs; (2) *polyneuritis*—due to degeneration of myelin sheaths reflexes related to kinesthesia are impaired, impairment of sense of touch, decreased intestinal motility, stunted growth in children, and poor appetite.
B₂ (riboflavin)	Not stored in large amounts in tissues. Most is excreted in urine. Small amounts supplied by bacteria of GI tract. Other sources include yeast, liver, beef, veal, lamb, eggs, whole-grain products, asparagus, peas, beets, peanuts.	Component of certain coenzymes (e.g., FAD) concerned with carbohydrate and protein metabolism, especially in cells of eye, integument, mucosa of intestine, blood.	Deficiency may lead to improper utilization of oxygen resulting in blurred vision, cataracts, and corneal ulcerations. Also dermatitis and cracking of skin, lesions of intestinal mucosa, and development of one type of anemia.
Niacin (nicotinamide)	Derived from amino acid tryptophan. Sources include yeast, meats, liver, fish, whole-grain products, peas, beans, nuts.	Essential component of coenzyme (NAD) concerned with energy-releasing reactions. In lipid metabolism, inhibits production of cholesterol and assists in fat breakdown.	Principal deficiency is *pellagra*, characterized by dermatitis, diarrhea, and psychological disturbances.
B₆ (pyridoxine)	Formed by bacteria of GI tract. Stored in liver, muscle, brain. Other sources include salmon, yeast, tomatoes, yellow corn, spinach, whole-grain products, liver, yogurt.	May function as coenzyme in fat metabolism. Essential coenzyme for normal amino acid metabolism. Assists production of circulating antibodies.	Most common deficiency symptom is dermatitis of eyes, nose, and mouth. Other symptoms are retarded growth and nausea.
B₁₂ (cyanocobalamin)	Only B vitamin not found in vegetables; only vitamin containing cobalt. Absorption from GI tract dependent on HCl and intrinsic factor secreted by gastric mucosa. Sources include liver, kidney, milk, eggs, cheese, meat.	Coenzyme necessary for red blood cell formation, formation of amino acid methionine, entrance of some amino acids into Krebs cycle, and manufacture of choline (chemical similar in function to acetylcholine).	Pernicious anemia and malfunction of nervous system due to degeneration of axons of spinal cord.
Pantothenic acid	Stored primarily in liver and kidneys. Some produced by bacteria of GI tract. Other sources include kidney, liver, yeast, green vegetables, cereal.	Constituent of coenzyme A essential for transfer of pyruvic acid into Krebs cycle, conversion of lipids and amino acids into glucose, and synthesis of cholesterol and steroid hormones.	Experimental deficiency tests indicate fatigue, muscle spasms, neuromuscular degeneration, insufficient production of adrenal steroid hormones.
Folic acid	Synthesized by bacteria of GI tract. Other sources include green leafy vegetables and liver.	Component of enzyme systems synthesizing purines and pyrimidines built into DNA and RNA. Essential for normal production of red and white blood cells.	Production of abnormally large red blood cells (macrocytic anemia).
Biotin	Synthesized by bacteria of GI tract. Other sources include yeast, liver, egg yolk, kidneys.	Essential coenzyme for conversion of pyruvic acid to oxaloacetic acid and synthesis of fatty acids and purines.	Mental depression, muscular pain, dermatitis, fatigue, nausea.
C (ascorbic acid)	Rapidly destroyed by heat. Some stored in glandular tissue and plasma. Sources include citrus fruits, tomatoes, green vegetables.	Exact role not understood. Promotes many metabolic reactions, particularly protein metabolism, including laying down of collagen in formation of connective tissue. As coenzyme may combine with poisons, rendering them harmless until excreted. Works with antibodies. Promotes wound healing.	Scurvy; anemia; many symptoms related to poor connective tissue growth and repair including tender swollen gums, loosening of teeth (alveolar processes also deteriorate), poor wound healing, bleeding (vessel walls fragile because of connective tissue degeneration), and retardation of growth.

radiation. Incidentally, the air temperature has no relationship to the radiation of heat to and from objects. Skiers can remove their shirts in bright sunshine, even though the air temperature is very low, because the radiant heat from the sun is adequate to warm them. In a room at 21°C (70°F), about 60 percent of heat loss is by radiation.

Conduction

Another method of heat transfer is **conduction.** In this process, body heat is transferred to a substance or object in contact with the body, such as chairs, clothing, or jewelry. About 3 percent of body heat is lost via conduction.

Convection

Convection is the transfer of heat by the movement of a fluid between areas of different temperature. When cool air makes contact with the body, it becomes warmed and is carried away by convection currents. Then more cool air makes contact with the body and is carried away. The faster the air moves, the faster the rate of conduction. About 15 percent of body heat is lost to the air by convection.

Evaporation

Evaporation is the conversion of a liquid to a vapor. Water has a high heat of evaporation. The *latent heat of evaporation* is the amount of heat necessary to evaporate 1 g of water at 30°C (86°F). Because of water's high latent heat of evaporation, every gram of water evaporating from the skin takes with it a great deal of heat—about 0.58 Calorie per gram of water. Under normal conditions, about 22 percent of heat loss occurs through evaporation. The evaporation of only 150 ml of water per hour is enough to remove all the heat produced by the body under basal conditions. Under extreme conditions, about 4 liters (1 gal) of perspiration is produced each hour and this volume can remove 2,000 Calories of heat from the body. This is approximately 32 times the basal level of heat production. The rate of evaporation is inversely related to relative humidity, the ratio of the actual amount of moisture in the air to the greatest amount it can hold at a given temperature. The higher the relative humidity, the lower the rate of evaporation.

A summary of heat-producing and heat-losing mechanisms is presented in Exhibit 25-5.

BODY TEMPERATURE REGULATION

If the amount of heat production equals the amount of heat loss, you maintain a constant body temperature near 37°C (98.6°F). If your heat-producing mechanisms generate more heat than is lost by your heat-losing mechanisms, your body temperature rises. If your heat-losing mechanisms give off more heat than is generated by heat-producing mechanisms, your temperature falls.

Hypothalamic Thermostat

Body temperature is regulated by mechanisms that attempt to keep heat production and heat loss in balance. A center of control of these mechanisms is found in the hypothalamus in a group of neurons in the anterior portion referred to as the *preoptic area.* If blood temperature increases, the neurons of the preoptic area fire impulses more rapidly. If something causes the blood's temperature to decrease, these neurons fire impulses more slowly. The preoptic area is adjusted to maintain normal body temperature and thus serves as your thermostat.

Impulses from the preoptic area are sent to other portions of the hypothalamus known as the heat-losing center and the heat-promoting center. The *heat-losing center,* when stimulated by the preoptic area, sets into operation a series of responses that lower body temperature. The *heat-promoting center,* when stimulated by the preoptic area, sets into operation a series of responses that raise body temperature. The heat-losing center is mainly parasympathetic in function; the heat-promoting center is primarily sympathetic.

Mechanisms of Heat Production

Suppose the environmental temperature is low or blood temperature falls below normal. Both stresses stimulate the preoptic area. The preoptic area, in turn, activates the heat-promoting center. In response, the heat-promoting center discharges impulses that automatically set into operation a number of responses designed to increase body heat and bring body temperature back up to normal.

● *Vasoconstriction* Impulses from the heat-promoting center stimulate sympathetic nerves that cause blood ves-

EXHIBIT 25-5 ▬▬▬▬
SUMMARY OF HEAT-PRODUCING AND HEAT-LOSING MECHANISMS

MECHANISM	COMMMENT
Heat Production	Most body heat is produced by the oxidation of foods, and the rate at which it is produced is called the metabolic rate. The rate is affected by exercise, strong sympathetic stimulation, hormones, and body temperature.
Heat Loss Radiation	Transfer of heat from the body to an object without physical contact. Example is losing heat to a cool object such as a floor.
Conduction	Transfer of heat from the body to any object in physical contact with the body such as clothing.
Convection	When cool air makes contact with the body, it is warmed and carried away by convection currents.
Evaporation	The conversion of a liquid to a vapor in which the evaporating substance (e.g., perspiration) removes heat from the body.

sels of the skin to constrict. The net effect of vasoconstriction is to decrease the flow of warm blood from the internal organs to the skin, thus decreasing the transfer of heat from the internal organs to the skin. This reduction in heat loss helps raise the internal body temperature.

● *Sympathetic Stimulation* Another response triggered by the heat-promoting center is the sympathetic stimulation of metabolism. The heat-promoting center stimulates sympathetic nerves leading to the adrenal medulla. This stimulation causes the medulla to secrete epinephrine and norepinephrine (NE) into the blood. The hormones, in turn, bring about an increase in cellular metabolism, a reaction that also increases heat production. This effect is called *chemical thermogenesis.*

● *Skeletal Muscles* Heat production is also increased by responses of skeletal muscles. For example, stimulation of the heat-promoting center causes stimulation of parts of the brain that increase muscle tone and hence heat production. As the muscle tone increases, the stretching of the agonist muscle initiates the stretch reflex and the muscle contracts. This contraction causes the antagonist muscle to stretch, and it too develops a stretch reflex. The repetitive cycle—called *shivering*—increases the rate of heat production by several hundred percent.

● *Thyroxine* Another body response that increases heat production is increased production of thyroxine. A cold environmental temperature causes the secretion of the regulating factor TRF produced by the preoptic area of the hypothalamus. The TRF in turn stimulates the anterior pituitary to secrete TSH, which causes the thyroid to release thyroxine into the blood. Since increased levels of thyroxine increase the metabolic rate, body temperature is increased.

Mechanisms of Heat Loss

Now suppose some stress raises body temperature above normal. The stress or higher temperature of the blood stimulates the preoptic area, which in turn stimulates the heat-losing center and inhibits the heat-promoting center. Instead of blood vessels in the skin constricting, they dilate. The skin becomes warm, and the excess heat is lost to the environment. At the same time, the metabolic rate and muscle tone are decreased. All these responses reverse the heat-promoting effects and bring body temperature down to normal.

When the body is subjected to high environmental temperatures or strenuous exercise, the high temperature of the blood signals the hypothalamus, which activates the heat-losing center. In response, impulses are sent out to the sweat glands of the skin, and they produce more perspiration. As the perspiration evaporates from the surface of the skin, the skin is cooled.

BODY TEMPERATURE ABNORMALITIES

A **fever** is an abnormally high body temperature. The most frequent cause of fever is infection from bacteria

(and their toxins) and viruses. Other causes are heart attacks, tumors, tissue destruction by x-rays or trauma, and reactions to vaccines. The mechanism of fever production is not completely understood. However, it is believed that foreign proteins, called *pyrogens* (*pyro* = fire), affect the hypothalamus by setting the thermostat at a higher temperature. It has been shown that 0.001 g of protein from the bacterium that causes typhoid fever can set the thermostat as high as 43.3°C (110°F) and the body temperature will continue to be regulated at this temperature until the protein is eliminated.

Suppose that disease-producing microbes set the thermostat at 39.4°C (103°F). Now the heat-promoting mechanisms (vasoconstriction, increased metabolism, shivering) are operating at full force. Thus even though body temperature is climbing higher than normal, say 38.3°C (101°F), the skin remains cold and shivering occurs. This condition, called a *chill,* is a definite sign that body temperature is rising. After several hours, body temperature reaches the setting of the thermostat and the chills disappear. But the body will continue to regulate temperature at 39.4°C (103°F) until the stress is removed. When the stress is removed, the thermostat is reset at normal—37°C (98.6°F). Since body temperature remains high in the beginning, the heat-losing mechanisms (vasodilation and sweating) go into operation to decrease body temperature. The skin becomes warm and the person begins to sweat. This phase of the fever is called the *crisis* and indicates that body temperature is falling.

Up to a point, fever is beneficial. The high body temperature is believed to inhibit the growth of some bacteria and viruses. Moreover, heat speeds up the rate of chemical reactions. This increase may help body cells to repair themselves more quickly during a disease. As a rule, death results if body temperature rises to 44.4 to 45.5°C (112 to 114°F). On the other end of the scale, death usually results when body temperature falls to 21.1 to 23.9°C (70 to 75°F).

In addition to fever, several other conditions are associated with heat regulation:

1. Heat cramp. This condition occurs as a result of profuse sweating, which removes water and NaCl from the body. Body temperature is not elevated. The salt loss causes painful contractions of muscles called heat cramp. The condition may be alleviated by taking salt tablets.

2. Sunstroke or **heatstroke.** This condition is brought about when the temperature and relative humidity are high, making it difficult for the body to lose heat by radiation or evaporation. As a result, there is a decreased flow of blood to the skin, perspiration is greatly reduced, and body temperature rises sharply. The skin is thus dry and hot—the temperature may reach 110°F. Brain cells are quickly affected and may be destroyed permanently. Treatment must be undertaken immediately and consists of cooling the body by immersing the victim in cool water and intravenous replacement of fluids and electrolytes.

3. Heat exhaustion or **heat prostration.** The body temperature is generally normal, or a little below, and the skin is cool and clammy (moist) due to profuse perspiration. Unlike heatstroke, in heat exhaustion the heart is generally at fault. Heat exhaustion is normally characterized by salt loss and results in muscle cramps, dizziness, vomiting, and fainting. Complete rest and salt tablets are recommended.

DISORDERS: HOMEOSTATIC IMBALANCES

Obesity

In affluent technological nations such as ours, the demand for physical work is steadily decreasing. This same environment provides a variety of appetizing, rich foods that are easily obtained. One consequence of these conditions is **obesity,** defined as a body weight 10 to 20 percent above a desirable standard as the result of an excessive accumulation of fat. There is little doubt that even moderate obesity is hazardous to health. The Framingham Study of the Public Health Service showed a high correlation between obesity and sudden death. Obesity seems solidly accepted as one of the "risk factors" in coronary artery disease.

Causes

Causes of obesity can be classified as regulatory or metabolic. Regulatory obesity seems to be far more common than metabolic obesity. People with **regulatory obesity** have no apparent metabolic abnormality that can account for the obesity. They simply ingest more high-energy-releasing foods than their bodies need. Causes include neurotic overeating, cultural dietary habits in otherwise normal people, and inactivity. Occasionally, regulatory obesity is caused by a disorder in the hypothalamus that reduces sensations of satiety. **Metabolic obesity** results primarily from a disorder that reduces the catabolism of carbohydrates and fats. An example is hyposecretion of thyroxine. It is not yet certain to what degree these metabolic disorders may be caused by changes in diet or physical activity.

In the clinical setting, obesity, whether regulatory or metabolic, seems more usefully categorized according to age of onset, degree of severity, or presence of an associated disorder such as diabetes, hypertension, osteoarthritis, and hyperlipidemia.

Treatment

Reduction of body weight involves keeping caloric intake well below energy expenditure. The goals during weight decrease are:

1. Loss of body fat with a minimal accompanying breakdown of lean tissue.

2. Maintenance of physical and emotional fitness during the reducing period.

3. Establishment of eating and exercise habits to maintain weight at the recommended level.

A number of factors must be considered in the formulation of a reducing diet. These include degree of overweight, age, state of physical fitness, normal level of physical activity, and the presence of related illnesses such as hypertension, coronary heart disease, diabetes mellitus, or gastrointestinal disorders. The presence of disorders in addition to obesity may require special dietary restrictions. Medical authorities agree that a sensible diet contains adequate amounts of essential nutrients and results in the loss of approximately 2 pounds a week and not much more.

The most drastic reducing diet is one that provides no Calories at all. Such diets must provide sufficient water, electrolytes (especially potassium), and vitamins to prevent dehydration, muscular weakness, mental confusion, heart malfunction, and other complications. Thus people treated with the "no-Calorie" diet should be hospitalized during the period of total Calorie starvation. The person remains at rest or is limited to light activity during the fasting period. During total Calorie restriction, fat is burned at an appreciable rate, but significant quantities of lean tissue also are broken down and lost.

If dietary restrictions and drugs still fail to control the patient's obesity, the only alternative may be surgery. Three surgical procedures for overweight are the jejunoileal (intestinal) bypass, gastroplasty(gastric stapling), and gastric banding. The suitable patients for these procedures are individuals whose weight is more than 100 pounds over the ideal body weight for height or 100 percent over the ideal.

The **jejunoileal (intestinal) bypass** involves creating a bypass connecting the first 35 cm (14 inches) of the jejunum to the last 10 cm (4 inches) of the ileum, thereby eliminating most of the absorptive surface of the small intestine. In some cases a second surgical procedure would be indicated to reconnect the intestine when weight returns to normal and the person has changed food intake patterns. However, jejunoileal bypass is not favored by most physicians because at least half of those who have the operation require hospital care for such problems as fluid and electrolyte imbalance, liver failure, kidney stones, renal failure, enteritis, malnutrition, bone disease, or intolerable anal and perineal irritation from diarrhea.

In **gastroplasty (gastric stapling),** a small upper stomach pouch (50 ml) is made with a small reinforced emptying orifice that provides an appropriate limitation of intake without bypassing any of the stomach. Gastroplasty allows ingested nutrients to pass through all portions of the digestive tract and allows iron and calcium to be exposed to the duodenum where they can be readily absorbed. Unlike jejunoileal bypass, it allows bile salts, fat-soluble vitamins, B_{12}, and protein to be absorbed in a normal manner.

Gastric banding involves fastening a lockable nylon band around the upper part of the stomach. This creates a pouch, about 50 ml in volume, that effectively limits the amount of food that can be ingested. Gastric banding is a variation of gastroplasty, but seems to involve fewer complications since it is less invasive.

Phenylketonuria (PKU)

By definition, **phenylketonuria** (fen'-il-kē'-tō-NOO-rē-a) or **PKU** is a genetic error of metabolism characterized by an elevation of the amino acid phenylalanine in the blood. It is frequently associated with mental retardation. The DNA of children with phenylketonuria lacks the gene that normally programs the manufacture of the enzyme phenylalanine hydroxylase. This enzyme is necessary for converting phenylalanine into the amino acid tryosine, an amino acid that enters the Krebs cycle. As a result, phenylalanine cannot be metabolized, and what is not used in protein synthesis builds up in the blood. High levels of phenylalanine are toxic to the brain during the early years of life when the brain is developing. Mental retardation can be prevented, when the condition is detected early, by restricting the child to a diet that supplies only the amount of phenylalanine necessary for growth.

Cystic Fibrosis

Cystic fibrosis is an inherited disease of the exocrine glands that affects the pancreas, respiratory system, and salivary and sweat glands. It is the most common lethal genetic disease of Caucasians—5 percent of the population are thought to be genetic carriers.

Among the most common signs and symptoms are pancreatic insufficiency, pulmonary involvement, and cirrhosis of the liver. It is characterized by the production of thick exocrine secretions that do not drain easily from the respiratory passageways. The buildup of the secretions leads to inflammation and replacement of injured cells with connective tissue that blocks these passageways. One of the prominent features is blockage of the pancreatic ducts so that the digestive enzymes canot reach the intestine. Since pancreatic juice contains the only fat-digestion enzyme, the person fails to absorb fats or fat-soluble vitamins and thus suffers from vitamin A, D, and K deficiency diseases. Calcium needs fat to be absorbed, so tetany also may result.

A child suffering from cystic fibrosis is given pancreatic extract and large doses of vitamins A, D, and K. The therapeutic diet is low, but not lacking, in fats and high in carbohydrates and proteins that can be used for energy and can also be converted by the liver into the lipids essential for life processes.

Celiac Disease

Celiac disease results in malabsorption by the intestinal mucosa due to the ingestion of gluten. *Gluten* is the water-insoluble protein fraction of wheat, rye, and oats. In susceptible persons, ingestion of gluten induces a morphological change in the intestinal mucosa accompanied by a variable amount of malabsorption.

The condition is easily remedied by administering a diet that excludes all cereal grains except rice and corn.

Kwashiorkor

Some dietary proteins are referred to as complete proteins, that is, they contain adequate amounts of essential amino acids. Sources of complete proteins are primarily animal products such as milk, meat, fish, poultry, and eggs. Incomplete proteins lack essential amino acids. An example is zein, the protein in corn, which lacks the essential amino acids tryptophan and lysine. The diet of many African natives consists largely of corn meal. As a result, many African children especially develop a protein deficiency disorder called **kwashiorkor**. It is characterized by hypoprotein edema of the abdomen, lethargy, failure to grow, and sometimes mental retardation.

STUDY OUTLINE

Metabolism (p. 630)

1. Nutrients are chemical substances in food that provide energy, act as building blocks in forming new body components, or assist in the functioning of various body processes.
2. There are six major classes of nutrients: carbohydrates, lipids, proteins, minerals, vitamins, and water.
3. Metabolism refers to all chemical reactions of the body and has two phases: catabolism and anabolism.
4. Catabolism is the term for decomposition reactions that provide energy.
5. Anabolism consists of a series of synthetic reactions whereby small molecules are built up into larger ones that form the body's structural and functional components. Anabolic reactions use energy.

Carbohydrate Metabolism (p. 630)

1. During digestion, polysaccharides and disaccharides are converted to monosaccharides, which are absorbed through capillaries in villi and transported to the liver via the hepatic portal vein.
2. Carbohydrate metabolism is primarily concerned with glucose metabolism.

Fate of Carbohydrates

1. Some glucose is oxidized by cells to provide energy; it moves into cells by facilitated diffusion and becomes phosphorylated to glucose-6-phosphate; insulin stimulates glucose movement into cells.
2. Excess glucose can be stored by the liver and skeletal muscles as glycogen or converted to fat.
3. Glucose excreted in the urine can produce glycosuria.

Glucose Catabolism

1. Glucose oxidation is also called cellular respiration.
2. The complete oxidation of glucose to CO_2 and H_2O involves glycolysis, the Krebs cycle, and the electron transport chain.

Glycolysis

1. Glycolysis refers to the breakdown of glucose into two molecules of pyruvic acid.
2. When oxygen is in short supply, pyruvic acid is converted to lactic acid; under aerobic conditions, pyruvic acid enters the Krebs cycle.
3. As a result of glycolysis, there is a net production of 2 molecules of ATP.

Krebs Cycle

1. Pyruvic acid is prepared for entrance into the Krebs cycle by conversion to a two-carbon compound (acetyl group) followed by the addition of coenzyme A to form acetyl coenzyme A.
2. The Krebs cycle involves decarboxylations and oxidations and reductions of various organic acids.
3. Each molecule of pyruvic acid that enters the Krebs cycle produces 3 molecules of CO_2, 4 molecules of $NADH_2$, 1 molecule of $FADH_2$, and 1 molecule of GTP.
4. The energy originally in glucose and then pyruvic acid is primarily in the reduced coenzymes $NADH_2$ and $FADH_2$.

Electron Transport Chain

1. The electron transport chain is a series of oxidation-reduction reactions in which the energy in $NADH_2$ and $FADH_2$ is liberated and transferred to ATP for storage.
2. The carrier molecules involved include FAD, coenzyme Q, and cytochromes.
3. The electron transport chain yields 32 molecules of ATP and H_2O.
4. The complete oxidation of glucose can be represented as follows:

$$C_6H_{12}O_6 + 6O_2 \rightarrow 36ATP + 6CO_2 + 6H_2O$$

Glucose Anabolism

1. The conversion of glucose to glycogen for storage in the liver and skeletal muscle is called glycogenesis. The process occurs in the liver and is stimulated by insulin.
2. The body can store about 500 g of glycogen.
3. The conversion of glycogen back to glucose is called glycogenolysis.
4. It occurs between meals and is stimulated by glucagon and epinephrine.
5. Gluconeogenesis is the conversion of fat and protein molecules into glucose. It is stimulated by cortisol, thyroxine, epinephrine, glucagon, and growth hormone (GH).
6. Glycerol may be converted to glyceraldehyde-3-phosphate and some amino acids may be converted to pyruvic acid.

Lipid Metabolism (p. 638)

1. During digestion, fats are ultimately broken down into fatty acids and monoglycerides.
2. Long-chain fatty acids and monoglycerides are carried in micelles for entrance into villi, digested to glycerol and fatty acids in epithelial cells, recombined to form triglycerides, and transported by chylomicrons through the lacteals of villi into the thoracic duct.

Fate of Lipids

1. Some fats may be oxidized to produce ATP.

2. Some fats are stored in adipose tissue.
3. Other lipids are used as structural molecules or to synthesize essential molecules. Examples include phospholipids of plasma membranes, lipoproteins that transport cholesterol, thromboplastin for blood clotting, and cholesterol used to synthesize bile salts and steroid hormones.

Fat Storage

1. Fats are stored in adipose tissue, mostly in the subcutaneous layer.
2. Adipose cells contain lipases that catalize the deposition of fats from chylomicrons and hydrolyze fats into fatty acids and glycerol.

Lipid Catabolism

1. Fat is released from depots and split into fatty acids and glycerol under the influence of growth hormone (GH).
2. Glycerol can be converted into glucose by conversion into glyceraldehyde-3-phosphate.
3. In beta oxidation, carbon atoms are removed in pairs from fatty acid chains; the resulting molecules of acetyl coenzyme A enters the Krebs cycle.
4. The formation of ketone bodies by the liver is a normal phase of fatty acid catabolism, but an excess of ketone bodies, called ketosis, may cause acidosis.

Lipid Anabolism: Lipogenesis

1. The conversion of glucose or amino acids into lipids is called lipogenesis. The process is stimulated by insulin.
2. The intermediary links in lipogenesis are glyceraldehyde-3-phospate and acetyl coenzyme A.

Protein Metabolism (p. 641)

1. During digestion, proteins are hydrolyzed into amino acids.
2. Amino acids are absorbed by the capillaries of villi and enter the liver via the hepatic portal vein.

Fate of Proteins

1. Amino acids, under the influence of growth hormone (GH) and insulin, enter body cells by active transport.
2. Inside cells, amino acids are synthesized into proteins that function as enzymes, hormones, structural elements, and so forth. Very little protein is used as a source of energy.

Protein Catabolism

1. Before amino acids can be catabolized, they must be converted to substances that can enter the Krebs cycle; these conversions involve deamination, decarboxylation, and hydrogenation.
2. Amino acids may also be converted into glucose, fatty acids, and ketone bodies.

Protein Anabolism

1. Protein synthesis is stimulated by growth hormone (GH), thyroxine, and insulin.
2. The process is directed by DNA and RNA and carried out on the ribosomes of cells.

Absorptive and Postabsorptive (Fasting) States (p. 642)

1. During the absorptive state, ingested nutrients enter the blood and lymph from the GI tract.
2. During the absorptive state, most blood glucose is used by body cells for oxidation. Glucose transported to the liver is converted to glycogen or fat. Most fat is stored in adipose tissue. Amino acids in liver cells are converted to carbohydrate, fats, and proteins.
3. During the postabsorptive (fasting) state, absorption is complete and the energy needs of the body are satisfied by nu-

trients already present in the body.
4. The major concern of the body during the postabsorptive state is to maintain normal blood glucose level. This involves conversion of liver and skeletal muscle glycogen into glucose, conversion of glycerol into glucose, and conversion of amino acids into glucose. The body also switches from glucose oxidation to fatty acid oxidation.

Regulation of Metabolism (p. 643)

1. Absorbed nutrients may be oxidized, stored, or converted, based on the needs of the body.
2. The pathway taken by a particular nutrient is enzymatically controlled and is regulated by hormones (see Exhibit 25-2).

Minerals (p. 645)

1. Minerals are inorganic substances that help regulate body processes.
2. Minerals known to perform essential functions are calcium, phosphorus, sodium, chlorine, potassium, magnesium, iron, sulfur, iodine, manganese, cobalt, copper, zinc, selenium, and chromium. Their functions are summarized in Exhibit 25-3.

Vitamins (p. 645)

1. Vitamins are organic nutrients that maintain growth and normal metabolism. Many function in enzyme systems.
2. Fat-soluble vitamins are absorbed with fats and include A, D, E, and K.
3. Water-soluble vitamins are absorbed with water and include the B vitamins and vitamin C.
4. The functions and deficiency disorders of the principal vitamins are summarized in Exhibit 25-4.

Metabolism and Body Heat (p. 646)

1. A Calorie is the amount of energy required to raise the temperature of 1,000 g of water 1°C from 14° to 15°C.
2. The Calorie is the unit of heat used to express the caloric value of foods and to measure the body's metabolic rate.
3. The apparatus used to determine the caloric value of foods is called a calorimeter.

Production of Body Heat

1. Most body heat is a result of oxidation of the food we eat. The rate at which this heat is produced is known as the metabolic rate.
2. Metabolic rate is affected by exercise, the nervous system, hormones, and body temperature.
3. Measurement of the metabolic rate under basal conditions is called the basal metabolic rate (BMR).

Loss of Body Heat

1. Radiation is the transfer of heat as infrared heat rays from one object to another without physical contact.
2. Conduction is the transfer of body heat to a substance or object in contact with the body.
3. Convection is the transfer of body heat by the movement of air that has been warmed by the body.
4. Evaporation is the conversion of a liquid to a vapor.

Body Temperature Regulation

1. A normal body temperature is maintained by a delicate balance between heat-production and heat-loss mechanisms.
2. The hypothalamic thermostat is the preoptic area.
3. Mechanisms that produce heat are vasoconstriction, sympathetic stimulation, skeletal muscle contraction, and thyroxine production.
4. Mechanisms of heat loss include vasodilation, decreased metabolic rate, decreased skeletal muscle contraction, and perspiration.

Body Temperature Abnormalities

1. Fever is an abnormally high body temperature caused by pyrogens; stages include chill and crisis.
2. Heat cramp is painful skeletal muscle contractions due to loss of salt and water.
3. Sunstroke results in decreased blood flow to skin, reduced perspiration, and high body temperature. Fluid therapy and body cooling are indicated.
4. Heat exhaustion results in a normal or below normal body temperature, profuse perspiration, nausea, cramps, and dizziness. Rest and salt tablets are indicated.

Disorders: Homeostatic Imbalances (p. 652)

1. Obesity is defined as a body weight 10 to 20 percent above desirable standard as the result of excessive accumulation of fat. Causes are regulatory or metabolic.
2. Phenylketonuria (PKU) is a genetic error of metabolism characterized by an elevation of phenylalanine in the blood.
3. Cystic fibrosis is a metabolic disease of the exocrine glands in which absorption of vitamins A, D, and K and calcium is inadequate.
4. Celiac disease is a condition in which the ingestion of gluten causes morphological changes in the small intestinal mucosa resulting in malabsorption.
5. Kwashiorkor is a protein deficiency disorder characterized by hypoprotein edema, lethargy, failure to grow, and sometimes mental retardation.

REVIEW QUESTIONS

1. Define a nutrient. List the six classes of nutrients and indicate the function of each.
2. What is metabolism? Distinguish between catabolism and anabolism and give examples of each.
3. How are carbohydrates absorbed and what are their fates in the body?
4. How does glucose move into body cells?
5. Define glycolysis. Describe its principal events and outcome.
6. What causes oxygen debt?
7. Describe how acetyl coenzyme A is formed.
8. Outline the principal events and outcomes of the Krebs cycle.
9. Explain what happens in the electron transport chain.
10. Summarize the outcomes of the complete oxidation of a molecule of glucose.
11. Define glycogenesis and glycogenolysis. Under what circumstances does each occur?
12. Why is gluconeogenesis important? Give specific examples to substantiate your answer.
13. How are fats absorbed and what are their fates in the body?
14. Where are fats stored in the body?
15. Explain the principal events of the catabolism of glycerol and fatty acids.
16. What are ketone bodies? What is ketosis?
17. Define lipogenesis and explain its importance.
18. How are proteins absorbed and what are their fates in the body?
19. Relate deamination to amino acid catabolism.
20. Summarize the major steps involved in protein synthesis.
21. Distinguish between essential and nonessential amino acids.
22. What is the absorptive state? Outline its principal events.
23. What is the postabsorptive (fasting) state? Outline its principal events.
24. Indicate the roles of the following hormones in the regulation of metabolism: insulin, glucagon, epinephrine, growth hormone (GH), thyroxine, cortisol, and testosterone.
25. What is a mineral? Briefly describe the functions of the following minerals: calcium, phosphorus, iron, iodine, copper, sodium, potassium, chlorine, magnesium, sulfur, zinc, fluorine, manganese, cobalt, chromium, and selenium.
26. Define a vitamin. Explain how we obtain vitamins. Distinguish between a fat-soluble and a water-soluble vitamin.
27. Compare avitaminosis and hypervitaminosis.
28. For each of the following vitamins, indicate its principal function and effect of deficiency: A, D, E, K, B_1, B_2, niacin, B_6, B_{12}, pantothenic acid, folic acid, biotin, and C.
29. Define a Calorie. How is the unit used?
30. How is the caloric value of foods determined?
31. What is metabolic rate? What factors affect it?
32. Define basal metabolic rate (BMR). How is it measured?
33. Define each of the following mechanisms of heat loss: radiation, conduction, convection, and evaporation.
34. Explain how body temperature is regulated by describing the mechanisms of heat production and heat loss.
35. Contrast fever, heat cramp, sunstroke, and heat exhaustion as body temperature abnormalities.
36. What is obesity? What are its causes? How is it treated?
37. Define phenylketonuria (PKU), cystic fibrosis, celiac disease, and kwashiorkor.

- Identify the external and internal gross anatomical features of the kidneys.
- Define the structural adaptations of a nephron for urine formation.
- Describe the blood and nerve supply to the kidneys.
- Describe the structure and function of the juxtaglomerular apparatus.
- Discuss the process of urine formation.
- Define glomerular filtration, tubular reabsorption, and tubular secretion.
- Compare the chemical composition of plasma, glomerular filtrate, and urine.
- Define the forces that support and oppose the filtration of blood in the kidneys.
- Explain the mechanism and importance of tubular reabsorption.
- Compare the obligatory and facultative reabsorption of water.
- Explain the mechanism and importance of tubular secretion.
- Explain the role of the countercurrent multiplier mechanism in producing dilute and concentrated urine.

- Compare the effects of blood pressure, blood concentration, temperature, diuretics, and emotions on urine volume.
- List and define the physical characteristics of urine.
- List the normal chemical constituents of urine.
- Define albuminuria, glycosuria, hematuria, pyuria, ketosis, bilirubinuria, urobilinogenuria, casts, and renal calculi.
- Discuss the structure and physiology of the ureters.
- Describe the structure and physiology of the urinary bladder.
- Explain the physiology of the micturition reflex.
- Compare the causes of incontinence, retention, and suppression.
- Explain the structure and physiology of the urethra.
- Discuss the causes of gout, glomerulonephritis, pyelitis, pyelonephritis, cystitis, nephrosis, and polycystic disease.
- Discuss the operational principle of hemodialysis.
- Define medical terminology associated with the urinary system.
- Explain the actions of selected drugs on the urinary system.

26 The Urinary System

The metabolism of nutrients results in the production of wastes by body cells, including carbon dioxide and excess water and heat. Protein catabolism produces toxic nitrogenous wastes such as ammonia and urea. In addition, many of the essential ions such as sodium, chloride, sulfate, phosphate, and hydrogen tend to accumulate in excess of the body's needs. All the toxic materials and the excess essential materials must be eliminated.

The primary function of the **urinary system** is to help keep the body in homeostasis by controlling the composition and volume of blood. It does so by removing and restoring selected amounts of water and solutes. Two kidneys, two ureters, one urinary bladder, and a single urethra make up the system (Figure 26-1). The kidneys regulate the composition and volume of the blood and remove wastes from the blood in the form of urine. They excrete selected amounts of various wastes, assume a role in erythropoiesis by forming renal erythropoietic factor, help to control blood pH, help to regulate blood pressure by secreting renin that activates the renin-angiotensin pathway, and participate in the activation of vitamin D. Urine is excreted from each kidney through its ureter and is stored in the urinary bladder until it is expelled from the body through the urethra. Other systems that aid in waste elimination are the respiratory, integumentary, and digestive systems (see Exhibit 26-3).

KIDNEYS

The paired **kidneys** are reddish organs that resemble kidney beans in shape. They are found just above the waist between the parietal peritoneum and the posterior wall of the abdomen. Since they are external to the peritoneal lining of the abdominal cavity, their placement is described as *retroperitoneal* (re'-trō-per-i-tō-NĒ-al). Other retroperitoneal structures include the ureters and suprarenal glands. Relative to the vertebral column, the kidneys are located between the levels of the last thoracic and third lumbar vertebrae and are partially protected by the eleventh and twelfth pairs of ribs. The right kidney is slightly lower than the left because of the large area occupied by the liver.

EXTERNAL ANATOMY

The average adult kidney measures about 10 to 12 cm (4 to 5 inches) long, 5.0 to 7.5 cm (2 to 3 inches) wide, and 2.5 cm (1 inch) thick. Its concave medial border faces the vertebral column. Near the center of the concave border is a notch called the **hilus,** through which the ureter leaves the kidney. Blood and lymph vessels and nerves also enter and exit the kidney through the hilus (Figure 26-2). The hilus is the entrance to a cavity in the kidney called the **renal sinus.**

Three layers of tissue surround each kidney. The innermost layer, the **renal capsule,** is a smooth, transparent, fibrous membrane that can easily be stripped off the kidney and is continuous with the outer coat of the ureter at the hilus. It serves as a barrier against trauma and the spread of infection to the kidney. The second layer, the **adipose capsule,** is a mass of fatty tissue surrounding the renal capsule. It also protects the kidney from trauma

and holds it firmly in place in the abdominal cavity. The outermost layer, the **renal fascia,** is a thin layer of fibrous connective tissue that anchors the kidney to its surrounding structures and to the abdominal wall.

CLINICAL APPLICATION
Floating kidney or **ptosis** (TŌ-sis) occurs when the kidney is no longer held in place securely by the adjacent organs or its covering of fat and slips from its normal position. Individuals, especially thin people, in whom either the adipose capsule or renal fascia is deficient may develop ptosis. It is dangerous because it may cause kinking of the ureter with reflux of urine and retrograde pressure. Pain occurs if the ureter is twisted. Also, if the kidneys drop below the rib cage, they become susceptible to blows and penetrating injuries.

INTERNAL ANATOMY

A coronal (frontal) section through a kidney reveals an outer, reddish area called the **cortex** and an inner, reddish-brown region called the **medulla** (Figure 26-2b, c). Within the medulla are 8 to 18 striated, triangular structures termed **renal (medullary) pyramids.** The striated appearance is due to the presence of straight tubules and blood vessels. The bases of the pyramids face the cortical area, and their apices, called **renal papillae,** are directed toward the center of the kidney. The cortex is the smooth-textured area extending from the renal capsule to the bases of the pyramids and into the spaces between them. It is thus divided into an outer cortical zone and an inner

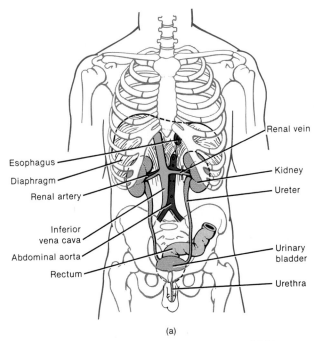

Esophagus
Diaphragm
Renal artery
Inferior vena cava
Abdominal aorta
Rectum
Renal vein
Kidney
Ureter
Urinary bladder
Urethra

(a)

FIGURE 26-1 Organs of the male urinary system. (a) Diagram.

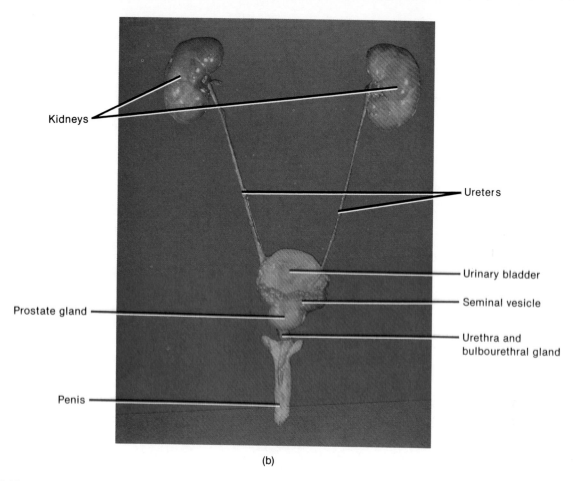

Kidneys

Ureters

Urinary bladder

Seminal vesicle

Urethra and
bulbourethral gland

Prostate gland

Penis

(b)

FIGURE 26-1 (*Continued*) Organs of the male urinary system. (b) Photograph. (Courtesy of C. Yokochi and J. W. Rohen, *Photographic Anatomy of the Human Body,* 2nd ed., 1979, IGAKU-SHOIN, Ltd., Tokyo, New York.)

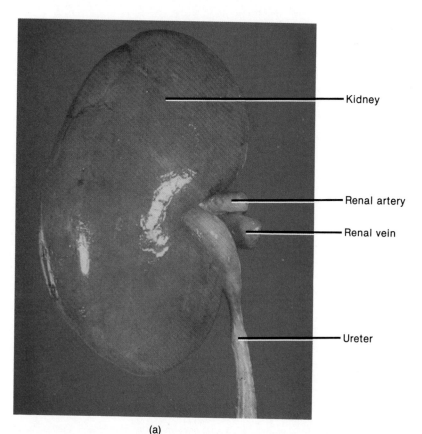

Kidney

Renal artery

Renal vein

Ureter

(a)

FIGURE 26-2 Kidney. (a) Photograph of the external view of the right kidney. (Courtesy of C. Yokochi and J. W. Rohen, *Photographic Anatomy of the Human Body,* 2nd ed., 1979, IGAKU-SHOIN, Ltd., Tokyo, New York.)

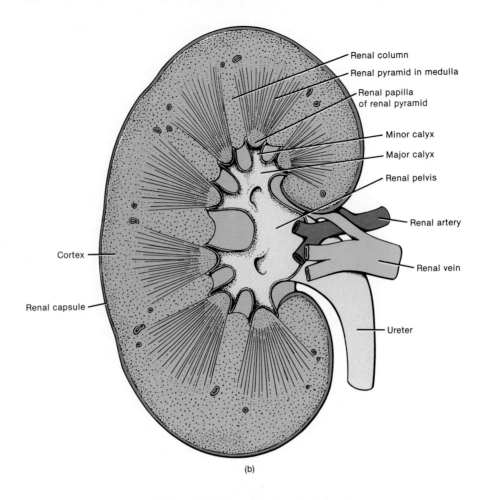

Renal column

Renal pyramid in medulla

Renal papilla
of renal pyramid

Minor calyx

Major calyx

Renal pelvis

Renal artery

Renal vein

Ureter

Cortex

Renal capsule

(b)

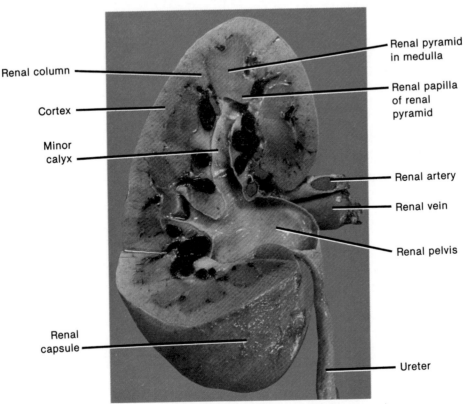

Renal column

Cortex

Minor
calyx

Renal
capsule

Renal pyramid
in medulla

Renal papilla
of renal
pyramid

Renal artery

Renal vein

Renal pelvis

Ureter

(c)

FIGURE 26-2 (*Continued*) Kidney. (b) Diagram of a coronal section of the right kidney illustrating the internal anatomy. (c) Photograph of a coronal and a cross section of the right kidney illustrating the internal anatomy. (Courtesy of C. Yokochi and J. W. Rohen, *Photographic Anatomy of the Human Body,* 2nd ed., 1979, IGAKU-SHOIN, Ltd., Tokyo, New York.)

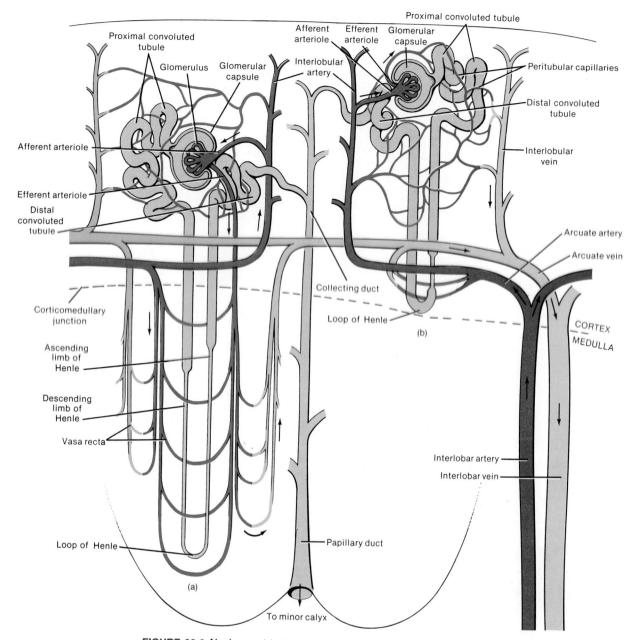

FIGURE 26-3 Nephrons. (a) Juxtamedullary nephron. (b) Cortical nephron.

juxtamedullary zone. The cortical substance between the renal pyramids forms the **renal columns.**

Together the cortex and renal pyramids constitute the parenchyma of the kidney. Structurally, the parenchyma of each kidney consists of approximately 1 million microscopic units called nephrons, collecting ducts, and their associated vascular supply. Nephrons are the functional units of the kidney. They help regulate blood composition and form urine.

In the renal sinus of the kidney is a large cavity called the **renal pelvis.** The edge of the pelvis contains cuplike extensions called **major** and **minor calyces** (KĀ-li-sēz).

There are 2 or 3 major calyces and 8 to 18 minor calyces. Each minor calyx collects urine from collecting ducts of the pyramids. From the major calyces, the urine drains into the pelvis and out through the ureter.

NEPHRON

The functional unit of the kidney is the **nephron** (NEF-ron) (Figure 26-3). Essentially, a nephron is a **renal tubule** and its vascular component. It begins as a double-walled cup, called the **glomerular (Bowman's) capsule,** lying in the cortex of the kidney. The outer wall, or *parietal layer,*

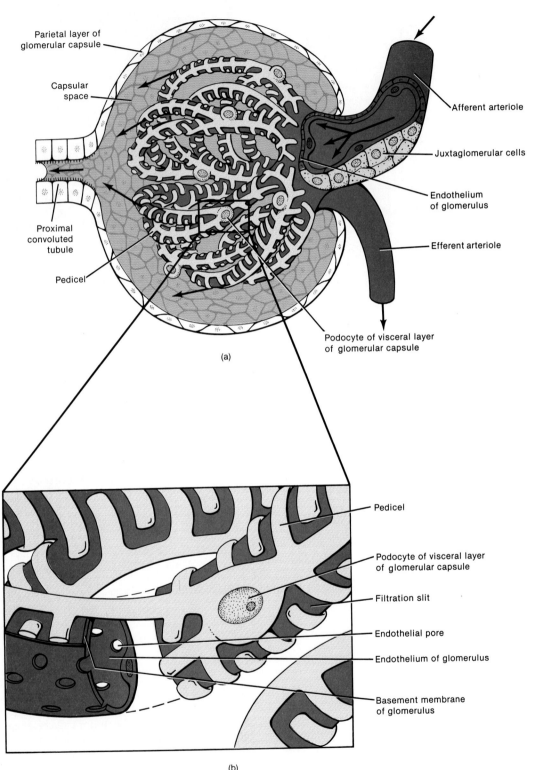

FIGURE 26-4 Endothelial-capsular membrane. (a) Parts of a renal corpuscle. (b) Enlarged aspect of a portion of the endothelial-capsular membrane.

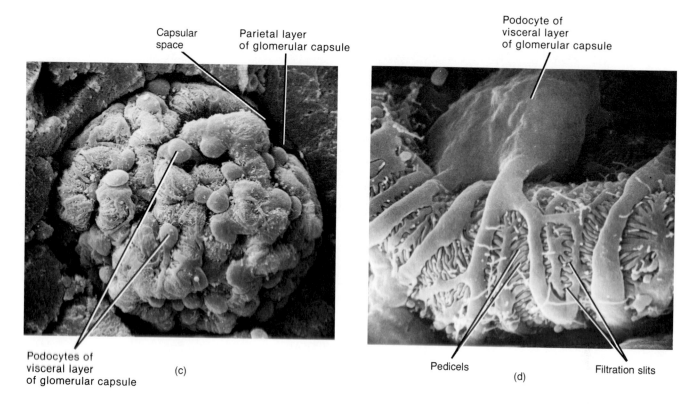

Capsular space   Parietal layer of glomerular capsule

Podocyte of visceral layer of glomerular capsule

Podocytes of visceral layer of glomerular capsule   (c)

Pedicels   (d)   Filtration slits

FIGURE 26-4 (Continued) Endothelial-capsular membrane. (c) Scanning electron micrograph of a renal corpuscle at a magnification of 1,230×. (Courtesy of Richard K. Kessel and Randy H. Kardon, *Tissues and Organs: A Text-Atlas of Scanning Electron Microscopy.* Copyright © 1979 by Scientific American, Inc.) (d) Scanning electron micrograph of a podocyte at a magnification of 7,800×. (Courtesy of Richard K. Kessel and Randy H. Kardon, *Tissues and Organs: A Text-Atlas of Scanning Electron Microscopy.* Copyright © 1979 by Scientific American, Inc.)

is composed of simple squamous epithelium (Figure 26-4). It is separated from the inner wall, known as the *visceral layer,* by the *capsular space.* The visceral layer consists of epithelial cells called podocytes. It surrounds a capillary network called the **glomerulus** (glō-MER-yoo-lus). Collectively, the glomerular capsule and its enclosed glomerulus constitute a **renal corpuscle** (KŌR-pus-sul).

The visceral layer of the glomerular capsule and the endothelium of the glomerulus form an **endothelial-capsular membrane.** This membrane consists of the following parts in the order in which substances filtered by the kidney must pass through.

1. Endothelium of the glomerulus. This single layer of endothelial cells has completely open pores (fenestrated) averaging 500 to 1,000 Å in diameter.

2. Basement membrane of the glomerulus. This extracellular membrane lies beneath the endothelium and contains no pores. It consists of fibrils in a glycoprotein matrix. It serves as the dialyzing membrane.

3. Epithelium of the visceral layer of the glomerular capsule. These epithelial cells, because of their peculiar shape, are called **podocytes.** The podocytes contain footlike structures called **pedicels** (PED-i-sels). The pedicels are arranged parallel to the circumference of the glomerulus and cover the basement membrane except for spaces between them called **filtration slits (slit pores).**

The endothelial-capsular membrane filters water and solutes in the blood. Large molecules, such as proteins, and the formed elements in blood do not normally pass

through it. The water and solutes that are filtered out of the blood pass into the capsular space between the visceral and parietal layers of the glomerular capsule and then into the renal tubule.

The glomerular capsule opens into the first section of the renal tubule, called the **proximal convoluted tubule,** which also lies in the cortex. Convoluted means the tubule is coiled rather than straight; proximal signifies that the glomerular capsule is the origin of the tubule. The wall of the proximal convoluted tubule consists of cuboidal epithelium with microvilli. These cytoplasmic extensions, like those of the small intestine, increase the surface area for reabsorption and secretion.

Nephrons are frequently classified into two kinds. A **cortical nephron** usually has its glomerulus in the outer cortical zone, and the remainder of the nephron rarely penetrates the medulla. A **juxtamedullary nephron** usually has its glomerulus close to the corticomedullary junction, and other parts of the nephron penetrate deeply into the medulla (Figure 26-3).

In the juxtamedullary nephron, the renal tubule straightens, becomes thinner, and dips into the medulla, where it is called the **descending limb of Henle** (HEN-lē). This section consists of squamous epithelium. The tubule then increases in diameter as it bends into a U-shaped structure called the **loop of Henle.** It then ascends toward the cortex as the **ascending limb of Henle,** which consists of cuboidal and low columnar epithelium.

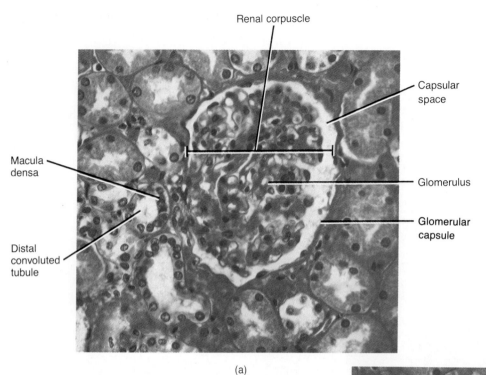

Renal corpuscle

Capsular space

Macula densa

Glomerulus

Glomerular capsule

Distal convoluted tubule

(a)

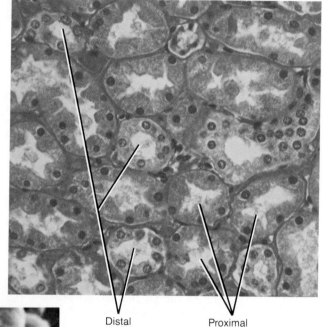

Distal convoluted tubules

Proximal convoluted tubules

(b)

FIGURE 26-5 Histology of a nephron. (a) Photomicrograph of the cortex of the kidney showing a renal corpuscle and surrounding renal tubules at a magnification of 400×. (© 1983 by Michael H. Ross. Used by permission.) (b) Photomicrograph of the cortex of the kidney showing renal tubules at a magnification of 400×. (© 1983 by Michael H. Ross. Used by permission.) (c) Scanning electron micrograph of several renal tubules at a magnification of 500×. (Courtesy of Fisher Scientific Company and S.T.E.M. Laboratories, Inc., Copyright, 1975.)

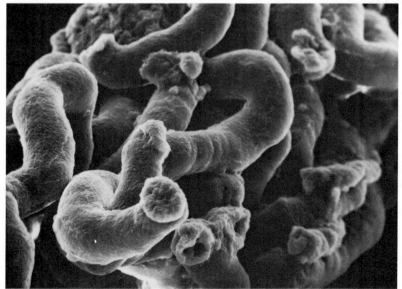

(c)

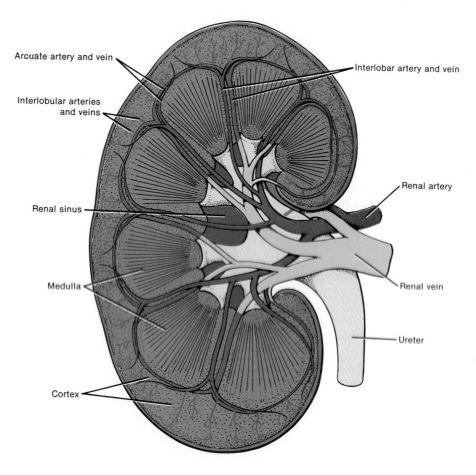

FIGURE 26-6 Blood supply of the right kidney seen in coronal section.

In the cortex, the tubule again becomes convoluted. Because of its distance from the point of origin at the glomerular capsule, this section is referred to as the **distal convoluted tubule.** The cells of the distal tubule, like those of the proximal tubule, are cuboidal. Unlike the cells of the proximal tubule, however, the cells of the distal tubule have few microvilli. In a cortical nephron, the proximal section runs into the distal tubule without the intervening limb of Henle. In both types of nephron, the distal tubule terminates by merging with a straight **collecting duct.**

In the medulla, the collecting ducts receive the distal tubules of several nephrons, pass through the renal pyramids, and open at the renal papillae into the minor calyces through a number of large **papillary ducts.** On the average, there are 30 papillary ducts per renal papilla. Cells of the collecting ducts are cuboidal; those of the papillary ducts are columnar.

The histology of a nephron and glomerulus is shown in Figure 26-5.

BLOOD AND NERVE SUPPLY

The nephrons are largely responsible for removing wastes from the blood and regulating its fluid and electrolyte content. Thus they are abundantly supplied with blood vessels. The right and left **renal arteries** transport about one-fourth the total cardiac output to the kidneys (Figure 26-6). Approximately 1,200 ml passes through the kidneys every minute.

Before or immediately after entering through the hilus, the renal artery divides into several branches that enter the parenchyma and pass as the **interlobar arteries** between the renal pyramids in the renal columns. At the base of the pyramids, the interlobar arteries arch between the medulla and cortex; here they are known as the **arcuate arteries.** Divisions of the arcuate arteries produce a series of **interlobular arteries,** which enter the cortex and divide into **afferent arterioles** (see Figure 26-3).

One afferent arteriole is distributed to each glomerular capsule, where the arteriole divides into the tangled capillary network termed the **glomerulus.** The glomerular capillaries then reunite to form an **efferent arteriole,** which leads away from the capsule and is smaller in diameter than the afferent arteriole. This variation in diameter helps raise the glomerular pressure. The afferent-efferent arteriole situation is unique because blood usually flows out of capillaries into venules and not into other arterioles.

Each efferent arteriole of a cortical nephron divides to form a network of capillaries, called the **peritubular capillaries,** around the convoluted tubules. The efferent arteriole of a juxtamedullary nephron also forms peritubular capillaries. In addition, it forms long loops of thin-walled vessels called **vasa recta** that dip down alongside the loop of Henle into the medullary region of the papilla.

The peritubular capillaries eventually reunite to form

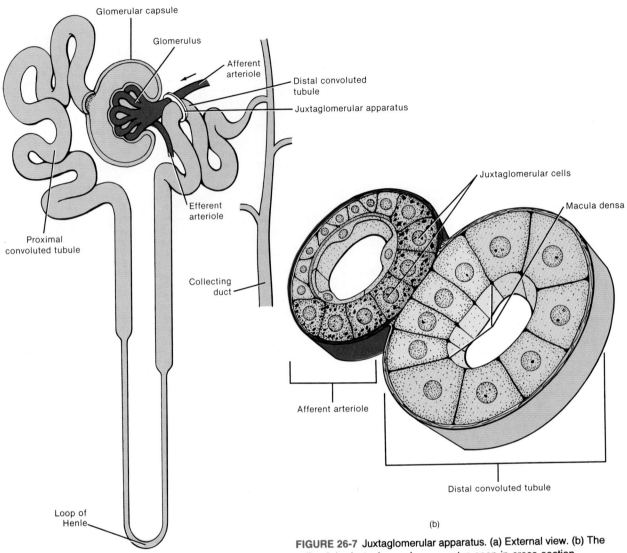

FIGURE 26-7 Juxtaglomerular apparatus. (a) External view. (b) The cells of the juxtaglomerular apparatus seen in cross section.

interlobular veins. The blood then drains through the **arcuate veins** to the **interlobar veins** running between the pyramids and leaves the kidney through a single **renal vein** that exits at the hilus. The vasa recta pass blood into the interlobular veins. From here, it goes to the arcuate veins, the interlobar veins, and then into the renal vein.

The nerve supply to the kidneys is derived from the **renal plexus** of the autonomic system. Nerves from the plexus accompany the renal arteries and their branches and are distributed to the vessels. Because the nerves are vasomotor, they regulate the circulation of blood in the kidney by regulating the diameters of the arterioles.

JUXTAGLOMERULAR APPARATUS

As the afferent arteriole approaches the renal corpuscle, the smooth muscle cells of the tunica media become modi-

fied. Their nuclei become rounded (instead of elongated), and their cytoplasm contains granules (instead of myofibrils). Such modified cells are called **juxtaglomerular cells.** The cells of the distal convoluted tubule adjacent to the afferent arteriole become considerably narrower. Collectively, these cells are known as the **macula densa.** Together with the modified cells of the afferent arteriole they constitute the **juxtaglomerular apparatus** (Figure 26-7). The juxtaglomerular apparatus helps to regulate renal blood pressure.

PHYSIOLOGY

The major work of the urinary system is done by the nephrons. The other parts of the system are primarily passageways and storage areas. Nephrons carry out three important functions. They control blood concentration

EXHIBIT 26-1

CHEMICALS IN PLASMA, FILTRATE, AND URINE DURING 24-HOUR PERIOD*

CHEMICAL	PLASMA	FILTRATE IMMEDIATELY AFTER GLOMERULAR CAPSULE	REABSORBED FROM FILTRATE	URINE
Water	180,000 ml	180,000 ml	178,000–179,000 ml	1,000–2,000 ml
Proteins	7,000–9,000	10–20	10–20	0[†]
Chloride (Cl⁻)	630	630	625	5
Sodium (Na⁺)	540	540	537	3
Bicarbonate (HCO₃⁻)	300	300	299.7	0.3
Glucose	180	180	180	0
Urea	53	53	28	25
Potassium (K⁺)	28	28	24	4
Uric acid	8.5	8.5	7.7	0.8
Creatinine	1.5	1.5	0	1.5

* All values, except for water, are expressed in grams. The chemicals are arranged in sequence from highest to lowest concentration in plasma.
† Although trace amounts of protein (170 to 250 mg) normally appear in urine, we will assume for purposes of discussion that all of it is reabsorbed from filtrate.

and volume by removing selected amounts of water and solutes. They help regulate blood pH. And they remove toxic wastes from the blood. As the nephrons go about these activities, they remove many materials from the blood, return the ones that the body requires, and eliminate the remainder. The eliminated materials are collectively called **urine.** The entire volume of blood in the body is filtered by the kidneys approximately 60 times a day. Urine formation requires three principal processes: glomerular filtration, tubular reabsorption, and tubular secretion.

Glomerular Filtration

The first step in the production of urine is called **glomerular filtration.** Filtration—the forcing of fluids and dissolved substances through a membrane by pressure—occurs in the renal corpuscle of the kidneys across the endothelial-capsular membrane. When blood enters the glomerulus, the blood pressure forces water and dissolved blood components (plasma) through the endothelial pores of the capillaries, basement membrane, and on through the filtration slits of the adjoining visceral wall of the glomerular capsule (see Figure 26-4). The resulting fluid is the **filtrate.** In a healthy person, the filtrate consists of all the materials present in the blood except for the formed elements and most proteins, which are too large to pass through the endothelial-capsular barrier. Exhibit 26-1 compares the constituents of plasma, glomerular filtrate, and urine during a 24-hour period. Although the values shown are typical, they vary considerably according to diet. The chemicals listed in plasma are those present in glomerular blood plasma before filtration. The chemicals listed in the filtrate immediately after the glo-

merular capsule are those that pass from the glomerular blood plasma through the endothelial-capsular membrane before reabsorption. The chemicals in the filtrate are the ones that have been filtered.

Renal corpuscles are especially structured for filtering blood. First, each capsule contains a tremendous length of highly coiled glomerular capillaries presenting a vast surface area for filtration. Second, the endothelial-capsular membrane is structurally adapted for filtration. Although the endothelial pores generally do not restrict the passage of substances, the basement membrane permits the passage of smaller molecules. Thus water, glucose, vitamins, amino acids, small proteins, nitrogenous wastes, and ions pass into the glomerular capsule. Large proteins and the formed elements in blood do not normally pass through the basement membrane. The filtration slits permit only the occasional passage of very small plasma proteins such as albumins. Third, the efferent arteriole is smaller in diameter than the afferent arteriole, so there is resistance to the outflow of blood from the glomerulus. Consequently, blood pressure is higher in the glomerular capillaries than in other capillaries. Glomerular blood pressure averages about 60 mm Hg, whereas the blood pressure of other capillaries averages only 30 mm Hg. Fourth, the endothelial-capsular membrane separating the blood from the space in the glomerular capsule is very thin (0.1 μm).

The filtering of the blood depends on a number of opposing pressures (Figure 26-8). The chief one is the **glomerular blood hydrostatic pressure.** *Hydrostatic (hydro* = water) *pressure* is the force that a fluid under pressure exerts against the walls of its container. Glomerular blood hydrostatic pressure means the blood pressure in the glomerulus. This pressure tends to move fluid out

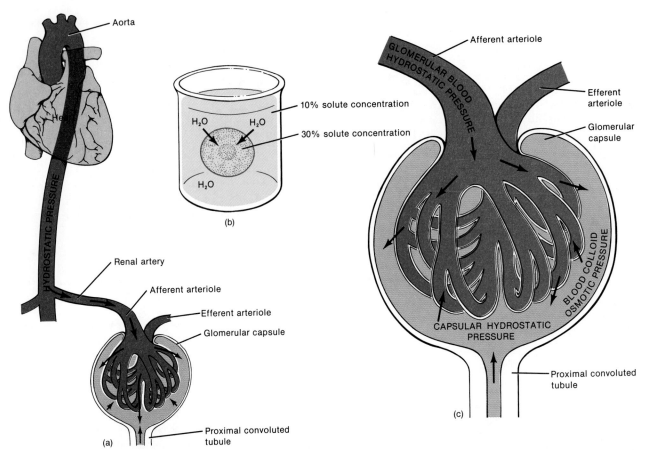

FIGURE 26-8 Glomerular filtration. (a) Heart action and resistance of walls of blood vessels provide the hydrostatic pressure for filtration. The hydrostatic pressure is counteracted by the wall of the glomerular capsule and the already present filtrate. (b) The development of osmotic pressure. When water moves into the cell, it swells as its osmotic pressure increases. (c) Application of hydrostatic pressure and osmotic pressure to glomerular filtration.

of the glomeruli at a force averaging about 60 mm Hg.

However, glomerular blood hydrostatic pressure is opposed by two other forces. The first of these, **capsular hydrostatic pressure,** develops in the following way. When the filtrate is forced into the space between the walls of the glomerular capsule, it meets with two forms of resistance: the walls of the capsule and the fluid that has already filled the renal tubule. As a result, some filtrate is pushed back into the capillary. The amount of "push" is the capsular hydrostatic pressure. It usually measures about 20 mm Hg (Figure 26-8c).

The second force opposing filtration into the glomerular capsule is the **blood colloid osmotic pressure.** *Osmotic pressure* is the pressure that develops because of water movement into a contained solution. Suppose we place a cell in a hypotonic solution. As water moves from the solution into the cell, the volume inside the cell increases, forcing the cell membrane outward (Figure 26-8b). Osmotic pressure always develops in the solution with the higher concentration of solutes. Hydrostatic pressure develops because of a force outside a solution. Osmotic pressure develops because of the concentration of the solution itself. Since the blood contains a much higher concentration of proteins than the filtrate does, water moves out

of the filtrate and back into the blood vessel. This blood osmotic pressure is normally about 30 mm Hg.

To determine how much filtration finally occurs, we have to subtract the forces that oppose filtration from the glomerular blood hydrostatic pressure. The net result is called the **effective filtration pressure,** which is abbreviated P_{eff}.

$$P_{eff} = \begin{pmatrix} \text{glomerular} \\ \text{blood} \\ \text{hydrostatic} \\ \text{pressure} \end{pmatrix} - \begin{pmatrix} \text{capsular} \\ \text{hydrostatic} + \\ \text{pressure} \end{pmatrix} \begin{pmatrix} \text{blood} \\ \text{colloid} \\ \text{osmotic} \\ \text{pressure} \end{pmatrix}$$

By substituting the values just discussed, a normal P_{eff} may be calculated as follows.

$$P_{eff} = (60 \text{ mm Hg}) - (20 \text{ mm Hg} + 30 \text{ mm Hg})$$
$$= (60 \text{ mm Hg}) - (50 \text{ mm Hg})$$
$$= 10 \text{ mm Hg}$$

This means that a pressure of about 10 mm Hg causes a normal amount of plasma to filter from the glomerulus into the glomerular capsule. This is about 125 ml of filtrate per minute.

Certain conditions may alter these pressures and thus the P_{eff}. In some forms of kidney disease, such as glomerulonephritis, glomerular capillaries become so permeable that the plasma proteins are able to pass from the blood into the filtrate. As a result, the capsular filtrate exerts an osmotic pressure that draws water out of the blood. Thus, if a capsular osmotic pressure develops, the P_{eff} will increase. At the same time, blood colloid osmotic pressure decreases, further increasing the P_{eff}.

The P_{eff} also is affected by changes in the general arterial blood pressure. Severe hemorrhaging produces a drop in general blood pressure, which also decreases the glomerular blood hydrostatic pressure. If the blood pressure falls to the point where the hydrostatic pressure in the glomeruli reaches 50 mm Hg, no filtration occurs, because the glomerular blood hydrostatic pressure equals the opposing forces. Such a condition is called renal suppression.

CLINICAL APPLICATION

Renal suppression or **anuria** (a-NOO-rē-a) is the failure of the kidneys to secrete urine. It may be caused by insufficient pressure to accomplish filtration or by inflammation of the glomeruli so that plasma is prevented from reaching the glomerulus.

A final factor that may affect the P_{eff} is the regulation of the size of the afferent and efferent arterioles. In this case, glomerular blood hydrostatic pressure is regulated separately from the general blood pressure. Sympathetic impulses and small doses of epinephrine cause constriction of both afferent and efferent arterioles. However, intense sympathetic impulses and large doses of epinephrine cause greater constriction of afferent than efferent arterioles. This intense stimulation results in a decrease in glomerular hydrostatic pressure even though blood pressure in other parts of the body may be normal or even higher than normal. Intense sympathetic stimulation is most likely to occur during the alarm reaction of the general adaptation syndrome. Blood may also be shunted away from the kidneys during hemorrhage.

Tubular Reabsorption

The amount of filtrate that flows out of all the renal corpuscles of both kidneys every minute is called the **glomerular filtration rate (GFR)**. In the normal adult, this rate is about 125 ml/min—about 180 liters (48 gal) a day. But as the filtrate passes through the renal tubules, it is reabsorbed into the blood at a rate of about 123 to 124 ml/min—about 178 to 179 liters/day. Thus only about 1 percent of the filtrate actually leaves the body—about 1 to 1.5 ml/min, or 1 to 2 liters/day. The movement of the filtrate back into the blood of the peritubular capillaries or vasa recta is called **tubular reabsorption.**

Tubular reabsorption is carried out by epithelial cells throughout the renal tubule. It is a very discriminating process. Only specific amounts of certain substances are reabsorbed, depending on the body's needs at the time. The maximum amount of a substance that can be reabsorbed under any condition is called the substance's **tubular maximum (Tm).** Materials that are reabsorbed include water, glucose, amino acids, and ions such as Na^+, K^+,

Ca^{2+}, Cl^-, HCO_3^-, and HPO_4^{2-}. Tubular reabsorption allows the body to retain most of its nutrients. Wastes such as urea are only partially reabsorbed. Exhibit 26-1 compares the values for the chemicals in the filtrate immediately after the glomerular capsule with those reabsorbed from the filtrate. It will give you an idea of how much of the various substances the kidneys reabsorb.

Reabsorption is carried out through both passive and active transport mechanisms. It is believed that glucose is reabsorbed by an active process involving a carrier system. The carrier, probably an enzyme, exists in the membranes of the tubular epithelial cells in a fixed and limited amount. Normally, all the glucose filtered by the glomeruli (125 mg/100 ml of filtrate/min) is reabsorbed by the tubules.

CLINICAL APPLICATION

The capacity of the glucose carrier system is limited. If the plasma concentration of glucose is significantly above normal, the glucose transport mechanism cannot reabsorb it all and the excess remains in the urine. If there is a malfunction in the tubular carrier mechanism, glucose appears in the urine even though the blood sugar level is normal. This condition is called **glycosuria.**

Sodium ions are actively transported (reabsorbed) mostly from the proximal and also from the distal convoluted tubules and collecting ducts. In the distal convoluted tubules and collecting ducts, Na^+ reabsorption varies with its concentration in extracellular fluid. When the Na^+ concentration of the blood is low, there is a drop in blood pressure, and the **renin-angiotensin pathway** goes into operation. The juxtaglomerular cells of the kidneys secrete an enzyme called renin, which converts angiotensin (synthesized by the liver) into angiotensin I. As it passes through the lungs, angiotensin I is converted into angiotensin II. Angiotensin II stimulates the adrenal zona glomerulosa of the cortex to produce aldosterone. Aldosterone brings about increased Na^+ and water reabsorption (Figure 26-9). Extracellular fluid volume increases and blood pressure is restored to normal.

When Na^+ ions move out of the tubules into the peritubular blood, the blood momentarily becomes more electropositive than the filtrate. Chloride, a negatively charged ion, follows the positive sodium ion out of the tubule by electrostatic attraction. Thus the movement of Na^+ ions influences the movement of Cl^- ions and other anions into the blood.

Cl^- ions, and not Na^+ ions, are actively transported from the ascending limb of Henle. In this portion of the nephron, then, Cl^- ions move actively and Na^+ ions follow passively. Whether Na^+ ions move actively and Cl^- ions follow passively or Cl^- ions move actively and Na^+ ions follow passively, the end result is the same.

Water reabsorption is controlled both by sodium transport and by water carrier molecules. As Na^+ ions are transported from the proximal convoluted tubules into the blood, the osmotic pressure of the blood becomes higher than that of the filtrate. Water follows the Na^+

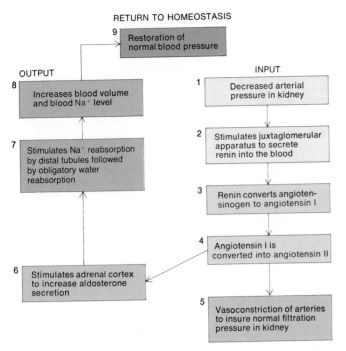

FIGURE 26-9 Role of the juxtaglomerular apparatus in maintaining normal blood volume.

ions into the blood in order to reestablish the osmotic equilibrium. About 80 percent of the water is reabsorbed by this method from the proximal convoluted tubule. This portion of reabsorbed water occurs as a function of the law of osmosis, so it is referred to as **obligatory reabsorption.** The proximal convoluted tubule has no control over osmosis because it is always permeable to water (Figure 26-10).

Passage of most of the remaining water in the filtrate can be regulated. The permeability of the cells of the distal and collecting tubules is controlled by the antidiuretic hormone (ADH), produced by the hypothalamus. When the blood-water concentration is low, the posterior pituitary releases ADH, a hormone that increases the

permeability of the plasma membranes of the distal tubule and collecting duct cells. As the membranes become more permeable, more water molecules pass into the cells. This type of absorption is called *facultative* (FAK-ul-tā'-tiv), meaning that it occurs under some conditions but not others. **Facultative reabsorption** is responsible for about 20 percent of the water in the filtrate. It is a major mechanism for controlling water content of the blood (Figure 26-10). Aldosterone has an effect similar to ADH.

Tubular Secretion

The third process involved in urine formation is **tubular secretion.** Whereas tubular reabsorption removes substances from the filtrate into the blood, tubular secretion adds materials to the filtrate from the blood. These secreted substances include potassium and hydrogen ions, ammonia, creatinine, and the drugs penicillin and para-aminohippuric acid. Tubular secretion has two principal effects. It rids the body of certain materials, and it controls the blood pH.

The body has to maintain normal blood pH (7.35 to 7.45) despite the fact that a normal diet provides more acid-producing foods than alkali-producing foods. To raise blood pH, the renal tubules secrete hydrogen and ammonium ions into the filtrate. Both these substances make the urine acidic.

Secretion of the hydrogen ions occurs through the formation of carbonic acid (Figure 26-11a). A certain amount of carbon dioxide normally diffuses from the peritubular blood into the cells of the distal tubules and collecting ducts. Once inside the epithelial cells, the CO_2 combines with water to form carbonic acid (H_2CO_3), which then dissociates into H^+ ions and bicarbonate ions (HCO_3^-). A low blood pH stimulates the cells to secrete the H^+ ion into the urine. As the H^+ enters the urine, it displaces another positive ion, usually Na^+, and forms a weak acid or a salt of the acid that is eliminated in urine. The displaced Na^+ or other positive ion diffuses from the urine into the tubule cell, where it combines with the bicarbonate ion to form sodium bicarbonate

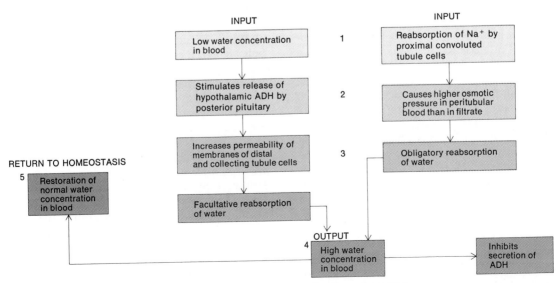

FIGURE 26-10 Factors that control water reabsorption.

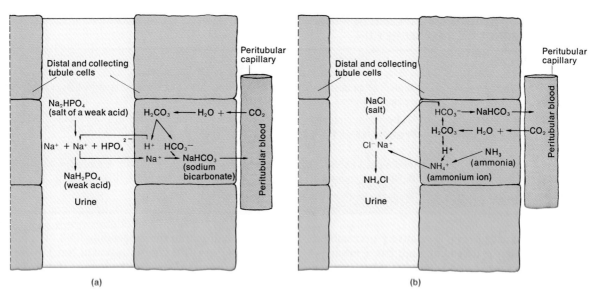

FIGURE 26-11 Role of the kidneys in maintaining blood pH. (a) Acidification of urine and conservation of sodium bicarbonate (NaHCO$_3$) by the elimination of hydrogen ions (H$^+$). (b) Acidification of urine and conservation of sodium bicarbonate (NaHCO$_3$) by the elimination of ammonium ions (NH$_4^+$).

(NaHCO$_3$), which then is absorbed into the blood. Not only is H$^+$ eliminated by the body and Na$^+$ conserved, but Na$^+$ is conserved in the form of NaHCO$_3$, which can buffer other H$^+$ ions in the blood.

A second mechanism for raising blood pH is secretion of the ammonium ion (Figure 26-11b). Ammonia, in certain concentrations, is a poisonous waste product derived from the deamination of amino acids. The liver converts much of the ammonia to a less toxic compound called urea. Urea and ammonia both become part of the glomerular filtrate and are subsequently expelled from the body. Any ammonia produced by the deamination of amino acids in the tubule is secreted into the urine. When ammonia (NH$_3$) forms in the distal and collecting tubule cells, it combines with H$^+$ to form the ammonium ion (NH$_4^+$). (The H$^+$ may come from the dissociation of H$_2$CO$_3$.) The cells secrete NH$_4^+$ into the filtrate, where it takes the place of a positive ion, usually Na$^+$, in a salt and is

eliminated. The displaced Na$^+$ diffuses into the renal cells and combines with HCO$_3^-$ to form NaHCO$_3$.

As a result of hydrogen and ammonium ion secretion, urine normally has an acidic pH of 6. The relationship of renal tubule ion excretion to blood pH level is summarized in Figure 26-12. Exhibit 26-2 summarizes filtration, reabsorption, and secretion in the nephrons.

Countercurrent Multiplier Mechanism

The various metabolic wastes in urine must be eliminated from the body regardless of its state of hydration. During dehydration, little urine is excreted, but it is concentrated. Such urine is said to be *hypertonic* (*hyperosmotic*) to the blood plasma. When there are large amounts of water in the body, more urine is excreted, but it is diluted. Such urine is said to be *hypotonic* (*hyposmotic*) to blood plasma. The ability of the kidneys to produce either hyperosmotic or hyposmotic urine depends in part on the **countercurrent multiplier mechanism.**

The countercurrent multipler mechanism is based on the anatomical arrangements of the juxtamedullary nephrons and vasa recta. Both begin in the cortex, dip deeply into the medulla, and return to the cortex. The normal concentration of the glomerular filtrate as it enters the proximal convoluted tubule is about 300 milliosmols (mOsm) per liter. A milliosmol represents the concentration of substances in the filtrate, mainly NaCl. It is a measure of the osmotic properties of a solution. This concentration of glomerular filtrate is isotonic to plasma. Figure 26-13 shows that the concentration of the interstitial fluid increases from 300 mOsm in the cortex to 1,200 mOsm in the inner medulla. As the filtrate passes through the ascending limb of Henle, Cl$^-$ is actively transported from the filtrate into the interstitial fluid. Na$^+$ follows passively and the concentration of NaCl becomes about four times greater at the bottom of the interstitial fluid than at the top.

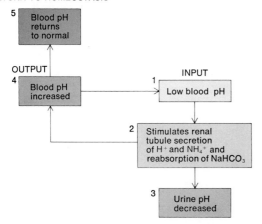

FIGURE 26-12 Summary of kidney mechanisms that maintain the homeostasis of blood pH.

EXHIBIT 26-2

FILTRATION, REABSORPTION, AND SECRETION

REGION OF NEPHRON	ACTIVITY
Renal corpuscle (endothelial-capsular membrane)	Filtration of glomerular blood under hydrostatic pressure results in formation of filtrate free of plasma proteins and cellular elements of blood.
Proximal convoluted tubule and descending and ascending limbs of Henle	Reabsorption of physiologically important solutes such as Na^+, K^+, Cl^-, HCO_3^-, and glucose. Obligatory reabsorption of water by osmosis (proximal convoluted tubule).
Distal convoluted tubule	Reabsorption of Na^+. Facultative reabsorption of water under control of ADH. Secretion of H^+, NH_3, K^+, creatinine, and certain drugs. Conservation of $NaHCO_3$.
Collecting duct	Facultative reabsorption of water under control of ADH.

The descending limb of Henle is relatively permeable to Na^+ and Cl^-, while the ascending limb has a mechanism that actively transports Cl^- from the filtrate into the interstitial fluid of the medulla. When the Cl^- is actively transported, Na^+ follows passively. Almost immediately they diffuse into the descending limb of Henle. This diffusion increases the concentration of NaCl in the filtrate that flows *downward* toward the loop of Henle. This continual active transport of Cl^- and passive movement of Na^+ out of the ascending limb of Henle into the interstitial

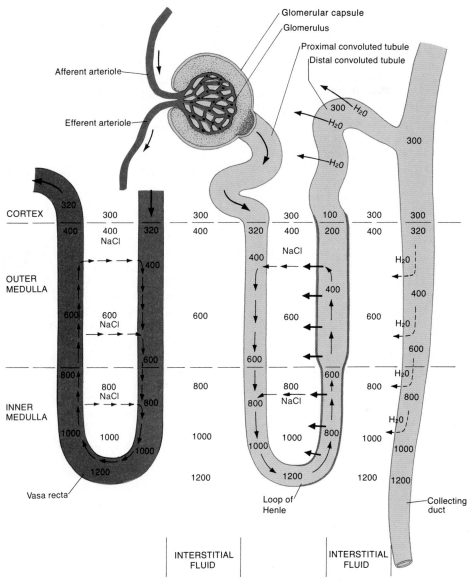

FIGURE 26-13 Countercurrent multiplier mechanism.

fluid of the medulla and then into the descending limb of Henle causes the high concentration of NaCl in the interstitial fluid of the medulla. Thus two parallel streams of liquid flowing opposite each other (*countercurrent*) result in an increased NaCl concentration in the medullary interstitial fluid (*multiplier*).

The high concentration of NaCl is also the result of the countercurrent flow of blood in the vasa recta. These vessels run parallel to the loop tubules and are freely permeable. As blood circulates through the descending limb of the vasa recta, Na^+ and Cl^- diffuse into it from the interstitial fluid of the medulla. The concentration of the blood reaches 1,200 mOsm at the tip of the vessel. As the blood flows through the ascending limb of the vasa recta, almost all of the NaCl diffuses back into the interstitial fluid of the medulla. By the time the blood in the vasa recta leaves the medulla, its concentration is only slightly higher than it was when it entered. Thus the countercurrent flow through the vasa recta permits most of the NaCl to remain in the interstitial fluid of the medulla.

The ascending limb of Henle is very impermeable to water. Yet large amounts of Cl^- are actively transported out of the ascending limb, and Na^+ follows passively into the interstitial fluid of the medulla. This means that the fluid in the ascending limb becomes more dilute as it approaches the cortex. In fact, the concentration of fluid decreases to about 100 mOsm. Thus the fluid in the ascending limb is hypotonic (hyposmotic) to the blood plasma. In the absence of ADH, no water is reabsorbed from the distal convoluted tubule and collecting duct as the fluid passes through. The fluid empties into the papillary ducts, calyces, and renal pelvis. The fluid thus entering the pelvis and passing through the ureter into the urinary bladder is very dilute. In the absence of ADH, the urine may be only one-fourth as concentrated as the blood plasma and glomerular filtrate.

In the presence of ADH, the epithelial pores of the distal convoluted tubule and collecting duct enlarge and they become permeable to water. As the dilute fluid passes through the distal convoluted tubule and collecting duct, most of the water is reabsorbed by osmosis. The loss of this water raises the concentration of the dilute fluid from 100 to 300 mOsm—the same concentration as the interstitial fluid of the cortex. As the fluid passes into the medulla through the collecting duct, it is exposed to the hyperosmotic interstitial fluid of the medulla. As a result, large amounts of water are absorbed by osmosis into the interstitial fluid of the medulla from the collecting duct. The highly concentrated interstitial fluid of the medulla causes the rapid osmosis of water when the epithelial pores are opened by ADH. Thus the concentration of fluid in the collecting duct approaches 1,200 mOsm—the concentration of the interstitial fluid of the medulla. The fluid that enters the papillary ducts, calyces, renal pelvis, and ureters and flows into the urinary bladder is a very concentrated urine. In the presence of ADH, it may be four times more concentrated than the blood plasma and glomerular filtrate.

HOMEOSTASIS

Excretion is one of the primary ways in which the volume, pH, and chemistry of the body fluids are kept in homeostasis. The kidneys assume a good deal of the burden for excretion, but they share this responsibility with several other organ systems.

OTHER EXCRETORY ORGANS

The lungs, integument, and alimentary canal all perform special excretory functions (Exhibit 26-3). A major responsibility for regulating body temperature through the excretion of water is assumed by sudoriferous (sweat) glands of the skin. The lungs maintain blood-gas homeostasis through the elimination of carbon dioxide. One way in which the kidneys maintain homeostasis is by coordinating their activities with other excretory organs. When the integument increases its excretion of water, the renal tubules increase their reabsorption of water, and blood volume is maintained. When the lungs fail to eliminate enough carbon dioxide, the kidneys attempt to compensate. They change some of the carbon dioxide into sodium bicarbonate, which becomes part of the blood buffer systems.

URINE

The by-product of the kidneys' activities is **urine.** In a healthy person its volume, pH, and solute concentration vary with the needs of the internal environment. During certain pathological conditions, the characteristics of urine may change drastically. An analysis of the volume and physical and chemical properties of urine tells us much about the state of the body.

Volume

The volume of urine eliminated per day in the normal adult varies between 1,000 and 2,000 ml (1 to 2 qt). Urine volume is influenced by a number of factors: blood pressure, blood concentration, diet, temperature, diuretics, mental state, and general health.

EXHIBIT 26-3
EXCRETORY ORGANS AND PRODUCTS ELIMINATED

EXCRETORY ORGANS	PRODUCTS ELIMINATED	
	PRIMARY	SECONDARY
Kidneys	Water, nitrogenous wastes from protein catabolism, and inorganic salts.	Heat and carbon dioxide.
Lungs	Carbon dioxide.	Heat and water.
Skin (sudoriferous glands)	Heat.	Carbon dioxide, water, salts, and urea.
Alimentary canal	Solid wastes and secretions.	Carbon dioxide, water, salts, and heat.

● **Blood Pressure**   The cells of the juxtaglomerular apparatus are particularly sensitive to changes in blood pressure. When renal blood pressure falls below normal, the juxtaglomerular apparatus secretes renin and the renin-angiotensin pathway is activated (see Figure 26-9). As a result, obligatory reabsorption of water increases, blood volume increases, and urine volume decreases. By raising the blood pressure, the juxtaglomerular apparatus ensures that the kidney cells receive enough oxygen and that the glomerular hydrostatic pressure is high enough to maintain a normal P_{eff}. The juxtaglomerular apparatus also regulates blood pressure throughout the body.

● **Blood Concentration**   The concentrations of water and solutes in the blood also affect urine volume. If you have gone without water all day and the water concentration of your blood becomes low, osmotic receptors in the hypothalamus stimulate the posterior pituitary to release ADH. The hormone stimulates the cells of the distal convoluted tubule and collecting duct to transport water out of the filtrate and into the blood by facultative reabsorption. Thus urine volume decreases and water is conserved.

If you have just drunk an excessive amount of liquid, urine volume may be increased through two mechanisms. First, the blood-water concentration increases above normal. This means that the osmoreceptors in the hypothalamus are no longer stimulated to secrete ADH and facultative water reabsorption stops. Second, the excess water causes the blood pressure to rise. In response, the renal vessels dilate, more blood is brought to the glomeruli, and the filtration rate increases.

The concentration of sodium ions in the blood also influences urine volume. Sodium concentration affects aldosterone secretion, which in turn affects both sodium reabsorption and the obligatory reabsorption of water.

● **Temperature**   When the internal or external temperature rises above normal, the cutaneous vessels dilate and fluid diffuses from the capillaries to the surface of the skin. As water volume decreases, ADH is secreted and facultative reabsorption increases. In addition, the increase in temperature stimulates the abdominal vessels to constrict, so the blood flow in the glomeruli and filtration decrease. Both mechanisms reduce the volume of urine.

If the body is exposed to low temperatures, the cutaneous vessels constrict and the abdominal vessels dilate. More blood is shunted to the glomeruli, glomerular blood hydrostatic pressure increases, and urine volume increases.

● **Diuretics**   Certain chemicals increase urine volume by inhibiting the facultative reabsorption of water. Such chemicals are called **diuretics,** and the abnormal increase in urine flow is called *diuresis.* Some diuretics act directly on the tubular epithelium as they are carried through the kidneys. Others act indirectly by inhibiting the secretion of ADH as they circulate through the brain. Coffee, tea, and alcoholic beverages are diuretics.

● **Emotions**   Some emotional states can affect urine volume. Extreme nervousness, for example, can cause an enormous discharge of urine because impulses from the brain cause an increase in blood pressure, resulting in an increased glomerular filtration rate.

Physical Characteristics

Normal urine is usually a yellow or amber-colored, transparent liquid with a characteristic odor. The color is caused by urochrome, pigments derived from the metabolism of bile. It varies considerably with the ratio of solutes to water in the urine. The less water there is, the darker the urine. Fever decreases urine volume in the same way that high environmental temperatures do, sometimes making the urine quite concentrated. It is not uncommon for a feverish person to have dark yellow or brown urine. The color of urine may also be affected by diet, such as a reddish color from beets, and by the presence of abnormal constituents, such as certain drugs. A red or brown to black color may indicate the presence of red blood cells or hemoglobin from bleeding in the urinary system.

Fresh urine is usually transparent. Turbid (cloudy) urine does not necessarily indicate a pathological condition, since turbidity may result from mucin secreted by the lining of the urinary tract. The presence of mucin above a critical level usually denotes an abnormality, however.

The odor of urine may vary. For example, the digestion of asparagus adds a substance called methyl mercaptan that gives urine a characteristic odor. In cases of diabetes, urine has a "sweetish" odor because of the presence of acetone. Stale urine develops the odor of ammonia due to ammonium carbonate formation as a result of urea decomposition.

Normal urine is slightly acid. It ranges between pH 5.0 and 7.8 and rarely becomes more acid than 4.5 or more alkaline than 8. Variations in urine pH are closely related to diet. These variations are due to differences in the end products of metabolism. Whereas a high-protein diet increases acidity, a diet composed largely of vegetables increases alkalinity. High altitude, fasting, and exercise also cause variations in urinary pH. Ammonium carbonate forms in standing urine. Since it can dissociate into ammonium ion and form a strong base, the presence

---

EXHIBIT 26-4
PHYSICAL CHARACTERISTICS OF NORMAL URINE

CHARACTERISTIC	DESCRIPTION
Volume	1–2 liters in 24 hours, but varies considerably.
Color	Yellow or amber, but varies with concentration and diet.
Turbidity	Transparent when freshly voided, but becomes turbid upon standing.
Odor	Aromatic, but becomes ammonialike upon standing.
pH	5.0–7.8, average 6.0; varies considerably with diet.
Specific gravity	1.008–1.030.

EXHIBIT 26-5

PRINCIPAL SOLUTES IN URINE OF ADULT MALE ON MIXED DIET

CONSTITUENT	AMOUNT (g)*	COMMENTS
ORGANIC		
Urea	25.0	Comprises 60–90 percent of all nitrogenous material. Derived primarily from deamination of proteins (ammonia combines with CO_2 to form urea).
Creatinine	1.5	Normal alkaline constituent of blood. Derived primarily from creatine (nitrogenous substance in muscle tissue).
Uric acid	0.8	Product of catabolism of nucleic acids derived from food or cellular destruction. Because of insolubility, tends to crystallize and is common component of kidney stones.
Hippuric acid	0.7	Form in which benzoic acid (toxic substance in fruits and vegetables) is believed to be eliminated from body. High-vegetable diets increase quantity of hippuric acid excreted.
Indican	0.01	Potassium salt of indole. Indole results from putrefaction of protein in large intestine and is carried by blood to liver, where it is probably changed to indican (less poisonous substance).
Ketone bodies	0.04	Also called acetone bodies. Normally found in small amounts. In cases of diabetes and acute starvation, ketone bodies appear in high concentrations.
Other substances	2.9	May be present in minute quantities depending on diet and general health. Include carbohydrates, pigments, fatty acids, mucin, enzymes, and hormones.
INORGANIC		
NaCl	15.0	Principal inorganic salt. Amount excreted varies with intake.
K^+	3.3	Occurs as chloride, sulfate, and phosphate salts.
Mg^{2+}	0.1	Occurs as chloride, sulfate, and phosphate salts.
Ca^{2+}	0.3	Occurs as chloride, sulfate, and phosphate salts.
SO_4^{2-}	2.5	Derived from amino acids.
PO_4^{3-}	2.5	Occurs as sodium compounds (monosodium and disodium phosphate) that serve as buffers in blood.
NH_4^+	0.7	Occurs as ammonium salts. Derived from protein catabolism and from glutamine in kidneys. Amount produced by kidney may vary with need of body for conserving Na^+ ions to offset acidity of blood and tissue fluids.

* Values are for a urine sample collected over 24 hours.

of ammonium carbonate tends to make urine more alkaline.

Specific gravity is the ratio of the weight of a volume of a substance to the weight of an equal volume of distilled water. Water has a specific gravity of 1.000. The specific gravity of urine depends on the amount of solid materials in solution and ranges from 1.008 to 1.030 in normal urine. The greater the concentration of solutes, the higher the specific gravity.

The physical characteristics of urine are summarized in Exhibit 26-4.

Chemical Composition

Water accounts for about 95 percent of the total volume of urine. The remaining 5 percent consists of solutes derived from cellular metabolism and outside sources such as drugs. The solutes are described in Exhibit 26-5.

CLINICAL APPLICATION

Several screening **tests for renal function** are available. The most commonly ordered test is the *blood urea nitrogen (BUN)* which measures blood levels of nitro-

gen in urea. Urea is an end product of protein metabolism produced in the liver and excreted in the liver. Although the BUN is relatively insensitive to mild degrees of renal dysfunction, it is a good clinical indicator of significant renal dysfunction.

The *serum creatinine test* measures the level of creatinine in blood. Creatinine is derived from the conversion of creatine phosphate in skeletal muscle tissue. Like the BUN, serum creatinine levels are elevated only when there is significant renal damage.

Renal clearance tests measure the volume of plasma that is cleared of a substance in a given period of time. The tests are used to assess glomerular filtration rates and renal blood flow. Clearance studies have been devised using injected substances such as inulin and naturally occurring substances such as creatinine and urea.

Abnormal Constituents

If the body's chemical processes are not operating efficiently, traces of substances not normally present may appear in the urine or normal constituents may appear in abnormal amounts. Analyzing the physical, chemical, and microscopic properties of urine often provides information that aids diagnosis. Such an analysis is called a **urinalysis (UA).**

● *Albumin*   Albumin is a normal constituent of plasma, but it usually appears in only very small amounts because the particles are too large to pass through the pores in the capillary walls. The presence of excessive albumin in the urine—**albuminuria**—indicates an increase in the permeability of the glomerular membrane. Conditions that lead to albuminuria include injury to the glomerular membrane as a result of disease, increased blood pressure, and irritation of kidney cells by substances such as bacterial toxins, ether, or heavy metals. Other proteins, such as globulin and fibrinogen, may also appear in the urine under certain conditions.

● *Glucose*   The presence of sugar in the urine is termed **glycosuria.** Normal urine contains such small amounts of glucose that clinically it may be considered absent. The most common cause of glycosuria is a high blood sugar level. Remember that glucose is filtered into the glomerular capsule. Later, in the proximal convoluted tubules, the tubule cells actively transport the glucose back into the blood. However, the number of glucose carrier molecules is limited. If more carbohydrates are ingested than can be used or stored as glycogen or fat, more sugar is filtered into the glomerular capsule than can be removed by the carriers. This condition, called **temporary (alimentary) glycosuria,** is not considered pathological. Another nonpathological cause is emotional stress. Stress can cause excessive amounts of epinephrine to be secreted. Epinephrine stimulates the breakdown of glycogen and the liberation of glucose from the liver. A **pathological glycosuria** results from diabetes mellitus. In this case, there is a frequent or continuous elimination of glucose because the pancreas fails to produce sufficient insulin. When glycosuria occurs with a normal blood sugar level, the problem lies in failure of the kidney tubular cells to reabsorb glucose.

● *Erythrocytes*   The appearance of red blood cells in the urine is called **hematuria.** Hematuria generally indicates a pathological condition. One cause is acute inflammation of the urinary organs as a result of disease or irritation from kidney stones. Whenever blood is found in the urine, additional tests are performed to ascertain the part of the urinary tract that is bleeding. One should also make sure the sample was not contaminated with menstrual blood from the vagina. Some long distance runners may also develop "red urine" due to damaged blood vessels in the bottom of the feet. As a result of the damage, blood appears in the urine.

● *Leucocytes*   The presence of leucocytes and other components of pus in the urine, referred to as **pyuria,** indicates infection in the kidney or other urinary organs. Again the source of the pus must be located, and care should be taken that the urine is not contaminated.

● *Ketone Bodies*   Ketone (acetone) bodies appear in normal urine in small amounts. Their appearance in high quantities, a condition called **ketosis (acetonuria),** may indicate abnormalities. It may be caused by diabetes mellitus, starvation, or simply too little carbohydrate in the diet. Whatever the cause, excessive quantities of fatty acids are oxidized in the liver and the ketone bodies are filtered from the plasma into the glomerular capsule.

● *Bilirubin*   As noted in Chapter 24, bilirubin is a product of the hemolysis of red blood cells and the breakdown of hemoglobin by reticuloendothelial cells. The pigment gives bile its major pigmentation. Bilirubin is carried to liver cells by an albumin molecule and is referred to as **unconjugated (free) bilirubin.** When it reaches the liver, the albumin is released. The bilirubin is then combined with certain substances in liver cells (glucuronic acid or sulfate) and is referred to as **conjugated bilirubin.** In this form it is secreted by the liver into the biliary system and then into the small intestine. Intestinal bacteria convert conjugated bilirubin into urobilogen, which gives feces its characteristic color. Above normal levels of bilirubin in urine are referred to as **bilirubinuria.** A routine urinalysis measures total bilirubin; it does not distinguish between unconjugated and conjugated. A normal level of total bilirubin rules out any significant liver dysfunction or excessive hemolysis of red blood cells. If the level is elevated, tests that differentiate between unconjugated and conjugated bilirubin may be performed. An increase in unconjugated bilirubin is more frequently associated with excessive hemolysis of red blood cells. An increase in conjugated bilirubin is more likely a result of liver dysfunction or obstruction of the biliary system.

● *Urobilinogen*   Part of the urobilinogen formed in the intestine is excreted with feces. Another portion is absorbed and returned to the liver where it is metabolized and excreted in bile. A small percentage of the urobilinogen that is absorbed escapes the liver and enters general circulation. The kidneys handle this amount of urobilinogen as though it were a foreign substance, that is, the tubular cells do not actively reabsorb the filtered urobilinogen. The presence of urobilinogen in urine is called **urobilinogenuria.** Thus, traces of urobilinogen in urine are normal. When urobilinogenuria is above normal, it indicates an increase in the production of bilirubin and inability of the liver to remove reabsorbed urobilinogen from the blood. Conditions that contribute to increased urobilinogen are hemolytic and pernicious anemia, infectious hepatitis, biliary obstruction, jaundice, cirrhosis, congestive heart failure, and infectious mononucleosis.

● *Casts*   Microscopic examination of urine may reveal **casts**—tiny masses of material that have hardened and assumed the shape of the lumens of the tubules and were then flushed out of the tubules by a buildup of filtrate behind them. Casts are named after the substances that compose them or for their appearance. There are white-blood-cell casts, red-blood-cell casts, epithelial casts that contain cells from the walls of the tubes, granular casts that contain decomposed cells which form granules, and fatty casts from cells that have become fatty.

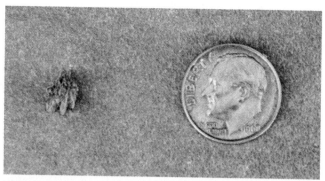

FIGURE 26-14 Photograph of a renal calculus adjacent to a dime for comparative size. (Courtesy of Matt Iacobino.)

● *Renal Calculi* Occasionally, the crystals of salts found in urine may solidify into insoluble stones called **renal calculi (kidney stones)** (Figure 26-14). They may be formed in any portion of the urinary tract from the kidney tubules to the external opening. Conditions leading to calculi formation include the ingestion of excessive mineral salts, a decrease in the amount of water, abnormally alkaline or acid urine, and overactivity of the parathyroid glands. Common constituents of stones are uric acid, calcium oxalate, and calcium phosphate crystals. The stones usually form in the pelvis of the kidney, where they cause pain, hematuria, and pyuria. A *staghorn calculus* may fill the entire collecting system; a *dendritic stone* involves some but not all of the calyces. Severe pain occurs when a stone passes through a ureter and stretches its walls. *Ureteral stones* are seldom completely obstructive because they are usually needle-shaped and urine can flow around them.

For kidney stones that become painful or obstructive, surgical removal is the typical alternative. Clinical trials are now underway to determine the effectiveness of removing kidney stones without surgery by using shock waves. In this procedure, a patient is placed in a water bath and is subjected to shock waves. Once the stones are shattered, the fragments are eliminated via the urine.

● *Microbes* In a properly collected and processed specimen the finding of bacteria may be of considerable importance. If bacteria are seen in a centrifuged specimen, but not in the unspun sample, it suggests a bacterial count of less than 10,000/ml. The presence of bacteria in an unspun sample suggests a count greater than 100,000/ml. The various bacteria in urine are identified by different microbiological tests. The number and type of bacteria vary with specific infections in the urinary tract. The most common fungus to appear in urine is *Candida albicans,* a common cause of vaginitis. The most frequent protozoan seen in urine is *Trichomonas vaginalis,* a cause of vaginitis in females and urethritis in males.

URETERS

Once urine is formed by the nephrons and collecting ducts, it drains through papillary ducts into the calyces surrounding the renal papillae. The minor calyces join to become the major calyces that unite to become the renal pelvis. From the pelvis, the urine drains into the ureters and is carried by peristalsis to the urinary bladder. From the urinary bladder, the urine is discharged from the body through the single urethra.

STRUCTURE

The body has two **ureters** (yoo-RĒ-ters)—one for each kidney. Each ureter is an extension of the pelvis of the kidney and extends 25 to 30 cm (10 to 12 inches) to the urinary bladder (see Figure 26-1). As the ureters descend, their thick walls increase in diameter, but at their widest point they measure less than 1.7 cm (½ inch) in diameter. Like the kidneys, the ureters are retroperitoneal in placement. The ureters enter the urinary bladder at the superior lateral angle of its base.

Although there are no anatomical valves at the openings of the ureters into the urinary bladder, there is a functional one that is quite effective. Since the ureters pass under the urinary bladder for several centimeters, pressure in the urinary bladder compresses the ureters and prevents backflow of urine when pressure builds up in the urinary bladder during urination. When this physiological valve is not operating, it is possible for cystitis (urinary bladder inflammation) to develop into kidney infection.

Three coats of tissue form the wall of the ureters. The inner coat, or mucosa, is mucous membrane with transitional epithelium. The solute concentration and pH of urine differ drastically from the internal environment of cells that form the wall of the ureters. Mucus secreted by the mucosa prevents the cells from coming in contact with urine. Throughout most of the length of the ureters, the second or middle coat, the muscularis, is composed of inner longitudinal and outer circular layers of smooth muscle. The muscularis of the proximal third of the ureters also contains a layer of outer longitudinal muscle. Peristalsis is the major function of the muscularis. The third, or external, coat of the ureters is a fibrous coat. Extensions of the fibrous coat anchor the ureters in place.

PHYSIOLOGY

The principal function of the ureters is to transport urine from the renal pelvis into the urinary bladder. Urine is carried through the ureters primarily by peristaltic contractions of the muscular walls of the ureters, but hydrostatic pressure and gravity also contribute. Peristaltic waves pass from the kidney to the urinary bladder, varying in rate from 1 to 5/min depending on the amount of urine formation.

URINARY BLADDER

The **urinary bladder** is a hollow muscular organ situated in the pelvic cavity posterior to the symphysis pubis. In the male, it is directly anterior to the rectum. In the female, it is anterior to the vagina and inferior to the uterus. It is a freely movable organ held in position by folds of the peritoneum. The shape of the urinary bladder depends on how much urine it contains. When empty

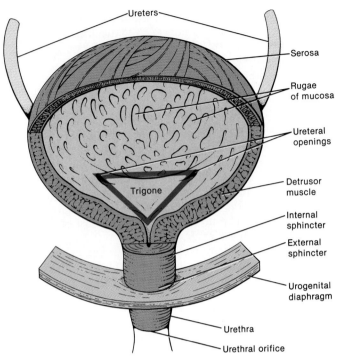

FIGURE 26-15 Urinary bladder and female urethra.

PHYSIOLOGY

Urine is expelled from the urinary bladder by an act called **micturition** (mik'-too-RISH-un), commonly known as urination or voiding. This response is brought about by a combination of involuntary and voluntary nervous impulses. The average capacity of the urinary bladder is 700 to 800 ml. When the amount of urine in the bladder exceeds 200 to 400 ml, stretch receptors in the urinary bladder wall transmit impulses to the lower portion of the spinal cord. These impulses initiate a conscious desire to expel urine and a subconscious reflex referred to as the **micturition reflex.** Parasympathetic impulses transmitted from the sacral area of the spinal cord reach the urinary bladder wall and internal urethral sphincter, bringing about contraction of the detrusor muscle of the urinary bladder and relaxation of the internal sphincter. Then the conscious portion of the brain sends impulses to the external sphincter, the sphincter relaxes, and urination takes place. Although emptying of the urinary bladder is controlled by reflex, it may be initiated voluntarily and stopped at will because of cerebral control of the external sphincter.

it looks like a deflated balloon. It becomes spherical when slightly distended. As urine volume increases, it becomes pear-shaped and rises into the abdominal cavity.

STRUCTURE

At the base of the urinary bladder is a small triangular area, the **trigone** (TRĪ-gōn), that points anteriorly (Figure 26-15). The opening to the urethra is found in the apex of this triangle. At the two points of the base, the ureters drain into the bladder. It is easily identified because the mucosa is firmly bound to the muscularis so that the trigone is typically smooth.

Four coats make up the wall of the urinary bladder (Figure 26-16). The mucosa, the innermost coat, is a mucous membrane containing transitional epithelium. Transitional epithelium is able to stretch—a marked advantage for an organ that must continually inflate and deflate. Rugae (folds in the mucosa) are also present. The second coat, the submucosa, is a layer of dense connective tissue that connects the mucosa and muscular coats. The third coat—a muscular one called the **detrusor** (de-TROO-ser) **muscle**—consists of three layers: inner longitudinal, middle circular, and outer longitudinal muscles. In the area around the opening to the urethra, the circular fibers form an **internal sphincter** muscle. Below the internal sphincter is the **external sphincter,** which is composed of skeletal muscle. The outermost coat, the serous coat, is formed by the peritoneum and covers only the superior surface of the organ.

URETHRA

The **urethra** is a small tube leading from the floor of the urinary bladder to the exterior of the body (see Figure 26-15). In females, it lies directly posterior to the symphysis pubis and is embedded in the anterior wall of the vagina. Its undilated diameter is about 6 mm (¼ inch), and its length is approximately 3.8 cm (1½ inch). The

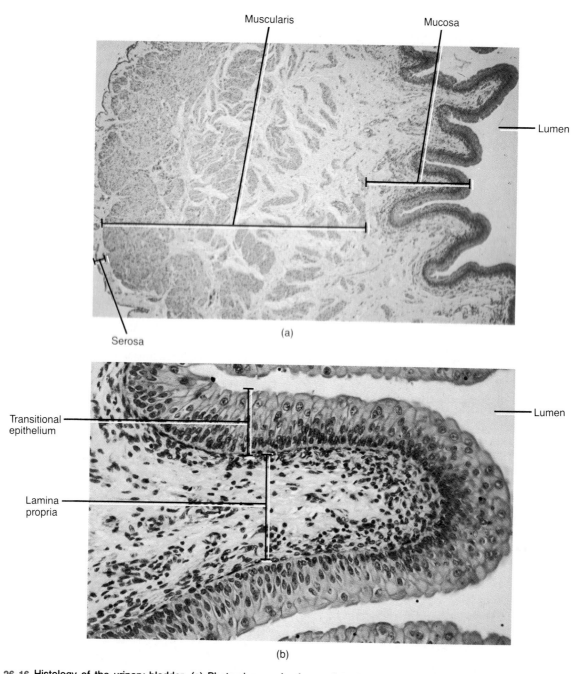

Muscularis   Mucosa

Lumen

Serosa

(a)

Transitional
epithelium

Lamina
propria

Lumen

(b)

FIGURE 26-16 Histology of the urinary bladder. (a) Photomicrograph of a portion of the wall of the urinary bladder at a magnification of 50×. (b) Photomicrograph of an enlarged aspect of the mucosa of the urinary bladder at a magnification of 200×. (© 1983 by Michael H. Ross. Used by permission.)

female urethra is directed obliquely inferiorly and anteriorly. The opening of the urethra to the exterior, the **urethral orifice** is located between the clitoris and vaginal opening.

In males, the urethra is about 20 cm (8 inches) long. Immediately below the urinary bladder it passes vertically through the prostate gland, then pierces the urogenital diaphragm, and finally pierces the penis and takes a curved course through its body (see Figures 28-1 and 28-8a).

STRUCTURE

The wall of the female urethra consists of three coats: an inner mucous coat that is continuous externally with that of the vulva, an intermediate thin layer of spongy tissue containing a plexus of veins, and an outer muscular coat that is continuous with that of the urinary bladder and consists of circularly arranged fibers of smooth muscle.

The male urethra is composed of two coats. An inner mucous membrane is continuous with the mucous membrane of the urinary bladder. An outer submucous tissue connects the urethra with the structures through which it passes.

PHYSIOLOGY

The urethra is the terminal portion of the urinary system. It serves as the passageway for discharging urine from the body. The male urethra also serves as the duct through which reproductive fluid (semen) is discharged from the body.

DISORDERS: HOMEOSTATIC IMBALANCES

Gout

Gout is a hereditary condition associated with an excessively high level of uric acid in the blood. When nucleic acids are catabolized, a certain amount of uric acid is produced as a waste. Some people seem to produce excessive amounts of uric acid, and others seem to have trouble excreting normal amounts. In either case, uric acid accumulates in the body and tends to solidify into crystals that are deposited in the joints and kidney tissue. When the crystals are deposited in the joints, the condition is called gouty arthritis. Gout is aggravated by excessive use of diuretics, dehydration, and starvation.

Glomerulonephritis (Bright's Disease)

Glomerulonephritis (Bright's disease) is an inflammation of the kidney that involves the glomeruli. One of the most common causes of glomerulonephritis is an allergic reaction to the toxins given off by streptococci bacteria that have recently infected another part of the body, especially the throat. The glomeruli become so inflamed, swollen, and engorged with blood that the glomerular membranes become highly permeable and allow blood cells and proteins to enter the filtrate. Thus the urine contains many erythrocytes and much protein. The glomeruli may be permanently changed, leading to chronic renal disease and renal failure.

Pyelitis and Pyelonephritis

Pyelitis is an inflammation of the kidney pelvis and its calyces. **Pyelonephritis** is the interstitial inflammation of one or both kidneys. It usually involves both the parenchyma and the renal pelvis and is due to bacterial invasion from the middle and lower urinary tracts or the bloodstream.

Cystitis is an inflammation of the urinary bladder involving principally the mucosa and submucosa. It may be caused by bacterial infection, chemicals, or mechanical injury.

Nephrosis

Nephrosis is a condition in which the glomerular membrane leaks, allowing large amounts of protein to escape from the blood into the urine. Water and sodium then accumulate in the body creating edema especially around the ankles and feet, abdomen, and eyes. Nephrosis is more common in children than adults, but occurs in all ages. Although it cannot always be cured, certain synthetic steroid hormones such as cortisone and prednisone, which are similar to the natural hormones secreted by the adrenal glands, can suppress some forms of it.

Polycystic Disease

Polycystic disease may be caused by a defect in the renal tubular system that deforms nephrons and results in cystlike dilations along their course. It is the most common inherited disorder of the kidneys. The kidney tissue is riddled with cysts, small holes, and fluid-filled bubbles ranging in size from a pinhead to the diameter of an egg. These cysts gradually increase until they squeeze out the normal tissue, interfering with kidney function and causing uremia. The chief symptom is weight gain.

The kidneys themselves may enlarge from their normal ½ lb to as much as 30 lb. Many people lead normal lives without ever knowing they have the disease, however, and the condition is sometimes discovered only after death. Although the disease is progressive, its advance can be slowed by diet, drugs, and fluid intake. Kidney failure as a result of polycystic disease seldom occurs before the mid-40s and frequently can be postponed until the 60s.

Hemodialysis Therapy

If the kidneys are so impaired by disease or injury that they are unable to excrete nitrogenous wastes and regulate pH and electrolyte concentration of the plasma, the blood must be filtered by an artificial device. Such filtering of the blood is called **hemodialysis.** *Dialysis* means using a semipermeable membrane to separate large nondiffusible particles from smaller diffusible ones. One of the best-known devices for accomplishing dialysis is the kidney machine (Figure 26-17). A tube connects it with the patient's radial artery. The blood is pumped from the artery through the tubes to one side of a semipermeable dialyzing membrane made of cellophane sheets. The other side of the membrane is continually washed with an artificial solution called the dialyzing solution. The blood that passes through the artificial kidney is treated with an anticoagulant. Only about 500 ml of the patient's blood is in the machine at a time. This volume is easily compensated for by vasoconstriction and increased cardiac output.

All substances (including wastes) in the blood except protein molecules and blood cells can diffuse back and forth across the semipermeable membrane. The electrolyte level of the plasma is controlled by keeping the dialyzing solution electrolytes at the same concentration found in normal plasma. Any excess plasma electrolytes move down the concentration gradient and into the dialyzing solution. If the plasma electrolyte

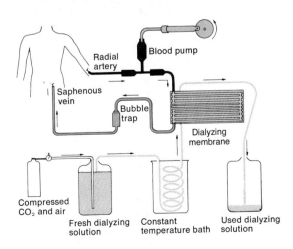

FIGURE 26-17 Operation of an artificial kidney. The blood route is indicated in red and blue. The route of the dialyzing solution is indicated in gold.

level is normal, it is in equilibrium with the dialyzing solution and no electrolytes are gained or lost. Since the dialyzing solution contains no wastes, substances such as urea move down the concentration gradient and into the dialyzing solution. Thus wastes are removed and normal electrolyte balance is maintained.

A great advantage of the kidney machine is that nutrition can be bolstered by placing large quantities of glucose in the dialyzing solution. While the blood gives up its wastes, the glucose diffuses into the blood. Thus the kidney machine beautifully accomplishes the principal function of the fundamental unit of the kidney—the nephron.

There are obvious drawbacks to the artificial kidney, however. Anticoagulants must be added to the blood during dialysis. A large amount of the patient's blood must flow through this apparatus to make the treatment effective, and so the slow rate at which the blood can be processed makes the treatment time-consuming. To date, no artificial kidney has been implanted permanently.

A recent development, called **continuous ambulatory peritoneal dialysis (CAPD),** promises to make hemodialysis more convenient and less time-consuming for many patients. CAPD uses the peritoneum instead of cellophane sheets as the dialyzing membrane. Since the peritoneum is a semipermeable membrane, it permits rapid bidirectional transfer of substances. A catheter is placed in the patient's peritoneal cavity and connected to a supply of dialyzing fluid. Gravity feeds the solution into the abdominal cavity from its plastic container. When the process is complete, the dialyzing fluid is returned from the abdominal cavity to the plastic container and then discarded.

MEDICAL TERMINOLOGY

Azotemia (*azo* = nitrogen-containing; *emia* = condition of blood) Presence of urea or other nitrogenous elements in the blood.

Cystocele (*cyst* = bladder; *cele* = cyst) Hernia of the urinary bladder.

Cystoscope (*skopein* = to examine) Instrument used to examine the urinary bladder.

Dysuria (*dys* = painful; *uria* = urine) Painful urination.

Enuresis (*enourein* = to void urine) Bed-wetting; may be due to faulty toilet training, to some psychological or emotional disturbance, or rarely to some physical disorder.

Intravenous pyelogram (*intra* = within; *veno* = vein; *pyelo* = pelvis of kidney; *gram* = written or recorded) X-ray film of the kidneys after injection of a dye.

Nephroblastoma (*neph* = kidney; *blastos* = germ or forming; *oma* = tumor) Embryonal carcinosarcoma; a malignant tumor arising from epithelial and connective tissue.

Nephrocele Hernia of the kidney.

Oliguria (*olig* = scanty) Scanty urine.

Polyuria (*poly* = much) Excessive urine.

Stricture Narrowing of the lumen of a canal or hollow organ, as the ureter or urethra.

Uremia (*emia* = condition of blood) Toxic levels of urea in the blood resulting from severe malfunction of the kidneys.

Urethritis Inflammation of the urethra, caused by highly acid urine, the presence of bacteria, or constriction of the urethral passage.

DRUGS ASSOCIATED WITH THE URINARY SYSTEM

Diuretics

Diuretics are drugs that stimulate the flow of urine. Many of these act on the renal tubule. One group inhibits active sodium transport. Examples include the benzothiadiazines like *chlorthiazide* (Diuril and Hygroton), *chlorthalidone* (Demi-Regroton), and *metolazone* (Diulo and Zaroxolyn). Another group of drugs inhibits active chloride transport; *ethacrynic acid* (Edecrin) and *furosemide* (Lasix) are examples. The third group includes *spironolactone* (Aldactone) which inhibits cellular binding of aldosterone, while *triamterene* (Dyrenium) inhibits sodium reabsorption and blocks potassium and hydrogen ion excretion.

Diuretics are used to control different types of edema, congestive heart failure, and hypertension.

STUDY OUTLINE

1. The primary function of the urinary system is to regulate the concentration and volume of blood by removing and restoring selected amounts of water and solutes. It also excretes wastes.
2. The organs of the urinary system are the kidneys, ureters, urinary bladder, and urethra.

Kidneys (p. 658)

External Anatomy; Internal Anatomy

1. The kidneys are retroperitoneal organs attached to the posterior abdominal wall.
2. Three layers of tissue surround the kidneys: renal capsule, adipose capsule, and renal fascia.
3. Internally, the kidneys consist of a cortex, medulla, pyramids, papillae, columns, calyces, and a pelvis.
4. The nephron is the functional unit of the kidneys.
5. Each juxtamedullary nephron consists of a glomerular capsule, glomerulus, proximal convoluted tubule, descending limb of Henle, loop of Henle, ascending limb of Henle, distal convoluted tubule, and collecting duct. The limb of Henle is absent in a cortical nephron.
6. The filtering unit of a nephron is the endothelial-capsular membrane. It consists of the glomerular endothelium, glomerular basement membrane, and epithelium (podocytes) of the visceral layer of the glomerular capsule.
7. The extensive flow of blood through the kidney begins in the renal artery and terminates in the renal vein.
8. The nerve supply to the kidney is derived from the renal plexus.
9. The juxtaglomerular apparatus consists of the juxtaglomerular cells of the afferent arteriole and the macula densa of the distal convoluted tubule.

Physiology

1. Nephrons are the functional units of the kidneys. They help form urine and regulate blood composition.
2. The nephrons form urine by glomerular filtration, tubular reabsorption, and tubular secretion.

3. The primary force behind glomerular filtration is hydrostatic pressure.
4. Filtration of blood depends on the force of glomerular blood hydrostatic pressure in relation to two opposing forces: capsular hydrostatic pressure and blood colloid osmotic pressure. This relationship is called effective filtration pressure (P_{eff}).
5. If glomerular blood hydrostatic pressure falls to 50 mm Hg, renal suppression occurs because the glomerular blood hydrostatic pressure exactly equals the opposing pressures.
6. Most substances in plasma are filtered by the glomerular capsule. Normally, blood cells and most proteins are not filtered.
7. Tubular reabsorption retains substances needed by the body, including water, glucose, amino acids, and ions. The maximum of a substance that can be absorbed is called tubular maximum.
8. About 80 percent of the reabsorbed water is returned by obligatory reabsorption, the rest by facultative reabsorption.
9. Chemicals not needed by the body are discharged into the urine by tubular secretion. Included are ions, nitrogenous wastes, and certain drugs.
10. The kidneys help maintain blood pH by excreting H^+ and NH_4^+ ions. In exchange, the kidneys conserve sodium bicarbonate.
11. The ability of the kidneys to produce either hyperosmotic or hyposmotic urine is based on the countercurrent multiplier mechanism.

Homeostasis (p. 673)

1. Besides the kidneys, the lungs, integument, and alimentary canal assume excretory functions.
2. Urine volume is influenced by blood pressure, blood concentration, temperature, diuretics, and emotions.
3. The physical characteristics of urine evaluated in a urinalysis (UA) are color, odor, turbidity, pH, and specific gravity.
4. Chemically, normal urine contains about 95 percent water water and 5 percent solutes. The solutes include urea, creatinine, uric acid, hippuric acid, indican, ketone bodies, salts, and ions.

5. Abnormal constituents diagnosed through urinalysis include albumin, glucose, erythrocytes, leucocytes, ketone bodies, bilirubin, urobilinogen, casts, renal calculi, and microbes.

Ureters (p. 677)

1. The ureters are retroperitoneal and consist of a mucosa, muscularis, and fibrous coat.
2. The ureters transport urine from the renal pelvis to the urinary bladder, primarily by peristalsis.

Urinary Bladder (p. 677)

1. The urinary bladder is posterior to the symphysis pubis. Its function is to store urine prior to micturition.
2. Histologically, the urinary bladder consists of a mucosa (with rugae), a muscularis (detrusor muscle), and a serous coat.
3. A lack of control over micturition is called incontinence; failure to void urine is referred to as retention.

Urethra (p. 678)

1. The urethra is a tube leading from the floor of the urinary bladder to the exterior.
2. Its function is to discharge urine from the body.

Disorders: Homeostatic Imbalances (p. 680)

1. Gout is a high level of uric acid in the blood.
2. Glomerulonephritis is an inflammation of the glomeruli of the kidney.
3. Pyelitis is an inflammation of the kidney pelvis and calyces; pyelonephritis is an interstitial inflammation of one or both kidneys.
4. Cystitis is an inflammation of the urinary bladder.
5. Nephrosis leads to protein in the urine due to glomerular membrane permeability.
6. Polycystic disease is an inherited kidney disease in which nephrons are deformed.
7. Filtering blood through an artificial device is called hemodialysis.
8. The kidney machine filters the blood of wastes and adds nutrients; a recent variation is called continuous ambulatory peritoneal dialysis (CAPD).

REVIEW QUESTIONS

1. What are the functions of the urinary system? What organs compose the system?
2. Describe the location of the kidneys. Why are they said to be retroperitoneal?
3. Prepare a labeled diagram that illustrates the principal external and internal features of the kidney.
4. What is a nephron? List and describe the parts of a nephron from the glomerular capsule to the collecting duct.
5. Distinguish between cortical and juxtaglomerular nephrons.
6. How are nephrons supplied with blood?
7. Describe the structure and importance of the juxtaglomerular apparatus.
8. What is glomerular filtration? Define filtrate.
9. Set up an equation to indicate how effective filtration pressure is calculated. What is the cause of renal suppression?
10. What are the major chemical differences among plasma, filtrate, and urine?
11. Define tubular reabsorption. Why is the process physiologically important? What is glomerular filtration rate?
12. What chemical substances are normally reabsorbed by the kidney? What is tubular maximum (Tm)?
13. Describe how glucose and sodium are reabsorbed by the kidneys. Where does the process occur?

14. How is chloride reabsorption related to sodium reabsorption?
15. Distinguish obligatory from facultative reabsorption of water. How is facultative reabsorption controlled?
16. Define tubular secretion. Why is it important? List some substances that are secreted.
17. Explain the mechanisms by which the kidneys help to control body pH.
18. Define the countercurrent multiplier mechanism. Why is it important?
19. Contrast the functions of the lungs, integument, and alimentary tract as excretory organs.
20. What is urine? Describe the effects of blood pressure, blood concentration, temperature, diuretics, and emotions on the volume of urine formed.
21. Describe the following physical characteristics of normal urine: color, turbidity, odor, pH, and specific gravity.
22. Describe the chemical composition of normal urine.
23. Define each of the following: albuminuria, glycosuria, hematuria, pyuria, ketosis, bilirubinuria, urobilinogenuria, casts, and renal calculi.
24. Describe the structure, histology, and function of the ureters.

25. How is the urinary bladder adapted to its storage function? What is micturition?
26. Contrast the causes of incontinence and retention.
27. Compare the location of the urethra in the male and female.
28. Define each of the following: gout, glomerulonephritis, pyelitis, pyelonephritis, cystitis, nephrosis, and polycystic disease.

29. What is hemodialysis? Briefly describe the operation of an artificial kidney. What is continuous ambulatory peritoneal dialysis (CAPD)?
30. Refer to the glossary of medical terminology associated with the urinary system. Be sure that you can define each term.
31. Describe the actions of selected drugs on the urinary system.

27 Fluid, Electrolyte, and Acid–Base Dynamics

The term **body fluid** refers to the body water and its dissolved substances. Fluid composes 45 to 75 percent of the body weight.

FLUID COMPARTMENTS AND FLUID BALANCE

About two-thirds of the fluid is located in cells and is termed **intracellular fluid (ICF).** The other third, called **extracellular fluid (ECF),** includes all the rest of the body fluids—interstitial fluid, plasma and lymph, cerebrospinal fluid, GI tract fluids, synovial fluid, the fluids of the eyes and ears, pleural, pericardial, and peritoneal fluids, and the glomerular filtrate.

Body fluids are separated into distinct compartments whose walls are the semipermeable membranes provided by the plasma membranes of cells. A compartment may be as small as the interior of a single cell or as large as the combined interiors of the heart and vessels. Fluids exist in compartments, but keep in mind that they are in constant motion from one compartment to another. In the healthy individual, the volume of fluid in each compartment remains stable—another example of homeostasis.

Water constitutes the bulk of all the body fluids. When we say that the body is in *fluid balance,* we mean it contains the required amount of water distributed to the various compartments according to their needs.

Osmosis is the primary way in which water moves in and out of body compartments. The concentration of solutes in the fluids is therefore a major determinant of fluid balance. Most solutes in body fluids are electrolytes—compounds that dissociate into ions. Fluid balance, then, means water balance, but it also implies electrolyte balance. The two are inseparable.

WATER

Water is by far the largest single constituent of the body, making up from 45 to 75 percent of the total body weight. The percentage varies from person to person and depends primarily on the amount of fat present and age. Since fat is basically water-free, lean people have a greater proportion of water to total body weight than fat people. Water proportion also decreases with age. An infant has the highest amount of water per body weight. In a normal adult male, water averages about 65 percent of the body weight. Females have more subcutaneous fat than males, and in a normal female, water averages about 55 percent.

FLUID INTAKE AND OUTPUT

The primary source of body fluid is water derived from ingested liquids (1,600 ml) and foods (700 ml) that has been absorbed from the alimentary canal. This water, called *preformed water,* amounts to about 2,300 ml/day. Another source of fluid is *metabolic water,* the water produced through catabolism. This amounts to about 200 ml/day. Thus total fluid input averages about 2,500 ml/day.

There are several avenues of fluid output. The kidneys on the average lose about 1,500 ml/day, the skin about 500 ml/day, the lungs about 300 ml/day, and the GI tract about 200 ml/day. Fluid output thus totals 2,500 ml/day. Under normal circumstances, fluid intake equals fluid output, so the body maintains a constant volume.

REGULATION OF INTAKE

Fluid intake is regulated by thirst. According to one theory, when water loss is greater than water intake, the resulting *dehydration* stimulates thirst through both local and general responses. Locally, it leads to a decrease in the flow of saliva that produces a dryness of the mucosa of the mouth and pharynx (Figure 27-1). Dryness is interpreted by the brain as a sensation of thirst. In addition, dehydration raises blood osmotic pressure. Receptors in the thirst center of the hypothalamus are stimulated by the increase in osmotic pressure and initiate impulses that also are interpreted as a sensation of thirst. The response, a desire to drink fluids, thus balances the fluid loss.

The initial quenching of thirst results from wetting the mucosa of the mouth and pharynx, but the major inhibition of thirst is believed to occur as a result of distension of the intestine and a decrease in osmotic pressure in fluids of the hypothalamus. Apparently, stimulated stretch receptors in the walls of the intestine send impulses that inhibit the thirst center in the hypothalamus.

REGULATION OF OUTPUT

Under normal circumstances, fluid output is adjusted by ADH and aldosterone, both of which regulate urine pro-

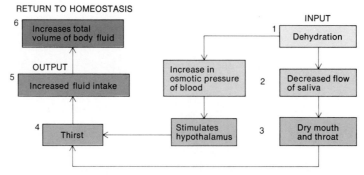

FIGURE 27-1 Regulation of fluid volume by the adjustment of intake to output.

duction (Chapter 26). Under abnormal conditions, other factors may influence output heavily. If the body is dehydrated, blood pressure falls, glomerular filtration rate decreases accordingly, and water is conserved. Conversely, excessive blood fluid results in an increase of blood pressure, glomerular filtration rate, and fluid output. Hypertension produces the same effect. Hyperventilation leads to increased fluid output through the loss of water vapor by the lungs. Vomiting and diarrhea result in fluid loss from the gastrointestinal tract. Finally, fever and destruction of extensive areas of the skin from burns bring about excessive water loss through the skin.

ELECTROLYTES

The body fluids contain a variety of dissolved chemicals. Some of these chemicals are compounds with covalent bonds, that is, the atoms that compose the molecule share electrons and do not form ions. Such compounds are called **nonelectrolytes.** Nonelectrolytes include most organic compounds, such as glucose, urea, and creatine. Other compounds, called **electrolytes,** have at least one ionic bond. When they dissolve in a body fluid, they dissociate into positive and negative ions. The positive ions are called cations; the negative ions are anions. Acids, bases, and salts are electrolytes. Most electrolytes are inorganic compounds, but a few are organic. For example, some proteins form ionic bonds. When the protein is put in solution, the ion detaches and the rest of the protein molecule carries the opposite charge.

Electrolytes serve three general functions in the body. First, many are essential minerals. Second, they control the osmosis of water between body compartments. Third, they help maintain the acid–base balance required for normal cellular activities.

CONCENTRATION

During osmosis, water moves to the area with the greater number of particles in solution. A particle may be a whole molecule or an ion. An electrolyte exerts a far greater effect on osmosis than a nonelectrolyte because an electrolyte molecule dissociates into at least two particles, both of them charged. Suppose the nonelectrolyte glucose and two different electrolytes are placed in solution:

$$C_6H_{12}O_6 \xrightarrow{\text{H}_2\text{O}} C_6H_{12}O_6$$
$$\text{Glucose}$$

$$NaCl \xrightarrow{\text{H}_2\text{O}} Na^+ + Cl^-$$
$$\text{Sodium chloride}$$

$$CaCl_2 \xrightarrow{\text{H}_2\text{O}} Ca^{2+} + Cl^- + Cl^-$$
$$\text{Calcium chloride}$$

Because glucose does not break apart when dissolved in water, a molecule of glucose contributes only one particle to the solution. Sodium chloride, on the other hand, contributes two ions, or particles, and calcium chloride contributes three. Thus calcium chloride has three times as great an effect on solute concentration as glucose.

Just as important, once the electrolyte dissociates, its ions can attract other ions of the opposite charge. If equal amounts of Ca^{2+} and Na^+ are placed in solution, the calcium ion will attract twice as many chloride ions to its area as the sodium ion.

To determine how much effect an electrolyte has on concentration, we must look at the concentrations of its individual ions. The concentration of an ion is commonly expressed in **milliequivalents per liter (meq/liter)**—the number of electrical charges in each liter of solution. The meq/liter equals the number of ions in solution times the number of charges the ions carries. Ion concentration can be calculated as shown in Exhibit 27–1.

DISTRIBUTION

Figure 27-2 compares the principal chemical constituents of plasma, interstitial fluid, and intracellular fluid. The chief difference between plasma and interstitial fluid is that plasma contains quite a few protein anions, whereas interstitial fluid has hardly any. Since normal capillary membranes are practically impermeable to protein, the protein stays in the plasma and does not move out of the blood into the interstitial fluid. Plasma also contains more Na^+ ions but fewer Cl^- ions than the interstitial fluid. In most other respects the two fluids are similar.

Intracellular fluid varies considerably from extracellular fluid, however. In extracellular fluid, the most abundant cation is Na^+ and the most abundant anion is Cl^-. In intracellular fluid, the most abundant cation is K^+ and the most abundant anion is HPO_4^{2-} (phosphate). Also, there are more protein anions in intracellular fluid than in extracellular fluid.

FUNCTIONS AND REGULATION

Sodium

Sodium (Na^+), the most abundant extracellular ion, represents about 90 percent of extracellular cations. Sodium is necessary for the transmission of impulses in nervous and muscle tissue. Its movement also plays a significant role in fluid and electrolyte balance.

> **CLINICAL APPLICATION**
> Sodium loss from the body may occur through excessive perspiration, certain diuretics, and burns. Such a loss can result in **hyponatremia** (hī'-pō-na-TRĒ-mē-a; *natrium* = sodium), a lower than normal blood sodium level. Hyponatremia is characterized by muscular weakness, headache, hypotension, tachycardia, and circulatory shock. Severe sodium loss can result in mental confusion, stupor, and coma.

The sodium level in the blood is controlled primarily by the hormone aldosterone from the adrenal cortex. Aldosterone acts on the distal convoluted tubules and collecting ducts of the kidneys and causes them to increase their reabsorption of sodium. The sodium thus moves from the filtrate back into the blood. Aldosterone is se-

EXHIBIT 27-1
CALCULATING CONCENTRATION OF IONS IN SOLUTION

The number of milliequivalents of an ion in each liter of solution is expressed by the following equation:

$$\text{meq/liter} = \frac{\text{milligrams of ion per liter of solution}}{\text{atomic weight}} \times \text{number of charges on one ion}$$

The atomic weight of an element is the number of protons and neutrons in an atom. Dividing the total weight of a solute by its atomic weight tells us how many ions there are in solution. The atomic weight of calcium is 40, whereas that of sodium is 23. Calcium is therefore a heavier element, and 100 g of calcium contains fewer atoms than does 100 g of sodium. The atomic weights of the elements can be found in a periodic table.

Using the preceding formula, we can calculate the milliequivalents per liter for calcium. In 1 liter of plasma there are normally 100 mg of calcium. Thus, by substituting this value in the formula, we arrive at:

$$\text{meq/liter} = \frac{100}{\text{atomic weight}} \times \text{number of charges}$$

The atomic weight of calcium is 40, and its number of charges is 2. By substituting these values we arrive at:

$$\text{meq/liter} = \frac{100}{40} \times 2 = 5$$

Let us now find the milliequivalents per liter of plasma for sodium.

Milligrams of ion per liter $= 3{,}300$

Number of charges $\quad = 1$

Atomic weight $\quad\quad\quad = 23$

meq/liter $\quad\quad\quad\quad = \dfrac{3{,}300}{23} \times 1$

$\quad\quad\quad\quad\quad\quad\quad = 143.0$

Even though calcium has a greater number of charges than sodium, the body retains many more sodium ions than calcium ions. Therefore, the milliequivalent for sodium in plasma is higher.

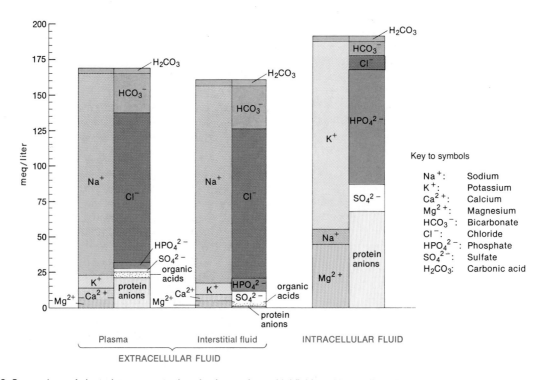

FIGURE 27-2 Comparison of electrolyte concentrations in plasma, interstitial fluid, and intracellular fluid. The height of each column represents the total electrolyte concentration.

creted in response to reduced blood volume or cardiac output, decreased extracellular sodium, increased extracellular potassium, and physical stress.

Chloride

Chloride (Cl⁻) is mainly an extracellular anion. However, it can easily diffuse between the extracellular and intracellular compartments. This movement makes chloride important in regulating osmotic pressure differences between compartments. In the gastric mucosal glands, chloride combines with hydrogen to form hydrochloric acid.

CLINICAL APPLICATION

An abnormally low level of chloride in the blood, called **hypochloremia** (hī'-pō-klō-RĒ-mē-a) may be caused by excessive vomiting, dehydration, and certain diuretics. Symptoms include muscle spasms, alkalosis, depressed respirations, and even coma.

The regulation of chloride is indirectly under control of aldosterone. Aldosterone regulates sodium reabsorption and chloride follows sodium passively.

Potassium

Potassium (K⁺) is primarily an intracellular electrolyte. The most abundant cation in intracellular fluid, it helps maintain the fluid volume in cells and control pH. Potassium also assumes a key role in the functioning of nervous and muscle tissue. When potassium ions move out of the cell, they are replaced by sodium ions and hydrogen ions. This shift of hydrogen ions helps regulate pH.

CLINICAL APPLICATION

A lower than normal level of potassium, called **hypokalemia** (hī'-pō-ka-LĒ-mē-a; *kalium* = potassium), may result from vomiting, diarrhea, high sodium intake, and kidney disease. Symptoms include cramps and fatigue, flaccid paralysis, mental confusion, increased urine output, shallow respirations, and changes in the electrocardiogram, including a lengthening of the Q-T interval and flattening of the T wave.

The blood level of potassium is under the control of mineralocorticoids, mainly aldosterone. The mechanism is exactly opposite that of sodium. When sodium concentration is low, aldosterone secretion increases and more sodium is reabsorbed. But when potassium concentration is high, more aldosterone is secreted and more potassium is excreted. This process occurs in the distal tubules and collecting ducts of the kidneys.

Calcium and Phosphate

Calcium (Ca²⁺) and phosphate (HPO₄²⁻) are stored in bones and teeth and released when needed. Calcium is principally an extracellular electrolyte; phosphate is principally an intracellular electrolyte. Calcium is a structural component of bones and teeth. It is also required for

blood clotting, chemical transmitter release, muscle contraction, and normal heartbeat. Phosphate is an important structural component of bones and teeth. In addition, it is necessary for the formation of nucleic acids (DNA and RNA), the synthesis of high-energy compounds (ATP and creatine phosphate), and buffering reactions.

Calcium and phosphate blood levels are regulated by several hormones. Parathyroid hormone (PTH) is released when the calcium blood level is low. PTH stimulates osteoclasts to release calcium and phosphate into the blood, increases the absorption of calcium from the gastrointestinal tract, and causes renal tubular cells to excrete phosphate. Calcitonin (CT), from the thyroid gland, decreases the blood level of calcium by stimulating osteoblasts and inhibiting osteoclasts. In the presence of calcitonin, osteoblasts remove calcium and phosphate from the blood and deposit them in bone.

Magnesium

Magnesium (Mg²⁺) is primarily an intracellular electrolyte. It is important for the sodium-potassium pump. It activates enzyme systems needed to produce cellular energy by the breakdown of ATP into ADP and enzyme systems involved in essential reactions in the liver and bone tissue.

CLINICAL APPLICATION

Symptoms of **hypomagnesemia (magnesium deficiency)** include increased neuromuscular and central nervous system irritability leading to tremor, tetany, and possibly convulsions. Diuretic-induced hypomagnesemia can result in cardiac arrhythmias. **Hypermagnesemia (magnesium excess)** may cause central nervous system depression, coma, and hypotension.

Magnesium level is regulated by aldosterone. When magnesium concentration is low, increased aldosterone secretion acts on the kidneys so that more magnesium is reabsorbed.

MOVEMENT OF BODY FLUIDS

BETWEEN PLASMA AND INTERSTITIAL COMPARTMENTS

The movement of fluid between plasma and interstitial compartments occurs across capillary membranes. This movement was discussed in detail in Chapter 21 (see Figure 21-8), but we will review it here. Basically, the fluid movement is dependent on four principal pressures: (1) blood hydrostatic pressure (BHP), (2) interstitial fluid hydrostatic pressure (IFHP), (3) blood osmotic pressure (BOP), and (4) interstitial fluid osmotic pressure (IFOP).

The difference between the two forces that move fluid out of plasma and the two forces that push it into plasma is the **effective filtration pressure** (P_{eff}). The P_{eff} at the arterial end of a capillary is 8.3 mm Hg, while that at the venous end is −6.7 mm Hg. Thus, at the arterial end of a capillary, fluid moves out (filtered) from plasma

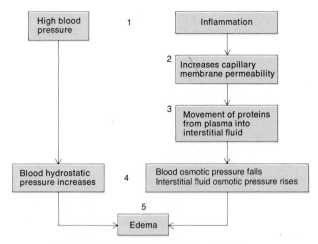

FIGURE 27-3 Conditions that produce edema.

into the interstitial compartment at a pressure of 8.3 mm Hg; at the venous end of a capillary, fluid moves in (reabsorbed) from the interstitial to plasma compartment at a pressure of −6.7 mm Hg. Thus, not all of the fluid filtered at one end of the capillary is reabsorbed at the other. The fluid not reabsorbed and any proteins that escape from capillaries pass into lymph capillaries. From here, the fluid (lymph) moves through lymphatics to the thoracic duct or right lymphatic duct for entrance into the cardiovascular system via the subclavian veins. Under normal conditions, there is a state of near equilibrium at the arterial and venous ends of a capillary in which filtered fluid and absorbed fluid, plus that picked up by the lymphatic system, are nearly equal. This near equilibrium is **Starling's law of the capillaries.**

CLINICAL APPLICATION

Occasionally the balance between interstitial fluid and plasma is disrupted. **Edema,** the abnormal increase in interstitial fluid resulting in tissue swelling, is an example of fluid imbalance (Figure 27-3). Hypertension may cause edema by raising the blood hydrostatic pressure. Another cause is inflammation. As part of the inflammatory response, capillaries become more permeable and allow proteins to leave the plasma and enter the interstitial fluid. Consequently, the blood osmotic pressure falls and the interstitial fluid osmotic pressure rises.

BETWEEN INTERSTITIAL AND INTRACELLULAR COMPARTMENTS

Intracellular fluid has a higher osmotic pressure than interstitial fluid. In addition, the principal cation inside the cell is K^+, whereas the principal cation outside is Na^+ (see Figure 27-2). Normally, the higher intracellular osmotic pressure is balanced by forces that move water out of the cell, so the amount of water inside the cell does not change. When a fluid imbalance between these two compartments occurs, it is usually caused by a change in the Na^+ or K^+ concentration.

Sodium balance in the body normally is controlled by aldosterone and ADH. ADH regulates extracellular

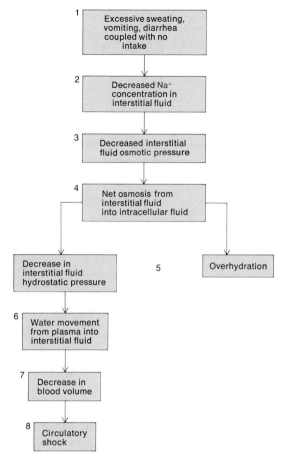

FIGURE 27-4 Interrelations between fluid imbalance and electrolyte imbalance.

fluid electrolyte concentration by adjusting the amount of water reabsorbed into the blood by the distal convoluted tubules and collecting ducts of the kidneys. Aldosterone regulates extracellular fluid volume by adjusting the amount of sodium reabsorbed by the blood from the kidneys. Certain conditions, however, may result in an eventual decrease in the sodium concentration in interstitial fluid. For instance, during sweating the skin excretes sodium as well as water. Sodium also may be lost through vomiting and diarrhea. Coupled with low sodium intake, these conditions can quickly produce a sodium deficit (Figure 27-4). The decrease in sodium concentration in the interstitial fluid lowers the interstitial fluid osmotic pressure and establishes an effective filtration pressure gradient between the interstitial fluid and the intracellular fluid. Water moves from the interstitial fluid into the cells, producing two results that can be quite serious.

The first result, an increase in intracellular water concentration, called **overhydration,** is particularly disruptive to nerve cell function. In fact, severe overhydration, or **water intoxication,** produces neurological symptoms ranging from disoriented behavior to convulsions, coma, and even death. The second result of the fluid shift is a loss of interstitial fluid volume that leads to a decrease in the interstitial fluid hydrostatic pressure. As the interstitial hydrostatic pressure drops, water moves out of the plasma, resulting in a loss of blood volume that may lead to circulatory shock.

ACID–BASE BALANCE

In addition to controlling water movement, electrolytes also help regulate the body's acid–base balance. The overall acid–base balance is maintained by controlling the hydrogen ion concentration of body fluids, particularly extracellular fluid. In a healthy person, the pH of the extracellular fluid is stabilized between 7.35 and 7.45. Homeostasis of this narrow range is essential to survival and depends on three major mechanisms: buffer systems, respirations, and kidney excretion.

BUFFER SYSTEMS

Most **buffer systems** of the body consist of a weak acid and a weak base, and they function to prevent drastic changes in the pH of a body fluid by changing strong acids and bases into weak acids and bases. Buffers work within fractions of a second. Recall that a strong acid dissociates into H^+ ions more easily than does a weak acid. Strong acids, therefore, lower pH more than weak ones because strong acids contribute more H^+ ions. Similarly, strong bases raise pH more than weak ones because strong bases dissociate more easily into OH^- ions. The principal buffer systems of the body fluids are the carbonic acid–bicarbonate system, the phosphate system, the hemoglobin-oxyhemoglobin system, and the protein system.

Carbonic Acid–Bicarbonate

The **carbonic acid–bicarbonate buffer system** is an important regulator of blood pH. This system is based on the weak acid, carbonic acid, and a weak base, primarily sodium bicarbonate. The following equations illustrate the mechanism.

$$HCl \;+\; NaHCO_3 \;\rightleftharpoons\; NaCl \;+\; H_2CO_3$$

Hydrochloric	Sodium	Sodium	Carbonic acid
acid	bicarbonate	chloride	(weak acid)
(strong acid)	(weak base)	(salt)	

$$NaOH \;+\; H_2CO_3 \;\rightleftharpoons\; H_2O \;+\; NaHCO_3$$

Sodium	Carbonic acid	Water	Sodium
hydroxide	(weak acid)		bicarbonate
(strong base)			(weak base)

Normal body processes tend to acidify the blood rather than make it more alkaline and the body needs more bicarbonate salt than it needs carbonic acid. In fact, when extracellular pH is normal (7.4), bicarbonate molecules outnumber carbonic acid by 20 to 1.

Phosphate

The **phosphate buffer system** acts in essentially the same manner as the bicarbonate buffer system. Its two components are sodium dihydrogen phosphate and sodium monohydrogen phosphate. The dihydrogen phosphate ion acts as the weak acid and is capable of buffering strong bases.

$$NaOH \;+\; NaH_2PO_4 \;\rightleftharpoons\; H_2O \;+\; Na_2HPO_4$$

Sodium	Sodium dihydrogen	Water	Sodium
hydroxide	phosphate		monohydrogen
(strong base)	(weak acid)		phosphate
			(weak base)

The monohydrogen phosphate ion acts as the weak base and is capable of buffering strong acids.

$$HCl \;+\; Na_2HPO_4 \;\rightleftharpoons\; NaCl \;+\; NaH_2PO_4$$

Hydrochloric	Sodium	Sodium	Sodium dihydrogen
acid	monohydrogen	chloride	phosphate
(strong	phosphate	(salt)	(weak acid)
acid)	(weak base)		

The phosphate buffer system is an important regulator of pH both in red blood cells and in the kidney tubular fluids. NaH_2PO_4 is formed when excess H^+ ions in the kidney tubules combine with Na_2HPO_4. In this reaction, the sodium released from Na_2HPO_4 forms sodium bicarbonate ($NaHCO_3$) and is passed into the blood. The H^+ ion that replaces sodium becomes part of the NaH_2PO_4 that is passed into the urine. This reaction is one of the mechanisms by which the kidneys help maintain pH by the acidification of urine.

Hemoglobin-Oxyhemoglobin

The **hemoglobin-oxyhemoglobin buffer system** is an effective method for buffering carbonic acid in the blood. When blood moves from the arterial end of a capillary to the venous end, the carbon dioxide given up by body cells enters the erythrocytes and combines with water to form carbonic acid (Figure 27-5). Simultaneously, oxyhemoglobin gives up its oxygen to the body cells, becomes reduced hemoglobin, and carries a negative charge. The hemoglobin anion attracts the hydrogen ion from the carbonic acid and becomes an acid that is even weaker than carbonic acid. When the hemoglobin-oxyhemoglobin system is active, the exchange reaction that occurs shows why the erythrocyte tends to give up its oxygen when pCO_2 is high.

Protein

The **protein buffer system** is the most abundant buffer in body cells and plasma. Proteins are composed of amino

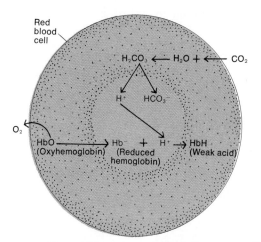

FIGURE 27-5 Hemoglobin-oxyhemoglobin buffer system. Oxyhemoglobin gives up its oxygen in an acid medium and buffers the acid. The bicarbonate (HCO_3^-) may remain in the cell and combine with potassium (K^+), or it may move out of the cell and combine with sodium (Na^+). In this way, much of the carbon dioxide is carried back to the lungs in the form of potassium bicarbonate ($KHCO_3$) or sodium bicarbonate ($NaHCO_3$).

acids. An amino acid is an organic compound that contains at least one carboxyl group (COOH) and at least one amine group (NH₂). The carboxyl group acts like an acid and can dissociate in this way.

$$NH_2-\overset{\overset{\textstyle R}{|}}{\underset{\underset{\textstyle H}{|}}{C}}-COO^-H^+$$

The hydrogen ion is then able to react with any excess hydroxide ion in the solution to form water.

The amine group has a tendency to act as a base.

$$COOH-\overset{\overset{\textstyle R}{|}}{\underset{\underset{\textstyle H}{|}}{C}}-NH_3^+OH^-$$

The hydroxide ion can dissociate, react with excess hydrogen ions, and also form water. Thus proteins act as both acidic and basic buffers.

RESPIRATIONS

Respirations also assume a role in maintaining the pH of the body. An increase in the carbon dioxide concentration in body fluids as a result of cellular respiration lowers the pH. This is illustrated by the following equation:

$$CO_2 + H_2O \rightleftharpoons H_2CO_3 \rightleftharpoons H^+ + HCO_3^-$$

Conversely, a decrease in the carbon dioxide concentration of body fluids raises the pH.

The pH of body fluids may be adjusted by a change in the rate of breathing, an adjustment that usually takes from 1 to 3 minutes. If the rate of breathing is increased, more carbon dioxide is exhaled and the blood pH rises. Slowing down the respiration rate means less carbon dioxide is exhaled, and the blood pH falls. Doubling the breathing rate increases the pH by about 0.23. Thus it can be increased from 7.4 to 7.63. Reducing the breathing rate to one-quarter its normal rate lowers the pH by 0.4. Thus it can be decreased from 7.4 to 7.0. If you consider that breathing rate can be altered up to eight times the normal rate, it should become obvious that alterations

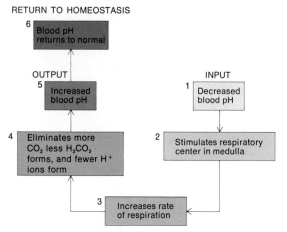

FIGURE 27-6 Relationship between pH and respirations.

in the pH of body fluids may be greatly influenced by respiration.

The pH of body fluids, in turn, affects the rate of breathing (Figure 27-6). If, for example, the blood becomes more acidic, the increase in hydrogen ions stimulates the respiratory center in the medulla and respirations increase. The same effect is achieved if the blood concentration of carbon dioxide increases. On the other hand, if the pH of the blood increases, the respiratory center is inhibited and respirations decrease. A decrease in the carbon dioxide concentration of blood has the same effect. The respiratory mechanism normally can eliminate more acid or base than can all the buffers combined.

KIDNEY EXCRETION

Since the role of the kidneys in maintaining pH has already been discussed in Chapter 26, we shall simply refer you to that section. Review Figure 26-11 very carefully.

A summary of the mechanisms that maintain body pH is presented in Exhibit 27-2.

ACID–BASE IMBALANCES

The normal blood pH range is 7.35 to 7.45. Any considerable deviation from this value falls under the category of acidosis or alkalosis. **Acidosis** is a condition in which blood pH ranges from 7.35 to 6.80. **Alkalosis** is a pH range of 7.45 to 8.00.

Respiratory acidosis is caused by hypoventilation. It occurs as a result of any condition that decreases the movement of carbon dioxide from the blood to the alveoli of the lungs and therefore causes a buildup of carbon dioxide, carbonic acid, and hydrogen ions. Such conditions include emphysema, pulmonary edema, injury to the respiratory center of the medulla, or disorders of the muscles involved in breathing. In uncompensated respiratory acidosis, the normal bicarbonate–carbonic acid ratio shifts from 20:1 to 10:1 or 8:1 and the blood pH decreases.

Respiratory alkalosis is caused by hyperventilation. It occurs as a result of any condition that stimulates the respiratory center. Such conditions include oxygen deficiency due to high altitude, severe anxiety, and aspirin overdose. In uncompensated respiratory alkalosis, the normal bicarbonate–carbonic acid ratio shifts from 20:1 to 20:0.5 and the blood pH increases.

Metabolic acidosis results from an abnormal increase in acidic metabolic products other than carbon dioxide and from the loss of bicarbonate ions from the body. Ketosis is a good example of metabolic acidosis brought on by an increase in the production of acidic metabolic products. Acidosis due to loss of bicarbonate may occur with diarrhea and renal tubular dysfunction. In uncompensated metabolic acidosis, the ratio of bicarbonate to carbonic acid is 12.5:1.

Metabolic alkalosis is caused by a nonrespiratory loss of acid by the body or excessive intake of alkaline drugs. Excessive vomiting of gastric contents results in a substantial loss of hydrochloric acid and is probably the most frequent cause of metabolic alkalosis. In uncompensated metabolic alkalosis, the ratio of bicarbonate to carbonic acid is 31.6:1.

The principal physiological effect of acidosis is depression of the central nervous system through depression of synaptic transmission. If the blood pH falls below 7, depression of the nervous system is so acute that the individual becomes disoriented and comatose. In fact, patients with severe acidosis usually die in a state of coma. On the other hand, the major physiological effect of alkalosis is overexcitability of the nervous system through facilitation of synaptic transmission. The overexcitability occurs both in the central nervous system and in peripheral nerves. Because of the overexcitability, nerves conduct impulses repetitively even when not stimulated by normal stimuli, resulting in nervousness, muscle spasms, and even convulsions.

CLINICAL APPLICATION

The primary **treatment for respiratory acidosis** is aimed at increasing the exhalation of carbon dioxide. Excessive secretions may be suctioned out of the respiratory tract, and artificial respiration may be given. **Treatment of metabolic acidosis** consists of intravenous solutions of sodium bicarbonate and correcting the cause of acidosis.

Treatment of respiratory alkalosis is aimed at increasing the level of carbon dioxide in the body. One corrective measure is to have the patient breathe into a paper bag and then rebreathe the exhaled mixture of CO_2 and oxygen from the bag. **Treatment for metabolic alkalosis** consists of giving a medication containing the chloride ion and correcting the cause of alkalosis.

Fluid Compartments and Fluid Balance (p. 686)
1. Body fluid is water and its dissolved substances.
2. About two-thirds of the body's fluid is located in cells and is called intracellular fluid (ICF).
3. The other third is called extracellular fluid (ECF). It includes interstitial fluid, plasma and lymph, cerebrospinal fluid, GI tract fluids, synovial fluid, and fluids of the eyes and ears, pleural, pericardial, and peritoneal fluids, and the glomerular filtrate.
4. Fluid balance means that the various body compartments contain the required amount of water.
5. Fluid balance and electrolyte balance are inseparable.

Water (p. 686)
1. Water is the largest single constituent in the body, varying from 45 to 75 percent of body weight depending on amount of fat present and age.
2. Primary sources of fluid intake are ingested liquids and foods and water produced by catabolism.
3. Avenues of fluid output are the kidneys, skin, lungs, and GI tract.
4. The stimulus for fluid intake is dehydration resulting in thirst sensations. Under normal conditions, fluid output is adjusted by aldosterone and ADH.

Electrolytes (p. 687)
1. Electrolytes are chemicals that dissolve in body fluids and dissociate into either cations (positive ions) or anions (negative ions).
2. Electrolyte concentration is expressed in milliequivalents per liter (meq/liter).
3. Electrolytes have a greater effect on osmosis than nonelectrolytes.
4. Plasma, interstitial fluid, and intracellular fluid contain varying kinds and amounts of electrolytes.
5. Electrolytes are needed for normal metabolism, proper fluid movement between compartments, and regulation of pH.
6. Sodium is the most abundant extracellular ion. It is involved in nerve impulse transmission, muscle contraction, and fluid and electrolyte balance. Its level is controlled by aldosterone.
7. Chloride is mainly an extracellular anion. It assumes a role in regulating osmotic pressure and forming HCl. Its level is controlled indirectly by aldosterone.
8. Potassium is the most abundant cation in intracellular fluid. It is involved in maintaining fluid volume, nerve impulse conduction, muscle contraction, and regulating pH. Its level is controlled by aldosterone.
9. Calcium is principally an extracellular ion that is a structural component of bones and teeth. It also functions in blood clotting, chemical transmitter release, muscle contraction, and heartbeat. Its level is controlled by parathyroid hormone (PTH) and calcitonin (CT).
10. Phosphate is principally an intracellular ion that is a structural component of bones and teeth. It is also required for the synthesis of nucleic acids and ATP and for buffer reactions. Its level is controlled by PTH and CT.
11. Magnesium is primarily an intracellular electrolyte that activates several enzyme systems. Its level is controlled by aldosterone.

Movement of Body Fluids (p. 689)
1. At the arterial end of a capillary, fluid moves from plasma into interstitial fluid. At the venous end, fluid moves in the opposite direction.
2. The state of near equilibrium at the arterial and venous ends

of a capillary between filtered fluid and absorbed fluid plus that picked up by the lymphatic system is referred to as Starling's law of the capillaries.
3. Fluid movement between interstitial and intracellular compartments depends on the movement of sodium and potassium and the secretion of aldosterone and ADH.
4. Fluid imbalance may lead to edema and overhydration (water intoxication).

Acid–Base Balance (p. 691)

1. The overall acid–base balance of the body is maintained by controlling the H^+ concentration of body fluids, especially extracellular fluid.
2. The normal pH of extracellular fluid is 7.35 to 7.45.
3. Homeostasis of pH is maintained by buffers, respirations, and kidney excretion.

4. The important buffer systems include: carbonic acid–bicarbonate, phosphate, hemoglobin-oxyhemoglobin, and protein.
5. An increase in rate of respirations, increases pH; a decrease in rate, decreases pH.

Acid–Base Imbalances (p. 692)

1. Acidosis is a blood pH between 7.35 and 6.80. Its principal effect is depression of the CNS.
2. Alkalosis is a blood pH between 7.45 and 8.00. Its principal effect is overexcitability of the CNS.
3. Respiratory acidosis is caused by hypoventilation; metabolic acidosis results from an abnormal increase in acid metabolic products (other than CO_2) and loss of bicarbonate.
4. Respiratory alkalosis is caused by hyperventilation; metabolic alkalosis results from nonrespiratory loss of acid or excess intake of alkaline drugs.

REVIEW QUESTIONS

1. Define body fluid. List several compartments and describe how they are separated.
2. What is meant by fluid balance? How are fluid balance and electrolyte balance related?
3. Describe the avenues of fluid intake and fluid output. Be sure to indicate volumes in each case.
4. Discuss the role of thirst in regulating fluid intake.
5. Explain how aldosterone and ADH adjust normal fluid output. What are some abnormal routes of fluid output?
6. Define a nonelectrolyte and an electrolyte. Give specific examples of each.
7. Distinguish between a cation and an anion. Give several examples of each.
8. How is the ionic concentration of a fluid expressed? Calculate the ionic concentration of sodium in a body fluid.
9. Describe the functions of electrolytes in the body.
10. Describe some of the major differences in the electrolytic concentrations of the three major fluid compartments in the body.
11. Name three important extracellular electrolytes and three important intracellular electrolytes.
12. Indicate the function and regulation of each of the following electrolytes: sodium, chloride, potassium, calcium, phos-

phate, and magnesium.
13. Describe the physiological effects of hyponatremia, hypochloremia, hypokalemia, hypomagnesemia, and hypermagnesemia.
14. Explain the forces involved in moving fluid between plasma and interstitial fluid. Summarize these forces by setting up an equation to express effective filtration pressure.
15. What is Starling's law of the capillaries?
16. Explain the factors involved in fluid movement between the interstitial fluid and the intracellular fluid.
17. Define edema. What are some of its causes?
18. Explain how the following buffer systems help to maintain the pH of body fluids: carbonic acid–bicarbonate, phosphate, hemoglobin-oxyhemoglobin, and protein.
19. Describe how respirations are related to the maintenance of pH.
20. Briefly discuss the role of the kidneys in maintaining pH.
21. Define acidosis and alkalosis. Distinguish between respiratory and metabolic acidosis and alkalosis.
22. What are the principal physiological effects of acidosis and alkalosis?
23. How are acidosis and alkalosis treated?

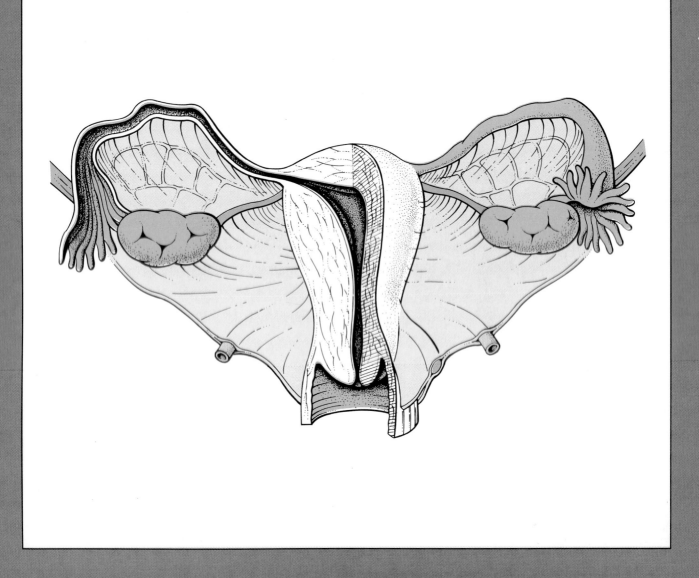

UNIT
V

This unit is designed to show you how the human organism is adapted for reproduction. It also traces the developmental sequence involved in pregnancy. Principles of inheritance and birth control conclude the unit.

Continuity

28 The Reproductive Systems

Reproduction is the mechanism by which the thread of life is sustained. It is the process by which a single cell duplicates its genetic material, allowing an organism to grow and repair itself. In this sense, reproduction maintains the life of the individual. But reproduction is also the process by which genetic material is passed from generation to generation. In this regard, reproduction maintains the continuation of the species.

The organs of the male and female reproductive systems may be grouped by function. The testes and ovaries, also called **gonads,** function in the production of gametes—sperm cells and ova, respectively. The gonads also secrete hormones. The production of gametes and their discharge into ducts classifies the gonads as exocrine glands, whereas their production of hormones classifies them as endocrine glands. The **ducts** transport, receive, and store gametes. Still other reproductive organs, called **accessory glands,** produce materials that support gametes.

MALE REPRODUCTIVE SYSTEM

The organs of the male reproductive system are the testes, or male gonads, which produce sperm; a number of ducts that either store or transport sperm to the exterior; accessory glands that add secretions constituting the semen;

and several supporting structures, including the penis (Figure 28-1).

SCROTUM

The **scrotum** is a cutaneous outpouching of the abdomen consisting of loose skin and superficial fascia. It is the supporting structure for the testes. Externally, it looks like a single pouch of skin separated into lateral portions by a median ridge called the **raphe** (RĀ-fē). Internally, it is divided by a septum into two sacs, each containing a single testis. The septum consists of superficial fascia and contractile tissue called the **dartos** (DAR-tōs), which consists of bundles of smooth muscle fibers. The dartos is also found in the subcutaneous tissue of the scrotum and is directly continuous with the subcutaneous tissue of the abdominal wall. The dartos causes wrinkling of the skin of the scrotum.

The location of the scrotum and the contraction of its muscle fibers regulate the temperature of the testes, organs that produce sperm and the hormones testosterone and inhibin. The production and survival of sperm require a temperature that is lower than body temperature. Because the scrotum is outside the body cavities, it supplies an environment about 3°F below body temperature. The

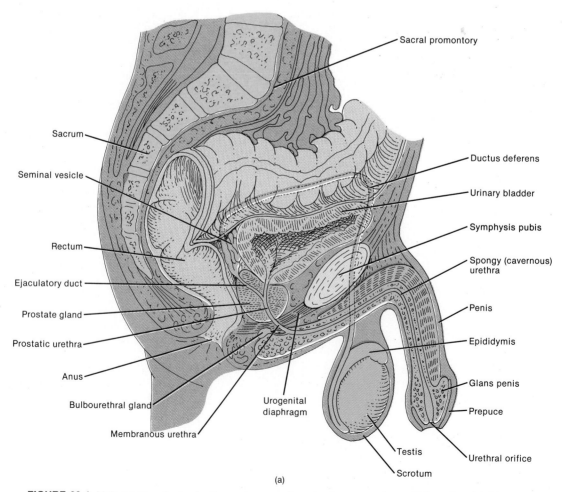

(a)

FIGURE 28-1 Male organs of reproduction and surrounding structures seen in sagittal section. (a) Diagram.

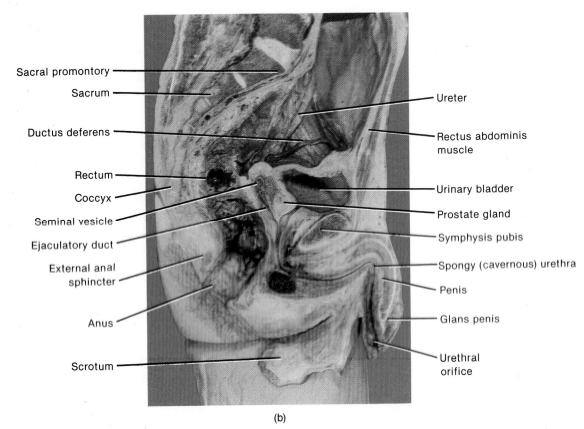

Sacral promontory

Sacrum

Ductus deferens

Rectum

Coccyx

Seminal vesicle

Ejaculatory duct

External anal sphincter

Anus

Scrotum

Ureter

Rectus abdominis muscle

Urinary bladder

Prostate gland

Symphysis pubis

Spongy (cavernous) urethra

Penis

Glans penis

Urethral orifice

(b)

FIGURE 28-1 (*Continued*) Male organs of reproduction and surrounding structures seen in sagittal section. (b) Photograph. (Courtesy of C. Yokochi and J. W. Rohen, *Photographic Anatomy of the Human Body*, 2nd ed., 1979, IGAKU-SHOIN, Ltd., Tokyo, New York.)

cremaster (krē-MAS-ter) **muscle,** a small, circular band of skeletal muscle, elevates the testes upon exposure to cold, moving them closer to the pelvic cavity where they can absorb body heat. Exposure to warmth reverses the process.

TESTES

The **testes** are paired oval glands measuring about 5 cm (2 inches) in length and 2.5 cm (1 inch) in diameter (Figure 28-2). They weigh between 10 and 15 g. The testes develop high on the embryo's posterior abdominal wall, and usually enter the scrotum by 32 weeks. Full descent is not complete until just prior to birth.

CLINICAL APPLICATION

When the testes do not descend, the condition is referred to as **cryptorchidism** (krip-TOR-ki-dizm). The condition occurs in about 3 percent of full-term infants and about 30 percent of premature infants. Cryptorchidism results in sterility because the cells involved in the initial development of sperm cells are destroyed by the higher body temperature of the pelvic cavity. Undescended testes can be placed in the scrotum by administering hormones or by surgical means prior to puberty without ill effects.

The testes are covered by a dense layer of white fibrous tissue, the **tunica albuginea** (al'-byoo-JIN-ē-a), that extends inward and divides each testis into a series of inter-

nal compartments called **lobules.** Each of the 200 to 300 lobules contains one to three tightly coiled tubules, the convoluted **seminiferous tubules,** that produce sperm by a process called **spermatogenesis.** This process is considered in detail in Chapter 29.

A cross section through a seminiferous tubule reveals that it is packed with sperm cells in various stages of development (Figure 28-3). The most immature cells, the **spermatogonia,** are located against the basement membrane. These cells collectively constitute the germinal epithelium. Toward the lumen of the tube, one can see layers of progressively more mature cells. In order of advancing maturity, these are primary spermatocytes, secondary spermatocytes, and spermatids. By the time a **sperm cell,** or **spermatozoon** (sper'-ma-tō-ZŌ-on), has reached full maturity, it is in the lumen of the tubule and begins to be moved through a series of ducts. Embedded between the developing sperm cells in the tubules are **sustentacular** (sus'-ten-TAK-yoo-lar) **cells.** These cells produce secretions that supply nutrients to the spermatozoa and secrete the hormone inhibin. Between the seminiferous tubules are clusters of **interstitial endocrinocytes.** These cells secrete the male hormone testosterone, the most important androgen.

Spermatozoa

Spermatozoa are produced or matured at the rate of about 300 million per day and, once ejaculated, have a life expectancy of about 48 hours within the female reproductive

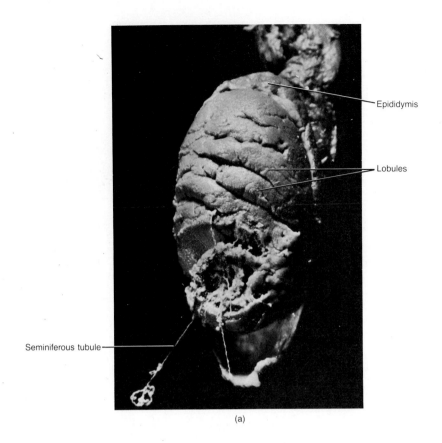

(a)

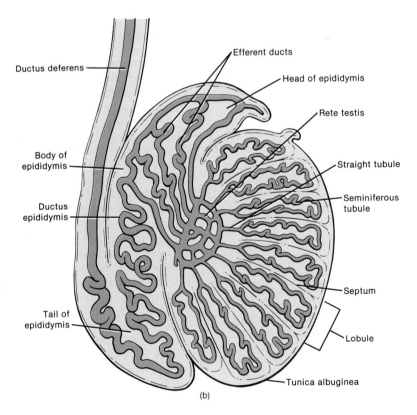

(b)

FIGURE 28-2 External and internal anatomy of testes. (a) Photograph of external view. (Courtesy of C. Yokochi and J. W. Rohen, *Photographic Anatomy of the Human Body,* 2nd ed., 1979, IGAKU-SHOIN, Tokyo, New York.) (b) Diagram of a sagittal section illustrating the internal anatomy.

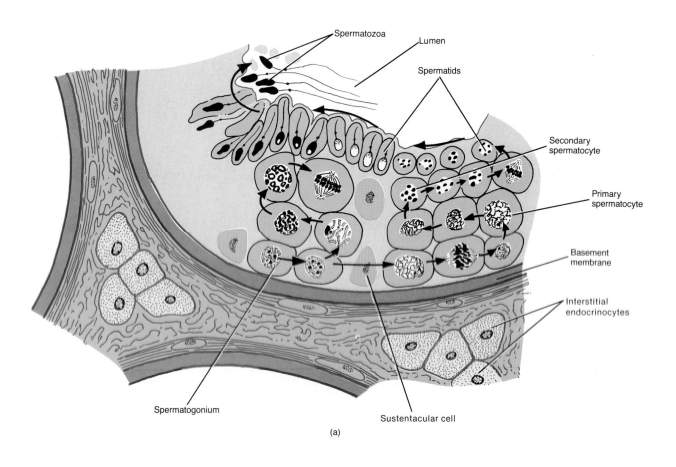

(a)

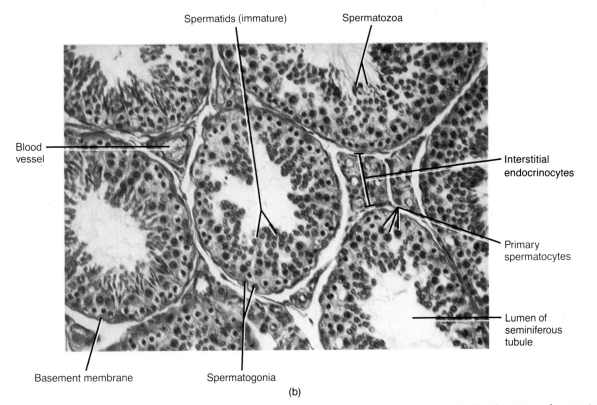

(b)

FIGURE 28-3 Histology of the testes. (a) Diagram of a cross section of a portion of a seminiferous tubule showing the stages of spermatogenesis. (b) Photomicrograph of an enlarged aspect of several seminiferous tubules at a magnification of 350×. (© 1983 by Michael H. Ross. Used by permission.)

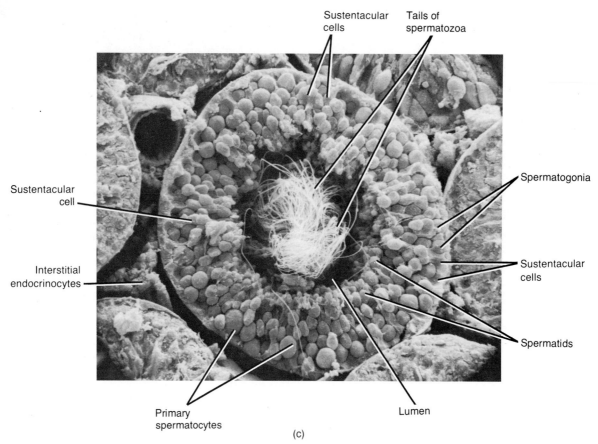

Sustentacular cells

Tails of spermatozoa

Sustentacular cell

Interstitial endocrinocytes

Spermatogonia

Sustentacular cells

Spermatids

Primary spermatocytes

Lumen

(c)

FIGURE 28-3 (*Continued*) Histology of the testes. (c) Scanning electron micrograph of a seminiferous tubule at a magnification of 663×. (Courtesy of Richard K. Kessel and Randy H. Kardon, *Tissues and Organs: A Text-Atlas of Scanning Electron Microscopy.* Copyright © 1979 by Scientific American, Inc.)

tract. A spermatozoon is highly adapted for reaching and penetrating a female ovum. It is composed of a head, a midpiece, and a tail (Figure 28-4). Within the **head** are the nuclear material and the *acrosome,* which contains enzymes (hyaluronidase and proteinases) that effect penetration of the sperm cell into the ovum. Numerous mitochondria in the **midpiece** carry on the metabolism that provides energy for locomotion. The **tail,** a typical flagellum, propels the sperm along its way.

Testosterone and Inhibin

Secretions of the anterior pituitary gland assume a major role in the developmental changes associated with puberty. At the onset of puberty the anterior pituitary starts to secrete gonadotropic hormones called follicle-stimulating hormone (FSH) and luteinizing hormone (LH). Their release is controlled from the hypothalamus by gonadotropin releasing factor (GnRF). Once secreted, the gonadotropic hormones have profound effects on male reproductive organs. FSH acts on the seminiferous tubules to initiate spermatogenesis and stimulate sustentacular cells. LH also assists the seminiferous tubules to develop mature sperm, but its chief function is to stimulate the interstitial endocrinocytes to secrete the hormone testosterone (tes-TOS-te-rōn).

Testosterone is synthesized from cholesterol or acetyl coenzyme A in the testes. It is the principal male hormone (androgen) and has a number of effects on the body. It controls the development, growth, and maintenance of the male sex organs. It also stimulates bone growth, protein anabolism, sexual behavior, final maturation of sperm, and the development of male secondary sex characteristics. These characteristics, which appear at puberty, include muscular and skeletal development resulting in wide shoulders and narrow hips; body hair patterns that include pubic hair, axillary and chest hair (within hereditary limits), facial hair, and temporal hairline recession; and enlargement of the thyroid cartilage of the larynx producing deepening of the voice. Testosterone also stimulates descent of the testes just prior to birth.

The interaction of LH with testosterone illustrates the operation of another negative feedback system (Figure 28-5). LH stimulates the production of testosterone. But once the testosterone concentration in the blood reaches a certain level, it inhibits the release of GnRF by the hypothalamus. This inhibition, in turn, inhibits the release of LH by the anterior pituitary. Thus testosterone production is decreased. However, once the testosterone concentration in the blood decreases to a certain level, GnRF is released by the hypothalamus. This release of LH stimulates the release of LH by the anterior pituitary, and

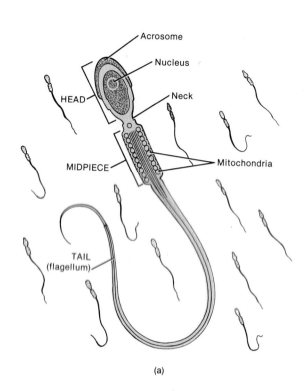

(a)

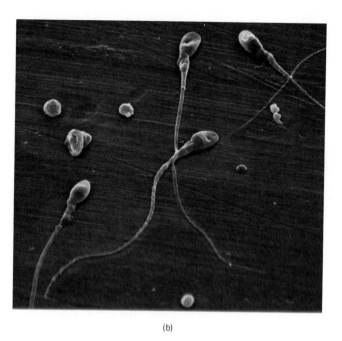

(b)

FIGURE 28-4 Spermatozoa. (a) Diagram of the parts of a spermato-zoon. (b) Scanning electron micrograph of several spermatozoa at a magnification of 2,000×. (Courtesy of Fisher Scientific Company and S.T.E.M. Laboratories, Inc., Copyright 1975.)

stimulates testosterone production. Thus the testosterone-LH cycle is complete.

Inhibin is a protein hormone that has a direct effect on the anterior pituitary by inhibiting the secretion of FSH. FSH brings about spermatogenesis and stimulates sustentacular cells. Once the degree of spermatogenesis required for male reproductive functions has been achieved, sustentacular cells secrete inhibin. The hormone feeds back negatively to the anterior pituitary to inhibit FSH and thus to decrease spermatogenesis (Figure

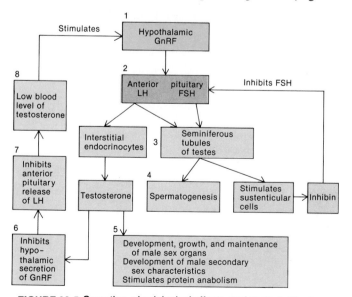

FIGURE 28-5 Secretion, physiological effects, and control of testos-terone and inhibin.

28-5). If spermatogenesis is proceeding too slowly, lack of inhibin production permits FSH secretion and an in-creased rate of spermatogenesis.

DUCTS

Ducts of the Testis

When the sperm mature, they are moved through the convoluted seminiferous tubules to the **straight tubules** (see Figure 28-2b). The straight tubules lead to a network of ducts in the testis called the **rete (RĒ-tē) testis.** Some of the cells lining the rete testis possess cilia that probably help move the sperm along. The sperm are next trans-ported out of the testis.

Epididymis

The sperm are transported out of the testis through a series of coiled **efferent ducts** in the epididymis that empty into a single tube called the ductus epididymis. At this point, the sperm are morphologically mature.

The two **epididymides** (ep'-i-DID-i-mi-dēz') are comma-shaped organs. Each lies along the posterior bor-der of the testis (see Figures 28-1 and 28-2) and consists mostly of a tightly coiled tube, the **ductus epididymis** (ep'-i-DID-i-mis). The larger, superior portion of the epi-didymis is known as the **head.** It consists of the efferent ducts that empty into the ductus epididymis. The **body** of the epididymis contains the ductus epididymis. The **tail** is the smaller, inferior portion. Within the tail, the ductus epididymis continues as the ductus deferens.

The ductus epididymis measures about 6 m (20 ft) in length and 1 mm in diameter. It is tightly packed within the epididymis, which measures only about 3.8 cm (1.5 inches). The ductus epididymis is lined with pseudostratified columnar epithelium, and its wall contains smooth muscle. The free surfaces of the columnar cells contain long, branching microvilli called *stereocilia* (Figure 28-6).

Functionally, the ductus epididymis is the site of sperm maturation. They require between 18 hours and 10 days to complete their maturation, that is, to become capable of fertilizing an ovum. The ductus epididymis also stores spermatozoa and propels them toward the urethra during ejaculation by peristaltic contraction of its smooth muscle. Spermatozoa may remain in storage in the ductus epididymis for up to four weeks. After that, they are reabsorbed.

Ductus (Vas) Deferens

Within the tail of the epididymis, the ductus epididymis becomes less convoluted, its diameter increases, and at this point it is referred to as the **ductus (vas) deferens** or **seminal duct** (see Figure 28-2). The ductus deferens, about 45 cm long (18 inches), ascends along the posterior border of the testis, penetrates the inguinal canal, and enters the pelvic cavity, where it loops over the side and down the posterior surface of the urinary bladder (see Figure 28-1). The dilated terminal portion of the ductus deferens is known as the **ampulla** (am-POOL-la). The ductus deferens is lined with pseudostratified epithelium and contains a heavy coat of three layers of muscle. Functionally, the ductus deferens stores sperm for up to several months and propels them toward the urethra during ejaculation by peristaltic contractions of the muscular coat.

CLINICAL APPLICATION

One method of sterilization of males is called **vasectomy,** in which a portion of each ductus deferens is removed. In the procedure, an incision is made in the scrotum, the ducts are located, each is tied in two places, and the portion between the ties is excised (see Figure 29-17a). Although sperm production continues in the testes, the sperm cannot reach the exterior because the ducts are cut and the sperm degenerate.

Traveling with the ductus deferens as it ascends in the scrotum are the testicular artery, autonomic nerves, veins that drain the testes, lymphatics, and the cremaster muscle. These structures constitute the **spermatic cord,** a supporting structure of the male reproductive system.

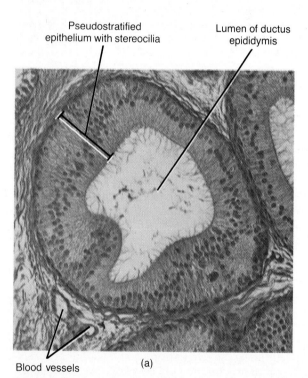

Pseudostratified epithelium with stereocilia

Lumen of ductus epididymis

Blood vessels

(a)

FIGURE 28-6 Histology of the ductus epididymis. (a) Photomicrograph of the ductus epididymis seen in cross section at a magnification of 160×. (b) Photomicrograph of an enlarged aspect of the mucosa of the ductus epididymis at a magnification of 350×. (© 1983 by Michael H. Ross. Used by permission.)

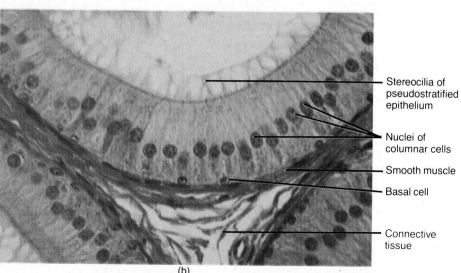

Stereocilia of pseudostratified epithelium

Nuclei of columnar cells

Smooth muscle

Basal cell

Connective tissue

(b)

The cremaster muscle, which also surrounds the testes, elevates the testes during sexual stimulation and exposure to cold. The spermatic cord passes through the **inguinal** (IN-gwin-al) **canal,** a slitlike passageway in the anterior abdominal wall just superior to the medial half of the inguinal ligament.

CLINICAL APPLICATION

The area of the inguinal canal is a weak spot in the abdominal wall. It is frequently the site of an **inguinal hernia**—a rupture or separation of a portion of the abdominal wall resulting in the protrusion of a part of an organ.

Ejaculatory Duct

Posterior to the urinary bladder are the **ejaculatory** (ē-JAK-yoo-la-tō'-rē) **ducts** (Figure 28-7). Each duct is about 2 cm (1 inch) long and is formed by the union of the duct from the seminal vesicle and ductus deferens. The ejaculatory ducts eject spermatozoa into the prostatic urethra.

Urethra

The **urethra** is the terminal duct of the system, serving as a passageway for spermatozoa or urine. In the male, the urethra passes through the prostate (PROS-tāt) gland, the urogenital diaphragm, and the penis. It measures about 20 cm (8 inches) in length and is subdivided into three parts (see Figures 28-1 and 28-8). The **prostatic urethra** is 2 to 3 cm (1 inch) long and passes through the prostate gland. It continues inferiorly and as it passes through the urogenital diaphragm, a muscular partition between the two ischiopubic rami, it is known as the **membranous** (MEM-bra-nus) **urethra.** The membranous portion is about 1 cm (½ inch) in length. As it passes through the corpus spongiosum of the penis, it is known as the **spongy (cavernous) urethra.** This portion is about 15 cm (6 inches) long. The spongy urethra enters the bulb of the penis and terminates at the external **urethral orifice.**

ACCESSORY GLANDS

Whereas the ducts of the male reproductive system store and transport sperm cells, the **accessory glands** secrete

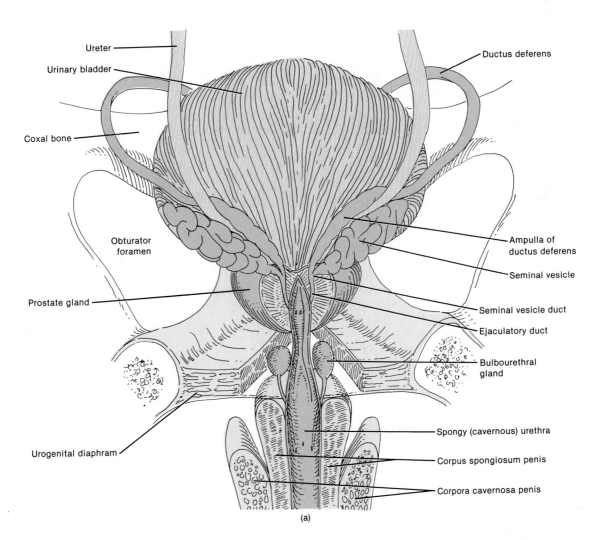

(a)

FIGURE 28-7 Male reproductive organs in relation to surrounding structures seen in posterior view. (a) Diagram.

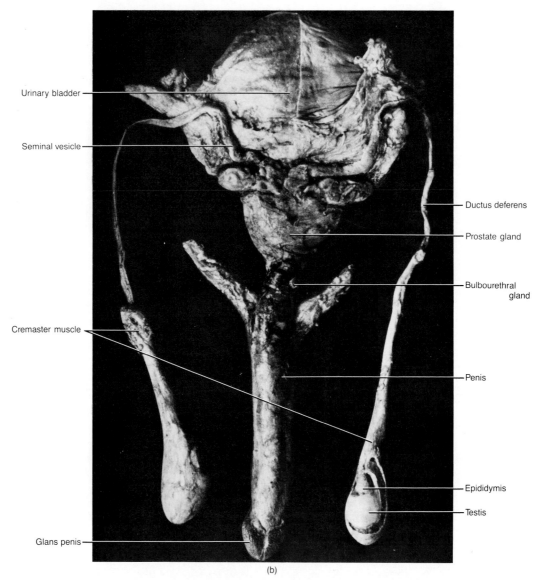

(b)

FIGURE 28-7 (*Continued*) Male reproductive organs in relation to surrounding structures seen in posterior view. (b) Photograph. (Courtesy of C. Yokochi and J. W. Rohen, *Photographic Anatomy of the Human Body,* 2nd ed., 1979, IGAKU-SHOIN, Ltd., Tokyo, New York.)

the liquid portion of semen. The paired **seminal vesicles** (VES-i-kuls) are convoluted pouchlike structures, about 5 cm (2 inches) in length, lying posterior to and at the base of the urinary bladder in front of the rectum (see Figure 28-7). They secrete an alkaline, viscous fluid, rich in the sugar fructose, and pass it into the ejaculatory duct. This secretion contributes to sperm viability. It constitutes about 60 percent of the volume of semen.

The **prostate gland** is a single, doughnut-shaped gland about the size of a chestnut (see Figure 28-7). It is inferior to the urinary bladder and surrounds the superior portion of the urethra. The prostate secretes an alkaline fluid into the prostatic urethra. The prostatic secretion constitutes 13 to 33 percent of the volume of semen and contributes to sperm motility.

The paired **bulbourethral** (bul'-bō-yoo-RĒ-thral) **glands** are about the size of peas. They are located beneath the prostate on either side of the membranous urethra (see Figure 28-7). The bulbourethral glands secrete mucus

for lubrication and a substance that neutralizes urine. Their ducts open into the spongy urethra.

SEMEN

Semen (seminal fluid) is a mixture of sperm and the secretions of the seminal vesicles, the prostate gland, and the bulbourethral glands. The average volume of semen for each ejaculation is 2.5 to 6 ml, and the average range of spermatozoa ejaculated is 50 to 100 million/ml. When the number of spermatozoa falls below 20 million/ml, the male is likely to be sterile. The very large number is required because only a small percentage eventually reach the ovum. And, although only a single spermatozoon fertilizes an ovum, fertilization seems to require the combined action at the ovum of a larger number of them. The intercellular material of the cells covering the ovum presents a barrier to the sperm. This barrier is digested by the hyaluronidase and proteinases secreted by the acrosomes of sperm. However, it appears that a single sperm does not produce enough of these enzymes to dissolve the barrier. A passageway through which one may enter can be created only by the action of many.

Semen has a slightly alkaline pH of 7.35 to 7.50. The prostatic secretion gives semen a milky appearance, and fluids from the seminal vesicles and bulbourethral glands give it a mucoid consistency. Semen provides spermatozoa with a transportation medium and nutrients. It neutralizes the acid environment of the male urethra and the female vagina. It also contains enzymes which activate sperm after ejaculation.

Semen has been shown to contain an antibiotic, called *seminalplasmin,* which has the ability to destroy a number of bacteria. Its antimicrobial activity has been described as similar to that currently exerted by penicillin, streptomycin, and tetracyclines. Since both semen and the lower female reproductive tract contain bacteria, seminalplasmin may keep these bacteria under control to help ensure fertilization.

Once ejaculated into the vagina, liquid semen coagu-lates rapidly because of a clotting enzyme produced by the prostate that acts on a substance produced by the seminal vesicle. This clot liquefies in a few minutes because of another enzyme produced by the prostate gland. It is not clear why semen coagulates and liquefies in this manner.

PENIS

The **penis** is used to introduce spermatozoa into the vagina (Figure 28-8). The distal end of the penis is a slightly enlarged region called the **glans,** which means shaped like an acorn. Covering the glans is the loosly fitting **prepuce** (PRĒ-pyoos), or **foreskin.**

CLINICAL APPLICATION

Circumcision (*circumcido* = to cut around) is a surgical procedure in which part or all of the prepuce is removed. It is usually performed by the third or fourth day after birth (or on the eighth day as part of a Jewish religious rite).

Internally, the penis is composed of three cylindrical masses of tissue bound together by fibrous tissue. The two dorsolateral masses are called the **corpora cavernosa penis.** The smaller midventral mass, the **corpus spongiosum penis,** contains the spongy urethra. All three masses of tissue are erectile and contain blood sinuses. Under the influence of sexual stimulation, the arteries supplying the penis dilate, and large quantities of blood enter the blood sinuses. Expansion of these spaces compresses the veins draining the penis so most entering blood is retained. These vascular changes result in an **erection,** a parasympathetic reflex. The penis returns to its flaccid state when the arteries constrict and pressure on the veins is relieved (Chapter 29). During ejaculation, a sympathetic reflex, the smooth muscle sphincter at the base of the urinary bladder is closed because of the higher pressure in the urethra caused by expansion of the corpus spongiosum penis. Thus urine is not expelled during ejaculation and semen does not enter the urinary bladder.

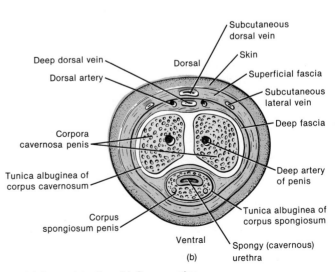

FIGURE 28-8 Internal structure of the penis. (a) Coronal section. (b) Cross section.

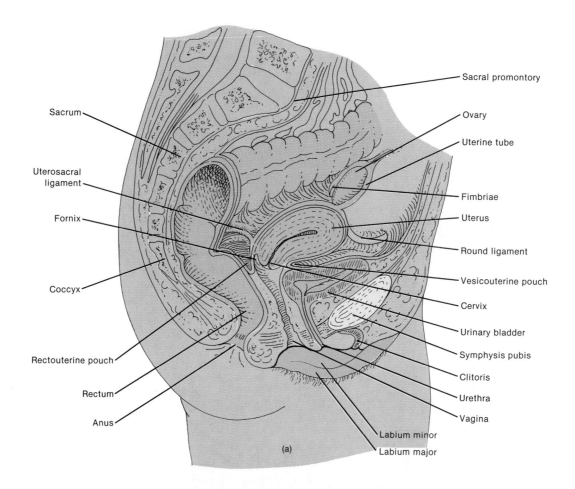

(a)

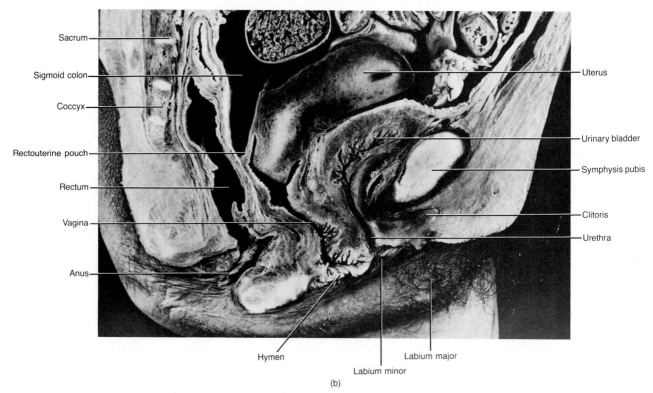

(b)

FIGURE 28-9 Female organs of reproduction and surrounding structures seen in sagittal section. (a) Diagram. (b) Photograph. (Courtesy of C. Yokochi and J. W. Rohen, *Photographic Anatomy of the Human Body,* 2nd ed., 1979, IGAKU-SHOIN, Ltd., Tokyo, New York.)

FEMALE REPRODUCTIVE SYSTEM

The female organs of reproduction include the ovaries, which produce ova (eggs); the uterine (fallopian) tubes, which transport the ova to the uterus (womb); the vagina; and external organs that constitute the vulva, or pudendum (Figure 28-9). The mammary glands also are considered part of the female reproductive system.

OVARIES

The **ovaries,** or female gonads, are paired glands resembling unshelled almonds in size and shape. They are positioned in the upper pelvic cavity, one on each side of the uterus. The ovaries are maintained in position by a series of ligaments (Figures 28-10 and 28-11). They are attached to the broad ligament of the uterus, which is itself part of the parietal peritoneum, by a double-layered fold of peritoneum called the **mesovarium** which surrounds the ovary and ovarian ligament. The ovaries are anchored to the uterus by the **ovarian ligament** and are attached to the pelvic wall by the **suspensory ligament.** Each ovary also contains a hilus, the point of entrance for blood vessels and nerves.

The microscope reveals that each ovary consists of the following parts (Figure 28-12).

1. Germinal epithelium. A layer of simple cuboidal epithelium that covers the free surface of the ovary and serves as a source of ovarian follicles.

2. Tunica albuginea. A capsule of collagenous connective tissue immediately deep to the germinal epithelium.

3. Stroma. A region of connective tissue deep to the tunica albuginea and composed of an outer, dense layer called the *cortex* and an inner, loose layer known as the *medulla.* The cortex contains ovarian follicles.

4. Ovarian follicles. Ova and their surrounding tissues in various stages of development.

5. Graafian (GRAF-ē-an) follicle. An endocrine gland made up of a mature ovum and its surrounding tissues. The Graafian follicle secretes hormones called estrogens.

6. Corpus luteum. Glandular body that develops from a Graafian follicle after extrusion of an ovum (ovulation). The corpus luteum produces the hormones progesterone, estrogens, and relaxin.

The ovaries produce ova, discharge ova (ovulation), and secrete the female sex hormones, progesterone, estrogens, and relaxin. They are analogous to the testes of the male reproductive system.

UTERINE TUBES

The female body contains two **uterine (fallopian) tubes** that extend laterally from the uterus and transport the ova from the ovaries to the uterus (see Figure 28-10). Measuring about 10 cm (4 inches) long, the tubes are positioned between the folds of the broad ligaments of the uterus. The funnel-shaped open distal end of each tube, called the **infundibulum,** lies close to the ovary but is not attached to it and is surrounded by a fringe of fingerlike projections called **fimbriae** (FIM-brē-ē). From the infundibulum the uterine tube extends medially and inferiorly and attaches to the superior lateral angle of

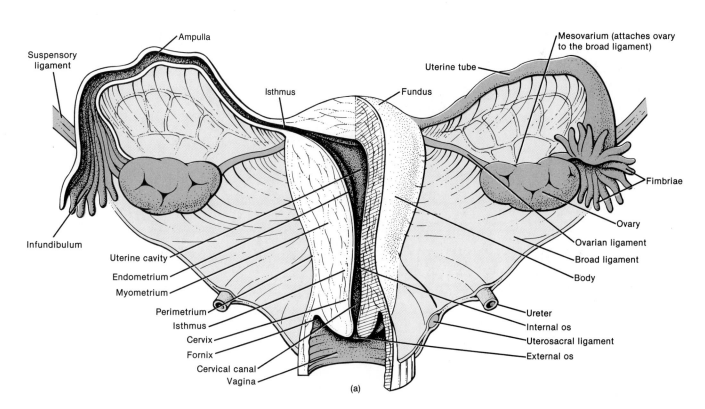

FIGURE 28-10 Uterus and associated structures seen in anterior view. (a) Diagram. The left side of the figure has been sectioned to show internal structures.

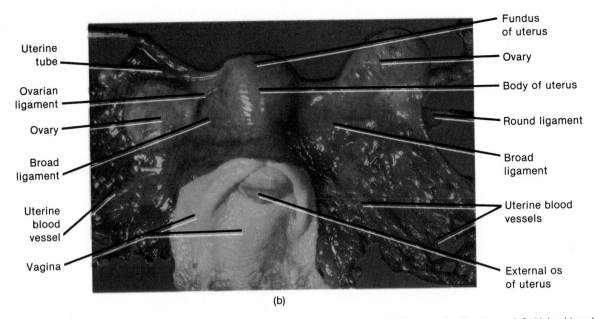

Uterine tube
Ovarian ligament
Ovary
Broad ligament
Uterine blood vessel
Vagina

Fundus of uterus
Ovary
Body of uterus
Round ligament
Broad ligament
Uterine blood vessels
External os of uterus

(b)

FIGURE 28-10 (*Continued*) Uterus and associated structures seen in anterior view. (b) Photograph. (Courtesy of C. Yokochi and J. W. Rohen, *Photographic Anatomy of the Human Body,* 2nd ed., 1979, IGAKU-SHOIN, Ltd., Tokyo, New York.)

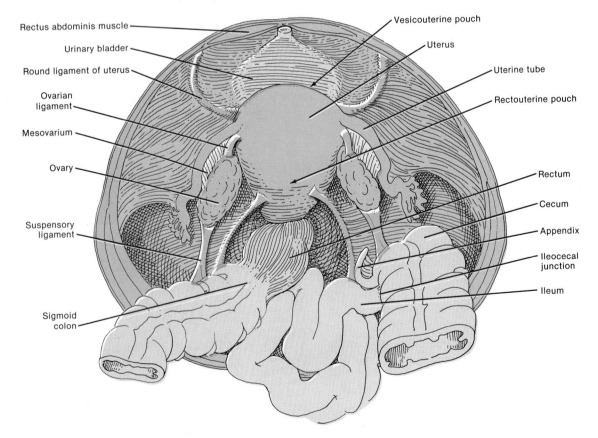

Rectus abdominis muscle
Urinary bladder
Round ligament of uterus
Ovarian ligament
Mesovarium
Ovary
Suspensory ligament
Sigmoid colon

Vesicouterine pouch
Uterus
Uterine tube
Rectouterine pouch
Rectum
Cecum
Appendix
Ileocecal junction
Ileum

FIGURE 28-11 Female pelvis viewed from above.

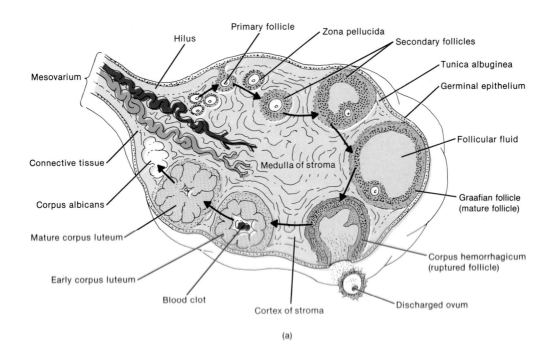

(a)

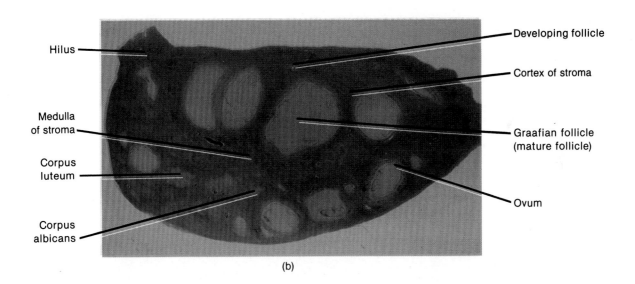

(b)

FIGURE 28-12 Histology of the ovary. (a) Diagram of the parts of an ovary seen in sectional view. The arrows indicate the sequence of developmental stages that occur as part of the ovarian cycle. (b) Photograph of an ovary in sectional view. (Courtesy of C. Yokochi and J. W. Rohen, *Photographic Anatomy of the Human Body,* 2nd ed., 1979, IGAKU-SHOIN, Ltd., Tokyo, New York.)

the uterus. The **ampulla** of the uterine tube is the widest, longest portion, making up about two-thirds of its length. The **isthmus** of the uterine tube is the short, narrow, thick-walled portion that joins the uterus.

Histologically, the uterine tubes are composed of three layers. The internal **mucosa** contains ciliated columnar cells and secretory cells, which are believed to aid the movement and nutrition of the ovum. The middle layer, the **muscularis,** is composed of a thick, circular region

of smooth muscle and an outer, thin, longitudinal region of smooth muscle. Peristaltic contractions of the muscularis and the ciliated action of the mucosa help move the ovum down into the uterus. The outer layer of the uterine tubes is a serous membrane, the **serosa.**

About once a month an immature ovum ruptures from the surface of the ovary near the infundibulum of the uterine tube, a process called **ovulation.** The ovum is swept into the tube by the ciliary action of the epithelium of

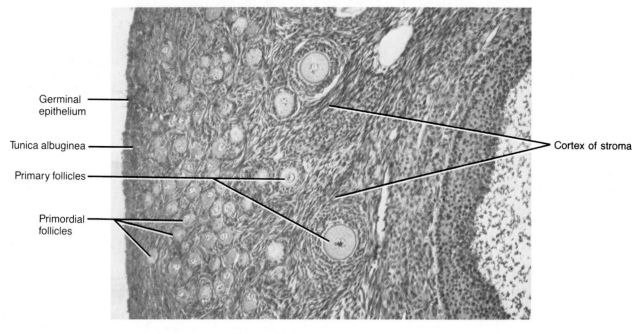

Germinal epithelium

Tunica albuginea

Primary follicles

Primordial follicles

Cortex of stroma

(c)

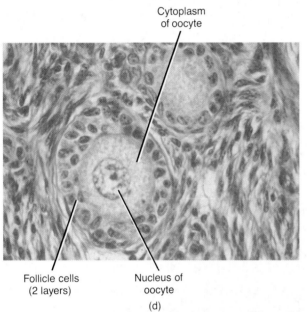

Cytoplasm of oocyte

Follicle cells (2 layers)

Nucleus of oocyte

(d)

FIGURE 28-12 (*Continued*) Histology of the ovary. (c) Photomicrograph of the cortex of the stroma of an ovary at a magnification of 60×. (d) Photomicrograph of an enlarged aspect of a primary follicle at a magnification of 300×. (e) Photomicrograph of an enlarged aspect of a secondary follicle at a magnification of 160×. The theca interna and theca externa are connective tissue coverings around the secondary follicle. (Photomicrographs © 1983 by Michael H. Ross. Used by permission.)

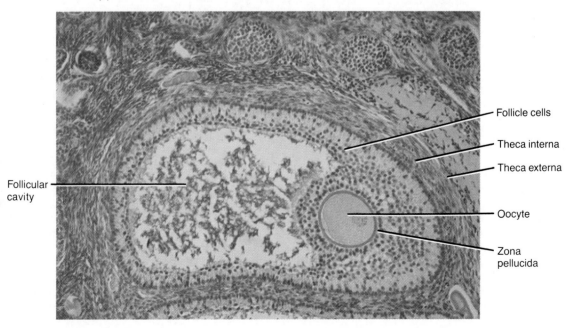

Follicle cells

Theca interna

Theca externa

Oocyte

Zona pellucida

Follicular cavity

(e)

the infundibulum. It is then moved along the tube by ciliary action supplemented by the peristaltic contractions of the muscularis. If the ovum is fertilized by a sperm cell, it usually occurs in the ampulla of the uterine tube. Fertilization may occur at any time up to about 24 hours following ovulation. The fertilized ovum, now referred to as a blastocyst, descends into the uterus within 7 days. An unfertilized ovum disintegrates.

CLINICAL APPLICATION

Fertilization may occur while the ovum is free in the pelvic cavity, and implantation may take place on one of the pelvic viscera. *Pelvic implantations* usually fail because the developing fertilized ovum does not make vascular connection with the maternal blood supply. On occasion, a fertilized ovum fails to descend to the uterus and implants in the uterine tube. In the case of a *tubular implantation,* the pregnancy must be terminated surgically before the tube ruptures. Both pelvic and tubular implantations are referred to as **ectopic** (ek-TOP-ik) **pregnancies.**

UTERUS

The site of menstruation, implantation of a fertilized ovum, development of the fetus during pregnancy, and labor is the **uterus.** Situated between the urinary bladder and the rectum, the uterus is shaped like an inverted pear (see Figures 28-9, 28-10, and 28-11). Before the first pregnancy, the adult uterus measures approximately 7.5 cm (3 inches) long, 5 cm (2 inches) wide, and 2.5 cm (1 inch) thick.

Anatomical subdivisions of the uterus include the dome-shaped portion above the uterine tubes called the **fundus,** the major tapering central portion called the **body,** and the inferior narrow portion opening into the vagina called the **cervix.** Between the body and the cervix is the **isthmus** (IS-mus), a constricted region about 1 cm (½ inch) long. The interior of the body of the uterus is called the **uterine cavity,** and the interior of the narrow cervix is called the **cervical canal.** The junction of the isthmus with the cervical canal is the **internal os.** The **external os** is the place where the cervix opens into the vagina.

Normally the uterus is flexed between the uterine body and the cervix. This is called **anteflexion.** In this position, the body of the uterus projects anteriorly and slightly superiorly over the urinary bladder, and the cervix projects inferiorly and posteriorly and enters the anterior wall of the vagina at nearly a right angle. Several structures that are either extensions of the parietal peritoneum or fibromuscular cords, referred to as ligaments, maintain the position of the uterus. The paired **broad ligaments** are double folds of parietal peritoneum attaching the uterus to either side of the pelvic cavity. Uterine blood vessels and nerves pass through the broad ligaments. The paired **uterosacral ligaments,** also peritoneal extensions, lie on either side of the rectum and connect the uterus to the sacrum. The **cardinal (lateral cervical) ligaments** extend below the bases of the broad ligaments between the pelvic wall and the cervix and vagina. These ligaments contain smooth muscle, uterine blood vessels, and nerves

and are the chief ligaments that maintain the position of the uterus and help to keep it from dropping down into the vagina. The **round ligaments** are bands of fibrous connective tissue between the layers of the broad ligament. They extend from a point on the uterus just below the uterine tubes to a portion of the external genitalia.

CLINICAL APPLICATION

Although the ligaments normally maintain the anteflexed position of the uterus, they also afford the uterine body some movement. As a result, the uterus may become malpositioned. A posterior tilting of the uterus is called **retroflexion.**

Histologically, the uterus consists of three layers of tissue. The outer layer, the **perimetrium** or **serosa,** is part of the visceral peritoneum. Laterally, it becomes the broad ligament. Anteriorly, it is reflected over the urinary bladder and forms a shallow pouch, the **vesicouterine** (ves'-i-kō-YOO-ter-in) **pouch** (see Figure 28-9). Posteriorly, it is reflected onto the rectum and forms a deep pouch, the **rectouterine** (rek-tō-YOO-ter-in) **pouch**—the lowest point in the pelvic cavity.

The middle layer of the uterus, the **myometrium,** forms the bulk of the uterine wall (Figure 28-13). This layer consists of three layers of smooth muscle fibers and is thickest in the fundus and thinnest in the cervix. During childbirth, coordinated contractions of the muscles help to expel the fetus from the body of the uterus.

The inner layer of the uterus, the **endometrium,** is a mucous membrane composed of two principal layers. The *stratum functionalis,* the layer closer to the uterine cavity, is shed during menstruation. The second layer, the *stratum basalis* (bā-SAL-is), is permanent and produces a new functionalis following menstruation. The endometrium contains numerous glands.

CLINICAL APPLICATION

More and more physicians are beginning to use **colposcopy** (kol-POS-ko-pē) in order to evaluate the status of the mucosa of the vagina and cervix. Colposcopy is the direct examination of the vaginal and cervical mucosa with a magnifying device (a colposcope) similar to a low-power binocular microscope. Various instruments that magnify the mucous membrane from about 6 to 40 times its actual size are commercially available. The application of a 3 percent solution of acetic acid removes mucus and enhances the appearance of mucosal columnar epithelium.

Blood is supplied to the uterus by branches of the internal iliac artery called *uterine arteries.* Branches called *arcuate arteries* are arranged in a circular fashion in the myometrium and give off *radial arteries* that penetrate deeply into the myometrium (Figure 28-14). Just before the branches enter the endometrium, they divide into two kinds of arterioles. The *straight arteriole* terminates in the basalis and supplies it with the materials necessary to regenerate the functionalis. The *spiral arteriole* penetrates the functionalis and changes markedly during the menstrual cycle. The uterus is drained by the *uterine veins.*

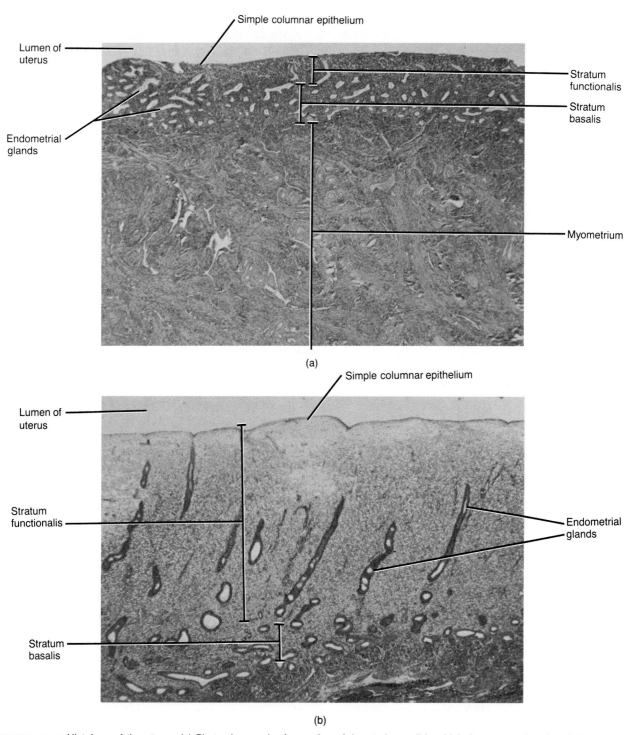

FIGURE 28-13 Histology of the uterus. (a) Photomicrograph of a portion of the uterine wall in which the stratum functionalis is in an early stage of proliferation at a magnification of 25×. (b) Photomicrograph of a portion of the uterine wall in which the stratum functionalis is in an advanced stage of proliferation at a magnification of 25×. Note the increased thickness of the stratum functionalis and the long, straight nature of the glands. (© 1983 by Michael H. Ross. Used by permission.)

ENDOCRINE RELATIONS: MENSTRUAL AND OVARIAN CYCLES

The principal events of the menstrual cycle can be correlated with those of the ovarian cycle and changes in the endometrium. All are hormonally controlled events.

The **menstrual cycle** is a series of changes in the endometrium of a nonpregnant female. Each month the endometrium is prepared to receive a fertilized ovum. An implanted fertilized ovum eventually develops into an embryo and then a fetus, which normally remains in the uterus until delivery. If no fertilization occurs, the stratum

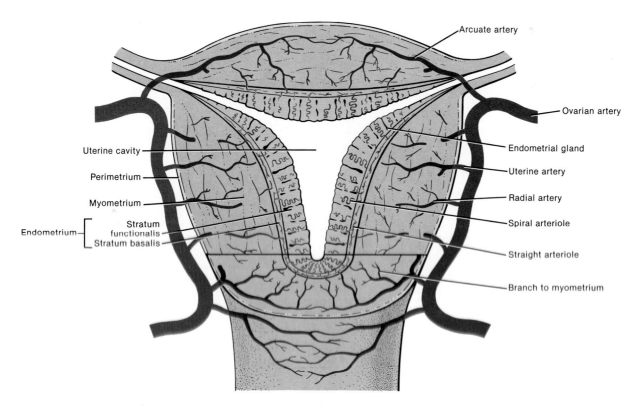

FIGURE 28-14 Blood supply of the uterus.

functionalis portion of the endometrium is shed. The **ovarian cycle** is a monthly series of events associated with the maturation of an ovum.

Hormonal Control

The menstrual cycle, ovarian cycle, and other changes associated with puberty in the female are controlled by a regulating factor from the hypothalamus called gonadotropin releasing factor (GnRF). Its influence is shown in Figure 28-15. GnRF stimulates the release of follicle-stimulating hormone (FSH) from the anterior pituitary. FSH stimulates the initial development of the ovarian follicles and the secretion of estrogens by the follicles. GnRF also stimulates the release of another anterior pituitary hormone—the luteinizing hormone (LH), which stimulates the further development of ovarian follicles, brings about ovulation, and stimulates the production of estrogens, progesterone, and relaxin by ovarian cells.

At least six different estrogens have been isolated from the plasma of human females. However, only three are present in significant quantities. These are *β-estradiol, estrone,* and *estriol.* Of these, *β-estradiol* exerts the major effect. It is synthesized from cholesterol or acetyl coenzyme A in the ovaries. As reference is made to estrogens in subsequent discussions, keep in mind that *β-estradiol* is the principal estrogen.

Estrogens, the hormones of growth, have three main functions. First is the development and maintenance of female reproductive structures, especially the endometrial lining of the uterus, secondary sex characteristics, and the breasts. The secondary sex characteristics include fat distribution to the breasts, abdomen, mons pubis, and hips; voice pitch; broad pelvis; and hair pattern. Second, they control fluid and electrolyte balance. Third, they increase protein anabolism. In this regard, estrogens are synergistic with growth hormone (GH). High levels of estrogens in the blood inhibit the release of GnRF by the hypothalamus, which in turn inhibits the secretion of FSH by the anterior pituitary gland. This inhibition provides the basis for the action of one kind of contraceptive pill.

Progesterone, the hormone of maturation, works with estrogens to prepare the endometrium for implantation of a fertilized ovum and the mammary glands for milk secretion. High levels of progesterone also inhibit GnRF. Progesterone, like estrogens, is synthesized from cholesterol or acetyl coenzyme A in the ovaries.

Relaxin goes into operation near the end of pregnancy. It relaxes the symphysis pubis and helps dilate the uterine cervix to facilitate delivery.

Menstrual Phase (Menstruation)

The duration of the menstrual cycle ranges from 24 to 35 days. For this discussion, we will assume an average duration of 28 days. Events occurring during the menstrual cycle may be divided into three phases: the menstrual phase, the preovulatory phase, and the postovulatory phase (Figure 28-16).

The **menstrual phase,** also called **menstruation** or the **menses,** is the periodic discharge of 25 to 65 ml of blood, tissue fluid, mucus, and epithelial cells. It is caused by a sudden reduction in estrogens and progesterone and

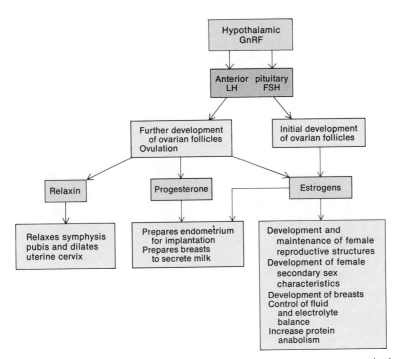

FIGURE 28-15 Secretion and physiological effects of estrogens, progesterone, and relaxin.

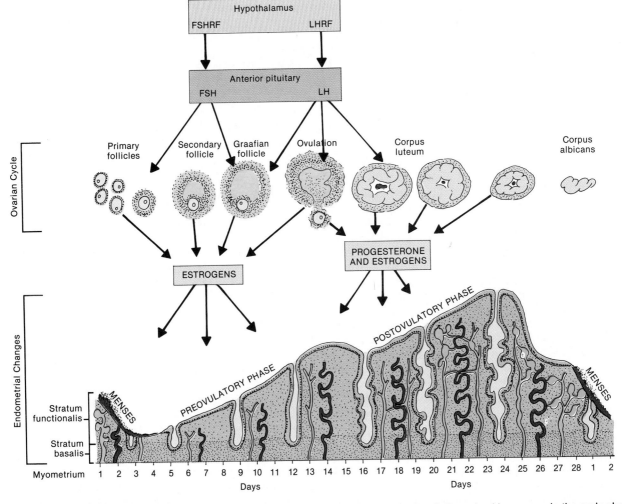

FIGURE 28-16 Correlation of menstrual and ovarian cycles with the hypothalamic and anterior pituitary gland hormones. In the cycle shown, fertilization and implantation have not occurred.

lasts for approximately the first 5 days of the cycle. The discharge is associated with endometrial changes in which the functionalis layer degenerates and patchy areas of bleeding develop. Small areas of the functionalis detach one at a time (total detachment would result in hemorrhage), the uterine glands discharge their contents and collapse, and tissue fluid is discharged. The menstrual flow passes from the uterine cavity to the cervix and through the vagina to the exterior. Generally the flow terminates by the fifth day of the cycle. At this time the entire functionalis has been shed and the endometrium is very thin because only the basalis remains.

During the menstrual phase, the ovarian cycle is also in operation. Ovarian follicles, called **primary follicles,** begin their development. At birth each ovary contains about 200,000 such follicles, each consisting of a potential ovum surrounded by a layer of cells. During the early part of each menstrual phase 20 to 25 primary follicles start to produce very low levels of estrogens. A clear membrane, the **zona pellucida,** also develops around the potential ova. Toward the end of the menstrual phase (days 4 to 5) about 20 of the primary follicles develop into **secondary follicles** as the cells of the surrounding layer increase in number and differentiate, secreting follicular (fō-LIK-yoo-lar) fluid. This fluid forces an immature ovum to the edges of the secondary follicles. The production of estrogens by the secondary follicles elevates the level of estrogens in the blood slightly. Ovarian follicle development is the result of GnRF secretion by the hypothalamus, which in turn stimulates FSH production by the anterior pituitary. During this part of the cycle, FSH secretion is relatively high. Although a number of follicles begin development each cycle, only one attains maturity. The others undergo atresia (death).

Preovulatory Phase

The **preovulatory phase,** the second phase of the menstrual cycle, is the time between menstruation and ovulation. This phase of the menstrual cycle is more variable in length than the other phases. It lasts from days 6 to 13 in a 28-day cycle.

FSH and LH stimulate the ovarian follicles to produce more estrogens, and this increase in estrogens stimulates the repair of the endometrium. Cells of the stratum basalis undergo mitosis and produce a new stratum functionalis. As the endometrium thickens, the short, straight endometrial glands develop and the arterioles coil and lengthen as they penetrate the functionalis. The thickness of the endometrium approximately doubles to about 4 to 6 mm. Because of the proliferation of endometrial cells, the preovulatory phase is also termed the **proliferative phase.** Still another name is the **follicular phase** because of increasing secretion of estrogens by the developing follicle. Functionally, estrogens are the dominant ovarian hormones during this phase of the menstrual cycle (Figure 28-17).

During the preovulatory phase, one of the secondary follicles in the ovary matures into a **Graafian follicle,** a follicle ready for ovulation. During the maturation process, the follicle increases its estrogen production. Early in the preovulatory phase, FSH is the dominant hormone

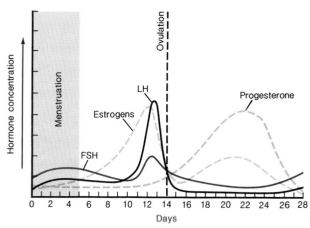

FIGURE 28-17 Relative concentrations of anterior pituitary hormones (FSH and LH) and ovarian hormones (estrogens and progesterone) during a normal menstrual cycle.

of the anterior pituitary, but close to the time of ovulation, LH is secreted in increasing quantities (Figure 28-17). Moreover, small amounts of progesterone may be produced by the Graafian follicle a day or two before ovulation.

Ovulation

Ovulation, the rupture of the Graafian follicle with release of the ovum into the pelvic cavity, occurs on day 14 in a 28-day cycle. Just prior to ovulation, the high level of estrogens that developed during the preovulatory phase inhibits GnRF production by the hypothalamus. This, in turn, inhibits FSH secretion by the anterior pituitary. Concurrently, LH secretion by the anterior pituitary is also inhibited by inhibition of GnRF. As a result of decreased levels of FSH and LH, ovulation occurs. Following ovulation, the Graafian follicle collapses and blood within it forms a clot called the **corpus hemorrhagicum.** The clot is eventually absorbed by the remaining follicular cells. In time, the follicular cells enlarge, change character, and form the **corpus luteum,** or yellow body.

CLINICAL APPLICATION

One method of **determining ovulation** involves recording basal temperature for about three or four consecutive months. The information is recorded on a basal temperature chart which plots days of the month against temperature in tenths of a degree Fahrenheit (Figure 28-18). The chart is begun with the first day of menstruation, at which time the date of menstruation and subsequent calendar dates are recorded on the top of the chart. The days of menstruation are marked with an X, and temperature is not recorded. When menstruation ceases, the temperature is taken immediately upon awakening each morning and marked on the chart. A sharp decrease in temperature followed by a sharp increase in temperature usually occurs about 14 days after the start of the last menstrual cycle. The 24 hours following this rise in temperature is the period immediately following ovulation and is generally considered the best time to become pregnant.

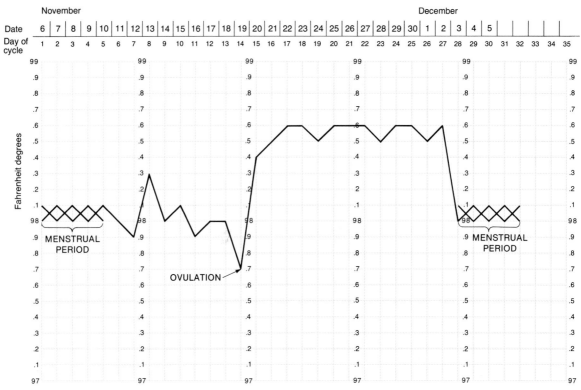

FIGURE 28-18 Basal temperature chart for determining ovulation.

The accuracy of the determination depends on many factors including individual variations, the accuracy of the temperature readings, and any factor other than the ovarian cycle that might affect body temperature.

Postovulatory Phase

The **postovulatory phase** of the menstrual cycle is the most constant in duration and lasts from days 15 to 28 in a 28-day cycle. It represents the time between ovulation and the onset of the next menses. Following ovulation, LH secretion stimulates the development of the corpus luteum. The corpus luteum then secretes increasing quantities of estrogens and progesterone. Progesterone is responsible for preparing the endometrium to receive a fertilized ovum. Preparatory activities include secretory activity of the endometrial glands that causes them to appear tortuously coiled, vascularization of the superficial endometrium, thickening of the endometrium, glycogen storage, and an increase in the amount of tissue fluid. These preparatory changes are maximal about 1 week after ovulation, and they correspond to the anticipated arrival of the fertilized ovum. During the postovulatory phase, FSH secretion again gradually increases and LH secretion decreases. The functionally dominant ovarian hormone during this phase is progesterone (see Figure 28-17). The relationship of progesterone to prostaglandins in causing painful menstruation will be considered at the end of the chapter.

If fertilization and implantation do not occur, the rising levels of progesterone and estrogens from the corpus luteum inhibit GnRF and LH secretion. As a result the corpus luteum degenerates and becomes the **corpus albicans.** The decreased secretion of progesterone and estrogens by the degenerating corpus luteum then initiates another menstrual period. In addition, the decreased levels of progesterone and estrogens in the blood bring about a new output of the anterior pituitary hormones—especially FSH in response to an increased output of GnRF by the hypothalamus. Thus a new ovarian cycle is initiated. A summary of these hormonal interactions is presented in Figure 28-19.

If, however, fertilization and implantation do occur, the corpus luteum is maintained throughout the pregnancy. For about 8 to 10 weeks it continues to secrete estrogens and progesterone. The corpus luteum is maintained by **human chorionic (kō-rē-ON-ik) gonadotropin (HCG),** a hormone produced by the developing placenta. The placenta itself secretes estrogens to support pregnancy and progesterone to support pregnancy and breast development for lactation. Once the placenta begins its secretion, the role of the corpus luteum becomes minor.

Menarche and Menopause

The menstrual cycle normally occurs once each month from **menarche** (me-NAR-kē), the first menses, to **menopause,** the last menses. The advent of menopause is signaled by the **climacteric** (klī-mak-TER-ik)—menstrual cycles become less frequent. The climacteric, which typically begins between ages 40 and 50, results from the failure of the ovaries to respond to the stimulation of gonadotropic hormones from the anterior pituitary. Some women experience hot flashes, copious sweating, head-

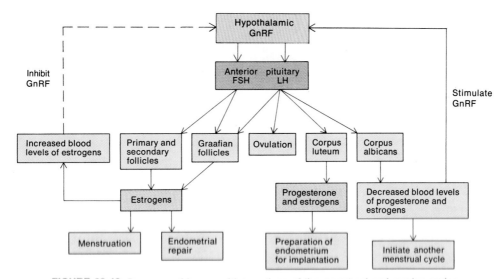

FIGURE 28-19 Summary of hormonal interactions of the menstrual and ovarian cycles.

ache, muscular pains, and emotional instability. In the postmenopausal woman there will be some atrophy of the ovaries, uterine tubes, uterus, vagina, external genitalia, and breasts.

The cause of menopause is related to a decreasing ability of aging ovaries to respond to FSH and LH. As a result there is a decrease in the production of estrogens by the ovaries. Throughout a woman's sexual life, some of the primary ovarian follicles grow into Graafian follicles with each sexual cycle, and eventually most of them degenerate. As a number of primary follicles diminishes, the production of estrogens by the ovary decreases.

VAGINA

The **vagina** serves as a passageway for the menstrual flow. It is also the receptacle for the penis during coitus, or sexual intercourse, and the lower portion of the birth canal. It is a muscular, tubular organ lined with mucous membrane and measures about 10 cm (4 inches) in length, extending from the cervix to the vestibule (see Figures 28-9 and 28-10). Situated between the urinary bladder and the rectum, it is directed superiorly and posteriorly, where it attaches to the uterus. A recess, called the **fornix,** surrounds the vaginal attachment to the cervix. The dorsal recess, called the posterior fornix, is deeper than the ventral and two lateral fornices. The fornices make possible the use of contraceptive diaphragms.

Histologically, the mucosa of the vagina is continuous with that of the uterus and consists of stratified squamous epithelium and connective tissue that lies in a series of transverse folds, the **rugae.** The muscularis is composed of longitudinal smooth muscle that can stretch considerably. This distension is important because the vagina receives the penis during sexual intercourse and serves as the lower portion of the birth canal. At the lower end of the vaginal opening, the **vaginal orifice,** is a thin fold of vascularized mucous membrane called the **hymen,** which forms a border around the orifice, partially closing it (see Figure 28-20).

The mucosa of the vagina contains large amounts of glycogen, which upon decomposition produces organic acids. These acids create a low pH environment that retards microbial growth. However, the acidity is also injurious to sperm cells. Semen neutralizes the acidity of the vagina to ensure survival of the sperm.

VULVA

The term **vulva (VUL-va), or pudendum** (pyoo-DEN-dum), is a collective designation for the external genitalia of the female (Figure 28-20). Its components are as follows.

The **mons pubis (veneris),** an elevation of adipose tissue covered by coarse pubic hair, is situated over the symphysis pubis. It lies anterior to the vaginal and urethral openings. From the mons pubis, two longitudinal folds of skin, the **labia majora** (LĀ-bē-a ma-JŌ-ra), extend inferiorly and posteriorly. The labia majora, the female homologue of the scrotum, contain an abundance of adipose tissue and sebaceous (oil) and sudoriferous (sweat) glands; they are covered by pubic hair on their upper outer surfaces. Medial to the labia majora are two folds of mucous membrane called the **labia minora** (MĪ-nō-ra). Unlike the labia majora, the labia minora are devoid of pubic hair and fat and have few sudoriferous glands. They do, however, contain numerous sebaceous glands.

The **clitoris** (KLI-to-ris) is a small, cylindrical mass of erectile tissue and nerves. It is located at the anterior junction of the labia minora. A layer of skin called the **prepuce** (foreskin) is formed at the point where the labia minora unite and covers the body of the clitoris. The exposed portion of the clitoris is the **glans.** The clitoris

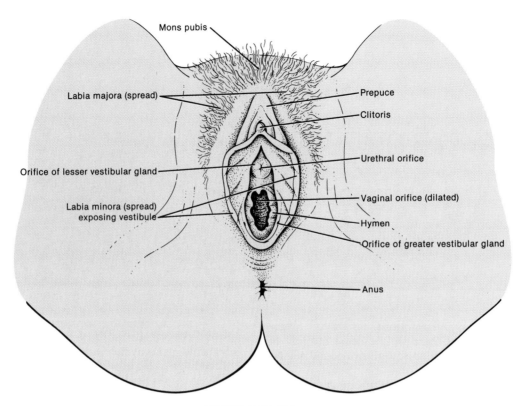

FIGURE 28-20 Vulva.

is homologous to the penis of the male in that it is capable of enlargement upon tactile stimulation and assumes a role in sexual excitement of the female.

The cleft between the labia minora is called the **vestibule.** Within the vestibule are the hymen, vaginal orifice, urethral orifice, and the openings of several ducts. The **vaginal orifice** occupies the greater poriton of the vestibule and is bordered by the hymen. Anterior to the vaginal orifice and posterior to the clitoris is the **urethral orifice.** Posterior and to either side of the urethral orifice are the openings of the ducts of the **lesser vestibular** (ves-TIB-yoo-lar) **glands.** These glands secrete mucus. On either side of the vaginal orifice itself are the **greater vestibular glands.** These small glands open by ducts into a groove between the hymen and labia minora and produce a mucoid secretion that supplements lubrication during sexual intercourse. The lesser vestibular glands are homologous to the male prostate. The greater vestibular glands are homologous to the male bulbourethral glands.

CLINICAL APPLICATION

One important sign in the **diagnosis of pregnancy** is a bluish discoloration of the vulva and vagina due to venous congestion. The discoloration appears at about the eighth to twelfth week and increases in intensity as the pregnancy progresses.

PERINEUM

The **perineum** (per'-i-NĒ-um) is the diamond-shaped area at the inferior end of the trunk between the thighs and buttocks of both males and females (Figure 28-21). It is bounded anteriorly by the symphysis pubis, laterally by the ischial tuberosities, and posteriorly by the coccyx. A transverse line drawn between the ischial tuberosities divides the perineum into an anterior **urogenital** (yoo'-rō-JEN-i-tal) **triangle** that contains the external genitalia and a posterior **anal triangle** that contains the anus.

CLINICAL APPLICATION

In the female, the region between the vagina and anus is known as the **clinical perineum.** If the vagina is too small to accommodate the head of an emerging fetus, the skin, vaginal epithelium, subcutaneous fat, and superficial transverse perineal muscle of the clinical perineum may tear. Moreover, the tissues of the rectum may be damaged. To avoid this, a small incision called an **episiotomy** (e-piz'-ē-OT-ō-mē) is made in the perineal skin and underlying tissues just prior to delivery. After delivery the episiotomy is sutured in layers.

MAMMARY GLANDS

The **mammary glands** are modified sweat glands (branched tubuloalveolar) that lie over the pectoralis major muscles and are attached to them by a layer of connective tissue (Figure 28-22). Internally, each mammary gland consists of 15 to 20 **lobes,** or compartments, separated by adipose tissue. The amount of adipose tissue determines the size of the breasts. However, breast size has nothing to do with the amount of milk produced. In each lobe are several smaller compartments called **lobules,** composed of connective tissue in which milk-secret-

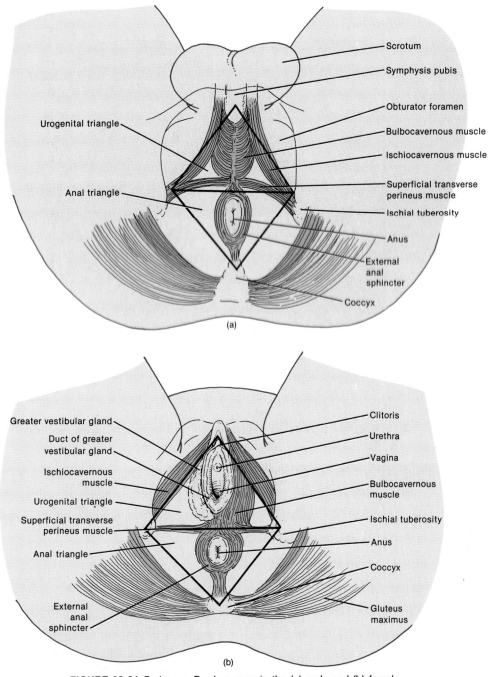

FIGURE 28-21 Perineum. Borders seen in the (a) male and (b) female.

ing cells referred to as **alveoli** are embedded (Figure 28-23). Alveoli are arranged in grapelike clusters. Between the lobules are strands of connective tissue called the **suspensory ligaments of the breast.** These ligaments run between the skin and deep fascia and support the breast. Alveoli convey the milk into a series of **secondary tubules.** From here the milk passes into the **mammary ducts.** As the mammary ducts approach the nipple, they expand to form sinuses called **ampullae** (am-POOL-ē), where milk may be stored. The ampullae continue as **lactiferous ducts** that terminate in the **nipple.** Each lactiferous duct conveys milk from one of the lobes to the exterior, although some may join before reaching the surface. The circular pigmented area of skin surrounding the nipple is called the

areola (a-RĒ-ō-la). It appears rough because it contains modified sebaceous glands.

At birth, both male and female mammary glands are undeveloped and appear as slight elevations on the chest. With the onset of puberty, the female breasts begin to develop—the ductile system matures, extensive fat deposition occurs, and the areola and nipple grow and become pigmented. These changes are correlated with an increased output of estrogens by the ovary. Further mammary development occurs at sexual maturity with the onset of ovulation and the formation of the corpus luteum. During adolescence, increased levels of progesterone cause the alveoli to proliferate, enlarge, and become secretory. Also, fat deposition continues, increasing the size

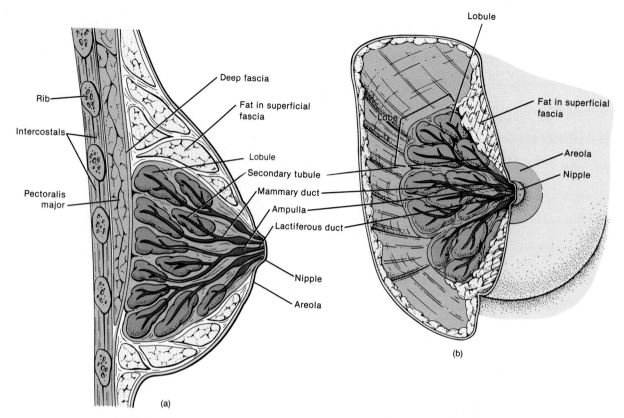

FIGURE 28-22 Mammary glands. (a) Sagittal section. (b) Anterior view, partially sectioned.

of the glands. Although the changes in mammary gland development are associated with estrogens and progesterone secretion by the ovaries, ovarian secretion is ultimately controlled by FSH, which is secreted in response to GnRF by the hypothalamus.

The essential function of the mammary glands is milk secretion and ejection, together called **lactation.** The secretion of milk is due largely to the hormone prolactin (PRL), with contributions from progesterone and estrogens. The ejection of milk occurs in the presence of oxytocin (OT). Lactation is considered in detail in Chapter 29.

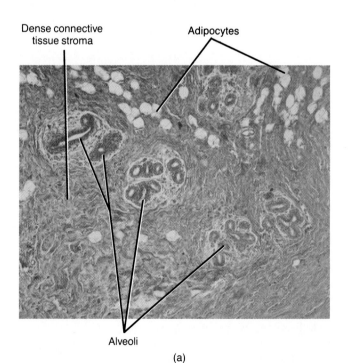

FIGURE 28-23 Histology of the mammary glands. (a) Photomicrograph of several alveoli in a resting (inactive) mammary gland at a magnification of 60×.

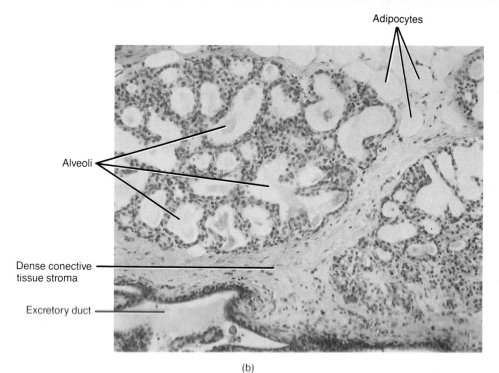

Adipocytes

Alveoli

Dense conective
tissue stroma

Excretory duct

(b)

FIGURE 28-23 (*Continued*) History of the mammary glands. (b) Photomicrograph of several alveoli during late pregnancy at a magnification of 160×. (© 1983 by Michael H. Ross. Used by permission.)

DISORDERS: HOMEOSTATIC IMBALANCES

Sexually Transmitted Diseases (STDs)

The general term **sexually transmitted disease (STD)** is applied to any of the large group of diseases that can be spread by sexual contact. The group includes conditions traditionally specified as **venereal diseases (VD)**, from Venus, goddess of love, such as gonorrhea, syphilis, and genital herpes, and several other conditions which are contracted sexually or may be contracted otherwise, but are then transmitted to a sexual partner.

Gonorrhea

Gonorrhea is an infectious sexually transmitted disease that affects primarily the mucous membrane of the urogenital tract, the rectum, and occasionally the eyes. The disease is caused by the bacterium *Neisseria gonorrhoeae.* Discharges from the involved mucous membranes are the source of infection, and the bacteria are transmitted by direct contact, usually sexual or during passage of a newborn through the birth canal.

Males usually suffer inflammation of the urethra with pus and painful urination. Fibrosis sometimes occurs in an advanced stage, causing stricture of the urethra. There also may be involvement of the epididymis and prostate gland. In females, infection may occur in the urethra, vagina, and cervix, and there may be a discharge of pus. However, infected females often harbor the disease without any symptoms until it has progressed to a more advanced stage. If the uterine tubes become involved, pelvic inflammation may follow. Peritonitis, or inflammation of the peritoneum, is a very serious disorder. The infection should be treated and controlled immediately because, if neglected, sterility or death may result. Although antibiotics have greatly reduced the mortality rate of acute peritonitis, it is estimated that between 50,000 and 80,000 women are made sterile by gonorrhea every year as a result of scar tissue formation that closes the uterine tubes. If the bacteria are transmitted to the eyes of a newborn in the birth canal, blindness can result.

Administration of a 1 percent silver nitrate solution or penicillin in the infant's eyes prevents infection. Penicillin is the drug of choice for the treatment of gonorrhea in adults.

Syphilis

Syphilis is a sexually transmitted disease caused by the bacterium *Treponema pallidum.* It is acquired through sexual contact or transmitted through the placenta to a fetus. The disease progresses through several stages: primary, secondary, latent, and sometimes tertiary. During the *primary stage,* the chief symptom is an open sore, called a *chancre,* at the point of contact. The chancre heals within 1 to 5 weeks. From 6 to 24 weeks later, symptoms such as a skin rash, fever, and aches in the joints and muscles usher in the *secondary stage.* These symptoms also eventually disappear (in about 4 to 12 weeks) and the disease ceases to be infectious, but a blood test for the presence of the bacteria generally remains positive. During this "symptomless" period, called the *latent stage,* the bacteria may invade body organs. When signs of organ degeneration appear, the disease is said to be in the *tertiary stage.*

If the syphilis bacteria attack the organs of the nervous system, the tertiary stage is called *neurosyphilis.* Neurosyphilis may take different forms, depending on the tissue involved. For instance, about two years after the onset of the disease, the bacteria may attack the meninges, producing meningitis. The blood vessels that supply the brain may also become infected. In this case, symptoms depend on the parts of the brain destroyed by oxygen and glucose starvation. Cerebellar damage is manifested by uncoordinated movements in such activities as writing. As the motor areas become extensively damaged, victims may be unable to control urine and bowel movements. Eventually, they may become bedridden, unable even to feed themselves. Damage to the cerebral cortex produces memory loss and personality changes that range from irritability to hallucinations.

Infection of the fetus with syphilis can occur after the fifth month. Infection of the mother is not necessarily followed by fetal infection, provided that the placenta remains intact. But once the bacteria gain access to fetal circulation, there is nothing to impede their growth and multiplication. As many as 80 percent of children born to untreated syphilitic mothers will be infected in the uterus if the fetus is exposed at the onset or in the early stages of the disease. About 25 percent of the fetuses will die within the uterus. Most of the survivors will arrive prematurely, but 30 percent will die shortly after birth. Of the infected and untreated children surviving infancy, about 40 percent will develop symptomatic syphilis during their lifetimes.

Syphilis can be treated with antibiotics (penicillin) during the primary, secondary, and latent periods. Certain forms of neurosyphilis may also be successfully treated, but the prognosis for others is very poor. Noticeable symptoms do not always appear during the first two stages of the disease. Syphilis, however, is usually diagnosed through a blood test whether noticeable symptoms appear or not. The importance of these blood tests and follow-up treatments cannot be overemphasized.

Genital Herpes

Another sexually transmitted disease, **genital herpes,** is now common in the United States. The sexual transmission of the herpes simplex virus is well established. Unlike syphilis and gonorrhea, genital herpes is incurable. Type I herpes simplex virus is the virus that causes typical cold sores. Type II herpes simplex virus causes painful genital blisters on the prepuce, glans penis, and penile shaft in males and on the vulva or sometimes high up in the vagina in females. The blisters disappear and reappear, but the disease itself remains in the body.

Genital herpes virus infection causes considerable discomfort, and there is an extraordinarily high rate of recurrence of the disease. Because genital herpes is statistically associated with later cervical carcinoma, all women with this infection should be reminded of the importance of annual Papanicolaou smears. For pregnant women with genital herpes, a Cesarean section will usually prevent complications in the child. Complications range from a mild asymptomatic infection to CNS damage to death.

Treatment of the symptoms involves pain medication, saline compresses, sexual abstinence for the duration of the eruption, and use of a drug called acyclovir. This drug interferes with viral DNA replication, but not host cell DNA replication. Although acyclovir speeds the healing and sometimes reduces the pain of initial genital herpes infections, it is not effective against recurrent infections because the genital herpes virus retreats into nerve cells in between attacks in the genital skin area.

Trichomoniasis

The microorganism *Trichomonas vaginalis,* a flagellated protozoan (one-celled animal), causes **trichomoniasis,** an inflammation of the mucous membrane of the vagina in females and the urethra in males. Symptoms include vaginal discharge and severe vaginal itch in women. Men can have it without symptoms, but can transmit it to women nonetheless. Sexual partners must be treated simultaneously.

Nongonococcal Urethritis (NGU)

The term **nongonococcal urethritis (NGU),** also known as **nonspecific urethritis (NSU),** refers to an inflammation of the urethra not caused by the bacterium *Neisseria gonorrhoeae.* The bacterium *Chlamydia trachomatis* is known to cause some cases of NGU. NGU affects both males and females. The urethra swells and narrows, impeding the flow of urine. Both urination and the urgency to urinate increase. Urination is accompanied by burning pain, and there may be a purulent (pus-containing) discharge.

Although nonmicrobial factors such as trauma (passage of a catheter) or chemical agents (alcohol and certain chemotherapeutic agents) can cause this condition, it is estimated that at least 40 percent of the cases of NGU are acquired sexually. In fact, NGU may be the most common sexually transmitted disease in the United States today. Although it is not a disease which must be reported to state departments of health and exact data are lacking, the Centers for Disease Control estimates that 4 to 9 million Americans have NGU. Since the symptoms are often mild in males and females are usually asymptomatic, many cases of infection go untreated. Complications are not common, but can be serious. Males may develop inflammation of the epididymis. In females, inflammation may cause sterility

by blocking the cervix or uterine tubes. As in gonorrhea, the bacteria can be passed from mother to infant during birth, infecting the eyes. NGU of chlamydial origin responds to treatment with tetracycline.

Male Disorders
Prostate

The prostate gland is susceptible to infection, enlargement, and benign and malignant tumors. Because the prostate surrounds the urethra, any of these disorders can obstruct the flow of urine. Prolonged obstruction may result in serious changes in the urinary bladder, ureters, and kidneys and may perpetuate urinary tract infections. Therefore, if the obstruction cannot be relieved by other means, surgical removal of part of or the entire gland is indicated. The surgical procedure is called **prostatectomy** (pros'-ta-TEK-tō-mē).

Acute and chronic infections of the prostate gland are common in postpubescent males, often in association with inflammation of the urethra. In **acute prostatis,** the prostate gland becomes swollen and tender. Appropriate antibiotic therapy, bed rest, and above-normal fluid intake are effective treatment.

Chronic prostatitis is one of the most common chronic infections in men of the middle and later years. On examination, the prostate gland feels enlarged, soft, and extremely tender. The surface outline is irregular and may be hard. This disease frequently produces no symptoms, but the prostate is believed to harbor infectious microorganisms responsible for some allergic conditions, arthritis, and inflammation of nerves (neuritis), muscles (myositis), and the iris (iritis).

An **enlarged prostate** gland, increasing to two to four times larger than normal, occurs in approximately one-third of all males over age 60. The cause is unknown, and the enlarged condition usually can be detected by rectal examination.

Tumors of the male reproductive system usually involve the prostate gland. Carcinoma of the prostate is the second leading cause of death from cancer in men in the United States and it is responsible for approximately 19,000 deaths annually. Its incidence is related to age, race, occupation, geography, and ethnic origin. Both benign and malignant growths are common in elderly men. Both types of tumors put pressure on the urethra, making urination painful and difficult. At times, the excessive back pressure destroys kidney tissue and gives rise to an increased susceptibility to infection. Therefore, even when the tumor is benign, surgery is indicated.

Sexual Functional Abnormalities

Impotence is the inability of an adult male to attain or hold an erection long enough for normal intercourse. Impotence could be the result of physical abnormalities of the penis, systemic disorders such as syphilis, vascular disturbances, neurological disorders, testosterone deficiency, or psychic factors such as fear of causing pregnancy, fear of sexually transmitted diseases, religious inhibitions, and emotional immaturity.

Infertility (sterility) is an inability to fertilize the ovum. It does not imply impotence. Male fertility requires production of adequate amounts of viable, normal spermatozoa by the testes, unobstructed transportation of sperm through the seminal tract, and satisfactory deposition in the vagina. The tubules of the testes are sensitive to many factors—x-rays, infections, toxins, malnutrition—that may cause degenerative changes and produce male sterility.

If inadequate spermatozoa production is suspected, a sperm analysis should be performed. Analysis includes measuring the volume of semen, counting the number of sperm per milliliter, evaluating sperm motility 4 hours after ejaculation, and determining the percentage of abnormal sperm forms (not to exceed 20 percent).

Female Disorders
Menstrual Abnormalities

Because menstruation reflects not only the health of the uterus, but also the health of the endocrine glands that control it, the ovaries and the pituitary gland, disorders of the female reproductive system frequently involve menstrual disorders.

Amenorrhea is the absence of menstruation. If a woman has never menstruated, the condition is called **primary amenorrhea.** Primary amenorrhea can be caused by endocrine disorders, most often in the pituitary gland and hypothalamus, or by genetically caused abnormal development of the ovaries or uterus. **Secondary amenorrhea,** the skipping of one or more periods, is commonly experienced by women sometime during their lives. Changes in body weight, either gains or losses, often cause amenorrhea. Obesity may disturb ovarian function, and similarly, the extreme weight loss which characterizes anorexia nervosa often leads to a suspension of menstrual flow. When amenorrhea is unrelated to weight, analysis of levels of estrogens often reveals deficiencies of pituitary and ovarian hormones.

Dysmenorrhea is painful menstruation caused by forceful contraction of the uterus. It is often accompanied by nausea, vomiting, diarrhea, headache, fatigue, and nervousness. Some cases are caused by pathological conditions such as uterine tumors, ovarian cysts, endometriosis, and pelvic inflammatory disease (PID). However, other cases of dysmenorrhea are not related to any pathologies. Although the cause of these cases is unknown, it appears that they may be triggered by an overproduction of prostaglandins by the uterus. Prostaglandins are known to stimulate uterine contractions, but they cannot do so in the presence of high levels of progesterone. As we have noted earlier, progesterone levels are high during the last half of the menstrual cycle. During this time, prostaglandins are apparently inhibited by progesterone from producing uterine contractions. However, if pregnancy does not occur, progesterone levels drop rapidly and prostaglandin production increases. This causes the uterus to contract and slough off its lining and may result in dysmenorrhea. The other symptoms of dysmenorrhea—nausea, vomiting, diarrhea, and headache—may be due to prostaglandin-stimulated contractions of the smooth muscle of the stomach, intestines, and blood vessels in the brain. Research is now underway to develop drugs to inhibit prostaglandin synthesis in order to treat dysmenorrhea.

Abnormal uterine bleeding includes menstruation of excessive duration or excessive amount, too frequent menstruation, intermenstrual bleeding, and postmenopausal bleeding. These abnormalities may be caused by disordered hormonal regulation, emotional factors, fibroid tumors of the uterus, and systemic diseases.

Premenstrual syndrome (PMS) is a term usually reserved for severe physical and emotional distress, occurring late in the postovulatory phase of the menstrual cycle and sometimes overlapping with menstruation. Signs and symptoms include edema, breast swelling and tenderness, abdominal distension, constipation, skin eruptions, fatigue and lethargy, greater need for sleep, depression or anxiety, irritability, headache, and cravings for sweet or salty foods. The basic cause of PMS is unknown. One postulated mechanism involves excessive levels of estrogens or inadequate levels of progesterone. Other proposed causes are vitamin B_6 deficiency and altered glucose metabolism (hypoglycemia).

Toxic Shock Syndrome (TSS)

Toxic shock syndrome (TSS) is primarily a disease of previously healthy, young, menstruating females who use tampons. It is also recognized in males, children, and nonmenstruating females. Clinically, TSS is characterized by high fever up to 40.6°C (105°F), sore throat or very tender mouth, headache, fatigue, irritability, muscle soreness and tenderness, conjunctivitis, diarrhea and vomiting, abdominal pain, vaginal irritation, and erythematosus rash. Other symptoms include lethargy, unresponsiveness, memory loss, hypotension, peripheral vasoconstriction, respiratory distress syndrome, intravascular coagulation, decreased platelet count, renal failure, circulatory shock, and liver involvement.

It is now clear that toxin-producing strains of the bacterium *Staphylococcus aureus* are necessary for development of the disease. Apparently, certain high absorbency tampons provide a substrate upon which the bacteria grow and produce toxin. Initial therapy is directed at correcting all homeostatic imbalances as quickly as possible. Antistaphylococcal antibiotics, such as penicillin or clindamycin, are also administered. In severe cases, high doses of corticosteroids are also administered.

Ovarian Cysts and Endometriosis

Ovarian cysts are fluid-containing tumors of the ovary. Follicular cysts may occur in the ovaries of elderly women, in ovaries that have inflammatory diseases, and in menstruating females. They have thin walls and contain a serous albuminous material. Cysts may also arise from the corpus luteum or the endometrium. The endometrium is the inner lining of the uterus that is sloughed off in menstruation.

Endometriosis is the growth of endometrial tissue outside the uterus. The tissue enters the pelvic cavity via the open uterine tubes and may be found in any of a dozen sites—on the ovaries, cervix, abdominal wall, and urinary bladder. Causes are unknown. Endometriosis is common in women 30 to 40 years of age. Symptoms include premenstrual pain or unusual menstrual pain. The unusual pain is caused by the displaced tissue sloughing off at the same time the normal uterine endometrium is being shed during menstruation. Infertility can be a consequence. Treatment usually consists of hormone therapy or surgery. Endometriosis disappears at menopause or when the ovaries are removed.

Infertility

Female infertility, or the inability to conceive, occurs in about 10 percent of married females in the United States. Once it is established that ovulation occurs regularly, the reproductive tract is examined for functional and anatomical disorders to determine the possibility of union of the sperm and the ovum in the uterine tube.

Disorders Involving the Breasts

The breasts of females are highly susceptible to cysts and tumors. Men are also susceptible to breast tumors, but certain breast cancers are 100 times more common in women.

In the female, the benign **fibroadenoma** is a common tumor of the breast. It occurs most frequently in young women. Fibroadenomas have a firm rubbery consistency and are easily moved about within the mammary tissue. The usual treatment is excision of the growth. The breast itself is not removed.

Breast cancer has the highest fatality rate of all cancers affecting women, but it is rare in men. In the female, breast cancer is rarely seen before age 30, and its occurrence rises rapidly after menopause. Breast cancer is generally not painful until it becomes quite advanced, so often it is not discovered early or, if noted, is ignored. Any lump, no matter how small, should be reported to a doctor at once. Treatment for breast cancer may involve hormone therapy, chemotherapy, a *modified* or *radical mastectomy,* or a combination of these. A radical mastectomy involves removal of the affected breast along with the underlying pectoral muscles and the axillary lymph nodes. Metastasis of cancerous cells is usually through the lymphatics or blood. Radiation treatment and chemotherapy may follow the surgery to ensure the destruction of any stray cancer cells.

The mortality from breast cancer has not improved significantly in the past 50 years. Early detection—especially by breast self-examination and mammography—is still the most promising method to increase the survival rate.

It is estimated that 95 percent of breast cancer is first detected by the women themselves. Each month after the menstrual period the breasts should be thoroughly examined for lumps, puckering of the skin, or discharge.

The most effective screening technique for routinely detecting tumors less than ½ inch (1.27 cm) in diameter is *mammography*. One reason that mammography is so useful is that it can detect small calcium deposits, called microcalcifications, in breast tissue. Such calcifications frequently indicate the presence of a tumor. The mammographic image, called a mammogram, can be obtained in either of two ways. In one procedure, x-rays are beamed onto a specially coated metal plate and the blue-on-white image produced provides the physician with detail of thicker as well as thinner portions of the breast. This procedure is called *xeroradiography*. In the other procedure, called *film-screen mammography*, the x-rays are beamed onto a fluorescent screen which exposes a film in contact with it. It uses a very low dose of radiation and can be adjusted to emphasize different kinds of tissue within the breast.

Another screening technique used is known as *thermography*. It measures and graphically records heat given off by various parts of the breast. Since tumors have high metabolic rates and rich vascular supplies, they give off more heat than normal breast tissue. Unfortunately, thermography cannot detect small tumors located deep within breast tissue.

One of the most recent breast cancer detecting procedures is *ultrasound*. The procedure is performed while the patient lies on her stomach in a specially designed hospital bed with her breasts immersed in a tank of water. Ultrasound produces images using a device that first emits a pulse of high-frequency sound and then records the echo on a monitor. Although ultrasound can neither detect microcalcifications or tumors less than 1 cm in diameter, it can be used to determine whether a lump is a benign cyst or a malignant tumor.

Another recent technique combines computed tomography with mammography (*CT/M*) and appears to overcome some of the limitations of mammography. The procedure is based upon the fact that breast carcinoma has an abnormal affinity for iodide. CT scans are made before and after the rapid intravenous infusion of an iodide contrast material. Comparison of the initial density of a suspected lesion with the density following infusion of the iodide gives an indication of the status of the tumor. CT/M affords definitive diagnostic help in instances where the mammographic and physical examinations are inconclusive and appears to be a significantly improved new method of breast cancer diagnosis.

Cervical Cancer

Another common disorder of the female reproductive system is cancer of the uterine cervix. It ranks third in frequency after breast and skin cancers. **Cervical cancer** starts with a change in the shape, growth, and number of the cervical cells called *cervical dysplasia* (dis-PLĀ-sē-a). Cervical dysplasia is not a cancer in itself, but the abnormal cells tend to become malignant.

Cervical cancer may be a sexually transmitted disease with a long incubation period. Inciting factors are not known, but type II herpes simplex virus has recently become suspect. Smegma and the DNA of spermatozoa have also been implicated. Cancer of the cervix (except for adenocarcinoma) rarely occurs in celibate women, and for unknown reasons it is rare in Jewish women.

Early diagnosis of cancer of the uterus is accomplished by the *Papanicolaou* (pap'-a-NIK-ō-la-oo) *test,* or Pap smear. In this generally painless procedure, a few cells from the vaginal fornix (that part of the vagina surrounding the cervix) and the cervix are removed with a swab and examined microscopically. Malignant cells have a characteristic appearance and indicate an early stage of cancer, even before symptoms occur. Estimates indicate that the Pap smear is more than 90 percent reliable in detecting cancer of the cervix.

To rule out invasive carcinoma a *cone biopsy* of the cervix is performed. A cone biopsy is a hospital procedure in which an inverted cone of tissue is excised. It requires an anesthetic and is usually done only when the field of dysplasia extends beyond the field of vision afforded by the colposcope or when abnormal cells have been detected. In another procedure, *punch biopsy* is combined with an *endocervical curettage* (ku'-re-TAZH) or *ECC;* this combination has a high degree of diagnostic accuracy. In a punch biopsy, a disc or segment of tissue is excised. Curettage is a procedure in which the cervix is dilated and the endometrium of the uterus is scraped with a spoon-shaped instrument called a curette. This procedure is commonly called a *D and C*. If the carcinoma has spread beyond the mucous membrane, treatment may involve complete or partial removal of the uterus, called a *hysterectomy,* or radiation treatment.

Pelvic Inflammatory Disease (PID)

Pelvic inflammatory disease (PID) is a collective term for any extensive bacterial infection of the pelvic organs, especially the uterus, uterine tubes, or ovaries. A vaginal or uterine infection may spread into the uterine tube (*salpingitis*) or even farther into the abdominal cavity, where it infects the peritoneum (*peritonitis*). PID is most commonly caused by gonorrhea, but any bacteria can trigger infection. Often the early symptoms of PID, which include increased vaginal discharge and pelvic pain, occur just after menstruation. As infection spreads, fever may develop in advanced cases along with painful abscesses of the reproductive organs. Early treatment with antibiotics can stop the spread of PID.

MEDICAL TERMINOLOGY

Leukorrhea (*leuco* = white; *rrhea* = discharge) A nonbloody vaginal discharge that may occur at any age and affects most women at some time.

Neoplasia (*neo* = new; *plas* = form, grow) A condition characterized by the presence of new growths (tumors).

Oophorectomy (*oophoro* = ovary; *ectomy* = removal of) Excision of an ovary. Bilateral oophorectomy refers to the removal of both ovaries.

Pruritus Itching.

Salpingectomy (*salpingo* = tube) Excision of a uterine tube.

Smegma (*smegma* = soap) The secretion, consisting principally of desquamated epithelial cells, found chiefly about the external genitalia and especially under the foreskin of the male.

Vaginitis Inflammation of the vagina.

STUDY OUTLINE

Male Reproductive System (p. 698)

1. Reproduction is the process by which genetic material is passed on from one generation to the next.
2. The organs of reproduction are grouped as: gonads (produce gametes), ducts (transport and store gametes), and accessory glands (produce materials that support gametes).
3. The male structures of reproduction include the testes, ductus epididymis, ductus deferens, ejaculatory duct, urethra, seminal vesicles, prostate gland, bulbourethral glands, and penis.

Scrotum

1. The scrotum is a cutaneous outpouching of the abdomen that supports the testes.
2. It regulates the temperature of the testes by contraction of the dartos to elevate them closer to the pelvic cavity.

Testes

1. The testes are oval-shaped glands (gonads) in the scrotum containing seminiferous tubules, in which sperm cells are made; sustentacular cells, which nourish sperm cells; and interstitial endocrinocytes, which produce the male sex hormone testosterone.
2. Failure of the testes to descend is called cryptorchidism.
3. Mature spermatozoa consist of a head, midpiece, and tail. Their function is to fertilize an ovum.
4. Spermatozoa are moved through the testes through the seminiferous tubules, straight tubules, rete testis, and efferent ducts.
5. At puberty GnRF stimulates anterior pituitary secretion of FSH and LH. FSH initiates spermatogenesis and LH assists spermatogenesis and stimulates production of testosterone.
6. Testosterone controls the growth, development, and maintenance of sex organs; stimulates bone growth, protein anabolism, and sperm maturation; and stimulates development of male secondary sex characteristics.
7. Inhibin is produced by sustentacular cells. Its inhibition of FSH helps to regulate the rate of spermatogenesis.

Ducts

1. The duct system of the testes includes the seminiferous tubules, straight tubules, and rete testis.
2. Sperm are transported out of the testes through the efferent ducts.
3. The ductus epididymis is lined by stereocilia and is the site of sperm maturation and storage.
4. The ductus deferens stores sperm and propels them toward the urethra during ejaculation.
5. Alteration of the ductus deferens to prevent fertilization is called vasectomy.
6. The ejaculatory ducts are formed by the union of the ducts from the seminal vesicles and ductus deferens and eject spermatozoa into the prostatic urethra.
7. The male urethra is subdivided into three portions: prostatic, membranous, and spongy (cavernous).

Accessory Glands

1. The seminal vesicles secrete an alkaline, viscous fluid that constitutes about 60 percent of the volume of semen and contributes to sperm viability.
2. The prostate gland secretes an alkaline fluid that constitutes about 13 to 33 percent of the volume of semen and contributes to sperm motility.
3. The bulbourethral glands secrete mucus for lubrication and a substance that neutralizes urine.
4. Semen (seminal fluid) is a mixture of spermatozoa and accessory gland secretions that provide the fluid in which spermatozoa are transported, provide nutrients, and neutralize the acidity of the male urethra and female vagina.

Penis

1. The penis is the male organ of copulation.
2. Expansion of its blood sinuses under the influence of sexual excitation is called erection.

Female Reproductive System (p. 709)

1. The female organs of reproduction include the ovaries (gonads), uterine tubes, uterus, vagina, and vulva.
2. The mammary glands are considered as part of the reproductive system.

Ovaries

1. The ovaries are female gonads located in the upper pelvic cavity, on either side of the uterus.
2. They produce ova, discharge ova (ovulation), and secrete female sex hormones (estrogens and progesterone).

Uterine Tubes

1. The uterine tubes transport ova from the ovaries to the uterus and are the normal sites of fertilization.
2. Implantation outside the uterus (pelvic or tubular) is called an ectopic pregnancy.

Uterus

1. The uterus is an inverted, pear-shaped organ that functions in menstruation, implantation of a fertilized ovum, development of a fetus during pregnancy, and labor.
2. The uterus is normally held in position by a series of ligaments.
3. Histologically, the uterus consists of an outer perimetrium, middle myometrium, and inner endometrium.

Endocrine Relations: Menstrual and Ovarian Cycles

1. The function of the menstrual cycle is to prepare the endometrium each month for the reception of a fertilized egg. The ovarian cycle is associated with the maturation of an ovum each month.
2. The menstrual and ovarian cycles are controlled by GnRF, which stimulates the release of FSH and LH.
3. FSH stimulates the initial development of ovarian follicles and secretion of estrogens by the ovaries. LH stimulates further development of ovarian follicles, ovulation, and the secretion of estrogens and progesterone by the ovaries.
4. Estrogens stimulate the growth, development, and maintenance of female reproductive structures; stimulate the development of secondary sex characteristics; regulate fluid and electrolyte balance; and stimulate protein anabolism.
5. Progesterone works with estrogens to prepare the endometrium for implantation and the mammary glands for milk secretion.
6. Relaxin relaxes the symphysis pubis and helps dilate the uterine cervix to facilitate delivery.
7. During the menstrual phase, the functionalis layer of the endometrium is shed with a discharge of blood, tissue fluid, mucus, and epithelial cells. Primary follicles develop into secondary follicles.
8. During the preovulatory phase, endometrial repair occurs. A secondary follicle develops into a Graafian follicle. Estrogens are the dominant ovarian hormones.
9. Ovulation is the rupture of a Graafian follicle and the release

of an ovum into the pelvic cavity brought about by inhibition of FSH and release of LH.

10. During the postovulatory phase, the endometrium thickens in anticipation of implantation. Progesterone is the dominant ovarian hormone.

11. If fertilization and implantation do not occur, the corpus luteum degenerates and low levels of estrogens and progesterone initiate another menstrual and ovarian cycle.

12. If fertilization and implantation do occur, the corpus luteum is maintained by placental HCG and the corpus luteum and placenta secrete estrogens and progesterone to support pregnancy and breast development for lactation.

13. The female climacteric is the time immediately before menopause, the cessation of the sexual cycles.

Vagina

1. The vagina is a passageway for the menstrual flow, the receptacle for the penis during sexual intercourse, and the lower portion of the birth canal.

2. It is capable of considerable distension to accomplish its functions.

Vulva

1. The vulva is a collective term for the external genitals of the female.

2. It consists of the mons veneris, labia majora, labia minora, clitoris, vestibule, vaginal and urethral orifices, and greater and lesser vestibular glands.

Perineum

1. The perineum is a diamond-shaped area at the inferior end of the trunk between the thighs and buttocks.

2. An incision in the perineal skin prior to delivery is called an episiotomy.

Mammary Glands

1. The mammary glands are modified sweat glands (branched tubuloalveolar) over the pectoralis major muscles. Their function is to secrete and eject milk (lactation).

2. Mammary gland development is dependent on estrogens and progesterone.

3. Milk secretion is due to mainly PRL and milk ejection is stimulated by OT.

Disorders: Homeostatic Imbalances (p. 723)

1. Sexually transmitted diseases (STDs) are diseases spread by sexual contact and include gonorrhea, syphilis, genital herpes, trichomoniasis, and nongonococcal urethritis (NGU).

2. Conditions that affect the prostate are prostatitis, enlarged prostate, and tumors.

3. Impotence is the inability of the male to attain or hold an erection long enough for intercourse.

4. Infertility is the inability of a male's sperm to fertilize an ovum.

5. Menstrual disorders include amenorrhea, dysmenorrhea, abnormal bleeding, and premenstrual syndrome (PMS).

6. Toxic shock syndrome (TSS) includes widespread homeostatic imbalances and is a reaction to toxins produced by *Staphylococcus aureus.*

7. Ovarian cysts are tumors that contain fluid.

8. Endometriosis refers to the growth of uterine tissue outside the uterus.

9. Female infertility is the inability of the female to conceive.

10. The mammary glands are susceptible to benign fibroadenomas and malignant tumors. The removal of a malignant breast, pectoral muscles, and lymph nodes is called a radical mastectomy.

11. Cervical cancer can be diagnosed by a Pap test.

12. Pelvic inflammatory disease (PID) refers to bacterial infection of pelvic organs.

REVIEW QUESTIONS

1. Define reproduction. Describe how the reproductive organs are classified and list the male and female organs of reproduction.

2. Describe the function of the scrotum in protecting the testes from temperature fluctuations. What is cryptorchidism?

3. Describe the internal structure of a testis. Where are the sperm cells made?

4. Identify the principal parts of a spermatozoon. List the function of each.

5. Explain the effects of FSH and LH on the male reproductive system. How are these hormones controlled by GnRF?

6. Describe the physiological effects of testosterone and inhibin on the male reproductive system.

7. How is testosterone level controlled?

8. Which ducts are involved in transporting sperm within the testes?

9. Describe the location, structure, and functions of the ductus epididymis, ductus deferens, and ejaculatory duct.

10. What is a vasectomy?

11. What is the spermatic cord? What is an inguinal hernia?

12. Give the location of the three subdivisions of the male urethra.

13. Trace the course of spermatozoa through the male system of ducts from the seminiferous tubules through the urethra.

14. Briefly explain the locations and functions of the seminal vesicles, prostate gland, and bulbourethral glands.

15. What is semen? What is its function?

16. How is the penis structurally adapted as an organ of copulation? What is circumcision?

17. How does an erection occur?

18. How are the ovaries held in position in the pelvic cavity? Describe the microscopic structure of an ovary.

19. What are the functions of the ovaries?

20. Where are the uterine tubes located? What is their function? What is an ectopic pregnancy?

21. Diagram the principal parts of the uterus.

22. Describe the arrangement of ligaments that hold the uterus in its normal position. What is retroflexion?

23. Discuss the blood supply to the uterus. Why is an abundant blood supply important?

24. Describe the histology of the uterus.

25. Define menstrual cycle and ovarian cycle. What is the function of each?

26. What is the function of each of the following in the menstrual and ovarian cycles: GnRF, FSH, LH, estrogens, progesterone, relaxin?

27. Briefly outline the major events of each phase of the menstrual cycle and correlate them with the events of the ovarian cycle.

28. Prepare a labeled diagram of the principal hormonal interactions involved in the menstrual and ovarian cycles.

29. What is menarche? Define female climacteric and menopause.

30. What is the function of the vagina? Describe its histology.

31. List the parts of the vulva and the functions of each part.
32. What is the perineum? Define episiotomy.
33. Describe the structure of the mammary glands. How are they supported?
34. Describe the passage of milk from the areolar cells of the mammary gland to the nipple.
35. Explain the roles of estrogens and progesterone in development of the mammary glands.
36. Define lactation. How is it controlled?
37. Define a sexually transmitted disease (STD). Describe the cause, clinical symptoms, and treatment of gonorrhea, syphilis, genital herpes, trichomoniasis, and nongonococcal urethritis (NGU).
38. Describe several disorders that affect the prostate gland.
39. Distinguish between impotence and infertility.

40. What are some of the causes of amenorrhea, dysmenorrhea, and abnormal uterine bleeding?
41. Describe the clinical symptoms of premenstrual syndrome (PMS) and toxic shock syndrome (TSS).
42. What are ovarian cysts? Define endometriosis.
43. Distinguish between a fibroadenoma and a malignant tumor of the breast.
44. What is a radical mastectomy? Describe the importance of mammography, thermography, ultrasound, and CT/M in detecting breast tumors.
45. What is a Pap smear? How is cervical tissue biopsied?
46. Define pelvic inflammatory disease (PID).
47. Refer to the glossary of medical terminology at the end of the chapter. Be sure that you can define each term.

- Define meiosis.

- Contrast the events and outcomes of spermatogenesis and oogenesis.

- Compare the role of the male and female in sexual intercourse.

- Explain the activities associated with fertilization, morula formation, blastocyst development, and implantation.

- Describe how external human fertilization is accomplished.

- Discuss the formation of the primary germ layers, embryonic membranes, placenta, and umbilical cord as the principal events of the embryonic period.

- List representative body structures produced by the primary germ layers.

- Discuss the function of the embryonic membranes.

- Compare the roles of the placenta and umbilical cord during embryonic and fetal growth.

- Discuss the principal body changes associated with fetal growth.

- Compare the sources and functions of the hormones secreted during pregnancy.

- Explain the three stages of labor.

- Explain the respiratory and cardiovascular adjustments that occur in an infant at birth.

- Discuss potential hazards to an embryo and fetus associated with chemicals and drugs, irradiation, alcohol, and cigarette smoking.

- Discuss the physiology and control of lactation.

- Explain the procedure of amniocentesis and its value in diagnosing diseases in the newborn.

- Contrast the various kinds of birth control and their effectiveness.

- Define inheritance and describe the inheritance of phenylketonuria (PKU), sex, and color blindness.

- Define medical terminology associated with development and inheritance.

29 Development and Inheritance

evelopment **anatomy** is the study of the sequence of events from the fertilization of an egg to the formation of an adult organism. As we look at the sequence from fertilization to birth, we will consider how reproductive cells are produced, the role of the male and female in sexual intercourse, and events associated with pregnancy, birth, and lactation. Finally, we will say a few words about birth control.

GAMETE FORMATION

DIPLOID AND HAPLOID CELLS

In sexual reproduction, each new organism is produced by the union and fusion of two different sex cells, one produced by each parent. The sex cells, called **gametes,** are the ovum produced in the female gonads (ovaries) and the sperm produced in the male gonads (testes). The union and fusion of gametes is called fertilization and the cell thus produced is known as a zygote. The zygote contains a mixture of chromosomes (DNA) from the two parents and, through its repeated mitotic division, develops into a new organism.

Gametes differ from all other body cells (somatic cells) with respect to the number of chromosomes in their nuclei. Somatic cells, such as brain cells, stomach cells, kidney cells, and all other uninucleated somatic cells, contain 46 chromosomes in their nuclei. Some somatic cells, such as skeletal muscle cells, are multinucleated and thus contain more than 46 chromosomes. However, since most somatic cells are uninucleated, these are the cells to which we will refer in the following discussion. Of the 46 chromosomes, 23 are a complete set that contain all the genes necessary for carrying out the activities of the cell. In a sense, the other 23 chromosomes are a duplicate set. The symbol n is used to designate the number of different chromosomes within the nucleus. Since somatic cells contain two sets of chromosomes, they are referred to as **diploid** (DIP-loyd) **cells** (di = two), symbolized as $2n$. In a diploid cell, two chromosomes that belong to a pair are called **homologous** (ho-MOL-o-gus) **chromosomes** or **homologues.**

If gametes had the same number of chromosomes as somatic cells, the zygote formed from their fusion would have double the number. The somatic cells of the resulting individual would have twice the number of chromosomes ($4n$) as the somatic cells of the parents, and with every succeeding generation, the number of chromosomes would double. The chromosome number does not double with each generation because of a special nuclear division called **meiosis.** Meiosis occurs only in the production of gametes. It causes a developing sperm or ovum to relinquish its duplicate set of chromosomes so that the mature gamete has only 23. Thus gametes are **haploid** (HAP-loyd) **cells,** meaning "one-half," and are symbolized n.

The formation in the testes of haploid spermatozoa by meiosis is called **spermatogenesis** (sper'-ma-tō-JEN-e-sis). The formation in the ovary of haploid ova by meiosis is referred to as **oogenesis** (ō'-ō-JEN-e-sis).

SPERMATOGENESIS

In humans, spermatogenesis takes about two to three weeks. The seminiferous tubules are lined with immature cells called **spermatogonia** (sper'-ma-tō-GŌ-ne-a) or sperm mother cells (Figure 29-1). Spermatogonia contain the diploid chromosome number and are the precursor cells for all the spermatozoa the male will produce. At puberty, when the anterior pituitary secretes FSH in response to GnRF from the hypothalamus, the spermatogonia embark on a lifetime of active mitotic division. Some of the daughter cells are pushed inward toward the lumen of the seminiferous tubule. These cells lose contact with the basement membrane of the seminiferous tubule, undergo certain developmental changes, and become known as **primary spermatocytes** (SPER-ma-tō-sītz'). Primary spermatocytes, like spermatogonia, are diploid, that is, they have 46 chromosomes. The other daughter cells formed by mitosis of the spermatogonia remain near the basement membrane and form a reservoir of precursor cells for future meiotic divisions.

Reduction Division

Each primary spermatocyte enlarges before dividing. Then two nuclear divisions take place as part of meiosis. In the first, DNA is replicated and 46 chromosomes (each made up of two chromatids) form and move toward the equatorial plane of the nucleus. There they line up by homologous pairs so that there are 23 pairs of duplicated chromosomes in the center of the nucleus. This pairing of homologous chromosomes is called **synapsis.** The four chromatids of each homologous pair then twist around each other to form a **tetrad.** In a tetrad, portions of one chromatid may be exchanged with portions of another. This process, called **crossing-over,** permits an exchange of genes among chromatids (Figure 29-2) that results in the recombination of genes. Thus the spermatozoa eventually produced may be genetically unlike each other and unlike the cell that produced them—hence the great variation among humans. Next the mitotic spindle forms and the chromosomal microtubules produced by the centromeres of the paired chromosomes extend toward the poles of the cell. As the pairs separate, one member of each pair migrates to opposite poles of the dividing nucleus. The cells formed by the first nuclear division (reduction division) are called **secondary spermatocytes.** Each cell has 23 chromosomes—the haploid number. Each chromosome of the secondary spermatocytes, however, is made up of two chromatids. Moreover, the genes of the chromosomes of secondary spermatocytes may be rearranged as a result of crossing-over.

Equatorial Division

The second nuclear division of meiosis is equatorial division. There is no replication of DNA. The chromosomes (each composed of two chromatids) line up in single file around the equatorial plane, and the chromatids of each chromosome separate from each other. The cells formed from the equatorial division are called **spermatids.** Each

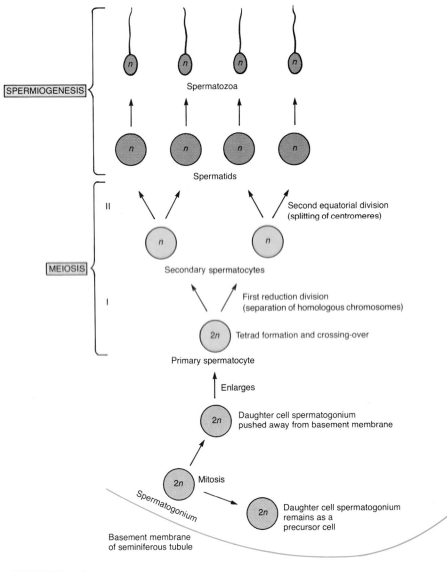

FIGURE 29-1 Spermatogenesis.

contains half the original chromosome number, or 23 chromosomes, and is haploid. Each primary spermatocyte therefore produces four spermatids by meiosis (reduction division and equatorial division). Spermatids lie close to the lumen of the seminiferous tubule.

Spermiogenesis

The final stage of spermatogenesis, called **spermiogenesis** (sper'-mē-ō-JEN-e-sis), involves the maturation of spermatids into spermatozoa. Each spermatid embeds in a sustentacular cell and develops a head with an acrosome and a flagellum (tail). Sustentacular cells extend from the basement membrane to the interior of the seminiferous tubule where they nourish the developing spermatids. Since there is no cell division in spermiogenesis, each spermatid develops into a single **spermatozoon (sperm cell).**

Spermatozoa enter the lumen of the seminiferous tubule and migrate to the ductus epididymis, where in 18

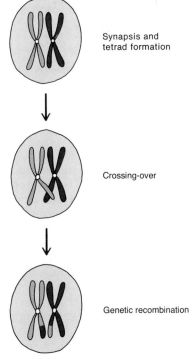

FIGURE 29-2 Crossing-over within a tetrad resulting in genetic recombination.

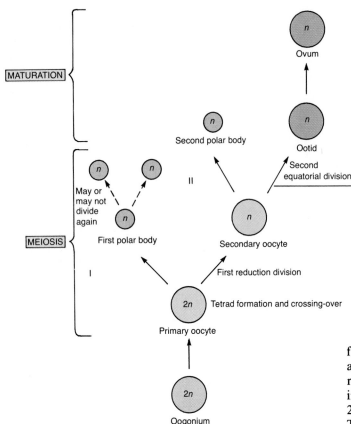

FIGURE 29-3 Oogenesis.

hours to 10 days they complete their maturation and become capable of fertilizing an ovum. Spermatozoa are also stored in the ductus deferens. Here, they can retain their fertility for up to several months.

OOGENESIS

The formation of a haploid ovum by meiosis in the ovary is referred to as oogenesis. With some exceptions, oogenesis occurs in essentially the same manner as spermatogenesis. It involves meiosis and maturation.

Reduction Division

The precursor cell in oogenesis is a diploid cell called the **oogonium** (ō'-o-GŌ-nē-um) or egg mother cell (Figure 29-3). Early in the prenatal development of the ovaries, oogonia proliferate by mitosis. But, as development continues, oogonia in the primary follicles lose their ability to carry on mitosis. At about the third month of prenatal development, oogonia develop into larger cells also containing the diploid number of chromosomes and called **primary oocytes** (Ō-o-sītz). They remain in this stage until their follicular cells respond to FSH from the anterior pituitary, which in turn has responded to GnRF from the hypothalamus.

Starting with puberty, several follicles respond each month to the rising level of FSH. As the preovulatory phase of the menstrual cycle proceeds and LH is secreted

from the anterior pituitary, one of the follicles reaches a stage in which the diploid primary oocyte undergoes reduction division. Synapsis, tetrad formation, and crossing-over occur, and two cells of unequal size, both with 23 chromosomes of two chromatids each, are produced. The smaller cell, called the **first polar body,** is essentially a packet of discarded nuclear material. The larger cell, known as the **secondary oocyte,** receives most of the cytoplasm.

Equatorial Division

At ovulation, the secondary oocyte with its polar body and surrounding supporting cells is discharged. The discharged secondary oocyte enters the uterine tube and, if spermatozoa are present and fertilization occurs, the second division takes place: the equatorial division.

Maturation

The secondary oocyte produces two cells of unequal size, both of them haploid. The larger cell is called an **ootid** and eventually develops into an **ovum,** or mature egg; the smaller is the **second polar body.**

The first polar body may undergo another division. If it does, meiosis of the primary oocyte results in a single haploid ovum and three polar bodies. In any event, all polar bodies disintegrate. Thus each oogonium produces a single ovum, whereas each spermatocyte produces four spermatozoa.

SEXUAL INTERCOURSE

Once spermatozoa and ova are developed through meiosis and maturation and the spermatozoa are deposited in the vagina, pregnancy can occur. **Sexual intercourse,** or **copulation** (in humans, called **coitus**), is the process by which spermatozoa are deposited in the vagina.

MALE SEXUAL ACT

Erection

The male role in the sexual act starts with **erection,** the enlargement and stiffening of the penis. An erection may be initiated in the cerebrum by stimuli such as anticipation, memory, and visual sensation, or it may be a reflex brought on by stimulation of the touch receptors in the penis, especially in the glans. In any case, parasympathetic impulses that pass from the sacral portion of the spinal cord to the penis cause dilation of the arteries of the penis, allowing blood to fill the cavernous spaces of the spongy bodies.

Lubrication

Parasympathetic impulses from the sacral cord also cause the bulbourethral glands to secrete mucus, which affords some **lubrication** for intercourse. The mucus flows through the urethra. The major portion of lubricating fluid is produced by the female. Without satisfactory lubrication, the male sexual act is difficult since unlubricated intercourse causes pain impulses that inhibit rather than promote coitus.

Orgasm

Tactile stimulation of the penis brings about emission and ejaculation. When sexual stimulation becomes intense, rhythmic sympathetic impulses leave the spinal cord at the levels of the first and second lumbar vertebrae and pass to the genital organs. These impulses cause peristaltic contractions of the ducts in the testes, epididymides, ductus deferens, and seminal ducts that propel spermatozoa into the urethra—a process called **emission.** Simultaneously, peristaltic contractions of the seminal vesicles and prostate expel seminal and prostatic fluid along with the spermatozoa. All these mix with the mucus of the bulbourethral glands, resulting in the fluid called semen. Other rhythmic impulses sent from the spinal cord at the levels of the first and second sacral vertebrae reach the skeletal muscles at the base of the penis, and the penis expels the semen from the urethra to the exterior. The propulsion of semen from the urethra to the exterior constitutes an **ejaculation.** A number of sensory and motor activities accompany ejaculation, including a rapid heart rate, an increase in blood pressure, an increase in respiration, and pleasurable sensations. These activities, together with the muscular events involved in ejaculation, are referred to as an **orgasm.**

FEMALE SEXUAL ACT

Erection

The female role in the sex act, like that of the male, also involves erection, lubrication, and orgasm. Stimulation of the female, as in the male, depends on both psychic and tactile responses. Under appropriate conditions, stimulation of the female genitalia, especially the clitoris, re-

sults in **erection** and widespread sexual arousal. This response is controlled by parasympathetic impulses sent from the sacral spinal cord to the external genitalia.

Lubrication

Parasympathetic impulses from the sacral spinal cord also pass to the greater vestibular glands, stimulating them to secrete mucus. This provides most of the **lubrication** during coitus. As was noted earlier, lack of sufficient lubrication results in pain impulses that inhibit rather than promote coitus.

Orgasm (Climax)

When tactile stimulation of the genitalia reaches maximum intensity, reflexes are initiated that cause the female **orgasm** or **climax.** Female orgasm is analogous to male ejaculation and may assume a role in fertilization of an ovum. The perineal muscles contract rhythmically from spinal reflexes similar to those that occur in the male ejaculation.

PREGNANCY

Pregnancy is a sequence of events that normally includes fertilization, implantation, embryonic growth, and fetal growth that terminates in birth. We will now examine the various events of pregnancy in detail.

FERTILIZATION AND IMPLANTATION

Fertilization

The term **fertilization** is applied to the penetration of an ovum by a spermatozoon and the subsequent union of the sperm nucleus and the nucleus of the ovum. Of the hundreds of millions of sperm cells introduced into the vagina, only about 2,000 arrive in the vicinity of the ovum. Fertilization normally occurs in the uterine tube when the ovum is about one-third of the way down the tube, usually within 24 hours after ovulation. Peristaltic contractions and the action of cilia transport the ovum through the uterine tube. The mechanism by which sperm reach the uterine tube is still unclear. Some believe that sperm swim up the female tract by means of whiplike movements of their flagella; others believe sperm are transported by muscular contractions of the uterus. Their motility is probably a combination of both.

Sperm must remain in the female genital tract 4 to 6 hours before they are capable of fertilizing an ovum. During this time, the enzymes hyaluronidase and proteinases are secreted by the acrosomes of the spermatozoa. They help dissolve the intercellular materials covering the ovum. The ovum is surrounded by a gelatinous covering called the **zona pellucida** (pe-LOO-si-da) and several layers of cells, the innermost of which are follicle cells, known as the **corona radiata** (Figure 29-4a). Normally, only one spermatozoon fertilizes an ovum, because once union is achieved, the electrical changes in the surface of the ovum block the entry of other sperm and enzymes produced

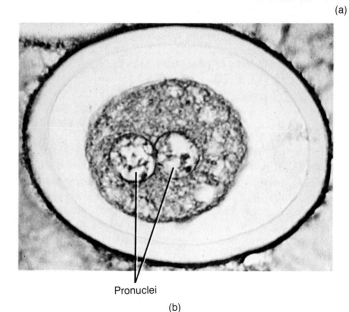

Ovum — Nucleus

Cytoplasm

Corona radiata

Second polar body

Spermatozoon

Zona pellucida

First polar body

(a)

Pronuclei

(b)

FIGURE 29-4 Fertilization and implantation. (a) Photomicrograph of a spermatozoon moving through the zona pellucida on its way to reach the nucleus of the ovum. (Courtesy of *The Rand McNally Atlas of the Body and Mind*, Rand McNally and Company, New York, Chicago, San Francisco, in association with Mitchell Beazley Publishers Limited, London, 1976.) (b) Photomicrograph showing male and female pronuclei. (Courtesy of Carolina Biological Supply Company.)

by the fertilized ovum alter receptor sites so that sperm already bound are detached and others are prevented from binding. In this way, polyspermy, the fertilization of an ovum by more than one spermatozoon, is prevented.

When a spermatozoon has entered the ovum, the tail is shed and the nucleus in the head develops into a structure called the **male pronucleus.** The nucleus of the ovum develops into a **female pronucleus** (Figure 29-4b). After the pronuclei are formed, they fuse to produce a **segmentation nucleus.** The segmentation nucleus contains 23 chromosomes from the male pronucleus and 23 chromosomes from the female pronucleus. Thus the fusion of the haploid pronuclei restores the diploid number. The fertilized ovum, consisting of a segmentation nucleus, cytoplasm, and enveloping membrane, is called a **zygote.**

CLINICAL APPLICATION

Dizygotic (fraternal) twins are produced from the independent release of two ova and the subsequent fertilization of each by different spermatozoa (Figure 29-5). They are the same age and are in the uterus at the same time, but they are genetically as dissimilar as any other siblings. They may or may not be the same sex. **Monozygotic (identical) twins** are derived from a single fertilized ovum that splits at an early stage in development. They contain the same genetic material and are always the same sex.

Monozygotic triplets are developed from a single fertilized ovum that splits twice; **dizygotic triplets** are derived from two ova from one or both ovaries, one of which splits after fertilization; and **trizygotic triplets** result from the fertilization of three ova from one or both ovaries.

Formation of the Morula

Immediately after fertilization, rapid cell division of the zygote takes place. This early division of the zygote is called **cleavage.** During this time the dividing cells are contained by the zona pellucida. Although cleavage in-

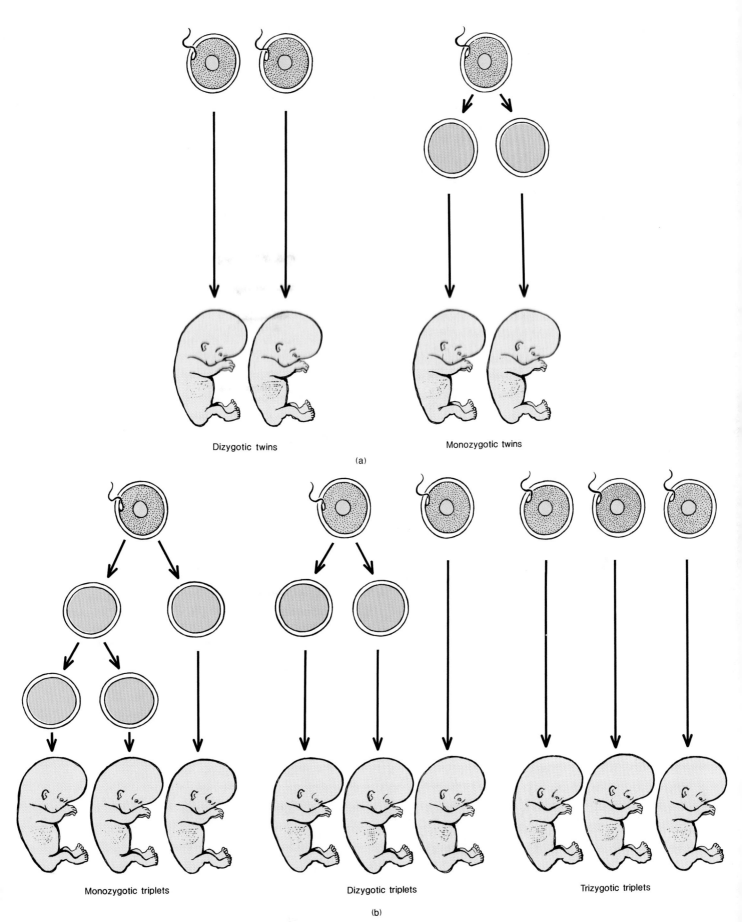

Dizygotic twins

Monozygotic twins

(a)

Monozygotic triplets

Dizygotic triplets

Trizygotic triplets

(b)

FIGURE 29-5 Possible embryonic origins of (a) twins and (b) triplets.

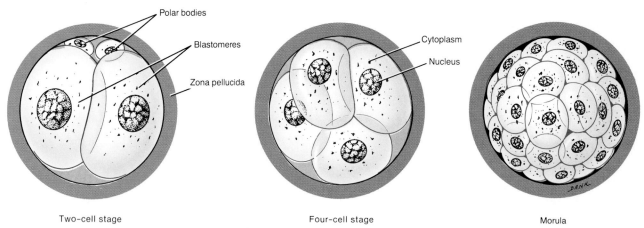

FIGURE 29-6 Formation of the morula.

Two-cell stage   Four-cell stage   Morula

creases the number of cells, it does not result in an increase in the size of the developing organism.

The first cleavage is completed after about 36 hours, and each succeeding division takes slightly less time (Figure 29-6). By the second day after conception, the second cleavage is completed. By the end of the third day there are 16 cells. The progressively smaller cells produced are called **blastomeres** (BLAS-tō-mērz). A few days after fertilization the successive cleavages have produced a solid mass of cells, the **morula** (MOR-yoo-la) or mulberry, which is about the same size as the original zygote.

Development of the Blastocyst

As the number of cells in the morula increases, it moves from the original site of fertilization down through the ciliated uterine tube toward the uterus and enters the uterine cavity. By this time, the dense cluster of cells is altered to form a hollow ball of cells. The mass is now referred to as a **blastocyst** (Figure 29-7).

The blastocyst is differentiated into an outer covering of cells called the **trophectoderm** (trō-FEK-tō-derm), an **inner cell mass,** and an internal fluid-filled cavity called the **blastocoel** (BLAS-tō-sēl). The trophectoderm ultimately forms part of the membranes composing the fetal portion of the placenta; the inner cell mass develops into the embryo.

Implantation

The attachment of the blastocyst to the endometrium occurs 7 to 8 days after fertilization and is called **implantation** (Figure 29-8). At this time, the endometrium is in its postovulatory phase. During implantation, the cells of the trophectoderm secrete an enzyme that enables the blastocyst to penetrate the uterine lining and become buried in the endometrium, usually on the posterior wall of the fundus or body of the uterus. The blastocyst, now only about $\frac{1}{100}$ of an inch in diameter, becomes oriented so that the inner cell mass is toward the endometrium. Implantation enables the blastocyst to absorb nutrients from the glands and blood vessels of the endometrium for its subsequent growth and development.

A summary of the principal events associated with fertilization and implantation is shown in Figure 29-9.

CLINICAL APPLICATION

In the early months of pregnancy, some females develop **hyperemesis gravidarum (morning sickness),** characterized by nausea and vomiting. Although the exact cause is unknown, it is possible that the degenerative products of digested portions of the endometrium during implantation may be responsible. Another possible cause is high levels of estrogens secreted by the placenta.

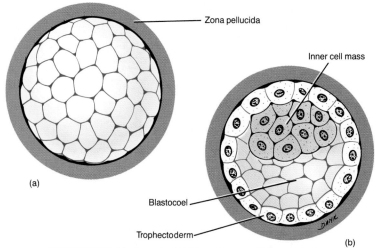

FIGURE 29-7 Blastocyst. (a) External view. (b) Internal view.

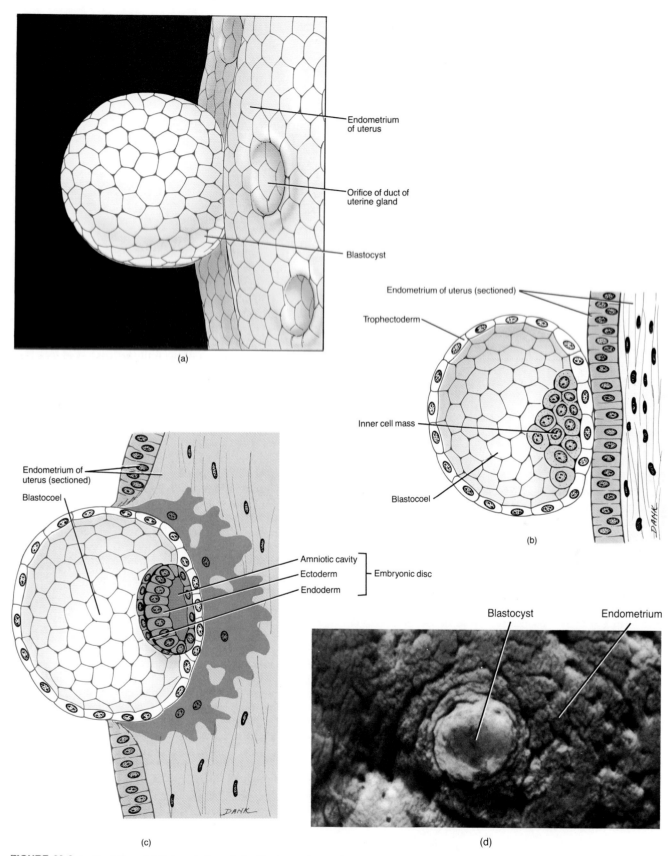

FIGURE 29-8 Implantation. (a) External view of the blastocyst in relation to the endometrium of the uterus about 5 days after fertilization. (b) Internal view of the blastocyst in relation to the endometrium about 6 days after fertilization. (c) Internal view of the blastocyst at implantation about 7 days after fertilization. (d) Photomicrograph of implantation. (Courtesy of Roberts Rugh and Landrum B. Shettles, M.D., with Richard N. Einhorn, *From Conception to Birth: The Drama of Life's Beginnings,* Harper & Row, Publishers, Inc., New York, 1971.)

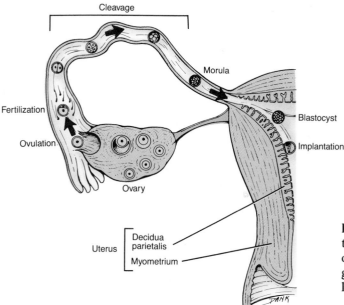

FIGURE 29-9 Summary of events associated with fertilization and implantation.

External Human Fertilization

On July 12, 1978, Louise Joy Brown was born near Manchester, England. Her birth was the first recorded case of **external human fertilization,** fertilization in a glass dish. The procedure developed for external human fertilization is believed to have been carried out as follows. The female is given FSH soon after menstruation, so that several ova, rather than the typical single one, will be produced (superovulation). Administration of LH may also ensure the maturation of the ova. Next, a small incision is made near the umbilicus, and the ova are aspirated from the follicles and placed in a medium that simulates the fluids in the female reproductive tract. The ova are then transferred to a solution of the male's sperm. Once fertilization has taken place, the fertilized ovum is put in another medium and is observed for cleavage. When the fertilized ovum reaches the 8-cell or 16-cell stage, it is introduced into the uterus for implantation and subsequent growth. The growth and development sequences that occur are similar to those in internal fertilization.

EMBRYONIC DEVELOPMENT

The first two months of development are generally considered the **embryonic period.** During this period the developing human is called an **embryo.** The months of development after the second month are considered the **fetal period,** and during this time the developing human is called a **fetus.** By the end of the embryonic period the rudiments of all the principal adult organs are present, the embryonic membranes are developed, and the placenta is functioning.

Beginnings of Organ Systems

Following implantation, the inner cell mass of the blastocyst begins to differentiate into the three **primary germ layers:** ectoderm, endoderm, and mesoderm. They are the embryonic tissues from which all tissues and organs of the body will develop. The various movements of cell groups leading to the establishment of the primary germ layers are referred to as **gastrulation.**

In the human, the germ layers form so quickly that it is difficult to determine the exact sequence of events. Before implantation, a layer of **ectoderm** (the trophectoderm) already has formed around the blastocoel (see Figure 29-8b). The trophectoderm will become part of the chorion, one of the fetal membranes. Within 8 days after fertilization, the inner cell mass moves downward so a space called the **amniotic cavity** lies between the inner cell mass and the trophectoderm. The bottom layer of the inner cell mass develops into an **endodermal** germ layer.

About the twelfth day after fertilization, striking changes appear (Figure 29-10a). A layer of cells from the inner cell mass has grown around the top of the amniotic cavity. These cells will become the amnion, another fetal membrane. The cells below the cavity are called the **embryonic disc.** They will form the embryo. At this stage, the embryonic disc contains ectodermal and endodermal cells; the mesodermal cells are scattered external to the disc. The cells of the endodermal layer have been dividing rapidly, so that groups of them now extend downward in a circle, forming the yolk sac, another fetal membrane. The **mesodermal** cells also have been dividing, and many have left the area of the embryonic disc and can be seen around the structures that are becoming fetal membranes.

About the fourteenth day, the cells of the embryonic disc differentiate into three distinct layers: the upper ectoderm, the middle mesoderm, and the lower endoderm (Figure 29-10b). At this time the two ends of the embryonic disc draw together, squeezing off the yolk sac. The resulting cavity inside the disc is the endoderm-lined **primitive gut.** The mesoderm in the disc soon splits into two layers, and the space between the layers becomes the **extraembryonic coelom.**

As the embryo develops (Figure 29-10c), the endoderm becomes the epithelial lining the digestive tract and respiratory tract and a number of other organs. The mesoderm forms the peritoneum, muscle, bone, and other connective tissue. The ectoderm develops into the skin and nervous system. Exhibit 29-1 provides more details about the fates of these primary germ layers.

EXHIBIT 29-1

STRUCTURES PRODUCED BY THE THREE PRIMARY GERM LAYERS

ENDODERM	MESODERM	ECTODERM
Epithelium of digestive tract (except the oral cavity and anal canal) and the epithelium of its glands.	All skeletal, most smooth, and all cardiac muscle.	All nervous tissue.
Epithelium of urinary bladder, gallbladder, and liver.	Cartilage, bone, and other connective tissues.	Epidermis of skin.
Epithelium of pharynx, auditory tube, tonsils, larynx, trachea, bronchi, and lungs.	Blood, bone marrow, and lymphoid tissue.	Hair follicles, arrector pili muscles, nails, and epithelium of skin glands (sebaceous and sudoriferous).
Epithelium of thyroid, parathyroid, pancreas, and thymus glands.	Endothelium of blood vessels and lymphatics.	Lens, cornea, and optic nerve of eye and internal eye muscles.
Epithelium of prostate and bulbourethral glands, vagina, vestibule, urethra, and associated glands such as the greater vestibular and lesser vestibular glands.	Dermis of skin.	Internal and external ear.
	Fibrous tunic and vascular tunic of eye.	Neuroepithelium of sense organs.
	Middle ear.	Epithelium of oral cavity, nasal cavity, paranasal sinuses, salivary glands, and anal canal.
	Mesothelium of coelomic and joint cavities.	
	Epithelium of kidneys and ureters.	Epithelium of pineal gland, hypophysis, and adrenal medulla.
	Epithelium of adrenal cortex.	
	Epithelium of gonads and genital ducts.	

Embryonic Membranes

During the embryonic period, the **embryonic membranes** form (Figure 29-11). These membranes lie outside the embryo and protect and nourish the embryo and later the fetus. The membranes are the yolk sac, amnion, chorion (KŌ-rē-on), and allantois (a-LAN-tō-is).

The **yolk sac** is an endoderm-lined membrane that, in many species, provides the primary or exclusive nutrient for the embryo. However, the human embryo re-

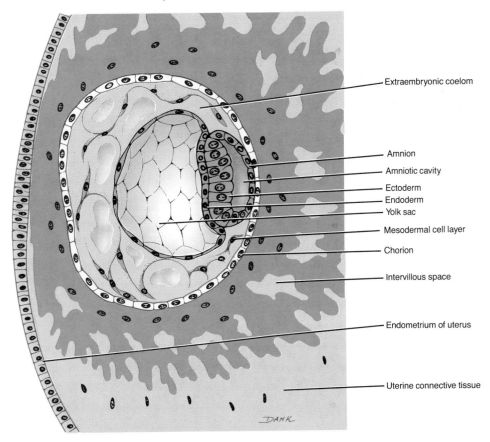

(a)

FIGURE 29-10 Formation of the primary germ layers and associated structures. (a) Internal view of the developing embryo about 12 days after fertilization.

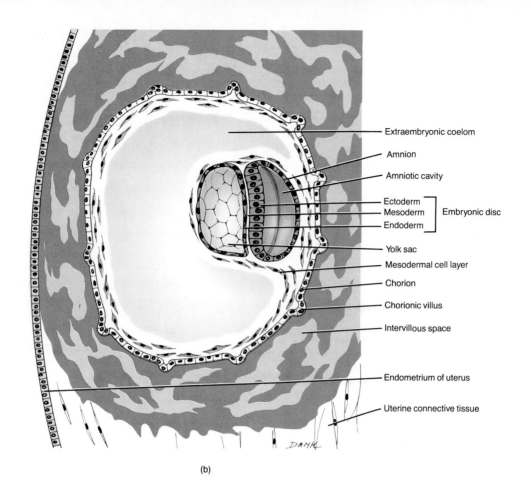

Extraembryonic coelom

Amnion

Amniotic cavity

Ectoderm ⎫
Mesoderm ⎬ Embryonic disc
Endoderm ⎭

Yolk sac

Mesodermal cell layer

Chorion

Chorionic villus

Intervillous space

Endometrium of uterus

Uterine connective tissue

(b)

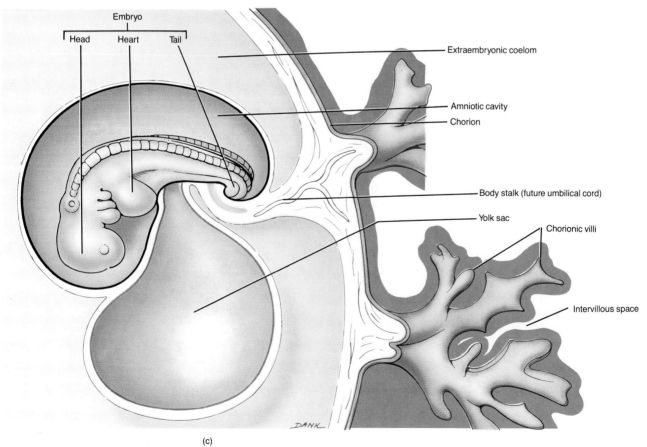

Embryo
Head   Heart   Tail

Extraembryonic coelom

Amniotic cavity

Chorion

Body stalk (future umbilical cord)

Yolk sac

Chorionic villi

Intervillous space

(c)

FIGURE 29-10 (*Continued*) Formation of the primary germ layers and associated structures. (b) Internal view of the developing embryo about 14 days after fertilization. (c) External view of the developing embryo about 25 days after fertilization.

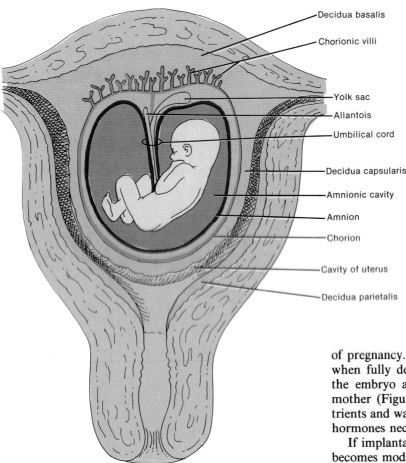

Decidua basalis

Chorionic villi

Yolk sac

Allantois

Umbilical cord

Decidua capsularis

Amnionic cavity

Amnion

Chorion

Cavity of uterus

Decidua parietalis

FIGURE 29-11 Embryonic membranes.

ceives its nourishment from the endometrium and the yolk sac remains small (see Figure 29-14b–d). During an early stage of development it becomes a nonfunctional part of the umbilical cord.

The **amnion** is a thin, protective membrane that initially overlies the embryonic disc and is formed by the eighth day following fertilization. As the embryo grows, the amnion entirely surrounds the embryo and becomes filled with **amniotic fluid** (see Figure 29-14c–g). Amniotic fluid serves as a shock absorber for the fetus. The amnion usually ruptures just before birth and with its fluid constitutes the "bag of waters."

The **chorion** is derived from the trophectoderm of the blastocyst and its associated mesoderm. It surrounds the embryo and, later, the fetus. Eventually the chorion becomes the principal part of the placenta, the structure through which materials are exchanged between mother and fetus. The amnion also surrounds the fetus and eventually fuses to the inner layer of the chorion.

The **allantois** is a small vascularized membrane. Later its blood vessels serve as connections in the placenta between mother and fetus. This connection is the umbilical cord.

Placenta and Umbilical Cord

Development of the placenta, the third major event of the embryonic period, is accomplished by the third month

of pregnancy. The **placenta** has the shape of a flat cake when fully developed and is formed by the chorion of the embryo and a portion of the endometrium of the mother (Figure 29-12). It provides an exchange of nutrients and wastes between fetus and mother and secretes hormones necessary to maintain pregnancy.

If implantation occurs, a portion of the endometrium becomes modified and is known as the **decidua** (dē-SID-yoo-a). The decidua includes all but the deepest layer of the endometrium and is shed when the fetus is delivered. Different regions of the decidua are named on the basis of their positions relative to the site of the implanted ovum (see Figures 29-11 and 29-13). The *decidua parietalis* (pa-rī-e-TAL-is) is the portion of the modified endometrium that lines the entire pregnant uterus, except for the area where the placenta is forming. The *decidua capsularis* is the portion of the endometrium between the embryo and the uterine cavity. The *decidua basalis* is the portion of the endometrium between the chorion and the muscularis of the uterus. The decidua basalis becomes the maternal part of the placenta.

During embryonic life, fingerlike projections of the chorion, called **chorionic villi** (kō'-rē-ON-ik VIL-ī), grow into the decidua basalis of the endometrium (see Figure 29-12a). These will contain fetal blood vessels of the allantois. They continue growing until they are bathed in maternal blood sinuses called **intervillous** (in-ter-VIL-us) **spaces.** Thus, maternal and fetal blood vessels are brought into proximity. It should be noted, however, that maternal and fetal blood do not normally mix. Oxygen and nutrients from the mother's blood diffuse into the capillaries of the villi. From the capillaries the nutrients circulate into the umbilical vein. Wastes leave the fetus through the umbilical arteries, pass into the capillaries of the villi, and diffuse into the maternal blood. The **umbilical cord** consists of an outer layer of amnion containing the umbilical arteries and umbilical vein, supported internally by mucous connective tissue from the allantois called Wharton's jelly (see Figure 29-14b–h).

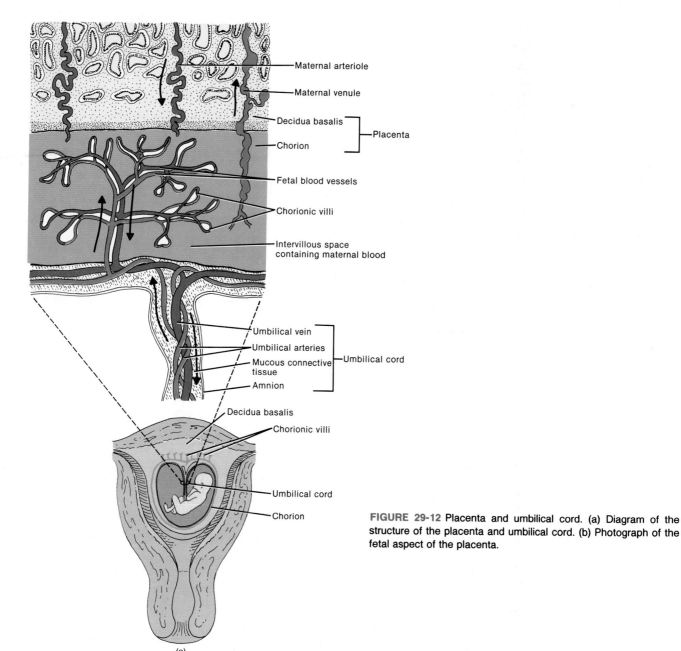

Maternal arteriole

Maternal venule

Decidua basalis

Chorion

Placenta

Fetal blood vessels

Chorionic villi

Intervillous space
containing maternal blood

Umbilical vein

Umbilical arteries

Mucous connective
tissue

Amnion

Umbilical cord

Decidua basalis

Chorionic villi

Umbilical cord

Chorion

(a)

FIGURE 29-12 Placenta and umbilical cord. (a) Diagram of the structure of the placenta and umbilical cord. (b) Photograph of the fetal aspect of the placenta.

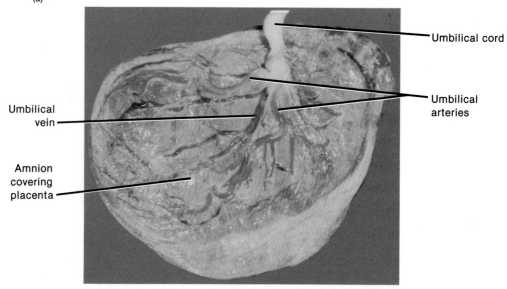

Umbilical cord

Umbilical
arteries

Umbilical
vein

Amnion
covering
placenta

(b)

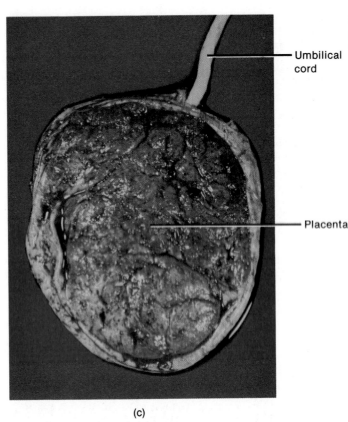

Umbilical cord

Placenta

(c)

FIGURE 29-12 (*Continued*) Placenta and umbilical cord. (c) Photograph of the maternal aspect of the placenta. (Courtesy of C. Yokochi and J. W. Rohen, *Photographic Anatomy of the Human Body,* 2nd ed., 1979, IGAKU-SHOIN, Ltd., Tokyo, New York.)

FETAL GROWTH

During the **fetal period,** organs established by the primary germ layers grow rapidly. The organism takes on a human appearance. A summary of changes associated with the fetal period is presented in Exhibit 29-2. Representative photos of embryos and fetuses are shown in various stages of development in Figure 29-14.

CLINICAL APPLICATION

At delivery, the placenta detaches from the uterus and is referred to as the **afterbirth.** At this time, the umbilical cord is severed, leaving the baby on its own. The scar that marks the site of the entry of the fetal umbilical cord into the abdomen is the **umbilicus (navel).**

Pharmaceutical houses use human placenta as a source of hormones, drugs, and blood. Portions of placentas are also used for burn coverage. The placental and umbilical cord veins are used in blood vessel grafts.

CLINICAL APPLICATION

Physicians can see into the uteri of pregnant women without exposing them to the known dangers of x-rays and without pain or intrusion. One such technique, called **ultrasound** (also **sonography** or **ultrasonography**), uses high frequency, inaudible sound waves, which are directed into the abdomen of the mother-to-be and then reflected back to a receiver. The reflected waves give a visual "echo" of what's inside the uterus. This echo is transformed electronically into an image on a screen. Another technique is the **Doppler Detector,** a fetal monitor that registers the baby's heart rate. It is used routinely in many hospitals during labor.

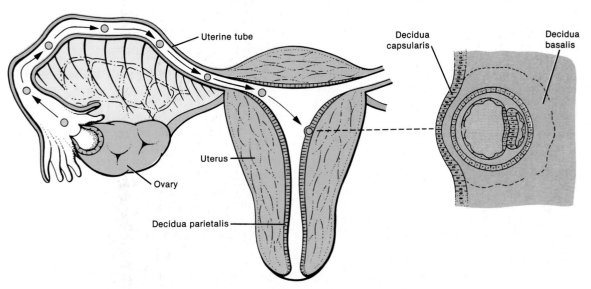

Uterine tube

Decidua capsularis

Decidua basalis

Uterus

Ovary

Decidua parietalis

FIGURE 29-13 Regions of the decidua.

EXHIBIT 29-2

CHANGES ASSOCIATED WITH FETAL GROWTH

END OF MONTH	APPROXIMATE SIZE AND WEIGHT	REPRESENTATIVE CHANGES
1	0.6 cm ($^3/_{16}$ inch)	Eyes, nose, and ears not yet visible. Backbone and vertebral canal form. Small buds that will develop into arms and legs form. Heart forms and starts beating. Body systems begin to form.
2	3 cm (1¼ inches) 1 g ($^1/_{30}$ oz)	Eyes far apart, eyelids fused, nose flat. Ossification begins. Limbs become distinct as arms and legs. Digits are well formed. Major blood vessels form. Many internal organs continue to develop.
3	7.5 cm (3 inches) 28 g (1 oz)	Eyes almost fully developed but eyelids still fused, nose develops bridge, and external ears are present. Ossification continues. Appendages are fully formed and nails develop. Heartbeat can be detected. Body systems continue to develop.
4	18 cm (6½–7 inches) 113 g (4 oz)	Head large in proportion to rest of body. Face takes on human features and hair appears on head. Skin bright pink. Many bones ossified, and joints begin to form. Continued development of body systems.
5	25–30 cm (10–12 inches) 227–454 g (½–1 lb)	Head less disproportionate to rest of body. Fine hair (lanugo) covers body. Skin still bright pink. Rapid development of body systems.
6	27–35 cm (11–14 inches) 567–681 g (1¼—1½ lb)	Head becomes even less disproportionate to rest of body. Eyelids separate and eyelashes form. Skin wrinkled and pink.
7	32–42 cm (13–17 inches) 1,135–1,362 g (2½–3 lb)	Head and body become more proportionate. Skin wrinkled and pink. Seven-month fetus (premature baby) is capable of survival.
8	41–45 cm (16½–18 inches) 2,043–2,270 g (4½–5 lb)	Subcutaneous fat deposited. Skin less wrinkled. Testes descend into scrotum. Bones of head are soft. Chances of survival much greater at end of eighth month.
9	50 cm (20 inches) 3,178–3,405 g (7–7½ lb)	Additional subcutaneous fat accumulates. Lanugo shed. Nails extend to tips of fingers and maybe even beyond.

In the United States today, most obstetricians prescribe ultrasound examinations only when there is some clinical question about the normal progress of the pregnancy. By far the most common use of diagnostic ultrasound is to determine true fetal age when the date of conception is unknown or mistaken by the mother.

Ultrasound has gained a wide application beyond obstetrics as well, in detecting the presence of tumors, gallstones, and other abnormal internal masses.

HORMONES OF PREGNANCY

The corpus luteum is maintained throughout pregnancy. For about 8 to 10 weeks after fertilization it continues to secrete **estrogens** and **progesterone.** Both these hormones maintain the lining of the uterus during pregnancy and prepare the mammary glands to secrete milk. The amount secreted by the corpus luteum, however, is only slightly more than that produced after ovulation in a normal menstrual cycle. The high levels of estrogens and progesterone needed to maintain pregnancy and develop the mammary glands for lactation are provided by the placenta.

The chorion of the placenta secretes **human chorionic gonadotropin (HCG).** The primary role of HCG seems to be to maintain the activity of the corpus luteum, especially with regard to continuous progesterone secretion— an activity necessary for the continued attachment of the fetus to the lining of the uterus (Figure 29-15).

CLINICAL APPLICATION
Human chorionic gonadotropin is excreted in the urine of pregnant women from about the eighth day of pregnancy reaching its peak of excretion about the eighth week of pregnancy. The HCG level decreases sharply during the fourth and fifth months and then levels off until childbirth. Excretion of HCG in the urine serves as the basis for most **pregnancy tests.**

The placenta begins to secrete estrogens and progesterone no later than the sixtieth day of pregnancy. They are secreted in increasing quantities until the time of birth. By the fourth month, when the placenta is established, the secretion of HCG is greatly reduced because the secretions of the corpus luteum are no longer essential. The placenta supplies the levels of estrogens and progesterone needed to maintain the pregnancy. The fetal hormones thus take over management of the mother's body in preparation for parturition (birth) and lactation. Following delivery, estrogens and progesterone in the blood decrease to normal levels.

Recent evidence indicates that the placenta produces a regulating factor called **placental luteotropic releasing factor (pLRF).** It is chemically similar to the luteotropic releasing factor (LRF) produced by the hypothalamus, which causes the synthesis and release of luteinizing hormone (LH) by the anterior pituitary. The proposed function of pLRF is to stimulate the secretion of HCG by the placenta.

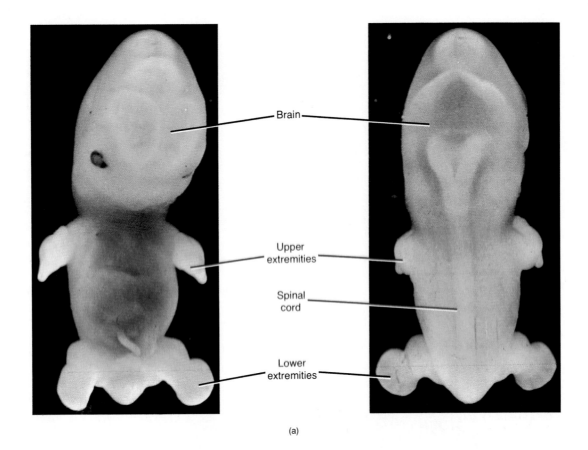

(a)

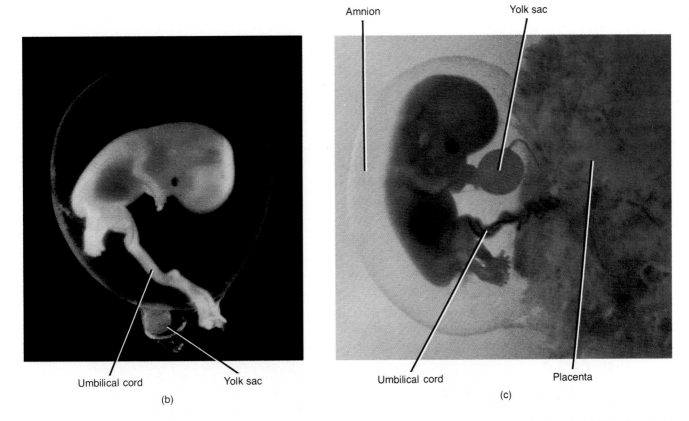

(b)

(c)

FIGURE 29-14 Human embryos and fetuses in various stages of development. (a) A 40-day embryo seen in anterior view (left) and posterior view (right). Note the brain, spinal cord, and upper and lower extremities. (b) Seven-week embryo with umbilical cord and yolk sac. (c) Eight-week embryo showing yolk sac, amnion, and placenta. (Courtesy of Roberts Rugh and Landrum B. Shettles, M.D., with Richard N. Einhorn, *From Conception to Birth: The Drama of Life's Beginnings,* Harper & Row, Publishers, Inc., New York, 1971.)

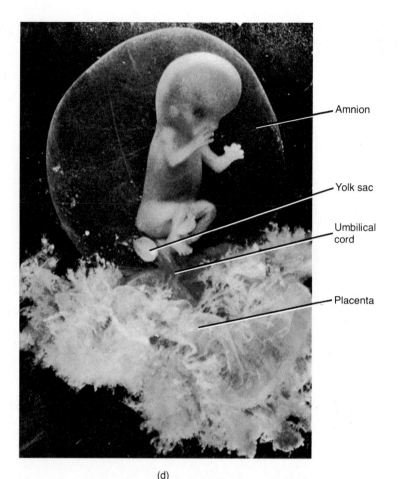

(d)

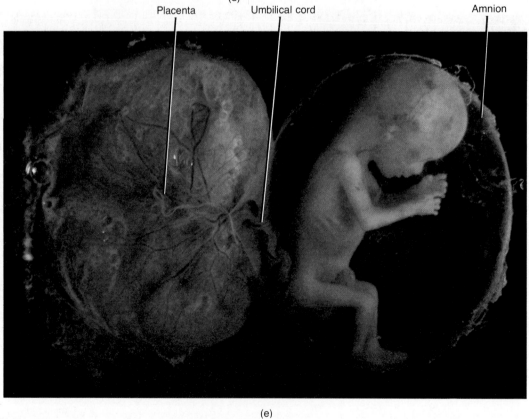

(e)

FIGURE 29-14 (*Continued*) Human embryos and fetuses in various stages of development. (d) Ten-week fetus. (e) Twelve-week fetus. Note the progressive development of the extremities. (Courtesy of Roberts Rugh and Landrum B. Shettles, M.D., with Richard N. Einhorn, *From Conception to Birth: The Drama of Life's Beginnings,* Harper & Row, Publishers, Inc., New York, 1971.)

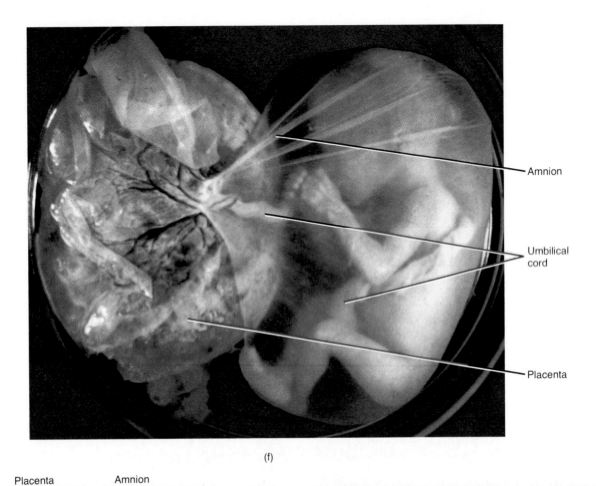

Amnion

Umbilical
cord

Placenta

(f)

Placenta   Amnion

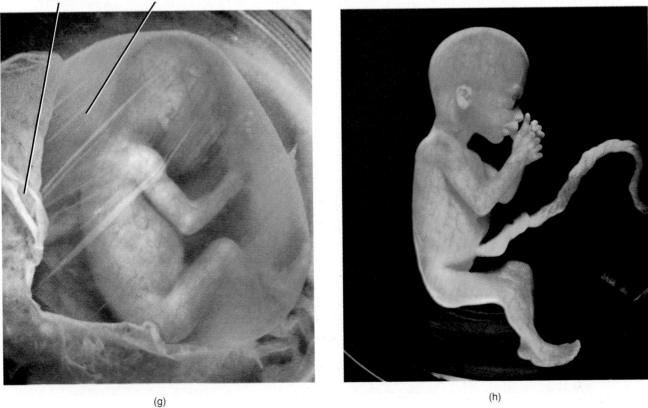

(g)

(h)

FIGURE 29-14 (*Continued*) Human embryos and fetuses in various stages of development. (f) Fourteen-week fetus. (g) Sixteen-week fetus. (h) Seventeen-week fetus sucking its thumb. (Courtesy of Roberts Rugh and Landrum B. Shettles, M.D., with Richard N. Einhorn, *From Conception to Birth: The Drama of Life's Beginnings,* Harper & Row, Publishers, Inc., New York, 1971.)

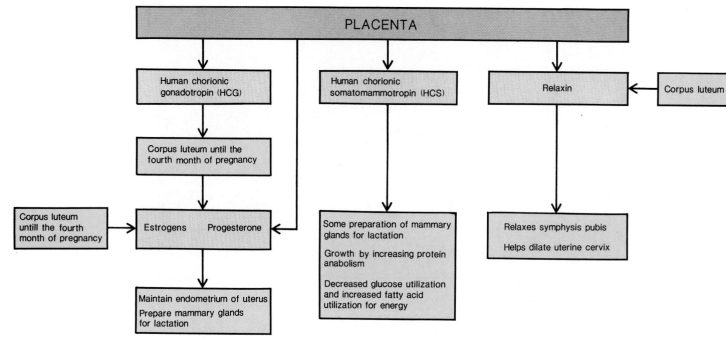

FIGURE 29-15 Hormones of pregnancy.

Another hormone produced by the chorion of the placenta is **human chorionic somatomammotropin (HCS).** Its secretion begins at about the same time as that of HCG, but its pattern of secretion is quite different. The rate of secretion of HCS increases in proportion to placental mass, reaching maximum levels after 32 weeks and remaining relatively constant after that. HCS is believed to stimulate some development of breast tissue for lactation, enhance growth by causing protein deposition in tissues, and regulate certain aspects of metabolism. For example, HCS causes decreased utilization of glucose by the mother, thus making more available for fetal metabolism. Also, HCS promotes the release of fatty acids from fat depots, providing an alternative source of energy for the mother's metabolism.

Relaxin is a hormone produced by the placenta and ovaries. Its physiological role is to relax the symphysis pubis and ligaments of the sacroiliac and sacrococcygeal joints and help dilate the uterine cervix toward the end of pregnancy. Both of these actions assist in delivery.

PARTURITION AND LABOR

The time the embryo or fetus is carried in the uterus is called **gestation** (jes-TĀ-shun). The total human gestation period is about 280 days from the beginning of the last menstrual period. The term **parturition** (par'-too-RISH-un) refers to birth. Parturition is accompanied by a sequence of events commonly called **labor.**

The **onset of labor** is apparently related to a complex interaction of many factors. Just prior to birth, the muscles of the uterus contract rhythmically and forcefully. Both placental and ovarian hormones seem to play a role in these contractions. Since progesterone inhibits uterine contractions, labor cannot take place until its effects are diminished. At the end of gestation, the level of estrogens

in the mother's blood is sufficient to overcome the inhibiting effects of progesterone and labor commences. It has been suggested that some factor released by the placenta, fetus, or mother rather suddenly overcomes the inhibiting effects of progesterone so that estrogens can exert their effect. Prostaglandins may also play a role in labor. Oxytocin (OT) from the posterior pituitary gland also stimulates uterine contractions. And, relaxin assists by relaxing the symphysis pubis and helping to dilate the uterine cervix.

Uterine contractions occur in waves, quite similar to peristaltic waves, that start at the top of the uterus and move downward. These waves expel the fetus. **True labor** begins when pains occur at regular intervals. The pains correspond to uterine contractions. As the interval between contractions shortens, the contractions intensify. Another sign of true labor in some females is localization of pain in the back, which is intensified by walking. A reliable indication of true labor is the "show" and dilation of the cervix. The "show" is a discharge of a blood-containing mucus that accumulates in the cervical canal during pregnancy. In **false labor,** pain is felt in the abdomen at irregular intervals. The pain does not intensify and is not altered significantly by walking. There is no "show" and no cervical dilation.

Labor can be divided into three stages (Figure 29-16).

1. The **stage of dilation** is the time from the onset of labor to the complete dilation of the cervix (Fig. 29-16). During this stage there are regular contractions of the uterus, usually a rupturing of the amniotic sac, and complete dilation (10 cm) of the cervix. If the amniotic sac does not rupture spontaneously, it is done artificially.

2. The **stage of expulsion** is the time from complete cervical dilation to delivery.

3. The **placental stage** is the time after delivery until the placenta or "afterbirth" is expelled by powerful uterine contrac-

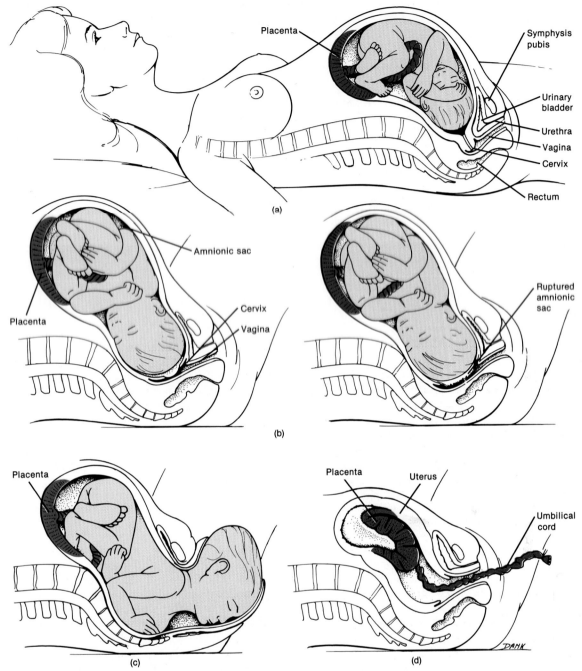

FIGURE 29-16 Parturition. (a) Fetal position prior to birth. (b) Dilation. Protrusion of amnionic sac through partly dilated cervix (left). Amnionic sac ruptured and complete dilation of cervix (right). (c) Stage of expulsion. (d) Placental stage.

tions. These contractions also constrict blood vessels that were torn during delivery. In this way, the possibility of hemorrhage is reduced.

CLINICAL APPLICATION
Pudendal (pyoo-DEN-dal) **nerve block** is used for procedures such as episiotomy. The primary innervation

to the skin and muscles of the perineum is the pudendal nerve. In the transvaginal approach, the needle is passed through the lateral vaginal wall to a point just medial to the ischial spine. The anesthesia results in loss of the anal reflex, relaxation of the muscles of the floor of the pelvis, and loss of sensation to the vulva and lower one-third of the vagina.

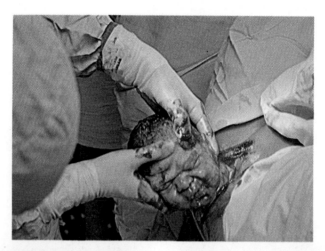

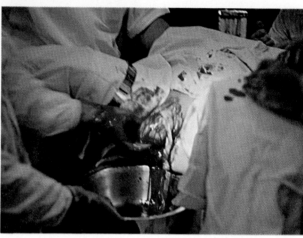

(e)

FIGURE 29-16 (*Continued*) Parturition. (e) Sequential photographs of parturition. Top: Emergence of the infant's head. (Courtesy of Kinne, Photo Researchers.) Bottom left: Delivery of the infant. (Courtesy of Thomas, Photo Researchers.) Bottom right: Delivery of the placenta. (Courtesy of McCartney, Photo Researchers.)

Various deformities of the female pelvis may be responsible for **dystocia**, (dis-TŌ-sē-a), that is, difficult labor. Pelvic deformities may be congenital or acquired from disease, fractures, or poor posture. Among other conditions associated with difficult labor are malposition of the fetus, malpresentation of the fetus, and premature rupture of the fetal membranes.

ADJUSTMENTS OF THE INFANT AT BIRTH

During pregnancy, the embryo and later the fetus is totally dependent on the mother for its existence. The mother supplies the fetus with oxygen and nutrients, eliminates its carbon dioxide and other wastes, and protects it against shocks, temperature changes, and certain harmful microbes. At birth the baby becomes self-supporting, and the newborn's body systems must make various adjustments. Following are some changes that occur in the respiratory and cardiovascular systems.

RESPIRATORY SYSTEM

The respiratory system is fairly well developed at least 2 months before birth as evidenced by the fact that prema-

ture babies delivered at 7 months are able to breathe and cry. The fetus depends entirely on the mother for obtaining oxygen and eliminating carbon dioxide. The fetal lungs are either collapsed or partially filled with amniotic fluid, which is absorbed at birth. After delivery the baby's supply of oxygen from the mother is stopped. Circulation in the baby continues and as the blood level of carbon dioxide increases, the respiratory center in the medulla is stimulated. This causes the respiratory muscles to contract and the baby draws its first breath. Since the first inspiration is unusually deep because the lungs contain no air, the baby exhales vigorously and naturally cries. A full-term baby may breathe 45 times a minute for the first two weeks after exposure to air. The rate is gradually reduced until it approaches a normal rate.

CARDIOVASCULAR SYSTEM

Following the first inspiration by the baby, the cardiovascular system must make several adjustments. The foramen ovale of the fetal heart between atria closes at the moment of birth. This diverts deoxygenated blood to the lungs for the first time. The foramen ovale is closed by two flaps of heart tissue that fold together and permanently fuse. The remnant of the foramen ovale is the fossa ovalis. Once the lungs begin to function, the ductus arteriosus is shut off by contractions of the muscles in its wall. The ductus arteriosus generally does not completely and irreversibly close for about 3 months following birth. Incomplete closing, as you already know, results in patent ductus arteriosus.

The ductus venosus of fetal circulation connects the umbilical vein directly with the inferior vena cava. It forces any remaining umbilical blood directly into the fetal liver and from there to the heart. When the umbilical cord is severed, all visceral blood of the fetus goes directly to the fetal heart via the inferior vena cava. This shunting of blood usually occurs within minutes after birth, but may take a week or two to complete. The ligamentum venosum, the remnant of the ductus venosus, is well established by the eighth postnatal week.

At birth, the infant's pulse may be from 120 to 160 per minute and may go as high as 180 following excitation. Several days after birth, there is a greater independent need for oxygen which stimulates an increase in the rate of erythrocyte and hemoglobin production. This increase usually lasts for only a few days. Moreover, the white blood cell count at birth is very high, sometimes as much as 45,000 cells per cubic millimeter, but this decreases rapidly by the seventh day.

Finally, the infant's liver may not be adjusted at birth to control the production of bile pigment. As a result of this and other complicating factors, a temporary jaundice may result in as many as 50 percent of normal newborns by the third or fourth day after birth.

POTENTIAL HAZARDS TO THE DEVELOPING EMBRYO AND FETUS

The developing fetus is susceptible to a number of potential hazards that can be transmitted from the mother. Such hazards include infectious microbes, chemicals and drugs, alcohol, and cigarette smoking. In addition, certain environmental conditions and pollutants can damage the fetus or even cause fetal death.

CHEMICALS AND DRUGS

Since the placenta is known to be an ineffective barrier between the maternal and fetal circulations, actually any drug or chemical dangerous to the infant may be considered potentially dangerous to the fetus when given to the mother. Many chemicals and drugs have been proven to be toxic and teratogenic to the developing embryo and fetus. A **teratogen** (*terato* = monster) is any agent or influence that causes physical defects in the developing embryo. Examples are pesticides, defoliants, industrial chemicals, some hormones, antibiotics, oral anticoagulants, anticonvulsants, antitumor agents, thyroid drugs, thalidomide, diethylstilbestrol (DES), LSD, and marijuana. In addition, accumulating evidence suggests that many agents assumed to be nonteratogenic can produce long-term adverse effects on reproductive ability and neurobehavioral development.

IRRADIATION

Ionizing radiations are potent teratogens. Treatment of pregnant mothers with large doses of x-rays and radium during the embryo's susceptible period of development may cause microcephaly (small size of head in relation to the rest of the body), mental retardation, and skeletal malformations. Caution is advised for diagnostic x-rays during the first trimester of pregnancy.

ALCOHOL

Alcohol has been a suspected teratogen for centuries, but only recently has a relationship been recognized between maternal alcohol intake and the characteristic pattern of malformations in the fetus. The term applied to the effects of intrauterine exposure to alcohol is **fetal alcohol syndrome (FAS)**. Studies to date have indicated that in the general population the incidence of FAS may exceed 1 per 1,000 live births, and may be by far the number one fetal teratogen. The symptoms shown by children may include slow growth before and after birth, small head, facial irregularities such as narrow eye slits and sunken nasal bridge, defective heart and other organs, malformed arms and legs, genital abnormalities, and mental retardation. There are also behavioral problems, such as hyperactivity, extreme nervousness, and a poor attention span.

CIGARETTE SMOKING

The latest evidence not only indicates a causal relationship between cigarette smoking during pregnancy and low infant birth weight, but also points to a strong probable association between smoking and a higher fetal and infant mortality. Infants nursing from smoking mothers have also been found to have an increased incidence of gastrointestinal disturbances. Other pathologies of infants of smoking mothers include an increased incidence of respiratory problems during the first year of life, including bronchitis and pneumonia. Cigarette smoking may be teratogenic and cause cardiac abnormalities and anencephaly (a developmental anomaly with absence of neural tissue in the cranium). Maternal smoking also appears to be a significant etiologic factor in the development of cleft lip and palate and has been tentatively linked with sudden infant death syndrome (SIDS).

LACTATION

The term **lactation** refers to the secretion and ejection of milk by the mammary glands. The major hormone in promoting lactation is **prolactin (PRL)** from the anterior pituitary gland. It is released in response to the hypothalamic prolactin releasing factor (PRF). Even though PRL levels increase as the pregnancy progresses, there is no milk secretion because estrogens and progesterone cause the hypothalamus to release prolactin-inhibiting factor (PIF). Following delivery, the levels of estrogens and progesterone in the mother's blood decrease and the inhibition is removed.

The principal stimulus in maintaining prolactin secretion during lactation is the sucking action of the infant. Sucking initiates impulses from receptors in the nipples to the hypothalamus. The impulses inhibit PIF production, and PRL is released by the anterior pituitary. The sucking action also initiates impulses to the posterior pituitary via the hypothalamus. These impulses stimulate the release of the hormone **oxytocin (OT)** by the posterior pituitary gland. This hormone induces certain cells surrounding the outer walls of the alveoli to contract, thereby compressing the alveoli. The compression moves milk from the alveoli of the mammary gland into the ducts, where it can be sucked. This process is referred to as **milk letdown.** Lactation often prevents the occurrence of female ovarian cycles for the first few months following delivery by inhibiting FSH and LH release by the anterior pituitary.

During late pregnancy and the first few days after birth, the mammary glands secrete a cloudy fluid called **colostrum.** Although it is not as nutritious as true milk, since it contains less lactose and virtually no fat, it serves adequately until the appearance of true milk on about the fourth day. Colostrum is thought to contain antibodies that protect the infant during the first few months of life.

Following birth of the infant, PRL level starts to return to the nonpregnant level, but each time the mother nurses the infant, nerve impulses from the nipples to the hypothalamus cause the release of PRF and the secretion of a tenfold increase in PRL by the anterior pituitary that lasts about an hour. PRL acts on the mammary glands to provide milk for the next nursing period. If this surge of PRL is blocked by injury or disease or if nursing is discontinued, the mammary glands lose their ability to secrete milk in a few days. However, milk secretion can continue for several years if the child continues to suckle. Milk secretion normally decreases considerably within seven to nine months.

BIRTH CONTROL

Methods of **birth control** include removal of the gonads and uterus, sterilization, and contraception.

REMOVAL OF GONADS AND UTERUS

Castration (removal of the testes), **hysterectomy** (removal of the uterus), and **oophorectomy** (ō'-of-ō-REK-tō-mē;

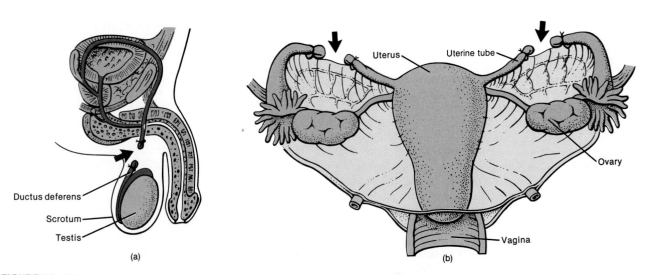

(a)

(b)

FIGURE 29-17 Sterilization. (a) Vasectomy. The ductus deferens of each testis is cut and tied after an incision is made into the scrotum. (b) Tubal ligation. Each uterine tube is cut and tied after an incision is made into the abdomen.

removal of the ovaries) are all absolute preventive methods. Once performed, these operations cannot be reversed and it is impossible to produce offspring. However, removal of the testes or ovaries has adverse effects because of the importance of these organs in the endocrine system. Generally these operations are performed only if the organs are diseased. Castration before puberty prevents the development of secondary sex characteristics.

STERILIZATION

One means of **sterilization** of males is **vasectomy**—a simple operation in which a portion of each ductus (vas) deferens is removed (Figure 29-17a). An incision is made in the scrotum, the tubes are located, and each is tied in two places. Then the portion between the ties is cut out. Sperm production can continue in the testes, but the sperm cannot reach the exterior.

Sterilization in females generally is achieved by performing a **tubal ligation** (lī-GĀ-shun). An incision is made into the abdominal cavity, the uterine tubes are squeezed, and a small loop called a knuckle is made (Figure 29-17b). A suture is tied tightly at the base of the knuckle and the knuckle is then cut. After four or five days the suture is digested by body fluids and the two severed ends of the tubes separate. The ovum thus is prevented from passing to the uterus, and the sperm cannot reach the ovum. Sterilization normally does not affect sexual performance or enjoyment.

Another method of sterilizing women is the **laparoscopic** (lap'-a-rō-SKŌ-pik) **technique**. After a woman receives local or general anesthesia, a harmless gas is introduced into her abdomen to create a gas bubble. The bubble expands the abdominal cavity and pushes the intestines away from the pelvic organs, permitting safe, easy access to the uterine tubes. The doctor makes a small incision at the lower rim of the umbilicus and inserts a laparoscope to view the inside of the abdominal cavity and the uterine tubes. The tubes can be closed with this instrument or a second incision can be made at the pubic hairline to insert a cautery forceps. Once the uterine tubes are sealed, the instrument is removed, the gas is released, and the incision is covered with a bandage. After a few hours the patient can usually go home.

CLINICAL APPLICATION

Microsurgery is any operation done with the aid of microscopes. Surgeons work with spectacle-mounted magnifiers or with microscopes. Laparoscopy is one such technique and several others have been mentioned throughout the text.

In the United States alone, more than 10 million people have been sterilized by artificial closures. For men who have had vasectomies, reversal through microsurgery has meant about a 90 percent chance of regaining a normal sperm count. Limited success has been attained in reversing women who have undergone sterilization surgery through microsurgery.

Microsurgery's impact on curing fertility problems has been nothing short of revolutionary. Using this technique, surgeons can remove a prolactin-secreting adenoma, a hormone-producing tumor of the pituitary gland that causes infertility in women, without damaging the pituitary gland itself. Microsurgery is also used to open blocked uterine tubes and spermatic ducts. Natural obstructions are a major cause of sterility in both sexes.

Neurosurgeons and ear and eye surgeons all rely on microsurgery, and thousands of people today already owe their health, children, limbs, and lives to microsurgery.

CONTRACEPTION

Contraception is the prevention of fertilization without destroying fertility by natural, mechanical, or chemical means.

Natural

The **natural methods** include complete or periodic abstinence. An example of periodic abstinence is the rhythm method, which takes advantage of the fact that a fertilizable ovum is available only during a period of three to five days in each menstrual cycle. During this time the couple refrains from intercourse. Its effectiveness is limited by the fact that few women have absolutely regular cycles. Moreover, some women occasionally ovulate during the "safe" times of the month, such as during menstruation.

Mechanical

Mechanical methods of contraception include the use of the condom by the male and the diaphragm by the female. The **condom** is a nonporous, elastic (rubber or similar material) covering placed over the penis that prevents deposition of sperm in the female reproductive tract. The **diaphragm** (Figure 29-18) is a dome-shaped structure that fits over the cervix and is generally used in conjunction with a sperm-killing chemical. The diaphragm stops the sperm from passing into the cervix. The chemical kills the sperm cells.

Another mechanical method of contraception is an **intrauterine device (IUD).** The device is a small object made of plastic, copper, or stainless steel and shaped like a loop, coil, T, or 7. It is inserted into the cavity of the uterus (Figure 29-19). It is not clear how IUDs operate. Some investigators believe they cause changes in the uterus lining that, in turn, produce a substance which destroys either the sperm or the fertilized ovum. Some IUDs release contraceptive agents in minute amounts; one new device secretes progesterone.

Chemical

Chemical methods of contraception include spermicidal and hormonal methods. Various foams, creams, jellies, suppositories, and douches make the vagina and cervix unfavorable for sperm survival.

The hormonal method, **oral contraception** (the pill), has found rapid and widespread use. Although several

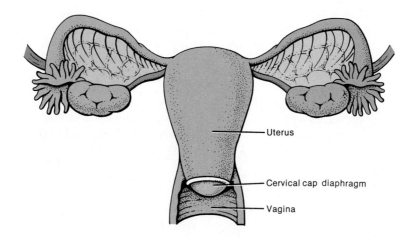

FIGURE 29-18 Cervical cap diaphragm. This particular type of diaphragm fits directly over the cervix of the uterus. The thin spring around the margin of the diaphragm opens outward, presses against the wall of the vagina, and stretches across the cervix.

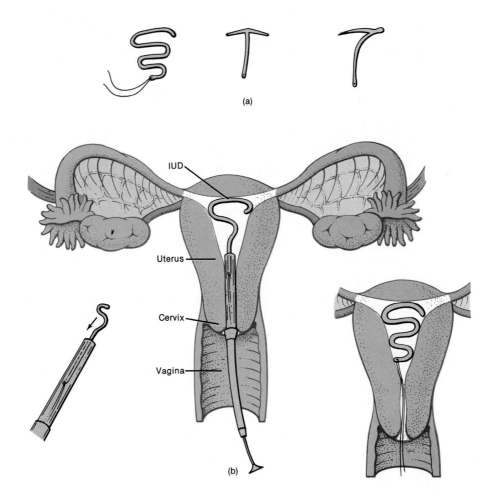

FIGURE 29-19 Intrauterine devices (IUDs). (a) Representative designs of three IUDs. (b) Procedure for insertion. The device is compressed to fit into a long, narrow-bore tube that is passed through the slightly dilated cervix. Once in position in the uterus, it spreads out to its former shape. IUDs have a thread or chain projecting into the vagina that is used as an indication that the device is still in place and for removal by the physician.

EXHIBIT 29-3
BIRTH CONTROL METHODS

METHOD	COMMENTS
Removal of gonads and uterus	Irreversible sterility. Generally performed if organs are diseased rather than as contraceptive method because of importance of hormones produced by gonads.
Sterilization	Procedure involving severing ductus (vas) deferens in males (vasectomy) and uterine tubes in females (tubal ligation and laparascopic technique).
Natural contraception	Abstinence from intercourse during time of month woman is fertile. Under ideal circumstances, effectiveness in women with regular menstrual cycles may approach that of mechanical and chemical contraceptives. Extremely difficult to determine fertile period. Effectiveness can be increased by recording body temperature each morning before getting up; a small rise in temperature indicates that ovulation has occurred one or two days earlier and ovum can no longer be fertilized.
Mechanical contraception **Condom**	Thin, strong sheath of rubber or similar material worn by male to prevent sperm from entering vagina. Failures caused by sheath tearing or slipping off after climax or not putting the sheath on soon enough. If used correctly and consistently, effectiveness similar to that of diaphragm.
Diaphragm	Flexible rubber dome inserted into vagina to cover cervix, providing barrier to sperm. Usually used with spermicidal cream or jelly. Must be left in place at least 6 hours after intercourse and may be left in place as long as 24 hours. Must be fitted by physician or other trained personnel and refitted every two years and after each pregnancy. Offers high level of protection if used with spermicide; rate of 2 to 3 pregnancies per 100 women per year is estimated for consistent users. If not used consistently, much higher pregnancy rates must be expected. Occasional failures caused by improper insertion or displacement during sexual intercourse.
Intrauterine device (IUD)	Small object (loop, coil, T, or 7) made of plastic, copper, or stainless steel and inserted into uterus by physician. May be left in place for long periods of time (some must be changed every two to three years). Does not require continued attention by user. Some women cannot use them because of expulsion, bleeding, or discomfort. Not recommended for women who have not had child because uterus is too small and cervical canal too narrow. Infrequently, insertion may lead to inflammation of pelvic organs.
Chemical contraception **Foams, creams, jellies, suppositories, vaginal douches**	Sperm-killing chemicals inserted into vagina to coat vaginal surfaces and cervical opening. Provide protection for about 1 hour. Effective when used alone, but significantly more effective when used with diaphragm or condom.
Oral contraceptive (OC)	Except for total abstinence or surgical sterilization, most effective contraceptive known. Side effects include nausea, occasional light bleeding between periods, breast tenderness or enlargement, fluid retention, and weight gain. Should not be used by women who have cardiovascular conditions (thromboembolic disorders, cerebrovascular disease, heart disease, hypertension), liver malfunction, cancer or neoplasia of breast or reproductive organs, or by women who smoke.

pills are available, the one most commonly used contains a high concentration of progesterone and a low concentration of estrogens. These two hormones act on the anterior pituitary to decrease the secretion of FSH and LH by inhibiting GnRF by the hypothalamus. The low levels of FSH and LH are not adequate to initiate follicle maturation or ovulation. In the absence of a mature ovum, pregnancy cannot occur.

Women for whom all oral contraceptives are contraindicated include those with a history of thromboembolic disorders (predisposition to blood clotting), cerebral blood vessel damage, hypertension, liver malfunction, heart disease, or cancer of the breast or reproductive system. About 40 percent of all pill users experience side effects— generally minor problems such as nausea, weight gain, headache, irregular menses, spotting between periods, and amenorrhea. The statistics on the life-threatening conditions associated with the pill such as blood clots, heart attacks, liver tumors, and gallbladder disease are somewhat more reassuring. For all the problems combined,

fewer than 2 deaths occurred per 100,000 users under age 30, 4 among those 30 to 35, 10 among those 35 to 39, and 18 among women over 40.

The major exception is that women who take the pill and smoke face far higher odds of developing heart attack and stroke than do nonsmoking pill users.

A summary of birth control methods is presented in Exhibit 29-3.

CLINICAL APPLICATION

The quest for an efficient male oral contraceptive has been disappointing until recently. An oral contraceptive, **gossypol**, which is derived from the cotton plant, has achieved an efficacy (power to produce effects) rate of 99.9 percent in clinical tests in more than 4,000 healthy males in China. Its action is to inhibit an enzyme required for spermatogenesis. Levels of blood luteinizing hormone (LH) and testosterone remained

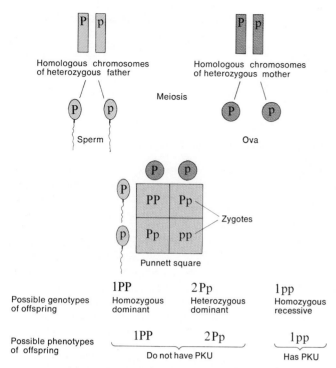

FIGURE 29-20 Inheritance of phenylketonuria (PKU).

---

unchanged and potency was not impaired. The number and morphology of spermatozoa gradually recover and fertility is restored to normal within three months after termination of therapy.

INHERITANCE

Inheritance is the passage of hereditary traits from one generation to another. It is the process by which you acquired your characteristics from your parents and will transmit your characteristics to your children. The branch of biology that deals with inheritance is called **genetics.**

GENOTYPE AND PHENOTYPE

The nuclei of all human cells except gametes contain 23 pairs of chromosomes—the diploid number. One chromosome from each pair comes from the mother, and the other comes from the father. Homologues, the two chromosomes in a pair, contain genes that control the same traits. If a chromosome contains a gene for height, its homologue will contain a gene for height.

The relationship of genes to heredity is illustrated admirably by the disorder called phenylketonuria or PKU (see Figure 29-20). People with PKU are unable to manufacture the enzyme phenylalanine hydroxylase. It is believed the PKU is brought about by an abnormal gene, which can be symbolized as p. The normal gene will be symbolized as P. The chromosome concerned with directions for phenylalanine hydroxylase production will have either p or P on it. Its homologue will also have p or P. Thus every individual will have one of the following genetic makeups, or **genotypes** (JĒ-nō-tīps): PP, Pp, or pp. Although people with genotypes of Pp have the abnor-

mal gene, only those with genotype pp suffer from the disorder because the normal gene, when present, dominates over and inhibits the abnormal one. A gene that dominates is called the **dominant gene,** and the trait expressed is said to be a dominant trait. The gene that is inhibited is called the **recessive gene,** and the trait it controls is called the recessive trait.

By tradition, we symbolize the dominant gene with a capital letter and the recessive one with a lowercase letter. An individual with the same genes on homologous chromosomes (for example, PP or pp) is said to be **homozygous** for the trait. An individual with different genes on homologous chromosomes (for example, Pp), is said to be **heterozygous** for the trait. **Phenotype** (FĒ-nō-tīp) refers to how the genetic makeup is expressed in the body. A person with Pp has a different genotype from one with PP, but both have the same phenotype—which in this case is normal production of phenylalanine hydroxylase.

To determine how gametes containing haploid chromosomes unite to form diploid fertilized eggs, special charts called **Punnett squares** are used. Usually, the male gametes (sperm cells) are placed at the side of the chart and the female gametes (ova) at the top (as in Figure 29-20). The four spaces on the chart represent the possible combinations of male and female gametes that could form fertilized eggs.

Exhibit 29-4 lists some of the variety of simple inherited structural and functional traits in humans. The classification of traits in this manner is rather arbitrary and somewhat artificial, since structure and function are intimately related. Nevertheless, some organizational scheme is

EXHIBIT 29-4
HEREDITARY TRAITS IN HUMANS

DOMINANT	RECESSIVE
Curly hair	Straight hair
Dark brown hair	All other colors
Coarse body hair	Fine body hair
Pattern baldness (dominant in males)	Baldness (recessive in females)
Normal skin pigmentation	Albinism
Brown eyes	Blue or gray eyes
Near or farsightedness	Normal vision
Normal hearing	Deafness
Normal color vision	Color blindness
Broad lips	Thin lips
Large eyes	Small eyes
Polydactylism (extra digits)	Normal digits
Brachydactylism (short digits)	Normal digits
Syndactylism (webbed digits)	Normal digits
Hypertension	Normal blood pressure
Diabetes insipidus	Normal excretion
Huntington's chorea	Normal nervous system
Normal mentality	Schizophrenia
Migraine headaches	Normal
Normal resistance to disease	Susceptibility to disease
Enlarged spleen	Normal spleen
Enlarged colon	Normal colon
A or B blood factor	O blood factor
Rh blood factor	No Rh blood factor

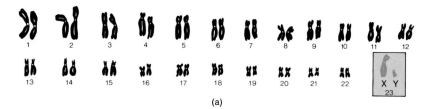

(a)

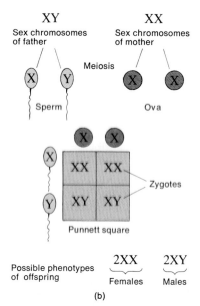

(b)

FIGURE 29-21 Inheritance of sex. (a) Normal human male chromosomes. Sex chromosomes are the 23rd pair indicated in the blue colored box. (b) Sex determination.

needed that gives an indication of the wide spectrum of inherited human characteristics. Many traits were at one time thought to be inherited in a simple Mendelian manner and were classified as "dominant" or "recessive." Then further research revealed that some traits were actually complex and that the genetic basis was correspondingly complicated. No wonder students of human genetics prefer to work with enzymes and antigens, in which the path from a gene to its product is much shorter, more direct, and less influenced by other genes and by environment. (If you would like to find out about the inheritance of some particular traits, the first place to look is McKusick's *Mendelian Inheritance in Man* [1971]).

Normal traits do not always dominate over abnormal ones, but genes for severe disorders are more frequently recessive than dominant. People who have severe disorders often do not live long enough to pass the abnormal gene on to the next generation. In this way, expression of the gene tends to be weeded out of the population.

CLINICAL APPLICATION

An exception is **Huntington's chorea**—a major disorder caused by a dominant gene and characterized by degeneration of nervous tissue, usually leading to mental disturbance and death. The first signs of Huntington's chorea do not occur until adulthood, very often after the person has already produced offspring.

INHERITANCE OF SEX

Microscopic examination of the chromosomes in cells reveals that one pair differs in males and in females (Figure 29-21a). In females, the pair consists of two rod-shaped chromosomes designated as X chromosomes. One X chromosome is present in males, but its mate is a hook-shaped structure called a Y chromosome. The XX pair in the female and the XY pair in the male are called the **sex chromosomes.** All other chromosomes are called **autosomes.**

The sex chromosomes are responsible for the sex of the individual (Figure 29-21b). When a spermatocyte undergoes meiosis to reduce its chromosome number, one daughter cell will contain the X chromosome and the other contain the Y chromosome. Oocytes have no Y chromosomes and produce only X-containing ova. If the ovum is subsequently fertilized by an X-bearing sperm, the offspring normally will be female (XX). Fertilization by a Y sperm normally produces a male (XY). Thus, sex is determined at fertilization. Although X and Y sperm are produced in equal amounts, more males are born than females. It is speculated that Y sperm swim faster than X sperm because X sperm contain more genetic material.

COLOR BLINDNESS AND X-LINKED INHERITANCE

The sex chromosomes also are responsible for the transmission of a number of nonsexual traits. Genes for these traits appear on X chromosomes, but many of these genes are absent from Y chromosomes. This feature produces a pattern of heredity that is different from the pattern described earlier. Let us consider color blindness. The gene for **color blindness** is a recessive one designated c. Normal color vision, designated C, dominates. The C/c genes are located on the X chromosome. The Y chromosome does not contain the segment of DNA that programs this aspect of vision. Thus the ability to see colors depends entirely on the X chromosomes. The genetic possibilities are:

X^CX^C	Normal female
X^CX^c	Normal female carrying the recessive gene
X^cX^c	Color-blind female
X^CY	Normal male
X^cY	Color-blind male

Only females who have two X^c chromosomes are color blind. In X^CX^c females the trait is inhibited by the normal, dominant gene. Males, on the other hand, do not have a second X chromosome that would inhibit the trait. Therefore all males with an X^c chromosome will be color blind. The inheritance of color blindness is illustrated in Figure 29-22.

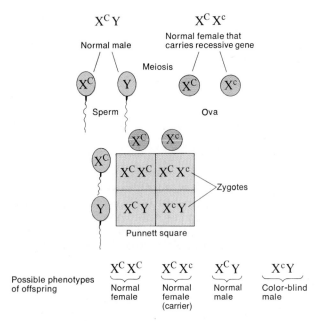

FIGURE 29-22 Inheritance of color blindness.

Traits inherited in the manner just described are called **X-linked traits.** Another X-linked trait is **hemophilia—** a condition in which the blood fails to clot or clots very slowly after an injury (Chapter 19). Like the trait for color blindness, hemophilia is caused by a recessive gene. If H represents normal clotting and h represents abnormal clotting, then X^hX^h females will have the disorder. Males with X^HY will be normal; males with X^hY will be hemophiliac. Actually, clotting time varies somewhat among hemophiliacs, so the condition may be affected by other genes as well.

A few other X-linked traits in humans are nonfunc-

tional sweat glands, certain forms of diabetes, some types of deafness, uncontrollable rolling of the eyeballs, absence of central incisors, night blindness, one form of cataract, white forelocks, juvenile glaucoma, and juvenile muscular dystrophy.

AMNIOCENTESIS

Amniocentesis is a technique of withdrawing some of the amniotic fluid that bathes the developing fetus to diagnose genetic disorders or to determine fetal maturity or well-being. The fluid is removed by hypodermic needle puncture of the uterus, usually 16 to 20 weeks after conception. Cells in the fluid are examined for biochemical defects and abnormalities in chromosome number or structure. Over 50 inheritable biochemical defects and close to 300 chromosomal disorders can be detected through amniocentesis, including hemophilia, certain muscular dystrophies, Tay-Sachs disease, myelocytic leukemia, Klinefelter's and Turner's syndromes, sickle cell anemia, thalassemia, and cystic fibrosis. When both parents are known or suspected to be genetic carriers of any one of these disorders, amniocentesis is advised.

CLINICAL APPLICATION

One chromosome disorder that may be diagnosed through amniocentesis is **Down's syndrome.** This disorder is characterized by mental retardation, retarded physical development (short stature and stubby fingers), distinctive facial structures (large tongue, broad skull, slanting eyes, and round head), and malformation of the heart, ears, hands, and feet. Sexual maturity is rarely attained. Individuals with the disorder usually have 47 chromosomes instead of the normal 46 (an extra chromosome in the twenty-first pair).

MEDICAL TERMINOLOGY

Abortion Premature expulsion from the uterus of the products of conception—embryo or nonviable fetus.

Cautery Application of a caustic (burning) substance or instrument for the destruction of tissue.

Cesarean section (*caedere* = to cut) Removal of the baby and placenta through an abdominal incision in the uterine wall.

Colpotomy (*colp* = vagina; *tome* = cutting) Incision of the vagina.

Culdoscopy (*skopein* = to examine) A procedure in which a culdoscope (endoscope) is used to view the pelvic cavity. The approach is through the vagina.

Hermaphroditism Presence of both male and female sex organs in one individual.

Karyotype (*karyon* = nucleus) The chromosomal elements typical of a cell, drawn in their true proportions, based

on the average of measurements determined in a number of cells. Useful in judging whether or not chromosomes are normal in number and structure.

Lethal gene A gene that, when expressed, results in death either in the embryonic state or shortly after birth.

Lochia The discharge from the birth canal consisting initially of blood and later of serous fluid occurring after childbirth. The discharge is derived from the former placental site and may last up to about a week and a half.

Mutation A permanent heritable change in a gene that causes it to have a different effect than it had previously.

Puerperal fever Infectious disease of childbirth, also called puerperal sepsis and childbed fever. The disease results from an infection originating in the birth canal and affects the endometrium.

STUDY OUTLINE

Gamete Formation (p. 732)
Diploid and Haploid Cells

1. Ova and sperm are collectively called gametes or sex cells and are produced in gonads.
2. Uninucleated somatic cells divide by mitosis, the process in which each daughter cell receives the full complement

of 23 chromosome pairs (46 chromosomes). Somatic cells are said to be diploid ($2n$).

3. Immature gametes divide by meiosis in which the pairs of chromosomes are split so that the mature gamete has only 23 chromosomes. It is said to be haploid (n).

Spermatogenesis

1. Spermatogenesis occurs in the testes. It results in the formation of four haploid spermatozoa.
2. The spermatogenesis sequence consists of reduction division, equatorial division, and spermiogenesis.

Oogenesis

1. Oogenesis occurs in the ovaries. It results in the formation of a single haploid ovum.
2. The oogenesis sequence consists of reduction division, equatorial division, and maturation.

Sexual Intercourse (p. 734)

1. The role of the male in the sex act involves erection, lubrication, and orgasm.
2. The female role also involves erection, lubrication, and orgasm (climax).

Pregnancy (p. 735)

1. Pregnancy is a sequence of events that includes fertilization, implantation, embryonic growth, fetal growth, and birth.
2. Its various events are hormonally controlled.

Fertilization and Implantation

1. Fertilization refers to the penetration of the ovum by a sperm cell and the subsequent union of the sperm and ovum nuclei to form a zygote.
2. Penetration is facilitated by hyaluronidase and proteinases produced by sperm.
3. Normally only one sperm fertilizes an ovum.
4. Early rapid cell division of a zygote is called cleavage, and the cells produced by cleavage are called blastomeres.
5. The solid mass of cells produced by cleavage is a morula.
6. The morula develops into a blastocyst, a hollow ball of cells differentiated into a trophectoderm (future embryonic membranes) and inner cell mass (future embryo).
7. The attachment of a blastocyst to the endometrium is called implantation.
8. It occurs by enzymatic degradation of the endometrium.

Embryonic Development

1. During embryonic growth, the primary germ layers and embryonic membranes are formed and the placenta is functioning.
2. The primary germ layers—ectoderm, mesoderm, and endoderm—form all tissues of the developing organism.
3. Embryonic membranes include the yolk sac, amnion, chorion, and allantois.
4. Fetal and maternal materials are exchanged through the placenta.
5. During the fetal period, organs established by the primary germ layers grow rapidly.

Hormones of Pregnancy

1. Pregnancy is maintained by human chorionic gonadotropin (HCG), estrogens, and progesterone.
2. Placental luteotropic releasing factor (pLRF) stimulates secretion of HCG.
3. Human chorionic somatomammotropin (HCS) assumes a role in breast development, protein anabolism, and glucose and fatty acid catabolism.
4. Relaxin relaxes the symphysis pubis and helps dilate the uterine cervix toward the end of pregnancy.

Parturition and Labor (p. 750)

1. The time an embryo or fetus is carried in the uterus is called gestation.
2. Parturition refers to birth and is accompanied by a sequence of events called labor.
3. The birth of a baby involves dilation of the cervix, expulsion of the fetus, and delivery of the placenta.

Adjustments of the Infant at Birth (p. 752)

1. The fetus depends on the mother for oxygen and nutrients, removal of wastes, and protection.
2. Following birth the respiratory and cardiovascular systems undergo changes in adjusting to self-supporting postnatal life.

Potential Hazards to the Developing Embryo and Fetus (p. 753)

1. The developing embryo and fetus is susceptible to many potential hazards that can be transmitted from the mother.
2. Examples are infections, microbes, chemicals and drugs, alcohol, and smoking.

Lactation (p. 753)

1. Lactation refers to the secretion and ejection of milk by the mammary glands.
2. Secretion is influenced by prolaction (PRL), estrogens, and progesterone.
3. Ejection is influenced by oxytocin (OT).

Birth Control (p. 754)

1. Methods include removal of gonads and uterus, sterilization (vasectomy, tubal ligation, laparascopic technique), and contraception (natural, mechanical, and chemical).
2. Contraceptive pills of the combination type contain estrogens and progesterone in concentrations that decrease the secretion of FSH and LH and thereby inhibit ovulation.

Inheritance (p. 758)

1. Inheritance is the passage of hereditary traits from one generation to another.
2. The genetic makeup of an organism is called its genotype. The traits expressed are called its phenotype.
3. Dominant genes control a particular trait; expression of recessive genes is inhibited by dominant genes.
4. Amniocentesis is the withdrawal of amniotic fluid. It can be used to diagnose inherited biochemical defects and chromosomal disorders, such as hemophilia, Tay-Sachs disease, sickle cell anemia, and Down's syndrome.
5. Down's syndrome is a chromosomal abnormality characterized by mental retardation and retarded physical development.
6. Sex is determined by the Y chromosome of the male at fertilization.
7. Color blindness and hemophilia primarily affect males because there are no counterbalancing dominant genes on the Y chromosomes.

REVIEW QUESTIONS

1. Define developmental anatomy.
2. Explain the terms haploid and diploid. What is the importance of meiosis?
3. Compare the events associated with spermatogenesis and oogenesis. How are the processes similar? How do they differ?
4. Explain the role of the male's erection, lubrication, and orgasm in the sex act. How do the female's erection, lubrica-

tion, and orgasm (climax) contribute to the sex act?

5. Define fertilization. Where does it normally occur? How is a morula formed?
6. Explain how twins, triplets, and quadruplets are produced.
7. Describe the components of a blastocyst.
8. What is implantation? How does the fertilized ovum implant itself? What causes morning sickness?
9. Describe the procedure for external human fertilization.
10. Define the embryonic period and the fetal period.
11. List several body structures formed by the endoderm, mesoderm, and ectoderm.
12. What is an embryonic membrane? Describe the functions of the four embryonic membranes.
13. Explain the importance of the placenta and umbilical cord to fetal growth.
14. Outline some of the major developmental changes during fetal growth.
15. List the hormones involved in pregnancy and describe the functions of each.
16. Define gestation and parturition.
17. Distinguish between false and true labor. Describe what happens during the stage of dilation, the stage of expulsion, and the placental stage of delivery.
18. Discuss the principal respiratory and cardiovascular adjustments made by an infant at birth.
19. Explain in detail some of the potential hazards for the developing embryo and fetus.
20. What is lactation? Name the hormones involved and their functions.
21. Briefly describe the following methods of birth control: removal of the gonads, sterilization, rhythm.
22. Distinguish between a condom, a diaphragm, and an IUD as methods of mechanical contraception.
23. List several examples and functions of chemical contraceptives.
24. Explain the operation of the combination contraceptive pill.
25. Define inheritance. What is genetics?
26. Define the following terms: genotype, phenotype, dominant, recessive, homozygous, heterozygous.
27. What is a Punnett square?
28. List several dominant and recessive traits inherited in humans.
29. Set up Punnett squares to show the inheritance of the following traits: sex, color blindness, hemophilia.
30. What is amniocentesis? What is its value?
31. What is X-linked inheritance?
32. Refer to the glossary of medical terminology associated with development and inheritance. Be sure that you can define each term.

Appendix A: Measurements

UNITS OF MEASUREMENT

When you measure something, you are comparing it with some standard scale to determine its *magnitude*. How long is it? How much does it weigh? How fast is it going? Some measurements are made directly by comparing the unknown quantity with the known unit of the same kind, for example, weighing a patient on a scale and taking the reading directly in pounds. Other measurements are indirect and are done by calculation, for example, counting a person's blood cells in a certain number of squares on a microscope slide and then calculating the total blood count.

Regardless of how a measurement is taken, it always requires two things: a *number* and a *unit*. When recording the weight of a patient, you would not just say 145. You have to give both the number (145) and the unit (pounds). When you count blood cells, you report the measurement as 10,000 (number) white blood cells per cubic millimeter of blood (unit).

All the units in use can be expressed in terms of one of three special units called *fundamental units*. These fundamental units have been established arbitrarily as length, mass, and time. Mass is perhaps an unfamiliar term to you. *Mass* is the amount of matter an object contains. The mass of this textbook is the same whether it is measured in a laboratory, under the sea, on top of a mountain, or even on the moon. No matter where you take it, it still has the same quantity of matter. *Weight*, on the other hand, is determined by the pull of gravity on an object. This textbook will not have the same weight on earth as on the moon because of the differences in gravitation. However, as long as we are dealing only with earthbound objects, weight and mass may be considered synonymous terms because the force of gravity on the surface of the earth is nearly constant. Thus weight remains nearly the same regardless of where the measurements are taken.

All units other than the fundamental ones are *derived units*—they can always be written as some combination of the three fundamental units. For example, units of volume are derived from units of length (the volume of a cube = length × width × height). Units of speed are combinations of distance and time (miles per hour).

Units are grouped into systems of measurement. The two principal systems of measurement commonly used in this country are the U.S. and the metric systems. The apothecary system is used by physicians and pharmacists.

U.S. SYSTEM

The **U.S. system** of measurement is used in everyday household work, industry, and some fields of engineering. The fundamental units in the U.S. system are the foot (length), the pound (mass), and the second (time).

The basic problem with the U.S. system is that there is no *uniform* progression from one unit to another. If you want to convert a measurement of 2½ yd to feet,

EXHIBIT A-1

U.S. UNITS OF MEASUREMENT

FUNDA-MENTAL OR DERIVED UNIT	UNITS AND U.S. EQUIVALENTS
Length	1 inch = 0.083 foot
	1 foot (ft) = 12 inches
	= 0.333 yard
	1 yard (yd) = 3 ft = 36 inches
	1 mile (mi) = 1,760 yd = 5,280 ft
Mass	1 grain (gr) = 0.002285 ounce
	1 dram (dr) = 27.34 gr
	= 0.063 ounce
	1 ounce (oz) = 16 dr = 437.5 gr
	1 pound (lb) = 16 oz = 7,000 gr
	1 ton = 2,000 lb
Time	1 second (sec) = 1/86,400 of a day
	1 minute (min) = 60 sec
	1 hour (hr) = 60 min = 3,600 sec
	1 day = 24 hr = 1,440 min
	= 86,400 sec
Volume	1 fluidram (fl dr) = 0.125 fluidounce
	1 fluidounce (fl oz) = 8 fl dr
	= 0.0625 quart = 0.008 gallon
	1 pint (pt) = 16 fl oz = 128 fl dr
	1 quart (qt) = 2 pt = 32 fl oz = 256 fl dr
	= 0.25 gallon
	1 gallon (gal) = 4 qt = 8 pt = 128 fl oz
	= 1,024 fl dr

you have to multiply it by 3 because there are 3 ft in a yard. If you want to convert the same length to inches, you have to multiply by 3 and then by 12 (or by 36) because there are 12 inches in a foot. In other words, to convert one unit of length to another, it is necessary to use *different* numbers each time. Conversions in the metric system are much easier since they are based on progressions of the number 10.

Exhibit A-1 lists U.S. units of measurement.

METRIC SYSTEM

The **metric system** introduced in France in 1790, is now used by all major countries except the United States. Scientific observations are almost universally expressed in metric units.

LENGTH

The standard of length in the metric system is the *meter* (m). It was originally defined in 1790 as one ten-millionth of the distance from the North Pole to the Equator. In 1889 it was redefined as the distance measured at 0°C between two lines on a bar of platinum-iridium kept at the International Bureau of Weights and Measures in France. Finally, in 1960 the meter was redefined as the length of 1,650,763.73 light waves emitted by atoms of the gas krypton under strictly specified conditions. The meter is equal to 39.37 inches.

A major advantage of the metric system is that units are related to one another by factors of 10. Thus 1 m is 10 decimeters (dm) or 100 centimeters (cm) or 1,000 millimeters (mm). Conversion from one unit to another is simple. Figure A-1 illustrates the differences between metric and U.S. conversions by comparing the meter stick and the yardstick. Exhibit A-2 lists the metric units of length with U.S. equivalents.

Since numbers with many zeros (very large numbers or very small fractions) are cumbersome to work with, they are expressed in *exponential form,* that is, as powers of 10. The form of exponential notation is

$$M \times 10^n$$

You can determine M and n in two steps. For example, how is 0.0000000001 written in exponential form? First, determine M by moving the decimal point so that only one nonzero digit is to the left of it:

0.0000000001.

The digit to the left of the decimal is 1; therefore $M = 1$. Second, determine n by counting the number of places you moved the decimal point. If you moved the point to the left, make the number positive; if you moved it to the right, it is negative. Since you moved the decimal point 10 places to the right, $n = -10$. Thus

$$0.0000000001 = 1 \times 10^{-10}$$

Now do a problem on your own. The wavelength of yellow light is about 0.000059 cm. Convert the centimeters to exponential form. If your answer is 5.9×10^{-5}, you are ready to continue. If you got the wrong answer, reread the discussion.

When we are working with a very large number, the same rules apply, but our exponential value will be positive rather than negative. Refer to Exhibit A-2. Note that 1 km equals 1,000 m. Even though 1,000 is not a cumbersome number, we can still convert it into exponential form. First, move the decimal point so there is only one nonzero digit to the left of it to determine M:

1.000.

Now, because the decimal has been moved three places to the left, n equals +3 or simply 3. Thus

$$1 \text{ km} = 1 \times 10^3 \text{ m}$$

EXHIBIT A-2
METRIC UNITS OF LENGTH AND SOME U.S. EQUIVALENTS

METRIC UNIT	MEANING OF PREFIX	METRIC EQUIVALENT	U.S. EQUIVALENT
1 kilometer (km)	kilo = 1,000	1,000 m	3,280.84 ft = 0.62 mi 1 mi = 1.61 km
1 hectometer (hm)	hecto = 100	100 m	328 ft
1 dekameter (dam)	deka = 10	10 m	32.8 ft
1 meter (m)	Standard unit of length		39.37 inches = 3.28 ft = 1.09 yd
1 decimeter (dm)	deci = 1/10	0.1 m	3.94 inches
1 centimeter (cm)	centi = 1/100	0.01 m	0.394 inch 1 inch = 2.54 cm
1 millimeter (mm)	milli = 1/1,000	0.001 m = 1/10 cm	0.0394 inches
1 micrometer (μm) [formerly micron (μ)]	micro = 1/1,000,000	0.000,001 m = 1/10,000 cm	3.94×10^{-5} inches
1 nanometer (nm) [formerly millimicron (mμ)]	nano = 1/1,000,000,000	0.000,000,001 m = 1/10,000,000 cm	3.94×10^{-8} inches
1 angstrom (Å)		0.000,000,000,1 m = 1/100,000,000 cm	3.94×10^{-9} inches

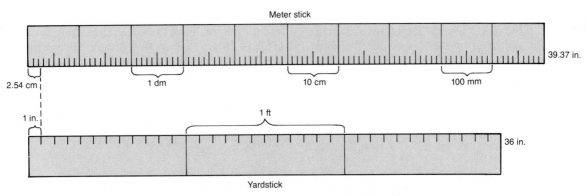

FIGURE A-1 Metric and U.S. units of length.

Do another problem on your own. The speed of light is about 30,000,000,000 cm/sec. Convert the centimeters to exponential form. Your answer should be 3×10^{10} cm.

Review Exhibit A-2 and note some common metric and U.S. equivalents. Note also the exponential forms.

MASS

Now let us look at the second fundamental unit of the metric system: mass. The standard unit of mass is the *kilogram* (kg). A kilogram is defined as the mass of a platinum-iridium cylinder kept at the International Bureau of Weights and Measures in France. The standard pound is defined in terms of the standard kilogram: 1 lb equals 0.4536 kg.

Exhibit A-3 lists metric units of mass and some U.S. equivalents.

EXHIBIT A-3
METRIC UNITS OF MASS AND SOME U.S. EQUIVALENTS

METRIC UNIT	METRIC EQUIVALENT	U.S. EQUIVALENT
1 kilogram (kg)	1,000 g	2.205 lb 1 ton = 907 kg
1 hectogram (hg)	100 g	
1 dekagram (dag)	10 g	0.0353 oz
1 gram (g)	1 g	1 lb = 453.6 g 1 oz = 28.35 g
1 decigram (dg)	0.1 g	
1 centigram (cg)	0.01 g	
1 milligram (mg)	0.001 g	0.015 gr
1 microgram (μg)	0.000,001 g	

TIME

The third fundamental unit of both the metric and the U.S. systems is time. The standard of time is the *second*. Formerly, the second was defined as 1/86,400 of a mean solar day. (A mean solar day is the average of the lengths of all days throughout the year.) Currently, the second is defined as the time required for 9,192,631,770 vibrations of cesium atoms when they are vibrating in a specific manner. Units of time are used in measuring pulse and heart rate, metabolic rate, x-ray exposure, and intervals between medications.

Exhibit A-1 lists the units of time.

VOLUME

Units of volume, or capacity, are derived units based on length. *Volume* in the U.S. system may be expressed as cubic feet (ft^3), cubic inches ($inch^3$), and cubic yards (yd^3) or as a unit of volume such as the quart. Volume in the metric system may be expressed in cubic units of length such as cubic centimeters (cm^3) or in terms of the basic unit of volume, the *liter*. A liter is defined as the volume occupied by 100 g of pure water at 4°C. Since 1 cm^3 of water at this temperature weighs 1 g, then 1,000 g of water occupies a volume of 1,000 cm^3. This means that a liter is equal to 1,000 cm^3 and 1 milliliter (ml) is equal to 1 cm^3. Because of this relationship, many volume-measuring devices, such as hypodermic needles, may be graduated in either milliliters or cubic centimeters.

Exhibit A-4 lists metric units of volume and some U.S. equivalents.

EXHIBIT A-4
METRIC UNITS OF VOLUME AND SOME EQUIVALENTS

METRIC UNIT	METRIC EQUIVALENT	U.S. EQUIVALENT
1 liter	1,000 ml	33.81 fl oz = 1.057 qt 946 ml = 1 qt
1 milliliter (ml)	0.001 liter	0.0338 fl oz 30 ml = 1 fl oz 5 ml = 1 teaspoon
1 cubic centimeter (cm³)	0.999972 ml	0.0338 fl oz

APOTHECARY SYSTEM

In addition to the U.S. and metric systems, there is the **apothecary system.** This system is commonly used by physicians prescribing medications and by pharmacists preparing them. Exhibit A-5 lists the important units and equivalents of the apothecary system. Note that the units of mass have the same names as in the U.S. system, but they are not equivalent (1 oz = 28.35 g; 1 oz ap = 30 g) and they do not have the same relationship to one another (1 oz = 16 dr; 1 oz ap = 8 dr ap). The units of volume are the same in both systems.

EXHIBIT A-5

APOTHECARY SYSTEM OF MASS AND VOLUME WITH METRIC EQUIVALENTS

FUNDAMENTAL UNIT	APOTHECARY UNIT AND CONVERSION	METRIC EQUIVALENT
Mass	1 grain (gr) = 0.002083 ounce	1 g = 15 gr
	1 dram (dr ap) = 60 gr	4 g = 1 dr
	1 ounce (oz ap) = 8 dr ap	30 g = 1 oz
	1 pound (lb ap) = 12 oz ap	1 kg = 32 oz
Volume	1 fluidram (fl dr) = 60 minims (min)	1 ml (or cm³) = 15 min
	1 fluidounce (fl oz) = 8 fl dr	4 ml (or cm³) = 1 fl dr
	1 pint (pt) = 16 fl oz	30 ml (or cm³) = 1 fl oz
		500 ml (or cm³) = 1 pt
		1,000 ml (or cm³) = 1 qt

FREQUENTLY USED CONVERSIONS BASED ON MILLIGRAM

1,000 mg (1 g)	= 15 grains
600 mg (0.6 g)	= 10 grains
300 mg (0.3 g)	= 5 grains
60 mg (0.06 g)	= 1 grain
30 mg (0.03 g)	= 0.50 (1/2) grain
20 mg (0.02 g)	= 0.33 (1/3) grain
10 mg (0.01 g)	= 0.166 (1/6) grain
5 mg (0.005 g)	= 0.083 (1/12) grain
4 mg (0.004 g)	= 0.66 (1/15) grain
1 mg (0.001 g)	= 0.016 (1/60) grain
0.5 mg (0.0005 g) =	0.0083 (1/120) grain
0.1 mg (0.0001 g) =	0.0016 (1/600) grain

Appendix B: Abbreviations

Many terms, especially medical terms, are commonly expressed in abbreviated form. In order to familiarize you with some of these abbreviations, an alphabetical list has been prepared. Most of the terms listed have been used in the book; some terms not referred to in the book have been included because of their frequent use.

AID	automatic implantable defibrillation; artificial insemination by donor
AIDS	acquired immune deficiency syndrome
AMI	acute myocardial infarction
ANS	autonomic nervous system
ARF	acute renal failure
AV	atrioventricular
BBB	blood–brain barrier
BEAM	brain electrical activity mapping
BMR	basal metabolic rate
BP	blood pressure
BS	blood sugar
BUN	blood urea nitrogen
CABG	coronary artery bypass grafting
CAC	cardioacceleratory center
CAD	coronary artery disease
CAPD	continuous ambulatory peritoneal dialysis
CBC	complete blood count
CCCC	closed-chest cardiac compression
CF	cystic fibrosis
CHD	coronary heart disease
CHF	congestive heart failure
CIC	cardioinhibitory center
CNS	central nervous system
CO	cardiac output; carbon monoxide
COAD	chronic obstructive airways disease
COPD	chronic obstructive pulmonary disease
CPR	cardiopulmonary resuscitation
C-section	cesarean section
CSF	cerebrospinal fluid
CT (CAT)	computed tomography (computed axial tomography)
CVA	cerebrovascular accident
CVP	central venous pressure
D & C	dilation and curettage
DBP	diastolic blood pressure
DES	diethylstilbestrol
DH	delayed hypersensitivity
DMD	Duchenne muscular dystrophy
DMSO	dimethyl sulfoxide
DSR	dynamic spatial reconstructor
DVT	deep vein thrombosis
EB virus	Epstein-Barr virus
ECF	extracellular fluid
ECG (EKG)	electrocardiogram
EDV	end-diastolic volume
EEG	electroencephalogram
EM	electron micrograph
EMG	electromyogram
EOG	electrooculogram
EPSP	excitatory postsynaptic potential
ER	endoplasmic reticulum
ESR	erythrocyte sedimentation rate
ESRD	end-stage renal disease
ESV	end-systolic volume
FAS	fetal alcohol syndrome
GFR	glomerular filtration rate
GI	gastrointestinal
Hb	hemoglobin
HBO	hyperbaric oxygenation
HBV	hepatitis B virus
HDCV	human diploid cell vaccine
HDL	high-density lipoprotein
HF	heart failure
HR	heart rate
HSV	herpes simplex virus
ICF	intracellular fluid
ID	intradermal
IM	intramuscular
INF	interferon
IPSP	inhibitory postsynaptic potential
IUCD	intrauterine contraceptive device
IUD	intrauterine device
IV	intravenous
LDL	low-density lipoprotein
LLQ	left lower quadrant
LUQ	left upper quadrant
LV	left ventricular
meq/liter	milliequivalents per liter
MG	myasthenia gravis
MI	myocardial infarction
mm³	cubic millimeter
mm Hg	millimeters of mercury
MS	multiple sclerosis
MVP	mitral valve prolapse
NGU	nongonococcal urethritis
NLMC	nocturnal leg muscle cramping
NMR	nuclear magnetic resonance
NPN	nonprotein nitrogen

NREM	nonrapid eye movement	**RUQ**	right upper quadrant
OC	oral contraceptive	**SA**	sinoatrial (sinuatrial)
OTC	over-the-counter	**SCA**	sickle cell anemia
PCT	positron computed tomography	**SIDS**	sudden infant death syndrome
PE	pulmonary embolism	**SLE**	systemic lupus erythematosus
P_{eff}	effective filtration pressure	**SNS**	somatic nervous system
PEMFs	pulsating electromagnetic fields	**SPF**	sun protection factor
PG	prostaglandin	**STD**	sexually transmitted disease
PID	pelvic inflammatory disease	**SV**	stroke volume
PKU	phenylketonuria	**TIA**	transient ischemic attack
PMNs	polymorphonuclear leucocytes	**Tm**	tubular maximum
PMS	premenstrual syndrome	**TM**	transcendental meditation
PNS	peripheral nervous system	**TM joint**	temporomandibular joint
PPNG	penicillinase-producing *Neisseria gonorrhea*	**TPN**	total parenteral nutrition
PTA	percutaneous transluminal angioplasty	**TSS**	toxic shock syndrome
RA	rheumatoid arthritis	**UA**	urinalysis
RAS	reticular activating system	**URI**	upper respiratory infection
RBC	red blood cell	**UTI**	urinary tract infection
RDS	respiratory distress syndrome	**UV**	ultraviolet
REM	rapid eye movement	**VD**	venereal disease
RES	reticuloendothelial system	**VF**	ventricular fibrillation
Rh	*Rhesus*	**VSD**	ventricular septal defect
RIG	rabies immune globulin	**VT**	ventricular tachycardia
RLQ	right lower quadrant	**WBC**	white blood cell

Glossary of Prefixes, Suffixes, and Combining Forms

Many medical terms are "compound" words, that is, they are made up of one or more word roots or combining forms of word roots with prefixes or suffixes. For example, *leucocyte* (white blood cell) is a combination of *leuco*, the combining form for the word root meaning "white," and *cyt*, the word root meaning "cell." Learning the medical meanings of the fundamental word parts will enable you to analyze many long, complicated terms.

The following list includes some of the most commonly used word roots, combining forms, prefixes, and suffixes used in making medical terms and an example for each.

Word Roots and Combining Forms

Acou-, Acu- hearing Acoustics (a-KOO-stiks), the science of sounds or hearing.

Acr-, Acro- extremity Acromegaly (ak'-rō-MEG-a-lē), hyperplasia of the nose, jaws, fingers, and toes.

Aden-, Adeno- gland Adenoma (ad-en-Ō-ma), a tumor with a glandlike structure.

Alg-, Algia- pain Neuralgia (nyoo-RAL-ja), pain along the course of a nerve.

Angi- vessel Angiocardiography (an'-jē-ō-kard-ē-OG-ra-fē), roentgenography of the great blood vessels and heart after intravenous injection of radiopaque fluid.

Arthr-, Arthro- joint Arthropathy (ar-THROP-a-thē), disease of a joint.

Aut-, Auto- self Autolysis (aw-TOL-i-sis), destruction of cells of the body by their own enzymes, even after death.

Bio- life, living Biopsy (BĪ-op-sē), examination of tissue removed from a living body.

Blast- germ, bud Blastocyte (BLAS-tō-sīt), an embryonic or undifferentiated cell.

Blephar- eyelid Blepharitis (blef-a-RĪt-is), inflammation of the eyelids.

Brachi- arm Brachialis (brā-kē-AL-is), muscle that flexes the forearm.

Bronch- trachea, windpipe Bronchoscopy (bron-KOS-kō-pē), direct visual examination of the bronchi.

Bucc-, cheek Buccocervical (bū-kō-SER-vi-kal), pertaining to the cheek and neck.

Capit- head Decapitate (dē-KAP-i-tāt), to remove the head.

Carcin- cancer Carcinogenic (kar-sin-ō-JEN-ik), causing cancer.

Cardi-, Cardia-, Cardio- heart Cardiogram (KARD-ē-o-gram), a recording of the force and form of the heart's movements.

Cephal- head Hydrocephalus (hī-drō-SEF-a-lus), enlargement of the head due to an abnormal accumulation of fluid.

Cerebro- brain Cerebrospinal (se-rē'-brō-SPĪN-al) fluid, fluid contained within the cranium and spinal canal.

Cheil- lip Cheilosis (kī-LŌ-sis), dry scaling of the lips.

Chole- bile, gall Cholecystogram (kō-lē-SIS-tō-gram), roentgenogram of the gallbladder.

Chondr-, Chondri-, Chondrio- cartilage Chondrocyte (KON-drō-sīt), a cartilage cell.

Chrom-, Chromat-, Chromato- color Hyperchromic (hī-per-KRŌ-mik), highly colored.

Crani- skull Craniotomy (krā-nē-OT-o-mē), surgical opening of the skull.

Cry-, Cryo- cold Cryosurgery (krī-ō-SERJ-e-rē), surgical procedure using a very cold liquid nitrogen probe.

Cut- skin Subcutaneous (sub-kyoo-TĀ-nē-us), under the skin.

Cysti-, Cysto- sac, bladder Cystoscope (SIS-tō-skōp), instrument for interior examination of the bladder.

Cyt-, Cyto-, Cyte- cell Cytology (sī-TOL-ō-jē), the study of cells.

Dactyl-, Dactylo- digits (usually fingers, but sometimes toes) Polydactylism (pol-ē-DAK-til-ism), above normal number of fingers or toes.

Derma-, Dermato- skin Dermatosis (der-ma-TŌ-sis), any skin disease.

Entero- intestine Enteritis (ent-e-RĪT-is), inflammation of the intestine.

Erythro- red Erythrocyte (e-RITH-rō-sīt), red blood cell.

Galacto- milk Galactose (ga-LAK-tose), a milk sugar.

Gastr- stomach Gastrointestinal (gas'-trō-in-TES-tin-al), pertaining to the stomach and intestine.

Gloss-, Glosso- tongue Hypoglossal (hī'-pō-GLOS-al), located under the tongue.

Glyco- sugar Glycosuria (glī'-kō-SHUR-ē-a), sugar in the urine.

Gyn-, Gyne-, Gynec- female, women Gynecology (gīn'-e-KOL-ō-jē), the medical specialty dealing with disorders of the female.

Hem-, Hemat- blood Hematoma (hē'-ma-TŌ-ma), a tumor or swelling filled with blood.

Hepar-, Hepato- liver Hepatitis (hep-a-TĪT-is), inflammation of the liver.

Hist-, Histio- tissue Histology (his-TOL-ō-jē), the study of tissues.

Hydr- water Hydrocele (HĪ-drō-sēl), accumulation of fluid in a saclike cavity.

Hyster- uterus Hysterectomy (his'-te-REK-tō-mē), surgical removal of the uterus.

Ileo- ileum Ileocecal valve (il'-ē-ō-SĒ-kal), folds at the opening between ileum and cecum.

Ilio- ilium Iliosacral (il'-ē-ō-SA-kral), pertaining to ilium and sacrum.

Lachry-, Lacri- tears Nasolacrimal (nā-zō-LAK-rim-al), pertaining to the nose and lacrimal apparatus.

Laparo- loin, flank, abdomen Laparoscopy (lap'-a-ROS-kō-pē), examination of the interior of the abdomen by means of a laparoscope.

Leuco-, Leuko- white Leucocyte (LYOO-kō-sīt), white blood cell.

Lip-, Lipo- fat Lipoma (lī-PŌ-ma), a fatty tumor.

Mamm- breast Mammography (ma-MOG-ra-fē), roentgenography of the mammary gland.

Mast- breast Mastitis (ma-STĪT-is), inflammation of the mammary gland.

Meningo- membrane Meningitis (men-in-JĪT-is), inflammation of the membranes of spinal cord and brain.

Metro- uterus Endometrium (en'-dō-MĒ-trē-um), lining of the uterus.

Morpho- form, shape Morphology (mor-FOL-o-jē), the study of form and structure of things.

Myelo- marrow, spinal cord Poliomyelitis (pō'-lē-ō-mī'-a-LĪT-is), inflammation of the gray matter of the spinal cord.

Myo- muscle Myocardium (mī'-ō-KARD-ē-um), heart muscle.

Necro- corpse, dead Necrosis (ne-KRŌ-sis), death of areas of tissue surrounded by healthy tissue.

Nephro- kidney Nephrosis (ne-FRŌ-sis), degeneration of kidney tissue.

Neuro- nerve Neuroblastoma (nyoor'-ō-blas-TŌ-ma), malignant tumor of the nervous system composed of embryonic nerve cells.

Oculo- eye Binocular (bī-NOK-yoo-lar), pertaining to the two eyes.

Odont- tooth Orthodontic (or-thō-DONT-ik), pertaining to the proper positioning and relationship of the teeth.

Oo- egg Oocyte (Ō-ō-sīt), original egg cell.

Ophthalm- eye Ophthalmology (of'-thal-MOL-o-jē), the study of the eye and its diseases.

Oss-, Osseo-, Osteo- bone Osteoma (os-tē-Ō-ma), bone tumor.

Oto- ear Otosclerosis (ō'-tō-skle-RŌ-sis), formation of bone in the labyrinth of the ear.

Patho- disease Pathogenic (path'-ō-JEN-ik), causing disease.

Ped- children Pediatrician (pēd'-ē-a-TRISH-an), medical specialist in the treatment of children.

Phag-, Phago- to eat Phagocytosis (fag'-ō-sī-TŌ-sis), the process by which cells ingest particulate matter.

Philic-, Philo- to like, have an affinity for Hydrophilic (hī'-drō-FIL-ik), having an affinity for water.

Phleb- vein Phlebitis (fli-BĪT-is), inflammation of the veins.

Pneumo- lung, air Pneumothorax (nyoo-mō-THŌR-aks), air in the thoracic cavity.

Pod- foot Podiatry (po-DĪ-a-trē), the diagnosis and treatment of foot disorders.

Procto- anus, rectum Proctoscopy (prok-TOS-kō-pē), instrumental examination of the rectum.

Psycho- soul, mind Psychiatry (sī-KĪ-a-trē), treatment of mental disorders.

Pyo- pus Pyuria (pī-YOOR-ē-a), pus in the urine.

Rhin- nose Rhinitis (rī-NĪT-is), inflammation of nasal mucosa.

Scler-, Sclero- hard Atherosclerosis (ath'-er-ō-skle-RŌ-sis), hardening of the arteries.

Sep-, Septic- toxic condition due to microorganisms Septicemia (sep'-ti-SĒ-mē-a), presence of bacterial toxins in the blood (blood poisoning).

Soma-, Somato- body Somatotropic (sō-mat-ō-TRŌ-pik), having a stimulating effect on body growth.

Sten- narrow Stenosis (ste-NŌ-sis), narrowing of a duct or canal.

Stasis-, Stat- stand still Homeostasis (hō'-mē-ō-STĀ-sis), achievement of a steady state.

Therm- heat Thermometer (ther-MOM-et-er), instrument used to measure and record heat.

Tox-, Toxic- poison Toxemia (tok-SĒ-mē-a), poisonous substances in the blood.

Trich- hair Trichosis (trik-Ō-sis), disease of the hair.

Viscer- organ Visceral (VIS-e-ral), pertaining to the abdominal organs.

Zoo- animal Zoology (zō-OL-o-jē), the study of animals.

Prefixes

A-, An- without, lack of, deficient Anesthesia (an'-es-THĒ-zha), without sensation.

Ab- away from, from Abnormal (ab-NOR-mal), away from normal.

Ad- to, near, toward Adduction (a-DUK-shun), movement of an extremity toward the axis of the body.

Ambi- both sides Ambidextrous (am'-bi-DEK-strus), able to use either hand.

Ante- before Antepartum (ant-ē-PAR-tum), before delivery of a baby.

Anti- against Anticoagulant (an-tī-kō-AG-yoo-lant), a substance that prevents coagulation of blood.

Bi- two, double, both Biceps (BĪ-seps), a muscle with two heads of origin.

Brachy- short Brachyesophagus (brā-kē-e-SOF-a-gus), short esophagus.

Brady- slow Bradycardia (brād'-ē-KARD-ē-a), abnormal slowness of the heartbeat.

Cata- down, lower, under, against Catabolism (ka-TAB-a-lizm), metabolic breakdown into simpler substances.

Circum- around Circumrenal (ser-kum-RĒN-al), around the kidney.

Col-, Con-, Com- with, together Congenital (kon-JEN-i-tal), existing at birth.

Contra- against, opposite Contraception (kon-tra-SEP-shun), the prevention of conception.

Crypt- hidden, concealed Cryptorchidism (krip-TOR-ka-dizm'), undescended or hidden testes.

Di-, Diplo- two Diploid (DIP-loyd), having double the haploid number of chromosomes.

Dis- separation, apart, away from Disarticulate (dis'-ar-TIK-yoo-lāt'), to separate at a joint.

Dys- painful, difficult Dyspnea (disp-NĒ-a), difficult breathing.

E-, Ec-, Ex- out from, out of Eccentric (ek-SEN-trik), not located at the center.

Ecto-, Exo- outside Ectopic pregnancy (ek-TOP-ik), gestation outside the uterine cavity.

Em-, En- in, on Empyema (em'-pī-Ē-ma), pus in a body cavity.

End-, Endo- inside Endocardium (en'-dō-KARD-ē-um), membrane lining the inner surface of the heart.

Epi- upon, on, above Epidermis (ep'-i-DER-mis), outermost layer of skin.

Eu- well Eupnea (YOOP-nē-a), normal breathing.

Ex-, Exo- out, away from Exocrine (EK-sō-krin), excreting outwardly or away from.

Extra- outside, beyond, in addition to Extracellular (ek'-stra-SEL-yoo-lar), outside of the cell.

Hemi- half Hemiplegia (hem-ē-PLĒ-jē-a), paralysis of only half of the body.

Heter-, Hetero- other, different Heterogeneous (het'-e-rō-JEN-ē-us), composed of different substances.

Homeo-, Homo- unchanging, the same, steady Homeostasis (hō'-mē-ō-STĀ-sis), achievement of a steady state.

Hyper- beyond, excessive Hyperglycemia (hī-per-glī-SĒ-mē-a), excessive amount of sugar in the blood.

Hypo- under, below, deficient Hypodermic (hī'-pō-DER-mik), below the skin or dermis.

Idio- self, one's own, separate Idiopathic (id'-ē-ō-PATH-ik), a disease without recognizable cause.

Inter- among, between Intercostal (int'-er-KOS-tal), between the ribs.

Intra- within, inside Intracellular (in'-tra-SEL-yoo-lar), inside the cell.

Iso- equal, like Isogenic (ī-sō-JEN-ik), alike in morphological development.

Macro- large, great Macrophage (MAK-rō-fāj), large phagocytic cell.

Mal- bad, abnormal Malnutrition (mal'-noo-TRISH-un), lack of necessary food substances.

Mega-, Megalo- great, large Megakaryocyte (meg'-a-KAR-ē-ō-sīt), giant cell of bone marrow.

Meta- after, beyond Metacarpus (met'-a-KAR-pus), the part of the hand between the wrist and fingers.

Micro- small Microtome (MĪ-krō-tōm), instrument for preparing very thin slices of tissue for microscopic examination.

Neo- new Neonatal (nē-ō-NĀT-al), pertaining to the first 4 weeks after birth.

Oligo- small, deficient Oliguria (ol-ig-YOO-rē-a), abnormally small amount of urine.

Ortho- straight, normal Orthopnea (or-thop-NĒ-a), inability to breathe in any position except when straight or erect.

Para- near, beyond, apart from, beside Paranasal (par-a-NĀ-zal), near the nose.

Per- through Percutaneous (per'-kyoo-TĀ-nē-us), through the skin.

Peri- around Pericardium (per'-i-KARD-ē-um), membrane or sac around the heart.

Poly- much, many Polycythemia (pol'-i-sī-THĒ-mē-a), an excess of red blood cells.

Post- after, behind Postnatal (pōst-NĀT-al), after birth.

Pre-, Pro- before, in front of Prenatal (prē-NĀT-al), before birth.

Pseud-, Pseudo- false Pseudoangina (soo'-dō-an-JĪ-na), false angina.

Retro- backward, located behind Retroperitoneal (re'-trō-per'-it-on-Ē-al), located behind the peritoneum.

Semi- half Semicircular (sem'-i-SER-kyoo-lar) canals, canals in the shape of a half circle.

Sub- under, beneath, below Submucosa (sub'-myoo-KŌ-sa), tissue layer under a mucous membrane.

Super- above, beyond Superficial (soo-per-FISH-al), confined to the surface.

Supra- above, over Suprarenal (soo-pra-RĒN-al), adrenal gland above the kidney.

Sym-, Syn- with, together, joined Syndrome (SIN-drōm), all the symptoms of a disease considered as a whole.

Tachy- rapid Tachycardia (tak'-i-KARD-ē-a), rapid heart action.

Trans- across, through, beyond Transudation (trans-yoo-DĀ-shun), oozing of a fluid through pores.

Tri- three Trigone (TRĪ-gon), a triangular space, as at the base of the bladder.

Suffixes

-able capable of, having ability to Viable (VĪ-a-bal), capable of living.

-ac, -al pertaining to Cardiac (KARD-ē-ak), pertaining to the heart.

-ary connected with Ciliary (SIL-ē-ar-ē), resembling any hairlike structure.

-asis, -asia, -esis, -osis condition or state of Hemostasis (hē-mō-STĀ-sis), stopping of bleeding or circulation.

-cel, -cele swelling, an enlarged space or cavity Meningocele (men-IN-gō-sēl), enlargement of the meninges.

-cid, -cide, -cis cut, kill, destroy Germicide (jer-mi-SĪD), a substance that kills germs.

-ectasia, -ectasis stretching, dilation Bronchiectasis (bron-kē-EK-ta-sis), dilation of a bronchus or bronchi.

-ectomize, ectomy excision of, removal of Thyroidectomy (thī-royd-EK-to-mē), surgical removal of thyroid gland.

-emia condition of blood Lipemia (lip-Ē-mē-a), abnormally high concentration of fat in the blood.

-ferent carry Efferent (EF-e-rent), carrying away from a center.

-form shape Fusiform (FYOO-zi-form), spindle-shaped.

-gen agent that produces or originates Pathogen (PATH-o-jen), microorganism or substance capable of producing a disease.

-genic produced from, producing Pyogenic (pī-ō-JEN-ik), producing pus.

-gram record, that which is recorded Electrocardiogram (e-lek'-trō-KARD-ē-ō-gram), record of heart action.

-graph instrument for recording Electroencephalograph (e-lek'-trō-en-SEF-a-lō-graf), instrument for recording electrical activity of the brain.

-ia state, condition Hypermetropia (hī'-per-me-TRŌ-pē-a), condition of farsightedness.

-iatrics, iatry medical practice specialties Pediatrics (pēd-ē-A-triks), medical science relating to care of children and treatment of their diseases.

-ism condition, state Rheumatism (ROO-ma-tizm), inflammation, especially of muscle and joints.

-itis inflammation Neuritis (nyoo-RĪT-is), inflammation of a nerve or nerves.

-logy, -ology the study or science of Physiology (fiz-ē-OL-o-jē), the study of function of body parts.

-lyso, -lysis solution, dissolve, loosening Hemolysis (hē-MOL-i-sis), dissolution of red blood cells.

-malacia softening Osteomalacia (os'-tē-ō-ma-LA-shē-a), softening of bone.

-oma tumor Fibroma (fi-BRŌ-ma), tumor composed mostly of fibrous tissue.

-ory pertaining to Sensory (SENS-o-rē), pertaining to sensation.

-ose full of Adipose (AD-i-pōz), characterized by presence of fat.

-pathy disease Neuropathy (nyoo-ROP-a-thē), disease of the peripheral nervous system.

-penia deficiency Thrombocytopenia (throm'-bō-sīt'-o-PĒ-nē-a), deficiency of thrombocytes in the blood.

-phobe, -phobia fear of, aversion to Hydrophobia (hī'-drō-FŌ-bē-a), fear of water.

-plasia, -plasty development, formation Rhinoplasty (RĪ-nō-plas-tē), surgical reconstruction of the nose.

-plegia, -plexy stroke, paralysis Apoplexy (AP-o-plek-sē), sudden loss of consciousness and paralysis.

-pnea to breathe Apnea (AP-nē-a), temporary absence of respiration, following a period of overbreathing.

-poiesis production Hematopoiesis (he-mat'-a-poy-Ē-sis), formation and development of red blood cells.

-rrhea flow, discharge Diarrhea (dī-a-RĒ-a), abnormal frequency of bowel evacuation, the stools with a more or less fluid consistency.

-scope instrument for viewing Bronchoscope (BRON-ko-skōp), instrument used to examine the interior of a bronchus.

-stomy creation of a mouth or artificial opening Tracheostomy (trā-kē-OST-o-mē), creation of an opening in the trachea.

-tomy cutting into, incision into Appendectomy (ap-en-DEK-to-mē), surgical removal of the appendix.

-trophy **state relating to nutrition or growth** Hypertrophy (hī-PER-trō-fē), excessive growth of an organ or part.

-tropic **turning toward, influencing, changing** Gonadotropic (gō-nad-a-TRŌ-pic), influencing the gonads.

-uria **urine** Polyuria (pol-ē-YOOR-ē-a), excessive secretion of urine.

Glossary of Terms

1. The strongest accented syllable appears in capital letters, for example, bilateral (bī-LAT-er-al) and diagnosis (dī-ag-NŌ-sis).
2. If there is a secondary accent, it is noted by a single quote mark ('), for example, constitution (kon'-sti-TOO-shun) and physiology (fiz'-ē-OL-ō-jē). Any additional secondary accents are also noted by a single quote mark, for example, decarboxylation (dē'-kar-bok'-si-LĀ-shun).
3. Vowels marked with a line above the letter are pronounced with the long sound as in the following common words:

   ā as in *māke*
   ē as in *bē*
   ī as in *īvy*
   ō as in *pōle*

4. Vowels not so marked are pronounced with the short sound, as in the following words:

   e as in *bet*
   i as in *sip*
   o as in *not*
   u as in *bud*

5. Other phonetic symbols are used to indicate the following sounds:

   a as in *above*
   oo as in *sue*
   yoo as in *cute*
   oy as in *oil*

Abatement (a-BĀT-ment) A decrease in the seriousness of a disorder or in the severity of pain or other symptoms.

Abdomen (ab-DŌ-men) The area between the diaphragm and pelvis.

Abdominal (ab-DŌM-i-nal) **cavity** Superior portion of the abdominopelvic cavity that contains the stomach, spleen, liver, gallbladder, pancreas, small intestine, most of the large intestine, kidneys, and ureters.

Abdominal thrust maneuver A first-aid procedure for choking. Employs a quick, upward thrust against the diaphragm that forces air out of the lungs with sufficient force to eject any lodged material. Also called the **Heimlich maneuver.**

Abdominopelvic (ab-dō-men-ō'-PEL-vic) **cavity** Inferior component of the ventral body cavity that is subdivided into an upper abdominal cavity and a lower pelvic cavity.

Abduction (ab-DUK-shun) Movement away from the axis or midline of the body or one of its parts.

Abortion (a-BOR-shun) The premature loss or removal of the embryo or nonviable fetus; any failure in the normal process of developing or maturing.

Abscess (AB-ses) A localized collection of pus and liquefied tissue in a cavity.

Absorption (ab-SORP-shun) The taking up of liquids by solids or of gases by solids or liquids; intake of fluids or other substances by cells of the skin or mucous membranes; the passage of digested foods from the GI tract into blood or lymph.

Absorptive state Metabolic state during which ingested nutrients are being absorbed by the blood or lymph from the GI tract.

Accommodation (a-kom-ō-DĀ-shun) A change in the curvature of the eye lens to adjust for vision at various distances; focusing.

Acetabulum (as'-e-TAB-yoo-lum) The rounded cavity on the external surface of the coxal bone that receives the head of the femur.

Acetylcholine (as'-ē-til-KŌ-lēn) **(ACh)** A chemical transmitter substance, liberated at synapses in the central nervous system, that stimulates skeletal muscle contraction.

Achilles (a-KIL-ēz) **tendon** The tendon of the soleus and gastrocnemius muscles at the back of the heel.

Acid (AS-id) A proton donor, or substance that dissociates into hydrogen ions (H^+) and anions, characterized by an excess of hydrogen ions and a pH less than 7.

Acidosis (as-i-DŌ-sis) A condition in which blood pH is between 7.35 and 6.80.

Acini (AS-i-nē) Masses of cells in the pancreas that secrete digestive enzymes.

Acne (AK-nē) Inflammation of sebaceous glands that usually begins at puberty; the basic acne lesions in order of increasing severity are comedones, papules, pustules, and cysts.

Acoustic (a-KOOS-tik) Pertaining to sound or the sense of hearing.

Acromegaly (ak'-rō-MEG-a-lē) Condition caused by hypersecretion of growth hormone (GH) during adulthood characterized by thickened bones and enlargement of other tissues.

Actin (AK-tin) The contractile protein that makes up thin myofilaments in muscle fiber.

Active transport The movement of substances, usually ions, across cell membranes, against a concentration gradient, requiring the expenditure of energy.

Acuity (a-KYOO-i-tē) Clearness or sharpness, usually of vision.

Acupuncture (AK-ū-punk'-chur) The insertion of a needle into a tissue for the purpose of drawing fluid or relieving pain.

It is also an ancient Chinese practice employed to cure illnesses by inserting needles into specific locations of the skin.

Acute (a-KYOOT) Having rapid onset, severe symptoms, and a short course; not chronic.

Adaptation (ad'-ap-TĀ-shun) The adjustment of the pupil of the eye to light variations. The property by which a neuron relays a decreased frequency of action potentials from a receptor even though the strength of the stimulus remains constant. The decrease in perception of a sensation over time while the stimulus is still present.

Addison's (AD-i-sonz) **disease** Disorder caused by hyposecretion of glucocorticoids characterized by muscular weakness, mental lethargy, weight loss, low blood pressure, and dehydration.

Adduction (ad-DUK-shun) Movement toward the axis or midline of the body or one of its parts.

Adenohypophysis (ad'-e-nō-hī-POF-i-sis) The anterior portion of the pituitary gland.

Adenosine triphosphate (a-DEN-ō-sēn trī-FOS-fāt) **(ATP)** The universal energy-carrying molecule manufactured in all living cells as a means of capturing and storing energy. It consists of the purine base *adenine* and the five-carbon sugar *ribose,* to which are added, in linear array, three *phosphate* molecules.

Adherence (ad-HĒR-ens) Firm contact between the plasma membrane of a phagocyte and an antigen.

Adhesion (ad-HĒ-zhun) Abnormal joining of parts to each other.

Adipocyte (AD-i-pō-sīt) Fat cell, derived from a fibroblast.

Adrenal (a-DRĒ-nal) **glands** Two glands located superior to each kidney. Also called the **suprarenal glands.**

Adrenal cortex (a-DRĒ-nal KOR-teks) The outer portion of an adrenal gland, divided into three zones, each of which has a different cellular arrangement and secretes different hormones.

Adrenal medulla (me-DUL-a) The inner portion of an adrenal gland, consisting of cells which secrete epinephrine and norepinephrine (NE) in response to the stimulation of preganglionic sympathetic neurons.

Adrenergic (ad'-ren-ER-jik) **fiber** A nerve fiber that, when stimulated, releases norepinephrine (noradrenaline) at a synapse.

Adrenocorticotropic (ad-rē'-nō-kor-ti-kō-TRŌP-ik) **hormone (ACTH)** A hormone produced by the adenohypophysis (anterior lobe) of the pituitary gland that influences the production and secretion of certain hormones of the adrenal cortex.

Adrenoglomerulotropin (a-drē'-nō-glō-mer'-yoo-lō-TRŌ-pin) A hormone secreted by the pineal gland that may stimulate aldosterone secretion.

Adventitia (ad-ven-TISH-yah) The outermost covering of a structure or organ.

Aerobic (air-Ō-bik) Requiring molecular oxygen.

Afferent neuron (NOO-ron) A neuron that carries an impulse toward the central nervous system. Also called a **sensory neuron.**

Afterimage Persistence of a sensation even though the stimulus has been removed.

Agglutination (a-gloot'-i-NĀ-shun) Clumping of microorganisms, blood corpuscles or particles; an immune response; an antigen-antibody reaction.

Agglutinin (a-GLOO-ti-nin) A specific principle or antibody in blood serum capable of causing the clumping of bacteria, blood corpuscles, or particles. Also called **isoantibody.**

Agglutinogen (a-GLOOT-in-ō-gen) A genetically determined antigen located on the surface of erythrocytes; basis for the ABO grouping and Rh system of blood classification. Also called **isoantigen.**

Aging Progressive failure of the body's homeostatic adaptive responses.

Agnosia (ag-NŌ-zē-a) A loss of the ability to recognize the meaning of stimuli from the various senses (visual, auditory, touch).

Agonist (AG-ō-nist) The prime mover—the muscle directly engaged in contraction as distinguished from muscles that are relaxing at the same time.

Agraphia (a-GRAF-ē-a) An inability to write.

Albinism (AL-bin-ism) Abnormal, nonpathological, partial or total absence of pigment in skin, hair, and eyes.

Albumin (al-BYOO-min) The most abundant (60 percent) and smallest of the plasma proteins, which functions primarily to regulate osmotic pressure of plasma.

Albuminuria (al-byoo'-min-UR-ēa) Presence of albumin in the urine.

Aldosterone (al-do-STĒR-ōn) A mineralocorticoid produced by the adrenal cortex that brings about sodium and water reabsorption and potassium excretion.

Aldosteronism (al'-do-STER-ōn-izm') Condition caused by hypersecretion of aldosterone characterized by muscular paralysis, high blood pressure, and edema.

Alkaline (AL-ka-līn) Containing more hydroxyl ions than hydrogen ions to produce a pH of more than 7.

Alkalosis (al-ka-LŌ-sis) A condition in which blood pH is between 7.45 and 8.00.

Allantois (a-LAN-tō-is) A small, vascularized membrane between the chorion and amnion of the fetus.

Allergen (AL-er-jen) An antigen that evokes a hypersensitivity reaction.

Allergic (a-LER-jik) Pertaining to or sensitive to an allergen.

All-or-none principle In muscle physiology, muscle fibers of a motor unit contract to their fullest extent or not at all. In neuron physiology, if a stimulus is strong enough to initiate an action potential, an impulse is transmitted along the entire neuron at a constant and maximum strength.

Alpha (AL-fa) **cell** A cell in the islets of Langerhans in the pancreas that secretes glucagon.

Altitude sickness Disorder caused by decreased levels of alveolar pO_2 as altitude increases and characterized by shortness of breath, nausea, and dizziness.

Alveolar-capillary (al-VĒ-ō-lar) **membrane** Structure in the lungs consisting of the alveolar wall and basement membrane and a capillary endothelium and basement membrane through which the diffusion of respiratory gases occurs. Also called the **respiratory membrane.**

Alveolar duct Branch of a respiratory bronchiole around which alveoli and alveolar sacs are arranged.

Alveolar macrophage (MAK-rō-fāj) Cell found in the alveolar walls of the lungs that is highly phagocytic. Also called a **dust cell.**

Alveolar sac A collection or cluster of alveoli that share a common opening.

Alveolus (al-VĒ-ō-lus) A small hollow or cavity; an air sac in the lungs; milk-secreting portion of a mammary gland.

Amenorrhea (ā-men-ō-RĒ-a) Absence of menstruation.

Amino acid An organic acid, containing a carboxyl group (COOH) and an amino group (NH₂), that is the basic unit from which proteins are formed.

Amnesia (am-NĒ-zē-a) A lack or loss of memory.

Amniocentesis (am'-nē-ō-sen-TĒ-sis) Removal of amniotic fluid by inserting a needle transabdominally into the amniotic cavity.

Amnion (AM-nē-on) The innermost fetal membranes; a thin transparent sac that holds the fetus suspended in amniotic fluid. Also called the **"bag of waters."**

Amphiarthrosis (am'-fē-ar-THRŌ-sis) Articulation midway between diarthrosis and synarthrosis, in which the articulating bony surfaces are separated by an elastic substance to which both are attached, so that the mobility is slight, but may be exerted in all directions.

Anabolism (a-NAB-ō-lizm) Synthetic reactions whereby small molecules are built up into larger ones.

Anaerobic (an-AIR-ō-bik) Not requiring molecular oxygen.

Anal (Ā-nal) **canal** The terminal 2 or 3 cm of the rectum; opens to the exterior at the anus.

Anal column A longitudinal fold in the mucous membrane of the anal canal that contains a network of arteries and veins.

Analgesia (an-al-JĒ-zē-a) Absence of normal sense of pain.

Anal triangle The subdivision of the male or female perineum that contains the anus.

Anamnestic (an'-am-NES-tik) **response** Accelerated, more intense production of antibodies upon a subsequent exposure to an antigen after the initial exposure.

Anaphase (AN-a-fāz) The third stage of mitosis in which the chromatids that have separated at the centromeres move to opposite poles.

Anaphylaxis (an'-a-fi-LAK-sis) A hypersensitivity reaction involving IgE antibodies, mast cells, and basophils; examples are hay fever, bronchial asthma, hives, and anaphylactic shock.

Anastomosis (a-nas-tō-MŌ-sis) An end-to-end union or joining together of blood vessels, lymphatics, or nerves.

Anatomical (an'-a-TOM-i-kal) **position** A position of the body universally used in anatomical descriptions in which the body is erect, facing the observer, the upper extremities are at the sides, and palms of the hands are facing forward.

Anatomy (a-NAT-ō-mē) The structure or study of structure of the body and the relationship of its parts to each other.

Androgen (AN-drō-jen) Substance producing or stimulating male characteristics, such as the male hormone testosterone.

Anemia (a-NĒ-mē-a) Condition of the blood in which the number of functional red blood cells or their hemoglobin content is below normal.

Anesthesia (an'-es-THĒ-zē-a) A total or partial loss of feeling or sensation, usually defined with respect to loss of pain sensation.

Aneurysm (AN-yoo-rizm) A saclike enlargement of a blood vessel caused by a weakening of the wall.

Angina pectoris (an-JĪ-na PEK-tō-ris) A pain in the chest related to reduced coronary circulation that may or may not involve heart or artery disease.

Angiography (an-jē-OG-ra-fē) X-ray examination of blood vessels after injection of a radiopaque substance into the common carotid or vertebral artery; used to demonstrate cerebral blood vessels and may detect brain tumors with specific vascular patterns.

Angiotensin (an-jē-ō-TEN-sin) Either of two forms of a protein associated with regulation of blood pressure. Angiotensin I is produced by the action of renin on angiotensinogen and is converted by the action of a plasma enzyme into angiotensin II, which controls aldosterone secretion by the adrenal cortex.

Ankyloglossia (ang'-ki-lō-GLOSS-ē-a) "Tongue-tied"; restriction of tongue movements by a short lingual frenulum.

Anomaly (a-NOM-a-lē) An abnormality that may be a developmental (congenital) defect; a variant from the usual standard.

Anopsia (an-OP-sē-a) A defect of vision.

Anorexia nervosa (an-ō-REK-sē-a ner-VŌ-sa) A disorder characterized by loss of appetite and bizarre patterns of eating.

Anosmia (an-OZ-mē-a) Loss of the sense of smell.

Anoxia (an-OK-sē-a) Deficiency of oxygen.

Antagonist (an-TAG-ō-nist) A muscle that has an action opposite that of the agonist and yields to the movement of the agonist.

Antepartum (an-tē-PAR-tum) Before delivery of the child; occurring (to the mother) before childbirth.

Anterior (an-TĒR-ē-or) Nearer to or at the front of the body. Also called **ventral.**

Anterior root The structure composed of axons of motor or efferent fibers that emerges from the anterior aspect of the spinal cord and extends laterally to join a posterior root, forming a spinal nerve. Also called a **ventral root.**

Antibiotic (an'-ti-bī-OT-ik) Literally, "antilife"; a chemical produced by a microorganism that is able to inhibit the growth of or kill other microorganisms.

Antibody (AN-ti-bod'-ē) A substance produced by certain cells in the presence of a specific antigen that combines with that antigen to neutralize, inhibit, or destroy it.

Anticoagulant (an-tī-cō-AG-yoo-lant) A substance that is able to delay, suppress, or prevent the clotting of blood.

Antidiuretic hormone (ADH) Hormone produced by neurosecretory cells in the paraventricular nucleus of the hypothalamus that stimulates water reabsorption from kidney cells into the blood and vasoconstriction of arterioles.

Antigen (AN-ti-jen) Any substance that when introduced into the tissues or blood induces the formation of antibodies or reacts with them.

Anulus fibrosus (AN-yoo-lus fī-BRŌ-sus) A ring of fibrous tissue and fibrocartilage that encircles the pulpy substance (nucleus pulposus) of an intervertebral disc.

Anus (Ā-nus) The distal end and outlet of the rectum.

Aorta (ā-OR-ta) The main systemic trunk of the arterial system of the body; emerges from the left ventricle.

Aortic (ā-OR-tik) **reflex** A reflex concerned with maintaining normal general systemic blood pressure.

Aphasia (a-FĀ-zē-a) Loss of ability to express oneself properly through speech or loss of verbal comprehension.

Apnea (ap-NĒ-a) Temporary cessation of breathing.

Apneustic (ap-NOO-stik) **area** Portion of the respiratory center in the pons that helps coordinate the transition between inspiration and expiration.

Apocrine (AP-ō-krin) **gland** A type of gland in which the secretory products gather at the free end of the secreting cell and are pinched off, along with some of the cytoplasm, to become the secretion, as in mammary glands.

Aponeurosis (ap'-ō-nyoo-RŌ-sis) A sheetlike layer of dense, regularly arranged connective tissue joining one muscle with another or with bone.

Appendicitis (a-pen-di-SĪ-tis) Inflammation of the vermiform appendix.

Appetite A desire for food stimulated by a psychological need; may not be related to hunger, which is a response to a nutritional or energy requirement.

Appositional growth Growth due to surface deposition of material, as in the growth in diameter of cartilage and bone. Also called **exogenous growth.**

Aqueous humor (ĀK-wē-us HYOO-mor) The watery fluid that fills the anterior cavity of the eye.

Arachnoid (a-RAK-noyd) The middle of the three coverings (meninges) of the brain.

Arachnoid villus (VIL-us) Berrylike tuft of arachnoid that protrudes into the superior sagittal sinus and through which the cerebrospinal fluid enters the bloodstream.

Arbor vitae (AR-bor VĒ-tē) The treelike appearance of the white matter tracts of the cerebellum when seen in midsagittal section. A series of branching ridges within the cervix of the uterus.

Arch of the aorta (ā-OR-ta) The most superior portion of the aorta, lying between the ascending and descending segments of the aorta.

Areflexia (a'-rē-FLEK-sē-a) Absence of reflexes.

Areola (a-RĒ-ō-la) Any tiny space in a tissue. The pigmented ring around the nipple of the breast.

Arm The portion of the upper extremity from the shoulder to the elbow.

Arrector pili (a-REK-tor PI-lē) Smooth muscles attached to hairs; contraction pulls the hairs into a more vertical position, resulting in "goose bumps."

Arrhythmia (a-RITH-mē-a) Irregular heart rhythm.

Arteriogram (ar-TĒR-ē-ō-gram) A roentgenogram of an artery after injection of a radiopaque substance into the blood.

Arteriole (ar-TĒ-rē-ōl) A small arterial branch that delivers blood to a capillary.

Artery (AR-ter-ē) A blood vessel that carries blood away from the heart.

Arthritis (ar-THRĪ-tis) Inflammation of a joint.

Arthrology (ar-THROL-ō-jē) The study or description of joints.

Arthroscopy (ar-THROS-co-pē) Surgical technique in which an arthroscope is inserted into a small incision, usually in the knee, to repair torn cartilage.

Articular (ar-TIK-yoo-lar) **capsule** Sleevelike structure around a synovial joint composed of a fibrous capsule and a synovial membrane.

Articular cartilage (KAR-ti-lij) Hyaline cartilage attached to articular bone surfaces.

Articular disc Fibrocartilage pad between articular surfaces of bones of some synovial joints. Also called a **meniscus** (men-IS-cus).

Articulation (ar-tik'-yoo-LĀ-shun) A joint.

Ascending colon (KŌ-lon) The portion of the large intestine that passes upward from the cecum to the lower edge of the liver where it bends at the right colic (hepatic) flexure to become the transverse colon.

Aseptic (ā-SEP-tik) Free from any infectious or septic material.

Asphyxia (as-FIX-ē-a) Unconsciousness due to interference with the oxygen supply of the blood.

Aspirate (AS-pir-āt) To remove by suction.

Association area A portion of the cerebral cortex connected by many motor and sensory fibers to other parts of the cortex. The association areas are concerned with motor patterns, memory, concepts of word-hearing and word-seeing, reasoning, will, judgment, and personality traits.

Association neuron (NOO-ron) A nerve cell lying completely within the central nervous system that carries impulses from sensory neurons to motor neurons. Also called an **internuncial** or **connecting neuron.**

Astereognosis (as-ter'-ē-ōg-NŌ-sis) Inability to recognize objects or forms by touch.

Astigmatism (a-STIG-ma-tizm) An irregularity of the lens or cornea of the eye causing the image to be out of focus and producing faulty vision.

Astrocyte (AS-trō-sīt) A neuroglial cell having a star shape.

Ataxia (a-TAK-sē-a) A lack of muscular coordination, lack of precision.

Atelectasis (at'-ē-LEK-ta-sis) A collapsed or airless state of all or part of the lung, which may be acute or chronic.

Atherosclerosis (ath'-er-ō-skle-RŌ-sis) A process in which fatty substances (cholesterol and triglycerides) are deposited in the walls of medium and large arteries in response to certain stimuli (hypertension, carbon monoxide, dietary cholesterol). Following endothelial damage, platelets release chemicals that stimulate uptake of cholesterol and proliferation of smooth muscle cells in the tunica media, resulting in formation of an atherosclerotic plaque that decreases the size of the arterial lumen.

Atom Unit of matter that comprises a chemical element; consists of a nucleus and electrons.

Atresia (a-TRĒ-zē-a) Abnormal closure of a passage, or absence of a normal body opening.

Atrial fibrillation (Ā-trē-al fib-ri-LĀ-shun) Asynchronous contraction of the atria that results in the cessation of atrial pumping.

Atrioventricular (AV) (ā'-trē-ō-ven-TRIK-yoo-lar) **bundle** The portion of the conduction system of the heart that begins at the atrioventricular node and passes through the cardiac skeleton separating the atria and the ventricles, then runs a short distance down the interventricular septum before splitting into right and left bundle branches.

Atrioventricular (AV) node The portion of the conduction system of the heart made up of a compact mass of conducting cells located near the orifice of the coronary sinus in the right atrial wall.

Atrioventricular (AV) valve A structure made up of membranous flaps or cusps that allows blood to flow in one direction only, from an atrium into a ventricle.

Atrium (Ā-trē-um) A superior chamber of the heart.

Atrophy (AT-rō-fē) Wasting away or decrease in size of a part, due to a failure, abnormality of nutrition, or lack of use.

Auditory ossicle (AW-di-tō-rē OS-si-kul) One of three small bones of the middle ear called the malleus, incus, and stapes.

Auditory tube The tube that connects the middle ear with the nose and nasopharynx of the throat. Also called the **Eustachian tube.**

Auricle (OR-i-kul) The flap or pinna of the ear. An appendage of an atrium of the heart.

Auscultation (aws-kul-TĀ-shun) Examination by listening to sounds in the body.

Autoimmunity An immunologic response against a person's own tissue antigens.

Autonomic ganglion (aw'-tō-NOM-ik GANG-lē-on) A cluster of sympathetic or parasympathetic cell bodies located outside the central nervous system.

Autonomic nervous system (ANS) Visceral efferent neurons, both sympathetic and parasympathetic, that transmit impulses from the central nervous system to smooth muscle, cardiac muscle, and glands; so named because this portion of the nervous system was thought to be self-governing or spontaneous.

Autonomic plexus (PLEK-sus) An extensive network of sympathetic and parasympathetic fibers; the cardiac, celiac, and pelvic plexuses are located in the thorax, abdomen, and pelvis, respectively.

Autopsy (AW-top-sē) The examination of the body after death; a postmortem study of the corpse.

Autoregulation (aw-tō-reg-yoo-LĀ-shun) A local, automatic adjustment of blood flow in a given region of the body in response to tissue needs.

Autosome Any chromosome other than the pair of sex chromosomes.

Avitaminosis (ā-vī-ta-min-Ō-sis) A deficiency of a vitamin in the diet.

Axilla (ak-SIL-a) The small hollow beneath the arm where it joins the body at the shoulders. Also called the **armpit.**

Axon (AK-son) The process of a nerve cell that carries an impulse away from the cell body.

Babinski (ba-BIN-skē) **sign** Extension of the great toe, with or without fanning of the other toes, in response to stimulation of the outer margin of the sole of the foot; normal up to 1½ years of age.

Back The posterior part of the body; the dorsum.

Bainbridge (BĀN-bridge) **reflex** The increased heart rate that follows increased pressure or distension of the right atrium.

Ball-and-socket joint   A synovial joint in which the rounded surface of one bone moves within a cup-shaped depression or fossa of another bone, as in the shoulder or hip joint. Also called a **spheroid joint.**

Basal metabolic (BĀ-sal met'-a-BOL-ik) **rate (BMR)**   The rate of metabolism measured under standard or basal conditions.

Basal ganglia (GANG-glē-a)   Paired clusters of cell bodies that make up the central gray matter in each cerebral hemisphere, including the caudate nucleus, lentiform nucleus, claustrum, and amygdaloid body. Also called **cerebral nuclei.**

Base   The broadest part of a pyramidal structure. A nonacid, or a proton acceptor, characterized by excess of hydroxide ions and a pH greater than 7. A ring-shaped, nitrogen-containing organic molecule which is one of the components of a nucleotide, for example, adenine, guanine, cytosine, thymine, and uracil.

Basement membrane   Thin, extracellular layer consisting of basal lamina secreted by epithelial cells and reticular lamina secreted by connective tissue cells.

Basilar (BAS-i-lar) **membrane**   A membrane in the cochlea of the inner ear that separates the cochlear duct from the scala tympani and on which the spiral organ rests.

Basophil (BĀ-sō-fil)   A type of white blood cell characterized by a pale nucleus and large granules which stain readily with basic dyes.

B cell   A lymphocyte that develops into a plasma cell that produces antibodies or a memory cell.

Belly   The abdomen. The gaster or prominent, fleshy part of a skeletal muscle.

Benign (bē-NĪN)   Not malignant.

Beta (BĀ-ta) **cell**   A cell in the islets of Langerhans in the pancreas that secretes insulin.

Bicuspid (bī-KUS-pid) **valve**   Atrioventricular valve on the left side of the heart. Also called the **mitral valve.**

Bilateral (bī-LAT-er-al)   Pertaining to two sides of the body.

Bile (bīl)   A secretion of the liver.

Biliary calculi (BIL-ē-er-ē CAL-kyoo-lē)   Gallstones formed by the crystallization of cholesterol in bile.

Bilirubin (bil-ē-ROO-bin)   A red pigment that is one of the end products of hemoglobin breakdown by the liver cells and is excreted as a waste material in the bile.

Bilirubinuria (bil-ē-roo-bi-NOO-rē-a)   The presence of above normal levels of bilirubin in urine.

Biliverdin (bil-ē-VER-din)   A green pigment that is one of the first products of hemoglobin breakdown in the liver cells and is converted to bilirubin or excreted as a waste material in bile.

Biofeedback   Process by which an individual gets constant signals (feedback) about various visceral body functions.

Biopsy (BĪ-op-sē)   Removal of tissue or other material from the living body for examination, usually microscopic.

Blastocoel (BLAS-tō-sēl)   The fluid-filled cavity within the blastocyst.

Blastocyst (BLAS-tō-sist)   In the development of an embryo, a hollow ball of cells that consists of a blastocoel (the internal cavity), trophectoderm (outer cells), and inner cell mass.

Blastomere (BLAS-tō-mēr)   One of the cells resulting from the cleavage of a fertilized ovum.

Bleeding time   The time required for the cessation of bleeding from a small skin puncture as a result of platelet disintegration and blood vessel constriction; ranges from 1 to 4 minutes.

Blind spot   Area in the retina at the end of the optic nerve in which there are no light receptor cells.

Blood   The fluid that circulates through the heart, arteries, capillaries, and veins and that constitutes the chief means of transport within the body.

Blood–brain barrier (BBB)   A special mechanism that prevents the passage of materials from the blood to the cerebrospinal fluid and brain.

Blood pressure (BP)   Force exerted by blood as it presses against and attempts to stretch blood vessels, especially arteries; clinically, a measure of the pressure in arteries during ventricular systole and ventricular diastole.

Body cavity   A space within the body that contains various internal organs.

Bolus (BŌ-lus)   A soft, rounded mass, usually food, that is swallowed.

Bony labyrinth (LAB-ī-rinth)   A series of cavities within the petrous portion of the temporal bone, forming the vestibule, cochlea, and semicircular canals of the inner ear.

Bradycardia (brad-ē-KAR-dē-a)   A slow heartbeat or pulse rate.

Brain   A mass of nerve tissue located in the cranial cavity.

Brain electrical activity mapping (BEAM)   Noninvasive procedure that measures and displays the electrical activity of the brain; used primarily to diagnose epilepsy.

Brain sand   Calcium deposits in the pineal gland that are laid down starting at puberty.

Brain stem   The portion of the brain immediately superior to the spinal cord, made up of the medulla oblongata, pons varolii, and midbrain.

Broca's (BRŌ-kaz) **area**   Motor area of the brain in the frontal lobe that translates thoughts into speech. Also called the **motor speech area.**

Bronchi (BRONG-kē)   Branches of the respiratory passageway including primary bronchi (the two divisions of the trachea), secondary or lobar bronchi (divisions of the primary that are distributed to the lobes of the lung), and tertiary or segmental bronchi (divisions of the secondary that are distributed to bronchopulmonary segments of the lung).

Bronchial asthma (BRONG-kē-al AZ-ma)   Usually allergic reaction characterized by smooth muscle spasms in bronchi resulting in wheezing and difficult breathing.

Bronchial tree   The trachea, bronchi, and their branching structures.

Bronchiole (BRONG-kē-ōl)   Branch of a tertiary bronchus further dividing into terminal bronchioles (distributed to lobules of the lung), which divide into respiratory bronchioles (distributed to alveolar sacs).

Bronchitis (BRONG-kī-tis)   Inflammation of the bronchi characterized by a productive cough.

Bronchogenic carcinoma (brong'-kō-JEN-ik kar'-si-NŌ-ma)   Cancer originating in the bronchi.

Bronchogram (BRONG-kō-gram)   A roentgenogram of the lungs and bronchi

Bronchopulmonary (brong'-kō-PUL-mō-ner'-ē) **segment**   One of the smaller divisions of a lobe of a lung supplied by its own branches of a bronchus.

Bronchoscope (BRONG-kō-skōp)   An instrument used to examine the interior of the bronchi of the lungs.

Buccal (BUK-al)   Pertaining to the cheek or mouth.

Buffer (BUF-er) **system**   A pair of chemicals, one a weak acid and one a weak base, that resists changes in pH.

Bulbourethral (bul'-bō-yoo-RĒ-thral) **gland**   One of a pair of glands located inferior to the prostate on either side of the urethra that secretes an alkaline fluid into the cavernous urethra. Also called a **Cowpers' gland.**

Bulimia (boo-LIM-ē-a)   A disorder characterized by uncontrollable overeating followed by forced vomiting or overdoses of laxatives.

Bundle branch   One of the two branches of the atrioventricular bundle, made up of specialized muscle fibers that transmit electrical impulses to the ventricles.

Bunion (BUN-yun) Inflammation and thickening of the bursa of a toe joint due to displacement of the great toe.

Burn An injury caused by heat (fire, steam), chemicals, electricity, or the ultraviolet rays of the sun.

Bursa (BUR-sa) A sac or pouch of synovial fluid located at friction points, especially about joints.

Bursitis (bur-SĪ-tis) Inflammation of a bursa.

Buttocks (BUT-oks) The two fleshy masses on the posterior aspect of the lower trunk, formed by the gluteal muscles.

Calcitonin (kal-si-TŌ-nin) **(CT)** A hormone produced by the thyroid gland that lowers the calcium and phosphate levels of the blood by inhibiting bone breakdown and accelerating calcium absorption by bones.

Calculus (KAL-kyoo-lus) A stone, or insoluble mass of crystallized salts or other material, formed within the body, as in the gallbladder, kidney, or urinary bladder.

Callus (KAL-lus) A growth of new bone tissue in and around a fractured area, ultimately replaced by mature bone. An acquired, localized thickening.

Calorie (KAL-ō-rē) A unit of heat. The small calorie (cal) is the standard unit and is the amount of heat necessary to raise 1 g of water 1°C from 14° to 15°C. The large Calorie (kilocalorie), used in metabolic and nutrition studies, is the amount of heat necessary to raise 1 kg of water 1°C, or 1,000 cal.

Calyx (KĀL-iks) Any cuplike division of the kidney pelvis. *Plural,* **calyces** (KĀ-li-sēz).

Canaliculus (kan'-a-LIK-yoo-lus) A small channel or canal, as in bones, where they connect the lacunae. *Plural,* **canaliculi** (kan'-a-LIK-yoo-lī).

Canal of Schlemm (shlem) A circular venous sinus located at the junction of the sclera and the cornea through which aqueous humor drains from the anterior chamber of the eyeball into the blood.

Cancer (KAN-ser) A malignant tumor of epithelial origin tending to infiltrate and give rise to new growths or metastases. Also called **carcinoma** (kar'-si-NŌ-ma).

Canthus (KAN-thus) The angular junction of the eyelids at either corner of the eyes.

Capillary (KAP-i-lar'-ē) A microscopic blood vessel located between an arteriole and venule through which materials are exchanged between blood and body cells.

Carbohydrate (kar'-bō-HĪ-drāt) An organic compound containing carbon, hydrogen, and oxygen in a particular amount and arrangement and comprised of sugar subunits; usually has the formula $(CH_2O)n$.

Carbon monoxide poisoning Hypoxia due to increased levels of carbon monoxide as a result of its preferential and tenacious combination with hemoglobin compared to oxygen.

Carcinogen (kar-SIN-ō-jen) Any substance that causes cancer.

Cardiac (KAR-dē-ak) **cycle** A complete heartbeat consisting of systole and diastole of both atria plus systole and diastole of both ventricles.

Cardiac muscle An organ specialized for contraction, composed of striated muscle cells, forming the walls of the heart, and stimulated by an intrinsic conduction system and visceral efferent neurons.

Cardiac notch An angular notch in the anterior border of the left lung.

Cardiac output (CO) The volume of blood pumped from one ventricle of the heart (usually measured from the left ventricle) in 1 min; about 5.2 liters/min under normal resting conditions.

Cardiac reserve The maximum percentage that cardiac output can increase above normal.

Cardioacceleratory (kar-dē-ō-ak-SEL-er-a-tō'-rē) **center** A group of neurons in the medulla from which cardiac nerves (sympathetic) arise; impulses along the nerves release epi-

nephrine that increases the rate and force of heartbeat.

Cardioinhibitory (kar-dē-ō-in-HIB-i-tō'-rē) **center** A group of neurons in the medulla from which parasympathetic fibers that reach the heart via the vagus (X) nerve arise; impulses along the nerves release acetylcholine that decreases the rate and force of heartbeat.

Cardiopulmonary resuscitation (rē-sus-i-TĀ-shun) **(CPR)** A technique employed to restore life or consciousness to a person apparently dead or dying; includes external respiration (exhaled air respiration) and external cardiac massage.

Caries (KĀ-rēz) Decay of a tooth or bone.

Carotid (ka-ROT-id) **body** Receptor in the carotid sinus that responds to alterations in blood levels of oxygen, carbon dioxide, and hydrogen ions.

Carotid sinus A dilated region of the internal carotid artery immediately above the bifurcation of the common carotid artery that contains receptors that monitor blood pressure.

Carotid sinus reflex A reflex concerned with maintaining normal blood pressure in the brain.

Carpus (KAR-pus) A collective term for the eight bones of the wrist.

Cartilage (KAR-ti-lij) A type of connective tissue consisting of chondrocytes in lacunae embedded in a dense network of collagenous and elastic fibers and a matrix of chondroitin sulfate.

Cartilaginous (kar'-ti-LAJ-i-nus) **joint** A joint without a joint cavity where the articulating bones are held tightly together by cartilage, allowing little or no movement.

Cast A small mass of hardened material formed within a cavity in the body and then discharged from the body; can originate in different areas and be composed of various materials.

Castration (kas-TRĀ-shun) The removal of the testes.

Catabolism (ka-TAB-ō-lizm) Decomposition processes that provide energy.

Cataract (KAT-a-rakt) Loss of transparency of the crystalline lens of the eye or its capsule or both.

Catheter (KATH-i-ter) A tube that can be inserted into a body cavity through a canal or into a blood vessel; used to remove fluids, such as urine or blood, and to introduce diagnostic materials or medication.

Cauda equina (KAW-da ē-KWĪ-na) A taillike collection of roots of spinal nerves at the inferior end of the spinal canal.

Cecum (SĒ-kum) A blind pouch at the proximal end of the large intestine to which the ileum is attached.

Cell The basic structural and functional unit of all organisms; the smallest structure capable of performing all the activities vital to life.

Cell inclusion A lifeless, often temporary, constituent in the cytoplasm of a cell as opposed to an organelle.

Cellular immunity That component of immunity in which specially sensitized lymphocytes (T cells) attach to antigens to destroy them.

Cementum (se-MEN-tum) Calcified tissue covering the root of a tooth.

Center of ossification (os'-i-fi-KĀ-shun) An area in the cartilage model of a future bone where the cartilage cells hypertrophy, secrete enzymes that result in the calcification of their matrix resulting in the death of the cartilage cells, followed by the invasion of the area by osteoblasts that then lay down bone.

Central canal A microscopic tube running the length of the spinal cord in the gray commissure.

Central fovea (FŌ-vē-a) A cuplike depression in the center of the macula lutea of the retina, containing cones only; the area of clearest vision.

Central nervous system (CNS) Brain and spinal cord.

Centrioles (SEN-trē-ōlz) Paired, cylindrical organelles within a centrosome, each consisting of a ring of microtubules and arranged at right angles to each other; function in cell division.

Centromere (SEN-trō-mēr) The clear, constricted portion of a chromosome where the two chromatids are joined; serves as the point of attachment for the chromosomal microtubules.

Centrosome (SEN-trō-sōm) A rather dense area of cytoplasm, near the nucleus of a cell, containing a pair of centrioles.

Cerebellar peduncle (ser-e-BEL-ar pe-DUNG-kul) A bundle of nerve fibers connecting the cerebellum with the brain stem.

Cerebellum (ser-e-BEL-um) The portion of the hindbrain lying posterior to the medulla and pons, concerned with coordination of movements.

Cerebral aqueduct (SER-e-bral AK-we-dukt) A channel through the midbrain connecting the third and fourth ventricles and containing cerebrospinal fluid.

Cerebral cortex The surface of the cerebral hemispheres, 2 to 4 mm thick, consisting of six layers of nerve cell bodies (gray matter) in most areas.

Cerebral palsy (PAL-zē) A group of nonprogressive motor disorders caused by damage to motor areas of the brain (cerebral cortex, basal ganglia, and cerebellum) during fetal life, birth, or infancy.

Cerebral peduncle (pe-DUNG-kul) One of a pair of nerve fiber bundles located on the ventral surface of the midbrain, conducting impulses between the pons and the cerebral hemispheres.

Cerebrospinal (se-rē'-brō-SPĪ-nal) **fluid (CSF)** A fluid produced in the choroid plexuses of the ventricles of the brain that circulates in the ventricles and the subarachnoid space around the brain and spinal cord.

Cerebrovascular (se-rē'-brō-VAS-kyoo-lar) **accident (CVA)** Destruction of brain tissue (infarction) resulting from disorders of blood vessels that supply the brain. Also called a **stroke.**

Cerebrum (SER-ē-brum) The two hemispheres of the forebrain, making up the largest part of the brain.

Ceruminous (se-ROO-mi-nus) **gland** A modified sudoriferous (sweat) gland in the external auditory meatus that secretes cerumen (ear wax).

Cervix (SER-viks) Neck; any constricted portion of an organ, especially the lower cylindrical part of the uterus.

Chemoreceptor (kē'-mō-rē-SEP-tor) Receptor that detects the presence of chemicals.

Chemotaxis (kē'-mō-TAK-sis) Attraction of phagocytes to a chemical stimulus.

Chemotherapy (kē'-mō-THER-a-pē) The treatment of illness or disease by chemicals.

Chiropractic (kī'-rō-PRAK-tik) A system of treating disease by using one's hands to manipulate body parts, mostly the vertebral column.

Chloride shift Diffusion of bicarbonate ions from red blood cells into plasma and of chloride ions from plasma into red blood cells that maintains ionic balance between red blood cells and plasma.

Cholecystectomy (kō'-lē-sis-TEK-tō-mē) Surgical removal of the gallbladder.

Cholesterol (kō-LES-te-rol) Classified as a lipid, the most abundant steroid in animal tissues; located in cell membranes and used for the synthesis of steroid hormones and bile salts.

Cholinergic (kō'-lin-ER-jik) **fiber** A nerve ending that liberates acetylcholine at a synapse.

Chondrocyte (KON-drō-sīt) Cell of mature cartilage.

Chondroitin (kon-DROY-tin) **sulfate** An amorphous matrix material found outside the cell.

Chordae tendineae (KOR-dē TEN-di-nē) Cords that connect the heart valves with the papillary muscles.

Chorion (KŌ-rē-on) The outermost fetal membrane; serves a protective and nutritive function.

Choroid (KŌ-royd) One of the vascular coats of the eyeball.

Choroid plexus (PLEK-sus) A vascular structure located in the roof of each of the four ventricles of the brain; produces cerebrospinal fluid.

Chromaffin (krō-MAF-in) **cell** Cell that has an affinity for chrome salts, due in part to the presence of the precursors of the chemical transmitter epinephrine; found, among other places, in the adrenal medulla.

Chromatid (KRŌ-ma-tid) One of a pair of identical connected nucleoprotein strands that are joined at the centromere and separate during cell division, each becoming a chromosome of one of the two daughter cells.

Chromatin (KRŌ-ma-tin) The threadlike mass of the genetic material consisting principally of DNA, which is present in the nucleus of a nondividing or interphase cell.

Chromosome (KRŌ-mō-sōm) One of the 46 small, dark-staining bodies that appear in the nucleus of a human diploid cell during cell division.

Chronic (KRON-ik) Long-term or frequently recurring; applied to a disease that is not acute.

Chylomicron (kī'-lō-MĪ-kron) Aggregate of triglycerides, phospholipids, and cholesterol coated with protein that enters a lacteal of a villus during absorption.

Chyme (kīm) The semifluid mixture of partly digested food and digestive secretions found in the stomach and small intestine during digestion of a meal.

Ciliary (SIL-ē-ar'-ē) **body** One of the three portions of the vascular tunic of the eyeball, the others being the choroid and the iris; includes the ciliary muscle and the ciliary processes.

Cilium (SIL-ē-um) A hair or hairlike process projecting from a cell that may be used to move the entire cell or to move substances along the surface of the cell.

Circadian (ser-KĀ-dē-an) **rhythm** A cycle of active and nonactive periods in organisms determined by internal mechanisms and repeating about every 24 hours.

Circle of Willis A ring of arteries forming an anastomosis at the base of the brain between the internal carotid and vertebral arteries and arteries supplying the brain.

Circulation time Time required for blood to pass from the right atrium, through pulmonary circulation, back to the left ventricle, through systemic circulation to the foot, and back again to the right atrium; normally about 1 min.

Circulatory (SER-kyoo-la-to'-rē) **shock** A condition in which body tissues, including critical organs, do not receive an adequate blood supply because of reduced cardiac output or reduced blood volume; signs include lowered blood pressure, feeble and rapid pulse, shallow and rapid breathing, paleness or cyanosis, and mental confusion or unconsciousness.

Circumcision (ser'-kum-SIZH-un) Removal of the foreskin, the fold over the glans penis.

Circumduction (ser'-kum-DUK-shun) A movement at a synovial joint in which the distal end of a bone moves in a circle while the proximal end remains relatively stable.

Circumvallate papilla (ser'-kum-VAL-āt pa-PIL-a) One of the circular projections which are arranged in an inverted V-shaped row at the posterior portion of the tongue; the largest of the elevations on the upper surface of the tongue and contains taste buds.

Cirrhosis (si-RŌ-sis) A liver disorder in which the parenchymal cells are destroyed and replaced by connective tissue.

Cisterna chyli (sis-TER-na KĪ-lē) The origin of the thoracic duct.

Cleavage The rapid mitotic divisions following the fertilization of an ovum, resulting in an increased number of progressively smaller cells, called blastomeres, so that the overall size of the zygote remains the same.

Cleft palate Condition in which the palative processes of the maxilla do not unite before birth; cleft lip, a split in the upper lip, is often associated with cleft palate.

Climacteric (klī-mak-TER-ik) Cessation of the reproductive function in the female or diminution of testicular activity in the male.

Climax The peak period or moments of greatest intensity during sexual excitement.

Clitoris (KLI-to-ris) An erectile organ of the female that is homologous to the male penis.

Clot The end result of a series of biochemical reactions that changes liquid plasma into a gelatinous mass; specifically, the conversion of fibrinogen into a tangle of polymerized fibrin molecules.

Clot retraction (rē-TRAK-shun) The consolidation of a fibrin clot to pull damaged tissue together.

Clotting time The time required for blood to coagulate as a result of platelet disintegration; ranges from 5 to 15 minutes.

Coagulation (cō-ag-yoo-LĀ-shun) Process by which a blood clot is formed.

Coccyx (KOK-six) The fused bones at the end of the vertebral column.

Cochlea (KŌK-lē-a) A winding, cone-shaped tube forming a portion of the inner ear and containing the spiral organ.

Cochlear duct The membranous cochlea consisting of a spirally arranged tube enclosed in the bony cochlea and lying along its outer wall. Also called the **scala media.**

Coenzyme A nonprotein substance that is associated with and activates an enzyme.

Coitus (KŌ-i-tus) Sexual intercourse. Also called **copulation.**

Collagen (KOL-a-jen) A protein that is the main organic constituent of connective tissue.

Collateral circulation The alternate route taken by blood through an anastomosis.

Colon The division of the large intestine consisting of ascending, transverse, descending, and sigmoid portions.

Color blindness Any deviation in the normal perception of colors, resulting from the lack of one or more of the photopigments of the cones.

Colostomy (kō-LOS-tō-mē) The surgical creation of a new opening from the colon to the body surface.

Colostrum (kō-LOS-trum) A thin, cloudy fluid secreted by the mammary glands a few days prior to or after delivery before true milk is secreted.

Colposcopy (kol-POS-kō-pē) Direct examination of the vaginal and cervical mucosa using a magnifying device.

Coma (KŌ-ma) Profound unconsciousness from which one cannot be roused.

Common bile duct A tube formed by the union of the common hepatic duct and the cystic duct that empties bile into the duodenum at the hepatopancreatic ampulla.

Complement A group of interacting plasma proteins that is activated by certain antigen-antibody reactions and follows a specific sequence of reactions on the surface of a cell on which the antibody is fixed so that the membrane is lysed.

Compliance The ease with which the lungs and thoracic wall can be expanded.

Computed tomography (tō-MOG-ra-fē) **(CT)** X-ray technique that provides a cross-sectional picture of any area of the body.

Concha (KONG-ka) A scroll-like bone found in the skull. *Plural,* **conchae** (KONG-kē).

Concussion (kon-KUSH-un) Traumatic injury to the brain that produces no visible bruising, but may result in abrupt, temporary loss of consciousness.

Cone The light-sensitive receptor in the retina concerned with color vision.

Congenital (kon-JEN-i-tal) Present at the time of birth.

Congestive heart failure (CHF) Chronic or acute state that results when the heart is not capable of supplying the oxygen demands of the body.

Conjunctiva (kon'-junk-TĪ-va) The delicate membrane covering the eyeball and lining the eyelids.

Conjunctivitis (kon-junk'-ti-VĪ-tis) Inflammation of the delicate membrane covering the eyeball and lining the eyelids.

Connective tissue The most abundant of the four tissue types in the body, performing the functions of binding and supporting; consists of relatively few cells in a great deal of intercellular substance.

Constipation (con-sti-PĀ-shun) Infrequent or difficult defecation caused by decreased motility of the intestines.

Contact inhibition Phenomenon by which migration of a growing cell is stopped when it makes contact with another cell of its own kind.

Contraception (kon'-tra-SEP-shun) The prevention of conception or impregnation without destroying fertility.

Contractility (kon'-trak-TIL-i-tē) The ability of muscle tissue to shorten.

Contralateral (kon'-tra-LAT-er-al) On the opposite side; affecting the opposite side of the body.

Conus medullaris (KŌ-nus med-yoo-LAR-is) The tapered portion of the spinal cord below the lumbar enlargement.

Convergence (con-VER-jens) An anatomical arrangement in which the synaptic knobs of several presynaptic neurons terminate on one postsynaptic neuron. The medial movement of the two eyeballs so that both are directed toward a near object being viewed in order to produce a single image.

Convulsion (con-VUL-shun) Violent, involuntary, tetanic contractions of an entire group of muscles.

Cornea (KŌR-nē-a) The transparent fibrous coat that covers the iris of the eye.

Corona radiata Innermost layer of follicle cells surrounding an ovum.

Coronary (KOR-ō-na-rē) **artery disease (CAD)** A condition in which the heart muscle receives inadequate blood due to an interruption of its blood supply.

Coronary artery spasm A condition in which the smooth muscle of a coronary artery undergoes a sudden contraction, resulting in vasoconstriction.

Coronary circulation The pathway followed by the blood from the ascending aorta through the blood vessels supplying the heart and returning to the right atrium. Also called **cardiac circulation.**

Coronary sinus (SĪ-nus) A wide venous channel on the posterior surface of the heart that collects the blood from the coronary circulation and returns it to the right atrium.

Corpora quadrigemina (KOR-por-a kwad-ri-JEM-in-a) Four small elevations on the dorsal region of the midbrain concerned with visual and auditory functions.

Corpus albicans (KOR-pus AL-bi-kanz) A white fibrous patch in the ovary that forms after the corpus luteum regresses.

Corpus callosum (ka-LŌ-sum) The great commissure of the brain between the cerebral hemispheres.

Corpus luteum (LOO-tē-um) A yellow endocrine gland in the ovary formed when a follicle has discharged its ovum; secretes estrogens, progesterone, and relaxin.

Cortex (KOR-teks) An outer layer of an organ. The convoluted layer of gray matter covering each cerebral hemisphere.

Costal cartilage (KOS-tal KAR-ti-lij) Hyaline cartilage that attaches a rib to the sternum.

Countercurrent multiplier mechanism The ability of the kidneys to produce either hyperosmotic or hyposmotic urine.

Cramp A spasmodic, especially a tonic, contraction of one or many muscles, usually painful.

Cranial (KRĀ-nē-al) **cavity** A subdivision of the dorsal body cavity formed by the cranial bones and containing the brain.

Cranial nerve   One of 12 pairs of nerves that leave the brain, pass through foramina in the skull, and supply the head, neck, and part of the trunk; each is designated by a Roman numeral and a name.

Craniosacral (krā-nē-ō-SĀ-kral) **outflow**   The fibers of parasympathetic preganglionic neurons, which have their cell bodies located in nuclei in the brain stem and in the lateral gray matter of the sacral portion of the spinal cord.

Craniotomy (krā'-nē-OT-ō-mē)   Any operation on the skull, as for surgery on the brain or decompression of the fetal head in difficult labor.

Cranium (KRĀ-nē-um)   The skeleton of the skull that protects the brain and the organs of sight, hearing, and balance; includes the frontal, parietal, temporal, occipital, sphenoid, and ethmoid bones.

Crenation (krē-NĀ-shun)   The shrinkage of red blood cells into knobbed, starry forms when placed in a hypertonic solution.

Cretinism (KRĒ-tin-izm)   Severe congenital thyroid deficiency during childhood leading to physical and mental retardation.

Crista (KRIS-ta)   A crest or ridged structure. A small elevation in the ampulla of each semicircular duct that serves as a receptor for dynamic equilibrium.

Cross-matching   A test performed prior to blood transfusion to check for compatibility of blood types. Red blood cells from donor are mixed with a sample of recipient's serum and then rbc's from recipient are mixed with a sample of donor's serum. If no agglutination occurs, the bloods are deemed compatible.

Cryosurgery (KRĪ-ō-ser-jer-ē)   The destruction of tissue by application of extreme cold.

Cryptorchidism (krip-TOR-ki-dizm)   The condition of undescended testes.

Cupula (KUP-yoo-la)   A mass of gelatinous material covering the hair cells of a crista, a receptor in the ampulla of a semicircular canal stimulated when the head moves.

Curvature (KUR-va-tūr)   A nonangular deviation of a straight line, as in the greater and lesser curvatures of the stomach. Abnormal curvatures of the vertebral column include kyphosis, lordosis, and scoliosis.

Cushing's syndrome   Condition caused by a hypersecretion of glucocorticoids characterized by spindly legs, "moon face," "buffalo hump," pendulous abdomen, flushed facial skin, and poor wound healing.

Cutaneous (kyoo-TĀ-nē-us)   Pertaining to the skin.

Cyanosis (sī'-a-NŌ-sis)   Slightly bluish or dark purple discoloration of the skin and the mucous membrane due to an oxygen deficiency.

Cyclic adenosine-3',5'-monophosphate (cyclic AMP)   Molecule formed from ATP by the action of the enzyme adenyl cyclase; serves as an intracellular messenger for some hormones.

Cystic (SIS-tik) **duct**   The duct that transports bile from the gallbladder to the common bile duct.

Cystitis (sis-TĪ-tis)   Inflammation of the urinary bladder.

Cystoscope (SIS-ti-skōp)   An instrument used to examine the inside of the urinary bladder.

Cytochrome (SĪ-tō-krōm)   A protein containing an iron portion capable of alternating between a reduced form and an oxidized form.

Cytokinesis (sī'-tō-ki-NĒ-sis)   Division of the cytoplasm.

Cytology (sī-TOL-o-jē)   The study of cells.

Cytoplasm (SĪ-tō-plazm)   Substance within a cell's plasma membrane and external to its nucleus. Also called **protoplasm.**

Cytoskeleton   Complex internal structure of cytoplasm consisting of microfilaments and microtubules.

Dead air volume   The volume of air that is inhaled but remains in spaces in the upper respiratory system and does not reach the alveoli for participation in gaseous exchange; about 150 ml.

Deafness   Lack of the sense of hearing or a significant hearing loss.

Debility (dē-BIL-i-tē)   Weakness of tonicity in functions or organs of the body.

Decibel (DES-i-bel) **(db)**   A unit that measures sound intensity (loudness).

Decidua (dē-SID-yoo-a)   That portion of the endometrium of the uterus (all but the deepest layer) that is modified for pregnancy and shed after childbirth.

Deciduous (dē-SID-yoo-us)   Falling off or being shed seasonally or at a particular stage of development. In the body, referring to the first set of teeth.

Decompression sickness   A condition characterized by joint pains and neurologic symptoms; follows from a too-rapid reduction of environmental pressure or decompression so that nitrogen that dissolved in body fluid under pressure comes out of solution as bubbles that form air emboli and occlude crucial blood vessels. Also called **caisson disease** or **bends.**

Decubitus (dē-KYOO-bi-tus) **ulcer**   Tissue destruction due to a constant deficiency of blood to tissues overlying a bony projection that has been subjected to prolonged pressure against an object such as a bed, cast, or splint. Also called **bedsore, pressure sore,** or **trophic ulcer.**

Decussation (dē'-ku-SĀ-shun)   A crossing over; usually refers to the crossing of most of the fibers in the large motor tracts to opposite sides in the medullary pyramids.

Deep   Away from the surface of the body.

Deep fascia (FASH-ē-a)   A sheet of connective tissue wrapped around a muscle to hold it in place.

Defecation (def-e-KĀ-shun)   The discharge of feces from the rectum.

Defibrillation (dē-fib-ri-LĀ-shun)   Delivery of a very strong electrical current to the heart in an attempt to stop ventricular fibrillation.

Deglutition (dē-gloo-TISH-un)   The act of swallowing.

Dehydration (dē-hī-DRĀ-shun)   Excessive loss of water from the body or its parts.

Delta cell   A cell in the islets of Langerhans in the pancreas that secretes somatostatin.

Demineralization (de-min'-er-al-i-ZĀ-shun)   Loss of calcium and phosphorus from bones.

Dendrite (DEN-drīt)   A nerve cell process carrying an impulse toward the cell body.

Dentin (DEN-tin)   The osseous tissues of a tooth enclosing the pulp cavity.

Dentition (den-TI-shun)   The eruption of teeth. The number, shape, and arrangement of teeth.

Deoxyribonucleic (dē-ok'-sē-ri'-bō-nyoo-KLĒ-ik) **acid (DNA)**   A nucleic acid in the shape of a double helix constructed of nucleotides consisting of one of four nitrogen bases (adenine, cytosine, guanine, or thymine), deoxyribose, and a phosphate group; encoded in the nucleotides is genetic information.

Depolarization (dē-pō-lar-i-ZĀ-shun)   Used in neurophysiology to describe the reduction of voltage across a cell membrane; expressed as a movement toward less negative (more positive) voltages on the interior side of the cell membrane.

Depression (dē-PRESS-shun)   Movement in which a part of the body moves downward.

Dermal papilla (pa-PILL-a)   Fingerlike projection of the papillary region of the dermis that may contain blood capillaries or Meissner's corpuscles, nerve endings sensitive to touch.

Dermatology (der-ma-TOL-ō-jē)   The medical specialty dealing with diseases of the skin.

Dermatome (DER-ma-tōm)   An instrument for incising the

skin or cutting thin transplants of skin. The cutaneous area developed from one embryonic spinal cord segment and receiving most of its innervation from one spinal nerve.

Dermis (DER-mis) A layer of dense connective tissue lying deep to the epidermis; the true skin or corium.

Descending colon (KŌ-lon) The part of the large intestine descending from the left colic (splenic) flexure to the level of the left iliac crest.

Developmental anatomy The study of development from the fertilized egg to the adult form. The branch of anatomy called embryology is generally restricted to the study of development from the fertilized egg through the eighth week in utero.

Diabetes insipidus (dī-a-BĒ-tēz in-SIP-i-dus) Condition caused by hyposecretion of antidiuretic hormone (ADH) and characterized by excretion of large amounts of urine and thirst.

Diabetes mellitus (MEL-i-tus) Condition caused by hyposecretion of insulin and characterized by hyperglycemia, increased urine production, excessive thirst, and excessive eating.

Diagnosis (dī-ag-NŌ-sis) Recognition of disease states from signs and symptoms by inspection, palpation, laboratory tests, and other means.

Dialysis (dī-AL-i-sis) The process of separating crystalloids (smaller particles) from colloids (larger particles) by the difference in their rates of diffusion through selectively permeable membrane.

Diapedesis (dī'-a-pe-DĒ-sis) The passage of white blood cells through intact blood vessel walls.

Diaphragm (DĪ-a-fram) Any partition that separates one area from another, especially the dome-shaped skeletal muscle between the thoracic and abdominal cavities.

Diaphysis (dī-AF-i-sis) The shaft of a long bone.

Diarrhea (dī-a-RĒ-a) Frequent defecation of liquid feces caused by increased motility of the intestines.

Diarthrosis (dī-'ar-THRŌ-sis) Articulation in which opposing bones move freely, as in a hinge joint.

Diastole (dī-AS-tō-lē) In the cardiac cycle, the phase of relaxation or dilation of the heart muscle, especially of the ventricles.

Diastolic (dī-as-TOL-ik) **blood pressure** The force exerted by blood on arterial walls during ventricular relaxation; the lowest blood pressure measured in the large arteries, about 80 mm Hg under normal conditions for a young, adult male.

Diencephalon (dī-'en-SEF-a-lon) A part of the brain consisting primarily of the thalamus and the hypothalamus.

Differential (dif-fer-EN-shal) **count** Determination of the number of each kind of white blood cell in a sample of 100 cells for diagnostic purposes.

Diffusion (dif-YOO-zhun) A passive process in which there is a net or greater movement of molecules or ions from a region of high concentration to a region of low concentration until equilibrium is reached.

Digestion (di-JES-chun) The mechanical and chemical breakdown of food to simple molecules that can be absorbed by the body.

Dilate (DĪ-lāte) To expand or swell.

Diploid (DIP-loyd) Having the number of chromosomes characteristically found in the somatic cells of an organism.

Diplopia (di-PLŌ-pē-a) Double vision.

Dislocation (dis-lō-KĀ-shun) Displacement of a bone from a joint with tearing of ligaments, tendons, and articular capsules. Also called **luxation** (luks-Ā-shun).

Dissect (DĪ-sekt) To separate tissues and parts of a cadaver (corpse) or an organ for anatomical study.

Dissociation (dis'-sō-sē-Ā-shun) Separation of inorganic acids, bases, and salts into ions when dissolved in water. Also called **ionization.**

Distal (DIS-tal) Farther from the attachment of an extremity to the trunk or a structure; farther from the point of origin.

Diuretic (dī-yoo-RET-ik) A chemical that increases urine volume by inhibiting facultative reabsorption of water.

Divergence (di-VER-jens) An anatomical arrangement in which the synaptic knobs of one presynaptic neuron terminate on several postsynaptic neurons.

Diverticulitis (dī-ver-tik-yoo-LĪ-tis) Inflammation of diverticula, saclike outpouchings of the colonic wall when the muscularis becomes weak.

Dominant gene A gene that is able to override the influence of the complementary gene on the homologous chromosome; the gene that is expressed.

Dorsal body cavity Cavity near the dorsal surface of the body that consists of a cranial cavity and vertebral canal.

Dorsiflexion (dor'-si-FLEK-shun) Flexion of the foot at the ankle.

Down's syndrome An inherited defect due to an extra copy of chromosome 21. Symptoms include mental retardation; a small skull, flattened front to back; a short, flat nose; short fingers; and a widened space between first two digits of the hand and foot. Also called **trisomy 21.**

Ductus arteriosus (DUK-tus ar-tē-rē-Ō-sus) A small vessel connecting the pulmonary trunk with the aorta; found only in the fetus.

Ductus deferens (DEF-er-ens) The duct that conducts spermatozoa from the epididymis to the ejaculatory duct. Also called the **vas deferens** or **seminal duct.**

Ductus venosus (ve-NŌ-sus) A small vessel in the fetus that helps the circulation bypass the liver.

Duodenal papilla (doo-ō-DĒ-nal pa-PILL-a) An elevation on the duodenal mucosa that receives the hepatopancreatic ampulla.

Duodenum (doo'-ō-DĒ-num) The first portion of the small intestine.

Dura mater (DYOO-ra MĀ-ter) The outer membrane (meninx) covering the brain and spinal cord.

Dynamic equilibrium (ē-kwi-LIB-rē-um) The maintenance of body position, mainly the head, in response to sudden movements such as rotation, acceleration, and deceleration.

Dynamic spatial reconstructor (DSR) An x-ray machine that has the ability to construct moving, three-dimensional, life size images of all or part of an internal organ from any view desired.

Dysfunction (dis-FUNK-shun) Absence of complete normal function.

Dyslexia (dis-LEK-sē-a) Impairment of ability to comprehend written language.

Dysmenorrhea (dis'-men-ō-RĒ-a) Painful menstruation.

Ectoderm The outermost of the three primary germ layers which gives rise to the nervous system and the epidermis of skin and its derivatives.

Edema (e-DĒ-ma) An abnormal accumulation of fluid in body tissues.

Effector (e-FEK-tor) The organ of the body, either a muscle or a gland, that responds to a motor neuron impulse.

Efferent (EF-er-ent) **ducts** A series of coiled tubes that transport spermatozoa from the rete testis to the epididymis.

Efferent neuron (NOO-ron) A neuron that conveys impulses from the brain and spinal cord to effectors which may be either muscles or glands. Also called a **motor neuron.**

Ejaculation (ē-jak-yoo-LĀ-shun) The reflex ejection or expulsion of semen from the penis.

Ejaculatory (e-JAK-yoo-la-tor'-ē) **duct** A tube that transports spermatozoa from the ductus deferens to the prostatic ure-thra.

Elasticity (e-las-TIS-i-tē)   The ability of tissue to return to its original shape after contraction or extension.

Electrocardiogram (e-lek'-trō-KAR-dē-ō-gram) **(ECG** or **EKG)**   A recording of the electrical changes that accompany the cardiac cycle.

Electroencephalogram (e-lek'-trō-en-SEF-a-lō-gram) **(EEG)**   A recording of the electrical impulses of the brain.

Electrolyte (ē-LEK-trō-līt)   Any compound that separates into ions when dissolved in water and is able to conduct electricity.

Electron transport chain   A series of oxidation-reduction reactions in the catabolism of glucose in which energy is released and transferred for storage to ATP.

Elevation (el-e-VĀ-shun)   Movement in which a part of the body moves upward.

Ellipsoidal (e-lip-SOY-dal) **joint**   A synovial joint structured so that an oval-shaped condyle of one bone fits into an elliptical cavity of another bone, permitting side-to-side and back-and-forth movements, as at the joint at the wrist between the radius and carpals. Also called a **condyloid joint.**

Embolism (EM-bō-lizm)   Obstruction or closure of a vessel by an embolus.

Embolus (EM-bō-lus)   A blood clot, bubble of air, fat from broken bones, mass of bacteria, or other debris or foreign material transported by the blood.

Embryo (EM-brē-ō)   The young of any organism in an early stage of development; in humans, the developing organism from fertilization to the end of the eighth week in utero.

Embryology (em'-brē-OL-ō-jē)   The study of development from the fertilized egg to the end of the eighth week in utero.

Emesis (EM-e-sis)   Vomiting.

Emmetropia (em'-e-TRŌ-pē-a)   The ideal optical condition of the eyes.

Emphysema (em'-fi-SĒ-ma)   A swelling or inflation of air passages due to loss of elasticity in the alveoli.

Emulsification (ē-mul'-si-fi-KĀ-shun)   The dispersion of large fat globules to smaller uniformly distributed particles.

Enamel (e-NAM-el)   The hard, white substance covering the crown of a tooth.

End-diastolic (dī-a-STOL-ik) **volume (EDV)**   The volume of blood that enters a ventricle during diastole.

Endocardium (en-dō-KAR-dē-um)   The layer of the heart wall, composed of endothelium and smooth muscle, that lines the inside of the heart and covers the valves and tendons that hold the valves open.

Endochondral ossification (en'-dō-KON-dral os'-i-fi-KĀ-shun)   The replacement of cartilage by bone. Also called **intracartilaginous ossification.**

Endocrine (EN-dō-krin) **gland**   A gland that secretes hormones into the blood; a ductless gland.

Endocytosis (en'-dō-sī-TŌ-sis)   The uptake into a cell of substances that are unable to penetrate the cell membrane; includes both phagocytosis and pinocytosis, processes in which the membrane first invaginates and then pinches off to enclose some of the surrounding medium.

Endoderm   The innermost of the three primary germ layers of the developing embryo which gives rise to the digestive tract, urinary bladder and urethra, and respiratory tract.

Endogenous (en-DOJ-e-nus)   Growing from or beginning within the organism.

Endolymph (EN-dō-lymf')   The fluid within the membranous labyrinth of the inner ear.

Endometriosis (en'-dō-MĒ-trē-ō'-sis)   The growth of endometrial tissue outside the uterus.

Endometrium (en'-dō-MĒ-trē-um)   The mucous membrane lining the uterus.

Endomysium (en'-dō-MIZ-ē-um)   Invaginations of the perimysium separating each individual muscle cell.

Endoneurium (en'-dō-NYOO-rē-um)   Connective tissue wrapping around individual nerve fibers.

Endoplasmic reticulum (en'-dō-PLAZ-mik re-TIK-yoo-lum) **(ER)**   A network of channels running through the cytoplasm of a cell that serves in intracellular transportation, support, storage, synthesis, and packaging of molecules. Portions of ER where ribosomes are attached to the outer surface are called granular or rough reticulum; portions that have no ribosomes are called agranular or smooth reticulum.

Endorphin (en-DOR-fin)   A peptide in the central nervous system that acts as a painkiller.

Endoscope (EN-dō-skōp')   An instrument used to look inside hollow organs such as the stomach (gastroscope) or urinary bladder (cystoscope).

Endosteum (en-DOS-tē-um)   The membrane that lines the medullary cavity of bones.

Endothelial-capsular (en-dō-THĒ-lē-al) **membrane**   A filtration membrane in a nephron of a kidney consisting of the endothelium and basement membrane of the glomerulus and the epithelium of the visceral layer of the glomerular capsule.

Endothelium (en'-dō-THĒ-lē-um)   The layer of simple squamous epithelium that lines the cavities of the heart and blood and lymphatic vessels.

End-systolic (si-STOL-ik) **volume (ESV)**   The volume of blood remaining in a ventricle following its systole.

Enkephalin (en-KEF-a-lin)   A peptide found in the central nervous system that acts as a painkiller.

Enteroendocrine (en-ter-ō-ENDO-krin) **cell**   A stomach cell that secretes the hormone stomach gastrin. Also called an **argentaffin cell.**

Enterogastric (en-te-rō-GAS-trik) **reflex**   A reflex that inhibits gastric secretion; initiated by food in the small intestine.

Enzyme (EN-zīm)   A substance that affects the speed of chemical changes; an organic catalyst, usually a protein.

Eosinophil (ē-ō-SIN-ō-fil)   A type of white blood cell characterized by granular cytoplasm readily stained by eosin.

Ependyma (e-PEN-de-ma)   Neuroglial cells that line ventricles of the brain.

Epicardium (ep-i-KAR-dē-um)   The thin outer layer of the heart wall, composed of serous tissue and mesothelium. Also called the **visceral pericardium.**

Epidemiology (ep'-i-DĒ-mē-ol'-ō-jē)   Medical science concerned with the occurrence and distribution of disease in human populations.

Epidermis (ep'-i-DERM-is)   The outermost layer of skin, composed of stratified squamous epithelium.

Epididymis (ep'-i-DID-i-mis)   An elongated tube along the posterior bladder of the testis in which sperm are stored. *Plural,* **epididymides** (ep'-i-DID-i-mi-dēz).

Epiglottis (ep'-i-GLOT-is)   A large, leaf-shaped piece of cartilage lying on top of the larynx, with its "stem" attached to the thyroid cartilage and its "leaf" portion unattached and free to move up and down to cover the glottis.

Epilepsy (EP-i-lep'-sē)   Neurological disorder characterized by short, periodic attacks of motor, sensory, or psychological malfunction.

Epimysium (ep'-i-MIZ-ē-um)   Fibrous connective tissue around muscles.

Epinephrine (ep-ē-NEF-rin)   Hormone secreted by the adrenal medulla that produces actions similar to those that result from sympathetic stimulation. Also called **adrenaline.**

Epineurium (ep'-i-NYOO-rē-um)   The outermost covering around the entire nerve.

Epiphyseal (ep'-i-FIZ-ē-al) **plate**   The plate between the epiphysis and diaphysis.

Epiphysis (ē-PIF-i-sis)   The end of a long bone, usually larger in diameter than the shaft (the diaphysis).

Episiotomy (ē-piz'-ē-OT-ō-mē)   Incision of the clinical perineum at the end of second stage of labor to avoid tearing.

Epistaxis (ep'-i-STAK-sis)   Hemorrhage from the nose; nosebleed.

Epithelial (ep'-i-THĒ-lē-al) **tissue**   The tissue that forms glands or the outer part of the skin and lines blood vessels, hollow organs, and passages that lead externally from the body.

Erection (ē-REK-shun)   The enlarged and stiff state of the penis (or clitoris) resulting from the engorgement of the spongy erectile tissue with blood.

Eructation (e-ruk'-TĀ-shun)   The forceful expulsion of gas from the stomach. Also called **belching.**

Erythema (er'-e-THĒ-ma)   Skin redness usually caused by engorgement of the capillaries in the lower layers of the skin.

Erythrocyte (e-RITH-rō-sīt)   Red blood cell.

Erythropoiesis (e-rith'-rō-poy-Ē-sis)   The process by which erythrocytes are formed.

Erythropoietin (ē-rith'-rō-POY-ē-tin)   A hormone formed from a plasma protein that stimulates red blood cell production.

Esophagus (e-SOF-a-gus)   A hollow muscular tube connecting the pharynx and the stomach.

Essential amino acids   Those 10 amino acids that cannot be synthesized by the human body at an adequate rate to meet its needs and therefore must be obtained from the diet.

Estrogens (ES-tro-jens)   Female sex hormones produced by the ovaries concerned with the development and maintenance of female reproductive structures and secondary sex characteristics, fluid and electrolyte balance, and protein anabolism. Examples are β-estradiol, estrone, and estriol.

Etiology (ē'-tē-OL-ō-jē)   The study of the causes of disease, including theories of origin and the organisms, if any, involved.

Euphoria (yoo-FŌR-ē-a)   A subjectively pleasant feeling of well-being marked by confidence and assurance.

Eupnea (yoop-NĒ-a)   Normal quiet breathing.

Euthanasia (yoo'-tha-NĀ-zē-ā)   The practice of ending a life in case of incurable disease.

Eversion (ē-VER-zhun)   The movement of the sole outward at the ankle joint.

Exacerbation (eg-zas'-er-BĀ-shun)   An increase in the severity of symptoms or of disease.

Excitability (ek-sīt'-a-BIL-i-tē)   The ability of muscle tissue to receive and respond to stimuli; the ability of nerve cells to respond to stimuli and convert them into impulses.

Excitatory postsynaptic potential (EPSP)   The slight decrease in negative voltage seen on the postsynaptic membrane when it is stimulated by a presynaptic terminal. The EPSP is a localized event that decreases in strength from the point of excitation.

Excrement (EKS-kre-ment)   Material cast out from the body as waste, especially fecal matter.

Excretion (eks-KRĒ-shun)   The process of eliminating waste products from a cell, tissue, or the entire body; or, the products excreted.

Exocrine (EK-sō-krin) **gland**   A gland that secretes substances into ducts that empty at covering or lining epithelium or directly onto a free surface.

Exocytosis (ex'-ō-sī-TŌ-sis)   A process of discharging cellular products too big to go through the membrane. Particles for export are enclosed by Golgi membranes when they are synthesized. Vesicles pinch off from the Golgi complex and carry the enclosed particles to the interior surface of the cell membrane where the vesicle membrane and cell membrane fuse and the contents of the vesicle are discharged.

Exogenous (ex-SOJ-e-nus)   Originating outside an organ or part.

Exophthalmic goiter (ek'-sof-THAL-mik GOY-ter)   Condition caused by hypersecretion of thyroid hormones characterized by protrusion of the eyeballs and an enlarged thyroid.

Expiration (ek-spi-RĀ-shun)   Breathing out; expelling air from the lungs into the atmosphere. Also called **exhalation.**

Expiratory (eks-PĪ-ra-tō-rē) **reserve volume**   The volume of air in excess of tidal volume that can be forcibly exhaled; about 1,200 ml.

Extensibility (ek-sten'-si-BIL-i-tē)   The ability of tissue to be stretched when pulled.

Extension (ek-STEN-shun)   An increase in the anterior angle between two bones, except in extension of the knee and toes, in which the posterior angle is involved; restoring a body part to its anatomical position after flexion.

External auditory (AW-di-tōr-ē) **canal** or **meatus** (mē-Ā-tus)   A canal in the temporal bone that leads to the middle ear.

External ear   The outer ear, consisting of the pinna, the external auditory canal, and the tympanic membrane or eardrum.

External nares (NA-rēz)   The external nostrils, or the openings into the nasal cavity on the exterior of the body.

Exteroceptor (eks'-ter-ō-SEP-tor)   A receptor adapted for the reception of stimuli from outside the body.

Extracellular fluid (ECF)   Fluid outside of body cells, such as interstitial fluid and plasma.

Extrinsic (ek-STRIN-sik)   Of external origin.

Extrinsic clotting pathway   Sequence of reactions leading to blood clotting that is initiated by the release of tissue thromboplastin by damaged tissue.

Exudate (EKS-yoo-dāt)   Escaping fluid or semifluid material that oozes from a space that may contain serum, pus, and cellular debris.

Face   The anterior aspect of the head.

Facilitated diffusion (fa-SIL-i-tā-ted dif-YOO-zhun)   Diffusion in which a substance not soluble by itself in lipids is transported across a semipermeable membrane by combining with a carrier substance.

Facilitation (fa-sil-i-TĀ-shun)   The process in which a nerve cell membrane is partially depolarized by a subliminal stimulus so that a subsequent subliminal stimulus can further depolarize the membrane to reach the threshold of impulse initiation.

Facultative (FAK-ul-tā-tive) **water reabsorption**   The absorption of water from distal convoluted tubules and collecting ducts of nephrons in response to antidiuretic hormone (ADH).

Falx cerebelli (falks ser'-e-BEL-lē)   A small triangular process of the dura mater attached to the occipital bone in the posterior cranial fossa and projecting inward between the two cerebellar hemispheres.

Falx cerebri (SER-e-brē)   A fold of the dura mater extending down into the longitudinal fissure between the two cerebral hemispheres.

Fascia (FASH-ē-a)   A fibrous membrane covering, supporting, and separating muscles.

Fascicle (FAS-i-kul)   A small bundle or cluster, especially of nerve or muscle fibers. Also called **fasciculus** (fa-SIK-yoo-lus); *plural,* **fasciculi** (fa-SIK-yoo-lī).

Fat   A lipid compound formed from one molecule of glycerol and three molecules of fatty acids; the body's most highly concentrated source of energy. Adipose tissue, composed of loose connective tissue and cells specialized for fat storage and present in the form of soft pads between various organs for support, protection, and insulation.

Fauces (FAW-sēs) The opening from the mouth into the pharynx.

Febrile (FĒ-bril) Feverish; pertaining to a fever.

Feces (FĒ-sēz) Material discharged from the rectum and made up of bacteria, excretions, and food residue. Also called **stool.**

Fenestra cochlea (fe-NES-tra KŌK-lē-a) A small opening between the middle and inner ear, directly below the oval window, covered by the second tympanic membrane. Also called the **round window.**

Fenestra vestibuli (ves-TIB-yoo-lī) A small opening between the middle and inner ear into which the footplate of the stapes fits. Also called the **oval window.**

Fertilization (fer'-ti-li-ZĀ-shun) Penetration of an ovum by a spermatozoon and subsequent union of the nuclei of the cells.

Fetal (FĒ-tal) **alcohol syndrome (FAS)** Term applied to the effects of intrauterine exposure to alcohol, such as slow growth, defective organs, and mental retardation.

Fetal circulation The circulatory system of the fetus, including the placenta and special blood vessels involved in the exchange of materials between fetus and mother.

Fetus (FĒ-tus) The latter stages of the developing young of an animal; in humans, the developing organism in utero from the beginning of the third month to birth.

Fever An elevation in body temperature above its normal temperature of 37°C (98.6°F).

Fibrillation (fi-bre-LĀ-shun) Irregular twitching of individual muscle cells (fibers) or small groups of muscle fibers preventing effective action by an organ or muscle.

Fibrin (FĪ-brin) An insoluble protein that is essential to blood clotting; formed from fibrinogen by action of thrombin.

Fibrinogen (fī-BRIN-ō-jen) A high-molecular-weight protein in the blood plasma that by the action of thrombin is converted to fibrin.

Fibrinolysis (fī-bri-NOL-i-sis) Dissolution of a blood clot by the action of a proteolytic enzyme that converts insoluble fibrin into a soluble substance.

Fibroblast (FĪ-brō-blast) A large, flat cell that forms collagenous and elastic fibers and the viscous ground substance of loose connective tissue.

Fibrocyte (FĪ-brō-sīt) A mature fibroblast that no longer produces fibers or matrix in connective tissue.

Fibromyositis (fi'-brō-mī-ō'-SĪ-tis) A group of symptoms including pain, tenderness, and stiffness of joints, muscles, or adjacent structures. Called **"charleyhorse"** when it involves the thigh.

Fibrosis (fī-BRŌ-sis) Abnormal formation of fibrous tissue.

Fibrous (FĪ-brus) **joint** A joint that allows little or no movement, such as a suture and syndesmosis.

Fibrous tunic (TOO-nik) The outer coat of the eyeball, made up of the posterior sclera and the anterior cornea.

Fight-or-flight response The effect of the stimulation of the sympathetic division of the autonomic nervous system.

Filiform papilla (FIL-i-form pa-PIL-a) One of the conical projections which are distributed in parallel rows over the anterior two-thirds of the tongue and contain no taste buds.

Filtration (fil-TRĀ-shun) The passage of a liquid through a filter or membrane that acts like a filter.

Filum terminale (FĪ-lum ter-mi-NAL-ē) Nonnervous fibrous tissue of the spinal cord that extends inferiorly from the conus medullaris to the coccyx.

Fimbriae (FIM-brē-ē) Fringelike structures, especially the lateral ends of the uterine tubes.

Fissure (FISH-ur) A groove, fold, or slit that may be normal or abnormal.

Fistula (FIS-choo-la) An abnormal passage between two organs or between an organ cavity and the outside.

Fixed macrophage (MAK-rō-fāj) Stationary phagocytic cell of the reticuloendothelial system (RES) found in the liver, lungs, brain, spleen, lymph nodes, subcutaneous tissue, and bone marrow. Also called a **histiocyte.**

Flaccid (FLAK-sid) Relaxed, flabby, or soft; lacking muscle tone.

Flagellum (fla-JEL-um) A hairlike, motile process on the extremity of a bacterium or protozoon. *Plural,* **flagella** (fla-JEL-a).

Flatfoot A condition in which the ligaments and tendons of the arches of the foot are weakened and the height of the longitudinal arch decreases.

Flatus (FLĀ-tus) Gas or air in the digestive tract; commonly used to denote passage of gas rectally.

Flexion (FLEK-shun) A folding movement in which there is a decrease in the angle between two bones anteriorly, except in flexion of the knee and toes, in which the bones are approximated posteriorly.

Flexor reflex A polysynaptic reflex arc that withdraws a part of the body from a harmful stimulus.

Fluoroscope (FLOOR-ō-skōp) An instrument for visual observation of the body by means of x-ray.

Follicle-stimulating (FOL-i-kul) **hormone (FSH)** Hormone secreted by the adenohypophysis (anterior lobe) of the pituitary gland that initiates development of ova and stimulates the ovaries to secrete estrogens in females and initiates sperm production in males.

Fontanel (fon'-ta-NEL) A membrane-covered spot where bone formation is not yet complete, especially between the cranial bones of an infant's skull.

Foot The terminal part of the lower extremity.

Foramen (fo-RĀ-men) A passage or opening; a communication between two cavities of an organ or a hole in a bone for passage of vessels or nerves.

Foramen ovale (ō-VAL-ē) An opening in the fetal heart in the septum between the right and left atria. A hole in the greater wing of the sphenoid bone that transmits the mandibular branch of the trigeminal (V) nerve.

Forearm (FOR-arm) The part of the upper extremity between the elbow and the wrist.

Fossa (FOS-a) A furrow or shallow depression.

Fourth ventricle (VEN-tri-kul) A cavity within the brain lying between the cerebellum and the medulla and pons.

Fracture (FRAK-chur) Any break in a bone.

Frontal plane A plane at a right angle to a midsagittal plane that divides the body or organs into anterior and posterior portions. Also called **coronal plane.**

Fulminate (FUL-mi-nāt') To occur suddenly with great intensity.

Functional residual (re-ZID-yoo-al) **volume** The sum of residual volume plus expiratory reserve volume; about 2,400 ml.

Fungiform papilla (FUN-ji-form pa-PILL-a) A mushroomlike elevation on the upper surface of the tongue appearing as a red dot; most contain taste buds.

Gallbladder A small pouch that stores bile, is located under the liver, and is filled and emptied of bile via the cystic duct.

Gamete (GAM-ēt) A male or female reproductive cell; the spermatozoon or ovum.

Ganglion (GANG-glē-on) A group of nerve cell bodies that lie outside the central nervous system. *Plural,* **ganglia** (GANG-glē-a).

Gangrene (GANG-rēn) Death and rotting of a considerable mass of tissue that usually is caused by interruption of blood supply followed by bacterial invasion (*Clostridium*).

Gastrointestinal (gas-trō-in-TES-ti-nal) **(GI) tract** A continuous tube running through the ventral body cavity extending from the mouth to the anus. Also called the **alimentary canal.**

Gavage (ga-VAZH) Feeding through a tube passed through the esophagus and into the stomach.

Gene (jēn) Biological unit of heredity; an ultramicroscopic, self-reproducing DNA particle located in a definite position on a particular chromosome.

General adaptation syndrome Wide ranging set of bodily changes triggered by a stressor that gears the body to meet an emergency.

Generator potential The graded depolarization that occurs in the region of an afferent neuron adjacent to a receptor cell.

Genetics The study of heredity.

Genital herpes (JEN-i-tal HER-pēz) A sexually transmitted disease caused by type II herpes simplex virus.

Genotype (JĒ-nō-tīp) The total hereditary information carried by an individual; the genetic makeup of an organism.

Germinal (JER-mi-nal) **epithelium** A layer of epithelial cells that covers the ovaries and produces ova and lines the seminiferous tubules of the testes and produces spermatozoa.

Gerontology (jer'-on-TOL-o-jē) The study of old age.

Gestation (jes-TĀ-shun) The period of intrauterine fetal development.

Giantism (GĪ-an-tizm) Condition caused by hypersecretion of growth hormone (GH) during childhood characterized by excessive bone growth and body size.

Gingivae (jin-JI-vē) Gums. They cover the alveolar processes of the mandible and maxilla and extend slightly into each socket.

Gingivitis (jin'-je-VĪ-tis) Inflammation of the gums.

Gland Single or group of specialized epithelial cells that secrete substances.

Glans penis (glanz PĒ-nis) The slightly enlarged region at the distal end of the penis.

Glaucoma (glaw-KŌ-ma) An eye disorder in which there is increased pressure due to an excess of fluid within the eye.

Gliding joint A synovial joint having articulating surfaces that are usually flat, permitting only side-to-side and back-and-forth movements, as between carpal bones, tarsal bones, and the scapula and clavicle. Also called an **arthrodial joint.**

Glomerular (glō-MER-yoo-lar) **capsule** A double-walled globe at the proximal end of a nephron that encloses the glomerulus. Also called **Bowman's capsule.**

Glomerular filtration The first step in urine formation in which substances in blood are filtered at the endothelial-capsular membrane and the filtrate enters the proximal convoluted tubule of a nephron.

Glomerular filtration rate (GFR) The total volume of fluid that enters all the glomerular capsules of the kidneys in 1 min; about 125 ml/min.

Glomerulonephritis (glō-mer-yoo-lō-nef-RĪ-tis) Inflammation of the glomeruli of the kidney. Also called **Bright's disease.**

Glomerulus (glō-MER-yoo-lus) A rounded mass of nerves or blood vessels, especially the microscopic tuft of capillaries that is surrounded by the glomerular capsule of each kidney tubule.

Glottis (GLOT-is) The air passageway between the vocal folds in the larynx.

Glucagon (GLOO-ka-gon) A hormone produced by the pancreas that increases the blood glucose level.

Glucocorticoids (gloo-kō-KOR-ti-koyds) A group of hormones of the adrenal cortex.

Gluconeogenesis (gloo'-kō-nē'-ō-JEN-e-sis) The conversion of a substance other than carbohydrate into glucose.

Glucose (GLOO-kōs) A six-carbon sugar, $C_6H_{12}O_6$; the major energy source for every cell type in the body. Its metabolism is possible by every known living cell for the production of ATP.

Glycogen (GLĪ-ko-jen) A highly branched polymer of glucose containing thousands of subunits; functions as a compact store of glucose molecules in liver and muscle cells.

Glycogenesis (glī'-kō-JEN-e-sis) The process by which many molecules of glucose combine to form a molecule called glycogen.

Glycogenolysis (glī'-kō-je-NOL-i-sis) The process of converting glycogen to glucose.

Glycolysis (glī-KŌL-i-sis) A series of chemical reactions that break down glucose into pyruvic acid, with a net gain of two molecules of ATP.

Glycosuria (glī'-kō-SOO-rē-a) The presence of glucose in the urine.

Gnostic (NOS-tik) Pertaining to the faculties of perceiving and recognizing.

Gnostic area Sensory area of the cerebral cortex that receives and integrates sensory input from various parts of the brain so that a common thought can be formed.

Goblet cell A goblet-shaped unicellular gland that secretes mucus.

Goiter (GOY-ter) An enlargement of the thyroid gland.

Golgi (GOL-jē) **complex** An organelle in the cytoplasm of cells consisting of four to eight flattened channels, stacked upon one another, with expanded areas at their ends; functions in packaging secreted proteins, lipid secretion, and carbohydrate synthesis.

Gomphosis (gom-FŌ-sis) A fibrous joint in which a cone-shaped peg fits into a socket.

Gonad (GŌ-nad) A gland that produces gametes and hormones; the ovary in the female and the testis in the male.

Gonadocorticoids (gō-na-dō-KOR-ti-koyd) Sex hormones secreted by the adrenal cortex.

Gonadotropic (go'-nad-ō-TRŌ-pik) **hormone** A hormone that regulates the functions of the gonads.

Gonorrhea (gon'-ō-RĒ-a) Infectious, sexually transmitted disease caused by the bacterium *Neisseria gonorrhoeae* and characterized by inflammation of the urogenital mucosa, discharge of pus, and painful urination.

Gout (gowt) Hereditary condition associated with excessive uric acid in the blood; the acid crystallizes and deposits in joints and the kidneys.

Graafian (GRAF-ē-an) **follicle** An endocrine gland consisting of a mature ovum and its surrounding tissue that secretes estrogens.

Gray matter Area in the central nervous system consisting of nonmyelinated nerve tissue.

Greater omentum (ō-MEN-tum) A large fold in the serosa of the stomach that hangs down like an apron over the front of the intestines.

Greater vestibular (ves-TIB-yoo-lar) **glands** A pair of glands on either side of the vaginal orifice that open by a duct into the space between the hymen and the labia minora. Also called **Bartholin's glands.**

Groin (groyn) The depression between the thigh and the trunk; the inguinal region.

Gross anatomy The branch of anatomy that deals with structures that can be studied without using a microscope. Also called **macroscopic anatomy.**

Growth hormone (GH) Hormone secreted by the adenohypophysis (anterior lobe) of the pituitary that brings about growth of body tissues, especially skeletal and muscular. Also known as **somatotropin** and **somatotropic hormone (STH).**

Gustatory (GUS-ta-tō'-rē) Pertaining to taste.

Gynecology (gī-ne-KOL-ō-jē) The branch of medicine dealing with the study and treatment of disorders of the female reproductive system.

Gyrus (JĪ-rus) One of the convolutions of the cerebral cortex of the brain. *Plural,* **gyri** (JĪ-rī).

Hair A threadlike structure produced by hair follicles that develops in the dermis. Also called **pilus.**

Hair follicle (FOL-li-kul) Structure composed of epithelium surrounding the root of a hair from which hair develops.

Hallucination (ha-loo'-sī-NĀ-shun) A sensory perception of something that does not really exist in the world, that is, a sensory experience created from within the brain.

Hand The terminal portion of an upper extremity, including the carpus, metacarpus, and phalanges.

Haploid (HAP-loyd) Having half the number of chromosomes characteristically found in the somatic cells of an organism; characteristic of mature gametes.

Hard palate (PAL-at) The anterior portion of the roof of the mouth, formed by the maxillae and palatine bones and lined by mucous membrane.

Haustra (HAWS-tra) The sacculated elevations of the colon.

Haversian (ha-VER-shun) **system** The basic unit of structure in adult compact bone, consisting of a Haversian canal with its concentrically arranged lamellae, lacunae, osteocytes, and canaliculi. Also called an **osteon.**

Head The superior part of a human, cephalic to the neck. The superior or proximal part of a structure.

Heart A hollow muscular organ lying slightly to the left of the midline of the chest that pumps the blood through the cardiovascular system.

Heart block An arrhythmia of the heart in which the atria and ventricles contract independently because of a blocking of electrical impulses through the heart at a critical point in the conduction system.

Heat exhaustion Condition characterized by cool, clammy skin, profuse perspiration, and salt loss that results in muscle cramps, dizziness, vomiting, and fainting. Also called **heat prostration.**

Hematocrit (hē'-MAT-ō-krit) An expression of the percentage of the volume occupied by blood cells. Usually calculated by centrifuging a blood sample in a graduated tube and then reading off the volumes of cells and total blood.

Hematology (hē'-ma-TOL-ō-jē) The study of the blood.

Hematoma (hē'-ma-TŌ-ma) A tumor or swelling filled with blood.

Hematopoiesis (hem'-a-tō-poy-Ē-sis) Blood cell production occurring in the red marrow of bones. Also called **hemopoiesis** (hē-mō-poy-Ē-sis).

Hematuria (hē'-ma-TOOR-ē-a) Blood in the urine.

Hemiballismus (hem'-i-ba-LIZ-mus) Violent muscular restlessness of half of the body, especially the upper extremity.

Hemocytoblast (hē'-mō-SĪ-tō-blast) Immature stem cell in bone marrow that develops along different lines into all the different mature blood cells.

Hemodialysis (hē'-mō-dī-AL-i-sis) Filtering of the blood by means of an artificial device so that certain substances are removed from the blood as a result of the difference in rates of their diffusion through a semipermeable membrane while the blood is being circulated outside the body.

Hemodynamics (hē-mō-dī-NA-miks) The study of factors and forces that govern the flow of blood through blood vessels.

Hemoglobin (hē'-mō-GLŌ-bin) **(Hb)** A substance in erythrocytes consisting of the protein globin and the iron-containing, red pigment heme and constituting about 33 percent of the cell volume; involved in the transport of oxygen and carbon dioxide.

Hemolysis (hē-MOL-i-sis) The escape of hemoglobin from the interior of the red blood cell into the surrounding medium; results from disruption of the integrity of the cell membrane by toxins or drugs, freezing or thawing, or hypotonic solutions.

Hemolytic disease of the newborn A hemolytic anemia of a newborn child that results from the destruction of the infant's rbc's by antibodies produced by the mother; usually the antibodies are due to an Rh blood type incompatibility. Also called **erythroblastosis fetalis.**

Hemophilia (hē'-mō-FĒL-ē-a) A hereditary blood disorder where there is a deficient production of certain factors involved in blood clotting, resulting in excessive bleeding into joints, deep tissues, and elsewhere.

Hemorrhage (HEM-or-rij) Bleeding; the escape of blood from blood vessels, especially when it is profuse.

Hemorrhoids (HEM-ō-royds) Dilated or varicosed blood vessels (usually veins) in the anal region. Also called **piles.**

Hemostasis (hē-MŌS-tā-sis) The stoppage of bleeding.

Hepatic (he-PAT-ik) Refers to the liver.

Hepatic duct A duct that receives bile from the bile capillaries. Small hepatic ducts merge to form the larger right and left hepatic ducts that unite to leave the liver as the common hepatic duct.

Hepatic portal circulation The flow of blood from the digestive organs to the liver before returning to the heart.

Hepatitis (hep-a-TĪ-tis) Inflammation of the liver due to a virus, drugs, and chemicals.

Hepatopancreatic (hep'-a-tō-pan'-krē-A-tik) **ampulla** A small, raised area in the duodenum where the combined common bile duct and main pancreatic duct empty into the duodenum. Also called the **ampulla of Vater.**

Hernia (HER-nē-a) The protrusion or projection of an organ or part of an organ through the wall of the cavity containing it.

Herniated (her'-nē-Ā-ted) **disc** A rupture of an intervertebral disc so that the nucleus pulposus protrudes into the vertebral cavity. Also called a **slipped disc.**

Heterozygous (he-ter-ō-ZĪ-gus) Possessing a pair of different genes on homologous chromosomes for a particular hereditary characteristic.

Hilus (HĪ-lus) An area, depression, or pit where blood vessels and nerves enter or leave the organ. Also called a **hilum.**

Hinge joint A synovial joint in which a convex surface of one bone fits into a concave surface of another bone, such as the elbow, knee, ankle, and interphalangeal joints. Also called a **ginglymus joint.**

Histamine (HIS-ta-mēn) Substance found in many cells, especially mast cells, basophils, and platelets, released when the cells are injured; results in vasodilation, increased permeability of blood vessels, and bronchiole constriction.

Histology (his-TOL-ō-jē) Microscopic study of the structure of tissues.

Holocrine (HŌL-ō-krin) **gland** A type of gland in which the entire secreting cell, along with its accumulated secretions, makes up the secretory product of the gland, as in the sebaceous (oil) glands.

Holter monitor Electrocardiograph worn by a person while going about everyday routines.

Homeostasis (hō'-mē-ō-STĀ-sis) The condition in which the body's internal environment remains relatively constant, within limits.

Homologous (hō-MOL-ō-gus) Similar in structure and origin, but not necessarily in function.

Homozygous (hō'-mō-ZĪ-gus) Possessing a pair of identical genes on homologous chromosomes for a particular hereditary characteristic.

Horizontal plane A plane that runs parallel to the ground and divides the body or organs into superior and inferior portions. Also called **transverse plane.**

Hormone (HOR-mōn) A secretion of an endocrine gland that alters the physiological activity of target cells of the body.

Human chorionic gonadotropin (kō-rē-ON-ik gō-nad-ō-TRŌ-pin) **(HCG)** A hormone produced by the developing placenta that maintains the corpus luteum.

Human chorionic somatomammotropin (sō-mat-ō-mam-ō-TRŌ-pin) **(HCS)** A hormone produced by the chorion of the placenta that may stimulate breast tissue for lactation, enhance body growth, and regulate metabolism.

Humoral (YOO-mor-al) **immunity** That component of immunity in which lymphocytes (B cells) develop into plasma cells that produce antibodies that destroy antigens.

Hunger A desire for food stimulated by a nutritional or energy requirement, as opposed to appetite, which is a response to a psychological need.

Hyaluronic (hī'-a-loo-RON-ik) **acid** An amorphous matrix material found outside the cell.

Hyaluronidase (hī'-a-loo-RON-i-dās) An enzyme that breaks down hyaluronic acid, increasing the permeability of connective tissues by dissolving the substances that hold body cells together.

Hydrocele (HĪ-drō-sēl) A fluid-containing sac or tumor. Specifically, a collection of fluid formed in the space along the spermatic cord and in the scrotum.

Hydrocephalus (hī-drō-SEF-a-lus) Abnormal accumulation of cerebrospinal fluid on the brain.

Hymen (HĪ-men) A thin fold of vascularized mucous membrane at the vaginal orifice.

Hyperbaric oxygenation (hī'-per-BA-rik ok'-sē-je-NĀ-shun) **(HBO)** Using pressure supplied by a hyperbaric chamber to cause more oxygen to dissolve in blood to treat patients infected with anaerobic bacteria (tetanus and gangrene bacteria). Also used to treat carbon monoxide poisoning, asphyxia, smoke inhalation, and certain heart disorders.

Hypercalcemia (hī'-per-kal-SĒ-mē-a) An excess of calcium in the blood.

Hypercapnia (hī'-per-KAP-nē-a) An abnormal increase in the amount of carbon dioxide in the blood.

Hyperemia (hī'-per-Ē-mē-a) An excess of blood in an area or part of the body.

Hyperextension (hī'-per-ek-STEN-shun) Continuation of extension beyond the anatomical position, as in bending the head backward.

Hyperglycemia (hī'-per-glī-SĒ-mē-a) An elevated blood sugar level.

Hypermetropia (hī'-per-mē-TRŌ-pē-a) A condition in which visual images are focused behind the retina with resulting defective vision of near objects; farsightedness.

Hyperplasia (hī'-per-PLĀ-zē-a) An abnormal increase in the number of normal cells in a tissue or organ, increasing its size.

Hyperpolarization (hī'-per-PŌL-a-ri-zā'-shun) Increase in the internal negativity across a cell membrane, thus increasing the voltage and moving it farther away from the threshold value.

Hypersecretion (hī'-per-se-KRĒ-shun) Overactivity of glands resulting in excessive secretion.

Hypersensitivity (hī'-per-sen-si-TI-vi-tē) Reaction to an antigen that results in pathological changes. Also called **allergy.**

Hypertension (hī'-per-TEN-shun) High blood pressure.

Hyperthermia (hī'-per-THERM-ē-a) An elevated body temperature.

Hypertonic (hī'-per-TON-ik) Having an osmotic pressure greater than that of a solution with which it is compared.

Hypertrophy (hī-PER-trō-fē) An excessive enlargement or overgrowth of tissue without cell division.

Hyperventilation (hī'-per-ven-ti-LĀ-shun) A rate of respiration higher than that required to maintain a normal level of plasma $p\text{CO}_2$.

Hypervitaminosis (hī'-per-vī-ta-min-Ō-sis) An excess of one or more vitamins.

Hypocalcemia (hī'-pō-kal-SĒ-mē-a) A below normal level of calcium in the blood.

Hypochloremia (hī'-pō-klō-RĒ-mē-a) Deficiency of chloride in the blood.

Hypoglycemia (hī'-pō-glī-SĒ-mē-a) An abnormally low concentration of glucose in the blood; can result from excess insulin (injected or secreted).

Hypokalemia (hī'-pō-kā-LĒ-mē-a) Deficiency of potassium in the blood.

Hypomagnesemia Deficiency of magnesium in the blood.

Hyponatremia (hī'-pō-na-TRĒ-mē-a) Deficiency of sodium in the blood.

Hyposecretion (hī'-pō-se-KRĒ-shun) Underactivity of glands resulting in diminished secretion.

Hypothalamic-hypophyseal (hī'-pō-thal-AM-ik hī'-po-FIZ-ē-al) **tract** A bundle of nerve processes made up of fibers that have their cell bodies in the hypothalamus but release their neurosecretions in the posterior pituitary gland or neurohypophysis.

Hypothalamus (hī'-pō-THAL-a-mus) A portion of the diencephalon, lying beneath the thalamus and forming the floor and part of the wall of the third ventricle.

Hypothermia (hī'-pō-THER-mē-a) A low body temperature.

Hypotonic (hī'-pō-TON-ik) Having an osmotic pressure lower than that of a solution with which it is compared.

Hypoventilation (hī-pō-ven-ti-LĀ-shun) A rate of respiration lower than that required to maintain a normal level of plasma $p\text{CO}_2$.

Hypoxia (hī-POKS-ē-a) Lack of adequate oxygen. Also called **anoxia.**

Hysterectomy (his-te-REK-to-mē) The surgical removal of the uterus.

Ileocecal (il'-ē-ō-SĒ-kal) **valve** A fold of mucous membrane that guards the opening from the ileum into the large intestine.

Ileum (IL-ē-um) The terminal portion of the small intestine.

Immunity (i-MYOON-i-tē) The state of being resistant to injury, particularly by poisons, foreign proteins, and invading parasites, due to the presence of antibodies.

Immunogenicity (im-yoo-nō-jen-IS-it-ē) Ability of an antigen to stimulate antibody production.

Immunoglobulin (im-yoo-nō-GLOB-yoo-lin) **(Ig)** An antibody synthesized by special lymphocytes (plasma cells) in response to the introduction of antigen. Immunoglobulins are divided into five kinds (IgG, IgM, IgA, IgD, IgE) based primarily on the larger protein component present in the immunoglobulin.

Immunologic (im'-yoo-nō-LOJ-ik) **tolerance** A state of specific nonresponsiveness to an antigen, such as a self antigen, following exposure to the antigen.

Immunosuppression (im'-yoo-nō-su-PRESH-un) Inhibition of the immune response.

Impetigo (im'-pe-TĪ-go) A contagious skin disorder characterized by pustular eruptions.

Implantation (im-plan-TĀ-shun) The insertion of a tissue or a part into the body. The attachment of the blastocyst to the lining of the uterus seven to eight days after fertilization.

Impotence (IM-pō-tens) Weakness; inability to copulate; failure to maintain an erection.

Incontinence (in-KON-ti-nens) Inability to retain urine, semen, or feces, through loss of spincter control.

Infant respiratory distress syndrome (RDS) A disease of newborn infants, especially premature ones, in which insufficient amounts of surfactant are produced and breathing is labored. Also called **hyaline membrane disease (HMD).**

Infarction (in-FARK-shun) The presence of a localized area of necrotic tissue, produced by inadequate oxygenation of the tissue.

Infectious mononucleosis (mon-ō-nook'-lē-Ō-sis) Contagious disease caused by a virus and characterized by an elevated mononucleocyte and lymphocyte count, fever, sore throat, stiff neck, cough, and malaise.

Inferior (in-FĒR-ē-or) Away from the head or toward the lower part of a structure. Also called **caudad.**

Inferior vena cava (VĒ-na CĀ-va) Large vein that collects blood from parts of the body inferior to the heart and returns it to the right atrium.

Infertility Inability to conceive or to cause conception. Also called **sterility.**

Inflammation (in'-fla-MĀ-shun) Localized, protective response to tissue injury designed to destroy, dilute, or wall off the infecting agent or injured tissue; characterized by redness, pain, heat, swelling, and sometimes loss of function.

Inflation reflex Reflex that prevents overinflation of the lungs. Also called the **Hering-Breuer reflex.**

Infundibulum (in'-fun-DIB-yoo-lum) The stalklike structure that attaches the pituitary gland (hypophysis) to the hypothalamus of the brain. The funnel-shaped, open, distal end of the uterine tube.

Ingestion (in-JES-chun) The taking in of food, liquids, or drugs, by mouth.

Inheritance The acquisition of body characteristics and qualities by transmission of genetic information from parents to offspring.

Inhibin A male sex hormone secreted by sustentacular cells that inhibits FSH release by the adenohypophysis (anterior pituitary) and thus spermatogenesis.

Inhibitory postsynaptic potential (IPSP) The increase in the internal negativity of the membrane potential so that the voltage moves further from the threshold value.

Insertion (in-SER-shun) The manner or place of attachment of a muscle to the bone that it moves.

Inspiration (in-spi-RĀ-shun) The act of drawing air into the lungs.

Inspiratory (in-SPĪ-ra-tō-rē) **capacity** The total inspiratory ability of the lungs; the sum of tidal volume and inspiratory reserve; about 3,600 ml.

Inspiratory reserve volume The volume of air in excess of tidal volume that can be inhaled by forced inspiration; about 3,100 ml.

Insula (IN-su-la) A triangular area of cerebral cortex that lies deep within the lateral cerebral fissure, under the parietal, frontal, and temporal lobes, and cannot be seen in an external view of the brain. Also called the **island** or **isle of Reil.**

Insulin (IN-su-lin) A hormone produced by the pancreas that decreases the blood glucose level.

Integumentary (in-teg'-yoo-MEN-tar-ē) Relating to the skin.

Intercalated (in-TER-ka-lāt-ed) **disc** An irregular transverse thickening of sarcolemma that separates cardiac muscle cells from each other.

Intercellular fluid That portion of extracellular fluid that bathes the cells of the body; the internal environment of the body. Also called **interstitial fluid.**

Interferon (in'-ter-FĒR-on) A protein naturally produced by virus-infected host cells that inhibits intracellular viral replication in uninfected host cells; artificially synthesized through recombinant DNA techniques.

Intermediate Between two structures, one of which is medial and one of which is lateral.

Internal capsule A thick sheet of white matter made up of myelinated fibers connecting various parts of the cerebral cortex and lying between the thalamus and the caudate and lentiform nuclei of the basal ganglia.

Internal ear The inner ear or labyrinth, lying inside the temporal bone, containing the organs of hearing and balance.

Internal nares (NA-rēz) The two openings posterior to the nasal cavities opening into the nasopharynx. Also called the **choanae.**

Interphase (IN-ter-fāz) The period during its life cycle when a cell is carrying on every life process except division; the stage between two mitotic divisions.

Interstitial (in'-ter-STISH-al) **endocrinocyte** A cell located in the connective tissue between seminiferous tubules in a mature testis that secretes testosterone. Also called an **interstitial cell of Leydig.**

Interstitial growth Growth from within, as in the growth of cartilage. Also called **endogenous growth.**

Intervertebral (in'-ter-VER-te-bral) **disc** A pad of fibrocartilage located between the bodies of two vertebrae.

Intracellular (in'-tra-SEL-yoo-lar) **fluid (ICF)** Fluid inside of body cells.

Intramembranous ossification (in'-tra-MEM-bra-nus os'-i-fĭ-KĀ-shun) The method of bone formation in which the bone is formed directly in membranous tissue.

Intraocular (in-tra-OC-yoo-lar) **pressure** Pressure in the eyeball, produced mainly by aqueous humor.

Intrapleural pressure Air pressure between the two pleural layers of the lungs, usually subatmospheric. Also called **intrathoracic pressure.**

Intrapulmonic pressure Air pressure within the lungs. Also called **intraalveolar pressure.**

Intrauterine device (IUD) A small metal or plastic object inserted into the uterus for the purpose of preventing pregnancy.

Intrinsic clotting pathway Sequence of reactions leading to blood clotting that is initiated by the release of chemicals by disintegrated platelets.

Intrinsic factor A glycoprotein synthesized and secreted by the gastric mucosa that facilitates vitamin B_{12} absorption.

Intubation (in'-too-BĀ-shun) Insertion of a tube through the nose or mouth into the larynx and trachea for entrance of air or to dilate a stricture.

In utero (YOO-ter-ō) Within the uterus.

Inversion (in-VER-zhun) The movement of the sole inward at the ankle joint.

In vitro (VĒ-trō) Literally, in glass; outside the living body and in an artificial environment such as a laboratory test tube.

In vivo (VĒ-vō) In the living body.

Ion (Ī-on) Any charged atom or group of atoms; usually formed when a substance, such as a salt, dissolves and dissociates.

Ipsilateral (ip'-si-LAT-er-al) On the same side, affecting the same side of the body.

Ischemia (is-KĒ-mē-a) A lack of sufficient blood to a part due to obstruction of circulation.

Islet of Langerhans (Ī-let of LAHNG-er-hanz) A cluster of endocrine gland cells in the pancreas that secretes insulin, glucagon, and somatostatin.

Isometric contraction A muscle contraction in which tension on the muscle increases, but there is minimal muscle shortening so that no movement is produced.

Isotonic (ī'-sō-TON-ik) Having equal tension or tone. Having equal osmotic pressure between two different solutions or between two elements in a solution.

Isotonic contraction A muscle contraction in which tension remains constant, but the muscle shortens and pulls on another structure to produce movement.

Isotope (Ī-sō-tōpe') A chemical element that has the same atomic number as another but a different atomic weight. Radioactive isotopes change into other elements with the emission of certain radiations.

Jaundice (JAWN-dis) A condition characterized by yellowness of skin, white of eyes, mucous membranes, and body fluids.

Jejunum (jē-JOO-num)   The middle portion of the small intestine.

Joint kinesthetic (kin'-es-THET-ik) **receptor**   A proprioceptive receptor located in a joint, stimulated by joint movement.

Juxtaglomerular (juks-ta-glō-MER-yoo-lar) **apparatus**   Consists of the macula densa (cells of the distal convoluted tubule adjacent to the afferent arteriole) and juxtaglomerular cells (modified cells of the afferent arteriole); secretes renin when blood pressure starts to fall.

Karyotype (KAR-ē-ō-tīp)   Chromosome characteristics of an individual or a group of cells.

Keratin (KER-a-tin)   An insoluble protein found in the hair, nails, and other keratinized tissues of the epidermis.

Ketone (KĒ-ton) **bodies**   Substances produced primarily during excessive fat metabolism, such as acetone, acetoacetic acid, and β-hydroxybutyric acid.

Ketosis (kē-TŌ-sis)   Abnormal condition marked by excessive production of ketone bodies.

Kidney (KID-nē)   One of the paired reddish organs located in the lumbar region that regulates the composition and volume of blood and produces urine.

Kinesthesia (kin-is-THĒ-sze-a)   Ability to perceive extent, direction, or weight of movement; muscle sense.

Krebs cycle   A series of energy-yielding chemical reactions in which carbon dioxide is formed and energy is transferred to carrier molecules for subsequent liberation. Also called the **citric acid cycle** and **tricarboxylic acid (TCA) cycle.**

Kyphosis (kī-FŌ-sis)   An exaggeration of the thoracic curve of the vertebral column, resulting in a "round-shouldered" or hunchback appearance.

Labial frenulum (LĀ-bē-al FREN-yoo-lum)   A medial fold of mucous membrane between the inner surface of the lip and the gums.

Labia majora (LĀ-bē-a ma-JOR-a)   Two longitudinal folds of skin extending downward and backward from the mons pubis of the female.

Labia minora (min-OR-a)   Two small folds of mucous membrane lying medial to the labia majora of the female.

Labium (LĀ-bē-um)   A lip. A liplike structure. *Plural*, **labia** (LĀ-bē-a).

Labyrinthine (lab-i-RIN-thēn) **disease**   Malfunction of the internal ear characterized by deafness, tinnitus, vertigo, nausea, and vomiting.

Laceration (las'-er-Ā-shun)   A torn, ragged, or mangled wound due to trauma.

Lacrimal (LAK-ri-mal) **canal**   A duct, one on each eyelid, commencing at the punctum at the medial margin of an eyelid and conveying the tears medially into the nasolacrimal sac.

Lacrimal gland   Secretory cells located at the superior lateral portion of each orbit that secrete tears into excretory ducts that open onto the surface of the conjunctiva.

Lacrimal sac   The superior expanded portion of the nasolacrimal duct that receives the tears from a lacrimal canal.

Lactation (lak-TĀ-shun)   The secretion and ejection of milk by the mammary glands.

Lacteal (LAK-tē-al)   One of many intestinal lymph vessels in villi that absorb fat from digested food.

Lacuna (la-KOO-na)   A small, hollow space, such as that found in bones in which the osteoblasts lie. *Plural*, **lacunae** (la-KOO-nē).

Lamellae (la-MEL-ē)   Concentric rings found in compact bone.

Lamina (LAM-i-na)   A thin, flat layer or membrane, as the flattened part of either side of the arch of a vertebra. *Plural*, **laminae** (LAM-i-nē).

Lamina propria (PRO-prē-a)   The connective tissue layer of a mucous membrane.

Lanugo (lan-YOO-gō)   Fine downy hairs that cover the fetus.

Large intestine   The portion of the digestive tract extending from the ileum of the small intestine to the anus, divided structurally into the cecum, colon, rectum, and anal canal.

Laryngitis (la-rin-JĪ-tis)   Inflammation of the mucous membrane lining the larynx.

Laryngopharynx (la-rin'-gō-FAR-inks)   The inferior portion of the pharynx, extending downward from the level of the hyoid bone to divide posteriorly into the esophagus and anteriorly into the larynx.

Larynx (LAR-inks)   The voice box, a short passageway that connects the pharynx with the trachea.

Lateral (LAT-er-al)   Farther from the midline of the body or a structure.

Lateral ventricle (VEN-tri-kul)   A cavity within a cerebral hemisphere that communicates with the lateral ventricle in the other cerebral hemisphere and with the third ventricle by way of the interventricular foramen.

Leg   The part of the lower extremity between the knee and the ankle.

Lens   A transparent organ lying posterior to the pupil and iris of the eyeball and anterior to the vitreous humor.

Lesion (LĒ-zhun)   Any localized, abnormal change in tissue formation.

Lesser omentum (ō-MEN-tum)   A fold of the peritoneum that extends from the liver to the lesser curvature of the stomach and the commencement of the duodenum.

Lesser vestibular (ves-TIB-yoo-lar) **gland**   One of the paired mucus-secreting glands that have ducts that open on either side of the urethral orifice in the vestibule of the female.

Leucocyte (LOO-kō-sīt)   A white blood cell.

Leucocytosis (loo-kō-sī-TŌ-sis)   An increase in the number of white blood cells, characteristic of many infections and other disorders.

Leucopenia (loo-kō-PĒ-nē-a)   A decrease of the number of white blood cells below 5,000/mm³.

Leukemia (loo-KĒ-mē-a)   A cancerlike disease of the blood-forming organs characterized by a rapid and abnormal increase in the number of white blood cells plus many immature cells in the circulating blood.

Leukoplakia (loo-kō-PLĀ-kē-a)   A disorder in which there are white patches in the mucous membranes of the tongue, gums, and cheeks.

Libido (li-BĒ-dō)   The sexual drive, conscious or unconscious.

Ligament (LIG-a-ment)   Dense, regularly arranged connective tissue that attaches bone to bone.

Limbic system   A portion of the forebrain, sometimes termed the visceral brain, concerned with various aspects of emotion and behavior, that includes the limbic lobe (hippocampus and associated areas of gray matter plus the cingulate gyrus), certain parts of the temporal and frontal cortex, some thalamic and hypothalamic nuclei, and parts of the basal ganglia.

Lingual frenulum (LIN-gwal FREN-yoo-lum)   A fold of mucous membrane that connects the tongue to the floor of the mouth.

Lipase (LĪ-pās)   A fat-splitting enzyme.

Lipid   An organic compound composed of carbon, hydrogen, and oxygen that is usually insoluble in water, but soluble in alcohol, ether, and chloroform; examples include fats, phospholipids, steroids, and prostaglandins.

Lipogenesis (li-pō-GEN-e-sis)   The synthesis of lipids from glucose or amino acids by liver cells.

Liver   Large gland under the diaphragm that occupies most of the right hypochondriac region and part of the epigastric region; functionally, it produces bile salts, heparin, and plasma proteins; converts one nutrient into another; detoxifies substances; stores glycogen, minerals, and vitamins; carries on phagocytosis of blood cells and bacteria; and helps to activate vitamin D.

Local   Pertaining to or restricted to one spot or part.

Lordosis (lor-DŌ-sis) An exaggeration of the lumbar curve of the vertebral column.

Lower extremity The appendage attached at the pelvic girdle, consisting of the thigh, knee, leg, ankle, foot, and toes.

Lumbar (LUM-bar) Region of the back and side between the ribs and pelvis; loins.

Lumen (LOO-men) The space within an artery, vein, intestine, or tube.

Lung One of the two main organs of respiration, lying on either side of the heart in the thoracic cavity.

Lunula (LOO-nyoo-la) The moon-shaped white area at the base of a nail.

Luteinizing (LOO-tē-in'-īz-ing) **hormone (LH)** A hormone secreted by the adenohypophysis (anterior lobe) of the pituitary gland that stimulates ovulation, progesterone secretion by the corpus luteum, and readies the mammary glands for milk secretion in females and stimulates testosterone secretion by the testes in males.

Lymph (limf) Fluid confined in lymphatic vessels and flowing through the lymphatic system to be returned to the blood.

Lymphangiography (lim-fan'-jē-OG-ra-fē) A procedure by which lymphatic vessels and lymph organs are filled with a radiopaque substance in order to be x-rayed.

Lymphatic (lim-FAT-ik) Pertaining to lymph. A large vessel that collects lymph from lymph capillaries and converges with other lymphatics to form the thoracic and right lymphatic ducts.

Lymph capillary Blind-ended microscopic lymph vessel that begins in spaces between cells and converges with other lymph capillaries to form lymphatics.

Lymph node An oval or bean-shaped structure located along the lymphatic vessels.

Lymphocyte (LIM-fō-sīt) A type of white blood cell, found in lymph nodes, associated with the immune system.

Lysosome (LĪ-sō-sōm) An organelle in the cytoplasm of a cell, enclosed by a single membrane and containing powerful digestive enzymes.

Lysozyme (LĪ-sō-zīm) A bactericidal enzyme found in tears, saliva, and perspiration.

Macula (MAK-yoo-la) A discolored spot or a colored area. A small, flat region on the wall of the utricle and saccule that serves as a receptor for static equilibrium.

Macula lutea (LOO-tē-a) The yellow spot in the center of the retina.

Malaise (ma-LĀYZ) Discomfort, uneasiness, and indisposition, often indicative of infection.

Malignant (ma-LIG-nant) Referring to diseases that tend to become worse and cause death; especially the invasion and spreading of cancer.

Mammary (MAM-ar-ē) **gland** Modified sweat gland of the female that secretes milk for the nourishment of the young.

Margination (mar'-ji-NĀ-shun) Accumulation and adhesion of neutrophils to the endothelium of blood vessels at the site of injury during the early stages of inflammation.

Marrow (MAR-ō) Soft, spongelike material in the cavities of bone. Red marrow produces blood cells; yellow marrow, formed mainly of fatty tissue, has no blood-producing function.

Mast cell A cell found in loose connective tissue along blood vessels that produces heparin, an anticoagulant. The name given to a basophil after it has left the bloodstream and entered the tissues.

Mastectomy (mas-TEK-tō-mē) Surgical removal of breast tissue.

Mastication (mas'-ti-KĀ-shun) Chewing.

Meatus (mē-Ā-tus) A passage or opening, especially the external portion of a canal.

Mechanoreceptor (me-KAN-ō-rē'-sep-tor) Receptor that detects mechanical deformation of the receptor itself or adjacent cells; stimuli so detected include touch, pressure, proprioception, hearing, and equilibrium.

Medial (MĒ-dē-al) Nearer the midline of the body or a structure.

Medial lemniscus (lem-NIS-kus) A flat band of myelinated nerve fibers extending through the medulla, pons, and midbrain and terminating in the thalamus on the opposite side. Second-order sensory neurons in this tract transmit impulses for discriminating touch, pressure, and vibration sensations.

Mediastinum (mē'-dē-as-TĪ-num) A mass of tissue found between the pleurae of the lungs that extends from the sternum to the vertebral column.

Medulla (me-DUL-la) An inner layer of an organ, such as the medulla of the kidneys.

Medulla oblongata (ob'-long-GA-ta) The most inferior part of the brain stem.

Medullary (MED-yoo-lar'-ē) **cavity** The space within the diaphysis of a bone that contains yellow marrow.

Medullary rhythmicity (rith-MIS-i-tē) **area** Portion of the respiratory center in the medulla that controls the basic rhythm of respiration.

Meiosis (mē-Ō-sis) A type of cell division restricted to sex-cell production involving two successive nuclear divisions which result in daughter cells with the haploid number of chromosomes.

Meissner's (MĪS-nerz) **corpuscle** The sensory receptor for the sensation of touch; found in the dermal papillae, especially in palms and soles.

Melanin (MEL-a-nin) A dark black, brown, or yellow pigment found in some parts of the body such as the skin.

Melanocyte (MEL-a-nō-sīt') A pigmented cell located between or beneath cells of the deepest layer of the epidermis that synthesizes melanin.

Melanocyte-stimulating hormone (MSH) A hormone secreted by the adenohypophysis (anterior lobe) of the pituitary gland that stimulates the dispersion of melanin granules in melanocytes.

Melanoma (mel'-a-NŌ-ma) A usually dark, malignant tumor of the skin containing melanin.

Melatonin (mel-a-TŌN-in) A hormone secreted by the pineal gland that may inhibit reproductive activities.

Membrane A thin, flexible sheet of tissue composed of an epithelial layer and an underlying connective tissue layer, as in an epithelial membrane, or of loose connective tissue only, as in a synovial membrane.

Membrane potential The voltage present, at any instant, across the cell membrane; measured with microelectrodes inside and outside the cell, it usually registers "resting" values of about −70 mV. Also called a **resting potential.**

Membranous labyrinth (mem-BRA-nus LAB-i-rinth) The portion of the labyrinth of the inner ear that is located inside the bony labyrinth and separated from it by the perilymph; made up of the membranous semicircular canals, the saccule and utricle, and the cochlear duct.

Memory The ability to recall thoughts; commonly classified as activated and long-term.

Menarche (me-NAR-kē) Beginning of the menstrual function.

Ménière's (men-YAIRZ) **syndrome** A type of labyrinthine disease characterized by fluctuating loss of hearing, vertigo, and tinnitus.

Meninges (me-NIN-jēz) Three membranes covering the brain and spinal cord, called the dura mater, arachnoid, and pia mater. *Singular,* **meninx** (MEN-inks).

Meningitis (men-in-JĪ-tis) Inflammation of the meninges, most commonly the pia mater and arachnoid.

Menopause (MEN-ō-pawz) The termination of the menstrual cycles.

Menstrual (MEN-stroo-al) **cycle** A series of changes in the endometrium of a nonpregnant female that prepares the lining of the uterus to receive a fertilized ovum.

Menstruation (men'-stroo-Ā-shun) Periodic discharge of blood, tissue fluid, mucus, and epithelial cells that usually lasts for 5 days; caused by a sudden reduction in estrogens and progesterone. Also called the **menstrual phase** or **menses.**

Merkel's (MER-kuls) **disc** An encapsulated, cutaneous receptor for touch located in deeper layers of epidermal cells.

Merocrine (MER-ō-krin) **gland** A secretory cell that remains intact throughout the process of formation and discharge of the secretory product, as in the salivary and pancreatic glands.

Mesenchyme (MEZ-en-kīm) An embryonic connective tissue from which all other connective tissues arise.

Mesentery (MEZ-en-ter'-ē) A fold of peritoneum attaching the small intestine to the posterior abdominal wall.

Mesocolon (mez'-ō-KŌ-lon) A fold of peritoneum attaching the colon to the posterior abdominal wall.

Mesoderm The middle of the three primary germ layers which gives rise to connective tissues, blood and blood vessels, and muscles.

Mesothelium (mez'-ō-THĒ-lē-um) The layer of simple squamous epithelium that lines serous cavities.

Metabolism (me-TAB-ō-lizm) The sum of all the biochemical reactions that occur within an organism, including the synthetic (anabolic) reactions and decomposition (catabolic) reactions.

Metacarpus (met'-a-KAR-pus) A collective term for the five bones that make up the palm of the hand.

Metaphase (MET-a-phāz) The second stage of mitosis in which chromatid pairs line up on the equatorial plane of the cell.

Metaphysis (me-TAF-i-sis) Growing portion of a bone.

Metastasis (me-TAS-ta-sis) The transfer of disease from one organ or part of the body to another.

Metatarsus (met'-a-TAR-sus) A collective term for the five bones located in the foot between the tarsals and the phalanges.

Micelle (mī-SEL) A spherical aggregate of bile salts that dissolves fatty acids and monosaccharides so that they can be transported into small intestinal epithelial cells.

Microcephalus (mi-kro-SEF-a-lus) An abnormally small head; premature closing of the anterior fontanel so that the brain has insufficient room for growth, resulting in mental retardation.

Microfilament (mī-krō-FIL-a-ment) Rodlike cytoplasmic structure, ranging in diameter from 30 Å to 120 Å; comprises contractile units in muscle cells and provides support, shape, and movement in nonmuscle cells.

Microglia (mi-krō-GLĒ-a) Neuroglial cells that carry on phagocytosis. Also called **brain macrophages.**

Microphage (MĪK-rō-fāj) Granular leucocyte that carries on phagocytosis, especially neutrophils and eosinophils.

Microtubule (mī-krō-TOOB-yool') Cylindrical cytoplasmic structure, ranging in diameter from 180 Å to 300 Å, consisting of the protein tubulin; provides support, structure, and transportation.

Microvilli (mī'-krō-VIL-ē) Microscopic, fingerlike projections of the cell membranes of small intestinal cells that increase surface area for absorption.

Micturition (mik'-too-RISH-un) The act of expelling urine from the bladder. Also called **urination.**

Midbrain The part of the brain between the pons and the diencephalon. Also called the **mesencephalon.**

Middle ear A small, epithelial-lined cavity hollowed out of the temporal bone, separated from the external ear by the eardrum and from the internal ear by a thin bony partition containing the oval and round windows; extending across the middle ear are the three auditory ossicles. Also called the **tympanic cavity.**

Midsagittal plane A plane through the midline of the body running vertical to the ground and dividing the body or organs into equal right and left sides. Also called a **median plane.**

Milking Contraction of skeletal muscle around veins causing venous valves to open and thus helping to drive blood back to the heart.

Milk letdown Contraction of alveolar cells to force milk into ducts of mammary glands, stimulated by oxytocin which is released from the posterior pituitary in response to suckling action.

Mineral Inorganic, homogeneous solid substance that may perform a function vital to life; examples include calcium, sodium, potassium, iron, phosphorus, and chlorine.

Mineralocorticoids (min'-er-al-ō-KOR-ti-koyds) A group of hormones of the adrenal cortex.

Minimal volume The volume of air in the lungs even after the thoracic cavity has been opened forcing out some of the residual volume.

Minute volume of respiration (MVR) Total volume of air taken into the lungs per minute; about 6,000/ml.

Mitochondrion (mi'-tō-KON-dre-on) A double-membraned organelle that plays a central role in the production of ATP; known as the powerhouse of the cell.

Mitosis (mī-TŌ-sis) The orderly division of the nucleus of a cell that ensures that each new daughter nucleus has the same number and kind of chromosomes as the original parent nucleus. The process includes the replication of chromosomes and the distribution of the two sets of chromosomes into two separate and equal nuclei.

Mitotic apparatus Collective term for continuous and chromosomal microtubules and centrioles; involved in cell division.

Modality (mō-DAL-i-tē) Any of the specific sensory entities, such as vision, smell, or taste.

Modiolus (mō-DĪ-ō'-lus) The central pillar or column of the cochlea.

Molecule (MOL-e-kyool) The chemical combination of two or more atoms.

Monoclonal antibody Antibody produced by in vitro clones of B cells hydridized with cancerous cells.

Monocyte (MON-ō-sīt') A type of white blood cell characterized by agranular cytoplasm; the largest of the leucocytes.

Mons pubis (monz PYOO-bis) The rounded, fatty prominence over the symphysis pubis, covered by coarse pubic hair.

Morbid (MOR-bid) Diseased; pertaining to disease.

Morula (MOR-yoo-la) A solid mass of cells produced by successive cleavages of a fertilized ovum a few days after fertilization.

Motor area The region of the cerebral cortex that governs muscular movement, particularly the precentral gyrus of the frontal lobe.

Motor end plate The portion of the muscle cell membrane directly under the termination of a motor neuron where the neuron releases its transmitter substance.

Motor unit A motor neuron together with the muscle cells it stimulates.

Mucous (MYOO-kus) **cell** A unicellular gland that secretes mucus. Also called a **goblet cell.**

Mucous membrane A membrane that lines a body cavity that opens to the exterior. Also called the **mucosa.**

Mucus The thick fluid secretion of the mucous glands and mucous membranes.

Multiple sclerosis (skler-Ō-sis) Progressive destruction of myelin sheaths of neurons in the central nervous system, short-circuiting conduction pathways.

Mumps Inflammation and enlargement of the parotid glands accompanied by fever and extreme pain during swallowing.

Murmur An unusual heart sound; may indicate a disorder such as a malfunctioning mitral valve or may have no clinical significance.

Muscle An organ composed of one of three types of muscle tissue (skeletal, cardiac, or visceral), specialized for contraction to produce voluntary or involuntary movement of parts of the body.

Muscle spindle An encapsulated receptor in a skeletal muscle, consisting of specialized muscle cells and nerve endings, stimulated by changes in length or tension of muscle cells; a proprioceptor.

Muscle tissue A tissue specialized to produce motion in response to nerve impulses by its qualities of contractility, extensibility, elasticity, and excitability.

Muscle tone A sustained, partial contraction of portions of a skeletal muscle in response to activation of stretch receptors.

Muscular dystrophy (DIS-trō-fē') Inherited myopathy characterized by degeneration of muscle cells that leads to progressive atrophy.

Muscularis (MUS-kyoo-la'-ris) A muscular layer or tunic of an organ.

Muscularis mucosae (myoo-KŌ-sē) A thin layer of smooth muscle cells located in the outermost layer of the mucosa of the alimentary canal, underlying the lamina propria of the mucosa.

Mutation (myoo-TĀ-shun) Any change in the sequence of bases in the DNA molecule resulting in a permanent alteration in some inheritable characteristic.

Myasthenia (mī-as-THĒ-nē-a) **gravis** Weakness of skeletal muscles caused by antibodies directed against acetylcholine receptors that inhibit muscle contraction.

Myelin (MĪ-e-lin) **sheath** A white, phospholipid, segmented covering, formed by Schwann cells, around the axons and dendrites of many peripheral neurons.

Myocardial infarction (mī'-ō-KAR-dē-al in-FARK-shun) Gross necrosis of myocardial tissue due to interrupted blood supply. Also called a **heart attack.**

Myocardium (mī'-ō-KAR-dē-um) The middle layer of the heart wall, made up of cardiac muscle, comprising the bulk of the heart, and lying between the epicardium and the endocardium.

Myofibril (mī'-ō-FĪ-bril) A threadlike structure, running longitudinally through a muscle cell, consisting mainly of thick myofilaments (myosin) and thin myofilaments (actin).

Myoglobin (mī-ō-GLŌ-bin) The oxygen-binding, iron-containing conjugated protein complex present in the sarcoplasm of muscle cells; contributes the red color to muscle.

Myogram (MĪ-ō-gram) The record or tracing produced by the myograph, the apparatus that measures and records the effects of muscular contractions.

Myology (mī-OL-ō-jē) The study of the muscles and their parts.

Myometrium (mī'-ō-MĒ-trē-um) The smooth muscle coat or tunic of the uterus.

Myopia (mī-Ō-pē-a) Defect in vision so that objects can be seen distinctly only when very close to the eyes; nearsightedness.

Myosin (MĪ-ō-sin) The contractile protein that makes up the thick myofilaments of muscle cells.

Myotonia (mī-ō-TO-nē-a) A continuous spasm of muscle; increased muscular irritability and tendency to contract and less ability to relax.

Myxedema (mix-e-DĒ-ma) Condition caused by hypothyroidism during the adult years characterized by swelling of facial tissues.

Nail A hard plate, composed largely of keratin, that develops from the epidermis of the skin to form a protective covering on the dorsal surface of the distal phalanges of the fingers and toes.

Nail matrix (MĀ-triks) The part of the nail beneath the body and root from which the nail is produced.

Nasal (NĀ-zal) **cavity** A mucosa-lined cavity on either side of the nasal septum that opens onto the face at an external naris and into the nasopharynx at an internal naris.

Nasal septum (SEP-tum) A vertical partition composed of bone and cartilage, covered with a mucous membrane, separating the nasal cavity into left and right sides.

Nasolacrimal (nā'-zō-LAK-ri-mal) **duct** A canal that transports the lacrimal secretion from the nasolacrimal sac into the nose.

Nasopharynx (nā'-zō-FAR-inks) The uppermost portion of the pharynx, lying posterior to the nose and extending down to the soft palate.

Neck The part of the body connecting the head and the trunk. A constricted portion of an organ such as the neck of a femur or the neck of the uterus.

Necrosis (ne-KRŌ-sis) Death of a cell or group of cells as a result of disease or injury.

Negative feedback The principle governing most control systems; a mechanism of response in which a stimulus initiates actions which reverse or reduce the stimulus.

Neoplasm (NĒ-ō-plazm) A mass of new, abnormal tissue; a tumor.

Nephritis (ne-FRĪT-is) Inflammation of the kidney.

Nephron (NEF-ron) The functional unit of the kidney.

Nephrosis (nef-RŌ-sis) A degenerative disease of the kidney in which glomerular damage results in leakage of protein into urine.

Nerve A cordlike bundle of nerve fibers and their associated connective tissue coursing together outside the central nervous system.

Nerve impulse A wave of negativity that sweeps along the outside of the membrane of a neuron. Also called an **action potential.**

Nervous tissue Tissue that initiates and transmits nerve impulses to coordinate homeostasis.

Neuritis (noo-RĪ-tis) Inflammation of a nerve.

Neuroeffector (noo-rō-e-FEK-tor) **junction** Cumulative term for neuromuscular and neuroglandular junctions.

Neurofibril (noo-rō-FĪ-bril) One of the delicate threads that forms a complicated network in the cytoplasm of the cell body and processes of a neuron.

Neuroglandular (noo-rō-GLAND-yoo-lar) **junction** Area of contact between a motor neuron and a gland.

Neuroglia (noo-ROG-lē-a) Cells of the nervous system that are specialized to perform the functions of connective tissue. The neuroglia of the central nervous system are the astrocytes, oligodendrocytes, microglia, and ependyma; neuroglia of the peripheral nervous system include the Schwann cells and the ganglion satellite cells. Also called **glial cells.**

Neurohypophysis (noo-rō-hī-POF-i-sis) The posterior lobe of the pituitary gland.

Neuromuscular (noo-rō-MUS-kyoo-lar) **junction** The area of contact between a motor neuron and a muscle fiber. Also called a **myoneural junction.**

Neuron (NOO-ron) A nerve cell, consisting of a cell body, dendrites, and an axon.

Neurosecretory (noo-rō-SĒC-re-tō-rē) **cell** A cell in a nucleus in the hypothalamus that produces oxytocin (**OT**) or antidiuretic hormone (ADH), hormones stored in the neurohypophysis of the pituitary gland.

Neutrophil (NOO-trō-fil) A type of white blood cell characterized by granular cytoplasm which stains as readily with acid or basic dyes. Also called a **polymorph.**

Night blindness Poor or no vision in dim light or at night, although good vision present during bright illumination; frequently caused by a deficiency of vitamin A.

Nipple A pigmented, wrinkled projection on the surface of the mammary gland which is the location of the openings of the lactiferous ducts for milk release.

Nissl (NIS-l) **bodies** Rough endoplasmic reticulum in the cell bodies of neurons that functions in protein synthesis.

Nociceptor (nō'-sē-SEP-tor) A receptor that detects pain.

Node of Ranvier (ron-vē-Ā) A space, along a myelinated nerve fiber, between the individual Schwann cells that form the myelin sheath and the neurilemma.

Nonessential amino acid An amino acid that can be synthesized by body cells through transamination, the transfer of an amino group from an amino acid to another substance.

Nongonococcal urethritis (NGU) A sexually transmitted disease characterized by inflammation of the urethra caused by a microorganism other than *Neisseria gonorrhoeae* or by nonmicrobial factors.

Norepinephrine (nor'-ep-ē-NEF-rin) **(NE)** A hormone secreted by the adrenal medulla that produces actions similar to those that result from sympathetic stimulation. Also called **noradrenaline.**

Nuclear magnetic resonance (NMR) A diagnostic procedure that focuses on the nuclei of atoms of a single element in a tissue, usually hydrogen, to determine if they behave normally in the presence of an external magnetic force; used to indicate the biochemical activity of a tissue.

Nucleic (noo-KLĒ-ic) **acid** An organic compound that is a long polymer of nucleotides, with each nucleotide containing a pentose sugar, a phosphate group, and one of four possible nitrogen bases (adenine, cytosine, guanine, and thymine or uracil).

Nucleosome (NOO-klē-ō-sōm) Elementary structural subunit of a chromosome consisting of histones and DNA.

Nucleus (NOO-klē-us) A spherical or oval organelle of a cell that contains the hereditary factors of the cell, called genes. A cluster of nerve cell bodies in the central nervous system. The central portion of an atom made up of protons and neutrons.

Nystagmus (nis-TAG-mus) Constant, involuntary, rhythmic movement of the eyeballs; horizontal, rotary, or vertical.

Obesity (ō-BĒS-i-tē) Body weight 10 to 20 percent over a desirable standard as a result of excessive accumulation of fat.

Obligatory water reabsorption The absorption of water from proximal convoluted tubules of nephrons as a function of osmosis.

Occlusion (ō-KLOO-zhun) The act of closure or state of being closed.

Olfactory (ōl-FAK-tō-rē) Pertaining to smell.

Olfactory bulb A mass of gray matter at the termination of an olfactory nerve, lying beneath the frontal lobe of the cerebrum on either side of the crista galli of the ethmoid bone.

Olfactory tract A bundle of axons that extends from the olfactory bulb posteriorly to the olfactory portion of the cortex.

Oligodendrocyte (o-lig-ō-DEN-drō-sīt) A neuroglial cell that supports neurons and produces a myelin sheath around axons of neurons of the central nervous system.

Oncogene (ONG-kō-jēn) Gene that has the ability to transform a normal cell into a cancerous cell.

Oncology (ong-KOL-ō-jē) The study of tumors.

Oogenesis (ō'-ō-JEN-e-sis) Formation and development of the ovum.

Oophorectomy (ō'-of-ō-REK-tō-mē) The surgical removal of the ovaries.

Opsonization (op-sō-ni-ZĀ-shun) The action of some antibodies that renders bacteria and other foreign cells more susceptible to phagocytosis. Also called **immune adherence.**

Optic (OP-tik) Refers to the eye, vision, or properties of light.

Optic chiasma (kī-AZ-ma) A crossing point of the optic nerves, anterior to the pituitary gland.

Optic disc A small area of the retina containing openings through which the fibers of the ganglion neurons emerge as the optic nerve. Also called the **blind spot.**

Optic tract A bundle of axons that transmits impulses from the retina of the eye between the optic chiasma and the thalamus.

Oral contraceptive (OC) A hormonal compound that is swallowed and that prevents ovulation, and thus pregnancy. Also called **"the pill."**

Orbit (OR-bit) The bony, pyramid-shaped cavity of the skull that holds the eyeball.

Organ A structure of definite form and function composed of two or more different kinds of tissues.

Organelle (or-gan-EL) A component or structure within a cell that is specialized to serve a specific function in cellular activities.

Organism (OR-ga-nizm) A total living form; one individual.

Orgasm (OR-gazm) Sensory and motor events involved in ejaculation for the male and involuntary contraction of the perineal muscles in the female at the climax of sexual intercourse.

Orifice (OR-i-fis) Any aperture or opening.

Origin (OR-i-jin) The place of attachment of a muscle to the more stationary bone, or the end opposite the insertion.

Oropharynx (or'-ō-FAR-inks) The second portion of the pharynx, lying posterior to the mouth and extending from the soft palate down to the hyoid bone.

Osmosis (os-MŌ-sis) The net movement of water molecules through a selectively permeable membrane from an area of high water concentration to an area of lower water concentration until an equilibrium is reached.

Osmotic pressure The force under which a solvent (water) moves through a selectively permeable membrane from higher to lower water concentration.

Ossicle (OS-si-kul) Small bone, as in the middle ear (malleus, incus, stapes).

Ossification (os'-i-fi-KĀ-shun) Formation of bone. Also called **osteogenesis.**

Osteoblast (OS-tē-ō-blast') A bone-forming cell.

Osteoclast (OS-tē-ō-clast') A large, multinuclear cell that destroys or resorbs bone tissue.

Osteocyte (OS-tē-ō-sīt') A mature osteoblast that has lost its ability to produce new bone tissue.

Osteology (os'-tē-OL-ō-jē) The study of bones.

Osteomalacia (os'-tē-ō-ma-LĀ-shē-a) A deficiency of vitamin D causing demineralization and softening of bone.

Osteomyelitis (os'-tē-ō-mī-i-LĪ-tis) Inflammation of bone marrow or of the bone and marrow.

Osteoporosis (os'-tē-ō-pō-RŌ-sis) Increased porosity of bone.

Otic (Ō-tik) Pertaining to the ear.

Otitis media (ō-TĪ-tus MĒ-dē-a) Acute infection of the middle ear cavity characterized by an inflamed tympanic membrane, subject to rupture.

Otolith (Ō-tō-lith) A particle of calcium carbonate embedded in the otolithic membrane that functions in maintaining static equilibrium.

Otolithic (ō-tō-LITH-ik) **membrane** Thick, gelatinous, glycoprotein layer located directly over hair cells of the macula in the saccule and utricle of the inner ear.

Ovarian (ō-VAR-ē-an) **cycle** A monthly series of events in the ovary designed to produce a mature ovum.

Ovarian follicle (FOL-i-kul) A general name for an ovum in any stage of development, along with its surrounding group of epithelial cells.

Ovary (Ō-var-ē) Female gonad that produces ova and the hormones estrogens, progesterone, and relaxin.

Ovulation (ō-vyoo-LĀ-shun) The rupture of a Graafian follicle with discharge of an ovum into the pelvic cavity.

Ovum (Ō-vum) The female reproductive or germ cell; an egg cell.

Oxidation (ok-si-DĀ-shun) The removal of electrons or the addition of oxygen to a molecule. The oxidation of glucose in the body is also called **cellular respiration.**

Oxygen debt The volume of oxygen required to oxidize the lactic acid produced by muscular exercise.

Oxyhemoglobin (ok'-sē-HĒ-mō-glō-bin) **(HbO₂)** Hemoglobin combined with oxygen.

Oxyphil cell A cell found in the parathyroid gland that secretes parathyroid hormone (PTH).

Oxytocin (ok'-sē-TŌ-sin) **(OT)** A hormone secreted by neurosecretory cells in the paraventricular nucleus of the hypothalamus that stimulates contraction of the smooth muscle cells in the pregnant uterus and contractile cells around the ducts of mammary glands.

Pacinian (pa-SIN-ē-an) **corpuscle** Oval pressure receptor located in subcutaneous tissue and consisting of concentric layers of connective tissue wrapped around an afferent nerve fiber.

Paget's (PAJ-ets) **disease** A disorder characterized by an irregular thickening and softening of the bones. The cause is unknown, but the bone-producing osteoblasts and the bone-destroying osteoclasts apparently become uncoordinated.

Palate (PAL-at) The horizontal structure separating the oral and the nasal cavities; the roof of the mouth.

Palliative (PAL-ē-a-tiv) Serving to relieve or alleviate without curing.

Palpate (PAL-pāt) To examine by touch; to feel.

Pancreas (PAN-krē-as) A soft, oblong organ lying along the greater curvature of the stomach and connected by a duct to the duodenum. It is both exocrine (secreting pancreatic juice) and endocrine (secreting insulin, glucagon, and somatostatin).

Pancreatic (pan'-krē-AT-ik) **duct** A single, large tube that drains pancreatic juice into the duodenum at the hepatopancreatic ampulla.

Papanicolaou (pap'-a-NIK-ō-la-oo) **test** A cytological staining test for the detection and diagnosis of premalignant and malignant conditions of the female genital tract. Cells scraped from the genital epithelium are smeared, fixed, stained, and examined microscopically. Also called a **Pap smear.**

Paralysis (pa-RAL-a-sis) Loss or impairment of motor function due to a lesion of nervous or muscular origin.

Paranasal sinus (par'-a-NĀ-zal SĪ-nus) A mucus-lined air cavity in a skull bone that communicates with the nasal cavity. Paranasal sinuses are located in the frontal, maxillary, ethmoid, and sphenoid bones.

Paraplegia (par-a-PLĒ-jē-a) Paralysis of the lower limbs only.

Parasympathetic (par'-a-sim-pa-THET-ik) **division** One of the two subdivisions of the autonomic nervous system, having cell bodies of preganglionic neurons in nuclei in the brain stem and in the lateral gray matter of the sacral portion of the spinal cord; primarily concerned with activities that restore and conserve body energy. Also called the **craniosacral division.**

Parathyroid (par'-a-THĪ-royd) **gland** One of four small endocrine glands embedded on the posterior surfaces of the lateral lobes of the thyroid.

Parathyroid hormone (PTH) A hormone secreted by the parathyroid glands that decreases blood phosphate level and increases blood calcium level.

Parenchyma (par-EN-ki-ma) The functional parts of any organ, as opposed to tissue that forms its ground substance or framework.

Parietal (pa-RĪ-e-tal) Pertaining to or forming the outer wall of a body cavity.

Parietal cell The secreting cell of the gastric glands that produces hydrochloric acid and intrinsic factor.

Parietal pleura (PLOO-ra) The outer layer of the serous pleural membrane that encloses and protects the lungs; the layer that is attached to the walls of the pleural cavity.

Parkinsonism (PARK-in-so-nizm') Progressive degeneration of basal ganglia of the cerebrum resulting in decreased production of dopamine that leads to tremor, slowing of voluntary movements, and muscle weakness. Also called **Parkinson's disease.**

Parotid (pa-ROT-id) **gland** One of the paired salivary glands located inferior and anterior to the ears connected to the oral cavity via a duct that opens into the inside of the cheek opposite the upper second molar tooth.

Paroxysm (PAR-ok-sizm) A sudden periodic attack or recurrence of symptoms of a disease.

Pars intermedia A small avascular zone between the adenohypophysis and neurohypophysis of the pituitary gland.

Parturition (par'-too-RISH-un) Act of giving birth to young; childbirth, delivery.

Patellar (pa-TELL-ar) **reflex** Extension of the lower leg by contraction of the quadriceps femoris muscle in response to tapping the patellar ligament.

Patent ductus arteriosus Congenital anatomical heart defect in which the fetal connection between the aorta and pulmonary trunk remains open instead of closing completely after birth.

Pathogenesis (path'-ō-JEN-e-sis) The development of disease or a morbid or pathological state.

Pathogen (PATH-ō-jen) A disease-producing organism.

Pathological (path'-ō-LOJ-i-kal) **anatomy** The study of structural changes caused by disease.

Pediatrician (pē'-dē-a-TRISH-un) A physician who specializes in the care and treatment of children and their illnesses.

Pedicel (PED-i-sel) Footlike structure, as on podocytes of a glomerular.

Pelvic (PEL-vik) **cavity** Inferior portion of the abdominopelvic cavity that contains the urinary bladder, sigmoid colon, rectum, and internal female and male reproductive structures.

Pelvic inflammatory disease (PID) Collective term for any extensive bacterial infection of the pelvic organs, especially the uterus, uterine tubes, and ovaries.

Pelvimetry (pel-VIM-e-trē) Measurement of the size of the inlet and outlet of the birth canal.

Pelvis The basinlike structure formed by the two pelvic bones, the sacrum, and the coccyx. The expanded, proximal portion of the ureter, lying within the kidney and into which the major calyces open.

Penis (PĒ-nis) The male copulary organ, used to introduce spermatozoa into the female vagina.

Pepsin Protein-digesting enzyme secreted by zymogenic (chief) cells of the stomach as the inactive form pepsinogen which is converted to active pepsin by hydrochloric acid.

Peptic ulcer An ulcer that develops in areas of the gastrointestinal tract exposed to hydrochloric acid; classified as a gastric ulcer if in the lesser curvature of the stomach and as a duodenal ulcer if in the first part of the duodenum.

Pericardial (per'-i-KAR-dē-al) **cavity** Small potential space between the visceral and parietal pericardium.

Pericardium (per'-i-KAR-dē-um) A loose-fitting serous membrane that encloses the heart, consisting of an outer fibrous layer and an internal serous layer. Also called **parietal pericardium, pericardial sac.**

Perichondrium (per'-i-KON-drē-um) The membrane that covers cartilage.

Perikaryon (per'-i-KAR-ē-on) The nerve cell body that contains the nucleus and other organelles.

Perilymph (PER-i-lymf') The fluid contained between the bony and membranous labyrinths of the inner ear.

Perimetrium (per-i-MĒ-trē-um) The serosa of the uterus.

Perimysium (per'-i-MIZ-ē-um) Invagination of the epimysium that divides muscles into bundles.

Perineum (per'-i-NĒ-um) The pelvic floor; the space between the anus and the scrotum in the male and between the anus and the vulva in the female.

Perineurium (per'-i-NYOO-rē-um) Connective tissue wrapping each muscle bundle.

Periodontal (per-ē-ō-DON-tal) **disease** A collective term for conditions characterized by degeneration of gingivae, alveolar bone, periodontal ligament, and cementum.

Periosteum (per'-ē-OS-tē-um) The membrane that covers bone and is essential for bone growth, repair, and nutrition.

Peripheral (pe-RIF-er-al) Located on the outer part or a surface of the body.

Peripheral nervous system (PNS) The part of the nervous system that lies outside the central nervous system—nerves and ganglia.

Peripheral resistance The hindrance to blood flow due to friction between blood and blood vessels; related to viscosity of blood and diameter and length of blood vessels.

Peristalsis (per'-i-STAL-sis) Successive muscular contractions along the wall of a hollow muscular structure.

Peritoneum (per'-i-tō-NĒ-um) The largest serous membrane of the body which lines the abdominal cavity and covers the viscera.

Peritonitis (per'-i-tō-NĪ-tis) Inflammation of the peritoneum.

Peroxisome (pe-ROKS-ī-sōm) Organelle similar in structure to a lysosome that contains enzymes related to hydrogen peroxide metabolism; abundant in liver cells.

Perspiration Substance produced by sudoriferous (sweat) glands containing water, salts, urea, uric acid, amino acids, ammonia, sugar, lactic acid, and ascorbic acid; helps maintain body temperature and eliminate wastes.

Peyer's (PĪ-erz) **patches** Aggregated lymph nodules that are most numerous in the ileum.

pH A symbol of the measure of the concentration of hydrogen ions in a solution. The pH scale extends from 0 to 14 with a value of 7 expressing neutrality, values lower than 7 expressing increasing acidity, and values higher than 7 expressing increased alkalinity.

Phagocytosis (fag'-ō-sī-TŌ-sis) The process by which cells ingest particulate matter; especially the ingestion and destruction of microbes, cell debris, and other foreign matter.

Phalanx (FĀ-lanks) The bone of a finger or toe. *Plural,* **phalanges** (fa-LAN-jēz).

Phantom pain A sensation of pain as originating in a limb that has been amputated.

Pharynx (FAR-inks) The throat; a tube that starts at the internal nares and runs partway down the neck where it opens into the esophagus posteriorly and into the larynx anteriorly.

Pheochromocytoma (fē-ō-krō'-mō-sī-TŌ-ma) Tumor of the chromaffin cells of the adrenal medulla.

Phenotype (FĒ-nō-tīp) The observable expression of genotype;

physical characteristics of an organism determined by genetic makeup and influenced by interaction between genes and internal and external environmental factors.

Phenylketonuria (fen'-il-kē'-tō-NOO-rē-a) **(PKU)** A disorder characterized by an elevation of the amino acid phenylalanine in the blood.

Photoreceptor Receptor that detects light on the retina of the eye. Also called **electromagnetic receptor.**

Physiology (fiz'-ē-OL-ō-jē) Science that deals with the functions of an organism or its parts.

Pia mater (PĒ-a MĀ-ter) The inner membrane (meninx) covering the brain and spinal cord.

Pineal (PIN-ē-al) **gland** The cone-shaped gland located in the roof of the third ventricle. Also called the **epiphysis cerebri.**

Pinealocyte (pin-ē-AL-o-sīt) Secretory cell of the pineal gland that produces hormones.

Pinna (PIN-na) The projecting part of the external ear. Also called the **auricle.**

Pinocytosis (pi'-nō-sī-TŌ-sis) The process by which cells ingest liquid.

Pituicyte (pi-TOO-i-sīt) Supporting cell of the posterior lobe of the pituitary gland.

Pituitary (pi-TOO-i-tar'-ē) **dwarfism** Condition caused by hyposecretion of growth hormone (GH) during the growth years and characterized by childlike physical traits in an adult.

Pituitary gland A small endocrine gland lying in the sella turcica of the sphenoid bone and attached to the hypothalamus by the infundibulum; nicknamed the "master gland." Also called the **hypophysis.**

Pivot joint A synovial joint in which a rounded, pointed, or conical surface of one bone articulates with a shallow depression of another bone, as in the joint between the atlas and axis and between the proximal ends of the radius and ulna. Also called a **trochoid joint.**

Placenta (pla-SEN-ta) The special structure through which the exchange of materials between fetal and maternal circulations occurs. Also called the **afterbirth.**

Plantar flexion (PLAN-tar FLEK-shun) Extension of the foot at the ankle joint.

Plaque (plak) A cholesterol-containing mass in the tunica media of arteries. A mass of bacterial cells, dextran (polysaccharide), and other debris that adheres to teeth.

Plasma (PLAZ-ma) The extracellular fluid found in blood vessels; blood minus the formed elements.

Plasma cell Cell that produces antibodies and develops from a B cell (lymphocyte).

Plasma (cell) membrane Outer, limiting membrane that separates the cell's internal parts from extracellular fluid and the external environment.

Platelet plug Aggregation of thrombocytes at a damaged blood vessel to prevent blood loss.

Pleura (PLOOR-a) The serous membrane that enfolds the lungs and lines the walls of the chest and diaphragm.

Pleural cavity Small potential space between the visceral and parietal pleurae.

Plexus (PLEK-sus) A network of nerves, veins, or lymphatic vessels.

Plicae circulares (PLĪ-kē SER-kyoo-lar-es) Permanent, deep, transverse folds in the mucosa and submucosa of the small intestine that increase the surface area for absorption.

Pneumonia (noo-MŌ-nē-a) Acute infection or inflammation of the alveoli of the lungs.

Pneumotaxic (noo-mō-TAK-sik) **area** Portion of the respiratory center in the pons that helps coordinate the transition between inspiration and expiration.

Podiatry (pō-DĪ-a-trē) The diagnosis and treatment of foot disorders.

Polar body The smaller cell resulting from the unequal division of cytoplasm during the meiotic divisions of an oocyte. The polar body has no function and is resorbed.

Polarized A condition in which opposite effects or states exist at the same time. In electrical contexts, having one portion negative and another positive; for example, a polarized nerve cell membrane has the outer surface positively charged and the inner surface negatively charged.

Poliomyelitis (pō'-lē-ō-mī-e-LĪ-tis) Viral infection marked by fever, headache, stiff neck and back, deep muscle pain and weakness, and loss of certain somatic reflexes; a serious form of the disease, **bulbar polio,** results in destruction of motor neurons in anterior horns of spinal nerves that leads to paralysis.

Polycythemia (pol'-ē-sī-THĒ-mē-a) An abnormal increase in the number of red blood cells.

Polyunsaturated fat A fat whose carbon atoms can still bond to two or more hydrogen atoms; found naturally in vegetable oils.

Polyuria (pol'-ē-YOO-rē-a) An excessive production of urine.

Pons varolii (ponz var-Ō-lē-ī) The portion of the brain stem that forms a "bridge" between the medulla and the midbrain, anterior to the cerebellum.

Positron emission tomography (PET) A type of radioactive scanning based on the release of gamma rays when positrons collide with negatively charged electrons in body tissues; it indicates where radioisotopes are used on the body.

Postabsorptive state Metabolic state during which absorption is complete and energy needs of the body must be satisfied by nutrients already in the body. Also called the **fasting state.**

Postcentral gyrus A gyrus immediately posterior to the central sulcus that contains the general sensory area of the cerebral cortex.

Posterior (pos-TĒR-ē-or) Nearer to or at the back of the body. Also called **dorsal.**

Posterior root The structure composed of afferent (sensory) fibers lying between a spinal nerve and the dorsolateral aspect of the spinal cord. Also called the **dorsal root** or **sensory root.**

Posterior root ganglion A group of cell bodies of sensory (afferent) neurons and their supporting cells located along the posterior root of a spinal nerve. Also called a **dorsal** or **sensory root ganglion.**

Postganglionic neuron (pōst'-gang-lē-ON-ik NOO-ron) The second visceral efferent neuron in an autonomic pathway, having its cell body and dendrites located in an autonomic ganglion and its unmyelinated axon ending at cardiac muscle, smooth muscle, or a gland.

Postpartum (pōst-PAR-tum) After parturition; occurring after the delivery of a baby.

Postsynaptic (pōst-sin-AP-tik) **neuron** The nerve cell that is activated by the release of a neurotransmitter substance from another neuron and carries action potentials away from the synapse.

Precentral gyrus A gyrus immediately anterior to the central sulcus that contains the primary motor area of the cerebral cortex.

Preganglionic (prē'-gang-lē-ON-ik) **neuron** The first visceral efferent neuron in an autonomic pathway, with its cell body and dendrites in the brain or spinal cord and its myelinated axon ending at an autonomic ganglion where it synapses with a postganglionic neuron.

Pregnancy Sequence of events including fertilization, implantation, embryonic growth, and fetal growth that normally terminates in birth.

Premenstrual syndrome (PMS) Severe physical and emotional stress occurring late in the postovulatory phase of the menstrual cycle and sometimes overlapping with menstruation.

Prepuce (PRĒ-pyoos) The loosely fitting skin covering the glans of the penis and clitoris. Also called the **foreskin.**

Presbyopia (prez-bē-Ō-pē-a) A loss of elasticity of the lens of the eye due to advancing age with resulting inability to focus clearly on near objects.

Pressoreceptor (press-ō-rē-SEP-tor) Nerve cells capable of responding to changes in blood pressure. Also called a **baroreceptor.**

Presynaptic (prē-sin-AP-tik) **neuron** A nerve cell that carries action potentials toward a synapse.

Prevertebral ganglion (prē-VERT-e-bral GANG-lē-on) A cluster of cell bodies of postganglionic sympathetic neurons anterior to the spinal column and close to large abdominal arteries. Also called a **collateral ganglion.**

Primary germ layer One of three layers of embryonic tissue called ectoderm, mesoderm, and endoderm, that give rise to all tissues and organs of the organism.

Principal cell Cell found in the parathyroid glands that secretes parathyroid hormone (PTH). Also called a **chief cell.**

Proctology (prok-TOL-ō-jē) The branch of medicine that treats the rectum and its disorders.

Progeny (PROJ-e-nē) Refers to offspring or descendants.

Progesterone (prō-JES-te-rōn) A female sex hormone produced by the ovaries that helps prepare the endometrium for implantation of a fertilized ovum and the mammary glands for milk secretion.

Prognosis (prog-NŌ-sis) A forecast of the probable results of a disorder; the outlook for recovery.

Projection (prō-JEK-shun) The process by which the brain refers sensations to their point of stimulation.

Prolactin (prō-LAK-tin) **(PRL)** A hormone secreted by the adenohypophysis (anterior lobe) of the pituitary gland that initiates and maintains milk secretion by the mammary glands.

Prolapse (PRŌ-laps) A dropping or falling down of an organ, especially the uterus or rectum.

Proliferation (pro-lif'-er-Ā-shun) Rapid and repeated reproduction of new parts, especially cells.

Pronation (prō-NĀ-shun) A movement of the flexed forearm in which the palm of the hand is turned posteriorly.

Properdin (prō-PER-din) A protein found in serum capable of destroying bacteria and viruses.

Prophase (PRŌ-fāz) The first stage in mitosis during which chromatid pairs are formed and aggregate around the equatorial plane region of the cell.

Proprioception (prō-prē-ō-SEP-shun) The receipt of information from muscles, tendons, and the labyrinth that enables the brain to determine movements and position of the body and its parts. Also called **kinesthesia.**

Proprioceptor (prō'-prē-ō-SEP-tor) A receptor located in muscles, tendons, or joints that provides information about body position and movements.

Prostaglandin (pros-ta-GLAN-din) **(PG)** A membrane-associated lipid composed of 20-carbon fatty acids with 5 carbon atoms joined to form a cyclopentane ring; synthesized in small quantities and basically mimics hormones in activities.

Prostatectomy (pros'-ta-TEK-tō-mē) The surgical removal of part of or the entire prostate gland.

Prostate (PROS-tāt) **gland** A muscular, doughnut-shaped gland inferior to the urinary bladder that surrounds the superior portion of the male urethra.

Prosthesis (pros-THĒ-sis) An artificial device to replace a missing body part.

Protein An organic compound consisting of carbon, hydrogen, oxygen, nitrogen, and sometimes sulfur and phosphorus and made up of amino acids linked by peptide bonds.

Prothrombin (prō-THROM-bin) An inactive protein synthesized by the liver, released into the blood, and converted to active thrombin in the process of blood clotting.

Prothrombin time A test to determine the amount of prothrombin in blood, based on the time required for a blood sample to clot after being treated with an anticoagulant and coagulation factors; normal time about 12 sec.

Protraction (prō-TRAK-shun) The movement of the mandible or clavicle forward on a plane parallel with the ground.

Proximal (PROK-si-mal) Nearer the attachment of an extremity to the trunk or a structure; nearer to the point of origin.

Pruritus (proo'-RĪ-tus) Itching.

Pseudopodia (soo'-dō-PŌ-dē-a) Temporary, protruding projections of cytoplasm.

Psoriasis (sō-RĪ-a-sis) Chronic skin disease characterized by reddish plaques or papules covered with scales.

Psychosomatic (sī'-kō-sō-MAT-ik) Pertaining to the relationship between mind and body. Commonly used to refer to those physiological disorders thought to be caused entirely or partly by emotional disturbances.

Ptosis (TŌ-sis) Drooping, as of the eyelid or the kidney.

Puberty (PYOO-ber-tē) The time of life during which the secondary sex characteristics begin to appear and the capability for sexual reproduction is possible; usually between the ages of 10 and 15.

Pudendum (pyoo-DEN-dum) A collective designation for the external genitalia of the female.

Pulmonary (PUL-mo-ner'-ē) Concerning or affected by the lungs.

Pulmonary circulation The flow of deoxygenated blood from the right ventricle to the lungs and the return of oxygenated blood from the lungs to the left atrium.

Pulmonary ventilation The inflow (inspiration) and outflow (expiration) of air between the atmosphere and the lungs. Also called **breathing.**

Pulp cavity A cavity within the crown and neck of a tooth, filled with pulp, a connective tissue containing blood vessels, nerves, and lymphatics.

Pulse The rhythmic expansion and recoil of the elastic arteries caused by the ejection of blood from the left ventricle. Pulse rate corresponds to the heart rate.

Pulse pressure The difference between the maximum (systolic) and minimum (diastolic) pressures; normally a value of about 40 mm Hg.

Pupil The hole in the center of the iris, the area through which light enters the posterior cavity of the eyeball.

Purkinje (pur-KIN-jē) **fibers** Muscle fibers in the subendocardial tissue of the heart specialized for conducting an impulse to the myocardium; part of the conduction system of the heart.

Pus The liquid product of inflammation containing leucocytes or their remains and debris of dead cells.

P wave The deflection wave of an electrocardiogram that records atrial depolarization.

Pyemia (pī-Ē-mē-a) Infection of the blood, with multiple abscesses, caused by pus-forming microorganisms.

Pyloric (pī-LOR-ik) **sphincter** A thickened ring of smooth muscle through which the pylorus of the stomach communicates with the duodenum. Also called the **pyloric valve.**

Pyogenesis (pi'-ō-JEN-e-sis) Formation of pus.

Pyramidal (pi-RAM-i-dal) **pathways** Collections of motor nerve fibers arising in the brain and passing down through the spinal cord to motor cells in the anterior horns.

Pyrexia (pī-REK-sē-a) A condition in which the temperature is above normal.

Pyuria (pī-YOO-rē-a) The presence of leucocytes and other components of pus in urine.

QRS wave The deflection wave of an electrocardiogram that records ventricular depolarization.

Quadriplegia (kwod'-ri-PLĒ-jē-a) Paralysis of all four limbs.

Quadrant (KWOD-rant) One of four parts.

Radiographic (rā'-dē-ō-GRAF-ic) **anatomy** Diagnostic branch of anatomy that includes the use of x-rays.

Rami communicantes (RĀ-mē kō-myoo-ni-KAN-tēz) Branches of a spinal nerve. *Singular,* **ramus communicans** (RĀ-mus ko-MYOO-ni-kans).

Rapid eye movement (REM) sleep A level of sleep characterized by symmetrical flutter of the eyes and eyelids and brain wave patterns similar to those of an awake person.

Reactivity (rē-ak-TI-vi-tē) Ability of an antigen to react specifically with the antibody whose formation it induced.

Receptor A specialized cell or a nerve cell terminal modified to respond to some specific sensory modality, such as touch, pressure, cold, light, or sound. A specific molecule or arrangement of molecules organized to accept only molecules with a complementary shape.

Recessive gene A gene that is not expressed in the presence of a dominant gene on the homologous chromosome.

Reciprocal innervation (re-SIP-rō-kal in-ner-VĀ-shun) The activation of those nerve fibers that specifically inhibit contractions of the antagonistic muscles in any muscle action.

Recombinant DNA Synthetic DNA, formed by joining a fragment of DNA from one source to a portion of DNA from another.

Recruitment (rē-KROOT-ment) The process of increasing the number of active motor units.

Rectum (REK-tum) The last 20 cm of the gastrointestinal tract, from the sigmoid colon to the anus.

Red nucleus A cluster of cell bodies in the midbrain, occupying a large portion of the tegmentum and sending fibers into the rubroreticular and rubrospinal tracts.

Reduction The addition of electrons and hydrogen ions to a molecule or, less commonly, the removal of oxygen from a molecule.

Referred pain Pain that is felt at a site remote from the place of origin.

Reflex Fast response to a change in the internal or external environment that attempts to restore homeostasis; passes over a reflex arc.

Reflex arc The most basic conduction pathway through the nervous system, connecting a receptor and an effector and consisting of a receptor, a sensory neuron, a center in the central nervous system for a synapse, a motor neuron, and an effector.

Refraction (rē-FRAK-shun) The bending of light as it passes from one medium to another.

Refractory (re-FRAK-to-rē) **period** A time during which an excitable cell cannot respond to a stimulus usually adequate to evoke an action potential.

Regimen (REJ-i-men) A strictly regulated scheme of diet, exercise, or activity designed to achieve certain ends.

Regional anatomy The division of anatomy dealing with a specific region of the body, such as the head, neck, chest, or abdomen.

Regulating factor Chemical secretion of the hypothalamus that can either stimulate or inhibit secretion of hormones of the adenohypophysis (anterior pituitary).

Regurgitation (rē-gur'-ji-TĀ-shun) Return of solids or fluids to the mouth from the stomach; flowing backward of blood through incompletely closed heart valves.

Relapse (rē-LAPS) The return of a disease weeks or months after its apparent cessation.

Relaxin A female hormone produced by the ovaries that relaxes the symphysis pubis and helps dilate the uterine cervix to facilitate delivery.

Remodeling Replacement of old bone by new bone tissue.

Renal (RĒ-nal) Pertaining to the kidney.

Renal corpuscle (KOR-pus'-l) A glomerular capsule and its enclosed glomerulus.

Renal erythropoietic (ē-rith'-rō-poy-Ē-tik) **factor** An enzyme released by the kidneys and liver in response to hypoxia that acts on a plasma protein to bring about the production of erythropoietin, which stimulates red blood cell production.

Renal pelvis A cavity in the center of the kidney formed by the expanded, proximal portion of the ureter, lying within the kidney, and into which the major calyces open.

Renal pyramid A triangular structure in the renal medulla composed of the straight segments of renal tubules.

Renal suppression The sudden stoppage of urine production and excretion. Also called **anuria.**

Renin (RĒ-nin) An enzyme released by the kidney into the plasma where it converts angiotensinogen into angiotensin I.

Renin-angiotensin (an'-jē-ō-TEN-sin) **pathway** A mechanism for the control of aldosterone secretion by angiotensin II, initiated by the secretion of renin by the kidney in response to low blood pressure.

Residual (re-ZID-yoo-al) **volume** The volume of air still contained in the lungs after a maximal expiration; about 1,200 ml.

Resistance Ability to ward off disease. The hindrance encountered by an electrical charge as it moves through a substance from one point to another. The hindrance encountered by blood as it flows through the vascular system or by air through respiratory passageways.

Respiration (res-pi-RĀ-shun) Overall exchange of gases between the atmosphere, blood, and body cells consisting of pulmonary ventilation, external respiration, and internal respiration.

Respiratory center Neurons in the reticular formation of the brain stem that regulate the rate of respiration.

Resting potential The voltage that exists between the inside and outside of a cell membrane when the cell is not responding to a stimulus; about −70 to −90 mV, with the inside of the cell negative.

Resuscitation (rē-sus'-i-TĀ-shun) Act of bringing a person back to full consciousness.

Retention (rē-TEN-shun) A failure to void urine due to obstruction, nervous contraction of urethra, or absence of sensation of desire to urinate.

Rete (RĒ-tē) **testis** The network of ducts in the testes.

Reticular (re-TIK-yoo-lar) **activating system (RAS)** An extensive network of branched nerve cells running through the core of the brain stem. When these cells are activated, a generalized alert or arousal behavior results.

Reticular formation A network of small groups of nerve cells scattered among bundles of fibers beginning in the medulla as a continuation of the spinal cord and extending upward through the central part of the brain stem.

Reticulocyte (rē-TIK-yoo-lō-sīt) An immature red blood cell.

Retina (RET-i-na) The inner coat of the eyeball, lying only in the posterior portion of the eye and consisting of nervous tissue and a pigmented layer comprised of epithelial cells lying in contact with the choroid. Also called the **nervous tunic.**

Retinene (re'-ti-NĒN) The pigment portion of the photopigment rhodopsin. Also called **visual yellow.**

Retraction (rē-TRAK-shun) The movement of a protracted part of the body backward on a plane parallel to the ground, as in pulling the lower jaw back in line with the upper jaw.

Retroflexion (re-trō-FLEK-shun) A malposition of the uterus in which it is tilted posteriorly.

Retroperitoneal (re'-trō-per-i-tō-NĒ-al) External to the peritoneal lining of the abdominal cavity.

Rheumatism (ROO-ma-tizm') Any painful state of the supporting structures of the body—bones, ligaments, joints, tendons, or muscles.

Rh factor An inherited agglutinogen on the surface of red blood cells.

Rhodopsin (rō-DOP-sin) A photosensitive, reddish-purple pigment in rods of the retina, consisting of the protein scotopsin plus retinene. Also called **visual purple.**

Ribonucleic (rī'-bō-nyoo-KLĒ-ik) **acid (RNA)** A single-stranded nucleic acid constructed of nucleotides consisting of one of four possible nitrogen bases (adenine, cytosine, guanine, or uracil), ribose, and a phosphate group; three types are messenger RNA (mRNA), transfer RNA (tRNA), and ribosomal RNA (rRNA), each of which cooperates with DNA for protein synthesis.

Ribosome (RĪ-bō-sōm) An organelle in the cytoplasm of cells, composed of ribosomal RNA and ribosomal proteins, that synthesizes proteins; nicknamed the "protein factory."

Rickets (RIK-ets) Condition affecting children characterized by soft and deformed bones resulting from inadequate calcium metabolism due to a vitamin D deficiency.

Right heart (atrial) reflex A reflex concerned with maintaining normal venous blood pressure.

Right lymphatic (lim-FAT-ik) **duct** A vessel of the lymphatic system that drains lymph from the upper right side of the body and empties it into the right subclavian vein.

Rigor mortis State of partial contraction of muscles following death due to lack of ATP that causes cross bridges of thick myofilaments to remain attached to thin myofilaments, thus preventing relaxation.

Rod A visual receptor in the retina of the eye that is specialized for vision in dim light.

Roentgenogram (RENT-gen-ō-gram) A photographic image.

Root canal A narrow extension of the pulp cavity lying within the root of a tooth.

Root hair plexus (PLEK-sus) A network of dendrites arranged around the root of a hair as free or naked nerve endings that are stimulated when a hair shaft is moved.

Rotation (rō-TĀ-shun) Moving a bone around its own axis, with no other movement.

Rugae (ROO-gē) Large folds in the mucosa of an empty hollow organ, such as the stomach and vagina.

Saccule (SAK-yool) The lower and smaller of the two chambers in the membranous labyrinth inside the vestibule of the inner ear containing a receptor organ for static equilibrium.

Sacral promontory (PROM-on-tor'-ē) The superior surface of the body of the first sacral vertebra that projects anteriorly into the pelvic cavity; a line from the sacral promontory to the superior border of the symphysis pubis divides the abdominal and pelvic cavities.

Saddle joint A synovial joint in which the articular surfaces of both of the bones are saddle-shaped or concave in one direction and convex in the other direction, as in the joint between the trapezium and the metacarpal of the thumb. Also called a **sellaris joint.**

Sagittal plane A plane parallel to a midsagittal plane that divides the body or organs into unequal left and right portions. Also called a **parasagittal plane.**

Saliva (sa-LĪ-va) A clear, alkaline, somewhat viscous secretion produced by the three pairs of salivary glands; contains various salts, mucin, lysozyme, and salivary amylase.

Salivary amylase (SAL-i-ver-ē AM-i-lās) An enzyme in saliva that initiates the chemical breakdown of starch, mostly in the mouth.

Salivary gland One of three pairs of glands that lie outside the mouth and pour their secretory product (called saliva) into ducts that empty into the oral cavity; the parotid, submandibular, and sublingual glands.

Salpingitis (sal'-pin-JĪ-tis)   Inflammation of the uterine or auditory tube.

Saltatory (sal-ta-TŌ-rē) **conduction**   The propagation of an action potential (nerve impulse) along the exposed portions of a myelinated nerve fiber. The action potential appears at successive nodes of Ranvier and therefore seems to jump or leap from node to node.

Sarcolemma (sar'-kō-LEM-ma)   The cell membrane of a muscle cell, especially of a skeletal muscle cell.

Sarcoma (sar-KŌ-ma)   A connective tissue tumor, often highly malignant.

Sarcomere (SAR-kō-mēr)   A contractile unit in a striated muscle cell extending from one Z line to the next Z line.

Sarcoplasm (SAR-kō-plazm)   The cytoplasm of a muscle cell.

Sarcoplasmic reticulum (sar'-kō-PLAZ-mik re-TIK-yoo-lum)   A network of saccules and tubes surrounding myofibrils of a muscle cell, comparable to endoplasmic reticulum; functions to reabsorb calcium ions during relaxation and to release them to cause contraction.

Satiety (sa-TĪ-e-tē)   Fullness or gratification, as of hunger or thirst.

Satiety center   A collection of nerve cells located in the medial portion of the hypothalamus, close to the third ventricle, that inhibit the feeding center.

Saturated fat   A fat whose carbon atoms are bonded to the maximum number of hydrogen atoms; found naturally in animal foods such as meat, milk, milk products, and eggs.

Scala tympani (SKA-la TIM-pan-ē)   The lower spiral-shaped channel of the bony cochlea, filled with perilymph.

Scala vestibuli (ves-TIB-u-lē)   The upper spiral-shaped channel of the bony cochlea, filled with perilymph.

Schwann (shwon) **cell**   A neuroglial cell of the peripheral nervous system that forms the myelin sheath and neurilemma of a nerve fiber by wrapping around a nerve fiber in a jelly-roll fashion.

Sciatica (sī-AT-i-ka)   Inflammation and pain along the sciatic nerve; felt at the back of the thigh running down the inside of the leg.

Sclera (SKLĒ-ra)   The white coat of fibrous tissue that forms the outer protective covering over the eyeball except in the area of the anterior cornea; the posterior portion of the fibrous tunic.

Sclerosis (skle-RŌ-sis)   A hardening with loss of elasticity of the tissues.

Scoliosis (skō'-lē-Ō-sis)   An abnormal lateral curvature from the normal vertical line of the spine.

Scotopsin (skō-TOP-sin)   The protein portion of the visual pigment rhodopsin found in rods of the retina.

Scrotum (SKRŌ-tum)   A skin-covered pouch that contains the testes and their accessory structures.

Sebaceous (se-BĀ-shus) **gland**   An exocrine gland in the dermis of the skin, almost always associated with a hair follicle, that secretes sebum. Also called an **oil gland.**

Sebum (SĒ-bum)   Secretion of sebaceous (oil) glands.

Secondary sex characteristic   A feature characteristic of the male or female body that develops at puberty under the stimulation of sex hormones, but is not directly involved in sexual reproduction, such as distribution of body hair, voice pitch, body shape, and muscle development.

Secretion (se-KRĒ-shun)   Production and release from a gland cell of a fluid, especially a functionally useful product as opposed to a waste product.

Selectively permeable membrane   A membrane that permits the passage of certain substances, but restricts the passage of others. Also called a **semipermeable membrane.**

Semen (SĒ-men)   A fluid discharge at ejaculation by a male that consists of a mixture of spermatozoa and the secretions of the seminal vesicles, the prostate gland, and the bulbourethral glands. Also called **seminal fluid.**

Semicircular canals   Three bony channels projecting superiorly and posteriorly from the vestibule of the inner ear, filled with perilymph, in which lie the membranous semicircular canals filled with endolymph. They contain receptors for equilibrium.

Semicircular ducts   The membranous semicircular canals filled with endolymph and floating in the perilymph of the bony semicircular canals. They contain cristae that are concerned with dynamic equilibrium.

Semilunar (sem'-ē-LOO-nar) **valve**   A valve guarding the entrance into the aorta or the pulmonary trunk from a ventricle of the heart.

Seminal vesicle (SEM-i-nal VES-i-kul)   One of a pair of convoluted, pouchlike structures, lying posterior and inferior to the urinary bladder and anterior to the rectum, that secrete a component of semen into the ejaculatory ducts.

Seminiferous tubule (sem'-i-NI-fer-us TOO-byool)   A tightly coiled duct, located in a lobule of the testis, where spermatozoa are produced.

Senescence (se-NES-ens)   The process of growing old; the period of old age.

Senility (se-NIL-i-tē)   A loss of mental or physical ability due to old age.

Sensation   A state of awareness of external or internal conditions of the body.

Sensory area   A region of the cerebral cortex concerned with the interpretation of sensory impulses.

Septal defect   An opening in the septum between the left and right sides of the heart.

Septum (SEP-tum)   A wall dividing two cavities.

Serosa (ser-Ō-sa)   Any serous membrane. The outermost layer or tunic of an organ formed by a serous membrane. The membrane that lines the pleural, pericardial, and peritoneal cavities.

Serous (SIR-us) **membrane**   A membrane that lines body cavities that does not open to the exterior. Also called the **serosa.**

Serum   Plasma minus its clotting proteins.

Sex chromosomes   The twenty-third pair of chromosomes, designated X and Y, which determine the genetic sex of an individual; in males, the pair is XY; in females, XX.

Sexual intercourse   The insertion of the erect penis of a male into the vagina of a female. Also called **coitus.**

Sexually transmitted disease (STD)   General term for any of a large number of diseases spread by sexual contact.

Shingles   Acute infection of the peripheral nervous system caused by a virus.

Shoulder   A synovial or diarthrotic joint where the humerus joins the scapula.

Sigmoid colon (SIG-moyd KŌ-lon)   The S-shaped portion of the large intestine that begins at the level of the left iliac crest, projects inward to the midline, and terminates at the rectum at about the level of the third sacral vertebra.

Sign   Any objective evidence of disease such as a lesion, swelling, or fever.

Sinoatrial (si-nō-Ā-trē-al) **(SA) node**   A compact mass of cardiac muscle cells specialized for conduction, located in the right atrium beneath the opening of the superior vena cava. Also called the **sinuatrial node** or **pacemaker.**

Sinus (SĪ-nus)   A hollow in a bone (paranasal sinus) or other tissue; a channel for blood; any cavity having a narrow opening.

Sinusitis (sīn-yoo-SĪT-is)   Inflammation of the mucous membrane of a paranasal sinus.

Sinusoid (SĪN-yoo-soyd)   A microscopic space or passage for blood in certain organs such as the liver or spleen.

Skeletal muscle   An organ specialized for contraction, composed of striated muscle cells, supported by connective tissue, attached to a bone by a tendon or an aponeurosis, and stimulated by somatic efferent neurons.

Skull The skeleton of the head consisting of the cranial and facial bones.

Sliding-filament theory The most commonly accepted explanation for muscle contraction in which actin and myosin myofilaments move into interdigitation with each other, decreasing the length of the sarcomeres.

Small intestine A long tube of the gastrointestinal tract that begins at the pyloric sphincter of the stomach, coils through the central and lower part of the abdominal cavity, and ends at the large intestine; divided into three segments: duodenum, jejunum, and ileum.

Sodium-potassium pump An active transport system located in the cell membrane that transports sodium ions out of the cell and potassium ions into the cell at the expense of cellular ATP. It functions to keep the ionic concentrations of these elements at physiological levels.

Soft palate (PAL-at) The posterior portion of the roof of the mouth, extending posteriorly from the palatine bones and ending at the uvula. It is a muscular partition lined with mucous membrane.

Solution A homogeneous molecular or ionic dispersion of one or more substances (solutes) in a (usually liquid) dissolving medium (solvent).

Somatic (sō-MAT-ik) **nervous system (SNS)** The portion of the peripheral nervous system made up of the somatic efferent fibers that run between the central nervous system and the skeletal muscles and skin.

Somesthetic (sō'-mes-THET-ik) Pertaining to sensations and sensory structures of the body.

Spasm (spazm) An involuntary, convulsive, muscular contraction.

Spermatic (sper-MAT-ik) **cord** A supporting structure of the male reproductive system, extending from a testis to the deep inguinal ring, that includes the ductus deferens, arteries, veins, lymphatics, nerves, cremaster muscle, and connective tissue.

Spermatogenesis (sper'-ma-tō-JEN-e-sis) The formation and development of the spermatozoa.

Spermatozoon (sper'-ma-tō-ZŌ-on) A mature sperm cell.

Spermicide (SPER-mi-sīd') An agent that kills spermatozoa.

Spermiogenesis (sper'-mē-ō-JEN-e-sis) The maturation of spermatids into spermatozoa.

Sphincter (SFINGK-ter) A circular muscle constricting an orifice.

Sphincter of the hepatopancreatic ampulla A circular muscle at the opening of the common bile and main pancreatic ducts in the duodenum. Also called the **sphincter** of **Oddi.**

Sphygmomanometer (sfig'-mō-ma-NOM-e-ter) An instrument for measuring arterial blood pressure.

Spina bifida (SPĪ-na BIF-i-da) A congenital defect of the vertebral column in which the two halves of the neural arch of a vertebra fail to fuse in midline.

Spinal (SPĪ-nal) **cord** A mass of nerve tissue located in the vertebral canal from which 31 pairs of spinal nerves originate.

Spinal (lumbar) puncture Withdrawal of some of the cerebrospinal fluid from the subarachnoid space in the lumbar region.

Spinal nerve One of the 31 pairs of nerves that originate on the spinal cord from posterior and anterior roots.

Spinous (SPĪ-nus) **process** A sharp or thornlike process or projection. Also called a **spine.** A sharp ridge running diagonally across the posterior surface of the scapula.

Spiral organ The organ of hearing, consisting of supporting cells and hair cells that rest on the basilar membrane and extend into the endolymph of the cochlear duct. Also called the **organ of Corti.**

Spirometer (spī-ROM-e-ter) An apparatus used to measure air capacity of the lungs. Also called a **respirometer.**

Spleen (SPLĒN) Large mass of lymphatic tissue between the fundus of the stomach and the diaphragm that functions in phagocytosis, production of lymphocytes, and blood storage.

Sprain Forcible wrenching or twisting of a joint with partial rupture or other injury to its attachments without dislocation.

Sputum (SPYOO-tum) Substance ejected from the mouth containing saliva and mucus.

Starling's law of the capillaries The movement of fluid between plasma and interstitial fluid is in a state of near equilibrium at the arterial and venous ends of a capillary, that is, filtered fluid and absorbed fluid plus that returned to the lymphatic system are nearly equal.

Starling's law of the heart The force of muscular contraction is determined by the length of the cardiac muscle fibers.

Stasis (STĀ-sis) Stagnation or halt of normal flow of fluids, as blood, urine, or of the intestinal mechanism.

Static equilibrium (ē-kwi-LIB-rē-um) The maintenance of posture in response to changes in the orientation of the body, mainly the head, relative to the ground.

Stellate reticuloendothelial (STEL-āte re-tik'-yoo-lō-en'-dō-THĒ-lē-al) **cell** Phagocytic cell that lines a sinusoid of the liver. Also called a **Kupffer cell.**

Stenosis (sten-Ō-sis) An abnormal narrowing or constriction of a duct or opening.

Sterile (STE-rīl) Free from any living microorganisms. Unable to conceive or produce offspring.

Sterilization (ster'-i-li-ZĀ-shun) Elimination of all living microorganisms. The rendering of an individual incapable of reproduction (e.g., castration, vasectomy, hysterectomy).

Stereocilia (ste'-rē-ō-SIL-ē-a) Groups of extremely long, slender, nonmotile microvilli projecting from epithelial cells lining the epididymis.

Stereognosis (ste'-rē-og-NŌ-sis) The ability to recognize the size, shape, and texture of an object by touch.

Sternal puncture Introduction of a wide-bore needle into the marrow cavity of the sternum for aspiration of a sample of red bone marrow.

Stimulus Any change in the environment, capable of initiating a response by the nervous system.

Stomach The J-shaped enlargement of the gastrointestinal tract directly under the diaphragm in the epigastric, umbilical, and left hypochondriac regions of the abdomen, between the esophagus and small intestine.

Strabismus (stra-BIZ-mus) A condition in which the visual axes of the two eyes differ, so that they do not both fix on the same object.

Straight tubule (TOO-byool) A duct in a testis leading from a convoluted seminiferous tubule to the rete testis.

Stratum basalis (STRĀ-tum ba-SAL-is) The outer layer of the endometrium, next to the myometrium, that is maintained during menstruation and gestation and produces a new functionalis following menstruation or parturition.

Stratum functionalis (funk'-shun-AL-is) The inner layer of the endometrium, the layer next to the uterine cavity, that is shed during menstruation and that forms the maternal portion of the placenta during gestation.

Stressor A stress that is extreme, unusual, or long-lasting and triggers the general adaptation syndrome.

Stretch receptor Receptor in the walls of bronchi, bronchioles, and lungs that sends impulses to the respiratory center that prevents overinflation of the lungs.

Stretch reflex A monosynaptic reflex triggered by a sudden stretch of a muscle and ending with a contraction of that same muscle.

Stricture (STRIK-cher) A local contraction of a tubular structure.

Stroke volume The volume of blood ejected by either ventricle in one systole; about 70 ml.

Stroma (STRŌ-ma)   The tissue that forms the ground substance, foundation, or framework of an organ, as opposed to its functional parts.

Subarachnoid (sub'-a-RAK-noyd) **space**   A space between the arachnoid and the pia mater that surrounds the brain and spinal cord and through which cerebrospinal fluid circulates.

Subcutaneous (sub'-kyoo-TĀ-nē-us)   Beneath the skin. Also called **hypodermic.**

Subcutaneous layer   A continuous sheet of loose connective tissue and adipose tissue between the dermis of the skin and the deep fascia of the muscles. Also called the **superficial fascia.**

Sublingual (sub-LING-gwal) **gland**   One of a pair of salivary glands situated in the floor of the mouth under the mucous membrane and to the side of the lingual frenulum, with a duct that opens into the floor of the mouth.

Submandibular gland   One of a pair of salivary glands found beneath the base of the tongue under the mucous membrane in the posterior part of the floor of the mouth, posterior to the sublingual glands, with a duct situated to the side of the lingual frenulum. Also called the **submaxillary gland.**

Submucosa (sub-myoo-KŌ-sa)   A layer of connective tissue located beneath a mucous membrane, as in the gastrointestinal tract or the urinary bladder where a submucosa connects the mucosa to the muscularis tunic.

Substrate   A substance with which an enzyme reacts.

Subthreshold stimulus   A stimulus of such weak intensity that it cannot initiate muscle contraction. Also called a **subliminal stimulus.**

Sudoriferous (soo'-dor-IF-er-us) **gland**   An apocrine or eccrine exocrine gland in the dermis or subcutaneous layer that produces perspiration. Also called a **sweat gland.**

Sulcus (SUL-kus)   A groove or depression between parts, especially between the convolutions of the brain. *Plural,* **sulci.**

Summation (sum-MĀ-shun)   The algebraic addition of the excitatory and inhibitory effects of many stimuli applied to a nerve cell body. The increased strength of muscle contraction that results when stimuli follow in rapid succession.

Sunstroke   Condition produced when the body cannot easily lose heat and characterized by reduced perspiration and elevated body temperature. Also called **heatstroke.**

Superficial (soo'-per-FISH-al)   Located on or near the surface of the body.

Superior (soo-PĒR-ē-or)   Toward the head or upper part of a structure. Also called **cephalad** or **craniad.**

Superior vena cava (VĒ-na CĀ-va)   Large vein that collects blood from parts of the body superior to the heart and returns it to the right atrium.

Supination (soo-pī-NĀ-shun)   A movement of the forearm in which the palm of the hand is turned anteriorly.

Suppuration (sup'-yoo-RĀ-shun)   Pus formation and discharge.

Surface anatomy   The study of the structures that can be identified from the outside of the body.

Surfactant (sur-FAK-tant)   A phospholipid substance produced by the lungs that decreases surface tension.

Susceptibility (sus-sep'-ti-BIL-i-tē)   Lack of resistance of a body to the deleterious or other effects of an agent such as pathogenic microorganisms.

Sustentacular (sus'-ten-TAK-yoo-lar) **cell**   A supporting cell of seminiferous tubules that produces secretions for supplying nutrients to spermatozoa and the hormone inhibin. Also called a **Sertoli cell.**

Suture (SOO-cher)   A fibrous joint, especially in the skull, where bone surfaces are closely united.

Sympathetic (sim'-pa-THET-ik) **division**   One of the two subdivisions of the autonomic nervous system, having cell bodies of preganglionic neurons in the lateral gray columns of the thoracic segment and first two or three lumbar segments of the spinal cord; primarily concerned with processes involving the expenditure of energy. Also called the **thoracolumbar division.**

Sympathomimetic (sim'-pa-thō-mi-MET-ik)   Producing effects that mimic those brought about by the sympathetic division of the autonomic nervous system.

Symphysis (SIM-fi-sis)   A line of union. A slightly movable cartilaginous joint such as the symphysis pubis between the anterior surfaces of the pubic bones.

Symptom (SIMP-tum)   An observable abnormality that indicates the presence of a disease or disorder of the body.

Synapse (SIN-aps)   The junction between the processes of two adjacent neurons; the place where the activity of one neuron affects the activity of another.

Synaptic (sin-AP-tik) **cleft**   The narrow gap that separates the membrane of an axon terminal of one nerve cell and the membrane of another nerve cell and across which the transmitter substance diffuses to affect the postsynaptic cell.

Synaptic delay   The length of time between the arrival of the action potential at the axon terminal and the membrane potential (IPSP or EPSP) change on the postsynaptic membrane; usually less than 1 msec.

Synaptic knob   Expanded distal end of a telodendrium that contains synaptic vesicles. Also called an **end foot.**

Synaptic vesicle   Membrane-enclosed sac in a synaptic knob that stores transmitter substances.

Synarthrosis (sin'-ar-THRŌ-sis)   An immovable joint.

Synchondrosis (sin'-kon-DRŌ-sis)   A cartilaginous joint in which the connecting material is hyaline cartilage.

Syndesmosis (sin'-dez-MŌ-sis)   A fibrous joint in which articulating bones are united by dense fibrous tissue.

Syndrome (SIN-drōm)   A group of signs and symptoms that occur together in a pattern characteristic of a particular disease or abnormal condition.

Syneresis (si-NER-e-sis)   The process of clot retraction.

Synergist (SIN-er-jist)   A muscle that assists the prime mover or agonist by reducing undesired action or unnecessary movement. Also called a **fixator.**

Synostosis (sin'-os-TŌ-sis)   A joint in which the dense fibrous connective tissue that unites bones at a suture has been replaced by bone, resulting in a complete fusion across the suture line.

Synovial (si-NŌ-vē-al) **cavity**   The space between the articulating bones of a synovial or diarthrotic joint, filled with synovial fluid.

Synovial fluid   Secretion of synovial membranes that lubricates joints and nourishes articular cartilage.

Synovial joint   A fully movable or diarthrotic joint in which a joint or synovial cavity is present between the two articulating bones.

Synovial membrane   The inner of the two layers of the articular capsule of a synovial joint, composed of loose connective tissue covered with epithelium that secretes synovial fluid into the joint cavity.

Syphilis (SIF-i-lis)   A sexually transmitted disease caused by the bacterium *Treponema pallidum.*

System   An association of organs that have a common function.

Systemic (sis-TEM-ik)   Affecting the whole body; generalized.

Systemic anatomy   The study of particular systems of the body, such as the skeletal, muscular, nervous, cardiovascular, or urinary systems.

Systemic circulation   The routes through which oxygenated blood flows from the left ventricle through the aorta to all the organs of the body and deoxygenated blood returns to the right atrium.

Systemic lupus erythematosus (er-i-them-a-TŌ-sus) **(SLE)**   An autoimmune, inflammatory disease that may affect every tissue of the body.

Systole (SIS-tō-lē) In the cardiac cycle, the phase of contraction of the heart muscle, especially of the ventricles.

Systolic (sis-TO-lik) **blood pressure** The force exerted by blood on arterial walls during ventricular contraction; the highest pressure measured in the large arteries, about 120 mm Hg under normal conditions for a young, adult male.

Tachycardia (tak'-i-KAR-dē-a) A rapid heartbeat or pulse rate.

Tactile (TAK-tīl) Pertaining to the sense of touch.

Taenia coli (TĒ-nē-a KŌ-lī) One of three flat bands of muscles running the length of the large intestine.

Target cell A cell whose activity is affected by a particular hormone.

Tarsus (TAR-sus) A collective term for the seven bones of the ankle.

T cell A lymphocyte that can differentiate into one of four kinds of cells—killer, helper, suppressor, or memory—all of which function in cellular immunity.

Tectorial (tek-TŌ-rē-al) **membrane** A gelatinous membrane projecting over and in contact with the hair cells of the spiral organ in the chochlear duct.

Telodendria (tel-ō-DEN-drē-a) The many small terminal branches of a neuron.

Telophase (TEL-ō-fāz) The final stage of mitosis in which the daughter nuclei become established.

Tendinitis (ten'-din-Ī-tis) Inflammation of a tendon and synovial membrane at a joint. Also called **tenosynovitis.**

Tendon (TEN-don) A white fibrous cord of dense, regularly arranged connective tissue that attaches muscle to bone.

Tendon organ A proprioceptive receptor found chiefly near the junction of tendons and muscles.

Tentorium cerebelli (ten-TŌ-rē-um ser'-e-BEL-ē) A transverse shelf of dura mater that forms a partition between the occipital part of the cerebral hemispheres and the cerebellum and that covers the cerebellum.

Teratogen (TER-a-tō-jen) Any agent or factor that causes physical defects in a developing embryo.

Testis (TES-tis) Male gonad that produces sperm and the hormones testosterone and inhibin.

Testosterone (tes-TOS-te-rōn) A male sex hormone (androgen) secreted by interstitial endocrinocytes of a mature testis; controls the growth and development of male sex organs, secondary sex characteristics, spermatozoa, and body growth.

Tetanus (TET-a-nus) An infectious disease caused by the toxin of *Clostridium tetani,* characterized by tonic muscle spasms and exaggerated reflexes, lockjaw, and arching of the back. A smooth, sustained contraction produced by a series of very rapid stimuli to a muscle.

Tetany (TET-a-nē) A nervous condition caused by hypoparathyroidism and characterized by intermittent or continuous tonic muscular contractions of the extremities.

Tetralogy of Fallot (tet-RAL-ō-jē of fal-Ō) A combination of four congenital heart defects: (1) constricted pulmonary semilunar valve, (2) interventricular septal opening, (3) emergence of aorta from both ventricles instead of from the left only, and (4) enlarged right ventricle.

Thalamus (THAL-a-mus) A large, oval structure located above the midbrain, consisting of two masses of gray matter covered by a thin layer of white matter.

Therapy (THER-a-pē) The treatment of a disease or disorder.

Thermoreceptor (THER-mō-rē-sep-tor) Receptor that detects changes in temperature.

Thigh The portion of the lower extremity between the hip and the knee.

Third ventricle (VEN-tri-kul) A slitlike cavity between the right and left halves of the thalamus and between the lateral ventricles.

Thoracic cavity Superior component of the ventral body cavity that contains two pleural cavities, the mediastinum, and the pericardial cavity.

Thoracic duct A lymphatic vessel that begins as a dilation called the cisterna chyli, receives lymph from the left side of the head, neck, and chest, the left arm, and the entire body below the ribs, and empties into the left subclavian vein. Also called the **left lymphatic duct.**

Thoracolumbar outflow The fibers of the sympathetic preganglionic neurons, which have their cell bodies in the lateral gray columns of the thoracic segment and first two or three lumbar segments of the spinal cord.

Thorax (THŌ-raks) The chest.

Threshold potential The membrane voltage that must be reached in order to trigger an action potential (nerve impulse).

Threshold stimulus Weakest stimulus from a neuron that is capable of initiating muscle contraction. Also called a **liminal stimulus.**

Thrombin (THROM-bin) The active enzyme formed from prothrombin which acts to convert fibrinogen to fibrin.

Thrombocyte (THROM-bō-sīt) A fragment of cytoplasm enclosed in a cell membrane and lacking a nucleus; found in the circulating blood; plays a role in blood clotting. Also called a **platelet.**

Thromboplastin (throm-bō-PLAS-tin) A factor, or collection of factors, whose appearance initiates the blood clotting process.

Thrombosis (throm-BŌ-sis) The formation of a clot in an unbroken blood vessel.

Thrombus A clot formed in an unbroken blood vessel.

Thymus (THĪ-mus) **gland** A bilobed organ, located in the upper mediastinum posterior to the sternum and between the lungs, that plays a role in the immunity mechanism of the body.

Thyroglobulin (thī-rō-GLŌ-byoo-lin) **(TGB)** A large glycoprotein molecule secreted by follicle cells of the thyroid gland in which iodine is combined with tyrosine to form thyroid hormones.

Thyroid cartilage (THĪ-royd KAR-ti-lij) The largest single cartilage of the larynx, consisting of two fused plates that form the anterior wall of the larynx. Also called the **Adam's apple.**

Thyroid colloid (KOL-loyd) A complex in thyroid follicles consisting of thyroglobulin and stored thyroid hormones.

Thyroid follicle (FOL-i-kul) Spherical sac that forms the parenchyma of the thyroid gland and consists of follicular cells that produce thyroxine (T_4) and triiodothyronine (T_3) and parafollicular cells that produce calcitonin (CT).

Thyroid gland An endocrine gland with right and left lateral lobes on either side of the trachea connected by an isthmus located in front of the trachea just below the cricoid cartilage.

Thyroid-stimulating hormone (TSH) A hormone secreted by the adenohypophysis (anterior lobe) of the pituitary gland that stimulates the synthesis and secretion of hormones produced by the thyroid gland.

Thyroxine (thī-ROK-sin) **(T_4)** A hormone secreted by the thyroid that regulates metabolism, growth and development, and the activity of the nervous system.

Tic Spasmodic, involuntary twitching by muscles that are ordinarily under voluntary control.

Tidal volume The volume of air breathed in and out in any one breath; about 500 ml in quiet, resting conditions.

Tinnitus (ti-NĪ-tus) A ringing or tingling sound in the ears.

Tissue A group of similar cells and their intercellular substance joined together to perform a specific function.

Tongue A large skeletal muscle on the floor of the oral cavity.

Tonsil (TON-sil) A mass of lymphoid tissue embedded in mucous membrane.

Torn cartilage A tearing of an articular disc in the knee.

Total lung capacity The sum of tidal volume, inspiratory reserve volume, expiratory reserve volume, and residual volume; about 6,000 ml.

Toxic (TOK-sik) Pertaining to poison; poisonous.

Toxic shock syndrome (TSS) A disease caused by the bacterium *Staphylococcus aureus*, occurring among menstruating females who use tampons and characterized by high fever, sore throat, headache, fatigue, irritability, and abdominal pain.

Trabecula (tra-BEK-yoo-la) Irregular latticework of thin plate of spongy bone. Fibrous cord of connective tissue serving as supporting fiber by forming a septum extending into an organ from its wall or capsule. *Plural*, **trabeculae** (tra-BEK-yoo-lē).

Trachea (TRĀ-kē-a) Tubular air passageway extending from the larynx to the fifth thoracic vertebra. Also called the **windpipe.**

Tracheostomy (trā-kē-OS-tō-mē) Creation of an opening into the trachea through the neck, with insertion of a tube to facilitate passage of air or evacuation of secretions.

Trachoma (tra-KŌ-ma) A chronic infectious disease of the conjunctiva and cornea of the eye caused by a strain of the bacterium *Chlamydia trachomatis*.

Tract A bundle of nerve fibers in the central nervous system.

Transcription (trans-KRIP-shun) The first step in the transfer of genetic information in which a single strand of a DNA molecule serves as a template for the formation of an RNA molecule.

Translation (trans-LĀ-shun) The construction of a new protein on the ribosome of a cell as dictated by the sequence of codons in mRNA.

Transmitter substance One of a variety of molecules synthesized within the nerve axon terminals, released into the synaptic cleft in response to an action potential, and affecting the membrane potential of the postsynaptic neuron. Also called a **neurotransmitter.**

Transplantation (trans-plan-TĀ-shun) The replacement of injured or diseased tissues or organs with natural ones.

Transverse colon (trans-VERS KŌ-lon) The portion of the large intestine extending across the abdomen from the right colic (hepatic) flexure to the left colic (splenic) flexure.

Trauma (TRAW-ma) An injury, either a physical wound or psychic disorder, caused by an external agent or force, such as a physical blow or emotional shock; the agent or force that causes the injury.

Treppe (TREP-ē) The gradual increase in the amount of contraction by a muscle caused by rapid, repeated stimuli of the same strength.

Triad (TRĪ-ad) A complex of three units in a muscle cell composed of a T tubule and the segments of sarcoplasmic reticulum on both sides of it.

Tricuspid (trī-KUS-pid) **valve** Atrioventricular valve on the right side of the heart.

Trigeminal neuralgia (trī-JEM-i-nal noo-RAL-jē-a) Pain in one or more of the branches of the trigeminal (V) nerve. Also called **tic douloureux.**

Trigone (TRĪ-gon) A triangular area at the base of the urinary bladder.

Triiodothyronine (trī-ī-od-ō-THĪ-rō-nēn) **(T₃)** A hormone produced by the thyroid gland that regulates metabolism, growth and development, and the activity of the nervous system.

Trophectoderm (trō-FEK-tō-derm) The outer covering of cells of the blastocyst.

Tropic (TRŌ-pik) **hormone** A hormone whose target is another endocrine gland.

Trunk The part of the body to which the upper and lower extremities are attached.

T tubules (TOOB-yools) **(transverse tubules)** Minute, cylindrical invaginations of the muscle cell membrane that carry the surface action potentials deep into the muscle cell.

Tubal ligation (li-GĀ-shun) A sterilization procedure in which the uterine tubes are tied and cut.

Tuberculosis (too-berk-yoo-LŌ-sis) An inflammation of the lungs and pleurae caused by *Mycobacterium tuberculosis* resulting in destruction of lung tissue and its replacement by fibrous connective tissue.

Tubular maximum (Tm) The maximum amount of a substance that can be reabsorbed by renal tubules under any condition.

Tubular reabsorption The movement of filtrate from renal tubules back into blood in response to the body's specific needs.

Tubular secretion The movement of substances in blood back into filtrate in response to the body's specific needs.

Tumor (TOO-mor) A growth of excess tissue due to an unusually rapid division of cells.

Tunica albuginea (TOO-ni-ka al'-byoo-JIN-ē-a) A dense layer of white fibrous tissue covering a testis or deep to the surface of an ovary.

Tunica externa (eks-TER-na) The outer coat of an artery or vein, composed mostly of elastic and collagenous fibers. Also called the **adventitia.**

Tunica interna (in-TER-na) The inner coat of an artery or vein, consisting of a lining of endothelium and its supporting layer of connective tissue. Also called the **tunica intima.**

Tunica media (MĒ-dē-a) The middle coat of an artery or vein, composed of smooth muscle and elastic fibers.

T wave The deflection wave of an electrocardiogram that records ventricular repolarization.

Twitch Rapid, jerky contraction of a muscle in response to a single stimulus.

Tympanic (tim-PAN-ik) **membrane** A thin, semitransparent partition of fibrous connective tissue between the external auditory meatus and the middle ear. Also called the **eardrum.**

Ulcer (UL-ser) An open lesion of the skin or a mucous membrane of the body with loss of substance and necrosis of the tissue.

Umbilical (um-BIL-i-kal) **cord** The long, ropelike structure, containing the umbilical arteries and vein, which connects the fetus to the placenta.

Umbilicus (um-BIL-i-kus) A small scar on the abdomen that marks the former attachment of the umbilical cord to the fetus. Also called the **navel.**

Unsaturated fat A fat whose carbon atoms can still bond to one more hydrogen atom; found naturally in vegetable oils.

Upper extremity The appendage attached at the shoulder girdle, consisting of the arm, forearm, wrist, hand, and fingers.

Uremia (yoo-RĒ-mē-a) Accumulation of toxic levels of urea and other nitrogenous waste products in the blood, usually resulting from severe kidney malfunction.

Ureter (yoo-RĒ-ter) One of two tubes that connect the kidney with the urinary bladder.

Urethra (yoo-RĒ-thra) The duct from the urinary bladder to the exterior of the body that conveys urine in females and urine and semen in males.

Urinalysis The physical, chemical, and microscopic analysis or examination of urine.

Urinary (YOO-ri-ner-ē) **bladder** A hollow, muscular organ situated in the pelvic cavity posterior to the symphysis pubis.

Urine The fluid produced by the kidneys that contains wastes or excess materials and excreted from the body through the urethra.

Urobilinogenuria The presence of urobilinogen in urine.

Urogenital (YOO'-rō-JEN-i-tal) **triangle** The region of the pelvic floor below the symphysis pubis, bounded by the symphysis pubis and the ischial tuberosities and containing the external genitalia.

Uterine (YOO-ter-in) **tube** Duct that transports ova from the ovary to the uterus. Also called the **fallopian tube** or **oviduct.**

Uterus (YOO-te-rus) The hollow, muscular organ in females that is the site of menstruation, implantation, development of the fetus, and labor. Also called the **womb.**

Utricle (YOO-tri-kul) The larger of the two divisions of the membranous labyrinth located inside the vestibule of the inner ear, containing a receptor organ for static equilibrium.

Uvea (YOO-vē-a) The three structures that make up the vascular tunic of the eye.

Uvula (YOO-vyoo-la) A soft, fleshy mass, especially the V-shaped pendant part, descending from the soft palate.

Vagina (va-JĪ-na) A muscular, tubular organ that leads from the uterus to the vestibule, situated between the urinary bladder and the rectum of the female.

Valvular stenosis (VAL-vyoo-lar STEN-Ō-sis) A narrowing of a heart valve, usually the mitral valve.

Varicose (VAR-i-kōs) Pertaining to an unnatural swelling, as in the case of a varicose vein.

Vasa vasorum (VĀ-sa va-SŌ-rum) Blood vessels supplying nutrients to the larger arteries and veins.

Vascular (VAS-kyoo-lar) Pertaining to or containing many blood vessels.

Vascular spasm Contraction of the smooth muscle in the wall of a damaged blood vessel to prevent blood loss.

Vascular tunic (TOO-nik) The middle layer of the eyeball, composed of the choroid, the ciliary body, and the iris.

Vasectomy (va-SEK-tō-mē) A means of sterilization of males in which a portion of each ductus deferens is removed.

Vasoconstriction (vāz-ō-kon-STRIK-shun) A decrease in the size of the lumen of a blood vessel caused by contraction of the smooth muscle in the wall of the vessel.

Vasodilation (vās'-ō-DĪ-lā-shun) An increase in the size of the lumen of a blood vessel caused by relaxation of the smooth muscle in the wall of the vessel.

Vasomotion (vāz-ō-MŌ-shun) Intermittent contraction and relaxation of the smooth muscle of metarterioles and precapillary sphincters that result in an intermittent blood flow.

Vasomotor (vā-sō-MŌ-tor) **center** A cluster of neurons in the medulla that controls the diameter of blood vessels, especially arteries.

Vein A blood vessel that conveys blood from tissues back to the heart.

Ventral (VEN-tral) **body cavity** Cavity near the ventral aspect of the body that contains viscera and consists of a superior thoracic cavity and an inferior abdominopelvic cavity.

Ventricle (VEN-tri-kul) A cavity in the brain or an inferior chamber of the heart.

Ventricular fibrillation (ven-TRIK-yoo-lar, fib-ri-LĀ-shun) Asynchronous ventricular contractions that result in circulatory failure.

Venule (VEN-yool) A small vein that collects blood from capillaries and delivers it to a vein.

Vermiform appendix (VER-mi-form a-PEN-diks) A twisted, coiled tube attached to the cecum.

Vermis (VER-mis) The central constricted area of the cerebellum that separates the two cerebellar hemispheres.

Vertebral (VER-te-bral) **canal** A cavity within the vertebral column formed by the vertebral foramina of all the vertebrae and containing the spinal cord. Also called the **spinal canal.**

Vertebral column The 26 vertebrae; encloses and protects the spinal cord and serves as a point of attachment for the ribs

and back muscles. Also called the **spine, spinal column,** or **backbone.**

Vertigo (VER-ti-go) Sensation of dizziness.

Vestibular (ves-TIB-yoo-lar) **membrane** The membrane that separates the cochlear duct from the scala vestibuli.

Vestibule (VES-ti-byool) A small space or cavity at the beginning of a canal, especially the inner ear, larynx, mouth, nose, and vagina.

Villus (VIL-lus) A projection of the intestinal mucosal cells containing connective tissue, blood vessels, and a lymphatic vessel; functions in the absorption of food. *Plural,* **villi** (VIL-ē).

Viscera (VIS-er-a) The organs inside the ventral body cavity. *Singular,* **viscus.**

Visceral (VIS-er-al) Pertaining to the organs or to the covering of an organ.

Visceral effector (e-FEK-tor) Cardiac muscle, smooth muscle, and glandular epithelium.

Visceral muscle An organ specialized for contraction, composed of smooth muscle cells, located in the walls of hollow internal structures, and stimulated by visceral efferent neurons.

Visceral pleura (PLOO-ra) The inner layer of the serous membrane that covers the lungs.

Visceroceptor (vis'-er-ō-SEP-tor) Receptor that provides information about the body's internal environment.

Vital capacity The sum of inspiratory reserve volume, tidal volume, and expiratory reserve volume; about 4,800 ml.

Vitamin An organic molecule necessary in trace amounts that acts as a catalyst in normal metabolic processes of the body.

Vitreous humor (VIT-rē-us HYOO-mor) A soft, jellylike substance that fills the posterior cavity of the eyeball, lying between the lens and the retina.

Vocal folds Pair of mucous membrane folds below the ventricular folds that function in voice production. Also called **true vocal cords.**

Volkmann's (FOLK-mans) **canal** A minute passageway by means of which blood vessels and nerves from the periosteum penetrate into compact bone.

Vomiting Forcible expulsion of the contents of the upper gastrointestinal tract through the mouth.

Vulva Collective designation for the external genitalia of the female. Also called **pudendum.**

Wallerian (wal-LE-rē-an) **degeneration** Degeneration of the distal portion of the axon and myelin sheath.

Wandering macrophage (MAK-rō-fāj) Phagocytic cell that develops from a monocyte, leaves the blood, and migrates to infected tissues.

Wart Generally benign tumor of epithelial skin cells caused by a virus.

White matter Aggregations or bundles of myelinated axons located in the brain and spinal cord.

Xiphoid (ZĪ-foyd) Sword-shaped. The lowest portion of the sternum.

Yolk sac An extraembryonic membrane that connects with the midgut during early embryonic development, but is nonfunctional in humans.

Zona fasciculata (ZŌ-na fa-sik'-yoo-LA-ta) The middle zone of the adrenal cortex that consists of cells arranged in long, straight cords and that secretes glucocorticoid hormones.

Zona glomerulosa (glo-mer'-yoo-LŌ-sa) The outer zone of the adrenal cortex, directly under the connective tissue covering, that consists of cells arranged in arched loops or round balls and that secretes mineralocorticoid hormones.

Zona pellucida (pe-LOO-si-da) Gelatinous covering that surrounds an ovum.

Zona reticularis (ret-ik'-yoo-LAR-is) The inner zone of the adrenal cortex, consisting of cords of branching cells that secrete sex hormones, chiefly androgens.

Zygote (ZĪ-gōt) The single cell resulting from the union of a male and female gamete; the fertilized ovum.

Zymogenic (zī'-mō-JEN-ik) **cell** A cell that secretes enzymes, for example, the zymogenic (chief) cells of the gastric glands that secrete pepsinogen.

Bibliography

UNIT I

Abston, S. "Burns in Children," *Clinical Symposia* 28, Ciba Pharmaceutical Company, 1976.

Allison, A. "Lysosomes and Disease," *Scientific American,* November 1967.

Artz, C. P. "Severe Burns: Current Concepts of Specialized Care," *Modern Medicine,* April 1973.

Bauer, W. R., et al. "Supercoiled DNA," *Scientific American,* July 1980.

Baxter, J. D. "Recombinant DNA and Medical Progress," *Hospital Practice,* February 1980.

Berkow, R. (Editor-in-Chief). *The Merck Manual,* 14th ed. Rahway, N.J.: Merck Sharp & Dohme, 1982.

Berlin, N. I. "Research Strategy in Cancer: Screening, Diagnosis, Prognosis," *Hospital Practice,* January 1975.

"Beyond Survival: Toward A Healthy View of Human Aging," *Modern Medicine,* 1 November 1976.

Bickers, D. R. "A Compendium of Topical Dermatological Agents," *Drug Therapy,* October 1979.

Bickers, D. R., and A. Kappas. "Metabolic and Pharmacologic Properties of the Skin," *Hospital Practice,* May 1974.

Bishop, M. J. "Oncogenes," *Scientific American,* March 1982.

Campbell, A. M. "How Viruses Insert Their DNA into the DNA of the Host Cell," *Scientific American,* December 1976.

Clemente, C. D. *Anatomy: A Regional Atlas of the Human Body.* Baltimore: Urban & Schwarzenberg, 1981.

Cohen, S. N., and James A. Shapiro. "Transposable Genetic Elements," *Scientific American,* February 1980.

Dahl, M. V. "Acne: How It Happens and How It's Treated," *Modern Medicine,* September 1982.

Dahl, M. V. "The Many Faces of Psoriasis: Recognizing and Treating Them," *Modern Medicine,* September 1981.

Daniels, F., Jr. "Saving Sun Worshippers from Their God," *Medical Opinion,* July 1976.

Davis, P. J. "Endocrines and Aging." *Hospital Practice,* September 1977.

de Duve, C. "Microbodies in the Living Cell," *Scientific American,* May 1983.

Dubois, E. L., and N. Talal. "Lupus: The New Great Imitator," *Medical World News,* 14 June 1976.

Fenske, N. A. "How Age Affects Skin," *Modern Medicine,* November 1982.

Geller, S. A. "Autopsy," *Scientific American,* March 1983.

Gilbert, W., and Lydia Villa-Komaroff. "Useful Proteins from Recombinant Bacteria," *Scientific American,* April 1980.

Green, M. "Viral Cell Transformation in Human Oncogenesis," *Hospital Practice,* September 1975.

Gunby, P. "PETT Scanners Probe Frontiers of Brain," *J. Amer. Med. Assoc.,* 13 August 1982.

Hayflick, L. "The Cell Biology of Human Aging," *Scientific American,* January 1980.

Henle, W., et al. "The Epstein-Barr Virus," *Scientific American,* July 1979.

Hoffman, J. F. "Ionic Transport across the Plasma Membrane," *Hospital Practice,* October 1974.

Holtzman, E. "The Biogenesis of Organelles," *Hospital Practice,* March 1974.

Howard-Flanders, P. "Inducible Repair of DNA," *Scientific American,* November 1981.

Human Physiology and the Environment in Health and Disease. p. 6: "Aging." Readings from *Scientific American.* San Francisco: Freeman, 1975.

Kaminester, L. H. "Sunlight, Skin Cancer, and Sunscreens," *J. Amer. Med. Assoc.,* 30 June 1975.

Keiffer, S. A., and E. R. Heitzman. *An Atlas of Cross-Sectional Anatomy.* New York: Harper & Row, 1979.

Kessel, R. G., and R. H. Kardon. *Tissues and Organs: A Text-Atlas of Scanning Electron Microscopy.* San Francisco: Freeman, 1979.

Koffler, D. "Systemic Lupus Erythematosus," *Scientific American,* July 1980.

Kornberg, R. D., and A. Klug. "The Nucleosome," *Scientific American,* February 1981.

Lake, J. A. "The Ribosome," *Scientific American,* August 1981.

Levine, N. "Eczema," *Modern Medicine,* August 1982.

————. "Sun Worship: Reducing the Ritual's Danger and Damage," *Modern Medicine,* June 1980.

————. "The Skin and Sports," *Modern Medicine,* April 1980.

McIver, R. T., Jr. "Chemical Reactions without Solvation," *Scientific American,* November 1980.

Miller, J. A. "Spelling Out a Cancer Gene," *Science News,* November 1982.

Moolten, S. E. "Bedsores: An Update," *Hospital Medicine,* August 1982.

Nicolson, G. H. "Cell Surfaces and Cancer Metastasis," *Hospital Practice,* August 1982.

Novick, R. P. "Plasmids," *Scientific American,* December 1980.

Oldendorf, W. H. "NMR Imaging: Its Potential Clinical Impact," *Hospital Practice,* September 1982.

Peterson, H. D. "Burn Sepsis—Management of the Severe Burn," *Forum on Infection,* September 1976.

Porter, K. R., and J. B. Tucker. "The Ground Substance of the Living Cell," *Scientific American,* March 1981.

Reith, E. M., and M. N. Ross. *Atlas of Descriptive Histology,* 3d ed. New York: Harper & Row, 1977.

Rubinstein, E. "Diseases Caused by Impaired Communication Among Cells," *Scientific American,* March 1980.

Sanders, B. B., Jr., and G. S. Stretcher. "Warts—Diagnosis and Treatment,"

J. Amer. Med. Assoc., 28 June 1976.

Satir, B. "The Final Steps in Secretion," *Scientific American,* October 1974.

Satir, P. "How Cilia Move," *Scientific American,* October 1974.

Schimke, R. T. "Gene Amplification and Drug Resistance," *Scientific American,* November 1980.

Sober, A. J., and T. B. Fitzpatrick. "Mela-noma: Be Ready for It," *Consultant,* September 1979.

Temin, H. M. "RNA-Directed Synthesis," *Scientific American,* January 1972.

Tortora, G. J. *Principles of Human Anatomy.* 3d ed. New York: Harper & Row, 1983.

Weaver, R. F. "The Cancer Puzzle," *National Geographic,* September 1976.

Wurtman, R. J. "The Effects of Light on the Human Body," *Scientific American,* July 1975.

Yokochi, C., and J. W. Rohen. *Photographic Anatomy of the Human Body,* 2d ed. Tokyo, New York: Igaku-shoin, Ltd., 1979.

UNIT II

Arehart-Treichel, J. "The Joint Destroyers," *Science News,* September 1982.

Avioli, L. A. "Management of Osteomalacia," *Hospital Practice,* January 1979.

Bassett, C. A., S. N. Mitchell, and S. R. Gaston. "Pulsing Electromagnetic Field Treatment in Ununited Fractures and Failed Arthrodeses," *J. Amer. Med. Assoc.,* 5 February 1982.

Blackshear, P. J. "Implantable Drug-Delivery Systems," *Scientific American,* December 1979.

Close, R. I. "Dynamic Properties of Mammalian Skeletal Muscles," *Physiol. Rev.,* 1972.

Cohen, C. "The Protein Switch of Muscle Contraction," *Scientific American,* November 1975.

DeLuca, H. F. "The Vitamin D Hormonal System: Implications for Bone Diseases," *Hospital Practice,* April 1980.

Elliott, J. "Electrical Stimulation of Bone Growth Wins Clinical Acceptance," *J. Amer. Med. Assoc.,* 3 April 1980.

Enis, J. E. "The Painful Knee," *Hospital Medicine,* December 1980.

Garn, S. "Bone-Loss and Aging," in *The Physiology and Pathology of Human Aging.* New York: Academic Press, 1975.

Hoyle, G. "How Is Muscle Turned On and Off?" *Scientific American,* April 1970.

Intramuscular Injections. New York: Wyeth Laboratories, 1972.

Johnson, G. T. (ed.) "A Primer on Arthritis Drugs," *The Harvard Medical School Health Letter,* September 1982.

———. "Arthroscopic Surgery on the Knee," *The Harvard Medical School Health Letter,* March 1982.

Lange, R. F. (ed.) "Chymopapain For Slipped Disc," *Internal Medicine Alert,* 13 December 1982.

Lisak, R. P. "Myasthenia Gravis: Mechanisms and Management," *Hospital Practice,* March 1983.

Margaria, R. "The Sources of Muscular Energy," *Scientific American,* March 1972.

Marx, J. L. "Osteoporosis: New Help For Thinning Bones," *Science,* February 1980.

Merz, B. "Try a Carbon Ribbon 'Round The Old Hurt Knee (and Shoulder)," *J. Amer. Med. Assoc.,* 8 October 1982.

Moskowitz, R. W. "Management of Osteoarthritis," *Hospital Practice,* July 1979.

Murray, J. M., and A. Weber. "The Cooperative Action of Muscle Proteins," *Scientific American,* February 1974.

"Prosthetic Knuckle Enables Deformed Hands to Grip and Pinch," *J. Amer. Med. Assoc.,* 22 September 1975.

Rodnan, G. P. (ed.). "Primer on the Rheumatic Diseases," *J. Amer. Med. Assoc.,* April 1973.

Rudd, E. M. "Living with Arthritis," *Drug Therapy,* March 1979.

Schumacher, H. R. "Osteoarthritis," *Modern Medicine,* November 30–December 15, 1978.

Smythe, H. "Nonsteroidal Therapy in Inflammatory Joint Disease," *Hospital Practice,* September 1975.

Steinbach, H. L., and R. H. Gold. "Pyogenic Infections of Bone: A Roentgenologic Guide," *Hospital Medicine,* July 1972.

Thompson, C. W. *Manual of Structural Kinesiology,* 9th ed. St. Louis: C. V. Mosby Co., 1981.

Thompson, G. R. "Diagnosis and Treatment of Septic Arthritis," *Hospital Medicine,* June 1976.

"Total Hip-Joint Replacement in the United States," *J. Amer. Med. Assoc.,* 15 October 1982.

Tronzo, R. G. "Bone: Self-repairing, Self-renewing," *Consultant,* April 1972.

Williams, R. C. "Rheumatoid Arthritis," *Hospital Practice,* June 1979.

Wilson, F. C. *The Musculoskeletal System: Basic Processes and Disorders,* 2d ed. Philadelphia: Lippincott 1983.

UNIT III

Arehart-Treichel, J. "Probing the Causes of MS," *Science News,* 30 January 1982.

Arnason, B. G. W. "Multiple Sclerosis: Current Concepts and Management," *Hospital Practice,* February 1982.

Basmajian, J. V. "Biofeedback: The Clinical Tool Behind The Catchword," *Modern Medicine,* 1 October 1976.

Binkley, S. "A Timekeeping Enzyme in the Pineal Gland," *Scientific American,* April 1979.

Blonde, L., and F. A. Riddick, Jr. "Hyperthyroidism: Etiology, Diagnosis, and Treatment," *Hospital Medicine,* March 1980.

Bloom, F. E. "Neuropeptides," *Scientific American,* October 1981.

Bluestone, C. D. "Otitis Media: Newer Aspects of Etiology Pathogenesis, Diagnosis, and Treatment," *Consultant,* December 1979.

"The Brain," entire issue devoted to the brain and the human nervous system, *Scientific American,* September 1979.

Cowart, V. S. "New Diagnostic Technique BEAMed at Dyslexia," *J. Amer. Med. Assoc.,* 16 April 1982.

Coyle, J. T., D. L. Price, and M. R. Delong. "Brain Mechanisms in Alzheimer's Disease," *Hospital Practice,* November 1982.

Dalessio, D. J. "Headache: Clinical Guide to Identifying the Cause," *Hospital Medicine,* September 1979.

Dole, V. P. "Addictive Behavior," *Scientific American,* December 1980.

Glasscock, M. E. "Ménière's Disease," *Hospital Medicine,* December 1979.

Gonzalez, E. R. "Parkinson's Disease: New Drugs on The Horizon But No Cure On The Horizon," *J. Amer. Med. Assoc.,* 3 July 1981.

Guillemin, R., and R. Burgus. "The Hormones of the Hypothalamus," *Scientific American,* November 1972.

Gunby, P. "New Focus on Spinal Cord Injury," *J. Amer. Med. Assoc.,* 27 March 1981.

Guyton, A. C. *Human Physiology and Mechanisms of Disease,* 3d ed. Philadelphia: Saunders, 1982.

Hayles, A. B. "Gigantism," *Pediatric Annals,* 9:4, April 1980.

Herbert, W. "Implanting an Electronic Earful," *Science News*, 27 November 1982.

"Hydrocephalus: Cranial Wrap Provides Hope," *Medical Tribune*, 18 July 1973.

Johnson, G. T. (ed.). "A Thyroid Tour," *The Harvard Medical School Health Letter*, November 1980.

———. "The Endorphins: The Body's Own Opiates," *The Harvard Medical School Health Letter*, February 1983.

———. "Middle Ear Infections," *The Harvard Medical School Health Letter*, May 1982.

———. "Senility," *The Harvard Medical School Health Letter*, May 1981.

Kaufman, H. E. "Corneal Transplantation: A Progress Report," *Hospital Practice*, July 1973.

Keynes, R. D. "Ion Channels in the Nerve-Cell Membrane," *Scientific American*, March 1979.

Klein, I., and G. S. Levey. "Thyroid Storm," *Hospital Medicine*, March 1982.

Krieger, D. T. "The Neuroendocrinology Series," *Hospital Practice*, April 1975.

Kupfer, D. J., and C. F. Reynolds III. "Sleep Disorders," *Hospital Practice*, February 1983.

Lange, R. F. (ed.). "Treatment For Multiple Sclerosis," *Internal Medicine Alert*, 14 February 1983.

Laros, R. K., Jr., et al. "Prostaglandins," *Amer. J. Nurs.*, June 1973.

Lester, H. A. "The Response to Acetylcholine," *Scientific American*, February 1977.

Levi-Montalcini, R., and P. Calissano. "The Nerve-Growth Factor," *Scientific American*, June 1979.

Livingston, S. "The Medical Treatment of Epilepsy," *Pediatrics Annals*, April 1979.

Llinas, R. R. "Calcium in Synaptic Transmission," *Scientific American*, October 1982.

McEwen, B. S. "Interactions between Hormones and Nerve Tissue," *Scientific American*, July 1976.

Mechner, F. "The Brain As a Target Organ of Endocrine Hormones," *Hospital Practice*, May 1975.

Melamed, M. A. "Cataracts: Recognition and Assessment," *Hospital Practice*, July 1982.

Morell, P., and W. T. Norton. "Myelin," *Scientific American*, May 1980.

Notkins, A. L. "The Causes of Diabetes," *Scientific American*, November 1979.

O'Malley, B. W. "Hormones, Genes, and Cancer," *Hospital Practice*, July 1975.

O'Malley, B. W., and W. T. Schrader. "The Receptors of Steroid Hormones," *Scientific American*, February 1976.

Pappenheimer, J. R. "The Sleep Factor," *Scientific American*, August 1976.

Parker, D. E. "The Vestibular Apparatus," *Scientific American*, November 1980.

Pines, M. "Memory: How It Can be Created and Destroyed," *Modern Medicine*, July 1974.

———. "The Memory Code: Is It Beyond Comprehension?" *Modern Medicine*, August 1974.

Regan, D. "Electrical Responses Evoked From the Human Brain," *Scientific American*, December 1979.

Rosenfield, D. B. "A Practical Neurological Screening Examination," *Hospital Medicine*, June 1982.

Rupp, W. M., et al. "The Use of An Implantable Insulin Pump in the Treatment of Type II Diabetes, *New England J. Med.*, July 1982.

Salans, L. B. "Diabetes Mellitus," *J. Amer. Med. Assoc.*, 5 February 1982.

Schwartz, J. H. "The Transport of Substances in Nerve Cells," *Scientific American*, April 1980.

Seyle, H. *Stress in Health and Disease*. London: Butterworths, 1976.

Smith, R. W. "Trigeminal Neuralgia," *Consultant*, October 1979.

Stein, J. M., and C. A. Warfield. "Phantom Limb Pain," *Hospital Practice*, February 1982.

Tolis, G. "Prolactin: Physiology and Pathology," *Hospital Practice*, February 1980.

Treichel, J. A. "Laser Therapy: Light Against Blindness, *Science News*, 15 May 1982.

Warfield, C. A., and Stein, J. M. "Pain Relief By Electrical Stimulation," *Hospital Practice*, March 1983.

Williams, R. *Textbook of Endocrinology*. 6th ed. Philadelphia: Saunders, 1981.

Wurtman, R. J. "The Pineal as a Neuroendocrine Transducer," *Hospital Practice*, January 1980.

Yahr, M. D. "Brain Tumors," *Hospital Medicine*, September 1973.

Yao, Y. "A Current View of Thyroid Function Tests," *Hospital Practice*, September 1981.

Young, W. R. "The Enduring Mystery of Dyslexia," *Reader's Digest*, 1976.

Zafar, S. "Diagnosis: Hypothyroidism," *Hospital Medicine*, January 1982.

Ziporyn, T. "Taste and Smell: The Neglected Senses," *J. Amer. Med. Assoc.*, 15 January 1982.

UNIT IV

Ackerman, S. "The Management of Obesity," *Hospital Practice*, March 1983.

Aledort, L. M. "Current Concepts in Diagnosis and Management of Hemophilia," *Hospital Practice*, October 1982.

Amsterdam, E. A. "A Realistic Guide to Changing Bad Habits," *Modern Medicine*, October 1981.

"Arrhythmias," *Hospital Practice*, May 1980.

Ash, S. R., et al. "Peritoneal Dialysis for Acute and Chronic Renal Failure: An Update," *Hospital Practice*, January 1983.

Baron, H. C. "Valvular Incompetence and Varicose Veins," *Hospital Medicine*, April 1976.

Bennett, W., and J. Gurin. "Do Diets Really Work?" *Science 82*, March 1982.

Bowen, W. H. "Prospects for the Prevention of Dental Caries," *Hospital Practice*, May 1974.

Buisseret, P. D. "Allergy," *Scientific American*, August 1982.

Burg, M. B. "The Nephron in Transport of Sodium, Amino Acids, and Glucose," *Hospital Practice*, October 1978.

Capra, J. D., and A. B. Edmundson. "The Antibody Combining Site," *Scientific American*, January 1977.

Carey, L. C. "Shock: Differential Diagnosis and Immediate Treatment," *Hospital Medicine*, May 1975.

Carpenter, C. B., and T. B. Strom. "Transplantation: Immunogenetic and Clinical Aspects," *Hospital Practice*, December 1982.

Carr, D. T. "Malignant Lung Disease," *Hospital Practice*, January 1981.

Chapman, M. L. "Axioms on Peptic Ulcer Disease," *Hospital Medicine*, January 1980.

Check, W. A. "Bad and Good News on Gastroplasty," *J. Amer. Med. Assoc.*, 16 July 1982.

Child, J., et al. "Blood Transfusions," *Amer. J. Nurs.*, September 1972.

Chlebowski, R. T. "Cancer of the Colon," *Hospital Medicine*, March 1982.

Coe, F. L. "Nephrolithiasis: Causes, Classification, and Management," *Hospital Practice*, April 1981.

Cooper, M. D., and A. R. Lawton, III. "The Development of the Immune System," *Scientific American*, November 1974.

Cowart, V. S. "Blood Substitues: Two Ways to Get There," *J. Amer. Med. Assoc.*, 14 January 1983.

Daughaday, C. C., and S. D. Douglas. "Phagocytes," *Pediatric Annals*, Immunology in Infancy and Childhood, June 1976.

Dickerson, E. "Cytochrome c and the Evolution of Energy Metabolism," *Scientific American,* March 1980.

Dolin, B. J. "Alcoholism and Its Treatment: 1983," *Hospital Medicine,* January 1983.

Doolittle, R. F. "Fibrinogen and Fibrin," *Scientific American,* December 1981.

"Dramatic Advances against Kidney Diseases," *Medical Tribune,* 23 June 1976.

Escher, D. J. W. "Use of Cardiac Pacemakers," *Hospital Practice,* September 1981.

Forget, B. G. "Hemolytic Anemias: Congenital and Acquired," *Hospital Practice,* April 1980.

Fozzard, H. A. "Electrophysiology of the Heart: The Effects of Ischemia," *Hospital Practice,* May 1980.

"Giving Up Smoking: How the Various Programs Work," *Medical World News,* 1 November 1976.

Goldstein, F. "Tracking and Treating Gallstones, *Modern Medicine,* May 15–30, 1979.

Goldstein, M. "The Aplastic Anemias," *Hospital Practice,* May 1980.

Gonzalez, E. R. "Implanted Pump Concentrates Chemotherapy for Liver Cancer," *J. Amer. Med. Assoc.,* August 1981.

Gotto, A. M. "How to Lower Lipids With Drugs and Diet," *Modern Medicine,* October 1981.

Grollman, A. "Body Fluids and Electrolytes," *Consultant,* May 1976.

Harrison, D. C. "Coronary Bypass: The First 10 Years," *Hospital Practice,* June 1981.

Heimlich, H. J. "A Life-Saving Maneuver to Prevent Food Choking," *J. Amer. Med. Assoc.,* 27 October 1975.

Herberman, R. B. "Natural Killer Cells," *Hospital Practice,* April 1982.

Herbert, V. "The Nutritional Anemias," *Hospital Practice,* March 1980.

Jarvik, R. K. "The Total Artificial Heart," *Scientific American,* January 1981.

Johansen, K. "Aneurysms," *Scientific American,* July 1982.

Johnson, G. T. (ed.). "The Cholesterol Controversy," *The Harvard Medical School Health Letter,* October 1981.

———. "Allergies—Parts I and II," *The Harvard Medical School Health Letter,* June and July 1981.

———. "The Tragedy of Anorexia Nervosa," *The Harvard Medical School Health Letter,* December 1981.

———. "Liver Ailments: Cirrhosis," *The Harvard Medical School Health Letter,* January 1981.

———. "Liver Ailments: Hepatitis," *The Harvard Medical School Health Letter,*

February 1981.

———. "Heart Arrhythmias," *The Harvard Medical School Health Letter,* January 1982.

———. "Heart Attacks and Counter Attacks," *The Harvard Medical School Health Letter,* February 1982.

———. "The Ups and Downs of Blood Pressure Numbers," *The Harvard Medical School Health Letter,* November 1982.

———. "Weight Control," *The Harvard Medical School Health Letter,* December 1980.

———. "Some Passing Thoughts on Kidney Stones," *The Harvard Medical School Health Letter,* March 1983.

———. "Urinary Tract Infections in Women," *The Harvard Medical School Health Letter,* December 1982.

Jones, J., et al. "Preventing Erythroblastosis Fetalis," *Current Prescribing,* November 1979.

Kassirer, J. P., and N. E. Madias. "Respiratory Acid–Base Disorders," *Hospital Practice,* December 1980.

Kaufman, C. E. "Essential Hypertension: Essentials of Management," *Hospital Medicine,* May 1982.

Kelly, D. H., and D. C. Shannon. "Sudden Infant Death Syndrome—Parts I and II," *Hospital Medicine,* October and November 1982.

Kohler, P. F., and J. Vaughan. "The Autoimmune Diseases," *J. Amer. Med. Assoc.,* 26 November 1982.

Kopp, U. C., and G. F. DiBona. "Diagnosis: Edema," *Hospital Medicine,* February 1982.

Lange, R. F. (ed.). "AIDS Marches On," *Internal Medicine Alert,* 31 January 1983.

———. "New Hope For Thalassemia and Sickle Cell Disease," *Internal Medicine Alert,* 27 December 1982.

Laragh, J. H. "Hypertension," *Drug Therapy,* January 1980.

Lawrence, S. "AIDS—No Relief in Sight," *Science News,"* 25 September 1982.

Leder, P. "The Genetics of Antibody Diversity," *Scientific American,* May 1982.

Luy, M. L. M. "The Roughage Rage: Would More Fiber in Our Diet Improve Health?" *Modern Medicine,* April 1976.

McCue, J. D. "The Pathogenesis of Infectious Mononucleosis," *Hospital Practice,* October 1982.

Mason, E. E. "Morbid Obesity: What Does Surgery Have to Offer?" *Consultant,* September 1979.

Mason, S. J., and S. Wolfson. "Angina Pectoris," *Hospital Medicine,* February 1980.

Milstein, C. "Monoclonal Antibodies," *Scientific American,* October 1980.

Moog, F. "The Lining of the Small Intestine," *Scientific American,* November 1981.

Naeye, R. L. "Sudden Infant Death," *Scientific American,* April 1980.

Newman, L. "More "Salt" Talks: Diet and Hypertension," *J. Amer. Med. Assoc.,* 10 December 1982.

Norman, P. S. "Asthma," *Drug Therapy,* January 1980.

Oberman, A. "Does Exercise Prevent Heart Disease?" *Modern Medicine,* October 1982.

Polish, E. "Jaundice: Guide to Diagnosis," *Hospital Medicine,* October 1976.

Race, G. J., and M. G. White. *Basic Urinalysis.* New York: Harper & Row 1979.

Rector, F. C., Jr., and M. G. Cogan. "The Renal Acidoses," *Hospital Practice,* April 1980.

Rees, M. K. "Cholesterol and the Heart," *Modern Medicine,* February 1982.

Reichel, J. "Pulmonary Emphysema," *Hospital Medicine,* February 1980.

Rose, N. R. "Autoimmune Diseases," *Scientific American,* February 1981.

Rosner, F. R., and H. W. Grunwald. "Acute Myelocytic Leukemia," *Hospital Medicine,* December 1982.

Schimke, R. N. "Metabolic Diseases," *Hospital Practice,* January 1983.

Schrier, R. W. "Acute Renal Failure," *J. Amer. Med. Assoc.,* 14 May 1982.

Schwartz, C. J., et al. "Atheroma: An Update," *Hospital Medicine,* December 1982.

Schwartz, R. S. "Therapeutic Uses of Immune Suppression and Enhancement," *Hospital Practice,* August 1981.

Shabetai, R. "Answers to Questions on Cardiac Catheterization," *Hospital Medicine,* August 1982.

Shaver, J. A. "Heart Murmurs: Innocent or Pathologic," *Hospital Medicine,* July 1982.

Snider, G. "Chronic Obstructive Pulmonary Disease: Advice on Dx and Rx," *Modern Medicine,* December 1982.

"Standards and Guidelines for Cardiopulmonary Resuscitation (CPR) and Emergency Cardiac Care (ECC)," *J. Amer. Med. Assoc.,* 244, August 1980.

Stobo, J. D. "Basic Mechanisms of Immunity," *Hospital Medicine,* July 1980.

Treichel, J. A. "Anorexia Nervosa: A Brain Shrinker?" *Science News,* 23 October 1982.

Van Itallie, T. B., and J. G. Kral. "The Dilemma of Morbid Obesity," *J. Amer. Med. Assoc.,* 28 August 1981.

Wallis, C. "Living on Borrowed Time," *Time,* December 1982.

West, S. "One Step Behind a Killer," *Science 83,* March 1983.

Westfall, U. A. "Electrical and Mechanical Events in the Cardiac Cycle," *Am. J. Nurs.,* February 1976.

Wilson, D. E., and H. Kaymakcalan. "Recent Trends in Duodenal Ulcer Disease," *Hospital Medicine,* August 1982.

Witzum, J. L. "Diagnosis and Treatment of Hyperlipidemia," *Hospital Medicine,* June 1978.

Zucker, M. B. "The Functioning of Blood Platelets," *Scientific American,* June 1980.

UNIT V

Abrams, J. "The Pap Test in 1980," *J. Med. Soc. of N.J.,* 77:11, October 1980.

Adams, M. M., et. al. "Down's Syndrome," *J. Amer. Med. Assoc.,* 14 August 1981.

Arehard-Treichel, J. "Fetal Ultrasound: How Safe?" *Science News,* 12 June 1982.

Begley, S., and J. Carey. "How Human Life Begins," *Newsweek,* 11 January 1982.

Burnett, L. S., et. al. "An Evaluation of Abortion: Techniques and Protocols," *Hospital Practice,* August 1975.

Carbone, P. P. "Options in Breast Cancer Therapy," *Hospital Practice,* February 1981.

Centerwall, W. R., and J. L. Murdoch. "Human Chromosome Analysis," *American Family Physician,* April 1975.

Cohen, R. J. "Breast Cancer: Guide to Early Diagnosis," *Hospital Medicine,* June 1980.

Corey, L. "The Diagnosis and Treatment of Genital Herpes," *J. Amer. Med. Assoc.,* 3 September 1982.

Crile, G., Jr. "Breast Examination: Step-by-Step Guide," *Hospital Medicine,* January 1980.

Davidson, J. M. "Hormones and Sexual Behavior in the Male," *Hospital Practice,* September 1975.

Federman, D. D. "Impotence: Etiology and Management," *Hospital Practice,* March 1982.

Fuchs, F. "Genetic Amniocentesis," *Scientific American,* June 1980.

"Genetic Counseling and Prevention of Birth Defects," *J. Amer. Med. Assoc.,* 9 July 1982.

Grobstein, C. "External Human Fertilization," *Scientific American,* June 1979.

Grumbach, M. M. "The Neuroendocrinology of Puberty," *Hospital Practice,* March 1980.

Hager, W. D. "Pelvic Inflammatory Disease," *Hospital Medicine,* April 1982.

Handsfield, H. H. "Sexually Transmitted Diseases," *Hospital Practice,* January 1982.

Hill, E. C. "Carcinoma of the Cervix: Diagnostic Guide," *Hospital Medicine,* May 1976.

Holmes, K. K. "Sexually Transmitted Diseases: An Overview and Perspectives on the Next Decade," *Infectious Diseases,* April 1980.

Johnson, G. T. (ed.). "Breast Cancer: The Challenge of Early Detection," *The Harvard Medical School Health Letter,* February 1983.

———. "Sexually Transmitted Diseases," *The Harvard Medical School Health Letter,* April 1981.

———. "Genital Herpes and Acyclovir," *The Harvard Medical School Health Letter,* September 1982.

———. "Cesarean Section," *The Harvard Medical School Health Letter,* June 1981.

———. "New Views on Feeding Babies," *The Harvard Medical School Health Letter,* August 1982.

Jones, O. W. "Where We Stand Today in Prenatal Diagnosis," *Medical Opinion,* May 1976.

Krauss, D. J. "The Physiological Basis of Male Sexual Dysfunction," *Hospital Practice,* February 1983.

Kruppel, R., et. al. "Medical Complications in Pregnancy—Parts 1 and 2," *Hospital Medicine,* September and October 1982.

Langer, A. "Practical Genetics in Office Practice," *Hospital Medicine,* August 1982.

"Menopause—The Normal Disorder," *Drug Therapy,* September 1976.

Michael, R. P. "Hormones and Sexual Behavior in the Female," *Hospital Practice,* December 1975.

Piver, M. S. "Chemotherapy for Gynecologic Malignancies," *Drug Therapy,* October 1976.

Resnick, M. I. "How to Find and Stage Prostatic Carcinoma," *Modern Medicine,* October 1976.

Rugh, R., and L. B. Shettles. *From Conception to Birth.* New York: Harper & Row, 1971.

Segal, S. J. "The Physiology of Human Reproduction," *Scientific American,* September 1974.

Shocket, B. R. "Medical Aspects of Sexual Dysfunction," *Drug Therapy,* June 1976.

Smith, D. W. "The Fetal Alcohol Syndrome," *Hospital Practice,* October 1979.

Sweet, R. L. "Toxic Shock: Accurate Diagnosis and Aggressive Treatment," *Modern Medicine,* February 1983.

Temple, M. J. "Chromosomal Syndromes," *Hospital Practice,* February 1983.

"The Breast Cancer Digest," NIH Publication No 80–1691, National Cancer Institute, Bethesda, Maryland 20205, May 1980.

"The Second-Generation IUD's—Progestasert," *Current Prescribing,* January 1976.

"Three Methods of Tubal Occlusion—And All Look Good," *Modern Medicine,* September 1976.

"Treating Menopausal Women and Climacteric Men," *Medical World News,* 28 June 1974.

Index